Caledonian Igneous Rocks of Great Britain

THE GEOLOGICAL CONSERVATION REVIEW SERIES

The comparatively small land area of Great Britain contains an unrivalled sequence of rocks, mineral and fossil deposits, and a variety of landforms that span much of the Earth's long history. Well-documented ancient volcanic episodes, famous fossil sites, and sedimentary rock sections used internationally as comparative standards, have given these islands an importance out of all proportion to their size. The long sequences of strata and their organic and inorganic contents have been studied by generations of leading geologists, thus giving Britain a unique status in the development of the science. Many of the divisions of geological time used throughout the world are named after British sites or areas; for instance, the Cambrian, Ordovician and Devonian systems, the Ludlow Series and the Kimmeridgian and Portlandian stages.

The Geological Conservation Review (GCR) was initiated by the Nature Conservancy Council in 1977 to assess and document the most important parts of this rich heritage. The GCR reviews the current state of knowledge of the key Earth science sites in Great Britain and provides a firm basis upon which site conservation can be founded in years to come. Each GCR title in the 42-volume series describes networks of sites of national or international importance in the context of a portion of the geological column, or a geological, palaeontological or mineralogical topic.

Within each individual volume, every GCR locality is described in detail in a self-contained account, consisting of an introduction (with a concise history of previous work), a description, an interpretation (assessing the fundamentals of the site's scientific interest and importance), and a conclusion (written in simpler terms for the non-specialist). Each site report is a justification of a particular scientific interest at a locality, of its importance in a British or international setting, and ultimately of its worthiness for conservation.

The aim of the Geological Conservation Review series is to provide a public record of the features of interest in sites being considered for notification as Sites of Special Scientific Interest (SSSIs). It is written to the highest scientific standards but in such a way that the assessment and conservation value of the site is clear. It is a public statement of the value placed on our geological and geomorphological heritage by the Earth science community that has participated in its production, and it will be used by the Joint Nature Conservation Committee, the Countryside Council for Wales, English Nature and Scottish Natural Heritage in carrying out their conservation functions. The three country agencies are also active in helping to establish sites of local and regional importance. Regionally Important Geological/Geomorphological Sites (RIGS) augment the SSSI coverage, with local groups identifying and conserving sites that have educational, historical, research or aesthetic value, enhancing the wider Earth heritage conservation perspective.

All the sites in this volume have been proposed for notification as SSSIs; the final decision to notify, or re-notify, sites lies with the governing councils of the appropriate country conservation agency.

Information about the GCR publication programme may be obtained from:

GCR Unit,
Joint Nature Conservation Committee,
Monkstone House,
City Road,
Peterborough PE1 1JY.

www.jncc.gov.uk/earthheritage

Titles in the series

1. **An Introduction to the Geological Conservation Review**
   N.V. Ellis (ed.), D.Q. Bowen, S. Campbell, J.L. Knill, A.P. McKirdy, C.D. Prosser, M.A. Vincent and R.C.L. Wilson

2. **Quaternary of Wales**
   S. Campbell and D.Q. Bowen

3. **Caledonian Structures in Britain South of the Midland Valley**
   Edited by J.E. Treagus

[handwritten note obscuring items 4–11:
The Old Red Sandstone £59.32
Caledonian Igneous £78.00
Tert Stratigraphy £59.13
Mass Movements £45.00
Coastal Geomorph 74.00
Total £315.45]

12. **Karst and Caves of Great Britain**
    A.C. Waltham, M.J. Simms, A.R. Farrant and H.S. Goldie

13. **Fluvial Geomorphology of Great Britain**
    Edited by K.J. Gregory

14. **Quaternary of South-West England**
    S. Campbell, C.O. Hunt, J.D. Scourse, D.H. Keen and N. Stephens

15. **British Tertiary Stratigraphy**
    B. Daley and P. Balson

16. **Fossil Fish of Great Britain**
    D.L. Dineley and S.J. Metcalf

17. **Caledonian Igneous Rocks of Great Britain**
    D. Stephenson, R.E. Bevins, D. Millward, A.J. Highton, I. Parsons, P. Stone and W.J. Wadsworth

Caledonian Igneous Rocks of Great Britain

Compiled and edited by

D. Stephenson
British Geological Survey, Edinburgh

R.E. Bevins
National Museum of Wales, Cardiff

D. Millward
British Geological Survey, Edinburgh

A.J. Highton
British Geological Survey, Edinburgh

I. Parsons
University of Edinburgh, Edinburgh

P. Stone
British Geological Survey, Edinburgh

and

W.J. Wadsworth
University of Manchester, Manchester

GCR Editor: **L.P. Thomas**

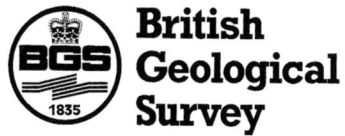

Published by the Joint Nature Conservation Committee, Monkstone House, City Road, Peterborough, PE1 1JY, UK

First edition 1999

© 1999 Joint Nature Conservation Committee

Typeset in 10/12pt Garamond ITC by JNCC
Printed in Great Britain by Hobbs the Printers Ltd. on 100 gsm Silverblade Matt.

ISBN 1 86107 471 9.

Apart from any fair dealing for the purposes of research or private study, or criticism or review, as permitted under the UK Copyright Designs and Patents Act, 1988, this publication may not be reproduced, stored, or transmitted, in any form or by any means, without the prior permission in writing of the publishers, or in the case of reprographic reproduction only in accordance with the terms of the licences issued by the Copyright Licensing Agency in the UK, or in accordance with the terms and licences issued by the appropriate Reproduction Rights Organization outside the UK. Enquiries concerning reproduction outside the terms stated here should be sent to GCR Team, JNCC.

The publisher makes no representation, express or implied, with regard to the accuracy of the information contained in this book and cannot accept any legal responsibility or liability for any errors or omissions that may be made.

British Geological Survey Copyright protected materials

1. The copyright of materials derived from the British Geological Survey's work is vested in the Natural Environment Research Council (NERC). No part of these materials (geological maps, charts, plans, diagrams, graphs, cross-sections, figures, sketch maps, tables, photographs) may be reproduced or transmitted in any form or by any means, or stored in a retrieval system of any nature, without the written permission of the copyright holder, in advance.

2. To ensure that copyright infringements do not arise, permission has to be obtained from the copyright owner. In the case of BGS maps this includes **both BGS and the Ordnance Survey.** Most BGS geological maps make use of Ordnance Survey topography (Crown Copyright), and this is acknowledged on BGS maps. Reproduction of Ordnance Survey materials may be independently permitted by the licences issued by Ordnance Survey to many users. Users who do not have an Ordnance Survey licence to reproduce the topography must make their own arrangments with the Ordnance Survey, Copyright Branch, Romsey Road, Southampton SO9 4DH (Tel. 01703 792913).

3. Permission to reproduce BGS materials must be sought in writing from the Intellectual Property Rights Manager, British Geological Survey, Kingsley Dunham Centre, Keyworth, Nottingham NG12 5GG (Tel. 0115 936 3100).

A catalogue record for this book is available from the British Library.

Contents

Contributors	xi
Acknowledgements	xiii
Access to the countryside	xv
Foreword P.E. Brown	xvii

1 Caledonian igneous rocks of Great Britain: an introduction   1

Introduction D. Stephenson	3
Site selection D. Stephenson	7
Tectonic setting and evolution D. Stephenson	9
Origin of the late Caledonian magmas D. Stephenson and A.J. Highton	17

2 Early Ordovician volcanic rocks and associated opholitic assemblages of Scotland   27

Introduction P. Stone and D. Flinn	29
The Shetland Ophiolite	36
The Punds to Wick of Hagdale D. Flinn	36
Skeo Taing to Clugan D. Flinn	41
Qui Ness to Pund Stacks D. Flinn	45
Ham Ness D. Flinn	48
Tressa Ness to Colbinstoft D. Flinn	51
Virva D. Flinn	54
The Highland Border Complex	58
Garron Point to Slug Head C.W. Thomas	58
Balmaha and Arrochymore Point J.R. Mendum	61
North Glen Sannox D.J. Fettes	65
The Ballantrae Complex	69
Byne Hill P. Stone	69
Slockenray Coast P. Stone	72
Knocklaugh P. Stone	78
Millenderdale P. Stone	81
Knockormal P. Stone	84
Games Loup P. Stone	87
Balcreuchan Port to Port Vad P. Stone	91
Bennane Lea P. Stone	96
Southern Uplands	100
Sgavoch Rock P. Stone	100

Contents

3 Mid-Ordovician intrusions of the North-east Grampian Highlands of Scotland — 105

Introduction *W.J. Wadsworth and D. Stephenson* — 107
Hill of Barra *W.J. Wadsworth* — 113
Bin Quarry *W.J. Wadsworth* — 115
Pitscurry and Legatesden quarries *W.J. Wadsworth* — 119
Hill of Johnston *W.J. Wadsworth* — 122
Hill of Creagdearg *W.J. Wadsworth* — 124
Balmedie Quarry *W.J. Wadsworth* — 127
Towie Wood *W.J. Wadsworth* — 130
Craig Hall *W.J. Wadsworth* — 132

4 Lake District and northern England — 135

Introduction *D. Millward* — 137
Eycott Hill *D. Millward* — 145
Falcon Crag *B. Beddoe-Stephens* — 149
Ray Crag and Crinkle Crags *M.J. Branney* — 153
Sour Milk Gill *M.J. Branney* — 160
Rosthwaite Fell *M.J. Branney* — 163
Langdale Pikes *M.J. Branney* — 167
Side Pike *M.J. Branney* — 171
Coniston *D. Millward* — 176
Pets Quarry *M.J. Branney* — 180
Stockdale Beck, Longsleddale *D. Millward* — 183
Bramcrag Quarry *S.C. Loughlin* — 187
Bowness Knott *D.J. Fettes* — 190
Beckfoot Quarry *B. Young* — 194
Waberthwaite Quarry *B. Young* — 196
Carrock Fell *D. Millward* — 198
Haweswater *D. Millward and B. Beddoe-Stephens* — 205
Grainsgill, Caldew Valley *S.C. Loughlin* — 207
Shap Fell Crags *S.C. Loughlin* — 211

5 Central England — 217

Introduction *D. Millward* — 219
Croft Hill *J.N. Carney and T.C. Pharaoh* — 221
Buddon Hill *J.N. Carney and T.C. Pharaoh* — 224
Griff Hollow *J.N. Carney and T.C. Pharaoh* — 227

6 Wales and adjacent areas — 231

Introduction *R.E. Bevins* — 233
Rhobell Fawr *R.E. Bevins* — 243
Pen Caer *R.E. Bevins* — 247
Aber Mawr to Porth Lleuog *R.E. Bevins* — 252
Castell Coch to Trwyncastell *R.E. Bevins* — 256
St David's Head *R.E. Bevins* — 259
Cadair Idris *D.G. Woodhall* — 264
Pared y Cefn-hir *D.G. Woodhall* — 269
Carneddau and Llanelwedd *D.G. Woodhall* — 273
Braich tu du *M. Smith* — 277

Contents

Llyn Dulyn *M. Smith*	281
Capel Curig *M. Smith*	283
Craig y Garn *M. Smith*	287
Moel Hebog to Moel yr Ogof *M. Smith*	290
Yr Arddu *M. Smith*	296
Snowdon Massif *M. Smith*	300
Cwm Idwal *M. Smith*	307
Curig Hill *M. Smith*	313
Sarnau *M. Smith*	316
Ffestiniog Granite Quarry *M. Smith*	319
Pandy *P.J. Brenchley*	322
Trwyn-y-Gorlech to Yr Eifl quarries *T.P. Young and W. Gibbons*	325
Penrhyn Bodeilas *T.P. Young and W. Gibbons*	327
Moelypenmaen *T.P. Young and W. Gibbons*	328
Llanbedrog *T.P. Young and W. Gibbons*	330
Foel Gron *T.P. Young and W. Gibbons*	332
Nanhoron Quarry *T.P. Young and W. Gibbons*	333
Mynydd Penarfynydd *T.P. Young and W. Gibbons*	334
Skomer Island *R.E. Bevins*	338
Deer Park *R.E. Bevins*	342

7 Late Ordovician to mid-Silurian alkaline intrusions of the North-west Highlands of Scotland — **345**

Introduction *I. Parsons*	347
Alkaline plutonic complexes	353
Loch Borralan Intrusion *I. Parsons*	353
Loch Ailsh Intrusion *I. Parsons*	366
Loch Loyal Syenite Complex *I. Parsons*	374
Alkaline minor intrusive rocks	379
'Grorudite' (peralkaline rhyolite, comendite)	380
Glen Oykel south *I. Parsons*	381
Creag na h-Innse Ruaidhe *I. Parsons*	381
The Canisp Porphyry (porphyritic quartz-microsyenite)	382
Beinn Garbh *I. Parsons*	383
The Laird's Pool, Lochinver *I. Parsons*	384
Cnoc an Leathaid Bhuidhe *I. Parsons*	384
'Hornblende porphyrite' (microdiorite, spessartite)	385
Cnoc an Droighinn *I. Parsons*	386
Luban Croma *I. Parsons*	386
Vogesite (hornblende-rich lamprophyre)	387
Allt nan Uamh *I. Parsons*	388
Glen Oykel north *I. Parsons*	389
'Nordmarkite' (quartz-microsyenite)	390
Allt na Cailliche *I. Parsons*	391
'Ledmorite' (melanite nepheline-microsyenite)	391
Camas Eilean Ghlais *I. Parsons*	392
an Fharaid Mhór *I. Parsons*	393

8 Late Silurian and Devonian granitic intrusions of Scotland — **395**

Introduction *A.J. Highton*	397
Loch Airighe Bheg *N.J. Soper*	405
Glen More *W.E. Stephens*	409

Contents

Loch Sunart *A.J. Highton*	412
Cnoc Mor to Rubh' Ardalanish *A.J. Highton*	417
Knockvologan to Eilean a'Chalmain *A.J. Highton*	421
Bonawe to Cadderlie Burn *A.J. Highton*	424
Cruachan Reservoir *A.J. Highton*	429
Red Craig *S. Robertson*	434
Forest Lodge *D. Stephenson*	438
Funtullich *W.E. Stephens*	444
Craig More *W.E. Stephens*	446
Garabal Hill to Lochan Strath Dubh-uisge *W.E. Stephens*	448
Loch Dee *W.E. Stephens*	452
Clatteringshaws Dam Quarry *W.E. Stephens*	456
Lea Larks *W.E. Stephens*	459
Lotus Quarries to Drungans Burn *W.E. Stephens*	460
Millour and Airdrie Hill *W.E. Stephens*	465
Ardsheal Hill and Peninsula *I.M. Platten*	468
Kentallen *I.M. Platten*	472

9 Late Silurian and Devonian volcanic rocks of Scotland — 479

Introduction *D. Stephenson*	481
South Kerrera *G. Durant*	489
Ben Nevis and Allt a'Mhuilinn *D.W. McGarvie*	492
The Glencoe Volcano – an introduction to the GCR sites *D.W. McGarvie*	497
Bidean nam Bian *D.W. McGarvie*	505
Stob Dearg and Cam Ghleann *D.W. McGarvie*	510
Buachaille Etive Beag *D.W. McGarvie*	513
Stob Mhic Mhartuin *D.W. McGarvie*	515
Loch Achtriochtan *D.W. McGarvie*	519
Crawton Bay *R.A. Smith*	522
Scurdie Ness to Usan Harbour *R.A. Smith*	525
Black Rock to East Comb *R.A. Smith*	528
Balmerino to Wormit *M.A.E. Browne*	531
Sheriffmuir Road to Menstrie Burn *M.A.E. Browne*	534
Craig Rossie *M.A.E. Browne*	537
Tillicoultry *M.A.E. Browne*	539
Port Schuchan to Dunure Castle *G. Durant*	542
Culzean Harbour *G. Durant*	546
Turnberry Lighthouse to Port Murray *G. Durant*	548
Pettico Wick to St Abb's Harbour *D. Stephenson*	552
Shoulder O'Craig *P. Stone*	556
Eshaness Coast *D. Stephenson*	559
Ness of Clousta to the Brigs *D. Stephenson*	565
Point of Ayre *N.W.A. Odling*	570
Too of the Head *N.W.A. Odling*	572

References	575
Glossary	619
Index	629

Contributors

Brett Beddoe-Stephens British Geological Survey, Murchison House, West Mains Road, Edinburgh EH9 3LA.

Richard E. Bevins Department of Geology, National Museum of Wales, Cathays Park, Cardiff CF1 3NP.

Michael J. Branney Department of Geology, University of Leicester, University Road, Leicester LE1 7RH.

Patrick J. Brenchley Department of Earth Sciences, University of Liverpool, The Jane Herdman Laboratories, Brownlow Street, Liverpool L69 3BX.

Michael A.E. Browne British Geological Survey, Murchison House, West Mains Road, Edinburgh EH9 3LA.

John N. Carney British Geological Survey, Kingsley Dunham Centre, Keyworth, Nottingham NG12 5GG.

Graham Durant Hunterian Museum, University of Glasgow, University Avenue, Glasgow G12 8QQ.

Douglas J. Fettes British Geological Survey, Murchison House, West Mains Road, Edinburgh EH9 3LA.

Derek Flinn Department of Earth Sciences, University of Liverpool, The Jane Herdman Laboratories, Brownlow Street, Liverpool L69 3BX.

Wes Gibbons Department of Earth Sciences, University College of Wales, Cardiff, PO Box 914, Cardiff CF1 3YE.

Andrew J. Highton British Geological Survey, Murchison House, West Mains Road, Edinburgh EH9 3LA.

Susan C. Loughlin British Geological Survey, Murchison House, West Mains Road, Edinburgh EH9 3LA.

Dave McGarvie The Open University, 2 Trevelyan Square, Boar Lane, Leeds LS1 6ED.

John R. Mendum British Geological Survey, Murchison House, West Mains Road, Edinburgh EH9 3LA.

David Millward British Geological Survey, Murchison House, West Mains Road, Edinburgh EH9 3LA.

Nicholas W.A. Odling Department of Geology and Geophysics, University of Edinburgh, The Grant Institute, West Mains Road, Edinburgh EH9 3JW.

Ian Parsons Department of Geology and Geophysics, University of Edinburgh, The Grant Institute, West Mains Road, Edinburgh EH9 3JW.

Timothy C. Pharaoh British Geological Survey, Kingsley Dunham Centre, Keyworth, Nottingham NG12 5GG.

Ian M. Platten School of Earth and Environmental Science, University of Greenwich, Medway Campus, Pembroke, Chatham Maritime, Kent ME4 4AW.

Contributors

Steven Robertson British Geological Survey, Murchison House, West Mains Road, Edinburgh EH9 3LA.

Martin Smith British Geological Survey, Murchison House, West Mains Road, Edinburgh EH9 3LA.

Richard A. Smith British Geological Survey, Murchison House, West Mains Road, Edinburgh EH9 3LA.

N. Jack Soper Gams Bank, Riverside, Threshfield, North Yorkshire BD23 4NP.

W. Edryd Stephens School of Geography and Geosciences, University of St Andrews, Purdie Building, North Haugh, St Andrews, Fife KY16 9ST.

David Stephenson British Geological Survey, Murchison House, West Mains Road, Edinburgh EH9 3LA.

Philip Stone British Geological Survey, Murchison House, West Mains Road, Edinburgh EH9 3LA.

Christopher W. Thomas British Geological Survey, Murchison House, West Mains Road, Edinburgh EH9 3LA.

W. John Wadsworth Department of Geology, University of Manchester, Manchester M13 9PL.

Derek G. Woodhall British Geological Survey, Murchison House, West Mains Road, Edinburgh EH9 3LA.

Brian Young British Geological Survey, Murchison House, West Mains Road, Edinburgh EH9 3LA.

Timothy P. Young Department of Earth Sciences, University College of Wales Cardiff, PO Box 914, Cardiff CF1 3YE.

Acknowledgements

This volume is the combined work of the 31 authors listed on pages xi–xii, most of whom, in addition to their own site descriptions, have made valuable comments on other aspects of the work. Individual chapters have been compiled by the authors of the relevant chapter introductions and have been edited by R.E. Bevins (Chapter 6), A.J. Highton (Chapter 8), D. Millward (chapters 1, 4, 5 and 9), D. Stephenson (chapters 3, 7, 8 and 9) and P. Stone (Chapter 2). Overall compilation and editing is by D. Stephenson. The GCR editor, L.P. Thomas provided guidance throughout the writing and compilation and the referee's comments of P.E. Brown were particularly welcome during the final stages of editing. The project has been managed by N.V. Ellis, N.K. Cousins and A.J. Carter for JNCC and D.I.J. Mallick and D.J. Fettes for BGS.

The initial site selection and site documentation for this volume was by R.E. Bevins, I.W. Croudace, B.P. Kokelaar, M.J. LeBas, D.T. Moffat, D. O'Halloran, I. Parsons, J.L. Roberts, W.E. Stephens, R.G. Thomas and W.J. Wadsworth. Since this initial exercise, much new mapping and refined interpretation has taken place, particularly in the Lake District and North Wales, and some revision of the original GCR site list has been necessary. M.J. Branney, M.F. Howells and B.P. Kokelaar have provided much helpful advice on this revision and the necessary amendments to the GCR documentation have been greatly facilitated by S. Campbell (for CCW), T. Moat (for EN) and R. Threadgould (for SNH).

Additional information and advice were provided by T.R. Astin (Shetland and Orkney sites), D. Gould (NE Grampian Highlands intrusions) and J.E.A. Marshall (Siluro-Devonian biostratigraphy). Diagrams were drafted by S.C. White and C.F. Pamplin (Xipress IT Solutions, Newmarket); photographs were scanned and prepared by T. Bain (BGS, Edinburgh). Photographs from the BGS collection are reproduced by kind permission of the Director, BGS ©NERC; all rights reserved (PR/23–27). The references were compiled by M.W. Kinnear, and the index was prepared by B.J. Amos.

Finally, on behalf of all of the site authors, we would like to record our thanks to the owners and managers of land and quarries who have allowed access to the sites, either during previous work or specifically for the GCR exercise.

Access to the countryside

This volume is not intended for use as a field guide. The description or mention of any site should not be taken as an indication that access to a site is open or that a right of way exists. Most sites described are in private ownership, and their inclusion herein is solely for the purpose of justifying their conservation. Their description or appearance on a map in this work should in no way be construed as an invitation to visit. Prior consent for visits should always be obtained from the landowner and/or occupier.

Information on conservation matters, including site ownership, relating to Sites of Special Scientific Interest (SSSIs) or National Nature Reserves (NNRs) in particular counties or districts may be obtained from the relevant country conservation agency headquarters listed below:

> Countryside Council for Wales,
> Plas Penrhos,
> Ffordd Penrhos,
> Bangor,
> Gwynedd LL57 2LQ.
>
> English Nature,
> Northminster House,
> Peterborough PE1 1UA.
>
> Scottish Natural Heritage,
> 12 Hope Terrace,
> Edinburgh EH9 2AS.

Foreword

Britain is exceptional in the continuity of geological history and variety of geological phenomena that are preserved within a comparatively small area. Since the early days of the geological sciences, the area has continued to provide outstanding contributions, theoretical and practical, to the understanding of Earth processes. No section of this long and distinguished history of scientific investigation is more noteworthy than that arising from the outstanding variety and preservation of the Caledonian igneous rocks. This volume describes localities that are regarded as representative of the long and complex evolution of the Caledonian igneous activity. Many of the sites listed have played a key role in interpretations marking major advances in geological thinking. One needs only to recall Hutton's deductions on the origin of granite from observations made in Glen Tilt over two hundred years ago, or the modern realization of the tectonic significance of Caledonian ophiolites. There are many problems remaining and new interpretations to be made, and the descriptions of key localities in this volume will, as well as the basic objective of conservation, provide both a tool and a stimulus for further research.

With regard to further research, one of the features that emerges from a review of the Caledonian igneous rocks and which is brought out in the introduction to the volume, is the breadth of interest these rocks have for different branches of the Earth sciences, including the petrologist looking for plate-tectonic models in explanation of the variety and spatial distribution of the igneous rocks, the structural geologist looking to the igneous rocks for support of his thoughts on ancient plate movements and the isotope geochemist endeavouring to provide a time framework for both. To all those interested in the comprehensive review presented, this GCR volume is potentially of great value in providing, as it does, summary access to both the detail and the broader picture of Caledonian igneous activity.

Accurate description and recording of field data is a fundamental aim of the Geological Conservation Review. Interpretations of the observations may vary over time but the role of the field geologist in providing the key data is paramount. In this review of the Caledonian igneous rocks the importance of detailed field observations is particularly well illustrated by the elegant modern interpretations of volcanological phenomena described at GCR sites in the Lake District and Glen Coe. These are outstanding examples of major advances resulting essentially from 'map and observation geology' (hammers nowadays tend to be rather frowned on, particularly at conservation sites). Detailed laboratory examination without ade-

Foreword

quate field support is always likely to lose much of its value, or at the worst the interpretations will be incorrect. The GCR review of the Caledonian igneous rocks is a welcome re-affirmation of the fundamental importance of field work.

The rocks described occur in Scotland, England and Wales and the variety and importance of the sites covered inevitably have made this a lengthy compilation. Individual site descriptions from thirty one contributors are organized into nine chapters under seven compilers. In most cases the sites have been described by acknowledged 'experts', many of whom have known and worked on the sites for many years. Some have been described by persons with no previous knowledge of the site, but with a background in related igneous rocks, and almost all have been visited by their author. The few exceptions that have not been visited had recent authoritative descriptions that could be summarized. Dr D. Stephenson and his team of co-authors are to be congratulated on the clarity achieved and also in preserving the individuality of presentation of the site descriptions whilst ensuring conformity with the overall aims and standard format of the Geological Conservation Review. The resulting volume will be valuable to both the amateur and professional for many years to come.

P. E. Brown FRSE
Professor Emeritus
University of St Andrews

Chapter 1

Caledonian igneous rocks of Great Britain: an introduction

Introduction

INTRODUCTION

D. Stephenson

Caledonian igneous rocks

This volume describes the igneous rocks of Scotland, England and Wales that were erupted, intruded or emplaced tectonically as a direct result of the Caledonian Orogeny (Figure 1.1). There is at present no agreed definition of the term 'Caledonian'. It has, for example, been used universally to describe the whole of the Caledonian mountain belt of Upper Proterozoic to middle Palaeozoic rocks (the Caledonides) which, prior to the more recent opening of the Atlantic Ocean, stretched continuously from the Appalachians, through Newfoundland and the British Isles to East Greenland and NW Scandinavia. In this volume the Caledonian Orogeny is taken to include all of the convergent tectonic and magmatic events arising from the closure of the 'proto-Atlantic', Iapetus Ocean in which many of the rocks of Late Proterozoic and Early Palaeozoic age had been deposited. It therefore includes subduction beneath the continental margins; the accretion or obduction of oceanic crust and island-arc material onto these margins; and ultimate collision of the continents, uplift and development of extensional molasse basins. Within this broad orogenic framework many separate 'events' are identified, several of which have commonly used specific names, most notably the mid-Ordovician peak of deformation and metamorphism in the Scottish Highlands, termed the Grampian Event by many authors, and the dominant Early Devonian deformation in northern England and Wales, which is now generally referred to as the Acadian. Many authors refer to these events as separate 'orogenies'. By this definition, most Caledonian igneous rocks of Britain range in age from about 500 Ma (earliest Ordovician) to around 390 Ma (end Early Devonian), with related activity continuing to around 360 Ma (end Late Devonian) in Orkney and Shetland (Figure 1.2). (The time-scale used throughout this volume is that of Harland *et al.* (1990), unless stated otherwise.)

The volume also includes volcanic rocks of Silurian age in southern Britain, which may be the result of the opening of a separate, younger ocean, but which have petrological features suggesting a mantle source modified by earlier Caledonian subduction.

Excluded from this volume are the volcanic rocks contemporaneous with Late Proterozoic sequences and intrusions that were emplaced during earlier (i.e. pre-Ordovician) phases of basin extension or compressive deformation. These rocks, regarded as 'Caledonian' by some authors (e.g. Read, 1961; Stephenson and Gould, 1995), will be discussed in the companion GCR volumes on *Lewisian, Torridonian and Moine rocks of Scotland*, *Dalradian rocks of Scotland* and *Precambrian rocks of England and Wales*. Devonian volcanic rocks in SW England were emplaced in extensional marine basins outside the area of Caledonian deformation. They were affected by the early deformation phases of the Variscan Orogeny and are described with other Variscan igneous rocks in the *Igneous Rocks of South-West England* GCR volume (Floyd *et al.*, 1993).

As a result of the various contrasting tectonic environments generated during the course of the orogeny, the Caledonian igneous rocks represent a wide range of compositions and modes of emplacement. Intense crustal shortening during the orogeny, followed by uplift and deep dissection since, has resulted in the exposure of numerous suites of oceanic crustal rocks, subaerial and submarine volcanic rocks, deep-seated plutons and minor intrusions, all now juxtaposed at the same erosion level. As in any orogenic province, geochemical affinities are dominantly calc-alkaline, although notable tholeiitic suites are present, particularly in rocks formed during local extensional events in the earlier stages of the orogeny, and there is one major suite of alkaline intrusions. Volcanic rocks with alkaline affinities are a minor component, occurring mostly as relics of Iapetus oceanic crust or as the products of localized crustal extension towards the end of the orogeny. Within each suite, compositions commonly range from basic to acid, and several intrusive suites include ultramafic lithologies. The alkaline intrusive suite of NW Scotland includes some highly evolved silica-undersaturated felsic rocks.

The contribution of Caledonian igneous rocks of Great Britain to igneous petrology and the understanding of igneous processes

All of the sites described in this volume are, by definition, of national importance and all have

Caledonian igneous rocks of Great Britain: an introduction

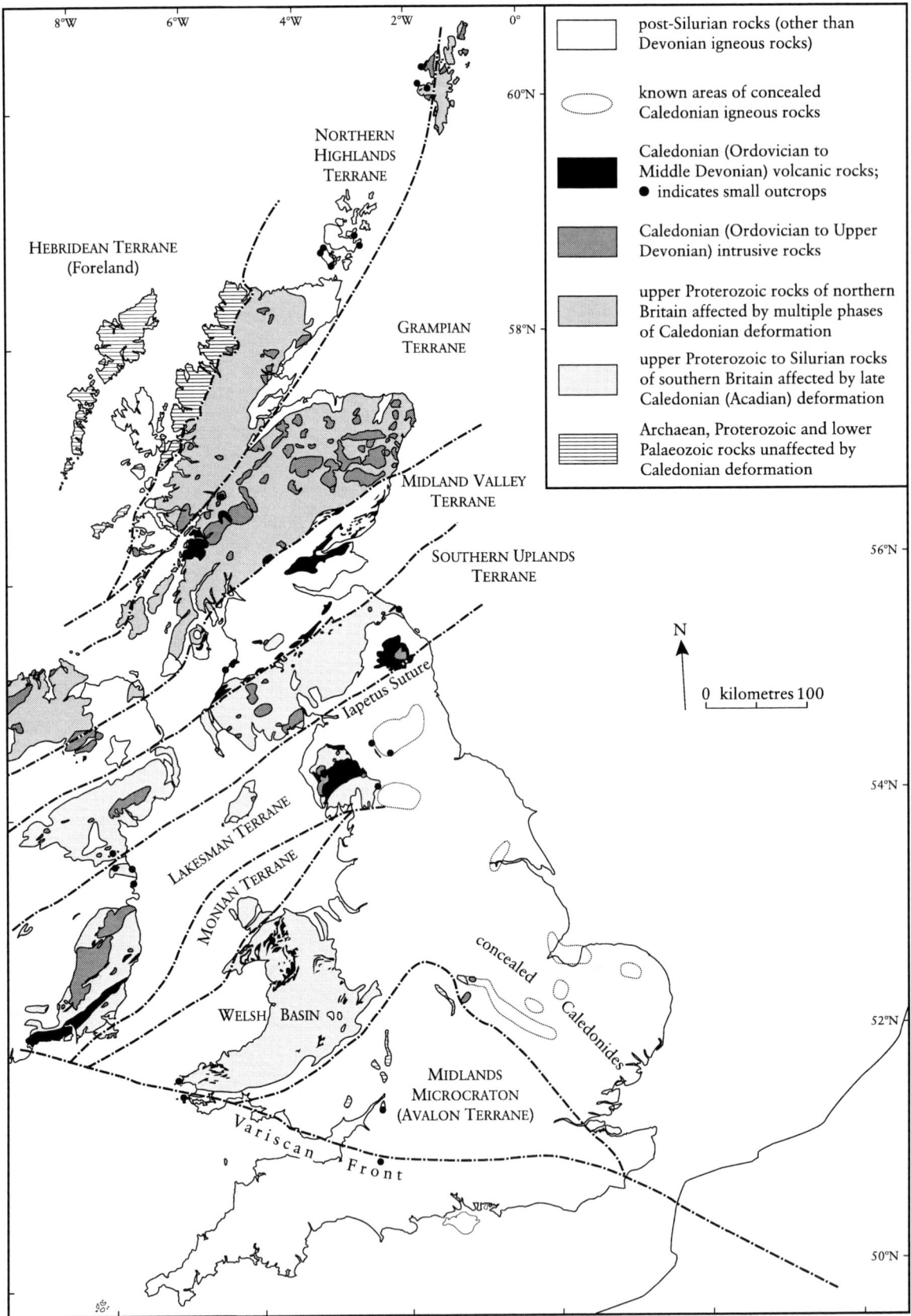

Figure 1.1 Location of Caledonian intrusive and extrusive igneous rocks of Great Britain relative to areas affected by Caledonian deformation and major terrane boundaries. Based on Brown *et al.* (1985). Nomenclature of terranes simplified after Gibbons and Gayer (1985), Bluck *et al.* (1992) and British Geological Survey (1996).

Introduction

provided evidence crucial to the mapping out and understanding of the British Caledonides. Additionally, because of their variety and their accessibility from major centres of early geological research, the Caledonian igneous rocks of Great Britain have played a major role in the initiation, testing and evolution of many theories of igneous processes. They have been, and no doubt will continue to be, the subjects of many truly seminal studies of major international significance.

Some of the first geological field investigations were conducted by James Hutton in the late eighteenth century in order to seek support for his 'Theory of the Earth', read in 1785 (Hutton, 1788). Among these were his excursions to Glen Tilt in the Grampian Highlands and the Fleet pluton in SW Scotland, where he deduced, from the presence of veins and other contact relationships, that granite had been intruded in a hot, fluid state (i.e. as a magma), surely one of the most fundamental geological concepts. By the early nineteenth century geologists were beginning to undertake detailed regional studies, such as those which led to the first accounts of Caledonian igneous rocks in Cumbria (1836) and North Wales (1843) by Adam Sedgwick and the description of the southern Grampians by James Nichol (1863). The Geological Survey was founded in 1835 and by the end of the century, many of the first editions of survey maps and memoirs of 'Caledonian' areas had been published. Based on work of this era, Archibald Geikie's *The Ancient Volcanoes of Great Britain*, published in two volumes in 1897, contains remarkably perceptive detailed accounts of most extrusive Caledonian rocks. The late nineteenth century also saw major developments in the science (and art) of descriptive petrography, exemplified by the work of J. Clifton Ward in the Lake District and Alfred Harker in both the Lake District and North Wales. This led to standard texts, drawing heavily on examples from the Caledonian province, such as Teall's *British Petrography* (1888), Hatch's *An Introduction to the Study of Petrology* (1891) and Harker's *Petrology for Students* (1895a). Subsequent editions of the last two were used to train generations of students into the late twentieth century.

The beginning of the twentieth century saw rapid developments in the understanding of igneous processes, building upon the systematic mapping and petrographic work of the previous half century and ongoing surveys. Between 1909 and 1916, C.T. Clough, H.B. Maufe and E.B. Bailey described their theory of cauldron subsidence, based upon Geological Survey mapping of the Caledonian volcanic rocks and intrusions at Glen Coe and Ben Nevis. This theory, linking surface volcanicity, subvolcanic processes and underlying plutons, was subsequently applied worldwide to both ancient and modern volcanoes and has been fundamental to the interpretation of ring intrusions and caldera structures. More recently, pioneering research on caldera development has once again focused on the British Caledonides, concentrating on the exhumed roots of such structures which are generally inaccessible in modern volcanoes. The work of M.F. Howells and co-workers in Snowdonia and of M.J. Branney and B.P. Kokelaar in the Lake District has been notable and most recently, attention has turned once again to a re-interpretation of the Glencoe volcano (Moore and Kokelaar, 1997, 1998).

The Caledonian volcanic rocks have provided numerous excellent examples of volcanological features and processes, many of which were the first to be recognized in ancient sequences; for example, the recognition of deposits from Peléan-type eruptions (i.e. incandescent ash flows or nuées ardentes) in Snowdonia by Greenly (in Dakyns and Greenly, 1905), which was later confirmed by Williams (1927) in a wider appraisal of the Ordovician volcanicity of Snowdonia. Welded ignimbrites (ash-flow tuffs) were subsequently recognized in the Lake District and Wales by Oliver (1954) and Rast *et al.* (1958) and at the time were the oldest known examples of such rocks. The later recognition, again in Snowdonia, that welding could take place in a submarine environment as well as subaerially (Francis and Howells, 1973; Howells *et al.*, 1973) led to a major re-appraisal of palaeogeographical reconstructions, not just in North Wales but in other volcanic sequences worldwide. In a further development, Kokelaar *et al.* (1985) recognized a silicic ash-flow tuff and related pyroclastic fall-out tuff on Ramsay Island in South Wales, which had been erupted, emplaced and welded, all in a submarine environment; this was the first record in the world of such ash-fall tuffs. It has become accepted that volcanic sequences commonly include shallow sills, many of which were emplaced within wet unconsolidated sediments. Rapid conversion of the water to steam fluidizes the sediment, which

Caledonian igneous rocks of Great Britain: an introduction

becomes intermixed with globules of magma; many examples have been documented in the Caledonian province and throughout the geological column. This is largely a result of a comprehensive account by Kokelaar (1982) which used case studies from the Siluro-Devonian of Ayrshire and the Ordovician of Snowdonia and Pembrokeshire.

Once established, Hutton's theory of the magmatic origin of granite remained virtually unchallenged in Great Britain for over a hundred years, until a growing body of opinion began to consider the possibility that at least some granites could have been generated *in situ* due to the 'transformation' of country rocks by fluid or gaseous 'fronts'. Support for this theory of 'granitization' reached a peak with studies of Caledonian and other granites in Ireland in the 1940s and 1950s. Although British Caledonian granites were not in the forefront of these investigations, several authors attributed the origin of the Loch Doon pluton to such a process (McIntyre, 1950; Rutledge, 1952; Higazy, 1954) and King's (1942) study of the progressive metasomatism of metasedimentary xenoliths in one of the Loch Loyal intrusions was widely quoted. In contrast, the Scottish plutons have prompted many studies of how large volumes of magma move through and become emplaced in the country rock. Read (1961) recognized that this 'space problem' could be overcome in a variety of ways, and made a widely adopted general classification of Caledonian plutons as 'forceful' or 'permitted'. More recent studies have produced more refined models of diapiric ('forceful') emplacement of plutons, such as that of Criffel in SW Scotland (Phillips *et al.*, 1981; Courrioux, 1987), but have also recognized the effects of intrusion during lateral shearing, as seen for example in the Strontian and Ratagain plutons of the NW Highlands (Hutton, 1987, 1988a, 1988b; Hutton and McErlean, 1991), in the Ballachulish, Ben Nevis, Etive, Glencoe, Rannoch Moor and Strath Ossian plutons of the SW Grampian Highlands (Jacques and Reavy, 1994) and the Fleet pluton of SW Scotland (Barnes *et al.*, 1995).

Several early attempts to integrate petrography and field observations with physical principals in order to understand the origin and evolution of igneous rocks were applied to Caledonian igneous rocks. These include the seminal studies by Alfred Harker on the Carrock Fell Complex (1894, 1895b), where in-situ differentiation processes were invoked to explain the lithological variation. Similar processes were also invoked by Teall (in Peach and Horne, 1899) to explain variations in the Loch Doon pluton, and the importance of the dominantly basic intrusions of the NE Grampians as an extreme fractionation series from peridotite to quartz-syenite was recognized by Read (1919). However, probably the most influential contribution was that of Nockolds (1941) on the Garabal Hill–Glen Fyne complex in the SW Grampian Highlands. This was one of the first geochemical studies of a plutonic complex and led to the concept of differentiation of a magma by the progressive removal of crystallized minerals from the remaining liquid (fractional crystallization), which dominated models of magmatic evolution for several decades. Ironically, more recent studies have shown that the evolution of this particular complex cannot be explained entirely by such a process (Mahmood, 1986), but the overall concept is still valid today, even though its application is more restricted. Other plutons of the Caledonian province have subsequently contributed further evidence to differentiation models, especially those in the Southern Uplands that exhibit strong concentric compositional zoning, such as Loch Doon (Gardiner and Reynolds, 1932; Ruddock, 1969; Brown *et al.*, 1979; Halliday *et al.*, 1980; Tindle and Pearce, 1981), Cairnsmore of Cairsphairn (Tindle *et al.*, 1988) and Criffel (Phillips, 1956; Halliday *et al.*, 1980; Stephens and Halliday, 1980; Stephens *et al.*, 1985; Holden *et al.*, 1987; Stephens, 1992). The Shap granite of Cumbria, in addition to being one of Britain's best known decorative stones, has been crucial in studies of metasomatism and crystal growth in igneous rocks (e.g. Vistelius, 1969) and continues to generate interest (e.g. Lee and Parsons, 1997). More-basic intrusions have contributed to other models to explain petrological diversity, such as magma mixing, hybridization and contamination. Studies by H.H. Read of the basic intrusions in the NE Grampians were among the first to recognize and document these processes (Read, 1919, 1923, 1935), and more recently the Appinite Suite of the Scottish Highlands has generated much discussion and speculation (e.g. Fowler, 1988a, 1988b; Platten, 1991; Fowler and Henney, 1996).

The alkaline igneous rocks of the NW Highlands of Scotland were among the first in the world to be recognized. Their distinctive chemistry and mineralogy was first noted by Heddle (1883a), with subsequent detailed descriptions by Teall (Horne and Teall, 1892; Teall, 1900), and they soon became a focus of international attention through the works of S.J. Shand and R.A. Daly who first introduced the crucial concept of silica saturation. For a period of over thirty years, from 1906–1939, the 'Daly–Shand hypothesis' of the origin of alkali rocks by 'desilication' of a granite magma by assimilation of limestone, was based on the common field association of the Assynt alkaline rocks with Cambro-Ordovician limestones and on similar associations worldwide. This hypothesis has now been rejected, but the NW Highland suite has continued to generate interest, not least because of its unusual tectonic setting. Most alkali provinces occur in an extensional, continental setting but the NW Highland rocks occur on the margin of a major orogenic belt. Recent, largely geochemical, studies have sought to relate them to subduction and to integrate theories of their origin with the origin of other, calc-alkaline, plutons of the Highlands (e.g. Thompson and Fowler, 1986; Halliday *et al.*, 1985, 1987; Thirlwall and Burnard, 1990; Fowler, 1992). The Siluro-Devonian volcanic rocks of northern Britain have also been attributed to subduction beneath the orogenic belt by Thirlwall (1981a, 1982) and hence, the whole magmatic province of the Highlands is a valuable complement to other continental margin provinces of the world, in which dominantly high-K calc-alkaline suites also include rare, highly alkaline derivatives.

On a broad tectonic scale, ophiolite complexes in Britain and throughout the orogenic belt are the main evidence for the former existence of the Iapetus Ocean and have enabled the position of the tectonic suture to be traced for thousands of kilometres. The study of these ophiolites has therefore made a major contribution to understanding the composition, structure and ultimate destruction of oceanic lithosphere, which has had implications far beyond the Caledonides since the initial, seminal review of Dewey (1974). In the northern part of the British Caledonides, much recent trace element and isotope work has been directed towards identifying the sources of magmas on the continental margin and, as a result, valuable information has been obtained on regional variations in the continental crust and subcontinental lithospheric mantle (e.g. Halliday, 1984; Stephens and Halliday, 1984; Thirlwall, 1989; Canning *et al.*, 1996). These variations have contributed to a division of the orogenic belt into distinct areas or 'terranes' and clearly have important regional implications in any reconstructions of the tectonic history of the Caledonides (see Tectonic setting and evolution section). Such studies, combined with similar work from the Appalachians, Newfoundland, Greenland and Scandinavia, make the Caledonides one of the most extensively studied and best understood orogenic belts in the world, acting as a stimulus and model for similar work elsewhere.

SITE SELECTION

D. Stephenson

The need for a strategy for the conservation of important geological sites is well illustrated by the words of James MacCulloch, writing in 1816 on his visit to the site in Glen Tilt where James Hutton had first demonstrated the magmatic origin of granite.

'Having blown up a considerable portion of the rock, I am enabled to say that it is of a laminated texture throughout, being a bed of which the alternate layers are limestone and that siliceous red rock which I consider as a modification of granite.'

Thankfully, despite the efforts of MacCulloch, the modern description of the Forest Lodge site in this volume can still be matched with the early accounts.

Igneous rocks and their contact relationships are on the whole less prone to damage than sedimentary rocks and fossil or mineral localities, but fine detail can be lost easily through injudicious hammering; minerals and delicate cavity features are subject to the attentions of collectors; and whole outcrops can be obscured by man-made constructions or removed by excavations. Indeed, their generally hard and resistant properties make igneous rocks an important source of construction materials and hence particularly vulnerable to large-scale commercial extraction. Whole igneous bodies can be lost in this way. Uses are many and varied; from the granite blocks that make up most of the old city of Aberdeen and many of the lighthouses and coastal defences of Britain, to the crushed dolerites that make excellent roadstone, and the

Caledonian igneous rocks of Great Britain: an introduction

multipurpose aggregates that can be derived from less resistant igneous rocks. As demand changes with time, new uses are constantly emerging, so that no igneous body can be considered safe from future exploitation. A prime example is the relatively small outcrop of the Shap granite in Cumbria, which was once prized as a decorative stone and worked on a moderate scale, but which is now worked extensively to be crushed and reconstituted to make concrete pipes.

The Geological Conservation Review (GCR) aims to identify the most important sites in order that the scientific case for the protection and conservation of each site is fully documented as a public record, with the ultimate aim of formal notification as a Site of Special Scientific Interest (SSSI). The notification of SSSIs under the National Parks and Access to the Countryside Act 1949 and subsequently under the Wildlife and Countryside Act 1981, is the main mechanism of legal protection in Great Britain. The origins, aims and operation of the review, together with comments on the law and practical considerations of Earth-science conservation, are explained fully in Volume 1 of the GCR series, *An Introduction to the Geological Conservation Review* (Ellis *et al.*, 1996). The GCR has identified three fundamental site-selection criteria; these are *international importance, presence of exceptional features* and *representativeness*. Each site must satisfy at least one of these criteria (Table 1.1). Many satisfy two and some fall into all three categories, such as the Ray Crag and Crinkle Crags site in the Lake District. The *international importance* of the British Caledonian igneous rocks has already been discussed, highlighting significant contributions to the understanding of igneous processes, many of which were identified, described or conceived for the first time from this province. Many of the sites that show *exceptional features* are also of international importance, and there are many others which provide excellent examples of features and phenomena that, although seen better elsewhere, are invaluable for research and/or teaching purposes. Good examples of the latter include the Side Pike site in the Lake District where the three main types of pyroclastic deposit can be demonstrated in close proximity; the many examples of ignimbrite in Wales, the Lake District, the Grampian Highlands and Shetland; and the layered basic intrusions of the NE Grampian Highlands, Carrock Fell in the Lake District and the Llŷn Peninsula and St David's in Wales.

The criterion of *representativeness* aims to ensure that all major stratigraphical, tectonic and petrological groupings of Caledonian igneous rocks are represented. With such a large outcrop area and wide geological and geographical range of Caledonian igneous rocks in Great Britain, it is difficult to do this while keeping the number of sites within reason. Hence there are some regionally important groups of rocks that are not represented, such as the voluminous Cairngorm Suite of granitic plutons or most of the numerous suites of minor intrusions throughout Great Britain. In many cases this is because there are no localities that show any exceptional features and there are none that exhibit the typical features of the suite any better than numerous other localities. Hence, for these suites, conservation is not a problem. However, it may be appropriate to designate 'Regionally Important Geological/Geomorphological Sites' (RIGS) to represent them so that, even though such status carries no legal protection, their importance is recognized and recorded. An attempt has been made in this volume to include, in an appropriate chapter introduction, a broad description of any group of rocks that is not represented by a GCR site, together with references to key publications. Hence, despite apparent gaps in the representativeness of the GCR site selection, the volume constitutes a complete review of all the Caledonian igneous rocks of Great Britain.

Some sites are important for more than just their igneous context. For example, the alkaline intrusions of the Assynt area (Chapter 7) provide vital structural and geochronological evidence for the complex timing of tectonic events in the NW Highlands, in particular the order and scale of movements on the various components of the Moine Thrust Zone. These aspects have been taken into account in the site selection and are relevant to discussions in the *Lewisian, Torridonian and Moine rocks of Scotland* GCR volume. Volcanic rocks within the stratigraphical column are important for radiometric dating and several of the sites in this volume have provided time-markers for regional sequences. Some of these have potential international significance in the construction of geological timescales, in particular the 'Old Red Sandstone' volcanic rocks (Chapter 9), many of which lie close to the Silurian–Devonian boundary on biostratigraphical evidence. Some 'igneous' sites in the Ballantrae ophiolite complex are important for their graptolite biostratigraphy and hence are also included in the *British Cambrian to Ordovician Stratigraphy* GCR volume (Rushton *et al.*, 1999). The Comrie pluton of the southern Grampian Highlands (Chapter 8) has been known for its metamorphic aureole since the days of Nichol (1863). It was the subject of a pioneering study by Tilley (1924) and continues to attract attention as a type example of low-pressure contact metamorphism (Pattison and Harte, 1985; Pattison and Tracy, 1991). At many sites, the igneous rocks have resulted in spectacular features of mountain or coastal geomorphology. In fact most of the more rugged mountains of mainland Britain owe their existence to Caledonian igneous rocks (e.g. those of Ben Nevis, Glen Coe, Glen Etive and most of the mountains of the Lake District and North Wales), while equally impressive coastal features occur within the GCR sites at Eshaness in Shetland, St Abb's Head in SE Scotland and the northern coast of Pembrokeshire in South Wales.

Features, events and processes that are fundamental to the understanding of the geological history, composition and structure of Britain are arranged for GCR purposes into subject 'blocks'. Three blocks comprise this volume, *Ordovician igneous rocks, Silurian and Devonian plutonic rocks,* and *Silurian and Devonian volcanic rocks*. Within each block, sites fall into natural groupings, termed 'networks', which are based upon petrological or tectonic affinites, age and geographical distribution. The ten networks in this volume contain 133 sites, which are listed in Tables 1.1a–c (pp. 19–26 herein), together with their principal reasons for selection. Some sites have features that fall within more than one network or block, for example the Ben Nevis and Allt a'Mhuilinn site, which encompasses an 'Old Red Sandstone' volcanic succession and a Late Caledonian pluton.

Site selection is inevitably subjective and some readers may feel that vital features or occurrences have been omitted or that others are over-represented. But the declared aim of the GCR is to identify *the minimum number and area of sites needed to demonstrate the current understanding* of the diversity and range of features within each block or network. To identify too many sites would not only make the whole exercise unwieldy and devalue the importance of the exceptional sites, but it would also make justification and defence of the legal protection afforded to these sites more difficult to maintain.

In this volume, most chapters represent a single GCR network as set out in Tables 1.1a–c, pp. 19–26. Continuity and descriptive convenience require two exceptions. Chapter 4 describes all the Caledonian igneous rocks of northern England, including those late Caledonian plutons which lie south of the Iapetus Suture, and Chapter 6 includes early Silurian volcanic rocks of South Wales and adjacent areas. Wherever possible, summaries of regional geology applicable to the whole network are given in the chapter introduction to avoid repetition (chapters 2, 3, 4, 5 and 7), but some networks span such a wide geographical area and range of geological settings that more specific accounts are necessary in the individual site descriptions (chapters 6, 8 and 9). In some cases, sections of general discussion apply to two or more sites (e.g. in the same pluton or group of intrusions) and in Chapter 7 this has necessitated a slight change of format to accommodate the many small sites that represent suites of minor intrusions. Space does not allow for more detailed accounts of country rock successions or structures and the reader is referred to *Geology of Scotland* (Craig, 1991), *Geology of England and Wales* (Duff and Smith, 1992) and volumes in the British Geological Survey's 'British Regional Geology' series.

TECTONIC SETTING AND EVOLUTION

D. Stephenson

The Iapetus Ocean was created in Late Proterozoic time by the rifting and pulling apart of a large supercontinent known as Rodinia. The opening started sometime around 650 million years (Ma) ago and by the beginning of Ordovician time, at 510 Ma, the ocean was at its widest development of possibly up to 5000 km. On one side of the ocean lay the supercontinent of Laurentia, which is represented today largely by the Precambrian basement rocks of North America, Greenland, the north of Ireland and the Scottish Highlands. On the opposite side lay the supercontinent of Gondwana, consisting of the basements of South America, Africa, India, Australia, East Antarctica and Western Europe (including south Ireland, England and Wales). A separate continent, Baltica (the basement of

Caledonian igneous rocks of Great Britain: an introduction

Scandinavia and Russia), was separated from Gondwana by an arm of the Iapetus Ocean, known as the Tornquist Sea (Figure 1.3). The wide separation is supported by palaeontological data, which show distinctly different faunal assemblages in the Lower Palaeozoic rocks of each former continent (Cocks and Fortey, 1982; McKerrow and Soper, 1989; McKerrow *et al.*, 1991) and by palaeomagnetic interpretations (Torsvik *et al.*, 1996). Upper Proterozoic and Cambrian sediments were deposited in extensional basins close to and on the passive margins of the flanking continents, and as turbidites that swept down the continental slopes onto the ocean floor. By analogy with modern oceans it can be assumed that new oceanic crust was generated at a mid-ocean ridge, with ocean-island type volcanicity along transform faults and above mantle plumes or 'hot-spots'.

The continental plates of Laurentia, Gondwana and Baltica started to converge during the early Ordovician, initiating new tectonic and magmatic processes that marked the start of the Caledonian Orogeny. The Iapetus oceanic crust was consumed in subduction zones beneath oceanic island arcs and beneath the continental margins and hence little is now preserved, except where slices have been sheared off and thrust over the continental rocks in the process known as obduction. Sedimentation continued in offshore, back-arc basins and in ocean trenches that developed above the subduction zones. These sediments were eventually scraped off and accreted against the Laurentian continental margin, where pre-existing rocks were undergoing deformation and metamorphism, resulting in considerable crustal shortening and thickening. On the margin of Gondwana, great thicknesses of sediment accumulated in marginal basins. Throughout all these events, new magma was being created by the melting of mantle and oceanic crustal material within and above the subduction zones and by melting within the thickened continental crust.

There have been many models to explain the relative movements of tectonic plates during the Caledonian Orogeny and, although at any one time there may be a broad consensus centred around one particular view, each has its drawbacks, and refinements are continually emerging with new evidence or new trains of thought. Early models postulating a straightforward convergence of two plates, Laurentia and Gondwana (Wilson, 1966; Dewey, 1969), were soon refined to encompass oblique closure (Phillips *et al.*, 1976; Watson, 1984). The recognition that sections of the Laurentian margin consist of distinct 'terranes' separated by strike-slip faults of great magnitude (e.g. Dewey, 1982; Gibbons and Gayer, 1985; Hutton, 1987; Kokelaar, 1988; Thirlwall, 1989; Bluck *et al.*, 1997), led to the realization that these terranes were probably not directly opposite each other as the Iapetus Ocean closed. It seems likely that Scotland lay on the margin of Laurentia, initially opposite Baltica, while England and Wales lay on the margin of Gondwana, opposite the Newfoundland sector of Laurentia. The terranes were only juxtaposed into their present relative positions (see Figure 1.3) by large sinistral (left-lateral) movements on the bounding faults during the later stages of the orogeny. The most recent models, which form the basis of the following synthesis, involve three converging continental plates, Laurentia, Baltica and Eastern Avalonia, the latter a microcontinent, including the Precambrian basement of England and Wales, which broke away from the margin of Gondwana in the early Ordovician and drifted towards Laurentia (e.g. Soper and Hutton, 1984; Pickering *et al.*, 1988; Soper *et al.*, 1992; Pickering and Smith, 1995; Torsvik *et al.*, 1996).

The exact sequence and timing of events as the three plates converged is the subject of much debate, in which the distribution, nature and timing of the igneous activity are crucial evidence. Figure 1.3 shows a simplified sequence that reflects a general consensus in which the key elements are as follows.

- Closure of the Tornquist Sea between Eastern Avalonia and Baltica, followed by strike-slip movement along the Tornquist Suture. Palaeontological evidence indicates that the two continents were close enough to share a common faunal assemblage by the late Ordovician, but final closure and suturing may have been later.
- Anticlockwise rotation of Baltica, followed by convergence with Laurentia, with subduction beneath the 'Scottish' sector of the Laurentian margin and closure during the early to mid-Silurian.
- Oblique convergence of Eastern Avalonia with Laurentia, with subduction beneath the Laurentian margin, resulting in closure by the early Silurian in the 'Irish' sector and

Tectonic setting and evolution

later, mid-Silurian closure in the 'Scottish' sector. The junction between the two fused plates passes through the Solway Firth in Britain and is known as the Iapetus Suture.
- Protracted continent–continent collision between Laurentia and Eastern Avalonia plus Baltica, with underthrusting beneath part of the Laurentian margin (?mid-Silurian to Mid-Devonian).
- Separation of a further microcontinent, Armorica, from the margin of Gondwana, which then collided with Eastern Avalonia during the Early Devonian (the Acadian Event).
- Sinistral re-alignment of terrane boundaries (?mid-Silurian to Mid-Devonian).

The main difference of opinion at the time of writing concerns the timing of the Baltica–Laurentia collision, which pre-dates or post-dates the Eastern Avalonia–Laurentia collision according to the model used. The timing and nature of the Eastern Avalonia–Baltica collision is also poorly constrained at present.

The tectonic history of the Caledonian Orogeny in Britain is summarized below, linking together coeval magmatic events in the various terranes across the whole orogenic belt (Figure 1.2). At times the magmatism was associated with distinct structural and metamorphic events, discussed in detail in the *Lewisian, Torridonian and Moine rocks of Scotland* and *Dalradian rocks of Scotland* GCR volumes. The most recent general references used in this compilation have been given above and more specific details will be found in the relevant chapter introductions. Although the histories of most individual terranes are now fairly well documented, a coherent overall model is difficult to achieve, particularly for the earlier parts of the orogeny when terranes were separated, possibly by hundreds of kilometres, prior to late-orogenic strike-slip re-alignment. Hence the timing of certain events may vary between terranes so that, for example, one sector of a continental margin may have been experiencing active subduction while another was still a passive margin.

Early Ordovician – Tremadoc and Arenig

In the early Ordovician much of the Iapetus Ocean was still in existence (Figure 1.3a). New oceanic crust was probably still being generated locally at ocean-ridge spreading centres and in oceanic island volcanoes, while elsewhere oceanic crust was being destroyed in intra-ocean subduction zones, generating widespread island-arc volcanism. The Highland Border Complex may have originated in a back-arc basin behind a volcanic arc on the Laurentian continental margin. Most of the Iapetus oceanic crust was destroyed during subsequent ocean closure, but vestiges remain in the form of ophiolite complexes of oceanic crust and upper mantle material that were accreted (welded) or obducted (thrust) onto the Laurentian margin during the earlier stages of closure (Chapter 2). Products of oceanic volcanism preserved in the ophiolites have been dated palaeontologically and/or radiometrically at mid-Tremadoc to Arenig and, in the case of the Highland Border Complex, Llanvirn. The date of accretion or obduction is more difficult to establish and was probably quite variable, since such processes must have continued throughout the early closure of the ocean. Estimates range from Early Cambrian (540 Ma) for the Highland Border Complex on the Isle of Bute to late Llanvirn or early Llandeilo for the complex elsewhere. The Shetland Ophiolite was probably obducted in the late Tremadoc, between 498 and 492 Ma, and the Ballantrae ophiolitic rocks were obducted before the deposition of Llanvirn shallow marine strata, probably in the late Arenig (c. 480–475 Ma).

On the opposite side of the Iapetus Ocean, ophiolites are preserved only in Norway, where they were obducted onto the margin of Baltica during the early Ordovician. On the margin of Gondwana, deep-marine turbidites (Skiddaw Group) were deposited on a passive margin in the Lakesman Terrane (cf. Lake District and Isle of Man) from Late Cambrian to early Llanvirn time. However, in the Wales sector, a passive margin gave way to active subduction of Iapetus oceanic crust during Tremadoc time. Subduction was associated with extension in the overlying continental crust and the development of the Welsh Basin, a marine marginal basin flanked by the Midlands Microcraton of the main continental mass and the periodically emergent Monian Terrane (cf. Irish Sea and Anglesey) adjacent to Iapetus. This basin dominated sedimentation and volcanism throughout the Caledonian Orogeny. Remnants of early localized calc-alkaline volcanism within the basin are preserved only at Rhobell Fawr in southern Snowdonia and

Caledonian igneous rocks of Great Britain: an introduction

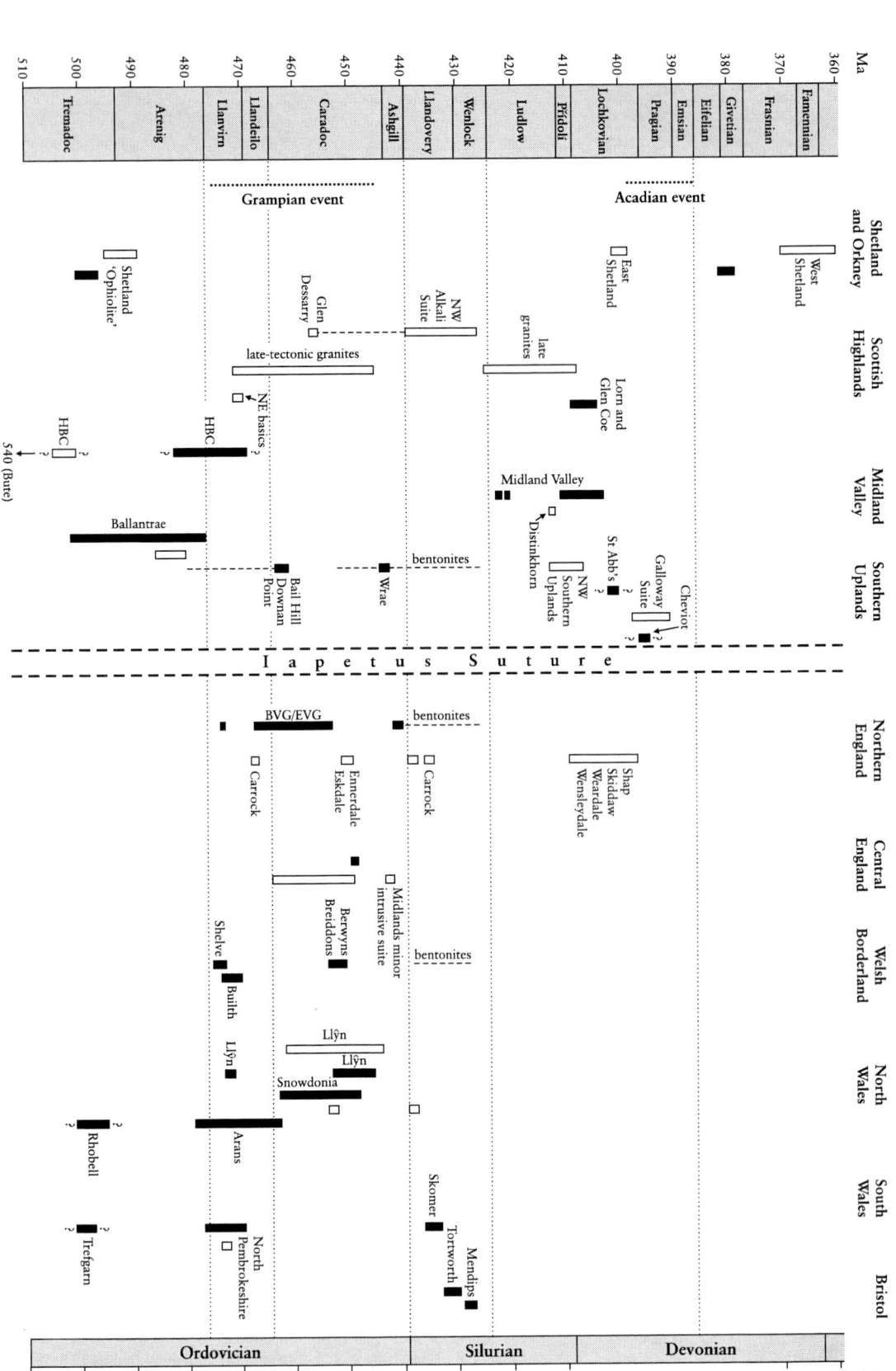

Figure 1.2 Stratigraphical distribution of Caledonian extrusive rocks (solid bars) and intrusions (open bars) in the various terranes of the British Caledonides. The time-scale is after Harland et al. (1990); horizontal lines indicate descriptive time units as used in the main text. Wherever possible the extrusive rocks are plotted according to their biostratigraphical range, as are the intrusions of Wales. Other intrusions have little or no biostratigraphical control and hence are plotted according to their currently accepted radiometric ages. This leads to some unavoidable discrepancies, in particular in the upper Silurian to Lower Devonian suites of Scotland, where intrusions and volcanic rocks in the same area are probably much closer in age than the diagram shows (see Chapter 9, Introduction). See individual chapter introductions for more detailed stratigraphical distribution charts.

at Treffgarne in north Pembrokeshire (Chapter 6). Both sequences are of late Tremadoc age and were followed by major uplift and local emergence; there is no evidence of further volcanic activity in the Welsh Basin until the late Arenig.

Mid-Ordovician – Llanvirn and Llandeilo

The mid-Ordovician saw the climax of the Caledonian Orogeny in the Scottish Highlands. This 'Grampian' Event is particularly well defined in the NE of the Grampian Terrane, where the main deformation episodes and the peak of regional metamorphism are dated by major tholeiitic basic intrusions that were emplaced at $c.$ 470 Ma (late Llanvirn), towards the end of the event (Chapter 3). In the central Grampian Highlands and the Northern Highlands Terrane, the comparable event may have been a little later ($c.$ 455 Ma). The tectonic cause of the Grampian Event is not clear, but current opinion favours collision of an intra-ocean island arc complex with the Laurentian margin, the remains of which may be present in the Highland Border and Ballantrae complexes. Crustal melting, which commenced during the peak of the Grampian Event, resulted in a suite of granite plutons which were intruded throughout late Llanvirn, Llandeilo and Caradoc time.

Volcanism continued in the Iapetus Ocean throughout Arenig to Caradoc time. The products of this activity, sparsely distributed within turbidite-facies sedimentary rocks, were accreted onto the Laurentian margin from Llandeilo time onwards in an imbricate thrust belt that now comprise the Southern Uplands Terrane (Chapter 2).

On the opposite side of Iapetus, the microcontinent of Eastern Avalonia began to drift away from Gondwana at about this time, creating the Rheic Ocean (Figure 1.3b). Areas of continental crust now occupied by the Lake District, central England and Wales were part of Eastern Avalonia, which now began to converge on Laurentia, consuming the Iapetus Ocean. The passive continental margin of the Lakesman Terrane first became unstable during the late Arenig and turbidites deposited during the Llanvirn contain volcanic debris. Widespread volcanism developed for the first time across the Welsh Basin with a mixture of extensional ridge-type magmas and subduction related products (Chapter 6). In north Pembrokeshire a wide range of effusive and pyroclastic rocks were all erupted in a submarine environment during latest Arenig to Llanvirn time. Basic lavas and a basic layered intrusion have tholeiitic affinities, although they do have some subduction-related calc-alkaline characteristics, which are more apparent in the related silicic rocks. In southern Snowdonia, the voluminous Aran Volcanic Group (late Arenig to earliest Caradoc) has transitional tholeiitic to calc-alkaline affinities, while both tholeiitic and calc-alkaline lavas are present in the Llanvirn sequence of the Builth Inlier.

Late Ordovician – Caradoc and Ashgill

The peak of deformation and metamorphism in the Northern Highlands Terrane and the central Grampian Highlands probably occurred during Caradoc time. Crustal melting continued to generate granite plutons and minor intrusions in both terranes (Figure 1.2). It is assumed that subduction continued throughout this period beneath the Laurentian margin, although subduction-related igneous rocks are conspicuously absent, apart from the small alkaline pluton of Glen Dessarry. The geochemistry of this pluton suggests a mantle source, similar to that of the later, more voluminous alkaline magmatism in the Northern Highlands Terrane.

Volcanic rocks, mainly with oceanic island characteristics, occur sporadically within the accreted turbidite sequences of the Southern Uplands Terrane, indicating the continued presence of Iapetus oceanic crust (Figure 1.3c), and calc-alkaline volcanic detritus suggests the presence of island arcs.

More significant igneous activity occurred at this time on the opposite side of the ocean, where active subduction was by now taking place beneath Eastern Avalonia. In the Lakesman Terrane regional uplift was closely followed by intense igneous activity. This resulted, in late Llandeilo and Caradoc time, in the extensive volcanic sequences of the Eycott and Borrowdale volcanic groups and emplacement of parts of the Lake District granitic batholith, early parts of the Carrock Fell Complex and numerous other intrusions (Chapter 4). The Eycott and early Carrock Fell magmas were continental-margin tholeiites, whereas the Borrowdale Volcanic Group originated from typical continental-margin calc-alkaline volcanoes with extensive

Caledonian igneous rocks of Great Britain: an introduction

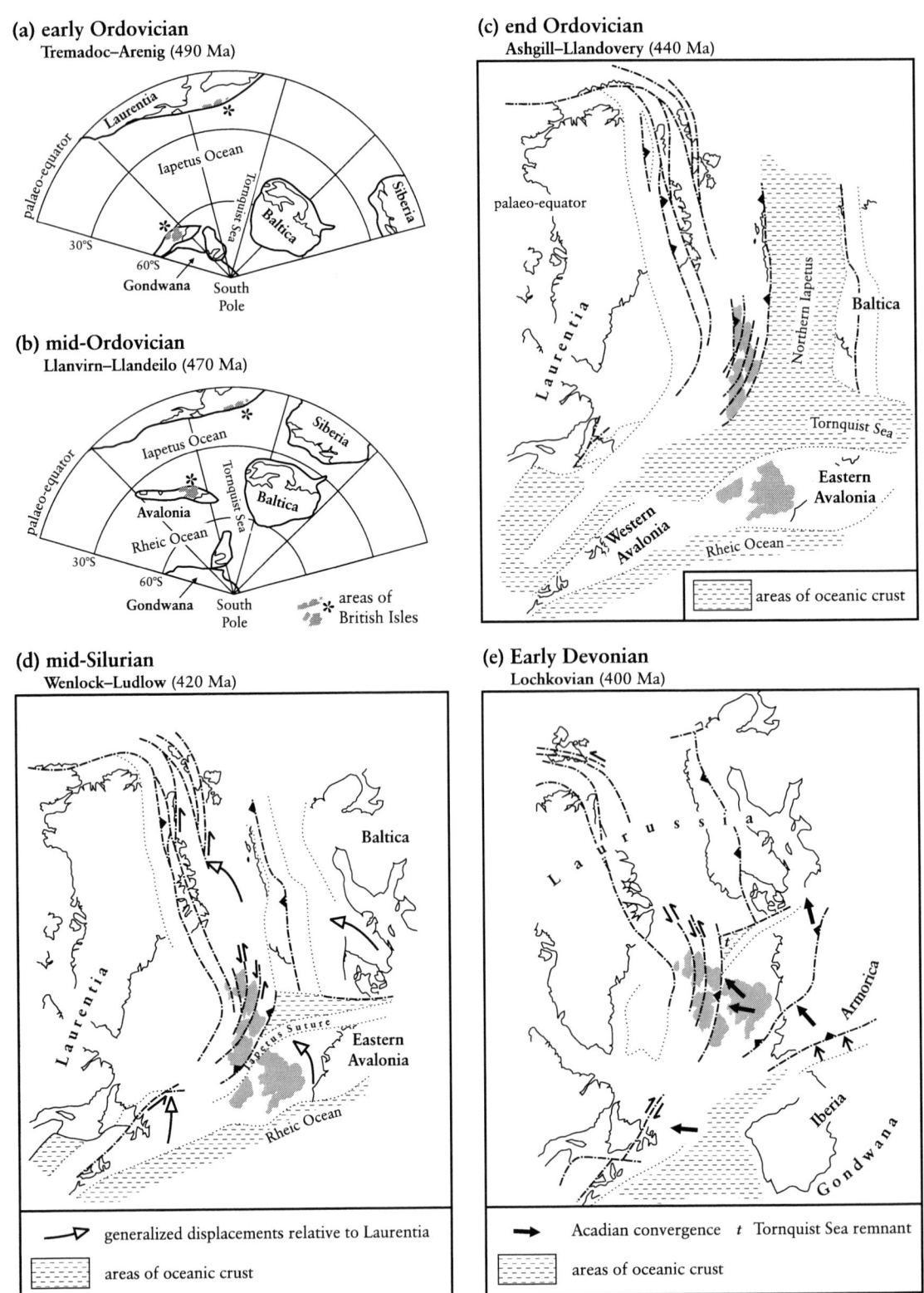

Figure 1.3 Reconstructions of the movements of continents bordering the Iapetus Ocean and Tornquist Sea during the Caledonian Orogeny. (a) and (b) are broad 'views' that are neccessary to encompass the great width of the ocean during the early stages of the orogeny. Note the separation of Avalonia from Gondwana during this time. Adapted from Torsvik *et al.* (1992) by Trench and Torsvik (1992). (c), (d) and (e) show the later stages of the orogeny in more detail, with progresssive narrowing of the oceanic areas, convergence of the continents and ultimate continent–continent collisions and strike-slip re-alignment of terranes. Adapted from Soper *et al.* (1992).

caldera development. The volcanoes were sub-aerial, of low relief, and were probably located in extensional basins, into which there was only one short-lived marine incursion, though subsidence at the end of the volcanic episodes allowed persistent marine conditions to become established. In Ashgill to early Llandovery time localized silicic volcanism occurred in a shallow marine setting and the later, tholeiitic, phases of the Carrock Fell Complex were emplaced.

In the Welsh Basin, the most intense and extensive episode of Caledonian volcanism occurred during Caradoc time, with major centres developed in Snowdonia, on the Llŷn Peninsula and in the Breiddon and Berwyn hills in the Welsh Borderland (Chapter 6). Both Snowdonia and Llŷn were also the focus for a number of major subvolcanic intrusions. Although extensive, the volcanic activity was relatively short lived in mid-Caradoc time, but the ages of intrusions range into the earliest Silurian. As with the earlier activity, there was a mixture of magma types, ranging from mid-ocean-ridge-type tholeiites, through low-K tholeiites to calc-alkaline, suggesting both extensional basin and subduction-related magmatism. In contrast to the Lake District, the volcanism was mostly submarine with some emergent centres, both within the basin in northern Snowdonia and in a shallow marine environment closer to the basin margins in the Welsh Borderland. Much of the volcanicity was controlled by contemporaneous faults, suggesting an extensional tectonic setting and major calderas were developed. This volcanic climax was the last Caledonian igneous activity to occur in North Wales and shows a geochemical transition towards a 'within-plate' environment. It has been suggested therefore that this marks the end of subduction of Iapetus crust beneath Eastern Avalonia.

In central England, calc-alkaline volcanic and plutonic rocks of Caradoc age occur mainly on the NE margin of the Midlands Microcraton (Chapter 5). These form part of a belt of arc-related rocks that extends eastwards into Belgium. Poor exposure means that their tectonic setting is not well understood, but their position on the eastern margin of Eastern Avalonia suggests that they may have been the product of subduction of oceanic lithosphere beneath the microcraton during closure of the Tornquist Sea (Figure 1.3c). It is, however, difficult to accommodate this subduction direction geometrically with the subduction of the Iapetus Ocean, which was occurring at the same time.

Early and mid-Silurian – Llandovery and Wenlock

It was during this period that the Iapetus Ocean finally closed along most of its length. Most current three-plate models show a triangular remnant of oceanic crust around the Laurentia–Baltica–Eastern Avalonia triple junction in mid-Llandovery time, but most authors agree that the ocean had closed completely by the end of Wenlock time, with the continents welded together along the lines of the Iapetus and Tornquist sutures (Figure 1.3d). Following the closure, a foreland basin developed, which propagated across the suture during Wenlock to Ludlow time and established for the first time continuity of sedimentation between the Southern Uplands and Lakesman terranes. There is little evidence for 'southward' active subduction beneath Eastern Avalonia after the end of the Ordovician and, away from the immediate proximity of the suture (i.e. the Lakesman Terrane), there is no igneous activity at all after Wenlock time. However, major igneous events on the Laurentian margin continued until well into the Early Devonian and all have subduction-related characteristics.

Both the Northern Highlands and the Grampian terranes were undergoing tectonic uplift during the early Silurian, but the only significant magmatic event was on the extreme 'north-western' edge of the Northern Highlands. Here, highly alkaline magmas were emplaced as plutons and minor intrusions, overlapping in time with large-scale thrust movements, which resulted in considerable crustal shortening and transposed the edge of the whole Caledonian 'mobile belt' over the foreland of Archaean and unmetamorphosed Proterozoic and Lower Palaeozoic rocks (Chapter 7). The timing and nature of these movements have much in common with the Scandian Event in the Norwegian sector of Baltica, so it may be that the Northern Highlands Terrane was still adjacent to Baltica at this time. Although alkaline magmatism is normally associated with crustal extension, it is now accepted that it can be generated by small degrees of partial melting in deep levels of a subduction zone or within mantle that contains relics of earlier subduction. Such an origin seems appropriate for the Caledonian alkaline rocks, which were emplaced farther from the continental margin (and hence the surface trace of the subduction zone) than any other igneous suite in the orogenic belt.

Evidence for contemporaneous volcanism is lacking in northern Britain. Calc-alkaline volcanic detritus in the Midland Valley and Southern Uplands terranes may have had a fairly local source. However, scattered metabentonites throughout Llandovery, Wenlock and early Ludlow sequences on the site of the former Iapetus Ocean probably represent ash-fall tuffs from very distant sources and it is impossible to speculate on their origin.

On the 'southern' margin of the Welsh Basin in south Pembrokeshire, localized volcanic activity occurred within a shallow marine sequence of Llandovery age (Chapter 6). This magmatism was more alkaline than that of earlier episodes and geochemical evidence suggests a within-plate oceanic mantle source that had been modified by the earlier subduction events. It is possible that this magmatism was generated during crustal extension on the margin of the Rheic Ocean, which was opening on the opposite side of Eastern Avalonia (Figure 1.3c, d), and hence may not be classed strictly as 'Caledonian'. On the southern edge of the Midlands Microcraton near Bristol, two local volcanic episodes of Llandovery and Wenlock age have a more definite calc-alkaline nature, but these too are very close to the inferred margin of the Rheic Ocean. Their tectonic affinities are uncertain, as are those of metabentonites in Silurian inliers of South Wales and the Welsh Borderland.

Late Silurian and Early Devonian – Ludlow to Emsian

By late Silurian time the continents had welded together along the Iapetus Suture to form the new supercontinent of Laurussia (Figure 1.3e). There is geophysical evidence that continental crust of Eastern Avalonia continued to be underthrust (?subducted) beneath Laurentia after the continent–continent collision, but only as far as the Moniaive shear zone in the centre of the Southern Uplands Terrane. However, calc-alkaline magmatism, with subduction-zone characteristics became widespread and voluminous throughout the former Laurentian terranes in the late Ludlow and continued throughout the Early Devonian. (The paradox of apparent subduction-related magmatism continuing after the cessation of active subduction is discussed in more detail in the next section.)

Large, essentially granitic, plutons were emplaced at all crustal levels in the Scottish Highland terranes and in eastern Shetland during early Ludlow to early Lochkovian time (Chapter 8). Their magmas were derived mainly from lower crustal sources, but with a recognizable mantle component. Granitic plutons with similar lower crustal and mantle characteristics were also emplaced at high crustal levels in late Ludlow to early Lochkovian time in the Midland Valley Terrane and in the NW part of the Southern Uplands Terrane. High-level granitic plutons and dyke-swarms were emplaced slightly later (Lochkovian to Pragian) in a broad zone that spans the projected position of the Iapetus Suture in the SE part of the Southern Uplands Terrane (Chapter 8) and in the Lakesman Terrane (Chapter 4). Of these, the youngest are those immediately NW of the suture, in the zone in which the Southern Uplands thrust belt was underthrust by Avalonian crust. These last plutons all have components with distinctive characteristics that have been attributed to derivation from the mid-crustal melting of sedimentary rocks, either Silurian greywackes similar to the country rocks or Late Cambrian to mid-Ordovician turbidites similar to those in the Lake District. Plutons SE of the suture have many geochemical similarities to those to the NW, but the influence of sedimentary rocks in their origin is more equivocal.

The late granitic plutons were emplaced during and immediately following the rapid crustal uplift which produced the Caledonian mountain chain. High-level crustal extension led to local fault-bound intermontane basins in the Grampian Highland and Southern Uplands terranes and the larger basins of the Midland Valley Terrane. Rapid erosion of the newly formed mountains resulted in the deposition of great thicknesses of continental molasse sediments in these basins during the latest Silurian and Early Devonian (the 'Old Red Sandstone'). The last major volcanic episode of the Caledonian Orogeny was coeval with the later stages of pluton emplacement throughout these terranes and resulted in great thicknesses of volcanic rocks within many of the Old Red Sandstone basins (Chapter 9). It is these high-K calc-alkaline volcanic rocks that provide the most compelling evidence for subduction-related magmatism in the late stages of the orogeny, after the closure of the Iapetus Ocean. In addition to their general arc-like geochemical and petrographical characteristics, these rocks exhibit marked spatial variations comparable with those of more recent continental-margin volcanic provinces. A more detailed discussion of possible explanations for this apparently anomalous style of magmatism is given in the next section.

To the south of the Iapetus Suture a major compressional deformation event produced folds, faults and a regional cleavage, which are seen particularly well in the 'slate belts' of the Lake District and Wales. North of the suture, broad folds were generated in the Lower Old Red Sandstone sequences of the Midland Valley Terrane. This event occurred in the Early Devonian with a probable climax in the Emsian and was coeval with the Acadian Orogeny in the Canadian Appalachians. Hence it is referred to in Great Britain as the Acadian Event (Soper *et al.*, 1987). The latest granite plutons of the Lakesman Terrane and the SE of the Southern Uplands Terrane were intruded during this time and the Shap, Skiddaw and Fleet plutons in particular have provided important evidence that emplacement was synchronous with cleavage development and lateral shearing. It has been suggested that the event was caused by further northward compression of Eastern Avalonia against Laurentia by the impact of Armorica, a microcontinent that had been drifting 'northwards' from the margin of Gondwana throughout the Silurian (e.g. Soper, 1986; Soper *et al.*, 1992) (Figure 1.3e).

Mid- and Late Devonian – Eifelian to Famennian

Plutons and dykes in eastern Shetland are of Early Devonian age and are comparable to the latest plutons elsewhere, but all other 'Caledonian' igneous activity in Orkney and Shetland is notably later than elsewhere in the British terranes. Volcanic activity occurred over a short period in the late Eifelian to early Givetian (Figure 1.2). By this time the orogeny had passed into an extensional phase and the 'Old Red Sandstone' sequences of mainly fluvial and lacustrine sediments were being deposited in fault-bound basins, which formed part of the wider Orcadian Basin. Most of the volcanic rocks have subduction-related characteristics but, in contrast to the late Caledonian volcanic rocks elsewhere in the province, many are transitional between calc-alkaline and tholeiitic in their chemistry. This has led to the suggestion that they acquired their characteristics from a much shallower subduction zone than the earlier rocks and hence at one time the area may have been close to a continental margin or trench. However, their tectonic relationship with the rest of the province is unclear. The lavas of the Isle of Hoy in the Orkneys are alkaline in nature and hence are more in keeping with the continental extensional setting; they may mark the start of the intra-plate alkaline volcanism that became widespread in northern Britain during the Early Carboniferous.

Plutons to the west of the Walls Boundary Fault in Shetland have yielded Frasnian to Famennian radiometric dates (Chapter 9). These probably represent minimum ages, but one pluton postdates early Givetian volcanic rocks. It is closely associated in space and time with a late phase of compressive deformation and metamorphism that post-dates the main extension responsible for the Orcadian Basin and is the last phase of Caledonian folding in Britain.

Given the unusual timing of the magmatic events and the unique structural sequence, it is tempting to suggest that Shetland must have originated in a separate terrane, well separated from the Northern and Grampian Highlands. The plutons are affected by movements on the Walls Boundary Fault (a possible continuation of the Great Glen Fault), but it is difficult to accommodate much lateral movement after the Mid-Devonian, by which time the Orcadian Basin covered Shetland, Orkney and parts of both the Northern Highlands and Grampian terranes, with a coherent pattern of sedimentation and common elements of stratigraphy throughout.

ORIGIN OF THE LATE CALEDONIAN MAGMAS

D. Stephenson and A.J. Highton

The source of the Caledonian magmas during the earlier parts of the orogeny, while the Iapetus Ocean was still in existence, is generally well constrained. Geochemical 'fingerprinting' enables suites of igneous rocks to be assigned to sources in various tectonic settings, although caution must be applied to most early Caledonian volcanic suites, which are commonly affected by varying degrees of of alteration and

Caledonian igneous rocks of Great Britain: an introduction

low-grade metamorphism. Most of the earlier suites have been attributed to mantle melting beneath spreading centres and within-plate oceanic islands, or to melting above island-arc subduction zones. All of these processes could have taken place either within the main ocean or in extensional marginal basins. The effects of subduction of oceanic crust beneath the continental margin of Avalonia can also be identified, by the presence of voluminous calc-alkaline igneous rocks. On the Laurentian margin, however, there is little evidence for subcontinental subduction during the early part of the orogeny, when the igneous activity was dominated by melting of mid- to upper-crustal late Proterozoic metasedimentary rocks.

Magmatism had more or less ceased in Eastern Avalonia at the end of the Ordovician, which implies that subduction had ceased on this margin of the Iapetus Ocean before final closure. However, magmatism on the Laurentian continental margin continued well after ocean closure and continent–continent suturing. Unlike the earlier magmatism, this later activity was markedly calc-alkaline; it has many subduction-related features and even shows spatial variations in geochemical parameters that are usually related to increasing depths of a subduction zone beneath a continental margin. The most compelling evidence for a subduction zone origin for the magmas is found in the geochemistry of the late Caledonian volcanic rocks (Groome and Hall, 1974; Thirlwall, 1981a, 1982; see Chapter 9: Introduction), although certain geochemical features of the plutonic rocks have led to similar conclusions (Stephens and Halliday, 1984; Thompson and Fowler, 1986; see Chapter 8: Introduction). Although the subduction model is attractive, in that it explains and unites many of the observed features of the late Caledonian igneous activity, it is beset by serious problems related to the overall timing of tectonic events and the distribution of some of the magmatism (Soper, 1986).

It is now generally accepted from sedimentological and structural evidence (e.g. Watson, 1984; Stone et al., 1987; Soper et al., 1992), palaeomagnetism (Trench and Torsvik, 1992) and geochemical studies (Stone et al., 1993), that the Iapetus Ocean had closed by the end of Wenlock time, that is before the onset of the late Caledonian volcanicity and indeed prior to the emplacement of most late Caledonian granitic plutons. There is no evidence for subsequent deformation in the Scottish Caledonides until the end of the Early Devonian, even though underthrusting of previously accreted oceanic sediments and Avalonian continental crust beneath the Laurentian margin must have continued for a while after the closure. Geophysical evidence and the sediment-derived isotope signatures of the plutons in the SE part of the Southern Uplands suggest that this underthrust material extends as far NW as the Moniaive shear zone, which overlies a geophysically defined basement discontinuity (Stone et al., 1987, 1997; Kimbell and Stone, 1995).

The continent–continent collision, marginal underthrusting and consequent crustal thickening probably resulted in rapid uplift of the Laurentian margin. The magmatism responsible for both the late Caledonian plutons and the volcanicity occurred during this continental uplift. The geochemistry of most of the plutons suggests that they are derived dominantly from the melting of a lower crustal, igneous source. In particular, their isotope characteristics (Halliday, 1984; Stephens and Halliday, 1984; Thirlwall, 1989) and the presence of inherited zircons from older crustal material, point to significant crustal recycling (O'Nions et al., 1983; Frost and O'Nions, 1985; Harmon et al., 1984). However, most authors are agreed that there is also a significant background input of magma from the subcontinental lithospheric mantle (Harmon et al., 1984; Stephens and Halliday, 1984; Tarney and Jones, 1994). This mantle-derived material is seen as mafic enclaves and appinitic rocks associated with many of the plutons and, possibly in its least modified form, in the calc-alkaline lamprophyres of the dyke swarms and in the near-contemporaneous lavas. These 'shoshonitic' basic to intermediate rocks, rich in potassium and other 'incompatible elements', are consistent with the melting of a hydrated K-rich mantle (Holden et al., 1987; Canning et al., 1996; Fowler and Henney, 1996), modified by mixing, both in the source region and on emplacement, with melts derived from the lower continental crust or subducted oceanic crust (Thirlwall, 1982, 1983b, 1986).

In an attempt to explain the apparently subduction-related geochemical signatures of the late Caledonian magmas, it has been suggested that the primary magmas may have originated by partial melting, during ultrametamorphism of a stationary slab of subducted oceanic crust, beneath the post-collision craton (Thirlwall,

Site selection criteria

1981a; Fitton *et al.*, 1982). Fitton *et al.* (1982) cited the Cascades of California as a modern example of active volcanism at a continental margin, possibly initiated by continued volatile loss from a now stationary slab. Watson (1984) favoured a two-stage model in which fluids expelled from a descending slab of oceanic crust rose into and metasomatically altered the overlying mantle wedge during active subduction. Partial melting of this modified mantle in response to lateral shearing and high-level movement on block faults after the end of subduction then gave rise to volcanic and plutonic rocks with subduction-related characteristics (Watson, 1984; Hutton and Reavy, 1992). A study of deep-seismic reflection profiles across the Iapetus Suture led Freeman *et al.* (1988) to suggest that the subcontinental mantle beneath the Avalonian crust became detached after continental collision and continued to be subducted, even though subduction of the continental crust had ceased. Zhou (1985) cited modern examples of continent–continent collision zones in Turkey, Iran and Tibet, where post-collision calc-alkaline magmatism is voluminous and Seber *et al.* (1996) presented geophysical evidence for the presence of detached slabs of subcontinental lithosphere, consequent upon collision and thickening of continental crust between Spain and Morocco. It therefore seems possible that the detachment and continued subduction or subsidence of a slab of metasomatized mantle and former oceanic crust may release residual melts or even generate new melts for some time after collision.

However, a further problem with the subduction model is that the volcanic rocks in the south-eastern Midland Valley are relatively close to the projected line of the Iapetus Suture. If the suture approximates to the original position of the trench, the inferred arc-trench gap of *c.* 60 km is substantially smaller than modern gaps (around 140 km minimum). This problem is only partly resolved by invoking significant crustal shortening as a result of deformation and underplating, or strike-slip movement during the later stages of the orogeny, juxtaposing terranes which were originally much more widely separated. The problem is even more acute in the Southern Uplands, where the significantly younger lavas and granite of the Cheviot Hills, for example, lie almost on the projected position of the suture. Analyses of volcanic rocks from the Cheviot Hills and St Abbs sequences do not fit into the spatial patterns seen NW of the Southern Upland Fault. Although these lavas, together with many of the dykes in the area (Rock *et al.*, 1986b), are undoubtedly calc-alkaline, their chemistries suggest derivation from a far deeper subduction-zone source than is possible given their position close to the suture. Therefore, any apparent subduction-related component must either derive from the mantle above a subsiding detached slab of lithosphere beneath the continental suture (cf. Zhou, 1985; Seber *et al.*, 1996), or be related to some other, post-Iapetus subduction zone yet to be identified (Soper, 1986); subduction of the Rheic Ocean beneath the southern margin of Eastern Avalonia is one possibility.

Table 1.1a Ordovician Igneous Rocks Block: networks and GCR site selection criteria

Volcanic Rocks and Ophiolites of Scotland Network, Chapter 2

Site name	GCR selection criteria
The Punds to Wick of Hagdale	Representative of lower part of Shetland Ophiolite, in particular the controversial intrusive relationship of dunite to mantle components. Internationally important in that it offers a rare section across the petrological Moho.
Skeo Taing to Clugan	Representative of lower part of Shetland Ophiolite, providing evidence for intrusive rather than layered cumulate relationships. Internationally important in that it offers a rare section across the geophysical Moho.
Qui Ness to Pund Stacks	Representative of upper part of Shetland Ophiolite, and illustrates relationships between dykes and underlying gabbro. Exceptional exposure of sheeted dyke complex, the clearest and most extensive in Britain.
Ham Ness	Representative of major structural relationships in Shetland Ophiolite with ultramafic rocks, gabbro and sheeted dykes brought into close proximity. Exceptional demonstration of emplacement of ultramafic nappe over sheeted dykes.

Caledonian igneous rocks of Great Britain: an introduction

Tressa Ness to Colbinstoft	Exceptional section in Shetland Ophiolite through base of ophiolitic nappe, illustrating tectonics of emplacement and enigmatic metasomatic relationships.
Virva	Representative of basal structures in Shetland Ophiolite with exceptional evidence pertaining to unusual intrusive relationships. Internationally important in terms of the tectonic emplacement mechanism of ophiolite complexes.
Garron Point to Slug Head	Representative of part of Highland Border Complex, containing a variety of ophiolitic igneous lithologies.
Balmaha and Arrochymore Point	Representative of part of the Highland Border Complex, providing evidence of the relationship of serpentinite to overlying clastic rocks.
North Glen Sannox	Exceptional section through pillow lavas of the Highland Border Complex, containing evidence for the tectonic relationship with adjacent Dalradian rocks.
Byne Hill	Representative of an important component of the Ballantrae Ophiolite. Exceptional illustration of a zoned gabbro–leucotonalite body intruded into ophiolitic serpentinite.
Slockenray Coast	Representative of several components of the Ballantrae Ophiolite. Exceptional features of upper part include ophiolitic mélange, mixing of coeval lava flows of different compositions and a lava-front delta. Lower part is an exceptional gabbro pegmatite contained within serpentinite cut by pyroxenite veins.
Knocklaugh	Representative of basal zone of Ballantrae Ophiolite. Internationally important section allowing interpretation of the metamorphic dynamothermal aureole at the base of an ophiolite in terms of its obduction while still hot.
Millenderdale	Unique representative within the Ballantrae Ophiolite of multiple dyke intrusion into gabbro. Exceptional development of unusual metamorphic and textural relationships.
Knockormal	Exceptional occurrences of blueschist and garnet-clinopyroxenite within the Ballantrae Ophiolite. Internationally important historically as a possible zone of very high pressure metamorphism.
Games Loup	Representative of interveining between ultramafic components of the Ballantrae Ophiolite and juxtaposition of ultramafic rock and spilitic pillow lavas by faulting.
Balcreuchan Port to Port Vad	Representative of Balcreuchan Group, the upper part of the Ballantrae Ophiolite. Exceptional example of structural imbrication of varied lava sequence, and the only unambiguous British example of boninitic lavas.
Bennane Lea	Representative of highest exposed part of Ballantrae Ophiolite, faulted against ultramafic rock. Exceptional illustration of relationships between deep-water chert, volcaniclastic sandstone, mass-flow conglomerate and submarine lava.
Sgavoch Rock	Representative of the earliest accreted component of the Southern Uplands thrust belt. Exceptional display of pillow lavas and associated volcanic features; arguably the finest in Britain.

Intrusions of the NE Grampian Highlands of Scotland Network, Chapter 3

Site name	GCR selection criteria
Hill of Barra	Representative of olivine-rich cumulates from lower part of Lower Zone in Insch intrusion.
Bin Quarry	Representative of troctolitic and gabbroic cumulates from upper part of Lower Zone in Huntly intrusion. Exceptional for small-scale layered structures.
Pitscurry and Legatesden quarries	Representative of cumulates from Middle Zone of Insch intrusion associated granular gabbros and later pegmatite sheets
Hill of Johnston	Representative of late-stage differentiates (ferromonzodiorites and quartz-syenites) of the Insch intrusion. Exceptional mineralogical and geochemical features.

Site selection criteria

Hill of Craigdearg	Representative example from Boganclogh of the quartz-biotite norites found in many of the 'Younger Basic' intrusions. Exceptionally fresh and Mg-rich ultramafic rocks, unlike the Lower Zone cumulates.
Balmedie Quarry	Exceptional examples in the Belhelvie intrusion of layered gabbros, sheared and crushed by post-magmatic tectonic events.
Towie Wood	Exceptional exposures in the Haddo House–Arnage intrusion of xenolithic complex and associated norites developed near the roof of a 'Younger Basic' intrusion.
Craig Hall	Representative example from Kennethmont granite–diorite complex of variety of rocks found in granitic intrusions broadly coeval with 'Younger basic' intrusions.

Lake District Network, Chapter 4

Site name	GCR selection criteria
Eycott Hill	Representative of Eycott Volcanic Group. Exceptional locality for 'Eycott-type' (orthopyroxene-plagioclase megaphyric) basaltic andesite.
Falcon Crags	Representative of pre-caldera volcanism in Borrowdale Volcanic Group. Internationally important example of dissected plateau-andesite province.
Ray Crag and Crinkle Crags	Internationally important for understanding 'piecemeal' caldera collapse. Representative type areas in Borrowdale Volcanic Group of stratified Scafell Caldera succession. Exceptional example of structures within an exhumed hydrovolcanic caldera and of welded ignimbrites.
Sour Milk Gill	Internationally important exposures of large-magnitude phreatoplinian ash-fall tuff, associated with development of 'piecemeal' caldera collapse.
Rosthwaite Fell	Exceptional illustration of variations in magmatic and hydromagmatic volcanism in internationally significant Scafell Caldera. Exceptional example of post-caldera lava, its vent and feeder.
Langdale Pikes	Exceptional examples of volcanotectonic faults. Internationally important example of caldera-lake sedimentary sequence and of subaqueous lag breccia associated with ignimbrite.
Side Pike	Exceptional exposures illustrating distinction between rocks of pyroclastic fall, flow and surge origin, and for rocks formed through magmatic, phreatomagmatic and phreatic processes. Representative of volcanic megabreccia within the internationally significant Scafell Caldera.
Coniston	Representative of post-Scafell Caldera volcanism and sedimentation in Borrowdale Volcanic Group.
Pets Quarry	Exceptional example of features of magma intrusion into wet sediment.
Stockdale Beck, Longsleddale	Representative of late Ordovician, post-Borrowdale Volcanic Group, volcanism in the north of England.
Bramcrag Quarry	Representative of Threlkeld microgranite.
Bowness Knott	Representative of Ennerdale granite.
Beckfoot Quarry	Representative of Eskdale granite.
Waberthwaite Quarry	Representative of Eskdale granodiorite.
Carrock Fell	Representative of Carrock Fell Complex. Internationally important for historical contributions to understanding of crystallization mechanisms.
Haweswater	Representative of Haweswater basic intrusions.

Central England Network, Chapter 5

Site name	GCR selection criteria
Croft Hill	Representative of South Leicestershire diorites.
Buddon Hill	Representative of Mountsorrel complex.
Griff Hollow	Representative of Midlands Minor Intrusive Suite.

Caledonian igneous rocks of Great Britain: an introduction

Wales Network, Chapter 6

Site name	GCR selection criteria
Rhobell Fawr	Representative of Rhobell Volcanic Group (Tremadoc), the earliest manifestation of Caledonian igneous activity in Britain south of the Iapetus Suture.
Pen Caer	Representative of Fishguard Volcanic Group (Llanvirn). Exceptional locality for products of major submarine basic–silicic volcanic complex. Internationally important for occurrence of silicic lava tubes.
Aber Mawr to Porth Lleuog	Internationally important for presence of silicic welded submarine ash-flow and ash-fall unit (Llanvirn), the first to be recognized worldwide.
Castell Coch to Trwyncastell	Representative of the youngest (Llanvirn) volcanic episode in north Pembrokeshire.
St David's Head	Exceptional composite intrusion showing evidence of multiple magma injection and in-situ fractional crystallization.
Cadair Idris	Representative of Aran Volcanic Group (Arenig–Caradoc), the most important volcanic episode in southern Snowdonia.
Pared y Cefn Hir	Representative of Aran Volcanic Group, with best exposed sequence of volcanic rocks of Arenig to Llanvirn age in North Wales.
Carneddau and Llanelwedd	Representative of Builth Volcanic Group (Llanvirn), the most important Ordovician volcanic episode in the Welsh Borderland.
Braich tu du	Representative of 1st Eruptive Cycle (Caradoc; Soudleyan) of Snowdon Centre.
Llyn Dulyn	Exceptional exposures of silicic ash-flow tuffs emplaced in subaerial environment, allowing palaeogeographical reconstruction of part of 1st Eruptive Cycle of Snowdon Centre. Complements Capel Curig.
Capel Curig	Exceptional exposures of silicic ash-flow tuffs emplaced in submarine environment, allowing palaeogeographical reconstruction of part of 1st Eruptive Cycle of Snowdon Centre. Complements Llyn Dulyn. Internationally important historically, for first recognition of welding in submarine ash-flow tuffs.
Craig y Garn	Representative site illustrating initiation of 2nd Eruptive Cycle (Caradoc; Soudleyan–Longvillian) of Snowdon Centre. Exceptional preservation of one of the thickest and most complete intra-caldera sequence of ash-flow tuffs in British Caledonides.
Moel Hebog to Moel yr Ogof	Representative of ash-flow tuffs of subaerial outflow facies from caldera at Craig y Garn GCR site, belonging to 2nd Eruptive Cycle of Snowdon Centre. Exceptional preservation of fault and subsidence related brecciation, sliding and widespread disruption of previously deposited ash-flow tuffs.
Yr Arddu	Representative of earliest activity from Snowdon Centre; ash-flow tuffs erupted from submarine fissure.
Snowdon Massif	Representative of main phases of intrusive and extrusive activity linked to evolution of major submarine caldera, of 2nd Eruptive Cycle of Snowdon Centre. Exceptional demonstration of complex inter-relationships, through time, between alternating basic–acid magmatism, changing styles of volcanic activity and effect on sedimentation.
Cwm Idwal	Exceptional illustration of thinned sequence representing outflow facies of major submarine caldera, linked to 2nd Eruptive Cycle of Snowdon Centre. Complements Snowdon Massif.
Curig Hill	Representative of lowest unit of final phase of magmatism related to 2nd Eruptive Cycle of Snowdon Centre.
Sarnau	Representative of middle and upper units of final phase of magmatism related to 2nd Eruptive Cycle of Snowdon Centre.
Ffestiniog Granite Quarry	Representative of sub-volcanic granitic intrusion linked to 2nd Eruptive Cycle of Snowdon Centre.
Pandy	Representative of Ordovician (Caradoc) igneous activity in the northern Welsh Borderland.
Trwyn-y-Gorlech to Yr Eifl	Representative of Garnfor multiple intrusion, a sub-volcanic intrusion related to the Upper Lodge Volcanic Group (Caradoc).
Penrhyn Bodeilas	Representative of Penrhyn Bodeilas Granodiorite, a sub-volcanic intrusion linked to Upper Lodge Volcanic Group (Caradoc).

Site selection criteria

Moelypenmaen	Representative of the Llanbedrog Volcanic Group (Caradoc).
Llanbedrog	Representative of high-level silicic intrusion associated with Llanbedrog Volcanic Group (Caradoc).
Foel Gron	Representative of most evolved member of suite of peralkaline intrusions associated with Llanbedrog Volcanic Group (Caradoc).
Nanhoron Quarry	Representative of least evolved member of suite of peralkaline intrusions associated with Llanbedrog Volcanic Group (Caradoc), preserving rare contact with lower Ordovician sedimentary rocks.
Mynydd Penarfynydd	Exceptional coastal exposures through layered basic sill, ranging from pictites through gabbros to intermediate compositions.

Table 1.1b Silurian and Devonian Plutonic Rocks Block: networks and GCR site selection criteria.

Alkaline Intrusions of the NW Highlands of Scotland Network, Chapter 7

Site name	GCR selection criteria
Loch Borralan Intrusion	Representative of the intrusion. Exceptional as only British examples of several rock types, including nepheline-syenite, pseudoleucite-syenite and carbonatite. Radiometric age and structural relationships important for timing of movements in Moine Thrust Zone. Internationally important for some of the most extreme potassium-rich igneous rocks found anywhere on Earth. Historically of great importance in development of hypotheses for evolution of igneous rocks.
Loch Ailsh Intrusion	Representative of the intrusion. Radiometric age and structural relationships important for timing of movements in Moine Thrust Zone. Internationally important as type-locality of alkali-feldspar-syenite 'perthosite', and because of unusually sodium-rich character of syenites.
Loch Loyal Syenite Complex	Representative of the complex and the only extensive British intrusion composed of peralkaline quartz-syenite (nordmarkite).
Glen Oykel south	Representative of 'grorudite' (peralkaline rhyolite) suite of dykes which are emplaced only in Ben More Nappe. Important structural relationship of dyke cutting Loch Ailsh intrusion establishes that the latter was emplaced prior to movements on Ben More Thrust.
Creag na h-Innse Ruaidhe	Representative of 'grorudite' suite of dykes in one of the outliers (klippen) of the Ben More Nappe, an important structural relationship.
Beinn Garbh	Representative and exceptional exposures of sills of 'Canisp Porphyry' (a striking feldspar-phyric quartz-microsyenite), the largest development of Caledonian magmatism in the Foreland.
The Lairds Pool, Lochinver	Representative of 'Canisp Porphyry' as a dyke cutting Lewisian basement, which indicates the western extent of this suite in the Foreland.
Cnoc an Leathaid Bhuidhe	Representative of Canisp Porphyry as a sill, close to, but not above the Sole Thrust, confirming the restriction of the suite to the Foreland.
Cnoc an Droighinn	Representatives of 'Hornblende Porphyrite' suite of sills in a setting of great structural complexity, in which the sills are repeated by imbrication.
Luban Croma	Representative of sills of 'Hornblende Porphyrite' suite, and others, illustrating range and variation of pre-deformational minor intrusive rocks in Assynt.
Allt nan Uamh	Representative of unaltered hornblende-rich lamprophyre (vogesite), an otherwise rare rock type which occurs widely in the Moine Thrust Zone of Assynt and Ullapool.
Glen Oykel north	Exceptional locality at which an enigmatic diatreme of brecciated dolomitic limestone in a fine-carbonate matrix is associated with a vogesite sill. May represent only example of transport by gas in Caledonian alkaline suite.
Allt na Cailliche	Representative of suite of quartz-syenite (nordmarkite) sills which occur only close to the Moine Thrust; the only igneous suite in Assynt whose emplacement was localized by the thrusts themselves.
Camas Eilean Ghlais	Representative of nepheline-syenite ('ledmorite') dykes, emplaced in the Foreland yet clearly trending towards the Loch Borralan Intrusion, with

Caledonian igneous rocks of Great Britain: an introduction

	implications for timing of thrust movements. Internationally important historically in demonstrating that alkaline magmatism did not involve reactions with limestone.
An Fharaid Mhór	Representative example of nepheline syenite ('ledmorite') dyke in the Foreland, trending towards the Loch Borralan intrusion.

Granitic Intrusions of Scotland Network, Chapter 8

Site name	GCR selection criteria
Loch Airighe Bheg	Representative of pluton within Rogart complex, Argyll and N. Highlands Suite. Exceptional examples of appinitic xenoliths exhibiting hybridization with host quartz-monzodiorite.
Glen More	Representative of Ratagain pluton, transitional alkaline member of Argyll and N. Highlands Suite. Exceptional for wide range of compositions, range of mantle and crustal sources, and extreme enrichment in Sr and Ba.
Loch Sunart	Representative of Strontian pluton, Argyll and N. Highlands Suite. Exceptional evidence for basic magmatism coeval with granodiorite emplacement. Internationally important for relationship to Great Glen Fault and deformation during emplacement and crystallization.
Cnoc Mor to Rubh' Ardalanish	Representative of eastern part of Ross of Mull pluton, Argyll and N. Highlands Suite, which shows reverse concentric zoning. Exceptional features of passive emplacement with stoping and assimilation of country rock.
Knockvologan to Eilean a' Chalmain	Representative of central part of Ross of Mull pluton. Exceptional examples of mafic enclaves, hybrid granitic rocks and internationally important example of 'ghost' stratigraphy in metasedimentary xenoliths.
Ben Nevis and Allt a'Mhuilinn (Chapter 9)	Representative of Ben Nevis pluton, Argyll and N. Highlands Suite. Internationaly important historically, for development of cauldron subsidence theory.
Bonawe to Cadderlie Burn	Representative of Etive pluton, Argyll and N. Highlands Suite and dyke swarm. Internationally important example of upper crustal, multiple pulse intrusion by a combination of block subsidence and diapirism within a shear-zone.
Cruachan Reservoir	Representative of marginal facies and hornfelsed envelope of Etive pluton, dyke swarm and screen of Lorn Plateau volcanic rocks.
Red Craig	Representative of Glen Doll diorite, South of Scotland Suite. Exceptional examples of assimilation of metasedimentary xenoliths with high-grade hornfelsing, local melting and hybridization.
Forest Lodge	Internationally important historically, as the site in Glen Tilt where Hutton first demonstrated the magmatic origin of granite in 1785.
Funtullich	Representative of Comrie pluton, South of Scotland Suite, a good example of a normally zoned, diorite to granite pluton. Exceptional internal contacts.
Craig More	Representative of Comrie pluton and aureole. Exceptional section across aureole, which has historical international importance.
Garabal Hill to Lochan Strath Dubh-uisge	Representative of Garabal Hill–Glen Fyne complex, South of Scotland Suite. Exceptional orderly sequence of intrusion from basic to acid. Internationally important historically, for studies of fractional crystallization.
Loch Dee	Representative of Loch Doon pluton, South of Scotland Suite, a fine example of a normally zoned pluton. Internationally important for studies of origin of compositional variation.
Clatteringshaws Dam Quarry	Representative of outer part of Fleet pluton, Galloway Suite, derived from melting of underthrust Lower Palaeozoic sedimentary rocks similar to those of Lake District.
Lea Larks	Representative of more evolved inner part of Fleet pluton, one of the most evolved late Caledonian granites. Internationally important for studies of extreme fractionation.
Lotus quarries to Drungans Burn	Representative of complete zonation of Criffel pluton. Internationally important for unusual transition from outer, mantle-derived rocks to inner granites derived from melting of sedimentary rocks.

Site selection criteria

Millour and Airdrie Hill	Representative of outer, mantle-derived part of Criffel pluton, Galloway Suite. Exceptional for mafic enclaves and foliation associated with emplacement. Internationally important for studies of diapirism.
Ardsheal Hill and peninsula	Representative and type area of Appinite Suite. Exceptional for range of ultramafic to acid compositions and for breccia-pipes. Internationally important for study of open system feeders to surface volcanism.
Kentallen	Representative example of appinitic intrusion. Exceptional Mg- and K-rich lithology, well-exposed contacts and complex age relationships.

Northern England Network, Chapter 4

Site name	GCR selection criteria
Grainsgill	Exceptional relationships of granite intrusion, greisen formation and mineralization in Skiddaw Granite.
Shap Fell Crags	Representative of Shap granite. Exceptional evidence for timing of Acadian deformation. Internationally important for study of K-feldspar megacrysts.

Table 1.1c Silurian and Devonian Volcanic Rocks Block: networks and GCR site selection criteria.

Scotland Network, Chapter 9

Site name	GCR selection criteria
South Kerrera	Representative of Lorn Plateau volcanic succession. Exceptional examples of subaerial lava features and interaction of magma with wet sediment.
Ben Nevis and Allt a'Mhuilinn	Representative of Ben Nevis volcanic succession. Exceptional intrusive tuffs. Internationally important as example of exhumed roots of caldera, and historically for development of cauldron subsidence theory.
Bidean nam Bian	Representative of entire succession of Glencoe volcanic rocks. Exceptional examples of ignimbrites, intra-caldera alluvial sediments and of sill complex intruded into unconsolidated sediments. Internationally important historically for development of cauldron subsidence theory and currently for evidence of graben-controlled volcanism.
Stob Dearg and Cam Ghleann	Representative of succession in eastern part of Glencoe caldera, including basal sedimentary rocks. Exceptional rhyolites, ignimbrites and intra-caldera sediments. Possible international importance for radiometric dating in conjunction with palaeontology close to Silurian/Devonian boundary.
Buachaille Etive Beag	Representative of Glencoe Ignimbrites. Exceptional exposures of pyroclastic flows separated by erosion surfaces and alluvial sediments.
Stob Mhic Mhartuin	Representative of Glencoe ring fracture and ring intrusion. Exceptional exposures of crush-rocks and intrusive tuff.
Loch Achtriochtan	Representative of Dalradian succession below Glencoe volcanic rocks. Exceptional topographic expression of ring fracture and ring intrusion.
Crawton Bay	Representative of Crawton Volcanic Formation.
Scurdie Ness to Usan Harbour	Representative of 'Ferryden lavas' and 'Usan lavas', comprising lower part of Montrose Volcanic Formation.
Black Rock to East Comb	Representative of 'Ethie lavas', comprising upper part of Montrose Volcanic Formation.
Balmerino to Wormit	Representative of eastern succession of Ochil Volcanic Formation. Possible international importance for radiometric dating in conjunction with palaeontology close to Silurian/Devonian boundary.
Sheriffmuir Road to Menstrie Burn	Representative of western succession of Ochil Volcanic Formation. Exceptional topographic expression of Ochil fault-scarp.
Craig Rossie	Representative of rare acid flow in upper part of Ochil Volcanic Formation.
Tillicoultry	Representative of diorite stocks, intruded into Ochil Volcanic Formation, sur-

Caledonian igneous rocks of Great Britain: an introduction

	rounded by thermal aureole, and cut by radial dyke swarm. Exceptional examples of diffuse contacts, due to metasomatism and contamination, with 'ghost' features inherited from country rock.
Port Schuchan to Dunure Castle	Representative of Carrick Hills volcanic succession. Exceptional features resulting from interaction of magma with wet sediment are of international importance.
Culzean Harbour	Representative of inlier of Carrick Hills volcanic succession. Exceptional features resulting from interaction of magma with wet sediment are of international importance.
Turnberry Lighthouse to Port Murray	Representative of most southerly inlier of Carrick Hills volcanic succession. Exceptional features resulting from interaction of magma with wet sediment are of international importance.
Pettico Wick to St Abb's Harbour	Representative of volcanic rocks in the SE Southern Uplands. Exceptional vent agglomerates, block lavas, flow tops and interflow high-energy volcaniclastic sediments.
Shoulder O'Craig	Representative of vent and minor intrusions in SW Southern Uplands.
Eshaness Coast	Representative of late Eifelian, Eshaness volcanic succession, NW Shetland. Exceptional exposures of ignimbrite, hydromagmatic tuffs, pyroclastic breccias, flow tops and magma–wet sediment interaction, all in spectacular coastal geomorphology.
Ness of Clousta to the Brigs	Representative of Givetian, Clousta volcanic rocks, Walls, Shetland, including phreatomagmatic deposits.
Point of Ayre	Representative of Givetian, Deerness Volcanic Member, mainland Orkney.
Too of the Head	Representative of Givetian, Hoy Volcanic Formation, Isle of Hoy, Orkney, unusual for alkaline character. Potential international importance as radiometric time marker in Mid-Devonian.

Wales Network, Chapter 6

Site name	GCR selection criteria
Skomer Island	Representative of most complete section through Skomer Volcanic Group (Llandovery), the most significant expression of late Caledonian volcanism in southern Britain.
Deer Park	Representative of Skomer Volcanic Group, providing critical biostratigraphical age constraints.

Chapter 2

Early Ordovician volcanic rocks and associated ophiolitic assemblages of Scotland

Introduction

INTRODUCTION

P. Stone and D. Flinn

At the beginning of the Palaeozoic Era the continental masses of Laurentia, Avalonia and Baltica were separated by large oceans. The Iapetus Ocean separated Laurentia from Eastern Avalonia and Baltica while the two latter continents were divided by the Tornquist Sea. Volcanic activity occurred within and at the margins of these oceans. The rocks described in this chapter all originated as varieties of Iapetus oceanic crust and upper mantle which were caught up in the orogenic fold belt produced as the ocean closed. This is the Caledonian–Appalachian Orogen which, prior to the Mesozoic and later opening of the Atlantic Ocean, formed a sinuous deformation belt extending from what is now northern Scandinavia, through Britain and Ireland, and on into maritime Canada, the eastern seaboard of the USA and possibly beyond. A pre-Atlantic continental reconstruction is shown in Figure 2.1.

The Iapetus Ocean probably reached its maximum width during the Late Cambrian or early Ordovician. Closure of the ocean then followed during the Ordovician and Silurian with continental suturing complete by the Early Devonian. Along the length of the orogen are preserved vestiges of the original ocean in the form of ophiolite complexes, assemblages variously comprising spilitic lava, sheeted mafic dykes, gabbros and ultramafic plutonic rocks; the ophiolite concept is discussed in more detail below. The distribution of the principal ophiolite complexes within the orogen is shown in Figure 2.1. Those now seen in Britain and Canada were obducted onto the continental margin of Laurentia, the Norwegian examples were obducted onto the margin of Baltica on the other side of the closing ocean. Despite their wide geographical range there is a surprising uniformity of age with most apparently being generated in the Late Cambrian or early Ordovician and then obducted shortly after. Of the Scottish ophiolites, the early Ordovician, Ballantrae Complex has much in common with those now exposed in Newfoundland, whereas

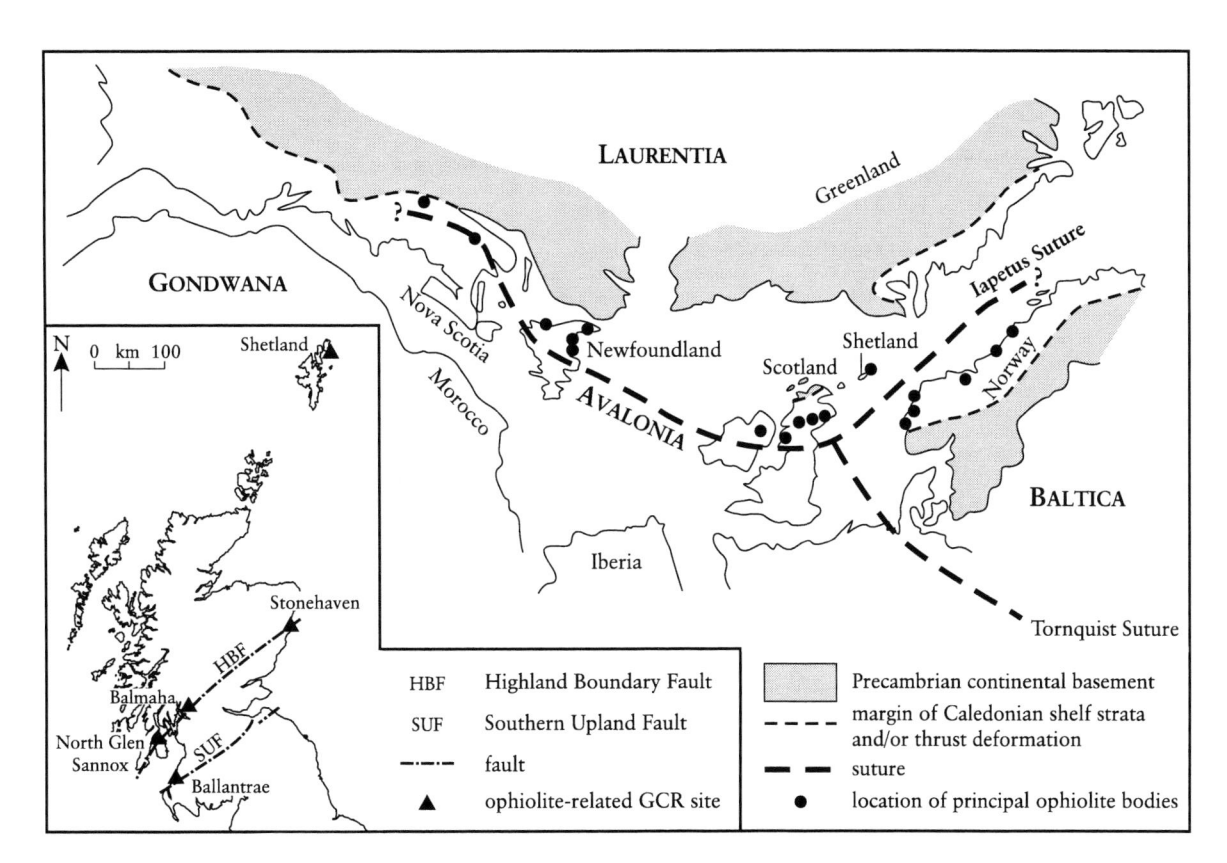

Figure 2.1 A pre-Atlantic reconstruction of the Caledonian Orogen showing the positions of the principal ophiolite complexes. A location map for the Scottish examples is shown inset.

the Shetland ophiolite is probably a little older with similar generation and obduction ages to some of the Norwegian examples. The broad coincidence of ages probably reflects the initiation of ocean closure when intra-oceanic subduction zones were formed adjacent to both continental margins. The volcanic products of these zones would have had greater obduction potential than the oceanic crust that was subsequently consumed by subduction at the continental margin. At that time the margins of the Iapetus Ocean were probably similar in character to the modern west Pacific Ocean with a complicated pattern of inter-related volcanic island arcs and back-arc basins.

The ophiolite model

The common association of serpentinite, gabbro, spilitic pillow lava and pelagic sedimentary rock has long been recognized by geologists and the idea of a kindred genetic relationship was introduced by Steinmann (1927). However, it was the later development of the plate tectonic theory that allowed a coherent interpretation as oceanic crust, generated at a constructive plate margin but subsequently accreted or thrust onto continental crust at a destructive margin. Until then the term *ophiolite* (from the Greek *ophi* meaning snake or serpent, hence serpentinite) had been used in a variety of ways but in 1972 the Geological Society of America convened a conference to formalize both its use and the underlying concept; the resulting proposals have been generally adopted. *Ophiolite*, as now defined (Anon, 1972), refers to a distinctive assemblage of mafic to ultramafic rocks which, in a complete development (Figure 2.2), consists of rock types in the following sequence, starting from the top and working down:

- *Mafic volcanic complex*, commonly of pillowed lava.
- *Mafic sheeted dyke complex*.
- *Gabbroic complex*, commonly with cumulus textures.
- *Ultramafic complex*, consisting of variable proportions of harzburgite, lherzolite and dunite (ultramafic rock names are defined in Figure 2.3), commonly serpentinized and with a tectonic fabric.

Associated rock types include: overlying pelagic sedimentary rocks such as chert, podi-

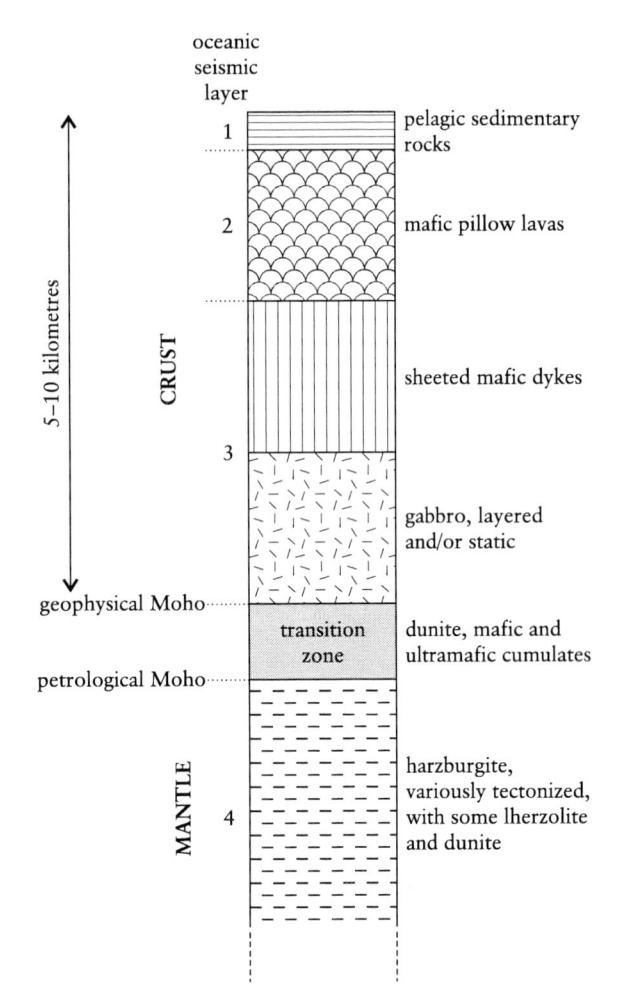

Figure 2.2 An idealized ophiolite succession compared to the seismic structure of oceanic crust and mantle (after Coleman, 1977). As an indication of scale, beneath the oceans the depth to the geophysical moho is between 5 and 10 km.

form chromite bodies in dunite, sodium-rich felsic intrusive rocks. The definition stresses that although *ophiolite* is generally interpreted to be oceanic crust and upper mantle, the use of the term should be independent of the supposed origin. As this terminology has been applied it has become common practice to refer to the whole assemblage as an *ophiolite complex*.

The comparison with oceanic crust relates the ultramafic rocks to the upper mantle and the gabbroic rocks and sheeted dykes to seismic layer 3 of the oceanic crust; the pillow lavas form

Introduction

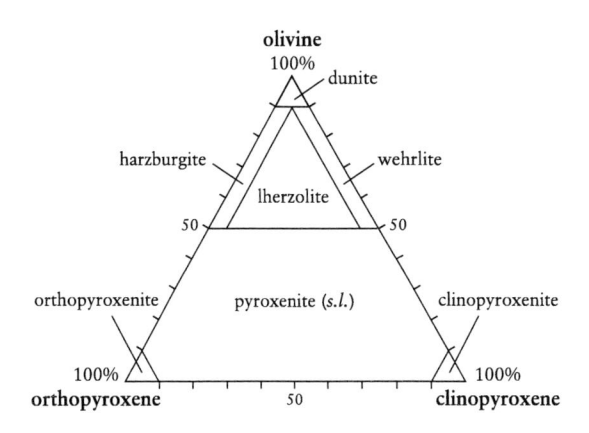

Figure 2.3 Descriptive nomenclature for ultramafic rocks in terms of their relative content of olivine and pyroxene.

layer 2 and the overlying pelagic strata form layer 1 (Figure 2.2). On this basis the geophysically defined Moho (occurring in the 5–10 km depth range beneath oceanic crust) coincides with the base of the gabbroic unit in an ophiolite complex. Beneath this level in many ophiolites, there is a transition zone of dunite and interlayered gabbroic and ultramafic lithologies passing down into harzburgite. The base of the transition zone has become known as the petrological Moho, which may lie up to several kilometres beneath the geophysical Moho. This zone is preserved within the Punds to Wick of Hagdale and Skeo Taing GCR sites on Shetland.

As work on ophiolites has progressed it has been established that many are in fact atypical of oceanic crust and many have been generated above subduction zones, i.e. at a destructive plate margin, rather than at a constructive mid-ocean ridge. Interpretation difficulties are also caused by the incomplete and fragmentary nature of most preserved ophiolites. The Scottish Caledonian examples of Shetland, the Highland Border and Ballantrae illustrate many of these problems and are of international importance in the overall assessment of the relationship of ophiolites to oceanic crust.

The Shetland Ophiolite

The Shetland Ophiolite forms a large part of the islands of Unst and Fetlar in the NE of Shetland (Gass *et al.*, 1982; Flinn, 1996). It is one of the series of ophiolite fragments distributed along the Caledonian orogenic belt from North America to Norway (Figure 2.1), most of which, including Shetland, were formed and obducted around 500 Ma or soon thereafter. Radiometric age data from Shetland suggests intrusive activity at 492 ± 3 Ma (U-Pb on zircon in a 'plagiogranite', *sensu* albite-granite, vein; Spray and Dunning, 1991) and the initiation of obduction at 498 ± 2 Ma (Ar-Ar on hornblende in amphibolite; Flinn *et al.*, 1991). The nearest to Shetland of the other Caledonian ophiolites is Karmoy in Norway, 400 km to the ESE across the orogen, while the nearest in the UK (excepting the small fragmentary vestiges of the Highland Border Complex) is Ballantrae, 600 km to the SSW along the orogen. The upper part (layers 1 and 2) of the characteristic ophiolite sequence (Figure 2.2) is not preserved in Shetland but the lower part is exceptionally well developed and exposed. Overall, the Shetland Ophiolite is among the finest and most accessible examples of ophiolitic rocks in Europe.

The ophiolite complex is composed of two tectonic units, the Upper and Lower nappes (Flinn, 1958), separated and underlain by imbricate zones composed of metasedimentary strata containing ophiolitic erosional debris from the nappes, acid and basic metavolcanic rocks and hornblende schist, all interleaved with tectonic slices of ophiolitic rocks. Figure 2.4a shows the distribution of these units together with that of the gneissose metasedimentary basement to the complex which crops out farther west; Figure 2.4b summarizes the lithological outcrop.

The *Upper Nappe* occurs as a series of widespread klippen composed largely of metaharzburgite of mantle origin; it rests either directly on the Lower Nappe or on the Middle Imbricate Zone separating the two nappes. The Upper Nappe is made of highly magnetic rocks, which enable it to be traced under the sea by means of positive aeromagnetic anomalies. The main Vord Hill Klippe occupies the centre of Fetlar and extends north under the sea to the island of Haaf Gruney. The nappe extends northwards to Hill of Clibberswick and westwards to the island of Sound Gruney via a series of small klippen, several of which also occur along the western edge of the complex in Unst. The Virva GCR site reveals that the Upper Nappe is underlain by hornblende schist derived from basic magma intruded into the thrust plane beneath the nappe and later cut into tectonic

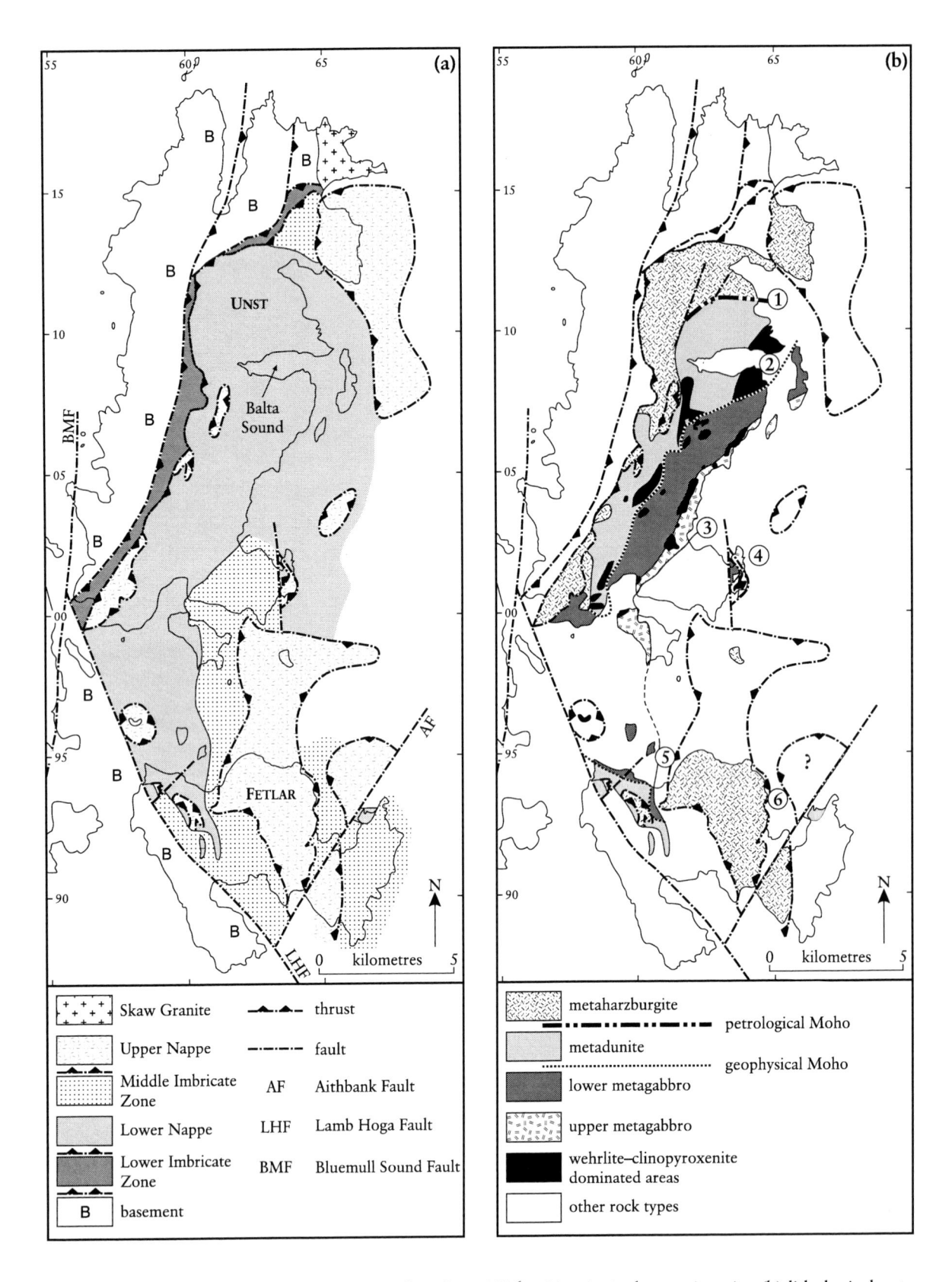

Figure 2.4 Maps of the Shetland Ophiolite (after Flinn, 1996): (a) principal tectonic units, (b) lithological outcrops. GCR sites: 1, The Punds to Wick of Hagdale; 2, Skeo Taing to Clugan; 3, Qui Ness to Pund Stacks; 4, Ham Ness; 5, Tressa Ness to Colbinstoft; 6, Virva.

Introduction

slices by the thrusting, while in the Tressa Ness to Colbinstoft GCR site the unusual lithology rodingite (a Ca-rich metasomatic rock) is present in the base of the Upper Nappe.

The *Lower Nappe* is widely exposed in Unst but is overlain by the Middle Imbricate Zone in the extreme north (Norwick) and the south-east (Muness). It is bound to the west by an eastward-dipping thrust and separated from the metamorphic basement farther west by the Lower Imbricate Zone. The Lower Nappe extends under the sea into the north of Fetlar on either side of the Vord Hill Klippe.

Unlike the Upper Nappe, the Lower Nappe is composed of a conformable series of vertically orientated layers each several kilometres thick. The layers represent much of the conventional ophiolitic pseudo-stratigraphical succession (Figure 2.2). A thick layer of metaharzburgite, very similar to that forming the Upper Nappe, forms the northern part of the Lower Nappe (Figure 2.4b). It is followed to the south (up the ophiolitic succession) by a thick metadunite layer. The Punds to Wick of Hagdale GCR site presents evidence that the metaharzburgite is serpentinized infertile mantle whereas the metadunite is a serpentinized intrusive mass overlying the mantle and separated from it by the petrological Moho (Figure 2.2). It corresponds to the ultramafic layered lower crustal unit (transition zone *sensu lato*) of the conventional ophiolite succession. The intrusive nature of the metadunite layer and the adjacent metagabbro layer to the south can be demonstrated in the Skeo Taing to Clugan GCR site by the presence in both layers of xenoliths and screens of banded wehrlite–clinopyroxenite. In the Qui Ness to Punds Stack GCR site the eastern boundary of the metagabbro layer is marked by a discontinuous layer of wehrlite–clinopyroxenite xenoliths and screens, beyond which is an upper metagabbro layer. The latter is characterized by fine-grained metagabbros intruded by a large number of basic, dyke-like sheets giving this layer a quasi-sheeted appearance. The Ham Ness GCR site presents the upper metagabbro layer overthrust by the Upper Nappe.

The Highland Border Complex

Within the Highland Boundary Fault Zone (Figure 2.1), a series of structurally isolated slivers contain various combinations of serpentinite, pillow lava, black shale, chert, limestone and sandstone, apparently ranging in age from Early Cambrian to late Ordovician. The igneous components of this lithological association are distinct from the Dalradian Supergroup rocks to the north and the whole assemblage contrasts with the Upper Palaeozoic strata of the Midland Valley Terrane to the south. All of the disparate elements are grouped together as the Highland Border Complex which remains largely enigmatic. The current consensus view of the regional development of the complex (see Robertson and Henderson, 1984; Curry *et al.*, 1984) envisages its formation in a small oceanic basin marginal to the Laurentian continent. However, there are aspects of the relationship between the Highland Border Complex and the Dalradian that remain poorly understood. In particular, some of the Early Cambrian sedimentary strata seem to lie in stratigraphical continuity with parts of the Dalradian succession and to have experienced a similar structural history (Tanner, 1995 and references therein). The problems are accentuated by the extreme tectonic disruption whereby the complex now exists as about ten fault-bound lenses, spread along the length of the Highland Boundary, from Stonehaven on the east coast of Scotland to Arran in the west (Figure 2.1).

Despite the tectonic disruption an overall succession has been established. The serpentinites are regarded as the oldest igneous component, of Late Cambrian or early Ordovician (Tremadoc?) age, and are overlain unconformably by conglomerates with serpentinite clasts; an associated limestone unit contains a mid-Arenig fauna. This part of the succession is well displayed at the Balmaha and Arrochymore Point GCR site and serpentinite is also present in the Garron Point to Slug Head GCR site. The relationship of the serpentinite to a slightly younger sequence dominated by basic pillow lava and black shale is uncertain but the lava-shale unit appears to range in age from the Arenig, through the Llanvirn and possibly into the Llandeilo. Associated with the lava and shale are tuff, chert and quartz-greywacke. A probable Arenig part of this unit, including an impressive array of pillow lava, is well exposed in the North Glen Sannox GCR site; the pillow lavas seen in the Garron Point to Slug Head GCR site may possibly be younger (Llanvirn–Llandeilo) in age although the evidence is meagre. The youngest part of the Highland Border Complex appears to be a Caradoc to possibly Ashgill sedimentary sequence of sandstone with ophiolitic, acid volcanic and quartz clasts, limestone and shale. Elements of this sedimentary association are seen in the Balmaha and Arrochymore Point and the North Glen Sannox GCR sites.

The closure of the Highland Border Complex basin (or basins) seems likely to have been a polyphase and protracted event involving oblique terrane amalgamation. A complicated and possibly protracted metamorphic history is suggested by stable isotope and petrographical evidence presented by Ikin and Harmon (1984). Indications of Cambrian tectonism come from the radiometric ages of around 540 Ma reported by Dempster and Bluck (1991) for an amphibolite associated with serpentinite on the island of Bute. This was taken as evidence for Cambrian initiation of ophiolite obduction but, elsewhere, Ordovician sedimentation clearly occurred at least during the Arenig–Llanvirn and Llandeilo–Caradoc intervals (Curry *et al.*, 1984). Structural evidence is similarly ambiguous. Henderson and Robertson (1982) thought that tectonic features suggest a similar deformation history for the Highland Border Complex and the Dalradian succession to the NW; they envisaged thrust emplacement of the complex towards the SE. Conversely, Curry *et al.* (1984) proposed that remnants of a marginal basin floor were obducted towards the NW onto the already deformed Dalradian. As a further complication, Harte *et al.* (1984) concluded that the later transcurrent faulting was largely responsible for the present lenticular disposition of the complex along the Highland Boundary Fault.

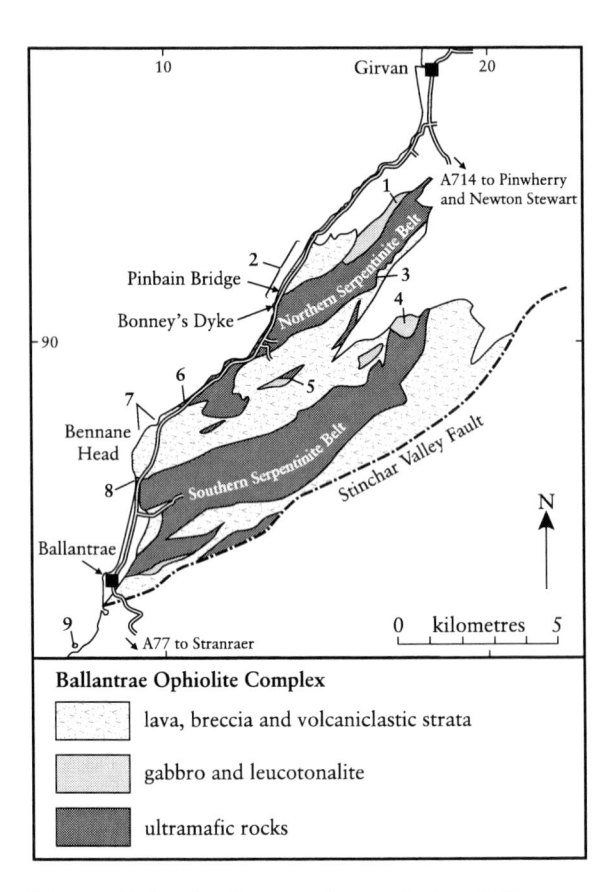

Figure 2.5 Outline geology of the Ballantrae Complex (after Stone and Smellie, 1988) showing the location of the GCR sites. 1, Byne Hill; 2, Slockenray coast; 3, Knocklaugh; 4, Millenderdale; 5, Knockormal; 6, Games Loup; 7, Balcreuchan Port to Port Vad; 8, Bennane Lea; 9, Sgavoch Rock.

The Ballantrae Complex

Between Girvan and Ballantrae, on the west coast of southern Scotland, a structurally complicated assemblage consisting mainly of serpentinized ultramafic rocks and tholeiitic lavas crops out over an area of about 75 km². This is the Ballantrae Complex, which has been widely interpreted as an obducted ophiolite (Church and Gayer, 1973; Bluck, 1978a) although there is little evidence of sheeted dykes and only fragmentary occurrences of layered gabbro. The outline geology and location of the GCR sites is shown in Figure 2.5. The internal structural relationships are complicated and all of the major lithological boundaries are faulted. The volcanic rocks form three fault-defined blocks trending NE–SW and separated by two blocks of ultramafic rock, the Northern and Southern serpentinite belts.

The lavas, together with abundant volcaniclastic breccia and sandstone, have been lithostratigraphically defined as the Balcreuchan Group (Stone and Smellie, 1988) aspects of which may be seen within the Slockenray Coast, Games Loup, Balcreuchan Port to Port Vad and Bennane Lea GCR sites. The lavas were erupted in a submarine environment with pillow structure widely developed. At a few localities, graptolites of Arenig (early Ordovician) age have been recovered from sedimentary interbeds (Stone and Rushton, 1983; Rushton *et al.*, 1986) while the tholeiitic basalt lavas have themselves given Sm-Nd radiometric ages of 501 ± 12 Ma and 476 ± 14 Ma (Thirlwall and Bluck, 1984). Eruption would therefore seem to have commenced

Introduction

during the Tremadoc and probably reached its peak during the Arenig. However, a number of studies of the lava geochemistry have concluded that eruption took place in disparate geotectonic environments with the Balcreuchan Group therefore containing a polygenetic structural assemblage (Wilkinson and Cann, 1974; Thirlwall and Bluck, 1984). The basis for this conclusion lies in the different trace element ratios and abundances characteristic of basalts erupted in either mid-ocean ridge, or within-plate, or island arc settings. A clear consensus has emerged that island-arc and within-plate lavas are both present but that there is very little evidence for a mid-ocean ridge volcanic component (Wilkinson and Cann, 1974; Lewis and Bloxam, 1977; Thirlwall and Bluck, 1984; Stone and Smellie, 1990). Further support for an island-arc involvement has come from the discovery of boninite lavas and breccias at the Balcreuchan Port to Port Vad GCR site (Smellie and Stone, 1992; Smellie *et al.*, 1995); these are unusual lithologies with modern examples known only from primitive, oceanic island arcs.

The ultramafic components of the Ballantrae Complex are now largely serpentinized; the protoliths were principally dunite and harzburgite with a small proportion of wehrlite and various pyroxenites. Ultramafic rocks are a particular feature of the Knockormal, Knocklaugh and Games Loup GCR sites. An origin within 'normal' (mid-ocean ridge) oceanic mantle was initially assumed in the ophiolite interpretation, and a geochemical study of a limited area of the Northern Serpentinite Belt by Jelínek *et al.* (1980) has supported this assumption. However, Stone and Smellie, (1990) considered that a range of features are atypical of ultramafic rock from mid-ocean ridge ophiolites but instead match those considered characteristic of supra-subduction zone (i.e. volcanic arc) environments by Pearce *et al.* (1984). Important aspects encouraging this conclusion are the overall lithological dominance of titanium-depleted harzburgite, the dominance of wehrlite over troctolite in the layered sequences, and the relative chromium enrichment of the accessory spinels over those found in a mid-oceanic setting. Thus the ultramafic rocks of the Ballantrae Complex are also indicative of an origin, for at least part of the complex, at an island arc above an oceanic subduction zone. The involvement of older oceanic mantle, perhaps forming the foundation to the arc development, is suggested by a Sm-Nd radiometric age of 576 ± 32 Ma reported by Hamilton *et al.* (1984) from a garnet clinopyroxenite body within the Knockormal GCR site, that was interpreted as a segregation in mantle harzburgite by Smellie and Stone (1984). A minimum age is provided by a U-Pb on zircon radiometric date of 483 ± 3 Ma reported by Bluck *et al.* (1980) from a leucotonalite, intruded into and chilled against the Northern Serpentinite Belt, within the Byne Hill GCR site.

The Byne Hill leucotonalite is gradational from gabbro but the gabbroic and sheeted dyke components of the ideal ophiolite assemblage are sparse and fragmentary within the Ballantrae Complex. Foliated gabbros cut by numerous dykes within the Millenderdale GCR site have been likened to a sheeted dyke association but with reservations (Bluck, 1978a). An exceptionally coarse pegmatitic gabbro segregation within harzburgite occurs within the Slockenray Coast GCR site.

The most likely setting for generation of the Ballantrae Complex would seem to be within a Tremadoc to Arenig oceanic island arc and associated marginal, back-arc basin. The youngest graptolite fauna recovered from the complex is of late Arenig age (about 480 Ma) (Stone and Rushton, 1983) while a number of radiometric dates overlap around an upper range of 475–485 Ma. Obduction of the ophiolite, perhaps caused by collision of the volcanic arc with the continental margin of Laurentia, could have followed soon after. Its initiation may well be dated by the K-Ar age of 478 ± 8 Ma reported by Bluck *et al.* (1980) from an amphibolite within the Knocklaugh GCR site. This amphibolite is part of a metamorphic assemblage thought to have formed in a dynamothermal aureole adjacent to the base of the Northern Serpentinite Belt when, as a slab of hot mantle material, it was thrust up through the crustal carapace of volcanic-arc lavas. A minimum age for obduction of the Ballantrae Complex is provided by the unconformably overlying, Llanvirn shallow-marine strata.

The Southern Uplands

The Southern Uplands Terrane formed as an accretionary thrust complex at the Laurentian continental margin during Ordovician and Silurian subduction of the Iapetus Ocean. The terrane has been widely interpreted as a fore-arc prism, formed above the active subduction zone

(Leggett *et al.*, 1979; Leggett, 1987). Alternatively, it may have developed from an Ordovician back-arc setting into a mid-Silurian foreland basin, the latter generated as the thrust front migrated onto the Avalonian continent following closure of the Iapetus Ocean (Stone *et al.*, 1987). In either model the resulting structural configuration is the same: a sequence of southward propagating imbricate thrusts separating lithostratigraphical tracts of steeply inclined strata that strike NE–SW. Internally the tracts have an overall sense of younging towards the north, whereas the minimum age of each tract decreases southwards. This phenomenon has been recently illustrated and discussed by Rushton *et al.* (1996 and references therein) while a full stratigraphy is given by Floyd (1996). The dominant lithology within each tract is turbidite-facies greywacke but this overlies a thin basal assemblage of black mudstone, chert and sporadic basic lava and hyaloclastite. The lava and hyaloclastite are found in the northern part of the terrane and appear to range in age from Arenig to Caradoc; recent reviews are provided by Phillips *et al.* (1995b) and Armstrong *et al.* (1996). The mudstone and chert range in age from the Caradoc well up into the mid-Silurian and throughout the succession are interbedded with subalkaline and mildly peralkaline metabentonite layers (Merriman and Roberts, 1990).

Phillips *et al.* (1995b, 1999) show that the oldest (Arenig) lavas include both alkaline within-plate basalts and tholeiitic varieties with probable mid-ocean ridge basalt characteristics. Other lavas of uncertain age within the Arenig–Caradoc range have compositions more suggestive of an island-arc source, while definite Caradoc lavas are of within-plate, ocean island affinity. Some of the latter are interbedded with the greywacke sequence rather than forming a base to the mudstone. Apart from these lava occurrences, three other examples of Ordovician volcanic rocks are worthy of note.

1. Peralkaline rhyolite and tuff (the Wrae or Tweeddale lavas), interbedded with greywacke of late Caradoc or early Ashgill age, may have been emplaced as a debris flow from a sea-mount volcano (Thirlwall, 1981b).
2. The Bail Hill Complex is a 1.8 km thickness of pyroclastic rock and lava, the latter comprising mugearite and hawaiite, thought to represent a sea-mount volcano enclosed by transgressive early Caradoc greywackes (Hepworth *et al.*, 1982; Phillips *et al.*, 1999).
3. The Downan Point Lava Formation (Stone and Smellie, 1988 and references therein) occupies a faulted wedge sandwiched between the northern margin of the Southern Uplands Terrane and the Ballantrae Complex. Tholeiitic basalts predominate and have within-plate, ocean island geochemical affinities. The formation has been associated historically with the Ballantrae Complex of Arenig age but a current consensus favours a younger, early Caradoc age and considers it to form the northernmost tract of the Southern Uplands. The coastal exposures present an extraordinary array of pillow structures, as seen at the Sgavoch Rock GCR site.

The full assemblage of Ordovician volcanic rocks in the northern part of the Southern Uplands Terrane is difficult to readily reconcile with either of the geotectonic models proposed for its development; fore-arc accretionary prism or back-arc (to Silurian foreland basin) thrust belt.

THE SHETLAND OPHIOLITE

THE PUNDS TO WICK OF HAGDALE (HP 647 103–644 113)

Derek Flinn

Introduction

The site lies on the east side of Unst within the Lower Nappe of the Shetland Ophiolite. On the east side is a continuously exposed low-lying cliff section which is of great importance in that it contains the uppermost kilometre of the mantle, the petrological Moho, and a kilometre of the overlying metadunite layer. Inland to the west abundant exposures, ideally weathered, ice-smoothed and lichen-free, reveal in unusual detail the petrography and mineralogy of these rocks. Hitherto, these two units of the ophiolite pseudo-stratigraphy in Unst have received the conventional labels of 'tectonized harzburgite' and 'cumulate dunite' and the latter has been interpreted as the result of the cumulate crystallization of olivine in the base of a magma chamber resting on tectonized mantle at an oceanic constructive plate margin (Gass *et al.*, 1982;

Prichard, 1985). However, Flinn (1996) has reached the contrary conclusion that the metadunite is an independent intrusive unit which did not form as a cumulate layer in a magma chamber and that the harzburgite is not tectonized. The site offers a magnificently exposed and easily accessible view of the mantle-lower crust junction.

Description

The rocks to the north of the petrological Moho in Figure 2.6, representing the uppermost mantle, are dominantly *metaharzburgite* with lesser amounts of metadunite. Both rocks, being uniformly and extensively altered to hydrous minerals, contain about 13 wt% of water. Lizardite-serpentine, which replaces olivine, weathers to a strong ochrous colour due to a relatively high iron content, while the hydrous minerals replacing pyroxenes weather white and stand slightly proud of the serpentine. Thus the two rocks are easily distinguished, even where intimately mixed. However, adjacent to some of the major shear zones the lizardite-serpentine has been recrystallized to antigorite-serpentine with magnetite and weathers white making it difficult there to distinguish between metaharzburgite and metadunite.

It is apparent that prior to serpentinization the metaharzburgite was composed dominantly of olivine together with 20–30% of orthopyroxene, and about 0.5% of accessory chromite. Thin sections commonly contain about 1% of clinopyroxene. The metadunite was originally almost entirely composed of olivine, with about 1% of accessory chromite. Clinopyroxene is present only very locally (see below). The grain size of the olivine and pyroxene is about 0.5 cm.

The metaharzburgite is rhythmically banded in places. The bands are sharply defined, rectilinear and dip steeply north. They are formed by alternations in olivine/pyroxene proportions, but are ungraded. Olivine-rich and pyroxene-rich pairs are about 30 cm thick, and are parallel to the petrological Moho towards which their persistence and frequency of occurrence increases. Generally, the bands can rarely be followed for more than a few metres, only rarely show signs of disruption or folding and are untectonized. The banding is very well displayed on the coast at The Viels (6444 1108) (Figure 2.7) and on the hillside to the west of the road.

The metaharzburgite is cut by many thin pyroxenite veins filling planar joint-like fractures. The veins are formed of pyroxenite with grain sizes up to one centimetre or more, and much less commonly of olivine-pyroxenite. They are always completely pseudomorphed by fine-grained hydrous minerals. They have been referred to as dykes (Prichard, 1985), but to avoid confusion with the basic dykes forming the quasi-sheeted dyke layer they are referred to here as veins. Many strike parallel to the rhythmic banding, where present, but rarely dip parallel to it; hence they cut the banding. In unbanded areas they have no preferred orientation. In a small number of localities they are seen to have been deformed by cross-cutting early shear zones. Pyroxenite veins are cut by metadunite bodies in the metaharzburgite. Since most orthopyroxene in the metaharzburgite is altered to the same secondary minerals as the pyroxenes in the veins, whereas clinopyroxene generally remains fresh, these were probably orthopyroxenite veins. An irregular sheet of fresh, olivine-free orthopyroxenite about 10 cm thick occurs at The Viels (6446 1108).

Metadunite within the metaharzburgite occurs as rare and sporadically distributed bodies that increase in frequency towards the petrological Moho within the northern part of Figure 2.6. Most are sub-equidimensional bodies or clots a metre or so in diameter with no detectable preferred orientation of their shapes; larger bodies up to a hundred metres or more across also occur and these are commonly of very irregular shape. Several of the larger metadunite bodies contain podiform chromite masses. Parallel-sided sheets of metadunite up to 20 cm thick and several metres in length occur with increasing frequency southward and, for two or three hundred metres north of the petrological Moho, the metadunite clots and sheets are particularly densely distributed, equalling in places the metaharzburgite in volume.

The metadunite sheets within the metaharzburgite adjacent to the petrological Moho are sub-parallel to the rhythmic banding and to the Moho; a high proportion are obliquely cross cutting by 10° or less, while a much smaller proportion cut at much greater angles or are exactly parallel. The clots of metadunite, especially those close to the petrological Moho often have invasive boundaries with thin wedge-like and sheet-like protuberances cutting the enclosing

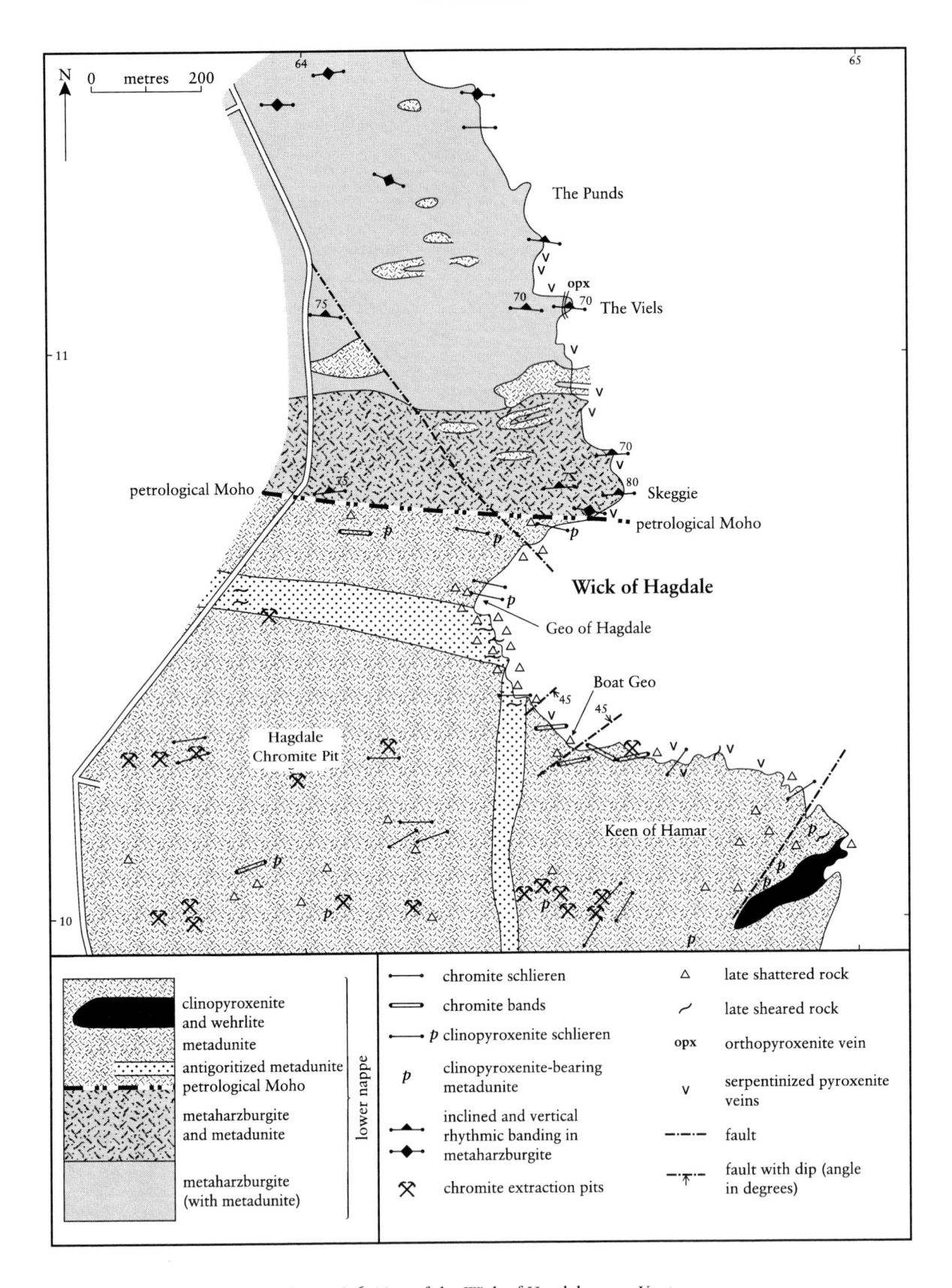

Figure 2.6 Map of the Wick of Hagdale area, Unst.

The Punds to Wick of Hagdale

Figure 2.7 Rhythmic banding in the metaharzburgite at The Viels (HP 6444 1108) close to the petrological Moho. The hammer is 37 cm long. (Photo: D. Flinn.)

metaharzburgite and giving them an intrusive appearance.

The *petrological Moho* is taken as the southern edge of the southernmost band of metaharzburgite. It has an E–W strike parallel to the rhythmic banding, and has a 70–80° dip to the north. It is exposed on the coast (Figure 2.6) and in several inland exposures.

The *metadunite* layer is dominantly composed of rock indistinguishable from the metadunite forming bodies and sheets within the metaharzburgite layer. Thin sections reveal equidimensional olivine grains about half a centimetre in size forming a triple-junction grain-boundary network, but which have been extensively replaced by iron-rich mesh-type serpentine so that the rock weathers to a bright ochrous colour, lacking the white-weathering altered pyroxenes characteristic of the metaharzburgite. The metadunite is lithologically uniform and texturally isotropic, except for the local presence of chromite and clinopyroxene.

Chromite, in both the metadunite layer and in the metadunite sheets and bodies within the metaharzburgite, occurs as disseminated accessory grains no more than 1 mm in size and as coarser grains several millimetres in diameter forming band- or vein-like concentrations that locally expand into podiform bodies. Some accessory chromite forms schlieren, typically several centimetres thick and ten times as long, in which the individual chromite grains are several millimetres apart. Schlieren are well exposed on the coast just south of the Moho in Figure 2.6 but occur more commonly in the metadunite to the west of the road. The bands formed of more concentrated coarsely crystalline chromite cannot be studied as extensively. They were used by prospectors as guides to the presence of podiform masses and have almost all been excavated, leaving pits, trenches and spoil heaps to mark their previous existence. The best remaining locality for the study of in-situ banded chromite occurs in the cliffs 100 m east of Boat Geo (6456 1030). The spoil heaps of some chromite pits are a good source of banded chromite, but the waste heaps of the famous Hagdale Quarry have been despoiled by collectors and carried away for infill. Specimens remaining in waste heaps elsewhere in Unst reveal bands of chromite cut by metadunite. The presence of angular fragments of massive chromite floating in metadunite and the folding of the chromite bands indicate that fracturing and deformation of the deposits took place while the dunite was still mobile. Disseminated accessory chromite is also present in the metaharzburgite, though much less abundantly than in the metadunite.

Clinopyroxene occurs locally and rarely as sub-millimetre, isolated, interstitial fresh grains. It may be either uniformly distributed or form schlieren in the metadunite layer but it never occurs in metadunite within the metaharzburgite. Both types of occurrence of clinopyroxene are to be found closely associated with similarly occurring chromite on the coast of the Wick of Hagdale in the 200–300 m to the south of the petrological Moho (Figure 2.6) and inland elsewhere. Interstitial clinopyroxene occurs most commonly in the top of the metadunite layer close to wehrlite–clinopyroxenite masses. An example is indicated in the SE corner of Figure 2.6 but this relationship is better displayed in the Skeo Taing to Clugan GCR site.

Interpretation

The metaharzburgite has been recognized as infertile mantle (Gass *et al.*, 1982; Prichard, 1985), on the basis of its composition and by analogy with other ophiolite complexes (Gass, 1980). This makes its boundary against the metadunite layer to the south the upper limit of the mantle and therefore the petrological Moho (Figure 2.2).

The metaharzburgite resembles that of many other ophiolites in the presence of the enclosed metadunite bodies, the pyroxenite veins, the rhythmic banding and the accessory chromite. Harzburgite in ophiolites is generally held to have been tectonized by convectional flow at a constructional plate margin before obduction and the tectonite nature of the harzburgite has been cited as evidence of the ophiolitic nature of the Shetland complex by the above mentioned authors. Bartholomew (1993) reported the occurrence throughout the metaharzburgite of two vertical foliations intersecting at high angles. These foliations were recognized in the field by the flattening of the accessory chromite grains and were confirmed in thin-section by a preferred orientation of grain shapes, interpreted by Bartholomew as arising from simple shear during mantle flow. However, Flinn (1996) found no trace of the two foliation directions and reported only that, in a number of places, a slight flattening of the chromite grains parallel to the plane of schlieren is present, but never in two widely different directions in the same area. He also noted that pyroxene grain-shapes are clearly etched out by weathering and show no grain-shape preferred orientation, even in rhythmically banded areas. The tectonite status of the metaharzburgite therefore remains controversial.

The conventionally accepted origin (Prichard, 1985) for the metadunite bodies within the metaharzburgite and the metadunite above the petrological Moho is that they crystallized out of basic magma, produced by adiabatic melting deep in the mantle, as it rose to form a magma chamber above the mantle. However, field evidence from this and the Skeo Taing to Clugan GCR site has been interpreted as showing that the metadunite layer above the mantle is an intrusive unit and not a layer accumulated on the floor of a magma chamber (Flinn, 1996).

In support of his interpretation Flinn cited Kelemen *et al.* (1995) who have suggested that MORB (mid-oceanic ridge basalt) magma produced by adiabatic melting deep in the mantle would be pyroxene undersaturated and would rise through the mantle by intergranular flow. Such magma would dissolve the pyroxene grains in the mantle and so become focused into conduits made more porous than the surroundings, the conduits thus becoming columns of dunite. It is possible that the enhanced porosity of the dunite in the MORB conduits (a MORB : dunite ratio of 1:10, according to Kelemen *et al.*, 1995) could so lower the cohesion of the olivine grains that the rising MORB would fluidize the dunite. The fluidized dunite would then flow up its conduits through the mantle giving rise to the observed invasive relationships of the dunite relative to the enclosing harzburgite, and including the sheet-like injections close to the Moho. The upward flowing dunite accumulated above the mantle as an intrusive layer several kilometres thick and when flow ceased the conduits within the mantle broke up into independent bodies.

Neither the conventional account of the origin of the metadunite bodies within the mantle (Prichard, 1985), nor that of Kelemen *et al.* (1995) provide an adequate explanation for the presence of podiform masses of chromite in some of the metadunite bodies within the mantle in Unst, whereas other, nearby and larger bodies contain only sparsely distributed accessory chromite grains. Nor do they account for the apparent random distribution of the chromite podiform masses in the metadunite layer above the mantle, or the fact that they are indistinguishable from those nearby within the mantle. Possibly the podiform masses were formed at depth in the mantle and carried up to their present positions in Unst in the mobilized dunite.

The origins of the clinopyroxene in the metadunite layer and of the wehrlite–clinopyroxenite masses in the upper part of the metadunite layer south of Figure 2.6 are more conveniently discussed in the context of the Skeo Taing to Clugan GCR site.

Conclusions

The Punds to Wick of Hagdale GCR site features an uninterrupted passage through the top kilometre of the mantle and the bottom kilometre of the overlying lower crustal succession. The profusion of glacially smoothed exposures makes the area ideal for the study of the environment of the petrological Moho, thus providing a rare opportunity of international importance in ophiolite geology. In particular, the site allows close investigation of the controversial intrusive relationship of the dunite to the mantle.

SKEO TAING TO CLUGAN (HP 646 067–639 084)

Derek Flinn

Introduction

The site lies on the east side of Unst within the Lower Nappe of the Shetland Ophiolite. It is bound to the east by a nearly continuously exposed low-lying cliff section which is of great importance in that it illustrates the uppermost kilometre of the metadunite layer, the geophysical Moho (not exposed), and a kilometre of the overlying lower metagabbro layer. Inland the site is not well exposed but presents evidence complementary to that in the cliffs. The metadunite layer has been interpreted conventionally as an ultramafic layer which, like overlying basic layers, accumulated in a magma chamber resting on the mantle below a constructive plate margin (Gass *et al.*, 1982; Prichard, 1985). However, much field evidence supports an alternative origin wherein the metadunite layer and the basic layers are separate intrusive bodies and not cumulate layers. Xenoliths of interbanded clinopyroxenite and wehrlite are prominent (Flinn, 1996). The site offers an easily accessible view of the passage from ultramafic to mafic rocks in the ophiolite succession and the very different form that this takes from the conventional ophiolite model.

Description

The geology of the Skeo Taing to Clugan site is summarized in Figure 2.8; note that the metadunite outcrop shown is a continuation (across strike) of that part of the same layer shown in Figure 2.6 and previously described at Wick of Hagdale (see The Punds to Wick of Hagdale GCR site). At Skeo Taing the top of the metadunite layer is exposed and typically is rich in wehrlite–clinopyroxenite xenoliths and screens, much more so than in the equivalent layer at Wick of Hagdale. Clinopyroxenite xenoliths too small to show on the map also occur in the Skeo Taing area; all are closely associated with interstitial clinopyroxene in the host metadunite.

In this account, metadunite containing interstitial clinopyroxene will be called clinopyroxene-bearing metadunite even where it contains sufficient clinopyroxene to be petrographically classed as wehrlite (Figure 2.3). This is done in order to distinguish it from the wehrlite associated with clinopyroxenite from which it is texturally distinct. In the metadunite small, fresh clinopyroxene grains occur interstitially to the larger, several-millimetre-sized serpentinized olivine grains whereas, in the wehrlite, the serpentinized small olivine grains are interstitial to the much larger, fresh clinopyroxene grains which in some samples are as much as a centimetre across.

The larger wehrlite–clinopyroxenite masses, those measured in tens of metres, form mound-like exposures rising a metre or two above the enclosing metadunite. The presence in them of abundant, fresh, large clinopyroxene grains makes the wehrlite–clinopyroxenite more resistant to erosion than the enclosing serpentinized dunite and clinopyroxene-bearing metadunite. As a result the wehrlite–clinopyroxenite masses

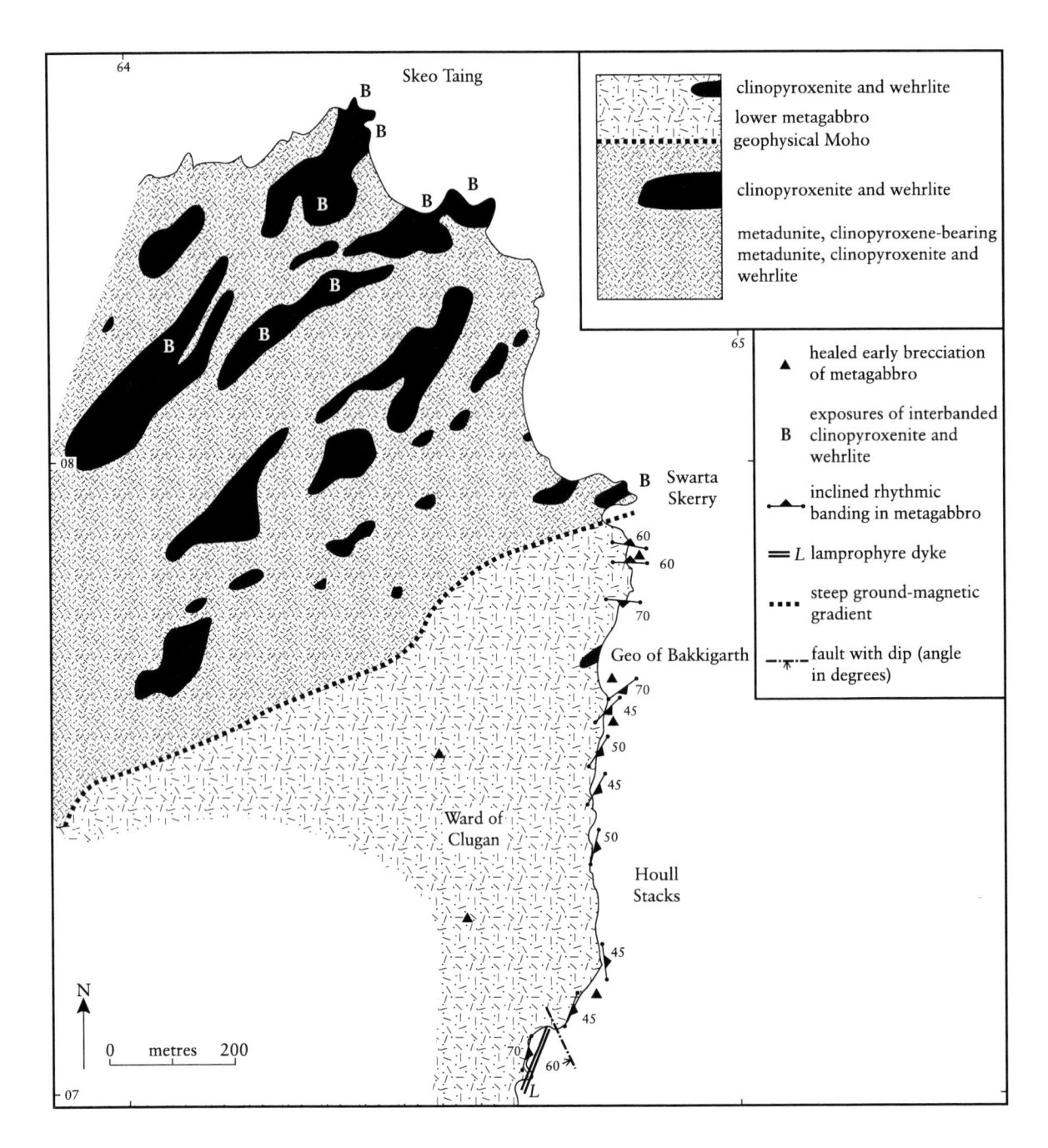

Figure 2.8 Map of the Skeo Taing area, Unst.

form the characteristic protuberant features that commonly carry an enhanced vegetation cover, probably due to the fertilizing effect of the Ca in the clinopyroxene. The low ground between the wehrlite–clinopyroxenite masses is occupied by their metadunite and clinopyroxene-bearing metadunite host rocks which, in the coastal exposures, can be seen to contain smaller wehrlite–clinopyroxenite xenoliths. The wehrlite–clinopyroxenite masses are only well exposed on the coast where it is apparent that they are formed of wehrlite, olivine-clinopyroxenite and clinopyroxenite complexly interbanded and intergraded on a scale ranging from several centimetres to several decimetres. The bands are much less rectilinear than those in the metaharzburgite and the metagabbro and can rarely be followed for more than a metre; some occurrences exhibit disruption and contortion of the banding.

Skeo Taing to Clugan

The contact between the metadunite layer and the lower metagabbro layer to the south is seen in only one small exposure in Unst and that is at 610 054, outside the area of Figure 2.8. However, the contact is closely confined by exposures on either side for considerable distances in several places and everywhere it can be followed by magnetometer since it gives rise to a very steep ground-magnetic gradient. The contact thus sharply defined lacks transitional lithologies and is curvilinear and unsheared. In the upward succession it marks the incoming of feldspar and the exit of olivine (fresh or serpentinized) except in xenoliths. This creates the most profound lithological change in the succession and is therefore considered to represent the geophysical Moho.

The lower metagabbro is as uniformly hydrated as the metaharzburgite and the metadunite (see The Punds to Wick of Hagdale GCR site). The clinopyroxenes have been replaced by single-crystal plates of colourless to pale-green amphibole and the feldspars have been altered to very low birefringence, very fine-grained pseudomorphs. As a result the rock weathers to a speckled pale-green and white pattern. Relict clinopyroxene is rare. Twinned, fresh albitic plagioclase occurs sporadically, associated with the altered plagioclase. The metagabbro contains no evidence of having ever contained olivine. Fresh interstitial grains of brown hornblende occur as a sparsely distributed original accessory. The early uniform hydration, as in the metaharzburgite, was followed by a later low-grade metamorphism closely related to shearing.

The metagabbro is characterized locally by regular alternations of pyroxene-rich and feldspar-rich ungraded bands, each several centimetres thick (Figure 2.9). South of the geophysical Moho in the coastal section (south of (648 076); Figure 2.8), this banding is seen to be repeated continuously for several hundred metres but most occurrences are small patches or even single bands. The attitude of the bands varies considerably but generally the dip direction is between east and south.

Large areas of the metagabbro have been intensively brecciated but subsequently healed. The metagabbro fragments, each no more than several centimetres across, have all been slightly rotated relative to each other but the rock is as resistant as the unbrecciated variety and the frac-

Figure 2.9 Rhythmic banding in metagabbro (HP 647 077). (Photo: D. Flinn.)

tures are not apparent in thin section. Such brecciation occurred earlier than the more common and widespread late shattering and shearing. This healed brecciation makes rhythmic banding difficult to detect.

At Geo of Bakkigarth (6475 0765), 200 m south of the geophysical Moho in Figure 2.8, there is a small xenolithic mass of wehrlite-dominated wehrlite–clinopyroxenite associated with some late shearing. This is one of a series of xenoliths, each up to several tens of metres across, composed of wehrlite–clinopyroxenite and clinopyroxene-bearing metadunite. These occur within the metagabbro close to its western boundary with the metadunite, but otherwise outside the area covered by Figure 2.8. Some are associated with swarms of pyroxenite fragments up to 10 cm across. Rhythmic banding in the metagabbro close to these blocks varies more widely than in the general mass of the metagabbro. All these wehrlite–clinopyroxenite masses are indistinguishable from those in the metadunite layer and like those are considered to be xenoliths.

Interpretation

In Shetland the succession above the petrological Moho takes the form of two distinct layers, each several kilometres thick, one of metadunite and the other of metagabbro. Both these layers contain fragments of a disrupted, coarsely crystalline and intensively banded wehrlite–clinopyroxenite layer. In the metadunite layer these xenoliths have been partly assimilated so that they are surrounded by metadunite containing isolated interstitial clinopyroxene grains.

The metadunite layer with its xenoliths of wehrlite–clinopyroxenite has been presented hitherto as ultramafic cumulates in which 'dunite grades into wehrlite and then pyroxenite'. This description is generally accompanied by idealized sections showing separate successive layers of dunite, wehrlite and clinopyroxenite (Gass *et al.*, 1982; Prichard, 1985; Lord *et al.*, 1994). While this representation is not wholly inappropriate, it hides the full sequence of relationships: wehrlite–clinopyroxenite xenoliths are enclosed in contamination aureoles of clinopyroxene-bearing metadunite suspended in an intrusive metadunite layer. Recently, the obvious absence in the field of continuous layers of wehrlite and clinopyroxenite has been explained as the result of tectonic disruption by NE-trending sinistral faults (Lord *et al.*, 1994). However, the absence of any evidence on the ground or on aerial photographs of such faults, together with the unfaulted continuity of the nearby metagabbro–metadunite boundary (as proved by magnetometery – see above), casts doubt on this interpretation.

Prior to the emplacement of the dunite layer it is possible that the wehrlite–clinopyroxenite masses formed a continuous unit that rested on the mantle and formed *in situ*, as cumulates from an early magma chamber, on the Gass (1980) model. This continuous unit was later disrupted and lifted piecemeal to a higher level, first by the arrival of the intrusive dunite layer and later by the gabbro.

The nature of the geophysical Moho, so clearly revealed in this area, is particularly relevant to the long-running volume problem in serpentinization. Serpentinization of an olivine-rich rock results in a decrease in density of about 20% and, according to petrographic calculations, either an increase in volume of about 45% (*c.* 13% linearly) or a loss of magnesia and silica of about 35% by weight (O'Hanley, 1992). In the past, serpentinization has been accepted as occurring with conservation of volume and loss of material to the surrounding rocks by metasomatism, or with conservation of material and volume increase taken up in bounding faults, with or without internal fracturing. Unst is unique in providing evidence that no expansion has taken place across the serpentinite–metagabbro contact and that no metasomatic transfer has occurred, despite the serpentinization of an olivine-rich rock. The problem remains an enigma.

Conclusions

The Skeo Taing to Clugan GCR site is of international importance in that it provides an accessible and almost continuously exposed coastal section through the geophysical Moho; a traverse across the boundary between the underlying ultramafic rocks and the overlying basic rocks in the ophiolite succession. However, the relationships revealed show that it is intrusive contacts that are involved and not compositional divisions within a continuous series of related rocks formed in a magma chamber. This controversial aspect adds to the overall value of the site in terms of ophiolite research.

QUI NESS TO PUND STACKS
(HP 622 032–622 035)

Derek Flinn

Introduction

The site presents a readily accessible view of the upper metagabbro, the highest exposed level of the pseudo-stratigraphy of the Shetland Ophiolite, within the Lower Nappe on the east side of Unst. It is bound to the east by a continuously exposed low-lying cliff section lying within the upper metagabbro. The site extends to the west for nearly a kilometre to include xenolithic screens of wehrlite–clinopyroxenite marking the boundary between the lower metagabbro layer and the upper metagabbro layer; it also includes the western limit of a swarm of basic 'sheeted dyke'-like intrusions (Flinn, 1996). The coastal section provides very good exposures of the latter and shows their relationship to rhythmic banding, minor fine-grained gabbroic intrusions, hornblende pegmatites, quartz-albite ('plagiogranite') veins and lamprophyres, all characteristic of the upper metagabbro layer. The 'sheeted dyke'-like intrusions, although normal to the present erosion surface, are emplaced obliquely to both the petrological and geophysical Mohos and to the metadunite and lower

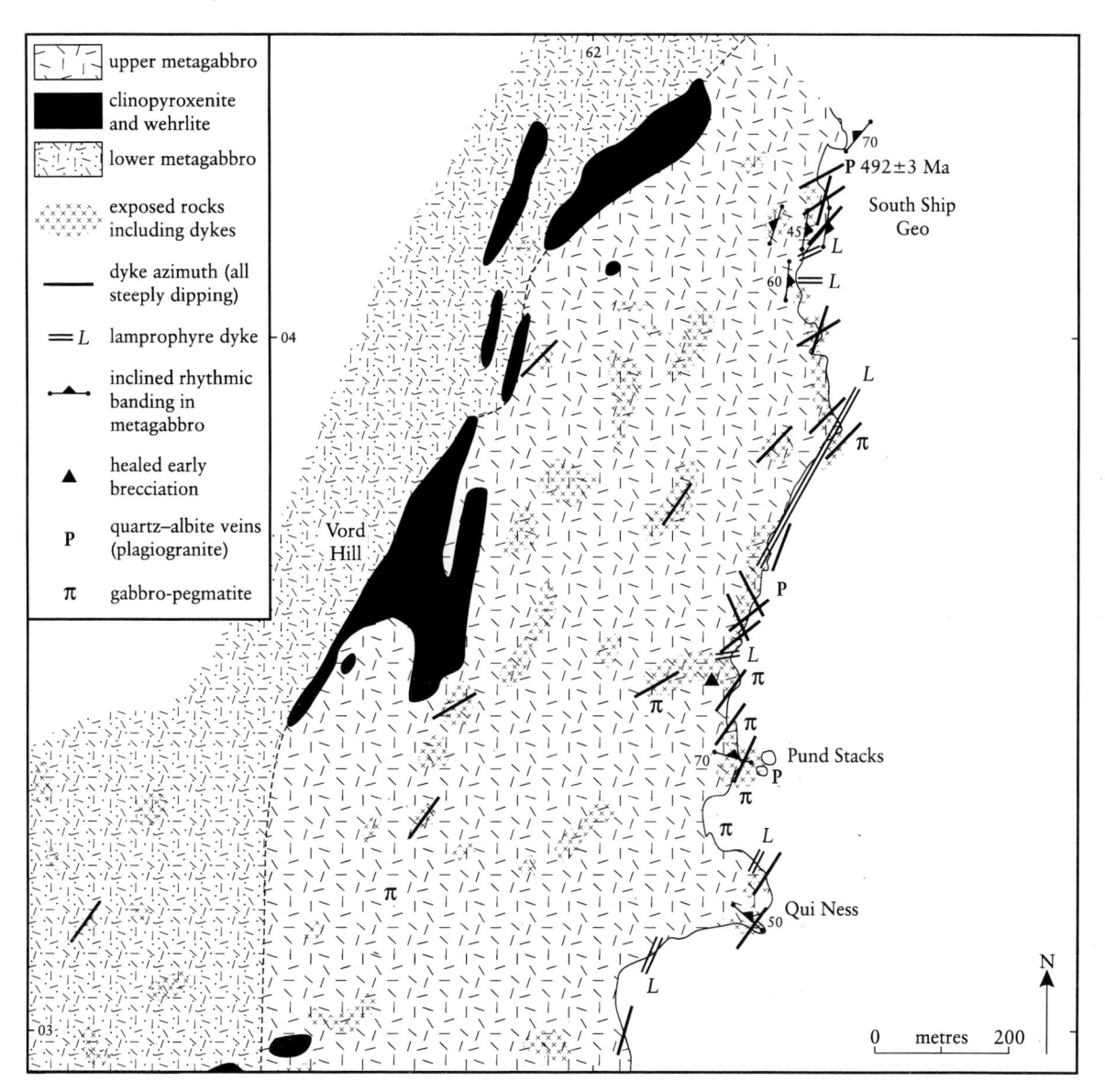

Figure 2.10 Map of the Qui Ness area, Unst.

Figure 2.11 Closely spaced, parallel metabasic sheets of the quasi-sheeted-dyke complex at Pund Stacks, Unst (HP 6218 0340). (Photo: D. Flinn.)

metagabbro layers underlying them (Flinn, 1996). In the Gass (1980) model, hitherto used to represent the Shetland Ophiolite (Gass *et al.*, 1982; Prichard, 1985), the dykes are shown as normal to the underlying layers.

Description

The boundary between the lower metagabbro layer and the upper metagabbro layer is not as clearly or sharply defined as the boundary between the lower metagabbro layer and the underlying metadunite layer (see Skeo Taing to Clugan GCR site). Along the east side of Unst it is marked by a discontinuous string of wehrlite–clinopyroxenite screens, lithologically indistinguishable from those above and below the metadunite–lower metagabbro contact, the geophysical Moho. The geological relationships are summarized in Figure 2.10.

The upper metagabbro layer is composed of a poorly contrasted assemblage of metagabbro, fine-grained metagabbro and 'sheeted-dyke'-like metabasic sheets, all subject to phyllitization and phacoidal shearing which tend to destroy lithological boundaries and to reduce the contrast between lithologies. The rocks of the upper metagabbro differ from the lower metagabbro in several respects: amphibole, where it occurs, is generally green rather than colourless and has feathery instead of granular grain boundaries; the plagioclase is commonly sodic and fresh or partly fresh instead of saussuritized; opaque grains with leucoxene/titanite rims are common.

Rhythmic banding occurs locally in the upper metagabbro, where it is texturally similar to the lower metagabbro, but it is of much more variable nature and band thickness than that seen in the lower metagabbro. Thin clinopyroxenitic bands occur locally but the greatest difference is seen at Qui Ness where there are several clearly defined bands of leucogabbro about 1 m thick.

The fine-grained metagabbros occur inextricably intermixed with coarser lower-metagabbro-like rock but, due to similar lithology and phyllitization, they are particularly difficult to distinguish from the metabasic sheets. The latter are best displayed at Pund Stacks (Figure 2.11), although they can be seen intermittently along the coast for 3 km to the north of that locality. Inland, the sheets are much more difficult to detect than on the coast due to lichen cover and generally poorer exposure. At most, the sheets occupy no more than 50% by volume of the outcrop. They are nearly vertical, have north to north-easterly strikes and are less than half a metre wide, discontinuous, and separated by narrow screens of host rock. They are oblique to any rhythmic banding developed in the host rock. None of the sheets cut the wehrlite–clinopyroxenite screens marking the western boundary of the upper metagabbro layer, even where these are well exposed, but west of the upper metagabbro boundary sheets do occur in the gaps between the screens, for example around 610 030.

The 'sheeted' dykes are invariably very fine-grained, highly altered by hydration, and often phyllitized so that chilled margins and rarely preserved traces of doleritic or basaltic texture are all that remain of their original state. The dykes range in composition from boninite to dacite. Due to the presence of groups of parallel sheets of slightly different strike it is possible, in places, to determine an order of emplacement from intersection relationships. Those dykes with chilled margins, poorly preserved traces of doleritic or basaltic texture or extreme fine grain were the last to be emplaced. The first emplaced tend to be darker coloured and to have a more northerly strike than general. Many of these early dykes also have a vertical penetrative schistosity, about 20° clockwise relative to their strike, which results in lenticular dismemberment. This makes them particularly difficult to distinguish from the similarly schistose and lenticularized fine-grained metagabbro components of the upper metagabbro.

Minor components of the upper metagabbro are hornblende pegmatite, lamprophyre and small quartz-albite ('plagiogranite') veins of hydrothermal or pegmatitic appearance. Elsewhere 'plagiogranite' occurs in the form of albite-granite (locally with granophyric texture) or quartz-albite porphyry. Pegmatites containing brown hornblende and altered plagioclase occur as very minor, irregular bodies in the area of Figure 2.10, although larger examples are found elsewhere. The lamprophyres can be seen in places along the coast forming more prominent, continuous sheets than those forming the metabasic, quasi-sheeted-dyke complex.

Interpretation

The upper metagabbro layer has been previously interpreted as the roof region of a high-level magma chamber at a constructive plate margin (Gass *et al.*, 1982; Prichard, 1985). Underplated gabbro was held to have been intruded by fine-grained gabbro and late felsic differentiates, such as hornblende pegmatites and 'plagiogranite'. The presence of swarms of basic sheets was considered to indicate a passage into an overlying sheeted-dyke layer. This formed the roof to the magma chamber, underlying the sea floor, and was either removed tectonically during obduction or was subsequently eroded. However, in terms of this model, it is difficult to explain the presence of xenolithic masses of wehrlite–clinopyroxenite in both the upper metagabbro and across the boundary between the base of the lower metagabbro and the top of the metadunite layer. These occurrences suggest that the metadunite and the metagabbro layers are separate intrusive bodies and not successive cumulate products of crystallization in a magma chamber as proposed previously.

Basic dykes are confined to the upper metagabbro and do not cut the screens of wehrlite–clinopyroxenite. They must, therefore, have been intruded separately from the two metagabbro layers, although geochemical analyses of the basic sheets and the metagabbro indicate that they are similar in composition (Gass *et al.*, 1982; Spray, 1988). Both the petrological Moho and the geophysical Moho in the Lower Nappe are steep to vertical and parallel to one another but the sheeted-dyke-like metabasic sheets in the upper metagabbro, instead of being normal to the Mohos as in the Cyprus model (Gass, 1980; Gass *et al.*, 1982) are more nearly parallel to them. There is no evidence to indicate whether the sheets were intruded before or after the rotation of the mantle, metadunite and lower metagabbro layers into the vertical position that they now occupy in the Lower Nappe in Unst. However, from the similarity of their metamorphic state (greenschist facies) to that of the ultramafic and basic lower layers, basic-sheet intrusion probably took place prior to the metamorphism of the ultramafic and basic layers. It is possible that the slice of crust and upper mantle forming the Lower Nappe slipped and rotated into a recumbent position on a listric fault at the constructive margin prior to the intrusion of the dykes.

Spray and Dunning (1991) obtained a U-Pb age of 492 ± 3 Ma from zircons extracted from a quartz-albite 'plagiogranite' vein at South Ship Geo (Figure 2.10). They considered that the vein was formed at high temperature, before the greenschist facies hydration of the metagabbro, whereas Flinn considers that it is a dacitic sheeted dyke. Either way, the zircon age obtained must be close to the age of formation of the ophiolite. This age is very close to that obtained from the Virva GCR site for the obduction of the Upper Nappe.

Conclusions

This site provides an easily accessible continuous section through the upper metagabbro and quasi-sheeted-dyke complex forming the uppermost exposed level in the Shetland Ophiolite. It reveals the lithologically and structurally complex nature of this layer and its separation from the underlying lower metagabbro by xenolithic screens of wehrlite–clinopyroxenite. This example of ophiolitic-sheeted dykes is the clearest and most extensive available in Britain.

HAM NESS
(HP 639 010–634 017)

Derek Flinn

Introduction

The site lies in the SE of Unst where continuous coastal sections and good inland exposure reveal a klippe of serpentinized ultramafic rock (the Mu Ness Klippe) of the Upper Nappe and its basal thrust resting on the Lower Nappe. The Lower Nappe is formed of two tectonically juxtaposed fragments. Rhythmically banded lower metagabbro lies to the south and a fine-grained facies of the upper metagabbro with quasi-sheeted-dyke-like metabasic sheets, quartz-albite ('plagiogranite') veins and lamprophyre dykes, lies to the north (Gass *et al.*, 1982; Prichard, 1985; Flinn, 1996). The site presents an easily accessible view of all these features characteristic of the highest level of the Shetland Ophiolite pseudo-stratigraphy and also illustrates the overthrust nature of the Upper Nappe.

Description

The Mu Ness Klippe is composed of highly sheared and serpentinized ultramafic rock forming the highest parts of the Mu Ness peninsula (Figure 2.12). The serpentinization is of antigorite type and has completely transformed the olivine in the rock so that the protolith of the serpentinite is not immediately obvious in the field. Thin-section examination shows that the klippe is composed of a mixture of metadunite, clinopyroxene-bearing dunite and wehrlite–clinopyroxenite. At the south end of the klippe are chromite prospecting pits and trenches, dug early in the 20th century, the waste from which shows the presence of banded coarse chromite in the dunite (see also The Punds to Wick of Hagdale GCR site report).

The klippe is separated from the underlying metagabbro by a thrust plane, which in many places is clearly exposed and easily accessible (Figure 2.13). At 638 009 it is flat lying and the overlying serpentinite is coarsely sheared parallel to the thrust, imparting a slate-like lamination enclosing sporadic fat, unsheared lenses. At 6388 0068 the shear zone includes a thin lens of blackish phyllite, which could be mistaken for a metasedimentary rock belonging to the Middle Imbricate Zone (Figure 2.4a). In the SE corner of the site, at 637 005, the thrust plane forms an upright, isoclinal synform in the cliffs with the closure only just above sea level. Here the shearing parallel to the thrust is very well displayed in both the serpentinite and the metagabbro.

The metagabbro mass on which the serpentinitic klippe rests is divided into two parts by a steeply dipping, structurally complex zone of gneissose or flaser gabbro. To the south of the flaser zone the metagabbro is of typical lower metagabbro type, both in its possession of the medium-grained, white and green speckled appearance and of the characteristic lower metagabbro rhythmic banding and the healed early brecciation (compare with the Skeo Taing to Clugan GCR site). North of the flaser zone, the rocks have many of the characteristics of the upper metagabbro (compare with the Qui Ness to Pund Stacks GCR site) with variable, fine- to medium-grain size and abundant metabasic sheets of quasi-sheeted-dyke type. The sheets range up to half a metre thick and alternate with screens of host metagabbro in equal volume. The intrusive sheets are particularly well displayed on the beach, at 6758 0160 and around 638 014, where they have been etched out by differential marine erosion, though large boulders interrupt the continuity. A reconstruction

Ham Ness

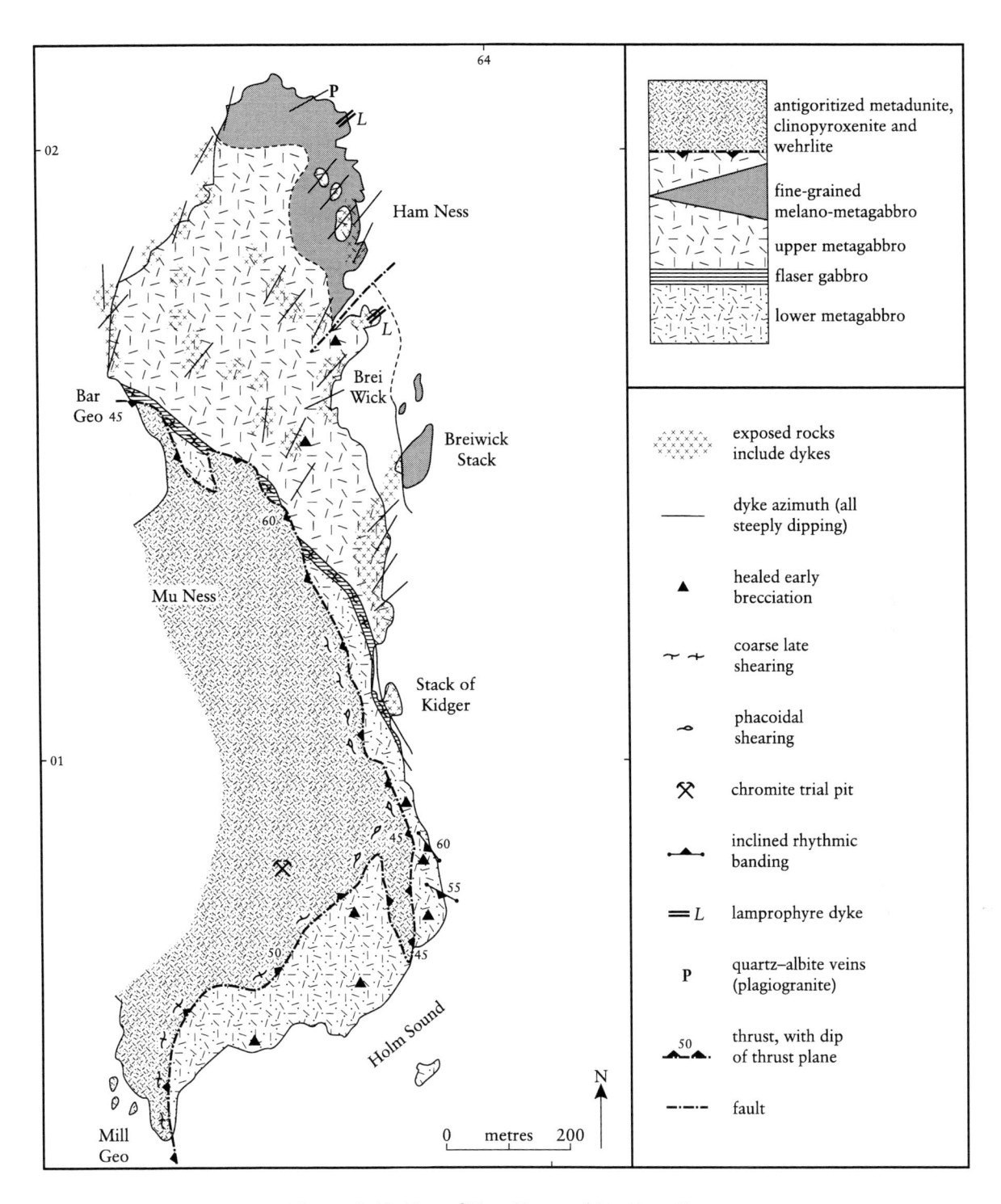

Figure 2.12 Map of Ham Ness and Mu Ness, Unst.

of the occurrence (at 638 014) has been published by Prichard (1985). Inland exposures also show metagabbro intruded by many metabasic sheets but these are more difficult to detect without the enhancing of visual contrast between the intruded rock and the host rock provided by marine erosion.

The north tip and the east side of the peninsula have been intruded by a fine-grained, epidote-hornblende gabbro. A lithologically similar rock forms The Vere, a half-tide rock in the sea 1.5 km to the NNE. This intrusion cuts the

Figure 2.13 The Mu Ness serpentinite klippe (dark coloured) resting on lower metagabbro of the lower nappe (light coloured) as seen from (HP 637 000). (Photo: D. Flinn.)

metabasic sheets, but is itself cut by several similar sheets of similar orientation. It is also cut by a quartz-albite vein ('plagiogranite') and by several lamprophyre dykes.

Interpretation

Within the square kilometre of Mu Ness and Ham Ness a thrust mass of ultramafic rock rests on metagabbro and on fine-grained metagabbro intruded by quasi-sheeted dykes. Three of the four main components to be expected in an ophiolite complex are thus clearly displayed in visible tectonic contact, in an easily accessible environment. The same components are also displayed elsewhere in the Lower Nappe in their natural (non-tectonic) relationship to each other (see Skeo Taing to Clugan and Qui Ness to Pund Stacks GCR sites).

Comparison of the two gabbroic ophiolite components represented in Mu Ness with the larger, in-situ bodies in the Lower Nappe reveals so little difference that the same interpretation may be applied. The serpentinite forming the Mu Ness Klippe is formed of metadunite, clinopyroxene-bearing metadunite and wehrlite –clinopyroxenite, similar to that described in the Skeo Taing to Clugan GCR site but here occurring at the base of the Upper Nappe. It is probably a tectonic slice cut from a level in the ophiolite above the mantle and tectonically emplaced immediately below the Upper Nappe, rather than forming a part of it. Elsewhere in the Shetland Ophiolite the island of Sound Gruney is lithologically and tectonically similar (Figure 2.4).

An upright isoclinal fold of the thrust plane below the Mu Ness Klippe, with a northerly axial trend and an amplitude of about 10 m, is clearly exposed in the cliff in the SE corner of Mu Ness. It is of interest in determining the history of obduction. No obduction can have taken place to either the east or the west in the presence of this synformal downfold as such movement would have had the effect of detaching it from the base of the klippe. The axis of the fold is approximately parallel to, and in line with, the axis of the major downfold of the thrust below the Vord Hill Upper Nappe klippe to the south in Fetlar (see the Tressa Ness to Colbinstoft and

Virva GCR site reports). Both these axes are approximately parallel to the lineations, fold axes, pebble elongations and L-tectonite fabrics of the metasedimentary rocks of the Middle Imbricate Zone between the Upper and Lower nappes (Flinn, 1958). Evidently at some time after the emplacement of the nappes by obduction they were constricted about a north to north-easterly axis, which resulted in folding on this axis and a pebble extension and lineation parallel to it (Flinn, 1958, 1992). The alternative interpretation, that thrusting during obduction took place in a northward direction parallel to the lineation and the fold axes (Cannat, 1989), was based largely on the conventional concept that lineations are parallel to transport directions, but this seems less plausible in the context of the Shetland Ophiolite.

Conclusions

The Ham Ness GCR site provides an easily accessible view of a detached thrust slice (klippe) of ultramafic rock from the Upper Nappe, resting on quasi-sheeted-dyke complex and lower metagabbro. There are abundant inland exposures and a continuous coastal section. This site is of particular importance for the Shetland Ophiolite in that it provides examples of the component ophiolitic units in tectonic juxtaposition and allows their structural relationships to be deduced. It is thus of major importance for any large-scale interpretations of the Caledonian Orogen.

TRESSA NESS TO COLBINSTOFT (HU 612 940–620 949)

Derek Flinn

Introduction

The site lies along the base of the Vord Hill Klippe of the Upper Nappe of the Shetland Ophiolite in Fetlar. The site is bound to the NW by a continuous low cliff exposing the base of the nappe, but not the thrust beneath it. Farther south the thrust is exposed in the coast and passes up the side of Stackaberg where it overlies thin slices of lower metagabbro which separate the nappe from the underlying metasedimentary phyllites of the Middle Imbricate Zone. The latter are also exposed in a window through both the klippe and the underlying metagabbro slices near the top of Stackaberg. Inland, to the east of the coastal section, the base of the klippe is seen to be intensely but crudely sheared, in many places in a phacoidal manner. This latter area is notable for the presence of a number of small rodingite (garnet-diopside-epidote-prehnite) bodies within unsheared metadunite lenses which form enclaves at the sheared base of the klippe (Phemister, 1964; Flinn, 1996).

Description

The site is bound to the west by accessible low cliffs of coarsely sheared serpentinite very close to the east-dipping basal thrust of the Vord Hill Klippe (Figure 2.14). The klippe is composed almost entirely of ochrous-weathering, strongly serpentinized harzburgite which forms the central part of Fetlar. The thrust plane beneath the klippe is associated not only with intense shearing of the base of the nappe but also with recrystallization of the serpentinized harzburgite to a white-weathering antigorite serpentinite (see also The Punds to Wick of Hagdale GCR site). Between Tressa Ness and Colbinstoft the zone of shearing and antigoritization extends about a quarter of a kilometre to the east of the coastal section and is very well exposed (Figure 2.15).

In the coastal section the shear fabric is continuous; it strikes parallel to the coastline and dips variably to the east. The shear fabric varies in nature along the coast from a crudely parallel-spaced cleavage to a coarsely anastomosing foliation enclosing close-packed lenticles of serpentinite. Inland the rocks present much the same appearance, except that larger (up to 100 m long) lenses of non-schistose and unsheared serpentinite are there enclosed in sheared serpentinite indistinguishable from that along the coast.

On the coast south of Easter Gors Geo the thrust marking the base of the klippe runs inland from under the sea. It is steeply dipping and separates the overlying metaharzburgite of the klippe from metagabbro of lower metagabbro type. The metagabbro, in turn, overlies phyllitic metasedimentary rocks of the Middle Imbricate Zone exposed along the coast to the west. The base of the klippe and the underlying metagabbro slices can be traced up the hillside to the summit area of Stackaberg where the thrust becomes horizontal. On the summit of Stackaberg the metaharzburgite forms a group of irregular, closely spaced erosional remnants of the klippe resting on a tectonic sheet of

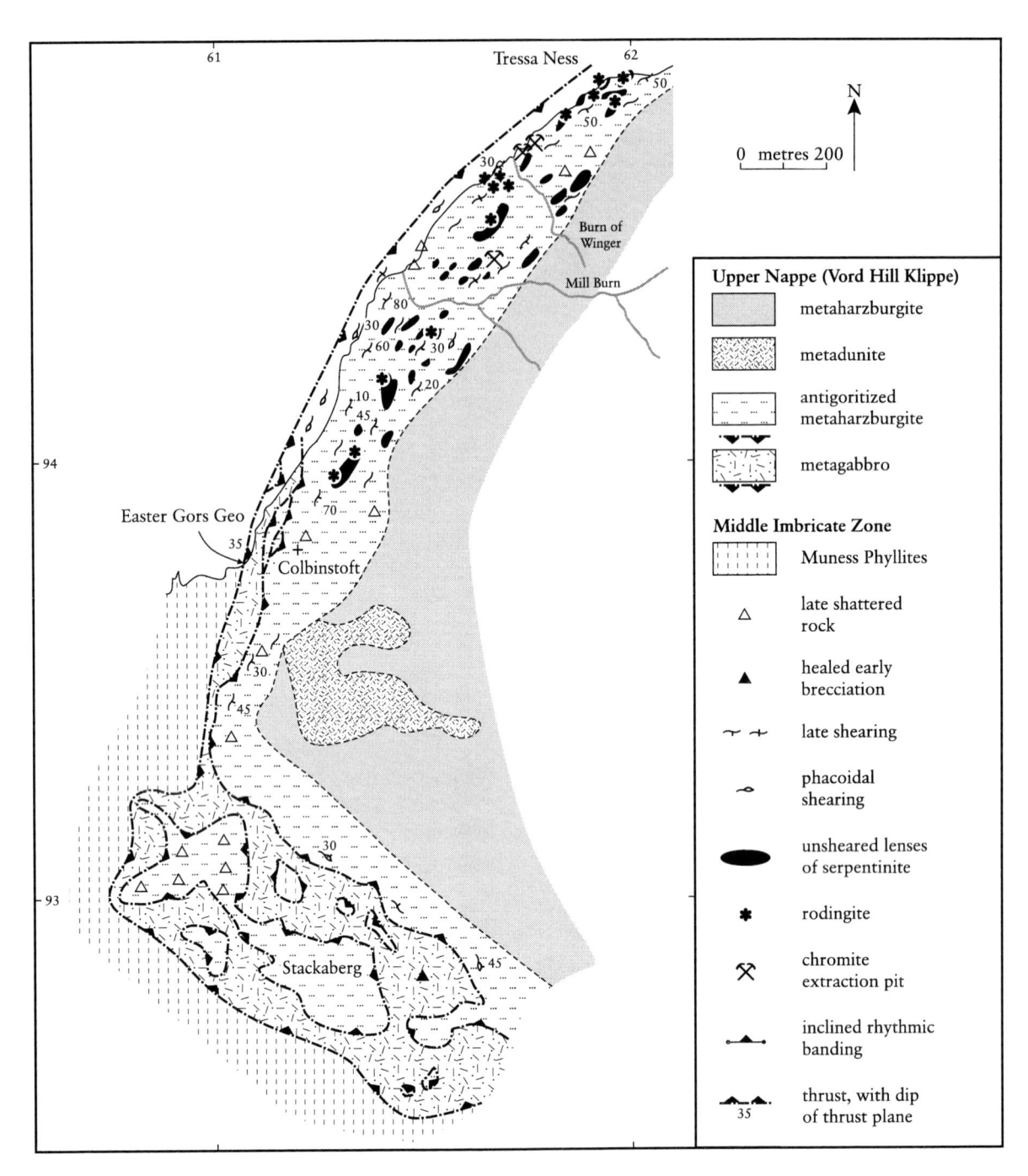

Early Ordovician volcanic rocks and ophiolites of Scotland

Figure 2.14 Map of the Stackaberg to Tressa Ness area, Fetlar.

metagabbro no more than a few tens of metres thick. In one place a window through the klippe and the underlying metagabbro reveals metasedimentary rocks of the Middle Imbricate Zone.

The most unusual lithological feature of the site is the presence of a number of rodingite bodies in the Colbinstoft to Tressa Ness area (Phemister, 1964). These are closely associated with large unsheared lenses of serpentinite enclosed within the schistosity or cleavage that dominates the base of the klippe. The rodingite occurrences are very irregular and ill-defined, vein-like bodies rarely more than a metre across in any direction. They grade continuously into the adjacent serpentinite, which has a similar greenish colour and surface texture. However, some, and especially the larger ones, have pink or even red cores due to the dominance of

Figure 2.15 Tressa Ness from the south showing the sheared antigoritized harzburgite (dark colour) and the unsheared lenses of antigoritized dunite (light colour). (Photo: D. Flinn.)

macroscopic garnet, for example at 6192 9487. The mineral content of the fine-grained, outer part of the veins can only be determined in thin section; diopside, epidote-type minerals, idocrase and chlorite are the principal constituents. The rodingite veins are not cut by the shearing associated with the base of the nappe. However, it is not clear if this is because the veins post-date the shearing or because they are protected by the unsheared lenses.

Interpretation

The unsheared lenses of serpentinite contained in the sheared base of the metaharzburgite klippe are formed of antigorite serpentinite derived from dunite (Phemister, 1964). Metadunite bodies of equidimensional, irregular and sheet-like form occur in the unsheared ochrous-weathering metaharzburgite nearby, but nowhere else in the Shetland Ophiolite are they more resistant to shearing than the enclosing metaharzburgite. The lenses weather white as a result of the complete antigoritization of the serpentine. In thin section they show no trace of serpentinized pyroxene, which is commonly visible in antigoritized harzburgite, and characteristically they contain more and slightly larger accessory chromite grains than would be expected of harzburgite.

According to Phemister (1964) the rodingite is the result of the metasomatic transformation of veins of gabbro (the protolith of the metagabbro) present within the ultramafic rock, by solutions from the cooling gabbro. He based this conclusion on the close proximity of the metagabbro with the serpentinite and the fact that in the whole of the Shetland Ophiolite rodingite only occurs where these two rocks are in contact. However, there are no 'veins of gabbro' present and the nearby metagabbro is in tectonic contact with the serpentinite, not in igneous contact as concluded by Phemister.

Rodingite occurs in Shetland only in the base of the Upper Nappe in Fetlar, Unst and the island of Sound Gruney to the west of Fetlar. The rodingite veins have the appearance in the field and in thin section of being hydrothermal and metasomatic. They contain diopsidic clinopyroxene (very little Fe or Al), garnet (grossular 50% and andradite 43%), epidote minerals including zoisite and clinozoisite, prehnite, idocrase, chlorite and titanite (Phemister, 1964). According to Phemister (1964) 'the transformation which resulted in the rodingites of Fetlar was induced by residual solutions from the gabbroic rock which were effective in carrying off Si, Al, and Na from the existing rock [gabbro] and possibly added Fe, Ca, and Mg'. Phemister's conclusions are similar to those in vogue today in other parts of the world for rodingite deposits associated with serpentinite (O'Hanley, 1996). However, it must be stressed that in Fetlar there is no evidence for the presence of gabbro veins in the base of the nappe and the gabbro was already consolidated and metamorphosed before it reached its present position below the serpentinite nappe, hence there is no clear gabbroic source of the metasomatizing solutions. Phemister (1964) also concluded that the rodingite veins were formed after the antigoritization (i.e. after the local, secondary serpentinization) and before the shearing of the base of the nappe. No definitive answer seems currently possible and the source of the Ca-rich metasomatizing solutions, the localization of the veins, and the timing of their formation are all problems waiting to be solved.

Conclusions

The Tressa Ness to Colbinstoft GCR site presents a remarkably well-exposed section through the base of the Upper Nappe. It illustrates the tectonics of emplacement of the nappe and the metamorphic and metasomatic processes that have taken place there to form rodingite deposits. The unusual characteristics of the latter provide one of the more enigmatic aspects of the Shetland Ophiolite. This adds to the value of the site since late-stage Ca-metasomatism is a widespread but poorly understood feature of many ophiolite complexes elsewhere in the world.

VIRVA
(HU 645 921)

Derek Flinn

Introduction

The site is bound to the east by a continuously exposed sea-cliff section revealing the base of the Vord Hill Klippe, a part of the Upper Nappe, the underlying thrust and the rocks below the thrust. Except at the south end of the site the cliffs are almost entirely composed of sheared and shattered serpentinite forming the cataclastic base of the Upper Nappe. In three places hornblendic rocks belonging to the underlying Middle Imbricate Zone are exposed. Both the hornblendic rocks and the overthrust ultramafic nappe are very well exposed and readily accessible. Hornblende schists are a characteristic feature of ophiolite complexes and are commonly found immediately beneath ophiolite nappes. They are conventionally interpreted as dynamothermal aureoles caused by the obduction of hot mantle nappes over basic volcanic rocks of the sea floor (Williams and Smyth, 1973). This interpretation has been applied to Fetlar by Gass *et al.* (1982), Prichard (1985) and Spray (1988), but has been rejected by Flinn on the basis of field evidence obtained from this site (Flinn *et al.*, 1991; Flinn, 1993).

Description

The sea cliffs northwards from Virva are formed of intensively shattered serpentinite lying immediately above the basal thrust of the Vord Hill Klippe of the Upper Nappe (Figure 2.16). The main body of the klippe to the west, underlying central Fetlar, is almost entirely formed of strongly ochrous-weathering metaharzburgite. However, one-quarter to one-half a kilometre west of the shoreline in Figure 2.16 the weathering colour changes eastward to white and the rocks become increasingly shattered and sheared. The change in weathering colour is due to the recrystallization of ochrous-weathering, relatively iron-rich lizardite serpentine to white-weathering, iron-poor antigorite serpentine and magnetite, a recrystallization that occurs along major shears in the metaharzburgite throughout the ophiolite. On the cliff face the serpentinite is seen to be intensely shattered and is coarsely sheared locally. Continuous fractures, many of which dip to the west, cut through this chaotically arranged assemblage. The condition of the serpentinite varies from a mass of fine powder to coherent blocks a metre or more across.

North of Virva the thrust beneath the serpentinite is submerged beneath the sea at the foot of the cliff and is exposed only at Demptster's Geo. At Virva (Figure 2.17) it is exposed as it passes inland to the south, where its course is marked by a very minor escarpment. The thrust plane is a thin zone of finely crushed and sheared rock a few centimetres thick. In the Virva area the base of the klippe adjacent to the thrust has been sheared into a crushed conglomeratic mass

Virva

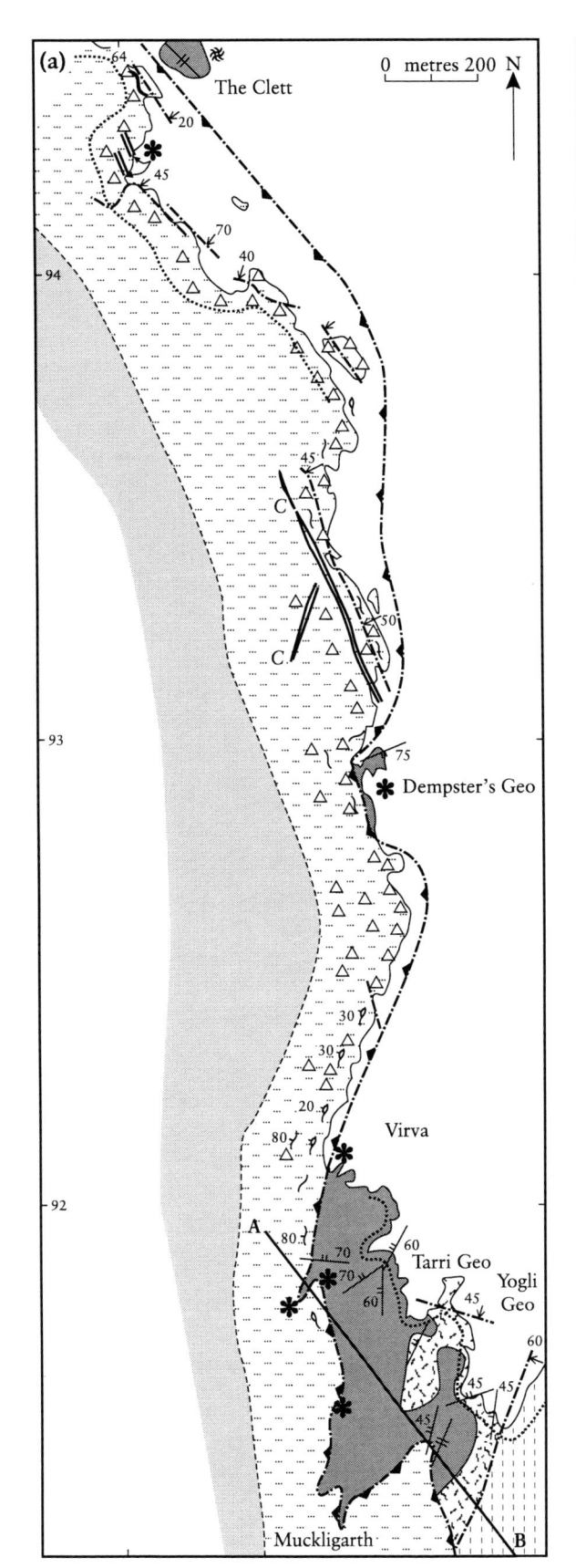

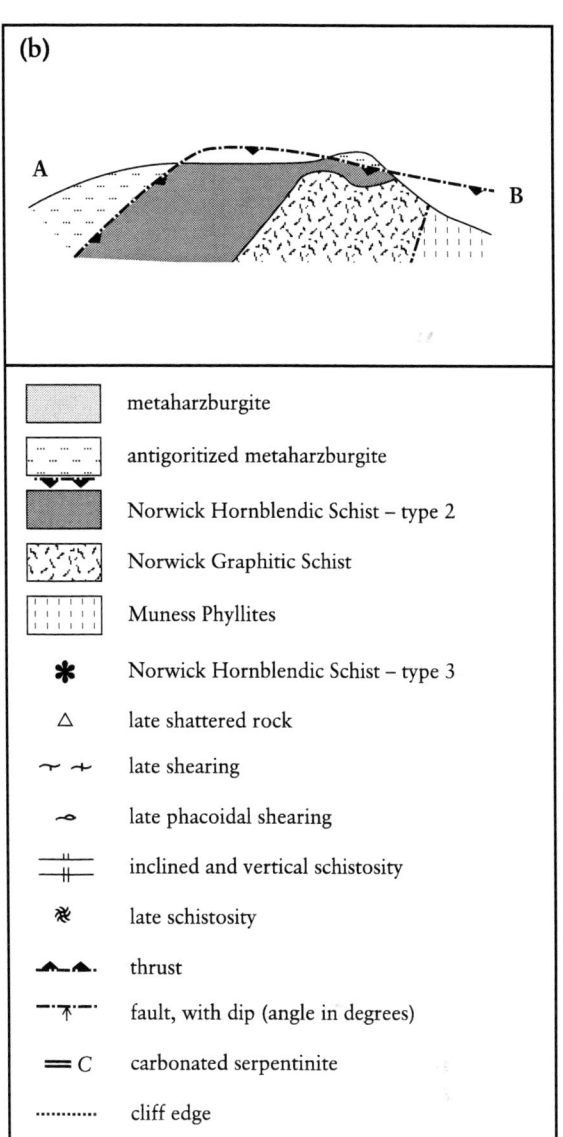

Figure 2.16 (a) Map of the Virva area, Fetlar. (b) cross section

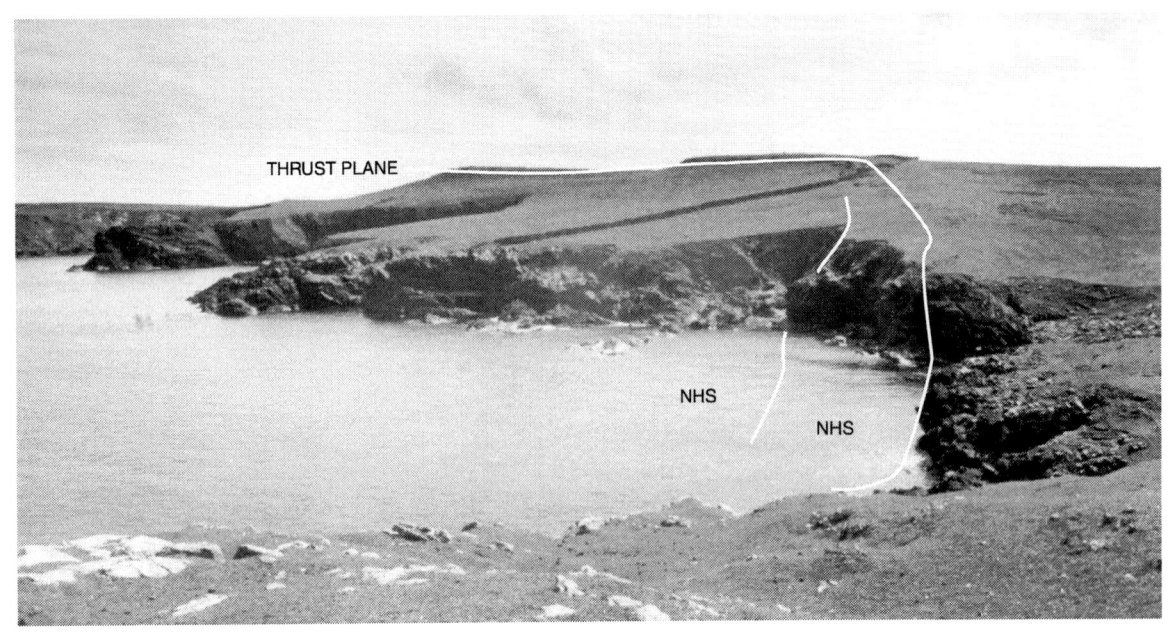

Figure 2.17 Virva from the north, showing the outcrop of the thrust below the Upper Nappe and the underlying hornblendic rocks. NHS, Norwick Hornblendic Schist (Photo: D. Flinn.)

some metres thick and composed of phacoidal serpentinite blocks ranging up to half a metre in diameter.

The hornblendic rocks beneath the nappe are equivalent to hornblendic rocks in Unst which occupy a similar tectonic position and were named the Norwick Hornblendic Schists by Read (1934). Three types of Norwick Hornblendic Schist have since been distinguished by Flinn (1993) of which two occur in the Virva site; these have been named Norwick Hornblendic Schist – type 2 and Norwick Hornblendic Schist – type 3 and are described below.

The Norwick Hornblendic Schist – type 2, here referred to as hornblende schist, is composed largely of green hornblende with the grains elongated parallel to the c-axes and arranged in the plane of the schistosity making them S-dominant tectonites with little sign of a lineation. The maximum grain size is rarely more than one millimetre. The schistosity is often emphasized by closely spaced, feldspar-rich, fine lenticular laminations. The schists contain epidote and chlorite in variable amounts, often associated with low-grade alteration. Titanite is a very common and often abundant accessory.

The Norwick Hornblendic Schist – type 3, here referred to as hornblende granofels, is a hornblende-clinopyroxene-garnet granofels with a grain size of several millimetres. It generally appears as black and homogeneous and the garnet and clinopyroxene grains can only be clearly discerned in the polished exposures near sea level. The hornblendes are brown and have a weak optical preferred orientation. The hornblende grain boundaries form a granoblastic polygonal mosaic indicative of recrystallization under isostatic conditions. The boundaries of the clinopyroxene grains cut across this hornblende mosaic showing that pyroxene replaced hornblende. Garnet (as grains generally about 2 mm in diameter), titanite, ilmenite, apatite, rutile and small amounts of quartz occur as accessories together with minor amounts of recrystallized (albitic) feldspar.

The relationship of the two types of Norwick Hornblendic Schist to the serpentinite nappe are best displayed at Virva. There, Gass *et al.* (1982) and Spray (1988) described a contact between the serpentinite and a clinopyroxene-garnet-hornblende rock (the granofels) which grades east (away from the serpentinite) through hornblende schist into greenschist. As described above, the mineral constituents of the granofels can only be identified in the field at the cliff-foot, where it has been scoured by the sea. Thin sections reveal that in fact the granofels grades to both east and west into chilled margins, a metre or more wide, against the adjacent rocks. In its outer part the chilled margin consists of fine-grained (less than 0.1 mm) aggregates of chlorite containing scattered opaque grains and very rare grains of clinozoisite 0.2–0.3 mm across. This contact facies passes into an inner zone dominated by closely packed anthophyllite needles in a matrix of chlorite but including rounded spots of pure chlorite about 0.3 mm across. This in turn passes through a random mixture of amphiboles, chlorite, epidotes, albite, ilmenite, and titanite, into the high-grade hornblende granofels.

At Virva (6445 9208) the hornblende-garnet-clinopyroxene granofels (Norwick Hornblendic Schist – type 3), with its chilled margins, extends along the coast for some 40 m from the thrust. The hornblende schist (Norwick Hornblendic Schist – type 2) cropping out immediately east of the granofels contains widely scattered lenticular streaks of clinopyroxenite, albite pegmatite, and quartzo-albitic leucosome; some tens of metres farther east it also contains two thin lenses of apparently psammitic, garnet-biotite gneiss. Several hundred metres east of the basal thrust the hornblende schist becomes almost unrecognizable due to crushing and late retrograde metamorphism.

Several other localities within the Virva site show important aspects of the granofels and schist. Chilled hornblende granofels can be seen welded to serpentinite in small exposures on the hillside south of Virva, at 6446 9157 and (6442 9185). At the latter exposure the granofels invades the serpentinite in a metre-wide dyke-like apophysis with a south-westerly trend. Hornblende schist forms the sea stack at Dempster's Geo and, between the stack and the serpentinite of the nappe in the adjacent cliff, fragments of chilled hornblende granofels occur among the shattered rock. Farther north, at the base of the cliff opposite to The Clett (640 943), two or three south-dipping intrusive sheets, less than a metre thick, consist of chilled hornblende granofels. At the southern margin of Figure 2.16, to the south of the coast at Virva, the basal thrust of the klippe can be seen to cut across the Norwick Hornblendic Schist into the underlying graphitic phyllite layer of the Middle Imbricate Zone.

Interpretation

Williams and Smyth (1973) showed that hornblende schists are a characteristic component of ophiolite complexes and used the Norwick Hornblendic Schists, as described by Read (1934), as one of a number of examples. Williams and Smyth interpreted the ophiolitic hornblendic rocks as a 'dynamothermal aureole' produced as hot harzburgitic mantle was obducted over basic rocks of the sea floor which were thus heated and thermally metamorphosed. Gass et al. (1982) applied this model to the Shetland Ophiolite. They showed that at Virva the grade of the hornblendic rocks decreases downwards from the basal thrust, stating that 'A thin zone of garnet-clinopyroxene amphibolite grades into amphibolite and then greenschist facies away from the contact'. However, farther west the base of the nappe overlies sheared, shattered and antigoritized serpentinite and to the east the greenschist, into which the amphibolite grades, is retrogressively metamorphosed amphibolite. These relationships do not fit readily with the simplistic interpretation of thermal metamorphism by an obducted, hot mantle slab.

Spray (1988) accepted the Gass et al. (1982) interpretation and drew further conclusions. He showed that the composition of the hornblendic rocks (supposedly the ocean floor overthrust by the ophiolite nappe) is significantly different both to that of the basic sheets of the quasi-sheeted-dyke complex and to that of their fine-grained upper metagabbro host (as exposed in Unst). On the basis of compositional comparisons with other ophiolite complexes he argued that the hornblendic rocks originated as true ocean floor whereas the Shetland Ophiolite was formed as the floor of a small basin marginal to that ocean. The marginal basin was envisaged as opening while the true ocean floor was subducted beneath it.

Flinn et al. (1991) and Flinn (1993) have argued that the Shetland ophiolitic nappes were already serpentinized when they were obducted and were thus too cool to have metamorphosed basic volcanic rocks of the ocean floor to form hornblende schist. It was also suggested that the protolith of the hornblende granofels was not basaltic oceanic floor but basic magma adiabatically melted from the mantle and segregated in the initial fracture arising from decoupling at the onset of obduction. The hornblende schist may have originated in a similar manner but somewhat earlier so that it served to lubricate the thrust during movement and thus became schistose as it cooled and crystallized.

Spray (1988) reported six K-Ar ages for hornblendes separated from four hornblendic lenses beneath the nappes. The oldest age obtained, 479 ± 6 Ma, was from the Virva occurrence (6445 9208) and the remainder ranged up to 465 ± 6 Ma. Spray interpreted this range of different ages as resulting from diachronous accretion of the hornblende schist during obduction. However, the widespread retrogressive metamorphism in the hornblende schists, leading to partial loss of radiogenic argon, makes these dates unreliable. A more dependable date is the ^{40}Ar-^{39}Ar step-heating age of 498 ± 2 Ma for hornblende separated from a Virva sample of the hornblende granofels (Flinn et al., 1991). In view of the likely low temperature of the nappe when obducted, the hornblende would not have taken long to cool to the closure temperature. This age is therefore regarded as a good estimate of the date of obduction of the Upper Nappe.

Conclusions

The coastal section north of Virva, Fetlar, provides an accessible and extensive view of the basal thrust of the Upper Nappe. This allows examination of the thrust's relationship to the underlying hornblendic granofels and schist. It is a crucial locality in the controversy surrounding the origin of these rocks and their relationship to the obduction of the Shetland Ophiolite and is of international significance. In particular, the site provides evidence that these hornblendic rocks cannot be a part of a thermal aureole formed as hot mantle was thrust over basic volcanic rocks of the ocean floor, following the conventional interpretation of hornblende schists found associated with ophiolites. Instead, a preferred interpretation is that the hornblendic rocks formed from basic magma which was intruded, from deep in the mantle, into the basal thrust of the ophiolite as it was obducted.

THE HIGHLAND BORDER COMPLEX

GARRON POINT TO SLUG HEAD (NO 893 877–886 873)

C. W. Thomas

Introduction

Mafic lavas, mafic and ultramafic intrusions and associated sedimentary rocks of the Highland Border Complex (HBC) are well exposed and readily accessible along the rugged, indented coastline between Slug Head and Garron Point, about 1 km NE of Stonehaven. At this site, the HBC crops out over about 1 km of strike, with an outcrop width of about 200 m, forming the most easterly and best exposed single occurrence of HBC rocks in Scotland. The complex is reverse-faulted against metagreywackes, grits and pelites of the Dalradian Southern Highland Group and is overlain unconformably by Old Red Sandstone facies, terrigenous clastic sediments of the Cowie Formation (Stonehaven Group), which are of Wenlock (Silurian) age.

The HBC rocks at Stonehaven were assigned a possible Arenig age on the first Geological Survey maps of the area (Sheet 67, Stonehaven), but Campbell (1911, 1913) provided the first detailed accounts of the rocks and described palaeontological evidence for their being Late Cambrian in age. The palaeontological evidence for the age of the rocks was based largely on brachiopod fragments found in the black shales intercalated with the lavas in Craigeven Bay. Although the faunal assemblages suggested a Late Cambrian age, Peach and others in the Geological Survey and Anderson (1947) considered the rocks to be Arenig or Caradocian in age, based on comparison with similar faunal assemblages in the Southern Uplands and the Stinchar Limestone of the Girvan district. Doubt was cast on the acceptability of the fossils from Craigeven Bay when Bulman and Rowell concluded that the identification of a comparable fauna from Aberfoyle was unreliable (Institute of Geological Sciences, 1963). However, more recent palaeontological work by Curry and colleagues (Curry et al., 1982; Curry et al., 1984; Ingham et al., 1985) has restored an Ordovician age for at least part of the HBC elsewhere.

Campbell (1913), in collaboration with Peach and others of the Geological Survey, identified the contact between the HBC and the Dalradian rocks as the Highland Boundary Fault, which previously had been placed at the boundary between the HBC rocks and the overlying Old Red Sandstone facies Silurian rocks. They recognized that the Silurian sedimentary rocks are unconformable on the HBC, despite local modification of the unconformity by faulting.

The petrogenesis of the pillow lavas and associated igneous rocks has been studied by Bloxam (1982).

Garron Point to Slug Head

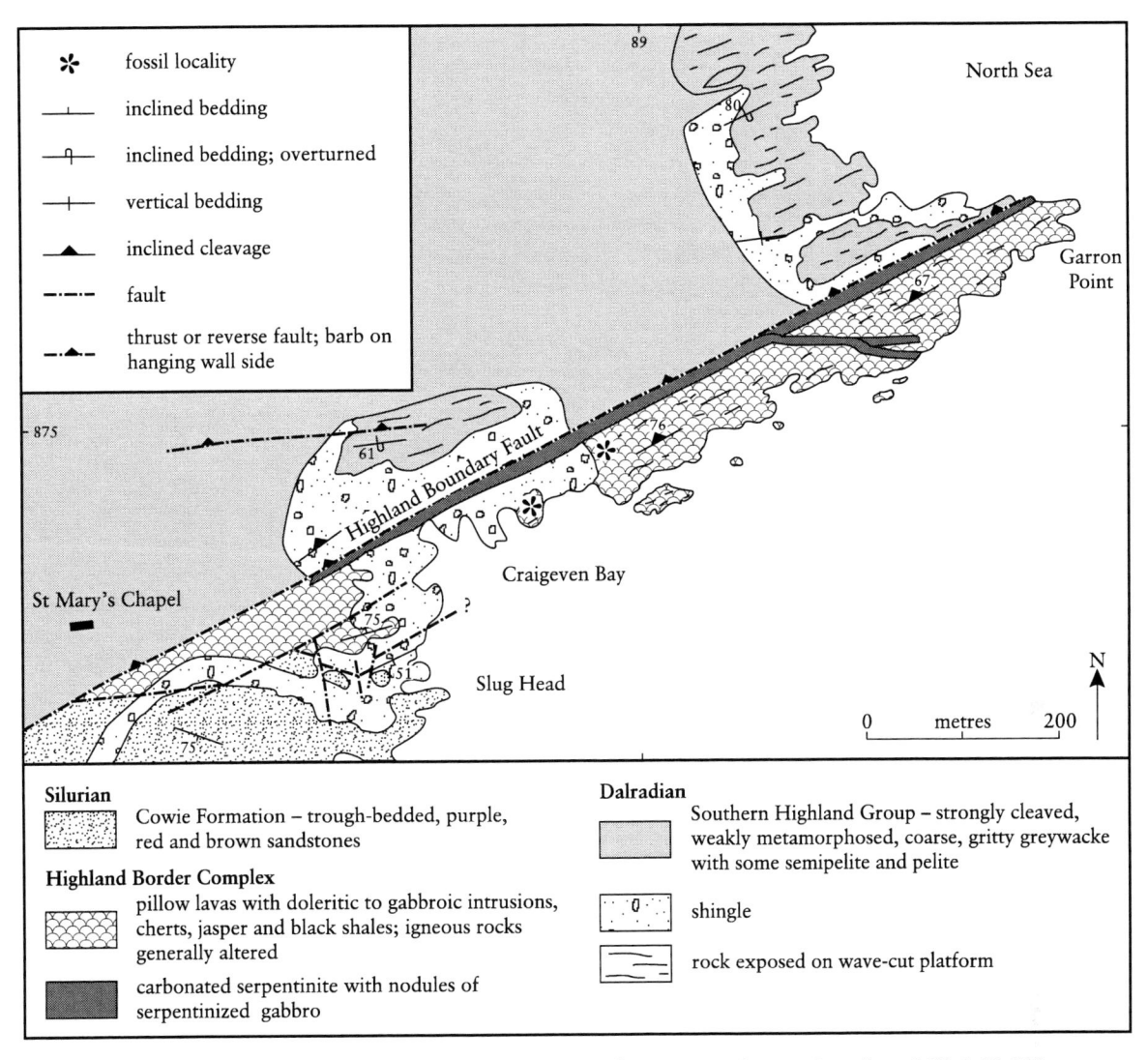

Figure 2.18 Map of the Garron Point to Slug Head area, Stonehaven. Based on BGS 1:10 000 Sheet NO88NE (1996).

Description

The Highland Border Complex is dominated largely by meta-igneous rocks at Stonehaven. Although metamorphosed to lower greenschist facies, extensively sheared and even phyllitic in places, spilitized mafic pillow lavas are generally well preserved (Campbell, 1913; Gillen and Trewin, 1987, plate 17) and clearly show the way-up, younging to the north. The rocks within the complex dip very steeply to the north and the pillow lavas become more intensely sheared and boudinaged northwards towards the tectonic contact with the Dalradian at the Highland Boundary Fault. The geological relationships are summarized in Figure 2.18 and illustrated in Figure 2.19.

The vesicular character of the pillow lavas is still recognizable and variolitic textures are commonly developed. Relict basaltic textures are preserved in places, although the development of epidote aggregates destroys most of the original igneous textures; fresh colourless calcic augite is occasionally preserved. Geochemical evidence from the pillow lavas suggests oceanic tholeiite affinities (Bloxam, 1982). Small, irregular doleritic and gabbroic intrusions in the pillow lavas are common on Garron Point.

A distinctive, orange-coloured, dolomitized and silicified serpentinite occurs along the Highland Boundary Fault and associated splays at Garron Point and is the most northerly unit of

Early Ordovician volcanic rocks and ophiolites of Scotland

Figure 2.19 The Highland Border Complex at Slug Head and Garron Point looking SW towards Stonehaven. (Photo: C.W. Thomas.)

the HBC at Stonehaven. The serpentinite is well-seen in Craigeven Bay and over the headland on to Garron Point and is up to 15 m thick. The conspicuous colouration is imparted by iron in the dolomite. The serpentinite also contains nodular inclusions of serpentinized gabbro. Bloxam (1982) records that the inclusions are difficult to distinguish from the serpentinite itself, being much more highly altered than the mafic intrusions outside of the serpentinite. The feldspars are now albite and the mafic minerals are largely replaced by hornblende, chlorite and serpentine.

Associated with the lavas and mafic intrusions are red and black cherts, including parallel-laminated black cherts with abundant pyrite, and siliceous siltstones and mudstones (Campbell, 1913; Henderson and Robertson, 1982). These occur as intercalations between flows or as lenticular masses, possibly filling the space between adjacent pillows. It is from these rocks that the faunal assemblage, dominated by brachiopods, was recovered. Grading indicates younging to the north, consistent with the younging indicated by the pillow lavas.

Interpretation

The lithological assemblage comprising the Highland Border Complex between Slug Head and Garron Point represents fragments of oceanic crust formed during the early Ordovician and tectonically emplaced against the Dalradian. The oceanic volcanotectonic setting for the rocks is clear from the lithological assemblage at Stonehaven. Although some of the mudstones and siltstones represent terrigenous material brought in by distal turbidites, most of the cherts were considered to be pelagic (Henderson and Robertson, 1982).

Bloxam (1982) suggested that the mafic inclusions in the serpentinite could be disrupted remnants of sheets intruded into the serpentinite or could have been incorporated tectonically from adjacent rocks. On the petrographical evidence, it appears that both the ultramafic protolith to the serpentinite and the mafic intrusions were serpentinized after the intrusion and albitization of the metamorphic rocks. By analogy with Ballantrae and similar rocks in California, Bloxam (1982) considered that the ultramafic protolith was probably emplaced as 'a low temperature 'mush' and/or solid tectonic slices caught up in the Highland Border *mélange*'. Henderson and Robertson (1982) considered that the whole of the HBC was emplaced tectonically.

The radiometric age of the HBC at Stonehaven remains undetermined. The faunal assemblages elsewhere in the HBC have Laurentian affinities and range in age from Arenig to possibly Llandeilo. However, the age of the HBC at Stonehaven is poorly constrained by the palaeontology and an Ordovician age has been assigned largely by analogy with the age derived from the faunal assemblage in the Dounans Limestone of the Highland Border Complex at Aberfoyle (e.g. Curry *et al.*, 1982; Ingham *et al.*, 1985).

Conclusions

The Stonehaven section provides one of the most extensive outcrops of igneous rocks in the Highland Border Complex, an understanding of which is an essential part of any tectonic interpretation of the Scottish sector of the Caledonian Orogen. This enigmatic assemblage represents remnants of ocean crust, consisting chiefly of metamorphosed tholeiitic pillow lavas with intercalated cherts, siltstones and mudstones. The complex also contains subordinate gabbroic and doleritic intrusions. Associated spatially with these rocks is a carbonated and silicified serpentinite with nodules of serpentinized gabbroic rocks; serpentinization occurred after the intrusion of the gabbros. The age of the HBC at Stonehaven is not determined, but is likely to be early- to mid-Ordovician, by analogy with HBC rocks elsewhere. There is general agreement that the HBC was tectonically emplaced into its current position. It is overthrust by Dalradian Southern Highland Group rocks to the north and is overlain unconformably to the SW by Old Red Sandstone facies rocks of Wenlock (Silurian) age.

BALMAHA AND ARROCHYMORE POINT
(NS 419 915, 416 913 AND 412 919)

J. R. Mendum

Introduction

The Balmaha and Arrochymore Point GCR site at the SE end of Loch Lomond contains readily accessible exposures of altered ultramafic members of the Highland Border Complex (HBC). They range from massive and sheared serpentinite, locally silicified, to fragmental serpentinite, typically highly carbonated. Related fragmental rocks with a limestone matrix contain chitinozoa that suggest an Arenig age. The HBC is bound to the NW by the Highland Boundary Fault and to the SE by the Gualann Fault (Bluck, 1992). Precambrian Dalradian metagreywackes and slates occur farther to the NW, and Lower Old Red Sandstone conglomerates to the SE. The ultramafic rocks form two belts, some 800 m apart, termed the Northern and Southern belts (Henderson and Fortey, 1982). They are separated by lithic arenites, part of the 'Loch Lomond Clastics' (Figure 2.20), that are interpreted as a younger component of the HBC. The HBC rocks are overlain unconformably by Upper Old Red Sandstone red-brown conglomerates and sandstones (Rosneath Conglomerate) and just NE of the site these beds overlap the Highland Boundary Fault.

The serpentinites are the oldest exposed components of the HBC, and are probably of early Ordovician (Tremadoc or Arenig) age. They range from massive altered peridotite through to sheared, fragmental, highly carbonated and highly silicified varieties. The adjacent metasedimentary 'Loch Lomond Clastics' show mineralogical and geochemical evidence of derivation from both ultramafic/mafic and quartzo-felspathic units.

The area was mapped by the Geological Survey in the late 19th century and was included

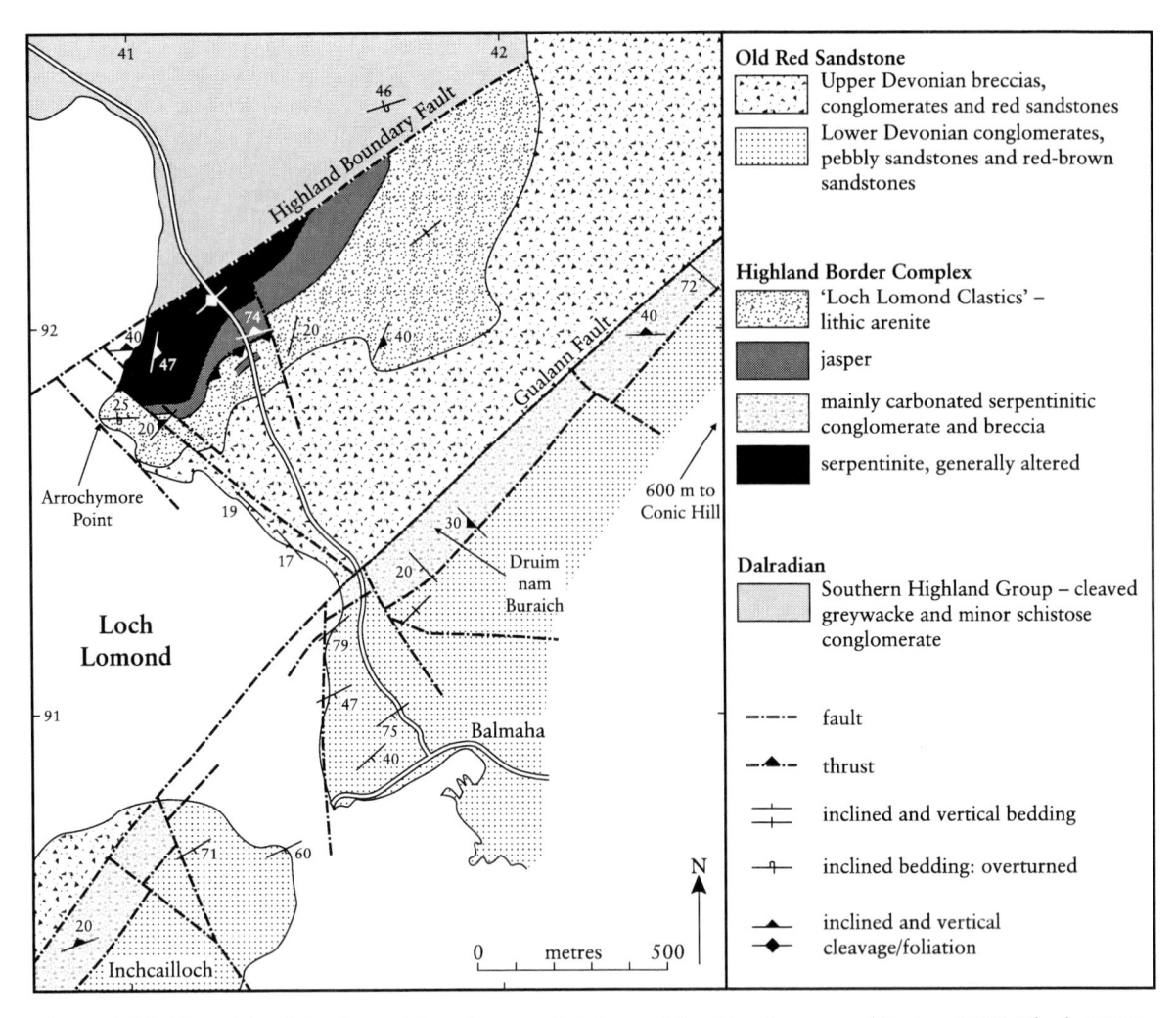

Figure 2.20 Map of the Balmaha and Arrochymore Point area (after Henderson and Fortey, 1982; Bluck, 1992).

in later studies of the HBC by Jehu and Campbell (1917) and by Anderson (1947). More recent work by Henderson and Fortey (1982), Henderson and Robertson (1982), Curry *et al.* (1982, 1984) and Bluck (1992) has described the nature of the HBC at Balmaha and in adjacent areas, notably in the Bofrishlie Burn section and the Lime Craig Quarry areas near Aberfoyle. These authors have discussed the relationships between the various component lithologies of the HBC and placed them in an overall stratigraphical and tectonic context.

Description

The southern serpentinite belt is exposed on Druim nam Buraich on the SW flank of Conic Hill (Figure 2.21). The serpentinite here is extensively altered, recrystallized, brecciated and veined by carbonate and silica. It is dominated by rounded to subangular pebble-like clasts typically 2–3 cm across and up to 15 cm long in places. In the least altered exposures (4254 9212) the clasts lie in a finer-grained (*c.* 2 mm diameter) matrix of altered serpentinite. Henderson and Fortey (1982) described these rocks as serpentinite conglomerate, derived by marine erosion from adjacent exposed intrusive serpentinite. Basic igneous rocks and quartz grains form minor clasts within the rock. Variations in fragment size and alignment of clasts is thought to reflect bedding with units typically about 50 cm thick. In other parts of the outcrop the texture of the rock is more variable and only a crude bedding structure is discernable. Only a weak gently dipping cleavage affects these lithologies. Thin-section studies (Henderson and Fortey, 1982) show that olivine 'mesh' and orthopyroxene 'bastite' textures are preserved in the larger clasts. Lizardite is the only serpentine mineral now present but tremolite, talc and chlorite are the dominant minerals and with further alteration they are replaced by ferroan dolomite and quartz. Chromite occurs both within the clasts and as separate detrital grains.

South-west along strike from Druim nam Buraich, exposures were noted by Bluck (1992) on the shore of Loch Lomond around 4156 9124. Normally below water-level, they comprise distinctive crystalline limestones and conglomerate containing clasts of dolerite, gabbro, spilite and well-rounded quartz arenite. The dolomitic limestone matrix has yielded chitinozoa suggesting an Arenig age. A similar lithology at Lime Craig Quarry near Aberfoyle (15 km to the NE) has yielded a silicified trilobite and ostracod fauna of early Arenig age from the Dounans Limestone. At this locality serpentinite is overlain by a serpentinite conglomerate with a dolomitic limestone matrix, that in turn passes upwards into the Dounans Limestone (Bluck *et al.*, 1984; Curry *et al.*, 1982; Ingham *et al.*, 1985). Similar serpentinite conglomerates are found on the Loch Lomond islands of Inchcailloch, at 4062 9010 adjacent to altered and brecciated basic volcanic rocks and serpentinite, and on Creinch (394 888). They also occur 90 km farther NE in the Highland Boundary Fault zone at Alyth, near Blairgowrie.

The northern serpentinite belt consists of variably foliated and altered serpentinite. Mesh, bastite and vein textures imply that it was initially a peridotite but the only non-serpentinized rock recorded, at a now-obscured locality near Arrochymore Point, is a pyroxenite composed mainly of chromian diopside (Henderson and Fortey, 1982). Antigorite, lizardite and chrysotile are present in the serpentinite seen elsewhere in the Arrochymore Point area. Deep-red altered chromite grains rimmed by opaque iron oxides and green, fine-grained, felted growths of chromium mica, chlorite and clay minerals are ubiquitous in the serpentinite. The foliation, defined mainly by ramifying sub-parallel fractures but locally by a mylonitic texture, is more marked adjacent to the northern faulted contact. The serpentinite was largely foliated prior to alteration to talc-magnesite and ferroan dolomite-quartz but, locally, subsequent deformation has caused the formation of cataclasite in the carbonate-quartz rocks.

Jasper, mainly dark red in colour, lies imme-

Figure 2.21 Looking SW from Conic Hill across Loch Lomond. Druim nam Buraich (foreground, right of centre) shows exposures of the southern serpentinite belt. (Photo: BGS no. D5406.)

diately SE of the serpentinite in the northern belt, but contacts are not exposed. Henderson and Fortey (1982) note that despite abundant brecciation and quartz veining relict pyroxene and olivine textures are visible, preserved by the fine-grained mosaic quartz. Early tectonic fabrics are also pseudomorphed by quartz and the jasper seems most likely to be the product of pervasive silicification of serpentinite. However, chert is commonly associated with black slates in other parts of the HBC and Bluck (1992) inferred that in these cases it is of sedimentary origin. In respect of the jasper, he commented particularly on the lack of interbedded red or black shales, the dearth of fossils, and the distinctive colour.

Immediately SE of the serpentinite and jasper forming the northern belt, cleaved gritty arenites show grading and immature grains indicative of deposition from turbiditic flows relatively proximal to their source area. The arenites are seen to be thrust over the jasper in the road cut at 4131 9194. Bluck (1992) reported that a soft green serpentinite conglomerate crops out just below the beach gravels on Arrochymore Point at 4095 9183, probably close to the arenite–serpentinite contact. This arenite unit, termed the Highland Border Grits by Tremlett (1973) and the 'Loch Lomond Clastics' by Henderson and Robertson (1982), consists of pink to grey, medium- to coarse-grained lithic arenite. In thin section Henderson and Fortey noted that it is composed of quartz, plagioclase feldspar, biotite, white mica and subsidiary chlorite, carbonate and accessory chromite. It also contains abundant grains and larger fragments of slate, chert, and ultramafic and mafic igneous rocks. Minor fine-grained acid volcanic clasts are also recorded. Development of kaolinite and extensive haematitic reddening was probably a late-stage alteration occurring in Early Devonian times. Cross-bedding is common and near Arrochymore Point, where the beds dip c. 25° to the NNW, it clearly shows that they are inverted. The cleavage in the arenites is marked by grain flattening, quartz pressure shadows and phengite mica trails. Henderson and Fortey (1982) noted that the angle between bedding and cleavage is very small so that the facing direction cannot be determined by that means. Bluck (1992) suggested that these rocks are Caradocian in age, correlating them with the similar Aberfoyle arenites from which chitinozoa have been obtained (Bluck *et al.*, 1984).

Interpretation

The Arrochymore Point outcrops are of massive, to foliated, serpentinized harzburgites and minor clinopyroxenites, commonly altered to talc-magnesite and ferroan dolomite-quartz assemblages. Pervasive silicification has altered part of the serpentinite to jasper. Farther south, the outcrops on Druim nam Buraich, near Balmaha, are resedimented serpentinite conglomerate, now almost entirely replaced by ferroan dolomite and quartz. Elsewhere these conglomerates apparently pass upwards into bedded limestones, although only limestone conglomerates are seen at Loch Lomond.

The serpentinite and serpentinite conglomerate outcrops are interpreted as the basal units of the HBC. They represent small fragments of sub-ocean-floor mantle and the detrital deposits formed by its erosion, under marine conditions, immediately following its thrust emplacment at the surface. The serpentinite assemblage is thought to be either Late Cambrian or early Ordovician in age. Basic intrusive rocks, now amphibolites, are associated with the ultramafic bodies near Aberfoyle and at Scalpsie Bay on Bute (Henderson and Robertson, 1982). They may represent original ocean-floor volcanic rocks subjected to dynamothermal metamorphism beneath the hot mantle slab as it was obducted. In this model the serpentinite conglomerates could have been generated by erosion at the active thrust front. Limestone deposition accompanied the later stages of serpentinite conglomerate formation and then continued under quieter marine conditions, in shallower water, as obduction was completed.

The overlying 'Loch Lomond Clastics', here represented only by lithic arenites, are a varied sequence of conglomerates, gritty arenites, shales and limestones. Robertson and Henderson (1984) used geochemical evidence to constrain the possible origin of the arenites. Analysis of a specimen of cleaved arenite from 200 m east of Arrochymore Point (4091 9178) showed high nickel (561 ppm) and chromium (1234 ppm) values in a rock with 77.7% silica. They concluded that the 'Loch Lomond Clastics' had been derived from erosion of ultramafic, mafic and quartzo-feldspathic rocks and hence deduced that they stratigraphically overlie the serpentinite, serpentinite conglomerate and succeeding limestone units. Indeed, they may be considerably younger than the serpentinite conglomerate and are possibly of Caradocian age.

Conclusions

The outcrops of the Highland Border Complex near Balmaha are of crucial importance in understanding its genesis. They provide examples of the oldest exposed parts of the complex, the serpentinite and serpentinite conglomerate, and reveal something of their relationships with overlying clastic strata. The serpentinite is generally regarded as of Late Cambrian or early Ordovician (Merioneth–Arenig?) age whereas the sedimentary rocks may range up to the late Ordovician (Caradoc?). Both the serpentinite bodies and adjacent clastic strata are fragments of the oceanic floor and sedimentary fill of small basins formed at the Laurentian continental margin. Much of the complex was emplaced both during collision of a large volcanic arc with the Laurentian margin in late Llanvirn times (*c.* 470 Ma), and by subsequent transcurrent faulting in the late Ordovician and Silurian. Some of the mafic and ultramafic elements may have been obducted during an earlier stage of this complex tectonic history.

NORTH GLEN SANNOX (NR 995465–995475)

D. J. Fettes

Introduction

The North Glen Sannox site presents an excellent section through the lava and black shale sequences of the Highland Border Complex (HBC) in north Arran. It illustrates one of the best and most varied developments of basic volcanic rocks within the complex. Downward-facing structures provide critical evidence for the study of the structural setting of the complex and its relationship to the Dalradian succession. The geology of the site is summarized in Figure 2.22.

Within this site the HBC outcrop forms a N–S strip that is about 2.5 km long, varies in width from 200 to 400 m and is broadly parallel to the regional strike. It is bound to the west by the turbiditic sandstones of the Dalradian Southern Highland Group and to the east by the Highland Boundary Fault and the Lower Old Red Sandstone succession (Figure 2.22). The strip is cut off to the south by the Palaeogene Northern Granite and to the north by the Highland Boundary Fault, which gradually transgresses northwards across the strike. The complex is composed predominantly of lavas with subordinate black schists, cherts and sandstones. Biostratigraphical control is poor but indicates an Arenig age.

The complex was first mapped by the Geological Survey around the beginning of the 20th century and was reported on in resultant memoirs by Gunn *et al.* (1903) and Tyrrell (1928). The area has been the subject of considerable subsequent study, notably by Anderson and Pringle (1944), Johnson and Harris (1967) and Henderson and Robertson (1982), most of whom were largely concerned with the structural setting of the HBC rocks and their relationship to the Dalradian succession.

Description

Excellent sections are provided through the Highland Border Complex outcrop by the North Sannox Burn (Figure 2.23) and by scattered exposures on the adjacent hillsides. In the burn, turbidites of the Southern Highland Group are exposed above the road-bridge and for about 130 m downstream to their contact with the HBC rocks. In the upstream part of this section massive pebbly sandstones predominate with quartz and feldspar clasts, but flaggy sandstones, siltstones and slates become more common downstream (towards the east). The bedding strikes *c.* NNE and is generally vertical or steeply dipping towards the east. Good evidence from graded beds indicates a consistent sense of younging, also to the east, in which direction the Dalradian strata are succeeded by the HBC assemblage. This may be divided into five units that comprise, from west to east:

1. black shales and cherts,
2. lower lava sequence,
3. black shales and cherts,
4. upper lava sequence,
5. green slates and sandstones.

Gunn *et al.* (1903) originally ascribed an Arenig age to these rocks on the basis of their similarity to the rocks at Ballantrae. Subsequently Anderson and Pringle (1944) reported a find of some poorly preserved brachipods from the central black shale unit. The shells were referred to *Acrotreta* and compared with *A. nicholsoni* described from Aberfoyle by Jehu and Campbell (1917). Curry *et al.* (1984), although unable to trace the original specimens, accepted the age albeit with some caution.

The contact between the Dalradian and HBC rocks is exposed on the south bank of the burn but is only readily examined when water levels are low. Anderson and Pringle (1944) placed the contact at the junction between fine-grained sandstones and black shales. They reported that there is no evidence of interbedding and, although the shales are disrupted, there is no sign of significant faulting; the disruption being ascribed to 'slight movement' at the contact during the regional folding. Eastwards of the contact the black shales continue for a few metres followed by a heavily brecciated and carbonated lava which may represent an altered autobrecciated flow. This is succeeded by a further 2–3 m of black shale with chert bands.

Above the black shales lies the lower lava unit, which is here over 100 m thick and forms the first major development of volcanic rocks in the section. The lavas are greenish-grey, spilitic, commonly pillowed and locally contain zones of autobrecciation; the latter were described by Henderson and Robertson (1982) as 'clast-supported lava rudites, composed of angular to subrounded clasts 0.05 to 0.25 m in diameter'. The pillows have chilled margins and concentric amygdaloidal zones. Johnson and Harris (1967) described almost undeformed pillows from Torr na Lair Brice with flat bottomed bun-shapes, pear-shapes and wedge-shapes all indicating younging to the east. Henderson and Robertson (1982) drew attention to the absence of intercalated sedimentary units in contrast to the basic volcanic rocks in the HBC in the North Esk and at Stonehaven.

The succeeding black shale unit is about 40 m thick and contains numerous chert beds and some beds of coarse-grained sandstone which may in part be volcaniclastic. It was from this unit, where it crops out in the Allt Carn Bhain (Figure 2.22), that Anderson and Pringle (1944) recovered the brachiopods.

The upper lava unit is up to 170 m thick and broadly similar to the lower unit although pillow structures are less well developed. Locally the rock is strongly cleaved and particularly towards its top the unit becomes so 'schistose' that Gunn

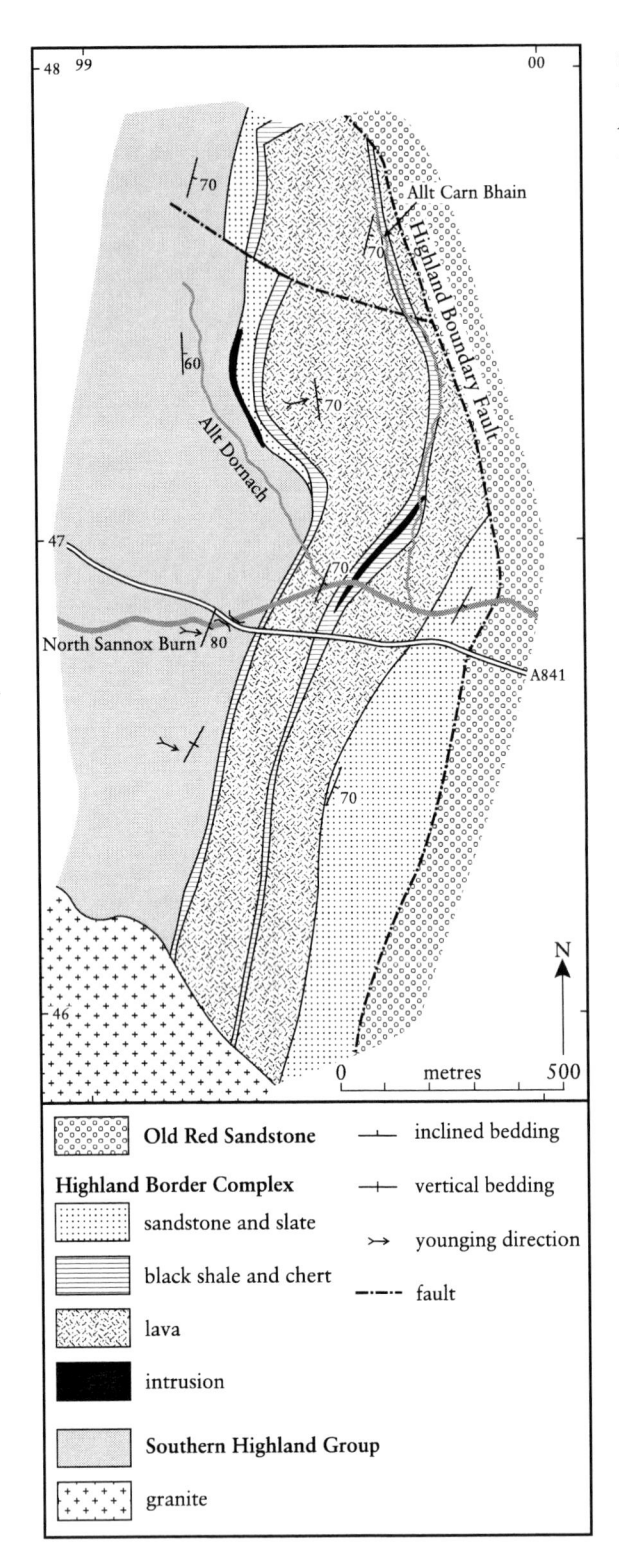

Figure 2.22 Map of the North Glen Sannox GCR site, in part after Johnson and Harris (1967) and Anderson and Pringle (1944).

Figure 2.23 North Sannox Burn below the road bridge, a section through the lower part of the Highland Border Complex. (Photo: Nature Conservancy Council.)

et al. (1903) compared its appearance to that of the Green Beds in the Southern Highland Group of the SW Highlands.

Above the lavas and forming the easternmost unit of the complex is a sandstone and slate unit broadly similar to the Southern Highland Group turbidites. In the west the unit is dominated by green slates and phyllites and the sequence becomes more sandstone rich towards the Highland Boundary Fault. Graded beds indicate an easterly sense of younging. Although Gunn *et al.* (1903) described apparent interbedding between slates and lavas at the westerly contact of the unit, this interpretation was strongly challenged by Anderson and Pringle (1944) who argued against a transitional contact in favour of a high-angle fault.

Gunn *et al.* (1903) reported intrusive basic rocks at two localities. The first lies at the base of the HBC succession *c.* 450 m north of the burn, the second crops out in the Allt Carn Bhain. Anderson and Pringle (1944) subsequently identified a third occurrence of a 2.5 m band within the central black shale unit, exposed on the south bank of the burn. They suggested that this outcrop may be an along-strike correlative of Gunn *et al.*'s (1903) second locality. The intrusive rocks appear to be altered hornblende gabbros which, although still retaining ophitic textures, now exhibit carbonated assemblages of uralitized amphibole and albitized plagioclase.

The petrology of the lavas and the associated intrusions has been studied by Ikin (1983) and Robertson and Henderson (1984). The lavas still display igneous textures although the mineralogy has been almost wholly altered with the original olivine-pyroxene-calcic plagioclase associations now represented by albite-actinolite-chlorite-epidote assemblages. The chemistry of the rocks has been significantly affected by spilitization with strong Na-enrichment, Ca-depletion and substantial movement of many mobile elements. A study of the immobile elements, however, allowed Ikin (1983) to classify the lavas. He recognized two groups: the first covers the greater part of the lower lava sequence and has affinites to alkali basalts characteristic of within plate settings; the second, which covers the uppermost part of the lower lava unit, the whole of the upper sequence and includes the intrusions, has tholeiitic affinities and mid-ocean-ridge basalt (MORB) characteristics. Robertson and Henderson (1984) noted that the MORB chemistry is characteristic of the basic volcanic rocks developed within the HBC along the Highland edge, whereas lavas with alkali basalt chemistry are only known from Arran and Aberfoyle.

The bedding in the Dalradian and HBC rocks is congruent and dips steeply to the east with a constant sense of easterly younging. In the Dalradian succession a penetrative cleavage (S1) strikes broadly parallel to bedding and dips to the east at moderate angles to give a downward sense of facing. Johnson and Harris (1967) reported that D1 folds are rare. They recorded D2 structures evidenced by the presence of fine crenulations in the slate bands; the crenulations plunging at moderately steep angles to the SSW. Within the HBC Johnson and Harris (1967) recorded a similar structural sequence to that in the Dalradian rocks with a penetrative downward-facing S1, tight or sub-isoclinal D1 minor folds and rare D2 crenulations.

Ikin (1983) noted the presence of two high-strain zones of 'mylonitized' lavas and associated breccias, approximately at the base of the two lava units. Henderson and Robertson (1982) also described the presence of high strain zones marked, in part, by mylonitized spilites; they noted that the high-strain fabrics post-date asymmetric folds at a locality to the north of the current site, but that the fabrics themselves are corrugated by minor folds.

Interpretation

Gunn et al. (1903) regarded the Highland Border Complex as an integral part of the 'metamorphic series of the Highlands', remarkable only for its unusual lithologies. This view was largely supported and expanded by Anderson and Pringle (1944) and Anderson (1947) who argued that the Dalradian and HBC rocks had been subjected to the same history of deformation and low-grade metamorphism. They regarded the junction between the Dalradian and HBC rocks as a modified 'disconformity or slight unconformity' and the eastern sandstone unit as a down-faulted block of Dalradian rock. They concluded that the Dalradian succession is of pre-Arenig age and that the folding and metamorphism was post-Arenig. They further argued that the lavas had been extruded on to a subsiding surface of Dalradian rocks and are broadly comparable with the ophiolitic sequences of the Alps and elsewhere.

Johnson and Harris (1967), influenced by the strong structural congruity between the Dalradian and HBC successions, concluded that 'such would appear to be conclusive proof that the ?Arenig has been involved in the same sequence of movements as the adjacent Dalradian'. However, Ikin (1983) and Henderson and Robertson (1982), impressed by the presence of high-strain zones within the HBC succession, argued for tectonic emplacement of the complex. The latter authors further argued for 'the HBC outcrop in Arran as a thrust-stack with zones of recrystallized Dalradian and schistose rock at the sole of each thrust sheet'. They stated that, 'because cleavages cannot be traced continuously from the HBC into the Dalradian, we cannot correlate the cleavages in the two units'. They noted the presence of downward-facing structures in the HBC rocks and argued that the thrusting occurred before the Dalradian D3 event, probably during D2. Robertson and Henderson (1984) were less categorical about the type of emplacement suggesting a general strike-slip regime. They regarded the eastern arenite-dominated sequence as part of the HBC succession rather than as a faulted block of Dalradian rocks (Robertson and Henderson, 1984, table 1). A current view of the regional development of the Highland Border Complex (for example, Robertson and Henderson, 1984; Curry et al., 1984) envisages the formation of the HBC in an oceanic basin marginal to a continental mass and subsequent tectonic juxtaposition with the Dalradian. Tanner (1995) however, in a discussion of the status of the HBC, emphasized the congruity of the younging evidence and minor structures between the Dalradian and HBC successions and concluded that the greater part of the HBC (at least including the Arenig rocks) was deposited in sequence with the Dalradian rocks.

On a regional scale the available biostratigraphical and radiometric age evidence for the formation of the HBC and the tectonism in the Dalradian rocks implies that the juxtaposition of the two successions could only have taken place after the initial Dalradian deformation events. It is these factors that are in conflict with the apparent congruity of the structural sequences in the Arran section. This dilemma has driven much of the recent research in the area and gives an added regional significance to the site.

Conclusions

The North Glen Sannox GCR site contains an excellent section through one of the major developments of basic volcanic rocks in the Highland Border Complex. It is of great importance in assessing the role of this enigmatic assemblage in terms of Caledonian orogenesis and terrane assembly. The lavas exhibit good pillow structures indicating eruption under water. Associated black shale and chert sequences have yielded fossils which suggest an Arenig age. These assemblages are characteristic of rocks generated on the ocean floor. Although early workers did not recognize a significant unconformity between the HBC and the adjacent Dalradian rocks, the subsequent recognition of highly sheared (mylonitic) zones within and at (or near the base of) the HBC, suggests that it was tectonically juxtaposed with the Dalradian early in the deformational history of the two assemblages. This may, however, be difficult to reconcile with regional evidence on the relative ages of the Dalradian and HBC deposition and deformation.

THE BALLANTRAE COMPLEX

BYNE HILL
(NX 180 945)

P. Stone

Introduction

Byne Hill is formed by an intrusive body ranging in composition from gabbro at the margins, through dioritic lithologies to a leucotonalite core; the gabbroic margin is chilled against the host serpentinized harzburgite. The leucotonalite, one of the key components in the interpretation of the Ballantrae Complex as an ophiolite, was originally described as an oceanic 'trondhjemite'.

The site lies at the northern margin of the Ballantrae Complex where a composite intrusive body cuts the Northern Serpentinite Belt and forms a ridge of high ground running SW for about 4 km from the northern flanks of Byne Hill (185 950) to the south-western spur of Grey Hill (158 924). The majority of the ridge is composed of gabbro which is chilled against the host ultramafic rock, mostly serpentinized harzburgite. This intrusive and chilled relationship can be seen at intervals along the SE side of the ridge; the NW margin of the intrusive mass has been faulted so that the igneous rocks are juxtaposed against either volcanosedimentary components of the Arenig ophiolite complex or younger Llandeilian boulder conglomerate (the Benan Conglomerate Formation of the Barr Group). The intrusive rock becomes progressively more leucocratic away from the chilled (SE) margin so that the gabbro passes transitionally into diorite, quartz-diorite and ultimately into 'trondhjemite', a sodium-rich leucotonalite consisting of quartz and plagioclase (now mainly albite but originally more calcic) with accessory hornblende and/or biotite. The most complete section through this transition is preserved on the SE flank of Byne Hill, the summit of which is composed of leucotonalite. A useful field guide to the area is provided by Bluck and Ingham (1992).

The definitive work on the Byne Hill igneous assemblage is the petrological and geochemical study by Bloxam (1968). However, the passage of gabbro into diorite and granite was first noted by Peach and Horne (1899) and the term 'trondhjemite' was first applied by Balsillie (1932) in recognition of the complete absence from the granite of K-feldspar. Following the detailed study by Bloxam, the presence of 'trondhjemite' in the Ballantrae Complex was identified by Church and Gayer (1973) as a key factor in its interpretation as an ophiolitic assemblage of obducted oceanic crust. The 'trondhjemite' has also been pivotal in dating the complex, providing a well-defined, U-Pb zircon age of 483 ± 3 Ma, taken by Bluck *et al.* (1980) as the time of crystallization of the leucotonalite magma. In modern usage leucotonalite is preferred to trondhjemite.

Description

The outline geology of the Byne Hill area is summarized in Figure 2.24. The host rock to the intrusive body is serpentinized harzburgite of the Northern Serpentinite Belt; along the SE flank of the hill the intrusive contact is approximately vertical but it is offset by numerous minor faults. A dark-green to black, massive appearance is characteristic of the serpentinite, which was originally mostly olivine, with sporadic, lustrous bronze-coloured relics after orthopyroxene. Within the serpentinite are pale

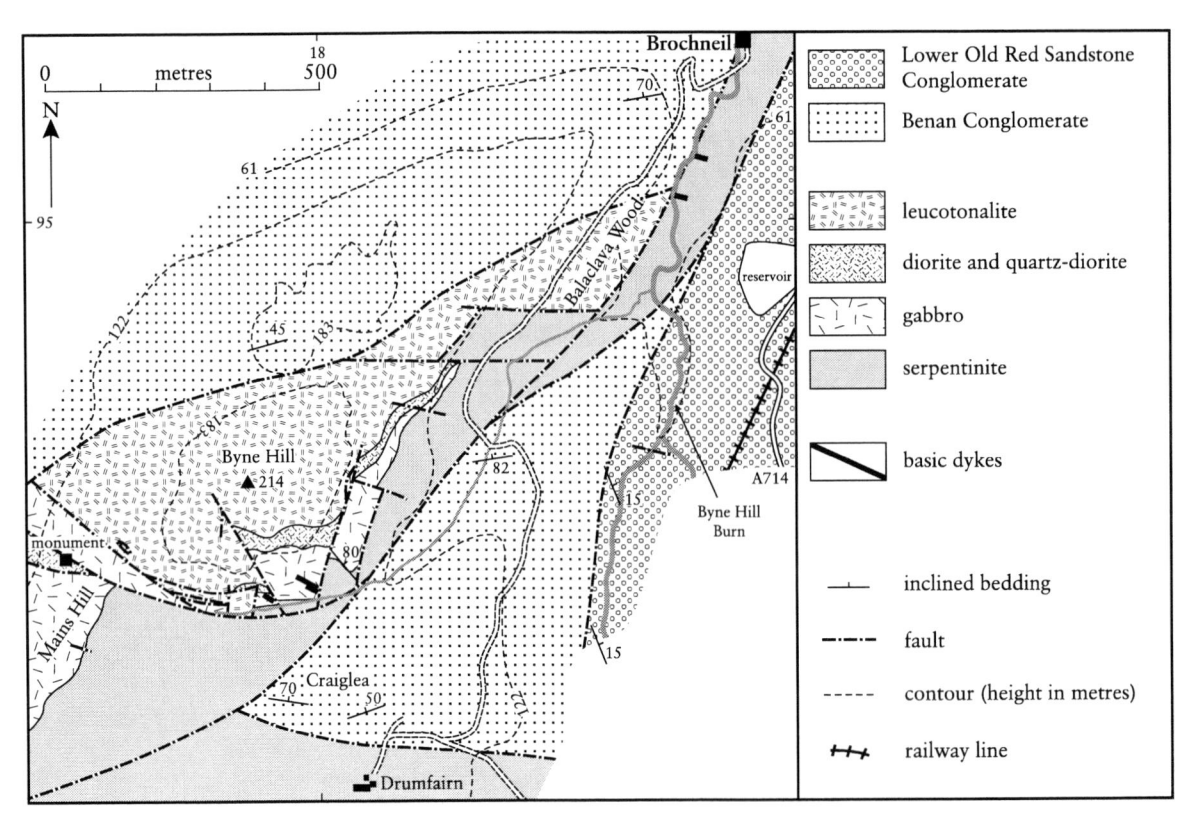

Figure 2.24 Map of the Byne Hill area (after Bloxam, 1968).

veins of calcium-rich secondary minerals such as hydrogrossular garnet and pectolite, believed to have been produced during the late stages of serpentinization. The veins cross the intrusive contact to penetrate the marginal gabbro establishing that intrusion was an early event preceding serpentinization. This deduction is reinforced by the local metasomatism of the marginal gabbro with the development of the same calcium-rich secondary minerals as seen in the veins.

The chilled margin of the intrusive body is doleritic but this coarsens to a gabbro within a few tens of centimetres of the contact. Olivine-gabbro forms the outermost facies but passes inwards into hornblende gabbro by the gradual loss of olivine and progressive mimetic replacement of clinopyroxene by brown hornblende. Parts of the gabbro are notably pegmatitic with individual crystals up to 5 cm across. The transition from gabbro, through diorite and into leucotonalite is relatively abrupt with the dioritic zone rarely wider than 10 m; the transitional lithologies contain distinctively lath-shaped albitized plagioclase, biotite, brown and green hornblende, with accessory quartz and apatite.

Bloxam (1968) stressed that gabbroidal minerals and textures survive as relict features in the dioritic transition zone. The inwards increase in quartz is thence marked and, over only 50 cm or so, quartz-diorite passes into the leucotonalite which contains macroscopic quartz and forms the summit area of Byne Hill. It is a pale-pink or very pale orange-coloured rock that weathers white. Quartz and plagioclase (with albitized rims to original oligoclase) together comprise about 90–95% of the leucotonalite with accessory green hornblende, biotite and zircon. The grain size is generally 1–2 mm, ranging up to 3 mm in places. Zircons from the summit area were used by Bluck *et al.* (1980) to produce their U-Pb age of 483 ± 4 Ma. At Byne Hill, the NW margin of the leucotonalite is faulted against the Benan Conglomerate (Figure 2.24) but towards that margin there does seem to be an increase in the proportion of hornblende in the rock. This rather slim evidence may imply that originally the leucotonalite may have formed the core of an elongate body with a symmetrical transition to gabbro towards both the SE and NW margins.

A characteristic feature of the leucotonalite

(and to a lesser extent of the quartz-diorite) is the pervasive network of cataclastic zones which traverse the rock. For the most part these are microfractures only clearly evident in thin section but locally they develop into mylonitized zones up to 2 cm across. A wide range of orientations is apparent but there is a preferential NE–SW trend to the macroscopic mylonite zones parallel to the long axis of the intrusive body. This trend is also parallel to the major fault forming the NW margin of the leucotonalite but there does not appear to be an increase in mylonitization towards that structure.

Interpretation

Byne Hill is formed by part of a composite, intrusive igneous body transitional from gabbro at its margins towards a leucotonalite core. The host rock is serpentinized harzburgite and the intrusive contacts are very steep or vertical. From the intrusive contact with serpentinite the transitional passage from gabbro passes through diorite and quartz-diorite into the leucotonalite with no sharp contacts and with gabbroidal minerals and textures persisting through the dioritic zone and possibly into the marginal hornblendic phase of the leucotonalite. The principal mineralogical changes are reported by Bloxam (1968) to be progressive amphibolitization and albitization of the gabbro accompanied by increases in SiO_2 and Na_2O content. Bloxam considered that these features, coupled with the textural evidence, suggest that the dioritic transitional facies contains hybrid rocks produced by reaction between crystalline gabbro and silicic, Na-rich solutions related to the leucotonalite. However, Bloxam also entertained the possibility that the leucotonalite could be entirely metasomatic in origin. Subsequently, Bluck *et al.* (1980) examined zircons from the leucotonalite and considered them to be of magmatic origin, their age therefore dating the time of crystallization.

The presence of leucotonalite or 'trondhjemite' within the Ballantrae Complex is one of the key lithological features supporting its interpretation as an ophiolite assemblage formed from obducted oceanic crust (Church and Gayer, 1973). The U-Pb age of 483 ± 4 Ma obtained from the zircons by Bluck *et al.* (1980) is thus of great importance in terms of the evolution of the oceanic crust which became the Ballantrae Complex. The zircon age is broadly early Arenig and intrusion was thus penecontemporaneous with eruption of the extensive lava sequences. At the margins of the intrusive body gabbro is chilled against harzburgite of the Northern Serpentinite Belt which must therefore have cooled by about 483 Ma. However, Bluck *et al.* (1980) also presented K-Ar age data from amphibolites associated with the metamorphic sole of the ophiolite produced at the base of the Northern Serpentinite Belt during its obduction (see Knocklaugh GCR site). These show that the metamorphic sole was cooling at 478 ± 8 Ma but in terms of error overlap there is no appreciable difference between the ages of lava eruption, ophiolite obduction and leucotonalite intrusion; all occurred around the early to middle part of the Arenig epoch. This relatively short time lapse between generation and obduction of the Ballantrae ophiolite encourages interpretation in terms of a back-arc or marginal basin formed close to the site of its eventual obduction onto the Laurentian continental margin. Supporting evidence comes from the chemistry of the gabbro with data presented by Stone and Smellie (1990) compatible with its generation in a relatively mature, oceanic island-arc system.

A final discussion point in the interpretation of the Byne Hill body has been its use in constraining the Arenig time-scale. The U-Pb zircon age of 483 ± 4 Ma is accepted as a reliable data point and has been extrapolated by Bluck *et al.* (1980) to clasts of leucotonalite contained in mélange interbeds within the volcanosedimentary succession of the Ballantrae Complex. Bluck *et al.* regarded these as of middle Arenig age but subsequent biostratigraphical work by Rushton *et al.* (1986) suggests that a lower Arenig age is more likely. Allowing some time for unroofing and erosion it seems most unlikely that lower Arenig strata could contain igneous material with a cooling age as young as *c.* 483 Ma. Nevertheless, the petrographical similarity between the Byne Hill leucotonalite and some of the mélange clasts is remarkable, particularly so in an example from the south of the Ballantrae Complex (the Craig Hill Breccia Formation of Stone and Smellie, 1988). Their relationship and its overall significance remains uncertain.

Conclusions

The Byne Hill site is of great importance in any assessment of the origin and tectonic role of the

Ballantrae Complex within the Caledonian Orogen. The gabbro–leucotonalite intrusion is one of the key lithological components supporting the interpretation of the complex as of ophiolitic origin. The intrusion has a marginal facies of gabbro, chilled against the serpentinized harzburgite host rock, with an inwards transition to leucotonalite through a narrow dioritic zone. Zircons have provided a reliable U-Pb date of 483 ± 4 Ma for cooling of the leucotonalite magma, thus constraining the age of the complex as a whole. The chemistry of the intrusive body, particularly the gabbro, is compatible with an origin at an oceanic volcanic island arc rather than within oceanic crust (*sensu stricto*).

SLOCKENRAY COAST
(NX 135 911 AND 136 914–143 924)

P. Stone

Introduction

The Slockenray Coast GCR site includes a part of the Northern Serpentinite Belt, its faulted margin with the Pinbain volcanosedimentary block, and an olistostrome, lavas and volcaniclastic strata within that block. The site falls into three parts each of which contains an important component of the Ballantrae Complex ophiolite. The northern sector forms the major part and provides an extensive section through an Arenig volcanosedimentary sequence of lava and intercalated volcaniclastic sedimentary rocks. The central part of the site is occupied by a structurally complex zone with a coarse conglomeratic olistostrome faulted against serpentinite. The isolated southern outlier contains a remarkable pegmatitic gabbro intruded into the serpentinite, which is also traversed by veins of coarse-grained pyroxenite. A general account of the geology is given by Stone and Smellie (1988).

The good accessibility and exposure have made it one of the most intensively studied parts of the Ballantrae Complex so that it has been central to several of the historical controversies which have raged over interpretation of the regional geology. Initially there was debate over a metamorphic versus an igneous intrusive origin for the complex and crucial evidence in support of the latter was obtained by Bonney (1878) from the pegmatitic gabbro at the southern end of the section. To mark this breakthrough the name 'Bonney's Dyke' was coined for the locality by Balsillie (1932) and it is generally known by this colloquial name in the geological literature. The gabbroic dyke has a striking black and white appearance, created by large plagioclase and altered clinopyroxene crystals; it displays a range of textures within an overall coarse pegmatitic style. The host rock is serpentinized harzburgite of the Northern Serpentinite Belt which at this locality contains numerous veins of coarse pyroxenite.

A short distance north from Bonney's Dyke, in the vicinity of Pinbain Bridge at the southern end of the main GCR sector, the Northern Serpentinite Belt is faulted against an unusual conglomeratic unit. This comprises a variety of clasts, both in terms of size and composition, contained in a black shaly matrix that is pervasively foliated in places. It was originally interpreted in volcanic terms but, with the development of the ophiolitic model, Church and Gayer (1973) noted its similarity to the extensive mélange deposits associated with the large-scale ophiolite bodies of Newfoundland. In this interpretation, masses of brecciated pillow lava, apparently interbedded with the conglomerate, may be rather viewed as large blocks within an original olistostrome.

The olistostrome is most pervasively foliated adjacent to its southern margin and the fabric becomes less intense towards the northern margin where there is another faulted sliver of serpentinite. This narrow zone provided the focus for much of the argument over a proposed intrusive origin for the serpentinite. The critical relationships were interpreted by Anderson (1936) in favour of intrusion of the serpentinite into the lava sequence while Balsillie (1937) was using information from elsewhere in the Ballantrae Complex to deduce that the serpentinite was much older than, and formed a basement to, the lavas (see particularly the Knockormal GCR site report). Bailey and McCallien (1957) re-examined the Pinbain locality, having been previously disposed towards the intrusive interpretation, and decided that the apparently intrusive veins are of secondary origin and are probably related to serpentinization of an ultramafic precursor. This conclusion is now generally accepted although Bailey and McCallien's broader interpretation, that the serpentinite originated as an ultramafic, submarine lava flow, has now been superseded by the ophiolite model.

Northwards from the olistostrome–serpentinite fault slice the coastal section comprises a relatively continuous volcanosedimentary succession. The beds are vertical or steeply inclined and young consistently northwards. Despite the abundant minor faulting this is the least structurally disturbed part of the Ballantrae Complex and many sedimentological and volcanic features are well displayed. Two aspects are of particular importance: a graptolite fauna collected from the southern end of the section and a remarkable lava-front delta assemblage in the central part at Slockenray itself. The graptolites were recovered and described by Rushton *et al.* (1986) who deduced an early Arenig age. The Slockenray relationships have been described in great detail by Bluck (1982) who emphasized their support for eruption of the lava sequence into relatively shallow water. The lack of spilitization of some of the Slockenray rocks was also stressed by Bluck following earlier debate over their exceptional lack of alteration. This was sufficiently striking that Bailey and McCallien (1957) admitted to briefly entertaining the idea that they were contained in a volcanic neck of Palaeogene age.

Description

The outline geology of the Slockenray–Pinbain section, which encompasses the Slockenray Coast GCR site, is shown in Figure 2.25. The Balcreuchan Group lithostratigraphy shown is based on that proposed informally by Bluck (1982).

At the southern extremity of the section, within the Northern Serpentinite Belt, a striking pegmatitic gabbro body (Bonney's Dyke) protrudes from the ultramafic rocks on the foreshore and just above the high-water mark. It is an elongate mass up to about 5 m wide, with an overall WNW trend made arcuate by the cumulative offsets of several minor sinistral faults (Figure 2.26). The gabbro is a pale-coloured rock formed almost exclusively of plagioclase and clinopyroxene, the latter in a rather fibrous variety of diopside and/or augite termed 'diallage' in the older literature. These minerals are now extensively altered, the plagioclase to a calcium-rich assemblage of pectolite, prehnite and hydrogrossular garnet and the pyroxene to an aggregate which includes actinolite and chlorite; some serpentine minerals are also present, which may indicate original olivine. Grain size is very variable (Figure 2.27) with fine-grained patches alternating irregularly with zones in which the plagioclase laths and pyroxene crystals range up to 2 cm. Xenoliths of serpentinite up to about 50 cm across are abundant locally. The margins of the gabbro body generally dip moderately to the south but three types of contact relationship are displayed: sharp contacts against serpentinite (including the xenoliths); more diffuse margins abutting coarse pyroxenite veins; and sheared margins that have recrystallized to a very fine-grained secondary mineral assemblage through calcium metasomatism. It is noteworthy that none of the gabbro margins are chilled unequivocally against the ultramafic rock.

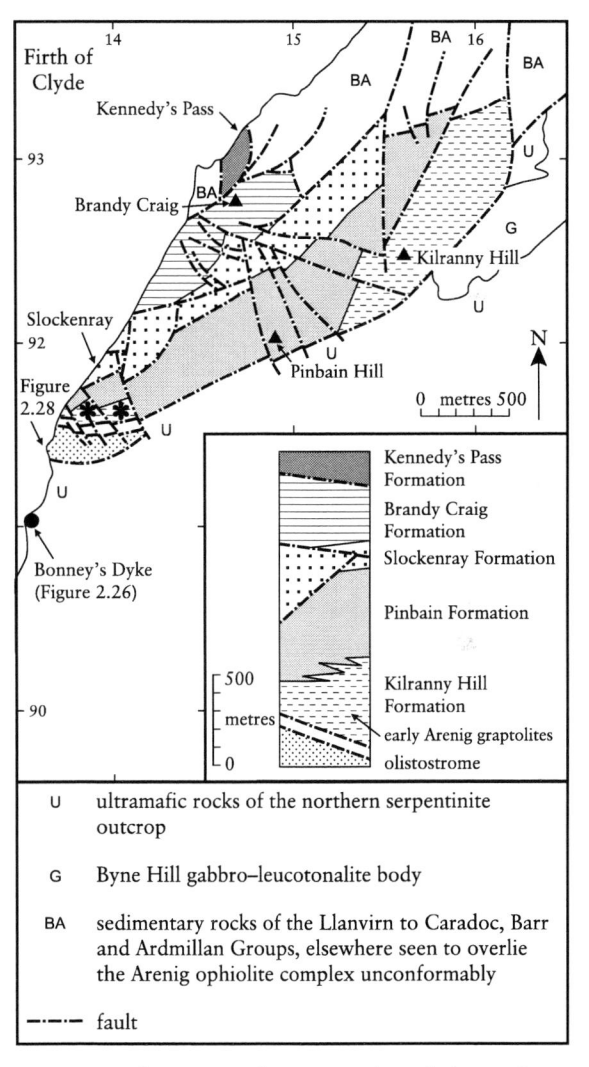

Figure 2.25 Map and stratigraphy of the Pinbain Block, the northernmost part of the Ballantrae Complex, after Bluck (1982) and Stone and Smellie (1988). * The area marked thus on the map was included within the Pinbain Formation by Bluck (1982), but is more closely related lithologically to the Kilranny Hill Formation.

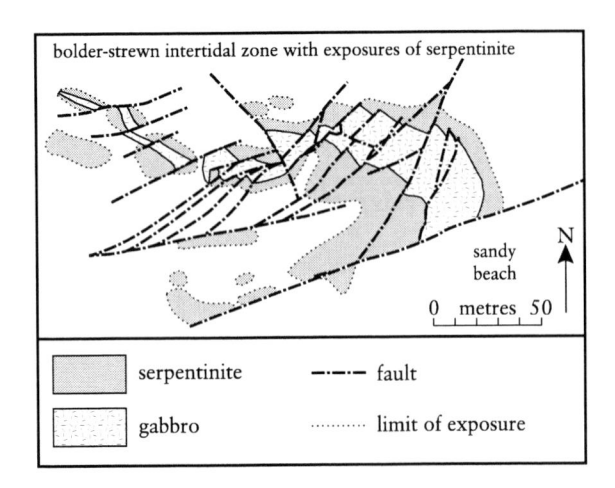

Figure 2.26 Map of Bonney's Dyke, after Bluck (1992). See Figure 2.25 for location.

The host rock to the pegmatitic gabbro is serpentinized harzburgite cut by abundant and anastomosing veins of coarse-grained pyroxenite. These stand out clearly as paler-green zones contrasting with the very dark background of the serpentinite. Some of the pyroxenite veins are spectacularly coarse with diallage crystals up to 10 cm long. Northwards from Bonney's Dyke the serpentinite is exposed sporadically towards the low-water mark but extensive rock outcrop does not recur until the Pinbain Bridge area (Figure 2.25). There, the faulted northern margin of the Northern Serpentinite Belt has been intruded by a dolerite dyke of Palaeogene age which has baked and hardened the adjacent ultramafic rock. Accordingly, this marginal zone of serpentinite has proved unusually resistant to erosion and now stands proud of the surrounding beach and rock platform.

Northwards, beyond the Palaeogene dyke, is an extensive outcrop of the Pinbain olistostrome. This has been described in great detail by Bluck (1978a, 1992) and is illustrated in Figure 2.28. The southern part of the olistostrome, adjacent to the serpentinite and dyke, consists of a mélange containing a great variety of clasts set in a pervasively foliated, black shaly matrix. The clasts range considerably in shape and size with some boulders several metres across. An equally large lithological range is present; basalt tends to form the larger clasts with some appearing to be intact pillows of lava, large pale clasts of carbonate may well be highly altered ultramafic rock but small clasts of dark-green serpentinite are also present. The other smaller clasts have a remarkably large lithological range with records of shale, tuff, chert, basalt, dolerite, gabbro, pyroxenite and, very rarely, glaucophane- and garnet-glaucophane schist (Bailey and McCallien, 1957). A large mass of brecciated pillow lava at the northern margin of the mélange may be an interbedded unit or may be an exceptionally large clast caught up in the olistostrome. Farther north still, the lithology becomes more conglomeratic with a more restricted clast size range and a generally unfoliated shale matrix. Granitic clasts, quartz, and fragments of quartz-cemented breccia appear at this level. The northern margin of the Pinbain olistostrome is formed by a complex fault zone introducing slivers of serpentinite and intruded by Palaeogene dolerite dykes. The olistostrome is thus structurally isolated from the volcano-sedimentary sequence forming the main part of the Slockenray section.

The Slockenray coastal section provides a cross-strike traverse through the Pinbain block, the northernmost and least structurally confused of the three principal volcanic units which together form the Balcreuchan Group within the Ballantrae Complex. Bluck (1982) divided the approximately 1500 m of lava and volcaniclastic sedimentary strata into five formations and this lithostratigraphical scheme, slightly modified by Stone and Smellie (1988, see especially table 4), is applied in Figure 2.25. The age of the sequence is fixed by the early Arenig graptolite

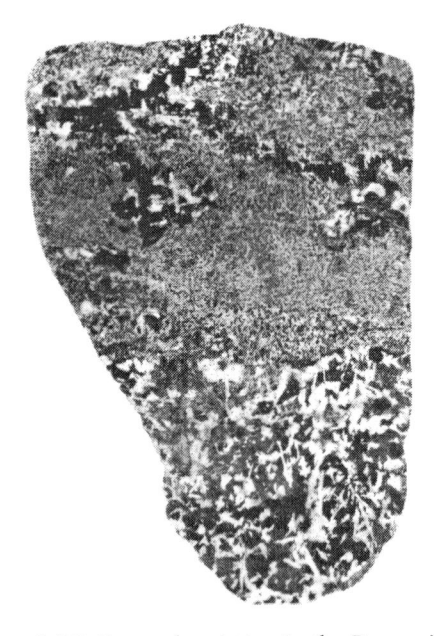

Figure 2.27 Textural variation in the Bonney's Dyke pegmatitic gabbro between plagioclase (pale) and pyroxene (dark). The long axis of the sample is 170 mm. (Photo: BGS no. MNS4007.)

Slockenray Coast

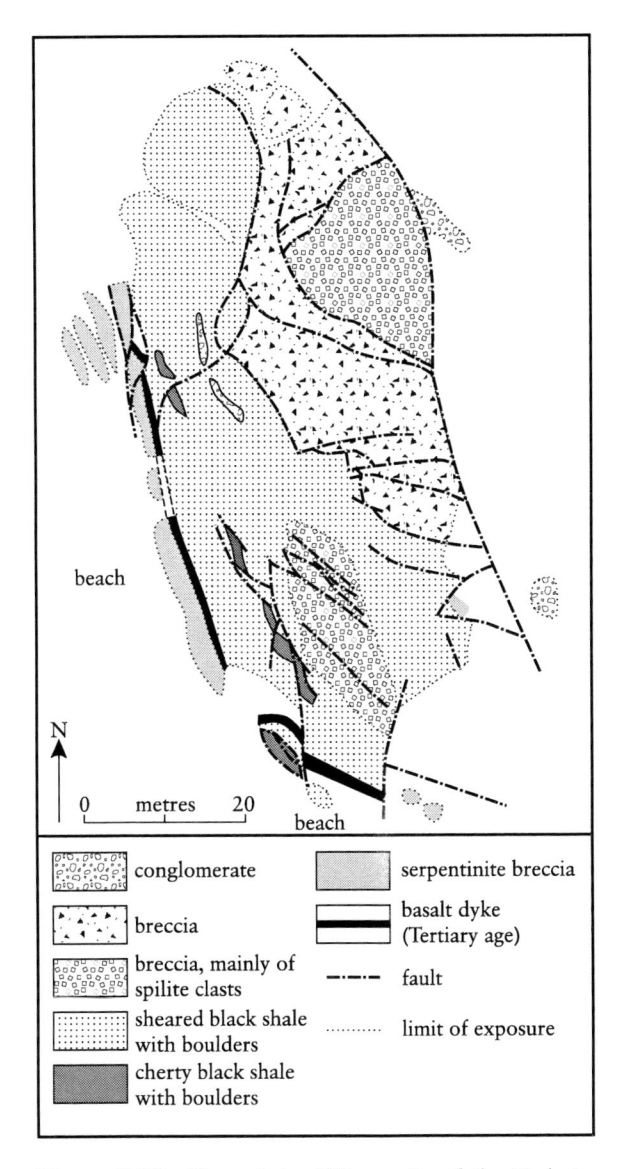

Figure 2.28 Map of the SW margin of the Pinbain olistostrome, after Bluck (1978a, 1992). See Figure 2.25 for location.

Plagioclase is the most common of the phenocryst minerals and there are also rare augite and olivine phenocrysts, all set in a matrix dominated by plagioclase and augite. Spilitization and low-grade metamorphism have created widespread secondary minerals including albite, chlorite, titanite, epidote and prehnite.

An area of particular interest is centred on Slockenray itself where there has apparently been mixing of two lava types, one aphyric and the other strikingly porphyritic (Figure 2.29). In the latter, large plagioclase laths are aligned in swirling patterns indicative of flow orientation and the zones of phenocrysts have relatively sharp margins against aphyric lava. However, apart from the presence or absence of phenocrysts there is no difference in the composition or appearance of the lava varieties and there are no signs of physical boundaries between them. It would appear that two different lavas, one aphyric and the other plagioclase-phyric, were erupted simultaneously and mixed together immiscibly during flow. Immediately to the south of the lavas, hyaloclastite breccia beds contain blocks of both lava types together with abundant detrital plagioclase crystals. Since the beds are sub-vertical in a north-younging sequence the breccias are older than the lavas which overlie them. The relationships at this locality have been described in detail by Bluck (1982) who considers the breccias to be lava-front accumulations ahead of an advancing flow which progressively over-rode its own frontal deposits.

In the northern part of the Slockenray coastal section features of interest include reddened, scoriaceous top surfaces of some lava flows and a bed of tuffaceous sandstone containing accretionary lapilli described by Smellie (1984). These features are of importance in any palaeoenvironmental analysis.

Interpretation

The coarse pegmatitic gabbro of Bonney's Dyke was quoted by Church and Gayer (1973) as representing one of the characteristic components of an ophiolite assemblage. Such lithologies may be found in the basal part of the gabbro unit or as a late intrusion into the upper part of the ophiolite 'stratigraphy'. However, while the lithology might be appropriate, the setting of the Bonney's Dyke example is incompatible with either of these styles of occurrence. If it origi-

fauna described from the southern extremity of its outcrop by Rushton *et al.* (1986). The fossils were recovered from thin shale laminae within volcaniclastic sandstone where the beds strike NE–SW and dip steeply to the NW. This is a fairly characteristic bedding attitude throughout the section and since younging is consistently to the NW the early Arenig age applies to the base of the sequence. A variety of volcaniclastic sedimentary rocks are interbedded with lavas and lava breccias. The lavas are variably porphyritic and aphyric, although even the aphyric lavas commonly contain scattered microphenocrysts.

Figure 2.29 Massive, porphyritic basalt lava at Slockenray (NX 1403 9197). The pale feldspar phenocrysts are tabular and range up to 2 cm in length. (Photo: BGS no. D4239.)

nated in the basal gabbro unit its present isolated situation would require interpretation as a clast or xenolith within the serpentinite. This is not supported by the marginal relationships or by the presence of serpentinite xenoliths in the pegmatite; these features all favour an intrusive origin. The intrusion is not into the upper levels of the ophiolite 'stratigraphy' as the model would predict but instead the pegmatite has been emplaced into mantle harzburgite, the lowest part of the ideal sequence. Thus, assuming that the pegmatite rose on intrusion, the implication of its current position is that the Ballantrae Complex ophiolite was structurally imbricated while the final stages of its igneous generation were still in progress. The calcium-metasomatism of the dyke margins shows that intrusion occurred before serpentinization of the ultramafic rock, further compressing the time-scale of ophiolite development, structural imbrication and alteration.

The olistostrome and breccia deposit exposed around the mouth of the Pinbain Burn is contained within a structurally isolated fault-bound block. It has been described and discussed by numerous authors, notably Bailey and McCallien (1957) and Bluck (1978a, 1992). The salient points of interpretation can be summarized as follows.

1. A wide variety of clast types are present, including the full range of ophiolitic rocks and some high-grade metamorphic lithologies.
2. Large clasts were slumped into relatively deep water where black shale was accumulating.
3. Intense shearing was focused into some parts of the unit during or immediately after its deposition to create a pervasively foliated mélange.
4. The slumped lithologies are intimately associated with tuff and pillow lava indicative of contemporaneous volcanicity.

In their original comparison of the Ballantrae Complex with the ophiolite model, Church and Gayer (1973) described the olistostrome as a syn-obduction deposit 'developed as a result of deformation beneath a forward moving thrust sheet of material previously produced by erosion from the leading edge of the self same nappe'. Bluck (1992) preferred an origin in an extensional environment, citing the volcanic activity in support of an association with normal rather than compressional faults. These brought the deeper levels of the ophiolite to the erosion level in fault scarps, the faults themselves probably rooting in the fissile shaly lithologies and there producing the localized intense ductile deformation. Such structures would almost certainly have been re-activated as thrusts during subsequent obduction-related basin inversion.

A mid- to late-Arenig graptolite was recorded from the olistostrome by Peach and Horne (1899, p. 442). Amphibolite *in situ* elsewhere in the Ballantrae Complex and lithologically identical to clasts in the olistostrome has been dated radiometrically at 478 ± 8 Ma. These relationships have been used by Bluck *et al.* (1980) to suggest a constraint on the Arenig time-scale. However, the graptolite was re-examined by Rushton *et al.* (1986) who concluded 'that it is quite undeterminable and even its graptolitic nature is questionable'. Stratigraphical deductions based on the Pinbain olistostrome should therefore be treated with caution.

The Pinbain volcanosedimentary sequence is approximately 1500 m thick, which suggests considerable synvolcanic subsidence during its accumulation. This point is emphasized by the abundant evidence for shallow-water sedimentation and possible subaerial eruption. Bluck (1982) described the Slockenray breccias in terms of an upward coarsening sequence built up as a hyalotuff delta, ahead of lava flows, in the shallow sub-tidal or intertidal zone. He summarized the situation as 'the lavas advanced over the sediments so a sequence was generated where these sediments have a source in lava flows which were eventually to overlie them'. Syn-eruption subsidence would seem implicit in this interpretation. Higher in the Pinbain Block succession the reddened tops of lava flows and the presence of accretionary lapilli (Smellie, 1984) are both indicative of subaerial eruption with the lapilli falling into shallow water.

Several cycles of lava advance have been described by Bluck (1992). Slow eruption rates produced a pillowed flow whereas rapid eruption produced a more massive lava; the lavas were continuously eroded into breccias and hyaloclastites and the sea may have transgressed the flow top between eruptions. The mixing of plagioclase-phyric and aphyric lavas, well displayed at Slockenray itself, requires the contemporaneous eruption of different magma types in close proximity. Flow segregation of the feldspar phenocrysts may be a possible alternative explanation but has not found much support from previous investigators.

The geochemistry of the Pinbain–Slockenray lavas has been generally interpreted in terms of within-plate, oceanic island eruption (Wilkinson and Cann, 1974; Thirlwall and Bluck, 1984; Stone and Smellie, 1990). This is most readily reconciled with the evidence for subsidence and shallow water in a marginal or back-arc basin. In this respect the sequence compares closely with that developed in the Bennane Head sector of the Balcreuchan Port to Port Vad GCR site but at Pinbain a geochemical curiosity occurs at the base of the succession. Here, volcaniclastic sandstones form the oldest stratigraphical level and are interbedded with graptolitic shales of early Arenig age (Rushton *et al.*, 1986). The sandstones are formed exclusively of volcanic material and on analysis prove to have the chemical characteristics of island-arc lavas (Stone and

Smellie, 1990; Smellie and Stone, 1992). If this association is accepted the back-arc model is strengthened since it provides an environment in which arc-derived sands could be overlain by within-plate lavas erupted during a phase of significant subsidence. Evidence of this sort, provided by the Slockenray Coast GCR site, is of fundamental importance in the interpretation of the Ballantrae Complex as a whole. A polygenetic source for the ophiolite seems increasingly likely with island-arc and within-plate components now structurally juxtaposed.

Conclusions

The Slockenray Coast GCR site contains three different geological components, each of which is fundamental to the ophiolitic interpretation of the Ballantrae Complex. Their characteristics are exceptionally well displayed in a section that has received much historical study.

At the southern end of the GCR site the coarse gabbro body forming Bonney's Dyke shows a spectacular development of very coarse-grained, pegmatitic texture. The lithology is an integral part of the ophiolite assemblage but its position here, intruded into serpentinite cut by pyroxenite veins, is somewhat anomalous. It suggests early structural disordering of the Ballantrae Complex while ophiolitic magmatism was still in progress.

The foliated conglomeratic (mélange) deposit exposed around the mouth of Pinbain Burn has been the subject of much debate. It contains clasts with a wide size and composition range, the latter including both ophiolitic lithologies and exotic, high-grade metamorphic rocks. An origin by large-scale slumping is now generally accepted.

The accessible and well-exposed Slockenray coastal section forms most of the GCR site; it contains the most complete and least structurally disturbed volcanosedimentary succession within the Ballantrae Complex. The lavas are believed to have been erupted in an oceanic island, within-plate environment and graptolite fossils from intercalated shale at the base of the sequence fix the age at early Arenig. Some of the lavas apparently resulted from the mixing of different varieties during flow and may have been erupted into relatively shallow water. At one locality the lava–sediment relationships have been interpreted in terms of a lava front delta deposited in an intertidal environment.

KNOCKLAUGH (NX 168 920)

P. Stone

Introduction

The Knocklaugh GCR site provides a section through a sequence of dynamothermal metamorphic rocks developed from shale and spilitic lava marginal to the Northern Serpentinite Belt of the Ballantrae Complex ophiolite. Lithologies range from chlorite and epidote schists, prograding into amphibolites towards the ultramafic belt, to the highest-grade metamorphic rock seen, which is a garnet metapyroxenite. However, since this lies just within the serpentinite it may be unrelated to the schist–amphibolite assemblage. The aureole of metamorphic rocks is generally about 40 to 50 m wide and can be traced discontinuously along the serpentinite margin for over 5 km.

The unusual metamorphic assemblage was first noted by Peach and Horne (1899, pp. 456–9) and subsequently described in some detail by Anderson (1936) who proposed an origin by tectonism during the intrusion of hot ultrabasic magma. Later, the development of plate tectonic concepts and the recognition of ophiolites as obducted oceanic lithosphere allowed the re-interpretation of the Ballantrae Complex in those terms. Church and Gayer (1973) considered the Knocklaugh rocks to be part of a metamorphic aureole formed at the base of an obducting slab of still-hot mantle material. They drew attention to the similarities with dynamothermal aureoles present beneath the large, well-preserved ophiolite complexes of Newfoundland and noted the inversion of the metamorphic succession, with the highest grade rocks at the top (closest to the over-riding hot slab) of an originally sub-horizontal metamorphic layer. More recent and detailed studies of the aureole rocks (Spray and Williams, 1980; Treloar *et al.*, 1980) have confirmed the broad interpretation while stressing the range of temperature and pressure conditions required for formation of the different components.

Description

Detailed descriptions of the metamorphic aureole rocks at Knocklaugh, in both petrographic and structural terms, are provided by Spray and

Knocklaugh

Williams (1980) and Treloar *et al.* (1980). A Sm-Nd isotopic age of 505 ± 11 Ma has been reported by Hamilton *et al.* (1984) for the high-grade metapyroxenite whereas Bluck *et al.* (1980) gave a K-Ar age of 478 ± 8 Ma for the lower-grade amphibolites. The geological setting is illustrated in Figure 2.30.

The Knocklaugh GCR site provides a section wherein the Arenig lavas and interbedded cherts and shales (oceanic upper crust) of the Balcreuchan Group become increasingly sheared and schistose towards the SE margin of the Northern Serpentinite Belt. Epidote schists are seen as slaty rocks with streaked epidote augen and granular boudins, up to a few millimetres across, dispersed within finer-grained, dark-green laminae of actinolitic hornblende, chlorite and albite. Lighter-coloured, yellow-green layers of fine-grained epidote and titanite are also present in places together with sporadic metasedimentary quartz-albite-epidote-muscovite-chlorite schists and small pods of recrystallized carbonate. Closer to the serpentinite the epidote schists pass abruptly into coarser-grained, dark-grey amphibolites composed largely of plagioclase and green or brown hornblende. The more highly foliated of the amphibolites may contain garnet, and pyroxene occurs as an accessory in amphibolites adjacent to the serpentinite margin. Hornblende-bearing garnet metapyroxenite also occurs within the contact zone but in some cases is separated from the amphibolites by thin slivers of serpentinite. This may be due to structural imbrication of the schistose aureole and adjacent serpentinite during the final stages of thrust movement. Alternatively, the garnet pyroxenite may have originated as segregations near the base of the ultrabasic precursor to the serpentinite and so would not be part of the dynamothermal aureole *sensu stricto*. The width of the aureole assemblage is generally 40 to 50 m. Locally this may reduce to 20 m and in places the entire aureole has been removed tectonically; conversely the epidote schist component may broaden locally to nearly 200 m.

Structurally, the epidote schists and amphibolites of the aureole are dominated by a slaty schistosity or fine gneissose foliation with a locally developed mineral aggregation lineation. These fabric elements dip or plunge steeply to the NNW. Within the schistose fabric tight and asymmetrical, reclined interfolial folds close both SW and NE. They are refolded by small,

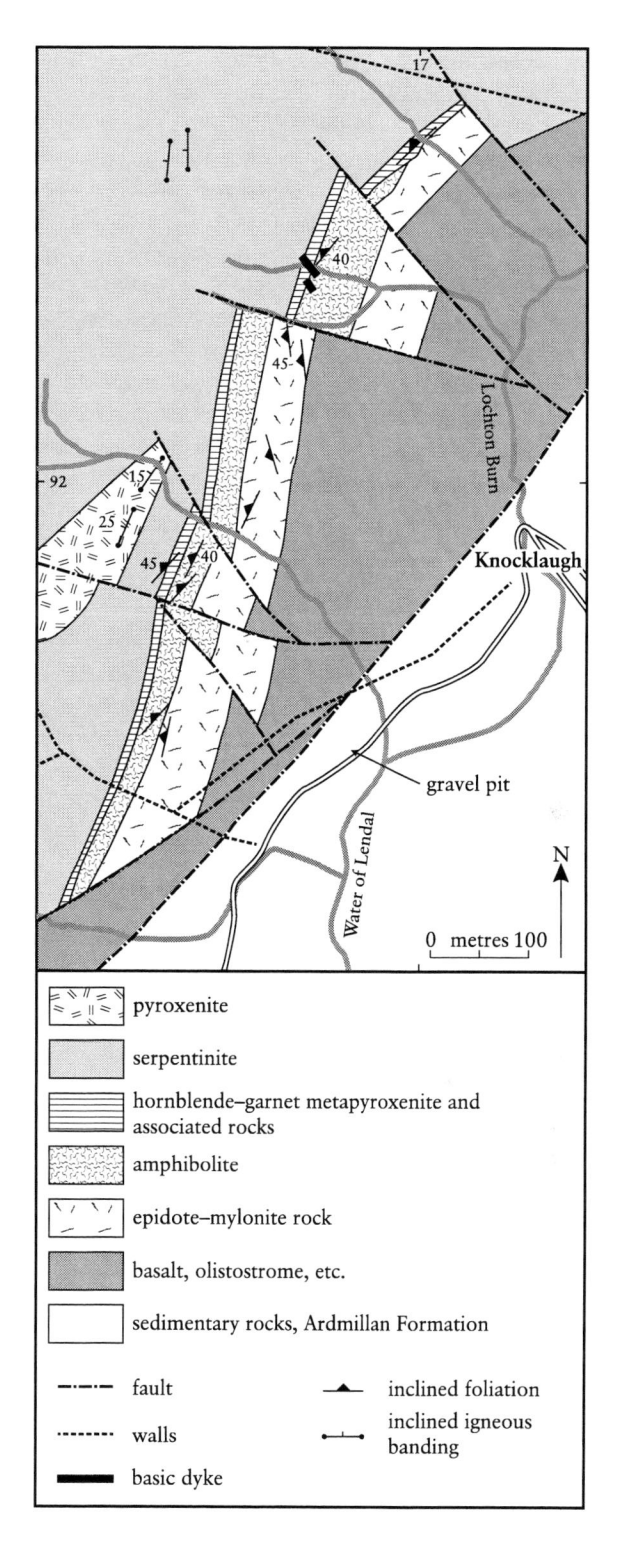

Figure 2.30 Map of the metamorphic aureole developed adjacent to serpentinite at Knocklaugh, after Treloar *et al.* (1980).

tight and angular, straight-limbed folds with an associated axial-planar crenulation cleavage. Such a history of polyphase deformation restricted to a narrow band of rocks is characteristic of a mylonite zone. The dominant fabric in the metapyroxenite is rather different, more compatible with deformation by prolonged laminar flow under granulite facies or perhaps upper mantle conditions. The relatively abrupt transitions between the different lithologies suggests that they have been juxtaposed by the tectonic slicing of an originally more extensive metamorphic aureole. All of these structural features are believed to have developed during obduction of the Ballantrae Complex since the aureole is cut across by unsheared doleritic dykes, which are themselves of probable Arenig age (Holub *et al.*, 1984).

Interpretation

The Knocklaugh GCR site provides a section through a narrow dynamothermal metamorphic aureole adjacent to one of the major serpentinite bodies of the Ballantrae Complex. Early interpretations of these relationships were contentious, with an intrusive origin for the serpentinite as a hot, ultrabasic magma generally preferred. This was either controlled by or caused the tectonic movements necessary to produce the localized shear fabrics (e.g. Anderson, 1936). Modern interpretations regard the aureole as a metamorphic 'sole' developed beneath a hot mantle slab as it was thrust over oceanic crust and ultimately obducted at a continental margin (Spray and Williams, 1980; Treloar *et al.*, 1980). Characteristically such 'soles' are inverted in that the highest-grade rocks are at the top, adjacent to the over-riding slab, with the metamorphic grade decreasing downwards. The interpretation of the metamorphic aureole in these terms thus identifies one of the key components in the recognition of the Ballantrae Complex as an obducted ophiolite.

The temperature–pressure conditions necessary for the generation of the various parts of the aureole are clearly varied. The highest-grade rocks have attracted the most interest with texture, mineralogy and mineral chemistry used to estimate the conditions of metamorphism. Treloar *et al.* (1980) considered the garnet metapyroxenite to have recrystallized under upper granulite facies or mantle conditions with $T = 900 \pm 70°C$ and $P = 10–15$ kbar. This lithology may not be part of the aureole *sensu stricto* but it rather formed as original mantle segregations where the same P–T conditions would apply. However, broadly comparable conditions were determined by Spray and Williams (1980) who considered that formation of the foliated garnet amphibolite required a minimum of $T = 850°C$ and $P = 7$ kbar. At these temperatures the adjacent ultramafic rock would have been unserpentinized suggesting that the onset of metamorphism was an early event in the assembly of the Ballantrae Complex, a deduction compatible with the available radiometric dates.

Various lines of evidence consistently indicate that the upper part of the metamorphic 'sole', i.e. that part closest to the serpentinite, formed under upper mantle conditions. Conversely, the lower-grade components of the aureole formed under very much less extreme conditions, perhaps as low as the greenschist facies, providing the characteristic inverted metamorphic profile. The full range of metamorphic lithologies are developed across only about 40 to 50 m, which represents an impossibly steep temperature gradient; further, the upper parts of the 'sole' were generated at much higher pressures than the lower parts. The limited radiometric age data suggests that the highest-grade rocks may have formed significantly earlier than the lower-grade components. Clearly the aureole assemblage as seen was not produced as a single entity and must have accreted piecemeal under decreasing pressure and temperature conditions. This would be in keeping with its interpretation as an obduction 'sole' wherein the metamorphic components would be both composite and tectonically transported. Treloar *et al.* (1980) envisaged 'a thin, transported and telescoped aureole made up of tectonic pieces of successively-formed metamorphic rocks that welded onto the base of an upward-moving and progressively-cooling peridotite slab'. Overall, it seems most likely that metamorphism was initiated in a deep thrust zone within oceanic lithosphere and continued as the thrust sheet moved upwards to shallower depths. The present array of aureole lithologies is thus the result of progressive incorporation at the base of the thrust sheet during its tectonic rise and emplacement.

Conclusions

Within the Ballantrae Complex ophiolite, at the

SE margin of the Northern Serpentinite Belt, a range of high- to low-grade metamorphic lithologies was produced during obduction. Epidote schist, amphibolite, garnet amphibolite and possibly garnet metapyroxenite form an intermittent aureole to the serpentinite body ranging in width from 20 m to almost 200 m. The garnet metapyroxenite occurs just within the serpentinite and is succeeded, in a south-easterly direction, first by foliated and garnet-bearing amphibolite and then by slaty, banded epidote schist. The regular arrangement of lithologies within the metamorphic aureole arose by the progressive addition of metamorphic slivers onto the base of a hot mantle peridotite body during its thrusting from oceanic lithosphere onto continental crust. This crucial evidence for the obduction of the Ballantrae Complex is well preserved in the Knocklaugh section.

MILLENDERDALE
(NX 177 905 AND 177 906)

P. Stone

Introduction

Mafic dykes rich in hornblende and pyroxene and with a granular texture are seen intruded into foliated gabbro in a series of rocky knolls about 500 m east of Millenderdale Farm. The foliation in the gabbro is generally steeply inclined with a variable, approximately NE–SW trend. The dykes range up to about 3 m across and comprise about 25% of the outcrop; for the most part they cross-cut the gabbro foliation but locally they swing into a parallel orientation. The relationships seen are unique within the Ballantrae Complex and are controversial in terms of its interpretation as an ophiolite assemblage.

The unusual characteristics of the gabbro–dyke assemblage were first described by Teall (in Peach and Horne, 1899) and discussed in some detail by Balsillie (1932) and Bloxam (1955) before its interpretation by Church and Gayer (1973) as representing a part of the sheeted-dyke component seen in an ideal ophiolite sequence. However, that interpretation has not been widely accepted and the problem has been summed up by Bluck (1978a) as follows.

'These outcrops have been interpreted as part of a sheeted dyke complex ... but the proportion of dykes to country rocks is considerably lower than is found in other ophiolites, e.g. 90–100% of the sheeted dyke complex is dyke rock at Troodos .... Dyke intrusion in a tectonically active regime where gabbroic rocks were forming and cooling might well fit the model of a spreading ridge, but on the exposed evidence the outcrop is merely gabbro cut by a few dykes ... a sheeted dyke complex is either poorly represented or not present at all in the Ballantrae ophiolite.'

Despite the sheeted-dyke controversy the Millenderdale gabbro–dyke assemblage preserves unusual textures and relationships that probably originated at an oceanic spreading ridge. Of particular interest is the granofels texture in the dykes and their apparent lack of chilling against the gabbro.

Description

Variably foliated and flasered gabbro, cut by mafic dykes, crops out to the east of Millenderdale Farm. The gabbro has been pervasively hornfelsed and in places is altered to a dark, granular-textured mafic rock consisting essentially of plagioclase, pyroxene and amphibole; the dykes were probably doleritic originally but have been almost entirely altered to a similar lithology. The Millenderdale gabbro lies at the NE end of the Southern Serpentinite Belt within the Ballantrae Complex but its relationship with the surrounding ultramafic rock is uncertain since no contacts are exposed; it may form a very large tectonic inclusion. The gabbro is a pale-coloured, medium- to very coarse-grained rock with extremely variable textures ranging from those produced by normal igneous crystallization to those developed during high-temperature metamorphism. That some of the apparent foliation must be a primary feature due to igneous crystallization differentiation is indicated by cryptic variation in pyroxene composition and the presence in some mafic bands of relict olivine.

The dykes are medium grained and dark-grey, locally weathering to a rusty-brown colour. Most have a granular, saccharoidal texture and are not chilled against the host gabbro, indicating that it was still hot at the time of dyke intrusion. The dykes have a very wide range of orientations and no dominant trend is evident; indeed, cross-cutting relationships between dykes are fairly common, perhaps indicating a protracted period of intrusion. Some dykes follow a meandering and anastomosing path and may enclose xenoliths of foliated gabbro, whereas others cut straight across the gabbro foliation (and earlier dykes) with no deviation.

Most dykes that lie parallel to the NE–SW gabbro foliation are foliated themselves, with augen structure developed in places, so that their margins appear gradational with the gabbro. The foliated dykes may then be cut by either of two further dyke varieties. Completely non-foliated dykes generally cross-cut the foliation at a high angle but a considerable range of intersections are seen and some non-foliated dykes lie parallel to the gabbro foliation; in some examples the gabbro foliation bends to become conformable with the dyke margin. A few of the dykes have a very faint internal foliation that bears no simple relationship to the main gabbro fabric; this variety may cut either foliated or non-foliated dykes but may in turn be cut by other non-foliated dykes.

The metamafic texture and composition of the dykes and some zones within the gabbro is a striking feature shared with numerous other doleritic enclaves within the Southern Serpentinite Belt. However, only at Millenderdale can the range of intrusive relationships be seen. The metamafic dyke rocks were described as beerbachites by Bloxam (1955) and this name has remained in common usage although beerbachite (*sensu stricto*) is a leucocratic hornfels dominated by plagioclase and containing little or no amphibole. The Millenderdale 'beerbachites' have a fine-grained (0.1–0.3 mm), equidimensional and polygonal granofels texture with only rare porphyroclasts of plagioclase or pyroxene, which results from the almost complete recrystallization of original igneous minerals (Figure 2.31a). They are high-grade, thermally metamorphosed rocks. Plagioclase, generally andesine, makes up between about 30% and 60% of the rock accompanied by variable quantities of pyroxene and brown amphibole, the latter forming by alteration of the former. The pyroxene is variously salite, diopside or hypersthene, the amphibole is Ti-rich ferroan pargasite, and olivine is a very rare accessory. A detailed mineralogical and chemical description is given by Jelinek *et al.* (1980). With the development of a foliation in the 'beerbachite' dykes the distribution of the different minerals becomes increasingly heterogeneous until the development of discrete mafic and leucocratic domains gives the rock a streaky appearance (Figure 2.31b).

Interpretation

Geological relationships preserved at Millenderdale are unique within the Ballantrae Complex and influence its interpretation as an ophiolite assemblage. It therefore forms an important complement to the GCR sites elsewhere in the Ballantrae district which illustrate other ophiolitic aspects. Unusual igneous and metamorphic features are also seen. The gabbro–dyke relationships require a protracted period of intrusion commencing while the gabbro was still in a relatively hot and plastic state. Development of foliations between phases of intrusion implies that the site of intrusion was tectonically active. The textural evidence indicates that much of the hornblende in the 'beerbachites' has replaced granoblastic pyroxene developed in original doleritic and gabbroic lithologies. This requires a post-magmatic metamorphism at about 900 to 1000°C (Jelinek *et al.*, 1980). Cooling from this event is probably recorded by the K-Ar age of 487 ± 8 Ma obtained from the flaser gabbro by Bluck *et al.* (1980) which is within error of most other indicators of the age of formation of the Ballantrae Complex. Geochemical data presented by Jelinek *et al.* (1980) for the gabbro and beerbachite dykes was interpreted in terms of a genetically related, ocean-floor tholeiite suite with the more evolved gabbros originating as the upper differentiates of a layered assemblage. In this context, although not representing a true sheeted-dyke complex, the Millenderdale gabbro–dyke association uniquely represents an important component of ophiolite 'stratigraphy'. However, data from the gabbro alone was thought by Stone and Smellie (1990) to be more indicative of within-plate, ocean-island magmatism and so the nature of the original intrusive setting remains controversial.

Quite apart from uncertainties over the geotectonic setting of intrusion there has also been debate over the circumstances and mechanics of incorporation of the metamafic rocks into the ultramafic belt. Various possibilities have been considered ranging from intrusion into a mass of hot peridotite with subsequent disruption and metamorphism as xenoliths, to metamorphism resulting from repeated intrusion in the same

Millenderdale

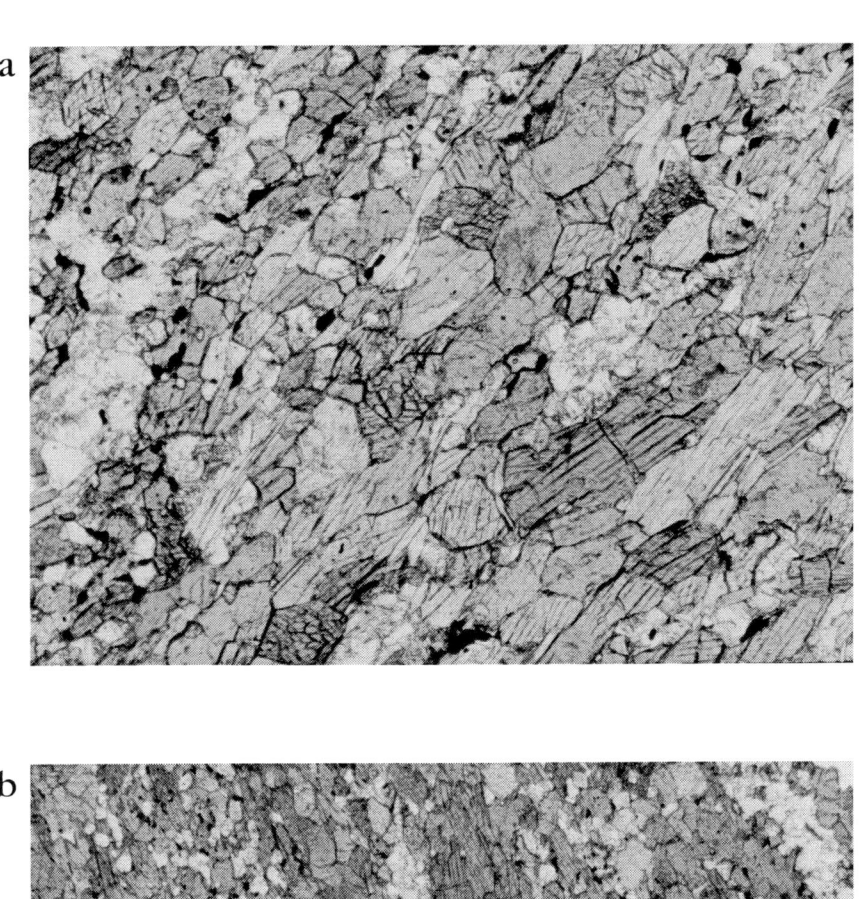

Figure 2.31 Photomicrographs illustrating textures seen in 'beerbachite' dykes near Millenderdale: (a) Polygonal, granoblastic texture developed between feldspar and brown amphibole. Plane polarized light, ×40. (Photo: BGS no. PMS469); (b) Development of mafic and leucocratic domains. Plane polarized light, ×20. (Photo: BGS no. PMS470.)

area with the gabbro–dyke blocks incorporated into the serpentinite during its cold, diapiric rise. Currently the most satisfactory explanation is probably the faulted spreading ridge model proposed by Jelinek *et al.* (1980). They envisaged a part of the ridge displaced by transform-fault movement such that spreading and fault movement in opposing directions combined to keep a ridge segment adjacent to the high heat-flow zone. Such an environment would allow both polyphase intrusion and high-temperature metamorphism in a tectonically active region.

Conclusions

The Millenderdale GCR site contains one of the more enigmatic components of the Ballantrae Complex, the interpretation of which has an important bearing on models for development of the complex as an ophiolite. A banded and foliated gabbro has been repeatedly intruded by doleritic dykes the earlier of which are themselves foliated. All of the dykes and parts of the gabbro have been subjected to high-temperature metamorphic recrystallization and are now granular-textured metamafic rocks ('beerbachites' *senso lato*) rich in pyroxene and amphibole. The explanation of these phenomena may lie in the development of the gabbro–dyke assemblage at or near an oceanic spreading centre; faulting is invoked to maintain the assemblage in the high heat-flow environment and allow thermal metamorphism. If such an interpretation is correct the Millenderdale rocks uniquely illustrate an aspect of ophiolite lithology not seen elsewhere in the Ballantrae Complex.

KNOCKORMAL
(NX 134 885, 137 890, 138 890 AND 143 891)

P. Stone

Introduction

That part of the Ballantrae Complex ophiolite exposed between Knockormal and Carleton Mains farms spans an anastomosing fault zone trending NE–SW through the central of the three volcanic belts (Figures 2.5 and 2.32). The area has long been famous for the reported occurrence within the fault zone of 'eclogite' (garnet clinopyroxenite with, in this case, accessory hornblende and spinel) and glaucophane-crossite schist. At its SW extremity this zone cuts out the metamorphic sole developed at the base of the Northern Serpentinite Belt and farther NE it introduces slivers of serpentinite into the upper Ordovician sedimentary sequence, which elsewhere overlies the lower Ordovician (Arenig) ophiolitic lithologies unconformably. A polyphase history of post-obduction movement is evident with cataclastic and brittle fabrics imposed locally.

The first extensive study of the Knockormal rocks was by Balsillie (1937) who identified the essential paradox; these are very high-grade metamorphic rocks requiring extremely high pressure for their formation, yet they are in close proximity to the extensive Ballantrae lava successions, which have experienced only very low-grade regional metamorphism. Balsillie's solution was to regard the high-grade rocks as relics of an ancient crystalline basement onto which the lavas were erupted, a controversial proposal at the time. However, a detailed study of mineralogy and mineral chemistry by Bloxam and Allen (1960) showed that the various glaucophane and other blue-amphibole rocks had been derived from spilitic lavas identical to those widely seen elsewhere in the Ballantrae Complex. Transitional lithologies were also discovered and an origin through in-situ faulting and shearing was favoured. The eclogite was also studied by Bloxam and Allen and described as a likely segregation within a host peridotite (now serpentinized), possibly forming as a primary component of a layered ultramafic body.

With the identification of the Ballantrae Complex as an obducted ophiolite, speculation was renewed and comparisons made with better exposed analogues elsewhere in the world. Church and Gayer (1973) introduced the possibility that all of the exotic lithologies at Knockormal could be contained within a large-scale, tectonized slump deposit formed during the obduction process. Bloxam (1980) suggested that the glaucophane schist could form tectonic inclusions contained within the serpentinite and derived from an older source, an echo of Balsillie's original proposal although differently effected. Nevertheless, it was the supposed association of eclogite with glaucophane schist and their joint implication of exceptionally high-pressure metamorphism that continued to attract most attention. Hence the inclusion of Ballantrae in that category on such regional compilations as the UNESCO-sponsored

Knockormal

Metamorphic Map of Europe (Fettes, 1978). With this background the Knockormal GCR site is of great importance both for interpretations of the Ballantrae Complex itself and for assessments of its large-scale regional relationships.

Description

The general distribution of lithologies in the Knockormal area is shown in Figure 2.32. The first descriptions of the blueschist and eclogitic lithologies were given by Balsillie (1937) who referred to 'a large knoll of foliated hornblendite and smaragdite-eclogite' with adjacent garnetiferous glaucophane schists 'derived from hornblendites'. Subsequently, in their extensive and detailed study, Bloxam and Allen (1960) located several zones of glaucophane schist in the vicinity of a low ridge capped by 'a prominent crag of eclogite'. Bloxam and Allen admitted the possibility that the eclogite 'crag' 'may not be precisely *in situ*' but concluded from their observation of other small exposures nearby that the material is indeed in place. The present appearance of the 'crag' clearly confirms it as a glacial erratic and there are currently no other exposures of the same lithology nearby. In 1975, a series of shallow boreholes was drilled in the vicinity of blueschist and eclogite 'exposures' and proved till with large boulders to a depth of at least 4 m, thus raising further doubt over the provenance of the schist and eclogite (J.E. Dixon, reported in Smellie and Stone, 1984). However, Smellie and Stone also reported in-situ material revealed in trenching operations: a drainage trench to the west of the Lendalfoot–Colmonell road (Trench 1, Figure 2.32) exposed a sequence of glaucophane-bearing schists while a garnet clinopyroxenite similar to the eclogite was recovered from a temporary excavation a short distance east of the crag/erratic (Trench 2, Figure 2.32).

The field relationships between the various ultramafic lithologies are uncertain but the contacts between the ultramafic rocks and the blueschists all appear to be faulted. The blueschists are disposed along the sheared and faulted margins of a complex tectonic lens, which contains dolerite and gabbro together with serpentinized ultramafic rock, wherlite and various pyroxenites including the 'eclogite'. The schistose foliation trends approximately parallel to the margins of the gabbro–ultramafic lens and appears to be consistent between outcrops. The schists have developed by progressive deformation of spilitic lavas which form the bulk of the outcrop in the Knockormal–Carleton Mains area. Locally, for example 400–500 m WSW from Knockormal farmhouse, the spilitic lavas are interbedded with black siliceous mudstone and recrystallized limestone. For the most part the

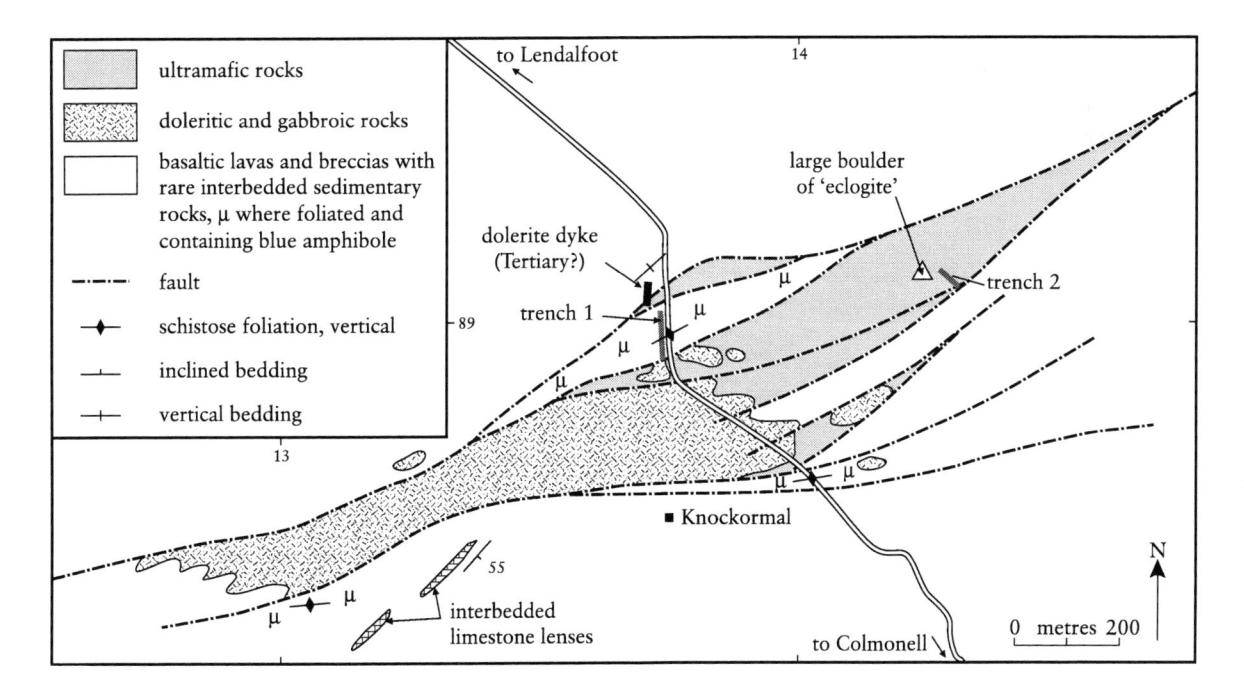

Figure 2.32 Map of the fault zone between Knockormal and Carleton Mains, after BGS (1988).

schists are of the greenschist facies but in some examples wisps of pale-blue glaucophane and/or crossite appear, apparently as replacements of actinolite or chlorite; rarely these may develop into macroscopic blue bands of fibrous crossite. These lithologies have been referred to as 'transitional glaucophane schists' by Bloxam and Allen (1960) and are the most common blueschist lithology. True glaucophane schist is rare and is restricted to narrow zones within the drainage trench and exposures 'on the crags near the Lendalfoot road' as reported by Bloxam and Allen (1960). In these examples the glaucophane forms a continuous foliation, apparently replacing chlorite and enclosing grains of epidote, titanite and albite with rare garnet present in places. In hand specimen the glaucophane schist is a hard, blue-grey rock with a silky lustre. Crossite amphibolites are interbanded with the glaucophane schists and are generally coarser grained with the crossite forming ragged fringes around crystals of green hornblende. A comprehensive account of the mineralogy and mineral chemistry of the Knockormal blueschists is provided by Bloxam and Allen (1960). All of the schistose lithologies show evidence of refolding, cataclasis and brittle deformation subsequent to the formation of the blue amphiboles.

Ultramafic lithologies exposed in the Knockormal area include serpentinized harzburgite, wherlite and clinopyroxenite. However, it is the reported presence of eclogite which gives this locality its unique importance. The eclogite was first reported by Balsillie (1937), together with the blueschists, as 'some of the most interesting rocks that have yet been discovered in the Ballantrae region'. Bloxam and Allen (1960) made a detailed study of material taken from the probable erratic boulder and described it as essentially amphibole (pargasite), clinopyroxene (fassaite) and garnet with accessory green spinel (ceylonite) and possibly zoisite. Smellie and Stone (1984) examined material from the same source and confirmed the earlier description although without the accessory spinel. The nearby excavation of a comparable lithology (Trench 2, Figure 2.32) was also reported by Smellie and Stone and is of importance in view of the uncertain origin of the originally described eclogite crag/boulder. A garnet clinopyroxenite from the NW end of Trench 2 consists of dark-green clinopyroxene (partially replaced by chlorite) and pale-pink garnet (generally fresh). Microprobe analyses presented by Smellie and Stone show that the mineral chemistry of the specimens from the erratic boulder and the trench is indistinguishable and suggests that the rocks had a common origin. The amphibole is tschermakitic hornblende, the pyroxene is fassaitic diopside and the garnet is almandine-rich pyrope. Small exposures close to the erratic boulder are of either serpentinized harzburgite or clinopyroxenite. A severely brecciated clinopyroxenite recovered from the SE end of Trench 2 consists of fresh, pale-green pyroxene occurring as large, ragged plates with strained and bent cleavage. This rock has apparently been affected by intense brittle fracture probably related to movement on the nearby fault (Figure 2.32).

Interpretation

The association of glaucophane schist and supposed eclogite has led to speculation that the Ballantrae Complex contains the remains of a high-pressure metamorphic belt (e.g. Fettes, 1978), with major implications for any interpretation of its origin. However, the ambiguous field relationships between the various lithologies and some unusual petrographical and chemical features have allowed the development of a number of alternative models.

The blueschist occurrences are all contained within an anastomosing fault zone and are marginal to a composite block of gabbro and ultramafic rock (Figure 2.32). If they formed *in situ* then an episode of intense shearing in a relatively low-temperature environment must be invoked, presumably during the obduction process. It seems doubtful that appropriate conditions could be achieved in this way since glaucophane is more characteristic of metamorphism during subduction when deep tectonic burial occurs in a zone of abnormally low heat flow. This paradox has led to alternative proposals whereby the blueschists were regarded as exotic lithologies brought into their current position by sedimentary or tectonic processes. Church and Gayer (1973) suggested that the blueschists (and the neighbouring 'eclogite') could be exotic clasts incorporated into a mélange which 'developed as a result of deformation beneath a forward moving thrust sheet of material previously produced by erosion from the leading edge of the self same nappe in an olistostrome environment.' These authors were particularly impressed by similarities between the Knockormal situation and more extensive analogues in the Newfoundland ophiolites. A different exotic origin was preferred by Bloxam (1980) who described the blueschists as tectonic inclusions contained within serpentinite; 'samples of Precambrian crust dismembered and caught-up in the serpentinite during its cold diapiric rise and tectonic emplacement.' The similarities of these proposals with the original deductions of Balsillie (1937) are striking and perhaps gain further support from the occurrence of blueschist clasts in undisputed mélange deposits elsewhere in the Ballantrae Complex (e.g. the Pinbain section within the Slockenray Coast GCR site).

The 'eclogite' has received even more attention than the blueschists and has generated a range of interpretations. These include: an ultramafic lens within serpentinized harzburgite (Bloxam and Allen, 1960), subsequently refined as a high-pressure phase of a layered peridotite sequence (Bloxam, 1980); part of a high-grade metamorphic belt (e.g. Fettes, 1978); a tectonically isolated sliver of a dismembered metamorphic sole to the ophiolite comparable to the in-situ examples seen at the Knocklaugh GCR site (Spray and Williams, 1980; Treloar *et al.*, 1980); an exotic clast derived from a mélange deposit (Church and Gayer, 1973) comparable, as with the blueschists, to fragments of garnet-clinopyroxene rock seen in other Ballantrae Complex mélanges. The detailed study by Smellie and Stone (1984) showed that the mineral paragenesis is consistent with an origin within the lowermost crust or upper mantle; the mineral chemistry is dissimilar to that of eclogites (*sensu stricto*) found in blueschist belts but is consistent with an ultramafic association. This led the latter authors to propose an origin as garnet pyroxenite formed by partial melting of a rising lherzolite diapir beneath an Arenig oceanic spreading centre. However, age dating by Hamilton *et al.* (1984) suggested that the Knockormal 'eclogite' is considerably older than Arenig (early Ordovician). Using material from the originally described crag/boulder (which they considered to be a clast within a mélange) these authors obtained a Sm-Nd garnet–pyroxene–whole rock isochron of 576 ± 32 Ma, about Early Cambrian. This is the oldest age yet determined for any component of the Ballantrae Complex and strengthens those interpretations that see the 'eclogite' as an exotic clast derived from pre-existing oceanic crust.

Despite a plethora of research the origin and relationships of the Knockormal blueschist-'eclogite' assemblage remain uncertain. The current consensus favours an exotic origin as clasts within a mélange deposit but the relative importance of sedimentary versus tectonic processes has not been established and the ultimate origin(s) of the clasts (if that is what they are) remains unknown. However, it does seem clear that there is no genetic link between the blueschists and the 'eclogite' and their present juxtaposition must therefore be regarded as fortuitous.

Conclusions

The 'eclogite' and glaucophane schist (blueschist) lithologies at Knockormal are internationally known as unusual components of the Ballantrae Complex. The 'eclogite' is most likely to have originated as a garnet(-hornblende) clinopyroxenite segregation within upper mantle harzburgite. It may now occur as a tectonic inclusion within serpentinite. It is the oldest rock so far identified within the complex (*c.* 576 Ma) and may represent a fragment of the oceanic crust which formed the basement to the younger (*c.* 500–490 Ma) components. The high-pressure blueschists occur at the sheared and faulted margins of a lensoid, composite block of gabbro and ultramafic lithologies that includes the 'eclogite'. Blueschist facies rocks appear to be transitional into lower-pressure greenschists derived from adjacent pillow lavas and there is continuity of the schistosity. Nevertheless, the blueschists are also widely considered to be exotic lithologies contained as blocks in a tectonized sedimentary slump deposit. Their ultimate origin is unclear but they may indicate a phase of subduction prior to the late-Arenig obduction of the Ballantrae Complex ophiolite. The combination of these unusual lithologies and the relationships between them, are among the most enigmatic problems in the interpretation of the ophiolite.

GAMES LOUP
(NX 103 880–107 882)

P. Stone

Introduction

At Games Loup a sub-vertical fault trends NE–SW

and juxtaposes two components of the Ballantrae Complex ophiolite: pillow lavas, with the geochemical characteristics of an oceanic island arc, and a serpentinite body in which red and green varieties show an interfingering relationship. The ultramafic rock lies to the NW of the fault and forms part of the Northern Serpentinite Belt whereas, to the SE, the sequence of pillow lavas is assigned to the Balcreuchan Group (Figure 2.33). A very large displacement is required to bring together these two units, which originated at very different structural levels – the serpentinite as mantle ultramafic rock and the pillow lavas as sea-floor volcanic flows. However, the fault exposed is a brittle structure that probably formed late in the assembly of the complex as part of a plexus of faults that cut through its central part. Faults within this zone affect Old Red Sandstone strata confirming a continuing history of movement well after the early to mid-Ordovician obduction of the ophiolite. In this respect the fault exposed at Games Loup, which forms the local margin to the Northern Serpentinite Belt, contrasts starkly with the structure seen in the Knocklaugh GCR site farther NE which also illustrates the marginal relationships of the Northern Serpentinite. At Knocklaugh the faulted margin preserves slivers of high-grade metamorphic lithologies produced during obduction; these have been removed by the later generation of faults as represented by the Games Loup example.

The serpentinite to the NW of the fault has a number of distinctive features: Balsillie (1937) first drew attention to the foliation seen at a number of localities, including Games Loup, while Stone and Smellie (1988) described the veined relationships between green and red varieties. On the SE side of the fault the lava sequence provides good examples of pillow structure with only local brecciation and very little intercalated sedimentary rock. The geochemistry of the Games Loup lavas has been extensively studied and quoted in support of polygenetic interpretations of the Ballantrae Complex. The consensus view is that the lavas were erupted at an oceanic island arc (Wilkinson and Cann, 1974; Lewis and Bloxam, 1977; Thirlwall and Bluck, 1984; Stone and Smellie, 1990). In addition, Thirlwall and Bluck gave a Sm-Nd radiometric age of 476 ± 14 Ma for the lavas, compatible with other evidence of an early Ordovician age for the complex as a whole. The outcrop of the Games Loup lava sequence continues southwards and its southern margin is included within the Balcreuchan Port to Port Vad GCR site.

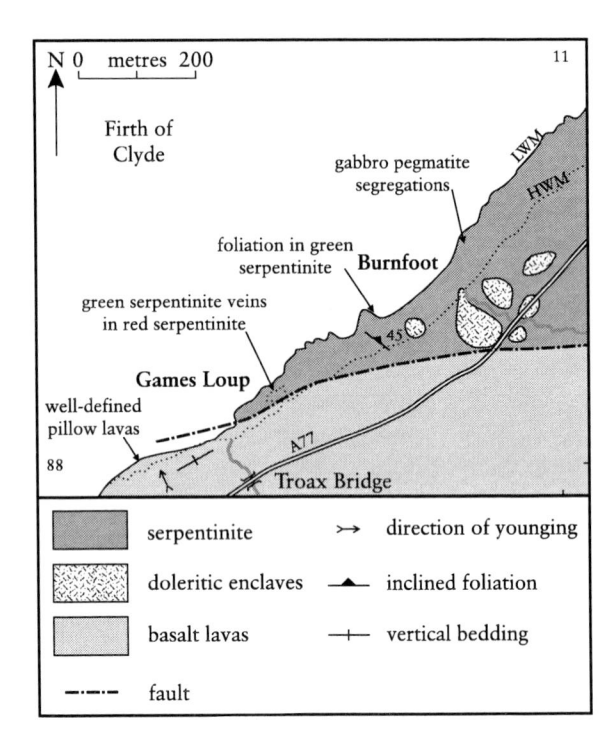

Figure 2.33 Map of the Games Loup area, after BGS 1:25 000 special sheet, Ballantrae (1988) and unpublished data.

Description

The serpentinite at Games Loup forms a series of low, rocky sea stacks and foreshore exposures on the NW side of the prominent fault gully. Adjacent to the fault the largest of the rocky areas consists mainly of red serpentinite but it is cut by prominent veins of green serpentinite up to about 15 cm across (Figure 2.34). The red serpentinite is slightly finer grained than the green variety but, that apart, the only difference is the dissemination of haematite in the serpentine crystal mesh, which imparts the red colouration. Both varieties were originally dunites consisting mainly of olivine with only a very sparse scattering of relict orthopyroxene and chrome-spinel grains. Running parallel to the vein margins are subordinate, very thin (mm-scale) veinlets containing fibrous chrysotile; these form an outer zone extending out for several centimetres from the principal vein margins (Figure 2.34).

Games Loup

Northwards, over about 50 m away from the fault, the proportion of green serpentinite increases until it is the red variety that appears to form sporadic veins. Other changes over the same interval include an increase in orthopyroxene relics (i.e. a more harzburgitic protolith) and the appearance of a faint and patchy foliation striking about NW and dipping moderately to the SW. Farther on, towards Burnfoot (108 882), highly altered doleritic enclaves several metres across occur within the green serpentinite, which there is also host to rare segregations of gabbroic pegmatite similar in lithology to that forming Bonney's Dyke within the Slockenray Coast GCR site.

There is no fabric development or change in the appearance of the serpentinite close to the fault zone, which forms a shingle-filled gully several metres across. The SE side is a substantial cliff of brecciated lava. Some of the brecciation might be of original autoclastic origin and some volcaniclastic material may be present but much of the texture seen probably arose as a brittle response to faulting. The fault seen here was probably initiated late in the structural history of the Ballantrae Complex and was not part of the deep-seated ductile fault system developed during ophiolite obduction.

A few metres south, away from the fault zone, most evidence of brecciation is lost and well-defined pillow shapes dominate the lava sequence. The pillows are generally relatively small with long axes ranging up to about 1 m. There are no sedimentary interbeds but the regular trend of the pillows defines an original horizontal layering, now steeply inclined, striking approximately NE and younging towards the NW. Vesicles are generally sparse. The lava is tholeiitic basalt, now extensively spilitized. A fine-grained matrix contains a scattering of small plagioclase and unusually fresh augite phenocrysts which, in some specimens, form as much as 15% of the rock. Sporadic pseudomorphs of chlorite and serpentinite, usually reddened by iron oxide, are probably the remains of original olivine phenocrysts. Other secondary minerals present include titanite, epidote and prehnite, probably generated by low-grade bur-

Figure 2.34 Green serpentinite veins cutting a red serpentinite host at Games Loup. Chrysotile veinlets are developed parallel to the vein margins. (Photo: BGS no. D3345.)

ial metamorphism. Since all of the margins of the Games Loup lava sequence are either faulted or unexposed it is not possible to estimate the original thickness; at least 250 m are now preserved.

Interpretation

The fault at Games Loup separates spilitic pillow lavas from ultramafic rock and as such forms the local margin of the Northern Serpentinite Belt within the Ballantrae Complex ophiolite. Elsewhere in the complex, as for example the Knocklaugh GCR site, this margin is marked by a zone of intense ductile faulting generated during the obduction of the ophiolite and the resulting juxtaposition of mantle and upper crustal lithologies. At Games Loup only brittle deformation is seen and this particular structure can best be regarded as a post-obduction fault locally cutting out the original structural boundary between the two Ballantrae Complex components.

The green serpentinized harzburgite and dunite seen in the Games Loup section are fairly typical of the common rock types within the Northern Serpentinite as a whole; the veined relationship with the red variety is, however, quite unusual. It may reflect original differences in the ultramafic protolith that have survived or been accentuated by serpentinization or alternatively it may be an entirely secondary effect developed during serpentinization. The subsidiary chrysotile veins might support the latter since their fibrous nature would require formation of the serpentine minerals in a dilational environment. The significance of the sporadic foliation is also uncertain. It is picked out by an elongation of the serpentine crystal lattice but it is not clear whether this reflects an original fabric in the protolith (hence perhaps a mantle tectonite) or whether subsequent deformation has been imposed following serpentinization.

The Games Loup lava sequence is exposed as a strike section south-westwards for about 750 m to the Balcreuchan Port area where it is terminated by a major fault zone. This part of the sequence contains an important lithological variation and falls within the Balcreuchan Port to Port Vad GCR site. Exposure is very poor inland but within 500 m of the coast volcaniclastic breccia and sandstone, containing only aphyric clasts, appear to be dominant. This transition may be stratigraphical, with the Games Loup lavas overlying an older clastic succession, or a fault may intervene. In a wider context the relative freshness of the lavas has allowed their use in geochemical characterization of geotectonic settings of eruption during the evolution of the Ballantrae Complex. Overall this has proved a controversial exercise but for the Games Loup lavas, almost uniquely, a succession of studies based on different element combinations has produced a consistent result (see the Balcreuchan Port to Port Vad GCR site report). Eruption of the lavas at an oceanic island arc was first proposed by Wilkinson and Cann (1974) based largely on Ti–Zr–Y–Nb relationships. A similar conclusion was then reached by Lewis and Bloxam (1977) using rare-earth element distributions. Thirlwall and Bluck (1984) used multi-element comparisons coupled with an assessment of Sr and Nd isotope ratios to interpret the lavas as primitive island-arc tholeiites similar to modern examples from the South Sandwich Islands. Data presented by Stone and Smellie (1990) are almost indistinguishable from those of Thirlwall and Bluck. Overall there is a selective enrichment in large ion lithophile elements (e.g. Rb, K, Ba, Sr) and a marked depletion in the high field strength elements (e.g. Y, Zr, Ti, Nb).

A Sm-Nd age of 476 ± 14 Ma was obtained by Thirlwall and Bluck (1984) from internal isochrons on the lavas and their separated clinopyroxene phenocrysts. The relatively large error overlaps with most other age determinations from the Ballantrae Complex and gives a time range compatible with its generally accepted early to mid-Arenig generation.

Conclusions

The Games Loup GCR site exposes several important geological features pertinent to interpretation of the Ballantrae Complex as an ophiolite. At Games Loup itself, the SE margin of the Northern Serpentinite Belt is a late brittle fault that juxtaposes the ultramafic rock against pillow lava. This structure has locally cut out the major ductile shear zone, which originally brought the two lithologies together during obduction of the ophiolite. To the north of the fault, serpentinized harzburgite and dunite occur as distinct red and green variants that have an interfingering relationship. In the largest exposures, adjacent to the fault, red serpentinite is dominant and is cut by veins of green serpentinite; to the north the proportion of the green variety increases sharply. South of the fault a sequence of basaltic pillow lavas has the tholeiitic geochemical characteristics of a primitive, oceanic island arc. The lavas have suffered relatively little alteration which has allowed their use in producing a Sm-Nd radiometric age of 476 ± 14 Ma, compatible with the Arenig age of the Ballantrae Complex as a whole.

BALCREUCHAN PORT TO PORT VAD (NX 100 878–093 869)

P. Stone

Introduction

Balcreuchan Port and Port Vad are both small embayments, backed by steep cliffs, on the northern flank of Bennane Head. The coastal section in this area affords magnificent exposure through part of the volcanosedimentary Balcreuchan Group in the central zone of the Ballantrae Complex ophiolite. The features preserved were fundamental in persuading Bonney (1878) that the rocks have a volcanic origin and an extensive description in those terms was provided by Peach and Horne (1899). A sequence of early Ordovician pillow lavas contains the very unusual lithology boninite, erupted only at primitive oceanic island arcs. This is the only unambiguous occurrence of boninite lava so far recorded from Britain. The island-arc lavas are faulted against a more disparate lava sequence believed to have originated at a within-plate, oceanic hot spot. Pillow structure is particularly well developed in the within-plate lavas, some of which are remarkably plagioclase-phyric. Interbedded sedimentary strata are sparsely fossiliferous, establishing an early to mid-Arenig age range and allowing structural imbrication of the sequence to be proven. Quite apart from its geological attractions Balcreuchan Port also features in Scottish legend as the lair of Sawney Bean and family, the infamous cannibal tribe finally brought to account and executed *en masse* in 1604 by personal command of King James VI.

The ophiolitic interpretation of the Ballantrae Complex (e.g. Church and Gayer, 1973) regarded the lavas as oceanic lithosphere and much subsequent work has been geochemically orientated in attempts to test this hypothesis. Wilkinson and Cann (1974) pioneered this approach, concluding that the Bennane Head section contains island-arc (at Balcreuchan Port) and ocean-island, hot-spot (south from Balcreuchan Port towards Port Vad) basalt lavas; a faulted relationship was inferred. Lewis and Bloxam (1977) also found a preponderance of island-arc lava types in the section but were also impressed by the apparent conformity of the sequence, placing it within a 4 km-thick lava pile. They considered this great thickness to be most compatible with an island-arc eruptive environment and regarded the hot-spot basalts as anomalous components within a single succession. A more sophisticated geochemical study by Thirlwall and Bluck (1984) confirmed the original conclusions of Wilkinson and Cann and re-established the fault juxtaposition of two very different lava sequences. The island-arc associations of the Balcreuchan Port rocks have also been confirmed by the discovery of characteristic boninite lavas (Smellie and Stone, 1992; Smellie *et al.*, 1995).

Graptolites had been recorded from sedimentary interbeds by Peach and Horne (1899) and assigned a generally mid-Arenig age. Further discoveries and a review of old collections allowed Stone and Rushton (1983) to establish biostratigraphical control on the lava sequence, which they showed to range from the early to the mid-Arenig. Despite the apparent conformity and uniform younging sense the biostratigraphy shows repetitions, and the structural imbrication of an originally relatively thin lava sequence seems the most likely explanation.

A detailed description of the Bennane Head section is given by Stone and Smellie (1988) and a field guide is provided by Stone (1996).

Description

The steep sea cliffs surrounding Balcreuchan Port expose an extensive array of pillow lavas intercalated with sporadic more massive lava flows. There is some autobrecciation of the pillows but intercalated sedimentary rocks are notable by their absence. The eastern margin of the bay coincides with a major N–S fault which, at sea level, juxtaposes the pillow lavas in the cliff against highly altered ultramafic rock forming the foreshore exposures. Alteration in the ultramafic rock lessens across the bay so that on the SW side serpentinized dunite and harzburgite are exposed in the intertidal zone. The cliff sections show no signs of ultramafic rock and a sub-horizontal structural contact is envisaged, between serpentinite below and pillow lava above, cut off by the N–S fault on the east side of the bay (Figure 2.35). An added point of interest in the foreshore exposures is a well-exposed Palaeogene dyke trending just east of north; it is about 50 to 60 cm wide with amygdales concentrated into zones parallel to the dyke margins.

The pillow lavas forming the Balcreuchan Port cliffs range up to about 1.5 m across but most are relatively small (less than 1 m). They are only sparsely vesicular. Small phenocrysts of plagioclase are much in evidence, accompanied by clinopyroxene phenocrysts (probably augite) and rare pseudomorphs after olivine; the matrix is dominated by altered glass. Spilitization has caused the alteration to albite of the originally more calcic plagioclase while low-grade metamorphism has produced the secondary mineral assemblage titanite-epidote-prehnite. The lavas are tholeiitic basalts forming a continuous sequence with those farther NE along the coast at the Games Loup GCR site. There, the relative freshness of the clinopyroxene phenocrysts allowed Thirlwall and Bluck (1984) to obtain a Sm-Nd age of 476 ± 14 Ma.

One very important lithological variation occurs in the sequence adjacent to the fault on the east side of Balcreuchan Port. Around the high-water mark a thickness of 5–10 m of lava shows only poorly defined pillows and is less porphyritic than most of the sequence. In thin-section the rock proves to have more glassy matrix and the scarcity of plagioclase microphenocrysts is confirmed. Conversely, clinopyroxene and olivine phenocrysts are slightly more abundant. Geochemical data show relatively high SiO_2, MgO, Cr and Ni but relatively low Al_2O_3 when compared to the normal tholeiites. These lavas are boninites (Smellie and Stone, 1992; Smellie *et al.*, 1995), a rare lithology that has no other unambiguous occurrence in Britain.

Westwards from Balcreuchan Port the lava sequence continues, with an increasing proportion of brecciated flows, as far as a second major N–S fault (Figure 2.35). Beyond this fault there is a marked change in the nature of the volcanic succession, which becomes lithologically much more variable. Adjacent to the fault about 30 m of sedimentary strata form the base of this varied sequence. Sandstone, shale, chert and conglomerate are all present and a graptolite fauna has been recovered that gives an early Arenig age (Stone and Rushton, 1983). The beds have a

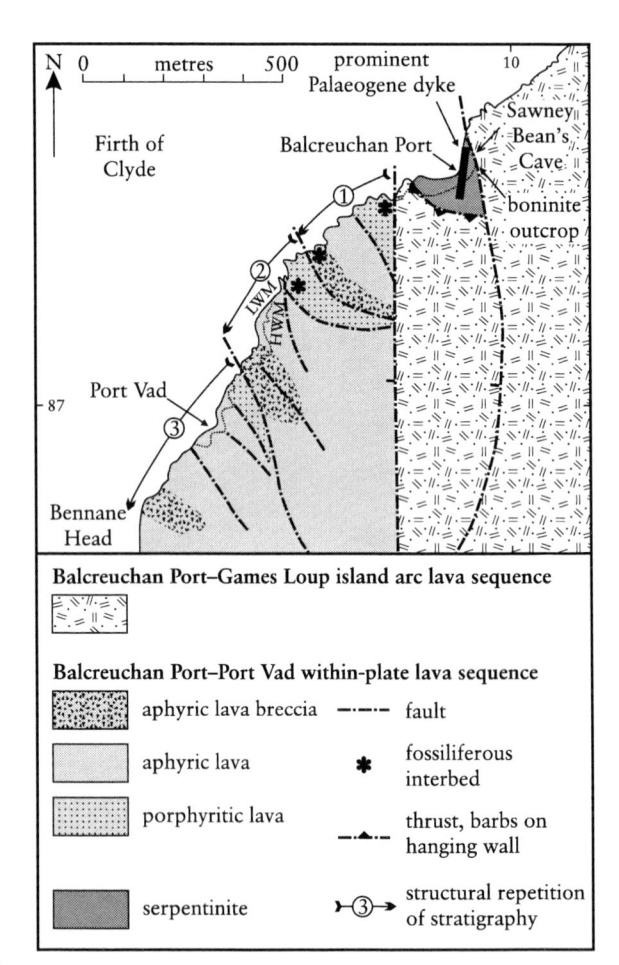

Figure 2.35 Map of the Balcreuchan Port to Port Vad area, after BGS 1:25 000 special sheet, Ballantrae (1988) and Stone and Smellie (1988).

N–S strike and are sub-vertical with sedimentary structures showing that they young towards the west. They are overlain conformably by exceptionally well-formed pillows of markedly porphyritic lava in which large plagioclase phenocrysts, up to 1 cm long and tabular in form, are contained in a pervasively reddened fine-grained matrix. The consistent asymmetry of the pillows (Figure 2.36), which range up to 2 m across, is consistent with the N–S strike, steep dip and westward younging deduced from the underlying sedimentary strata. The red, feldspar-phyric lavas form a unit about 150 m thick that is cut in places by thin dykes (less than 50 cm) of unreddened, aphyric basalt. These are the feeders for the overlying unit, which consists of about 150 m of dark grey-green aphyric lava in well-formed pillows generally smaller than those of the reddened lava type. Despite the presence of numerous minor faults in the section a conformable relationship can be seen between the two lava units, with aphyric pillows resting directly on reddened feldspar-phyric pillows. There is no intervening sedimentary material. Up sequence the aphyric pillows become increasingly more brecciated and interbeds of clastic rock appear. For one of these sedimentary intervals, just SW from the mouth of the Bennane Burn, a graptolite fauna indicates a mid-Arenig age (Stone and Rushton, 1983).

In the Bennane Head sector the succession described above (sedimentary beds (early Arenig)–red feldspar-phyric pillow lavas–aphyric pillow lavas–aphyric lava breccia–sedimentary interbeds (mid-Arenig)) is then terminated by faulting. However, biostratigraphical and lithostratigraphical comparisons suggest that it is repeated to the SW in a series of imbricate fault slices (Stone and Rushton, 1983). Throughout the imbricated succession the overall strike and dip and the younging direction remain constant. One repetition extends from red feldspar-phyric pillow lava just SW of Bennane Burn to brecciated aphyric pillow lava NE from Port Vad. At this point faulting re-introduces the red feldspar-phyric lava and another repetition of the sequence continues thence around Bennane Head (Figure 2.35). From Port Vad this highest structural repetition continues farther up the sequence with a much greater thickness of aphyric lava breccia preserved. It is this lithology (Figure 2.37) that forms the steep cliffs of Bennane Head itself. Ultimately the stratigraphy continues southwards to the Bennane Lea GCR site where the youngest part of the succession is exposed.

Interpretation

The lava sequence in the Balcreuchan Port to Port Vad area can be divided into two parts on lithostratigraphical grounds. The north-eastern sector consists exclusively of pillow lava and lava breccia with little obvious lithological variation and a complete absence of sedimentary interbeds. In contrast, the south-western sector contains an alternation of aphyric and reddened feldspar-phyric pillow lavas with abundant sedimentary intercalations. From the available evidence their ages are indistinguishable: the north-eastern lavas have a Sm-Nd radiometric age of 476 ± 14 Ma, the south-western lavas are early to mid-Arenig on biostratigraphical grounds. They are divided by a major N–S fault.

In a series of geochemical studies the Balcreuchan Port to Games Loup (north-eastern) lavas have consistently been interpreted as the

Figure 2.36 Large, well-formed pillows of reddened, plagioclase-phyric basalt exposed SW of Balcreuchan Port. (Photo: BGS no. D3585.)

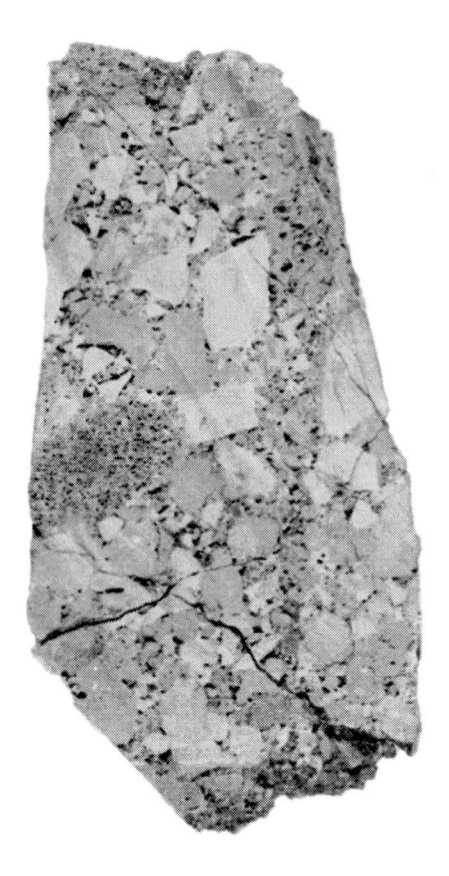

Early Ordovician volcanic rocks and ophiolites of Scotland

Figure 2.37 Volcaniclastic breccia of aphyric and vesicular lava clasts from Bennane Head. The long axis of the sample is 165 mm. (Photo: BGS no. MNS3838.)

products of eruption at an oceanic island arc (Wilkinson and Cann, 1974; Lewis and Bloxam, 1977; Thirlwall and Bluck, 1984). Most of the analyses published by these authors were derived from the Games Loup end of the section, now designated as a separate GCR site. More recent results from the Balcreuchan Port area (Stone and Smellie, 1990; Smellie and Stone, 1992; Smellie *et al.*, 1995) have confirmed the bulk of the lava sequence as of low-Ti tholeiitic character, and have also established the presence of the boninitic lavas. Examples of the geochemical discriminations used, based on abundances and ratios of trace elements such as Ti, Y and Zr, are shown in Figure 2.38a.

Despite their relative scarcity the boninites are of great importance in interpretation of the Ballantrae Complex because of the unique combination of circumstances involved in their petrogenesis. Important factors include: they are found exclusively in the intra-oceanic realm; their major element contents resemble those of primary magmas derived by partial melting of strongly depleted sources; the partial melting may occur at abnormally high geothermal gradients at relatively shallow levels. In most modern examples a general association with supra-subduction zone extension is apparent and in this respect the interbedding of the boninites with 'normal' island-arc tholeiites is important. Hence, arc-splitting during the initiation of a back-arc basin was invoked by Smellie and Stone (1992). After more detailed analytical work Smellie *et al.* (1995) noted that the low-Ti tholeiitic rocks, an unusual suite in their own right, shared some of the boninitic characteristics, such as the high Cr content (Figure 2.38b) and could not be definitively distinguished from basalts erupted in back-arc basins. A solution was proposed whereby the distinctive features of the combined sequence arose from mantle source heterogeneity caused by metasomatism above a subducting slab. Whatever the detailed mechanism the association with an intra-oceanic island arc seems unequivocal.

The more varied volcanosedimentary succession extending SW from Balcreuchan Port towards Port Vad has also been included in a number of geochemical studies (Wilkinson and Cann, 1974; Lewis and Bloxam, 1977; Thirlwall and Bluck, 1984; Stone and Smellie, 1990). The geochemical results have been consistently indicative of eruption in a within-plate environment above a mantle hot spot with the lavas forming a Hawaiian-type ocean island (Figure 2.38a). Controversy was introduced into the interpretation by Lewis and Bloxam who regarded the whole Ballantrae Complex succession as essentially conformable and abnormally thick, over 4 km and dominated by island-arc lavas. This apparently great thickness could be best accommodated in an island arc environment and so the Balcreuchan Port to Port Vad results were taken to be anomalous. The dilemma was resolved by recognition of structural imbrication of the within-plate sequence (Stone and Rushton, 1983). The nature of both the intercalated sedimentary rocks and those farther south at the top of the sequence (within the Bennane Lea GCR site) suggests deposition in fairly deep water and the reddening of the porphyritic lavas may have been caused by lengthy sea-floor weathering. This contrasts with the shallow-water environment deduced for the volcanosedimentary succession forming the Slockenray Coast GCR site, which contains geochemically indistinguishable (but largely unreddened) lavas

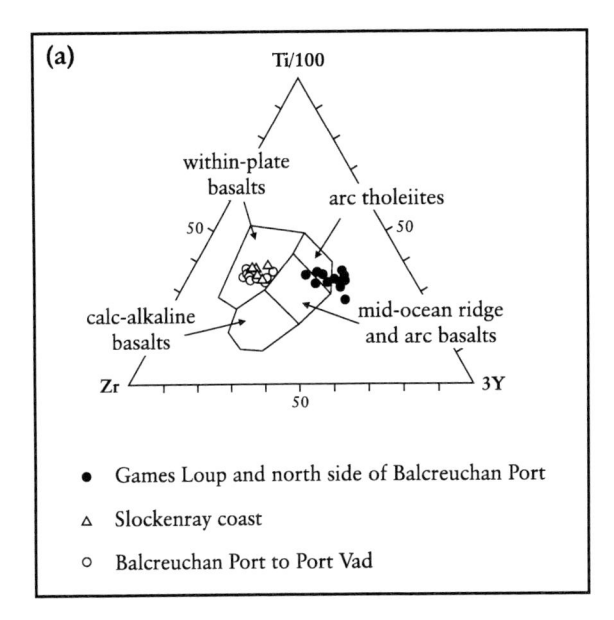

Figure 2.38 Geochemical discrimination diagrams for Ballantrae Complex basalt lavas:
(a) Ti–Zr–Y, fields from Pearce and Cann (1973).
(b) Cr–Y, showing comparison with representative boninites after Smellie and Stone (1992); fields from Pearce (1982).

of the same early Arenig age. It would seem reasonable to regard these two successions as facies variants formed coevally at the same ocean island.

Between Balcreuchan Port and Port Vad the N–S fault zone separating the two lithologically and geochemically distinct lava sequences (Figures 2.35 and 2.38) intersects the coast at 0971 8751, about 100 m west of the SW corner of Balcreuchan Port. It has had a polyphase history of movement but is unlikely to be the original structure formed during the imbrication of the lava sequence, which was probably a low-angle thrust. Some trace of this may be preserved to the immediate east of the main fault in a small and complex subsidiary fault-block. On the west side of the main fault a pronounced swing of strike suggests a sinistral movement laterally. However, in regional terms the fault is part of a plexus forming the eastern margin of a largely offshore Permo-Triassic basin and so has a significant downthrow to the west. Taking this latter movement alone, the island-arc sequence at Balcreuchan Port may have originally been at a greater structural depth within the obducted ophiolite thrust stack than the juxtaposed within-plate sequence forming Bennane Head.

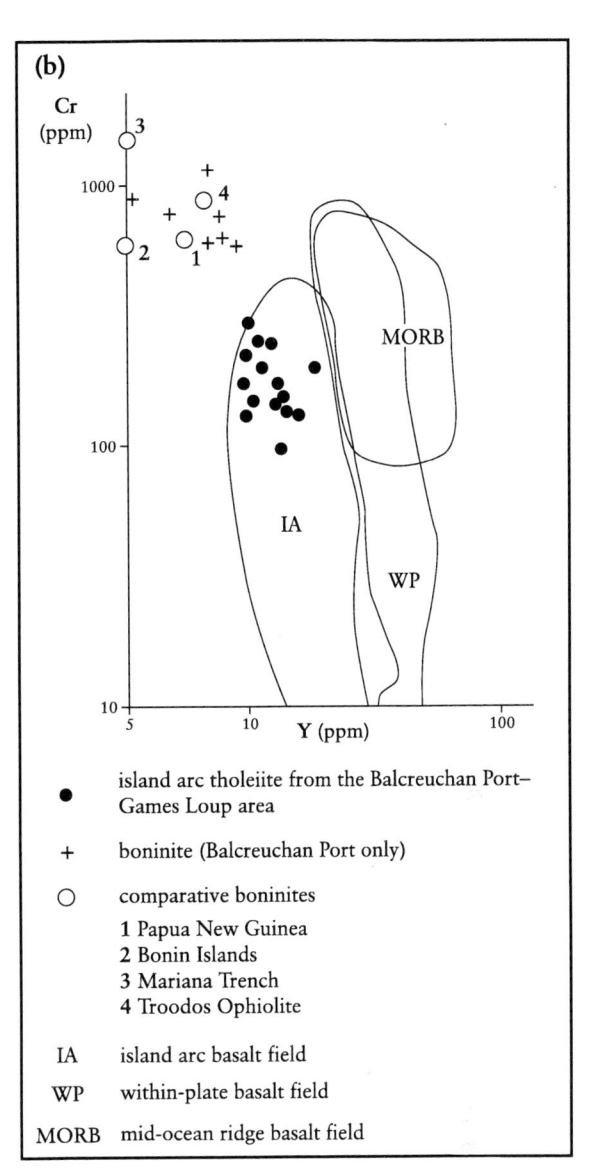

Conclusions

The Balcreuchan Port to Port Vad sector of the Ballantrae Complex provides crucial data for its interpretation and is important internationally for the more general assessment of ophiolite generation and obduction models. Within the site, two dissimilar lava sequences are separated by a major N–S fault. To the east of the fault, at Balcreuchan Port, pillow lavas and lava breccias of uniform tholeiitic basalt were generated in an oceanic island arc. Rare intercalated boninite lavas are definitive of island-arc eruptions; they are a rare but petrogenetically important lithology and no other unambiguous examples are currently known from Britain. To the west of the fault, towards Port Vad, a varied assemblage of basalts were erupted in a within-plate, ocean-island environment; pillow-forms are well developed and sedimentary interbeds of deep-water facies contain graptolites of early to mid-Arenig age. The rare combination of biostratigraphy and distinctive lithostratigraphy proves the structural repetition of the within-plate sequence.

BENNANE LEA (NX 091 861)

P. Stone

Introduction

At Bennane Lea two major components of the Ballantrae Complex ophiolite, the central sector of the volcanosedimentary Balcreuchan Group and the Southern Serpentinite Belt, are separated by a major fault. The locality lies on the south side of Bennane Head and the massive basalt breccias seen there continue southwards forming steep sea cliffs. Higher in the succession, towards Bennane Lea, pillowed and massive lava flows appear intercalated with the breccias near sea level while the main cliffs are composed of breccia, here rather finer grained and with more sandstone and chert interbeds. In the foreshore exposures the breccia and lava are succeeded southwards and up-sequence by a sedimentary assemblage, initially of black mudstone followed by chert and coarse conglomerate; tuffaceous sandstone interbeds occur sporadically throughout the succession. The major fault, trending approximately E–W, cuts out the higher sedimentary beds and juxtaposes serpentinite against the chert and conglomerate. Coinciding with this abrupt lithological change is an equally abrupt change in topography. The softer serpentinite has been more readily eroded so that the steep sea cliffs and restricted foreshore typical of the volcanic rocks are replaced by a broad sandy beach and an extensive raised beach backed by relict sea cliffs cut in glacial till. Within the broad, sandy foreshore small exposures show that here Permian red sandstone (the fringe of a major, mainly offshore basin) overlies the serpentinite unconformably.

Since the coastal and foreshore exposures at Bennane Lea contain such a wide variety of lithologies with complex structural relationships, it has been the focus for much geological investigation. Detailed sketch maps have been published by Peach and Horne (1899), Bailey and McCallien (1957) and Bluck (1978a, 1992) and the area has featured in most of the controversies over the origin of the Ballantrae Complex. Peach and Horne regarded the ultramafic rock, now serpentinized, as intrusive into the volcanic sequence; a position defended by Anderson (1936), using some evidence from Bennane Lea, against the then-current consensus that it formed an older basement to the lavas. Bailey and McCallien (1957) proposed the radical alternative that the serpentinite originated as a submarine lava interbedded with the volcanic rocks. The interpretation of the Ballantrae Complex in ophiolitic terms (Church and Gayer, 1973) provided an integrated model for the igneous rocks, with the sedimentary assemblage regarded as a deep-water, slumped olistostrome. Detailed work by Bluck (1978a) built on this hypothesis, with evidence from the Bennane Lea conglomerates supporting his proposition that the complex had suffered considerable structural disruption prior to its final obduction. Bluck envisaged an island-arc–marginal basin environment for its original generation.

Description

Basalt breccias form the steep sea cliffs of Bennane Head and from there extend south towards Bennane Lea where they are succeeded by a mixed assemblage mainly composed of chert and conglomerate. A detailed geological sketch map of the Bennane Lea area is shown in Figure 2.39.

The Bennane Head breccias overlie aphyric pillow lavas seen at the southern margin of the Balcreuchan Port to Port Vad GCR site and extend that volcanosedimentary sequence southwards. A thickness of between 200 and 300 m is present and, for the most part, the breccias are very coarse with clasts up to 2 m across and commonly in the 2–10 cm range. They are composed exclusively of aphyric basalt although there is considerable difference in the vesicularity, with some clasts completely devoid of vesicles whereas others are scoriaceous (Figure 2.37). Although some clasts are partially reddened there are no clasts of the distinctive, reddened porphyritic basalt seen lower in the sequence towards Balcreuchan Port. The clasts are generally angular although a small proportion always show some degree of rounding and this proportion increases slightly up-sequence, towards the south. Bedding is difficult to detect in the massive breccia units but sporadic interbeds of volcaniclastic sandstone define an approximately NW–SE strike and a very steep dip; grading in the sandstone confirms younging towards the SW, continuing the trend established farther north between Balcreuchan Port and Port Vad.

Lava flows reappear at the top of the breccia sequence but the flows are relatively thin and pillows are only developed locally. The lava is aphyric basalt, vesicular in places, with interbeds of breccia, sandstone or siliceous mudstone between the flows. The proportion of intercalated sedimentary rock increases up-sequence, southwards, and the clast size of the breccias reduces. Bluck (1992) reported that the composition of the breccias also becomes more variable with the inclusion of acid volcanic lithologies. Near the top of this lava-dominated unit, which is about 100 m thick, black siliceous mudstone is particularly well developed and contains a mid-Arenig graptolite fauna (Peach and Horne, 1899; Stone and Rushton, 1983). The mudstone contains chalcopyrite and fracture surfaces within it are coated by green, secondary copper minerals.

Minor faulting complicates this part of the sequence and across the faulted zone there is a change of strike to almost E–W, the dip remaining steep. However, the sedimentary interval continues and any stratigraphical break is thought to be fairly small. It is also possible that there is a minor intra-formational unconformity at this point but the evidence is obscure and inconclusive. Immediately south of the fault pale-green volcaniclastic sandstone is interbedded with chert and the relationship between the two lithologies suggests that the sandstone remained fluid after the chert had become partially lithified. The resulting soft-sediment deformation structures were first described and illustrated by Bailey and McCallien (1957).

Southwards and up-sequence the proportion of chert increases sharply. It is reddish-brown in colour and individual beds rarely exceed 10 cm in thickness; radiolaria were described by Aitchison (1998) and Bluck (1992) reported that locally the chert may contain glass shards. Boudinage and other features characteristic of soft-sediment deformation are widely present and increase in both size and frequency towards the south. This trend towards increased deformation is made more complex by a concomitant increase in tectonic folding with several large fold hinges, plunging steeply seawards, affecting the cherts on the foreshore. The old sea cliffs behind the raised beach provide a more extensive and spectacular view of these structures. The range of deformation features makes it difficult to assess the thickness of the chert unit but around 30 to 40 m seems likely.

Towards the top of the chert sequence coarse conglomerate makes an abrupt appearance, overlying and interdigitating with the chert beds. Clast size ranges up to about 1 m. In the vicinity of the conglomerate the soft-state deformation of the chert reaches a maximum with slump folds showing bulbous thickening of their hinge zones (Figure 2.40). The clasts in the conglomerate are mainly of either aphyric basalt (generally rounded) or chert (generally angular), both with an obvious local provenance, but rarer clast types include a pinkish, coarse-grained syenitic lithology, massive pyrite and carbonate. The latter is of some interest in that Bailey and McCallien (1957) reported dissolving fragments in hot acid and recovering grains of chrome-spinel; the carbonate was therefore regarded as a highly altered ultramafic rock. In addition to the above Bluck (1992) reported clasts of acid volcanic rock. The conglomerate does not extend inland from the foreshore exposures but Bluck described it as increasing in proportion towards and beyond the low-tide mark. The shoreline would therefore seem to coincide with an interfingering zone between two distinct lithofacies.

The southernmost of the conglomerate beds is only a few metres short of the major fault juxtaposing this volcanosedimentary succession and ultramafic rock of the Southern Serpentinite Belt. In this interval lies an enigmatic lithology that has some characteristics of an intrusive dolerite and some of a volcaniclastic sediment. Its northern margin appears to be in conformable contact with chert but its origin remains uncertain; Bluck (1992) refered to it as 'dolerite-tuff' and speculated that it might be a high-level sill, despite an absence of peperitic margins. The southern margin of this body is faulted against serpentinite with a thin zone of quartz-carbonate alteration along the contact. The serpentinite itself is pervasively reddened and contains large enclaves of altered gabbroic rock. These form small bosses between which expo-

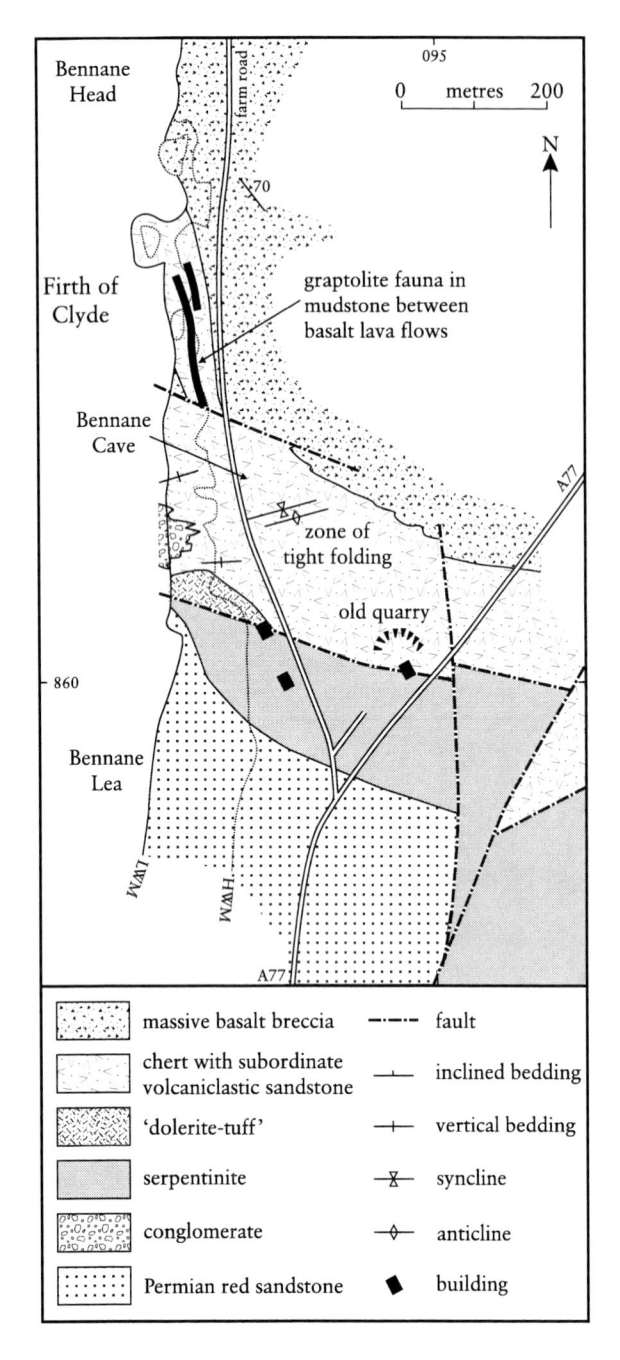

Early Ordovician volcanic rocks and ophiolites of Scotland

Figure 2.39 Map of the Bennane Lea area, after BGS 1:25 000 special sheet, Ballantrae (1988) and Stone and Smellie (1988).

sure of the serpentinite is only sparse. A few metres farther south the outcrop of ultramafic rock ends at the unconformable contact with Permian red sandstone.

Interpretation

The Bennane Lea section illustrates the dominantly sedimentary and volcaniclastic upper part of a Balcreuchan Group lava sequence. Coarse, oligomictic basalt breccia in the north of the section probably accumulated as flow-front talus and in terms of their petrography and geochemistry the clasts are identical to the subjacent lavas seen farther north at Port Vad. The presence of interbeds of graded volcaniclastic sandstones suggests deposition in relatively deep water but the rounding of some lava clasts probably arose through wave action in shallow water. The most likely combination of events would seem to be the eruption of aphyric basalt lava into shallow water, autobrecciation of the lava front (and perhaps some wave erosion) with partial rounding of some fragments by wave action, followed by avalanching of the talus accumulation into deeper water to cover basalts erupted earlier in the same eruptive episode. A few lava flows then extended across the accumulated breccia.

The association of these later lava flows with graptolitic mudstone is also indicative of a deep-water depositional environment but the evidence of the overlying chert–conglomerate sequence is ambiguous. In straightforward lithological terms the chert might be regarded as a deep-sea deposit but its extensive soft sediment deformation requires slumping to have occurred, presumably on an unstable slope. The interdigitating conglomerates with their rounded clasts appear to have slumped from a shallow-water environment and the glass shards in the chert indicate relatively proximal and probably subaerial, contemporaneous volcanicity. Cobbles and boulders in the conglomerate are derived from ophiolitic lithologies, including altered ultramafic rock, which suggests that obduction was already in progress when this part of the Ballantrae complex was deposited. Overall, a tectonically unstable environment is indicated, interpreted by Bluck (1978a) in terms of obduction within an active marginal (back- or intra-arc) basin. Developing this theme, Bluck (1992), emphasized the evidence for acid volcanicity and the abrupt facies changes, concluding that the Bennane Lea succession represents the rift facies developed during the splitting of a volcanic arc. The lavas to the north were regarded as part of the original arc structure. This interpretation was also influenced by the thickness of the succession which, at over 2 km, Bluck thought anomalous for anything but an island-arc environment. However, this figure was reached by including within the same suc-

Bennane Lea

Figure 2.40 Slump fold in chert interbedded with conglomerate and volcaniclastic sandstone at Bennane Lea. (Photo: BGS no. D3333.)

cession the whole coastal outcrop between Bennane Lea and Games Loup, which a series of geochemical studies has shown to be polygenetic (Wilkinson and Cann, 1974; Thirlwall and Bluck, 1984; Stone and Smellie, 1990). On the geochemical evidence (Figure 2.38) the Bennane Lea strata overlie within-plate, ocean-island basalts with both the breccia clasts and the lavas above the breccia showing the same geochemical characteristics; on this basis none of these lavas can have any direct connection with an island arc. Further, when the evidence for structural imbrication presented by Stone and Rushton (1983) is taken fully into account the exposed thickness of the succession seen in part at Bennane Lea does not exceed 900 m (Stone and Smellie, 1988). The dilemma remains unresolved but the problem cannot be addressed by evidence from the Bennane Lea GCR site taken in isolation; comparison with the contiguous outcrop, as represented by the Balcreuchan Port to Port Vad and Games Loup GCR sites is important. It is also useful to compare the Bennane Lea succession with the within-plate, ocean-island volcanosedimentary rocks of the Slockenray Coast GCR site, which show some remarkable similarities.

Whatever the correct interpretation of the Bennane Lea volcanosedimentary sequence, at its southern margin it is faulted against serpentinized ultramafic rock. The fault is one of the major structures of the Ballantrae Complex juxtaposing mantle rocks of the Southern Serpentinite Belt and upper crust of the Balcreuchan Group, represented here by the chert–conglomerate–volcaniclastic assemblage. Movement on this scale seems unlikely to be achieved in a single phase of faulting and a history of periodic re-activation is probable. However, in this context the narrow zone of quartz-carbonate alteration adjacent to the fault is significant. This developed as a side-effect of serpentinization at a relatively early stage in the history of the complex. Its presence suggests that the Bennane Lea Fault was also an early-formed feature.

Conclusions

The Bennane Lea GCR site reveals several unique features of great importance in the interpretation of the Ballantrae Complex as an ophiolite. Some aspects remain controversial. The exposed strata form the highest preserved beds of a volcanosedimentary sequence dominated elsewhere, in its lower part, by basalt lava. These lavas have the geochemical characteristics of within-plate, ocean-island eruption, a feature shared by large clasts within breccias and sporadic lava flows within the Bennane Lea sector. However, the unusual chert–conglomerate–volcaniclastic rock assemblage has been interpreted, on mainly sedimentological grounds, as being more compatible with a volcanic arc environment. The compositional range of clasts in the conglomerate suggests that ophiolite obduction was already in progress when it was deposited; graptolites in subjacent mudstones give deposition a maximum age of mid-Arenig. At the southern margin of the site, volcanosedimentary upper crustal rocks are faulted against serpentinized mantle rocks, across one of the major lithological breaks within the Ballantrae Complex. Bennane Lea provides rare exposure across a fault of this magnitude.

SOUTHERN UPLANDS

SGAVOCH ROCK (NX 075 810)

P. Stone

Introduction

The coastal sections on the mainland opposite the offshore Sgavoch Rock provide the finest array of pillow lavas to be seen in Britain. Individual pillow shapes range from almost spherical, through elliptical 'bolster' shapes into sheet flows; lava tubes can also be clearly identified with well-preserved pillow buds at their margins. The lava sequence is structurally confined between the Stinchar Valley Fault to the north and the Dove Cove Fault to the south (Figure 2.41). Extensive coastal exposure is continuous from the Sgavoch Rock area southwards to Downan Point but the outcrop narrows inland as the two faults converge towards the NE. These lavas have traditionally been associated with the Arenig (early Ordovician) Ballantrae ophiolite complex with which they are juxtaposed across the Stinchar Valley Fault. However, recent interpretations have given more weight to relationships at the southern margin of the Sgavoch–Downan lavas where the faulted base of the sequence is intimately associated with chert and shale containing a *gracilis* Biozone (Caradoc, mid- to late Ordovician) graptolite fauna. On this basis the lavas should be regarded as the earliest accreted unit within the Southern Uplands imbricate thrust belt. A poorly constrained radiometric (Sm-Nd) age of 468 ± 22 Ma (Thirlwall and Bluck, 1984) does not differentiate between the two alternatives. A summary of the debate is given by Stone and Smellie (1988) who introduced the lithostratigraphical name Downan Point Lava Formation.

The spectacular pillow structures within the Downan Point Lava Formation (DPF) were first noted and illustrated by Peach and Horne (1899) and subsequently discussed in more detail by Bloxam (1960). A more recent description of the Sgavoch Rock locality was given by Bluck (1992). The pillow lavas exposed preserve a range of features characteristic of submarine eruption and are arguably the best British examples of their kind; they also occupy an important position in terms of the regional geology of southern Scotland. To the north, the ophiolitic Ballantrae Complex was generated and obducted onto the Laurentian continental margin during the early Ordovician. To the south the Southern Uplands Terrane was sequentially accreted at the Laurentian margin during the late Ordovician and early Silurian. The DPF represents renewed volcanicity within or marginal to the Iapetus Ocean immediately prior to the initiation of Southern Uplands accretion and in this context the DPF represents the most extensive volcanic fragment preserved within the Southern Uplands Terrane.

Description

The geological setting of the Sgavoch Rock area is shown in Figure 2.41. The lavas and the associated breccias, cherts and shales accumulated in a submarine environment and it is the process of eruption under water that produces the characteristic pillow shapes. The pillows form as lava is squeezed out from points of weakness in the walls of lava tubes, a phenomenon known as budding. Each emerging tongue of lava is rapidly chilled by contact with the water and by the time it has grown to pillow size the hardening skin prevents further growth. The pillow may then break free of the lava front or be overtaken by new tongues, which similarly swell into pillows. Since the pillows accumulate in a semi-solid state they settle and mould around each other, finally solidifying into interlocking patterns. In an ideal example the top surface of each pillow is a convex dome whereas the base may be flat or irregular, moulded to the shape of the underlying surface; vesicles are typically concentrated towards the top of the pillow and in many of the examples seen they form prominent concentric zones. These features are illustrated in Figure 2.42.

The full array of pillow lava features is most spectacularly displayed on the small rocky headland due east from the Sgavoch Rock; note that the latter feature is covered at high tide. On the headland the pillow attitude shows bedding to be steeply dipping or vertical with an approximately NE–SW strike; the asymmetry of the pillows establishes that the original top of the lava pile lies towards the NW. Since the lavas were erupted onto a probably sub-horizontal sea floor they have therefore been rotated through 90° and now become sequentially younger seaward. The Sgavoch (DPF) pillows range from slightly elliptical with long axes of 20–50 cm through to larger, elongate 'bolster' shapes up to 2 m across. Within the pillow lava sequence there are other sheet-like lava bodies which probably represent original lava tubes carrying magma forward to the eruptive front. In some cases lava pillows can be seen budding from the extremities and top surfaces of the sheets in a remarkable illustration of their mode of formation. The proportion of unpillowed sheet-flows in the succession increases southwards towards Downan Point.

Several types of pelagic sediment are intimately associated with the lavas. Lenses of laminated black chert, up to 3 m long and 30 cm thick, occupy what were probably hollows in the lava pile topography and indicate local breaks in lava accumulation. Red chert or dark siliceous mudstone may fill gaps left between pillows and in some cases the partial draining of pillows has left spaces that are now filled with either chert or laminated carbonate. Euhedral calcite crystals are seen growing from the pillow margins into the inter-pillow sediment and appear to have done so while the sediment was still soft, displacing sedimentary laminae. Carbonate is also the most common filling for the original vesicles.

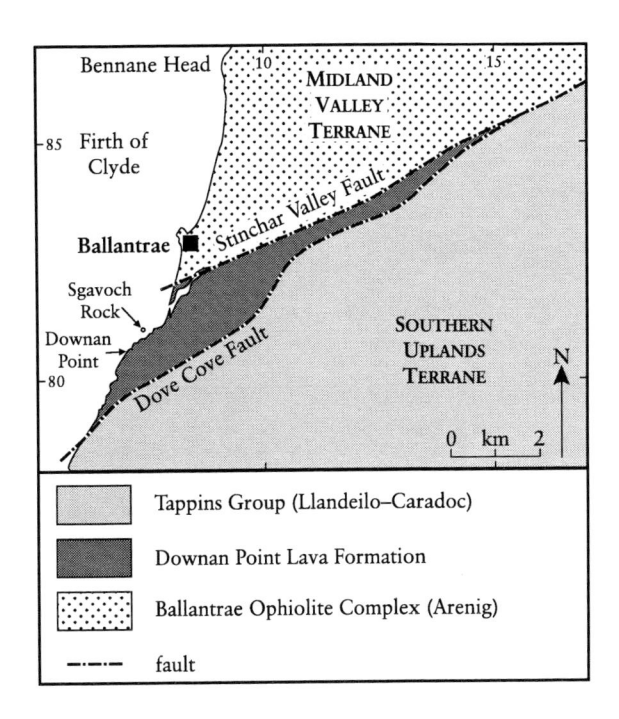

Figure 2.41 The position of the Sgavoch Rock GCR site within the Downan Point Lava Formation and its relationship to the Ballantrae Complex and the Southern Uplands Terrane.

Lava breccias are interspersed in the volcanic sequence and are usually dominated by small pillows as little as 10 cm in diameter. These show intact chilled margins and so are not clasts resulting from the disintegration of larger pillows although such clasts are also present. The breccias are commonly rich in fine-grained matrix and may have been emplaced as debris flows from the lava front. However, they are also pervasively bleached to a pale yellowish-green colour, probably by hydrothermal alteration, and the restriction of this alteration to the clastic zones means that an origin as intrusion breccia cannot be entirely ruled out.

The pillows all have green chloritic rims, which are the result of alteration of the original glassy chilled margins; the basalt forming the pillow cores is tholeiitic, fine grained, mostly aphyric and generally vesicular. Plagioclase, clinopyroxene and chlorite form the matrix, which encloses rare, but very locally abundant, small plagioclase phenocrysts.

Interpretation

The magnificent array of pillow lavas exposed in the Sgavoch section provides a rare opportunity to examine in detail the products of submarine volcanic eruptions. The age of this particular eruption remains uncertain: the lavas could be of Arenig age and a part of the Ballantrae Complex ophiolite that crops out immediately to the north; alternatively, and perhaps more likely, the lavas may be of Caradoc age and form the oldest accreted unit within the Southern Uplands imbricate thrust terrane. Several GCR sites within the Ballantrae Complex show comparable (although less spectacular) pillow lava developments but there are only comparatively meagre equivalents within the Southern Uplands. The DPF lava pile at Sgavoch has been rotated to a sub-vertical attitude but has otherwise escaped deformation. There is no penetrative cleavage and no evidence that the pillow shapes have been modified tectonically. The consistent westward younging of the steeply inclined flows militates against structural complexity introduced by folding or faulting. The metamorphic grade is very low. This fortunate state of affairs contributes to the value of the Sgavoch section as an ideal site for the investigation of submarine volcanism.

The submarine environment of eruption is not in dispute and broadly similar lava flow morphologies have been observed forming at mid-ocean ridges and around Pacific islands. The alternation of pillowed and sheeted flows reflects varying rates of extrusion on an unstable sea-floor, the more rapid the advance of the lava

Figure 2.42 A spectacular array of pillow lavas from the Downan Point Lava Formation exposed on the coast adjacent to the offshore Sgavoch Rock. The lavas are steeply inclined and slightly overturned. (Photo: BGS no. D1572.)

front the less opportunity there is for the production of pillows. Submarine slopes are indicated by a preponderance of jumbled, elongate pillows with the steeper slopes producing breccias as lava pillows and fragments cascaded downwards. The interpillow chert may represent either background pelagic sediment or be an essentially hydrothermal deposit; the laminated chert lenses are most likely to have a sedimentary origin and may even indicate intermittent turbidity current activity. A fairly deep water setting seems probable.

Several geochemical studies of the DPF pillow basalts have been carried out and the results were summarized and assessed by Thirlwall and Bluck (1984). A consensus view considers that they are most closely comparable to modern oceanic island, 'Hawaiian-type' basalts and thus a within-plate geotectonic setting has been proposed for their eruption. The source of the magma erupted at Sgavoch was therefore a mantle plume of some sort. Lavas of closely similar composition are found within the Ballantrae Complex at the Slockenray Coast and Balcreuchan Port to Port Vad GCR sites but are also known from small, scattered outcrops contained within fault zones defining the northern tracts of the Southern Uplands thrust belt. The age and structural association of the DPF basalts at Sgavoch therefore remains uncertain.

Conclusions

The Sgavoch Rock GCR site contains arguably the finest array of submarine pillow lavas to be seen in Britain. It is a spectacular locality of major volcanological interest. The lavas are of Ordovician age but may be associated either with the Ballantrae Complex ophiolite (early Ordovician) or with the slightly later (mid-Ordovician) initiation of the Southern Uplands accretionary thrust belt. The environment of eruption was in fairly deep water adjacent to a 'Hawaiian-type' oceanic island. A remarkable variety of features is present. The overall appearance of the pillow pile is particularly striking but a wealth of detail is also preserved: variable pillow shape clearly indicating the original top of the sequence, pillows budding from lava tubes, drained pillow cores filled with sediment, marked zonation by vesicles (relict gas bubbles) and interbedded lava breccias to list only the more obvious.

Chapter 3

Mid-Ordovician intrusions of the North-east Grampian Highlands of Scotland

Introduction

INTRODUCTION

W. J. Wadsworth and D. Stephenson

The Caledonian Orogeny in the Grampian Highlands reached a climax during the mid-Ordovician, with the formation of major folds and the peak of the Barrovian regional metamorphism (the Grampian Event). At this time the region formed the continental margin of Laurentia and the closure of the Iapetus Ocean was well underway. The precise reason for the timing of the orogenic climax is not clear; final closure of the ocean, with continent–continent collision and suturing on the southern margin of Laurentia, did not occur until the late Silurian, although Soper and Hutton (1984) suggested that westward subduction of Baltica beneath the eastern margin may have occured in the Ordovician. From evidence in Ireland, Dewey and Shackleton (1984) and Ryan and Dewey (1991) have suggested that the climax arises from the collision of the Laurentian margin with an island-arc complex related to a southward-dipping intra-ocean subduction zone. Given the evidence for the obduction of early Ordovician oceanic and island-arc material at Ballantrae, the Highland Border and Shetland (see Chapter 2), an extension of this model to include the Grampian Highlands seems plausible.

Large intrusions of tholeiitic basic and ultramafic rock in the NE Grampian Highlands are believed to have been emplaced into the Dalradian metasedimentary succession at about the same time as, or shortly after, the regional metamorphism reached its climax. Until recently this magmatism was believed to have occurred at around 490 Ma ago, but more precise U-Pb zircon age determinations now suggest 470 Ma (Rogers *et al.*, 1994). The basic intrusions are closely associated spatially with the Buchan-style (high temperature/low pressure) metamorphism and with migmatites and S-type granites in the NE Grampian area, also dated at around 470 Ma. The generation and upward migration of substantial volumes of basic magma clearly implies a regime of high heat flow. Whether the emplacement of large bodies of basic magma was directly responsible for the crustal melting that produced the granites and for the Buchan metamorphism, or whether these are all a general effect of the high heat flow, is still a matter of debate. However, the tholeiitic nature of the basic magmatism and the need to create space for its emplacement as large tabular intrusions, strongly suggests that all of this occurred in an extensional tectonic setting. This means that a period of crustal extension, triggering a major thermal perturbation, must have occurred immediately following the major crustal shortening. Although the effects of this extension were widespread in the NE Grampians, with voluminous magma generation, the event must have been short-lived, since the intrusions were affected by the subsequent D3 compressional deformation while they were still hot (see below). The tholeiitic magmatism and the implied short-lived crustal extension are unique to the NE of the Grampian Terrane; Caledonian tholeiitic intrusions occur nowhere else in the British or Irish sectors of the Laurentian continental margin.

The basic intrusions were divided by Read (1919) into 'Older' and 'Younger' groups, on the basis that the principal phase of folding and metamorphism intervened between the two magmatic events. This distinction has subsequently been shown to be too simple, and it is now clear that the 'Younger Basic' intrusions have been considerably modified by later Caledonian orogenic activity. This has involved large-scale folding and disruption of the original igneous bodies during the regional D3 deformation phase. Much of this disruption is associated with major shear zones (Ashcroft *et al.*, 1984; Fettes *et al.*, 1991) and considerable movement on these zones is indicated in places by contrasts in thermal history (Beddoe-Stephens, 1990), disruption of Dalradian stratigraphy and the removal of thermal aureoles from the margins of some basic intrusions. The shearing has resulted in substantial retrograde metamorphism, converting the peridotites and gabbros into serpentinites and amphibolites respectively (Mongkoltip and Ashworth, 1986). There has also been considerable deformation, resulting in the formation of schistose metagabbros and mylonites (see the Balmedie Quarry GCR site report). As a result, some of the 'Younger Basic' rocks have acquired characteristics previously regarded as hallmarks of the 'Older Basic' rocks. Careful reinvestigation of the apparently 'Older' mafic bodies has revealed that many of these have relict primary features linking them to the 'Younger Basic' intrusions. This is especially the case along the Portsoy Lineament (Figure 3.1), which effectively marks the western margin of the 'Younger Basic' province (Fettes *et al.*,

Mid-Ordovician intrusions of NE Scotland

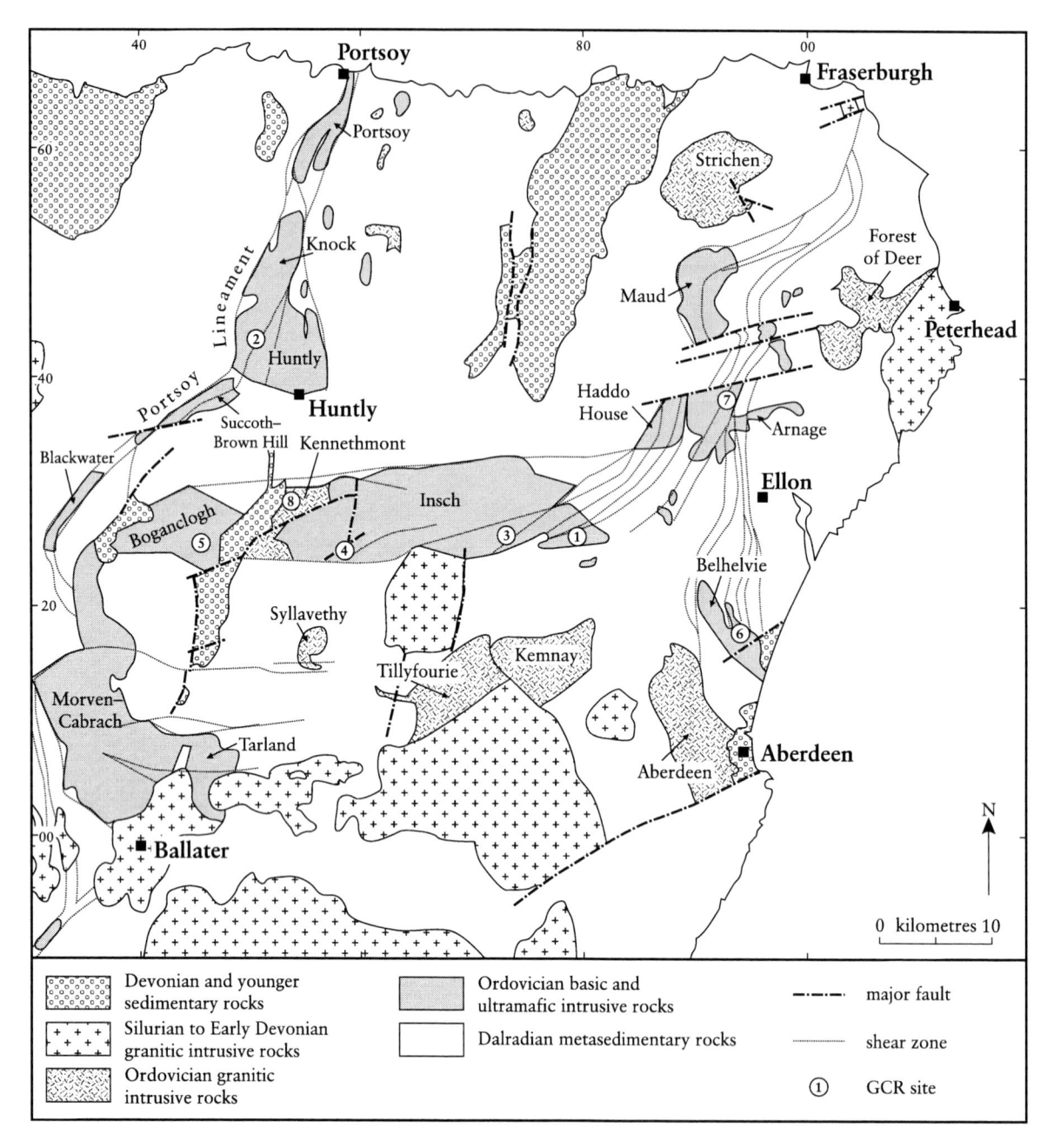

Figure 3.1 Location of basic intrusions and late Caledonian granitic intrusions in the NE Grampian Highlands, modified after Ashcroft *et al.* (1984) by Gould (1997). GCR sites: 1, Hill of Barra; 2, Bin Quarry; 3, Pitscurry and Legatesden quarries; 4, Hill of Johnston; 5, Hill of Craigdearg; 6, Balmedie Quarry; 7, Towie Wood; 8, Craig Hall.

1991). Here, for example, the Blackwater intrusion, which has been thoroughly deformed and metamorphosed, is now thought to be part of the 'Younger Basic' suite (Fettes and Munro, 1989). On the other hand, the nearby Succoth–Brown Hill mafic complex may well be a genuinely 'Older Basic' intrusion, and has more calc-alkaline, arc-like geochemical and mineralogical features (Gunn *et al.*, 1996). Other evidence that Read's original distinction is still relevant, although based on different criteria now, is discussed elsewhere in this chapter (see the Hill of Creagdearg GCR site report).

The precise nature of the 'Younger Basic' intrusive event is not known, because of the subsequent tectonic modifications, but Read (1923) suggested the formation of a large sheet-like body initially, and this view has generally been

Introduction

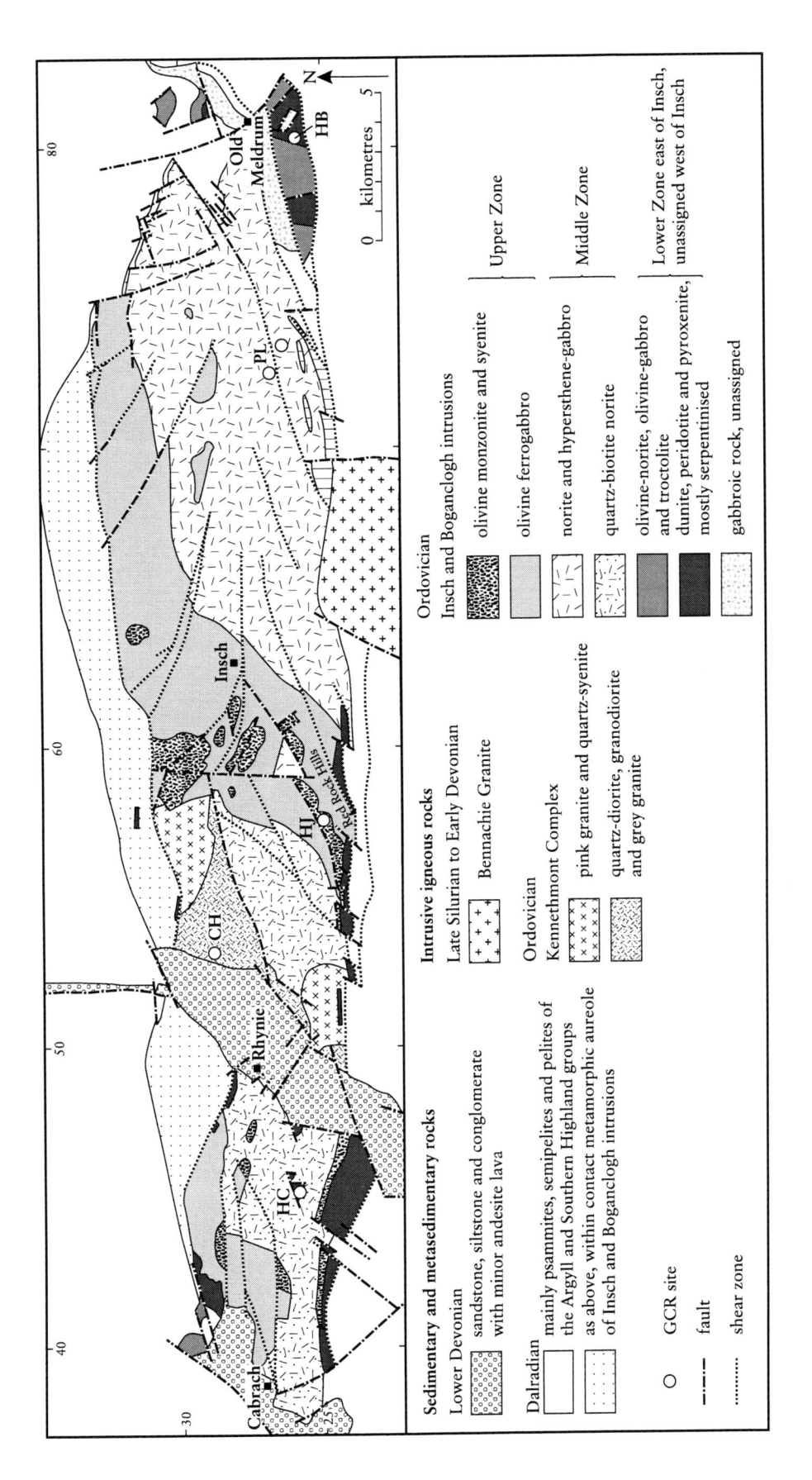

Figure 3.2 Map of the Insch, Boganclogh and Kennethmont intrusions, adapted from Gould (1997). GCR sites: CH Craig Hall; HB Hill of Barra; HC Hill of Craigdearg; HJ Hill of Johnston; PL Pitscurry and Legatesden quarries.

Figure 3.3 Typical landscape of the Insch intrusion. View WNW from Candle Hill, Oyne towards the 'Red Rock Hills'. Foreground rocks are norites of the Middle Zone. (Photo: W.J. Wadsworth.)

reinforced by subsequent investigations (see Wadsworth, 1982). However, there are enough differences in detail between the individual intrusions to indicate some degree of independent development, perhaps in fairly distinct 'compartments' within a single complex body, and the possibility of completely separate intrusions cannot be ruled out. This uncertainty does not invalidate the general conclusion that the 'Younger Basic' intrusions represent a single, coherent magmatic episode.

Six principal 'Younger Basic' intrusions are generally recognized, although some of these are composite, in that they comprise two or three adjacent, but apparently detached, masses at the present level of erosion (Figure 3.1):

1. Belhelvie
2. Insch–Boganclogh (Figures 3.2 and 3.3)
3. Huntly–Knock
4. Morven–Cabrach and Tarland
5. Haddo House–Arnage
6. Maud

The 'Younger Basic' rocks clearly have tholeiitic geochemical affinities. The parental magma is inferred to have been basaltic, with normative *hypersthene*, and the cumulate succession follows a typically tholeiitic differentiation trend. This is expressed mineralogically by the widespread occurrence of orthopyroxene (norites are common), by the absence of olivine in the intermediate fractionation stages, and by the appearance of quartz in the most evolved rocks, which are also distinguished by high levels of such elements as phosphorous, barium and zirconium. Broadly similar tholeiitic fractionation trends are encountered in the classic layered sequences of the Skaergaard (East Greenland) and Bushveld (South Africa) intrusions.

The most distinctive characteristic of the 'Younger Basic' masses, viewed collectively, is the diversity of rock types represented. Of these, the most significant is the fractionation series comprising all gradations from peridotites, through gabbros, ferrogabbros and syenodiorites, to extreme quartz-syenites. Many of the rocks in this sequence are texturally cumulates, and in some cases they exhibit small-scale layering (see the Bin Quarry GCR site report).

Introduction

Zone		Rock type	Mineral composition							Estimated thickness of unit (m)
			%Fo Olivine	%En Orthopyroxene	#Mg Clinopyroxene	%An Plagioclase	Or/Ab/Cn Alkali feldspar	#Mg Amphibole	#Mg Biotite	
Upper Zone	c	syenite			6	28 31 30	80/20/0 81/18/1 79/19/3	19	8	50
	b	olivine monzonite	6 13	20 24 29	40	49	76/19/5 63/24/13	17	18	50
	a	olivine-ferrogabbro	18 47	34 IP {47 {52 58	46 54 63	52 50–56		31 53	31 58	200
Middle Zone		norite, granular gabbro, quartz-biotite norite		IP {44 {53 71	55 78	47–52 69–74		43 60	42 70	2000+
					hiatus					
Lower Zone		olivine-norite, troctolite, dunite	77 87	79 87	82 88	70–76 78–84				1800+
					hiatus					
Bogan-clogh		dunite, harzburgite, rare wehrlite	89 92	91 95	89 92 95					unknown

——— cumulus phase   - - - - - intercumulus only   #Mg = 100Mg/(Mg + Fe)   IP = inverted pigeonite

Figure 3.4 Petrographical divisions and variation of mineral compositions in the Insch, Boganclogh and Belhelvie intrusions, from Gould (1997). Data from Ashcroft and Munro (1978), Gould (1997), Styles (1994) and Wadsworth (1986, 1988, 1991).

They are believed to represent gravitative accumulation of crystals precipitated from a basalt parental magma, which progressively evolved towards more felsic compositions. Similar cumulate successions are characteristic of major layered intrusions such as the Bushveld Complex in South Africa (Eales and Cawthorne, 1996), and these form the basis of the classification scheme now applied to the 'Younger Basic' differentiation series (Figure 3.4).

Three main 'stratigraphical' units are recognized, namely the Lower Zone (LZ), comprising peridotites, troctolites and olivine-gabbros; the Middle Zone (MZ), comprising olivine-free two-pyroxene gabbros; and the Upper Zone (UZ) comprising olivine-ferrogabbros, syenodiorites (see the Hill of Johnston GCR site report) and quartz-syenites. Both the LZ and UZ are further divided into three subzones (a, b and c). The cumulate sequence reaches its maximum development in the Insch intrusion, where all three zones are represented (Wadsworth, 1986, 1988; Gould, 1997), but where the LZ is poorly exposed (see the Hill of Barra GCR site report). However, the Belhelvie intrusion appears to provide a more complete duplicate of this part of the succession (Wadsworth, 1991). The Morven–Cabrach (Allan, 1970) and Huntly–Knock (Munro, 1984) intrusions both contain cumulates, although the Huntly rocks, which are broadly equivalent to the LZ elsewhere, are significantly different in some details (see the Bin Quarry GCR site report).

The intermediate stage of the main fractionation series (MZ) which is only fully represented in the Insch intrusion (see the Pitcaple GCR site report), consists essentially of olivine-free two-pyroxene gabbros ('hypersthene-gabbros' of Read *et al.*, 1965). These were originally interpreted as a separate intrusion, unrelated to the neighbouring LZ and UZ rocks at Insch, but are now regarded as an integral part of the succession (Wadsworth, 1988). However, not all the MZ rocks are normal cumulates. There is an intricate association of cumulates and fine-grained granular gabbros, some of which are porphyritic, containing prominent plagioclase phenocrysts. These granular rocks have been variously explained as thermally metamorphosed gabbros (Whittle, 1936), as the products of late-stage basic magma injections into the MZ cumulates (Clarke and Wadsworth, 1970) and, most recently, as an integral component of the MZ, but possibly representing a gabbroic 'roof facies', parts of which foundered periodically into the cumulate pile (Wadsworth, 1988). Rather similar granular gabbros also occur in the Huntly intrusion (Fletcher and Rice, 1989).

Another variety of olivine-free rock of broadly gabbroic composition is found in many of the 'Younger Basic' intrusions. This is generally described as quartz-biotite norite and appears to form distinct bodies of homogeneous basic rocks, without obvious cumulate textures. These are particularly important in the Insch–Boganclogh intrusion (especially in the Boganclogh area) and the Morven–Cabrach intrusion, and also occur at Huntly, Haddo House–Arnage and Maud. They are usually taken to represent 'Younger Basic' magma which has crystallized *in situ* under relatively hydrous conditions.

Further diversity among the 'Younger Basic' intrusions is associated with their margins, where these represent igneous rather than tectonic contacts, and especially in what was probably the roof region of the original bodies. The characteristic assemblage of these occurrences is predominantly xenolithic, with a wide variety of hornfels fragments in a cordierite norite (grading to tonalite) matrix (see the Towie Wood site report). The fragments are generally interpreted as the refractory residues (restites) resulting from partial melting of Dalradian country rock, that was heated by the 'Younger Basic' magma. They fall into three categories, namely quartz-rich, silica-deficient (aluminous), and (rarer) calc-silicate, which were derived from psammitic, pelitic and calcareous metasedimentary rocks respectively. The matrix material probably represents the partial melt component, which may have mixed locally with the basic magma. These marginal xenolithic complexes are most prominent in the Haddo House–Arnage intrusion (Read, 1923, 1935; Gribble, 1968), but are also an important component of the Huntly–Knock intrusion (Dalrymple, 1995), and occur locally at Insch (Read, 1966). Contact metamorphism of the Dalradian country rock without partial melting is also well-documented by Droop and Charnley (1985). From a study of hornfelsed pelites adjacent to the basic intrusions, they have estimated temperatures of 700–850°C and pressures of 4–5 kbar, implying an emplacement depth of 15–18.5 km.

The relationships of the mafic igneous rocks (both 'Older' and 'Younger') of the area, to the more widespread and better-known Caledonian granites is by no means clear-cut. Read (1961), following Barrow (in Barrow and Cunningham Craig, 1912), recognized two principal periods of granite emplacement. His 'Older Granites' (often migmatitic) were believed to be younger than the 'Older Basic' bodies and the main episodes of Caledonian folding, and probably coeval with the Buchan regional metamorphism. He described the 'Younger Basic' intrusions as representing a 'basic interlude' between the 'Older Granites' and the main late Caledonian granites, which he termed 'Newer Granites', and which included a distinct sub-group termed the 'Last Granites'. Subsequently, the 'Newer Granites' were divided into three groups, based mainly upon radiometric ages by Pankhurst and Sutherland (1982), the youngest of which was broadly equivalent to Read's 'Last Granites'. Many of the radiometric ages used in this grouping are now considered to be unreliable and the 'Newer Granites' are now more simply classified as either late tectonic or post-tectonic, depending more on their relationships to the tectonic history of the surrounding rocks or, where this is not known, by their petrological and geochemical affinities (Stephenson and Gould, 1995; see Chapter 8).

Syntectonic granites (equivalent to Read's 'Older Granites') in the Scottish Highlands probably originated during many different events over a long time-span, ranging from about 800 Ma to the peak of Caledonian regional metamorphism and deformation at 470 Ma. In the NE Grampian Highlands they are represented, in areas of middle amphibolite facies metamorphism and above, by migmatitic segregations and elsewhere by small, sheet-like masses of deformed muscovite-biotite granite, most notably those at Portsoy, Windyhills, Keith and Muldearie. There are no published modern age determinations (only a Rb-Sr isochron age of 655 ± 17 Ma for the Portsoy granite; Pankhurst, 1974). Most are highly sheared, with a strong augen texture, but their sheet-like form suggests that their emplacement was controlled by pre-existing shear zones. They could be similar in age to the Ben Vuirich Granite of Perthshire (590 Ma), but equally they could be coeval with the 'Newer Basics' and not much older than the D3 shearing. No sites have yet been selected for the GCR to represent the syntectonic granites.

Most of the granites which are spatially associated with the 'Younger Basics' can be classified as late tectonic (Figure 3.1). Two suites are recognized. One consists of diorites, tonalites and granodiorites, with characteristics of a mantle or lower crustal igneous source (I-type); some are foliated and many are highly xenolithic. They include the Kennethmont granite–diorite complex, which is located at the western end of the Insch 'Younger Basic' mass (Figure 3.2), but which is regarded as a result of a slightly younger separate magmatic event (see the Craig Hall GCR site report). It also includes the Syllavethy, Corrennie and Tillyfourie intrusions (Harrison, 1987b; Gould, 1997), which all lie within the area of an extensive tonalite–granodiorite vein complex to the south of Insch. The other late tectonic suite consists of biotite-muscovite granites, which are commonly foliated and garnet-bearing. Contacts with country rock are gradational and compositions are compatible with an S-type origin by partial melting of mid- to upper- crustal metasedimentary rocks. They post-date the D3 folding, the regional migmatization and the basic intrusions, but their most likely age of intrusion, based on U-Pb monazite determinations is comparable to these events at around 470 Ma. They include large bodies such as the granites of Strichen, Forest of Deer, Kemnay (Gould, 1997) and Aberdeen (Munro, 1986; Kneller and Aftalion, 1987), and many smaller bodies. Many of the minor granite and pegmatite sheets that cut the 'Younger Basic' masses may be coeval with this suite. No sites have yet been selected for the GCR to represent this suite.

The remaining NE Grampian granites (Figure 3.1), including those of the East Grampian Batholith (which can be taken as marking the southern limit of the area) and large bodies west of Aberdeen (e.g. Bennachie, Hill of Fare, Crathes, Cromar) and to the north (Peterhead) are all post-tectonic I-type or transitional to alkaline (A-type) granites. They have emplacement ages in the range 420 to 395 Ma and, like the late Caledonian granitic intrusions as a whole (see Chapter 8: Introduction), can be related to late-orogenic uplift.

HILL OF BARRA (NJ 803 257)

W. J. Wadsworth

Introduction

Olivine-rich ultramafic rocks believed to represent the Lower Zone of the 'Younger Basic'

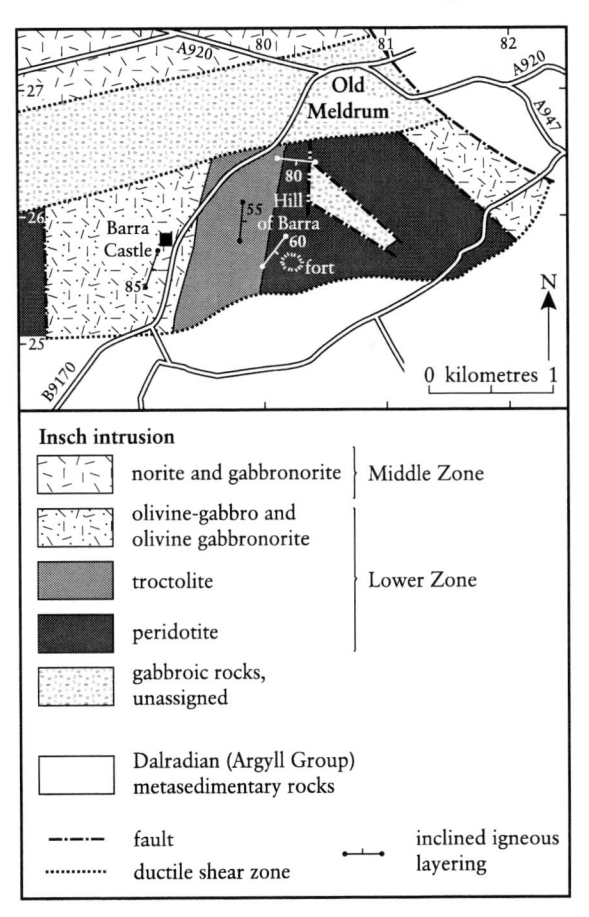

Figure 3.5 Geological map of the area around the Hill of Barra GCR site, Insch intrusion, from Ashcroft and Munro (1978) and BGS 1:10 000 sheets NJ72NE (1989) and NJ82NW (1989).

layered sequence can be recognized in both the Belhelvie and Insch intrusions, but natural exposures are generally very poor because the rocks are heavily serpentinized. However, Hill of Barra, at the eastern end of the Insch intrusion, is exceptional in forming a positive topographic feature and in providing relatively good exposures, despite the degree of serpentinization (Gould, 1997) (Figure 3.5).

In terms of the original layered sequence, the Hill of Barra peridotites are referred to the basal subdivision of the Lower Zone (LZa) as defined by Wadsworth (1982) and are classified as olivine cumulates. Although the eastern end of the Insch intrusion is structurally complicated, geophysical evidence (mainly magnetic anomalies) and borehole sampling have shown that there is a relatively continuous sequence from LZa peridotites (olivine cumulates), to LZb troctolites (plagioclase-olivine cumulates) and LZc norites (plagioclase-orthopyroxene-olivine cumulates) along the southern edge of the intrusion, between Old Meldrum and Cuttlecraigs (Ashcroft and Munro, 1978). In this segment, the succession appears to 'young' from east to west, with the observed layered structures mostly striking approximately N–S and generally dipping steeply eastwards, indicating that the rocks have been overturned (Figure 3.5). An apparently identical sequence of Lower Zone cumulates has been described from the Belhelvie intrusion (Wadsworth, 1991).

Description

The principal outcrops on Hill of Barra form the west-facing ramparts of an ancient fort. They comprise dark-weathering serpentinized peridotite, which displays fairly well-developed jointing; one set of joints dips steeply eastwards, and another set steeply westwards. There is also evidence of very faint compositional layering, which appears to be the result of slight variations in the amount of interstitial feldspar in the original cumulate. This rudimentary layering is dipping at angles between 50° and 60° towards the ESE, approximately parallel to the eastward-dipping joints. The upper slopes of Hill of Barra are also littered with blocks of layered peridotite and troctolite. The freshly exposed rock surfaces are dominated by dull-black serpentinized olivine, but scattered poikilitic crystals of intercumulus pyroxene can be discerned, and at least a trace of interstitial plagioclase (now substantially altered to secondary minerals), can be seen on weathered surfaces, since it is slightly more resistant to weathering than the serpentine. The plagioclase content may approach 10% by volume in the relatively feldspathic layers. Westwards from Hill of Barra, towards Barra Castle, plagioclase gradually increases in abundance, and eventually occurs as cumulus grains, giving rise to the troctolitic (plagioclase-olivine) cumulates of LZb.

Although the ultramafic rocks of LZa are heavily serpentinized, remnants of fresh olivine occur locally and have compositions in the range of Fo_{87-86} (Wadsworth, 1991). The serpentine forms a distinctive mesh-structure, and is associated with granular aggregates and stringers of magnetite. The intercumulus pyroxene generally occurs as large poikilitic crystals, and although augite is the most obvious variety of pyroxene, because it is relatively unaltered, orthopyroxene has also been recorded. The original intercumulus plagioclase has been almost entirely replaced by turbid, isotropic material, possibly hydrogrossular, but rare patches with relict multiple twinning can be distinguished. The plagioclase composition is rather variable, but is generally in the bytownite–labradorite range. Corona structures are sometimes developed at the contact between original olivine and plagioclase, and these consist typically of a zone of granular orthopyroxene, immediately adjacent to the olivine, surrounded by a zone of fibrous amphibole in symplectic intergrowth with turbid isotropic material (probably altered feldspar). These coronas are believed to be the result of reaction between olivine and plagioclase under metamorphic conditions (Mongkoltip and Ashworth, 1983).

Interpretation

The significance of the ultramafic rocks at the eastern end of the Insch intrusion as the most primitive members of an extreme fractionation series (culminating in the 'syenitic' rock compositions found farther west; see the Hill of Johnston GCR site report), was first recognized by Read, Sadashivaiah and Haq (1961). They interpreted the layered peridotites and troctolites as basal cumulates, formed by gravitative differentiation of gabbroic magma, and recognized that the original sub-horizontal layering has been thoroughly disturbed by subsequent tectonic events. The cumulate theme was developed and refined by Clarke and Wadsworth (1970), who recognized that the Hill of Barra peridotites are olivine cumulates, and that cumulus plagioclase and pyroxenes (both clinopyroxene and orthopyroxene) only appear higher in the succession (i.e. farther west) to give rise to cumulate troctolites and gabbros. A more detailed structural study of the eastern end of the Insch intrusion by Ashcroft and Munro (1978) identified a number of separate fault blocks of LZ cumulates, with the Hill of Barra rocks forming part of a steeply dipping cumulate sequence, younging from east to west, and locally overturned. It is on this basis that the Hill of Barra rocks are referred to LZa of the complete layered sequence (Wadsworth, 1982, 1991). It is now thought that these ultramafic cumulates are not the same as, or even closely related to, the serpentinized peridotites found along the southern margin of the Insch–Boganclogh intrusion (see the Creag Dearg GCR site report).

Conclusions

The rocks of the Hill of Barra GCR site are typical of the layered ultramafic unit found in the 'Younger Basic' masses (Insch and Belhelvie), and believed to represent the early-formed olivine-rich cumulates. The original peridotite has been highly serpentinized, but there is some evidence of rudimentary layering, dipping steeply eastwards and indicating considerable post-depositional tectonic disturbance. Exposures of these ultramafic rocks are generally very poor; this highlights the significance of the western slopes of Hill of Barra, which provide relatively good outcrops.

BIN QUARRY (NJ 498 431)

W. J. Wadsworth

Introduction

Although the 'Younger Basic' igneous bodies are generally interpreted as layered intrusions, because of their large-scale compositional variations, they rarely display convincing small-scale layered structures. However, the Bin Quarry, which is located near the western margin of the Huntly–Knock intrusion (Figure 3.6), is exceptional in this respect, and exposes a sequence of spectacularly layered cumulates.

The Bin Quarry rocks are broadly troctolitic (plagioclase-olivine cumulates) to gabbroic (plagioclase-olivine-augite cumulates) in composition, but there are considerable modal variations from layer to layer (generally on a scale of centimetres or tens of centimetres), producing mafic (olivine-rich) and felsic (plagioclase-rich) lithologies locally. These layered rocks have moderately steep dips towards the west, but since they are believed to represent the lower part (LZ) of the Huntly–Knock cumulate sequence, which generally 'youngs' from west to east in this area, (Munro 1984), they must have been overturned tectonically in the vicinity of the Bin Quarry. This view is supported by the evidence of small-scale 'sedimentological' features such as graded bedding which are clearly displayed in the quarry (Shackleton, 1948), and by progressive variations in the mineral compositions in this part of the Huntly intrusion.

Description

The main face of the quarry is 130 m in length, and exposes an apparently continuous succession of layered cumulates dipping at angles between 40° and 60° (averaging 50°) towards the WNW; the exposed stratigraphical thickness is therefore approximately 100 m. The layering is exceptionally well-developed and is generally very regular (Figures 3.7, 3.8). It consists of small-scale (centimetres to tens of centimetres) lithological variations from peridotite (olivine cumulate) to troctolite (plagioclase-olivine cumulate) and olivine-gabbro (plagioclase-augite-olivine cumulate), with considerable variations in grain size and texture as well. The olivine cumulate layers are best observed towards the western end of the main face (Figure 3.7) where they occur as thin (2–25 cm) units of dark, highly serpentinized peridotite, with scattered poikilitic crystals of pyroxene. Some of these layers display obvious grading into the adjacent troctolites. This graded bedding is consistent in direction and provides clear evidence that the stratigraphical succession youngs from west to east (Shackleton, 1948).

Although most of the layering is regular, examples of laterally impersistent, wispy layering and local cross-bedding are found, notably in the large loose blocks at the foot of the main face towards the western end of the quarry. These have probably fallen from the exceptionally well-

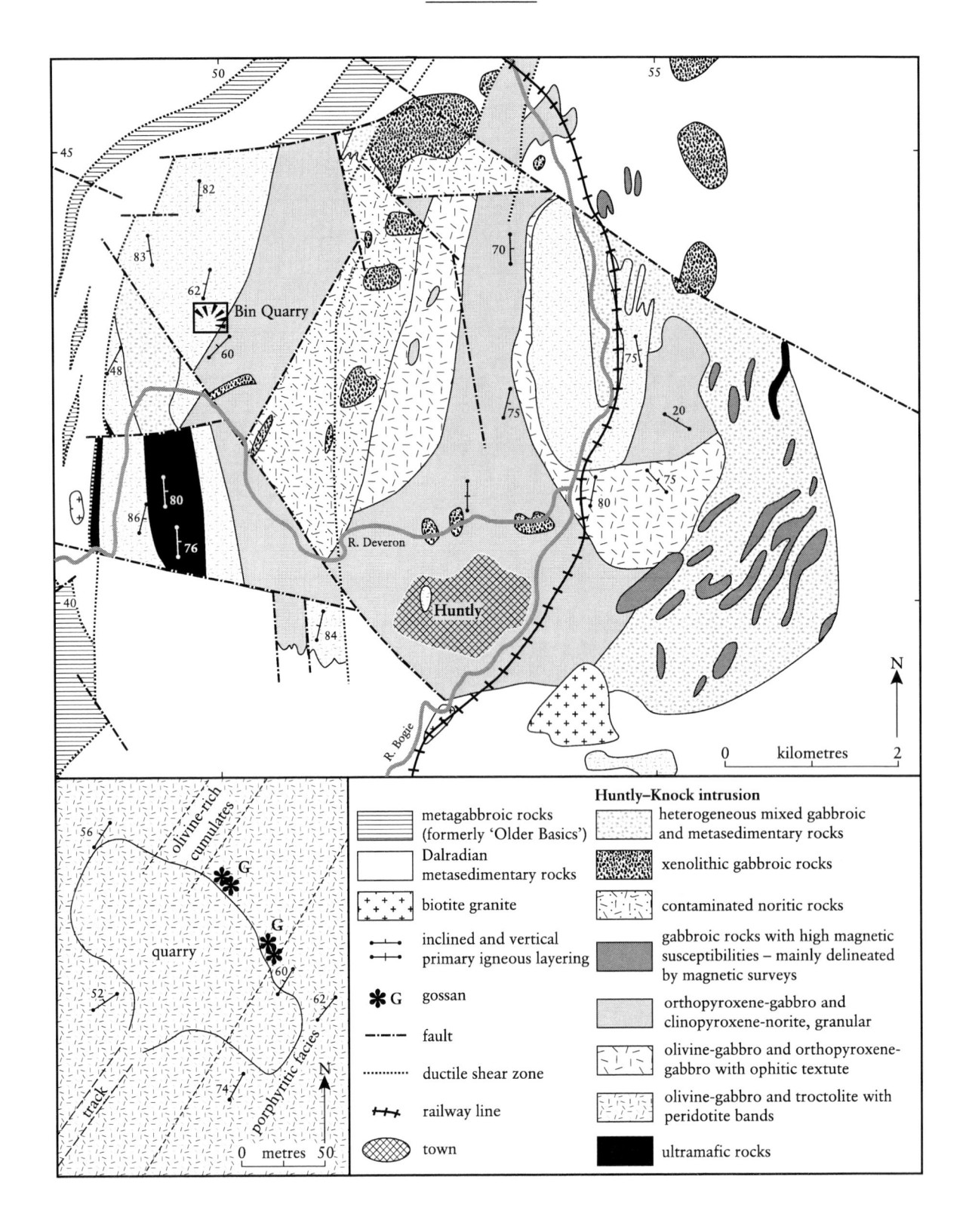

Figure 3.6 Map of the southern part of the Huntly–Knock intrusion, from BGS 1:50 000 Sheet 86W (in press), with details of the Bin Quarry GCR site, from Gunn and Shaw (1992).

Bin Quarry

Figure 3.7 Layered olivine-gabbro cumulates of the Huntly intrusion Lower Zone in the Bin Quarry. Layering dips at 50° to the NW, but modal layering, 'sedimentary' structures and variations in mineral composition show that the sequence 'youngs' to the SE and hence is inverted. (Photo: BGS no. D4122.)

Figure 3.8 Block of layered olivine-gabbro cumulate of the Huntly intrusion Lower Zone in the Bin Quarry. The layering, which is inverted in this photograph, reflects both modal and mineral compositional variation, ranging from peridotite, through mafic gabbro and troctolite to anorthosite. (Photo: BGS no. D4121.)

layered material which can be seen near the top of the face in this area. There is considerable variation in grain size between layers, and there are also lenses of very coarse-grained pegmatitic gabbro which are associated with significant sulphide mineralization (Fletcher, 1989; Gunn and Shaw, 1992). Two areas of gossan seen in the main quarry face (Figure 3.6) mark the location of irregular pegmatitic pyroxenite sheets. The troctolites and gabbros are well jointed, with the principal joint surfaces dipping at moderate angles (20–25°) towards the east. They are mostly well spaced, but are closer together (and slightly steeper) in the central part of the main quarry face.

The troctolitic and gabbroic cumulates are distinctively speckled black and white rocks, with variable proportions of cumulus plagioclase to cumulus olivine, and typically displaying a preferred orientation of the tabular plagioclase crystals parallel to the lithological layering (a feature normally referred to as 'igneous lamination'). The olivines sometimes display corona structures consisting of granular orthopyroxene and fibrous amphibole and presumably of metamorphic origin. Serpentinization of the olivines has resulted in the development of distinctive expansion cracks, radiating from the altered olivines out into the surrounding feldspars. The troctolites usually contain scattered crystals of intercumulus augite, and they are interlayered with more obviously gabbroic cumulates, in which the augite is of cumulus habit. Orthopyroxene is not present as a primary mineral (either cumulus or intercumulus) in these rocks. Cumulus mineral compositions in the troctolites and gabbros have been determined as follows: plagioclase (An_{74}), olivine (Fo_{80}) and augite ($Ca_{48}Mg_{44}Fe_8$) (Munro, 1984). There is also evidence of a slight but progressive change in olivine composition from west to east within the quarry (Fo_{81} to Fo_{78}), and this fits well with the graded layering structures in indicating eastward 'younging' of the succession despite the westward dips.

Thin sheets of pegmatitic gabbro, consisting of quartz, feldspar and biotite, occur in the eastern part of the main face. They are generally sub-horizontal in attitude, but have some steeper offshoots. They appear to have caused extensive amphibolitization of the mafic minerals in the immediately adjacent cumulates. Some vein-like areas of alteration contain radiating needles of xonotlite and botryoidal prehnite (Gillen, 1987).

Interpretation

The recognition that the Bin Quarry provides a classic example of cumulate rocks formed at a relatively early stage of gabbroic magma crystallization is based on a combination of lithological features. These include: a) the small-scale layering itself, which is predominantly the result of frequent changes in the cumulus mineral assemblage, but also involves textural features, such as the development of well-laminated plagioclase-rich layers; b) the occurrence of distinctive textures comprising both cumulus and intercumulus components; c) the cumulus mineral compositions (relatively magnesian olivine and augite, relatively calcic plagioclase) and their slight but systematic variation with stratigraphical position. The recognition of specifically sedimentational aspects (especially graded bedding, and its use as a 'way-up' criterion) was first applied to the 'Younger Basics' by Shackleton (1948), as the results of observations at the Bin Quarry, and this approach has been substantiated by most other studies in the area (e.g. Stewart and Johnson, 1960; Read, 1961).

In general terms, the predominantly troctolitic cumulates of the Huntly–Knock area are regarded as equivalent to Lower Zone rocks at Belhelvie and Insch, but from higher levels than LZa (as seen at the Hill of Barra GCR site in the Insch mass, or on the western edge of the Belhelvie mass). They are therefore referred to LZb/LZc, since they contain cumulus plagioclase and augite (Wadsworth, 1982; Munro, 1984; Fletcher, 1989). However, there is now considerable evidence that the lower part (LZ) of the Huntly–Knock succession is not identical to the Insch and Belhelvie equivalents. For example, there is no evidence of a distinct ultramafic (LZa) unit in the Huntly–Knock mass, although olivine-rich cumulate layers occur locally, as in the Bin Quarry sequence. Shallow drilling of the poorly exposed western margin (Munro, 1984) suggests that troctolitic and gabbroic rocks persist from the quarry area as far as the contact with Dalradian country rocks. In addition, there is the scarcity of orthopyroxene in the Huntly–Knock LZ cumulates, and the compositions of the co-existing cumulus minerals (olivine, plagioclase and augite) show small but significant differences between Huntly–Knock and Belhelvie–Insch (Munro, 1984; Wadsworth, 1991). These features lend support to the view that the regionally available 'Younger Basic' gabbro magma underwent progressive fractional crystallization in at least two separate magma chambers (or distinct compartments of a single complex chamber) with resultant slight differences in the respective crystallization sequences, as advocated by Weedon (1970), Ashcroft and Munro (1978) and Munro (1984). This is in contrast to the proposal by Wadsworth (1970, 1982) that the 'Younger Basics' represent a single-layered intrusion, subsequently disrupted by tectonic events.

Conclusions

The troctolitic and gabbroic cumulates of the Bin Quarry GCR site, with their magnesian olivines and pyroxenes, and calcic feldspars, are excellent examples of a relatively early stage of fractionation in the Lower Zone (LZb/LZc) of the 'Younger Basic' layered sequence, although they are not quite as primitive as the ultramafic (LZa) cumulates at the Hill of Barra GCR site. Various features of the rocks are of particular significance in the context of the 'Younger Basic' intrusions. Most prominent is the small-scale layering, with its associated 'sedimentary' structures, providing convincing evidence of gravity accumulation of crystals. The steep dips indicate post-depositional tectonic disturbance, and the graded (olivine-rich to plagioclase-rich) layers show that the sequence has been overturned, at least in the quarry area.

PITSCURRY AND LEGATESDEN QUARRIES
(NJ 728 267 AND 737 263)

W. J. Wadsworth

Introduction

The intermediate fractionation stages of the 'Younger Basic' magmatic event, which are collectively termed the Middle Zone (MZ), are best represented in the Insch intrusion, especially in the area around Pitcaple. The Insch MZ rocks are mainly olivine-free, two-pyroxene gabbros, with mineral compositions broadly intermediate between those of the Lower Zone (LZ) and Upper Zone (UZ) cumulates respectively, but displaying much greater textural diversity and structural complexity than either. Two principal textural variants are found throughout the Insch MZ, associated in approximately equal abundance, namely gabbroic cumulates and relatively fine-grained granular gabbros (FGG). However, their precise distribution and relationships are difficult to define, partly because of generally poor natural exposures, and partly because this area of the Insch intrusion lies within a major shear-belt (Read, 1956; Ashcroft *et al.*, 1984; Kneller and Leslie, 1984) so that the original rocks have been substantially modified. The Insch MZ rocks were originally described as a separate intrusion of hypersthene-gabbro, apparently unrelated to the main differentiation series (Read *et al.*, 1965) but Clarke and Wadsworth (1970) recognized a distinct cumulate element that partly bridges the gap between LZ and UZ, and all the MZ gabbros are now interpreted as integral components of the Insch sequence (Wadsworth, 1988; Gould, 1997).

The most important exposures of the Insch MZ gabbros are found in quarries, especially in the Pitcaple area (Figure 3.9). Two separate, but neighbouring quarries (Pitscurry and Legatesden) provide complementary information about the different MZ components and their relationships. Because both quarries lie within one of the main shear-belts affecting the Insch intrusion, much of the gabbroic material has been deformed and amphibolized, so that the

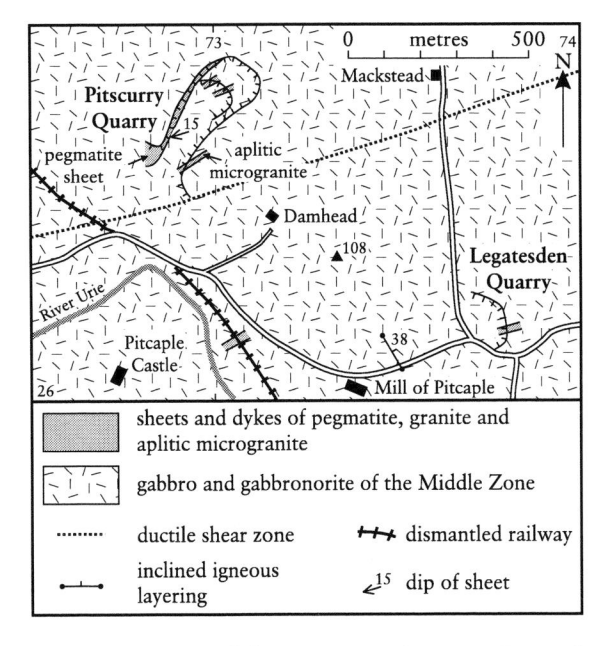

Figure 3.9 Map of the area around Legatesden and Pitscurry quarries, Insch intrusion, from BGS 1:10 000 Sheet NJ72NW (1989).

primary mineralogy is often difficult to decipher. However, this is an intrinsic part of the petrological variety of the area, and the localities selected are representative of the whole range of MZ gabbro types, including exceptionally fresh samples of the original gabbros as well as a complete spectrum of secondary modifications.

Description

Pitscurry Quarry

This is a large working quarry, which provides extremely fresh material from three texturally distinct varieties of gabbro; the relatively coarse-grained MZ cumulates and two types of fine-grained gabbro of characteristically granular (?recrystallized) appearance. Most of the granular gabbros are aphyric (fine-grained granular gabbro or FGG), but some contain abundant plagioclase phenocrysts (porphyritic granular gabbros or PGG). Despite the continuous exposure, the field relationships between these different gabbro components are difficult to decipher, largely because the face is too steep to be readily accessible, but partly because of the combination of prominent jointing and local deuteric alteration along the joint planes.

In general terms, the relatively coarse-grained gabbros, assumed to be MZ cumulates, are found at the western end of the working face. They consist of cumulus orthopyroxene (En_{47}), augite ($Ca_{45}Mg_{32}Fe_{23}$) and plagioclase (An_{60}), and appear to be unlayered.

The rest of the working face comprises members of the fine-grained granular gabbro suite. In the central part of the face the rocks are olivine-bearing, which is unusual, but their textural features are typical of the more commonly encountered olivine-free types of FGG. These gabbros consist of olivine (Fo_{63}), augite ($Ca_{45}Mg_{41}Fe_{14}$) and plagioclase (An_{65}) and are also exceptionally fresh. Farther east they pass into PGG with an abundance of large plagioclase phenocrysts (An_{80} zoned to An_{65}) in a groundmass virtually identical to the olivine-bearing FGG described above.

These varieties of granular gabbro (normal FGG, olivine-bearing FGG and PGG) are also encountered in the same relative positions in the newly-developed quarry area above the main working face, close to Pitscurry Wood, but even here there is as yet no direct evidence of their age relationship, only negative features in the sense that there are no obvious chilled margins, intrusive veins, or xenoliths of one rock type in another.

The more southerly part of the quarry area consists mostly of coarse-grained gabbroic rocks, with textural features similar to the MZ cumulates elsewhere, but they have been thoroughly

Figure 3.10 Norite of the Middle Zone, Insch intrusion, intruded by a 10 m-thick sheet of pegmatitic granite with narrow veins branching off the main sheet, Pitscurry Quarry, Pitcaple. (Photo: BGS no. D4332.)

amphibolitized. Pitscurry Quarry also contains examples of the later pegmatitic granite sheets. These are best seen in the western face, where they form a 10 m-thick, approximately horizontal sheet, with minor offshoots (Figure 3.10). Smaller inclined or vertical sheets are seen in the northern face. In addition to feldspar, quartz and micas, these pegmatites contain garnet, black tourmaline (schorl) and rare beryl (Leslie, 1987).

Legatesden Quarry

This small quarry (no longer worked) is entirely within MZ cumulates, but displays gradations from fresh material at the NW end of the exposure into moderately deformed and amphibolitized rocks elsewhere. The fresh gabbros consist of cumulus plagioclase (An_{60}) and orthopyroxene (En_{55}), together with scattered subhedral grains of opaque oxide, and a small amount of interstitial augite and biotite.

Just to the SE of the central part of the main quarry face, close to a 2 m-wide sheet of pegmatitic granite, the cumulates exhibit well-developed layering, consisting of an alternation of felsic and mafic units on a relatively small scale (centimetres to tens of centimetres). Some of the mafic layers are rather wispy and laterally impersistent, and there is also an indication of upward grading from the principal mafic layer in this outcrop. The base of this layer is also remarkably uneven in a way that is reminiscent of loading structures in sediments and clearly implies a considerable degree of post-cumulus instability. Unfortunately these layered cumulates have been thoroughly altered in proximity to the pegmatite sheet, and now consist predominantly of chlorite, moderately sodic plagioclase, which is rather strained and locally recrystallized, and epidote. As well as this local modification of the cumulates, there is also a more general increase in degree of shearing and alteration from NW to SE in the quarry. The earlier stages appear to involve plagioclase deformation and the replacement of the original pyroxene by colourless amphibole. More advanced alteration results in the recrystallization of plagioclase (and formation of epidote) and the development of chlorite at the expense of the secondary amphibole. Detailed discussion of the textural and mineralogical modification to gabbros involved in shear zones is given by Kneller and Leslie (1984).

Interpretation

Although the Insch MZ has obvious geographical coherence, lying between the LZ to the east and the UZ to the west and NW, and displays broadly intermediate petrological characteristics, in detail it turns out to be unexpectedly complicated. This is seen not only in the intricate association of MZ cumulates and granular gabbros (FGG and PGG), but also in the absence of a simple cumulate stratigraphical sequence from SE to NW (Wadsworth, 1988). Both of these features imply that there was considerable disruption during and after the formation of the Insch MZ, and there have been many different interpretations of this unit, both in terms of its internal complexity and its relationship to the adjacent LZ and UZ cumulates. It is hoped that investigations in the Pitcaple area, in particular in the large working quarry at Pitscurry, will eventually resolve the situation.

Whittle (1936) was the first to investigate the Insch (MZ) hypersthene-gabbros and, on textural grounds, concluded that the granular gabbros are older than the coarse-grained gabbros (now regarded as cumulates) and have been thermally metamorphosed by them. Read *et al.* (1965) were more concerned with the significance of the hypersthene-gabbro unit as a whole, rather than with the internal textural features, and decided that it represents a distinct intrusion, invading the Insch cumulates (LZ and UZ) and not directly related to them. To Read and his colleagues, the main significance of the hypersthene-gabbros lies in their lack of olivine (and relative abundance of orthopyroxene), which they took to indicate large-scale contamination of the regionally available 'Younger Basic' magma by argillaceous sedimentary material.

Clarke and Wadsworth (1970) re-interpreted the coarse-grained hypersthene-gabbros as an integral part of the Insch cumulate succession, thus defining the MZ stage of differentiation. However, they believed the associated granular gabbros to be slightly younger than the cumulates, and to represent invasion by pulses of the parental 'Younger Basic' magma. Wadsworth (1988) was persuaded by the mineralogical evidence that the MZ cumulates and the granular gabbros are both part of a coherent, intermediate fractionation stage. He suggested that the FGG and PGG represent material which crystallized near the intrusion margins (probably the roof) and subsequently foundered into the contemporary cumulate pile from time to time as large 'rafts' of essentially solid material. Such a mechanism would not only explain the intimate association of MZ cumulates and FGG/PGG, but might also account for some of the structural complexity of the cumulate succession. One important line of evidence is the occurrence of abundant small FGG xenoliths in a MZ cumulate matrix at Candle Hill (662 265), between Pitcaple and Insch (Wadsworth, 1988).

However, it must be emphasized that although Pitscurry quarry provides excellent exposures of the various MZ components, their precise relationships are not immediately evident, and await a more thorough investigation.

Conclusions

Pitscurry and Legatesden quarries are representative of the Middle Zone (MZ) of the 'Younger Basic' layered sequence. Between them they provide access to the great variety of Insch MZ rocks, both primary and secondary (shear-belt modification), as well as later pegmatitic granite sheets. Pitscurry is particularly important in terms of the close association of unusually fresh MZ cumulates and granular gabbros, whereas Legatesden is significant in displaying small-scale layering in MZ cumulates (rarely seen elsewhere), with evidence of post-cumulus instability.

HILL OF JOHNSTON (NJ 575 250)

W. J. Wadsworth

Introduction

One of the most distinctive features of the 'Younger Basics' is the development of a pronounced fractionation trend towards iron-rich, felsic residual material, which appears to be represented by the final stages of the Insch cumulate succession from Lower Zone (LZ), Middle Zone (MZ) and eventually through to Upper Zone (UZ) stages. Read *et al.* (1961) were the first to describe these late-stage differentiates, which they recognized as forming a series of small hills (referred to as the 'Red Rock Hills', because of the reddish 'syenitic' rocks found at their summits), trending from SW to NE, to the west of Insch town (Figure 3.11). Clarke and Wadsworth (1970) formally grouped all the olivine-bearing rocks (and closely related felsic material) lying to the west and NW of the predominantly olivine-free MZ gabbros (hypersthene-gabbros of Read *et al.*, 1965) as comprising the Insch UZ, which they subdivided into three sub-zones (UZa, b and c) according to their detailed mineralogy. This classification was further refined by Wadsworth (1986). The UZa rocks are widespread throughout this part of the Insch intrusion, and are ferrogabbros (Fe-rich olivine-plagioclase-pyroxene cumulates). The ferrogabbros are overlain locally (in the 'Red Rock Hills') by UZb, which consists of ferromonzodiorites (monzonites locally) (similar to UZa, but with cumulus alkali feldspar), and then by UZc, which is always heavily altered but is approximately quartz-syenitic in composition.

This type of sequence is represented to varying degrees on all the 'Red Rock Hills', but exposures are generally very poor (Figure 3.12). The most complete sequence (both in terms of exposure and variety of rock types) is found at Hill of Johnston, the most south-westerly of the 'Red Rock Hills', where UZb has been quarried for roadstone and where UZc is at least seen *in situ*, although by no means well exposed.

Description

Although UZa rocks are not exposed in the immediate vicinity of Hill of Johnston, excellent examples of very fresh material from this sub-zone occur elsewhere in the north-western part of the Insch intrusion, e.g. at Brankston (589 308), and clearly represent the immediate precursors to the more extreme differentiates at this locality. They are essentially gabbroic cumulates, comprising Fe-rich olivine, two pyroxenes (Fe-rich orthopyroxene and ferroaugite) and plagioclase (approximately An_{50}).

The 'stratigraphically' lowest rocks of the sequence at Hill of Johnston are exposed in the roadside quarry at the SW foot of the hill, and in another small quarry at Mill of Johnston, 200 m to the SE.

The rocks from the lower part of the main quarry (no longer worked) are cumulates, although there is no small-scale layering visible at this locality. They are mineralogically complex, consisting mainly of cumulus plagioclase (An_{45}), alkali feldspar, olivine (Fo_9) and ferroaugite ($Ca_{42}Mg_{21}Fe_{37}$), with relatively abundant apatite and zircon, both of which may be cumulus phases, and intercumulus hornblende

Hill of Johnston

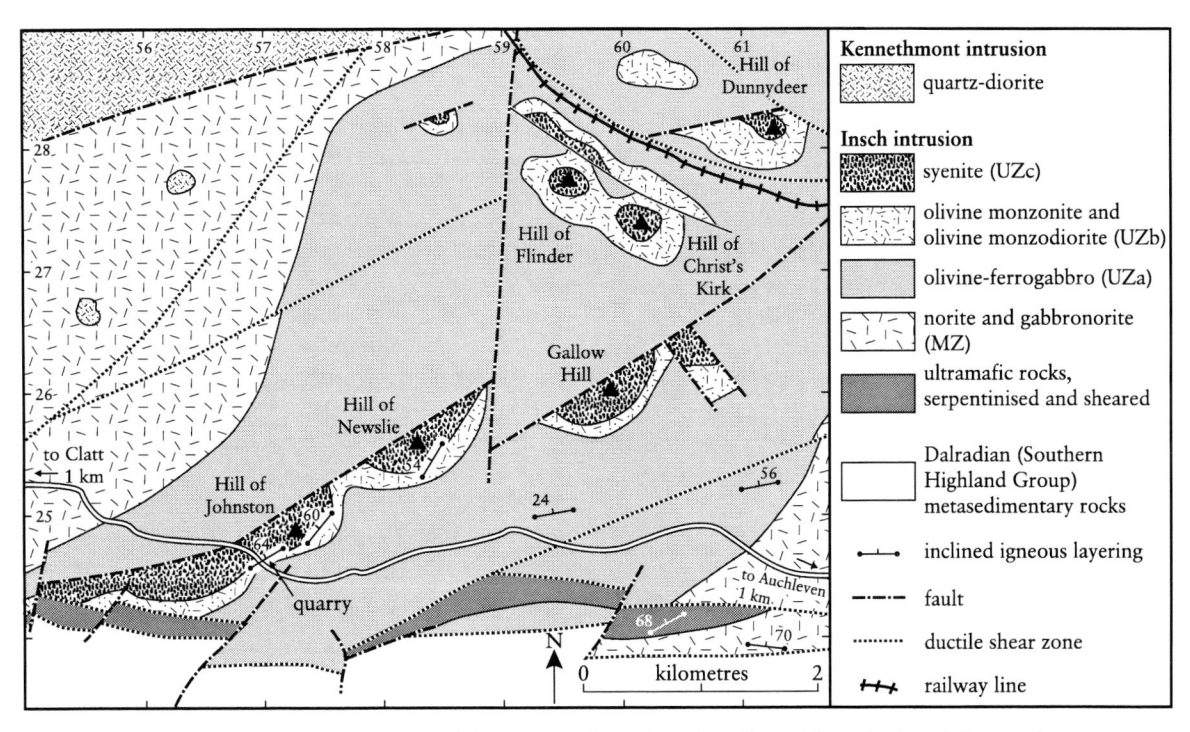

Figure 3.11 Map showing the location of the principal 'Red Rock Hills' (UZb and UZc of the Insch intrusion), west of Insch, from BGS 1:50 000 Sheet 76W (1993).

and biotite. Orthopyroxene (En_{24}) occurs in some rocks and interstitial quartz is generally present. The cumulus alkali feldspar, which is microperthitic orthoclase, is notable for its high Ba content ($Ab_{19}Or_{71}Cn_{10}$). These rocks were termed syenogabbros by Read *et al.* (1961) but are probably more accurately described as olivine ferromonzodiorites (or ferromonzonites) and have been interpreted as representing UZb (Wadsworth, 1986). They are well jointed, and tend to weather spheroidally.

At higher levels in the quarry the rocks are slightly less mafic, and also contain more alkali feldspar ($Ab_{19}Or_{76}Cn_5$) relative to plagioclase (An_{37}). The ferromagnesian minerals tend to be more altered, with olivine (Fo_6) almost completely serpentinized, and ferroaugite ($Ca_{42}Mg_{14}Fe_{44}$) occurring as relict cores in amphibole. Zircon and apatite are still abundant, and are accompanied by interstitial biotite and quartz. These relatively felsic rocks also occur as near-vertical veins, up to 5 cm across,

Figure 3.12 The 'Red Rock Hills': Hill of Christ's Kirk (left distance) and Hill of Dunnideer (centre distance, with ruined castle) from near Auchleven. The hills are composed of syenite and olivine monzonite and the foreground is underlain by olivine-ferrogabbros, all of the Upper Zone, Insch intrusion. (Photo: BGS no. D4542.)

cutting the more mafic UZ material.

Above the quarry, there are numerous small natural outcrops on the SW slopes of Hill of Johnston. The rocks are distinctly reddish in colour, hence the term 'Red Rock Hills', and they are generally rather altered. Some examples are fresh enough to indicate that they are very rich in alkali feldspar, and approach syenitic compositions as described by Read et al. (1961). There is no clear textural evidence that they are cumulates, but they have been referred to UZc by Wadsworth (1986). The alkali feldspar is not Ba-rich (approximately $Ab_{18}Or_{81}Cn_1$) at this level in the intrusion, and it is always heavily sericitized. The mafic minerals, mainly amphibole and biotite have been largely replaced by chlorite. Apatite, zircon and interstitial quartz are also present. Some of these rocks appear to have been silicified.

Despite the absence of overt layering in these UZ rocks, the general occurrence of the more differentiated UZc material towards the summit of Hill of Johnston suggests that the cumulate succession is approximately horizontal in this area of the Insch intrusion. Elsewhere in the 'Red Rock Hills' sub-horizontal layering occurs at Hill of Dunideer and Hill of Christ's Kirk, but dips of 50° to the NNW have been recorded at Hill of Newleslie (Gould, 1997).

Interpretation

The Hill of Johnston outcrops provide the clearest evidence available that the parental magma of the 'Younger Basics' was capable of evolving towards extremely felsic and iron-rich compositions, as represented by the rocks of UZb and UZc. Read et al. (1961) demonstrated the essential coherence of the Insch olivine-gabbros and associated 'syenogabbros' and 'syenites' of the 'Red Rock Hills' on the basis of geological, petrological and chemical characteristics. They also hinted at a broader association between these rocks (now referred to the Insch UZ) and the peridotites and troctolites at the eastern end of the Insch intrusion (now referred to the LZ).

Clarke and Wadsworth (1970) developed this theme, and extended it to include the hypersthene-gabbros, as representing the intermediate stage (MZ) of the complete fractionation sequence. They identified the bulk of the rocks in the intrusion as cumulates, on textural grounds, and confirmed the general progression of mineral compositions expected in such a situation. Wadsworth (1986) continued this approach for the Insch UZ, and was able to construct a detailed cumulate succession, emphasizing the trend towards extreme iron enrichment, and comparing this with broadly similar trends in the Bushveld (South Africa), Skaergaard (Greenland) and Fongen–Hyllingen (Norway) layered intrusions. The Insch UZb rocks, particularly well displayed at Hill of Johnston, are probably the uppermost true cumulates in the succession, and are noteworthy for the large number of cumulus minerals represented (olivine, orthopyroxene, clinopyroxene, plagioclase, alkali feldspar, apatite, zircon and Fe-Ti oxide). The UZc rocks are generally interpreted as having crystallized from the residual magma after significant crystal settling had ceased (Read et al., 1961; Clarke and Wadsworth, 1970; Wadsworth, 1986). This is also indicated by the veins of broadly similar quartz-syenite found locally within UZb.

Conclusions

The Hill of Johnston GCR site is particularly important in providing information about the later stages (UZ) of crystallization in the 'Younger Basics' in general, and the Insch intrusion in particular. In this way, it is complementary to the LZ (Hill of Barra and Bin Quarry) and MZ (Pitscurry and Legatesden) GCR sites. The most significant geochemical aspects of the Hill of Johnston rocks are their pronounced iron, barium and zirconium enrichment, as indicated by the olivine and pyroxenes (UZb and c), alkali feldspar (UZb) and cumulus zircon (UZb), respectively. Similar features are known to result from extreme fractionation of tholeiitic basic magma in other layered intrusions worldwide, e.g. Bushveld, Skaergaard and Fongen–Hyllingen, but the Hill of Johnston is the only example in Britain.

HILL OF CREAGDEARG (NJ 453 259)

W. J. Wadsworth

Introduction

In addition to the main cumulate fractionation sequence of the 'Younger Basic' intrusions, as seen in the Insch, Belhelvie and Huntly intrusions, there are other igneous rocks that are spatially associated with the cumulates and may be genetically related to them. These are particularly characteristic of the western end of the Insch intrusion and of the Boganclogh intrusion, which is the westward continuation of the Insch intrusion (Busrewil *et al.*, 1973; Gould, 1997). The Boganclogh area contains three principal igneous rock types: a northern strip of ferrodiorites, a central region of quartz-biotite norites, and a southern belt of heavily serpentinized ultramafic rocks (Figure 3.2). The ferrodiorites are broadly equivalent to the Insch Upper Zone; the quartz-biotite norites are also represented in the Insch mass, but are very poorly exposed there; serpentinites occur sporadically along the western part of the southern margin of the Insch mass (Read, 1956), but reach their maximum development in the Boganclogh area (Blyth, 1969).

The central belt of quartz-biotite norites at Boganclogh contains smaller areas of the other components. Two such areas occur at Hill of Creagdearg and at nearby Red Craig (Figure 3.13). In each case, well-exposed ultramafic rocks (peridotites) with an unusually high proportion of fresh olivine, are surrounded by typical quartz-biotite norites. The names of the two hills are derived from the characteristic reddish-brown colour of the weathered surfaces of the peridotite (Figure 3.14) which stand out in marked contrast to the grey-weathering norites.

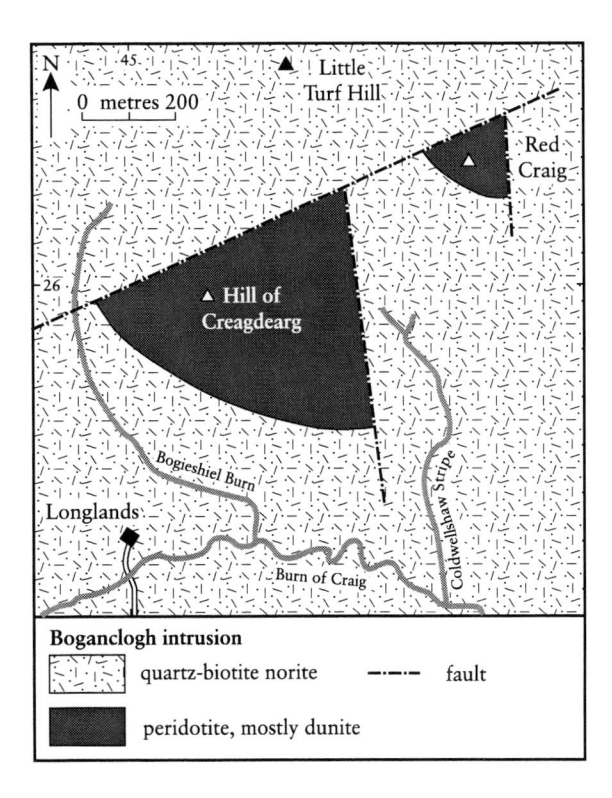

Figure 3.13 Map of the Hill of Creagdearg and Red Craig area, Boganclogh intrusion, from BGS 1:10 000 sheets NJ42NW (1991) and NJ42NE (1991).

Description

The two areas of peridotite (Hill of Creagdearg and Red Craig) can be clearly delineated, because of the excellent exposure and the obvious colour contrast between the main rock types in the field. The northern and eastern margins of both peridotites appear to be steep and faulted, but towards the SW the junction with the norites is believed to be gently dipping and is probably an original igneous contact (Gould, 1997). Despite this, there is no unambiguous evidence of relative age. The peridotites appear to be totally unaffected in proximity to the norites, and there is no evidence of inclusions or veining of one rock type by the other. The only hint of age relationships is a slight reduction of grain size in the quartz-biotite norite towards its contact with peridotite, although this is not very pronounced and may be unrelated to marginal chilling, since the norite shows considerable variations in grain size throughout its outcrop.

The peridotites appear to be essentially massive and structureless in the field, with no evidence of small-scale layering such as is seen in the Insch Lower Zone cumulates at the Hill of Barra GCR site. They consist of a high proportion (80 to 95%) of olivine, together with minor spinel and scattered orthopyroxene crystals. The degree of serpentinization is variable, but in general these rocks are remarkably fresh compared with the ultramafic rocks along the southern margin of the Insch–Boganclogh mass.

In thin section, the olivine crystals average approximately 2 mm in length, with polygonal boundaries showing 120° angles at triple junctions. The spinel is chromite, with translucent red-brown margins, and occurs both within and between the olivine crystals. The orthopyroxene occurs as sparse grains, similar in size and shape to the olivines, and in places displaying exsolution lamellae of augite. Olivine and orthopyroxene are altered to antigorite and bastite respectively. The olivine (Fo_{91}) and orthopyroxene (En_{91}) are significantly more magnesian than in

Figure 3.14 Fresh peridotite (dunite) with brown-weathering crust, Red Craig, Boganclogh intrusion. (Photo: BGS no. D4532.)

the LZa cumulates at Insch or Belhelvie, where olivine is typically Fo_{87-86}.

The adjacent quartz-biotite norites are easily distinguished from the ultramafic rocks, not only by the grey colour of their weathered surfaces, but by the prominent sub-horizontal jointing, giving rise to distinctive 'slabby', tor-like outcrops. The norites consist of plagioclase (An_{60}), augite ($Ca_{45}Mg_{37}Fe_{18}$), orthopyroxene (En_{55}), poikilitic biotite (up to 20% by volume), hornblende, and interstitial quartz (2 to 3% by volume). Ilmenite and magnetite are typically present and are sometimes accompanied by traces of pyrrhotite.

Interpretation

Although the quartz-biotite norite is clearly part of the 'Younger Basic' magmatic spectrum, being represented in many of the individual intrusions, the status of the ultramafic rocks at Boganclogh is more contentious. From the detailed investigations of Blyth (1969), it is clear that the peridotites at Hill of Creagdearg and Red Craig are essentially part of the main ultramafic belt which lies along the southern margin of the Boganclogh intrusion, and extends eastwards as a series of discontinuous lenses along the southern edge of the Insch intrusion (Read, 1956). It is also clear that these ultramafic rocks are associated with major shear zones, and that the southern boundary of the Insch–Boganclogh mass is tectonic. This accounts for the highly serpentinized nature of these rocks generally, but inevitably tends to obscure evidence of their primary origin. For this reason, the relatively unserpentinized peridotites from Hill of Creagdearg and Red Craig are particularly significant.

Hinxman and Wilson (1890) suggested a possible correlation between the serpentinites of the southern marginal belt, and the broadly similar rocks at the eastern end of the Insch intrusion (now referred to as LZa cumulates). Both Read (1956) and Blyth (1969) discounted this correlation, emphasizing certain differences between the two ultramafic associations. Blyth, in particular, stressed the absence of any gradations to more gabbroic lithologies in the Boganclogh peridotites and serpentinites. However, neither Read nor Blyth speculated further about the source of the strongly tectonized serpentinites, except to imply that they came 'from depth'.

Busrewil *et al.* (1973), on the other hand,

believed that the mineralogical and textural evidence, especially from the Hill of Creagdearg peridotites, supports the link between the LZ cumulates and the marginal serpentinites. The olivine compositions, in particular, appear to show overlap between the two occurrences. However, more recent investigations have shown that the Hill of Creagdearg peridotites contain distinctly more magnesian olivines (Fo_{91}) than in the LZa cumulates (Fo_{87-86}). This, together with the lack of feldspar, and the absence of convincing cumulate textures, now re-inforce Blyth's view that the Boganclogh ultramafic rocks are not related to the cumulate succession. Further, the olivine (and orthopyroxene) compositions suggest a close connection with mantle peridotites, either in the form of tectonically emplaced mantle fragments, or as the early crystallization products of very primitive, mantle-derived, magma, which have subsequently been involved in major tectonic disturbance.

In this connection, it is of interest to note the occurrence of similar, highly magnesian olivines in the Succoth–Brown Hill intrusion (S–BH), to the NW of Boganclogh (Gunn *et al.*, 1996), although the crystallization sequence olivine–clinopyroxene–plagioclase at S–BH is in marked contrast to the early appearance of orthopyroxene in the Boganclogh peridotites, suggesting a fundamentally different, possibly more calc-alkaline magma (Styles, 1994). Like the southern margin of the Insch–Boganclogh mass, the S–BH intrusion is also associated with a major shear zone (within the Portsoy Lineament) and the latter is believed to indicate a significant magmatic event earlier than the 'Younger Basic' intrusions. On this basis the S–BH intrusion, and possibly the Boganclogh ultramafic rocks, provide support for Read's original (1919) idea that two main episodes ('Older' and 'Younger') of basic magmatism are represented in NE Scotland.

The quartz-biotite norites are certainly part of the 'Younger Basic' activity, and therefore are probably significantly younger than the peridotites at Hill of Creagdearg, although the age relationships are not convincingly displayed in the field. The quartz-biotite norites are generally interpreted as samples of basic magma similar to the parental magma of the cumulate succession, but which crystallized under relatively hydrous conditions (Wadsworth, 1988). They probably represent in-situ crystallization without significant crystal settling.

Conclusions

The Hill of Creagdearg GCR site provides evidence of primitive mafic magma associated with a 'Younger Basic' intrusion. The unusually fresh peridotites are believed to be quite distinct from, and possibly older than, the olivine-rich Lower Zone cumulates of the 'Younger Basic' suite, found at the eastern end of the Insch intrusion and at Belhelvie. Associated with the peridotites are quartz-biotite norites which are a significant component of the 'Younger Basic' activity in general, but are particularly widespread at Boganclogh.

BALMEDIE QUARRY (NJ 944 182)

W. J. Wadsworth

Introduction

In addition to the wide range of igneous components found in the 'Younger Basic' intrusions, other variations have been imposed by later tectonic events. Late Caledonian (D3) folding was responsible for the variable, and often steep, dips in the layered cumulates, and may have caused some disruption of the original igneous complex into smaller bodies. Another aspect of the post-intrusion structural disturbance was the development of major shear zones, which may have played an important part in the final emplacement and configuration of the 'Younger Basic' masses (Ashcroft *et al.*, 1984). These shear zones are associated particularly with the present margins of the intrusions, notably the Insch–Boganclogh mass (Figure 3.2), but they also produced significant modifications internally. These modifications may be dominantly mineralogical, involving the progressive amphibolitization of the gabbros, or textural, involving the formation of gabbro mylonites, or a combination of both effects.

One of the most prominent shear zones runs approximately N–S from near Fraserburgh southwards towards Aberdeen and, near its southern end, it intersects the Belhelvie intrusion (Figure 3.15). Although enough of the intrusion is unaffected for it to be possible to establish a detailed cumulate stratigraphy (Munro, 1986; Wadsworth, 1991), substantial areas have been thoroughly amphibolitized (Stewart, 1946) and, locally at least, show evidence of pronounced textural changes with the production of schistose (or flaser) gabbros and mylonites. Excellent examples of both types of modification associated with the shear zones are seen in Balmedie Quarry, just north of Belhelvie village, and have been fully documented. (Boyd and Munro, 1978; Munro, 1986).

Description

Balmedie Quarry is situated in the centre of an approximately 1 km-wide shear-zone stretching northwards from Belhelvie village to the eastern margin of the Belhelvie intrusion (Figure 3.15). The original cumulates are believed to represent the upper part (LZc) of the Belhelvie succession (Wadsworth, 1991) and consist of cumulus olivine (Fo_{77}), orthopyroxene (En_{79}), augite ($Ca_{45}Mg_{45}Fe_{10}$) and plagioclase (An_{75}). Primary layering in the least-deformed examples shows a wide variety of modal variations, with all graduations between mafic (peridotite) and felsic (anorthosite) layers, although most of the these cumulates are gabbroic or noritic.

Deformation affects the whole area of the quarry, but the intensity varies considerably, even over a scale of a few centimetres. No systematic pattern of variation is recognizable within the area studied. In the initial stages of modification the original mineralogy is retained, but the texture is modified to the extent of producing localized strain effects in individual crystals, as well as some degree of marginal granulation and minor recrystallization. Where the deformation was more intense, the textural changes are accompanied by the development of aggregates of secondary amphibole crystals in place of the original pyroxene. With increasing degree of deformation the original igneous textures are obliterated, and the resultant rock consists of lensoid clusters of amphibole, biotite and opaque minerals (representing the original ferromagnesian minerals) and irregular plagioclase porphyroclasts, in a fine-grained aggregate of mafic material.

The most intense deformation is restricted to

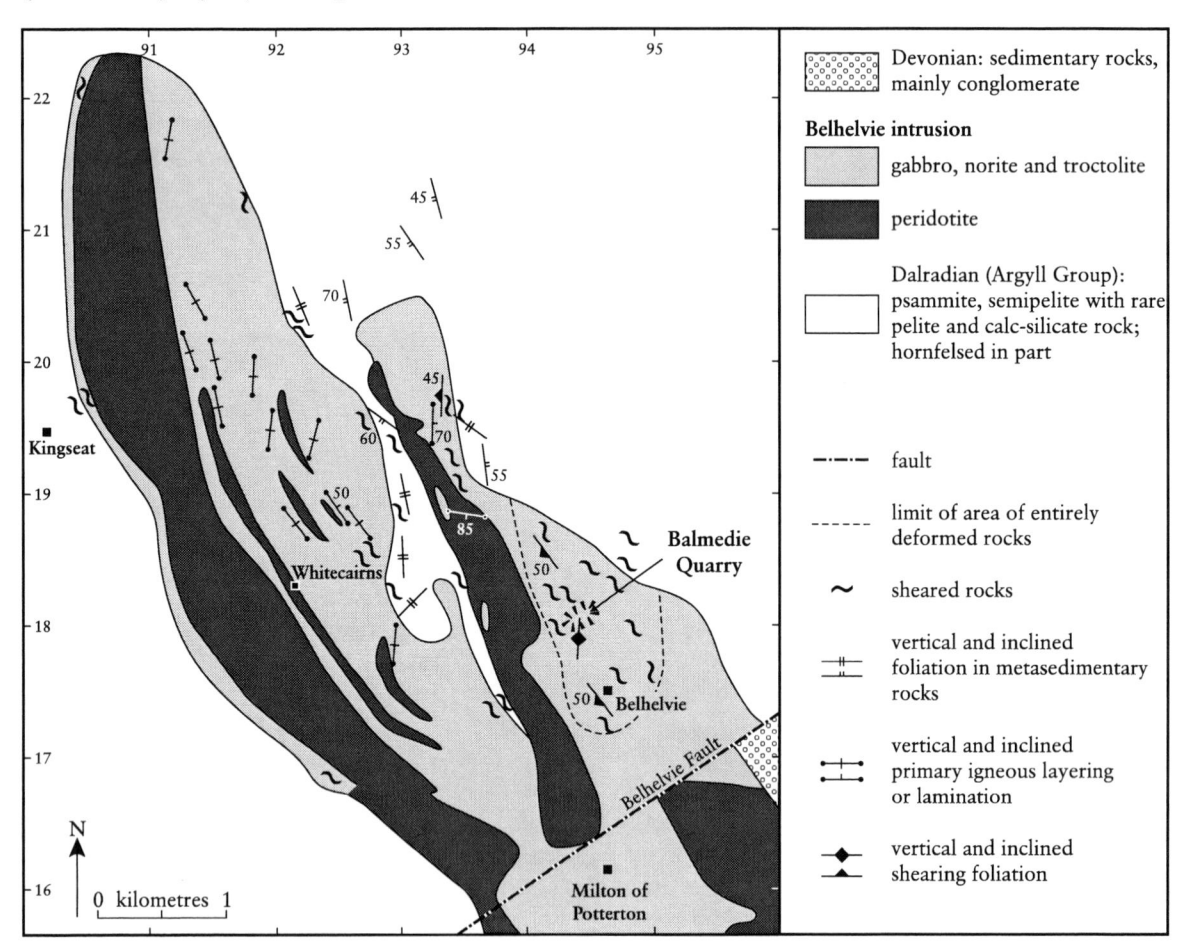

Figure 3.15 Map of the northern part of the Belhelvie intrusion, showing the position of the Balmedie Quarry GCR site in relation to a major shear zone, after Boyd and Munro (1978).

Balmedie Quarry

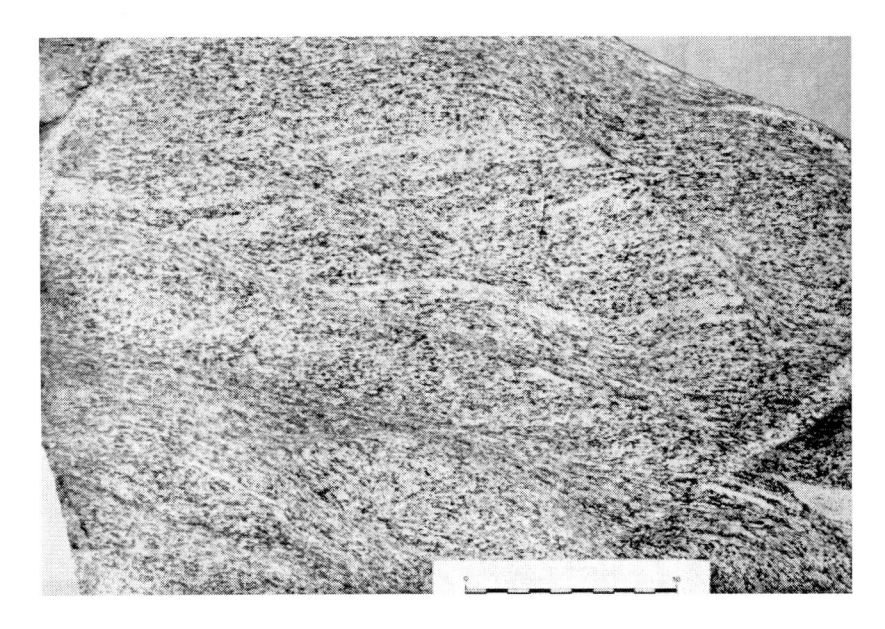

Figure 3.16 Balmeddie Quarry; block of deformed mafic rock, cut by narrow zones of mylonite that in part conform with and in part transgress an earlier foliation. Scale in centimetres. (Photo: from Boyd and Munro, 1978, plate 1a.)

narrow zones (generally less than 1 cm wide), which cut the more typically foliated gabbros in sinuous fashion (Figure 3.16). The rocks in these zones are essentially mylonites, and consist of porphyroclasts of frayed plagioclase (and more rarely amphibole or pyroxene) in a groundmass of amphibole, biotite, opaque minerals and plagioclase. The larger feldspar fragments typically retain their original composition (An_{80-75}) in their cores, but have strongly zoned margins (An_{45-40}). Large crystals of quartz occur locally, and there is considerable evidence that there has been an episode of late-stage silicification. Although the detailed attitude of both the general foliation, and the intensely mylonitized zones, is extremely variable, there is an overall tendency for these structures to strike approximately N–S, and to display steep dips. Some of the more sinuous mylonite zones form minor folds with steeply plunging axes.

In addition to the deformed gabbroic rocks, there is a small strip of hornfelsed and mylonitized metasedimentary rocks (10–15 m wide) in the eastern part of the quarry, and there are also some granitic minor intrusions, which are generally medium grained, but include tourmaline-bearing pegmatitic types. Most of these granites are undeformed, and appear to post-date the foliation in the mafic rocks but some of them have been affected by shearing, at least locally.

Interpretation

In general, the shear zones appear to represent localized regions of strong mechanical distortion and dislocation, accompanied by low pressure (2–3 kbar), amphibolite facies (600°C) metamorphism, which occurred soon after the intrusion and crystallization of the 'Younger Basic' igneous bodies approximately 470 Ma ago (Ashcroft *et al.*, 1984). In Balmedie Quarry, the shear zone is itself cut by granitic minor intrusions, one of which has been radiometrically dated at 462 ± 5 Ma (Pankhurst, 1982). Most of these granitic intrusions are undeformed, but some of them show evidence of marginal crushing and shearing (Munro, 1986). This suggests that there may have been more than one episode of granite veining, or that deformation continued, at least on a reduced scale, after the 'Younger Basic' igneous activity had ceased.

The textural features of the modified gabbros in Balmedie Quarry suggest that a pervasive foliation was developed at an early stage in the deformation history, resulting in relatively limited shearing, granulation and recrystallization. This then seems to have been followed by an episode of more intense deformation, restricted to the narrow mylonite zones. The broadly similar structural trends of the foliation and the mylonite zones suggests that the two types of deformation were related, and represent a continuum of structural disturbance rather than discrete events. From the mineralogical evidence, the initial stages of deformation appear to have been simply cataclastic, but the later stages were characterized by recrystallization in the presence of volatile components, resulting in the formation of amphibole, biotite and relatively sodic plagioclase at the expense of the original pyroxenes and calcic plagioclase (Kneller and Leslie, 1984). It is also evident that additional silica was added during this stage (Boyd and Munro, 1978).

Conclusions

Balmedie Quarry is important for the remarkably clear evidence that it exhibits of the effects of post-magmatic tectonic events on the original layered gabbros and norites of the 'Younger Basic' intrusions. These effects are partly mechanical, resulting in locally intense shearing and crushing, to give rise to flaser gabbros and mylonites. Mineralogical changes are also represented; these involve metamorphism of the original high-temperature igneous assemblage (olivine, pyroxenes, calcic plagioclase) to a lower grade (amphibolite facies) assemblage of amphibole, biotite and sodic plagioclase, together with additional silica. The main period of deformation was followed by intrusion of minor granite sheets, some of which are pegmatitic.

TOWIE WOOD (NJ 933 383)
POTENTIAL GCR SITE

W. J. Wadsworth

Introduction

One of the most distinctive aspects of the 'Younger Basic' intrusions is the local occurrence of a marginal facies of cordierite norites rich in xenolithic material derived from the Dalradian country rock. These are known from both the Insch and Huntly intrusions, but reach their maximum development in the Haddo House–Arnage intrusion (Read, 1923, 1935; Read and Farquhar, 1952; Gribble, 1968), notably in the classic, but poorly exposed, areas of Craigmuir Wood and Wood of Schivas, in the Ythan Valley. Here the least modified component of the 'Younger Basic' intrusions is quartz-norite, although olivine-norite occurs locally.

The xenolithic complexes comprise abundant small fragments of hornfels in a matrix of igneous aspect texturally, but with a distinctive mineralogy involving various proportions of plagioclase, cordierite, biotite, orthopyroxene and quartz, sometimes accompanied by garnet and alkali feldspar. These rocks are generally referred to as cordierite norites although some of them are quartz-rich and would be more accurately described as tonalites. The xenoliths represent a wide range of compositions, but fall into two principal categories, namely silica-deficient, aluminous types and silica-rich types. These are believed to represent argillaceous and quartz-rich Dalradian metasedimentary rocks, respectively. Local concentrations of calc-silicate hornfels xenoliths (e.g. from Craigmuir Wood, near Haddo House) were presumably derived from calcareous layers in the Dalradian (Read, 1935). The cordierite norites and associated xenoliths have generally been taken to represent some form of interaction between 'Younger Basic' magmas and the adjacent country rock, especially in the roof region of the original intrusions, but there has been considerable disagreement about the precise nature of this interaction.

Description

The Towie Wood site lies towards the northern extremity of the eastern or Arnage area of the Haddo House–Arnage intrusion (Figure 3.17). Most of the exposures are found on, or close to, the former railway line and in the adjacent Ebrie Burn. At Towie Wood, the old railway cutting provides exposures of quartz-norite, while a small, but compact and well-exposed area of xenolithic material occurs just to the south, on the eastern side of the railway trackbed, where it forms a distinct knoll (Munro and Leslie, 1987).

The quartz-norite, which is essentially homogenous and xenolith-free, consists of plagioclase (An_{60-50}), orthopyroxene (En_{68-55}), clinopyroxene, hornblende, biotite and quartz. It is of medium- to coarse-grain size, with conspicuous biotite crystals (Munro and Leslie, 1987).

The xenolithic complex has an igneous-textured matrix of rather variable grain size and modal proportions, in which the principal minerals are plagioclase, cordierite, biotite, orthopyroxene, garnet and quartz. This cordierite norite is intimately associated with xenolithic material, although the relationship between the two components is by no means constant. Locally the xenoliths are dominant, and appear to represent relatively undisturbed country rock invaded by veins and stringers of norite. This arrangement was taken by Read (1923) to indicate proximity to the roof of the intrusion. Elsewhere, xenoliths and matrix are approximately equal in abundance, with an apparently chaotic mixture of xenolith types in random orientation. This type of assemblage grades into more homogeneous cordierite norites, with fewer relict xenoliths.

Two principal types of xenolith are present. One type is quartz-rich, presumably representing more psammitic Dalradian material. These tend to be angular in shape, and show all size gradations from centimetres to tens of centimetres across; some display relict bedding structures. In thin section they are seen to be plagioclase-quartz-biotite hornfelses, with sporadic garnet, cordierite or orthopyroxene. The other type comprises compact, homogeneous blue-grey hornfelses derived from pelitic Dalradian metasediments. They tend to be smaller (generally less than 5 cm across) and more rounded than the siliceous xenoliths. Mineralogically, they are silica deficient, consisting of variable proportions of cordierite, plagioclase, spinel, sillimanite and garnet, accompanied by corundum, orthopyroxene or biotite. Some of the argillaceous xenoliths are composed entirely of cordierite. In addition to the xenoliths and cordierite norite matrix, quartz-rich veins and stringers are quite widespread.

Interpretation

Although there has been broad agreement that the cordierite norites and associated xenolithic assemblages are genetically related, and were developed as part of the 'Younger Basic' magmatic event, various explanations of their precise relationship and significance have been presented.

Read (1923) considered the cordierite norites to represent basic magma which had been contaminated by assimilation of country rock. He suggested that Si, Ca, Na and K had been selectively extracted from Dalradian pelites by the magma, leaving a residue of silica-deficient, aluminous xenoliths. He also regarded the more homogeneous quartz-norites of the Haddo House–Arnage area, as mildly contaminated intermediates between the cordierite norites

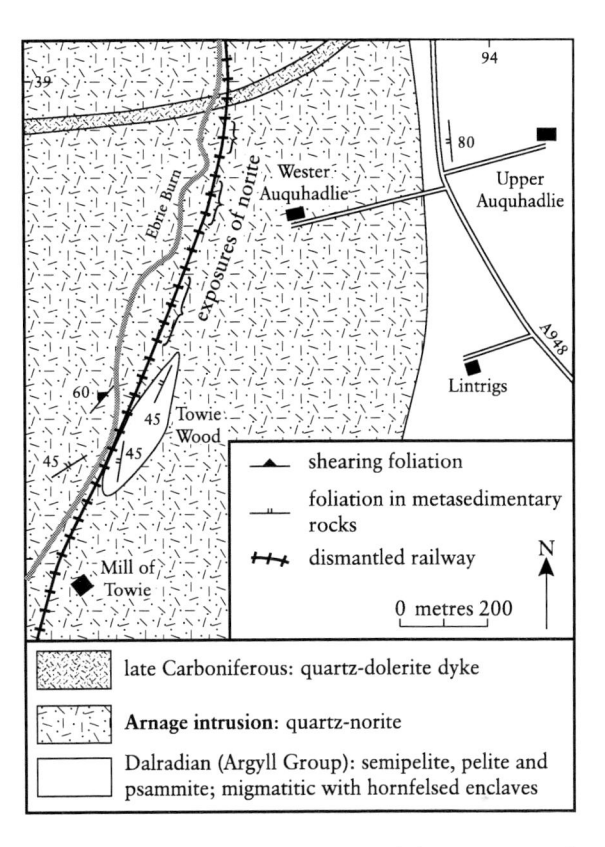

Figure 3.17 Map of the area around the Towie Wood GCR site, Arnage–Haddo intrusion, from BGS 1:10 000 Sheet NJ93NW (1986) by W. Ashcroft and M. Munro, Aberdeen University.

and the olivine-gabbros more typical of the 'Younger Basic' activity in general. Read subsequently re-interpreted these xenolithic rocks as parts of an older migmatite complex implying that the associated 'Younger Basic' norites were not actively involved in their formation (Read and Farquhar, 1952). However, he continued to invoke large-scale regional contamination of 'Younger Basic' magma with Dalradian pelitic material to explain the apparently anomalous hypersthene-gabbro unit (now interpreted as Middle Zone cumulates) in the Insch intrusion (Read et al., 1965), and appeared to accept a contamination origin for xenolithic cordierite norites occurring locally at the margin of this intrusion.

Gribble (1968) re-investigated the xenolithic complexes at Haddo House–Arnage, and concluded that the 'Younger Basic' intrusions had played a vital part in their formation by raising temperatures in the adjacent country rocks, so that they began to melt. The cordierite norites were believed to represent partial melts of the

Dalradian material, and the xenoliths as the refractory residues (restites) after melt extraction had occurred. This conclusion is supported by the strontium isotope data of Pankhurst (1969), which shows no significant overlap between the initial $^{86}Sr/^{87}Sr$ ratios for the normal norites (0.706 to 0.715) and the cordierite norite matrix material (0.720 to 0.731), whereas the matrix and xenolith ratios overlap each other.

Conclusions

The Towie Wood site is the most compact and well-exposed locality to display the characteristic features of the xenolithic complexes associated with the 'Younger Basic' intrusions. The xenolithic fragments are particularly abundant, and most of the main Dalradian country rock types are represented, together with a variety of cordierite norite matrix material; the relationship between these components is clearly displayed. More normal, homogeneous norites occur close to the xenolithic complex, although the contact between the two is not exposed.

CRAIG HALL (NJ 530 292)

W. J. Wadsworth

Introduction

Forming the western margin of the main Insch intrusion, the Kennethmont granite–diorite complex comprises homogeneous granites (both pink and grey varieties) and diorites, together with a group of xenolithic rocks consisting of dioritic fragments in a granitic matrix (Gould, 1997). The complex as a whole is poorly exposed, but the principal components were mapped and described by Sadashivaiah (1954), who suggested that they are essentially unrelated to the 'Younger Basic' magmatic event. He concluded that they represent a later invasion of granitic magma which interacted locally with already consolidated basic rocks, similar to hypersthene-gabbros of the Insch mass, to produce hybrid xenolithic diorites. Read and Haq (1965) provided geochemical data to support these conclusions. Busrewil *et al.* (1975) re-investigated the Kennethmont granite–diorite complex, and recognized that the situation is more complicated than indicated by Sadashivaiah. They showed that two distinct magmas (granitic and dioritic) were responsible for the formation of the xenolithic rocks, and that the 'Younger Basic' rocks were not involved at all. They also realized that the pink granite member of the complex is unrelated to the other rock types, although it is probably of similar age. It has been dated at 453 ± 4 Ma (Pankhurst, 1982), and is probably one of the late tectonic granites.

Because of the poor exposure, and the way in which the principal components of the complex are distributed, it is impossible to identify a compact site in which they are all present. However, the Craig Hall area (Figure 3.18) is important in that the xenolithic assemblages, with diorite fragments in various stages of assimilation in a grey granitic matrix, are well represented. Although exposures are scarce, there is abundant loose material that is believed to be of very local derivation.

Description

All gradations occur between rocks consisting of relatively coherent dioritic xenoliths in apparently unmodified grey granite matrix, to more granodioritic material containing small (millimetre to centimetre) mafic inclusions, with rounded or irregular shapes; these are believed to represent residual, partly digested, xenoliths. The grey granite comprises quartz, microcline, oligoclase and biotite (with minor titanite and apatite), and appears to show all gradations to granodioritic compositions, as plagioclase becomes the dominant feldspar. The dioritic xenoliths are relatively mafic, with pyroxene (mostly clinopyroxene, but some orthopyroxene) hornblende, biotite and titanite all present, although biotite is usually the dominant mineral. Plagioclase (An_{35}) is the principal felsic constituent, although small amounts of quartz occur in some samples. The mafic inclusions in the granodiorite consist of biotite and hornblende. The inclusions are typically finer grained than the acid matrix, and they show gradations from an igneous texture, with a few relict phenocrysts of plagioclase, to a distinctly granular metamorphic texture. Chemical data (Busrewil *et al.*, 1975) show gradational relationships from the xenolithic basic material, through 'contaminated' rocks, with partly digested xenoliths, to the grey granitic matrix.

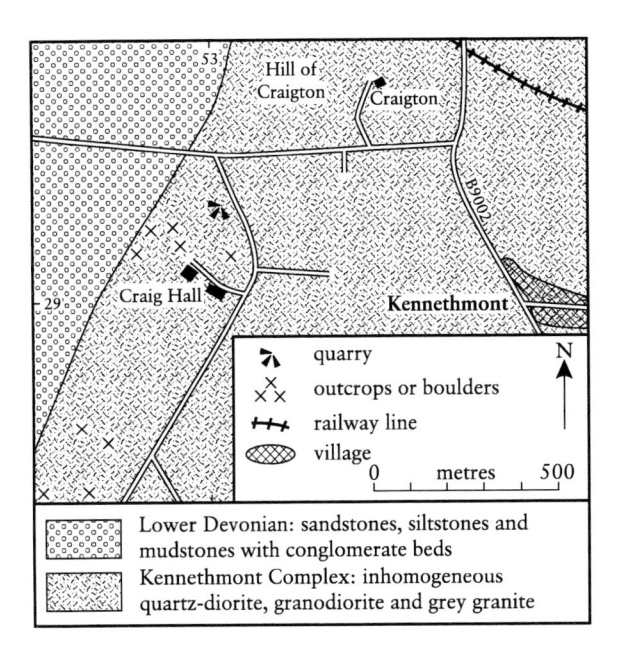

Figure 3.18 Map of the area around the Craig Hall GCR site, Kennethmont, from unpublished BGS maps.

Interpretation

The xenolith-bearing granitic rocks of the Kennethmont granite–diorite complex clearly imply interaction between acid magma and a solidified basic component, but the origin of the latter, and the extent to which hybridization was capable of producing relatively homogeneous intermediate rock types (diorites), has been a controversial issue. Sadashivaiah (1954), followed by Read and Haq (1965) identified the 'Younger Basic' intrusions as the most likely source of the mafic xenoliths, since a variety of suitable compositions was locally available in the Insch intrusion. Of these the hypersthene-gabbros were regarded as the most appropriate protolith. Further, it was implied that the Kennethmont diorites represent the end-products of the hybridization process.

Busrewil *et al.* (1975) presented detailed chemical evidence (especially rare-earth element data) which effectively excluded any of the 'Younger Basic' components as the source of the xenoliths. Instead, they showed that there is considerable geochemical coherence between the dioritic xenoliths and the more homogeneous diorites found in the vicinity. They concluded that both diorite and granite magma were emplaced at about the same time, approximately 465 Ma ago (i.e. perhaps 5 Ma after the 'Younger Basic' event), and that locally the diorite crystallized first, to be invaded, disrupted and partly assimilated by the grey granite. A possible genetic connection between the diorite and grey granite component of the Kennethmont area was envisaged, but the pink granite component was shown to be unrelated, although of much the same age.

Conclusions

The Craig Hall area is of interest in that it provides evidence of a slightly younger magmatic event (the Kennethmont granite–diorite complex), spatially associated with the Insch intrusion, but apparently otherwise unrelated to it. In particular, it is of significance in the occurrence of xenolithic material, superficially resembling the 'Younger Basic' xenolithic complexes (see the Towie Wood site report), but of quite different origin (and age). The xenolithic assemblage of the Kennethmont complex essentially involved the interaction of two magmas (dioritic and granitic), whereas the typical 'Younger Basic' xenolithic association represents local partial melting of adjacent country rock.

Chapter 4

Lake District and northern England

Introduction

INTRODUCTION

D. Millward

In Ordovician times the Lake District and northern England formed part of the microcontinent of Eastern Avalonia, that comprised southern Britain and adjacent parts of continental Europe. Faunal and palaeomagnetic evidence show that in the early Ordovician, Eastern Avalonia was attached to the supercontinent of Gondwana in a high southern latitude. Avalonia–Gondwana was separated from Laurentia to the north by the Iapetus Ocean (Figure 2.1). Avalonia then rifted from the supercontinent in the mid-Ordovician, drifted northwards during closure of the Iapetus and eventually collided with Laurentia in tropical latitudes during the Silurian. The line of closure of the ocean now lies beneath the Solway Firth (Figure 4.1).

During the Palaeozoic, from Late Cambrian to early Llanvirn times, deep-marine turbidites of the Skiddaw Group were deposited on the passive margin of Eastern Avalonia (Cooper *et al.*, 1995). Regional uplift from this non-volcanic, deep oceanic environment to a subaerial one was the prelude to volcanism. Subsequently, two episodes of intense igneous activity occurred in the Lake District. Firstly, during the Ordovician (Llandeilo–Ashgill), basaltic, andesitic and rhyolitic lavas and voluminous pyroclastic rocks were erupted at the continental margin of Avalonia, above a subduction zone in which the Iapetus Ocean was being consumed (Fitton *et al.*, 1982). The Eycott Volcanic Group (EVG) and the Borrowdale Volcanic Group (BVG) represent two contemporaneous volcanic fields (Figure 4.2). This phase of magmatism culminated in the emplacement of much of the Lake District granitic batholith (Hughes *et al.*, 1996). Secondly, in late Silurian to Early Devonian times, additional granitic plutons were emplaced in the Lake District and northern England subsequent to the closure of the Iapetus.

The volcanoes were subaerial (Branney, 1988a), of low relief and probably located within extensional basins, a feature illustrated by the single marine incursion in the upper BVG. The earliest known terrestrial trace fossils in the world occur in volcaniclastic sedimentary rocks from the BVG (Johnson *et al.*, 1996). The oldest

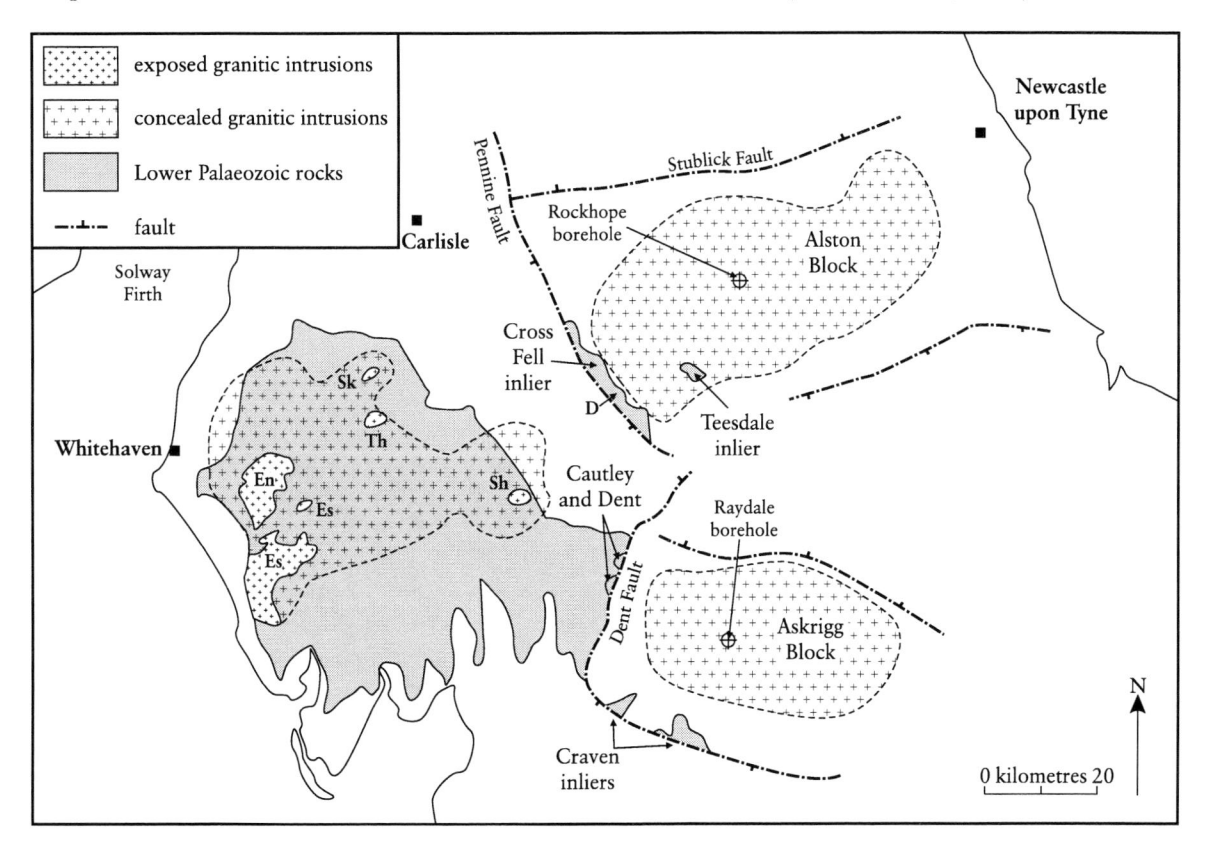

Figure 4.1 Lower Palaeozoic inliers of northern England and locations of the major, buried batholiths. Exposed granitic intrusions: En, Ennerdale; Es, Eskdale; Sh, Shap; Sk, Skiddaw; Th, Threlkeld; D Dufton.

Lake District and northern England

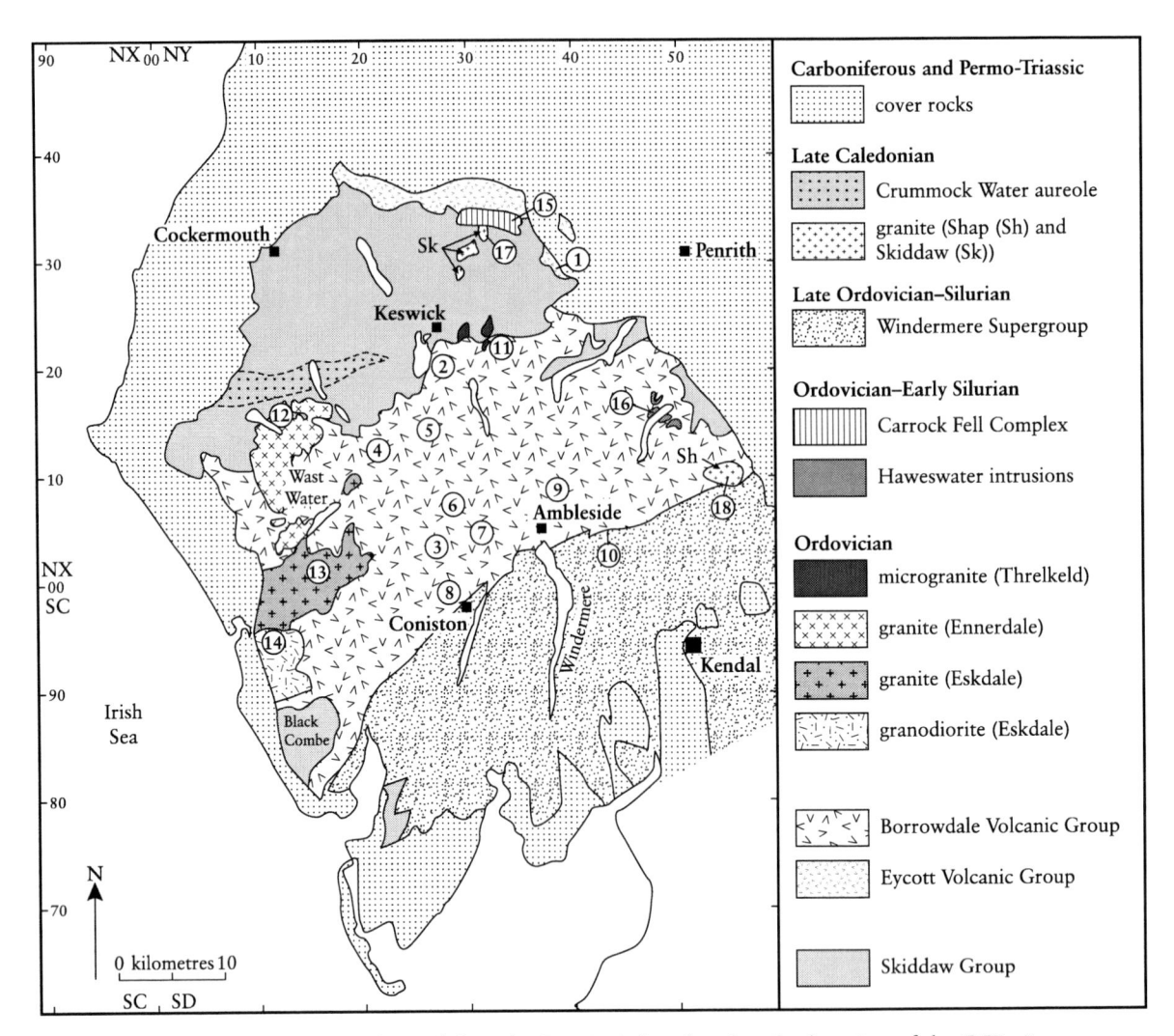

Figure 4.2 Lower Palaeozoic geology of the Lake District inlier showing the location of the GCR sites. 1, Eycott Hill; 2, Falcon Crags; 3, Ray Crag and Crinkle Crags; 4, Sour Milk Gill; 5, Rosthwaite Fell; 6, Langdale Pikes; 7, Side Pike; 8, Coniston; 9, Pets Quarry; 10, Stockdale Beck, Longsleddale; 11, Bramcrag Quarry; 12, Bowness Knott; 13, Beckfoot Quarry; 14, Waberthwaite Quarry; 15, Carrock Fell; 16, Haweswater; 17, Grainsgill; 18, Shap Fell Crags.

record of volcanism is in the western Lake District where phreatomagmatic tuff cones were eroded and redeposited. The main part of the lower BVG is dominated by andesite block lavas that built up sub-horizontal plateaux (Petterson *et al.*, 1992). The lavas emanated from many vent sites and pyroclastic deposits and volcanic mudflows locally filled valleys; unconsolidated volcaniclastic deposits were reworked by rivers and sheet floods. At the end of the effusive phase the terrain had a relative relief of no more than 110 m. Though of subtly different geochemical composition, the EVG had an evolution comparable to this lower part of the BVG.

A widespread voluminous andesitic ash-fall deposit heralded a major change to explosive volcanism that dominated the later part of the BVG episode. The succession of lavas, overlain by thick, ponded silicic ignimbrites and followed by lava domes and lacustrine sedimentary strata is typical of a caldera cycle (Branney and Soper, 1988). In the central Lake District the recently discovered Scafell Caldera contains a thick pyroclastic fill (Branney and Kokelaar, 1994a). Caldera collapse developed through incremental, piecemeal subsidence with many closely spaced faults. The waning stage of the collapse phase comprised small-volume pyroclastic flows

Introduction

Figure 4.3 The Scafell Caldera: within the caldera, thick sheets of welded silicic ignimbrite are overlain by caldera-lake sedimentary rocks. These are well displayed here in the Scafell Syncline, a structure that formed by Early Devonian tectonic compression of the Ordovician downsag caldera. In the crags on the left of the skyline the rocks dip to the right, and on the right the lacustrine rocks exposed in the pointed peak of Bowfell dip to the left. The bedded rocks in the foreground are breccias avalanched from local volcanotectonic faults during caldera collapse. (Photo: BGS no. D4031.)

and lava domes and the broad depression that marked the site of the caldera was then filled with lake sediments. The low-relief crustal sag that resulted from caldera collapse following the paroxysmal eruption of voluminous pyroclastic flows is seen today tectonically tightened into the Scafell Syncline (Figure 4.3).

In the SW Lake District repeated cycles of pyroclastic eruptions and volcaniclastic sedimentation succeed the Scafell Caldera succession in the Duddon Basin, the volcanotectonic sag that was later deformed as the Ulpha Syncline (Millward *et al.*, in press). Calderas were probably formed, though the geometry of these is less well understood than in the Scafell Caldera. Also, regional subsidence and extensional faulting may have contributed to development of the basin and marine conditions were established for a short time. After the events of the Duddon Basin succession, fluvial and lacustrine sedimentation became widespread throughout the Lake District. Contemporaneous pyroclastic activity continued, producing ash-fall tuffs and generating sediment-gravity flows that contained juvenile material. Basaltic andesite and andesite sills were emplaced into unconsolidated, wet sediments. Further extensive silicic pyroclastic eruptions periodically interrupted the dominantly sedimentary regime.

As the Eycott and Borrowdale volcanic episodes waned, marine conditions became established across the eroded and thermally subsiding volcanic pile and lasted for more than 40 million years from late Ordovician to the Early

Devonian. The earliest Ashgill rocks are shallow marine and carbonate rich. Silicic volcanic eruptions occurred in the east of the Lake District and, in the SW, tuff-turbidites were deposited. A marked increase in subsidence and sedimentation rate during the Ludlow epoch has been associated with a foreland basin migrating southward across the Lake District during the final stages in the closure of the Iapetus Ocean (Kneller, 1991). This was initiated when the northern margin of Eastern Avalonia collided with the margin of Laurentia producing a flexural basin ahead of a SE-propagating thrust sequence (Kneller *et al.*, 1993a). Final inversion of the foreland basin and development of typical 'slate belt' structures, folds, cleavage and associated faults appears to have climaxed during the Acadian Event in the Early Devonian. Cleavage formation was synchronous with emplacement of the Early Devonian Shap and Skiddaw granites and associated dykes (Soper and Kneller, 1990).

The sequence of igneous events in the Lake District and northern England is summarized in Figure 4.4 and the sites selected to represent the magmatic evolution of this area are located on Figure 4.2.

Volcanic rocks in the Skiddaw Group

The Skiddaw Group contains few volcanic rocks. The middle part of the Llanvirn, Tarn Moor Formation comprises mudstone with up to 5% volcaniclastic beds, including bentonite and tuffaceous, turbiditic sandstone (Cooper *et al.*, 1995). Several igneous sheets in the succession, for many years regarded as lavas and considered to represent the earliest record of volcanism in the Lake District, were recently shown to be sills and probably related to the later, more substantial, subduction-related volcanic episodes described below (Hughes and Kokelaar, 1993).

Eycott Volcanic Group (EVG)

In the northern part of the Lower Palaeozoic inlier the Skiddaw Group is overlain unconformably by the EVG, comprising at least 3200 m of subaerial medium-K, continental margin, tholeiitic volcanic rocks (Cooper *et al.*, 1993). The succession comprises basaltic andesite, andesite and dacite lavas and sills along with interbedded tuff, lapilli-tuff and volcaniclastic sedimentary rocks. Massive intermediate and acidic lithic-rich and vitric lapilli-tuff comprise the uppermost 800 m of the group. The type section for the EVG is described in the Eycott Hill GCR site report, which also contains examples of the distinctive plagioclase-megaphyric basaltic andesite. Volcaniclastic sedimentary rocks at the base of the EVG contain a marine microflora indicating that the volcanic rocks are not older than late Llanvirn and possibly of Llandeilo–Caradoc age (Millward and Molyneux, 1992).

Borrowdale Volcanic Group (BVG)

The BVG forms the high fells of the central Lake District and occurs in the Cross Fell and Teesdale inliers (Figure 4.1). The succession comprises about 8 km of subaerial, calc-alkaline continental-margin lavas, sills, pyroclastic and volcaniclastic rocks associated with caldera development (Millward *et al.*, 1978; Branney and Soper, 1988). The rocks are unusual among volcanic suites worldwide because of the presence of almandine–pyrope garnet phenocrysts that occur in andesite, dacite and rhyolite (Fitton, 1972). The BVG is generally considered to be Llandeilo or early Caradoc (Wadge, 1978). The Holehouse Gill Formation, within the upper part of the BVG, is probably Caradoc (Harnagian–Soudleyan) on the basis of its marine microflora (Molyneux, 1988). The radiometric date of 457 ± 4 Ma for the BVG (Sm-Nd on garnet–whole-rock pairs; Thirlwall and Fitton, 1983) is compatible with the Caradoc biostratigraphical age.

Subaerial volcanic successions, such as the BVG, are rare in the geological record. Volcanic landforms, particularly those constructed with abundant pyroclastic deposits, are not generally preserved because of explosive disintegration, weathering and erosion; most are represented by volcaniclastic and tuffaceous sedimentary rocks within subaqueous sequences. However, the extensional tectonic regime and successive episodes of caldera collapse ensured preservation of the BVG. Many primary volcanic features are exquisitely preserved on the weather-worn exposures in the fells. The combination, therefore, of volcanic processes, tectonism, uplift and erosion during the last 450 Ma provides a unique insight into the deeper levels of, and processes that occur within, this type of volcanic province.

Introduction

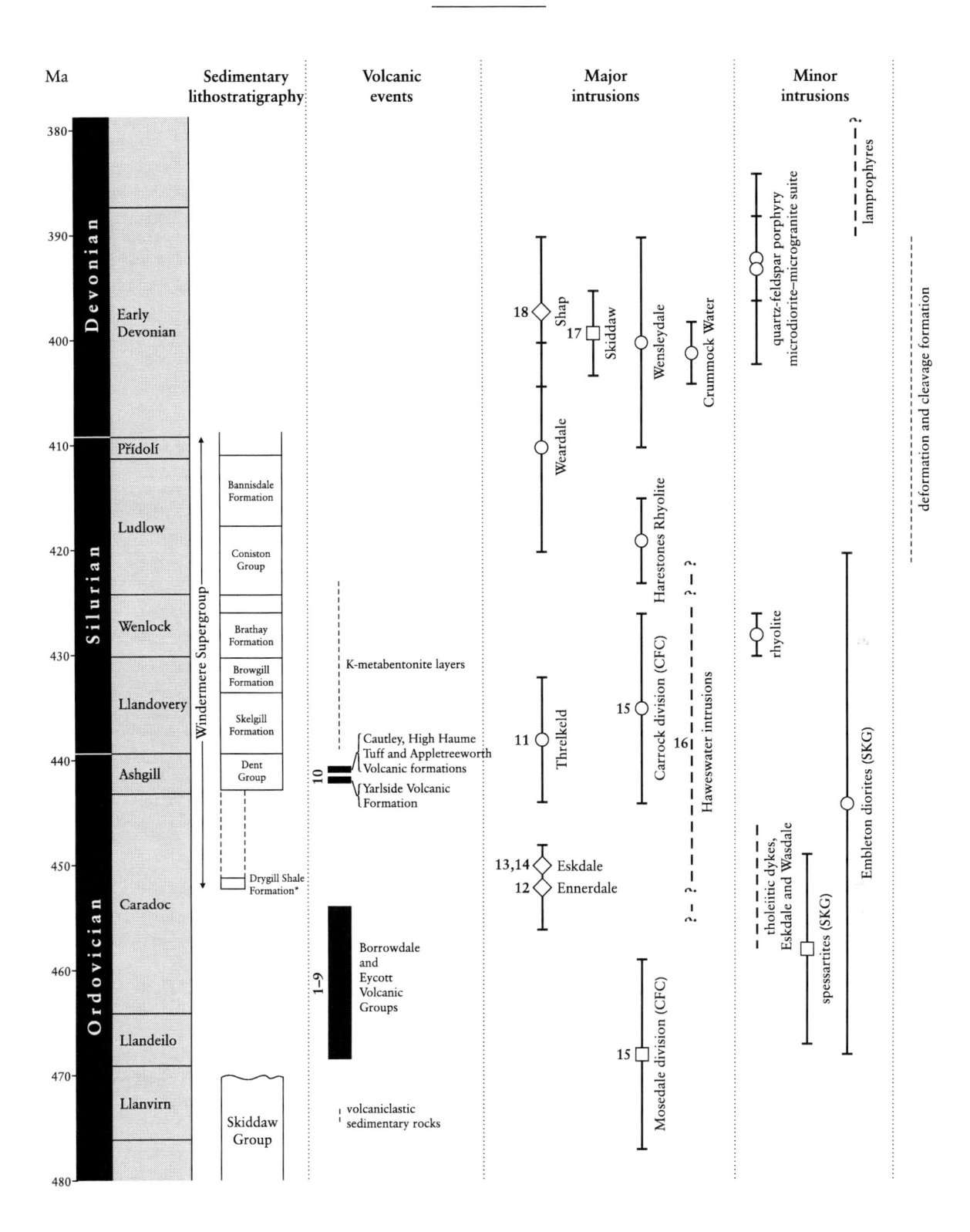

Figure 4.4 Summary chart of Lake District Caledonian igneous rocks. Bold numbers are GCR sites (numbers as per Figure 4.2). Radiometric dates: diamonds U-Pb; circles Rb-Sr; squares K-Ar; see text for sources. 1σ error bars are shown. Period of deformation and cleavage formation from Merriman *et al.* (1995). CFC Carrock Fell Complex; SKG Skiddaw Group.

*Note: In the northern Lake District the base of the Windermere Supergroup is at the base of the Drygill Shale Formation; in the central/southern Lake District the base of the supergroup coincides with the base of the Dent Group (Kneller *et al.*, 1994).

Lake District and northern England

Up to 2.8 km of andesite block lavas with subordinate basalt, dacitic and rhyolitic lavas and sills comprise the lower part of the BVG (Petterson *et al.*, 1992). Andesitic, dacitic and rhyolitic ash-fall tuff, pyroclastic surge deposits and ignimbrites constitute small parts of the succession and provide important stratigraphical markers. Interbedded sedimentary rocks mostly comprise tephra which has been reworked, redeposited and lithified to form volcaniclastic siltstone, sandstone, conglomerate and breccia. The more resistant andesites typically form traplike topography locally, and an excellent example is well illustrated in the Falcon Crag GCR site. Extensive tabular lavas emanating from many eruption centres are characteristic of this plateau-andesite province. Fine examples of plateau-basalt sequences occur in the British Tertiary Igneous Province, but the BVG is the best-preserved example of a calc-alkaline, Palaeozoic plateau-andesite province. More recent examples are known from the USA, Mexico, South America and New Zealand, but these are largely concealed by intracaldera successions. The superb exposure of the lower BVG provides a three-dimensional picture of this type of volcanic province.

The upper part of the BVG developed through intermediate and silicic explosive volcanism and caldera formation (Branney and Soper, 1988), the products of which, together with extensive intercalated volcaniclastic sedimentary rocks, form a succession two to five kilometres thick. Though volcaniclastic rocks comprise most of the upper BVG, sills are very abundant locally. A number of volcaniclastic successions, probably associated with major eruption centres, are recognized in the western Lake District, but correlation of these with sequences of similar lithologies in the east has yet to be made. The oldest of the successions occurs in the Scafell area in the central Lake District. In the SW this is overlain by the pyroclastic and volcaniclastic sedimentary succession in the Duddon Basin which is up to 3 km thick. The youngest part of the BVG overlies both the Scafell Caldera and Duddon Basin successions, and predominantly comprises volcaniclastic sedimentary rocks with intercalated thick ignimbrite sheets (Millward *et al.*, in press). A significant and dominantly andesitic welded ignimbrite succession is largely concealed beneath Permo-Triassic cover rocks in west Cumbria (Millward *et al.*, 1994).

The best known of the pyroclastic successions is that in the central Lake District fells. Pyroclastic rocks there were described as 'streaky' by Walker (1904) and Green (1915a). Marr (1916) included these rocks within his Sty Head Garnetiferous Group, but was uncertain whether many were extrusive or intrusive. Walker (1904) postulated an extrusive origin for the 'streaky' rocks of Rosthwaite Fell in Borrowdale (see the Rosthwaite Fell GCR site report). The occurrences in the Crinkle Crags and Langdale area were interpreted as intrusive rhyolites by Hartley (1932) (see the Ray Crag and Crinkle Crags GCR site report). One of the most significant contributions to the understanding of the BVG was the interpretation of the 'streaky' rocks as welded tuff by Oliver (1954). He postulated that these rocks, which constituted his Airy's Bridge Group and Lincomb Tarns Formation (Oliver, 1961), were formed from nuée ardentes or incandescent ash-flows (now known as pyroclastic flows) and compared them with welded tuff or ignimbrite from his native New Zealand. At the time these Lower Palaeozoic welded tuffs from the Lake District, along with similar rocks from Wales, were the oldest known examples of this rock type.

The complex sequence of garnet-bearing andesitic to rhyolitic pyroclastic rocks in the central Lake District was recently interpreted to be part of the Scafell Caldera, a unique example of an exhumed hydrovolcanic caldera (Branney and Kokelaar, 1994a). This structure is of international interest because the level of erosion provides a superb insight into its well-layered caldera-fill succession and the mechanisms of piecemeal caldera collapse. It is also important because of the repeated involvement of water during eruption, causing abrupt alternations between magmatic and hydromagmatic explosions. In contrast to most of the massive caldera-fill successions that have been studied elsewhere, the dramatic lateral variations in the intracaldera strata of the Scafell sequence show how different parts of the caldera subsided at different times and at different rates. The piecemeal nature of the collapse contrasts with popular models of simple, piston-like collapse (Lipman, 1984; c.f. Ben Nevis and Allt a' Mhuilinn and five Glencoe GCR sites, this volume).

The piecemeal caldera model with its layered silicic caldera-fill pyroclastic rocks, the progressive deformation of the welded tuffs and the formation of volcanotectonic faults is developed in

Introduction

the Ray Crag and Crinkle Crags GCR site. The other sites within the central Lake District illustrate significant variations in different parts of the caldera. A proximal volcanic lake environment for the initial phase of the volcanism occurs in the Sour Milk Gill GCR site. The eruption of silicic lava during the waning stage of this eruptive cycle is a major topic of the Rosthwaite Fell GCR site, and the Langdale Pikes GCR site superbly illustrates the post-collapse caldera lake succession. Though the Side Pike GCR site illustrates the volcanic megabreccia associated with the Scafell Caldera, it also provides probably the best locality in Britain where the characteristics of rocks with pyroclastic fall, flow and surge origin may be contrasted.

An example of the alternation of pyroclastic and volcaniclastic sedimentary rocks from the Duddon Basin is given by the Coniston GCR site. In the past, andesite in the BVG has been assumed to be largely lava, but the Pets Quarry GCR site illustrates that many may be high-level sills.

Volcanic rocks in the Windermere Supergroup

Two episodes of acid volcanism are preserved in the Dent Group (Figure 4.4). The first is the major silicic, early Cautleyan, Yarlside Volcanic Formation, north of Kendal (Stockdale Beck, Longsleddale GCR site). The second includes thin, widespread volcaniclastic successions of late Rawtheyan age, represented by the High Haume Tuff and Appletreeworth Volcanic formations in the south and SW of the Lake District respectively, and the Cautley Volcanic Member in the Cautley and Dent inliers. Contemporaneous volcanism from unknown sources is recorded by thin K-metabentonite beds in the Skelgill, Browgill and Brathay formations of early Llandovery to Wenlock age in the southern Lake District and in the Cautley and Dent inliers (Fortey *et al.*, 1996). Though turbidite sedimentation hindered preservation of ash-fall layers in the upper part of the Windermere Supergroup, early Ludlow K-metabentonites are recorded from the Ribblesdale (Craven) inlier (Figure 4.1; Romano and Spears, 1991).

Threlkeld intrusions

East of Keswick, microgranite outcrops on Low Rigg, Threlkeld Knotts and Bramcrag are sufficiently similar to suggest that they are part of a single, irregular body intruded into the Skiddaw Group and the base of the Borrowdale Volcanic Group (BVG). Geophysical modelling shows that it is a laccolith (Lee, 1989). The calc-alkaline intrusion is geochemically similar to acidic BVG rocks and is distinct from the other Lake District granites (O'Brien *et al.*, 1985). The Rb-Sr isochron age of 438 ± 6 Ma (Rundle, 1981) is Ordovician. The microgranite and its relationships with the Skiddaw Group and BVG are only exposed in the Bramcrag Quarry GCR site.

Lake District batholith

An elongate multicomponent batholith occurs at shallow depths beneath the central Lake District (Bott, 1974; Lee, 1986). This was probably emplaced as a set of stacked laccoliths (Evans *et al.*, 1993). The batholith is exposed in the west as the Eskdale and Ennerdale intrusions. Geological relationships (Branney and Soper, 1988) and their Caradoc age (452–450 Ma; U-Pb on zircon; Hughes *et al.*, 1996) suggest that these were subvolcanic, though they are geochemically distinct from the volcanic suites (O'Brien *et al.*, 1985).

The Ennerdale intrusion is well illustrated by the Bowness Knott GCR site, and lies to the west and NW of Wast Water (Figure 4.2). The intrusion is mostly in contact with the Skiddaw Group and BVG, but at the southern end of Wast Water it is possibly faulted against the Eskdale granite. It is a relatively thin tabular body, less than 2 km thick (Lee, 1989; Evans *et al.*, 1993). Granophyric-textured porphyritic granite is dominant, but dolerite, dioritic, and hybridized dioritic, granodioritic and melanocratic granitic rocks occur locally, adjacent to the margin of the intrusion.

The Eskdale intrusion comprises the Eskdale granite and the Eskdale granodiorite. The southern part of the granite is at the base, or within the lowest part, of the BVG, but northwards the contact rises to within a few hundred metres of the base of the Scafell Caldera succession. The granodiorite is a discordant intrusion that cuts through from the base of the BVG to the Duddon Basin succession. The Eskdale granite consists of medium-grained muscovite granite, aphyric and megacrystic microgranite, and coarse- to very coarse-grained granite.

Microgranite is most common in the northern part where the low-dipping contacts and inliers of hornfelsed volcanic rocks suggest that the roof zone of the intrusion is exposed. Xenoliths of country rock are extremely rare in the Eskdale granite and a thin zone of microgranite occurs at the margin. The main features of the intrusion are illustrated by the Beckfoot Quarry GCR site. Metasomatic recrystallization of the granite to quartz-white mica greisen occurs locally within all but the coarse facies of the Eskdale granite (Young *et al.*, 1988). Topaz is abundant in some greisens with accessory fluorite. A distinctive quartz-andalusite rock is associated with topaz greisen adjacent to the contact with the Skiddaw Group near Devoke Water (SD 1529 9733). The Eskdale granodiorite is typically medium grained, with hornblende and locally abundant biotite; it lacks muscovite. A marginal microgranodiorite is developed along the contact. Almandine garnet is locally present, particularly in rock exposed in the Waberthwaite Quarry GCR site.

Carrock Fell Complex

The multiple intrusions of the dyke-like Carrock Fell Complex were emplaced at the boundary between the Skiddaw Group and EVG in the north of the Lake District (Figure 4.4). It is the largest mafic intrusion in the Lake District; the characteristics of this layered intrusion are superbly illustrated in the Carrock Fell GCR site, but also feature in the Grainsgill GCR site. The layered cumulate gabbros of the Mosedale division are genetically related to the EVG (Hunter, 1980). Microgabbro, apatite-bearing ferrodioritic rocks and granophyric microgranite forming the Carrock division are related by fractional crystallization of a tholeiitic parent magma and were emplaced as a later dyke complex (Hunter, 1980). In the NW, later silicic bodies include the Harestones Rhyolite and lenticular micrographic microgranite intrusions emplaced along the Roughton Gill Fault. Radiometric dates of 468 ± 9 Ma (K-Ar, whole rock) and 435 ± 9 Ma (Rb-Sr, whole rock) for the Mosedale and Carrock divisions respectively probably indicate an Ordovician emplacement age (Rundle, 1979). The Silurian isochron for the Harestones Rhyolite (419 ± 4 Ma; Rb-Sr, whole rock) may be reset.

Haweswater basic intrusions

Gabbro, dolerite and dioritic rocks, cut by aplitic veins, and forming small outcrops totalling 2.6 km² are spread out over 19 km² on both sides of Haweswater (Figure 4.2; Haweswater GCR site). The intrusions were emplaced into the BVG. No radiometric date has been obtained, but cleavage in the rocks indicates an Early Palaeozoic emplacement age.

Late Caledonian intrusions

In the northern part of the Lake District the roof zone of the broadly cylindrical Skiddaw granite is exposed in the River Caldew, Grainsgill Beck and Sinen Gill (Lee, 1986). It is a coarse-grained biotite granite. At the northern end of the Grainsgill outcrop the granite passes into greisen, associated with mineralization that cuts both the granite and the adjacent Carrock Fell Complex. This significant feature is described in the Grainsgill GCR site report. The granite was emplaced into the Skiddaw Group and is surrounded by a classic, concentrically zoned aureole that grades outwards from cordierite-andalusite hornfels through cordierite-chiastolite and cordierite hornfels zones to spotted slates.

The sub-cylindrically shaped Shap granite cuts the BVG and adjacent Windermere Supergroup in the eastern Lake District (Figure 4.2). An extensive subcrop is present to the north of the outcrop (Lee, 1986). The pink and grey granite contains orthoclase megacrysts that are a distinctive and important petrogenetic feature (Shap Fell Crags GCR site). The intrusion is late Caledonian (397 ± 7 Ma; K-Ar; Rundle, 1992).

To the north of the Ennerdale intrusion an ENE-trending elongate zone of bleached and recrystallized Skiddaw Group rocks defines the Crummock Water aureole (Cooper *et al.*, 1988; Figures 4.2, 4.4). The metasomatic event has been dated at *c.* 400 Ma. The aureole is believed to be associated with a buried, highly evolved granite intruded along the northern margin of the Lake District batholith.

In the northern Pennines, the Lower Palaeozoic basement of the Alston Block is underpinned by the buried Weardale granite (Figure 4.1). The existence of this batholith was detected by geophysical surveys and confirmed by the Rookhope borehole (Dunham *et al.*, 1965; Bott, 1967). It is an aphyric two-mica granite, with pegmatitic and aplitic facies, and a sub-horizontal foliation. It is peraluminous and geochemically similar to the Skiddaw granite, a correlation that is supported by the Rb-Sr whole-rock isochron age of 410 ± 10 Ma (Holland and Lambert, 1970). The batholith is 60 × 25 km in extent, with cupolas such as the pink Dufton microgranite of the Cross Fell inlier (Hudson, 1937), rising from it. A further negative gravity anomaly over the Askrigg Block and the Raydale borehole confirmed the presence of the Wensleydale intrusion comprising pink, medium-grained granite (Figure 4.1; 400 ± 10 Ma; Rb-Sr, whole rock; Dunham and Wilson, 1985).

Minor intrusions

Suites of minor intrusions of diverse composition cut the Lower Palaeozoic rocks. Many of these are spatially and genetically linked with the major intrusions, but others illustrate a significantly greater range of magmatic compositions than is represented by the larger bodies. High-level intrusions associated with the Eycott and Borrowdale volcanic groups are intimately part of those groups. A brief description of the other suites is given below but no GCR sites specifically relate to the minor intrusions. Minor igneous intrusive bodies that crop out within GCR sites are described where appropriate.

Throughout the Skiddaw Group there are dykes, sheets and small plutons of porphyritic hornblende diorite, including augite spessartite, microdiorite, dolerite and olivine-augite hornblendite. These are calc-alkaline and are geochemically similar to the BVG; the K-Ar hornblende age of 458 ± 9 Ma supports the connection (Rundle, 1979). East of Cockermouth a suite of sills and stock-like intrusions of aphyric basalt, andesite and microgranodiorite is compositionally similar to the EVG and the Mosedale division of the Carrock Fell Complex. The Embleton Diorite from this suite has been dated at 444 ± 24 Ma by Rb-Sr (Rundle, 1979).

Aphyric basalt and dolerite dykes are abundant within the BVG around the Eskdale granite in Eskdale and Wasdale. The suite comprises high-Fe-Ti tholeiitic, and calc-alkaline, groups (Macdonald *et al.*, 1988). The latter may be associated with basalt lavas in the BVG, but the other is unique. No radiometric ages have been determined, but they are believed to span emplacement of the Eskdale granite.

Aphyric and sparsely microporphyritic rhyolite dykes are locally abundant within the BVG near the Eskdale and Ennerdale intrusions. The rocks are fine grained to cryptocrystalline and spherulitic. The dykes are genetically linked with the Ennerdale intrusion and the Rb-Sr isochron ages of 436–428 Ma are probably reset (Al Jawadi, 1987; Rundle, 1992).

Distinctive quartz-feldspar porphyry dykes occur sporadically throughout the Lake District. A geochemically variable suite of microdiorite–microgranite porphyry minor intrusions crop out in the Scafell area, in the Duddon valley and in the Windermere Supergroup. These may be linked with the Skiddaw–Shap intrusive episode as indicated by their geochemistry and by similar Rb-Sr isochron ages (Al Jawadi, 1987; Rundle, 1992).

Sparse, uncleaved, lamprophyre dykes cut all lithostratigraphical units within the Lake District and Cross Fell inliers. They are probably Early Devonian (Macdonald *et al.*, 1985). The Sale Fell minette (Eastwood *et al.*, 1968) is part of this suite.

EYCOTT HILL
(NY 382 283–397 305)

D. Millward

Introduction

Located in the NE of the Lake District Lower Palaeozoic inlier, the Eycott Hill GCR site demonstrates a complete succession of lavas and interbedded volcaniclastic rocks of the Eycott Volcanic Group (EVG) (Figure 4.5) and may be regarded as the type section. The volcanic rocks are well exposed in typical trap topography of successive scarp and dip slopes. Ward (1876, 1877) first described the geology of Eycott Hill and included the sequence on a geological cross section. A detailed geological map and description of the succession is included in the Geological Survey memoir for the Cockermouth district (Eastwood *et al.*, 1968, pp. 70–72). The area is included in Geological Survey sheets 23 (1997) and 29 (1999).

For many years the volcanic rocks in the northern part of the Lake District were considered to be an outlier of, and were included within, the Borrowdale Volcanic Group (BVG). It was not until the geochemical work of Fitton (1971) that the northern volcanic rocks, including those of Eycott Hill, were shown to be dis-

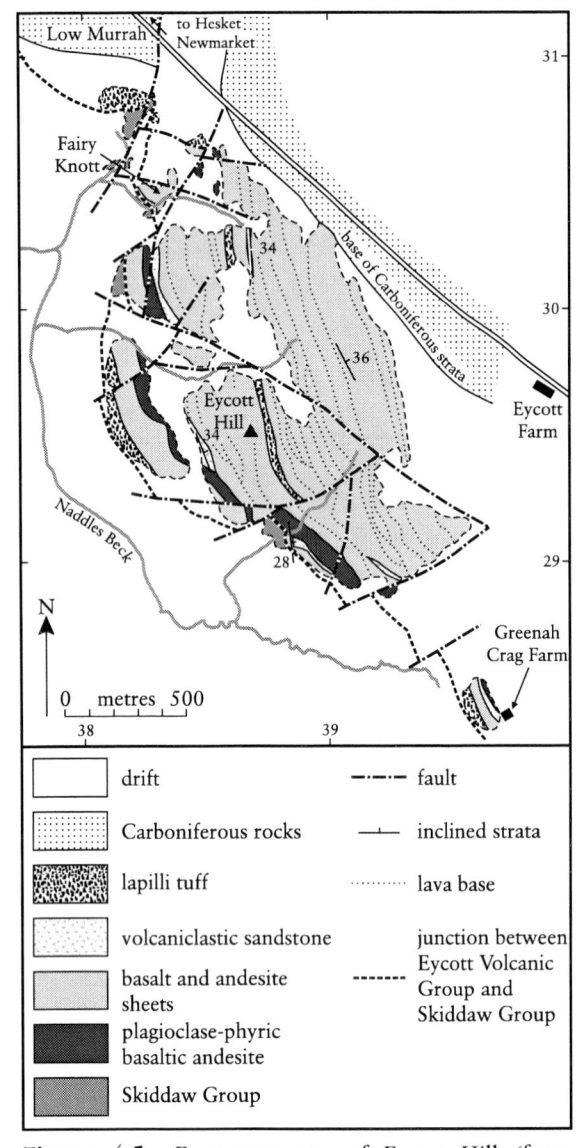

Figure 4.5 Exposure map of Eycott Hill (from Millward and Molyneux, 1992).

tinct from the larger outcrop in the central Lake District. A short while later the northern volcanic rocks were interpreted to be earliest Llanvirn in age and to overlie the Skiddaw Group conformably; they were then defined formally as the EVG (Downie and Soper, 1972). However, the age and basal relationship of the EVG have been challenged recently by Millward and Molyneux (1992) who mapped an unconformity at the base of the group and suggested that the Eycott and Borrowdale volcanic groups may have been contemporaneous. The site provides evidence crucial to the current understanding of the base of the EVG.

The site is also of historical interest because it is the type area for the distinctive orthopyroxene-plagioclase-megaphyric basaltic andesite, given the local name of 'Eycott-type' basaltic andesite by Eastwood *et al.* (1968) (Figure 4.6). Ward (1875, 1876, 1877) first described and illustrated these coarsely porphyritic rocks that contain feldspar crystals locally more than 2 cm long, and Teall (1888, pp. 225–228), in his classic work on the petrography of British rocks, described them in detail as 'labradorite-pyroxene-porphyrite'.

Description

The EVG in the GCR site is generally well exposed between Low Murrah and Greenah Crag Farm (Figure 4.5), consisting of an 800 m-thick succession of continental margin-type tholeiitic lavas and interbedded volcaniclastic rocks dipping eastwards at 30–40°. The volcanic rocks unconformably overlie Skiddaw Group mudstone of possible Cambrian age (Millward and Molyneux, 1992); the angular truncation of the uppermost lavas by the overlying basal beds of the Carboniferous cover rocks is well seen in the north of the GCR site (Figure 4.7).

More than 20 aphyric and highly porphyritic basaltic andesite and andesite lavas, varying from 20–90 m thick, constitute most of the sequence; several lava margins are present. Two 'Eycott-type' basaltic andesite lavas, each about 25 m thick, and one rhyolite crop out near the base of the sequence. Simple lavas dominate, having massive central zones, and becoming increasingly amygdaloidal towards the top and bottom; flow-banding and a fine-scale platy jointing parallel to the base are typically present in the lower part of the lavas. Rubbly flow-breccia indicates that most are aa-lavas. The lowest lava is heterogeneous with repeated alternations of massive and amygdaloidal, clinkery material suggesting that it is compound.

The lavas are typically porphyritic, containing up to 46% phenocrysts and glomerocrysts set in a fine- to very fine-grained groundmass of stumpy plagioclase laths, intergranular clinopyroxene, opaques and interstitial chlorite or dark-brown mesostasis, presumably after glass. A small number of lavas are aphyric or nearly so. Most of the phenocrysts are labradorite euhedra and scattered glomerocrysts; in the 'Eycott-type' basaltic andesite these may be up to 5 cm, but are mostly 1–1.5 cm. The plagioclase is fresh to turbid, typically with multiple and compound zoning. Small inclusions of chlorite or brown

Eycott Hill

Figure 4.6 Pyroxene-plagioclase-megaphyric ('Eycott-type') basaltic andesite, Eycott Volcanic Group. The coin is 25 mm diameter. (Photo: BGS no. A6605.)

mesostasis are scattered throughout some crystals or in rings. Small 'rounded' anhedra of fresh clinopyroxene are present only in the lowest lavas on Eycott Hill. Chlorite pseudomorphs after orthopyroxene occur in basaltic andesite and andesite. Though Eastwood *et al.* (1968) reported pseudomorphs after olivine from elsewhere in the EVG, this mineral has not been found in rocks from Eycott Hill.

On the southern side of the inlier, less than 10 m of volcaniclastic rocks underlie the 'Eycott-type' basaltic andesite lavas, thickening to 90 m in the north. Thin interbeds, up to 5 m thick, are also present between andesites at higher lev-

Figure 4.7 Eycott Hill: craggy scarp and dip-slope topography of the Eycott Volcanic Group in the foreground, contrasted with smooth, regular scarps in the overlying Carboniferous Limestone in the background. View looking NE. (Photo: BGS no. A6616.)

els in this sequence. These sedimentary rocks are generally parallel-bedded, fine- to coarse-grained volcaniclastic sandstone, though rock fragments in the soil on the west side of Eycott Hill (3829 3013) comprise graded, very coarse-grained sandstone and pebbly layers with mudstone clasts, presumably derived from the Skiddaw Group.

The basal unit of the succession west of Eycott Hill and at Greenah Crag Farm is a weakly bedded, poorly sorted lapilli-tuff comprising closely packed fragments of intensely amygdaloidal or wispy scoria along with angular to subrounded clasts of non-amygdaloidal, variously textured basalt, andesite and rhyolite; rare fragments of gabbro are present. These beds were probably deposited from mixed scoria-fall and hydroclastic eruptions. By contrast the basal, unbedded, massive, poorly sorted, matrix supported lapilli-tuff north of Fairy Knott is probably an ignimbrite.

A pinkish weathered welded dacitic lapilli-tuff, about 600 m in outcrop length and 20 m thick is present approximately 370 m above the base of the succession. This ignimbrite contains fiamme that are generally chloritized and contain coarse quartz-feldspar spherulitic devitrification. Crystal content comprises 16% plagioclase, 2.5% pseudomorphs after mafic minerals, and less than 1% granular opaque. Angular to subrounded, non-vesicular andesite clasts comprise the lithic component (4%).

Interpretation

The division of the middle Ordovician volcanic rocks in the Lake District Lower Palaeozoic inlier into separate lithostratigraphical units during the early 1970s marked a major change in the understanding of the history of the Lake District magmatic province, and arose from separate geochemical and biostratigraphical studies. Firstly, Fitton and Hughes (1970), and Fitton (1971) demonstrated clearly that the volcanic rocks of Binsey, the Caldbeck Fells and Eycott Hill show some tholeiitic characteristics compared with the calc-alkaline rocks of the central Lake District. Secondly, examination of microfloras from siltstone at the base of the volcanic succession in the Binsey area, west of the GCR site, led Downie and Soper (1972) to infer that the volcanic rocks are of earliest Llanvirn age. Thus, the northern volcanic rocks became known as the Eycott Volcanic Group, which was recognized as a precursor to the BVG. Wadge (1978) shortened the name to Eycott Group, but also included within it pelitic rocks in the eastern Lake District that were considered to be of similar age. Neither Moseley (1984) nor Millward and Molyneux (1992) followed this chronostratigraphical approach.

The relationship of the volcanic rocks to the underlying Skiddaw Group in the Lake District has been much in contention for many years (see Wadge, 1978, for summary). Eastwood et al. (1968) and Downie and Soper (1972) described passage beds comprising interbedded tuffaceous sedimentary rocks and andesite sheets at the junction in the west of the outcrop. However, Millward and Molyneux (1992, fig. 7) demonstrated a marked angular unconformity beneath these passage beds at the Chapel House Reservoir (2582 3551); biostratigraphical support showed that the Skiddaw Group beneath the mapped unconformity ranges in age from possible Late Cambrian to early Llanvirn. The andesite sheets in the passage beds were interpreted as sills and the sedimentary rocks designated as the Overwater Formation.

Ward (1876) showed a faulted contact on Eycott Hill and Eastwood et al. (1968) mapped a conformable base. A more complex basal relationship was discussed by Millward and Molyneux (1992). The characteristics of the lowest volcanic deposits vary along strike. North of Fairy Knott dark-grey, water-laid, laminated siltstone and silty claystone (Skiddaw Group) are overlain by a coarse, heterolithic, lapilli-tuff and tuff-breccia. In the stream just to the south of Fairy Knott about 4 m of tuffaceous sandstone overlie the Skiddaw Group and this bed can be mapped southwards to lie at least 120 m above the base of the volcanic sequence, with intervening pyroxene andesite and ash-fall lapilli-tuff. These lateral facies changes indicate emplacement of the volcanic rocks onto an existing topography.

The unconformable base and a succession predominantly of simple porphyritic lavas rather than volcaniclastic rocks demonstrate that the EVG has much in common with the Birker Fell Formation, the basal formation of the BVG. The sheeted andesite lava complex of the calc-alkaline BVG has been interpreted by Petterson et al. (1992) to have formed volcanoes with very shallow sides that constructed a plateau succession, probably within a graben-like structure. The geochemically distinct EVG may thus be interpreted as the product of similar constructions in a penecontemporaneous but separate volcanic field (Millward and Molyneux, 1992).

Conclusions

The Eycott Hill GCR site is significant as the type section for the Eycott Volcanic Group and for the well-known coarsely porphyritic 'Eycott-type' basaltic andesites. Crucial evidence for the onset of volcanism in the Lake District is present. More than 20 basaltic andesite and andesite aa-lavas, along with thin intercalations of pyroclastic and sedimentary rocks are well displayed by the crag and dip-slope topography. The volcanic rocks have some tholeiitic characteristics and were erupted onto an eroded landscape of Skiddaw Group rocks at the continental margin of Eastern Avalonia during the closure of the Iapetus Ocean. Despite geochemical differences with the Borrowdale Volcanic Group, the Eycott Volcanic Group exhibits a remarkably similar style of volcanism to the lower part of the former.

FALCON CRAG
(NY 272 206–274 199)

B. Beddoe-Stephens

Introduction

Falcon Crag and Brown Knotts overlook the eastern shores of Derwent Water and provide a well-exposed, 650 m-thick succession of lavas and interbedded volcaniclastic rocks within the Birker Fell Formation, the basal part of the Borrowdale Volcanic Group (BVG). The volcanic succession dips gently to the east and forms a series of prominent terrace features, particularly on Brown Knotts, developed as trap topography (Figure 4.8). This GCR site is one of only three or four superb examples of this landform in the Lake District.

The GCR site (Figure 4.9) is an excellent and characteristic example of the largely subaerial, plateau-andesite field that is exposed extensively in the south-western, western and northern parts of the BVG outcrop and which developed as a precursor to large-scale ignimbrite volcanism and caldera collapse of the upper parts of the succession (Petterson *et al.*, 1992). Features that are typical of block- and aa-lavas are displayed by flows that are generally basaltic andesite to andesite in composition. In addition, the occurrence of some high-level, co-magmatic sills emplaced into wet volcaniclastic sediment are indicated by the presence of peperitic margins (Suthren, 1977; Branney and Suthren, 1988). The inter-relationships of primary and reworked pyroclastic rocks within the largely lava-dominated succession are also illustrated and of particular interest are volcaniclastic rocks formed by hydrovolcanic processes.

This area was first mapped and described by the Geological Survey (Ward, 1876). Remapping and re-interpretation of the rocks in terms of volcanic environments and sedimentary facies was undertaken by Suthren (1977). Extensive fieldwork in the area, and to the south by the Geological Survey (Petterson *et al.*, 1992) has shown that within the andesite-dominated lower part of the BVG the whole succession is considered best as one formation, with distinctive units defined as members.

Description

A generalized vertical section of the BVG rocks within the GCR site is shown in Figure 7.10. A major fault forms the eastern shore of Derwent Water and juxtaposes Skiddaw and Borrowdale Volcanic group rocks. However, the unconformity between the BVG and underlying Skiddaw Group is exposed nearby in Cat Gill, close to Kettlewell car park (268 194), and farther south near Troutdale Cottages (261 173). In the south of the GCR site, below Brown Knotts, the lowest unit in the succession is an aphyric andesite, very fine grained and typified by a conspicuous fluidal pilotaxitic texture in thin section. It commonly contains zones of internal brecciation, suggesting that it is probably a compound flow. In places it has a prominent laminar flow fabric. This andesite, up to 200 m thick, has been mapped extensively in Borrowdale and southwestwards to Dale Head. To the north, in the lower reaches of Cat Gill (270 210), a distinctive breccia appears to underlie the aphyric andesite and probably represents the basal bed of the BVG in this area. The breccia is reddish purple, massive and very poorly sorted with a polymict assemblage of typically highly angular volcanic clasts up to block size including pale-coloured, flow-laminated felsic rocks and darker andesitic scoria. Former pumiceous clasts and fragments are also present, as are rarer clasts of (Skiddaw Group) sedimentary lithologies.

Figure 4.8 Falcon Crag (left) and Brown Knotts showing the terraced, trap-like topography of lavas and interbedded volcaniclastic rocks. (Photo: B. Beddoe-Stephens.)

Falcon Crag

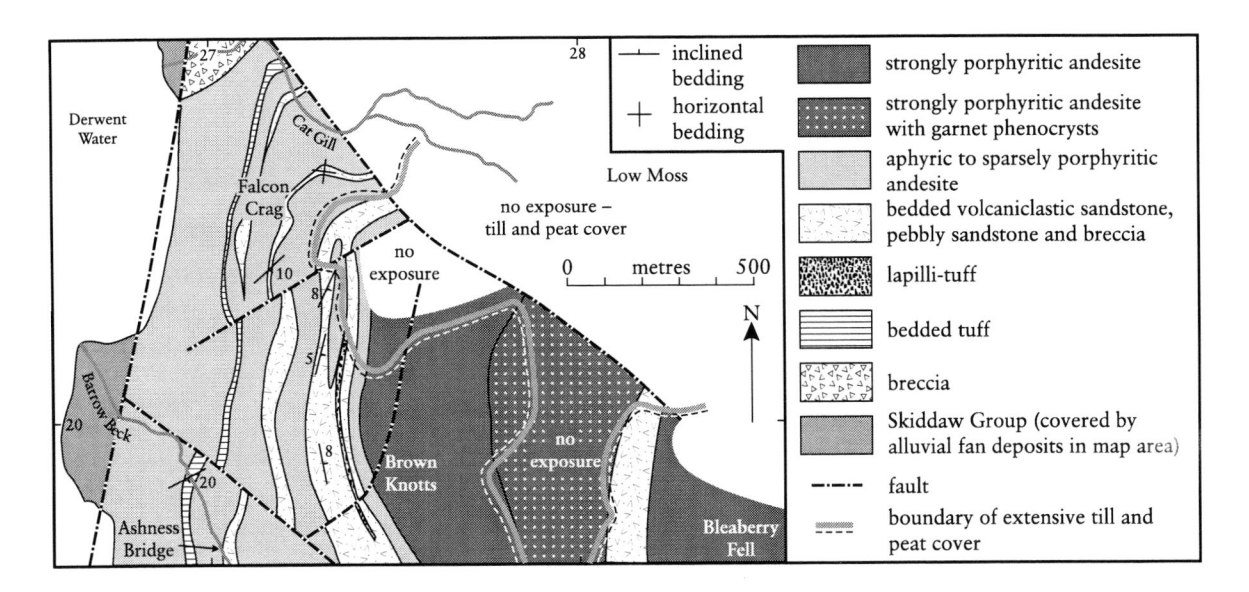

Figure 4.9 Map of the Falcon Crag GCR site.

The lower half of the succession (Figure 4.10) comprises generally sparsely porphyritic to aphyric fine-grained lavas with intercalated, bedded volcaniclastic units up to several tens of metres thick. The upper part is, by contrast, dominated by moderately to strongly porphyritic andesite. The lavas vary in thickness from 10–100 m (Figure 4.10), but in many cases have a low aspect ratio (thickness/extent ratio). This is particularly evident in the lower part of the section where a conspicuous trap-like topography forms the crags of Falcon Crag and Brown Knotts (Figure 4.8). Higher in the sequence the lavas are generally thicker and probably have higher aspect ratios.

The top, and locally the base, of the lavas are typically flow-brecciated, but the interiors are massive. Flow-breccias typically consist of angular to subangular, commonly amygdaloidal andesite blocks, and are clast supported. In the absence of volcaniclastic sandstone beds between lavas it is not always easy to establish whether a particular flow-breccia represents the top of one lava or the base of another, and in some cases a mixed breccia may form. The interstices of flow-breccias are commonly filled with finely laminated sediment washed in by percolating water. Autobreccia, in which solid blocks or crusts are re-incorporated into fluid lava, is present in the thick garnet-bearing andesite.

Within the central parts of some lavas, planar flow-laminae and jointing may be developed, usually parallel or sub-parallel to the base. In aphyric lavas at the base of the succession, flow-laminae or fine-scale banding form pervasively with superimposed platy jointing. A more discontinuous, irregular but sub-parallel flow-joint set is typically formed in the garnet-bearing andesite and the overlying andesite, which forms the summit of Bleaberry Fell.

The porphyritic andesites contain predominately plagioclase phenocrysts up to 2 or 3 mm with subordinate pyroxene (or pseudomorphs thereof). A thick lava that can be traced around the west and south sides of Bleaberry Fell, and forms the prominent knoll near the sheepfold (278 201), is distinctive for its conspicuous red garnet phenocrysts up to 5 mm across. In thin section these commonly have irregular, corroded margins. This same lava carries an accessory population of dark-coloured, irregular, rounded fine-grained diorite or dolerite xenoliths up to several centimetres across; on weathered surfaces these form recesses.

Though most of the andesites in the succession can be satisfactorily interpreted as lavas, there is evidence that some sheets were intrusive into volcaniclastic deposits. One thin, highly amygdaloidal and autobrecciated sheet terminates laterally in bedded sandstone indicating intrusion (2735 2015). Of more diagnostic value is the presence of peperitic upper margins, as described by Suthren (1977) and Branney and Suthren (1988), for the thin aphyric andesite on Brown Knotts (2744 2005).

Most of the volcaniclastic rocks are reworked,

Bedded ash-fall tuffs are preserved locally, particularly in the c. 15 m-thick volcaniclastic unit near the top of the section (Figure 4.10). Though the beds are locally channelled and disrupted, planar beds of lapilli-tuff and tuff dominate and drape irregular surfaces. The beds are both normally and reversely graded, and contain common angular and scoriaceous clasts. Interbedded cross-laminated sandstone beds appear to represent the flushing out of fines from the ash deposits.

Lapilli-tuff occurs at several horizons in the succession. A massive to weakly stratified fine lapilli-tuff is well exposed at the base of the lower main face of Falcon Crag (2706 2043). It comprises abundant, highly angular juvenile clasts and sporadic paler lithic fragments. In thin section the clasts are predominantly devitrified glass; blocky to vesicular forms are present. An unbedded lapilli-tuff is exposed for a distance of about 1 km on Brown Knotts. It is about 10 m thick, occurs towards the top of the thickest volcaniclastic unit and locally overlies a thin sill (2737 2015). This lapilli-tuff is rather fissile because of a prominent low-angle foliation produced by the alignment of abundant, dark fiamme-like fragments. It also contains common angular, andesitic to felsitic lithic lapilli in a fine-grained matrix, which includes siltstone fragments.

Interpretation

The GCR site illustrates most of the features that are critical to the recent interpretation by Petterson et al. (1992) of the Birker Fell Formation, the lowest unit of the BVG. The conspicuous trap-like topography of dominantly andesite lava interbedded with volcaniclastic rocks is well illustrated in the crags of Falcon Crag and Brown Knotts. In his analysis of the volcanic environment in which this succession was deposited, Suthren (1977) noted the lack of evidence for any significant topographic relief, such as might be expected on a steep-sided stratocone. The lavas form tabular sheets broadly concordant with the intercalated, laterally persistent, bedded volcaniclastic rocks that must have been deposited sub-horizontally. Thus, lava extrusion occurred over a subdued, flat-lying landscape. Using evidence from the western part of the BVG, Petterson et al. (1992) concluded that extensive areas of the Birker Fell Formation aggraded as a subaerial, flat-lying or

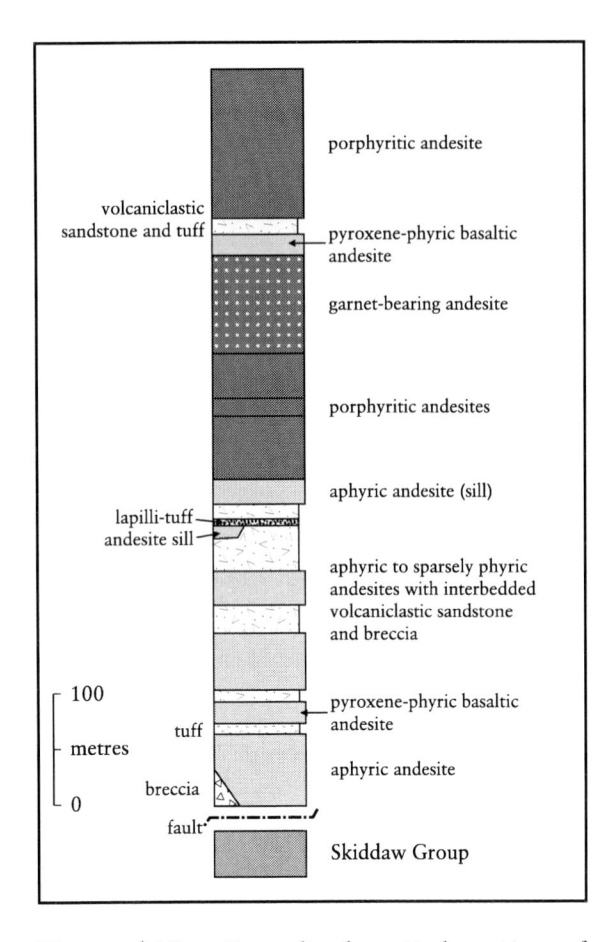

Figure 4.10 Generalized vertical section of Borrowdale Volcanic Group rocks between Brown Knotts and Bleaberry Fell; Falcon Crag GCR site.

comprising sandstone, pebbly sandstone and breccia. On the bench forming the top surface to the lower main face of Falcon Crag (271 205) are massive to weakly stratified pebbly sandstone and fine breccia with andesite clasts up to 20 cm suspended in a finer matrix. Clasts vary from subrounded to subangular. Structures typical of deposition by water are common, with abundant cross-bedding, sporadic ripple sets and rip-up clasts; erosional channels are common and are filled with coarser, gravelly lags. Silt- to sand-grade, planar bedded rocks also occur and these are affected by small-scale synsedimentary deformation locally. Examples of these features can be seen either side of the footpath from Ashness Bridge and traversing Brown Knotts (273 202). Boulder conglomerates, composed dominantly of andesitic lithologies, occur in lenticular masses or within eroded channels (e.g. at 2716 1993) and commonly contain lenses of bedded sandstone.

shield-like plateau-andesite field with lavas erupted from many centres.

The development of the thick (up to 2 km or more), subaerial, flat-lying andesite pile that comprises the Birker Fell Formation suggests accumulation within an actively subsiding basin or rift zone developed within a continental arc (Petterson *et al.*, 1992). Analogies in active orogenic arcs of this type of environment are to be found in Japan, New Zealand, the western USA and Central America.

The volcaniclastic rocks are also important in the interpretation of the volcanism and several features are illustrated in the succession of the GCR site. Suthren (1977) interpreted the local basal breccia as a mass flow deposit. The high degree of fragmentation, form and vesicularity of the juvenile fragments and presence of substrate material suggest an explosive hydromagmatic origin, though subsequent reworking of such a deposit to form debris flows is likely and cannot be excluded. A further example of rocks probably formed by hydroclastic processes is the massive to weakly stratified fine-grained lapilli-tuff, containing predominantly devitrified glass, that is well exposed at the base of the lower main face of Falcon Crag (2706 2043).

The distinctive, unbedded lapilli-tuff on Brown Knotts illustrates the problems inherent in the interpretation of ancient pyroclastic rocks. Suthren (1977) interpreted the lapilli-tuff as a welded ignimbrite, because of the eutaxitic-like fabric. Many of its characteristics are consistent with its origin as a lithic-rich ignimbrite. However, the fabric may not have been the result of welding. Alteration of pumice to clay minerals may occur rapidly after deposition, causing loss of strength within the clast (Branney and Sparks, 1990). Collapse of the vitroclasts may then occur through subsequent loading and resulting in a fabric that resembles the effects of welding. Alternatively, this rock may have had a phreatomagmatic origin.

In common with much of the BVG, there is no evidence (faunal or lithological) that volcanism was submarine (Branney, 1988a). However, the presence of hydroclastic tuffs towards the base of the sequence in this GCR site possibly indicates that periodically subaqueous conditions were present. Other volcaniclastic beds record the extensive reworking by runoff (transient fluvial systems or sheetwash) of ash-fall deposits and re-deposition as planar- and cross-bedded sandstone, or turbidite-like mass flows into ephemeral lakes. Locally, coarse debris-flow deposits and conglomerates record occasional high-energy conditions, with much of the coarse debris representing material stripped from the blocky carapaces of lavas.

Conclusions

The Falcon Crags GCR site illustrates the typical characteristics of the Birker Fell Formation, the thick, lowest part of the Borrowdale Volcanic Group. This rare calc-alkaline plateau-andesite lava field of Ordovician age is of international interest. The well exposed and readily accessible succession of lavas and bedded volcaniclastic rocks seen in the trap-like topography of Brown Knotts is an excellent example of the plateau-andesite lava field that dominated the early stage of volcanism. However, the dominantly reworked material in the volcaniclastic rocks was probably derived from the reworking of unconsolidated pyroclastic deposits, and illustrates the low-preservation potential of the latter within the geological record; pyroclastic and hydroclastic eruptions were far more common during the volcanic episode represented by the Birker Fell Formation than is recorded specifically in the succession.

RAY CRAG AND CRINKLE CRAGS (NY 241 038–NY 265 055)

M. J. Branney

Introduction

West of Great Langdale, the Ray Crag and Crinkle Crags GCR site provides exceptional exposures of the Borrowdale Volcanic Group (BVG) succession within the Scafell Caldera (Figures 4.11, 4.12). The volcanic rocks record the eruptive history of a caldera-collapse cycle that fluctuated from magmatic to phreatomagmatic as vent regions were periodically flooded during the subsidence (Branney and Kokelaar, 1994a). The phreatomagmatic eruptions were large in scale; one of the ash-fall layers records the largest magnitude phreatoplinian eruption yet documented (Branney, 1991), and evidence for its origin by subaerial fallout and for its subsequent seismically induced deformation when the caldera subsided is superbly exposed. The exhumed internal structure of the Scafell Caldera shows how it fractured into many fragments as it collapsed;

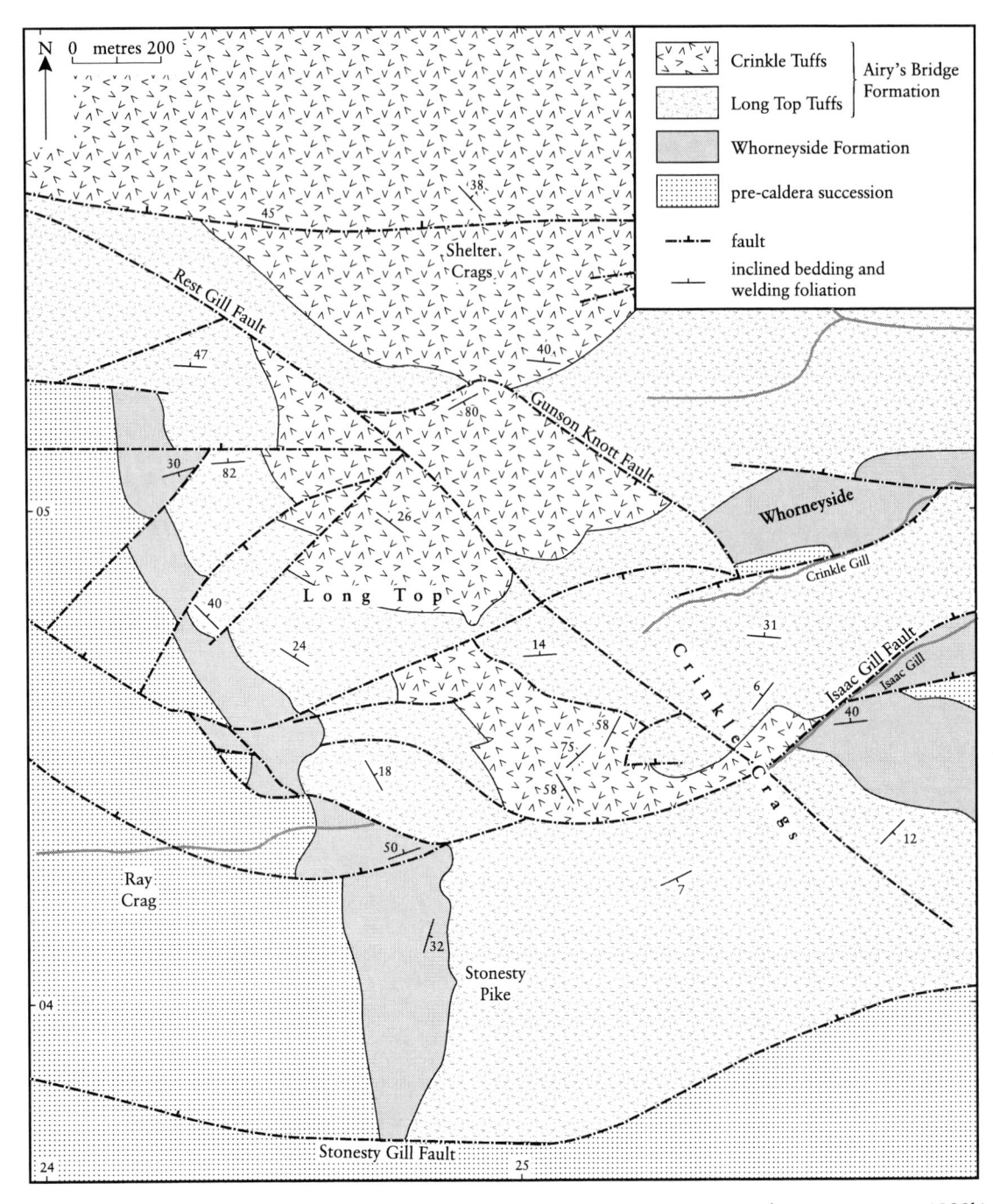

Figure 4.11 Simplified map of the Ray Crag and Crinkle Crags GCR site (mapping by M. J. Branney, 1988b).

this has become known as 'piecemeal' caldera collapse. Criteria to distinguish volcanotectonic faults and to determine movement histories are exhibited in cross section on Crinkle Crags. Many of the faults show complex movement histories, and the fault scarps they produced shed rock avalanches into the subsiding caldera. Tuff-filled fissures and a graben at Crinkle Crags record localized extension caused by downsag-ging of the caldera floor during subsidence (Branney, 1995).

The geology of Crinkle Crags was described by Green (1913), Hartley (1932), and Oliver (1961), but it was not until the late 1980s that the 15 km-diameter Scafell Caldera was recognized (Branney and Soper, 1988; Branney, 1988b; Davis, 1989). The GCR site is of international importance because it provides unique information about the architecture of large hydrovolcanic calderas, and shows how their eruptions and collapse geometries evolve during subsidence. Type localities of many of the formations and members that comprise the caldera-fill succession are within the site. Fuller descriptions of the many geological features exhibited are given in Branney (1988b, 1990a, 1991, 1995), Davis (1989), Branney *et al.* (1992) and Branney and Kokelaar (1994b, 1994c); geological maps of the Scafell Caldera are published by the British Geological Survey as 1 : 50 000 sheets 29 (1999) and 38 (1996), and in Branney and Kokelaar (1994a).

Description

The oldest rocks within the GCR site are lavas that pre-date the onset of the major explosive volcanism associated with the Scafell Caldera (Branney, 1988b, 1990a). On Ray Crag (243 042), west of Crinkle Crags, thin subaerial, amygdaloidal pahoehoe, aa, and blocky basaltic-andesite lavas are displayed, probably to the best advantage of any locality in Britain (Figure 4.11). The lavas are interstratified with thin beds of basic tuff and volcaniclastic sedimentary rock. These belong to the Lingcove Formation of Branney and Soper (1988) and Branney and Kokelaar (1994a) and Ray Crags is the type locality for the formation. Petterson *et al.* (1992) and Millward *et al.* (in press) included these rocks as part of the Throstle Garth Member of the Birker Fell Formation.

Spectacular, continuous exposures around the southern part of Crinkle Crags are the type area of the Whorneyside Formation, the product of the first major explosive phase of the Scafell Caldera; the type area is named after Whorneyside Force (2615 0535) (Branney, 1991). The formation is composed of an andesitic ignimbrite that passes up into an approximately 30 m-thick, thinly bedded andesitic ash-fall tuff with small accretionary lapilli and ash pellets. Ash-shower fallout onto a land surface is indicated by the size grading and stratification, by fallout layers that drape small steep-sided erosional rills cut by ephemeral streams, and by an absence of subaqueous facies (Branney, 1991). Impact structures occur at the base of some larger fallout lithic clasts (2488 0406), the maximum size of which increases systematically northwards, indicating a vent position to the north (see the Sour Milk Gill GCR site report for comparative description of subaqueous facies of the Whorneyside Formation). However, the average grain size remains constant and fine grained with distance from the source. This is a typical feature of phreatoplinian ashes. The tuff is extensively and spectacularly deformed, with multiple soft-state thrusts in some places (e.g. at Stonesty Pike, 2488 0402) and extensional faults elsewhere. These probably record surficial gravity sliding and lateral spreading during caldera collapse. A superbly displayed angular unconformity north of Stonesty Pike (at 2487 0420) shows white silicic tuff of the Airy's Bridge Formation draping the irregular top of the deformed andesitic tuff.

An andesite sheet south of Rest Gill (at 242 054) may record a late phase of the Whorneyside eruption. It is instructive because at outcrop, it closely resembles an autobrecciated, flow-banded block-lava buried by the overlying Airy's Bridge Formation tuffs, but mapping shows that it is a sill intruded into the phreatomagmatic tuff. In contrast to other high-level sills like the one described in the Pets Quarry GCR site report, this sill does not show textural features that are diagnostic of an intrusive origin, such as spalled hydroclasts and local steam-vesiculation of the host deposit; it would probably be taken to be a lava but for its locally discordant contacts.

The overlying Airy's Bridge Formation, the type area of which is on Crinkle Crags, is divided into two. The lower part, the Long Top Tuffs, is superbly exposed on Long Top (246 047), and comprises approximately 120 m of welded silicic ignimbrites interstratified with thin units of non-welded phreatomagmatic tuff (Figure 4.12). The silicic ignimbrites vary from bedded to massive and most are intensely welded. The lowest phreatomagmatic tuff layer is the Stonesty Tuff, a bedded, white, flinty, formerly vitric, tuff with abundant accretionary lapilli (type locality: Stonesty Pike, 2488 0403; Figure 4.11). Its thickness variation suggests that the source vent lay to the south. The next two phreatomagmatic beds, the Cam Spout Tuff (2445 0474) and the Hanging Stone Tuff (2841 0442), are both extensive, cream-coloured accretionary lapilli-bearing, cross-bedded pyroclastic surge and ash-fall layers, that record phreatomagmatic eruptions from vents to the north of Crinkle Crags.

The upper part of the Airy's Bridge Formation, the Crinkle Tuffs, comprises silicic ignimbrites that are more intensely welded than

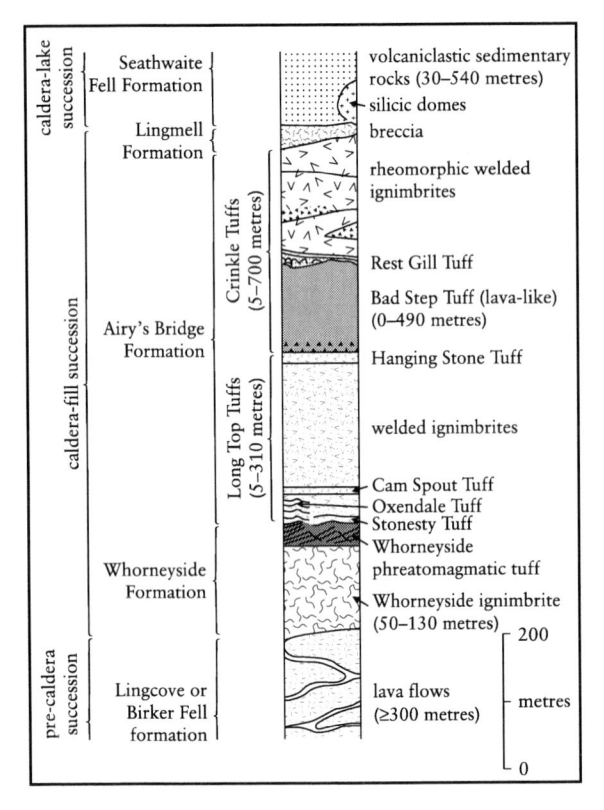

Figure 4.12 Generalized lithostratigraphy of the Scafell Caldera succession (after Branney and Kokelaar, 1994a).

the underlying Long Top Tuffs, and contain superb rheomorphic folds and lineations. The lowest of these ignimbrites is the garnetiferous Bad Step Tuff (type area: Bad Step, 2487 0481, near the summit of Crinkle Crags). This unit is of particular interest because it shows extreme welding. It is flow-banded, flow-folded, autobrecciated and lava-like, and is indistinguishable from a rhyolite block lava, except near its base where, instead of having a widespread basal autobreccia that is typical of block-lavas, it is commonly unbrecciated and has well-preserved pyroclastic textures. Around Bad Step, and on the eastern slopes of Shelter Crags (2530 0553), the basal part of the Bad Step Tuff is a poorly stratified pyroclastic breccia with a welded tuff matrix. The fiamme-bearing eutaxitic matrix has siliceous nodules and grades upward into lava-like rock with siliceous nodules, indicating that the entire unit was deposited by a single hot pyroclastic flow (Branney et al., 1992). The breccia is a rare UK example of a co-ignimbrite lag breccia (Walker, 1985), commonly used to indicate proximity to ignimbrite source vents.

The Bad Step Tuff is separated from overlying rheomorphic ignimbrites of the Crinkle Tuffs by the Rest Gill Tuff, a distinctive, laminated, turquoise fine to coarse tuff, less than 2 m thick, and thought to be the product of an extremely wet phreatomagmatic eruption. It is named after Rest Gill (2446 0558) where it is well displayed, showing evidence of contemporaneous local reworking by surface water. The overlying thick, massive, garnetiferous welded ignimbrites have eutaxitic and parataxitic welding fabrics,

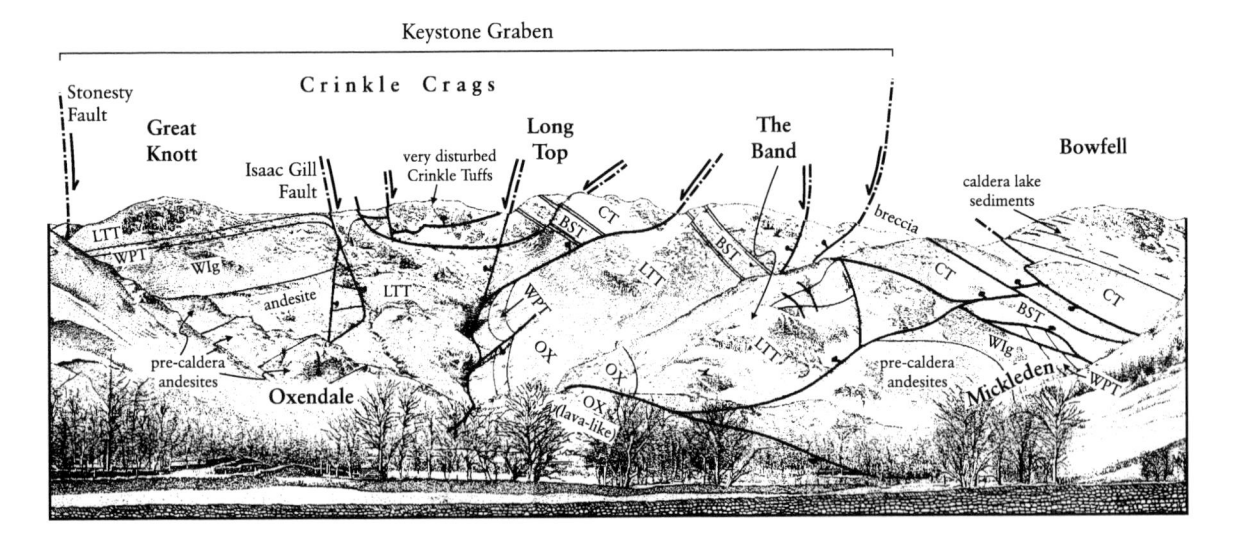

Figure 4.13 Field sketch of Oxendale and Crinkle Crags, viewed from the east, showing complex extensional faulting in an area separating caldera-floor rocks in the centre and right of the picture from sub-horizontal pre-caldera strata on the left (from Branney, 1995). BST, Bad Step Tuff; CT, Crinkle Tuffs; LTT, Long Top Tuffs; OX, Oxendale Tuff; WIg, Whorneyside ignimbrite; WPT, Whorneyside phreatomagmatic tuff.

Ray Crag and Crinkle Crags

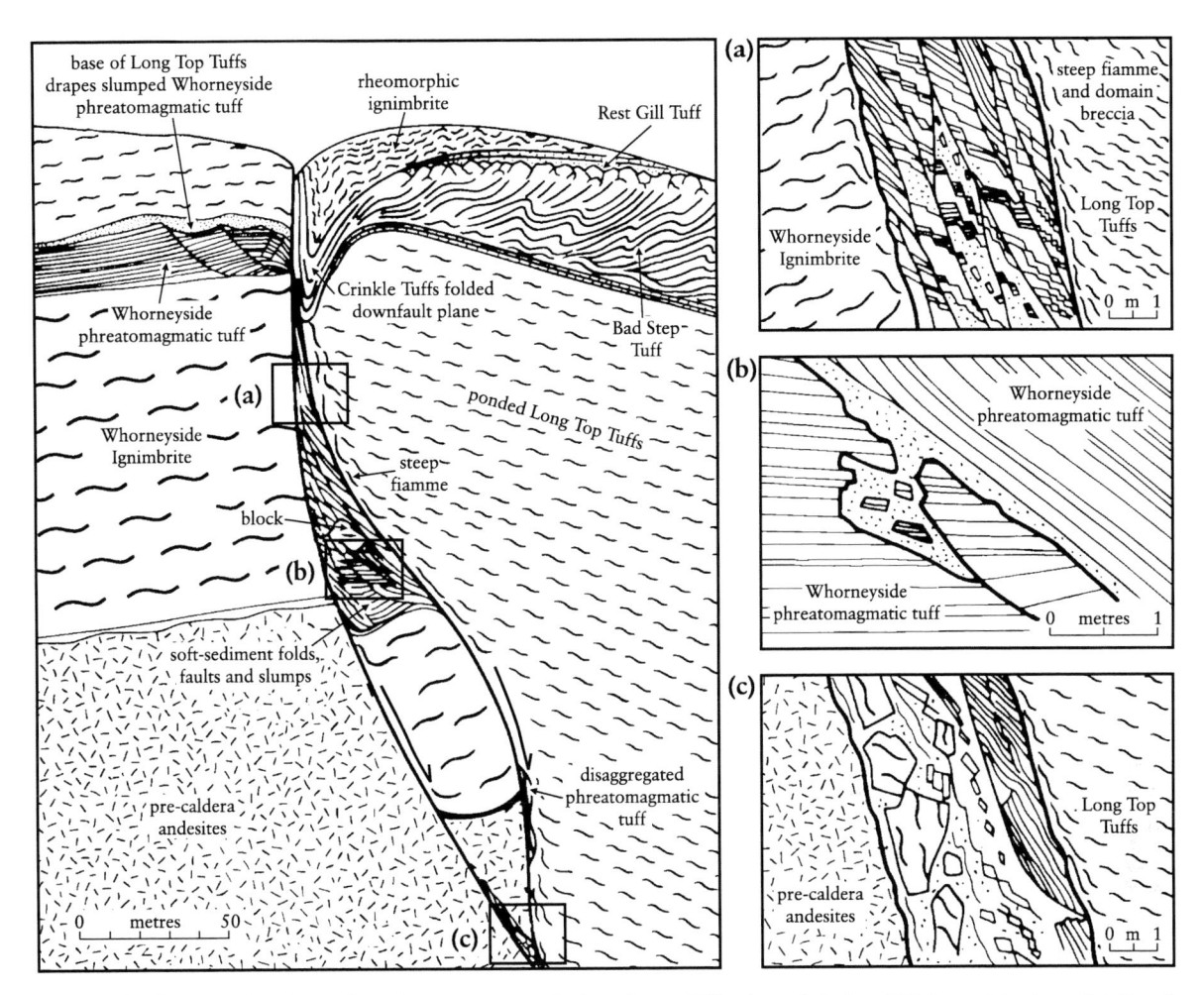

Figure 4.14 Field sketches of a volcanotectonic fault at Isaac Gill, viewed facing SW. Inset (a) gives details of disrupted bedded tuff of the Whorneyside Formation within the fault zone. Inset (b) shows homogenized tuff along one of the many slide surfaces in the bedded tuff with footwall pull-apart structures and locally derived intraclasts. Slumped bedded tuff dips steeply towards the downthrow side of the main fault. Inset (c) shows mixing of blocks of ignimbrite and homogenized tuff (from Branney and Kokelaar, 1994a).

the latter with a rheomorphic lineation; locally in high-strain zones the fabric is flow-banded. The ignimbrites contain numerous lenses of mesobreccia, formed from rock avalanches produced by unstable, growing caldera-fault scarps. Some mesobreccia has a welded tuff matrix (e.g. at 2487 0559) and fiamme have been deformed plastically around rigid blocks during compaction while the deposit was still hot.

The internal structure of the Scafell Caldera is well displayed on Crinkle Crags. Many closely spaced volcanotectonic faults are exposed in plan and in section (Figures 4.11, 4.13). For example, the Isaac Gill Fault (Figure 4.14) is a superb example of a caldera-floor fault that moved prior to the lithification of wet phreatomagmatic tuff, and before some of the ignimbrites cooled down and became brittle (Branney, 1988b). On the east side of Crinkle Crags, the Whorneyside phreatomagmatic tuff has slumped down the fault plane resulting in steeply inclined bedding, intense soft-state disruption, disaggregation, and local incorporation of angular blocks of Whorneyside ignimbrite. Welded silicic ignimbrites show evidence of hot ductile shear and folding adjacent to the volcanotectonic faults (Figure 4.14). The thickness of successive pyroclastic units changes abruptly across the faults, which allows the complex movement history of individual faults to be constrained (Branney and Kokelaar, 1994a).

Acid and basic dykes cut the succession at Crinkle Crags, for example on Stonesty Pike (2491 0385), Great Cove (2515 0470) and Rest Gill (2448 0520). They include porphyritic rhyolite and altered basalt; composite dykes have

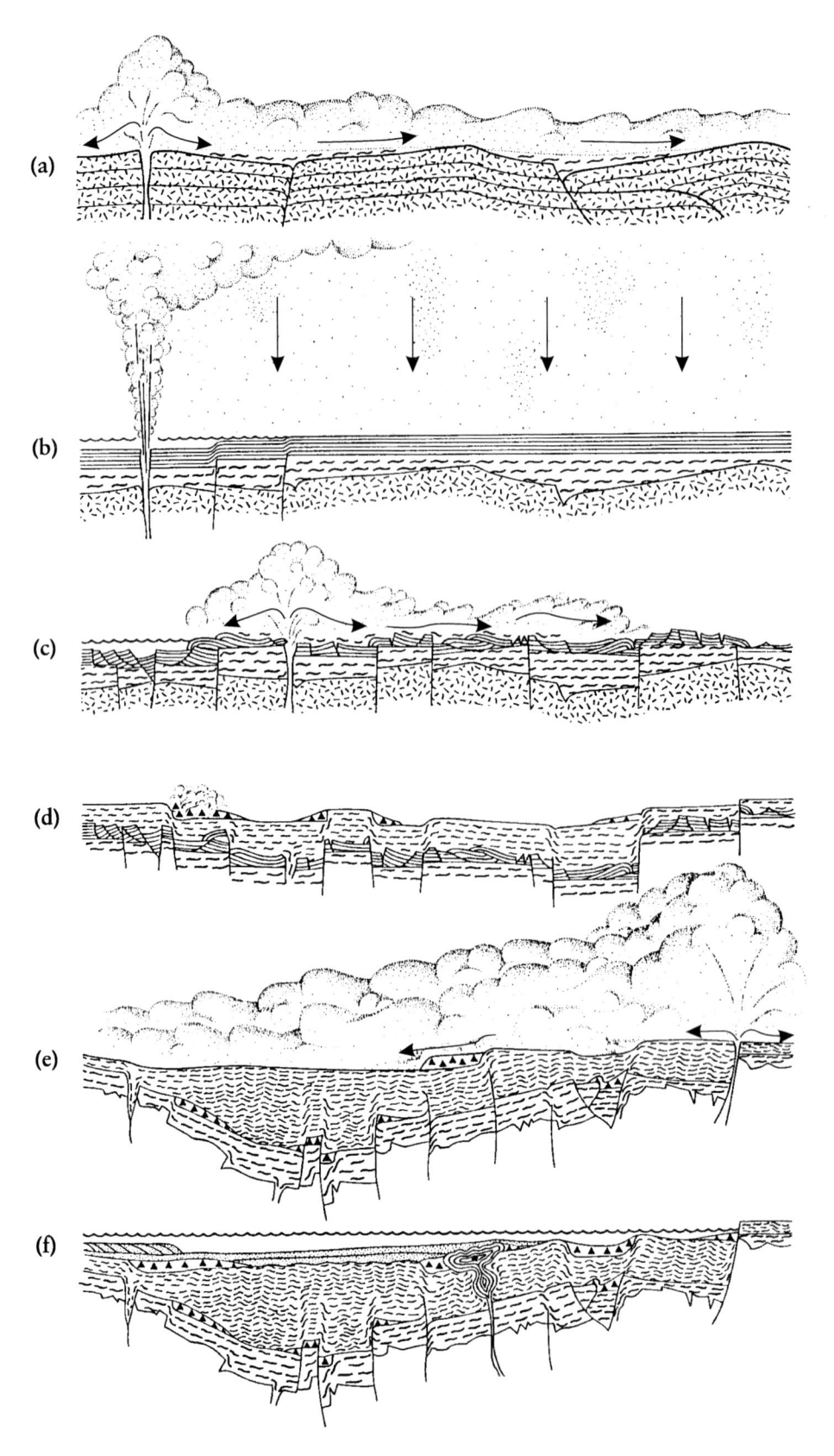

Figure 4.15 Schematic summary of the evolution of the Scafell Caldera by piecemeal collapse. The cartoons are simplified and not to scale. (a) Emplacement of the Whorneyside ignimbrite across low-profile andesite volcanoes. (b) Proximal subsidence facilitates aqueous inundation of the vent, changing the eruption style to phreatoplinian (see the Sour Milk Gill GCR site report). (c) Onset of widespread piecemeal subsidence, deformation of Whorneyside phreatomagmatic tuff deposits and burial under hot silicic ignimbrites of the Long Top Tuffs erupted from new vents. (d) Continued subsidence with ductile deformation of hot ignimbrites and collapse of growing fault scarps. (e) Final stages of the paroxysmal eruption of the Crinkle Tuffs. (f) Caldera lake formation with deposition of subaqueous volcaniclastic sediments, along with post-collapse silicic dome emplacement (see the Rosthwaite Fell GCR site report) (from Branney and Kokelaar, 1994a).

columnar jointed garnetiferous silicic centres and basic margins. Other silicic dykes contain mafic xenoliths. Epidotized, amygdaloidal basic dykes have characteristic irregular, chilled margins.

Interpretation

The inferred sequence of events that occurred during the collapse of the Scafell Caldera and seen within the Crinkle Crags GCR site is summarized in Figure 4.15 (Branney and Kokelaar, 1994a). The pre-existing plateau or shield-like andesite volcanoes, with a surface topography of less than about 110 m, were completely buried by the andesitic Whorneyside ignimbrite (50–130 m thick after welding compaction) during the first major explosive eruption of the Scafell Caldera (Figure 4.15a). The ignimbrite was sourced from the north of the Lake District, where proximal subsidence caused inundation of the vents, and a change in the style of eruption from magmatic to phreatomagmatic (Whorneyside phreatomagmatic tuff; Figure 4.15b; Branney, 1991). The rapid evacuation of large volumes of andesitic magma caused widespread subsidence, with block faulting, tilting, and seismic shock. This produced soft-state sliding and spreading of the wet, unconsolidated fallout tuff (Figure 4.15c). It also caused a change in vent position and the onset of silicic explosive ignimbrite eruptions, that rapidly buried and preserved the slumped Whorneyside phreatomagmatic tuff (Figure 4.15c). There followed a fluctuation between explosive magmatic eruptions producing hot ignimbrite and violent phreatomagmatic eruptions (Long Top Tuffs). Piecemeal caldera subsidence continued, and the hot ignimbrites deformed in a ductile manner at the numerous active volcanotectonic faults (Figure 4.15d). The fault scarps collapsed locally, generating hot rock avalanches. During the climax of the eruption, thick high-grade rheomorphic ignimbrites were emplaced and localized subsidence accelerated (Figure 4.15e). Finally, the caldera became flooded, forming a large caldera lake (Figure 4.15f). The subaqueous sedimentary rocks and tuffs that buried the caldera-fill ignimbrites, and the post-collapse silicic domes are considered in the Langdale Pikes and Rosthwaite Fell GCR site reports. The succession was subsequently intruded by dykes and pervasively altered hydrothermally, a typical feature of modern caldera fills. The area is underlain at shallow depths by the Ordovician Eskdale granite, thought by Branney (1988a, b) to be the evolved remnant of the subvolcanic magma chamber.

Structurally, the GCR site has the form of a complex 'keystone graben', formed by extension at the margin of the caldera as it subsided (Figure 4.13). During subsidence the centre of the caldera sagged, and the resultant monoclinal flexing at the caldera margin (Crinkle Crags) caused brittle extension, and the development of tuff-filled fissures (Figure 4.14) and the keystone graben. Dips of strata at Crinkle Crags are thought to record block rotations during caldera collapse, later tightened during Acadian tectonic compression (Branney and Soper, 1988). Similar keystone-type extensional structures occur around margins of young calderas (Branney, 1995), but are rarely seen in section.

Conclusions

The Ray Crag and Crinkle Crags GCR site is of international importance because uplift and glacial dissection have revealed the internal structure and pyroclastic succession of the Scafell Caldera. It is possibly the best-exposed exhumed example of a large explosive andesite–rhyolite caldera known, with an abundance of varied deposits produced by the explosive disintegration of magma by water. Most modern calderas of this type are inaccessible because they are flooded or buried by their erupted products. Therefore, an understanding of the Scafell Caldera provides important insights into the evolution and possible internal structure of modern, hazardous, flooded caldera volcanoes, like Taal (Philippines), Santorini (Greece), Kikae (Japan) and Rabaul (Indonesia).

Crinkle Crags exhibits a 'piecemeal' caldera structure resulting from the differential foundering of adjacent caldera-floor fault blocks. Such complexity was not widely envisaged for calderas prior to the 1980s. Several continuously exposed faults can be traced down through the caldera stratigraphy into the caldera floor. These faults grew in a complex way during the eruption, and they display features that readily distinguish them from other types of faults. The stratified caldera-fill succession allows reconstruction of how a caldera subsides progressively with time, and how the corresponding eruptive activity varies. Most caldera fills studied previously are non-stratified, preventing detailed reconstruction of caldera evolution during collapse.

Crinkle Crags is the type locality of the Whorneyside Formation, which records the largest-magnitude explosive eruption yet documented that was produced by large volumes of water coming into contact with magma. The GCR site is also the type area for the Airy's Bridge Formation, which includes some of the best examples in Britain of cross-bedded pyroclastic surge deposits, garnet-bearing welded ignimbrites, avalanche deposits, and lag breccia associated with ignimbrite. The very densely welded ignimbrites, including one that is so intensely welded that it resembles a flow-banded, flow-folded lava, are unusual among calc-alkaline ignimbrites. Finally, on Ray Crag are excellent examples of pahoehoe, aa and blocky basic lavas that were erupted prior to the major explosive phase of the Scafell Caldera.

SOUR MILK GILL
(NY 235 122)

M. J. Branney

Introduction

The Sour Milk Gill GCR site contains a superb record of volcanic and sedimentary processes that occurred during a major explosive eruption in a volcanic lake that was proximal to the vent (Figure 4.16). The rocks were produced by the first major explosive eruption associated with the Scafell Caldera (Figure 4.12; Branney and Kokelaar, 1994a). They belong to the Whorneyside Formation, which comprises an ignimbrite overlain by a bedded phreatomagmatic tuff (Figure 4.12). Substantially more than 100 km^3 of magma were erupted, burying the western Lake District beneath more than 30 m of fallout ash (Branney, 1991). This ash-fall layer is believed to record the largest magnitude phreatoplinian eruption yet documented (Branney, 1991). Facies associations, and the size of impacted lithic clasts in the tuff indicate that the volcanic vent lay just to the NW of Sour Milk Gill, where subsidence had created a large volcanic lake, possibly connected to the sea. Though the recent interpretation of the bedded tuff succession in the Whorneyside Formation is the main reason for the selection of this GCR site, the overlying Airy's Bridge Formation also contains important features that are additional to those seen in the Ray Crag and Crinkle Crags and Rosthwaite Fell GCR sites.

Oliver (1961) did not distinguish the succession at Sour Milk Gill from the dominantly

Figure 4.16 Sour Mill Gill from Seathwaite Farm, Borrowdale. Southward-dipping proximal lacustrine Whorneyside phreatomagmatic tuff (centre of the picture), intruded by andesite sills (right) is overlain by silicic ignimbrites of the Airy's Bridge Formation (left). (Photo: M. J. Branney.)

andesitic lower part of the Borrowdale Volcanic Group (BVG). The rocks at Sour Milk Gill were described in detail first by Suthren (1977) and Suthren and Furnes (1980), who recognized the lacustrine character. Branney (1988b, 1991) presented evidence that these rocks are the proximal facies of the Whorneyside phreatoplinian eruption. Davis (1989) interpreted the overlying Airy's Bridge Formation and a further account of the area is by Suthren and Davis (1990). The following details are derived mainly from the most comprehensive account by Kokelaar and Branney (1999). Arthropod tracks have been recorded from a loose block of bedded volcaniclastic rocks from this site (Johnson et al., 1994).

Description

Strata at Sour Milk Gill dip 50° S, towards the centre of the Scafell Caldera. On the north side of Sour Milk Gill a stack of andesite sills, more than 300 m thick, dips beneath the 160 m-thick stratified upper part of the Whorneyside Formation (Figures 4.12, 4.17). The sills were intruded into wet ash and sediment during, and shortly after, the Whorneyside eruption, and their upper contacts locally show discordant, invasive, apophyses of peperite (see the Pets Quarry GCR site report for discussion of peperite). The Whorneyside ignimbrite crops out farther NE, where it underlies the sills.

To the SE of Sour Milk Gill, for example around Seathwaite Farm, the Whorneyside phreatomagmatic tuff is subaerial. Its pyroclastic origin is indicated by abundant accretionary lapilli, ballistic lithic blocks with impact structures, and draping of topography. The parallel thin stratification records unsteady ash-shower fallout from a vast umbrella cloud (Branney, 1991). Rainfall during the eruption produced surface water that eroded minor rills into the subaerial ash. The 'V'-shaped fluvial rills draped by succeeding fallout ash layers (2422 1222) are probably the best examples seen in Britain.

At Sour Milk Gill the ash fell into shallow, standing water. Volcaniclastic lithofacies have been divided into six categories (Kokelaar and Branney, 1999), listed here in decreasing order of abundance:

1. Parallel-stratified very fine to very coarse tuff, interpreted as lithified fallout ash, exhibits rare impact structures, and can be subdivided into water-settled and subaerial varieties, based on the lamination and grading patterns, and on occurrences of loading structures and polygonal desiccation cracks.

2. Massive to laminated, fine-grained to pebbly sandstone, in places with matrix-supported intraclasts and/or dewatering structures, is thought to have been deposited by shallow-water turbidity currents.

3. Cross-laminated, scoured and rippled siltstone and sandstone, is interpreted to represent wave and current reworking of fallout ash in shallow water.

4. Lenticular beds, up to 4 m thick, of very poor-

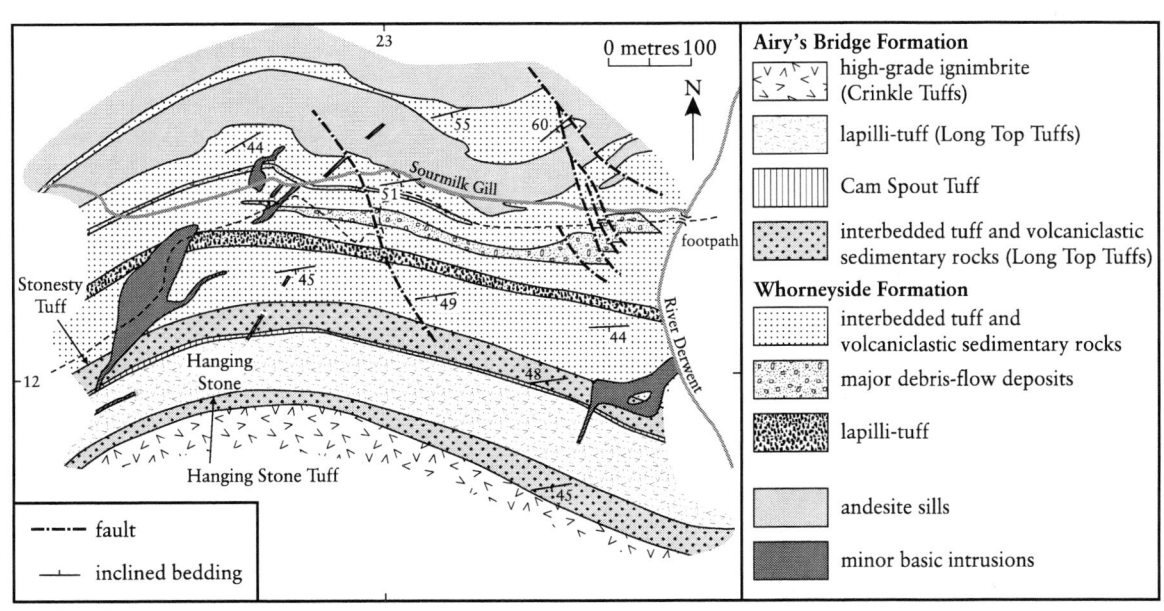

Figure 4.17 Simplified map of Sour Milk Gill (after Kokelaar, in Kokelaar and Branney, 1999).

ly sorted, block-rich sandstone and breccia are interpreted as debris-flow deposits. Most blocks are of andesite and are probably derived from proximal ballistic fallout and/or high-level intrusions that emerged onto the lake floor. One bed contains numerous blocks of subaerial tuff containing accretionary lapilli, and must have been derived from the lake shore.

5. Rare cross-stratified, medium- to very coarse-grained sandstone and gravel conglomerate is thought to have been deposited from dilute stream-flow currents that caused migration of sediment sheets and dunes.

6. Poorly sorted, parallel-stratified medium- to very coarse-grained sandstone and gravel conglomerate, with some low-angle truncations, is interpreted to represent deposition from laminar hyperconcentrated currents. A unit of massive, andesitic lapilli-tuff, 11 m thick, is considered to be a non-welded ignimbrite, and is overlain by a probable co-ignimbrite fallout tuff with accretionary lapilli.

The succession passes up into the Long Top Tuffs, which here comprise interbedded and mixed andesitic and silicic tuffs, and volcaniclastic sedimentary rocks. Some beds grade from andesitic to silicic, and contemporaneous fallout from two sources is indicated. Several of the silicic beds contain chlorite- and epidote-rich fiamme in a laminated or massive, fine-grained silicic matrix. These fiamme probably represent pumice lapilli that collapsed during diagenesis and burial (Davis, 1989; Branney and Sparks, 1990). Though this mechanism of fiamme formation is best seen in the Pets Quarry GCR site, Sour Milk Gill is also a good place to examine such features, and to compare them with fiamme formed by welding of hot tuff, as seen in silicic eutaxitic ignimbrites of the Crinkle Tuffs exposed higher in the section.

In the Airy's Bridge Formation, the Cam Spout Tuff is well exposed on the south side of Sour Milk Gill (228 121). It is a spectacular cross-bedded tuff, 6 m thick, with abundant large accretionary lapilli and lapilli-impact structures and has been traced around the Scafell Caldera to Cam Spout in Eskdale and southwards to Wrynose (Branney and Kokelaar, 1994a). The turquoise, laminated Rest Gill Tuff is the lowest recognizable unit of the Crinkle Tuffs at Sour Milk Gill, and probably represents a particularly wet, large-scale phreatomagmatic eruption. Overlying the Rest Gill Tuff is a thick silicic ignimbrite with superb columnar polygonal cooling joints (at Hanging Stone; 228 120). Thin-sections show intense welding of the shards, with perlitic cracks and pseudomorphs after spherulites. Ignimbrites of the Crinkle Tuffs reach 500 m thick in this part of the Scafell Caldera.

Graphite, which gave rise to the Keswick pencil industry, can be found on disused spoil tips 500 m north of Sour Milk Gill (231 128). The mineral occurs in irregular pipe-like bodies within altered basic intrusions into the Whorneyside Formation (Firman, 1978a). The origin of the graphite is enigmatic, and may be related to underlying Skiddaw Group rocks (see the *Mineralization of Great Britain* GCR volume).

Interpretation

Rocks within the Whorneyside Formation in the Sour Milk Gill GCR site record the evolution of a dynamic, near-vent lake environment during a major explosive eruption accompanied by rapid differential volcanic subsidence. Beautifully preserved sedimentary structures record rapid ash fallout into shallow water, with sediment reworking by wave and current action, and the deposition of block- and ash-rich slurries. The spectacular soft-state deformation, Neptunian dykes, and dewatering structures resulted from rapid burial, shaking and tilting of watery lake sediment during the eruption. The accumulating ash and sediment at Sour Milk Gill were intruded by coeval andesitic magma, forming shallow, peperitic intrusions. Blocks, ash, sediment, and peperitic sills are all broadly of the same andesitic composition and may be co-magmatic. Primary volcanic aggradation by fallout from pyroclastic flows and from reworked equivalents of these produced about 69% of the succession, whereas secondary input from a variety of ash- and breccia-laden flows contributed the remaining 31% (Kokelaar and Branney, 1999). Sedimentation was demonstrably rapid, despite the deceptive fine grain-size and intricate stratification, because the 160 m-thick succession is a time correlative of *c.* 30 m of subaerial fallout ash to the south and was deposited during a single eruption, possibly lasting a few months or years. However, despite the rapid aggradation, water depth generally seems to have kept up with aggradation, because throughout the succession very small wave ripples indicate very shallow water and periodic emergence is indicated by polygonal mud cracks. The SE margin of the subsiding proximal basin was probably controlled by an active fault through Seathwaite (Kokelaar and Branney, 1999).

Elsewhere in the Lake District, the Whorneyside phreatomagmatic tuff is widely overlain by the Airy's Bridge Formation with an angular unconformity (see the Ray Crag and Crinkle Crags GCR site report) but at Sour Milk Gill the contact is conformable and gradational. Andesitic and silicic layers are interstratified, showing that the Airy's Bridge silicic eruption started before the andesitic Whorneyside eruption had ceased. Therefore, the widespread unconformity at the base of the Airy's Bridge Formation elsewhere must record a geologically instantaneous event, namely rapid caldera collapse, associated soft-state deformation and burial. This is in contrast to a protracted period of uplift and erosion. During the Airy's Bridge eruption the centre of subsidence shifted southwards with time from Sour Milk Gill, which became persistently emergent. This is illustrated by the Cam Spout Tuff (Figures 4.12, 4.17) which represents another important phreatomagmatic phase in the formation of the Scafell Caldera. Its presence at Sour Milk Gill shows that the area had become emergent, because pyroclastic surges are gaseous and too buoyant to invade standing water. This southward migration of subsidence means that the upper, subaerial part of the succession at Sour Milk Gill is thinner than ponded equivalents to the south. For example, the Bad Step Tuff is absent at Sour Milk Gill, but is over 400 m thick in Langdale (see the Langdale Pikes GCR site report), indicating the development of highly irregular caldera-floor topography as it subsided.

Conclusions

The Sour Milk Gill GCR site is internationally important because the exposed rocks were deposited in a volcanic lake situated near the vent of an exceptionally large-magnitude eruption caused by the explosive interaction of water and andesitic magma. The volcaniclastic rocks record particularly energetic explosions, rapid unsteady (pulsatory) ash fallout near to the vent, and contemporaneous reworking. Near-vent deposits of major phreatomagmatic eruptions at modern volcanoes, such as Lake Taal in the Philippines, and Lake Taupo in New Zealand, are mostly submerged and inaccessible. Therefore, this GCR site provides a rare view of the accumulation of these deposits, subsidence near to the vent, and how such watery basins are invaded by magma during the eruption to form stacks of sills. The site is particularly instructive because the volcanological context is well known from the continuity of outcrop around the Scafell Caldera, and the quality of the exposures allows individual beds to be traced for hundreds of metres.

Within the overlying Airy's Bridge Formation the Cam Spout Tuff at Sour Milk Gill is possibly the best example of a cross-bedded pyroclastic surge deposit in Britain, and the overlying welded tuffs show superb welding textures and polygonal columnar joints. Marked differences between the Sour Milk Gill succession and those of the Rosthwaite Fell and Ray Crag and Crinkle Crags GCR sites demonstrate that caldera subsidence can occur in a complex, piecemeal manner with adjacent fault blocks subsiding at different rates and being flooded at different times. This evidence has been highly influential in the way ideas about caldera collapse have developed in recent years.

ROSTHWAITE FELL (NY 258 122)

M. J. Branney

Introduction

Almost continuous exposure on the slopes of the Rosthwaite Fell GCR site, Borrowdale, provides a further section through the caldera-fill succession in the northern part of the Scafell Caldera (see the Ray Crag and Crinkle Crags GCR site report; Figures 4.12, 4.18). However, the site has been selected principally for two important aspects. Firstly, the pre-caldera lavas and an overlying ignimbrite are overlain by a pile of andesite sheets, which represents shallow-level ponding of andesite magma into accumulating ash in a proximal, subsiding volcanotectonic basin during the phreatomagmatic phase of the Whorneyside eruption. Secondly, post-caldera collapse magmatism is represented by the Rosthwaite Rhyolite, a coulée whose fault-controlled intrusive feeder is exposed in cross section. Silicic tuffs and ignimbrites of the Airy's Bridge and Lingmell formations, and caldera lake sedimentary rocks of the Seathwaite Fell

Figure 4.18 Rosthwaite Fell, from the village of Rosthwaite, Borrowdale. (Photo: D. Millward.)

Formation are also well displayed (Figure 4.19). The area has been described by Oliver (1954, 1961) and Millward (1976). The most recent accounts are by Davis (1989), Branney *et al.* (1993), Kneller *et al.* (1993b), Kneller and McConnell (1993) and Branney and Kokelaar (1994a); the resurvey associated with this work is included in the Geological Survey 1:50 000 Sheet 29 (1999). During the resurvey parts of the 'Birker Fell Andesite Group' of Oliver (1961) were re-assigned to the Lingcove Formation of Branney *et al.* (1990) and sills within the Whorneyside Formation.

Description

The lowest units exposed at Rosthwaite Fell are autobrecciated andesite sheets that pre-date the Scafell Caldera eruptions. They belong to the Lingcove Formation of Branney *et al.* (1990) and to the Birker Fell Formation of Petterson *et al.* (1992). These rocks are overlain by the Whorneyside ignimbrite, which was the first phase of the Whorneyside eruption and marks the start of the major explosive episode of the Scafell Caldera (Branney, 1991). The ignimbrite is coarser grained than elsewhere in the caldera (it contains abundant blocks) and this may indicate relative proximity to source. The ignimbrite varies abruptly in thickness, from 60–120 m, and this is thought to represent ponding in the underlying lava topography. The ignimbrite is overlain by more than 690 m of andesite sheets intercalated with thin beds of parallel-bedded andesitic tuff. This subaerial fallout tuff is phreatomagmatic and is interstratified with some debris-flow breccias and reworked layers. The andesite sheets are individually up to 380 m thick and have flow-banded and flow-folded central parts, and marginal autobreccias. Some are sills, and locally the upper contacts are peperitic. The origin of others is equivocal and lavas may be present. The Whorneyside Formation on Rosthwaite Fell is about 700 m thick, much thicker than on the south side of the Scafell Caldera (130 m).

The lower part of the Airy's Bridge Formation (Long Top Tuffs) comprises thin, bedded, welded silicic ignimbrites and subordinate pyroclastic surge and fall deposits. This part of the succession is thinner (75–120 m) than in the southern part of the caldera (over 200 m thick), and it thins further towards the NE (Kokelaar, in Branney *et al.*, 1993). By contrast, the upper

Rosthwaite Fell

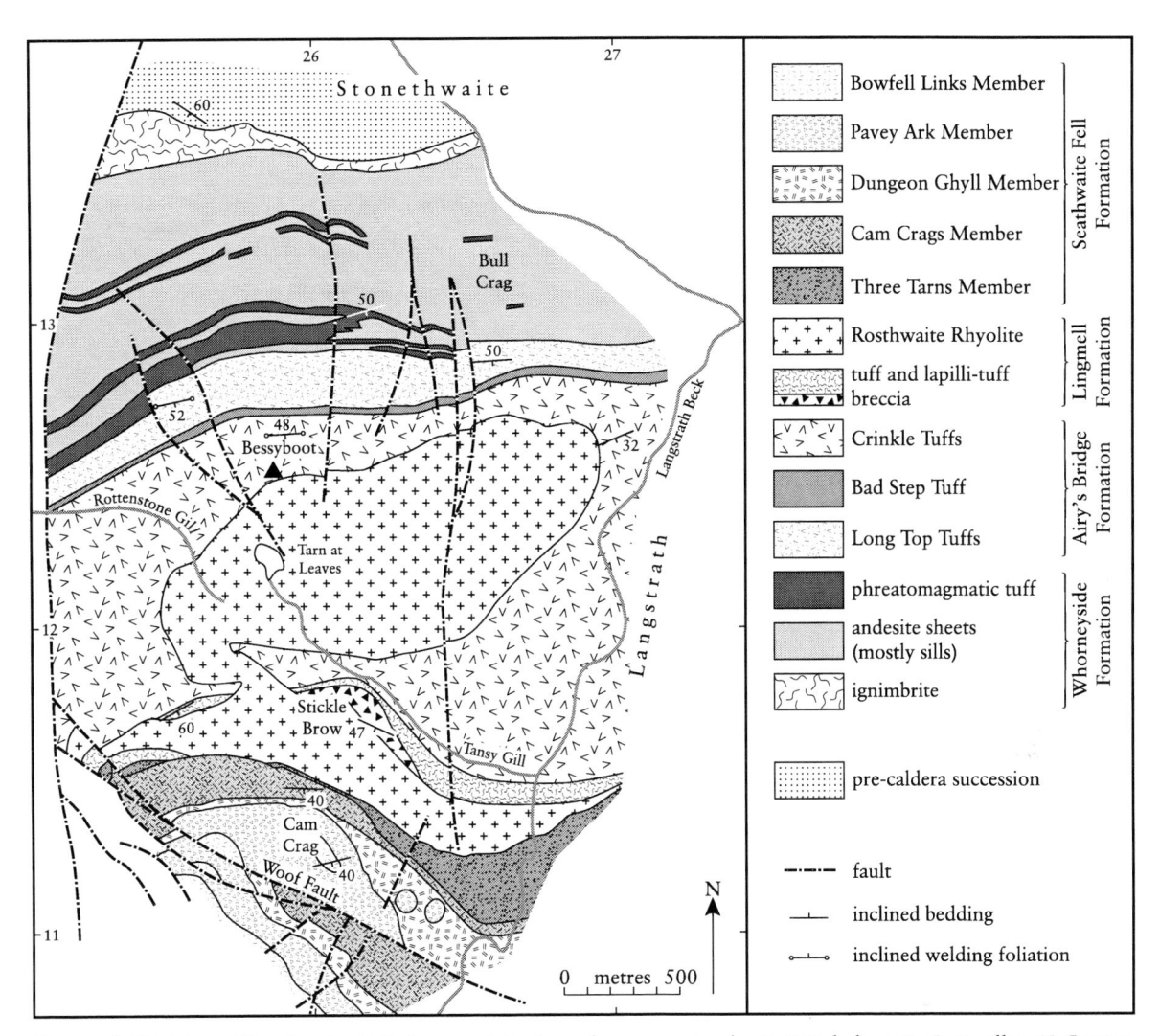

Figure 4.19 Map of Rosthwaite Fell, Borrowdale (based on mapping by B. P. Kokelaar, B. C. Kneller, N. Davies and M. J. Branney, for British Geological Survey).

member of the Airy's Bridge Formation (Crinkle Tuffs) is relatively thick on Rosthwaite Fell (c. 660 m). Ignimbrites of the Crinkle Tuffs are massive, intensely welded and commonly rheomorphic. Parataxitic fabrics, lineations and large-scale folds, all caused by rheomorphism, are best developed in the middle part of the Crinkle Tuffs. In a 10 m-wide zone adjacent to the Rosthwaite Rhyolite and a smaller rhyolite intrusion near Langstrath Beck, welding fabrics in the Crinkle Tuffs are deflected into concordance with the intrusive contacts. Columnar jointing is also present locally. The lava-like Bad Step Tuff (Branney et al., 1992) is absent from the Rosthwaite Fell succession, but a massive eutaxitic lapilli-tuff, 3–15 m thick with a fine-grained top on Bessyboot (between 2543 1254 and 2679 1282) may represent a distal co-ignimbrite correlative (Branney et al., 1993).

Subaerial, thinly stratified, clast-supported and locally eutaxitic, lapilli-tuffs and tuffs of the Lingmell Formation unconformably overlie the Crinkle Tuffs, and thicken westwards from the Langstrath towards Stickle Brow (261 118). A lens of clast-supported breccia containing blocks of welded Crinkle Tuffs thickens to 75 m towards Stickle Brow over a distance of about 300 m, possibly due to ponding in a half-graben, or formation of an apron of blocks along a volcanotectonic fault scarp (Kneller et al., 1993b). The breccias are avalanche deposits, possibly with some pyroclastic breccias related to the extrusion of an overlying post-caldera-collapse rhyolite coulée (called the Rosthwaite Rhyolite; Millward, 1976; Davis, 1989). The rhyolite is crystal poor, flow-folded and perlitic, and has upper and lower autobreccias. It is of particular interest because it has both intrusive and extrusive parts connected by a short rhyolite-filled conduit and vent (257 118), now exposed in diagonal section. Its intrusive part is well exposed for a distance of 1.5 km (264 122), and the extrusive part is 130 m thick and continuously exposed from vent to lateral terminations, 1.9 km apart (Figure 4.19). It may have risen along the fault whose scarp generated the avalanche breccias.

The Rosthwaite Rhyolite is onlapped by coarsening-upwards laminated pumiceous sedimentary rocks that represent turbidites, and/or water-laid pyroclastic deposits (Three Tarns Member, Seathwaite Fell Formation), and by overlying beds of deltaic pebbly volcaniclastic sandstone and pebble conglomerate derived from the north (Cam Crags Member) (Kneller and McConnell, 1993). The latter unit thickens markedly westwards across Rosthwaite Fell from the Langstrath (262 113), and the overlying sedimentary unit, the Dungeon Gill Member, correspondingly thins in this direction, because of ponding against the delta topography. The Dungeon Gill Member comprises fine-grained sandstone and siltstone with disrupted, nebulous or chaotic bedding, patches of coarse-grained sandstone, and large pods of breccia derived from the overlying Pavey Ark Member and injected with flames of sandstone. These are sediment gravity-flow deposits with intense soft-sediment disruption involving liquefaction, probably caused by sedimentary loading, slumping, and seismic shock. The overlying Pavey Ark Member, exposed on both sides of the WNW-trending fault along Woof Gill (258 112), is a breccia, *c.* 25 m thick, that grades up into 40 m of massive fine-grained sandstone (Raine, 1998). The breccia has layers containing abundant andesite blocks (2598 1090) whose shapes indicate that they were once hot and fluidal, as with basaltic volcanic bombs.

Interpretation

The volcanic succession of Rosthwaite Fell records the evolution of the northern part of the Scafell Caldera. It fits the generalized sequence of events inferred by Branney and Kokelaar (1994a; Figure 4.15). Marked differences in the succession between Rosthwaite Fell and other parts of the Scafell Caldera reflect proximities to former vents and very localized subsidence. For example, during the later stages of the Whorneyside eruption, the northern part of the Scafell Caldera (including the Rosthwaite Fell and Sour Milk Gill GCR sites) lay within an actively subsiding northern depocentre, and the deposits are thicker with more aqueously reworked components and sills than farther south in the caldera, where the Whorneyside phreatomagmatic tuff was deposited by fallout on to a flat subaerial ignimbrite plain (Branney, 1988b; 1991). However, subsidence at Rosthwaite Fell then declined during the Long Top Tuff eruptions: the ignimbrites are thin, and lowest units of the succeeding Crinkle Tuffs, the lava-like Bad Step Tuff (Branney et al., 1992) and the overlying, phreatomagmatic, Rest Gill Tuff (Branney and Kokelaar, 1994a), are not recognised in the Rosthwaite Fell succession. By contrast, other parts of the caldera (e.g. in Langdale; see the Langdale Pikes GCR site report) were undergoing dramatic subsidence at this time. Subsidence at Rosthwaite Fell then increased during the late climactic phase of the eruption, resulting in the thickly ponded uppermost Crinkle Tuffs. Thus, at Rosthwaite Fell most of the subsidence during the Airy's Bridge eruptions post-dated the Rest Gill Tuff; this is later than that farther south.

Active volcanotectonic faults produced ephemeral scarps which shed rock avalanches, and provided pathways for the ascent of the Rosthwaite Rhyolite. The entire area was then inundated and buried with caldera-lake sediments. A delta advanced from the north, and its toe was obstructed by the extant Rosthwaite Rhyolite coulée (Kneller and McConnell, 1993). The Pavey Ark Member is found extensively within the Scafell Caldera, and represents a catastrophic eruption-generated subaqueous gravity flow. It may be an intracaldera equivalent of spatter-rich co-ignimbrite lag breccias deposited from voluminous proximal pyroclastic flows, such as those that occur on rims of modern flooded explosive calderas, such as Santorini (Mellors and Sparks, 1991). If so, it is the only intracaldera example recorded worldwide.

Conclusions

This GCR site is important because is provides remarkably continuous exposure through the caldera-collapse cycle within the internationally significant Scafell Caldera. The site illustrates variations in the nature of alternations between explosive eruption produced by the release of gas from magma, and those driven by explosive vaporization of water on contact with magma. It also shows how different parts of the caldera subsided at different times and at different rates. It is thus complementary to the other GCR sites within the Scafell Caldera. Post-collapse magmatism is a principal feature of this site, which provides a rare and beautifully exposed cross section through the Rosthwaite Rhyolite along with its vent and feeder, centred on a fault that was active during the volcanism.

LANGDALE PIKES
(NY 271 063–NY 300 082)

M. J. Branney

Introduction

The imposing Langdale Pikes have long fascinated geologists, and many crags in the area, such as Gimmer, Raven, and Whitegill are classic climbing localities (Figure 4.20). The Langdale Pikes GCR site illustrates further features of the Scafell Caldera of the Borrowdale Volcanic Group. Variations within the ignimbrites contrast markedly with correlatives in the neighbouring sites, such as the Ray Crag and Crinkle Crags, and Rosthwaite Fell GCR sites, to provide evidence that the Scafell Caldera collapsed in a chaotic, piecemeal fashion, with numerous localized graben that ponded thick tuffs. However, the Langdale Pikes GCR site has been selected principally for its continuous sections through the Scafell Caldera lake succession, arguably the best example of its kind worldwide. Dramatic thickness variations of individual units in the succession indicate continued differential subsidence of the caldera after it was flooded. There is also a unique example of a subaqueously emplaced intracaldera pyroclastic lag breccia containing scoria. Neolithic stone-axe factories within the area exploited silicified fine-grained volcaniclastic rocks of the caldera-lake facies and represent one of Britain's oldest stone industries.

Previous work in Langdale was by Hartley (1932), Millward (1976), and Moseley and Millward (1982). Recent mapping, which involved tracing distinctive thin tuff layers as stratigraphical markers, has revealed the presence of a major reverse fault, the Langdale Fault, which runs along the northern slopes of Great Langdale (Branney and Kokelaar, 1994a) and this has led to a substantial change in the inter-

Figure 4.20 The Langdale Pikes, viewed from the south. Welded ignimbrites within the Scafell Caldera form the lower crags and Gimmer Crag (top left of centre). Pyroclastic and sedimentary rocks of the caldera-lake succession are well exposed on Pike of Stickle (far left) and Harrison Stickle (right of centre). (Photo: D. Millward.)

pretation of the stratigraphy. The revised stratigraphy is simplified and thinner, because units near the valley floor are repeated up slope by the fault (Figures 4.12 and 4.21). Recent descriptions are given in Kokelaar *et al.* (1990), Branney *et al.* (1992), Kneller *et al.* (1993b), Kneller and McConnell (1993) and Branney and Kokelaar (1994a); these accounts are summarized in Millward *et al.* (in press) and the results are incorporated in the Geological Survey 1:50 000 Sheet 38 (1996).

Description

The Langdale Pikes GCR site extends from Grave Gill (277 065) to Whitegill Crag (298 071), and from the summits of Pike of Stickle (274 073), Harrison Stickle (298 073) and Pavey Ark (284 078) to the lowest exposures on the north side of Great Langdale. The lowest unit present is a single exposure of the Long Top Tuffs (Airy's Bridge Formation) at Grave Gill. More generally, the lowermost exposures are of the Bad Step Tuff, the basal unit of the Crinkle Tuffs (Figure 4.12), because this was thickly ponded in a small, ephemeral intracaldera graben in this part of Great Langdale. In this GCR site the Bad Step Tuff is over 400 m thick, but its base is not exposed.

The Bad Step Tuff is an ignimbrite that closely resembles a flow-folded rhyolite lava. Its pyroclastic origin is inferred on the basis of evidence elsewhere, where it is thinner (see the Ray Crag and Crinkle Crags GCR site report). Its uppermost 10 m are autobrecciated and interstices between the blocks are infilled with a fine-grained silicic tuff of possible co-ignimbrite ash-fall origin. It is overlain by the Rest Gill Tuff, a turquoise laminated silt- to fine sand-grade unit, which exceeds 3 m thick locally on the downthrow (south) side of the Langdale Fault, but is only a few centimetres thick on the upthrow side. This is overlain by massive eutaxitic ignimbrites of the Crinkle Tuffs, which locally display superb examples of small- and medium-scale rheomorphic folds. They thicken markedly towards the NW. Intercalated with them, and overlying them, are layers and lenses of breccia, up to 20 m thick, with angular, framework-supported blocks (up to 2 m across) of Bad Step Tuff and eutaxitic Crinkle Tuffs ignimbrites. These pass up into subaqueously deposited sedimentary rocks. The breccias overlying the uppermost ignimbrite of the Crinkle Tuffs, and the transitional beds into the lacustrine facies, have been grouped somewhat arbitrarily together as the Lingmell Formation (Figure 4.21).

The overlying lacustrine rocks belong to the Seathwaite Fell Formation (Figure 4.21). The total thickness of the caldera lake succession is approximately 540 m, but individual units vary dramatically in thickness. The lowest unit of this formation is the Three Tarns Member, which varies laterally from 3 m of ripple cross-laminated sandstone and siltstone above Whitegill Crag, to 80 m of laminated silicic mudstone, siltstone and fine-grained sandstone at Pike of Stickle. It commonly shows parallel and wavy lamination and slump structures. The mudstone is flinty, with a conchoidal fracture, and was worked for stone axes around Harrison Stickle and Pike of Stickle in Neolithic times. Epidote- and chlorite-rich fiamme are concentrated towards the tops of some beds and are inferred to represent pumice lapilli flattened during burial as a result of their diagenetic alteration to clays (cf. Branney and Sparks, 1990).

Two prominent dark bands seen in crags just below the summits of Pike of Stickle and Harrison Stickle form the Harrison Stickle Member (Figure 4.21). These are massive breccias and are 30 m thick on Harrison Stickle. They are overlain by the Dungeon Ghyll Member, about 60 m of intensely disturbed siltstone and epidotized sandstone (Kneller and McConnell, 1993). Some beds contain fiamme.

The succeeding Pavey Ark Member, named after Pavey Ark (285 078), comprises a massive, heterolithic coarse breccia, with an upwards-fining top that exhibits cross-bedding. The breccia is about 200 m thick in the east and thins westwards. South of the Langdale Fault it is about 10 m thick, and locally occurs as discontinuous pods. Its grain size varies gradationally. The breccia contains blocks of lapilli-tuff, rhyolite, sandstone, and concentrations of ragged amygdaloidal andesite clasts, whose embayed and folded shapes, sometimes draped and deformed around lithic clasts, indicate they were hot and plastic (Figure 4.22). These juvenile bombs commonly contain small angular rhyolite clasts. The Pavey Ark Member is overlain on Thunacar Knott by siltstone, sandstone and tuff of the upper part of the Seathwaite Fell Formation.

Interpretation

The dramatic thickness variations in the Crinkle

Langdale Pikes

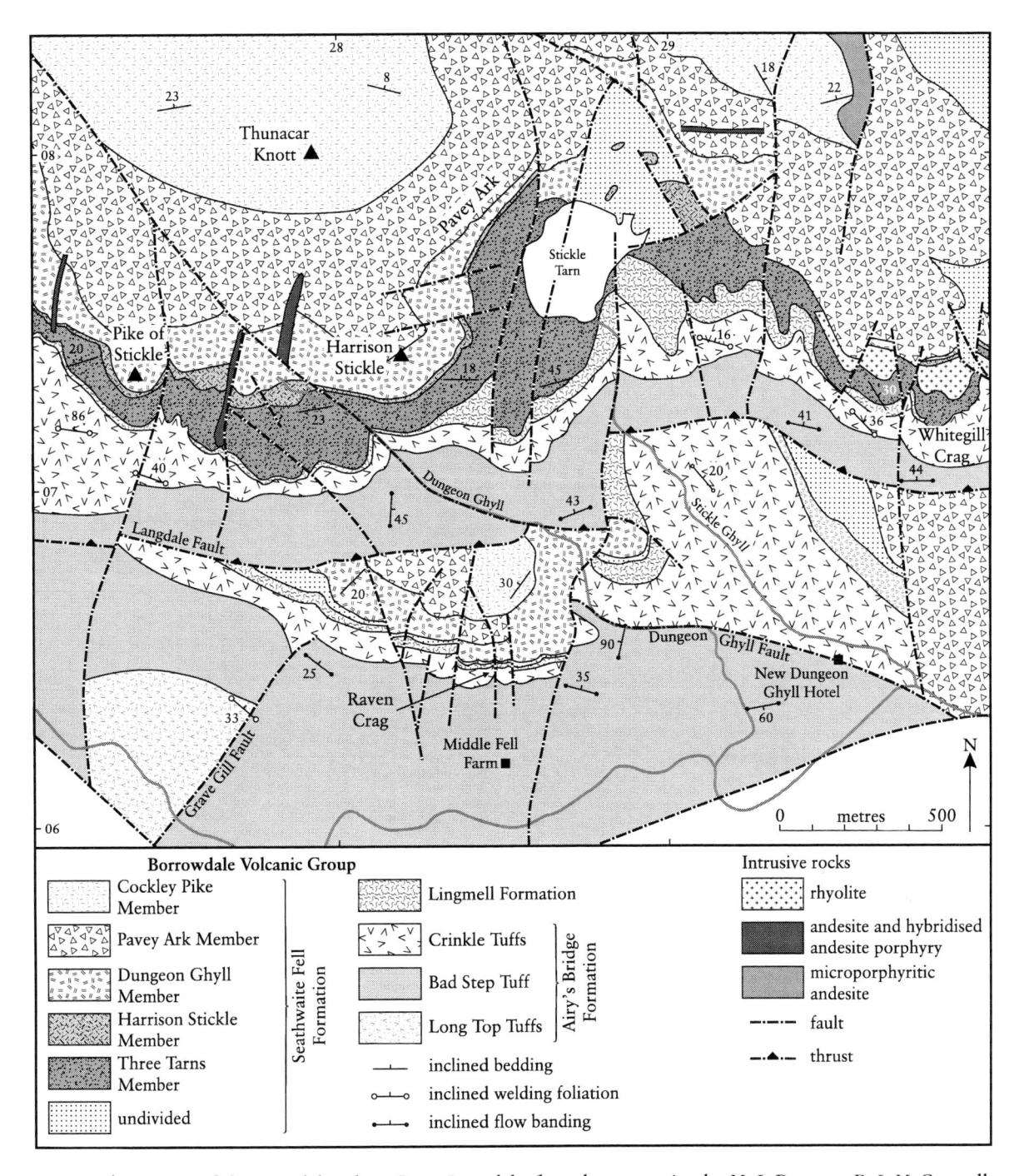

Figure 4.21 Map of the Langdale Pikes, Great Langdale (based on mapping by M. J. Branney, B. J. McConnell and B. C. Kneller, for British Geological Survey).

Tuffs ignimbrites and overlying caldera lake sedimentary rocks indicate a complex pattern of caldera subsidence, in which different areas subsided rapidly at different rates. The thickness of individual units in the caldera lake succession varies dramatically, indicating continued differential subsidence of the caldera after it was flooded. Volcanotectonic faults that bound the areas of differential subsidence are exposed in cross section. Zones of vertical and highly attenuated welding fabrics along the fault planes indicate ductile shear rather than brittle fracture and show that the faults moved when the ignimbrites were still hot (Branney and Kokelaar, 1994a).

Figure 4.22 Pavey Ark breccia from the top of Pavey Ark. Rag-shaped clasts with fluidal outlines indicate that they were hot bombs and spatter incorporated into a pyroclastic density current during a large explosive eruption. The rock is interpreted as a subaqueous scoria-rich co-ignimbrite lag breccia, similar to subaerial phreatomagmatic scoria-rich deposits around Taal caldera lake in the Philippines. (Photo: M. J. Branney.)

Good examples are the Dungeon Ghyll Fault (290 066) and the Grave Gill Fault (277 065), where 10 m-wide shear zones with vertical fiamme are well exposed. As the caldera progressively subsided, these faults were re-activated, sometimes with dip-slip throws in the opposite direction to the previous displacements. For example, thickness variations across the Grave Gill Fault (277 065) show that an early easterly downthrow, which ponded the Bad Step Tuff, was reversed during the next phase of the eruption (see Branney and Kokelaar, 1994a). The net displacement rate of these faults is much more rapid than has been recorded from tectonic faults. Contrasting successions on either side of the Langdale Fault indicate that it moved during caldera collapse and formed an unstable topographic scarp on the caldera floor that shed rock avalanches containing blocks over 2 m across. Elsewhere, strata geometries and thickness changes indicate that fault blocks rotated during caldera collapse. For example, well-exposed ignimbrites thicken gradually from 40–200 m (278 070–273 070) below Pike of Stickle, and indicate a sudden fault-block rotation of at least 20° just after the Bad Step Tuff eruption (Branney and Kokelaar, 1994a).

A transition from subaerial to lacustrine deposition as a result of flooding of the Scafell Caldera is recorded by the Lingmell Formation, and the upwards-fining succession may be a result of water gaining access to vent regions, causing an increase in explosivity (phreatomagmatism). The basal deposits of the Seathwaite Fell Formation are probably turbidites and ash falling directly into lake water, with wave and current reworking. The Harrison Stickle and Dungeon Ghyll members were rapidly emplaced as sediment-gravity flows, possibly derived from pyroclastic eruptions, and their soft-state disruption may record slumping, liquefaction due to sudden loading, and seismic shock. The Pavey Ark Member records another, larger volume sediment-gravity flow. Clasts within it closely resemble mafic scoria 'rags' found in association with proximal lag breccias at flooded calderas like Santorini, Greece (Mellors and Sparks, 1991). Away from the Langdale Pikes, the Pavey Ark Member fines upwards and laterally into sandstone and locally, eutaxitic tuff (Kneller and McConnell, 1993). The presence of hot magma spatter in the gravity flows suggests that the flows were generated by a pyroclastic eruption. The breccias probably represent intracaldera, lacustrine scoriaceous co-ignimbrite lag breccias from a major pyroclastic flow eruption that occurred in the flooded caldera. The fine-grained component could represent water-settled or trapped ash from the pyroclastic flows, or from the disturbed lake bed. This interpretation makes the Pavey Ark Member unique worldwide, because all other documented scoria-bearing co-ignimbrite lag breccias were subaerially deposited on caldera rims, and their intracaldera equivalents are inaccessible beneath caldera lakes.

Conclusions

The geology of the Langdale Pikes is remarkable because it records the final stages of collapse of a large caldera, and its inundation by water. It exposes the entire caldera lake succession (Seathwaite Fell Formation) of the Scafell Caldera, arguably the best-exposed succession of its type worldwide. Sedimentary structures recording subaqueous pyroclastic and sedimentary processes, and soft-sediment deformation are beautifully picked out by recent weathering of the glaciated rock surfaces. The site includes the type localities of the Harrison Stickle and the Pavey Ark members of the Seathwaite Fell Formation. The latter is the only example recorded in the world of subaqueous spatter-bearing breccia. It has international significance because it complements interpretations of subaerial deposits formed in similar eruptions at the Santorini, Taal, and Rabaul calderas.

Continuous sections through the uppermost caldera ignimbrites display superb plastic folds and both abrupt and gradual thickness variations, indicating fault-block subsidence and fault-block rotations within a subsiding caldera floor. The site was a fault-controlled depositional centre for the Bad Step Tuff, the ponding of which is localized within the Langdale part of the caldera. The site exposes internationally important examples of exhumed caldera-floor faults that have complex re-activation histories, and vertical welding fabrics along them. The Langdale Fault was originally a normal fault, active during the volcanic eruptions, but it was subsequently re-activated as a thrust, repeating the Langdale stratigraphy. It formed a scarp that shed avalanches, and the resultant coarse breccias exposed on the Langdale Pikes are among Britain's finest examples of caldera-collapse breccias.

SIDE PIKE
(NY 293 053)

M. J. Branney

Introduction

Well-preserved subaerial pyroclastic successions in the ancient geological record are rare worldwide, largely because they are lost by erosion. The Side Pike GCR site contains possibly the best example in Britain of an ancient subaerial volcanic succession that exhibits in close association the three main categories of pyroclastic deposit: surge, flow and fallout deposits. It also records the three principal types of volcanic explosion, magmatic, phreatomagmatic and phreatic, and it also includes features that are interpreted to have resulted from a rootless explosion caused by the interaction of surface water with hot welded ignimbrite. Post-eruption volcaniclastic sedimentary deposits are represented by debris-flow breccias and aqueous volcaniclastic siltstones and sandstones, belonging to the Seathwaite Fell Formation.

Evidence for volcanotectonic faulting closely associated with caldera volcanism is also well preserved in this site. Large vertical syn-eruptive displacements are indicated by structures on fault planes and by abrupt thickness changes in pyroclastic and sedimentary units across the faults. The most intensely fractured part of the Scafell Caldera broke into blocks 10–1000 m in size; the resulting chaotic megabreccia covers an area of more than 5 km², and is known as the Side Pike Complex. Megabreccia is a characteristic feature of large calderas throughout the world (e.g. in the San Juan mountains, Colorado; Lipman, 1984), and represents caldera floor and/or wall rocks that fragmented as a result of caldera collapse. Side Pike forms a megablock, 500 m across, that lies near the eastern margin of the megabreccia.

The rocks of Side Pike were described first by

Branney (1988a, 1988b, 1990b), and Branney and Kokelaar (1994a).

Description

Side Pike is a small glaciated peak on the south side of Great Langdale. Strata generally dip about 25° to the east and are intensely faulted (Figure 4.23). The lowest unit (the Lingmoor Tuff; A of Figure 4.24) is exposed on the south and west flanks and consists of fine tuff with thin parallel stratification, low-angle cross-stratification and abundant accretionary lapilli, many of which have several concentric laminations. It was deposited by ash fallout and from pyroclastic surges (Figure 4.25a). Above it lie approximately 6 m of pale-weathered silicic tuffs in which stratification becomes more diffuse and subtle with height. They grade up into a pink, massive eutaxitic lapilli-tuff, which is about 30 m thick and is interpreted as a welded ignimbrite (B of Figure 4.24; Figure 4.25b). The size and degree of flattening of the fiamme change with height in the ignimbrite; the most flattened fiamme occur toward the centre. Locally (2902 0516), most of the thickness of the welded ignimbrite has been brecciated into angular, jigsaw-fitting blocks, with interstitial fine silicic tuff. The breccia grades laterally into coherent, unbrecciated ignimbrite, and the brecciation clearly occurred *in situ*. Where coherent, the ignimbrite is overlain by a 20 cm-thick layer of cream-coloured, very fine (formerly vitric) tuff with abundant 1 cm-diameter accretionary lapilli (C of Figure 4.24). The tuff layer exhibits little stratification, and probably records suspension fallout of fine ash after the passage of the pyroclastic density current that deposited the underlying ignimbrite (Branney 1988a, 1988b). It is therefore a 'co-ignimbrite ash-fall' deposit.

The vitric ash at the top of the ignimbrite forms a prominent grassy ledge. Above this lie 1–2 m of cross-bedded, fine to coarse tuff (D of Figure 4.24), in which sorting, undulatory sand-wave cross-stratification and abundant accretionary lapilli indicate deposition from phreatomagmatic pyroclastic surges (Figure 4.25c). The surge deposit is incised by the irregular base of a monolithological breccia (E of Figure 4.24) that contains angular blocks of eutaxitic lapilli-tuff, closely similar to the underlying in-situ welded ignimbrite. The breccia's geometry, poor sorting and lack of internal organization indicate emplacement from a debris flow.

A diverse succession of fallout deposits, ignimbrites and aqueously deposited bedded volcaniclastic sedimentary rocks (F of Figure 4.24) overlies the debris-flow breccia. The sedimentary rocks exhibit spectacular soft-state deformation structures, and are locally overturned (e.g. at 2907 0537). Such localized and intense deformation is characteristic of the Side Pike Complex. Two amygdaloidal andesite sheets with marginal autobreccias lie within the bedded succession (Figure 4.23).

Between the andesite sheets lies the Lingmoor Tuff, a thinly bedded fine tuff with abundant accretionary lapilli (Figure 4.23). The tuff is also exposed in the saddle between Side Pike and Lingmoor Fell, and on the eastern end of Lingmoor Fell (2985 0510). It reaches 10 m thick and includes ash-fall layers with accretionary lapilli, and thicker, massive ignimbrite layers with matrix-supported fiamme. The top of the tuff is cut by gullies filled with sediment that contains locally derived intraclasts of accretionary lapilli-bearing Lingmoor Tuff. This indicates post-eruption reworking, possibly by ephemeral streams.

Vertical and deformed fiamme occur in eutaxitic lapilli-tuffs in the immediate vicinity of NNE-trending faults in the saddle between Side Pike and Lingmoor Fell (Figure 4.23). For example, vertical fiamme occur in a welded silicic ignimbrite, more than 40 cm thick, and plastered on to the faulted face of a massive andesite at 2938 0530. Many of the fiamme are sub-parallel to the fault plane and indicate fault displacement while the ignimbrite was still hot and able to shear in a ductile manner (Branney and Kokelaar, 1994a). The thickness of the massive silicic ignimbrite underlying the Seathwaite Fell Formation on Lingmoor Fell and Side Pike varies dramatically in thickness across the faults, as does the thicknesses of lacustrine sedimentary rocks between the ignimbrite and the Lingmoor Tuff (see cross section in Figure 4.23).

Interpretation

The lowest part of the succession in the GCR site illustrates the distinction between the varieties of pyroclastic deposit and also the characteristic types of volcanic eruption. The silicic tuff (B of Figure 4.24) that grades up from stratified into massive ignimbrite records the prolonged passage of a pyroclastic density current in which the

Side Pike

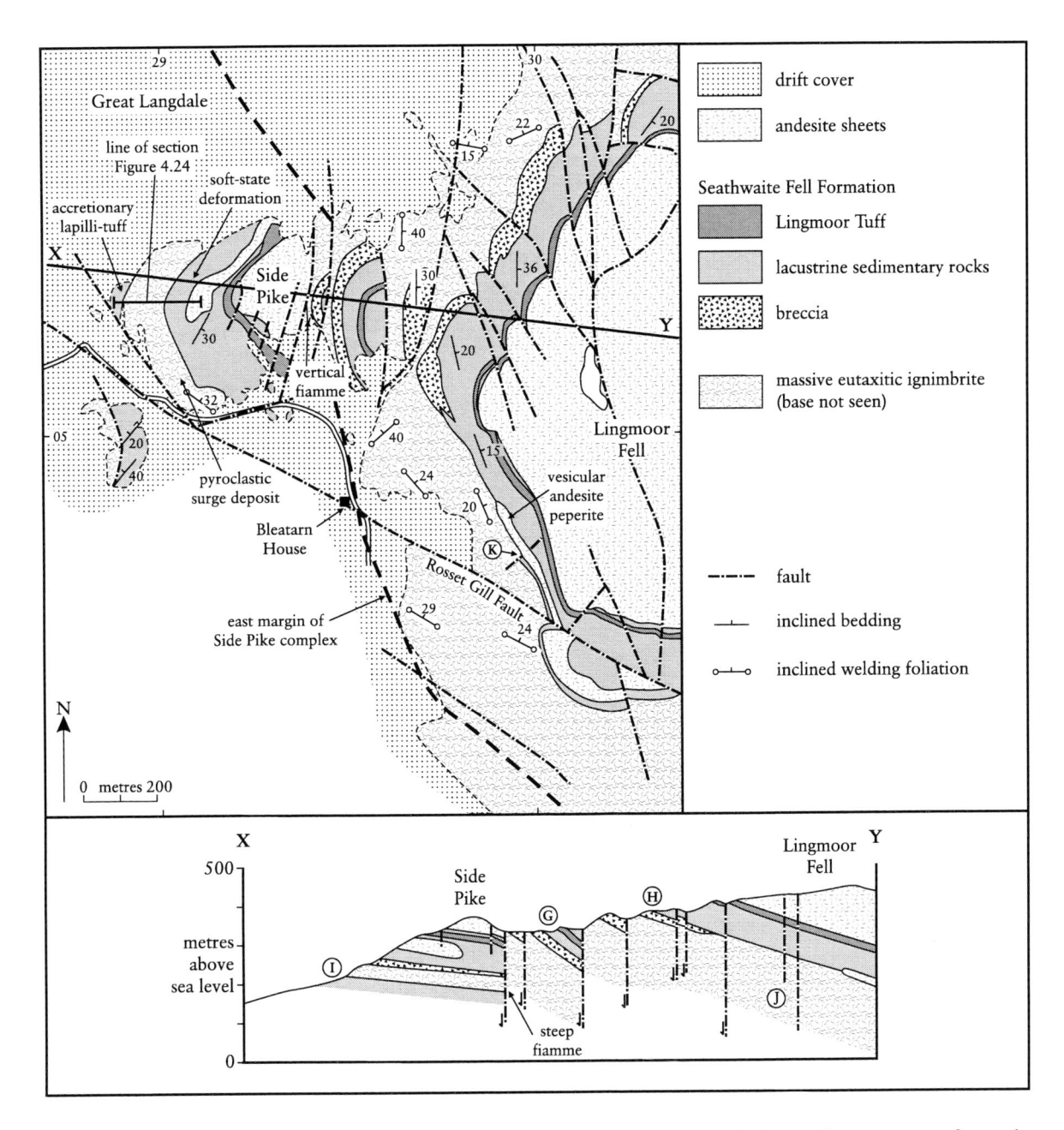

Figure 4.23 Map and true scale cross section (X–Y) of Side Pike, to show thickness changes across formerly eastward-downthrowing volcanotectonic faults, which have since been re-activated in the opposite sense. Note the change in thickness of lacustrine sedimentary rocks (between G and H) and of ignimbrite (between I and J), and the steep fabrics at two of the faults that record hot deformation of ignimbrite. A peperitic sill cuts a fault at K indicating that the fault pre-dates dewatering of the sediments. Localities G to K are described in the text. (Mapping by M. J. Branney and E. W. Johnson.)

concentration of particles increased with time. The eruption was magmatic and the welding indicates a high temperature. Gentle fallout of fine ash from a dilute co-ignimbrite ash cloud left in the wake of the density current gave rise to the thin co-ignimbrite ash-fall layer on the top of the ignimbrite (C of Figure 4.24). Co-ignimbrite clouds are known to loft high into the atmosphere, and it is probable that the thin vitric ash layer once covered an area of several hundreds of square kilometres. During and after its deposition, the underlying hot ignimbrite was undergoing welding compaction during cooling. Shortly after the eruption, surface water, possibly a small water-course, penetrated down into the hot ignimbrite and caused a violent rootless phreatic (steam) explosion, blasting a crater into the ignimbrite and shattering parts into blocks. The pyroclastic surge deposit (D of Figure 4.24) may be derived from this explosion or from similar contemporary explosive eruptions. Blocks of ballistic ejecta around the rootless vent sloughed away as debris flows (lahars; E of Figure 4.24) and these were locally deposited into channels cut into the unconsolidated pyroclastic surge deposit. These events occurred in rapid succession because the unconsolidated, thin co-ignimbrite ash layer beneath must have been protected from erosion by rapid burial.

Further pyroclastic eruptions and aqueous reworking were followed by the phreatomagmatic eruption of the Lingmoor Tuff. This is important stratigraphically because it has been correlated widely around the western part of the Borrowdale Volcanic Group outcrop. It lies within the upper part of the Seathwaite Fell Formation. The facies association resembles that of the Neapolitan Yellow Tuff of the Campanian region of Italy, which was erupted from beneath the Bay of Naples (Scarpati *et al.*, 1993), and a broadly similar style of eruption is possible. Abrupt lateral facies variations across the NNE-trending volcanotectonic faults east of Side Pike indicate that the faults were active and influencing the local palaeogeography at the time of the eruption.

At Side Pike there is little evidence diagnostic of either an intrusive or an extrusive origin for the andesite sheets that occur within the bedded succession. However, their stratigraphical positions coincide with two andesite sheets on Lingmoor Fell (Figure 4.23) where there is evidence that the lower one is a high-level peperitic sill indicating intrusion into a wet substrate (see the Pets Quarry GCR site report for details of the mechanisms of intrusion). It cuts a contemporaneous volcanotectonic fault (K on Figure 4.23) and has fluidized and vesiculated suprajacent bedded sediment (Branney and Suthren, 1988). The critical top contact of the upper andesite sheet is not exposed.

The precise origin of the large-scale brecciation of the Side Pike Complex is not known. There appears to have been more than one phase of early fracturing, characterized by soft-

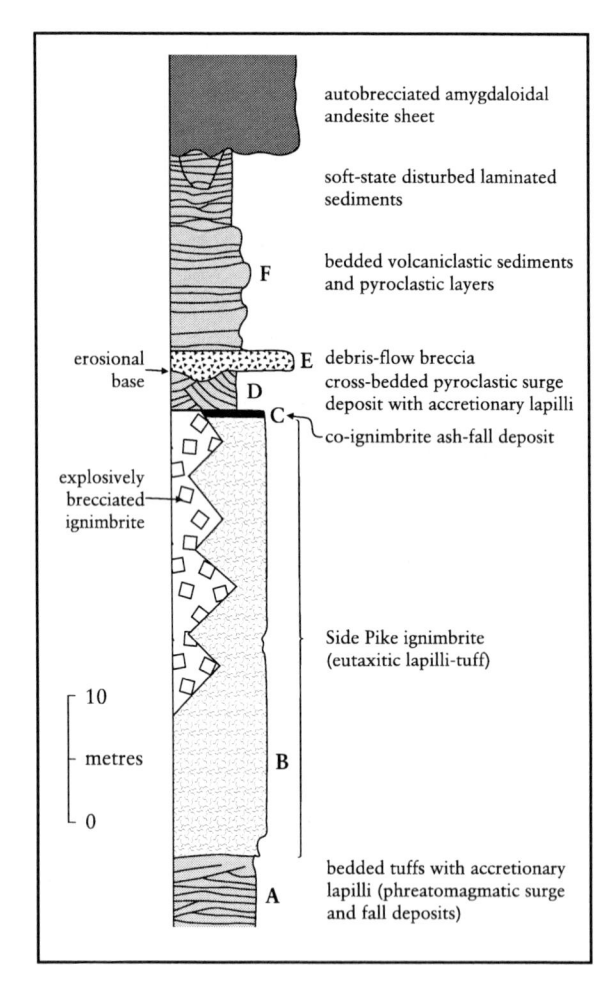

Figure 4.24 Generalized vertical section through part of the pyroclastic succession on the west side of Side Pike; see Figure 4.23 for the location and the text for explanation of units referred to as A to F. After Branney (1988b and 1990b).

state styles of deformation. The megablocks contain coherent to intensely deformed successions, some of which correlate with the Scafell caldera-floor and caldera-fill successions. However, several megablocks have 'exotic' volcanic sequences of unknown provenance, even though these may, in some cases, be correlated from one megablock to another (Branney, 1988b). The subaerial pyroclastic succession on Side Pike correlates with the successions seen in several other megablocks on Wrynose Fell, just to the SW, but this succession has not been recognized outside the Side Pike Complex; thus its precise stratigraphical position in the Borrowdale Volcanic Group remains uncertain. However, uppermost units in the Side Pike megablock are thought to correlate with the

Side Pike

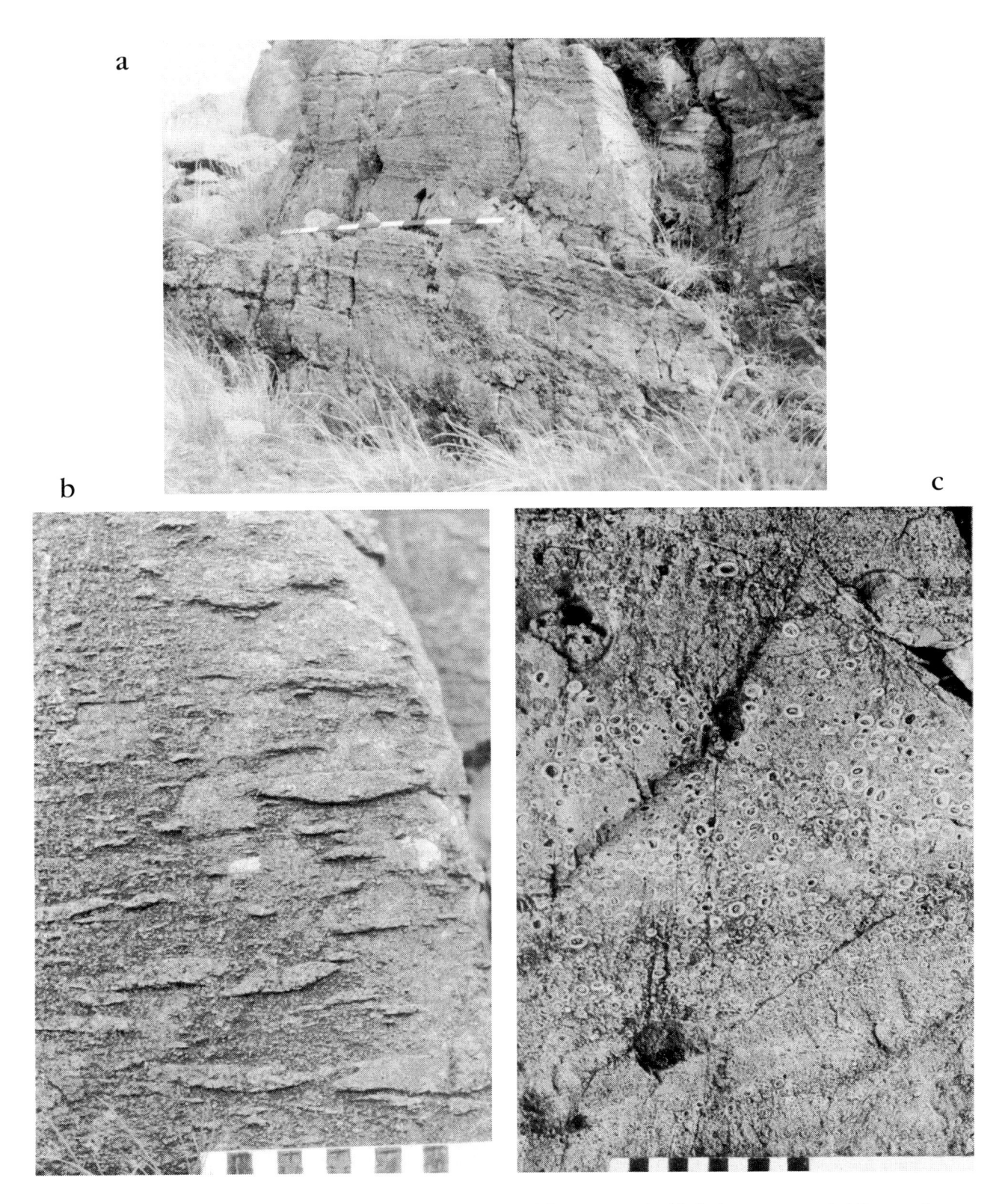

Figure 4.25 Subaerial pyroclastic rocks at Side Pike, Langdale: (a) Cross-bedded phreatomagmatic fallout and surge deposits, which underlie the Side Pike ignimbrite. (b) Rhyodacitic welded ignimbrite (the Side Pike ignimbrite). (c) Accretionary lapilli-tuff in pyroclastic surge deposit that overlies the Side Pike ignimbrite. (Photos: M. J. Branney.)

upper part of the Seathwaite Fell Formation, which is the record of a lake that filled the Scafell Caldera after it subsided (see Langdale Pikes GCR site).

Conclusions

The Side Pike GCR site is part of a breccia that is exposed over more than 5 km². The breccia comprises enormous blocks and is associated with the formation of the Scafell Caldera. The facies association at Side Pike is diagnostic of a subaerial environment and its discovery was influential in determining the overall non-marine character of the Borrowdale Volcanic Group. This is in contrast, for example, with the island volcano setting of Caradoc rocks in North Wales (see Chapter 6). Side Pike is a rare site in the ancient geological record that exhibits the three main categories of pyroclastic rock: fallout, surge and flow deposits in close association. It also includes superb, rare examples of an ash-fall deposit associated with ignimbrite and the record of rootless steam explosions that occurred shortly after an ignimbrite eruption, while the ash deposits were still hot. Though secondary explosions of this type are well known at modern volcanoes, such as following the recent ignimbrite eruptions of Mount St Helens (USA) and Mount Pinatubo (Philippines), such clear evidence from Lower Palaeozoic rocks is rare. Side Pike also provides excellent evidence for differential ground subsidence along faults generated during volcanism which are now exposed in cross section.

CONISTON
(SD 290 978–303 990)

D. Millward

Introduction

North-west of Coniston village, Mouldry Bank and Long Crag provide a well-exposed and readily accessible example of a section through the uppermost formations of the Borrowdale Volcanic Group (BVG) in the SW Lake District (Figure 4.26). About 1800 m of steeply dipping pyroclastic and volcaniclastic sedimentary rocks are intruded by basaltic andesite sills (Figure 4.27). The section is a typical example of the upper part of the BVG succession in which fluvial and lacustrine volcaniclastic sedimentary deposits are intercalated with the products of voluminous silicic pyroclastic flow eruptions. These events post-date the Scafell Caldera ignimbrite succession. On Long Crag are spectacular examples of columnar joints in welded ignimbrite. Just to the west of this site are the Coniston Copper Mines, the location of a significant British mining industry during the last century, and represented in the 'Mineralogy of the Lake District' site network of the GCR. The sedimentary rocks are strongly cleaved and have been quarried extensively for slate.

Figure 4.26 View of Long Crag from Coniston. The crags have been sculpted out of the ignimbrite of the Lincomb Tarns Formation and the low ground exposes Windermere Supergroup rocks, unconformably overlying the volcanic rocks. (Photo: D. Millward.)

Coniston

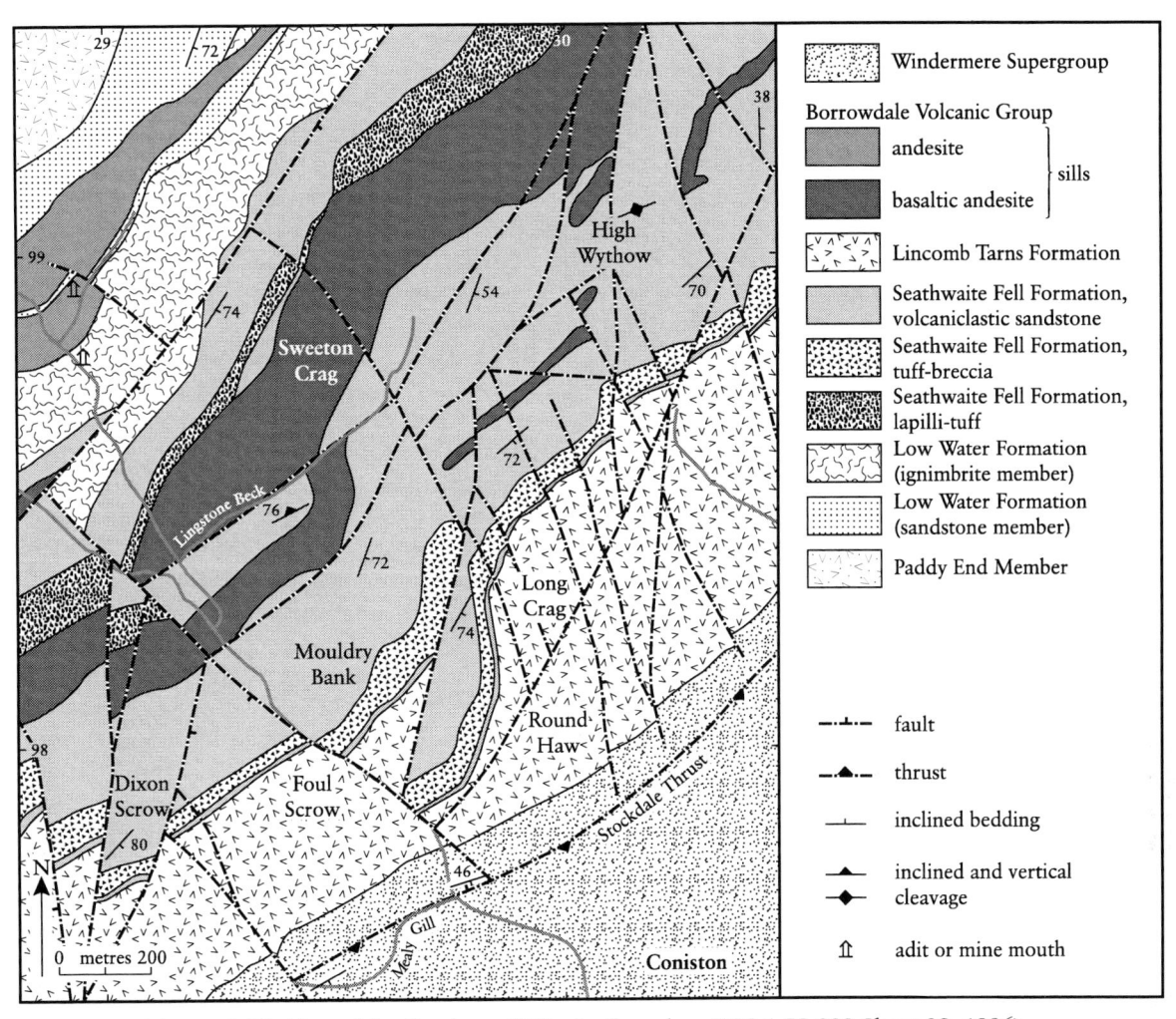

Figure 4.27 Map of the Coniston GCR site (based on BGS 1:50 000 Sheet 38, 1996).

The Coniston Fells were first mapped by the Geological Survey in 1882, but it was Mitchell (1940), who established the volcanic stratigraphy and structure of the area. The repetition of volcaniclastic sedimentary and pyroclastic rocks, both at Coniston and elsewhere in the BVG, was interpreted by Mitchell and others (e.g. Green, 1920) in the first half of the twentieth century as a set of tight folds with steeply dipping axial planes. This structural model for the BVG was subsequently abandoned by Mitchell (e.g. 1956) as studies of the sequence progressed. The GCR site description is based on work by Millward (1980), Millward *et al.* (in press) and the Geological Survey 1:50 000 Sheet 38 (1996). Re-interpretation of the stratigraphy and structure of the central Lake District has resulted in revision of the lithostratigraphical nomenclature from Mitchell (1940). His Upper Tilberthwaite Tuffs are divided between the Low Water and Seathwaite Fell formations; the Yewdale Bedded Tuffs become part of the Seathwaite Fell Formation; and the Yewdale Breccia is correlated with the Lincomb Tarns Formation of the central Lake District. The Wrengill Andesites are re-interpreted as sills (Branney and Suthren, 1988). Watson (1984) considered brecciated rocks at Colt Crag (280 980) to the west of the GCR site to lie within a possible vent for the BVG, but these rocks have been interpreted subsequently as autobrecciated andesite sills. Part of the area is covered by a field guide (Moseley, 1990).

Description

In the Coniston area the volcanic rocks dip steeply to the SE, within the steep limb of the monocline that is the principal structure of the area encompassing the uppermost BVG and Windermere Supergroup SW of Ambleside. The volcanic sequence in the Coniston area postdates the pyroclastic fill to the Scafell Caldera (Figure 4.12). The volcanic rocks are overlain unconformably by the basal beds of the Dent Group (formerly Coniston Limestone and basal unit of the Windermere Supergroup), but these rocks are poorly exposed in the site. NE-trending faults (Figure 4.27) are associated with backthrusts within the monocline.

The lowest unit in the succession, the Paddy End Member of the Lickle Formation crops out in the NW of the site, has the appearance of a rhyolite and is 150–170 m thick (equivalent to the Paddy End Rhyolite of Mitchell, 1940). The Paddy End Member comprises a single bed of homogeneous, white to pale-pink weathered, splintery, devitrified felsite, having a fine-scale foliation parallel to the base. Relict vitroclastic textures are present; the crystal content is low and there are few lithic clasts. The base and top of the unit are autobrecciated.

The Paddy End Member is overlain unconformably by the Low Water Formation, consisting of about 600 m of welded dacitic lapilli-tuff and intercalated volcaniclastic rocks. The basal 220 m comprise thinly to thickly parallel-bedded sandstone and pebbly sandstone, intercalated with coarse tuff and fine lapilli-tuff. Some of these deposits were probably water-laid, as is indicated by some dark-grey laminae of non-volcanic detritus and soft-sediment deformation. The succeeding two sheets of unbedded dacitic lapilli-tuff have a well-developed eutaxitic texture. Fiamme comprise up to 20% of the rock, which is also rich in angular, non-vesicular lithic lapilli. A lithic-rich basal zone up to 10 m thick contains blocks up to 70 cm across and forms a very poorly sorted, clast-supported breccia.

The Seathwaite Fell Formation mainly consists of up to 850 m of bedded volcaniclastic siltstone and sandstone locally intercalated with thin units of pyroclastic rock, particularly near the top of the formation. Though the base is conformable in the type area for the formation in the central Lake District (see the Langdale Pikes GCR site report), in the Coniston Fells an unconformity marks the base of the formation and the basal beds are probably markedly diachronous (Millward *et al.*, in press).

The basal lithofacies comprises up to 150 m of greenish-grey thinly bedded and laminated fine- and medium-grained sandstone and siltstone with sparse interbeds of pebbly coarse-grained sandstone intraclasts. Bedding and lamination are predominantly planar, but wavy bed forms, ripple cross-lamination and climbing ripples occur in places. Soft-state deformation structures occur throughout the succession. Dark-grey and brownish-grey beds occur locally near the base of the succession, and contain rare and poorly preserved acritarchs. None of the genera is demonstrably indigenous or diagnostic of age and they may have been derived from older marine strata, such as the Skiddaw Group. A 60–135 m-thick bed of massive poorly sorted dacitic lapilli-tuff, containing abundant lithic lapilli overlies these rocks.

Most of the formation in the area comprises a coarse lithofacies association of massive and thickly bedded, coarse-grained and pebbly sandstone intercalated with well-bedded and laminated, medium- to coarse-grained sandstone and minor siltstone. Bedding is generally poorly defined and predominantly planar, though trough and ripple cross-bedding are common locally. Sedimentary structures that are associated with rapid rates of deposition occur throughout the succession and include convolute laminae, flames, ball and pillow structures and dewatering pipes; soft-state syndepositional deformation and microfaults are also common. Towards the top of the succession, scours and channels, some containing pebble and cobble conglomerate, indicate fluvial deposition.

Though beds of tuff and lapilli-tuff occur throughout the formation, the most prominent that crops out throughout the GCR site is a coarse volcaniclastic rock, 25–80 m thick, occurring within the uppermost 150 m of the formation (Figure 4.27). Lithic blocks are concentrated in a basal 10 m-thick tuff-breccia, which passes gradationally upwards into overlying massive, lithic-rich eutaxitic lapilli-tuff; the uppermost 2 m are stratified. Characteristic pock marks on the rock surface show where pumice has been removed by weathering. This is overlain locally by the Glaramara Tuff, a white-weathered, splintery, bedded fine tuff, locally with accretionary lapilli and interbedded with thin beds of eutaxitic lapilli-tuff (see also the Side Pike GCR site report). Up to 30 m of parallel-laminated and cross-laminated, fine- to coarse-grained sandstone separates the tuff from the overlying Lincomb Tarns Formation.

The Lincomb Tarns Formation is a lithic-rich eutaxitic lapilli-tuff, up to 350 m thick. It is part of an extensive ignimbrite that crops out from the Coniston area, through Ambleside and

Grasmere to the Scafell area. It forms the rocky fells of Long Crag and Foul Scrow that overlook Coniston. Particularly conspicuous are abundant pink, angular, welded tuff and non-vesicular rhyolite fragments, commonly up to 2 cm. The lithic population is bimodal, also including many fine-grained basaltic andesite pyroclasts, in places with cumulose margins. Crystals, dominantly plagioclase with subordinate pseudomorphs after pyroxene and an opaque, form 4–14% of the rock. Fiamme are sparse to abundant (less than 5% to more than 35%) and some are identifiable pieces of long-tube pumice, altered to chlorite and an opaque mineral. A relict vitroclastic texture is preserved locally.

A characteristic feature of the formation in this area is columnar cooling joints, spectacularly displayed on Foul Scrow and Long Crag (Millward, 1980). At Long Crag at least two columnar zones have average column dimensions of 11 cm and 5.5 cm respectively. At Long Crag there is local small-scale variation in the inclination of the columns, which Millward attributed to unevenness in the cooling surfaces, possibly of separate eruptive units. The columnar zones have sharp bases and tops, and the column diameters remain constant along their length. However, the columnar zones have very restricted lateral extents.

Andesite and basaltic andesite sills were emplaced into the sedimentary rocks. These are variably massive to autobrecciated and locally intensely amygdaloidal. The contacts cut across bedding and the marginal breccias are intimately mixed with sandstone that in places is amygdaloidal.

Interpretation

Along with most of the early workers, Mitchell (1940) considered the volcaniclastic rocks in the BVG to be dominantly pyroclastic and the others, including the Paddy End Member and columnar-jointed parts of the Lincomb Tarns Formation to be lavas. Radical re-interpretation of the volcanic rocks of the Coniston area has occurred with the resurvey by the British Geological Survey, such that the new facies model shows a sequence dominated by volcaniclastic sedimentary deposits.

Abrupt lateral facies variations and faulting have caused many problems in correlating the volcanic succession across the Lake District and in particular between the Scafell and Coniston areas (Mitchell, 1956; Millward *et al.*, 1978). These problems have been resolved during recent work by the British Geological Survey (1:50 000 Sheet 38 and Millward *et al.*, in press). Within the Scafell Caldera, the pyroclastic succession of the Airy's Bridge and Lingmell formations (Figure 4.12) is overlain conformably by the dominantly sedimentary Seathwaite Fell Formation. By contrast, in the Coniston area the Airy's Bridge Formation (exposed to the NW of the GCR site) is succeeded by other welded ignimbrites and then overlain unconformably by the Seathwaite Fell Formation. The intervening ignimbrites of the Paddy End Member and the Low Water Formation are part of a succession that is developed more fully in a depositional centre in the SW of the Lake District. Detailed correlation of units within the Seathwaite Fell Formation in the Scafell and Coniston areas shows that only the uppermost part of the formation is present in the latter (Millward *et al.*, in press). Thus, the base of the Seathwaite Fell Formation is strongly diachronous.

Bedded volcaniclastic rocks are a significant feature of this GCR site. Their petrographical characteristics, bed forms and abundant sedimentary structures are indicative of deposition in fluvial and lacustrine regimes (Millward *et al.*, in press). Thin beds of accretionary lapilli-tuff and andesitic tuff represent small-scale subaerial ash-fall eruptions that occurred periodically during the major periods of relative quiescence represented by parts of the Low Water Formation and by most of the Seathwaite Fell Formation. However, the thin pyroclastic deposits are probably under-representative of the volcanic activity during deposition of the sedimentary formations, because of the poor preservation potential of unconsolidated tephra in the subaerial environment. Also, the presence of grey beds near the base of the Seathwaite Fell Formation, compared with the generally green colour of most of the sedimentary rocks is indicative of an influx of non-volcanic detritus from outside the volcanic basin, possibly associated with an interruption in the supply of volcanic detritus.

Interpretation of pyroclastic rocks in the Lake District underwent major change because of Oliver's (1954) recognition of many of the thick and extensive massive, eutaxitic lapilli-tuffs as ignimbrite. An extrusive origin for the Paddy End Member is confirmed from the basal field relationships and from the inclusion of felsite blocks in the overlying formations. The massive, lava-like felsite of the Coniston area passes south-westwards into welded lapilli-tuff (Millward *et al.*, in press), indicating that it is interpreted best as a rheomorphic ignimbrite. Mitchell (1940) mapped the columnar jointed parts of the Lincomb Tarns Formation (his Yewdale Breccia) as separate columnar lavas, but Millward (1980) demonstrated the pyroclastic nature of these rocks and included the columnar parts as part of a compound ignimbrite comprising multiple flow units. The bases to columnar zones may thus coincide with flow-unit boundaries, though there is little other evidence to support this.

The Wrengill Andesites were interpreted as lavas by Mitchell (1940). Branney and Suthren (1988) critically examined the contact zones of a number of andesite sheets at different levels in the BVG, including one NE of Church Beck. They included the intimate mixing of marginal andesite breccia with sandstone, sandstone amygdales in the andesite blocks and injection of sandstone into the andesite among extensive criteria diagnostic of intrusion of the andesite bodies into wet sediment. These andesite bodies also cut across the stratigraphy. Thus, there are no lavas within the succession seen in the GCR site.

Conclusions

The Coniston GCR site is an excellent and well-exposed representative example of the rocks deposited during the latest stages of mid-Ordovician volcanism in the Lake District. Fluvial and lacustrine sedimentation following major episodes of caldera collapse associated with the Scafell Caldera was interrupted by the emplacement of further voluminous dacitic ignimbrites from other centres. One such ignimbrite has spectacular developments of columnar cooling joints. Basaltic andesite and andesite magma was intruded into the water-saturated sediment pile.

PETS QUARRY
(NY 392 073)

M. J. Branney

Introduction

Pets Quarry lies 300 m west of Kirkstone Pass.

The quarry is cut into volcaniclastic sedimentary rocks of the Seathwaite Fell Formation (Figure 4.12) and the currently active faces are changeable. The excellent exposures of the contact relationships of a high-level sill illustrate features diagnostic of an intrusive origin into wet sediment. The sedimentary rocks are of considerable interest in their own right, because they record catastrophic syn-eruptive lacustrine sedimentation of hydroclasts by sustained turbidity currents. The site has been described and interpreted by Branney (1988b) and Branney and Sparks (1990).

The Borrowdale Volcanic Group (BVG) contains abundant sub-concordant igneous sheets throughout which, because they are readily distinguishable from volcaniclastic lithologies, have long been used as a basis for defining lithostratigraphical formations. However, it was established recently that sheets at many stratigraphical levels are intrusive (Branney and Suthren, 1988). The proportion of sills within the BVG remains unclear, but the recognition of sills that resemble blocky lavas has thrown into question the general practice of using the presence or absence of andesite sheets within local successions to define and to correlate lithostratigraphical units.

Early workers (e.g. Marr, 1916) compared andesite sheets in the BVG with modern autobrecciated lavas. Green (1913, 1915b) contended that many were sills, but subsequent workers (see Moseley and Millward, 1982 and references therein) concurred with the earlier interpretation, considering that the general concordance, brecciation, and lack of baking at upper contacts indicate an extrusive origin.

Recognition of high-level sills in the BVG (Branney and Suthren, 1988) followed work elsewhere (e.g. Kokelaar, 1982; Hanson and Schweickert, 1986), which had shown that sills intruded into wet sediment commonly do not bake the sedimentary host, for three reasons (Kokelaar, 1982). Firstly, steam generated at the magma contact insulates the host. Secondly, sediment immediately adjacent to the advancing margin of the intrusion is explosively disaggregated and excavated by steam, and is rapidly transported away along the magma–sediment contact in a fluidized state. And thirdly, an envelope of steam surrounding the invading magma can prevent the intrusion exerting directed stress on to the host to deform it. The removal of steam-fluidized sediment from the site of

Pets Quarry

intrusion can give rise to strange contact geometries with an apparent 'space problem'. Well-preserved or only slightly deformed bedding in the sedimentary rock is sharply truncated by sill contacts, indicating that substantial volumes of host sediment have been removed with little trace. Contact relationships may be complicated further by differential burial compaction of bedding around irregular sill margins. This can produce structures that resemble draped or mantle bedding, similar to that which characterizes fall-out ash. Perhaps understandably, many such sills have been mistaken for lavas whose auto-brecciated tops have been draped or infilled, and then buried by ash or sediment. The origin of many andesite sheets in the BVG whose upper contacts are not particularly well exposed remains equivocal. The non-genetic term 'sheets' has been advocated for these (Branney and Suthren, 1988).

Description

The upper part of an irregularly shaped andesite sill is exposed in the quarry (Figure 4.28). The uppermost 8 m of the sill are brecciated; its base is not seen in the quarry, but is exposed a few metres below. Closely packed jigsaw-fit breccia grades upwards into an open framework-supported texture, and some of the uppermost blocks are apparently supported by sedimentary rock. At one place, loosely packed blocks form a 4 m-high vertical face at the top of the breccia. This is steeper than the repose angle, and it is unlikely that this could have been maintained without support of the sediment. However, there is no evidence that andesite debris from this body was reworked into the immediately adjacent sediments. Interstices in the breccia are occupied by andesitic sedimentary rock. In places this exhibits undisturbed lamination sub-parallel to the local dip. Elsewhere it comprises wispy discontinuous contorted laminations with soft-sediment shears and dislocations, or it is thoroughly homogenized. Bedding commonly abuts directly against the andesite blocks, though in many places there is a contact zone, 5 mm to 5 cm wide, of homogenized (formerly fluidized), pale, fine-grained sediment. Clouds of in-situ peperite, comprising locally spalled and quenched small hydroclasts set in homogenized sediment, occur adjacent to complex, highly irregular andesite block margins, particularly within zones of sediment between andesite blocks. Ellipsoidal chlorite amygdales up to 10 mm in diameter occur within the peperitic sedimentary matrix, and also occur in sediment that infills small vesicles in the andesite. There is ubiquitous penetration of fine-grained sediment into narrow fissures in the andesite and in the overlying sedimentary rocks. Many of the small vesicles in andesite adjacent to the fissures, or near the margins of the separated blocks, are completely filled with fine-grained sediment. However, vesicles in andesite away from the block margins contain no sediment.

The host rocks are cleaved volcaniclastic siltstone, sandstone, and breccia, and bedding dips about 20° to the NE. Soft-state faults, and loading and dewatering structures are common. Some beds contain chloritic fiamme, whose shape indicates pre-cleavage burial compaction. Other beds contain angular blocks, up to 40 cm across, of vesicular andesite identical to those of the brecciated sill. The silty matrix of one bed contains vertical trails of carbonate-filled vesicles trails rising from andesite blocks, and carbonate-filled (steam?) geopetal cavities on the underside of andesite blocks.

Interpretation

The volcaniclastic rocks exposed in the quarry are correlated with the upper part of the Seathwaite Fell Formation, below the Glaramara Tuff (see Figure 4.23 and the Side Pike GCR site report). Chloritic fiamme in some of the volcaniclastic sedimentary rocks are interpreted to record waterlogged pumice or scoria clasts that compacted in subaqueous sediment by burial during diagenesis (Branney and Sparks, 1990). They cannot have formed by welding, because the rock is lacustrine and sedimentary rather than pyroclastic, and it could not have been hotter than 100°C. The sedimentary lithofacies include high-density turbidites and debris-flow deposits. The turbidites do not exhibit Bouma sequences. This is because the turbidity currents were prolonged and high density rather than dilute, single-surge events dominated by waning flow, and so they deposited disordered sequences of divisions, including variously graded, massive or stratified layers and scour-and-fill structures (Branney and Suthren, 1988; Branney et al., 1990).

The following features seen in the andesite are diagnostic of high-level intrusion into wet sediment (Branney and Suthren, 1988):

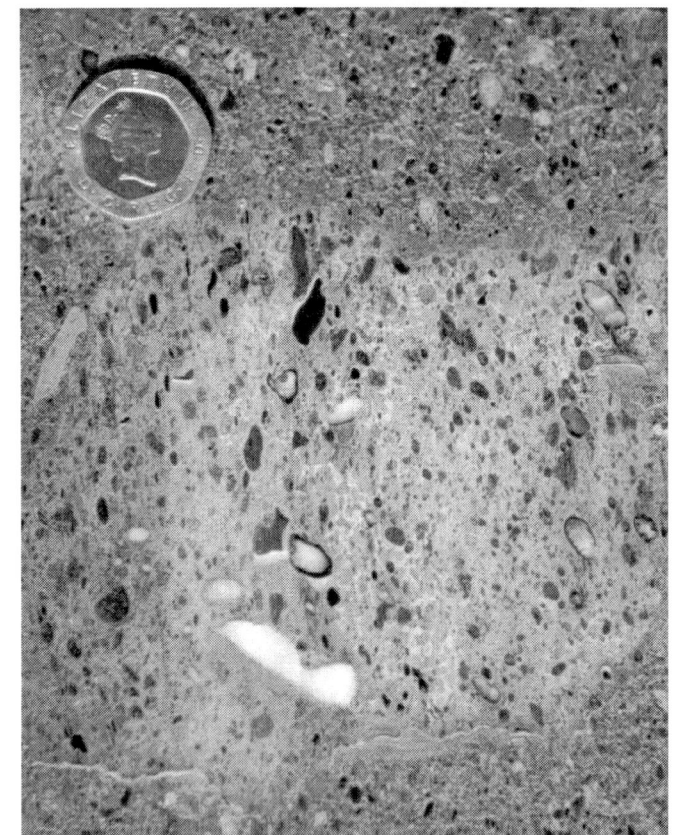

Figure 4.28 Details of peperitic andesite intrusions in the Seathwaite Fell Formation at Pets Quarry, Kirkstone Pass. (a) Andesite blocks have intruded lacustrine volcaniclastic sands and silts (pale coloured); patches of angular hydroclasts surround some block margins and sediment has been injected into cracks between the blocks. (b) Breccia formed by reworking of hot peperite on the lake floor. Geopetal sand partially infills vesicles later filled with carbonate (white) and chlorite (black) in the large andesite block. White carbonate beneath the block preserves a cavity that probably formed when the hot block heated water in the sediment matrix after its emplacement in a debris flow. The coin is 22 mm across. (Photos: M. J. Branney.)

1. Localized vesiculation of sediment above the andesite sheet suggests that the overlying sediment was already present at the time of andesite emplacement and was heated by the andesite.
2. Matrix-support of some andesite blocks also indicates that the sediment above the sheet was in place before the andesite was introduced.
3. Localized clouds of peperite around block margins indicate that block margins were undergoing in-situ hydroclastic decrepitation.
4. Sporadic, pale, faintly laminated fine-grained sediment rims around some andesite blocks are inferred to be remnants of sediment left behind after removal of disaggregated sediment by steam fluidization.
5. A lack of evidence of sedimentary reworking of the in-situ peperite is consistent with accumulation of the sediment before emplacement of the andesite.
6. Ubiquitous penetration of sediment into cracks indicates that the injected sediment was highly mobile, and probably water-fluidized and/or steam-fluidized.
7. Considering the volume of breccia, deformation that can be ascribed to sill emplacement is minimal. This suggests that sediment had been excavated by fluidization from sites now occupied by the andesite blocks.

Unequivocal criteria demonstrating intrusion, such as vesicles in the sedimentary rock, are not clearly visible where the upper intrusive contact of the sill in Pets Quarry is traced away from the fresh quarry face. It is also interesting that the diagnostic features do not occur everywhere along the contact. Movement of warm pore water through sediment is likely to have occurred as the intrusion cooled, and some of the host deformation may have been patchy post-emplacement dewatering and subsequent burial compaction. The andesite sill is autobrecciated in a similar manner to a block lava. This indicates that it had a similar rheology to a viscous block lava, and that it was in direct contact with only a steam carapace, so that the enclosing country rock was not able to exert significant mechanical constraint on the magma flow.

The sedimentary beds containing angular andesite blocks are also significant. The carbonate-filled geopetal cavities on the underside of blocks, and vesicles rising from their tops suggest that some of the andesite blocks remained sufficiently hot to vaporize the pore water of the debris-flow deposit after debris flow had ceased. The general facies association indicates that these beds were emplaced rapidly from unstable extrusive or unroofed parts of contemporaneous high-level sills that became emergent on the lake floor.

Conclusions

The Pets Quarry GCR site is perhaps the best and most accessible location in the Lake District that illustrates the processes of magma intrusion into near surface, wet sediment. Superb exposures show how sediment immediately adjacent to hot magma is mobilized by steam fluidization. Sediment only a little distance away from the contact is neither significantly heated nor disturbed, because a steam carapace around the invading magma effectively insulates the host, both mechanically and thermally, from the hot magma. The superficial similarity of this intrusion to a block-lava is instructive, and emphasizes the need for caution when interpreting the origin, and stratigraphical importance, of andesite sheets elsewhere.

The volcaniclastic sedimentary rocks show that sustained turbidity currents and debris flows may be generated during volcanic eruptions in lakes, and how the sedimentary facies produced in this way can be completely different from the much better-known sedimentary facies characteristic of non-volcanic turbidite settings. The site also exhibits pumice that has been flattened by low-temperature diagenesis and burial compaction. This closely resembles a lenticular type of disc (fiamme) formed in hot ignimbrite due to welding compaction (see also the Sour Milk Gill GCR site report). This alternative origin for such similar fragments is highly significant to the interpretation of volcaniclastic successions worldwide.

STOCKDALE BECK, LONGSLEDDALE (NY 477 049–493 060)

D. Millward

Introduction

Pyroclastic rocks and an enigmatic felsite seen within the Stockdale Beck, Longsleddale GCR site were probably erupted about 10 Ma after the end of the Borrowdale Volcanic Group (BVG) activity (Figure 4.29). These rocks are important in understanding the latest stages of Early Palaeozoic volcanism in northern England. Furthermore, the felsite is a good example of an extensive lava-like body of silicic composition with characteristics that have been taken to indicate that it is either a lava or a rheomorphic ignimbrite. The GCR site contains the type section of the Yarlside Volcanic Formation (Kneller et al., 1994), the new name for the suite of rocks formerly known as the 'Yarlside Rhyolite' (Marr, 1892; Ingham et al., 1978), 'Stockdale Rhyolite' (Gale et al., 1979) and 'Stockdale Rhyolite Member' (Millward and Lawrence, 1985; Lawrence et al., 1986). The formation is the thickest of several minor volcanic successions within the Dent Group of the Lake District and neighbouring northern Pennines. The felsite was distinguished first by Sedgwick (1836), but was described comprehensively only recently (Millward and Lawrence, 1985). The Rb-Sr isochron age of 421 ± 3 Ma has been used by Gale et al. (1979) to revise calibration of the Palaeozoic time-scale.

Description

The Yarlside Volcanic Formation crops out from Stile End (471 047) to near Shap Wells (577 096) in the eastern Lake District, but is in places extensively covered by Quaternary deposits. The formation overlies fine-grained sandstone and conglomerate of the Stile End Formation and locally, between the River Sprint and Stockdale Beck, volcanic rocks fill small-scale depressions in the top of the underlying formation (Figure 4.29). The uppermost beds of the Stile End Formation may have been thermally metamorphosed by the volcanic rocks (Ingham et al., 1978). The volcanic rocks are overlain by a pebble-conglomerate comprising felsite fragments at the base of the Kirkley Bank Formation (formerly Applethwaite Member of Lawrence et al., 1986). About 60 m of volcanic rocks are preserved near Sadgill Wood, thickening to 180 m east of Mere Crag around Stockdale Beck; the formation thins out west of the GCR site towards Stile End.

Most of the succession consists of a single bed of pink to pale-grey and greyish-green, splintery, massive to intensely fractured felsite that is platy jointed, flow-banded and flow-folded. It is high-silica rhyolite in composition. In Stockdale Beck, the type section, an almost completely exposed section through 180 m of felsitic rock comprises a single unit in which the lowest 45 m are strongly flow-folded and the uppermost

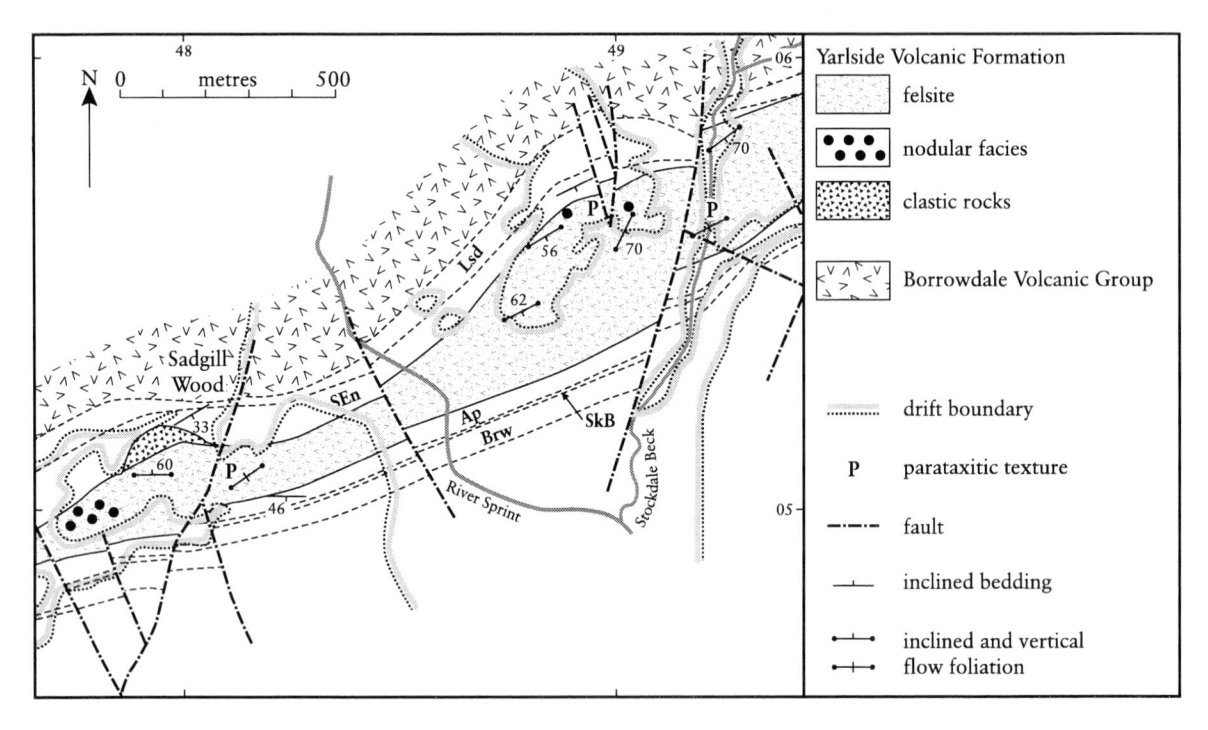

Figure 4.29 Map of the Stockdale Beck, Longsleddale GCR site (after Millward and Lawrence, 1985). Windermere Supergroup abbreviations: Ap, Kirkley Bank and Ashgill formations; Brw, Browgill Formation; Lsd, Longsleddale Member; SEn, Stile End Formation; SkB, Skelgill Formation.

Stockdale Beck, Longsleddale

30 m are massive with a vitroclastic-like texture and a devitrified fabric overprinting perlitic cracking; the central part comprises a mixture of these two facies. Abundant small, subangular to subrounded felsite clasts are present throughout. In the lower part of the unit, flow-banding is generally concordant with bedding in the underlying sedimentary rocks; in the upper part dips are generally steeper than the regional dip suggesting that ramp-like structures may be present. Flow folds range from small-scale open undulations to isoclinal, intrafolial structures, with amplitudes of a few centimetres to several metres. No autobreccia is associated with the felsitic rock.

The abundant small spherulites and perlitic cracking testify to the original glassy state. Devitrification textures include sutured fine-grained mosaics of quartz and feldspar, snowflake texture and elongate axiolitic structures. Recrystallization of the felsic rock has produced a fine- to medium-grained mosaic of anhedral quartz and subhedral albite. Locally the rock is riddled with small veins of quartz. Concentric perlitic cracks locally provide nucleation points for recrystallization, the fractures marked by chlorite and the intervening areas by partial, spherulite-like clusters of quartz and feldspar.

A nodular facies, referred to as 'agate-ball' structure by Sedgwick (1836), is well developed within the site (Figure 4.30). Most of the locally abundant nodules are single or intergrown expanded spherulites, up to 20 cm diameter, in which the original radiating quartz–feldspar fibres commonly have been ghosted by an overprinted snowflake texture. The central star-like cavities were filled subsequently by quartz with subordinate sericite and carbonate. The basal part of the unit west of Stockdale Beck (489 056) contains another type of nodule, up to 40 cm, without central cavities and with a concentric recrystallization fabric overprinting the undisturbed flow-banding.

In the west of the GCR site approximately 10 m of medium-bedded eutaxitic-textured lapilli-tuff occur at the base of the formation. Small lapilli-sized chloritized fiamme and pink felsite clasts occur within a vitroclastic matrix (Millward and Lawrence, 1985, fig. 4A). In the upper part of the formation in Stockdale Beck, and immediately to the east, grey, unbedded tuff contains subangular to subrounded felsite clasts set in a microcrystalline siliceous groundmass, which in places has a recrystallization fabric overprinting perlitic cracking. Strata up to 40 cm thick, local-

Figure 4.30 The Yarlside Volcanic Formation, approximately 450 m NW of Stockdale, showing flow-banded felsite containing large nodules formed by intense silicification. (Photo: BGS no. L3143.)

ly showing evidence of reworking, occur at the top of the formation east of Stockdale Beck (498 061) and comprise devitrified glass shards, fragments of pumice and sparse rhyolite clasts, but crystals are notably absent.

Interpretation

The Yarlside Volcanic Formation is probably the most voluminous post-BVG Lower Palaeozoic volcanic deposit in the Lake District and adjacent areas (Ingham et al., 1978). The felsite is the only post-BVG lava-like volcanic rock, because the rest are clearly volcaniclastic. After the end of the major volcanic episode represented by the BVG, the Lake District underwent erosion and thermal subsidence that allowed clastic sediments derived from the volcanic massif to accumulate in a shore-face or beach environment, and this gave way subsequently to shallow-water carbonate shelf conditions. The similar outcrop distributions of the Yarlside Volcanic Formation and the underlying Stile End Formation suggest that they filled a coastal embayment (Kneller et al., 1994). Rocks of the Yarlside Volcanic Formation were erupted into this environment, probably from a source to the south of the outcrop (Ingham et al., 1978). Eruption of the felsite probably caused temporary emergence and the rocks were reworked into the base of the overlying Kirkley Bank Formation (Millward and Lawrence, 1985).

Rocks of the Yarlside Volcanic Formation have been interpreted in most accounts as extrusive and probably lava (Rutley, 1885a; Green, 1915b; Marr, 1916; Mitchell, 1934, 1956; Gale et al., 1979). Gale et al. (1979) interpreted the three separately exposed parts of the outcrop as evidence for three lavas. However, Millward and Lawrence (1985) proved by detailed mapping that a single continuous outcrop is present. They described characteristics of the felsite that, in their opinion, are not typical of felsic lava. These include the facies association with thin pyroclastic beds in a marine and otherwise non-volcanic environment, the absence of autobreccia, and the local presence of welded-tuff-like textures. They suggested that the felsite is interpreted best as a rheomorphic ignimbrite. During the last ten years there has been considerable debate about whether lava-like felsic bodies elsewhere are rheomorphic ignimbrites or true lavas (see Manley, 1996). The Bad Step Tuff in the BVG of the central Lake District described by Branney et al. (1992) is an excellent example of a lava-like ignimbrite (see the Ray Crag and Crinkle Crags GCR site report). However, textures seen in the felsite of the Yarlside Volcanic Formation have been described from probable lavas by Manley (1996), casting doubt on the interpretation by Millward and Lawrence (1985) and it remains possible that the felsite is an extensive lava.

The probable short time span represented by the volcanic episode and the close biostratigraphical control from fossiliferous beds above and below make the felsite a potential control point for calibration of the geological time-scale. Gale et al. (1979) used the Rb-Sr isochron age of 421 ± 3 Ma for the felsite as the date for the Ashgill. However, the same numerical age was obtained subsequently for the Laidlaw Volcanics near Canberra, Australia, which are also well constrained biostratigraphically, as early Ludlow (Wyborn et al., 1982). Compston et al. (1982) re-examined the data presented by Gale et al. (1979) and concluded that emplacement of the felsite took place at least 430 Ma ago and that there was a net loss of Sr during hydrothermal circulation at around 412 Ma. The biostratigraphical age for the Yarlside Volcanic Formation is unequivocal and the time-scale of Harland et al. (1990), which does not use the Yarlside Volcanic Formation date, suggests an age of about 445 Ma.

The debate on the radiometric age of the felsite in the Yarlside Volcanic Formation has wider implications in Lake District research. Rundle (1979) defined a c. 420 Ma magmatic event on the basis of similar Rb-Sr isochron ages obtained for the Stockdale Rhyolite, the Ennerdale and Carrock granites, and the Harestones Rhyolite. In addition to the earlier biostratigraphical age for the Yarlside Volcanic Formation, U-Pb determinations on zircons from the Ennerdale intrusion have indicated a Caradoc age (452 ± 4 Ma, Hughes et al., 1996), considerably older than its Rb-Sr date. Other dating methods have not yet been applied to the Carrock microgranite nor to the Harestones Rhyolite, but it seems unlikely that the Rb-Sr dates for these represent the age of intrusion. If the c. 420 Ma date is not the age of emplacement then what is the significance of this early Ludlow event? Hughes et al. (1996) suggested that resetting may be related to water–rock interaction caused by tectonic events at the onset of basin inversion which was associated with foreland basin and mountain front development in the Lake District upon closure of the Iapetus Ocean (Kneller *et al.*, 1993a).

Conclusions

The Stockdale Beck, Longsleddale GCR site contains the type section of the Yarlside Volcanic Formation, which comprises an extensive lava-like felsite and locally preserved pyroclastic and reworked pyroclastic rocks that were erupted about 10 Ma after the main phase of volcanism in the Lake District had ceased. These rocks were erupted into a shallow-marine environment, probably causing emergence locally. Lava and very intensely welded ignimbrite are possible interpretations of these rocks. The site is important because it contains the only Early Palaeozoic example of a lava-like felsite that post-dates the BVG. The felsite is also probably the most voluminous volcanic rock that post-dates the BVG. The intercalation of the felsite within a biostratigraphically well-constrained marine sedimentary succession gives this site potential value for the calibration of the geological time-scale.

BRAMCRAG QUARRY (NY 320 220)

S.C. Loughlin

Introduction

The Threlkeld microgranite is exposed in three main outcrops about 4 km east of Keswick (Figure 4.31). The outcrops on Low Rigg, Threlkeld Knotts and in Bramcrag Quarry are interpreted to represent a single irregular laccolith (Firman, 1978b). Prior to the excavation of Bramcrag Quarry in the early 1970s, the age and relationship of the Threlkeld microgranite to the Ordovician country rocks was difficult to ascertain (Hadfield and Whiteside, 1936; Rastall, 1940). Wadge (1972) suggested that the microgranite was emplaced prior to deposition of the Borrowdale Volcanic Group (BVG). However, the section exposed in Bramcrag Quarry clearly shows that the Threlkeld microgranite cuts both the Skiddaw Group and the lowest parts of the BVG. The petrography and geochemistry of the Threlkeld microgranite were described by Caunt (1984) who proposed that assimilation of Skiddaw Group material facilitated the low-pressure crystallization of garnet. The Threlkeld microgranite is now thought to be a high-level intrusion, contemporaneous with the thick succession of ignimbrites in the upper part of the BVG, though this theory is not entirely supported by isotopic dating (Wadge *et al.*, 1974; Rundle, 1981, 1992).

The Bramcrag Quarry GCR site is one of the best exposures of the Threlkeld microgranite and thus provides critical evidence for the timing of mid-Ordovician intrusive activity in the Lake District. In addition, Bramcrag Quarry illustrates the unconformable relationship between the Skiddaw Group and the overlying BVG (Figure 4.32).

Description

The Threlkeld microgranite, previously referred to as 'adamellite', is medium grained, light-grey and contains abundant feldspar phenocrysts (up to 5 mm in size) together with striking rounded

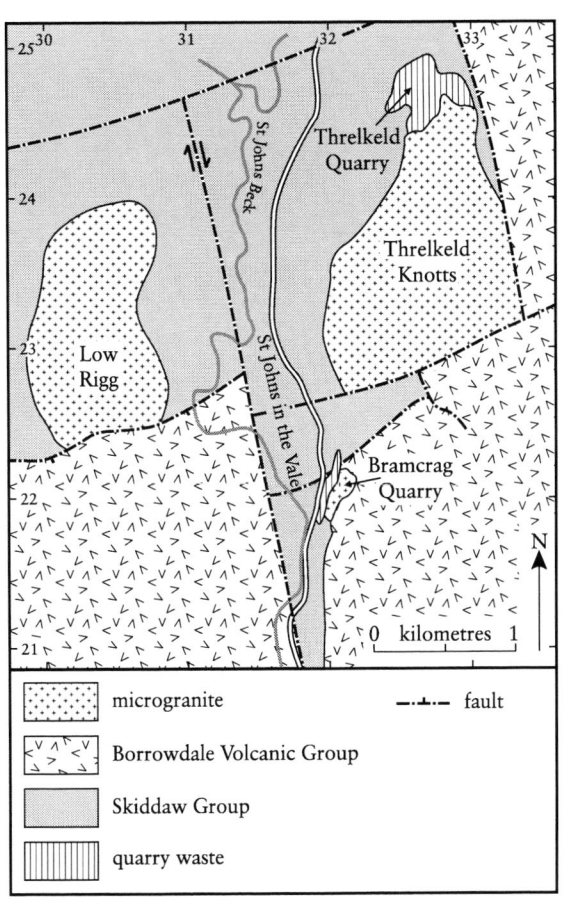

Figure 4.31 Map of the Threlkeld microgranite showing the location of Bramcrag Quarry.

quartz crystals. The feldspar phenocrysts are predominantly euhedral to subhedral albite with some pink orthoclase; they show no preferred orientation and may form glomeroporphyritic clusters. The feldspar phenocrysts show varying degrees of alteration to sericite, calcite, chlorite and quartz. Subhedral, elongate chlorite pseudomorphs intergrown with iron oxides may be after amphibole (Campbell, 1995) or biotite (Caunt, 1984). Sporadic anhedral garnet with ragged margins is also present. The groundmass is mainly composed of fine-grained granular quartz and sericitized plagioclase with some iron oxide and elongate chlorite pseudomorphs, perhaps after amphibole; accessory minerals include zircon and apatite.

The upper, sub-horizontal contact of the microgranite is exposed high along an inaccessible, worked face in the eastern part of the quarry, where it is overlain by mudstone, siltstone and tuffaceous sandstone belonging to the lowest part of the BVG (Figure 4.32). The sedimentary rocks are weakly hornfelsed and show thermal spotting, but this contact metamorphism typically extends for only a few centimetres from the contact and rarely to more than one metre.

The microgranite has narrow chilled margins (less than 1 cm) from which phenocrysts are absent. The contact is generally sharp and planar; in the northern part of the quarry, a tongue of microgranite has penetrated the overlying sedimentary rocks with clear discordance. Debris on the quarry floor illustrates this type of intrusive relationship; there are also blocks of distinctive pyroxene-phyric basic lavas from higher levels in the BVG.

In the southern part of the quarry, the microgranite contact with the BVG dips at 5–10° to the SE. Some slickensided quartz veining can be seen along the contact suggesting that shearing has occurred. There is also some evidence that minor shearing occurred within the microgranite. Stepped slickensides and quartz slickencrysts found on selected joints within the microgranite indicate an oblique dextral-reverse movement (Campbell, 1995).

At the southern end of Bramcrag Quarry, a near-vertical intrusive contact of the microgranite cuts through the dark-grey mudstone and siltstone of the Skiddaw Group. Several metres above the Skiddaw Group rocks, which dip steeply to the SW, are sub-horizontal volcaniclas-

Figure 4.32 Bramcrag Quarry: the Threlkeld microgranite is overlain by lavas and volcaniclastic sedimentary rocks of the basal part of the Borrowdale Volcanic Group. The contact slopes from top right to lower left. (Photo: BGS no. L2041.)

tic sedimentary rocks and lavas of the BVG. Unfortunately, the plane of the unconformity is obscured by scree.

Xenoliths of Skiddaw Group lithologies are common, particularly near the margins of the Threlkeld microgranite, and range in size from a few millimetres to over one metre in diameter. These xenoliths are commonly spotted by contact metamorphism and some show intense folding within them (Rastall, 1940). Rare xenoliths of BVG rocks are also present. A near-vertical ENE-trending cleavage occurs throughout the quarry and post-dates the microgranite.

Interpretation

Hadfield and Whiteside (1936) reported whole-rock geochemical analyses on the Threlkeld microgranite which showed that, despite the abundance of xenoliths in certain places, the overall composition of the granite seemed to be largely unaffected by their assimilation. The compositional uniformity of the Threlkeld microgranite was also recognized by O'Brien et al. (1985) who showed that, along with other Lake District granites, it follows a calc-alkaline trend on an AFM diagram. The Threlkeld microgranite has a small negative Eu anomaly indicating that plagioclase fractionation may have occurred before intrusion (O'Brien et al., 1985). There are some significant similarities in trace element geochemistry between published BVG compositions (e.g. Fitton et al., 1982) and the Threlkeld microgranite, but there are also some differences, particularly with respect to P, Nb, Zr and Y, suggesting that they either had different fractionating phases, or different sources (O'Brien et al., 1985). Low initial $^{87}Sr/^{86}Sr$ ratios of 0.7055 (Wadge et al., 1974) attest to the I-type nature of the Threlkeld microgranite.

Garnets in the Threlkeld microgranite show a wide range of compositions, in contrast to the restricted compositions of those in the host rock (Caunt, 1984). There is a compositional overlap between garnets from the Threlkeld microgranite and garnets from the BVG. The Threlkeld microgranite garnets may be magmatic in origin (Oliver, 1956a, 1956b) or they may be xenocrystic (Rastall, 1940; Fitton, 1972). Caunt (1984) suggested that incorporation of Skiddaw Group material enriched the microgranite magma in Al_2O_3 and Fe_2O_3 allowing moderately low-pressure crystallization of garnet (cf. Fitton et al., 1982 for garnet crystallization in BVG rocks).

The age of the Lake District intrusions, including the Threlkeld microgranite, has been a major debate since the beginning of the 20th century (e.g. Marr, 1900; Harker, 1902; Rastall and Wilcockson, 1915). The relationship of the Threlkeld microgranite to the contact between the Skiddaw and Borrowdale Volcanic groups is of crucial importance in determining the timing of its intrusion. Green (1917) suggested that the Threlkeld microgranite is a sill intruded into a conformable contact between the Skiddaw and Borrowdale Volcanic groups. Taking into account cleavage in the microgranite he suggested an age that was post-lower BVG but pre-Devonian. Hadfield and Whiteside (1936) considered the microgranite to be a laccolith intruded below the BVG. They found xenoliths of BVG material in the microgranite and also proposed that the intrusion post-dates the lower BVG.

A different view was expressed by Rastall (1940). Based on his discovery of a xenolith of sharply folded Skiddaw Group material within the microgranite, Rastall (1940) argued that because folding of the Skiddaw Group was Caledonian, the microgranite must be Caledonian or younger. Wadge (1972) interpreted the contact between the Skiddaw Group and the BVG as an angular unconformity, suggested that the stock-like Threlkeld microgranite was intruded into the Skiddaw Group and that it was then partially eroded before deposition of the overlying BVG. Subsequently, the excavation at Bramcrag Quarry exposed field relationships that prove clearly that the Threlkeld microgranite is younger than the lowest strata of the BVG (Wadge et al., 1974).

Gravity data presented by Bott (1974) show an anomaly overlying both the Skiddaw granite and the Threlkeld microgranite, but a more recent interpretation of the gravity data by Lee (1986) revealed that each intrusion has a separate Bouguer anomaly. An anomaly beneath the NE margin of the Threlkeld microgranite (Lee, 1986) was interpreted as a separate granite cupola corroborating Firman's (1978b) interpretation of the outcrops of the Threlkeld microgranite as a single irregular laccolith.

The Threlkeld microgranite is considered to be a subvolcanic intrusion related to the extrusive rocks of the BVG (Rundle, 1992), and it must therefore be Ordovician in age (older than c. 440 Ma). The Rb-Sr isochron age of 445 ± 15 Ma for the Threlkeld microgranite reported by Wadge *et al.* (1974) was recalculated by Rundle (1981) as 438 ± 6 Ma. Rundle (1992) acknowledged that this date could represent either the emplacement age of the microgranite or a subsequent 'resetting' event. He suggested that the emplacement of the Eskdale granite, which he had dated at about 430 Ma using Rb-Sr and K-Ar methods, may have been responsible for the resetting event. However, recent U-Pb analyses of zircons gave an age of 450 ± 3 Ma for the emplacement of the Eskdale granite and 452 ± 4 Ma for the Ennerdale granite (previously 420 ± 4 Ma using a Rb-Sr isochron; Rundle, 1992) suggesting that the Rb-Sr whole rock and K-Ar mineral methods do not correspond to the emplacement of the intrusions, but to a resetting event (Hughes *et al.*, 1996). This in turn casts doubt on the accuracy of the Rb-Sr isochron age as the date of emplacement for the Threlkeld microgranite. Hughes *et al.* (1996) suggested that there were only two phases of acid magmatism relating to the exposed parts of the Lake District batholith: the Eskdale, Ennerdale and Threlkeld intrusions belong to a subduction-related Caradoc phase whereas the Shap and Skiddaw intrusions belong to a later, Devonian phase.

The significance of the sheared contact between the Threlkeld microgranite and the lower BVG is uncertain. The presence of apophyses of microgranite within the overlying sedimentary rocks suggests that absolute movement along this contact was minimal (Campbell, 1995).

Conclusion

The age of the Threlkeld microgranite and its relationship with the Ordovician country rock and other Lake District intrusions has been the subject of debate for many years. The relationships exposed at Bramcrag Quarry are crucial in constraining the further interpretation of the isotopic dates. The Bramcrag Quarry site is the only locality that demonstrates that the Threlkeld microgranite intrudes both the Skiddaw Group and the lowest part of the Borrowdale Volcanic Group. The intrusion is considered, on petrographical and geochemical evidence, to be contemporaneous with the thick ignimbrites of the upper part of the Borrowdale Volcanic Group. The available Rb-Sr age of 438 ± 6 Ma post-dates the cessation of volcanism and may represent either the age of emplacement of the Threlkeld microgranite or a resetting event. Accurate U-Pb dates on zircons may offer a resolution to this problem.

BOWNESS KNOTT (NY 112 156)

D. J. Fettes

Introduction

Bowness Knott is bisected by the nearly NE-trending contact between the Ennerdale intrusion to the east and the Skiddaw Group rocks to the west (Figure 4.33). The GCR site provides excellent sections through this contact and allows an examination of the relationship of the intrusion to the regional deformational events.

The Ennerdale intrusion (previously referred to as the Ennerdale 'Granophyre') is one of the major igneous bodies of the Lake District and as such it is a surface expression of the Lake District batholith (Bott, 1974). The intrusion crops out between Buttermere in the north and Wasdale in the south, a distance of *c.* 14 km with an average width of *c.* 4 km. It consists predominantly of a relatively uniform pink or grey medium- to fine-grained granite. There are also a number of localized zones of more basic and related hybrid rocks, notably at Burtness Combe above Buttermere, Bowness Knott, the Bleng Valley and at Mecklin Wood in Wasdale.

The body was first mapped by the early surveyors of the Geological Survey (Ward, 1876). The first comprehensive account was given by Rastall (1906). Subsequent detailed accounts were given for those parts of the mass within the Gosforth (37) and Whitehaven (28) geological sheets by Trotter *et al.* (1937) and Eastwood *et al.* (1931) respectively. More recently, the mass and its regional setting were described by Clark (1963) who included petrochemical and petrogenetic discussion.

There has been a number of studies on the radiometric age of the intrusion (Brown *et al.*, 1964; Rundle, 1979; Hughes *et al.*, 1996) and on the geophysical characteristics (Bott, 1974; Lee, 1986; Evans *et al.*, 1994). Many of these studies have focused on the regional setting of the mass, and its age relative to the other igneous complexes and the regional tectonism of the Lake District.

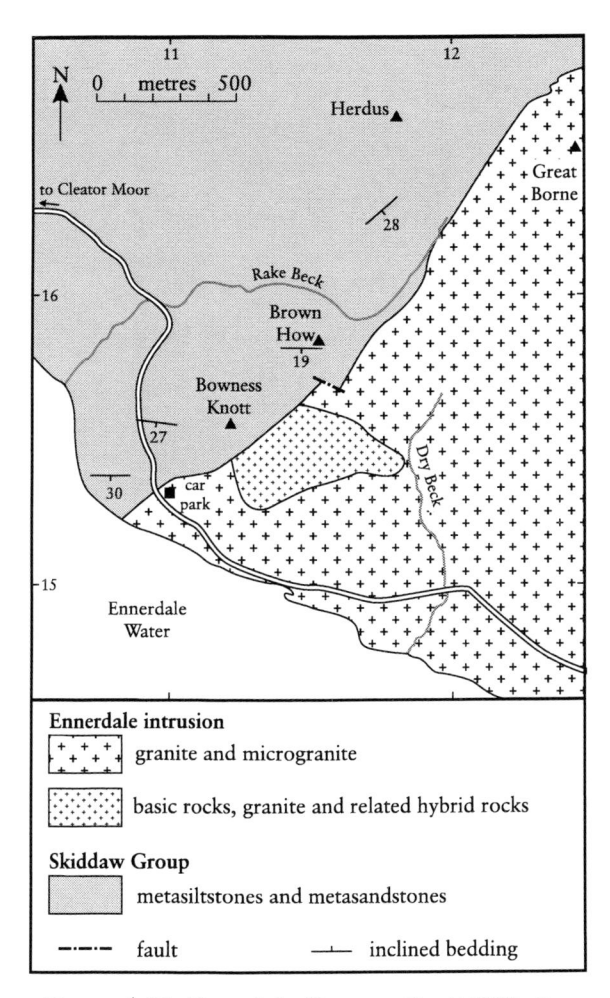

Figure 4.33 Map of the Bowness Knott GCR site.

Description

Bowness Knott (333 m) lies on the north side of Ennerdale Water guarding the entrance to the main valley (Figure 4.34). It has steep crag and scree-covered slopes to the west and south, falling off more gently to the east, where available exposure is masked by recent afforestation. To the north the ground flattens off with the minor top of Brown How before rising steeply to Herdus and Great Borne. Rake Beck, which rises between these last two hills, flows steeply down to Brown How where it turns through a right angle and continues down the north side of Bowness Knott.

The contact between the Ennerdale intrusion and the country rocks is steep or near vertical in this area. It runs up a gully near the eastern end of the crags on the south side of Bowness Knott, across the shoulder east of the top, east of Brown How and up the east side of Rake Beck. Over most of its length the contact can be localized to a few metres, though it is only exposed on the south slope of Great Borne. Its trend is parallel to one of the principal joint directions in the intrusion and the country rock.

The country rocks belong to the Buttermere Formation of the Skiddaw Group and are believed to be of late Arenig or early Llanvirn age (Cooper *et al.*, 1995). They are predominantly composed of finely laminated siltstone with subordinate sandstone and mudstone. The bedding strikes around E–W, with low dips to the south; to the north of the GCR site the strike swings nearer to NE–SW. The slump folds and soft-sediment deformational structures characteristic of this formation elsewhere are not significantly developed in this area though small slump folds are cut by a granite vein at one locality (1116 1539). Tectonic structures are also sparse and the regional cleavage is generally absent or only weakly developed; where present it strikes broadly parallel to bedding but with steeper dips. Within the GCR site the rocks are all thermally metamorphosed by the intrusion, developing a hard splintery texture. The hornfels has a greenish-grey colour but as the main contact is approached alteration on joints marked by pale-pink or red staining may pervade the whole rock giving it a pale bleached look. Hughes and Fettes (1994) recorded the presence of biotite and incipient spots of cordierite and/or andalusite close to the contact. Quartz veins are widely developed within these rocks, both as thin laminae parallel to the bedding and as larger cross-cutting structures. These veins are themselves cut by the granitic veins.

Marginal granitic veining within the Skiddaw Group is generally sparse though where present it may be relatively abundant, for example on the crags above the Bowness Knott car park (109 155). Most of the veins persist for only a few tens of metres away from the main contact. The veins are irregular and branching and range from a few centimetres to two or three metres in width. They may be markedly cross-cutting in respect of the sedimentary bedding and have sharp margins and no sign of chilling. Locally, (for example at 1118 1540), the marginal veins may become so numerous near the main contact that they isolate large blocks of the country rock. However, such effects are uncommon, as are true xenoliths.

The main mass of the Ennerdale intrusion is

Lake District and northern England

Figure 4.34 Bowness Knott from the west with the low summit of Brown How to the left. (Photo: D. J. Fettes.)

composed of a relatively uniform pinkish medium- to fine-grained granite, commonly with chloritic clots and white feldspar phenocrysts. Locally, particularly towards the margins, the granite may become more felsitic. Patches of light-coloured microgranite occur within the mass and locally these have transitional contacts with the main granite, though elsewhere they appear to be later; they probably relate to a series of late-stage minor acid intrusions within the mass including microgranites, fine-grained granophyric microgranites and aplitic microgranites (Eastwood et al., 1931; Clark, 1963). The mass is well jointed with two or three near-vertical sets and one sub-horizontal set. The rock consists of quartz, plagioclase, potash feldspar, biotite and chlorite; accessories include epidote, iron oxides, titanite, apatite and zircon. Rastall (1906) showed that granophyric textures are generally absent from the margins of the body and increase towards the centre where, the '... intergrowth becomes continuously finer in texture, and of an increasingly perfect micropegmatitic structure'. Chemical analyses of the granite (Clark, 1963; O'Brien et al., 1985; Millward et al., in press) show the body to be chemically coherent but with anomalously high soda and correspondingly low potash values, suggesting some form of metasomatic exchange. The chemical characteristics are consistent with an I-type granite generated within a volcanic-arc environment.

Basic rocks occur in a zone near the margin of the main mass on the eastern shoulder of Bowness Knott. The zone covers an area of c. 500 × 200 m, though the transitional nature of its margins make exact definitions difficult. Within the zone there is a complex inter-relationship of rock types. Three main varieties may be recognized. The first and most basic variety is a dark compact dark doleritic rock that contains some interstitial potash feldspar (Clark, 1963); the second is a dark-grey dioritic rock which, in part, grades into the third variety comprising the common granite of the mass. On crags at 114 155 veins of granite cut the more basic varieties and blebs of granitic material lie within more basic varieties. The more basic varieties are most numerous near the main contact and to the east the amount of ferromagnesian minerals in the intermediate types falls eventually to give way to the uniform granite of the mass. Clark (1963) described dioritic rocks cut by a vein of granite that is itself cut by a vein of finer-grained granite. Rastall (1906) described a peculiar form of intermediate rock, which he termed 'needle rock', characterized by long acicular crystals of uralitized augite.

Minor intrusions are found throughout the GCR site. Clark (1963) described a series of microdiorite dykes on and around Bowness Knott. They vary from 0.2–4 m in width and have SE and SW trends. They are composed of plagioclase and secondary actinolitic amphibole, possibly after pyroxene. Acid dykes also occur, many having a splintery felsitic appearance (Eastwood *et al.*, 1931). A notable example is provided by a 3 m-thick spherulitic felsite which trends ENE and crosses Rake Beck (1136 1603). The spherulites, which are up to 15 mm across, are regular and arranged in bands parallel to the margins. Other examples of major acid veins may be examined at 110 156 where a 2 m vein cuts obliquely across the bedding lamination and at 1143 1602 where a 2 m vein trending at 135° runs along a major joint and is itself finely jointed at its margins.

Interpretation

Rastall (1906) regarded the Ennerdale intrusion as a laccolith or series of laccoliths that pre-dated the regional cleavage-forming event, a view supported by Green (1917). Eastwood *et al.* (1931) and subsequently Clark (1963) argued against these views. They noted the near vertical attitude of the contact of the granite mass across Ennerdale and the fact that it flattens out to the NE across Starling Dodd and suggested that this indicates a stock-like geometry. They further argued that the mass was affected by regional-scale thrusting and therefore must pre-date that event. They noted that the granophyre cuts folds in the metasedimentary rocks and tentatively suggested that the veins must post-date the cleavage, assuming that the latter is genetically related to the folds. However, if the folds are accepted as slump structures the basis of the argument fails. The early radiometric dates on the granite seemed to support the late Caledonian age; Brown *et al.* (1964) derived a whole-rock K-Ar age of 370 ± 20 Ma for the main granite, and Rundle (1979) a whole-rock Rb-Sr age of 420 ± 4 Ma. However, Hughes and Fettes (1994) concluded from thin-section studies and variations in the nature of the cleavage across the aureole that the cleavage has been imposed on the hornfels and that in consequence the intrusion pre-dates the regional deformation. This view was supported by Hughes *et al.* (1996) who presented a U-Pb zircon age of 452 ± 4 Ma, indicating a late Caradoc age of intrusion.

Lee (1989) interpreted the Ennerdale mass as a shallow laccolith *c.* 1–2 km thick overlying a denser granitic mass. Evans *et al.* (1994) refined this model and suggested that the mass consists of a 1100 m-thick laccolith underlain by further laccolith-style members of the Lake District batholith.

Much discussion has taken place on the basic and acid complexes (Rastall, 1906; Eastwood *et al.*, 1931; Clark, 1963). It is generally accepted, on the basis of the complex age relationships, the transitional rock types and the confinement of the basic rocks to the margins of the mass, that there has been some form of hybridization between an early basic magmatic phase and the later granitic magma, probably by some form of metasomatic exchange. Both Clark (1963) and Millward *et al.* (in press) argued that the early basic phase was dioritic and that the doleritic rocks represent some form of cumulate or early differentiate.

The Ennerdale intrusion is most probably a subvolcanic intrusion emplaced in a supra-subduction zone setting as a series of two or three magmatic pulses. Locally, the host rocks were metamorphosed and hardened, and this carapace largely resisted the subsequent regional cleavage-forming events.

Conclusions

The Bowness Knott GCR site provides excellent sections through the contact of the Ennerdale intrusion and the host Skiddaw Group rocks and allows a study of the nature of the intrusion and, in particular, its age relative to the regional events.

The Ennerdale intrusion, one of the surface expressions of the Lake District batholith, is a relatively uniform, pink granite, characterized for the greater part by granophyric intergrowths and feldspar phenocrysts. Diorite phases occur within the main mass and as minor intrusions around Bowness Knott; they have locally hybridized with the slightly later granite. The hybrid rocks contain a range of rock types including the famous 'needle rock' well seen on the eastern shoulder of Bowness Knott.

The regional cleavage is only incipiently developed in the aureole of the intrusion and overprints the aureole minerals indicating that its development post-dates the intrusion. Radiometric dating suggests a late Caradoc age for the intrusion linking it with the formation of the Borrowdale Volcanic Group. Chemical characteristics of the mass are consistent with a subvolcanic setting for its intrusion.

BECKFOOT QUARRY (NY 164 003)

B. Young

Introduction

Beckfoot Quarry is located in the central part of the Eskdale pluton, the westernmost exposed portion of the largely concealed Lake District batholith. The pluton consists of two major components: a well-exposed northern granite and a generally poorly exposed southern granodiorite. The geology and petrography of the Eskdale pluton has been described by Dwerryhouse (1909), Simpson (1934), Trotter *et al.* (1937), Firman (1978b), Ansari (1983), Young (1985), Young *et al.* (1988) and Millward *et al.* (in press). The geochemistry of the granites has been discussed by Ansari (1983) and O'Brien *et al.* (1985). Within the northern part of the pluton three main facies of granitic rocks may be distinguished: medium-grained aphyric muscovite granite (the so-called 'normal' granite of several authors); a series of microgranites that vary from aphyric to markedly porphyritic and megacrystic; and a local development of a coarse- to very coarse-grained granite. The distribution of these facies is shown on recent Geological Survey maps of the area (1:25 000 special sheet, 1991; 1:50 000 Sheet 38, 1996).

Previously held views that the Eskdale pluton is a late Caledonian intrusion, similar in age to the granites of Shap and Skiddaw, were dispelled by the Rb-Sr age of 429 ± 4 Ma (Rundle, 1979) and the discovery of cleavage within the Eskdale granite (Allen, 1987). More recently, Hughes *et al.* (1996) have published a U-Pb age of 452 ± 4 Ma, which confirms the late Ordovician age of this subvolcanic intrusion.

The Beckfoot Quarry GCR site (Figure 4.35) is the only site selected for the GCR within the granite of the Eskdale pluton; the contrasting granodiorite that comprises the southern portion of the Eskdale pluton is described in the Waberthwaite Quarry GCR site report (see below). Good examples of the 'normal' granite and its relationships with the microgranite facies are exposed (Figure 4.35). Beckfoot Quarry was chosen for sample collection for U-Pb dating on zircon because the granite there is xenolith free and largely unaffected by vein mineralization (Hughes *et al.*, 1996). Therefore, it is a key site in understanding the magmatic history of the Lake District.

Description

Beckfoot Quarry is an abandoned quarry within the medium-grained 'normal' granite and microgranite facies of the Eskdale pluton (Figure 4.35). Microgranites are common within the granite, especially in the NE part of the intrusion. The commonly complex field relationship of the microgranites with the 'normal' granite has been described by Young (1985).

Medium-grained 'normal' granite forms much of the eastern face of the quarry. This rock is typically a pale-pink equigranular granite composed of perthitic feldspar, sodic plagioclase and quartz with some muscovite and biotite. Accessory minerals are generally scarce but zircon is common in altered biotite. Tourmaline is found locally as clusters of radiating black crystals coating joint surfaces (Jones, 1915). Much of the central and western parts of the quarry expose a variety of microgranite lithologies. These are mineralogically identical to the granite, but commonly exhibit various textures even within a single small exposure. Aphyric and porphyritic variants occur, the latter with abundant quartz or feldspar phenocrysts, commonly in clustered aggregates. The relationship of the microgranite to the granite within the quarry is not clear, though in the immediate neighbourhood of the quarry the microgranites appear to lie beneath the granite.

Both the granite and microgranite are cut by a roughly WSW–ENE vein composed of microgranite fragments cemented by quartz and pale-fawn dolomite. Some haematite staining occurs locally on joint surfaces within the microgranite.

Interpretation

The identical mineralogical composition of the 'normal' granite and associated microgranites within the Eskdale pluton are consistent with a co-magmatic origin. Moreover, the complex and often intricate relationships between these facies have been cited as the result of partial mixing of two or more pulses of the same, or extremely similar, magma in a partially crystallized or plastic state (Young *et al.*, 1988). Clusters of quartz and feldspar phenocrysts within the microgranites have been interpreted as partially absorbed xenoliths of coarsely crystallized granite

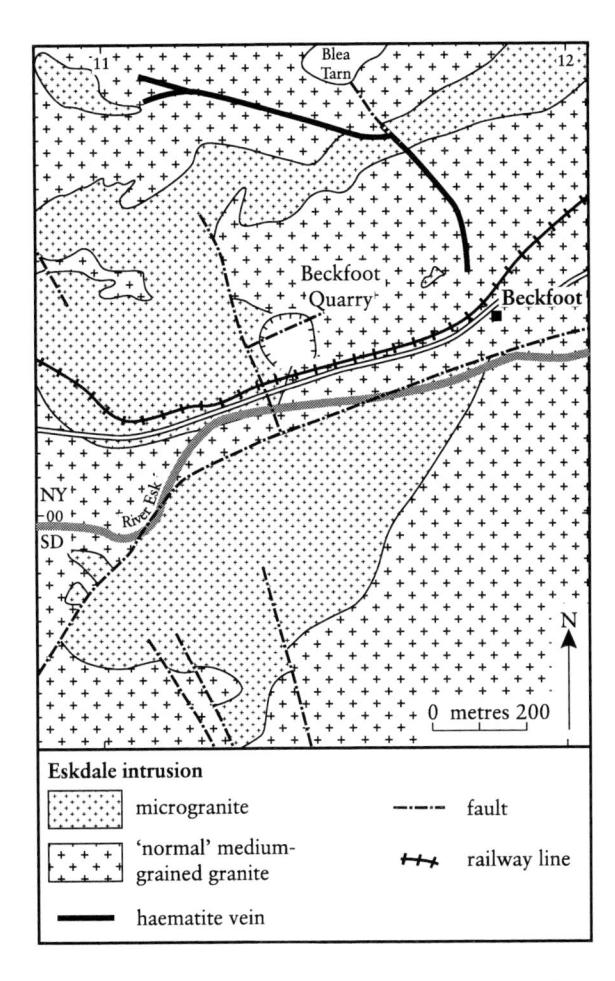

Figure 4.35 Map of the area around Beckfoot Quarry.

(Millward *et al.*, in press).

Mapping suggests that the Eskdale granite exposed in the central and upper parts of Eskdale lies within a few hundred metres of the roof of the intrusion. Outcrops of hornfelsed Borrowdale Volcanic Group rocks overlie granite NE of Blea Tarn (168 012) and NE of Boot (181 014; 185 016).

Ansari (1983) and O'Brien *et al.* (1985) noted the differences in geochemistry between the granite and granodiorite in the Eskdale pluton. Analytical data for the lithologies emphasize the more evolved nature of the former and show a compositional hiatus between them. Ansari (1983) concluded on this basis that there is no genetic relationship between the granite and granodiorite, and that they are effectively separate intrusions. However, O'Brien *et al.* (1985) suggested that the granodiorite and granite are petrogenetically linked by crystal–liquid fractionation of a common parental magma intermediate in composition to the granite and granodiorite. This is supported by similar initial $^{87}Sr/^{86}Sr$ ratios of 0.7076 ± 0.0005 and 0.7073 ± 0.0007 for the granite and granodiorite, respectively (Rundle, 1979).

Compositional differences are present between the three principal granite lithologies (Millward *et al.*, in press). Early, coarse-grained, predominantly primary textured granites show consistently lower levels of differentiation compared with 'normal' granite. By contrast, the microgranites, which record a spectrum of two-phase crystallization textures, span or exceed the compositional range of the coarse and normal granites. Crystallization of early magma, represented by the coarse granite, produced residual, more evolved melts that locally infiltrated, invaded and disrupted to varying degrees zones of partly crystallized granite, producing the more chemically variable mixed or hybridized compositions.

An end-Silurian or Devonian age for the granite was inferred for many years and only Green (1917) favoured an Ordovician age. The first radiometric date on the Eskdale granite was a Rb-Sr isochron age of 429 ± 4 Ma (Rundle, 1979), thus clearly establishing an Early Palaeozoic age and a link with the magmatic episode typified by the subduction-related volcanism of the Borrodale Volcanic Group. Allen (1987) noted the presence of cleavage locally within parts of the Eskdale granite though this cleavage is not seen in Beckfoot Quarry. The U-Pb age of 452 ± 4 Ma published by Hughes *et al.* (1996) from zircons obtained from the 'normal' granite of Beckfoot Quarry confirms the Eskdale pluton as late Ordovician and probably subvolcanic.

Conclusions

The Beckfoot Quarry GCR site contains good examples typical of both the medium-grained ('normal') and microgranite facies of the Eskdale pluton. The Eskdale granite is the western part of the largely concealed Lake District batholith and has been shown to be a late Ordovician subvolcanic intrusion. The quarry provided the sample from which the late Ordovician U-Pb zircon age of 452 ± 4 Ma was obtained.

WABERTHWAITE QUARRY (SD 112 944)

B. Young

Introduction

Waberthwaite Quarry, located between Broad Oak and Waberthwaite in west Cumbria, provides one of the very few exposures of the biotite granodiorite facies within the generally poorly exposed southern granodiorite of the Eskdale pluton (Figure 4.36). The northern well-exposed granite has been described in the Beckfoot Quarry GCR site. Important descriptions of these rocks include those by Dwerryhouse (1909), Simpson (1934), Trotter *et al.* (1937), Firman (1978b), Ansari (1983), Young (1985), Young *et al.* (1988) and Millward *et al.* (in press). The biotite granodiorite facies is shown separately from the granodiorite on the Geological Survey 1:50 000 Sheet 38 (1996), though this distinction is not made on the 1:25 000 special sheet (1991). The presence within the granodiorite at Waberthwaite of almandine-rich garnet in the granodiorite and associated aplitic microgranites, and within some of the xenoliths is noteworthy.

Description

Though very poorly exposed the Eskdale granodiorite includes a number of lithologies (Figure 4.36). Microgranodiorite is commonly developed adjacent to the margin of the intrusion. The main mass of the body appears to be composed of a pink medium-grained granodiorite in which some biotite and amphibole are generally present. Both of these rock types are commonly altered with extensive sericitization and saussuritization of feldspar, and chloritization of mafic constituents. In the vicinity of Waberthwaite Quarry a distinctive grey, biotite-rich granodiorite occurs, though the relationship of this to the main body of granodiorite is not clear. This rock is best seen in Waberthwaite Quarry. Alteration is less intense in this rock than in much of the granodiorite body.

Though Waberthwaite Quarry has effectively been disused for many years (Figure 4.37), small quantities of rock have been extracted in recent years for use as dimension stone. These have been marketed as 'Broad Oak Granite'. Despite these small-scale workings, the quarry faces at

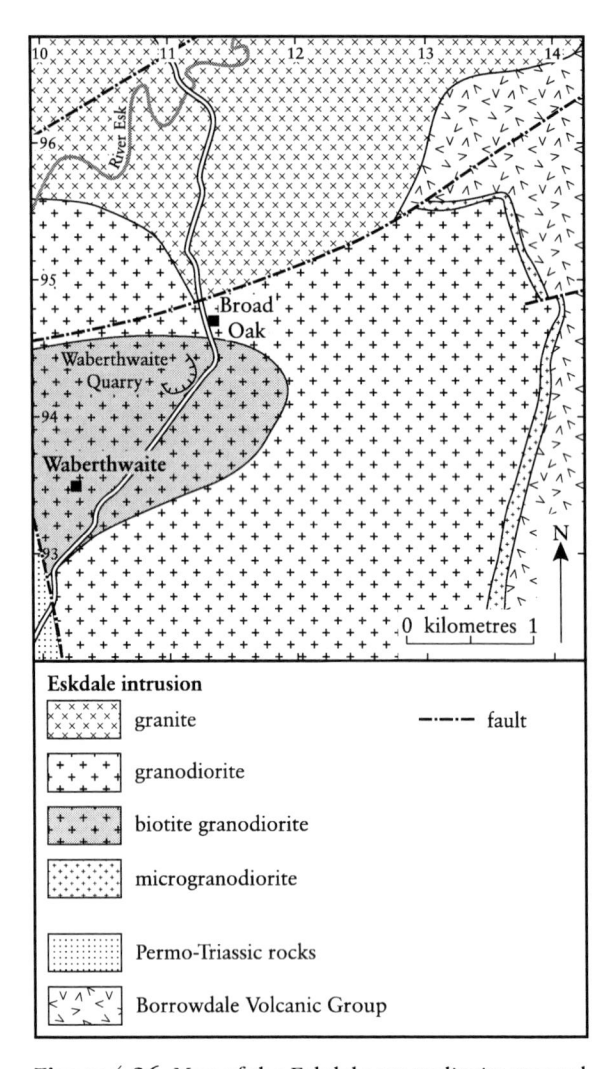

Figure 4.36 Map of the Eskdale granodiorite around Waberthwaite Quarry.

the time of writing are rather weathered and the site is considerably overgrown. The granodiorite exposed in the quarry is a grey medium-grained biotite-rich facies. Mafic xenoliths are common and a few narrow (up to 15 cm) aplitic veins are present.

Typically the biotite-rich granodiorite consists of plagioclase, orthoclase, some perthite and quartz with abundant, commonly fresh, biotite. Almandine garnet is locally conspicuous, both in the main mass of the granodiorite and in some biotite-rich xenoliths and aplitic veins. Ansari (1983) described its presence in equant crystals 1–2 mm in diameter intergrown with quartz and feldspar. Firman (1978b) noted that the garnets range up to 5 mm in diameter and that they commonly appear to be broken fragments of

Waberthwaite Quarry

Figure 4.37 Photograph of Waberthwaite Quarry taken in 1935. (Photo: BGS no. A6707.)

euhedral crystals. A little amphibole is also present and tourmaline occurs both in the granodiorite and as clusters in aplitic veins. Accessory minerals include apatite, zircon and opaque oxides. Saussuritic alteration of plagioclase is common and perthite is extensively sericitized. Biotite is in places chloritized with the development of some secondary titanite and anatase.

The aplitic rocks are typically fine grained, equigranular and contain chloritized biotite together with more perthite than plagioclase. Almandine garnet and tourmaline are locally present.

Interpretation

The relative abundance of almandine garnet within the biotite-rich granodiorite has long been known. Simpson (1934) suggested that the garnet may be the result of a late-stage concentration effect in the crystallization of the granodiorite. Trotter *et al.* (1937) favoured assimilation of Borrowdale Volcanic Group rocks in the formation of the granodiorite, though they discounted the possibility of the garnets being derived from a garnetiferous lava. Firman (1978b) noted that many of the garnets appear to be broken fragments of euhedral crystals and their xenocrystic origin was advocated by Ansari (1983).

In his description of the Eskdale pluton Simpson (1934) referred to the presence of abundant xenoliths in the rock exposed in the higher parts of Waberthwaite Quarry, apparently overlying 'grey granite'. He interpreted the abundance there of xenoliths as evidence for the nearness of the roof of the intrusion. The exposures available today do not support Simpson's suggestion of an upper xenolith-rich zone. However, evidence from recent mapping of the Eskdale pluton by the British Geological Survey (Millward *et al.*, in press) is consistent with much of the exposed part of this body lying close to the original roof zone of the intrusion.

A geochemical comparison of the granite and granodiorite parts of the Eskdale pluton has been summarized in the Beckfoot Quarry GCR site (Ansari, 1983; O'Brien *et al.*, 1985; Millward *et al.*, in press). The Eskdale intrusion was originally linked, on petrogenetic grounds, with the late Caledonian Skiddaw and Shap granites (e.g. Firman, 1978b). The recognition of cleavage within the Eskdale granite (Allen, 1987) together with a Rb-Sr age of 429 ± 4 Ma (Rundle, 1979) suggested an early, pre-deformation date of emplacement. Though less precise, the isochron age of 429 ± 22 Ma obtained by Rundle (1979) for the granodiorite indicates that this part of the pluton may be of closely similar age. Field evidence of the age relationships is unclear though Young (1985) suggested that the granodiorite may be the earlier intrusion. The U-Pb zircon age of 452 ± 4 Ma for the Eskdale granite (Hughes *et al.*, 1996) has been discussed in the Beckfoot Quarry GCR site report, but an accurate age for the granodiorite is not available. However, if the very limited evidence of field relationships suggested by Young (1985) is accepted, the granodiorite must pre-date this slightly.

Conclusions

Waberthwaite Quarry is important in providing the best available exposures of the garnet-bearing biotite granodiorite facies of the Eskdale pluton. The quarry offers a unique opportunity to study mafic xenoliths in the intrusion, possibly derived from Borrowdale Volcanic Group. The comparative abundance of almandine garnet, at least some of which may be derived from country rocks, is noteworthy.

CARROCK FELL
(NY 340 320–360 342)

D. Millward

Introduction

The Carrock Fell GCR site lies at the eastern extent of the Carrock Fell Complex, the largest mafic intrusion in the Lake District (Figure 4.38). The two principal sheet-like units of the complex, the Mosedale and Carrock divisions, are petrographically and geochemically distinct. The former is equivalent to the Carrock Fell Gabbro of Harker (1894, 1895b) and Eastwood *et al.* (1968), the Carrock Fell Gabbro Series of Harris and Dagger (1987) and the Mosedale series of Hunter and Bowden (1990). The 'diabase' of earlier workers, along with the 'granophyres' of Carrock Fell and Rae Crags form the Carrock division, equivalent to the Carrock series of Hunter and Bowden (1990). 'Division' is used on the Geological Survey 1:50 000 Sheet 23 (1997) for units within the complex to avoid confusion with the chronostratigraphical usage of 'series'. At its western extent the complex also contains three later felsic intrusions located along the Roughton Gill and Drygill fault systems: the Iron Crag and Red Covercloth microgranites and the Harestones rhyolite (Figure 4.38).

Mafic rocks within the complex show many of the features typical of small- to medium-size layered gabbroic intrusions. This type of igneous body is atypical among the Caledonian intrusions of England and Wales. The GCR site is the best-exposed section through the complex and includes good examples of those features that have been the subject of much research interest for more than a century. Geochemical data show that the Mosedale division is probably cogenetic with the continental-margin tholeiitic rocks of the Eycott Volcanic Group (Fitton, 1971; Hunter, 1980), and that the Carrock division was formed by crystal fractionation of an evolved, low-Mg tholeiitic basaltic magma (Hunter, 1980). The co-existence of the latter magma type with continental-margin volcanic suites makes this site of major importance in understanding the evolution of the Lake District magmatic province.

The complex was described first by Ward (1876), but Harker's (1894, 1895b) seminal works were a landmark in geology because, for the first time, physical principals were applied to field observations and the interpretations of in-situ differentiation processes within igneous intrusions. Later accounts of the complex include those by Eastwood *et al.* (1968), Skillen (1973), and Harris and Dagger (1987). However, the most important recent work is by Hunter (1980) whose study of the detailed petrology, mineralogy and geochemistry of these rocks resulted in a modern understanding of the crystallization and emplacement mechanisms. The account by Hollingworth (1937) of a field visit to Carrock Fell is a reminder of the popularity of the GCR site for excursions and there is an excellent field guide (Hunter and Bowden, 1990).

Description

The east-facing crags of the eastern part of

Carrock Fell

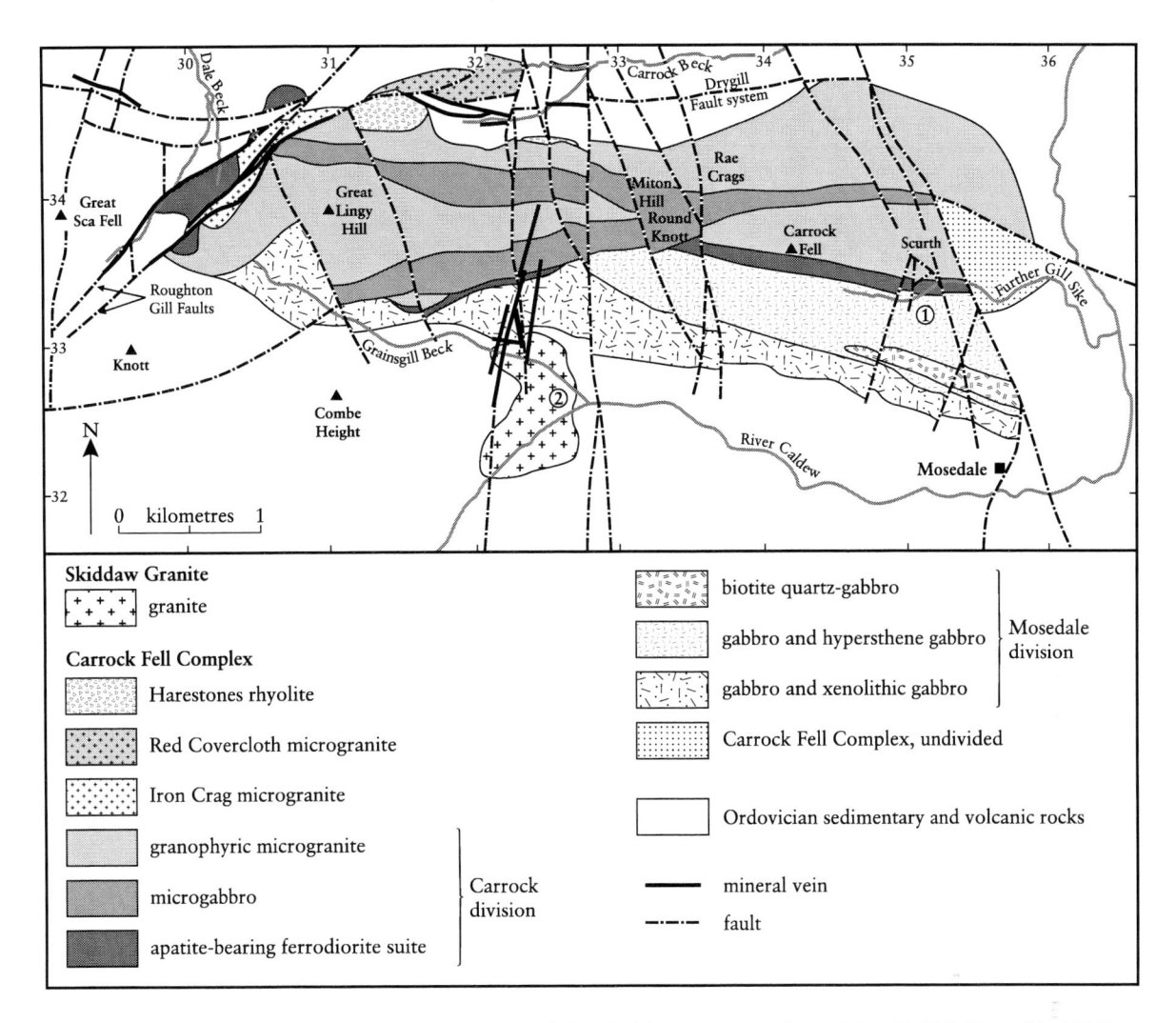

Figure 4.38 Map of the Carrock Fell Complex and the Skiddaw granite (after BGS 1:50 000 Sheet 23, 1997).

Carrock Fell rise from the Caldew river valley at about 220 m above OD to the summit of Carrock Fell at 660 m (Figure 4.39). The NNW-trending line of crags is formed by the Carrock End Fault, which juxtaposes the Carrock Fell Complex on the west against Skiddaw Group mudstones to the east (Figure 4.40); the latter are deeply eroded and extensively drift covered. Skiddaw Group rocks are exposed in the south of the GCR site.

The Mosedale division crops out in the southern part of the complex and consists dominantly of layered ortho- and mesocumulate gabbros. A broad symmetry of melagabbro at the margins passing progressively inwards into gabbro and leucogabbro, with a central mesocratic gabbro, has been recognized since Harker's classic work. Three sheet-like units are defined (Figure 4.40) amalgamating some of the divisions of Eastwood *et al.* (1968). The first unit crops out along the contact with the Skiddaw Group, and comprises gabbro and hornfelsed contact gabbro. This is succeeded to the north by poikilitic leucogabbro, gabbro and hypersthene-gabbro. Intruding the second unit in the east of the division is a westerly thinning wedge of biotite quartz-gabbro. Modal layering, mineral lamination and cryptic variation are characteristic features of the Mosedale division.

On the southern flanks of Carrock Fell, relatively homogeneous, medium-grained gabbro is in contact with folded mudrocks of the Skiddaw Group. The irregular sharp contact dips very steeply to the south. The mudrocks are hornfelsed and some pelitic beds contain garnet. The locally chilled gabbro contains many xenoliths

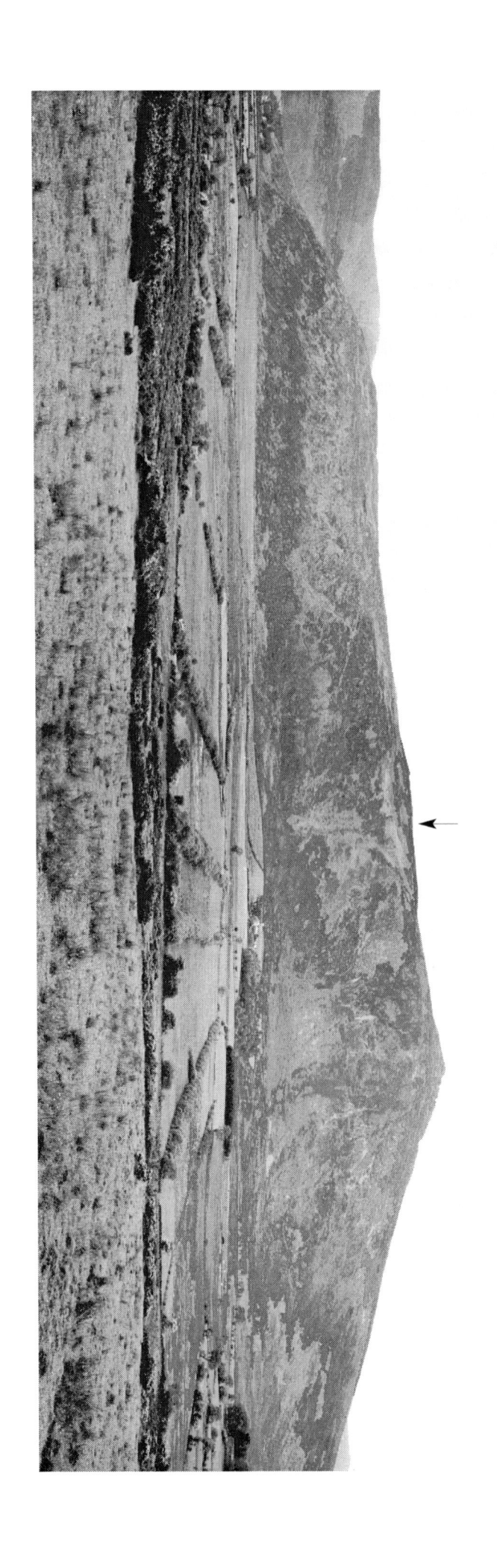

Figure 4.39 Panoramic view of the eastern end of Carrock Fell from the east side of Mosedale. The Caldew valley and the village of Mosedale are located far left. The stream just left of centre (arrowed) is Further Gill Syke and the amphitheatre-like landslip scar to the right of this, beneath the summit, is the Scurth. (Photomosaic: BGS nos. A6751 and A6752)

and screens of country rock, including large tabular masses 150 m or more in length. Though Skiddaw Group xenoliths occur near to the contact, hornfelsed basalt and basaltic andesite from the Eycott Volcanic Group are more widespread through the unit. The gabbros comprise plagioclase, Ca-rich pyroxene and Fe-Ti oxides; olivine is absent. Though magnetite is present, conspicuous chains of large ilmenite grains are more abundant. The dark colour of these rocks results from abundant secondary amphibole. Biotite is relatively common in rocks close to the contact and surrounding the large xenoliths; it was produced during late-stage magmatic reactions. Layering in the gabbros, typically in 1–20 cm-thick units dipping steeply to the NNE, is shown by variations in the proportions of plagioclase, clinopyroxene and Fe-Ti oxide primocrysts; individual layers may be divided by differences in the abundance of poikilitic clinopyroxene (Figure 4.41).

The marginal gabbros pass northward into a unit comprising poikilitic leucogabbro, gabbro and hypersthene-gabbro. The contact is diffuse over a few centimetres and the normally massive, coarse leucogabbro becomes noticeably finer grained in the 2–3 m adjacent to the junction. Northwards, the massive leucogabbro is succeeded by two zones, each comprising leucogabbro grading into modally layered and then laminated gabbros. The northern zone is more mafic than the other and corresponds to the 'fluxion gabbro' of Eastwood *et al.* (1968). In these rocks concentrations of augite, hypersthene, plagioclase and oxides mark igneous lamination, which generally dips steeply north. Adjacent to the contact with the Carrock division are layered and laminated gabbros, particularly rich in ilmenite and titanomagnetite. These rocks have a more melanocratic, and locally mottled, appearance caused by hydrothermal alteration during emplacement of the Carrock division.

The mesocratic biotite quartz-gabbro is finer grained than the leucogabbros which it intrudes. The gabbro contains abundant quartz, along with augite, hypersthene and inverted pigeonite, and is the most evolved member of the Mosedale division. Though poorly exposed, the southern contact is gradational over 2–3 m, and at the northern margin biotite quartz-gabbro passes into coarser poikilitic leucogabbro.

Mosedale division rocks are variably affected by hydrothermal alteration with the replacement of plagioclase by sericite, pyroxene by amphibole and/or chlorite, and titanomagnetite by titanite, haematite or leucoxene. Quartz and alkali feldspar are more abundant in the altered rocks and there are interstitial aggregates and cross-cutting veins of calcite, prehnite, apatite and epidote.

Rocks of the Carrock division crop out in the northern part of the GCR site (Figures 4.38, 4.40). Three units are recognized: a narrow marginal zone adjacent to the Mosedale division comprising an apatite-bearing ferrodioritic suite; the main masses of micrographic microgranite of Rae Crags and Carrock Fell; and the microgabbro of Round Knott and Miton Hill. The primary mineralogy of the Carrock division is Ca-rich pyroxene, plagioclase and Fe-Ti oxides; accessory apatite, amphibole and zircon become conspicuous in some of the ferrodioritic rocks. Alkali feldspar and quartz occur as micrographic intergrowths. Olivine is absent and Ca-poor pyroxene occurs only as exsolution lamellae in Ca-rich pyroxene. Alteration of these rocks varies considerably but is rarely complete. Feldspar is sericitized whereas pyroxene is replaced by secondary amphibole, primary amphibole by biotite and secondary amphibole, and Fe-Ti oxides by haematite, leucoxene, rutile and titanite.

The orthocumulate ferrodioritic suite is well exposed in, and to the north of, Further Gill Sike (351 333) (Figure 4.40). There is complete gradation in the suite from ferrogabbro through ferrodiorite, ferromonzodiorite to ferromicrogranite; pegmatites also occur. The ferrodiorite is laminated in places. Ferrogabbro contains abundant Fe-Ti oxides and conspicuous acicular apatite up to 10 mm. The latter enables ready field distinction from oxide-rich gabbroic cumulates of the Mosedale division nearby. Locally, the ferrogabbros coarsen and pass into granophyric pegmatite. Intrusion relationships are highly variable from gradational transitions to sharp contacts; chilling is not present within the marginal zone.

The contact between the marginal ferrodioritic rocks and the Mosedale division is not seen. Crescumulate pyroxene and plagioclase is developed locally normal to the mapped margin of the Carrock division. To the north of the upper reaches of Further Gill Sike, porphyritic ferromicrogranite is chilled against the main mass of microgranite. A 5 m-wide zone of hornfelsed microgranite occurs within the ferromicrogran-

places containing dendritic and platy ferroaugite in spectacular growths up to 30 cm across. Locally, there are numerous easterly trending dykes of ferroandesite and dacite which may represent chilled cogenetic magmas.

The microgabbro (previously the 'diabase') is best seen on Round Knott (334 336), west of Carrock Fell summit (Figure 4.38). The layered augite-plagioclase cumulates there are altered. Fine-grained aplitic back-veining into the rocks south of Round Knott suggests that the microgabbro was intruded into the microgranite, and locally remelted it (Hunter and Bowden, 1990).

The main masses of reddish brown to reddish and pale pinkish-grey micrographic microgranite are best exposed in the Crags of the Scurth (348 337). Typically, zoned euhedral albite–oligoclase crystals are engulfed in micrographic intergrowths of quartz and feldspar; spherulitic intergrowths and some patches with microcrystalline felsitic texture are also present.

Interpretation

Ward (1876) considered the Carrock Fell Complex to be metamorphosed volcanic rocks. However, Harker (1894) proposed a primary igneous origin, explaining the decrease in the colour index of the gabbros within the Mosedale division as the result of the concentration, during crystallization, of more mafic magma towards the cooler margins of the intrusion. Eastwood *et al.* (1968) discussed some of the reasons why this mechanism was not favoured by petrologists and proposed that successive injections of crystal mush were derived from a deeper, unconsolidated intrusion. They also regarded the microgranite as the product of partial fusion of sialic crust caused by heat from the intrusion of mafic magma.

Later, the Mosedale division was interpreted by Hunter (1980) and Hunter and Bowden (1990) as a multiple-layered gabbroic intrusion, with replenishment events giving rise to later leucogabbro–gabbro–hypersthene-gabbro and biotite quartz-gabbro. A significant proportion of the Mosedale division may have been lost through faulting before emplacement of the Carrock division (Hunter and Bowden, 1990). Fitton (1971) and Hunter (1980) demonstrated the geochemical similarities and the probable cogenetic relationships between the Mosedale division and the Eycott Volcanic Group.

Emplacement of the Carrock division has

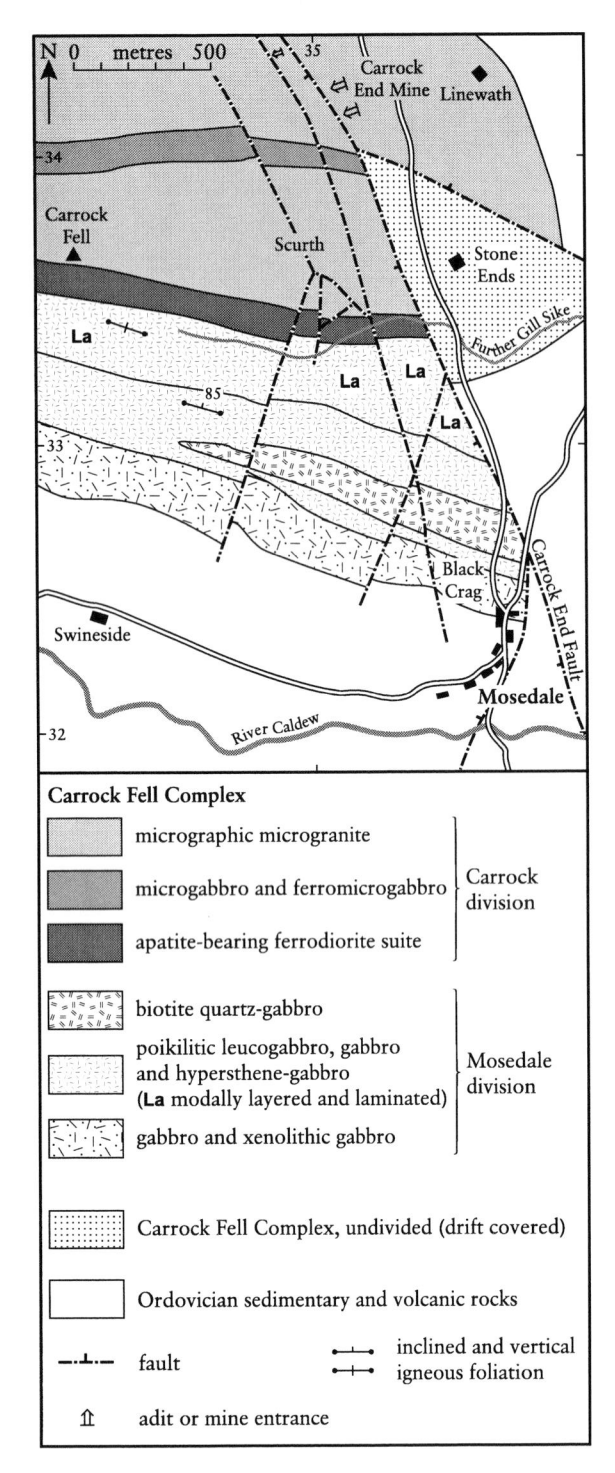

Figure 4.40 Map of the Carrock Fell GCR site.

ite and is possibly a stoped block from the main mass.

In the topographically highest parts of the intrusion, WNW of Further Gill Sike (349 334) there are drusy, granophyric pegmatitic rocks, in

Carrock Fell

Figure 4.41 Modally layered gabbros from the Mosedale division. Alternation of leucocratic feldspathic and melanocratic mafic layers. (Photo: BGS no. A6743.)

been discussed also by Hunter (1980) and Hunter and Bowden (1990) who concluded that a low-Mg, basaltic magma fractionated along a tholeiitic liquid line of descent. The main masses of microgranite crystallized as a single entity and subsided from the uppermost and most evolved part of a zoned magma body. The ferrodioritic marginal rocks, considered previously to be hybrids (Harker, 1894, 1895b; Eastwood *et al.*, 1968), were clearly demonstrated to have crystallized along the side-walls, possibly from boundary layer flows that were feeding the evolved roof zone. Crystal fractionation was enhanced by rapid crystallization and subsequent filter pressing of interstitial liquid. Direct evidence for the proximity of the roof of the complex is not present within the GCR site, though to the west on Balliway Rigg (299 338), there is a sub-horizontal contact where ferrogabbro is overlain by andesite of the Eycott Volcanic Group (Figure 4.38). However, within the GCR site pegmatitic rocks of the Further Gill Sike area crystallized from water-saturated melt trapped beneath the roof of the intrusion (Hunter and Bowden, 1990).

The present dyke-like shape of the complex led Harker (1894) to propose that these rocks were emplaced nearly vertically. This has remained largely unchallenged, particularly for the Carrock division. Palaeomagnetic data have been used in support of near-vertical emplacement, though the Mosedale and Carrock divisions were not treated separately (Briden *et al.*, 1973; Faller and Briden, 1978). However, Harris and Dagger (1987) contended that emplacement of the Mosedale division as a sub-horizontal sheet-like body was consistent with the field and petrographical data and this is also entirely consistent with a genetic association with the Eycott Volcanic Group. A palaeomagnetic study (Piper, 1997) strongly supports emplacement of the Mosedale division sub-horizontally, followed by deformation and subsequent vertical intrusion of the Carrock division.

A wide range of ages for the complex has been proposed. Harker (1895b) concluded that the Carrock Fell Complex post-dates deformation of the Skiddaw and Eycott Volcanic groups and, from the petrographical similarities with some of the intrusions of the Hebrides, suggested a Palaeogene age (Harker, 1902). An Ordovician emplacement age for the complex was first argued by Green (1917). However, Eastwood *et al.* (1968) favoured a post-Caledonian, pre-Carboniferous age. The K-Ar whole-rock age of 356 ± 20 Ma of biotite hornfels from the contact aureole of the complex determined by Brown *et al.* (1964) is younger than 399 ± 6 Ma obtained for the Skiddaw granite by Miller (1961) and does not accord with field and petrographical evidence that the gabbros are earlier than the hydrothermal alteration and mineralization associated with the Skiddaw granite (Eastwood *et al.*, 1968). The anomalous radiometric date was probably caused by argon loss (Brown *et al.*, 1964).

An Ordovician age for the Mosedale division seems probable. The cogenetic relationship of these rocks with the Eycott Volcanic Group implies a similar age; the volcanic rocks have a late Llanvirn to Caradoc biostratigraphical age range (Downie and Soper, 1972; Millward and Molyneux, 1992). A minimum emplacement K-Ar age of 468 ± 10 Ma on biotite from the Mosedale division gabbro (Rundle, 1979) is late Llanvirn on the time-scale of Harland *et al.* (1990).

The Carrock division microgranite from Rae Crags post-dates the Caradoc (Longvillian) Drygill Shale Formation (Figure 4.38). It has a Rb-Sr isochron age of 416 ± 20 Ma (Rundle, 1979). Though it had long been accepted from field evidence that rocks of the Carrock division are younger than those of the Mosedale division (Harker, 1894; Eastwood *et al.*, 1968; Skillen, 1973), the differing radiometric ages obtained by Rundle were the first indication that there may have been a considerable time gap between emplacement of the two divisions. A radiometric age similar to the Carrock microgranite was also determined for the Harestones rhyolite (Rundle, 1979), but with the uncertainty in the significance of the many dates of about 420 Ma in the Lake District (see discussion in the Stockdale Beck, Longsleddale GCR site report), 416 Ma may be a resetting event.

The Carrock Fell Complex has been cited as evidence in the lengthy debate about the possibility of a late Ordovician orogeny in the Lake District. This idea was proposed originally by Green (1920) to explain the truncation of major fold-like structures such as the Ulpha Syncline by the Windermere Supergroup. The near-vertical emplacement of the Carrock Fell Complex and in particular the Carrock division, into steeply dipping rocks has been used as support for such an orogenic episode (Briden and Morris, 1973; Briden *et al.*, 1973; Faller and Briden, 1978; Piper, 1997; Piper *et al.*, 1997). However, the reinterpretation recently of the Ulpha Syncline and other related folds in the Borrowdale Volcanic Group as volcanotectonic structures (Branney and Soper, 1988) removes the need for an episode of large-scale folding in the late Ordovician. Tilting of the Skiddaw and Eycott Volcanic groups before intrusion of the Carrock division need not have resulted from an orogenic event as concluded by Piper *et al.* (1997), but from synvolcanic extensional faulting.

Conclusions

The Carrock Fell Complex is a multiple dyke-like mafic–felsic intrusion containing layered gabbroic rocks that was emplaced at the junction between the Skiddaw and Eycott Volcanic groups. It is unique in the Lake District and is atypical among the intrusive rocks of England and Wales. Components within the complex are mineralogically and geochemically distinct. The gabbros of the earliest, Mosedale division are considered to be associated with the continental-margin tholeiitic rocks of the Eycott Volcanic Group and are thus of Ordovician age. These are cut by the dyke-like Carrock division, comprising a gabbroic–ferrodioritic–microgranitic suite related by fractional crystallization of a tholeiitic basalt magma and dated at 416 ± 20 Ma. The GCR site best illustrates the main features of, and the relationships between, these divisions. To the west of the site the complex also contains intrusions of feldspar-phyric micrographic microgranite and feldspar-quartz-phyric rhyolite that post-date, and are probably unrelated geochemically to, the Carrock division. The GCR site is an important focus of research into crystallization mechanisms in layered igneous rocks and is internationally significant. Though the rocks are petrographically, mineralogically and geochemically well characterized, the age of intrusion of the components remains to be understood fully.

Haweswater

HAWESWATER
(NY 480 140–500 167)

D. Millward and B. Beddoe-Stephens

Introduction

The group of cleaved and presumably related, mainly mafic, intrusive rocks that crop out within an area of about 19 km² around the northern part of the Haweswater Reservoir collectively form the Haweswater 'Complex' of Nutt (1966, 1970, 1979). This is the largest group of mafic intrusions within the outcrop of the Borrowdale Volcanic Group (BVG) and the GCR site includes excellent examples of the principal rock types, including layered dolerite and gabbro (Figure 4.42). Marginal intrusive breccias are a significant feature. The presence of dolerite in the Haweswater area was reported by Dakyns *et al.* (1897) and Walker (1904), and the rocks were described by Green (1915b), Hancox (1934) and Nutt (1979). Nutt (1979) interpreted the intrusions as a subvolcanic magma chamber and the focus of an eruptive centre for the BVG.

Description

Geological maps and descriptions of the Haweswater intrusions by Hancox (1934) and Nutt (1970, 1979) show a group of dolerite and related rocks cropping out around the Haweswater Reservoir, encompassing part of Bampton Common and Naddle Forest from Willdale Beck in the north to Whelter Knotts (472 134) in the SW and Harper Hills in the SE. The intrusions comprise a combined outcrop area of about 2.6 km² within a total area of 19 km². The host rocks are lava and pyroclastic formations within the 3200 m-thick succession of the BVG in the eastern Lake District (Nutt, 1979). The major fault along Haweswater, interpreted by Nutt (1970) as separating intrusions within the lower part of the BVG succession to

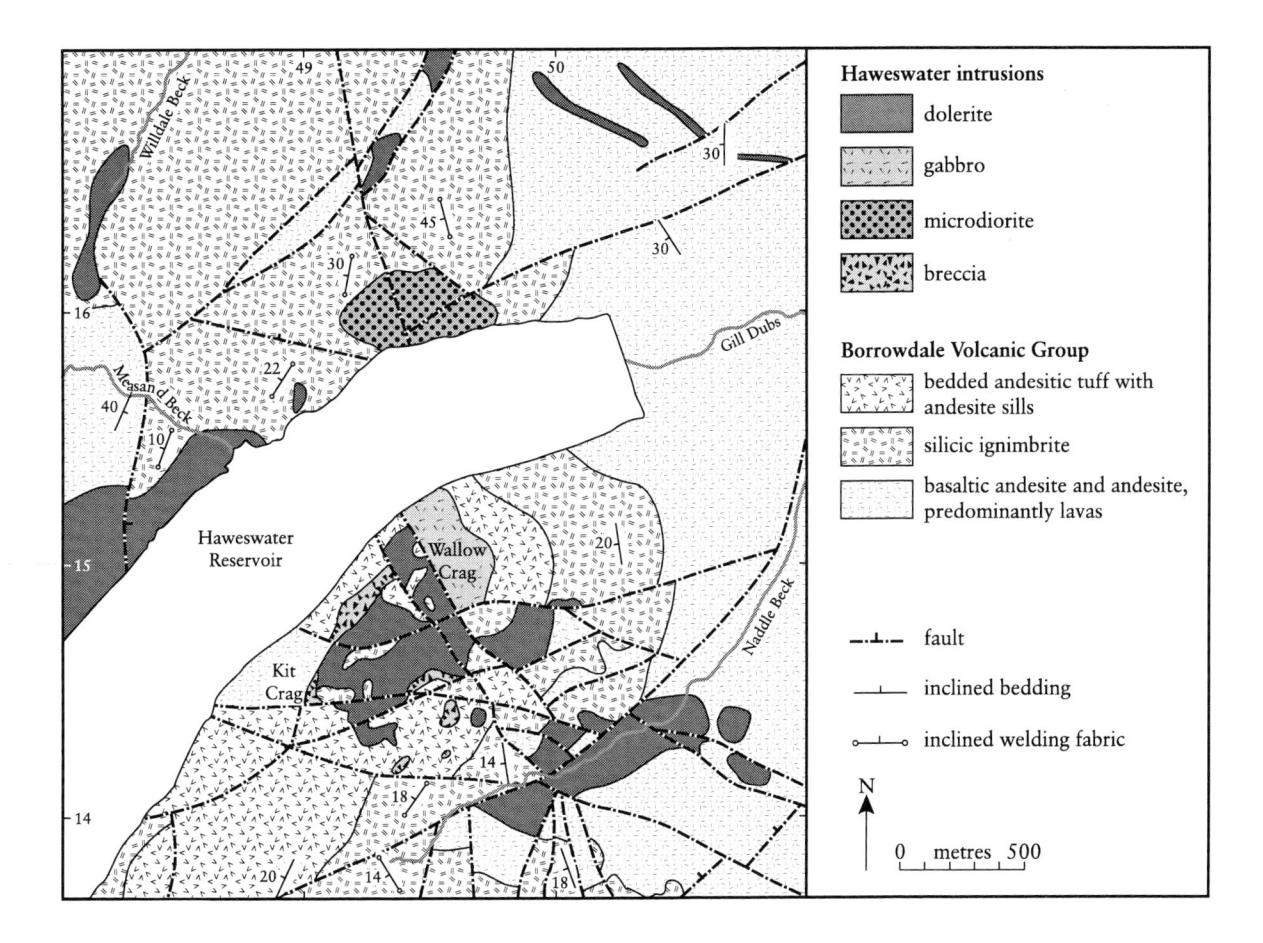

Figure 4.42 Map of the Haweswater intrusions, based on unpublished British Geological Survey maps by D. Millward and B. Beddoe-Stephens.

the NW from those within the upper part to the SE, is not supported by recent mapping of the area by the British Geological Survey (Figure 4.42); the same volcanic formations are present in both areas.

The intrusions are dominantly dolerite, with subordinate fine-grained gabbro, microdiorite, intrusive breccia and locally abundant aplitic veins. The dolerite ranges from leucocratic to melanocratic, with the more mafic rocks occurring NW of Haweswater. Leucocratic microgabbro crops out, for example, on Wallow Crag (495 151), and dolerite and gabbro on and below Wallow Crag are compositionally layered with vertical bands 2–30 cm thick. Most contacts of the intrusion with the host rocks are near vertical or vertical. Exceptions include horizontal contacts east and NE (490 145) of Kit Crag. Xenoliths are present locally within the contact zone.

North-west of Haweswater, some dolerite masses are associated with andesite along a significant fault-zone. Both NW and SE of the reservoir, marginal zones to the dolerite locally exhibit a microporphyritic texture. Microdiorite ('augite-porphyrite' of Hancox, 1934), interpreted by Nutt (1979) to have formed from the alteration of dolerite, is a minor component associated with dolerite SE of the reservoir, but forms a large body on the NW shore, south of Great Birkhouse Hill. Small garnets are a rare accessory mineral in the microdiorite.

Mineralogical alteration is moderate to intense, though original textures are preserved: most rocks are typically subophitic, intersertal or intergranular, and commonly porphyritic; locally, ophitic texture and ophimottling are present. Plagioclase is generally replaced by albite and/or mats of sericite, epidote and chlorite. Fresh clinopyroxene is present in some areas, though this is more typically replaced epitaxially by amphibole or chlorite and fibrous amphibole. Unaltered clinopyroxene in gabbros is accompanied by pseudomorphs of chlorite and/or fibrous amphibole, possibly after orthopyroxene. Interstitial quartz is common in thin-section; locally intersertal orthoclase and microperthite are present associated with micrographic intergrowths with quartz. Accessory minerals include opaque oxide, apatite, biotite and tourmaline.

Thin microcrystalline veins are abundant locally. One type comprises cloudy albite laths with amphibole, chlorite and minor secondary quartz ('doleritic aplite' of Hancox, 1934 and Nutt, 1979). Other veins consist of quartz, perthite, minor albite and accessory tourmaline ('granitic aplite' of Nutt, 1979). Hancox (1934) also described fibrous chloritic veins.

Intrusive breccia forms the marginal rocks of the faulted intrusion south of Wallow Crag (Figure 4.42). Also present is a narrow dyke on the NW shore of Haweswater (476 140). The breccia comprises fragments of dolerite and gabbro, as well as wall-rocks of andesite, rhyolite and devitrified welded tuff. The intensely altered matrix is feldspathic and andesitic. The vertical pipe of andesite breccia cutting dolerite by Low Goat Gill (506 142), SE of Naddle Beck, described by Nutt (1979, p. 729), has been re-interpreted during the recent mapping as autobrecciated andesite within the BVG country rock, capping an irregular top to the intrusion. Clasts within the breccia are wholly of andesite.

Interpretation

The age of the intrusions is poorly constrained and no radiometric age determinations have been published. Green (1915b) stated that the rocks are not cleaved and are thus implicitly post-Acadian. Later, he reported having seen samples that are cleaved, thus establishing the Early Palaeozoic age (Green, 1917). The cleavage, the regional association, and similarities in alteration styles and geochemistry with the volcanic rocks, indicated to Hancox (1934) and Nutt (1979) that the intrusions were emplaced before the regional cleavage-forming event and are Ordovician in age.

Though the Haweswater intrusions were not distinguished by the primary geological survey, Dakyns et al. (1897, pp. 20–21) described a dolerite, locally porphyritic and containing much augite, forming Wallow Crag and compared this with similar rocks from Measand and Colby (now submerged beneath the reservoir at c. 493 157), on the north side of Haweswater (Figure 4.42). Though they felt that the Wallow Crag mass is intrusive, the Measand rock was interpreted as lava.

Walker (1904) described the Haweswater mafic rocks as 'quartz-diabase' and interpreted them as intrusions; he also concluded that they are part of a large body, and similar to the lavas of Eycott Hill, which are now interpreted geochemically as tholeiitic (Fitton, 1971). Green

(1915b) used 'hypersthene-dolerite' for the Haweswater rocks, and he and Hancox (1934) both commented on the presence of orthopyroxene along with interstitial quartz and K-feldspar, characteristics that might suggest a tholeiitic affinity. However, Fitton (1971) and Nutt (1979) concluded from geochemical analyses of the intrusions that they are indistinguishable from, and thus compatible with, the calc-alkaline BVG. This conclusion does not accord with the preliminary interpretation of new, and as yet unpublished, geochemical data for the Haweswater rocks (University of Lancaster: R. Macdonald, pers. comm.), which indicate that a tholeiitic affinity is possible. If this is substantiated by further work then the dolerites cannot have been a magma chamber and vent site for the BVG. However, the Haweswater rocks do not have the high-Ti values of the tholeiitic rocks of the Eycott Volcanic Group and Carrock Fell Complex in the north (Hunter, 1980), and pre-cleavage dykes cutting the BVG and Eskdale granite in the west (Macdonald *et al.*, 1988).

Surface contact exposures indicate steep-sided bodies and previous workers concurred that the widespread group of intrusions is linked at depth. Density values for representative rock types from the intrusions fall within the range for the BVG and Lee (1986) found that the gravity anomalies over the Haweswater area can be interpreted best if the mafic intrusions are underlain, at depths of as little as 1 km, by low density material, probably a granitic mass associated with the Shap granite. It is thus unlikely that a substantial body of mafic composition is present at depth in this area.

Conclusions

The Haweswater intrusions are a unique group of coarse-grained mafic and intermediate masses within the outcrop of the Borrowdale Volcanic Group and are probably Ordovician in age. The rocks are best exposed within the GCR site. The dolerite and associated rocks have been considered as a subvolcanic magma body, but this is not supported by geophysical evidence. Moreover, if studies in progress are substantiated, suggesting that the complex may be tholeiitic, then these intrusions are further examples in the Lake District magmatic province of the association between calc-alkaline volcanic and tholeiitic intrusive rocks.

GRAINSGILL, CALDEW VALLEY (NY 327 327)

S.C. Loughlin

Introduction

The Grainsgill GCR site, located within the Caldew valley in the northern Lake District (Figure 4.43), is an area of diverse geology that has been of interest to geologists for more than a century. The northernmost of only three small outcrops of the Skiddaw granite is exposed here and the contact between the granite and the highly deformed Skiddaw Group into which it is intruded is clearly demonstrated. Locally, the northern part of the granite outcrop is extensively altered to greisen. Only 100 m north of the granite margin, the Carrock Fell Complex is intruded at the boundary between the Skiddaw Group and Eycott Volcanic Group (see also the Carrock Fell GCR site report). The greisen, the hornfelsed Skiddaw Group and the Carrock Fell Complex are cut by arsenic- and tungsten-bearing veins, the largest of which have been worked from the now abandoned Carrock Mine.

The area was first mapped by the Geological Survey (Ward, 1876) and this was followed by several accounts of specific aspects of the geology (e.g. Harker, 1895b; Finlayson, 1910). Hitchen (1934) concluded that the granite, greisen and mineralization are related to one igneous episode. Field evidence also led to the deduction that the greisen formed by metasomatism of the granite and this was confirmed by subsequent studies (Thimmaiah, 1956; Ewart, 1962; Eastwood *et al.*, 1968). Shepherd *et al.* (1976) used fluid inclusion and isotope techniques to suggest a genetic link between mineralization and alteration of the granite.

This GCR site provides a very important insight into the relationship between igneous intrusion, metasomatism and mineralization. In addition it contains important evidence of the timing of emplacement of the Skiddaw granite, one of only two late Caledonian intrusions in the Lake District. The mineral veins at Grainsgill will also be described in the *Mineralization of Great Britain* GCR volume.

Description

The oldest rocks in the Grainsgill GCR site are

very strongly folded metapelites of the Skiddaw Group. Fold interference patterns suggest that several generations of deformation are present. On the sides of Carrock Fell, north of the River Caldew, the axial planes of folds are almost vertical and trend N–S, perpendicular to the contact between the Skiddaw Group and the Carrock Fell Complex. Contact metamorphism has altered rocks of the Skiddaw Group to biotite-cordierite hornfels within a classically concentric aureole 10 km in diameter. As the greisen is approached cordierite is altered to chlorite and biotite is replaced by muscovite.

The Carrock Fell Complex is a multiple dyke-like intrusion, emplaced between the Skiddaw Group and the Eycott Volcanic Group and is exposed in Brandy Gill, 100 m to the north of Grainsgill (see the Carrock Fell GCR site report for detailed description). In Brandy Gill, the southern part of the Carrock Fell Complex comprises gabbro and xenolithic gabbro of the Mosedale division, cut upstream by apatite-bearing ferromicrodiorite, ferromicrogabbro and micrographic ferromicrogranite sheets belonging to the Carrock division.

The Skiddaw granite is exposed in Grainsgill Beck and in the River Caldew (Figure 4.43); the intrusion appears to be a steep-sided dome with its long axis approximately N–S (Hitchen, 1934). It is laterally extensive at shallow depths (Lee, 1986). The contact between the intrusion and the hornfelsed sedimentary rocks is well exposed in both Grainsgill Beck and the River Caldew just upstream from their confluence. On Coombe Height, between the two rivers, Skiddaw Group rocks overlie the flat roof of the granite.

The Skiddaw intrusion is a medium-grained biotite granite comprising orthoclase, oligoclase, quartz and biotite. Accessory minerals include zircon, apatite, ilmenite, pyrrhotite, pyrite, epidote, titanite, anatase, brookite, and rutile (Rastall and Wilcockson, 1915). In the River Caldew, feldspars are partially replaced by muscovite as a result of incipient greisen formation, and the biotite is partially replaced by chlorite. The degree of greisen formation increases northwards to Grainsgill where the original biotite granite is pervasively converted to a muscovite-quartz greisen. Apatite, rutile, pyrite and arsenopyrite also occur in the greisen.

The principal rocks within the GCR site are transected by NNE-trending tungsten-bearing veins ranging in width from less than 1 cm to

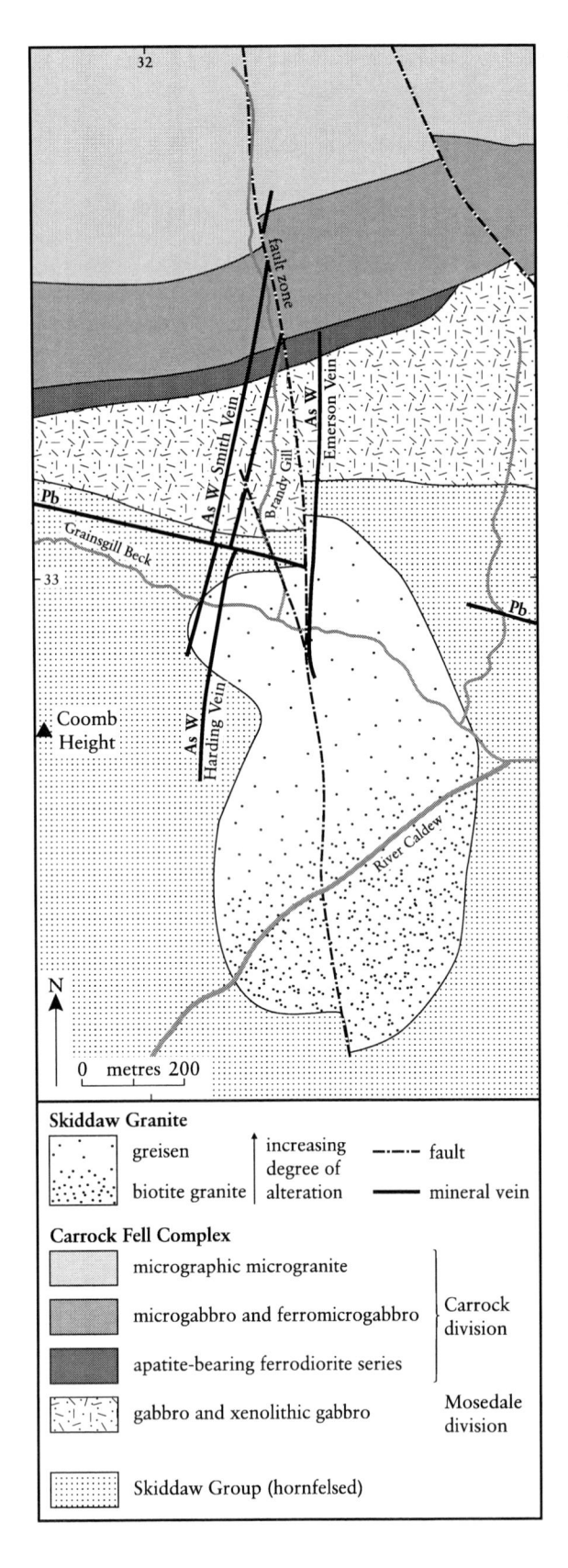

Figure 4.43 Map of the Grainsgill GCR site (from Geological Survey maps).

1.5 m and dipping steeply to the west. Within about 60 cm of large veins and less for smaller veins, the granite is pervasively altered to greisen comprising quartz, muscovite, pyrite, arsenopyrite and apatite. The veins have also altered the cordierite-andalusite-biotite hornfels of the Skiddaw Group. Biotite is altered to chlorite and andalusite to sericite as a vein is approached and within 20–30 cm of the contact, the chlorite and sericite are altered to muscovite; pyrite and tourmaline occur in small amounts in this new assemblage. Alteration is also noticeable in the gabbros of the Carrock Fell Complex within 50 cm of the veins. Within 15 cm of a vein the gabbro is completely altered to quartz, biotite, sericite and some muscovite, and immediately adjacent to the vein the original mineralogy is replaced by arsenopyrite, pyrite, quartz, calcite and muscovite.

The largest veins worked at Carrock Mine are, from west to east, the Smith, Harding and Emerson (Figure 4.43). The dominantly quartz veins contain local pockets of wolframite, scheelite, arsenopyrite, pyrite, pyrrhotite, sphalerite and ankerite along with accessory molybdenite, chalcopyrite, bismuthinite, apatite, dolomite, calcite, fluorite, and muscovite with rare joseite, cosalite, cassiterite and gold (Finlayson, 1910; Hitchen, 1934; Ewart, 1962; Shepherd *et al.*, 1976; Young, 1987). Despite variable relative mineral proportions, all the veins have a similar paragenetic sequence (Hitchen, 1934; Shepherd *et al.*, 1976).

The two near-vertical, quartz- and ankerite-filled NNW-trending fault zones in the upper and lower parts of Brandy Gill do not appear to have hydrothermally altered the country rocks (Ewart, 1962). Narrow E–W veins carrying quartz, galena and sphalerite slightly displace the tungsten-bearing veins and are considered to be related to the extensive E–W lead and zinc veins found in the Caldbeck Fells to the north of the GCR site (Shepherd *et al.*, 1976). Two veins of 'greisen' were reported by Hitchen (1934): one of these is about 1 m wide and cuts the Carrock Fell Complex in Brandy Gill and the other cuts the hornfelsed Skiddaw Group in the upper reaches of Grainsgill. Ewart (1962) found several more of these veins underground, and suggested that they may be metasomatized aplitic veins.

Interpretation

Ward (1876) described a 'very quartzo-micaceous granite' at Grainsgill and suggested that this is part of the Skiddaw granite, because the contact between it and the biotite granite exposed in the River Caldew is gradational. Harker (1895b) recognized that the granite had undergone some metasomatism, but suggested that it formed primarily as a late-stage acid melt squeezed from the crystallizing Skiddaw granite by northerly directed pressure. Finlayson (1910) related the ore minerals genetically to the greisen. Rastall and Wilcockson (1915) noted that accessory minerals of the granite are relatively scarce in the greisen, and that minerals such as arsenopyrite are more common.

Hitchen (1934) proposed that the granite, greisen and mineralization are related to one igneous episode. He described the gradual change from biotite granite to greisen, involving the replacement of feldspars by muscovite. Where the greisen is entirely quartz and muscovite, the original outlines of feldspar crystals can still be discerned, even in hand specimen. He suggested further that the large number of mineral veins in the greisen, and the increase in abundance of arsenopyrite, indicate that intense localized hydrothermal activity had occurred. The scarcity of accessory minerals was attributed to their removal by hydrothermal fluids. Hitchen concluded that, following the consolidation of the granite, greisen was formed as a result of hydrothermal activity and aqueous solutions deposited minerals in fissures and cracks.

Subsequent work confirmed the metasomatic origin of the greisen (Thimmaiah, 1956; Ewart, 1962; Eastwood *et al.*, 1968). Thimmaiah (1956) and Ewart (1962) considered that the only true greisen in the area is the thoroughly altered granite near the quartz-tungsten veins, whereas Eastwood *et al.* (1968) also described the altered hornfels adjacent to the veins as greisen. However, geochemical studies by Roberts (1983) showed that, despite their similar appearance in the field, the metasomatized hornfels has a very different chemistry to the greisen.

A genetic link between alteration of the granite and mineralization was proposed by Shepherd *et al.* (1976) using fluid inclusion and oxygen isotope data. They found that mineral deposition and greisen formation indicate equilibration with fluids of similar isotopic compositions under similar P–T conditions. K-Ar dating could not distinguish between cooling of the

Skiddaw granite, greisen formation and tungsten mineralization (minimum age of granite intrusion, 392 ± 4 Ma; mean age of mineralization alteration, 385 ± 4 Ma) suggesting that they occurred within a relatively short space of time. Fluid inclusions in vein quartz contain NaCl brines and high tungsten concentrations occur in the altered granite of the mineralized area. Based on their data they suggested that mineralizing fluids were moderately saline with high Na/K, enriched in tungsten and periodically charged with CO_2; compared with magmatic fluids the mineralizing fluids were depleted in $\delta^{18}O$. Shepherd *et al.* (1976) produced a model of ore genesis whereby meteoric water was drawn convectively into the northern part of the Skiddaw granite through the adjacent relatively permeable igneous rocks of the Carrock Fell Complex. Chemical exchange increased the salinity of these fluids before they reached the granite where they became mixed with hot magmatic fluid enriched in tungsten. The granite was metasomatized to form greisen and the circulating fluids became enriched in silica due to the breakdown of primary igneous silicates. These fluids convected and mixed with more non-magmatic fluids along the N–S faults which cut the complex, producing the mineral veins.

Roberts (1983) proposed that leaching of the greisen and metasomatized hornfels by the circulating fluids provided the Na for the NaCl brines found in the fluid inclusions. Therefore, the initial fluid did not necessarily have high Na/K ratios. He favoured a model in which brines enriched in K due to feldspar alteration were drawn down to the intrusion and then resulted in the formation of greisen and K-metasomatism.

A recent model proposed by E. S. Burden (Derby University, pers. comm. 1996) combined fluid inclusion studies with mineralogical data and mass-balance calculations to show that greisen can form from granite simply by the addition of H_2O. The process involves a series of linked ionic sub-reactions and the alteration of K-feldspar to muscovite releases K-ions which are necessary to convert plagioclase to muscovite.

Geochemical data for the Skiddaw granite show clear magmatic trends consistent with either in-situ fractionation or derivation from a deeper fractionating magma chamber (O'Brien *et al.*, 1985). The Skiddaw granite has highly fractionated rare earth element (REE) patterns and, like the Shap granite, is thought to be derived from a mafic source containing residual garnet. It also contains elevated levels of the heat-producing radioactive elements and the geothermal potential of the granite has been assessed (Webb and Brown, 1984a; Wheildon *et al.*, 1984).

The age of emplacement of the Skiddaw granite has been discussed by several authors (Miller, 1961; Brown *et al.*, 1964; Shepherd and Darbyshire, 1981) and K-Ar dating of fresh biotite gives an age of 399 ± 8 Ma (recalculated from Shepherd *et al.*, 1976), very similar to the Shap granite (Rundle, 1982). Eastwood *et al.* (1968) showed that the hydrothermal alteration associated with the granite post-dates the Carrock Fell Complex. In the outer parts of the Skiddaw granite aureole, andalusite clearly overgrows the main cleavage indicating that contact metamorphism post-dates the Acadian deformation. However, in the inner hornfels zone of the aureole, the cleavage weakly wraps around andalusite porphyroblasts. The implication is that the granite was emplaced during the late stages of the Acadian Event, perhaps during a period of stress relaxation between successive cleavage-forming events (Soper and Roberts, 1971; Soper and Kneller, 1990; see also the Shap Fell Crags GCR site report).

Conclusions

The late Caledonian Skiddaw granite shows a clear northwards gradation from biotite granite to greisen and it can be shown that hydrothermal alteration associated with emplacement of the granite was also responsible for the greisen formation and the deposition of some economically important mineral veins. The nearby intrusive Carrock Fell Complex provided a permeable pathway for fluids to circulate down towards the granite and explains why metasomatism and mineral deposition are concentrated only in the northern part of the Skiddaw granite. The Grainsgill GCR site is significant because it provides an important insight into the relationship of a late Caledonian intrusion to greisen formation and mineralization. The site also provides important evidence regarding the timing of emplacement of late Caledonian intrusions in the Lake District.

SHAP FELL CRAGS
(NY 555 084)

S.C. Loughlin

Introduction

The Shap Fell Crags GCR site comprises the Shap Pink Quarry and is the principal exposure of the Shap granite, renowned as a distinctive, decorative building stone. The igneous processes that can be demonstrated within the quarry make this an internationally significant site. The site (Figure 4.44) has been a popular field locality for undergraduate students and amateur geologists for many years. The granite and its associated metasomatism and mineralization have been described by many authors (e.g. Harker and Marr, 1891, 1893; Rastall and Wilcockson, 1915; Grantham, 1928; Firman, 1957, 1978a, 1978b) and it is featured in numerous classic geological textbooks (e.g. Teall, 1888; Hatch *et al.*, 1971; Holmes, 1993).

The origin of the abundant pink Carlsbad-twinned K-feldspar megacrysts, which are the trademark of the Shap granite, has been a source of much controversy. The megacrysts occur not only in the granite but also within microgranular mafic enclaves within the granite, leading to the suggestion that the megacrysts are porphyroblasts resulting from late-stage crystallization from potassium-rich metasomatic fluids (Vistelius, 1969; Firman, 1978b; Le Bas, 1982a). However, recent work favours a phenocrystic origin for the megacrysts (e.g. Vernon, 1986; Lee *et al.*, 1995; Cox *et al.*, 1996; Lee and Parsons, 1997).

The intrusion is steep sided, and was emplaced close to the boundary between the Borrowdale Volcanic Group and the Windermere Supergroup (Locke and Brown, 1978). Radiometric dates obtained by a number of methods indicate an Early Devonian age (Pidgeon and Aftalion, 1978; Wadge *et al.*, 1978; Rundle, 1992). The granite was originally thought to post-date Caledonian deformation (Boulter and Soper, 1973), but recent field evidence proves that it was emplaced during, not after, the late Caledonian Acadian Event (Soper and Kneller, 1990). Thus, the granite is crucial in dating Caledonian deformation events within the Lake District.

A geochemical study of the Shap granite was made by O'Brien *et al.* (1985). Elevated levels of the radioactive elements U, Th, Rb and K explain the high present-day heat production of the granite which has been investigated as a potential source of geothermal energy (Wheildon *et al.*, 1984).

Description

Grantham (1928) recognized that the Shap granite was formed in three stages. To the west of the quarry, near Wasdale Head Farm (549 081), is a grey granite ('stage' I) which grades up into, and is transgressed by, the familiar pink, coarse-grained granite ('stage' II) which occupies up to 90% of the intrusion. The 'stage' III granite is less common and cuts the earlier mass in dyke-like bodies between 1 cm and 1 m wide.

The 'stage' I granite contains about 15% pink, Carlsbad-twinned orthoclase-perthite megacrysts up to 5 cm in length. The groundmass is composed of orthoclase, plagioclase zoned from andesine to albite, quartz, and biotite. By contrast, the 'stage' II granite contains less biotite and up to 30% pink K-feldspar megacrysts which commonly show a preferred alignment. The 'stage' III granite is very similar in appearance to the 'stage' II granite, but contains up to 60% pink K-feldspar megacrysts and has even less biotite. Accessory minerals include titanite, apatite, magnetite, zircon, fluorite, monazite, allanite, amphibole and pyrite (Firman, 1978b). Complex microtextures in the megacrysts have been described by Lee *et al.* (1995), Cox *et al.* (1996) and Lee and Parsons (1997). The megacrysts also contain numerous inclusions of plagioclase, biotite and quartz that increase in size and abundance towards the rim.

The 'stage' I and 'stage' II granites contain abundant enclaves (Figure 4.45). Most common are rounded, microdioritic enclaves comprising fine- to medium-grained aggregates of plagioclase, quartz, biotite and K-feldspar. Slightly rounded, pink K-feldspar megacrysts with oligoclase rims may constitute 5–8% of the enclaves; some of these megacrysts occur partly within the enclave and partly within the granite host. Clots of granite matrix up to 1.5 cm across have been reported from within some enclaves (Cox *et al.*, 1996). In addition, the 'stage' II granite contains enclaves of 'stage' I granite that range from only a few centimetres across to rafts with dimensions of $36 \times 30 \times 6$ m (Grantham, 1928). Identifiable xenoliths of hornfelsed andesitic

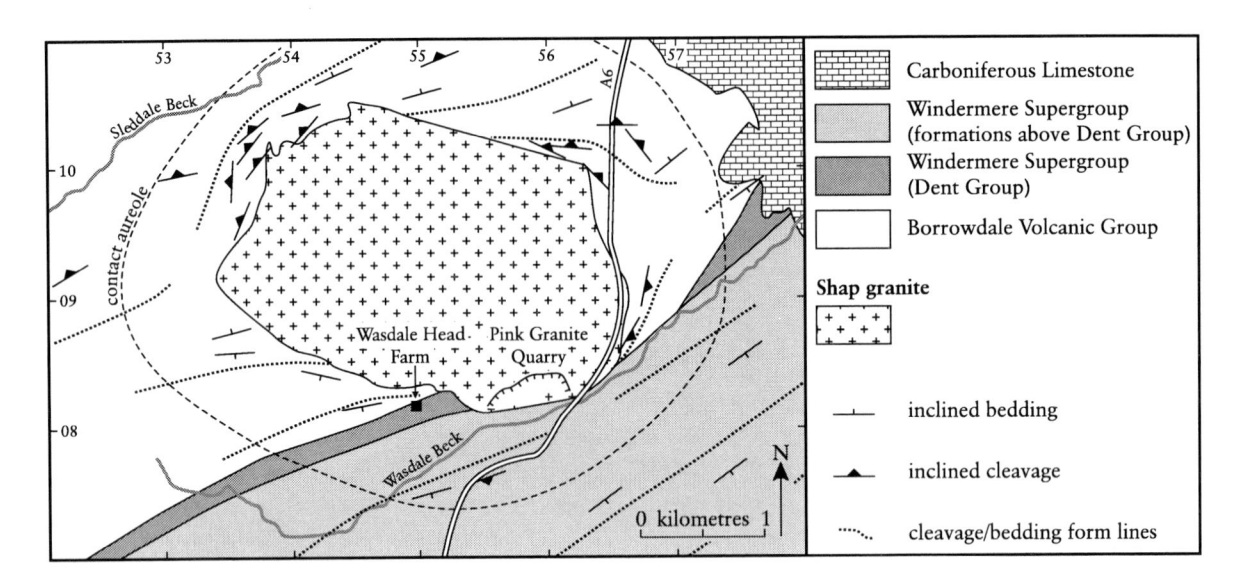

Figure 4.44 Map of the Shap Fell Crags GCR site (after Soper and Kneller, 1990).

rocks from the Borrowdale Volcanic Group occur near the margins of the intrusion. Xenoliths of Dent Group lithologies (formerly the Coniston Limestone) were abundant near the southern contact of the intrusion, but are now rare because quarrying has advanced away from this contact. Harker and Marr (1891) noted that K-feldspar megacrysts do not occur in 'xenoliths formed from solid rock fragments'.

Aplitic veins are common within the granite and country rocks, and pegmatitic rocks are common locally; both contain quartz, K-feldspar and a little plagioclase. Molybdenite and hydrothermal vein minerals coat many surfaces within a pervasive, blocky joint system. Two distinct vein assemblages are seen: the first contains quartz, calcite, bismuthinite and chalcopyrite, and the later veins contain quartz, calcite, haematite, fluorite and baryte. These veins commonly penetrate the centre of 'stage' III granite bodies giving the orthoclase on either side of the vein a deeper pink colour. Where such veins cut the 'stage' II granite, darkening of the feldspars may occur up to 5 m from the vein giving rise to the 'Dark Shap', a highly prized variety of the Shap granite that is difficult to extract. 'Light Shap' refers to the normal unaltered pink variety of granite.

Though the Shap granite outcrop is small (c. 5 km^2), gravity data (Lee, 1986) and contact metamorphism suggest that it extends at shallow depths at least 10 km farther to the NW. Andesite and tuff within the metamorphic aureole have been converted to biotite hornfels with some amphibole in places. The more aluminous sedimentary rocks of the Windermere Supergroup may contain sillimanite, andalusite and cordierite. The country rocks are folded and faulted and a near-vertical cleavage is deflected around the Shap granite (Figure 4.44; Boulter and Soper, 1973; Soper and Kneller, 1990). At several localities in the volcanic rocks of the metamorphic aureole, biotite has clearly overgrown the cleavage (Boulter and Soper, 1973). The granite has an associated dyke swarm which is crudely orientated NE–SW, sub-parallel to the Acadian folding and faulting.

Interpretation

Early researchers proposed that the K-feldspar megacrysts in the Shap granite are phenocrysts (Harker and Marr, 1891; Grantham, 1928), though others preferred a porphyroblastic origin (Vistelius, 1969; Le Bas, 1982a). However, modern microtextural, geochemical and isotopic analysis favours a phenocrystic origin (Vernon, 1986; Lee et al., 1995; Cox et al., 1996; Lee and Parsons, 1997). Features cited as evidence for this theory include the following:

1. The apparent alignment of some megacrysts.
2. The higher Ba content in K-feldspar megacrysts than in K-feldspar in the groundmass suggesting earlier crystallization.

Figure 4.45 Pink granite with xenoliths containing K-feldspar megacrysts. (Photo: D. Millward.)

3. Chemical and isotopic zoning of the megacrysts consistent with growth in a magma body.
4. Inclusions in the megacrysts which increase in size and frequency towards the rim suggesting that the megacryst and inclusions were growing simultaneously in a magmatic environment.
5. The typically euhedral inclusions exhibit zonal alignment implying growth in a magmatic environment.

Evidence for a porphyroblastic origin is the occurrence of megacrysts within the microdioritic enclaves and, more importantly, in some examples transgressing the margins of enclaves. The scarcity of K-feldspar in the groundmass and clear evidence of metasomatism prompted several authors to suggest that late-stage potassic metasomatic fluids circulated through the granite causing porphyroblastic growth of orthoclase (Vistelius, 1969; Firman, 1978b; Le Bas, 1982a).

The microdioritic enclaves were termed the 'Early Basic Granite' by Grantham (1928), who proposed that they represent an early peripheral hybrid granite. Grantham considered xenoliths of hornfelsed Borrowdale Volcanic Group rocks to be very rare except at the margins of the intrusion, whereas Firman (1978b) suggested that most enclaves represent material from the Borrowdale Volcanic Group at varying stages of recrystallization and assimilation. Recent work suggests that the enclaves formed during mixing between a dioritic magma and the host granite, and that some megacrysts may have been incorporated into the dioritic magma during that process (Vernon, 1986; Cox et al., 1996). This mechanism may account for the slightly rounded appearance of the megacrysts and the oligoclase rims, which are absent from megacrysts in the granite. Some K-feldspar megacrysts at the boundary between granite and microdiorite enclaves contain shells of inclusions that are difficult to explain if the megacrysts are porphyroblasts (Vernon, 1986). In discussion of this problem Vernon (1986) also cited the rare occurrence in granites elsewhere of xenoliths as well as megacrysts transgressing the boundary of an enclave. He considered this as evidence of their incorporation before solidification.

Based on geochemistry, O'Brien et al. (1985) showed that the Shap granite is unlikely to have been generated from a sedimentary source such as the Skiddaw Group, despite having moderately high initial $^{87}Sr/^{86}Sr$ values of 0.707 (Wadge et al., 1978) and high $\delta^{18}O$ of +11.0‰ (Harmon and Halliday, 1980). The Skiddaw Group sedimentary rocks contain particularly high concentrations of boron and on mantle-normalized plots show a pronounced negative phosphorous anomaly; neither of these features is a characteristic of the Shap granite. It is calc-alkaline with a restricted composition. High levels of large-ion lithophile (LIL) elements in the Shap granite suggest that it is magmatically evolved, but compatible elements such as Mg, Cr, Ni, Ti, V, Sr and Ba show no evidence for an extended history of crystal fractionation. Changes in fluid content or pressure, and partial hybridization appear to have dominated the geochemical evolution of these rocks. It is unclear to what extent metasomatism might have fundamentally increased LIL concentrations, though the mobility of K and Rb during metasomatism has been recognized for a long time (e.g. Vistelius, 1969). O'Brien et al. (1985) proposed that the Shap granite was the site of discharge of high-temperature fluids from depth thus explaining high concentrations of Li (Farrand, 1960) and other incompatible elements such as Rb.

O'Brien et al. (1985) also showed that the Lake District granites as a whole have increasingly fractionated rare earth element patterns with time. The Shap granite has the most fractionated REE pattern and also the highest La/Yb suggesting that it was derived from a mafic source containing residual garnet. Ordovician (Caradoc) Lake District granites with lower La/Yb were derived from a source without residual garnet, perhaps reflecting changing thermal conditions during magma generation (O'Brien et al., 1985). The younger, Early Devonian granites of Shap and Skiddaw contain higher abundances of radioactive elements than the other Lake District granites. Spears (1961) recognized that 95% of the alpha radioactivity in the Shap granite comes from accessory minerals (e.g. titanite, monazite and zircon).

It has long been assumed that emplacement of the Shap granite post-dated the main Acadian deformation event, because contact metamorphic minerals have overgrown cleavage within the metamorphic aureole of the granite (Boulter and Soper, 1973). However, some felsitic microgranite dykes within the Shap dyke-swarm, which cut folded and cleaved Silurian sedimentary rocks, are themselves weakly cleaved, implying that the dykes were emplaced during cleavage formation (Soper and Kneller, 1990). Some of these cleaved dykes cut the Shap granite and therefore the granite itself may also have been emplaced during the cleavage-forming event. Boulter and Soper (1973) interpreted the deflection of the cleavage around the Shap granite as having resulted from the forcible injection of the intrusion. However, it is also possible that the cleavage 'wraps' around the granite because shortening continued after the granite was emplaced (Figure 4.44; Soper and Kneller, 1990). Soper and Kneller inferred from this evidence that the cleavage formed incrementally, with periods of stress relaxation allowing the injection of dykes and possibly the emplacement of the granite itself.

The age of the Shap granite is therefore crucial in determining the timing of cleavage formation. Using K-Ar and Rb-Sr methods on biotite, several early authors dated the Shap granite as late Silurian or Early Devonian (Kulp et al., 1960; Lambert and Mills, 1961; Dodson et al., 1961). In a study of these and other available isotopic dates, Brown et al. (1964) recognized that the Shap granite is coeval with the Skiddaw granite. More recently, three separate radiometric dating methods have produced similar dates: U-Pb on zircons gave an age of 390 ± 6 Ma (Pidgeon and Aftalion, 1978); 394 ± 3 Ma was obtained using the mineral–whole-rock Rb-Sr isochron (Wadge et al., 1978), but it is the K-Ar age of 397 ± 7 Ma on fresh biotite that is taken to indicate the age of emplacement of the Shap granite (Rundle, 1992). These conclusions are consistent with the age of cleavage formation of between 418 ± 3 Ma and 397 ± 7 Ma recently obtained on mica concentrates from the Foredale metabentonite in the Ribblesdale inlier (Merriman et al., 1995). Thus, Acadian deformation in northern England occurred in the Early Devonian, some 23 Ma after Silurian deformation in the Southern Belt of the Southern Uplands (c. 420 Ma, Barnes et al., 1989).

Conclusions

The Shap granite is a widely used decorative building stone whose principal outcrop is within the Shap Fell Crags GCR site. This internationally significant site illustrates several important features that are crucial to the continuing debate concerning the magmatic evolution of granites and their large K-feldspar crystals. The 397 Ma Shap granite, its metamorphic aureole and dyke-swarm provide crucial evidence in the determination of the sequence and timing of the late Caledonian, Acadian Event in the Lake District. Field evidence suggests that the deformation was episodic and that the granite and the associated dyke swarm were emplaced during periods of stress relaxation.

Chapter 5

Central England

Introduction

INTRODUCTION

D. Millward

Occurrences of Caledonian intrusive and volcanic rocks in central England comprise a relatively small number of exposures within Lower Palaeozoic and Precambrian inliers, and provings from deep boreholes (Figure 5.1). Many of the localities and borehole sites are on the NE margin of the Midlands Microcraton, though some occur within it. For many years these igneous rocks were considered as Precambrian, along with those of Charnwood Forest. However, Le Bas (1972) proposed a Caledonian age for the igneous rocks at Warboys, Mountsorrel, South Leicestershire, Nuneaton and Dost Hill on the basis that they all could be inferred to intrude Lower Palaeozoic rocks. Recently, geochemical data and an increasing number of accurate U-Pb zircon dates have confirmed that igneous events occurred in this area during the late Proterozoic (at *c.* 615 Ma) and during the Ordovician (Noble *et al.*, 1993; Pharaoh *et al.*, 1993). The Precambrian occurrences are described in the *Precambrian of England and Wales* GCR volume.

Caledonian volcanic rocks in central England are almost entirely concealed, with the exception of the Barnt Green Volcanic Formation (Old *et al.*, 1991). The lower Silurian volcanic rocks of the Mendips and Tortworth area are described in Chapter 6 along with rocks of Skomer with which they are probably associated. Volcanic rocks are known to be widespread from well-documented borehole records (Pharaoh *et al.*, 1991, 1993; Figure 5.1). The volcanic rocks are calc-alkaline, arc-related and have been distinguished from the late Proterozoic rocks on their trace element abundances and isotopic compositions (Pharaoh *et al.*, 1991, 1993; Noble *et al.*, 1993). Though the compositional range includes basaltic andesite, andesite and rhyolite, felsic tuff predominates; no dacite is recorded. An Ordovician age has been determined for felsic tuff from the North Creake Borehole (449 ± 13 Ma, U-Pb zircon, Noble *et al.*, 1993).

The Ordovician plutons are also calc-alkaline (Le Bas, 1972, 1982b) and one group is aligned broadly along a NW–SE belt of crust with strongly positive aeromagnetic anomalies (Allsop, 1987; Pharaoh *et al.*, 1991, 1993). The recent summary of the occurrences and geochemistry of the igneous rocks of Central England by Pharaoh *et al.* (1993) has shown that the largely Triassic cover rocks conceal a substantial magmatic province of late Ordovician age. Pharaoh *et al.* (1993) speculated that the magnetic anomaly belt may mark the magmatic core of a continental calc-alkaline volcanic and plutonic province. Le Bas (1972, 1982b) and Pharaoh *et al.* (1991) have suggested that this belt of arc-related rocks extends eastwards into Belgium. The development of the putative arc at the eastern margin of Avalonia may have been the product of the subduction of oceanic lithosphere during closure of the Tornquist Sea (Pharaoh *et al.*, 1993). The lithologies, geochemistry and timing of Caledonian magmatism have considerable similarities with the Lake District magmatic province, but the tectonic relationships are yet to be understood.

Plutonic rocks

Three groups of plutonic rocks are recognized (Figure 5.1). The first of these, known as the 'South Leicestershire diorites', crops out on the margin of the Midlands Microcraton SW of Leicester. The second group is associated with a belt of positive aeromagnetic anomalies up to 10 km wide and extending about 125 km from Hathern, near Derby to St Ives in Huntingdonshire (Allsop, 1987); examples of these plutonic rocks include the exposed Mountsorrel complex and occurrences in the Rempstone, Kirby Lane and Warboys 1 boreholes. The xenolithic granodiorite from Rempstone is petrographically similar to Mountsorrel. Farther NE, in south Lincolnshire, cleaved and altered granophyric microgranite in the Claxby borehole has been dated at 457 ± 20 Ma (Noble *et al.*, 1993). The microgranite is geochemically similar to Ordovician intrusions of the Lake District (Pharoah *et al.*, 1997). While it is part of a third group of silicic plutons interpreted from geophysical anomalies in the area around the Wash (Busby *et al.*, 1993), it is not considered to be a major component of that batholith (Pharoah *et al.*, 1997).

The *South Leicestershire diorites* comprise diorite, tonalite and microtonalite and are, or were formerly, exposed at Stoney Stanton (SP 490 950), Croft (SP 510 967), Coal Pit Lane Quarry (SP 542 992), Enderby (SP 542 992), Red Hill Quarry, Narborough (SP 532 975) and Narborough Quarry (SP 525 975); the Countesthorpe borehole also penetrated these rocks

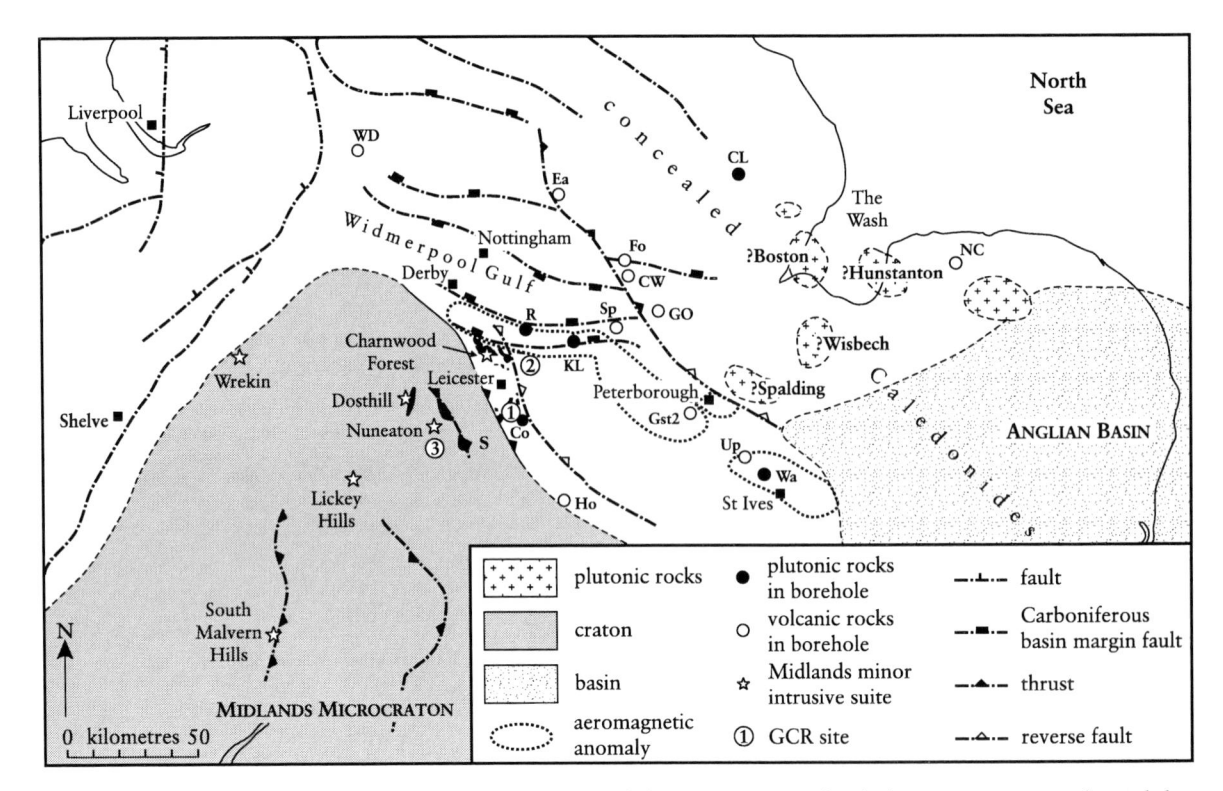

Figure 5.1 Map of central England showing locations of the occurrences of Caledonian igneous rocks and the GCR sites (after Pharaoh *et al.*, 1993). GCR sites: 1, Croft Hill; 2, Buddon Hill (Mountsorrel); 3, Griff Hollow. Occurrences of plutonic rocks: Cl, Claxby; Co, Countesthorpe; KL, Kirby Lane; R, Rempstone; S, South Leicestershire diorites; Wa, Warboys 1. Occurrences of volcanic rocks: CW, Coxs Walk; EA, Eakring 146; Fo, Foston; GO, Great Osgrove Wood 1; Gst2, Gas Stamford 2; Ho, Hollowel; NC, North Creake 1; Sp, Sproxton; Up, Upwood 1; WD, Woo Dale 1.

(Figure 5.1). The diorite is quartz bearing and contains hornblende and sparse augite. The tonalite is described in detail from the Croft Hill GCR site. Le Bas (1972) considered that these occurrences form a composite pluton about 14 km wide, intruding the Cambrian Stockingford Shale Group. The age of the intrusions is taken at 449 ± 18 Ma (U-Pb; Pidgeon and Aftalion, 1978; recalculated by Noble *et al.*, 1993).

The *Mountsorrel complex*, about 10 km north of Leicester, comprises gabbro, diorite and granodiorite. The gabbro is exposed only on a small island in Swithland Reservoir and is composed of labradorite and ophitic brown hornblende enclosing relict augite; pseudomorphs after olivine may be present. The diorite is similar to the South Leicestershire suite (above) but contains more biotite. However, the most extensive rock type is a biotite granodiorite well illustrated by the Buddon Hill GCR site near Mountsorrel. The age of the complex is taken at 463 ± 32 Ma (U-Pb; Pidgeon and Aftalion, 1978; recalculated by Noble *et al.*, 1993).

Midlands Minor Intrusive Suite

The Cambrian and Tremadoc rocks within, and at the margin of, the Midlands Microcraton are intruded by lamprophyre (spessartite) and diorite dykes and sills. The field occurrence, petrography and geochemistry are detailed in Carney *et al.* (1992), Thorpe *et al.* (1993a) and Bridge *et al.* (1998). Exposures of these rocks are in quarries close to Nuneaton, near the Wrekin in Shropshire and in the Malvern Hills (Figure 5.1). The suite is typified by the 50 m-thick composite sill of spessartite, hornblende diorite and hornblende meladiorite described from the Griff Hollow GCR site; that sill has a U-Pb emplacement age of 442 ± 3 Ma (Noble *et al.*, 1993). In the Tremadoc Shineton Shales of the Wrekin area a single lenticular mass is exposed in an old quarry (SJ 645 087). The mineralogy of this occurrence differs from those at Nuneaton in containing no magmatic amphibole and up to 20% clinopyroxene. The presence of olivine and absence of quartz distinguishes this group from the other Caledonian igneous rocks of central England.

A small, faulted inlier at the southern end of the Lickey Hills, south of Birmingham (Figure 5.1) comprises water-laid crystal and crystal-lithic tuffs together with other volcaniclastic sedimentary rocks of the Tremadoc Barnt Green Volcanic Formation, intruded by aphyric microdiorite intrusions (Old *et al.*, 1991). Geochemical comparisons led Carney *et al.* (1992) to suggest that these rocks may belong to an early, extrusive phase related to the Midlands Minor Intrusive Suite, thus implying that this magma type was available over a substantial time-span.

CROFT HILL
(SP 510 967)

J. N. Carney and T. C. Pharaoh

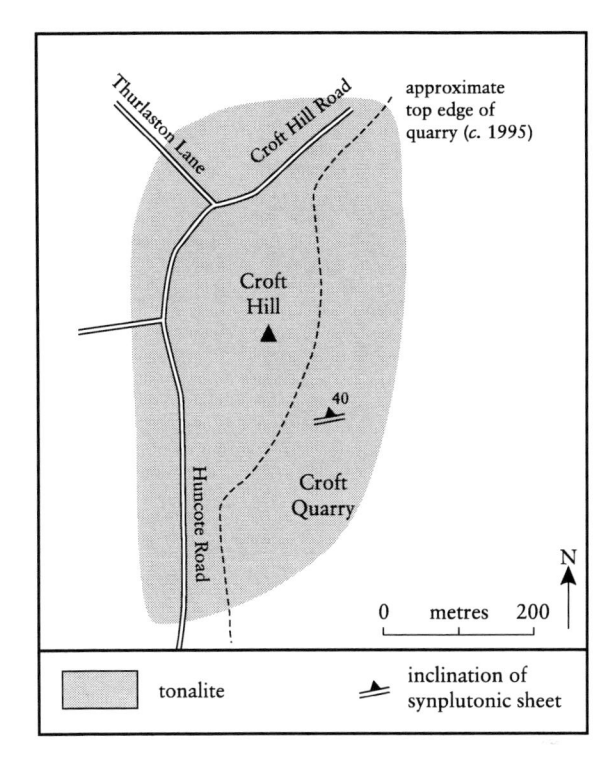

Figure 5.2 Map of the Croft Hill GCR site.

Introduction

The Croft Hill site exhibits coarse-grained plutonic rocks as small crags and pavements on the summit and flanks of the 123 m-high Croft Hill, and in the face of Croft Quarry, which is excavated into the SE side of the hill (Figure 5.2). These rocks have been studied petrographically over a long period of time; they were originally described as syenite (Hill and Bonney, 1878) before Whitehead (in Eastwood *et al.*, 1923) suggested that, as no alkali feldspar is present, they should be classified as quartz-diorite or tonalite. The Croft pluton is now assigned to an assemblage of calc-alkaline intrusive rocks which, because of their geographical distribution, are collectively termed the 'South Leicestershire diorites' (Le Bas, 1968). These bodies are mostly hidden beneath Triassic strata, but have magnetic properties that enable them to be traced at depth as a series of small batholiths. The Croft Hill exposure is part of a composite pluton about 14 km wide, which is linked at depth with similar plutonic rocks cropping out around Stoney Stanton to the west and Enderby to the east (Allsop and Arthur, 1983).

The significance of these rocks to the geology of central England is, in part, their age. A U-Pb date of 449 ± 18 Ma was obtained on zircon from the tonalitic rocks exposed at Enderby NE of Croft Hill (Pidgeon and Aftalion, 1978; recalculated by Noble *et al.*, 1993), and is the currently accepted emplacement age of the South Leicestershire diorites. The plutons therefore belong to an Ordovician (late Caradoc) intrusive event, contemporaneous with the subduction-related magmatism of central Wales and the Lake District (Pharaoh *et al.*, 1993). Their presence confirms that an extension of the Caledonian orogenic belt lies beneath much of central and eastern England (Le Bas, 1972; Pharaoh *et al.*, 1987).

The Croft Hill site with its adjacent quarry offers extensive exposures demonstrating the petrology and internal intrusive history of a typical South Leicestershire pluton. It is also the location for an analcime-molybdenite style of mineralization, which is here more intensively developed than elsewhere in the pluton.

Description

On the NW flank of Croft Hill small crags of inequigranular tonalite display abundant small white plagioclase phenocrysts set in a crumbly yellow or brown, medium-grained weathered base. At the summit, pavements of the same

Figure 5.3 View of Croft Hill, showing the NW face of Croft Quarry. To the left of the picture, the dark-grey parallel lineaments dipping from top left to bottom right represent a swarm of synplutonic intrusive sheets. (Photo: J. N. Carney.)

rock type are transected by an orthogonal fracture system, the principal trends of which are 360°, 260° and 240°.

Below the summit of Croft Hill, in the NW face of Croft Quarry (Figure 5.3), is exposed massive, pale-grey, inequigranular medium- to coarse-grained tonalite. This is characterized by common large crystals of white euhedral plagioclase, up to 6 mm long, within a pink, medium-grained quartzo-feldspathic base studded with black oxide granules. The tonalite has a hypidiomorphic, inequigranular texture, with abundant euhedral plagioclase crystals, some with labradorite cores and rimmed by grainy, inclusion-filled albite. Surrounding these crystals are aggregates of smaller, inclusion-filled sodic plagioclase crystals which are in part idiomorphic and in part form an interlocking granular intergrowth with quartz. Clinopyroxene forms sporadic euhedra and aggregates largely altered to chloritic minerals; small tatters of biotite are similarly altered. Plagioclase is pervasively replaced by patches and veinlets of albite. Pumpellyite occurs interstitially, and Webb and Brown (1989) noted radial prehnite infilling cavities and zeolites occupying veins.

A major feature of the intrusion occurs lower down in the same quarry face, where several sheets of darker-grey tonalite, 1 m to 3 m wide, extend up the face for tens of metres and form a well-spaced 'swarm' dipping at about 40° to the NNW. The contact between these sheets and the host tonalite is sharp and irregular; slivers of the sheets are incorporated as xenoliths in the host. Neither the sheet nor the host is chilled, but the latter has intimately permeated the sheets as diffusely margined stringers of pale-grey tonalite which divide the sheet into rectangular to ovoid segments. The grey tonalite is mineralogically similar to the host, but is medium grained and non-porphyritic. Elsewhere in the quarry augite-bearing microdiorite xenoliths are common in the tonalite (Le Bas, 1968).

Prominent joint systems seen in the quarry below Croft Hill comprise a master set dipping 25° NNE and a subordinate set at right angles to this. In the SW quarry face, a prominent discontinuity outlines what appears to be a gentle dome-shaped structure, dividing intrusive rocks with different joint orientations. The nature of this structure cannot be determined as this part of the quarry is inaccessible.

The mineralization in Croft Quarry principally comprises the replacement of feldspar minerals by albite and analcime, producing the characteristic pink colour of the tonalites (King, 1968). The final stages of this alteration is known by local miners as 'rammel', which occurs in layers up to 12 m thick of completely disaggregated rock, commonly with cavities lined by crystals of analcime, calcite and quartz; prehnite, datolite, laumontite and dolomite are accessory minerals of this assemblage. The occurrence of minor amounts of molybdenite in this association is unusual because it normally belongs to a high-temperature environment, whereas analcime characterizes mesothermal and lower temperature environments (King, 1968).

Triassic strata mantle the southern part of the Croft tonalite and the highly irregular contact is well exposed just outside the limits of the site (Le Bas, 1993).

Interpretation

The rocks of Croft Hill constitute one of the few exposures of Caledonian igneous 'basement' in central England. Research during the 20th century has emphasized the unusual mode of occurrence of these rocks, as a pinnacle or inselberg buried beneath Triassic strata (Bosworth, 1912). Recent geophysical investigations have demonstrated that a much larger parent body is present at depth (Allsop and Arthur, 1983). Thus, the composition and variation seen within the Croft intrusion should be considered together with the very similar lithologies exposed at Stoney Stanton and Enderby, and encountered in the Countesthorpe borehole (Le Bas, 1972).

The Croft tonalitic intrusion is evidently part of a small, zoned batholith, with diorites to the west and microtonalite to the east (Le Bas, 1972). The Croft Quarry exposures suggest that even on a small scale, the batholith may have multiple phases of intrusion. The first phase comprised pale-grey inequigranular coarse-grained tonalite, which constitutes most of the northern face below Croft Hill. Parallel sheets of a darker-grey and more evenly grained tonalite were then emplaced into the main body, probably as a series of synplutonic intrusions because neither these sheets nor the host show chilling. Subsequent minor remobilization of the host resulted in brecciation of the synplutonic sheets. They were then extensively invaded by stringers emanating from the host, and in part stoped-out to form xenoliths within the latter.

The South Leicestershire diorites exhibit a strongly calc-alkaline geochemical signature (Le Bas, 1972; Webb and Brown, 1989). Pharaoh *et al.* (1993) noted that the Croft rocks show moderate enrichments of large-ion lithophile elements (K, Rb and Ba), Th and Ce, and relative depletion of Nb and Ta, which are patterns typical of calc-alkaline magmas arising within a volcanic arc founded on continental crust. The subduction zone above which the magmas were generated may have been situated to the east of central England in late Ordovician times, its activity related to the phase of plate convergence that closed the Tornquist Sea (Noble *et al.*, 1993).

The widespread extent of albite and analcime replacement suggests pervasive deuteric alteration of the Croft rocks, a process possibly enhanced by the relatively complex emplacement history of the pluton evidenced by its zonation on a regional scale. The occurrence of molybdenite in association with these secondary minerals is unusual and not understood fully.

Conclusions

The tonalites of the Croft Hill GCR site belong to the suite of late Ordovician intrusions known as the 'South Leicestershire diorites'. They represent one of the few exposures of the central zone of a small batholith, the larger part of which lies hidden beneath Triassic rocks. On a regional scale, the Croft body shows compositional zoning, indicative of a complex intrusive history. This is demonstrated at the smaller scale in Croft Quarry, where several parallel sheets of equigranular tonalite cut the host inequigranular coarse-grained tonalite, but were then brecciated and invaded during subsequent mobilization of the latter. Pervasive post-emplacement alteration of the tonalites involved the conversion of feldspar to albite, analcime and other zeolites, and was accompanied by minor molybdenite mineralization. The age and calc-alkaline geochemistry of the Croft rocks show that they were generated within a SE extension of the Caledonian magmatic belt and were contemporaneous with the volcanic and intrusive rocks of central Wales and the Lake District.

Central England

BUDDON HILL
(SK 562 154)

J. N. Carney and T. C. Pharaoh

Introduction

The Buddon Hill GCR site comprises isolated rocky knolls in woodland to the north and east of the Swithland Reservoir and includes most of the central and western parts of the Mountsorrel roadstone quarry (Figure 5.4). Features that underline the importance of the site to studies of central England Caledonian magmatism include: the considerable extent of exposure, the variety of igneous rock types and contact relationships that they portray, and the more silicic nature of the Mountsorrel complex (Le Bas, 1968, 1972) compared with the otherwise compositionally similar Croft pluton (see the Croft Hill GCR site report). Geophysical studies indicate that the complex extends beneath younger cover rocks to the village of Thrussington, about 8 km to the east (Hallimond, 1930; McLintock and Phemister, 1931). Borehole provings also suggest that comparable granodioritic rocks are regionally developed (Pharaoh *et al.*, 1993).

Early workers thought the Mountsorrel rocks were part of the Precambrian Charnwood Forest massif (e.g. Hill and Bonney, 1878). Lowe (1926) drew attention to the predominantly granodioritic composition, whereas Jones (1927) suggested, on the basis of jointing, that they represent an intrusion of 'post-Charnian' age. Meanwhile, Watts (prior to publication in 1947) had concluded that the rocks have petrological affinities with the igneous 'Caledonian group' of the Lake District. This view was seemingly confirmed by Meneisy and Miller (1963), who obtained a K-Ar age of 379 ± 17 Ma on biotite from a diorite from Brazil Wood, to the SW of Buddon Hill. However, the emplacement age is constrained better by the U-Pb determination of 463 ± 32 Ma (early Caradoc) on zircon from the Mountsorrel complex (Pidgeon and Aftalion, 1978; recalculated by Noble *et al.*, 1993). The Caledonian age of the Mountsorrel rocks is in keeping with their strongly calc-alkaline geochemistry, comparable to some intrusions of the Brabant Massif of Belgium and the English Lake District (Le Bas, 1972; Pharaoh *et al.*, 1993). These rocks therefore provide evidence for Ordovician calc-alkaline magmatism in eastern England, probably associated with subduction of

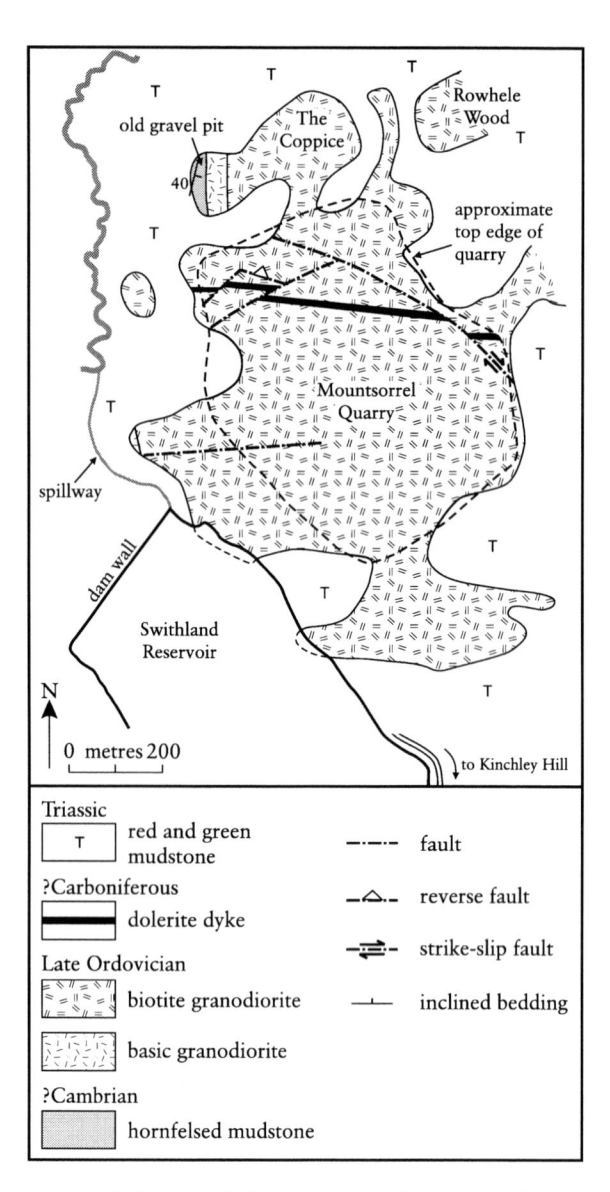

Figure 5.4 Map of the Buddon Hill GCR site, based in part on BGS 1:10 560 scale mapping.

oceanic lithosphere from an eastern Tornquist branch of the Iapetus Ocean (Pharaoh *et al.*, 1993).

The site contains one of the few exposures of contact relationships between these plutonic rocks and cleaved, hornfelsed mudstone. Le Bas (1968) noted that the metasedimentary country rocks are more iron-rich, and less siliceous, than the Swithland Slates of the Precambrian Charnian Supergroup, and may instead correlate with Cambrian strata similar to the Stockingford Shale Group exposed around Nuneaton (e.g. Taylor and Rushton, 1971). However, there is no direct evidence to support either hypothesis.

A basic dyke, thought to belong to the late Carboniferous alkaline cycle of magmatism (Le Bas, 1968), and the highly irregular unconformable base of the Triassic strata are displayed within the Mountsorrel Quarry.

Description

The slopes above the NE shore of Swithland Reservoir (5607 1461) expose pink-weathered, inequigranular coarse-grained biotite granodiorite which represents the voluminous, main phase of the Mountsorrel complex. Laths and euhedral plates of plagioclase (about 60%) are surrounded by hypidiomorphic-granular aggregates of albite, inclusion-filled perthitic alkali feldspar and clear quartz; the mafic minerals are mainly brown biotite and yellow to green hornblende. The granodiorite encloses sporadic pink to grey, rounded xenoliths of equigranular diorite averaging 50 mm across. It is cut by sheets of fine- to medium-grained aplitic microgranite between 10 mm and 0.15 m wide: similar late sheets are visible in exposures northwards along this slope, up to the wall of the dam.

North-east of the dam the granodiorite is coarser grained and more inequigranular, a texture that is well displayed at the summit of the prominent knoll overlooking the spillway (5575 1501). This exposure exhibits two phases of aplitic intrusion (Figure 5.5); the first comprises a vertical sheet 0.45 m thick striking NNE and the second, a cross-cutting sheet dipping 40° to the NE. At the base of this knoll, to the west, in a small quarry is a pale-grey, mafic-rich 'basic granodiorite' (Le Bas, 1968), suggesting proximity to the margin of the pluton.

Contact relationships between granodiorite and country rock were described by Watts (1947) at the 'Old Gravel Pit' (5582 1555), about 670 m NNE of the dam. At this locality, now much overgrown, purple, micaceous, hornfelsed mudstone is traversed by sinuous veinlets of pink granodiorite, and by sheets several centimetres wide of pink to grey, medium-grained equigranular 'basic granodiorite'. According to Watts (1947), the hornfels contains garnet and cordierite.

Granodiorite forming the eastern part of Mountsorrel Quarry appears relatively devoid of aplitic microgranite sheets and contains only sporadic, small rounded xenoliths of dark-grey medium-grained diorite. Farther west, aplitic sheets and xenoliths are more common, and two phases of intrusion can be recognized, the second accompanied by marginal chilling of the sheets. The fresh granodiorite is a grey, inequigranular, coarse-grained rock speckled with biotite. High-temperature alteration, which has caused reddening of feldspars and replacement of the mafic minerals, occurs in association with wide zones of closely spaced E–W orientated joints, the surfaces of which show slickenfibre development. Within the altered zones, King (1968) determined a lower temperature mineral assemblage of dolomite, epidote, chlorite, quartz, chalcopyrite and pyrite together with minor amounts of galena, calcite, haematite and baryte. The style of this mineralization is comparable to that seen at Shap in the Lake District (see the Shap Fell Crags GCR site report, Chapter 4).

A sub-vertical dolerite dyke 2–3 m wide crosses the quarry, following an approximately E–W course related to the trend of the principal fracture systems in the granodiorite. It is dark-green, medium grained and reminiscent in appearance to Carboniferous minor intrusions found in the region; the precise age is uncertain. In the eastern quarry face the dyke shows an apparent dextral offset along the NW fault which crosses the quarry. The western face exhibits an offset along an arcuate reverse fault dipping steeply to the NE. A second mineralizing event has filled joints parallel to the dolerite and consists of cavernous veinlets with dolomite, calcite, bitumen, a clay mineral and pyrite. A hydrothermal origin for the bitumen was suggested by Sylvester-Bradley and King (1963), and Ponnamperuma and Pering (1966).

In the eastern face of the quarry an excellent example of a Triassic valley eroded into the granodiorite is preserved. The lower part of the valley is occupied by a breccia of granodiorite fragments, overlain by red and green parallel-bedded mudstone and siltstone.

Interpretation

The site exposes a variety of plutonic rocks near to the western margin of the Mountsorrel complex. Close to the contact, the exposures show that the plutonic rocks have veined the country rock, with the growth of garnet and cordierite in hornfelsed mudstone of possible Cambrian age.

The earliest phase of the Mountsorrel complex may be the dioritic xenoliths. These are

Figure 5.5 Exposure of granodiorite at Buddon Hill showing two phases of aplite intrusion. (Photo: T. C. Pharaoh.)

possibly cognate inclusions derived from a basic precursor that was disrupted during intrusion of the main-phase biotite granodiorite, though other explanations are possible. The persistence of basic granodiorite close to the western contact of the pluton was attributed by Taylor (1934) to reaction between granodiorite and pre-existing diorite. A different interpretation was proposed by Le Bas (1968), who suggested that at Kinchley Hill, 450 m south of the site, the intrusive relationships indicate that granodiorite and diorite were contemporaneous, facilitating processes of hybridization that basified the former and acidified the latter. The main-phase biotite granodiorite was subsequently veined by at least two generations of aplitic microgranite sheets. These were interpreted by Taylor (1934) to be the relatives of a sodic 'aplogranite magma' which, after hybridizing at depth with gabbro or diorite, had given rise to the main body of granodiorite. Alternatively, the aplitic rocks could be differentiates of the granodiorite injected back into the partially cooled pluton. This late magmatic environment favoured the molybdenite mineralization seen in association with aplitic and pegmatitic rocks in other parts of the Mountsorrel complex (King, 1959, 1968).

The locally extensive low-temperature alteration of the Mountsorrel body was attributed to deuteric mineralization by King (1968), though it may also be related to the formation of E–W joint systems, implying an underlying tectonic control.

Major element geochemical data discussed by Le Bas (1972) demonstrate the calc-alkaline magmatic lineage of the Mountsorrel complex.

Studies of trace elements show enrichment of the large ion lithophile elements Th and Ce with respect to high field-strength elements such as Nb, Zr and Y, further confirming the calc-alkaline arc affinities of the complex and its similarity to the South Leicestershire diorites (Pharaoh *et al.*, 1993). All of these rocks are therefore interpreted as products of late Ordovician subduction-related magmatism.

The final magmatic event, uniquely exposed in Mountsorrel Quarry, was the emplacement of the E–W-trending dolerite dyke, possibly in Carboniferous times. Hydrothermal fluid circulation caused the low-grade mineralization that is spatially associated with the dyke. The combination of dextral strike-slip and reverse faulting that affected the dyke suggests a significant component of Variscan transpression which may have contributed to the structural complexity of the Mountsorrel pluton.

Conclusions

The Buddon Hill GCR site represents the only extensive exposures of the Mountsorrel complex of late Ordovician calc-alkaline plutonic rocks. It clearly demonstrates a sequence of intrusion commencing with diorite and biotite granodiorite and closing with aplitic microgranite. Alteration of these phases by water-rich fluids derived from the magmas, to assemblages that include albite may have been, at least in part, structurally controlled. One of the main features of the site is an exposure of the western contact of the complex. This shows granodiorite veining fine-grained metasedimentary rocks which have been converted to a garnet-cordierite hornfels. A dolerite dyke crossing the Mountsorrel Quarry may be Carboniferous in age; it has acted as a passive marker, illustrating a possible Variscan phase of deformation. The highly irregular unconformity at the base of the Triassic strata is preserved around the margin of Mountsorrel Quarry.

GRIFF HOLLOW (SP 361 895)

J. N. Carney and T. C. Pharaoh

Introduction

The Griff Hollow GCR site is situated within the Nuneaton inlier of pre-Devonian rocks (Bridge *et al.*, 1998). It is largely occupied by the Griff No. 4 aggregate quarry (Figure 5.6). The site has been selected because it exposes the type example of a composite hornblende diorite sill, one of a swarm of concordant bodies collectively termed the Midlands Minor Intrusive Suite (Carney *et al.*, 1992). The sill, about 50 m thick, is intruded into Upper Cambrian mudstone belonging to the Outwoods Shale Formation of the Stockingford Shale Group (Taylor and Rushton, 1971). Its emplacement age has been constrained at 442 ± 3 Ma by a U-Pb determination on baddeleyite from a pegmatitic segregation (Noble *et al.*, 1993). This late Ordovician (earliest Ashgill) age is similar to that determined for the South Leicestershire diorites exposed farther east.

The Midlands Minor Intrusive Suite differs from the South Leicestershire diorites in a number of important respects. The Griff Hollow composite sill is intimately associated with narrow sheets of fine-grained lamprophyre, a rock that is not associated with the South Leicestershire diorites. Petrographical studies further support the assertion made by Allport (1879) that the Nuneaton sills contain olivine and are thus devoid of essential quartz. Other significant geochemical differences exist between the two Ordovician intrusive suites (Bridge *et al.*, 1998). Therefore the site is important in comparative studies aimed at resolving the causes of both regional and local petrological variations within the late Ordovician (Caledonian) magmatic rocks of central England.

The NW corner of the quarry contains an exposure of the unconformity between the sill and overlying Coal Measures sandstones, providing field evidence for the pre-Carboniferous (Westphalian) age of the Midlands Minor Intrusive Suite.

Description

Igneous layering within the Griff Hollow sill dips at about 20° to the SW, parallel to the base of the sill and concordant with bedding in the Stockingford Shale Group. The host strata contain slumped bedding, but were evidently well consolidated at the time of intrusion; they are not spotted, but are flinty and hard for several centimetres from the contact.

The sole of the intrusion consists of a 20 cm-thick chilled zone of fine-grained spessartite

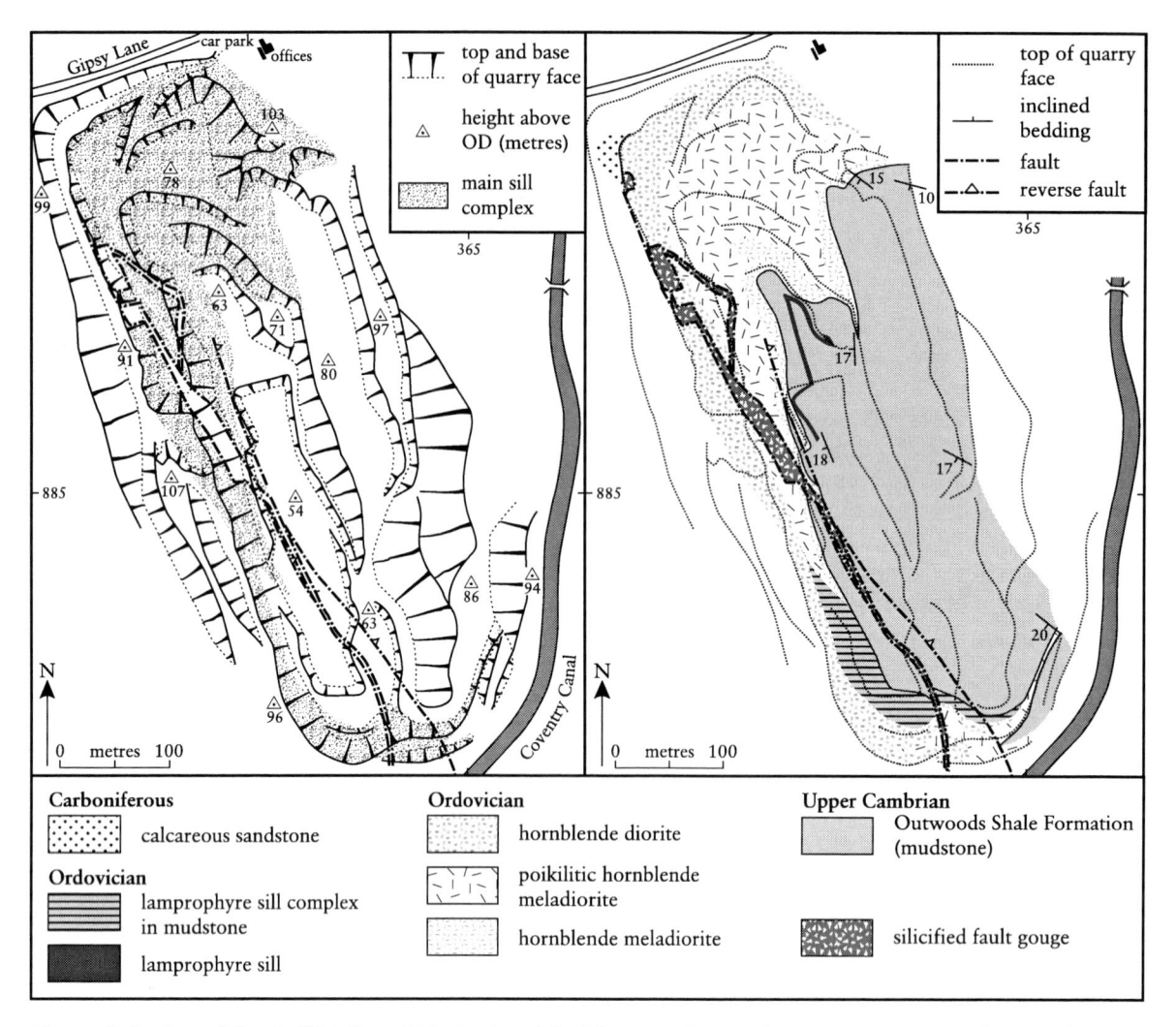

Figure 5.6 Map of the Griff Hollow GCR site (modified from Bridge *et al.*, 1998). Positions of quarry faces in 1990 are shown.

lamprophyre. Abundant flow-aligned plagioclase laths are accompanied by a few per cent of small plagioclase and chloritized mafic phenocrysts. The base is much altered with interstitial chlorite and carbonate, which also fill small vugs. The lowest part of the sill in the SE of the quarry is replaced by a sheeted complex consisting of *c.* 1 m-wide lamprophyre sills interfingered with the host mudstone. These sills are texturally and mineralogically similar to the basal facies of the main part of the composite sill and to discrete lamprophyre sheets cutting the rest of the Griff Hollow intrusion.

Overlying the basal lamprophyre are 4–5 m of dark-grey, medium-grained hornblende meladiorite characterized by a thickly developed planar foliation, parallel to the basal contact of the sill, and consisting of dark-grey diorite interlayered with a paler-grey, more feldspathic variety. The meladiorite is highly magnetic due to the presence of several per cent of iron-titanium oxides in the rock. Mafic minerals, comprising some 40% of the rock, consist of pale-yellow to brown euhedral hornblende, altered olivine and pyroxene. Euhedral plagioclase forms the remainder of the rock. The importance of volatiles in the late-stage crystallization of the intrusion is indicated by the occurrence of much interstitial chlorite, carbonate and pyrite.

Sharply succeeding the hornblende meladiorite is a layer of poikilitic hornblende meladiorite, about 22 m thick (Figure 5.7), which represents the most mafic-enriched part of the sill. In its lower part it is a highly distinctive black, pyritous, coarse-grained rock with hornblende crystals up to 25 mm long. The rock is dominated by areas in which large interlocking plates of dark reddish-brown hornblende poikilitically enclose highly altered olivine, clinopyroxene and plagioclase; the accessory minerals are iron-titanium oxides, apatite, and secondary actinolite, chlorite, pyrite and pumpellyite. This facies passes up into a more plagioclase-rich meladiorite in which the clinopyroxene is fresher. In the middle to lower parts the poikilitic meladiorite layer contains sporadic elliptical pegmatitic segregations. These show cores enriched in pink albite intergrown with large crystals and aggregates of white mica, pyrite and carbonate. In the south-eastern part of the sill (Figure 5.7), the poikilitic hornblende meladiorite forms metre-long pods within the hornblende diorite.

The topmost facies of the sill, about 24 m thick, consists of pale-grey, pyritous, coarse-grained hornblende diorite with prominent white plagioclase and black hornblende laths.

Interpretation

The Griff Hollow intrusion contains the complete assemblage of lithologies found in composite diorite sills of the late Ordovician Midlands Minor Intrusive Suite. Lamprophyres that occur along the sole and within the main body of the intrusion are interpreted as the chilled equivalents of the magmas which formed the diorite sills.

Previously, all of the Nuneaton sills were described as camptonites (e.g. Lapworth, 1898; Le Bas, 1968), but Hawkes (in Taylor and Rushton, 1971) noted that the diopsidic nature of the pyroxene, sodic composition of the plagioclase and absence of any obvious alkaline mineralogy is more in keeping with a classification as the spessartite variety of lamprophyre. However, Hawkes was incorrect to suggest that olivine is absent. The presence of olivine is compatible with the recent classification of spessartite as advocated by Rock (1987). Recent studies have shown that hornblende compositions in the Griff sill range from edenitic to pargasitic (Bridge *et al.*, 1998), and these are also compatible with a spessartite lineage for the Midlands Minor Intrusive Suite magmas.

Vertical mineralogical variations within the sill indicate that the amount of olivine and pyroxene remains approximately constant between the hornblende meladiorite and the succeeding

Figure 5.7 View of the western face of Griff No.4 Quarry. The base of the sill is at the foot of the lowest face, behind the stockpile in the centre of the photograph. The middle face, dark-grey in tone, exposes the poikilitic hornblende meladiorite layer and the upper face is in pale-grey hornblende diorite. The regular bedding above the latter represents the Coal Measures unconformably overlying the sill. (Photo: J. N. Carney.)

poikilitic hornblende meladiorite layer. The latter contains correspondingly larger amounts of interstitial hornblende, which suggests crystallization of a magma enriched in iron, magnesium and related elements. The coarse-grained, in places pegmatitic, texture of the poikilitic hornblende meladiorite further suggests relatively slow crystallization of hornblende in the presence of a volatile phase. These relationships show that in-situ fractionation involving mafic crystal accumulation is unlikely to have caused the compositional layering, and suggest that the intrusion formed by the multiple injection of related magma batches (Thorpe *et al.*, 1993). The field observations generally support multiple intrusion of these sills (e.g. Le Bas, 1968).

Geochemical data relevant to the petrological processes that contributed to the compositional diversity of the Griff Hollow sill are discussed in Thorpe *et al.* (1993) and Henney (in Bridge *et al.*, 1998). Henney showed that only one of the Griff Hollow rocks shows Eu enrichment, indicating that feldspar fractionation is unlikely to have occurred. Similarly, the values of MgO, Ni and Cr in most cases show a limited range indicating that they have undergone only minor fractionation, and in some rocks their abundances are comparable with those of unfractionated, primitive mantle-derived melts. Thorpe *et al.* (1993) also favoured a model involving low-degree, mantle-derived partial melting under volatile-rich conditions, with varying degrees of crystal fractionation of these liquids at the base of the crust or during ascent.

The tectonic setting of magma generation can be inferred from the geochemical data. Thorpe *et al.* (1993) noted that on the total alkali–silica diagram, the diorites and lamprophyres from Griff Hollow have an alkalic trend, with compositional affinities that span the basalt–trachyandesite fields and including types with normative olivine and hypersthene or nepheline. The rocks have high TiO_2 values relative to the South Leicestershire diorites, and Zr/Y and Zr concentrations appropriate to magmas generated in within-plate tectonic settings. However, the high La/Ta and Th/Ta ratios (Thorpe *et al.*, 1993), and high Ba/Ta (Henney, in Bridge *et al.*, 1998) also suggest the involvement of a subduction zone component in their genesis. It is considered probable that the Midlands Minor Intrusive Suite originated from lithospheric mantle previously enriched during Ordovician subduction, and subjected to low-degree partial melting during cessation of subduction. This occurred at a time when the accretion of Caledonian orogenic terranes had enlarged the Midlands cratonic crust (Thorpe *et al.*, 1993).

Conclusions

The Griff Hollow GCR site exposes diorites and lamprophyres belonging to the late Ordovician Midlands Minor Intrusive Suite, and emplaced into the Upper Cambrian Stockingford Shale Group. The site contains a typical example of a composite layered hornblende diorite sill, consisting of a lamprophyric sole, passing upwards into a hornblende-enriched facies and pale hornblende diorite. The diversity of rock types has not arisen from in-situ crystal fractionation, but from processes involving partial melting and possibly fractionation at deep levels within the crust or upper mantle, followed by the multiple intrusion of genetically related magma batches. The Griff Hollow sill belongs to essentially the same late Ordovician (Caledonian) magmatic episode as the quartz-bearing South Leicestershire diorites, but differs petrographically and geochemically, and may have been emplaced slightly later, when the tectonic setting of this region was undergoing transition from a subduction-controlled to a within-plate type of regime.

Chapter 6

Wales and adjacent areas

Introduction

INTRODUCTION

R. E. Bevins

Geological setting

In early Ordovician times Wales lay at a latitude of approximately 60°, in the southern hemisphere on the margin of a major continent, Gondwana (Cocks and Fortey, 1982; Torsvik and Trench, 1991). This continent was separated from two other major continents, Laurentia and Baltica, by the Iapetus Ocean and Tornquist Sea respectively. Active subduction of Iapetus oceanic lithosphere beneath the continental margin of Gondwana was associated with extension of the overlying crust and the generation of abundant magmatism (Bevins *et al.*, 1984; Kokelaar *et al.*, 1984b). This extension also led to the development of the Welsh Basin, a marine basin up to 150 km wide, which was to dominate the palaeogeography of Wales for over 100 million years. During mid-Ordovician times, Eastern Avalonia, a continental fragment or microcontinent, split away from Gondwana and drifted northwards, linked to continued subduction. Collision with Laurentia may have occured as early as Caradoc times in this sector of Iapetus. Wales was a part of this microcontinent (Soper, 1986) and the character and timing of igneous activity across the Welsh region was dictated by this subduction-dominated tectonic evolution, from the earliest activity in southern Snowdonia and north Pembrokeshire in Tremadoc times through to the last recorded major activity in southern Pembrokeshire, of Llandovery age.

Sedimentation throughout the region was largely of marine character; for long periods of time, black muds accumulated in deep waters, interrupted periodically by incursions of coarser sediment, transported by gravity flows. At the margins of the Welsh Basin shallower water conditions prevailed, and periodically the Midland Platform to the SE of the basin was emergent. Sedimentation was also strongly influenced by relatively short-lived volcanic episodes, when large volumes of coarser detritus were available for reworking in the marine environment (Kokelaar *et al.*, 1984b).

Kokelaar (1988) has argued convincingly that magmas in the Ordovician Welsh Basin were channelled up major fractures that cut through the crust, in an overall sinistrally transtensional tectonic environment. Major volcanism was restricted to narrow graben-like zones above the most important fractures, leading to the development of thick local accumulations. In some cases, as in central Snowdonia, extension was linked to caldera development (Howells *et al.*, 1991).

The geochemistry of the various Caledonian igneous rocks across Wales has been studied in considerable detail. The earliest lavas show a strong subduction-zone influence, being low-K tholeiitic to calc-alkaline in character (Kokelaar *et al.*, 1984b). Variations from basic to silicic composition have been identified, resulting from low pressure fractional crystallization (Kokelaar, 1986). Subsequent Ordovician events were characterized by bimodal sequences, showing a range of chemical affinities, from Mid-Ocean-Ridge Basalt (MORB) types through low-K tholeiites to calc-alkaline, some lava sequences showing more than one chemical type (Bevins, 1982; Bevins *et al.*, 1984; Smith and Huang, 1995). The origin of the silicic rocks has been a matter of considerable debate, linked variably to melting of the crust as the hot mafic magmas rose towards the surface (Kokelaar *et al.*, 1984b), to fractional crystallization of the mafic magmas (Bevins *et al.*, 1991), or to a combination of these processes (Leat *et al.*, 1986). Studies of the chemical variations seen within a single intrusion showing extreme compositional variation in South Wales (Bevins *et al.*, 1994), and also from detailed studies of a closely related suite of minor silicic intrusions and ash-flow tuffs in North Wales (Campbell *et al.*, 1987; Thorpe *et al.*, 1993b) favour the bulk of the silicic rocks being derived from more mafic compositions through fractional crystallization.

The chemistry of late Ordovician igneous rocks from Snowdonia suggest that subduction beneath Eastern Avalonia ceased during Caradoc times (Leat and Thorpe, 1989), although the effects of the subduction are still to be seen in later igneous rocks. An episode of igneous activity in South Wales during Llandovery times was more alkaline in composition than earlier episodes, with the eruption of basalts, hawaiites and mugearites, as well as peralkaline rhyolitic rocks (Thorpe *et al.*, 1989). These lavas still, however, show a chemical signature related to the modification of mantle-derived magmas most probably by fluids driven off from subducted oceanic crust.

Wales and adjacent areas

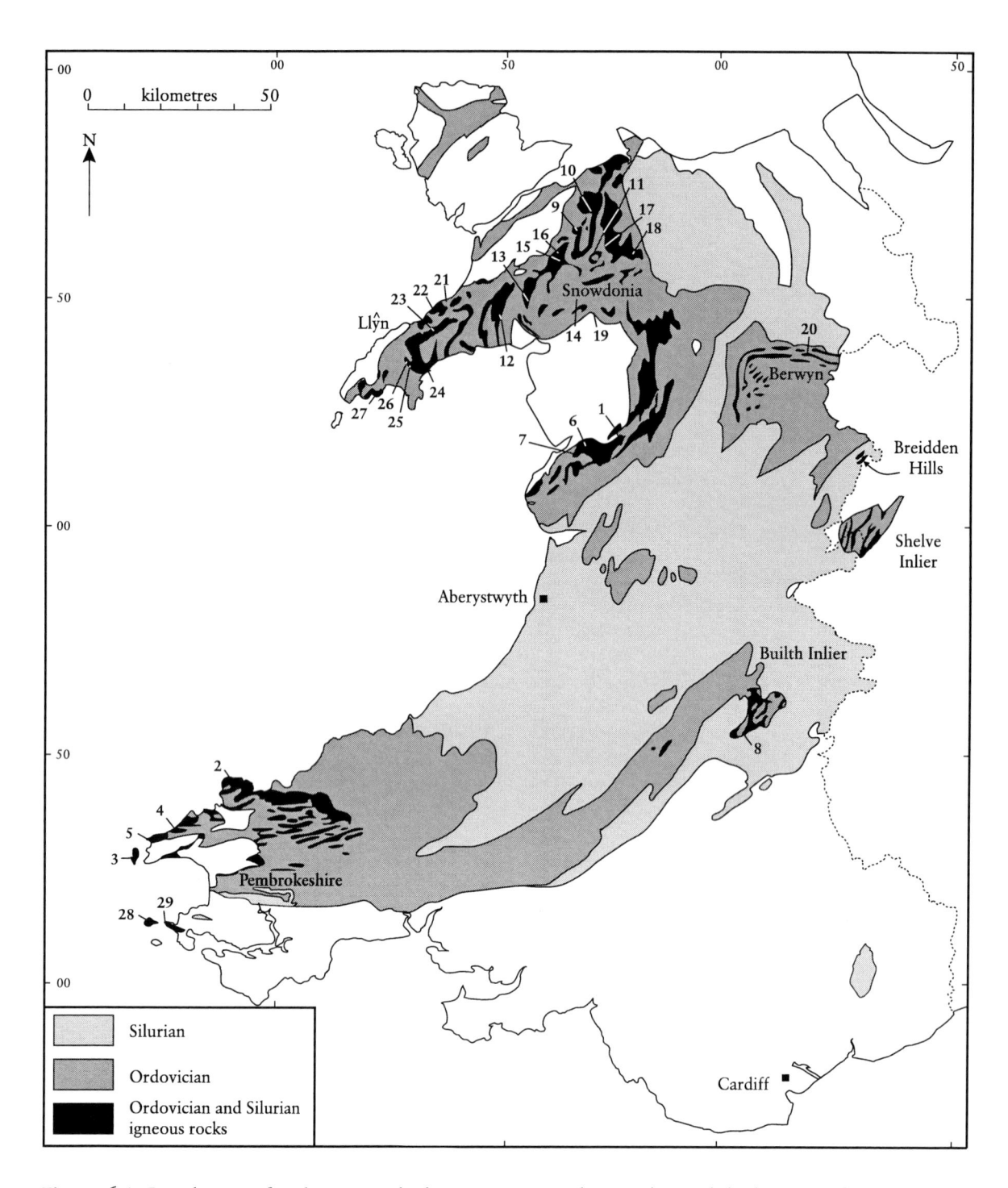

Figure 6.1 Distribution of Ordovician and Silurian igneous rocks in Wales, and the location of GCR sites. 1, Rhobell Fawr; 2, Pen Caer; 3, Aber Mawr to Porth Lleuog; 4, Castell Coch to Trwyncastell; 5, St David's Head; 6, Cadair Idris; 7, Pared y Cefn-hir; 8, Carneddau and Llanelwedd; 9, Braich tu du; 10, Llyn Dulyn; 11, Capel Curig; 12, Craig y Garn; 13, Moel Hebog to Moel yr Ogof; 14, Yr Arddu; 15, Snowdon massif; 16, Cwm Idwal; 17, Curig Hill; 18, Sarnau; 19, Ffestiniog granite quarry; 20, Pandy; 21, Trwyn-y-Gorlech to Yr Eifl quarries; 22, Penrhyn Bodeilas; 23, Moelypenmaen; 24, Llanbedrog; 25, Foel Gron; 26, Nanhoron quarry; 27, Mynydd Penarfynydd; 28, Skomer Island; 29, Deer Park.

Introduction

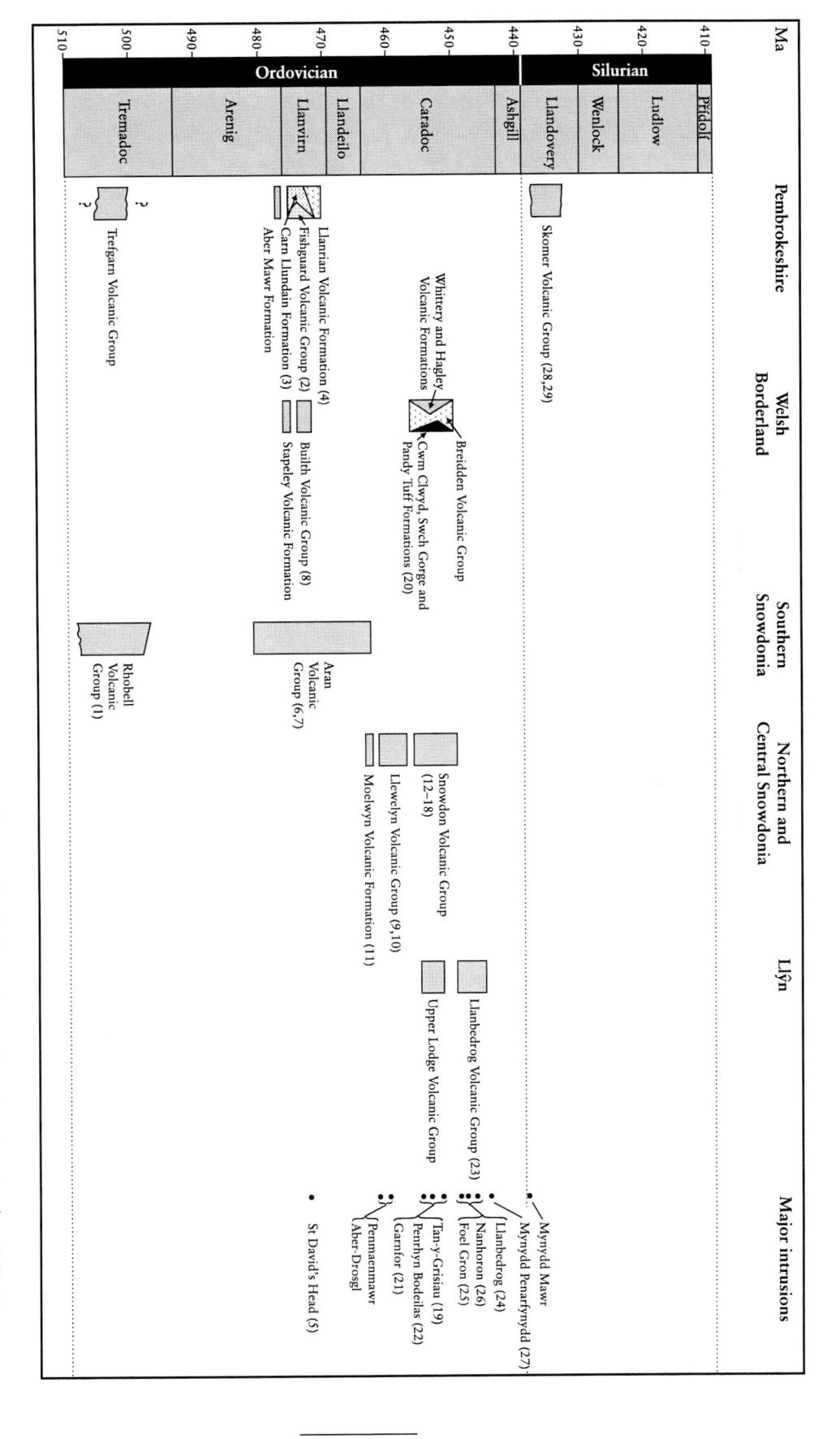

Figure 6.2 Generalized stratigraphical successions of the Lower Palaeozoic sequences of Wales, highlighting the major volcanic episodes. GCR sites are numbered and listed on Figure 6.1.

Wales and adjacent areas

The igneous sequence

The sites selected to represent the Caledonian igneous history of Wales are located on Figure 6.1. A general stratigraphical representation of the main volcanic events is presented in Figure 6.2. An overall review is presented in this section, followed by individual site reports.

Tremadoc

The earliest expressions of Caledonian igneous activity in the Welsh region are represented by the Treffgarne Volcanic Group, exposed in north Pembrokeshire (Bevins *et al.*, 1984) (not formally represented in the GCR at the time of writing), and the Rhobell Volcanic Complex, exposed in southern Snowdonia (Kokelaar, 1979, 1986) at the Rhobell Fawr GCR site. Here, basic lavas and related high-level basic, intermediate and silicic intrusions are thought to represent the eroded remnants of a calc-alkaline volcano linked to the first stages of subduction of Iapetus oceanic crust. Rare cumulate blocks, unique in the British Caledonide region, provide critical evidence for the petrogenesis of the various rock types present. The Rhobell igneous rocks have low-K tholeiitic to calc-alkaline geochemical affinities.

Arenig to Llandeilo

Following the localized igneous activity in Tremadoc times, Arenig to Llanvirn times, in contrast, saw widespread volcanism across Wales, from Pembrokeshire in the south, through southern Snowdonia, to Llŷn in the north, with the first activity also seen in the Welsh Borderland.

In north Pembrokeshire (Figure 6.3), a major volcanic centre developed during latest Arenig to Llanvirn times in the vicinity of Fishguard (Bevins, 1979; Bevins and Roach, 1979a), characterized by the predominantly bimodal basic–silicic volcanic rocks exposed at the Pen Caer GCR site. Here, up to 1800 m of basaltic lavas and relatively minor rhyolitic lavas and ash-flow tuffs, which make up the Fishguard Volcanic Group, represent the most important Ordovician volcanic centre in South Wales. The sequence developed entirely in a submarine environment and the sequence of pillow lavas exposed around the Pen Caer coast is among the finest seen anywhere in Great Britain. At the base of the succession, in the Porth Maen Melyn area, dacitic to rhyodacitic lavas exhibit elongate flow tubes developed at a steep flow front (Bevins and Roach, 1979b); flow phenomena of this type are rarely developed in lavas of such

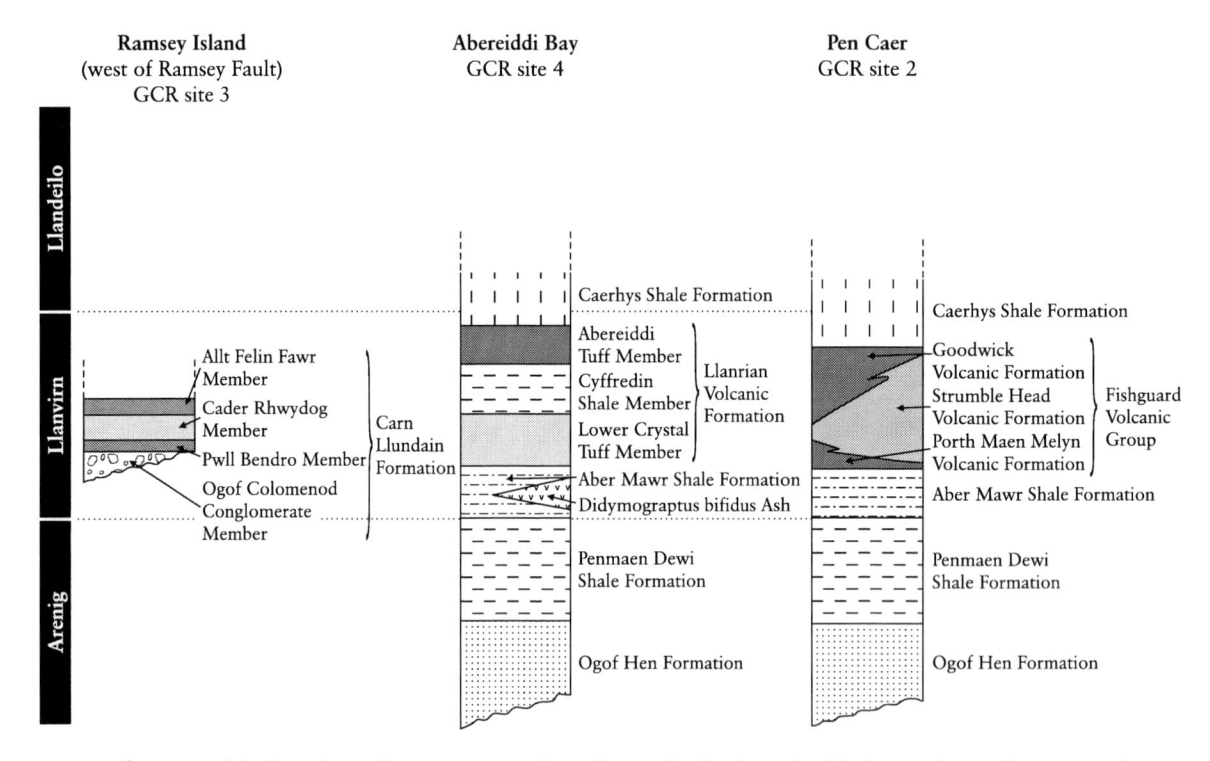

Figure 6.3 Simplified geological successions of north Pembrokeshire, highlighting the Ordovician volcanic sequences. GCR site numbers are listed on Figure 6.1.

Introduction

silicic compositions and hence the features are of international significance. The basic lavas show a tholeiitic chemistry of N-type (normal) MORB character, with a minor subduction zone influence (Bevins, 1982), although the dacitic lavas and high-level intrusions at the base of the sequence have calc-alkaline affinities (Bevins *et al.*, 1992).

Farther to the west in north Pembrokeshire, a major submarine volcanic centre developed in the vicinity of Ramsey Island (Kokelaar *et al.*, 1985), the products comprising the Carn Llundain Formation, exposed in the Aber Mawr to Porth Lleuog GCR site. An extremely well-exposed coastal section shows Cambrian conglomerates and sandstones overlain by a thick Arenig to Llanvirn silicic pyroclastic and volcaniclastic sequence. The succession comprises abundant tuff turbidites of the Pwll Bendro Member, ash-flow tuffs of the Cader Rhwydog Member, and at the top of the pile an autobrecciated rhyolitic flow forming the Allt Felin Fawr Member. Of major significance is the development of a thick welded submarine ash-flow and ash-fall unit in the Cader Rhwydog Member, which was the first of such ash-fall tuffs in the world to be described.

The youngest volcanic episode in north Pembrokeshire is demonstrated by basaltic tuffs of *Didymograptus murchisoni* Biozone age (Bevins and Roach, 1979a; Kokelaar *et al.*, 1984b), exposed in the Castell Coch to Trwyncastell GCR site area. The tuffs represent distal deposits derived by slumping of loose ash and lapilli on the flanks of a submarine volcano. In the vicinity of Trwyncastell, fine-grained distal silicic ash-flow tuffs of slightly earlier Llanvirn (*Didymograptus bifidus* Biozone) age are also exposed. Although precise correlation is not possible, these ash-flow tuffs most probably represent the distal deposits from the kind of volcanic centre developed to the west, on Ramsey Island (Aber Mawr to Porth Lleuog GCR site), or to the east, at the Pen Caer GCR site.

The origin of the silicic igneous rocks in the Caledonide region of Wales has been a matter of contention for many years. Recent detailed geochemical studies of the St David's Head Intrusion (Bevins *et al.*, 1994), in north Pembrokeshire (St David's Head GCR site), have indicated that in this case the silicic rocks were generated as a result of low-pressure fractional crystallization from the more basic compositions, providing a model for the generation of some of the Ordovician rhyolitic rocks associated with tholeiitic basaltic lavas (for example at the Pen Caer GCR site).

Another major volcanic centre of early Ordovician age was located in southern Snowdonia, to the SW of Dolgellau. The Cadair Idris GCR site is of importance in representing the most complete, well-exposed sequence through the Aran Volcanic Group, which ranges in age from Arenig to earliest Caradoc (Pratt *et al.*, 1995). Like the Pen Caer GCR site, the igneous episode was bimodal in character, with predominantly rhyolitic ash-flow tuffs associated with an interval of eruption of basaltic lavas that are commonly pillowed. The basalts show transitional tholeiitic–calc-alkaline affinities (Kemp and Merriman, 1994), contrasting with the more strongly tholeiitic lavas of the Pen Caer GCR site. The nearby Pared y Cefn-hir GCR site provides an excellent section through basic and acid tuffs of the lower part of the Aran Volcanic Group, and represents the best-exposed section of volcanic rocks of Arenig to Llanvirn age in North Wales.

In the Welsh Borderland, volcanic rocks of Llanvirn to Llandeilo age crop out in the Builth and Shelve inliers (Figure 6.4), as well as farther south around Llandeilo. The thickest sequence lies at the southern end of the Builth Inlier, where the Carneddau and Llanelwedd GCR site exposes mainly basic and silicic tuffs, as well as basic, intermediate and silicic lavas of Llanvirn age (Bevins *et al.*, 1984; Kokelaar *et al.*, 1984b; Bevins and Metcalfe, 1993). In contrast to the Pen Caer and the Cadair Idris GCR sites, lavas and intrusive rocks of intermediate composition are also present. Of particular interest is that both tholeiitic and calc-alkaline lavas are present in the lava sequence exposed at the Carneddau and Llanelwedd GCR site (Smith and Huang, 1995). The site also has historical importance in being the area in which one of the first palaeogeographical reconstructions of a volcanic environment was based (Jones and Pugh, 1949).

Caradoc

During Caradoc times igneous activity shifted to North Wales, with centres located in Snowdonia, to the west on Llŷn, and to the east in the Berwyn and Breidden Hills in the Welsh Borderland (Figure 6.4). Snowdonia, however, was the focus of the most important activity in the Welsh Caledonide region and has for many

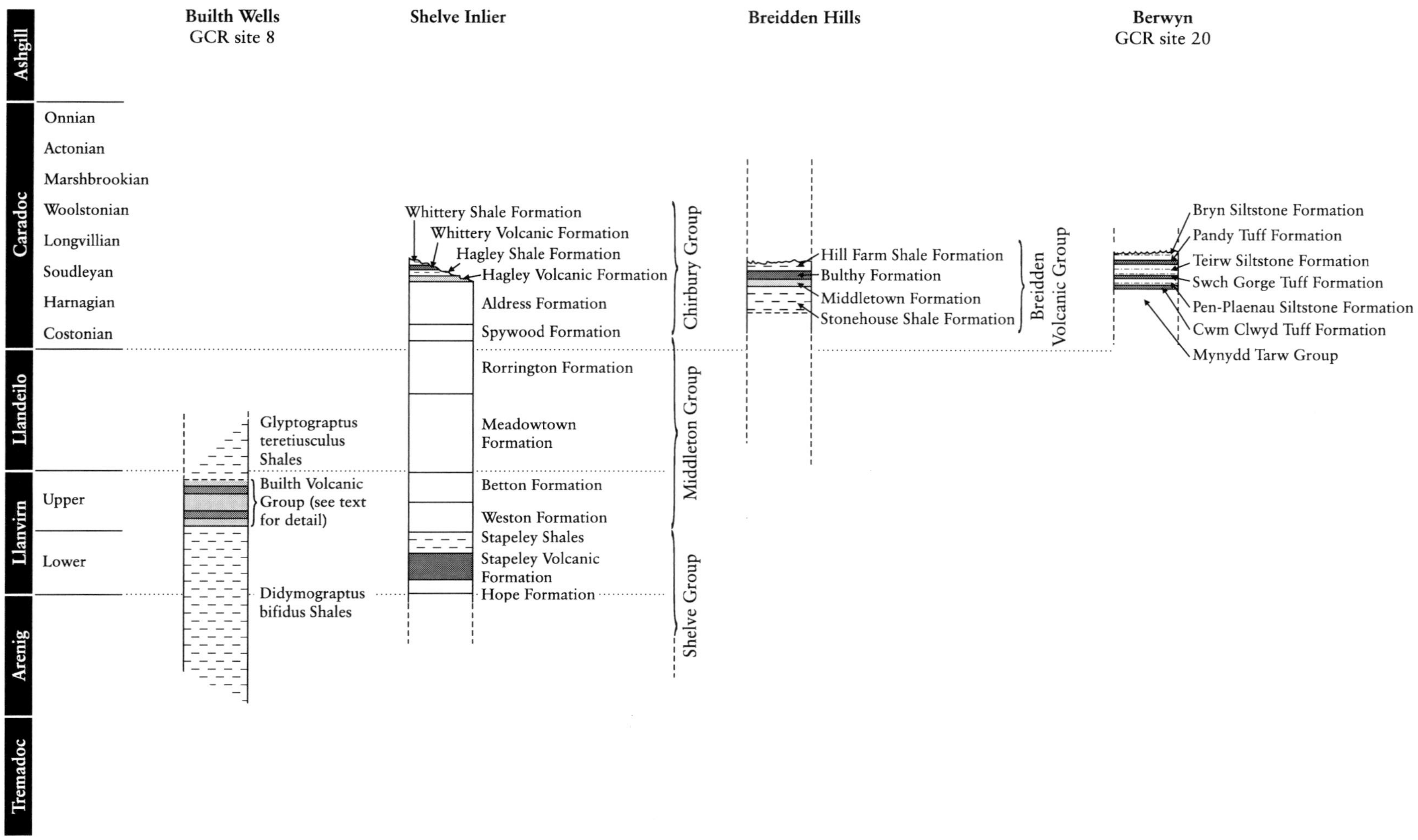

Figure 6.4 Simplified geological successions of the Welsh Borderland, highlighting the Ordovician volcanic sequences. GCR site numbers are listed on Figure 6.1.

Introduction

years played a crucial part in the understanding of igneous processes, in particular the mechanisms of ash-flow tuff eruption and emplacement.

The first regional investigation of Snowdonia was instigated as early as 1846 by the Geological Survey, being completed in 1852, with later publication of a sheet explanation (Ramsay, 1866) and a memoir (Ramsay, 1881). Harker (1889) published results of the first detailed petrographical investigations of the Snowdon Ordovician volcanic rocks and recognized that a number of different volcanic centres had been in existence. A critical observation, which went largely unrecognized for many years, was that by Greenly (in Dakyns and Greenly, 1905), namely, that some of the silicic rocks had textures similar to those of Peléan deposits. This point was reinforced by Williams (1927) in an influential paper reviewing the form, petrography and structure of the Ordovician volcanic rocks of central Snowdonia. A crucial stimulus to the investigation of the volcanic history of Snowdonia, however, was provided by Oliver (1954) and later by Rast et al. (1958), who recognized that many of the silicic volcanic rocks were in fact welded ash-flow tuffs. At the time it was assumed that such tuffs could only be generated in subaerial environments and hence palaeogeographical reconstructions were revised to take account of substantial, periodic rises and falls in sea level, with the environment periodically changing from subaerial to submarine. Consideration of these environmental changes led to the proposal of a major subaerial volcano and related caldera structure (Bromley, 1969; Rast, 1969; Beavon, 1980).

This palaeogeographical reconstruction for Snowdonia was challenged and substantially revised following the crucial identification in Snowdonia of ash-flow tuffs that had welded in the submarine environment (Francis and Howells, 1973; Howells *et al.*, 1973). This work, of major international significance, stimulated a thorough reinvestigation of the Ordovician volcanic rocks of northern and central Snowdonia, based on detailed field mapping by officers of the British Geological Survey. Most recently the area has been the subject of a comprehensive, multi-disciplinary investigation of the Caradoc volcanism, co-ordinated by the British Geological Survey but also involving researchers from other institutions. The culmination of this project was the publication of a definitive memoir of the igneous history of this part of the Caledonide region (Howells *et al.*, 1991).

Although the volcanic successions are voluminous in central and northern Snowdonia, accounting for approximately half of the Ordovician succession in those areas, the volcanic activity was relatively short-lived, being restricted to two chronostratigraphical stages, namely the Soudleyan and the Longvillian. A lack of adequately preserved faunas precludes determination of an accurate biostratigraphical age for the volcanic episode, but it appears to have occurred partly during *Diplograptus multidens* Biozone times and partly during *Dicranograptus clingani* Biozone times.

Two eruptive cycles have been determined in northern and central Snowdonia, separated by a period of quiescence and deposition of siliciclastic sediments. The earlier, 1st Eruptive Cycle comprises the Llewelyn Volcanic Group, while the later, 2nd Eruptive Cycle comprises the Snowdon Volcanic Group (Figure 6.5).

The Llewelyn Volcanic Group comprises five formations that are exposed in a NE-trending tract of country across northern Snowdonia, with distal facies locally reaching central Snowdonia. The first four formations represent an earlier phase of activity, the formations being in part contemporaneous. The eruptions occurred from four different centres, with deposition being strongly controlled by contemporaneous faults; hence thicknesses vary rapidly. The most northerly formation, the Conwy Rhyolite Formation, is exposed to the SW of Conwy, and comprises flow-banded rhyolitic lavas and ash-flow tuffs (Howells *et al.*, 1991). To the SW, the Foel Fras Volcanic Complex is composed chiefly of trachyandesitic lavas and ash-flow tuffs and associated high-level intrusions (Howells *et al.*, 1983; Ball and Merriman, 1989), with the maximum thickness being developed within a caldera structure at the centre of eruption. The Foel Grach Basalt Formation is exposed farther to the SW again, with deposition of massive and pillowed basalts and hyaloclastite breccias in two separate basins which, considering thickness variations, had the form of half-graben. Magma ascent most probably took place up the bounding faults (Howells *et al.*, 1991). The most south-westerly formation of the early phase of the 1st Eruptive Cycle is the Braich tu du Formation, composed mainly of rhyolitic lavas and tuffs. The Braich tu du GCR site exposes sequences that demonstrate the volcanic evolu-

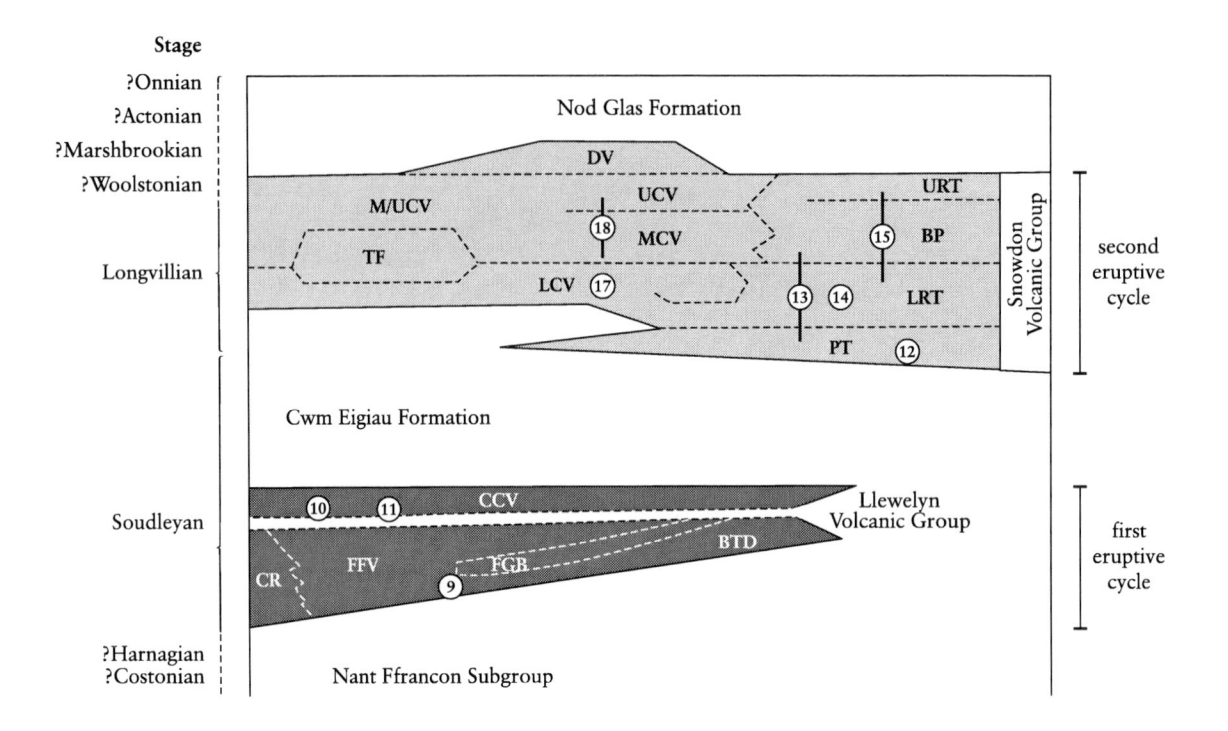

Figure 6.5 Simplified geological successions of northern Snowdonia, highlighting the Ordovician volcanic sequences, in particular the Llewelyn Volcanic Group of the 1st Eruptive Cycle, and the Snowdon Volcanic Group of the 2nd Eruptive Cycle. GCR site numbers are listed on Figure 6.1. BP, Bedded Pyroclastic Formation; BTD, Braich tu du Volcanic Formation; CCV, Capel Curig Volcanic Formation; CR, Conwy Rhyolite Formation; DV, Dolgarrog Volcanic Formation; FFV, Foel Fras Volcanic Complex; FGB, Foel Grach Basalt Formation; LCV, Lower Crafnant Volcanic Formation; LRT, Lower Rhyolitic Tuff Formation; MCV, Middle Crafnant Volcanic Formation; PT, Pitts Head Tuff Formation; TF, Tal y Fan Volcanic Formation; UCV, Upper Crafnant Volcanic Formation; URT, Upper Rhyolitic Tuff Formation.

tion of the three most south-westerly formations described above.

The final volcanic episode of the 1st Eruptive Cycle is represented by the Capel Curig Volcanic Formation, exposed across northern and eastern Snowdonia (Howells *et al.*, 1979; Howells and Leveridge, 1980). The formation chiefly comprises ash-flow tuffs, both welded and non-welded in character. These tuffs were erupted from three volcanic centres; two located in the north were in a subaerial environment, the third in the south in a submarine environment. Howells and Leveridge (1980) demonstrated, on the basis of the various facies of the tuffs, that the ash-flows passed from the subaerial to the submarine environment. The Llyn Dulyn GCR site illustrates the subaerial character of the Capel Curig Volcanic Formation ash-flow tuffs while the Capel Curig GCR site, in contrast, is of importance in being representative of the submarine facies of the ash-flow tuffs. Critically, the tuffs are welded in close contact with the adjacent sedimentary rocks (Francis and Howells, 1973), and this is interpreted as representing welding of the ash-flow tuffs in a submarine environment. This site is of international importance in representing the first location where the welding of tuffs in a submarine situation was identified.

The Snowdon Volcanic Group, representing the 2nd Eruptive Cycle, comprises a complex sequence of silicic ash-flow tuffs, rhyolitic and basaltic lava flows and hyaloclastites emplaced in a shallow to offshore marine environment. Contemporaneous high-level silicic intrusions are abundant. The group crops out across Snowdonia, trending from the NE to the SW over a distance of some 45 km. Three centres of activity have been defined (Howells *et al.*, 1991), the Llwyd Mawr Centre in the SW, the Snowdon

Introduction

Centre around the Snowdon Massif, and the Crafnant Centre in the NE. The development of caldera structures is a feature of each of these centres.

The Llwyd Mawr Centre comprises silicic ash-flow tuffs of the Pitts Head Tuff Formation. The site of eruption is considered to have been close to Llwyd Mawr (Roberts, 1969). Two facies are identified, an intracaldera facies, where up to 700 m of welded ash-flow tuffs occur owing to ponding within the caldera (Reedman *et al.*, 1987) which most probably developed in a sub-aerial setting. This intracaldera facies of the Pitts Head Tuff is exposed at the Craig y Garn GCR site. The Pitts Head Tuff Formation can be traced to the east into the Moel Hebog syncline. The formation is represented by two, relatively thin ash-flow tuffs (up to 90 m in total), which can be traced crossing from a subaerial to a submarine environment. In both environments the tuffs are welded. These tuffs are thought to represent the outflow facies from the Llwyd Mawr Centre (Reedman *et al.*, 1987), and are exposed in the Moel Hebog to Moel yr Ogof GCR site area.

Activity linked to the Snowdon Centre was dominated by the eruption of voluminous acidic ash-flow tuffs linked to major caldera collapse. The earliest activity from this centre, which occurred prior to development of the caldera, led to the accumulation of 180 m of welded ash-flow tuffs, the Yr Arddu Tuffs, erupted along a NNE-trending fissure (Howells *et al.*, 1987). The deposits from this phase of activity are exposed in the Yr Arddu GCR site area.

Following eruption of the Yr Arddu Tuffs, a major caldera-forming event in central Snowdonia occurred, associated with the eruption of a huge volume of acidic ash-flow tuffs, forming the Lower Rhyolitic Tuff Formation (Howells *et al.*, 1986). The formation is up to 600 m thick and volume estimates range from 30 km^3 in the central area, to 20 km^3 away from this area. Marked variations in thickness define intracaldera and outflow facies. On the basis of such thickness and facies variations, the caldera has been calculated as being up to 12 km across. Intrusive rhyolites are particularly abundant within the caldera area, with five geochemically distinct groups being determined (Campbell *et al.*, 1987). The form of the caldera was strongly influenced by four NE- to NNE-trending basement fractures. The various lithologies and the different facies of this caldera-forming event are seen across central Snowdonia, in the Moel Hebog to Moel yr Ogof, Snowdon Massif, and Cwm Idwal GCR site areas.

The caldera-forming event was followed by an episode of predominantly basic volcanic activity, and the Bedded Pyroclastic Formation is dominated by basic tuffaceous sedimentary rocks, basaltic lavas, hyaloclastites, and high-level basic intrusive rocks (Howells *et al.*, 1991). Kokelaar (1992) demonstrated the strong tectonic influence in the timing, location and style of the basaltic volcanism associated with development of the Bedded Pyroclastic Formation. The formation is exposed around the Snowdon Massif and the sequences are represented in the Moel Hebog to Moel yr Ogof, Snowdon Massif, and Cwm Idwal GCR site areas.

The final activity of the Snowdon Centre was the eruption of further silicic ash-flow tuffs of the Upper Rhyolitic Tuff Formation, possibly linked to the generation of another caldera structure. Eruption of the tuffs was associated with the development of high-level rhyolite intrusions, all of the silicic rocks being of peralkaline composition. The Upper Rhyolitic Tuff is best exposed in the Snowdon Massif GCR site area.

Following eruption of the Upper Rhyolitic Tuffs of the Snowdon Centre, the focus of activity shifted to eastern and north-eastern Snowdonia, with eruptions from the Crafnant Centre. Activity was dominated by the eruption and emplacement of silicic ash-flow tuffs in a relatively deep marine environment. The earliest events were linked to the generation of three primary, non-welded tuffs, derived from a local source, which form the Lower Crafnant Volcanic Formation (Howells *et al.*, 1973), exposed at the Curig Hill GCR site area. Later activity shifted to the east, deposits from which are outflow tuffs of the Middle and Upper Crafnant volcanic formations (Howells *et al.*, 1978), seen in the Sarnau GCR site area.

Most acidic intrusions in Snowdonia are closely related to the centres of eruption of the silicic ash-flow tuffs and the caldera structures. Two major acidic intrusions, however, fall outside of the caldera areas, namely the Mynydd Mawr and the Tan y Grisiau granitic intrusions. The character of the latter intrusion is demonstrated at the Ffestiniog Granite Quarry GCR site.

Volcanism during Caradoc times was not solely restricted to Snowdonia. To the east, in the

Berwyn, explosive silicic volcanism occurred contemporaneous with the caldera-forming events in Snowdonia. At the Pandy GCR site, pumiceous ash-fall, ash-flow and pyroclastic surge deposits testify to short-lived subaerial volcanoes, linked to uplift associated with the volcanism, and demonstrate the character of activity across NE Wales.

To the west of Snowdonia, the products of Caradoc-age volcanism are also seen on Llŷn (Figure 6.6). The products from two distinct centres have been determined, although distal ash-flow tuffs of the Pitts Head Tuff Formation erupted from the Llwyd Mawr Centre (see Craig y Garn GCR site) are exposed in the east of the peninsula. On Llŷn, an earlier ?Soudleyan–Longvillian event centred in the Nefyn district was responsible for the generation of volcanic rocks that include the Upper Lodge Formation (broadly equivalent to the 'Upper Lodge Group' of Matley and Heard, 1930) and the Allt Fawr Rhyolitic Tuff Formation, while a younger event, of Woolstonian age, was linked to generation of the Llanbedrog Volcanic Group (Young *et al.*, in press). Both centres are characterized by the development of subvolcanic high-level intrusions. These intrusions were previously thought to be of Devonian age (Tremlett, 1962). Later studies by Croudace (1982) and Young *et al.* (in press) have demonstrated convincingly that the volcanism and intrusions were contemporaneous.

Subvolcanic intermediate to silicic intrusions linked to the Upper Lodge volcanic rocks are exposed in northern Llŷn. The Trwyn-y-Gorlech to Yr Eifl GCR site provides excellent exposures through the Garnfor Multiple Intrusion, demonstrating a basic to acid evolution with time, possibly linked to the evacuation of a fractionating, layered magma chamber at depth. The nearby Penrhyn Bodeilas GCR site shows intermediate to silicic intrusive components of the contemporaneous Penrhyn Bodeilas Intrusion which contains co-magmatic mafic enclaves, also suggestive of high-level fractionation processes.

A major volcanic centre of Woolstonian age, linked to the generation of a caldera collapse structure, developed farther to the west on Llŷn. Basic to intermediate lavas were erupted while explosive activity produced a series of ash-flow tuffs. The high proportion of intermediate lavas seen in the group, for example at the Moelypenmaen GCR site, contrasts with the predominantly bimodal basic–silicic activity of the

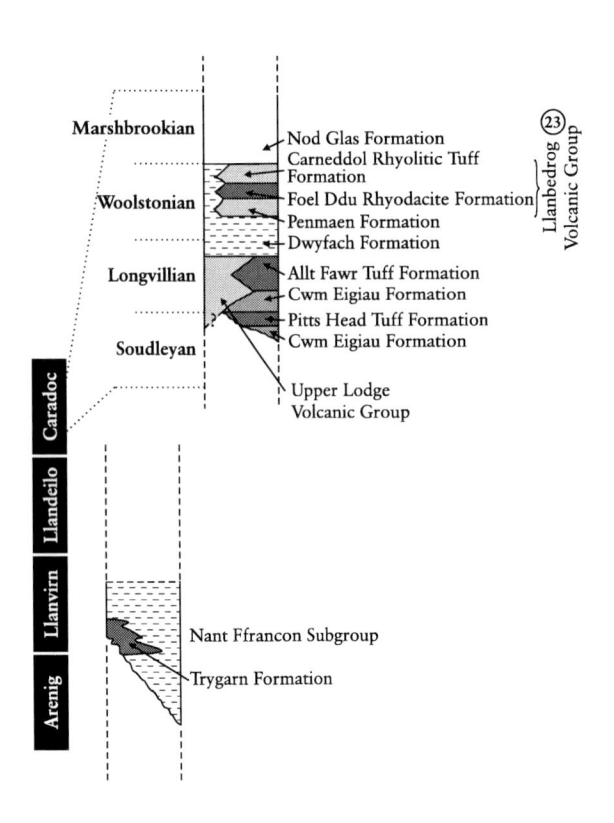

Figure 6.6 Simplified geological successions of Llŷn, highlighting the Ordovician volcanic sequences. GCR site numbers are listed on Figure 6.1.

central Snowdonia centres (e.g. at the Cwm Idwal GCR site). Related high-level silicic intrusions again developed, and are well exposed at a number of sites in western Llŷn. At the Llanbedrog GCR site the Mynydd Tir-y-cwmwd Porphyritic Granophyric Microgranite is thought to be a subvolcanic intrusion associated with the Carneddol Rhyolitic Tuff Formation (of the Llanbedrog Volcanic Group). Other intrusions linked to the Llanbedrog group, and occurring along its western margin, are of particular interest in being peralkaline in character, such as the Foel Gron Granophyric Microgranite (exposed at the Foel Gron GCR site) and the Nanhoron Granophyric Microgranite (exposed at the Nanhoron Quarry GCR site).

The final phase of volcanic activity on Llŷn is seen in the west of the peninsula, and occurred during Nod Glas (late Caradoc) times. Young *et al.* (in press) have linked the Mynydd Penarfynydd Intrusion, one of the most spectacular layered mafic intrusions in the whole of the British Isles, to this episode of volcanism; previously it was considered to be of earlier Ordovician (Llanvirn) age. The intrusion is magnificently exposed in the coastal cliffs of the Mynydd Penarfynydd GCR site.

Llandovery and later Silurian events

The youngest major igneous activity in the Welsh Caledonide region was located in southern Pembrokeshire (Ziegler et al., 1969; Thorpe et al., 1989). Basic, intermediate and silicic lavas, tuffs and high-level intrusions of the Llandovery-age, Skomer Volcanic Group, are exposed at the Skomer Island and Deer Park GCR sites. The major part of the sequence was probably emplaced in a subaerial environment, with the eruption of relatively thin, basic to intermediate flows of considerable lateral extent, which are excellently exposed in cliffs along the western coast of the Skomer Island GCR site area. The local development of pillowed flows, however, shows that subaqueous conditions prevailed intermittently. Rhyolitic rocks form thick flows or domes of relatively limited extent, although rhyolitic ash-flow tuffs are more extensive and provide critical stratigraphical markers.

Geochemically the lavas have alkaline characteristics; however, they still show subduction zone signatures, reflecting the dominant process in the tectonic evolution of this part of the British Caledonides from Tremadoc through to Llandovery times. Two contrasting rhyolite groups in the Skomer Volcanic Group were previously considered not to be related geochemically (Thorpe et al., 1989). Recent work, however, has suggested that they are linked through low-pressure fractional crystallization, with the contrasting trace element concentrations resulting from the precipitation of different minor mineral assemblages.

The Deer Park GCR site is located on the mainland adjacent to the Skomer Island GCR site. Although exposures are more difficult to access, the Deer Park GCR site is crucial in providing a diagnostic age constraint. Faunas within adjacent sedimentary rocks at Anvil Bay are indicative of an early Upper Llandovery (C_{1-2}) age for the Skomer Volcanic Group (Walmsley and Bassett, 1976; Bassett, 1982).

Silurian volcanic rocks in areas of England close to South Wales, include basaltic lavas of Llandovery age that crop out in the Tortworth area (Van de Kamp, 1969; Cave, 1977), and andesites and andesitic tuffs of Wenlock age exposed in the Mendip Hills (Van de Kamp, 1969; Hancock, 1982). Also of some interest is the widespread occurrence of bentonites throughout the Welsh Borderland, in particular in sequences of Wenlock and Ludlow age (Teale and Spears, 1986). A major volcanic event is recorded in the Townsend Tuff unit, of Přídolí age, which can be traced across a wide area of southern Wales and the Welsh Borderland (Allen and Williams, 1981). Whether these various ash-fall tuffs are truly Caledonian, however, remains uncertain as the source of volcanism is not known.

RHOBELL FAWR (SH 787 257)

R. E. Bevins

Introduction

Early Ordovician basic volcanic rocks and associated high-level basic, intermediate and silicic intrusions, comprising the Rhobell Volcanic Complex, are exposed in a remote tract of country on the eastern side of the Harlech Dome, in southern Snowdonia, centred around Rhobell Fawr (Figure 6.7). The site is of importance as a representative of the only substantial remnant of the early arc igneous episode in Wales, linked to subduction of Iapetus oceanic lithosphere beneath the Welsh Basin. Rare cumulate blocks in the basaltic lavas provide critical evidence for petrogenesis of the various igneous rocks, related to processes occurring in a thermally and compositionally stratified magma chamber. This is the only record of such cumulate blocks in the Caledonides of the British Isles.

The first detailed description of the volcanic sequences exposed around Rhobell Fawr was provided by Wells (1925). In the mid to late 1970s there was a resurgence of interest in the various Ordovician volcanic sequences across Wales, following the work of Fitton and Hughes (1970) who linked their origin to subduction of oceanic crust beneath the Welsh Basin. The area of Rhobell Fawr was studied in detail by Kokelaar (1977), including a detailed investigation of the petrology, geochemistry and petrogenesis (see Kokelaar, 1986).

The Rhobell Volcanic Group lies with marked unconformity on folded sedimentary rocks of Cambrian and earliest Tremadoc age. A maximum vertical thickness of 260 m of lavas, chiefly

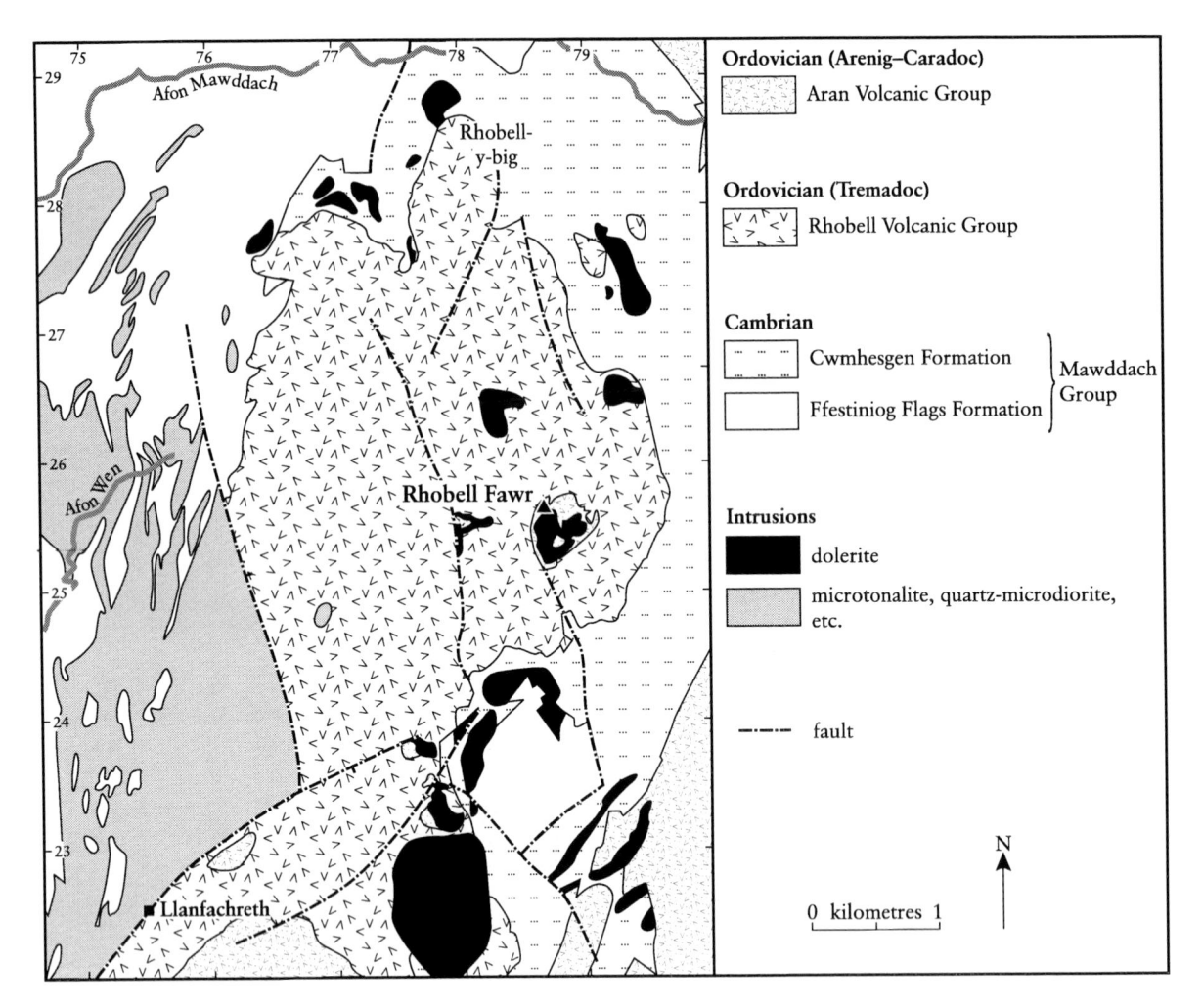

Figure 6.7 Map of the Rhobell Fawr GCR site, adapted from Kokelaar (1977).

plagioclase-clinopyroxene-phyric basalts, is exposed around Rhobell Fawr, although Kokelaar (1986) estimated, from structural considerations, that originally up to 2 km of basalts were erupted. K-Ar age determinations on pargasites from basalt lavas of the complex give an age of 508 ± 11 Ma (Kokelaar *et al.*, 1982).

A N–S fault zone, termed the Rhobell Fracture, appears to have strongly influenced development of the volcanic pile, as well as defining the eastern margin of an upfaulted horst of Cambrian rocks, known as the Harlech Dome. Indeed, further movements along the Rhobell Fracture in late Tremadoc times led to folding, faulting and erosion of igneous rocks of the Rhobell Volcanic Complex, such that today they are overlain by lowest Arenig strata with marked unconformity.

Description

Kokelaar (1977) divided the Rhobell Volcanic Group into four formations:

1. Ffridd Graig-wen Formation (youngest)
2. Eglwys Rhobell Formation
3. Rhobell Ganol Formation
4. Blaen-y-Glyn Formation (oldest)

These formations successively overlap to the east. The four formations are composed almost entirely (99%) of plagioclase-clinopyroxene-phyric lavas. Rare pargasite-bearing lavas are found chiefly in the Blaen-y-Glyn Formation (Figure 6.8), while porphyritic basalts are present mostly in the Eglwys Rhobell Formation. Variations that are present in the succession are considered by Kokelaar (1986) to reflect changes

Figure 6.8 Porphyritic pargasite-bearing basalt of the Rhobell Volcanic Complex, Rhobell Fawr. (Photo: BGS no. L 1274.)

in effusion rates. Breccias occurring in intimate association with the lavas are thought to be autoclastic in origin, while minor volcaniclastic units are possibly water-reworked deposits.

The plagioclase-clinopyroxene-phyric basalts are extensively altered, with a range of secondary minerals characteristic of the prehnite-pumpellyite and lower greenschist facies. Euhedral plagioclase phenocrysts, almost always albitized, form up to 40% of the mode, and commonly show evidence of normal or oscillatory zoning, or contain concentric zones of inclusions towards their margins. Clinopyroxenes are more commonly fresh, and have augitic compositions. These also show normal and oscillatory zoning, and inclusion-rich zones occur towards the crystal margins. Groundmass in the basalts is invariably altered.

Pargasite-bearing basalts form a minor component of the Rhobell lavas. Typically pargasite phenocrysts form 1–5% of the mode, and reach up to 4 cm in length. Both oscillatory zoning and zones at crystal margins rich in inclusions are present, sometimes both being sharply truncated by later crystal growth. Clinopyroxenes range up to 4% of the mode, and reach up to 7 mm in diameter. Both normal and oscillatory zoning are present, and margins to crystals are inclusion-rich. Clinopyroxenes are augitic in composition. Plagioclase phenocrysts, up to 2 mm in length, form up to 20% of the mode; typically they are altered. The groundmass is fine grained and altered.

Locally the lavas are cumulo-phyric, with cumulus ferromagnesian phenocrysts forming up to 50% of the mode. In the cumulo-phyric basalts pargasite forms up to 47% of the mode, while clinopyroxenes form up to 10%, and are diopsidic in composition. Plagioclase is typically absent from the cumulo-phyric basalts.

Rarely cumulo-phyric lavas of the Eglwys Rhobell Formation contain cognate cumulate blocks which reach up to 10 cm in diameter (Figure 6.9). Kokelaar (1986) reported the presence of two classes of cumulate blocks, namely pargasite mesocumulates and adcumulates, and pargasite-salite (augite) mesocumulates and adcumulates. The former class of block, which is the most common type found, contains pargasite, up to 1.5 cm in diameter, as the only cumulus phase, with variable amounts of post-cumulus material, possibly showing fine-scale grain size and textural lamination. Small inclusions are present in many of the pargasite crystals. The pargasite-augite cumulate blocks are similar to the pargasite cumulate blocks but contain up to 20% modal augite. Some of the augites contain minor chlorite inclusions, possibly after

Figure 6.9 Cognate cumulate block in basalt lava of the Rhobell Volcanic Complex, Rhobell Fawr. (Photo: R.E. Bevins.)

glass.

Intrusions associated with the Rhobell Volcanic Group comprise basic, intermediate and silicic varieties. Although sheet-like with respect to the Cambrian strata within which they are contained, they are dyke-like with respect to the base of the volcanic pile. Basic varieties are doleritic, typically dominated by clinopyroxene and plagioclase; leucodoleritic varieties contain only sparse clinopyroxene. Intrusions of intermediate composition (not exposed in the GCR site area) are represented by porphyritic microdiorites, which contain plagioclase and hornblende phenocrysts (up to 40% and 3% of the mode respectively), the former showing normal and oscillatory zoning. In the more silicic (microtonalitic) intrusions, the modal proportion of hornblende decreases (to *c.* 1% of the mode) while that of quartz, present both as phenocrysts and in the groundmass, increases. The phenocrysts reach up to 0.4 mm in diameter, and form up to 5% of the mode.

Interpretation

Basaltic rocks of the Rhobell Volcanic Complex are thought to reflect a subaerial sequence of lavas which produced a volcanic pile up to 2 km thick around a fissure zone, now represented by an intense swarm of dykes. In the central zone of the swarm, intervening screens of sedimentary rocks are absent, and up to 1 km of E–W dilation across the Rhobell Fracture Zone has been estimated (Kokelaar, 1986). Activity was focused along this zone for up to 24 km in a N–S orientation, most probably with other volcanic piles developing locally. The abundant clasts of porphyritic igneous rock and the feldspar-rich nature of many sandstones in the Arenig of this part of the Welsh Basin support such a contention.

Textures in the cumulate blocks, in particular the inclusion-free and inclusion-rich zones, have been interpreted by Kokelaar (1986) as reflecting varying conditions of crystal growth, especially variations in temperature within a magma chamber, coupled possibly with variations in composition. Crystals that grew and accumulated in this magma chamber were periodically disturbed, as is shown by the presence of oscillatory zoning and by the sharp terminations to growth patterns in crystals. Disrupted crystals or blocks were incorporated into erupted magmas.

Kokelaar (1986) provided extensive whole-rock and mineral geochemistry for samples from the Rhobell Volcanic Complex and combined these data to present a petrogenetic model to explain the diverse rock types present. Geochemical and mineralogical constraints suggest that the variety of igneous rocks present in the Rhobell Volcanic Complex was derived by fractional crystallization from a basic parent. The fractionation process was heavily influenced by pargasite in the early stages, along with clinopyroxene. Geochemically, the rocks show calc-alkaline affinities, characteristic of destructive plate margins, which supports the suggestion that these rocks were generated as a result of the subduction of Iapetus oceanic lithosphere beneath the Welsh Basin in early Ordovician times (Bevins *et al.*, 1984; Kokelaar *et al.*, 1984b).

Conclusions

Basic, intermediate and silicic rocks of the Rhobell Volcanic Complex represent the extrusive and high-level intrusive products of a calc-alkaline volcanic episode which developed over a major fracture, the Rhobell Fracture, in early Ordovician times. Dyke rocks were emplaced as a swarm along the central zone of the fracture, reflecting up to 1 km of crustal extension in an E–W direction.

The majority of the lavas are basaltic, rarely containing amphibole phenocrysts and blocks of amphibole-pyroxene-rich rock that are unique in the British Caledonides and which provide important constraints on the origin and evolution of the magmas. Textures in the blocks imply crystal accumulation, under varying temperature conditions, in a compositionally varied magma chamber. Periodically this chamber was disturbed, releasing crystals and blocks into the erupting magmas. The various igneous rocks were generated from a basic parental magma, by crystal fractionation, dominated by the removal of the amphibole crystals from the magma.

The rocks show calc-alkaline affinities, in keeping with suggestions that these Ordovician volcanic rocks were derived as a result of the subduction of Iapetus oceanic lithosphere beneath Wales. Indeed, they represent the only substantial remnant of an early volcanic-arc episode in Wales. Following eruption of the Rhobell lavas, the character of magmatism changed, with the later eruption of magmas more typical of a back-arc basin environment (see, for example, the Pen Caer GCR site report).

PEN CAER
(SM 887 391–909 409, 937 404–952 402)

R. E. Bevins

Introduction

The Pen Caer Peninsula, located to the west of Fishguard in north Pembrokeshire, largely comprises Ordovician lavas and related intrusive rocks, chiefly of bimodal basic and silicic compositions, but rarely of intermediate composition. These are associated with pyroclastic and sedimentary volcaniclastic rocks and argillites, the latter reflecting the background sedimentation that was punctuated by the relatively short-lived volcanic inputs.

The site is of importance in view of the excellent coastal exposures, which demonstrate the nature of subaqueous extrusive and related high-level intrusive igneous processes in the southern part of the Welsh Basin in early Ordovician times. In addition, the geochemistry of the igneous rocks has been used to establish the ensialic marginal arc-basin setting in which these various extrusive and intrusive rocks were emplaced (Bevins *et al.*, 1984; Kokelaar *et al.*, 1984b).

The extrusive volcanic rocks, up to 1800 m thick, belong to the Fishguard Volcanic Group of Llanvirn age, and were emplaced entirely in a submarine environment. They show a range of forms related to their mode of eruption and emplacement, which included the quiet effusion of basic sheets and pillowed flows and silicic dome-producing extrusions, the dramatic effects of thermal shattering and breccia production, and the explosive eruption associated with the generation of hyaloclastites and silicic ash-flow tuffs.

Basic intrusive sheets show a variety of forms, directly related to depth of emplacement. Those emplaced at some depth in the sequence show parallel tops and bases and relatively coarse-grained textures. At higher levels, however, the intrusions invaded still-wet sediments, their contacts are irregular, sometimes with flame structures at the sediment–igneous contact, and textures are those characteristic of rapid quenching.

The extrusive products were subject to rapid reworking, and volcaniclastic rocks were generated. For the most part, however, the lava pile must have been well-below wave-base, as such rocks are relatively rare. More common are mudstone intercalations, which reflect temporary cessations in the volcanic activity and a return to 'normal' background sedimentation.

The first detailed description of the igneous rocks of the Pen Caer Peninsula was by Reed (1895). This study was extended subsequently by Cox (1930), and later by Thomas and Thomas (1956). The most recent investigations are those of Bevins (1979) and Bevins and Roach (1979a, 1979b). Bevins (1982) described aspects of the petrology and geochemistry of the volcanic and related intrusive rocks of the Fishguard Volcanic Complex. A detailed description of parts of the site area appears in the field guide of Kokelaar *et al.* (1984a).

Description

The Fishguard Volcanic Group in the Pen Caer area comprises three formations, namely:

1. Goodwick Volcanic Formation (youngest)
2. Strumble Head Volcanic Formation
3. Porth Maen Melyn Volcanic Formation (oldest)

The Porth Maen Melyn Volcanic Formation is exposed to the north of the bay known as Porth Maen Melyn (Figure 6.10), to the south of Strumble Head. The lowermost unit of the formation comprises 10 m of extensively recrystallized bedded rhyolitic tuffs that show evidence of shardic textures. The tuffs appear to be non-welded, silicic ash-flow tuffs, emplaced subaqueously. This unit is overlain by two units, 35 m thick in total, of lithic-crystal-vitric breccias and crystal-vitric tuffs. The lower unit shows a fining upwards through the lowermost 50 cm. Clasts in the breccias are chiefly of angular to sub-rounded rhyolitic lava, sometimes with a perlitic texture, although in the lower part of the sequence basic lava and dolerite are common. Crystals, chiefly bi-pyramidal quartz and plagioclase, are prominent in the fine-grained basal part of the sequence. Pumice clasts are represented by weathered-out streaks in the tuffs, while shardic fragments are seen in thin section. These tuffs are thought to be sediment-gravity flow deposits, derived from the reworking of primary silicic eruptive volcanic detritus. The basic clasts were most probably eroded from wall rocks in the volcanic edifice during eruption. Overlying these breccias and tuffs is a rhyodacitic lava flow up to 40 m thick; to the east it is massive, while traced westwards it becomes pillowed, showing elongate flow tube structures, associated with inter-pillow breccias. The flow is quenched and shows a perlitic texture, reflecting the recrystallization of a glass. The top of the flow is autobrecciated. The tubes and pillows are thought to result from the rapid effusion of hot magma at the steep front of a lava flow, as described by Bevins and Roach (1979b).

The Strumble Head Volcanic Formation con-

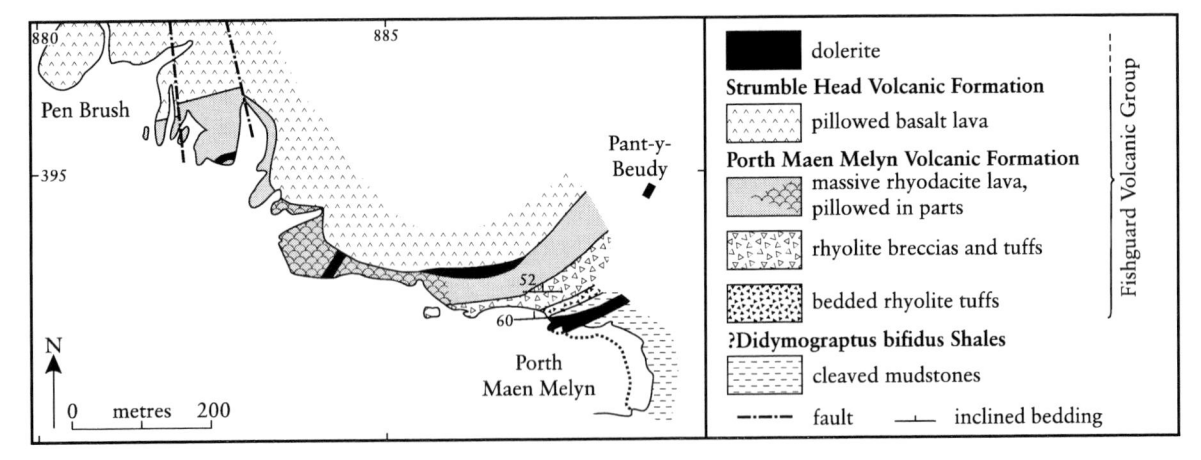

Figure 6.10 Map of the Porth Maen Melyn area, Pen Caer GCR site (after Kokelaar *et al.*, 1984a).

Pen Caer

Figure 6.11 Pillow lava from the Fishguard Volcanic Group, Strumble Head. The pillow in the centre of the photograph is 75 cm across (long axis). (Photo: R.E. Bevins.)

formably overlies the Porth Maen Melyn Volcanic Formation, a contact well exposed in the steep cliffs on the north side of the bay. Immediately overlying the contact are classic pillowed basaltic lavas, individual pillows typically reaching up to 40 cm in diameter (Figure 6.11). Such lavas form the bulk of the succession exposed from Porth Maen Melyn northwards to Strumble Head and then eastwards to the extremity of the outcrop of the Strumble Head Volcanic Formation at Carnfathach. The pillowed flows commonly show well-developed inter-pillow breccias, while intercalations of isolated pillow breccias and hyaloclastites are sporadically developed. Locally, elongate, steeply inclined lava tubes are developed, as are classic necking structures. Intercalated with the pillowed lavas are massive, commonly lensoid lava sheets, as well as high-level intrusive basaltic sheets. Some intrusive sheets show pillowed forms at their base, while the thicker sheets commonly possess well-developed columnar jointing.

The top of the chiefly basaltic Strumble Head Volcanic Formation and the base of the overlying chiefly silicic Goodwick Volcanic Formation is marked by a complex interdigitation of lavas and high-level intrusions (Figure 6.12). The various facies of a thick rhyolite flow or dome at the base of the Goodwick Volcanic Formation are excellently exposed on Penfathach. The lowermost facies, a breccia, comprises 4–5 cm-diameter clasts of flow-banded rhyolite showing little to no post-brecciation rotation. This passes upwards into rhyolites with excellent large- to small-scale flow-folding, and locally with well-developed columnar jointing. Locally, perlitic textures are preserved. Contact with the overlying autobrecciated carapace is gradational. To the east, basaltic tuffs are exposed at the foot of cliffs at Porth Maen. These tuffs are generally parallel-bedded, and show both normal and inverse grading. Clasts are typically 2–3 mm in diameter, reaching up to *c.* 1.5 cm. Offsets to bedding probably relate to soft sediment faulting. These tuffs are thought to be sediment-gravity flows slumping off the flanks of a submarine, possibly shallow-water volcano. Farther to the east, and up-section, coarse lithic-rich ash-flow tuffs are exposed. These are overlain by fine-grained silicic tuffaceous sedimentary rocks that have been intruded by a basic sill. This sill shows bulbous protrusions at its base, indicative

Wales and adjacent areas

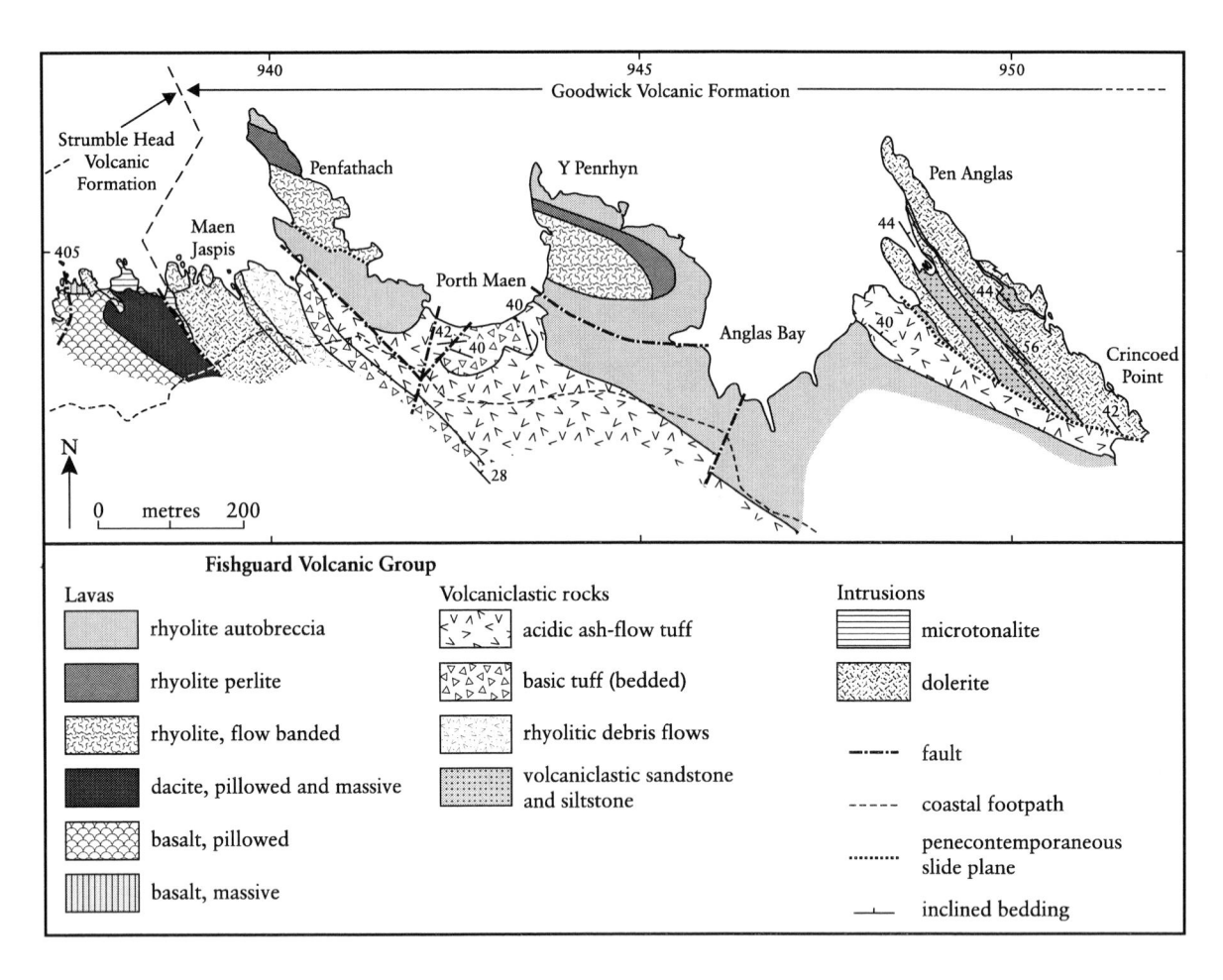

Figure 6.12 Map of the north-eastern part of the Pen Caer GCR site (after Kokelaar *et al.*, 1984a).

of wet sediment–magma interactions. The top of a higher basic intrusion, exposed on Pen Anglas, has particularly well-developed columnar jointing, yet is pillowed at its top, again indicative of intrusion of magma into wet sediment (Figure 6.13).

Interpretation

This site demonstrates the complex facies developed in extrusive and high-level intrusive rocks associated with the generation and evolution of a submarine volcanic complex, with compositions ranging from basic through intermediate to silicic.

Classic pillowed and sheet-like forms are developed in the basic lavas, along with inter-pillow and isolated-pillow breccias. In critical sections, elongate, steeply dipping lava tubes can be seen. Unconsolidated tuffs, probably generated in relatively shallow-water environments, were reworked locally and led to the generation of bedded basaltic volcaniclastic sediments.

Some basic magmas failed to reach the seawater–sediment interface, and were intruded at a high level. At deeper levels in the sequence the intrusions are relatively thick, typically coarse-grained, and show sharp concordant contacts. At shallower levels, the intrusions are thinner, finer grained, even quenched, and show irregular, sometimes flamed or pillowed, contacts.

More silicic, dacitic, magmas produced a 40 m-thick flow that erupted onto the sea floor. Part of the exposure reveals a massive, central facies with an autobrecciated flow top; at the steep front of the flow, however, elongate flow tubes developed, associated with inter-pillow breccias. The flow appears to have been glassy throughout. Such flow features in dacitic lavas are extremely rare, making this site of interna-

Figure 6.13 Magma-wet sediment relationships at the base of a high-level basic intrusion, Pen Anglas. Magma has injected and loaded down into unlithified tuffaceous sediment, while the wet sediment has locally flamed up into the magma. Lens cap for scale. (Photo: R.E. Bevins.)

tional significance in terms of understanding eruptive igneous processes.

Silicic magmas were emplaced as extrusive domes on to the sea floor, with well-developed autobrecciated carapaces and showing magnificent flow-folding phenomena. In addition, explosive eruptions produced deposits of ash, pumice, glass shards and lithic fragments; some deposits were derived from primary ash-flows, in other cases they represent the slumping of previously deposited debris on the unstable flanks of submarine volcanoes and domes.

Bevins (1982) investigated the geochemistry of the basaltic rocks of the Fishguard Volcanic Group. He reported that they show close similarities to N-type (normal) Mid-Ocean-Ridge Basalt (MORB), but were slightly enriched in the light rare earth elements and Th, coupled with a marked depletion in Nb. Such features are thought to reflect a minor subduction zone influence in their genesis and are characteristic of basalts erupted in back-arc or marginal basins. This is consistent with the Welsh Basin occupying a tectonic position at the margin of Gondwana during early Ordovician times, and with subduction of the Iapetus Ocean crust beneath the continental margin.

Conclusions

This site provides excellent coastal exposures through a submarine volcanic pile, showing a range of extrusive and intrusive igneous phenomena. The products provide critical evidence for the nature of submarine volcanism and the processes involved in the development of a submarine volcanic complex. It is one of the classic sites in the British Isles for pillowed basaltic lavas and related tuffs and breccias. Of international note, however, are exposures through a dacite lava that shows a transition from a massive centre to a flow front with elongate flow tubes; such phenomena are extremely rare in lavas of such silicic composition. Also, the nature of the growth and development of rhyolitic domes on the sea floor is demonstrated, associated with the products of more explosive activity and the subsequent reworking of loose debris in the submarine environment. Finally, the geochemistry of the basalts indicates that the lavas were erupted in a marginal basin environment linked to subduction of Iapetus Ocean crust beneath Gondwana in Ordovician times.

ABER MAWR TO PORTH LLEUOG (SM 698 240–700 231)

R. E. Bevins

Introduction

The volcanic and intrusive rocks of Ramsey Island form part of a belt of bimodal basic–acidic Ordovician igneous rocks in north Pembrokeshire, interpreted as having been emplaced in a marginal basin related to closure of the Iapetus Ocean. The site is of importance in providing important exposures of silicic volcanic and related rocks in the southern part of the Welsh Basin, and in a wider context it provides critical insight into the nature of submarine explosive eruptions and the emplacement of welded ash-flow and ash-fall tuffs.

Ramsey Island is composed of a sequence of rocks ranging from mid-Cambrian through to Ordovician (Llanvirn) in age, in part sedimentary but mainly igneous. The igneous rocks are entirely Ordovician, and are chiefly silicic pyroclastic and volcaniclastic rocks, associated with relatively minor silicic extrusive and intrusive equivalents. The rugged coastline provides excellent exposures and presents detailed sections through the succession.

Early studies of the geology of Ramsey Island concentrated on establishing the age and stratigraphical relationships of the various lithological units. Those by Pringle (1914, 1915, 1930) are particularly important in providing the first detailed account and geological map of the island. Later investigations by Bevins and Roach (1979a), Kokelaar (1982), and Kokelaar *et al.* (1985) presented an interpretation of the various pyroclastic and volcaniclastic rocks. In particular, the studies of Kokelaar *et al.* (1985) have provided an insight into the processes related to the submarine eruption of silicic magmas and their subsequent reworking.

A major N–S fault divides Ramsey Island into two, with contrasting stratigraphies and geological evolutions. The GCR site covers the area to the west of the Ramsey Fault.

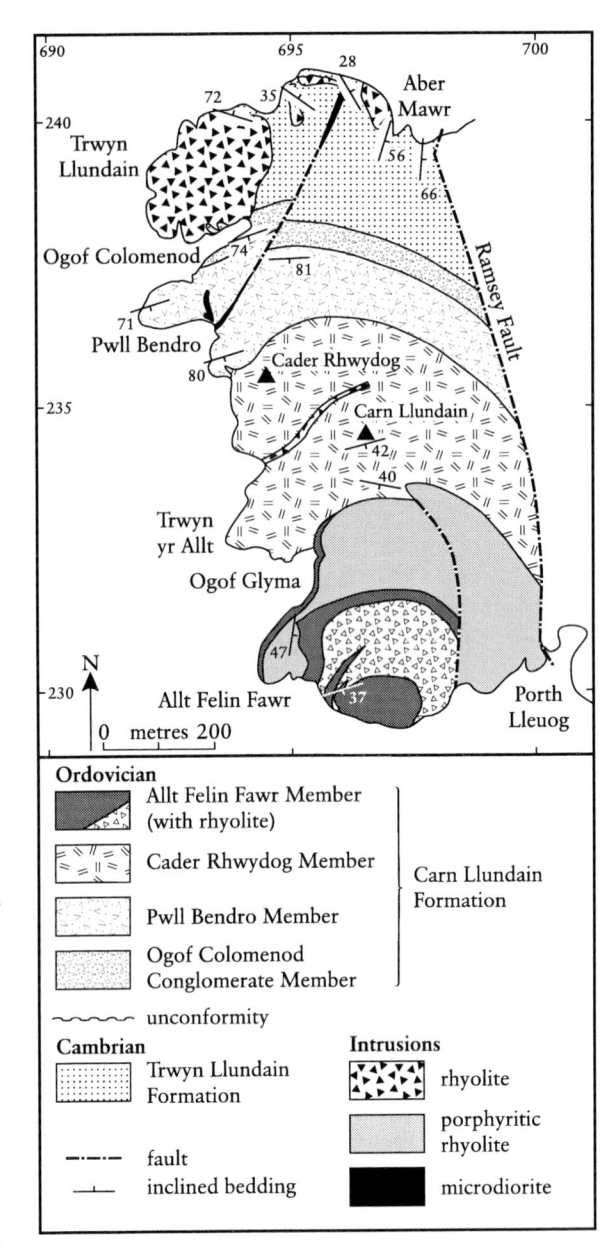

Figure 6.14 Map of the Aber Mawr to Porth Lleuog GCR site, Ramsey Island (after Kokelaar *et al.*, 1985).

Description

The succession to the west of the Ramsey Fault is dominated by silicic pyroclastic rocks of Llanvirn age, comprising the Carn Llundain Formation (Figure 6.14). The lowermost rocks, buff-coloured sandstones, are thought to be of mid-Cambrian age. A spectacular 35 m-thick, poorly bedded conglomerate, the Ogof

Aber Mawr to Porth Lleuog

Figure 6.15 Ragged, elongate pumice fragments up to 12 cm in length in the Cader Rhwydog Tuff, Cader Rhwydog, Ramsey Island. (Photo: R.E. Bevins.)

Colomenod Conglomerate Member, overlies these sandstones with marked unconformity and represents the lowermost member of the Carn Llundain Formation. The conglomerate is poorly sorted, with rounded cobbles and pebbles of rhyolite, set in a finer matrix of rhyolite and rhyolitic sandstone.

The Pwll Bendro Member, comprising 165 m of silicic lapilli-tuffs and fine-grained typically massive tuffs, overlies and grades up from the underlying conglomerates. These tuffs are poorly sorted and poorly bedded, with individual tuff beds ranging from a few centimetres to 20 m in thickness, although typically they are in the range 1–2 m. They chiefly comprise lapilli of rhyolite, tube pumices, plagioclase and quartz crystals, and glass shard fragments. The tuffs are non-welded. A sequence of three ash-flow tuffs, the Cader Rhwydog Tuff (CRT), the Trwyn yr Allt Tuff (TAT) and the Ogof Glyma Tuff (OGT), comprise the overlying Cader Rhwydog Member, totalling some 223 m in thickness. The lowermost tuff, which is 186 m thick, dominates the succession, the other tuffs being 22 and 15 m thick respectively. The CRT is excellently exposed about Carn Llundain. The base is slightly unconformable in relation to the underlying tuffs and shows slight down cutting. A deeply incised 'palaeo' gulley is also seen. Near the base the CRT includes rounded pebbles and granules of rhyolite, angular clasts of rhyolite, and mudstone clasts. The main part of the CRT, however, contains ragged, commonly elongate, tube pumices (Figure 6.15), up to 32 cm, along with rhyolite fragments, and whole and broken quartz and plagioclase crystals. The pumice fragments are commonly flattened and in places are moulded around rhyolite clasts. Flattened

Figure 6.16 Wet sediment disturbance in thinly laminated turbiditic tuffs at the top of the Trwyn yr Allt Tuff, Cader Rhwydog Member, Carn Llundain, Ramsey Island. (Photo: R.E. Bevins.)

porphyritic rhyolite fiamme are also present. Poorly developed columnar jointing is seen extending upwards from the base for around 100 m. The uppermost 25 m or so of the CRT shows a marked fining, passing eventually into very fine-grained vitric tuffs. The TAT succeeds the CRT conformably, and comprises massive rhyolitic lapilli-tuffs with pumice clasts and quartz and plagioclase crystals. At the top these pass into thinly laminated turbiditic tuffs showing evidence of wet-sediment disturbance (Figure 6.16). The overlying OGT is similar to the TAT, but it also contains streaky, flattened pumices.

The Cader Rhwydog Member is succeeded conformably by several metres of thinly bedded and laminated tuffs of the Allt Felin Fawr Member, although the succession is complicated by a slightly discordant, cross-cutting porphyritic rhyolite intrusion, which shows irregular bulbous and peperitic margins. A maximum of 25 m is exposed beneath the sill with a further 8 m above the sill. The tuffs are silicic, containing rhyolite and pumice lithic clasts and quartz and plagioclase in a fine-grained siliceous matrix. The tuffs are well bedded, fine upwards and show loading structures at the base of individual beds. In the finer-grained tuffs, ripple drift cross-lamination, soft-sediment disruption and convolute lamination are all seen. This sequence is partly repeated by sliding. A large channel structure, up to 20 m wide and 6 m deep, is seen on a small headland to the SE of Allt Felin Fawr. Above this lies a poorly sorted heterolithic deposit, 3 m thick, which contains clasts of pumice, porphyritic rhyolite, rhyolite, fine-grained silicic tuff, and crystals set in a fine-grained silicic matrix. This unit is followed by a thin (0.5 m) conglomerate bed, poorly sorted heterolithic deposits (1.5 m), and then 2 m of laminated and thinly bedded silicic tuffs.

Overlying these various units is an autobrecciated rhyolite exposed to the east and south of Allt Felin Fawr. In the east the lava reaches up to 35 m in thickness, but to the SW the flow becomes compound, as two tuffaceous wedges come in. In the lowermost tuffaceous wedge a thin sequence of turbiditic tuffs occurs, which is overlain by an ash-flow tuff containing rhyolite and pumice lapilli in a vitric matrix with quartz and plagioclase crystals. The top grades sharply into fine tuffs. Towards the SW a flattening fabric is seen and the tuff contains randomly orientated slabs (1 m thick and 10 m in length) of laminated pumice- and crystal-bearing tuffs, in addition to an individual block (2.1 m in diameter) of porphyritic rhyolite with peperitic and lobate margins. A 0.5 m-thick, chaotic unit, containing rhyolite clasts and crystals in a mudstone matrix, lies between the ash-flow tuff below and a further ash-flow tuff above. Similar to the lower ash-flow tuff, this upper unit contains further contorted slabs of laminated fine tuffs. In places this ash-flow disrupts the thin chaotic unit below, with flames of muddy material penetrating up to 2 m into the ash-flow tuff. To the SW the ash-flow tuff is completely broken up into sac-like masses separated by thin veneers of the chaotic muddy deposit.

The upper wedge in the rhyolite lava flow comprises 20 m of medium- to fine-grained silicic tuffs, which wedge out against the rhyolite. The coarser tuffs show erosive bases, normal grading of the lithic component, and loading structures. The uppermost unit of these tuffs comprises the youngest strata exposed of the Carn Llundain Formation.

Interpretation

The entire sequence of Ordovician volcanic rocks exposed on Ramsey Island is considered to have been erupted and emplaced in a submarine environment, and those of the GCR site are considered to be of proximal origin. Uplift prior to the onset of deposition of the Carn Llundain Formation is represented by the marked unconformable relationship at the base of that formation and is probably related to intrusion at a high level of significant volumes of silicic magma.

The lowest conglomerates of the Ogof Colomenod Member are thought to result from a series of sediment-gravity flows, generated perhaps by volcanotectonic instability and derived from a littoral or supralittoral environment, possibly linked to a rhyolitic volcanic island, but emplaced in deeper water.

Tuffs of the Pwll Bendro Member are considered to have been generated essentially from cold high-density turbidity currents, with the thinner-bedded units resulting from less dense turbidity currents. Eruption of the primary ashes must have been rapid because of the lack of any intercalated background sediments. However it is difficult to establish whether this was in a subaerial or a submarine environment, although Kokelaar *et al.* (1985) favoured a submarine eruptive column as the source. In comparison with tuffs higher in the sequence this source seems to have been at some distance from the site area. They are most probably derived from the slumping of debris immediately following accumulation on the flank of a submarine volcano.

The overlying tuffs of the CRT show crucial evidence for heat retention following emplacement, including the flattening of pumices and the moulding of pumices around lithic clasts. This tuff is interpreted as having been erupted from a major, single, entirely submarine eruption, emplaced as a hot ash-flow tuff with associated ash-fall. A totally subaqueous environment is supported by the presence of soft-sediment convolutions in the overlying fine tuffs. These tuffs are thought to be of proximal origin.

The TAT largely comprises turbiditic tuffs, again thought to be derived from the reworking of unconsolidated ash deposits derived from explosive, most probably submarine, rhyolitic volcanism. Tuffs of the OGT, however, show welded shardic fragments, implying the welding of hot juvenile fragments, suggestive of a very proximal environment.

The Allt Felin Fawr Member is characterized by deposits derived from turbidity currents, which are invaded by a rhyolitic intrusion showing evidence of emplacement into wet, poorly consolidated tuffs, associated with fluidization. This had a catastrophic effect on the overlying tuffs, which were slumped away, exposing the rhyolite at the sea bed, to suffer later reworking into younger debris-flow deposits. Slumping of the tuffs resulted in their repetition in the stratigraphical succession.

The rhyolite appears to have been an autobrecciated lava erupted on to the sea floor, with very shallow ($c.$ 6°) slopes. The two wedges intercalated with the rhyolite in the west of the outcrop were derived from the slumping of primary rhyolitic ash-flow tuffs, the extrusive rhyolite providing a topographic restriction to the distribution of the reworked tuffs. The contained contorted slabs are thought to represent turbiditic and ash-fall deposits genetically associated with the ash-flow tuffs but incorporated later, following slumping and mass-gravity flow. The contained block of porphyritic lava from the immediately subjacent sill is considered to reflect the contemporaneous nature of the intrusive activity, the explosive silicic volcanism and the quiet effusion of rhyolitic lavas, all in a submarine environment.

Conclusions

The rugged coastline and crags of Ramsey Island provide magnificent exposures through rhyolitic volcanic rocks that were erupted and emplaced close to their source, entirely in a submarine environment. A variety of processes are demonstrated, including the eruption of rhyolitic lava on to the sea floor, and the emplacement and reworking of silicic ash-flow and ash-fall deposits.

These exposures provide one of the most crucial sites in the British Isles for the interpretation of submarine silicic volcanic processes, and indeed are of international importance in demonstrating the submarine emplacement and welding of a silicic ash-flow tuff and related ash-fall derived from a major submarine explosive eruption. This account is the first record of such welded ash-fall tuffs in the world.

CASTELL COCH TO TRWYNCASTELL (SM 775 303–796 316)

R. E. Bevins

Introduction

Volcanic rocks are exposed extensively in the north Pembrokeshire region. The major activity, for example around Fishguard, appears to have occurred during Llanvirn times, chiefly during *Didymograptus bifidus* Biozone times. In the Abereiddi area, volcanic rocks of the Llanrian Volcanic Formation (of Hughes *et al.*, 1982) are exposed on the north and south sides of the bay, on opposing limbs of the Llanrian Syncline. On the south side of the bay, between Castell Coch in the west and Melin Abereiddi in the east, basaltic tuffs of the Didymograptus Murchisoni Ash of Cox (1915), or the Abereiddi Tuff Member of Hughes *et al.* (1982), are of *Didymograptus murchisoni* Biozone age. On the north side of Abereiddi Bay, silicic tuffs of *Didymograptus bifidus* Biozone age, are exposed; these are the Llanrian Volcanic Group of Cox (1915), or the Lower Crystal Tuff Member of the Llanrian Volcanic Formation of Hughes *et al.* (1982).

The first important description of these tuffs was by Cox (1915). Bevins and Roach (1979b) provided an initial interpretation of the environment and the depositional processes responsible for the genesis of the various tuffs. This was expanded upon by Kokelaar *et al.* (1984a). The geochemistry of tuffs belonging to the Abereiddi Tuff Member was described by Bevins *et al.* (1992).

Tuffs belonging to the Abereiddi Tuff Member represent the youngest Ordovician volcanic rocks exposed in the southern part of the Welsh Basin.

Description

The Ordovician sequence in the area around Abereiddi contains significant volcanic and volcaniclastic rocks, exposed in two E–W coastal sections. These rocks are interbedded with strongly cleaved mudstones that reflect the character of the background sedimentation which was punctuated periodically by short-lived volcanic events.

At the western end of the cliffs forming the south side of Abereiddi Bay (Figure 6.17), the Abereiddi Tuff Member is well exposed to the NE of Aber Creigwyr. The tuffs dip to the north at around 45°, and are represented by two fining-upwards sequences, 95 m thick in total. The upper tuff unit lies conformably on the lower, and the base of each unit is sharp and planar.

The lower tuff unit is poorly to moderately bedded, with successively higher beds becoming overall of finer grade and showing more clearly defined coarse-grained bases (coarse-tail grading). The upper unit, in contrast, is not obviously bedded, although it does show evidence of grading. The tuffs are composed of angular, scoriaceous basalt lapilli and sparsely vesicular blocks and bombs, up to 40 cm in diameter, set in a fine-grained tuffaceous matrix (Figure 6.18).

To the east, in the area around Melin Abereiddi, the tuffs are thinner (around 60 m). The tuffs of the lower unit rest conformably on tuffaceous mudstones, and are poorly sorted, with crude coarse-tail grading and ill-defined bedding. Upwards the tuffs of the lower unit become more thinly bedded and finer grained. The upper unit, which has a distinctive lapilli-tuff at its base, shows a coarser grain-size. These

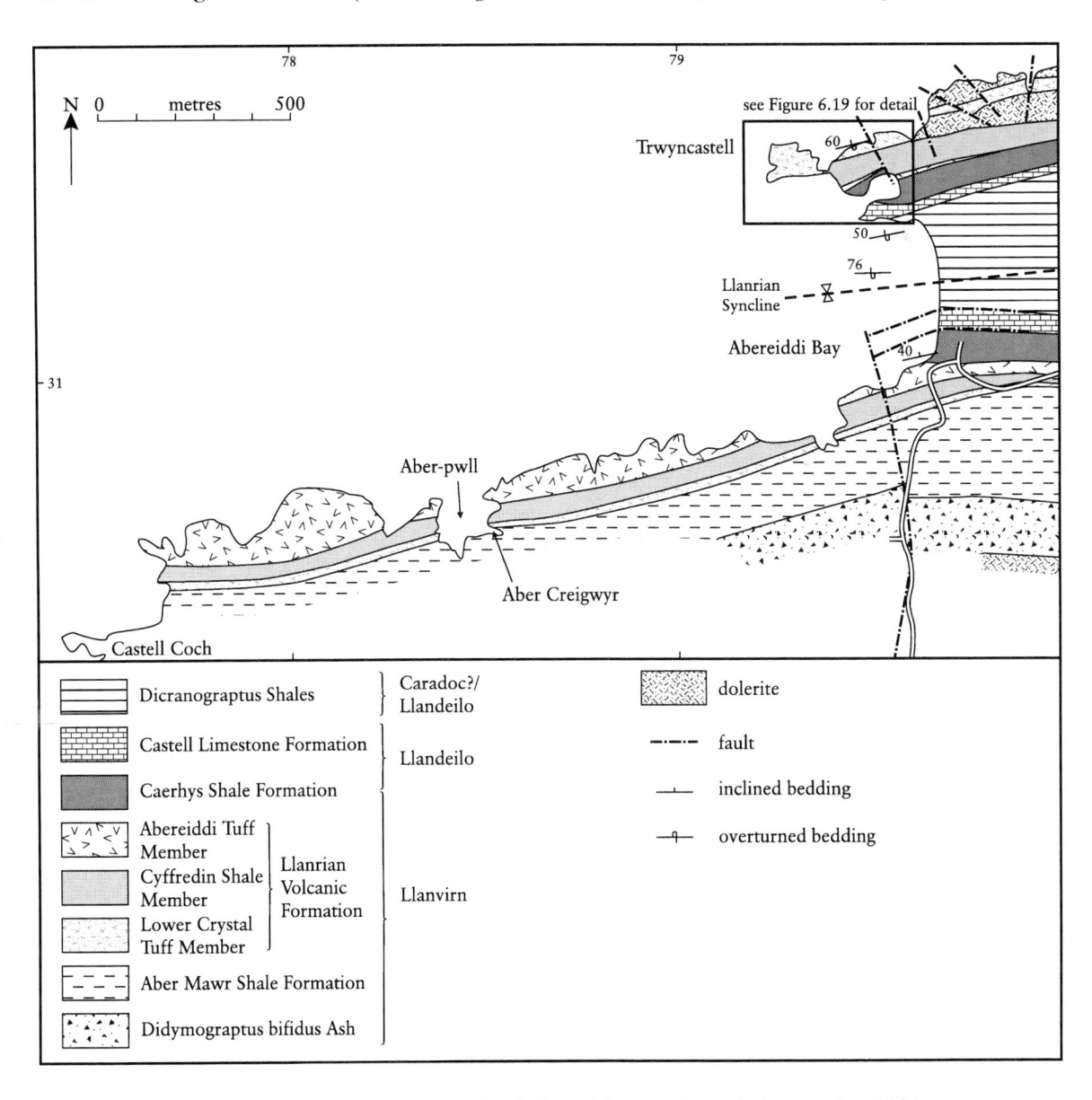

Figure 6.17 Map of the south side of Abereiddi Bay (after Kokelaar *et al.*, 1984a).

Figure 6.18 Coarse lapilli-tuff of the Abereiddi Tuff Member, south side of Abereiddi Bay. (Photo: R.E. Bevins.)

tuffs are well bedded, with bed thickness ranging from 3–25 cm; some beds showing normal grading, although most show little to no evidence of grading. Parallel and ripple drift lamination are present. Bed bases tend to be planar, although locally erosive, down-cutting bases are also seen. Upwards, these tuffs pass conformably into tuffaceous mudstones and mudstones.

On the north side of Abereiddi Bay (Figure 6.19) the Abereiddi Tuff Member is represented by a thin sequence of tuffs exposed in the disused quarry. Structurally above but stratigraphically below this unit, the Lower Crystal Tuff Member, up to 100 m thick, is chiefly represented by crystal-rich volcaniclastic sandstones that show clear evidence of normal grading. On the promontory of Trwyncastell the group comprises a sequence of fine-grained silicic tuffs, considered by Cox (1915) to be rhyolitic lavas.

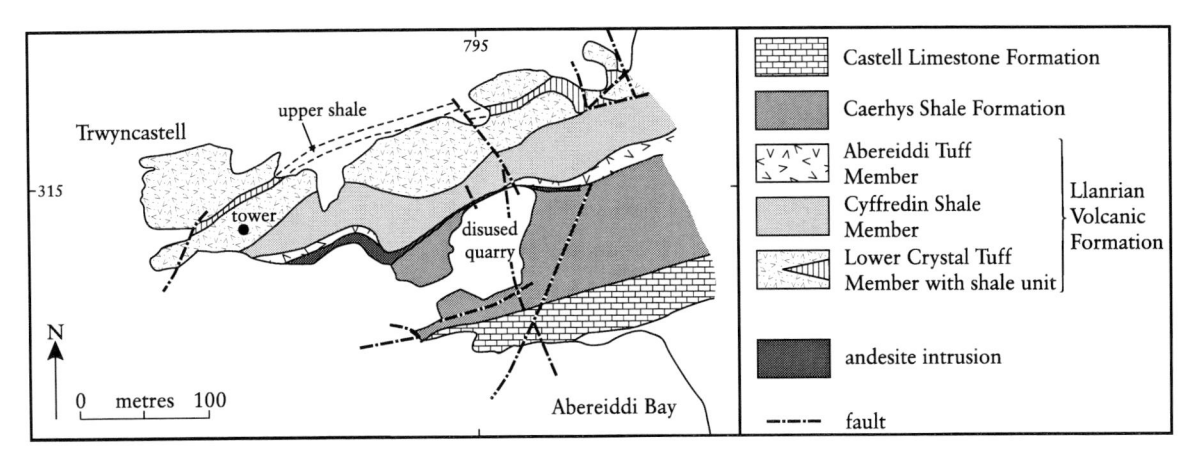

Figure 6.19 Map of the north side of Abereiddi Bay (adapted from Hughes *et al.*, 1982).

Subsequent studies (Bevins and Roach, 1979a), however, have shown that they contain shards and are fine-grained, silicic tuffs.

Interpretation

The silicic rocks of the Llanrian Volcanic Formation were interpreted by Cox (1915) as being rhyolitic lavas. However, they contain shardic fragments, along with crystals and are interpreted as ash-flow tuffs, emplaced subaqueously and later extensively recrystallized.

The tuffs of the Abereiddi Tuff Member were deposited entirely in a submarine environment and are interpreted as deposits chiefly from low- and high-density turbidity currents and from debris flows. They become thicker towards the west, as well as becoming coarser grained and less well bedded. The sequence is thought to have been generated by the periodic slumping of unstable basaltic debris down the flanks of a submarine volcano. The increase in bedding features and the fining of grain-size upwards through the sequence possibly reflects a decrease in the rate of activity with time, such that the input of slumped material and coarse-grained debris decreased progressively. Periodic increases in activity are reflected, for example, by the lapilli-tuffs at the base of the upper unit of the sequence.

Conclusions

Ordovician tuffs exposed to the north and south of Abereiddi Bay reflect subaqueous silicic and basaltic volcanism. The silicic tuffs were probably generated as ash-flow tuffs from primary explosive eruptions and developed broadly contemporaneous with similar tuffs to the east around Fishguard (at the Pen Caer GCR site) and to the west in the vicinity of Ramsey Island (at the Aber Mawr to Porth Lleuog GCR site). Direct correlations are not possible due to lack of outcrop continuity, but they clearly form part of the major bimodal basic–silicic volcanism which characterized this part of the Welsh Basin in early Ordovician times.

Basaltic tuffs appear to have been generated on the flanks of a submarine volcano; loose ash and lapilli were reworked due to slumping, leading to the accumulation of a sequence of turbidity deposits. These tuffs are the youngest Ordovician volcanic deposits in the southern part of the Welsh Basin.

ST DAVID'S HEAD
(SM 733 275–747 288)

R. E. Bevins

Introduction

The St David's Head area in north Pembrokeshire, comprises chiefly Ordovician gabbroic and closely related rocks, forming the St David's Head Intrusion. The intrusion is broadly sheet-like in form and, as a result of Caledonian folding, occurs as two linear, near continuous outcrops, up to 2 km in length, on opposing limbs of a tight NE-trending syncline. Where present, an igneous layering dips steeply (50–80°) to the SE and NW on opposing limbs of the syncline. A maximum thickness of 570 m is

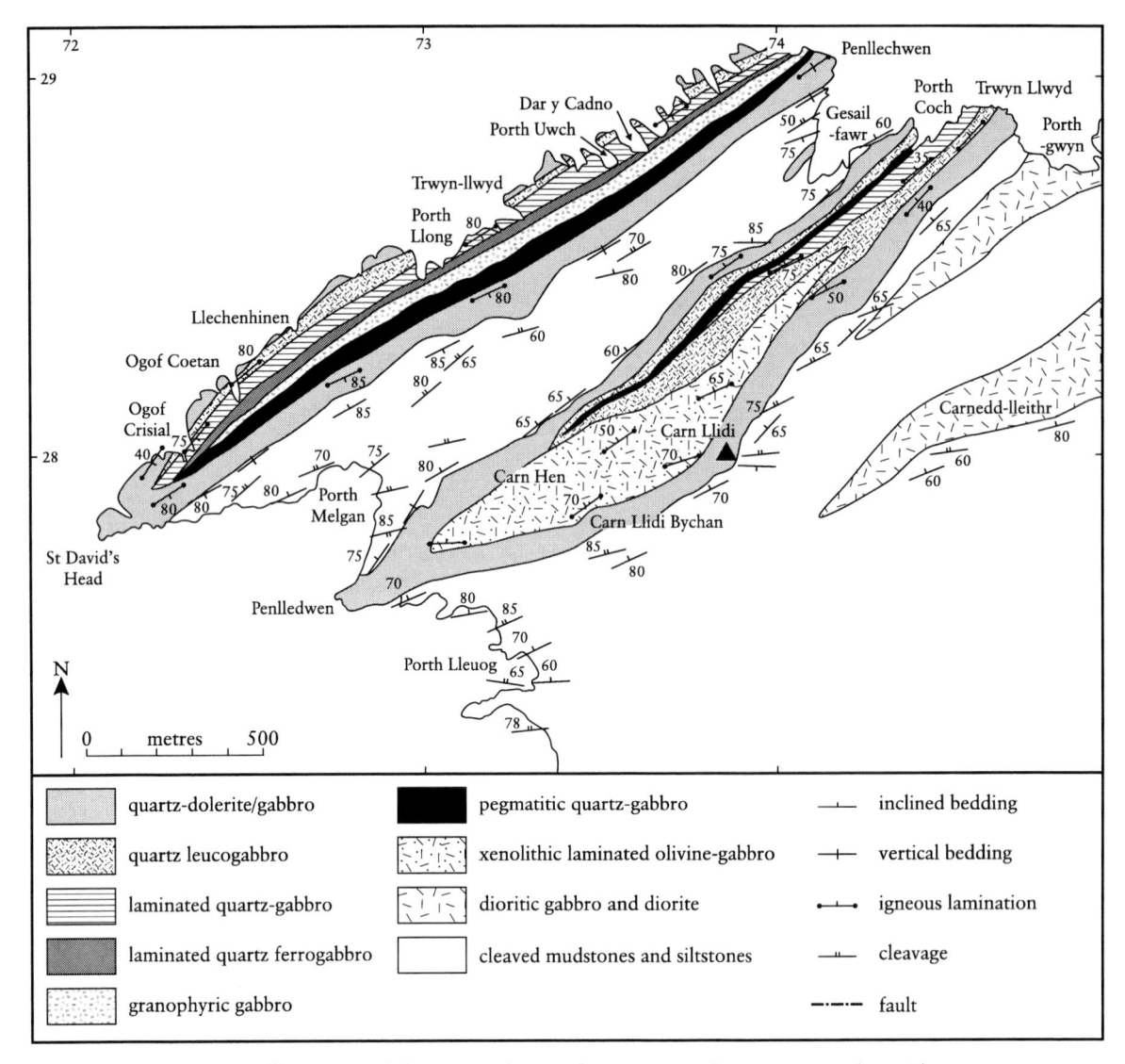

Figure 6.20 Map of the St David's Head Intrusion (after Bevins *et al.*, 1994).

seen on the south-eastern (Carn Llidi) limb, while only *c.* 385 m is exposed on the north-western (St David's Head) limb.

The intrusion invaded sedimentary rocks (mudstones to fine-grained sandstones) of Ordovician, probable Arenig, age prior to Caledonian (end-Silurian) folding. Contact metamorphic effects are seen in manganiferous siltstones, with the development of spessartine and cordierite porphyroblasts (now pseudomorphed).

The earliest studies on the St David's Head Intrusion were those by Elsden (1905, 1908), who described the presence of various norites, quartz-norites, and enstatite diorites, noting interbanding between and a regular distribution of the different rock types. Roach (1969) elaborated on the earlier work, identifying seven major petrological types in the intrusion orientated parallel to the contacts, in addition to minor, cross-cutting aplite veins. Roach (1969) considered that the two outcrops represent two separate sheets, while Bevins and Roach (1982) suggested that the two sheets are, in fact, opposing limbs of a synclinal structure. Bevins *et al.* (1991) presented geochemical data, highlighting the petrogenetic link between the basic and silicic rocks. Most recently, Bevins *et al.* (1994) provided a detailed petrological and geochemical description of the major rock types present in the intrusion, arguing for a complex evolution, linked to multiple magma injection and in-

St David's Head

Figure 6.21 Macrorhythmic layering between laminated quartz-gabbro (lighter bands) and laminated quartz-ferrogabbro (darker bands), looking to the NE across Porth Llong, St David's Head Intrusion. (Photo: R.A. Roach.)

situ crystallization.

This is one of only two layered intrusions in the Caledonian sequences of Wales, and illustrates the complex nature of processes involved in the development of a layered intrusion, as well as providing critical evidence for the origin of contemporaneous, closely related silicic lavas and pyroclastic rocks.

Description

The St David's Head Intrusion is magnificently exposed in sea cliffs and adjacent exposures in the area to the north of Whitesand Bay. It occurs as two NE-trending outcrops, the north-western forming the coastal section from St David's Head to Penllechwen, the south-eastern forming the less well-exposed section between Penlledwen and Trwyn Llwyd, best exposed on the crags of Carn Llidi. Seven major petrological types are recognized in the St David's Head Intrusion (Figure 6.20), in addition to the relatively minor, cross-cutting aplite veins.

Quartz-dolerite and quartz-gabbro

Quartz-dolerite and quartz-gabbro form an outer unit to the intrusion, up to 100 m thick and best exposed in cliffs at the south-western extremity of the St David's Head outcrop. Here, intermittent centimetre-scale microrhythmic felsic–mafic segregations occur, passing in places into irregular pegmatitic patches. Mineralogically, the gabbros are dominated by plagioclase (max An_{65}) and clinopyroxene (cores $Ca_{40}Mg_{49}Fe_{11}$ to $Ca_{40}Mg_{43}Fe_{17}$; rims $Ca_{40}Mg_{47}Fe_{13}$ to $Ca_{40}Mg_{36}Fe_{24}$) along with minor altered olivine, orthopyroxene and ilmenite, and a quartz-feldspar mesostasis. Texturally, the gabbros are subophitic.

Xenolithic laminated olivine-gabbro

The xenolithic laminated olivine-gabbro unit is up to 200 m thick, being exposed only on the south-eastern limb of the intrusion and with the best exposures present in the steep crags of Carn Llidi. This unit is characterized by the pres-

Figure 6.22 Granophyric gabbro from the St David's Head Intrusion, showing slight variations in felsic components, east of St David's Head. (Photo: R.A. Roach.)

ence of mafic and felsic cognate xenoliths up to 1 m in length. The host gabbro is poorly laminated and is characterized by plagioclase, clinopyroxene, orthopyroxene (typically pseudomorphed), altered olivine and minor Ti-rich biotite and hastingsitic amphibole. The mafic xenoliths are predominantly composed of clinopyroxene, pseudomorphed orthopyroxene and pseudomorphed olivine. Clinopyroxenes are the most Mg-rich in the intrusion, with compositions in the range $Ca_{42}Mg_{47}Fe_{11}$ to $Ca_{40}Mg_{45}Fe_{15}$; orthopyroxenes show very restricted compositions, from $Ca_4Mg_{72}Fe_{24}$ to $Ca_4Mg_{68}Fe_{28}$. Plagioclase compositions reach An_{71}, the most Ca-rich feldspars in the intrusion.

Quartz-leucogabbro

Quartz-leucogabbro units are present in both outcrops of the intrusion. These gabbros are characterized by relatively high modal proportions of plagioclase relative to the mafic minerals. Clinopyroxenes are similar in composition to those in the xenolithic laminated olivine-gabbro unit, while plagioclases reach An_{61}, although generally they are albitized. Minor ilmenite and quartz are also present.

Laminated quartz-gabbro

Quartz-gabbros showing a pronounced mineral lamination are best exposed in the vicinity of Ogof Crisial. Locally, these gabbros are interlayered with laminated quartz-ferrogabbros (Figure 6.21). The lamination is due to the alignment of the major mineral phases, namely tabular plagioclase, elongate, prismatic clinopyroxene (and rarer altered orthopyroxene) and elongate ilmenites. Clinopyroxenes show a restricted range of compositions from $Ca_{43}Mg_{39}Fe_{18}$ to $Ca_{42}Mg_{33}Fe_{25}$. Plagioclase compositions reach a maximum calcic component of An_{58}. Commonly, the plagioclases show strong compositional zoning.

Laminated quartz-ferrogabbro

A relatively thin unit of laminated quartz-ferrogabbro in the St David's Head outcrop is best exposed on the north-eastern side of Porth

Llong. Towards the south-western end of the outcrop laminated ferrogabbros, up to 1 m thick, are spectacularly interlayered on a macrorhythmic scale with laminated quartz-gabbros. The lamination is similar to that in the laminated quartz-gabbros, except that ilmenite is more abundant in the ferrogabbros. Mineral compositions are nearly identical to those in the laminated quartz-gabbros.

Granophyric gabbro

The principal exposures of granophyric gabbro occur in the north-western outcrop of the St David's Head Intrusion, for example in the clifftop crags above Porth Llong, although coarse pegmatitic to granophyric gabbros are also present in the Carn Llidi outcrop. These gabbros are isotropic, dominated by plagioclase (altered), with relatively minor clinopyroxene (the most Fe-rich in the St David's Head Intrusion, reaching $Ca_{42}Mg_{27}Fe_{31}$), orthopyroxene, and ilmenite. Apatite, as stout prisms up to 2 cm long, is abundant and of probable cumulus origin. These gabbros are characterized by interstitial quartz-alkali feldspar granophyric intergrowths (Figure 6.22).

Pegmatitic quartz-gabbro

Distinctive pegmatitic quartz-gabbros are present in both outcrops, towards the top of the intrusion. Geochemical evidence presented by Bevins *et al.* (1994) suggests that these gabbros are in fact coarse-grained equivalents of either quartz-gabbro, leucogabbro or granophyric gabbro. Distinctive in these pegmatitic quartz-gabbros are prismatic clinopyroxene crystals up to 12 cm in length.

In addition to the seven main petrological units, thin (up to 30 cm wide) cross-cutting aplite veins are particularly well developed in the vicinity of Ogof Crisial. The aplites are dominated by albite and quartz, associated with minor ilmenite, amphibole, apatite, titanite and zircon.

Contact relationships and layering

Contacts between the petrological units vary from sharp to gradational, while certain units are interbanded. Relationships are described in detail by Bevins *et al.* (1994). The nature of the contacts is crucial in establishing the history of the intrusive events. In addition, the St David's Head Intrusion shows three types of mineral and compositional layering within the individual petrological units, namely:

1. Mineral layering in the more evolved laminated quartz-gabbros and laminated quartz-ferrogabbros, related to the parallel alignment of tabular plagioclase, pyroxene and ilmenite crystals.
2. Macrorhythmic modal layering, up to 1 m thick, related to an alternation of laminated quartz-gabbro and laminated quartz-ferrogabbro units.
3. Centimetre-scale, felsic-mafic, microrhythmic modal layering in the quartz-gabbro unit.

Geochemistry

Bevins *et al.* (1994) provided the first detailed geochemistry for the full range of gabbroic and related rocks of the St David's Head Intrusion. Roach (1969) had previously presented major element analyses for a small number of samples from the quartz-dolerites and quartz-gabbros, while Bevins *et al.* (1991) had used the geochemistry of the silicic (aplitic) derivatives of the St David's Head rocks to speculate on the origin of silicic eruptive rocks associated with the neighbouring Ordovician volcanic centres (e.g. the Pen Caer GCR site). The apparent link by fractional crystallization between the St David's Head aplites and the more basic gabbros led Bevins *et al.* (1991) to suggest that many of the rhyolitic lavas and ignimbrites in the adjacent sequences were similarly derived from more basic compositions by fractional crystallization.

Interpretation

The petrological varieties present in the St David's Head Intrusion and their contact relationships reveal a complex origin, thought to relate in part to in-situ fractionation and in part to multiple events of magma injection. For example, interbanding relationships between leucogabbro and laminated quartz-gabbro at Ogof Crisial imply that the leucogabbro was intruded later than the crystal accumulation that gave rise to the laminated quartz-gabbro, indicating separate intrusive events. In contrast, the lamination in the laminated quartz-gabbros and the laminated quartz-ferrogabbros is thought to be of cumulus origin, resulting from the periodic sedimentation and accumulation of ilmenite crystals. The presence of mafic (olivine +orthopyroxene+clinopyroxene) cognate xenoliths in the xenolithic gabbros has been taken as evidence for the existence of high-level magma chambers in which these mafic minerals accumulated, prior to incorporation in later basic magmas and transport to higher crustal levels.

The geochemical data set of Bevins *et al.* (1994) confirmed the earlier proposals of Bevins *et al.* (1991) that the various petrological types are related through crystal fractionation. In particular, strong correlations between highly incompatible elements demonstrate that all the different rock types present in the intrusion, from basic through to silicic compositions, are petrogenetically linked, while plots of highly compatible elements versus incompatible elements illustrate the role of clinopyroxene and olivine in the fractionation process. Not all of the chemical variations determined, however, can be explained by crystal fractionation. Certain major and minor element variations suggest the importance of cumulate processes, in particular accumulation of olivine-orthopyroxene, ilmenite, and of apatite.

Conclusions

The St David's Head Intrusion is one of only two layered intrusions in the Caledonides of Wales, and provides evidence for a variety of high-level igneous processes. A great variety of rock types are present, resulting in a number of different rock units, ranging from basic through to silicic composition. Evidence from the field relationships between these different units, afforded by magnificent coastal exposures, combined with geochemical evidence, suggests that the units are all related to each other magmatically, but that a variety of igneous processes have operated, leading to the variety of compositions exposed today. These include in-situ fractional crystallization, the incorporation of crystalline material (cognate xenoliths) from an underlying high-level magma chamber, and the injection of different magma batches of contrasting compositions at slightly different times. The geochemical variation present in the intrusion provides critical evidence for the origin of rhyolitic lavas and ash-flow tuffs exposed elsewhere in the Ordovician sequences of Pembrokeshire.

CADAIR IDRIS (SH 750 149, 667 133, 712 148–711 100)

D. G. Woodhall

Introduction

The Cadair Idris GCR site comprises a well-exposed succession of acid and basic volcanic rocks of Llanvirn to basal Caradoc age, belonging to the Aran Volcanic Group (Figure 6.23). The latter comprises up to 2 km of volcanic and sedimentary rocks of Arenig to Caradoc age, that crop out around the southern and eastern parts of the Harlech Dome. At the Cadair Idris GCR site, subaqueously emplaced silicic ash-flow tuffs in the lower and upper parts of the succession, are separated by a thick series of submarine basalt lavas with interbedded tuffs and mudstones. A number of dolerite sills, and a thick microgranite sheet, are also well exposed. The whole succession dips to the south or southeast.

The first detailed descriptions of the igneous rocks of Cadair Idris were given by Cox (1925) and Cox and Wells (1927) in their accounts of the stratigraphy, tectonics and intrusive igneous rocks of the Dolgellau area. Some of the intrusions within the site have been the subject of detailed structural, petrographical and geochemical studies. Lake and Reynolds (1912) described the structure of the Mynydd y Gader dolerite, which lies in the northern part of the site area, while Davies (1955, 1956, 1959) described the petrology, geochemistry and contact metamorphism associated with the Cadair Idris granophyre (now classed as a microgranite) and the Pen-y-gader dolerite, both of which crop out extensively. The site area was included in a recent resurvey of the Cadair Idris district by the British Geological Survey (Pratt *et al.*, 1995), which involved a geochemical study of the igneous rocks by Kemp and Merriman (1994).

Exposures of the Aran Volcanic Group in the Cadair Idris GCR site area represent products of the most voluminous early Ordovician igneous episode in southern Snowdonia.

Description

The stratigraphical succession established most recently by the British Geological Survey (Pratt *et al.*, 1995) is presented below (Table 6.1), along

Cadair Idris

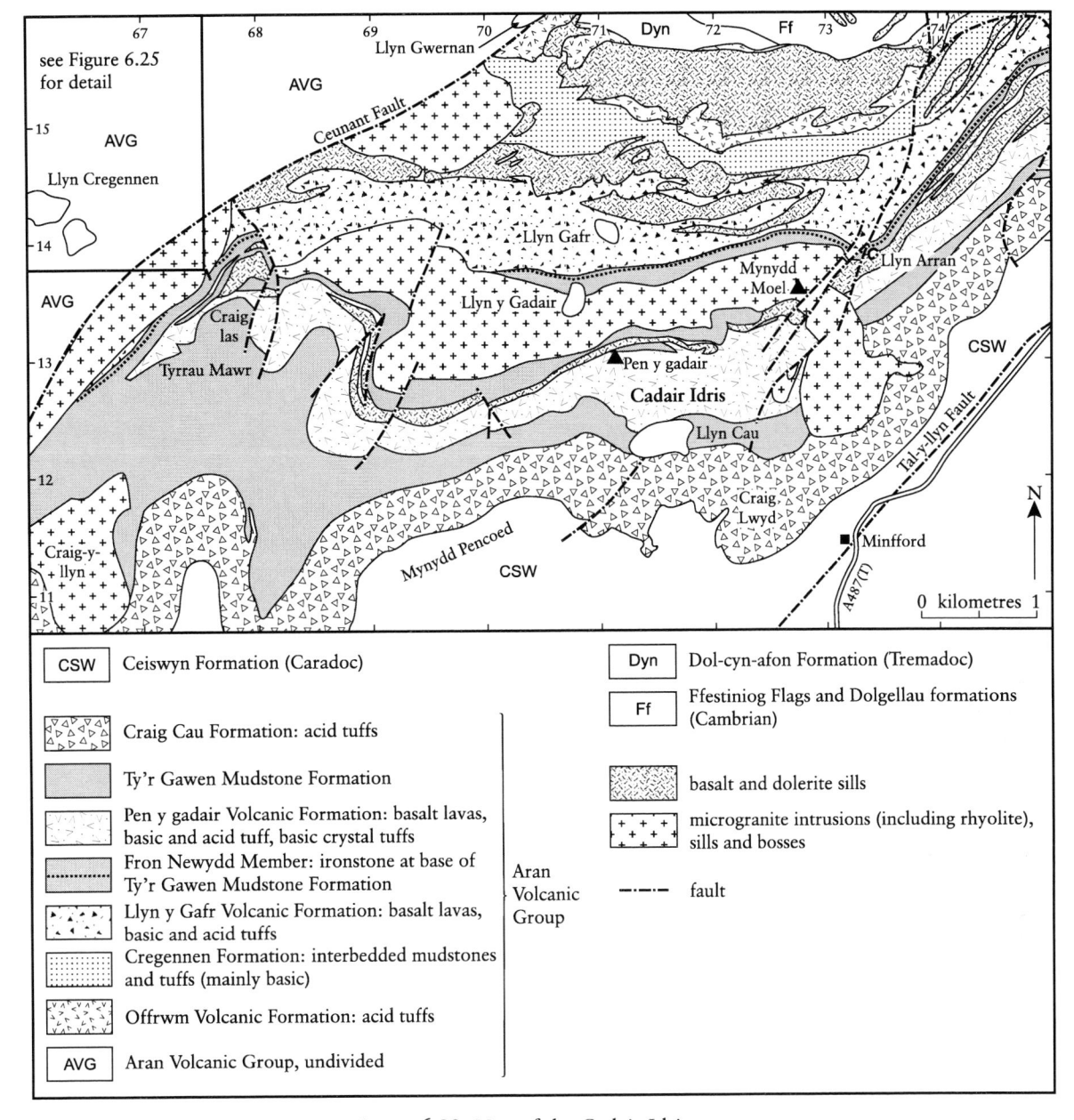

Figure 6.23 Map of the Cadair Idris area.

with the earlier terminology of Cox (1925) and Cox and Wells (1927).

Silicic ash-flow tuffs of the Offrwm Volcanic Formation crop out on Mynydd y Gader to the north of the site area, where they are intruded by the Mynydd y Gader dolerite sill (Lake and Reynolds, 1912). In contrast to the Pared y Cefn-hir GCR site, at Cadair Idris there are no interbedded mudstones, although the tuffs are petrographically similar and display microscopic evidence of welding. Bedded tuffs up to 3 m thick occur locally at the top of the formation.

The overlying Cregennen Formation consists of interbedded tuffs and mudstones, which crop out along the south side of Mynydd y Gader. In the west of the area, relationships are complex as a result of slumping and later intrusions, and therefore correlation with the Pared y Cefn-hir GCR site has proved difficult (Pratt *et al.*, 1995). The upper part of the formation, in the vicinity of the Penrhyn-gwyn slate quarries, includes a silicic ash-flow tuff up to 25 m thick, which is possibly equivalent to a similar tuff seen on Pared y Cefn-hir itself. This tuff is composed of

Wales and adjacent areas

Table 6.1 Stratigraphy of the Cadair Idris area, showing correlations with earlier nomenclature.

Pratt *et al.* (1995)	Cox (1925) and Cox and Wells (1927)	Thickness (m)
Craig Cau Formation	Upper Acid Group	> 400
Ty'r Gawen Mudstone Formation	Llyn Cau Mudstone	150
Penygadair Volcanic Formation	Upper Basic Group	200
Ty'r Gawen Mudstone Formation	Llyn y Gadair Mudstones and ash	200
Llyn y Gafr Volcanic Formation	Llyn y Gafr Volcanic Formation	360
Cregennen Formation		160
Offrwm Volcanic Formation	Lower Acid Group	80

microscopic pumice fragments and glass shards that occur in a fine-grained quartzose recrystallized matrix (probably formerly vitric dust). Farther east, the formation is relatively undisturbed and consists of a coarse-grained basic tuff, 60 m thick and overlain by up to 100 m of mudstones within which there are impersistent basic turbiditic tuffs. The coarse-grained basic tuffs are massive and poorly sorted and are made up mainly of clasts up to 15 cm across of basic and acidic volcanic rocks. At the base there are clasts, up to 0.5 m across, of contorted laminated siltstone and the uppermost 10–15 m are bedded and finer grained. These basic tuffs are interpreted as debris flow deposits.

The Llyn y Gafr Volcanic Formation crops out in the relatively low ground between Mynydd y Gader and Cadair Idris. At Llyn y Gafr, the type section of the formation, massive basalt lavas in the lower part of the formation are up to 60–70 m thick and have pillowed tops up to 10 m thick. There are a few intercalations, up to 40 m thick, of basic tuffs and mudstones, along with coarse-grained debris flows of basic tuffs which resemble those in the underlying Cregennen Formation. The upper part of the formation consists of massive, vesicular and pillowed basalt lava, 70 m thick, which is overlain by coarse-grained breccia up to 75 m thick. The breccia, probably of pyroclastic origin but redeposited by a debris flow, is composed of clast-supported angular blocks of highly vesicular basalt. These blocks are mainly up to 20 cm across, although at the base massive basalt and

Figure 6.24 Pillowed basalts from the Penygadair Volcanic Formation, SW of Penygadair summit. (Photo: D.G. Woodhall.)

pillow fragments are up to 1 m across. West of Llyn y Gafr the lavas of the lower part of the formation are poorly exposed, but the basaltic breccia described above increases in thickness to nearly 150 m. East of Llyn y Gafr the proportion of basalt lava, particularly that which is massive, decreases in relation to basic tuff and mudstone, and the breccia in the upper part of the formation wedges out.

The overlying Penygadair Volcanic and Ty'r Gawen Mudstone formations are intercalated. At the base of the lowest mudstone, which crops out adjacent to Llyn y Gadair and Llyn Arran, is a 25 m-thick unit of black pyritous mudstone with numerous phosphatic nodules and with lenticular oolitic and pisolitic ironstone up to 2 m thick. The mudstone has yielded fossils indicative of either the *Nemagraptus gracilis* Biozone (Llandeilo) or the lowest part of the *Diplograptus multidens* Biozone (basal Caradoc) (Pratt *et al.*, 1995). The ironstone has been worked as a source of low grade ore, with small excavations marking its outcrop.

The lowest part of the Penygadair Volcanic Formation consists of 10 m of basic feldspar crystal-rich tuffs, emplaced as debris flow deposits and turbidites, which crop out east of Llyn y Gadair. An intercalation of the Ty'r Gawen Mudstone Formation and a large microgranite intrusion separate these tuffs from a 200 m-thick succession dominated by basalt lavas. Pillow lavas dominate exposures at the summit of Cadair Idris (Penygadair) and eastwards as far as Gau Graig (Figure 6.24). At Penygadair, the type area of the formation, individual flows up to 15 m thick are locally apparent where they are separated by thin intervening tuffs and/or mudstones. The thicker flows are massive at the base, with some incipient columnar jointing, but are pillowed at the top. The pillows are closely packed with little or no interpillow sediment and tend to decrease in size upwards within a flow from about 2 m across to about 0.2 m. They display radial and concentric fractures and most have quartz-filled amygdales. There are rare lava tubes up to 10 m across infilled with flow-banded basalt, and locally hyaloclastite occupies depressions on flow surfaces. At Penygadair there are two interbedded silicic ash-flow tuffs each about 15 m thick and composed of quartz, recrystallized pumice fragments, and glass shards along with feldspar crystals. There is microscopic evidence for welding. The lower of the two silicic ash-flow tuffs is the most persistent, extending from Tyrrau Mawr in the west to Gau Graig in the east. Between Penygadair and Tyrrau Mawr the basalt lavas wedge out and the formation consists of silicic ash-flow tuffs with interbedded tuffaceous sandstones and mudstones, altogether 65 m thick, overlain by a lenticular basaltic (?)pyroclastic breccia, 3 m thick. The breccia is composed of angular blocks of vesicular basalt up to 0.2 m across in a matrix of finer grained scoriaceous basalt fragments, which grade into glass shards.

The highest intercalation of the Ty'r Gawen Mudstone Formation crops out south of Penygadair, where part of the outcrop is concealed by Llyn Cau. In the steep slopes between Llyn Cau and Craig Cau the mudstones incorporate an olistostrome, 200 m long and 20 m thick, made up of blocks of pillowed basalt up to 4 m across in a matrix of tuffaceous mudstone with a variable amount of volcanic rock fragments and feldspar crystals. It also contains irregular contorted clasts of green tuff, up to 0.3 m across, which were probably incorporated and deformed while unlithified.

The lower part of the Craig Cau Formation, in exposures at Craig Cau and Craig Cwm Amarch, consists of a basal blocky tuff, 20–40 m thick, overlain by a disrupted sequence of tuffs and tuffaceous mudstones with lenticular mudstones and basalt lavas, altogether 200 m thick. The basal tuff is possibly an ash flow but it has incorporated many blocks of vesicular basalt (some resembling pillow fragments), dolerite, basic tuff and more rarely rhyolite, along with feldspar crystals, in a dark-green muddy matrix. The blocks decrease in size and frequency upwards and the top is relatively block-free and more silicic in appearance. The upper part of the formation consists of an ash-flow tuff, 180 m thick, with up to 10 m of bedded tuffs at the top. The ash-flow tuff is non-welded and feldspar crystal-rich in the basal 2–3 m, although above the tuff is strongly welded with a contorted welding fabric indicative of rheomorphic folding. This becomes less evident upwards as the attitude of the welding fabric becomes approximately bedding parallel. The overlying bedded tuffs are fine grained and include laminated tuffs and thin intercalations of mudstone.

Most of the igneous intrusions within the site area are dolerite sills, up to 200 m thick, of which the Mynydd y Gader dolerite (Lake and Reynolds, 1912) and the Penygadair dolerite (Davies, 1956) are the most extensive. The

Cadair Idris microgranite, previously referred to as 'granophyre' (Davies, 1959), is for much of its outcrop a sill up to 600 m thick. It forms a prominent escarpment immediately north of Cyfrwy, Penygadair and Mynydd Moel, with columnar jointing evident in the escarpment cliffs. At Mynydd Moel the sill becomes a highly discordant boss, within which the microgranite becomes fine grained and grades into rhyolite. Whereas the sill is emplaced into the lowest part of the Ty'r Gawen Mudstone Formation, the boss has intruded overlying formations as high as the upper part of the Craig Cau Formation. Here it loses its identity as a result of the lithological similarity between recrystallized ash-flow tuff and rhyolite (Davies, 1959; Pratt et al., 1995).

Interpretation

The tuffs of the Offrwm Volcanic and Cregennen formations are interpreted as the subaqueous products of contemporaneous acid and basic explosive volcanism at both the Cadair Idris and the Pared y Cefn-hir GCR sites. However, the absence of interbedded mudstone in the Offrwm Volcanic Formation on Mynydd y Gader suggests that the formation is represented here by a single thick ash-flow tuff, which was possibly emplaced into shallower water than the tuffs of this formation at the Pared y Cefn-hir GCR site. The scarcity of mudstone at the base of the Cregennen Formation along the central part of Mynydd y Gader indicates that shallow-water conditions persisted during the emplacement of the lowest basic tuffs by one or more debris flows. However, the mudstone-dominated succession above indicates the establishment of deeper water conditions in which relatively fine-grained turbiditic tuffs were deposited.

The basalt lavas of the Llyn y Gafr and Penygadair volcanic formations are the products of effusive volcanism from an unknown source area. The basalt breccias possibly represent coarse pyroclastic material that accumulated close to a vent, but it is possible that the breccias represent accumulations of pyroclastic material redeposited by debris flows at greater distances from the source vents. The mudstone intercalation that separates the two formations indicates a hiatus in volcanism. The pyritous mudstones and ironstones are interpreted as the result of low sedimentation rates during this hiatus (Pratt et al., 1995).

The presence of the few silicic ash-flow tuffs in the Cregennen and Penygadair Volcanic formations indicates that explosive silicic volcanism was only intermittent. However, during the emplacement of the Craig Cau Formation the situation was very different. The basaltic material incorporated in the ash-flow tuffs and other deposits in the lower part of the formation is consistent with sporadic basaltic volcanism. The lack of such material higher up in the formation suggests that basic volcanism had died out before the subaqueous emplacement of the thick silicic ash-flow tuffs that dominate the upper part of the formation. It has been suggested that the boss of the Cadair Idris microgranite lies close to the source of the silicic tuffs of the Craig Cau Formation (Cox and Wells, 1927; Davies, 1956, 1959; Pratt et al., 1995).

Geochemical analyses (Kemp and Merriman, 1994) confirm the existence of bimodal basaltic and rhyolitic lavas and tuffs, and suggest that the associated dolerite and microgranite intrusions are probably cogenetic. The rocks evolved largely by crystal fractionation of transitional tholeiitic/calc-alkaline magmas emplaced into thinned continental lithosphere, probably in an extensional marginal basin. The magmas are believed to have been generated in the mantle by the melting of subduction-modified N-type (normal) Mid-Ocean-Ridge Basalt (MORB).

Conclusions

The Cadair Idris GCR site represents the most important episode of Ordovician volcanism in southern Snowdonia. The well-exposed volcanic and associated intrusive rocks of the site area include examples of a subaqueously emplaced volcanic succession dominated by basaltic pillow lavas and silicic ash-flow tuffs.

Evidence for welding in the silicic tuffs at the base and top of the volcanic succession indicates that they were emplaced as hot ash-flow tuffs contemporaneous with explosive volcanism. However, the mode of emplacement of the basic tuffs and their association with contemporaneous volcanism is less clear. The existence of well-exposed pillow lavas provides clear evidence that effusive basaltic volcanism took place and it is likely that there was associated explosive volcanism. Many of the basic tuffs display evidence of mass-flow emplacement, suggesting that the products of this explosive volcanism were dispersed in debris flows and turbidity currents.

Pared y Cefn-hir

Some of the numerous dolerite sills display evidence of intrusion contemporaneous with the basaltic volcanism and it also seems likely that the silicic magma that produced the Cadair Idris microgranite was closely related to that which generated the silicic ash-flow tuffs of the Craig Cau Formation.

PARED Y CEFN-HIR
(SH 666 152)

D. G. Woodhall

Introduction

The Pared y Cefn-hir GCR site (Figure 6.25) incorporates the best-exposed succession of volcanic rocks of Arenig to Llanvirn age in North Wales. The succession forms the lowest part of the Aran Volcanic Group.

A notable aspect of the site is the strong topographical expression of certain volcanic units and igneous intrusions. These form a series of NE-trending ridges, of which Pared y Cefn-hir itself is the most prominent (Figure 6.26). Intervening depressions are formed of less-resistant grey mudstones. The volcanic rocks consist mostly of acid and basic tuffs, all emplaced subaqueously, some as ash-flow tuffs. There are subordinate basaltic pillowed lava flows. The igneous intrusions consist of sills of dolerite and microgranite, some with well-exposed chilled margins and contacts with country rocks.

The first detailed description and map of the igneous rocks of the Pared y Cefn-hir area was that of Cox and Wells (1921), in their account of the stratigraphy, structure and intrusive igneous rocks of the district between Arthog and Dolgellau. The petrography of certain intrusive igneous rocks was also described. The site was re-examined during the recent resurvey of the Cadair Idris district by the British Geological Survey (Pratt *et al.*, 1995), which included a petrological and geochemical study of the igneous rocks by Kemp and Merriman (1994) (see the Cadair Idris GCR site report).

Description

The stratigraphy of the Pared y Cefn-hir area presented below (Table 6.2) is based on the recent

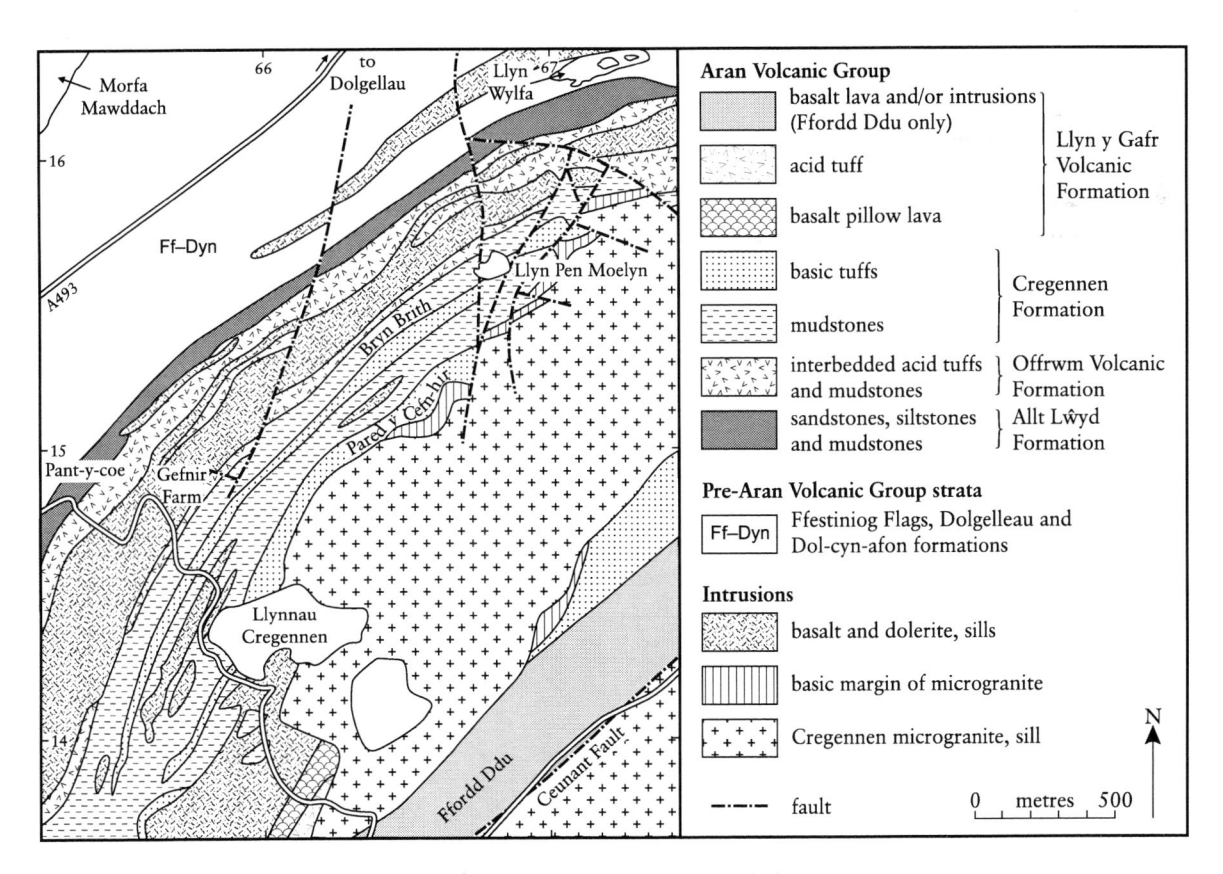

Figure 6.25 Map of the Pared y Cefn-hir area.

Wales and adjacent areas

Figure 6.26 View of the Pared y Cefn-hir area from the SW. The prominent ridge is formed by the Cefn-hir Member and the rocky slopes to the right are in the Cregennen microgranite. (Photo: D.G. Woodhall.)

resurvey by the British Geological Survey (Pratt *et al.*, 1995), with the earlier terminology of Cox and Wells (1921) for comparison.

The Allt Lŵyd Formation consists of sandstones, siltstones and mudstones which crop out and are moderately well exposed in the vicinity of Gefnir Farm and Llyn Wylfa, but which wedge out immediately south of Llyn Wylfa. Petrographical analyses of the sandstones show that the majority are of volcanic provenance, composed of feldspar crystals and basalt fragments, while few, typically in the lower part of the formation, are quartzose sandstones of non-volcanic provenance. Acritarch floras from interbedded mudstones indicate an Arenig age.

The Offrwm Volcanic Formation rests conformably on the Allt Lŵyd Formation and is best exposed immediately north of Gefnir Farm. Here, it consists of units 1–15 m thick of silicic ash-flow and turbiditic tuff, most of which are separated by intervals of dark-grey mudstone up to 25 m thick. There are subordinate tuffaceous sandstones, probably also deposited as turbidites. Typically, the silicic tuffs are pale yellowish-grey weathering and fine-grained. Several tuff units are planar bedded at the top and some contain clasts of either fine-grained silicic tuff or contorted grey mudstone. A few contain siliceous nodules and several display a bedding-parallel welding foliation, best seen in the highest tuff unit, which is the most easily identifiable and persistent. It increases in thickness, from 8 m at Gefnir Farm to 15 m farther to the NE. Tuff units lower in the formation are

Table 6.2 Stratigraphy of the Pared y Cefn-hir area, showing correlations with earlier nomenclature.

Pratt *et al.* (1995)	Cox and Wells (1921)	Thickness (m)
Llyn y Gafr Volcanic Formation	Lower Basic Volcanic Series	> 100
Cregennen Formation	Moelyn, Crogenen and Bifidus slates	225
Cefn-hir Member	Cefn Hir Ashes	45
Bryn Brith Member	Bryn Brith Beds	55
Offrwm Volcanic Formation	Lower Acid Volcanic Series	90
Allt Lŵyd Formation	Basement Series	80

probably also persistent but are difficult to distinguish from each other owing to intermittent exposure and numerous dolerite sills that complicate the succession. Thin sections show abundant glass shards and minute pumice fragments intensely altered to, and in many instances greatly obscured by, a microcrystalline quartzo-feldspathic aggregate. Feldspar crystals are abundant in some of the lowest tuffs. The interbedded mudstones are dark-grey and strongly cleaved. Graptolite faunas from a number of localities within the site area indicate the *Didymograptus artus* Biozone (early Llanvirn) (Pratt *et al.*, 1995).

The Cregennen Formation comprises 60 m of mudstone at the base, which crops out, but is poorly exposed, immediately NW of Bryn Brith where it rests sharply on the upper acid tuff of the underlying Offrwm Volcanic Formation. Basic tuffs in the middle of the Cregennen Formation constitute the Bryn Brith Member and are particularly well exposed on Bryn Brith. Here a single unit of massive, coarse-grained, poorly sorted basic tuff, 55 m thick and probably emplaced from debris flows, passes upwards into 10 m of finer grained, planar-bedded turbiditic tuff. The massive tuff is composed chiefly of abundant ragged fragments, up to 5 mm across, of vesicular, altered basalt, but it also contains numerous subangular blocks of bedded basic tuff up to 30 cm across, the bedding of which is highly contorted, indicating incorporation in an unlithified state. The Bryn Brith Member is overlain by 65 m of mudstone, which is poorly exposed in the low ground between Bryn Brith and Pared y Cefn-hir. The Cefn-hir Member lies at the top of the formation and is well exposed on the central and southern parts of the prominent ridge of Pared y Cefn-hir. On the southern part of the ridge a 2 m-thick bed of silicic ash-flow tuff lies in the middle of the member but underlying beds of basic tuff, 1–15 m thick, wedge out north-eastwards. Consequently, on the central part of the ridge the silicic tuff lies at the base of the member. Here, it is overlain by 2–3 m of mudstone followed by 25 m of massive, coarse-grained, blocky and poorly sorted basic tuffs, which form a series of debris flow units 1–10 m thick. Each unit has finer grained, planar bedded, turbiditic tuff at the top, in some cases up to 2 m thick. The basic tuffs are petrographically similar to those of the Bryn Brith Member but feldspar crystals are more common, and blocks of bedded basic tuff are accompanied by those of basalt, silty mudstone and acid volcanic rock. Basaltic pillow lavas, up to 5 m thick, occur locally near the top of the member at the SW end of the ridge. The mudstones of the Cregennen Formation have yielded graptolites suggestive of

Figure 6.27 Basic xenoliths in the margin of the Cregennen microgranite, south side of Pared y Cefn-hir (6651 1506). (Photo: D.G. Woodhall.)

the *Didymograptus artus* Biozone, and fragmentary trilobites obtained from mudstone within the Bryn Brith Member SW of Bryn Brith suggest a similar early Llanvirn age (Pratt *et al.*, 1995). Basaltic pillow lavas of the overlying Llyn y Gafr Volcanic Formation are locally well exposed immediately south of Llynnau Cregennen.

The igneous intrusions are mostly dolerite sills which range in thickness from a few metres to *c.* 100 m. They occur typically within mudstones, which most probably facilitated intrusion. Dolerite exposures near Llyn Wylfa and Gefnir Farm display narrow (up to 0.5 m) chilled margins against mudstone country rocks. In each case mudstone 'flames' penetrate from 2 cm to as much as 1 m into the sill, and indicate that the sediment was unlithified at the time of intrusion. The dolerites are typically composed of plagioclase, clinopyroxene, iron-oxide and accessory apatite. Extensive alteration has produced a range of secondary minerals; albite replacing plagioclase, actinolite (along with chlorite) replacing pyroxene, titanite replacing iron-oxide, and intergrowths of chlorite, epidote, quartz and stilpnomelane replacing the groundmass of porphyritic rocks. In spite of the alteration, primary subophitic and ophitic igneous textures are preserved. The chilled margins tend to be vesicular and porphyritic with feldspar phenocrysts in a groundmass of feldspar microlites and chlorite.

The Cregennnen microgranite sill, which is 500 m thick, crops out extensively in the eastern part of the site. It has intruded the Cregennen and Llyn y Gafr formations. It is particularly well exposed on the eastern side of Pared y Cefn-hir where there is a distinct lower marginal facies developed up to 10 m from the contact. The microgranite proper is characterized by granophyric intergrowths of feldspar and quartz, but there are some alkali feldspar phenocrysts and accessory amounts of apatite and zircon. The feldspars have been altered to epidote and pumpellyite, and late stage stilpnomelane crystals are common. The marginal facies has abundant hornblende and biotite, but geochemically it is similar in composition to the microgranite (Kemp and Merriman, 1994). The contact between the basic margin and microgranite proper is marked by a conspicuous xenolithic zone several metres thick (Figure 6.27). This zone is interpreted as earlier, partly crystalline magma that was disrupted by the intrusion of the main part of the microgranite (Pratt *et al.*, 1995).

Interpretation

The sandstones of the Allt Lŵyd Formation are interpreted as shallow-marine deposits (Pratt *et al.*, 1995), with a volcaniclastic component derived from the contemporaneous erosion of pre-existing volcanic rocks, most probably of the Rhobell Volcanic Group (Kokelaar, 1979).

The absence of sandstones and shallow-water bedforms in the Offrwm and Cregennen formations suggests deeper-water conditions, possibly established by marked local subsidence. The tuffs of these formations are the products of contemporaneous explosive acidic and basaltic volcanism. This is clearly indicated by the presence of welding fabrics in silicic ash-flow tuffs in the Offrwm Volcanic Formation, while the presence of unabraded glass shards and/or pumice fragments in both acid and basic tuffs is regarded as being further evidence for contemporaneous volcanism, even though they are present in debris flow deposits and turbidites. The latter formed by the resedimentation of pyroclastic material during or soon after explosive volcanism. The fact that these tuffs are interbedded with mudstones clearly indicates subaqueous emplacement. The source of this volcanism has not been identified (Pratt *et al.*, 1995), but the occurrence of basaltic pillow lavas near the top of the Cefn-hir Member and in the overlying Llyn y Gafr Formation suggests that the effusive basaltic volcanism took place at an unknown source possibly closer to the site area.

Conclusions

The well-exposed volcanic and intrusive igneous rocks of the Pared y Cefn-hir GCR site are of national importance as they represent the best exposures of the Aran Volcanic Group of Arenig to Llanvirn age. They are also of importance as a succession of subaqueously emplaced volcanic rocks and for the associated igneous intrusions. The exposures are easily accessible for educational purposes and the site complements the Cadair Idris GCR site.

The volcanic rocks consist of acid and basic tuffs that were emplaced subaqueously, some as ash-flows and others as debris flow deposits and turbidites. Ash-flow tuffs are only distinguishable where evidence of welding fabrics can be seen, with some difficulty and uncertainty, in some of the acid tuffs. The presence of abundant angular glass fragments in both the acid and basic tuffs suggests that they were a product of explosive volcanism. These deposits contrast with the sandstones derived from reworked volcanic material which dominate the Allt Lŵyd Formation. The site includes a number of dolerite sills, some showing evidence for emplacement into unlithified sediment, which in turn is evidence that intrusive activity was approximately contemporaneous with the volcanism. Silicic magmatism represented by the Cregennen microgranite sill however, is clearly later than that indicated by the silicic tuffs of the Offrwm, Cregennen and Llyn y Gafr formations.

CARNEDDAU AND LLANELWEDD (SO 050 520–075 549)

D. G. Woodhall

Introduction

The Carneddau and Llanelwedd GCR site lies at the southern end of the Builth Inlier of Ordovician (Llanvirn to Llandeilo) volcanic and sedimentary rocks (Figure 6.28). The volcanic rocks consist mainly of rhyolitic and basaltic tuffs, but there are basaltic, andesitic and dacitic lavas in various parts of the succession, most particularly at the top. Doleritic and dacitic intrusions also occur.

The history of research on the Builth Inlier extends back to general observations made by Murchison (1833, 1839, 1867, 1872) who noted that the largest 'trap' district in Radnorshire extends southwards from Llandegley to Builth Wells. The stratigraphy of the inlier was described by Elles (1940) but most emphasis in this work was on the sedimentary rocks. The detailed survey of Jones and Pugh (1941, 1949) was used by the Geological Survey in the compilation of a 1:25 000 scale geological map (1977). The whole of the Builth Inlier has recently been resurveyed by the British Geological Survey (BGS).

Petrological and geochemical studies of the volcanic rocks by Furnes (1978) and Smith and Huang (1995) indicated a range in composition, from basalt through andesite and dacite to rhyolite, which is wider than that which occurs gen-

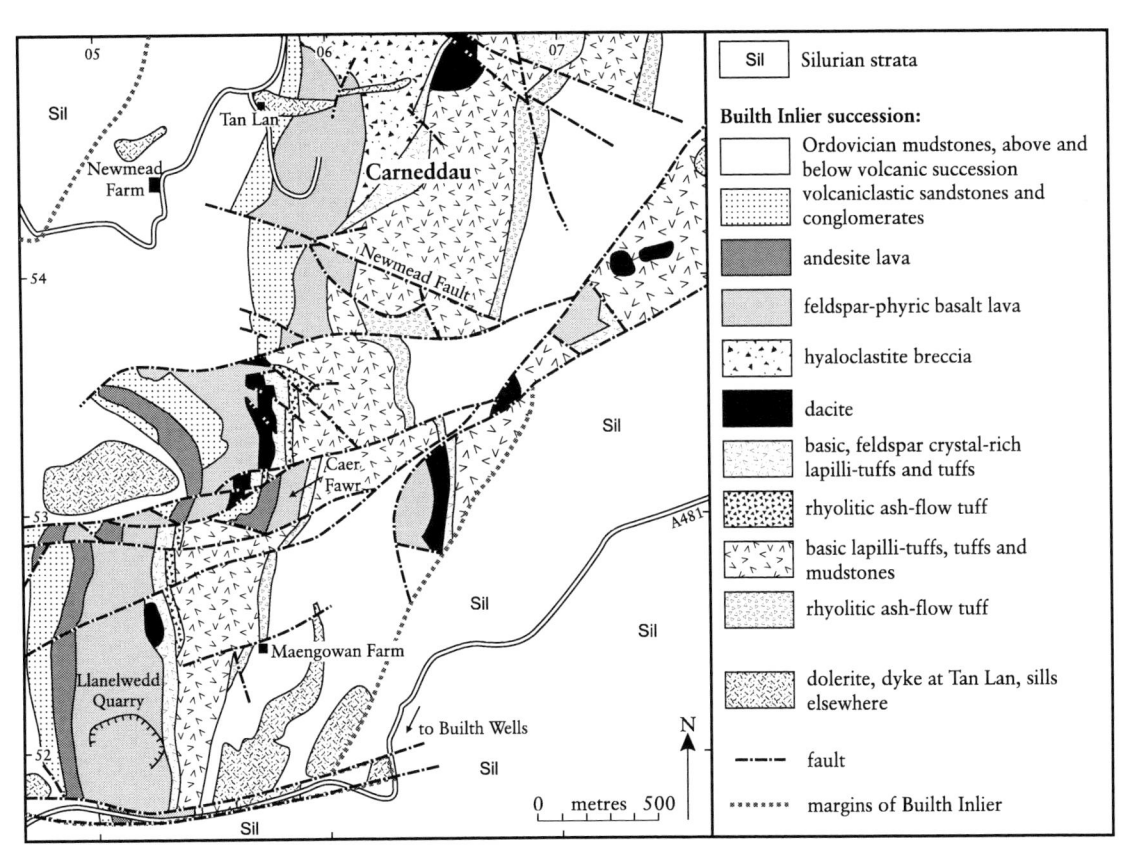

Figure 6.28 Map of the southern part of the Builth Inlier.

erally in North Wales. The volcanic rocks appear to have mainly calc-alkaline geochemical affinities, although some tholeiitic flows are also present.

Description

The volcanic succession given in Table 6.3 is based on the recent BGS resurvey of the Builth Inlier, with the stratigraphical units of Jones and Pugh (1949) for comparison.

The volcanic succession is underlain and overlain by mudstone-dominated sedimentary successions which crop out outside of the site area. The entire succession lies within the *Didymograptus murchisoni* Biozone.

The silicic ash-flow tuff at the base of the volcanic succession crops out along the eastern side of the site where it forms a subdued east-facing escarpment and is repeatedly offset by major E–W faults. It is composed of pumice fragments up to 3 mm across, along with glass shards and feldspar crystals, which can still be seen under the microscope despite intense quartz- and/or chlorite-dominated alteration. Much of the tuff appears to be bedded, but this is interpreted as a secondary feature resulting from the development of massive, resistant layers formed by intense silicification, and fissile, less-resistant layers produced by chloritization. True primary bedding is evident locally, however, where the uppermost *c.* 5 m of tuff is reworked. Erosion of the ash-flow tuff generated tuffaceous sands, composed of rhyolitic rock fragments, up to 2 m of which are locally preserved in erosional hollows at the top of the tuff.

The silicic tuff is sharply overlain by up to 100 m of massive lapilli-tuffs, emplaced as a series of debris-flow deposits. The lapilli-tuffs are composed of abundant poorly sorted fragments of basic volcanic rock up to 3 cm across, with scattered rhyolite fragments up to 15 cm across. Individual debris flow units are difficult to distinguish owing to poor exposure but they appear to be separated either by bedded tuffs, or by interbeds of dark-grey siltstone less than 10 cm thick. The massive lapilli-tuffs pass upwards into a series of turbidites, altogether about 50 m thick, each of which consist of normally graded beds of lapilli-tuff and/or tuff and which are well exposed near Maengowan Farm. Individual turbidites are sharp-based and range in thickness from 0.5 m to at least 4 m. The turbidites are overlain by up to 65 m of mudstones within which there are a number of interbedded tuff turbidites each less than 1 m thick. Within the lapilli-tuffs, tuffs and mudstones there are scattered pods of dacite lava (keratophyres of the Geological Survey 1:25 000 map, 1977) some of which are associated with hyaloclastite breccias.

An impersistent silicic ash-flow tuff occurs in the middle of the volcanic succession and is overlain by more basic lapilli-tuffs and tuffs. These were also emplaced as debris-flow

Table 6.3 Stratigraphy and lithologies of volcanic rocks of the Builth Inlier.

Lithology	Stratigraphy (after Jones and Pugh, 1949)	Thickness (m
Silicic ash-flow tuff	Rhyolitic ash and ashy mudstones of the Cwmamliw Series	35
Sandstones and conglomerates of volcanic provenance	Sandstones of the Newmead Series, including the boulder beds	65
Feldspar-phyric basalt and andesite lavas, passing laterally into hyaloclastite breccia	Spilites, keratophyres and bouldery spilitic ash of the Builth Volcanic Series	250
Feldspar crystal-rich basic lapilli-tuffs and tuffs	Pebbly feldspar ash of the Builth Volcanic Series	50
Silicic ash-flow tuff		0–35
Basic lapilli-tuffs, tuffs and mudstones, with subordinate dacite and hyaloclastite	Red agglomerate, ash and shales of the Builth Volcanic Series	> 200
Silicic ash-flow tuff	Rhyolitic ash of the Llandrindod Volcanic Series	50

Carneddau and Llanelwedd

Figure 6.29 Llanelwedd quarries, Builth Wells, viewed from the south. Westerly-dipping basic lavas, belonging to the Builth Volcanic Group, comprise much of the quarry area. Other volcanic units form the prominent features in the hills behind the quarry, the slack ground being eroded into softer shales. (Photo: R.E. Bevins.)

deposits and turbidites, but they differ from those lower in the volcanic succession in that many individual flow units contain abundant feldspar crystals. Some of the tuffs are well bedded and display hummocky cross-stratification. An impersistent flow-banded dacite lava, 20 m thick, is well exposed immediately north of Caer Fawr.

The tuffs and dacite lava described above are sharply overlain by a series of feldspar-phyric basalt lavas, with subordinate andesite lava, lapilli-tuffs and tuffs. The lavas reach up to 250 m thick at Llanelwedd where they are well exposed in a series of quarries (Figure 6.29), but they are only 150 m thick immediately to the north of the quarries. The lavas in the lowest 30 m are microporphyritic and the basal flow can be seen resting sharply on tuffs in a small disused quarry immediately SE of the main working quarry. Individual flows are 10–20 m thick and include up to 5 m of brown-weathering, clast-supported flow surface breccia. These lavas wedge out immediately north of the main quarry. In the eastern part of the main quarry the microporphyritic lavas are overlain by a series of feldspar-phyric basalt lavas, 40 m thick, with some individual flows up to 12 m thick. Flow surfaces are either brecciated or highly vesicular, but a few flows are separated by several metres of lapilli-tuff. In the NW and western parts of the main quarry massive feldspar-phyric basalt displays no obvious flow surface features and may therefore be intrusive. Two such sheets, 35–50 m thick, are separated by up to 8 m of blocky lapilli-tuff. This deposit, probably emplaced as an ash-flow tuff, contains abundant blocks of feldspar-phyric basalt along with many of gabbro and rhyolite. In the extreme west of the main quarry feldspar-phyric basalt, considered to be extrusive, locally displays pillow-like structures up to 2 m across defined by concentric layers of amygdales. Similar structures are seen locally where the lavas crop out north of the quarries. The highest lava, 35 m thick, crops out immediately west of the main quarry and is andesitic in composition (Furnes, 1978). It is locally flow-banded and distinctly brownish-yellow weathering at the top. In the NW part of the site area the feldspar-phyric basalt lavas overlie, and interdigitate with, massive poorly sorted hyaloclastite breccia composed of angular blocks of massive and vesicular feldspar-phyric basalt up to 0.5 m across. The finest-grained material consists of basalt fragments and feldspar crystals;

the coarsest blocks are up to 2 m across.

The volcanic provenance of the sandstones and conglomerates at the top of the volcanic succession is indicated by the abundance of feldspar crystals and basalt clasts. The sandstones rest with marked unconformity on feldspar-phyric basalt and andesite lava. This unconformity was described in detail by Jones and Pugh (1949) who documented features such as fossil sea cliffs, stacks, wave-worn surfaces and screes, which they attributed to an ancient shoreline. Mapping of the unconformity clearly indicates that it has considerable relief. This may be at least 50 m in places, indicated by thickness variations across faults, notably the E–W Newmead Fault where the sandstones thin abruptly northwards from 85 m to 25 m. This may, however, be a result of contemporaneous faulting as well as relief on the unconformity. The lack of lavas or tuffs within the sandstones indicates that they are a product of post-volcanic sedimentation brought about by the erosion of pre-existing volcanic rocks.

There are few igneous intrusions within the site area. Many of the keratophyres shown on the 1:25 000 geological map (1977) have been re-interpreted during the recent BGS resurvey as dacite lavas, although the distinction between lavas and high-level intrusions has been difficult. A prominent E–W dolerite dyke adjacent to Tanlan cuts across feldspar-phyric basalt lavas and sandstones, and feeds a sill emplaced at the base of the overlying mudstones. The dyke increases in thickness to 80 m as it approaches the sill. The latter is exposed in a disused quarry at Tanlan where a concordant contact with the mudstones is seen.

Interpretation

The silicic ash-flow tuff at the base of the sequence was produced during a violently explosive volcanic eruption, and the ash-flow probably travelled many tens of kilometres from its source. The location of the source is not known, but a coarse-grained facies found in northern parts of the Builth Inlier (Davies *et al.*, 1996) is consistent with a source still farther to the north. Evidence for subaqueous emplacement, including contacts with underlying mudstones and fossiliferous reworked tuffs at the top, is present mainly in parts of the inlier outside of the site area. It is considered that emplacement took place into an open-shelf environment and brought about marked shoaling following the emplacement of up to 50 m of ash-flow tuff. This led to the establishment of a shoreface zone possibly no more than 20 m deep within which the top of the tuff was reworked.

The basic lapilli-tuffs and tuffs also represent the products of explosive volcanism but this was probably less violent. The emplacement of the fragmented (pyroclastic) material as debris flow deposits and turbidites was probably contemporaneous with the volcanism as suggested by the angularity of the fragments and preservation of delicate glass shard structures. The increasing frequency upwards of finer-grained tuffs and interbedded mudstones suggests that there was a gradual reduction in frequency and duration of successive volcanic eruptions. The dacite pods are interpreted as viscous magma bodies emplaced within unlithified mud and ash, which in at least some instances broke through to the surface. The associated hyaloclastites formed by non-explosive quench fragmentation of magma in wet sediment and/or water.

Some of the mudstones mentioned above possibly accumulated during a short hiatus in volcanism that was terminated by the emplacement of another subaqueous silicic ash-flow. The impersistence of this tuff within the site area is interpreted as a result of emplacement in a shallow-water high-energy environment, and it is thought likely that immediately following its emplacement the tuff was eroded by strong (?)tidal currents; the remaining occurrences are the result of preservation in sheltered sea-floor depressions. The overlying feldspar crystal-rich lapilli-tuffs and tuffs represent the products of further contemporaneous basic explosive volcanism. However the shallow-water bedforms in some bedded tuffs suggest shoaling brought about by the continued accumulation of debris-flow deposits and turbidites. Dacite magmatism persisted.

The shoreface environment established as a result of shoaling was displaced by the northward spread of feldspar-phyric basalt lavas. Pre-existing dacite lavas probably formed local topographical barriers limiting the extent of the earliest lavas. This is suggested by the abrupt termination of the basal microporphyritic lavas immediately north of the main quarry in proximity to a dacite lava. The feldspar-phyric basalt lavas have been interpreted as subaqueously emplaced (Jones and Pugh, 1949; Nicholls, 1958; Baker and Hughes, 1979; Metcalfe, 1990; Bevins and Metcalfe, 1993) on the basis of localized occurrences of interbedded black shales and pillows. However, they have also been interpreted as subaerial (Furnes, 1978), and this interpretation is preferred because of the nature of flow surfaces; lavas with brecciated surfaces are interpreted as aa flows, whereas those that are highly amygdaloidal at the top are interpreted as pahoehoe flows. In addition, the pillows are re-interpreted as cross sections through lobate pahoehoe lava on the basis of the presence of concentric zones of amygdales and the absence of prominent chilled margins and radiating fracture patterns. The massive feldspar-phyric basalt sheets which lack flow surface features are interpreted as high-level synvolcanic intrusions (sills). The hyaloclastite breccias in the north of the site area are thought to be the products of quench fragmentation of the lavas as they encountered relatively deep water; however, the poorly sorted nature of the breccia is consistent with debris-flow deposition. The fact that the breccia is first encountered immediately beneath feldspar-phyric basalt lavas suggests that the earliest breccias were over-ridden by later lavas.

The immediate post-volcanic period in the site area was marked by marine erosion, which was accompanied by sandstone deposition and possibly also by faulting. A combination of erosion and faulting may have produced a cliffed coastline, the ultimate burial of which by sands could account for the relief on the unconformity, most evident across the Newmead Fault. The latest volcanism to affect the Builth Inlier did not leave any deposits within the site area. This volcanism resulted in the subaqueous emplacement of a further silicic ash-flow tuff, comprising the Cwmamliw Series of Jones and Pugh (1949). The pile of feldspar-phyric basalt lavas appears to have acted as a barrier preventing the southward spread of this ash-flow tuff into the Carneddau and Llanelwedd site area.

Furnes (1978) investigated the geochemistry of the volcanic rocks of the site area, although this work remains unpublished. Kokelaar *et al.* (1984b) reported that the lavas have calc-alkaline affinities, while the more detailed investigations of Metcalfe (1990) and Smith and Huang (1995) identified the presence of tholeiitic lavas interbedded with those of calc-alkaline affinity.

Conclusions

The volcanic rocks of the Builth Inlier, most of which are seen within the Carneddau and Llanelwedd GCR site, are of national importance for their value in reconstructing the plate tectonic history of both Wales and the British Isles during Ordovician times. The extensive outcrops provide detailed successions through the thickest sequence of Ordovician volcanic rocks at the margin of the Welsh Basin. The sequence has mainly calc-alkaline affinities, in contrast to volcanic rocks of Llanvirn age in north Pembrokeshire (see the Pen Caer GCR site report), which are chiefly tholeiitic. Silicic ash-flow tuffs at the base and near the middle of the volcanic succession represent the onset of separate volcanic cycles characterized by more basic explosive activity and finally basaltic lavas. The best exposures are those of the late-stage lavas seen in the Llanelwedd quarries. The continued working of the main quarry, leading to constantly changing exposures, has resulted in divergent views as to the subaqueous or subaerial emplacement of the lavas.

Finally, the site is of historical interest in that it is in part the area studied in detail by Jones and Pugh (1949), which led to the presentation of one of the first detailed palaeogeographical reconstructions of an ancient volcanic environment.

BRAICH TU DU (SH 650 606–648 630)

M. Smith

Introduction

The steep western slopes of the ridge of Braich tu du, which form the eastern side of the Nant Ffrancon Pass, exhibit a classic condensed section up through a heterogeneous sequence of acid ash-flow tuffs, intrusions and marine sedimentary rocks. The section includes representatives of both the 1st and 2nd eruptive cycles of Caradoc caldera activity in North Wales as well as evidence for the background sedimentation, which provides valuable information on the general palaeoenvironment.

The site (Figure 6.30) lies on the north-western limb of a major synclinal structure, the Idwal Syncline. Moderate to gentle south-easterly dips

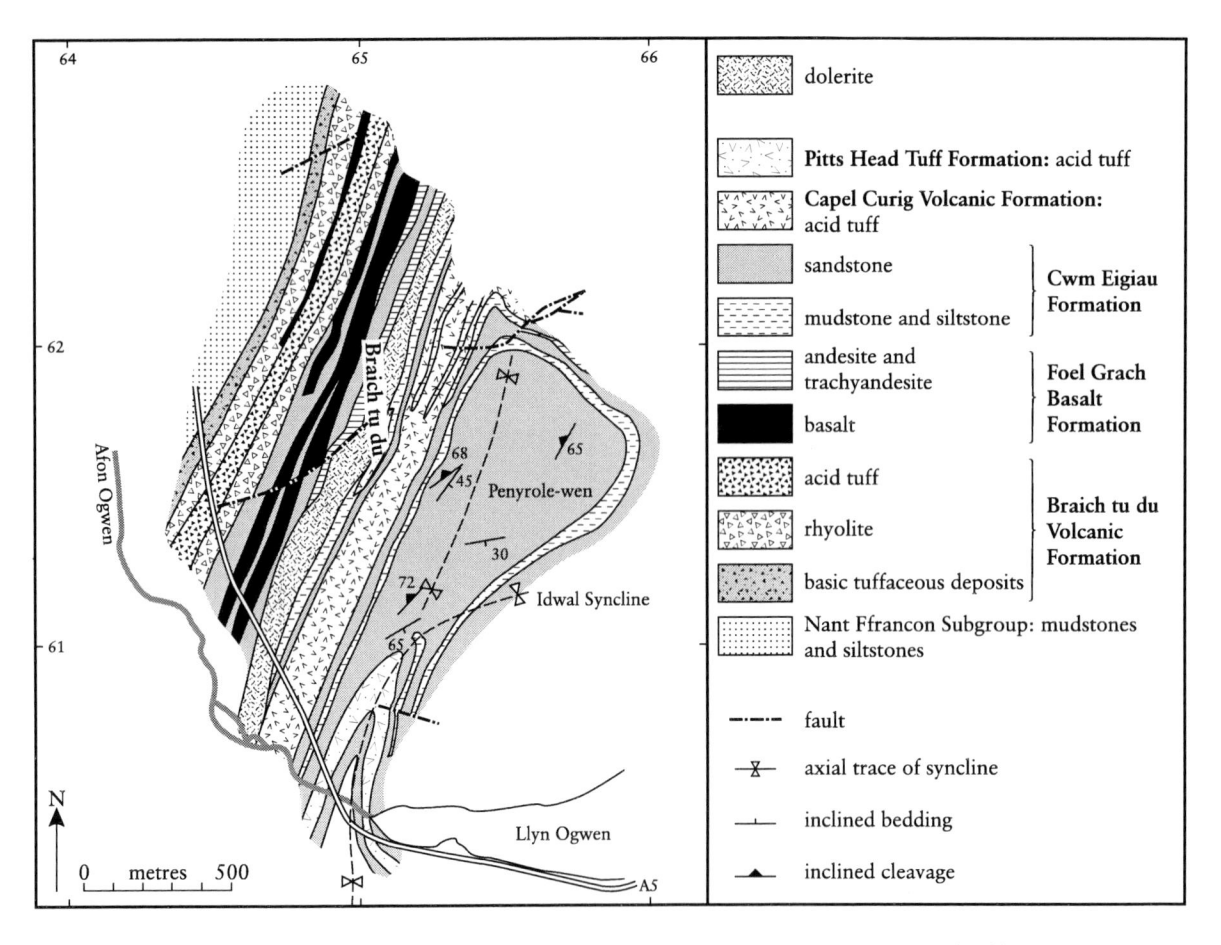

Figure 6.30 Map of the Braich tu du area. Adapted from BGS 1:25 000 Sheet 65/66 (1985).

expose up to 1000 m of section younging from NW to SE.

The area was included in the original primary geological survey completed in 1852 (Ramsay, 1881) and was resurveyed by the Geological Survey in 1968. It is included in the 1:50 000 scale Geological Sheet 106 (Bangor) (1985), although no detailed descriptions are available in the literature.

Description

The lower beds exposed at the north-western end of the GCR site comprise cleaved grey siltstones of the Nant Ffrancon Subgroup, which pass conformably into the overlying volcanic rocks of the 1st Eruptive Cycle belonging to the Llewelyn Volcanic Group (Soudleyan). This group comprises five formations, four of which are exposed in this section.

The Braich tu du Volcanic Formation is a heterogeneous sequence, up to 280 m thick, of rhyolitic flows and acid ash-flow tuffs, locally with basalt and basic tuffs. On the ridge of Braich tu du, the type locality for the formation, the sequence is dominated by two thick rhyolite flows separated by a welded ash-flow tuff with basalt lavas, basic tuffs and sedimentary rocks. The lower rhyolite, 60 m thick, is overlain by an intensely welded ash-flow tuff, up to 90 m thick (Figure 6.31). The tuff, characterized by small prismatic albite phenocrysts, has a weakly welded basal zone rich in lithic and cognate clasts. At 6–7 m above the base, welding is intense and chloritic clasts (up to 10 cm in length), representing flattened fiamme, are accentuated by quartzose recrystallization. Internally, the central part of the flow is characterized by highly contorted rheomorphic flow-folding and brecciation. The top is pervasively autobrecciated. The upper rhyolite is 45 m thick and, like the lower flow, displays excellent flow-banding and flow-folding with prominent columnar jointing and autobrecciation along the upper and lower

Braich tu du

contacts. Petrographically, the rhyolites contain up to 25% phenocrysts of sodic plagioclase and cryptoperthitic feldspar set in a devitrified groundmass of spherulitic intergrowths of quartz and feldspar.

Thin basalt lava flows and water-lain basaltic tuffs, associated with sandstones and siltstones, interdigitate throughout the sequence. The basalts and basic tuffs, originally considered by Howells *et al.* (1983) as part of the formation, have been re-interpreted as belonging to the Foel Grach Basalt Formation and indicate contemporaneity of the two formations (Howells *et al.*, 1991).

The Foel Grach Basalt Formation generally overlies the Braich tu du tuffs along strike; interdigitation of the two formations shows them to be essentially contemporaneous. At Braich tu du up to 180 m of basalt lavas and tuffs are exposed, interbedded with marine sandstones and siltstones, but the formation wedges out rapidly to the south. The basalt flows are plagioclase-phyric, amygdaloidal and tend to have massive columnar-jointed cores with blocky brecciated carapaces. Primary flow alignment of phenocrysts and feldspar microlites is evident in

Figure 6.31 The lower rhyolite (R) and ash-flow tuff (T) members of the Braich tu du Volcanic Formation separated by a thin sequence of marine siltstone, sandstone and basic tuffaceous sedimentary rocks (S) on the NE slopes of Nant Ffrancon (SH 648 621). The columnar joints in the welded tuff (T) are perpendicular to its base, which dips steeply to the right. Reproduced from Howells *et al.* (1991).

thin section. The rocks are metamorphosed to lower greenschist facies grade with plagioclase phenocrysts altered to albite ± carbonate ± epidote ± white mica and hornblende to chlorite ± clinozoisite ± actinolite ± carbonate.

Two thin ash-flow tuffs, interlayered with sandstones and siltstones, comprise the Foel Fras Volcanic Formation at Braich tu du, and represent the outflow facies from the Foel Fras Volcanic Complex (Howells *et al.*, 1991). The tuffs, of trachyandesite composition, are bedded and non-welded, with extensive reworking of the upper contacts. They comprise fragmentary feldspar crystals and lithic clasts in a matrix of albite + quartz + sericite + calcite + anatase. Relict shards replaced by chlorite and cryptocrystalline silica are common. The lithic clasts are mainly of trachyte lava and microdiorite with variable amounts of sandstone and silty mudstone. A thick dolerite sill separates the formation from the overlying Capel Curig Volcanic Formation.

Three distinct acid ash-flow tuffs, interlayered with sedimentary rocks and belonging to the Capel Curig Volcanic Formation, are well exposed between Bwlch yr Ole Wen (6530 6206) and Clogwyn Llys (6500 6140). These tuffs, originally mapped as two separate units (the 1st and 4th members) by Howells and Leveridge (1980), are now considered to be entirely within the 4th member (Howells *et al.*, 1991). In marked contrast to the primary character of the Capel Curig tuffs in the Capel Curig and Llyn Dulyn districts (see the Capel Curig and Llyn Dulyn GCR site reports), the tuffs here show no primary characteristics and are composed entirely of slumped tuff, block-and-ash tuffs, debris-flow deposits, accretionary lapilli-tuffs and thin primary ash-flow tuffs. The lowermost unit is dominated by well-bedded, thin tuffaceous sedimentary rock and is underlain by up to 20 m of coarse sandstone, which wedges out along strike against a thick dolerite sill. Within the sequence, interlayered sandstones and siltstones contain a sparse shelly fauna dominated by the brachiopod *Dinorthis*.

Cessation of volcanic activity at the end of the 1st Eruptive Cycle saw a period of shallow to offshore marine sedimentation of the Cwm Eigiau Formation. Above the prominent ridge and summit area around Penyrole-wen (6534 6148), these strata are composed of well-bedded medium-grained fossiliferous sandstones with thin interlayers of siltstone and mudstone. These sedimentary rocks are described more fully in the Cwm Idwal GCR site account.

The 2nd Eruptive Cycle commenced with eruptive activity from the Llwyd Mawr Centre, with emplacement of acid ash-flow tuffs of the Pitts Head Tuff Formation. Along the lower slopes, north of Pont Pen-y-benglog, the northernmost expression of the Pitts Head Tuff Formation is exposed in the core of the Idwal Syncline. In contrast to the sections on Moel Hebog (see the Moel Hebog to Moel yr Ogof GCR site report) and Craig y Garn (see site report), the tuffs here are underlain and overlain by marine sedimentary rocks and deposition clearly took place in a submarine environment. The lower tuff, 30–40 m thick, has a thin non-welded base that grades up into white-weathering eutaxitically welded tuff with dark chloritic fiamme. Irregular zones of siliceous nodules are scattered throughout. More detailed descriptions of the formation and its confining sedimentary strata are provided in the description for the Cwm Idwal GCR site, covering the area located immediately to the south.

Interpretation

The Braich tu du Volcanic Formation has been interpreted as representing the outflow facies from small-scale caldera-like structures that were active prior to the main 1st Eruptive Cycle centres in northern Snowdonia. The laterally restricted, but locally thick, accummulations of the Braich tu du tuffs south of their eruptive centre at Foel Fras, and their interdigitation with the Foel Grach Basalt Formation are interpreted by Howells *et al.* (1991) to reflect topographically controlled deposition within a series of small fault-controlled troughs. Within the troughs, subsidence kept pace with accumulation of the volcanic deposits and basalt effusion was controlled by fissures located along the trough margins.

The overlying Capel Curig tuffs were emplaced and reworked in a submarine environment. The tuffs contain accretionary lapilli indicating that they are the products of subaerial eruptions probably from a centre located to the south (Howells *et al.*, 1991).

The sedimentary rocks above and below the Pitts Head tuff have been studied in detail by Orton (1988). Sedimentary features and bedform analysis suggest that the underlying sedimentary rocks reflect a shallow shelf environment with shelf–ridge sands and interbar silts and muds, with the thicker, coarser sand bodies marking interbar storm events. Above the tuffs there is a fining-up sandstone-dominated sequence with trough cross-bedding, parallel lamination and winnowed fossiliferous beds, which is interpreted as accumulating in a transgressive non-barred wave-influenced shoreline (Reedman *et al.*, 1987). The contained shelly faunas, dominated by *Dinorthis–Macrocoelia* communities, suggest water depths of less than 25 m (Pickerill and Brenchley, 1979). Thus, the Pitts Head tuffs were probably emplaced within a marine mid- to outer storm-dominated shelf environment. The tuffs clearly retained sufficient heat to weld upon emplacement and the streaming off of volatiles from the base of the tuffs facilitated the formation of irregular gas cavities and ductile deformation within the main body of the tuffs.

Conclusions

The GCR site at Braich tu du preserves an impressive section through a wide time-span of volcanic activity and sedimentation within the evolving Snowdon Graben. It contains representatives of most of the 1st Eruptive Cycle and the first major ash-flow tuff-forming eruptions related to the 2nd Eruptive Cycle. It is the type locality for the Braich tu du Volcanic Formation with excellent exposed examples of welded acid ash-flow tuffs.

LLYN DULYN
(SH 703 661)

M. Smith

Introduction

The Llyn Dulyn GCR site lies within the main outcrop of volcanic strata of Caradoc (Soudleyan) age which forms a narrow strip trending NE–SW across northern Snowdonia. Originally mapped and described by the Geological Survey (Ramsay, 1881), the area was remapped by the Geological Survey in 1968–71 and is described in Howells *et al.* (1981). The site (Figure 6.32) provides excellent exposures of a part of the 1st Eruptive Cycle of early Caradoc volcanic activity and preserves the critical contact relationships between the ash-flow tuffs and their confining sedimentary rocks.

Three acid ash-flow tuffs, belonging to the Capel Curig Volcanic Formation, are distinguished and represent the emanations from a series of subaerial volcanic centres in North Wales (Howells *et al.*, 1991). These tuffs are in part correlatives to the three members described at the Capel Curig GCR site but with some important differences; the tuffs here have unwelded bases, contain no interlayered sediments and show no evidence for magma–water interaction. These features, in conjunction with the character of the adjacent sedimentary strata, indicate a subaerial emplacement of hot pyroclastic flow deposits and provide an important contrast with the subaqueous environment at Capel Curig. Recognition of these contrasting environments by Francis and Howells (1973) and Howells *et al.* (1973) led to a radical reinterpretation of Lower Palaeozoic palaeoenvironments in North Wales.

Description

The crags forming the south-western backwall to Llyn Dulyn (Figure 6.33) (between 7002 6632 and 7030 6631), expose a near-complete dip section through the Capel Curig Volcanic Formation and into the overlying sedimentary rocks of the Cwm Eigiau Formation. Three of the four members of the formation are present, in upward succession informally termed the 1st, 2nd, and 3rd members. They have a cumulative thickness of up to 180 m and characteristically comprise well-bedded, pale-grey, blocky and jointed fine-grained tuffs.

The lowest (1st) member, exposed in the western half of the site, is equivalent to the Garth Tuff farther south. It comprises a single cooling unit, 18 m thick, of a primary welded ash-flow tuff. The basal non-welded zone has an even, planar base with scattered tabular feldspar crystals, often pseudomorphed by sericite, and passes up into a welded central part with a eutaxitic foliation accentuated by concordant segregations of secondary quartz. The topmost 5 m is composed predominantly of non-welded shards and a well-defined zone of thin-walled siliceous nodules.

The 2nd or middle member, correlated with the Racks Tuff at Capel Curig, has a basal zone marked by a non-welded lithic crystal tuff and is overlain by strongly welded ash-flow tuff for 87 m. This thick unit comprises two separate cooling units. The lower tuff, 34 m thick, is rich

Wales and adjacent areas

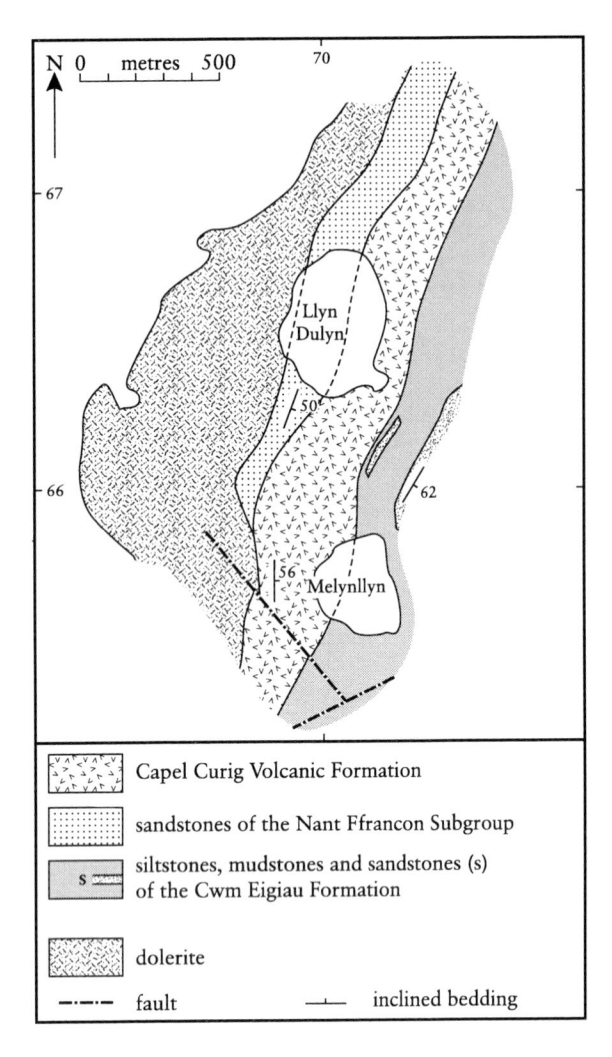

Figure 6.32 Map of Llyn Dulyn, after BGS 1:25 000 sheets 65/66 (1985) and 76 (1981).

in lithic clasts including mudstone, welded tuff, altered acidic and basic intrusive rock and feldspar crystals. In thin section, the crystals are of albite–oligoclase with rounded and resorbed textures and commonly are altered to sericite and carbonate. Delicate shards are scattered throughout, set in a matrix of sericite aggregates. The prominent eutaxitic foliation is defined by aligned, dark-green chloritic fiamme with ragged terminations consisting of chlorite flakes and sericite, which are interpreted as recrystallized pumice. Above 34 m from the base of the unit, feldspar crystals are notably less common and lithic clasts are absent. At 72 m, within the upper cooling unit, siliceous recrystallization along the foliation is distinctive and tends to obscure the eutaxitic foliation. The top part is characterized by a prominent bedding plane that can be traced throughout the immediate area.

The 3rd member, which has no representative in the Capel Curig area, comprises up to 75 m of thinly banded welded tuff with distinctive layers of collapsed pumice and evidence of upward grading. Originally interpreted as a single unit by Howells *et al.* (1981), the fine-grained banding is interpreted here to represent a sequence of thin welded primary ash-flow tuffs. With thicknesses up to 4 m, the tuffs contain pumice-rich layers and locally graded concentrations of feldspar crystals towards the top. They are recrystallized with a platy quartzo-feldspathic mosaic obscuring the original shardic fabric. In the upper parts of the member, feldspar crystals up to 3.5 mm in diameter are common along

Figure 6.33 Moderately dipping welded ash-flow tuffs of the Capel Curig Volcanic Formation, SW side of Llyn Dulyn. (Photo: BGS no L1501.)

with segregations of quartz, sericite and chlorite. The uppermost beds are very fine grained with a pronounced penetrative cleavage.

The underlying sedimentary strata exposed in the extreme western part of the site comprise a heterolithic sequence of sandstones, conglomerates and siltstones which form part of the Nant Ffrancon Subgroup. The lower beds are coarse-grained sandstones and conglomerates in units up to 8 m thick and grading up into 2 m-thick cross-bedded sandstones. These coarse-grained beds are interlayered with thinner beds and lenses of laminated tuffaceous siltstone, which may also occur as irregularly orientated blocks. The upper beds are characterized by medium-grained graded sandstones and laminated tuffaceous siltstones. The sandstones are tabular and cross-laminated with symmetrical ripple marks.

The sedimentary rocks overlying the welded tuffs represent the local basal strata of the Cwm Eigiau Formation in the area. They crop out across the rounded shoulder SE of Llyn Dulyn and extend southwards to Melynllyn (Figure 6.32). Up to 12 m of grey siltstone are present immediately above the tuffs, passing upwards into fine-grained, cross-bedded sandstones.

Interpretation

Analysis of the regional thickness, lithological and textural variations, and volcanic features within the tuffs and adjacent sedimentary rocks of the Capel Curig Volcanic Formation were used by Howells and Leveridge (1980) and Howells *et al.* (1991) to identify the likely source areas for the individual eruptive events and the environments of deposition (Figure 6.34). The individual members of the Capel Curig Formation exposed at the Llyn Dulyn site are interpreted as massive acid ash-flows erupted from subaerial vents to the north (1st and 2nd members) and west (3rd member), travelling southwards and eastwards down the volcano flanks and onto a broad plain. The exceptional thin development (only 20 m) of the 1st member is interpreted by Howells and Leveridge (1980) to represent emplacement on a locally developed topographical high within this plain.

Features within the tuffs, including the presence of non-welded bases, normally graded lithic clasts and crystals, and the intense siliceous segregation accentuating the eutaxitic foliation, are all indicative of subaerial ash-flow emplacement. When combined with the sedimentary features in the underlying Nant Ffrancon Subgroup strata, which indicate shallow, lacustrine conditions succeeded by high-energy alluvial braided streams, they provide persuasive evidence for emplacement in a subaerial environment. This interpretation contrasts markedly with that of Capel Curig GCR site where the tuffs have extended into a shallow-marine environment (Figure 6.34).

Conclusions

Llyn Dulyn provides one of the best-exposed sections through the Capel Curig Volcanic Formation belonging to the 1st Eruptive Cycle during Caradoc times in northern Snowdonia. It is a classic site in which to demonstrate the emplacement of ash-flow tuffs in a subaerial environment. The tuffs, derived from eruptive centres located to the north and west, represent a series of hot rhyolitic ash-flows and display textbook examples of welding and cooling textures. Environmental interpretations of the surrounding sedimentary rocks and the lack of evidence for interaction between magma and wet sediment emphasize the subaerial environment. This contrasts markedly with the submarine environment proposed for the tuffs exposed at the Capel Curig GCR site and enhances reconstructions of Lower Palaeozoic environments in North Wales.

CAPEL CURIG (SH 700 575–707 565)

M. Smith

Introduction

During Caradoc (Soudleyan) times, volcanicity in North Wales was characterized by a series of climactic acid ash-flow eruptions sourced from a number of subaerial volcanic centres in northern Snowdonia. These represent part of the 1st Eruptive Cycle of Howells *et al.* (1991). These ash-flow deposits, and their bounding sedimentary strata, provide important evidence for palaeoenvironment reconstructions and record a transgression from a subaerial environment across a shoreline into a subaqueous environment. Within this framework the exposures west of Capel Curig village are of international importance, as it was here that welded submarine ash-flow tuffs were first identified in ancient

Wales and adjacent areas

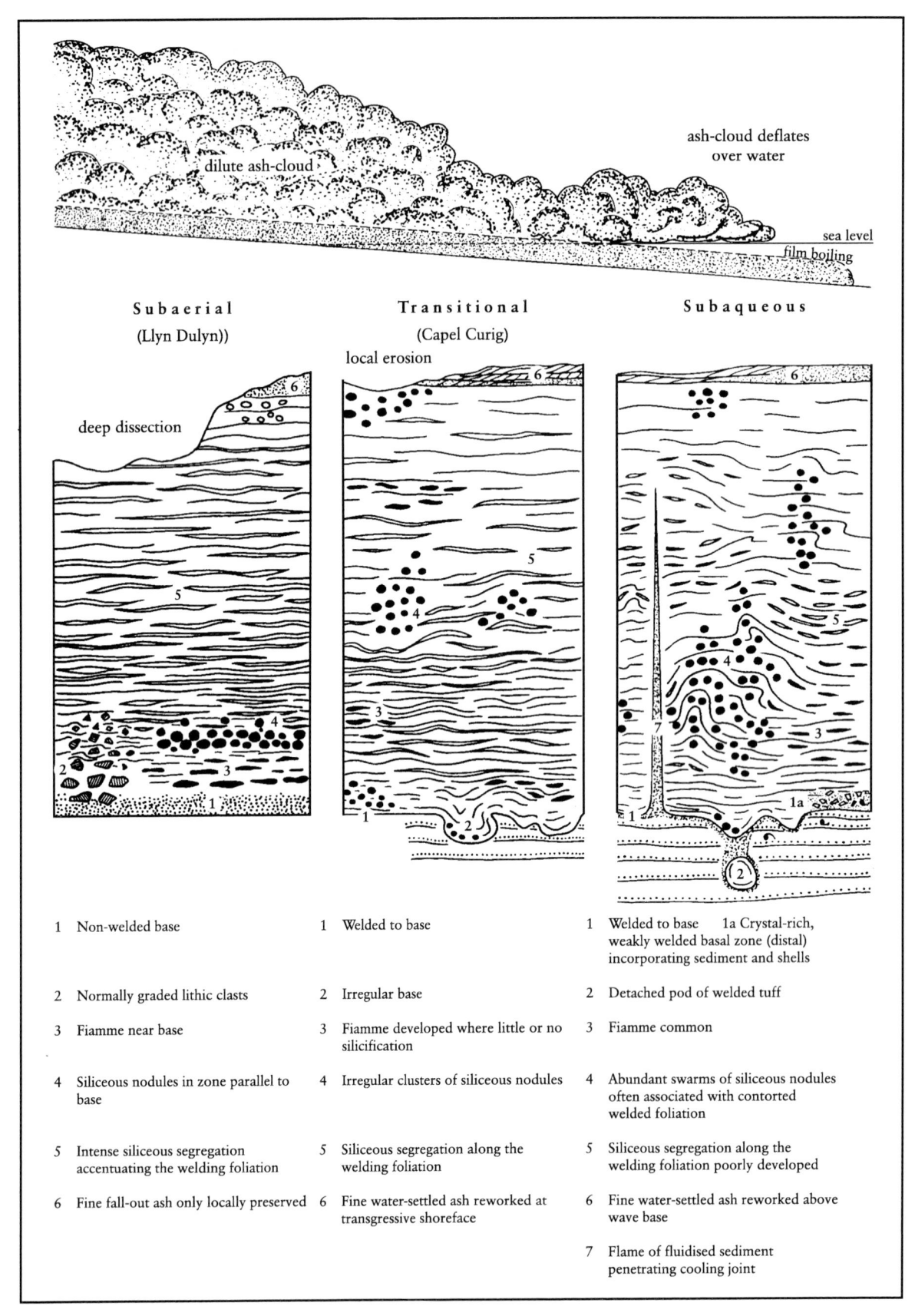

Figure 6.34 Model of ash-flow emplacement, and of contact and internal facies relations of welded ash-flow tuffs with respect to environment of deposition (after Howells *et al.*, 1991).

rocks by Francis and Howells (1973) and Howells *et al.* (1973). The sections also provide classic examples of magma–sediment interaction and are pertinent to the continuing debate on subaqueous welding of ash-flow tuffs (see for example Cas and Wright, 1987, 1991; McPhie *et al.*, 1993).

The GCR site includes volcanic rocks and interlayered marine sedimentary rocks that form the type area for the Capel Curig Volcanic Formation. They lie within the core of the Capel Curig Anticline and are well exposed in the craggy ground to the NW and SE of Llynnau Mymbyr. Originally mapped and described by Ramsay (1881), the type area was remapped by Williams (1922) who interpreted most of the volcanic rocks as rhyolite lavas emplaced in a marine environment. Recognition by Oliver (1954) and Rast *et al.* (1958) that many of the 'rhyolite lavas' of North Wales are in fact welded tuff or ignimbrite radically changed views of the Caradoc palaeoenvironments. The latter authors considered that all of the ignimbrites were erupted and emplaced subaerially. Subsequent detailed mapping by Francis and Howells (1973) in the Capel Curig area convincingly demonstrated that some of the tuffs were emplaced in a submarine environment, work which provided a stimulus for further investigation and led to more realistic assessments of the palaeogeography (Howells *et al.*, 1979; Howells and Leveridge, 1980; Orton, 1988; Howells *et al.*, 1991). This site is complemented by the Llyn Dulyn GCR site, which demonstrates the lateral equivalents of these ash-flow tuffs emplaced in a subaerial environment (Figure 6.35).

Of the four volcanic members distinguished within the Capel Curig Volcanic Formation, three are present at the Capel Curig GCR site and in upward succession are the Garth Tuff, the Racks Tuff and the Dyffryn Mymbyr Tuff. Geochemical investigations indicate that the tuffs are rhyolitic to rhyodacitic in composition and are spatially and geochemically related to a series of subvolcanic intrusions (Howells *et al.*, 1991). The members may be distinguished by their trace element compositions and can be related by fractional crystallization, with the oldest (the Garth Tuff) being the least evolved.

Description

The Garth Tuff, forming the lower crags north of Llynnau Mymbyr (Figure 6.36), is a massive, cream to white, well-jointed unit, generally 10 m thick but increasing up to 40 m in the western part of the Capel Curig Anticline. It is massive and welded in the lower and middle parts, grading up through a zone with faint bedding planes into a reworked upper part, up to 20 m thick, with current bedding and ripples. A prominent eutaxitic foliation is always parallel to the regional dip except at the margins to the tuff. The top is concordant with the overlying sandstones. Lithic clasts, including devitrified perlitic glass and welded tuff, recrystallized shards, siliceous nodules, and isolated and fragmented albite phenocrysts are scattered throughout the unit. The fine-grained matrix is composed of sericite, chlorite, quartz and feldspar. On the south-eastern limb of the anticline, sections exposed above the A4086 road and in forestry cuttings (e.g. 7137 5732), show that the bedded top also contains accretionary lapilli. Here, the base is remarkably discordant, with flames of the underlying sediment penetrating deeply into the tuff. Tuff–sediment contacts at 90° to the regional bedding are not uncommon. The adjacent sedimentary strata are often highly disturbed and are interpreted as reconstituted bedded sandstones. Numerous small tuff apophyses penetrate the sediment and comprise admixtures of tuff and sediment, (e.g. at 7091 5665). Locally, these may be detached and resemble tuff-pipes at outcrop. Large detached bodies of tuff, up to 100 m × 250 m in plan, and surrounded by the underlying sandstones, are sporadically preserved. These bodies of tuff are lithologically identical to the main tuff.

The overlying strata, between the Garth and Racks tuffs, form part of the Cwm Eigiau Formation and comprise fine-grained pale-green sandstones with thin, impersistent, often disrupted, grey, cleaved mudstones and siltstones in the upper part. Poorly preserved fossiliferous bands are present in crags at 7095 5789 and a disused quarry at 7094 5784. Shelly forms dominate, and include *Dalmanella* sp., *Dinorthis* cf. *berwynensis*, *Howellites* sp. and *Macrocoelia* sp. (Howells *et al.*, 1978).

The Racks Tuff, lithologically comparable to the Garth Tuff, is well exposed on the south-eastern limb of the anticline, where it forms a well-bedded and non-welded unit up to 30 m thick. In contrast, on the northern crags (around 706 579) the tuff is thinner, massive and welded throughout with patches of siliceous nodules. The outcrop is discontinuous and podiform, with the tuff locally wedged out. The lower contact is highly irregular with large flames of mudstone transgressing the upper and lower tuff contacts (e.g. 704 578). The adjacent mudstones and sandstones are highly contorted and are penetrated by thin apophyses of tuff.

The Dyffryn Mymbyr Tuff is only present on the north-western limb of the anticline and thins markedly to the NE, grading from a coarse-grained lithic tuff containing accretionary lapilli to a tuffaceous mudstone.

Supplementary sites in the Lledr Valley (at 7656 5436 and at Rolwyd (765 512) to the SW of Capel Curig) provide additional evidence for the emplacement of isolated pods of tuff within unconsolidated marine sediments (Francis and Howells, 1973). A possible mechanism for the formation of these pods is described in Howells *et al.* (1991, fig. 27).

Interpretation

The presence of eutaxitic welding fabrics and relict shardic textures within the tuffs of the Capel Curig Formation, exposed on the flanks of the Capel Curig Anticline, are indicative of their formation as hot pyroclastic flow deposits. Regional lithological and geochemical studies show that these tuffs (the Garth and Racks tuffs) were erupted from subaerial centres in northern Snowdonia during Caradoc times and transported southwards. Changes in emplacement and cooling textures within the tuffs, combined with complex sediment–tuff relationships and studies of the adjacent sediment, indicate that the environment of deposition changed from subaerial

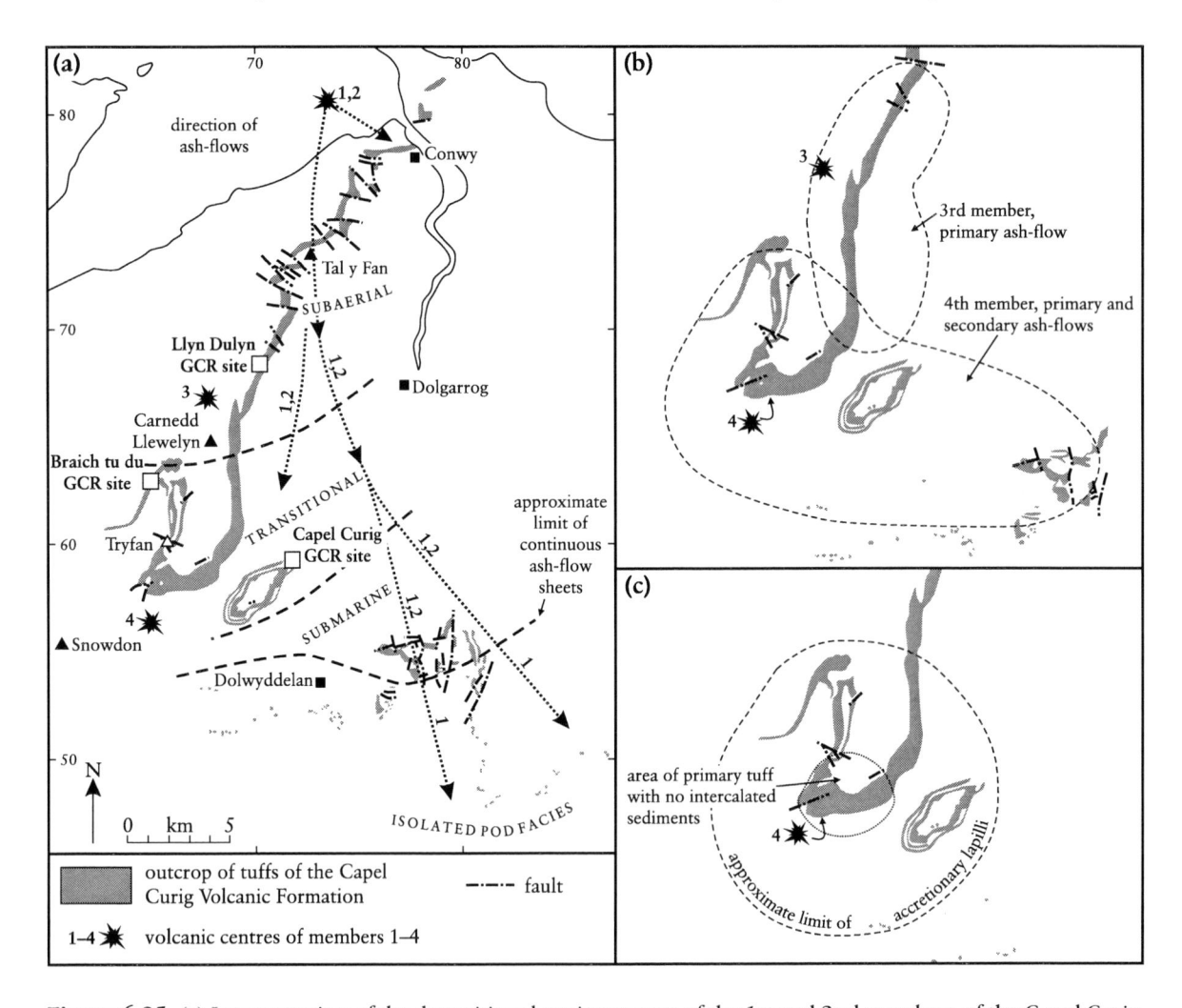

Figure 6.35 (a) Interpretation of the depositional environments of the 1st and 2nd members of the Capel Curig Volcanic Formation, showing flow directions and the distribution of isolated pods of the 1st Member. (b), (c) Distribution of the 3rd and 4th members of the Capel Curig Volcanic Formation. After Howells and Leveridge (1980).

Craig y Garn

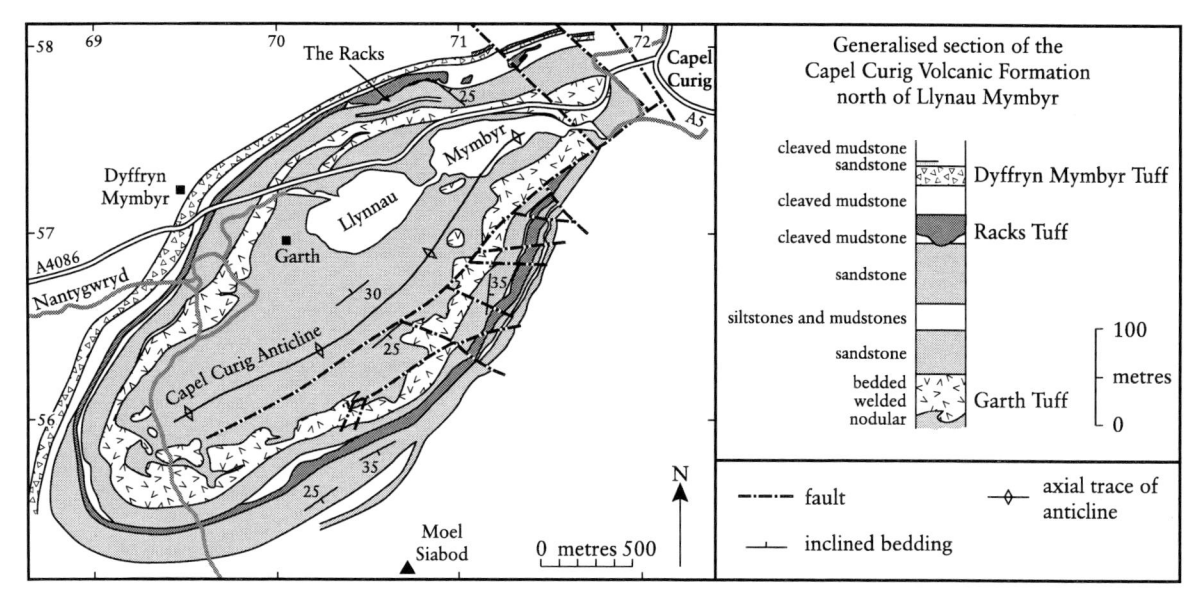

Figure 6.36 Map and vertical section of the Capel Curig Volcanic Formation in the Capel Curig Anticline (after Francis and Howells, 1973).

to submarine (Howells and Leveridge, 1980; Howells *et al.*, 1991). The outcrops around Capel Curig represent one of the key areas in this reconstruction, marking the shallow-water transition zone between a shoreline located just north of Capel Curig, and deeper basinal conditions farther south. This interpretation, supported by the sedimentary features and shelly faunas of the enclosing sediment and evidence for reworking of the upper parts of individual tuffs, contrasts markedly with the subaerial conditions at the Llyn Dulyn GCR site.

As the tuffs transgressed the shoreline into the submarine environment, the hot, gaseous and dense ash-flows interacted with the semi-lithified and water-saturated sediments on the sea bed. The remarkable irregularities of the lower tuff contacts may be attributed to the disturbance of the underlying wet unlithified sediments by the rapid emplacement of hot ash-flows on an uneven surface or to seismic shocks attendant on eruption (Francis and Howells, 1973). Either mechanism would induce the sediments to deform thixotropically and the ash-flows to collapse downwards by unequal loading to form irregular lobes and pipe-like masses, which in extreme cases may have become completely detached. Fluidization and magma–water interaction at the tuff–sediment contact probably facilitated these processes. Although still a matter of debate (see Howells *et al.*, 1991, pp. 163–5) water depths are generally assumed to be less than the thickness of the pyroclastic flow deposits (McPhie *et al.*, 1993).

Conclusions

The Capel Curig GCR site provides classic exposures exemplifying the delivery of subaerial pyroclastic flow deposits into a shallow-marine environment. As one of the first documented examples of this process in an ancient environment, the site is of historical as well as international scientific interest. The presence of a strong foliation within the tuffs and the interaction with the underlying fossiliferous marine sediments have been central to the arguments that the ash-flows crossed the shoreline hot and intact and continued across the sea bed, retaining sufficient heat to become welded when they stopped moving.

CRAIG Y GARN (SH 510 440–504 466)

M. Smith

Introduction

The Craig y Garn GCR site represents an important dip section through one of the main eruptive centres of the 2nd Eruptive Cycle of Caradoc volcanic activity in Snowdonia. It lies within the eastern half of an elongate synformal outlier of the main outcrop of the Snowdon Volcanic Group. This outlier is interpreted as the site of a volcanotectonic collapse structure whose location and development was in part influenced by north- to NE-trending fractures in the underlying basement.

Formerly known as the Llwyd Mawr Ignimbrite, the strata, which were originally described by Sedgwick (1843) and later by Ramsay (1881) and Harker (1889), were regarded as rhyolitic lava flows and tentatively correlated with similar lavas on Moel Hebog. This account draws on the work of Roberts (1969) whose detailed mapping and studies of petrography and deformation led to their re-interpretation as ash-flow tuffs (ignimbrites). Roberts also confirmed the earlier suggestion of Shackleton (1959) that within this thick (over 700 m) intra-caldera sequence there is no evidence for any appreciable subdivision. Later deformation studies by Roberts and Siddans (1971) used variations in the compactional strain, as seen in lithic clasts and pumice, to identify two separate eruptive pulses. However, a more recent geochemical study by Howells *et al.* (1991) shows little evidence for trace element compositional variation throughout the sequence and reconfirms the original suggestion that Llwyd Mawr represents one of the thickest accumulations of welded ash-flow tuffs related to an individual volcanic centre in Britain.

The Craig y Garn site includes the basal contact of the tuff sequence with marine mudstones of Llanvirn age. Elsewhere the tuff sequence is overlain by Longvillian age strata and thus an upper Soudleyan to Longvillian age is likely. In the northern half of the site intrusive rhyolite domes and a possible vent breccia may represent resurgent activity within the caldera. Lithological and geochemical studies (Howells *et al.*, 1991) include the Craig y Garn rocks within the Pitts Head Tuff Formation and confirm a correlation with the outflow facies on Moel Hebog (Reedman *et al.*, 1987), as described in the Moel Hebog to Moel yr Ogof GCR site report.

The site is partly included in the 1:50 000 scale Geological Sheet 119 (Snowdon) (1997) but has not been resurveyed in detail.

Description

The site encompasses some 3 km² along the western side of Cwm Pennant and includes the minor hills of Craig y Garn and Llywd Mawr (Figure 6.37). Scattered exposures extending eastwards and north-eastwards from the slate quarries at Hendre-ddu (5180 4442) show many of the features typical of a major ash-flow tuff.

The lower beds rest abruptly, but concordantly on dark bluish-grey micaceous mudstones and silty mudstones of the Nant Ffrancon Subgroup. Graptolites recovered from these strata are characterized by *D. murchisoni* (Shackleton, 1959; Howells and Smith, 1997) indicating the Llanvirn *D. murchisoni* Biozone. In contrast to the strata exposed beneath the ash-flow tuff on Moel Hebog there is no evidence for the Llandeilo or Caradoc stages, thus indicating the existence of a major volcanotectonic break at the base of the tuffs.

Throughout most of its eastern outcrop the base of the tuff sequence is underlain by a rhyolitic sill. The sill, up to 75 m thick, is exposed around the eastern flanks of Craig y Garn and comprises flow-banded and flow-folded rhyolite, locally autobrecciated and spherulitic at its lower contact. Contact metamorphism and alteration related to the intrusion of the sill has baked the overlying tuffs and protected them from cleavage development.

The sequence up through the tuff pile as exposed on Craig y Garn commences with a compact, silicified, blue-grey non-welded crystal-lithic-vitric tuff up to 3 m thick. The tuff contains a variety of clast types including rounded to weakly flattened pumice clasts, up to 10 cm in diameter, mudstone clasts, up 8 mm in length, and rare angular rhyolitic clasts, less than 2.5 cm, set in a devitrified matrix of glass shards, chloritized feldspar (mainly plagioclase and rarer anorthoclase), quartz crystals and dust. The mudstone clasts are common only in the basal 30 cm. The shards are typically well preserved and in thin section show Y-shaped or four-sided morphologies.

At about 4–5 m above the base the glass shards show increasing distortion and incipient welding, and a planar eutaxitic foliation is developed, although much of the foliation is obscured by spherulitic recrystallization. Quartz phenocrysts are rare above this level. The foliation dips consistently westwards and is concordant throughout. Above 10 m, spherulitic recrystallization reduces in intensity and welded textures are again evident. Shard distortion and collapse and flattening of pumice clasts and lapilli thereafter increases logarithmically upwards and at around 17 m above the base a strong parataxitic foliation defined by flattened shards is present.

Craig y Garn

Locally perlitic fracturing may be observed. Layers of radial and concentric siliceous nodules are commonly developed immediately below the parataxitic zone. The parataxitic foliation extends for the next 45 m. At 62 m, the tuff is strongly recrystallized; shards are flattened and outlines are completely destroyed. The textures continue to the top of the section.

The changes in the degree of deformation and recrystallization are matched by variations in joint style. Above the base, the joints are crudely perpendicular to the lower contact, and at about 18 m become crudely columnar with rectangular cross sections. At 18 m to about 58 m, the joints are platy and above 58 m polygonal forms dominate, with cross-sectional diameters increasing from 12 cm to 50 cm at 75 m above the base.

Immediately north of Llywd Mawr, along the northern margin of the site, an intrusive rhyolite dome can be seen cutting the tuffs. The dome is composed of flow-banded rhyolite with feldspar phenocrysts, and in places is fringed by an autobrecciated facies comprising blocks (less than 2 m in length) of flow-banded and flow-folded rhyolite and zones of siliceous nodules. The dome and its breccia carapace are surrounded by a non-welded vitroclastic tuff, locally agglomeratic with blocks of flow-banded rhyolite, welded tuff and rare mudstone.

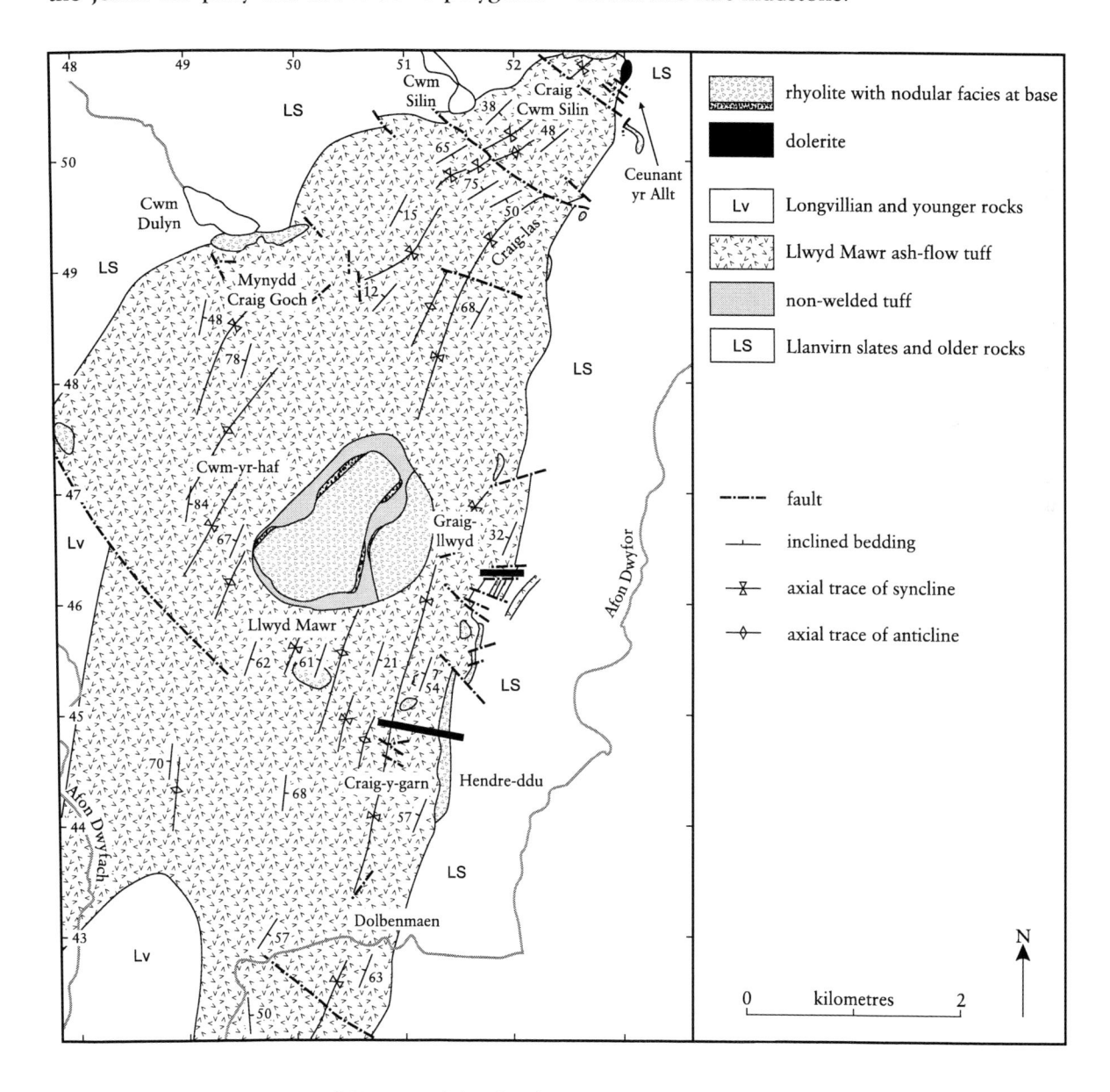

Figure 6.37 Map of the Llwyd Mawr Centre (after Roberts, 1969).

Interpretation

The textural and lithological characteristics of the Pitts Head Tuff Formation on Craig y Garn and the general absence of interaction with the enveloping marine sediments indicate that this impressive thick sequence of rhyolitic tuff was emplaced subaerially, probably within a subsiding volcanotectonic depression or caldera. A volcanotectonic break, of unknown magnitude, is indicated by the absence of Caradoc strata in the east and compares with a full sequence on the eastern side of Cwm Pennant at Moel Hebog. This implies uplift and erosion in the vicinity of the caldera prior to collapse (Howells *et al.*, 1991). The lack of a decrease in the intensity of welding and general absence of non-welded tuff led Roberts (1969) to infer that a large volume of tuff may have been removed by erosion and thus the 700 m thickness estimate must be regarded as a minimum. Lateral equivalents of the Pitts Head tuffs are exposed in the Moel Hebog area, where they represent the outflow facies from the caldera (see the Moel Hebog to Moel yr Ogof GCR site report).

The progressive changes recorded by the degree of flattening, foliation development and joint style are considered to result from the cooling of a single unit, albeit composed of more than one ash-flow, subsequently intensely recrystallized and altered. The widespread development of spherulitic recrystallization beneath the main zone of development of a parataxitic fabric probably resulted from the trapping of volatiles exsolving from the basal non-welded tuff.

The latter stages of caldera evolution were marked by the forceful emplacement of rhyolitic domes and sills along the caldera margin and within feeder pipes or vents already choked with an agglomeratic vitric-clastic tuff in the centre of the caldera. Geochemically and petrographically comparable to the tuffs, these domes and sills may represent the late degassed equivalent of the tuff magma.

Conclusions

The Craig y Garn GCR site preserves one of the thickest and most complete sections through a Lower Palaeozoic caldera fill in Britain. In excess of 700 m of welded rhyolitic ash-flow tuff was ponded or entrapped within a major volcanic depression that formed within the Snowdon Graben and marked the initiation of the 2nd Eruptive Cycle across Snowdonia in Caradoc time. The site is important for the preservation of textures typical of welded ash-flow tuffs and for its correlation with tuffs present on Moel Hebog, which are considered to have emanated from the same caldera (see the Moel Hebog to Moel yr Ogof GCR site report).

MOEL HEBOG TO MOEL YR OGOF (SH 568 464–557 483)

M. Smith

Introduction

The Moel Hebog to Moel yr Ogof GCR site is one of two GCR sites that lie at or near the margin of one of the major eruptive caldera centres defined in the Caradoc rocks of North Wales. The Snowdon Centre, belonging to the 2nd Eruptive Cycle of Howells *et al.* (1991), is interpreted as a marine island caldera complex whose margins may be recognized by one or more of the following features:

1. Localized shallowing and emergence within a regionally subsiding marine environment.
2. Ponding of ash-flow tuff deposits.
3. Large-scale disruption of the volcanosedimentary sequences.
4. The emplacement of a series of rhyolitic domes and sills and basaltic magmas.

The complex geology displayed on Moel Hebog and Moel yr Ogof (Figure 6.38) shows many of the above features, including excellently exposed examples of subaerially emplaced acid ash-flow tuffs, primary and reworked intra-caldera tuffs, basic pillow lavas and spectacular large-scale disruption and slumping of blocks off the rim of the caldera.

The exposed strata comprise the Snowdon Volcanic Group which is divided into four formations. Only the three lowest formations are exposed at the Moel Hebog to Moel yr Ogof GCR site; the upper formation is seen at the Snowdon Massif GCR site. The lowest part of the succession crops out on the eastern side of Moel Hebog and includes marine sandstones of the Cwm Eigiau Formation overlain by acid ash-flow tuffs of the Pitts Head Tuff Formation, representing the outflow facies from the adjacent Llywd Mawr Centre (see the Craig y Garn GCR site report). These are in turn overlain by ash-flow tuffs representing the intracaldera facies of the Lower Rhyolitic Tuff Formation. The Bedded Pyroclastic Formation crops out at Moel yr Ogof and is interpreted as the site of an eruptive basic vent subsequently capped and intruded by rhyolite domes and sills.

The site area was originally described by Williams (1927) and Shackleton (1959), both of whom regarded the tuffs as extrusive rhyolites. They were subsequently shown to be welded ash-flow tuffs by Rast *et al.* (1958) and were remapped in detail at the 1:10 000 scale by the British Geological Survey between 1984 and 1985. Detailed descriptions are provided by Reedman *et al.* (1987) and Howells *et al.* (1991) and the site is included on the 1:50 000 scale Geological Sheet 119 (Snowdon) (1997).

Description

Sedimentary rocks of the Cwm Eigiau Formation underlying the Pitts Head Tuff Formation crop out along the south-eastern part of the site, east of South Buttress (around 5695 4671) and along the main summit path from Beddgelert (at 5694 4729) (Figure 6.38). Described and logged in detail by Orton (1988), they comprise a lower succession of interlayered siltstones and mudstones passing up into medium- to coarse-grained tabular sheets of pebbly sandstones with trough cross-bedding. Thin interlayers of white-weathering vitric tuff or tuffaceous sedimentary rock are developed sporadically. Debris flows and rapid variations in grain size characterize the uppermost units although bedding features are generally destroyed within 0.5 m of the overlying tuff.

The 2nd Eruptive Cycle of Caradoc volcanicity in Snowdonia, represented on Moel Hebog by the Pitts Head Tuff Formation, commenced with the eruption of ash-flow tuffs from the Llywd Mawr Centre. The intracaldera facies is described in the Craig y Garn GCR site report and is lithologically identical to the outflow facies on Moel Hebog, which is represented by two distinct layers of ash-flow tuff.

The lower tuff, up to 90 m thick, forms the lower parts of the crags immediately east of the summit of Moel Hebog (Figure 6.39), and is composed of welded and non-welded crystal-rich tuff. The basal contact appears conformable on the underlying sandstones. Locally, a distinctive unit, approximately 1 m thick, of thinly bedded non-welded vitroclastic tuff with polygonal jointing, intervenes between the main tuff and the underlying sandstones. The base of the main tuff unit is non-welded but grades rapidly up into strongly jointed welded tuff with chloritic fiamme and distinctive zones of siliceous nodules (Figure 6.40). These nodular zones are 1–3 m thick and individual nodules up to 40 cm in diameter are not uncommon; microscopically they comprise a quartz mosaic. Planar concordant siliceous segregations accentuate the strong parataxitic fabric in the remainder of the overlying tuff. The top of the tuff is irregular and eroded but locally a fine-grained top is preserved. The tuffs are devitrified and recrystallized and in thin section comprise aggregates of sericite, quartz, feldspar and chlorite. Shards are well defined and euhedral phenocrysts of albite–oligoclase feldspar and perlitic fracturing may be seen in the basal welding zone.

A prominent feature of this tuff unit is the development of areas of autobrecciation. Brecciated tuffs, composed of rotated angular clasts of welded tuff, occur as thin discontinuous zones or along joints but elsewhere may occupy the entire thickness of the tuff. Reedman *et al.* (1987) noted that the crystal-rich or weakly welded basal portions of the tuff thin when traced laterally to areas of pervasive brecciation, and develop lobate protrusions into the underlying sediment or in places are completely absent. These changes are matched by a reduction in the development of siliceous nodules close to the zones of brecciation. Where the breccia occupies the complete tuff it is markedly discordant with the underlying sediment. Locally, the upper parts of the brecciated tuff are disrupted and may occur as detached rafts and fragments within ash-flow tuff deposits of the overlying Lower Rhyolitic Tuff Formation (Figure 6.39). A subsidiary site located *c.* 800 m north of Moel Hebog, around Y Braich (5660 4785), displays more complex relationships between the Pitts Head and Lower Rhyolitic tuffs with large overturned rafts forming a volcanic megabreccia.

The upper tuff, up to 70 m thick, occupies the middle part of the Ladder Buttress (567 467) and differs from the lower flow in the absence of a basal nodular zone. It wedges out to the NE and is not present north of the summit of Moel Hebog. Geochemically, the upper tuff is distinguished by its lower TiO_2 content and enrichment in Nb relative to the lower tuff and the intracaldera tuffs on Llywd Mawr (Howells *et al.*, 1991).

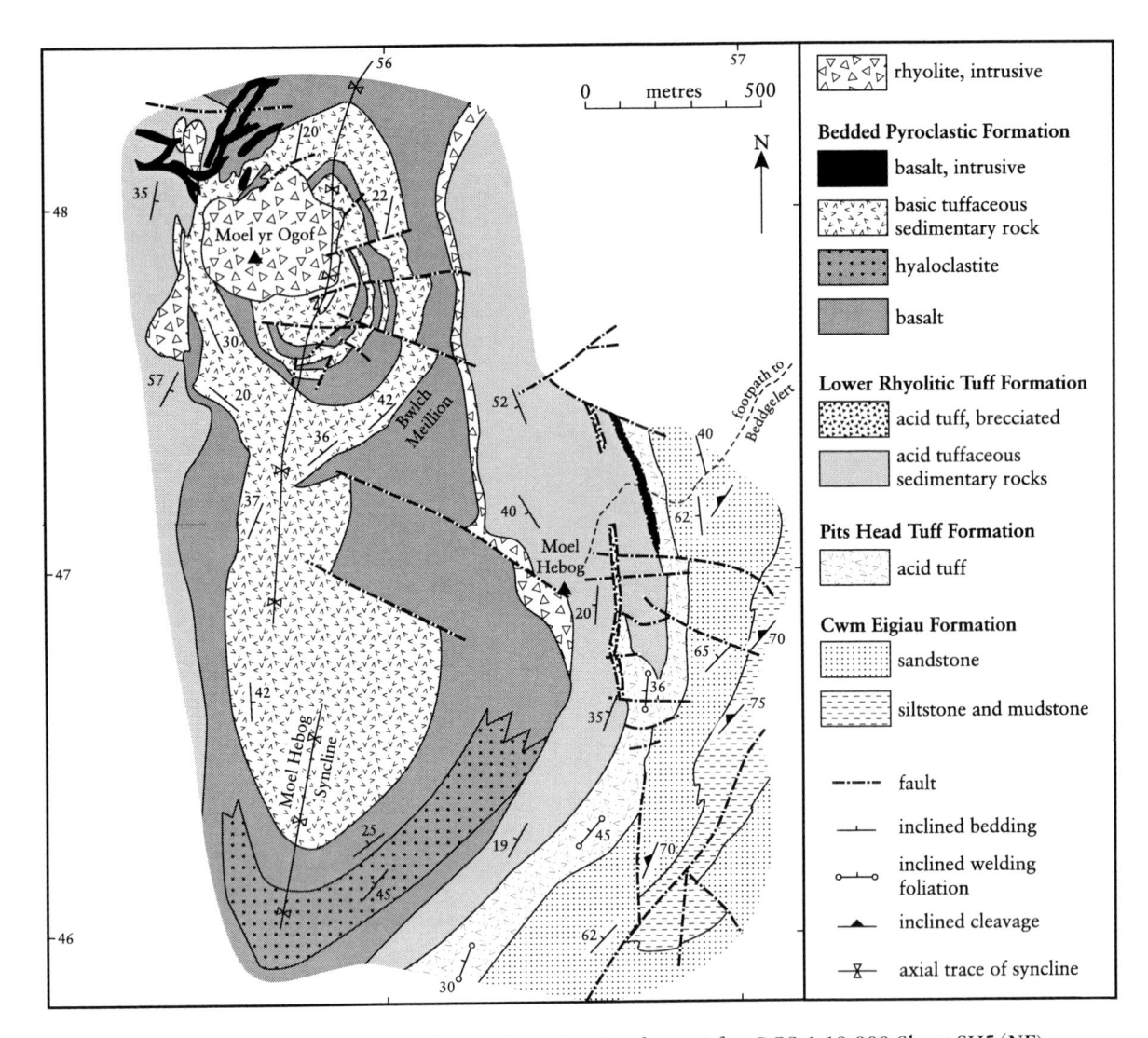

Figure 6.38 Map of the Moel Hebog and Moel yr Ogof area (after BGS 1:10 000 Sheet SH54NE).

The upper parts of the crags immediately east of the summit area are formed of primary and reworked ash-flow tuffs of the Lower Rhyolitic Tuff Formation. These tuffs rest unconformably on the underlying upper or lower units of the Pitts Head tuff. The lower, basal unit comprises up to *c.* 20 m of massive, brown-weathering primary welded lapilli-tuff with a well-developed eutaxitic foliation and is overlain by massive non-welded tuff and reworked tuffaceous sedimentary rocks. The reworked tuffs comprise planar-bedded tuffaceous siltstones and sandstones with 20–50 cm-thick tuff layers and contain numerous zones of hummocky cross-stratification and coarse-grained debris-flow conglomerates (Fritz *et al.*, 1990). The upper tuff is well bedded and displays trough cross-bedding. It is interlayered with tuffaceous sandstones and rare current-rippled vitric tuffs. Locally, the base is marked by a 1 m-thick clast-supported breccia containing pumice blocks and bombs and thin layers of parallel laminated tuff. Loading and irregular basal contacts to individual beds indicate deposition onto a semi-lithified substrate.

The later stages in the evolution of the main caldera phase at the Snowdon Centre were marked by an episode of basaltic volcanic activity. The deposits from this activity, known as the Bedded Pyroclastic Formation (Howells *et al.*, 1983), are widely dispersed across Snowdonia. In the north-western part of the GCR site the formation is superbly exposed in the crags surrounding the southern, eastern and northern flanks of Moel yr Ogof (5583 4767), which are

Moel Hebog to Moel yr Ogof

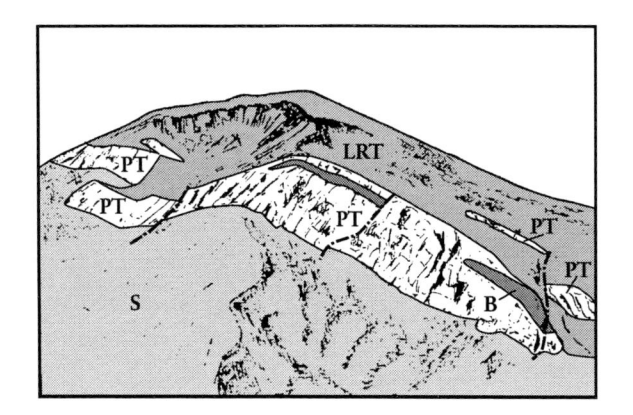

Figure 6.39 Moel Hebog, viewed from the NE, showing primary and reworked tuffs of the Lower Rhyolitic Tuff Formation (LRT) overlying and enclosing disrupted rafts and blocks of the Pitts Head Tuff Formation (PT) near the southern margin of the Lower Rhyolitic Tuff Formation caldera. The Pitts Head Tuff Formation is underlain by sediments (S) and intruded by basalt (B). Reproduced from Howells *et al.* (1991).

located within the core of the Moel Hebog Syncline. Well-featured ground rising up from Bwlch Meillionen (Figure 6.41) shows up to 230 m of extrusive basalts, variably pillowed with associated pillow breccias and hyaloclastites.

The lower beds within Bwlch Meillionen comprise a crudely bedded basaltic breccia with bombs and blocks of basalt, up to 50 cm in diameter, and lapilli in a matrix of basaltic tuff. These are interlayered with basic tuffs and thin rhyolitic vitric tuffs. Above, and to the north, the main crags show blocky basaltic lavas with pillow forms up to 1 m across. The pillowed basalt lavas can be traced laterally into pillow breccia deposits, hyaloclastites and well-bedded basaltic tuffs. With a decreasing frequency of basalt lava flows, the succession grades up into reworked basaltic tuffaceous and volcaniclastic sandstones with two prominent basalt lava flows. The tuffaceous rocks are generally well bedded, 1–10 cm thick with parallel- and cross-lamination. Transgressive basaltic sills and dykes occur within the succession and are particularly numerous to the north of Moel yr Ogof. These sills and dykes can be traced downward into a basalt/dolerite dyke feeder system exposed immediately to the west of the GCR site and

Wales and adjacent areas

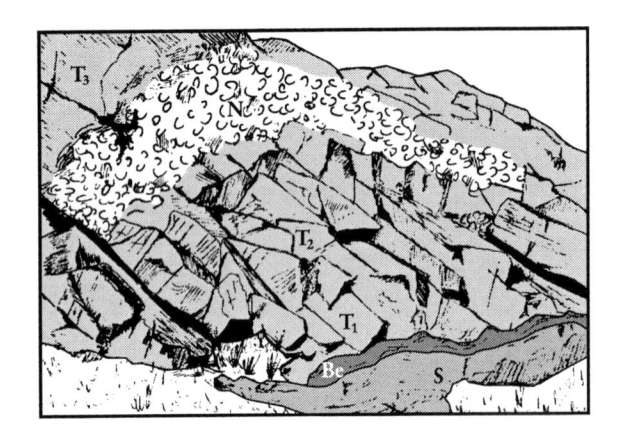

Figure 6.40 The lower outflow tuff of the Pitts Head Tuff Formation, Moel Hebog. The ash-flow tuff overlies coarse-grained sandstones (S) and bedded tuffs (Be). The base of the ash-flow tuff comprises non-welded tuff (T1) and is overlain by columnar jointed welded tuff (T2). A prominent zone of siliceous nodules (N) is overlain by densely welded tuff (T3) with a conspicuous, silicified, welding foliation (SH 5684 4694). Reproduced from Howells *et al.* (1991). (Photo: BGS no. A14658.)

down through the sedimentary rocks below the Pitts Head Tuff Formation to a massive dolerite sill.

Immediately south and west of the summit of Moel Hebog and on Moel yr Ogof, flow-banded and flow-folded autobrecciated rhyolite sills and domes cap the succession. These acid intrusive rocks form part of the second phase of rhyolite intrusions in Snowdonia (Campbell *et al.*, 1987) post-dating caldera subsidence, and were associated with resurgent activity broadly contemporaneous with the Bedded Pyroclastic Formation.

Interpretation

The bleached weathered appearance of the Pitts Head Tuff Formation makes it one of the most distinctive tuff sequences in Snowdonia. On Moel Hebog the formation is represented by two primary rhyolitic ash-flow tuffs. Lithological and geochemical similarities with the 700 m-thick intracaldera tuff sequence on Llywd Mawr support the interpretation that the Pitts Head tuffs are the outflow facies from the Llywd Mawr Centre, although detailed correlations with the

Moel Hebog to Moel yr Ogof

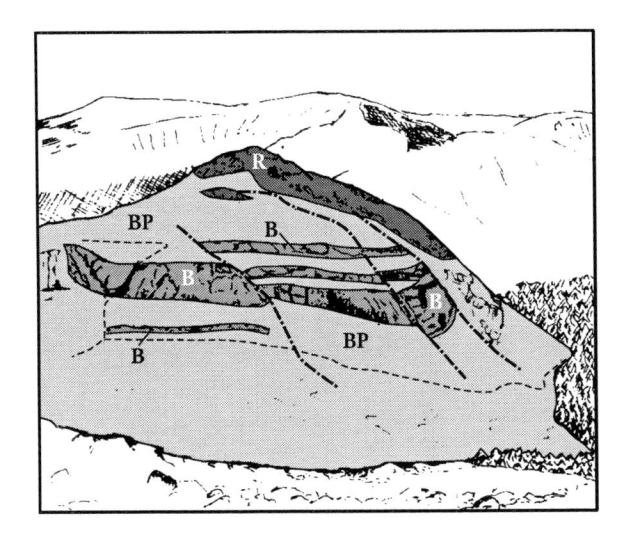

Figure 6.41 View, generally northwards, from Moel Hebog, showing broad features of geology on Moel yr Ogof (SH 556 478). Basaltic tuffs, hyaloclastites and volcaniclastic sediments (BP) and pillowed or massive basalts (B) of the Bedded Pyroclastic Formation are intruded by rhyolite (R). Reproduced from Howells *et al.* (1991). (Photo: BGS no. A14659.)

tuff sequence on Craig y Garn remain uncertain. Evidence from the south side of Moel Hebog and to the north indicates a considerable time gap between the upper and lower tuffs and therefore they must represent two distinct eruptive events. Facies and bedform analyses of the underlying sedimentary rocks have been interpreted by Orton (1988) to indicate alluvial plain and fan environments dominated by braided stream deposits derived from the W or SW.

These studies, combined with post-emplacement textures, thus support a subaerial emplacement for both tuffs. The non-welded, locally bedded base and the conspicuous zones of nodules marking the transition from non-welded to welded tuff probably resulted from the entrapment of volatiles near the zone of intense welding and growth during compactional welding and cooling, an interpretation supported by textural relationships. Above, the intense silicified planar parataxitic fabrics, with collapsed pumice clasts replaced by silica, indicate post-emplacement welding and compaction. In contrast, sections in the Pitts Head tuff farther north towards Snowdon show no siliceous nodule development and welding fabrics extend to the base of the flows. These changes correspond with a progressive north-eastward change to a submarine environment.

The evidence for post-emplacement brecciation and ductile flow of the Pitts Head tuff in a still plastic state and its presence as isolated rafts and blocks within tuffs of the Lower Rhyolitic Tuff Formation are distinctive and important features of the geology along the east side of Moel Hebog. These features have been interpreted as indicating instability and slope generation, probably by contemporary fault movements, post-Pitts Head tuff emplacement and during Lower Rhyolitic Tuff Formation times (Reedman *et al.*, 1987; Howells *et al.*, 1991). When this evidence is combined with the restriction of the upper Pitts Head tuff to the southern part of Moel Hebog and localization of acid and basic intrusive activity, it supports a model of fault-controlled topography along a caldera margin with the mass movement of welded tuff to form megabreccia deposits within a subsiding caldera.

The developing caldera was then infilled by the Lower Rhyolitic Tuff Formation with the widespread ponding and reworking of ash-flow tuffs and related mass-flow deposits within a shallow-marine environment. The basal welded unit is a primary ash-flow tuff possibly erupted from a vent to the east around Beddgelert. The presence of welding at the upper contact suggests that the upper part of this flow was eroded prior to the deposition of the overlying reworked tuff and breccia deposits.

The complex lateral facies relationships within the Bedded Pyroclastic Formation on Moel yr Ogof, with basaltic dykes and sills feeding up into basaltic pillow piles and lavas, suggest the formation of a basic vent and renewed activity along the caldera structure. The formation of pillows is inferred to reflect continued subsidence within the caldera and the continuation of marine conditions.

Conclusions

The Moel Hebog to Moel yr Ogof GCR site is a key site for the interpretation of the textures and emplacement mechanisms of acid ash-flow tuffs related to the 2nd Eruptive Cycle of Caradoc volcanic activity in Snowdonia, and in the identification of faulting and renewed volcanism along the caldera margin of an ancient submarine volcano. The Pitts Head tuffs represent the outflow from the caldera of the Llwyd Mawr Centre (see the Craig y Garn GCR site report). Their subaerial emplacement within an alluvial fan grading offshore into a shallow shelf provides an important contrast with marine conditions farther north. Subsequent fault-related activity along the south-western caldera margin of the Snowdon Centre and continued subsidence during the main phase of volcanic activity is graphically displayed in the brecciation, sliding and widespread disruption of previously emplaced ash-flow tuffs. Renewed basaltic activity and the intrusion of rhyolite domes serve to emphasize the importance of the caldera margin fracture in the channelling of magma to shallow crustal levels during later phases of resurgent activity.

YR ARDDU
(SH 621 452–631 472)

M. Smith

Introduction

The present-day mountainous glaciated terrain of Snowdonia represents the deeply eroded roots to a series of large-scale (15–20 km in diameter) caldera-forming eruptive centres. These centres, of Caradoc age, have been largely identified from detailed geological mapping and studies of the internal facies and thickness variations and the emplacement environments of the bedded volcaniclastic deposits (Howells *et al.*, 1991). Within this environment, smaller-scale eruptive centres are surprisingly scarce. One of the best exposed and clearly defined of these smaller centres is that of Yr Arddu, interpreted as one of the earliest phases of activity related to the 2nd Eruptive Cycle of volcanic activity in Snowdonia in Caradoc times (Howells *et al.*, 1991). The well-exposed acid ash-flow tuffs, intrusive lava domes and breccias, and associated sedimentary rocks provide an important example of emplacement mechanisms and volcanic processes proximal to a fissure-controlled eruptive vent.

Originally mapped by the Geological Survey in 1851 as 'contemporaneous felsite', Yr Arddu was not described in detail until the work of Beavon in 1963. Beavon subdivided the tuff sequence on Yr Arddu into the Lower, Middle and Upper lapilli-tuffs, which he correlated with various outflow tuffs of the Lower Rhyolitic Tuff Formation on Moel Hebog and Snowdon. Shelly faunas in the subjacent sandstones were ascribed a Soudleyan age (Williams and Harper, in Beavon, 1963). Yr Arddu was later remapped by the British Geological Survey and was described by Howells *et al.* (1987). This work refuted the correlations proposed by Beavon, and showed that, geochemically, the Yr Arddu Tuffs form a distinctive group within the Lower Rhyolitic Tuff Formation with significantly lower Nb/Th ratios than other ash-flow tuffs of the formation (Howells *et al.*, 1991, figs 54 and 55).

The GCR site, which includes all of the main mass of Yr Arddu, covers an area of some 3 km² and includes the outcrop of the Yr Arddu Tuffs and the immediate underlying sedimentary strata (Figure 6.42). It is included in the geological 1:25 000 scale Sheet SH 64 and 65 (Snowdon) (1989) and the 1:50 000 scale Sheet 119 (Snowdon) (1997). The description given below is based largely on Howells *et al.* (1987) and is presented in stratigraphical order.

Description

Yr Arddu forms an elongate synclinal outlier situated along the site of a deep-seated fracture on the south-eastern margin of the Snowdon eruptive centre (Howells *et al.*, 1991). The site includes sandstones and siltstones of the Cwm Eigiau Formation (interlayered with rare, thin, acid tuffs), which are overlain by the Yr Arddu Tuffs, one of the oldest units within the Lower Rhyolitic Tuff Formation. The Yr Arddu Tuffs form a pile more than 180 m thick, intruded by later rhyolite domes, a distinctive breccia dyke, and sills of dolerite.

Exposure on Yr Arddu is excellent and all of the main lithologies can be studied in a traverse from immediately south of Gareg Bengam at 6182 4520 to the vicinity of Llynnau Cerrig-y-myllt at 6330 4722 (Figure 6.42).

The lower beds comprise interlayered siltstones and mudstones coarsening up into blue-grey, well-bedded sandstones and siltstones. The sandstones are flaggy to massive and locally conglomeratic with common cross-bedding and channelized pebbly sandstones. The presence of plagioclase crystals and lithic-tuff fragments indicates that the coarser sandstones are probably volcaniclastic. The siltstones are grey and homogeneous and may include thin layers of laminated sandstone up to 2 cm thick. Thin (up 3 m thick), pale weathering beds representing fine-grained reworked air-fall and/or primary non-welded rhyolitic ash-flow tuff with reworked tops are commonly developed. The primary ash-flow deposits comprise delicate cuspate and bubble shards and a few feldspar crystals. Also present are tuffaceous sedimentary rocks, thin coarse-grained lithic tuffs and debris-flow deposits with acid tuff clasts in a silt matrix (e.g. at 6181 4512). The tuffaceous sedimentary rocks, 1–20 cm thick, commonly display trough cross-bedding and abundant minor syndepositional faults.

Shelly faunas, dominated by disarticulated brachiopods and fragmentary trilobites, occur typically in the coarser sandstone layers and have been collected mainly from the south-eastern and north-eastern margins of Yr Arddu, (for example at 6224 4519 and 6360 4701). Originally interpreted as indicative of a Soudleyan age, the presence of *Kloucekia apiculata*, *Flexicalymene planimarginata* and *Broeggerolithus nicholsoni* (see plate 9 in Howells and Smith (1997) and plate 4 in Howells *et al.* (1991) for examples) indicate a Longvillian age. Detailed collecting has established the Longvillian–Soudleyan boundary on the eastern side of Yr Arddu.

Upslope, outcrops from Gareg Bengam to the summit of Yr Arddu are dominated by massive white-weathering, welded and non-welded rhy-

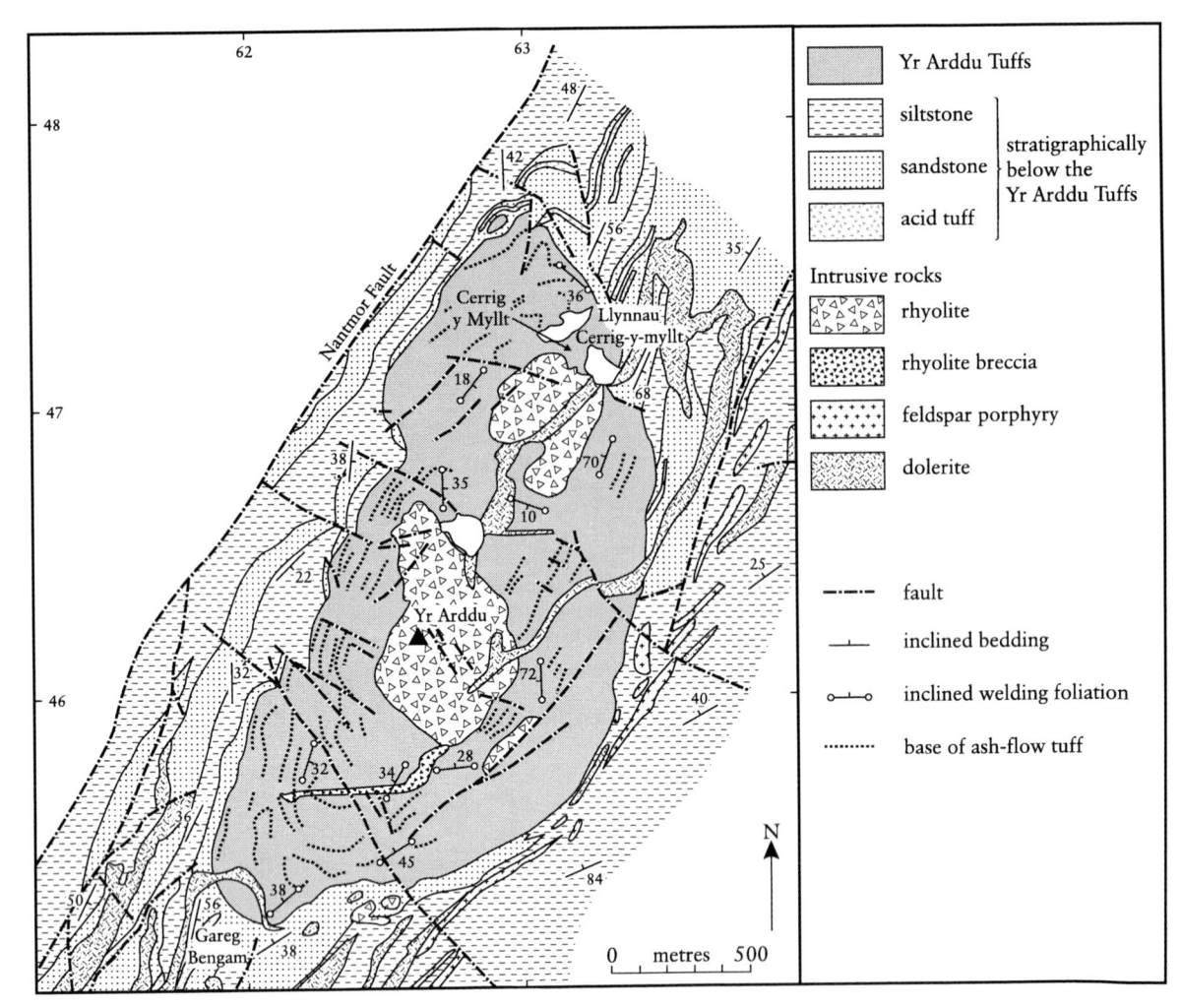

Figure 6.42 Map showing the Yr Arddu Tuffs, subjacent sedimentary rocks and associated intrusions (after Howells *et al.*, 1987).

olitic pumice-lapilli tuffs interlayered with pyroclastic breccia deposits. They form a series of distinctive scarp features and intervening depressions, reflecting the primary stratification in the tuffs, and define a broad synclinal structure. Resting discordantly on the underlying sandstones, the tuffs are variably cleaved and eutaxitic fabrics, defined by chloritic segregations and fiamme, subdivide the sequence above Gareg Bengam into a series of welded and non-welded ash-flows. They are devitrified, locally crystal rich, and siliceous nodules, up to 40 cm in diameter, are often concentrated near the bases and tops of flows. Petrographical descriptions of these tuffs are given in Howells *et al.* (1987). Reworking of the tops of the tuff beds is common near the base of the sequence as indicated by the presence of cross-lamination and shelly debris, including disarticulated brachiopods and crinoid fragments (e.g. at 6219 4524). The tuffs grade into block-and-ash tuffs and pyroclastic breccia deposits, dominated by pumice fragments up to 35 cm in diameter and blocks of acid tuff and rhyolite up to 1.5 m. Minor clast compositions include siltstone, sandstone, basalt and dolerite. A weak eutaxitic foliation is present and is commonly moulded around the blocks.

Midway up the slope to the summit and near the axis of the synform, a dyke-like apophysis of rhyolitic breccia cross-cuts the tuffs. This apophysis, up to 10 m wide, comprises blocks of flow-banded and flow-folded rhyolite up to 2 m in diameter set in a matrix of lapilli-tuff.

Immediately south of the main summit crags, the largest of the two rhyolite domes on Yr Arddu cuts through the tuffs. The contact zone is marked by the spectacular development of

Yr Arddu

Figure 6.43 Siliceous nodules at the top of an acid ash-flow tuff, Lower Rhyolitic Tuff Formation, Yr Arddu. (Photo: BGS no. A14435.)

siliceous nodules locally up to the size of footballs (Figure 6.43). Typically the rhyolites are sparsely porphyritic, and flow-banded. The second and smaller dome is centred about Cerig y Myllt where it is cut by a dolerite sill.

Interpretation

By Caradoc times Snowdonia had undergone regional subsidence and was the site of a major NE-trending incipient rift or graben structure that became the focus for the 2nd Eruptive Cycle of magmatism in North Wales. The bedforms and faunal assemblages of the background sediment within the graben reflect the development of a moderate- to high-energy subtidal shallow-marine environment and indicate local uplift and temporary emergence prior to the onset of volcanic activity.

Within this incipient graben structure, and close to the eastern margin of the future Lower Rhyolitic Tuff Formation caldera, the Yr Arddu Tuffs represent a significant local accumulation of primary welded ash-flow deposits with lesser volumes of block- and ash-flow tuff and pyroclastic breccia deposits (Howells *et al.*, 1987). In contrast to the more internally uniform outflow facies of the Lower Rhyolitic Tuff Formation their heterogeneity suggests proximity to an eruptive source, and a linear NE-trending fissure is thought to underlie the site. The general absence of grading, the concentration of blocks near the bases of individual flows, the limited evidence for reworking and erosion, and the lack of interbedded sediments all suggest that volcanic activity from this vent was largely uninterrupted and dominated by suppressed eruptive columns. Geochemical correlations with the rhyolite domes suggest that they were probably emplaced into the fissure during the waning stages of activity.

The inward-dipping (centroclinal) configuration and the locally large discordance at the base of the tuff pile have been interpreted by Howells *et al.* (1987) as volcanotectonic features subsequently modified by tectonism. The outflow facies from this fissure-controlled centre crops out as a series of welded ash-flow tuffs on Moel y Dyniewyd immediately to the NW of Yr Arddu.

Conclusions

The Yr Arddu GCR site provides a magnificent section through a minor fissure-controlled eruptive centre formed in a shallow-marine environment. Significant volumes of tuff and breccia accumulated close to their source, within an elongate depression and were later intruded by rhyolite domes representing resurgent activity along an underlying fissure. This fissure line represents an important early axis of magmatic activity parallel to the trend of the main graben structure and later fissures within the developing Lower Rhyolitic Tuff Formation caldera.

SNOWDON MASSIF
(SH 622 562–615 524)

M. Smith

Introduction

During Caradoc times, volcanic activity in Snowdonia migrated spatially within a large graben-like structure, termed the Snowdon Trough (Campbell *et al.*, 1988; Kokelaar, 1988) which was marked by the formation of a series of large caldera centres within a predominantly marine environment. Deep-seated NW-trending fractures influenced both the formation of this trough and the tectonic evolution of individual caldera structures during distinct phases of collapse and resurgence. Of these caldera structures, the Snowdon Centre is the largest and most clearly defined (Howells *et al.*, 1991) and has been the subject of detailed investigation by numerous workers over the last two decades of the 20th century. The GCR site lies within this centre and preserves a thick volcanic succession, recording developments within the northern, deepest part of the caldera, which developed during the 2nd Eruptive Cycle of Howells *et al.* (1991).

The geology of Snowdon was originally described in detail by Williams (1927) and was remapped by the Geological Survey between 1970 and 1983. It is included in the 1:25 000 scale Geological Sheets SH64/65 (Snowdon) (1989) and SH65/66 (Passes of Nant Ffrancon and Llanberis) (1985) and the 1:50 000 scale Geological Sheet 119 (Snowdon) (1997). General field guides are provided by Roberts (1979) and Howells *et al.* (1981) and detailed descriptions for the various parts of the succession, including geochemical analyses, are given by Howells *et al.* (1986, 1991) and Kokelaar (1992).

The succession within the GCR site is contained in three major rock basins or cwms separated by narrow serrated ridges, and comprises three formations belonging to the Snowdon Volcanic Group (Figure 6.44). The lower unit, best exposed in Cwm Llan, Cwm Tregalan, around Llyn Llydaw and Lliwedd, comprises a thick accumulation of acid ash-flow deposits known as the Lower Rhyolitic Tuff Formation (LRTF). This is succeeded by basaltic activity represented by the Bedded Pyroclastic Formation (BPF) which crops out extensively within the north-facing Cwm Glas and Cwm Uchaf, west of Glaslyn, the summit area of Snowdon and the north-eastern flanks of Lliwedd. The youngest strata, the Upper Rhyolitic Tuff Formation (URTF), mark a return to acidic volcanism, possibly related to resurgent caldera activity, and are preserved only around the northern cwms, particularly on Clogwyn y Person and Snowdon summit itself. Numerous rhyolitic sills and domes dominate the northern half of the site and show complex intrusive and extrusive relationships with the above strata.

Volcanogenic quartz–sulphide mineralization is important throughout the Snowdon Massif and has been related to hydrothermal alteration by mineralizing fluids during the waning stages of caldera activity (Reedman *et al.*, 1985). At a regional scale, the rocks are buckled into a series of open NE-trending synclinal and anticlinal fold structures representing the imprint of the Caledonian Orogeny in the area.

Description

The site area exposes strata of the Cwm Eigiau Formation, the three formations of the Snowdon Volcanic Group, and related high-level intrusions (Figure 6.44).

Fine-grained siltstones and mudstones of the Cwm Eigiau Formation are exposed in a number of small quarries in Cwm Llan (6134 5250) and represent the oldest strata in the GCR site area.

Snowdon Massif

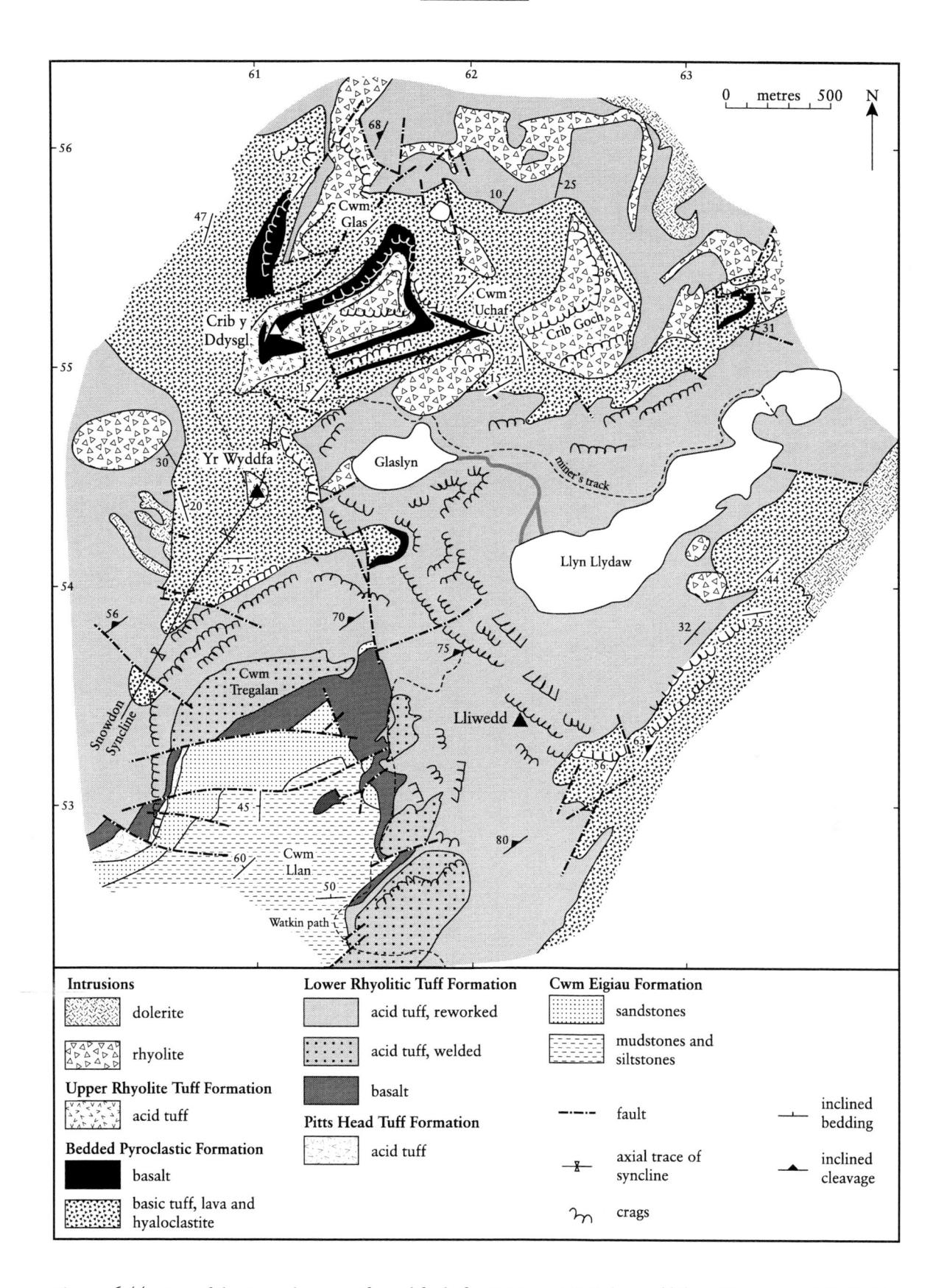

Figure 6.44 Map of the Snowdon massif, modified after BGS 1:25 000 sheets 64/65 (1989) and 65/66 (1985).

Wales and adjacent areas

They pass upwards into sandstones, locally pebbly, with wave-washed, reworked concentrates of detrital magnetite and ilmenite.

The Lower Rhyolitic Tuff Formation (LRTF) generally rests with sharp conformity on the lower Pitts Head tuffs or sedimentary rocks of the Cwm Eigiau Formation. However, in places, up to 100 m of intrusive and extrusive basaltic sheets, associated with pillow breccias, hyaloclastites and basic tuffaceous sandstones rest with marked discordance on the underlying Pitts Head tuffs. Southwards, they rapidly cut down through to the underlying sandstones. Well exposed in the west wall of Cwm Tregalan, these basic rocks are referred to as the sub-LRTF basalts (Howells *et al.*, 1991) and are comparable to the sub-LRTF basalts in the east limb of the Idwal Syncline.

The basal unit of the LRTF, which crops out around Cwm Llan, comprises a white-weathered, intensely jointed, recrystallized and foliated, welded ash-flow tuff. Immediately south of Cwm Tregalan, at the southern margin of the site (around 618 528), the basal unit passes laterally into more impersistently welded tuffs with large pods of silicified welded tuff. Finely recrystallized, the basal tuff is seen to be dominated, in thin section, by aggregates of quartz, sericite and chlorite with isolated altered feldspar phenocrysts preserved as remnants of the original fabric.

The basal tuff is overlain by one of the thickest sequences of non-welded intracaldera ash-flow tuffs in central Snowdonia. Magnificently exposed on the north face of Lliwedd (Figure 6.45), the greater part of the formation comprises up to *c.* 500 m of uniform, massive, unbedded, non-welded rhyolitic pumice-lapilli ash-flow tuff with small clasts, up to 4 mm, of tubular pumice. The base of this sequence is exposed farther north in the Pass of Llanberis (Howells and Smith, 1997) and the upper contact with the overlying BPF can be traced around Glaslyn and the SE side of Crib Goch. Petrographically, the tuffs are dominated by varying admixtures of shards and feldspar crystals set in a matrix of sericite and chlorite (see Howells *et al.*, 1986 for further details).

The non-welded tuffs are overlain by 38 m of reworked tuffs that represent the uppermost part of the LRTF and have been described in detail by Fritz *et al.* (1990) and Howells *et al.* (1986). These beds crop out around the Snowdon Massif, but are best exposed along the western shore of Llyn Gwynant immediately to the SE of the GCR site. They comprise coarse-grained tuffaceous sandstones, interlayered with lesser amounts of laminated tuffaceous fine-grained sandstones and mudstones. Sedimentary structures include dune trough cross-stratification, wave ripples, and hummocky cross-stratification. Large concretionary nodules occur as isolated pods near the base of the section and are interpreted by Fritz *et al.* (1990) as

Figure 6.45 Lliwedd from Miner's Track showing the contact between the Lower Rhyolitic Tuff Formation and the Bedded Pyroclastic Formation near the centre of the ridge. (Photo: BGS no. A14391.)

early diagenetic features. Contorted bedding and small sedimentary dykes indicate soft-sediment deformation, possibly in response to rapid depositional rates. At 15 m above the base of the reworked tuffs, there is a prominent 12 m-thick bed of acid ash-flow tuff, which in turn is overlain by coarse-grained tuffaceous sandstones with abundant sedimentary structures, including herringbone cross-beds, horizontal lamination and trough cross-stratification; it is extensively bioturbated in the uppermost 3 m.

The Bedded Pyroclastic Formation (BPF) is preserved across Snowdonia, mainly within a series of synclinal inliers, and records the shallow-marine accumulation of basaltic pillow and sheet lavas, breccias, hyaloclastites and basic tuffs from a series of vents. These deposits show complex internal relationships and interdigitate with well-bedded tuffaceous sediments. Around Snowdon, the BPF is preserved high up in the glaciated cwms of Cwm Glas and Cwm Uchaf, in the steep cliffs above Glaslyn, and around the upper flanks of Snowdon summit (Figure 6.46). The complex geological history contained in these sections was described in detail by Kokelaar (1992) and Kokelaar *et al.* (1994) and was summarized by Howells *et al.* (1991). Here, only a brief account of the main lithologies and their geological features is presented and the reader is referred to the above accounts for further information.

The basal units of the BPF, exposed in the NE face of Snowdon above Glaslyn (Figure 6.46), comprise up to 95 m of basaltic tuffs, breccias and hyaloclastites. The contact with the underlying LRTF is marked by 5–6 m of thinly bedded tuff-turbidites and cobble conglomerates with vesicular scoria and glassy shard fragments. Two distinct agglomerate vents or necks, up to 280 m in diameter, have been distinguished by their markedly discordant relationships to the LRTF and the lower tuffs and turbidites (Kokelaar, 1992). The vents consist of subangular to rounded basic lapilli and blocks of basalt, up to 20 cm across, set in a rather indeterminate fine-grained basaltic matrix. The basal beds and the vents are then, in part, cut out by the overlying sequence, which consists of turbidites, conglomerates and breccia deposits. In Cwm Glas this break is marked by a 6 m-thick sequence of flow-banded rhyolite lava and the emplacement of large rhyolite intrusions within the LRTF. The overlying sedimentary strata have a total thickness of 75 m and comprise reworked turbidites, lithic-vitric breccias, and conglomerates. Beds are dominated by basaltic clasts including contorted spatter and bombs, but also include rhyolites and shelly debris.

The above strata are overlain by 190 m of heterolithic sedimentary rock. A basal unit, marked by cross-stratified matrix-supported conglomerates, is succeeded by up to 50 m of coarse- to fine-grained turbiditic sandstones, granule conglomerates and siltstones. Sedimentary structures are abundant and include planar and trough cross-bedding, cross-lamination, hummocky cross-stratification and wave ripples. Separating these beds from the overlying turbidites is a distinctive marker bed, some 4 m thick, of fine- to medium-grained altered sandstones and acid tuffaceous beds with carbonate nodules and brachiopod and crinoid debris. The uppermost beds, which form the crags above Glaslyn and the upper south face of Crib y Ddysgl, are composed of 140 m of massive thickly bedded turbiditic sandstones passing up into more thinly bedded, finer-grained sandstones.

In Cwm Glas, the overlying beds indicate a return to basaltic activity with up to 85 m of basaltic tuffs and lavas interleaved with turbiditic sandstones. The lavas are vesiculated, plagioclase-phyric, pillowed and often form columnar-jointed sheets up to 4 m thick. Detailed mapping has traced these flows to the vicinity of the earlier vents in Glaslyn (Kokelaar, 1992). Finally, the complete sequence is overlain by up to 6 m of pebbly and turbiditic sandstones, and silicic siltstones. Rich in basalt scoria fragments, these beds also contain a rich derived shelly fauna with brachiopods of a *Dinorthis* assemblage suggesting water depths of less than 10 m.

The final activity of the Snowdon Centre is represented by the Upper Rhyolitic Tuff Formation (URTF) which is restricted in its outcrop to a series of small outliers within central Snowdonia. The formation includes peralkaline acidic ash-flow tuffs, bedded tuffs and tuffaceous sedimentary rocks and rare basaltic beds and can be related compositionally to the last phase of rhyolite intrusion (Howells *et al.*, 1991). Within the GCR site, the formation is superbly exposed on Clogwyn y Person (Figure 6.47) and Crib y Ddysgl, where up to 100 m are preserved and rest with gentle unconformity on the BPF. Pebbly sandstones, with both rhyolitic and basaltic clasts, locally mark the base, but laterally these beds are overstepped by the main ash-flow tuff. The main tuff, up to 35 m thick, has a

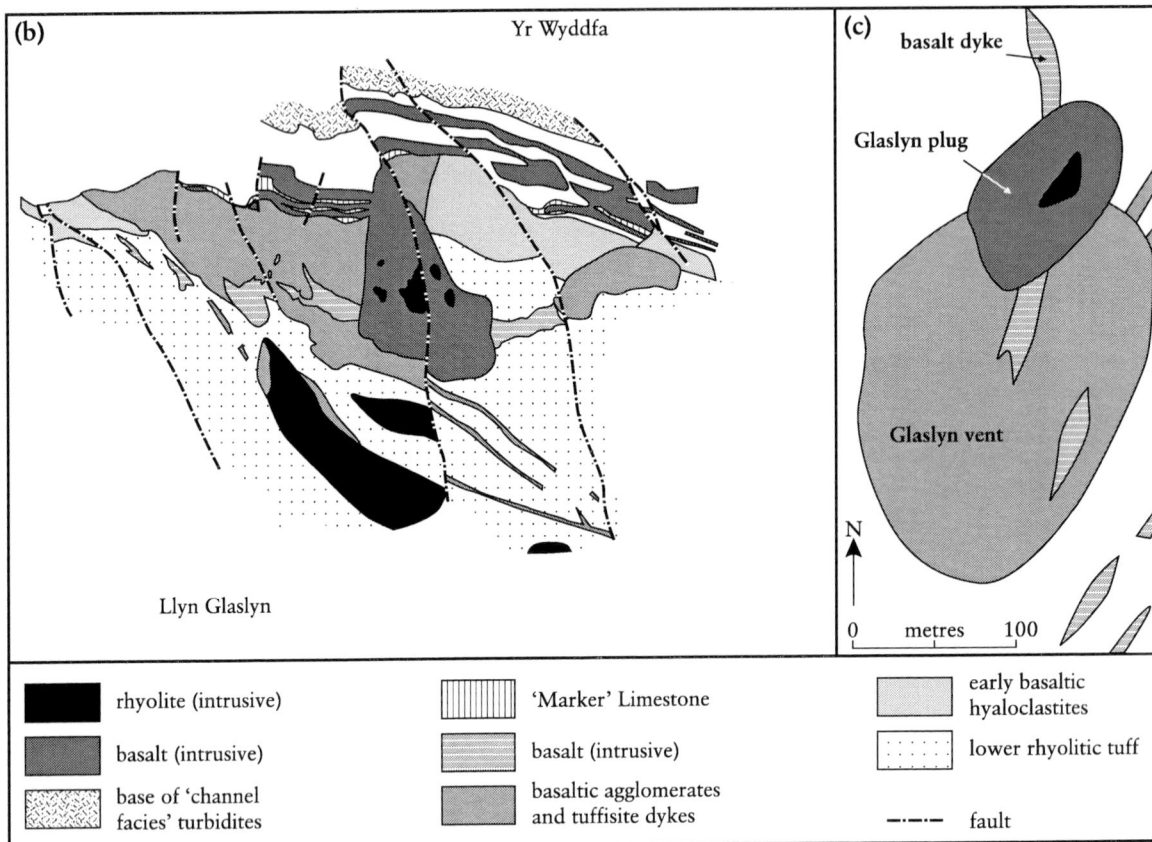

Figure 6.46 Details of the Bedded Pyroclastic Formation cropping out on the NE face of Yr Wyddfa, Snowdon above Llyn Glaslyn. (a) Photograph of NE face with geological boundaries added. (b) Key to the geological units exposed in a. (c) Sketch map of the Glaslyn Vent Complex. Reproduced from Howells *et al.* (1991).

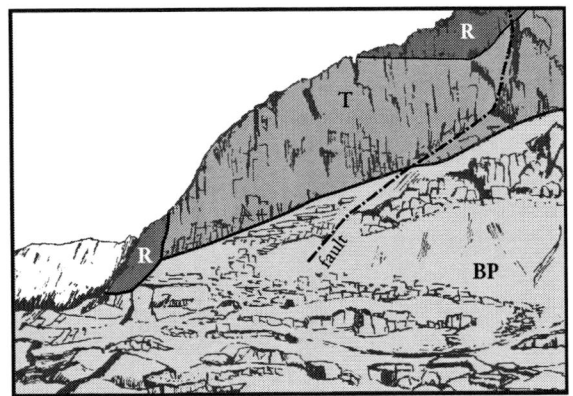

Figure 6.47 The ridge of Clogwyn y Person (SH 615 554) viewed from Cwm Glas. The ridge comprises well-jointed, acidic ash-flow tuff (T) at the base of the Upper Rhyolitic Tuff Formation. Below lie bedded basaltic sediments, basalt and hyaloclastite of the Bedded Pyroclastic Formation (BP), and above an intrusive rhyolite (R). Reproduced from Howells *et al.* (1991).

distinctive, bleached weathered surface with lithic clasts and carbonate nodules, and grades up into fine-grained silicified tuff near the top of the section. Petrographically, the URTF is heterogeneous (Howells *et al.*, 1991) and comprises quartz, feldspar, sericite and chlorite with dispersed shards and a few lithic clasts of acid tuff, perlitic glass and chloritized basaltic fragments. On Crib y Ddysgl, the welded tuff is overlain by up to 40 m of fine-grained, bedded tuffs, tuffaceous siltstones and thin intercalations of basaltic tuff. The siltstones are silicified, with planar and low-angle cross-lamination. Snowdon summit is composed of a small outlier, some 25 m thick, of flaggy, acid tuffs and tuffaceous siltstones which are assigned to the URTF.

Both intrusive and extrusive rhyolite bodies are intimately associated with the LRTF and were emplaced prior to, during and after the deposition of the ash-flow tuffs. They are clearly exposed on the northern flanks of Snowdon, typically forming the serrated ridges south of the Llanberis Pass, such as Crib Goch. The intrusions cut the LRTF and are generally overlain by the basic tuffs and lavas of the BPF. On Clogwyn y Person, a rhyolite dyke that intrudes the BPF can be traced up into a dome overlying the main ash-flow tuff of the URTF, indicating late-stage activity. Typically pale-weathering, the rhyolites are strongly jointed, flow-banded and sparsely porphyritic; perlitic fracturing and autobrecciation are common.

Interpretation

The lower strata of the Cwm Eigiau Formation form part of the substrate onto which the 2nd phase of volcanic activity linked to the Snowdon Centre, here represented by the Pitts Head tuffs, was emplaced. Regional studies have inferred the presence of a NE-dipping palaeoslope with prograding alluvial fans in the south passing northwards into delta-front deposits (Reedman et al., 1987; Howells et al., 1991). In Cwm Llan, the presence of bands of heavy mineral concentrates indicates wave activity and the progradation of a wave-influenced beach. Into this environment, volcanic activity, within the evolving Snowdon caldera, commenced with the localized eruption of basaltic lavas and tuffs from a series of small vents which were probably controlled by deep-seated NE-trending fractures. The marked unconformity at the base of these basalts indicates uplift and erosion, and Howells et al. (1991) suggested that this, in part, reflects the upward propagation of pre-existing basement faults, probably triggered by the emplacement of magma into the cover sequence and the formation of a periclinal anticline within the sediments underlying the LRTF. This structure, termed the Beddgelert Pericline, is unique in the Caradoc strata of Snowdonia and heralds the main phase of ash-flow eruption; it later became the focus for volcanic activity, faulting and mineralization.

The LRTF represents a major period of ash-flow eruption and caldera collapse. From the evidence of facies and thickness variations, basal contacts and the distribution of associated rhyolite intrusions, Howells et al. (1986) inferred the presence of a caldera structure some 15 km in diameter. The section of tuffs within the Snowdon GCR site represents the infill to this caldera. Early eruptions, represented by the basal welded unit, are interpreted as reflecting a distinct eruptive event, possibly located on vents aligned along the NE-trending crest of the Beddgelert Pericline. In Cwm Tregalan, this unit rests conformably on reworked and subaqueously erupted basalts, indicating emplacement in a marine environment. This contrasts with the predominantly subaerial emplacement farther south.

The overlying main phase tuffs were ponded within the Snowdon caldera and, although uniform in appearance, geochemical data indicate that their accumulation involved at least two eruptive phases (Howells et al., 1991). As the caldera progressively subsided, marine incursions began to rework the accumulating tuff pile. Sections through the upper part of the LRTF represent a progressively shallowing marine environment with rapidly fluctuating water depths and local topographical highs within the caldera. Sedimentation in a tidally influenced beach environment is indicated by the herringbone cross-beds, re-activation surfaces and bioturbation (Fritz et al., 1990).

Although the subsidence is greatest in the northern part of the LRTF caldera, as indicated by the thick pile of intracaldera tuffs, on the evidence of the extrusion of rhyolite domes, and the subsequent complex interplay between basaltic magmatism and water depth, it is considered that the general subsidence was interrupted by resurgent uplift. This uplift has been attributed to the emplacement of acid magma at depth and is represented by the intrusion of a series of rhyolites both along NE-trending fractures and along the caldera margins (Campbell et al., 1987; Howells et al., 1991). Following resurgence and reworking, basic magmatism represented by the BPF occurred across Snowdonia. The detailed work of Kokelaar (1992) in the Snowdon Massif recognized a complex history with repeated uplift, emergence and subaerial erosion of a series of basalt island volcanoes. A total uplift of more than 336 m and subsidence of more than 500 m has been calculated by Kokelaar (1992.).

A return to acid volcanism is marked by the URTF, which is intimately associated with high-level rhyolite intrusion and final activity within the Snowdon caldera.

Conclusions

The Snowdon Massif GCR site provides important sections detailing the main phases of extrusive and intrusive volcanic activity related to the Snowdon Centre, a major caldera that developed in a predominantly marine setting. The spectacular ice-sculpted cwms and hanging valleys of Snowdon offer an unrivalled opportunity to study the complex three-dimensional geological relationships within part of this caldera

Cwm Idwal

Figure 6.48 The Idwal Syncline viewed along the axis, across Llyn Ogwen towards Cwm Idwal and the Devil's Kitchen. (Photo: BGS no. L2390)

structure. Numerous studies have revealed the complex inter-relationships, through time, between alternating acid and basic magmatism, changing styles of volcanic activity and the background sedimentation. These relationships, most clearly expressed within the later stages of basaltic activity, represented by the Bedded Pyroclastic Formation, provide valuable insights into the ancient environments of Snowdonia during Caradoc times.

CWM IDWAL
(SH 646 606–640 583)

M. Smith

Introduction

Cwm Idwal is a National Nature Reserve of outstanding geological, geomorphological and botanical interest that is easily accessible from the A5 trunk road near Llyn Ogwen. The geology is varied and complex and includes features of volcanological, sedimentological, and structural importance that are clearly displayed in the eastern and southern cliffs above Llyn Idwal and in the lower ground NW of Ogwen Cottage (Figure 6.48).

The GCR site encompasses a heterogeneous sequence of rock types ranging from rhyolitic ash-flow tuffs, basic tuffs and lavas, to intrusive rhyolites, all interlayered with volcaniclastic marine sedimentary rocks (Figure 6.49). It includes representatives of the two main eruptive cycles related to major caldera activity within central Snowdonia during Caradoc times. Outflow tuffs from both the 1st Eruptive Cycle, related to the Llwyd Mawr Centre, and the 2nd Eruptive Cycle, related to the Snowdon Centre are present. Of particular interest are the sections through the Lower Rhyolitic Tuff Formation (LRTF), of the 2nd Eruptive Cycle, which record the deposition of caldera-sourced pyroclastic breccias and welded tuffs, passing up into reworked tuffs and turbidites. The shelly faunas contained within the sedimentary rocks indicate an age range of Soudleyan to Longvillian (mid-Caradoc).

The primary survey of the area was completed in 1852 (Ramsay, 1881) and later the area was partly described by Williams (1930). Detailed remapping by the Geological Survey at the

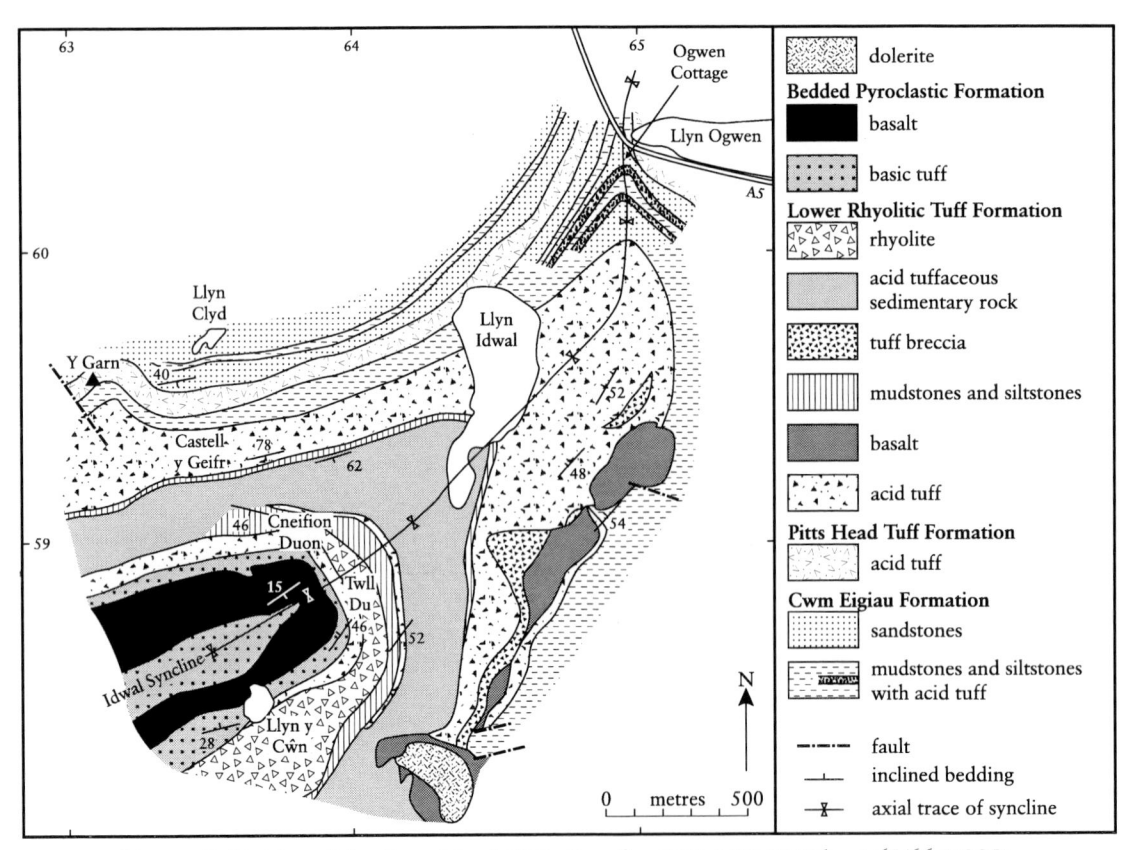

Figure 6.49 Map of the Cwm Idwal GCR site, after BGS 1:25 000 Sheet 65/66 (1985).

1:10 560 scale was completed in 1977 and incorporated into the 1:25 000 scale geological Sheet SH65/66 (Passes of Nant Ffrancon and Llanberis) (1985). General descriptions of Cwm Idwal were given by Roberts (1979) and Howells *et al.* (1981), with detailed descriptions of parts of the succession in Reedman *et al.* (1987), Fritz *et al.* (1990), Howells *et al.* (1991) and Kokelaar *et al.* (1994). Geochemical data were presented by Howells *et al.* (1991).

In ascending order the succession includes the Cwm Eigiau Formation, the Pitts Head Tuff Formation, the Lower Rhyolitic Tuff Formation (LRTF) and the Bedded Pyroclastic Formation (BPF). The strata are deformed in a classic open symmetrical synclinal fold termed the Idwal Syncline (Fitches, 1992). This NE-trending structure, and the complementary Tryfan Anticline to the east, are major fold structures formed during Caledonian orogenesis.

Description

In the site area, strata occurring immediately below the Snowdon Volcanic Group (the Cwm Eigiau Formation), and three formations belonging to the Snowdon Volcanic Group are magnificently exposed.

The lower strata of the Cwm Eigiau Formation, which crop out in the western part of the site (Figure 6.49) and form the crags west of Llyn Idwal, can be traced both south-westwards to Y Garn and northwards into the Braich tu du GCR site. They overlie the Capel Curig Volcanic Formation conformably and comprise steeply dipping very fine-grained structureless sandstones interlayered with siltstones and thin mudstones. Beds containing disarticulated brachiopod shells are common and are well exposed around 6487 6041, NW of Ogwen Cottage. By comparison with the Berwyn district, these faunal associations have been assigned to the *Dinorthis–Macrocoelia* community (Pickerill and Brenchley, 1979) and provide important information on the palaeoenvironment prior to the main volcanic events. A prominent knoll at the north-western end of Llyn Idwal (6430 5981) shows excellent exposures of cross-bedded sandstones overlain by contorted, wavy siltstones and fine-grained sandstones. The overlying welded tuff also contains a raft of contorted sandstone. Similar features, with lobate and

Cwm Idwal

Figure 6.50 The Idwal Slabs, on the eastern limb of the Idwal Syncline, composed of acidic ash-flow tuffs of the Lower Rhyolitic Tuff Formation. (Photo: BGS no. L2636.)

flame structures, are also preserved in quarry sections west and NW of Ogwen Cottage, (for example at 6492 6082), and together these features indicate syn-emplacement deformation of semi-lithified sediments immediately beneath the Pitts Head tuffs.

The overlying acidic ash-flow tuff is the most northerly representative of the Pitts Head Tuff Formation in Snowdonia and is the distal outflow facies from a major caldera centre located some 25 km SW at Llywd Mawr (see the Craig y Garn GCR site report). The tuff varies from 30–50 m in thickness and is well exposed in numerous sections and cuttings immediately south and north of the A5 road, (for example at 649 606), and in the lower western crags rising up across the ridge of Castell y Geifr to Y Garn. The base is generally concordant on the underlying sandstones.

Three main sub-units are recognized in the Pitts Head Tuff at this locality: a basal non-welded zone 1–2 m thick and rich in feldspar crystals with a vitroclastic texture; this grades up into 25–45 m of white, siliceous, welded tuff with a prominent eutaxitic foliation; and an upper fine-grained tuff. Distinctive ragged, flattened and streaky fiamme, up to 5 cm in length, are present throughout the central part of the tuff. Some 5 m above the base, cooling and contraction joints are well developed and fiamme increase markedly in size. The welding foliation is accentuated by siliceous segregations locally reaching 30 cm in length, although these are less well developed compared with the southern outcrops around Moel Hebog. Zones of siliceous nodules 1–3 cm in diameter occur irregularly through the tuff and rheomorphic flow textures, indicated by the variable orientation of the fiamme with respect to the margins of the flow, are common. Locally, (for example at 6476 6024), strong linear fabrics may be observed, and are interpreted as having formed by extreme extension in the flowing and compacting tuff (Howells *et al.*, 1991). The upper sub-unit comprises cleaved vitric dust-rich tuff, which is variably welded and locally reworked.

The Pitts Head tuffs, in turn, are sharply overlain by up to 200 m of sandstones representing a continuation of the Cwm Eigiau Formation with fining-upward sequences and rare thin mudstones. These strata form the striated and ice-sculpted ridges, crags, and waterfalls immediately east of Llyn Ogwen and have been described in detail by Orton (1988) in Cwm

Bochlwyd, immediately east of the GCR site. Sedimentary features are common, including trough cross-bedding, swaley cross-stratification, and low-angle cross-lamination to horizontal lamination. Upwards, the sandstones become planar bedded with thin layers of tuff-turbidite and are characterized by careous-weathering lenses of winnowed, disarticulated brachiopod shells, coquina-filled scours and rare, low to moderately dipping cross-beds with shells dispersed along the foresets.

The acidic ash-flow tuffs, intrusive rhyolites, breccias and interbedded sedimentary rocks which comprise the overlying Lower Rhyolitic Tuff Formation (LRTF) dominate the well-known cliff sections and prominent Idwal Slabs around the eastern and southern parts of the site (Figure 6.50).

The LRTF is a heterogeneous unit up to 110 m thick, which rests with marked disconformity on the underlying sedimentary rocks. This discordance increases to the south to an unconformity (Howells *et al.*, 1986). In the west, the LRTF cuts down through easterly-dipping sandstones, locally cutting out thin tuffaceous units, and comprises welded primary ash-flow tuffs with no intercalated sediments. In contrast, on the eastern limb of the Idwal Syncline the basal relationships are complicated by intrusive basalts, lavas and the development of a pyroclastic breccia facies at the base of the LRTF.

The basal sections of the LRTF are best seen at the foot of a wall immediately SW of two streams that drain into Llyn Idwal at 6470 5894. Here, the basal beds include vesicular, massive basaltic pillow lavas, heterolithic basaltic debris-flow deposits and thin turbiditic tuffs. The debris-flow deposits contain subrounded basalt blocks up to 1 m in diameter, in a fine-grained matrix of basic tuff with dispersed feldspar crystals. The overlying pyroclastic breccia, up to 12 m thick, comprises a coarse lithic breccia occurring as layers or lenses which thin and pass laterally into the host matrix-supported ash-flow tuff. Individual blocks, 0.1–0.7 m in length, range in composition from basalt, acid tuff and rhyolite to rare sandstone and siltstone. A crude stratification can be discerned by variations in grain and clast size, and lithology.

The main part of the LRTF is a thick sequence of stratified welded lapilli ash-flow tuffs with beds up to 1.5 m thick and a prominent eutaxitic foliation. Thin breccia layers similar to the basal breccia have been recorded up to 42 m above the base (Kokelaar *et al.*, 1994). In thin section, the fiamme are chloritic and are set in a matrix of undeformed devitrified shards and fine-grained dust. Several units are indicated on the 1:25 000 scale map and are separated by laterally impersistent layers of siltstone. The highest beds are composed of well-bedded and upwardly graded reworked tuffs. A dark porcellaneous laminated fine-grained tuff marks the top of the sequence.

The overlying sedimentary rocks, up to 100 m in thickness, have been described in detail by Fritz *et al.* (1990). The lowest bed is a 2 m-thick laminated pyritic mudstone, passing up into tuffaceous siltstones with hummocky and swaley cross-stratification and abundant large carbonate concretions. Above, are up to 75 m of brown-weathering, greenish coarse-grained volcaniclastic sandstones with interbeds of pale tuffaceous siltstone and rare impersistent ash-flow tuffs. Individual bed thicknesses range between 10–50 cm and show a progressive upward decrease in the volcaniclastic component. The sandstones, up to 7 cm thick, are massive and planar bedded with flat, locally scoured bases and reworked hummocky tops. Sedimentary features include cross-lamination, ripples, grading, washouts, intraformational unconformities, and slump folds with contorted bedding indicating deformation of semi-lithified sediment (for example at 6402 5904).

The upper parts of the backwall of Cwm Idwal are dominated by a thick columnar-jointed rhyolitic lava flow, which can be traced across the core of the synclinal structure. The rhyolite lava is dark blue-grey, flinty and finely banded with locally developed perlitic fracturing. The upper surface, exposed along the footpath between Cneifion Duon and Y Garn, is brecciated, and in places, hollows and depressions are infilled with coarse volcanic detritus. The zones of brecciation contain classic jig-saw breccia fabrics and can be traced laterally into flow-banded rhyolite indicating in-situ autobrecciation. The rhyolite lava is in places separated from the overlying BPF by up to 35 m of planar and cross-laminated rhyolitic tuffaceous sandstone representing the top of the LRTF.

The Bedded Pyroclastic Formation (BPF) crops out to the SW of Twll Du (or Devil's Kitchen) and around Llyn y Cwn, immediately south of the GCR site. Up to 24 m of flaggy-bedded, greenish basic tuffaceous sedimentary rocks and coarse-grained tuffaceous sediments

Cwm Idwal

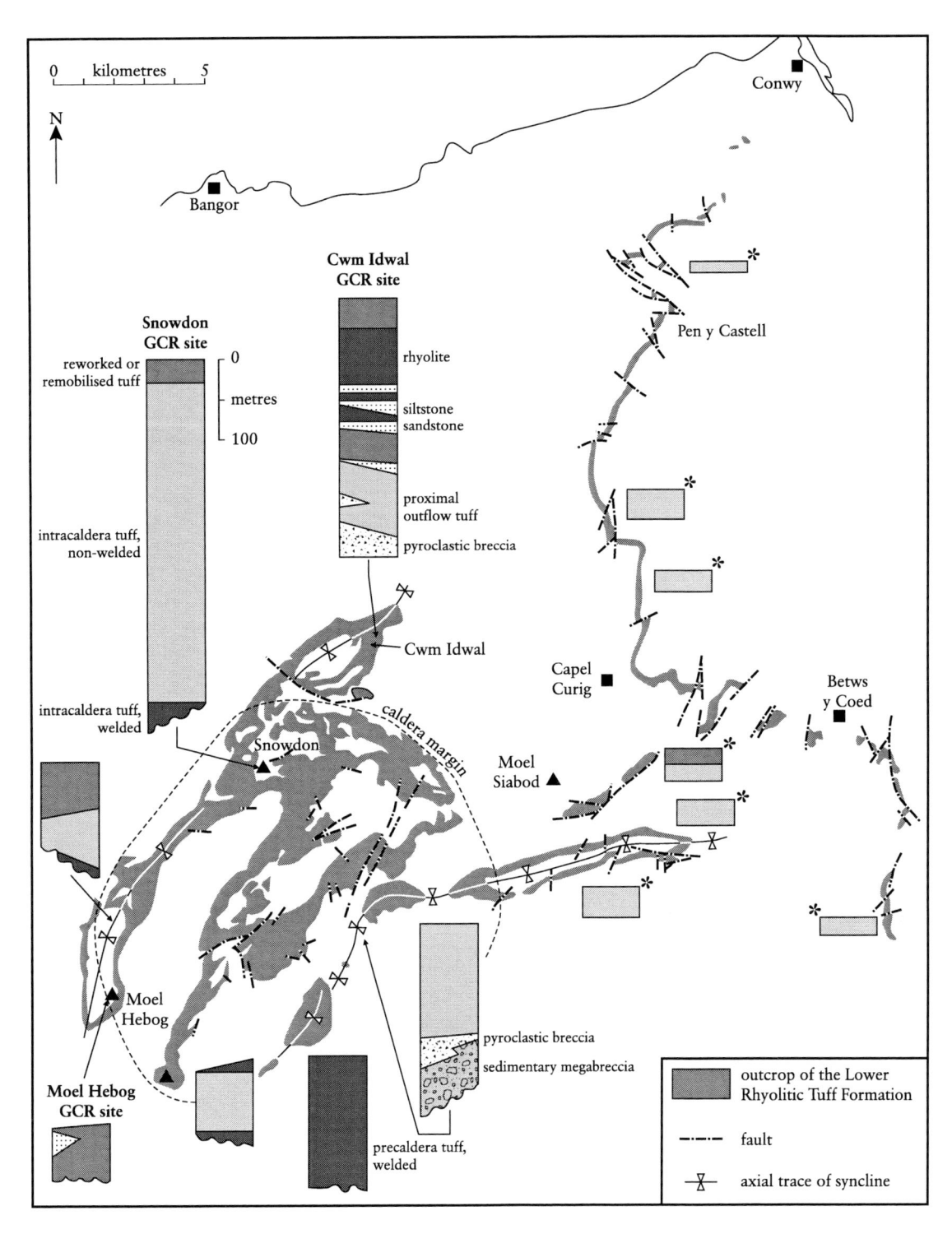

Figure 6.51 Outcrop and measured sections of the Lower Rhyolitic Tuff Formation. Asterisks indicate distal outflow tuff sections. After Howells *et al.* (1991).

including block- and lapilli-rich beds are exposed. Sedimentary structures, including cross-lamination and wave-rippled surfaces, are common. Elsewhere, these beds grade into volcaniclastic sedimentary rocks with a sparse shelly fauna, interpreted as indicating a probable upper Longvillian age (Howells *et al*., 1991). The latter are overlain by up to 70 m of autobrecciated non-vesicular basaltic lavas with sparse plagioclase phenocrysts and variably developed pillow forms. Where well developed, the pillows reach 1.5 m in diameter. At the very top of the section the basalt lavas are overlain by basic tuffaceous sedimentary rocks displaying cross-lamination.

Interpretation

The geology of Cwm Idwal and the surrounding area provides an unrivalled opportunity in Snowdonia to assess the changes in sedimentation in the marine environment caused by the introduction of large volumes of hot ash-flow deposits proximal to a large caldera structure. The lower sedimentary rocks and contained faunal assemblages of the Cwm Eigiau Formation suggest a shallow-marine environment with water depths of less than *c*. 25 m. The interlayered sandstones and siltstones are interpreted as shelf-ridge sands with interbar silts and muds, within which the more massive and thicker sandstones may represent discrete storm events (Howells *et al*., 1991).

Into this environment the Pitts Head tuffs were deposited as hot gas-charged pyroclastic flows. The submarine emplacement and post-emplacement features of the Pitts Head tuffs in Cwm Idwal contrast with the subaerial environment at the Moel Hebog to Moel yr Ogof GCR site and imply a north-easterly dipping palaeoslope (Reedman *et al*., 1987). The lack of disruption along the basal contact and the internal fabrics suggest that the tuff appears to have ingested little water, and retained sufficient heat to weld on emplacement. The overlying, less dense gas-rich cloud above the tuff is thought to have travelled across the water surface, eventually settled, and is represented by the upper vitric dust tuff (Howells *et al*., 1991). Estimates of water depth, often problematical in shallow-marine settings, is considered to be less than 20 m, and therefore it would seem unlikely that the tuffs would have been completely submerged.

The overlying sandstones, with their abundant sedimentary features and transported faunal debris, represent the continuation of marine conditions with high-energy regimes on a mid- to outer storm-dominated shelf (Orton, 1988). With time, deeper water conditions prevailed; the upper turbiditic sandstones and siltstones, with a lack of coarse-grained detritus, record subsidence prior to the next period of volcanic activity.

The following cycle of volcanic activity recorded by the LRTF commenced with localized basic magmatism and the formation of a distinctive suite of pyroclastic breccia deposits. These breccias, identical to intracaldera lag breccias south of Snowdon, are only found in the Cwm Idwal area up to 4 km north of the margin of the LRT caldera (Figure 6.51). From their lithology and nature, they are interpreted to be co-ignimbritic lag breccias (Howells *et al*., 1986). The overlying main part of the LRTF shows a marked absence of any compositional or fabric variations both between or within individual beds, suggesting repeated pulses of ash-flows from a single eruptive phase (Howells *et al*., 1986). This is supported by their trace element geochemistry, which shows consistent relative abundances of the elements Zr, Nb, Th and TiO_2. Vertical variations in trace element profiles suggest a break near the top of the main body of pyroclastic breccias.

The primary tuffs were overlain by remobilized pyroclastic debris and sediments, interpreted to represent deposition on a pyroclastic apron that formed along the northern margin of the caldera (Orton, 1988; Fritz *et al*., 1990). A rapidly shallowing sequence from deep, non-volcanically influenced sedimentation represented by the black mudstones to above storm-wave base with water depths of *c*. 100 m is indicated by the hummocky and cross-stratified sandstones. The increase in grain size and extensive reworking of the overlying sandstones may represent turbidity-current deposition and progradation of the apron, fed by sediment from the caldera margin, to within and above storm-wave base (Fritz *et al*., 1990). Howells *et al*. (1991) noted that the overall volume of reworked material exposed around the northern edge of the caldera is small compared to the infill. This suggests that the edifice of the Snowdon caldera had a limited subaerial expression and ponding of eruptive products occurred within a shallow-marine depression.

A thick rhyolite flow was intruded into the upper part of the LRTF; the presence of an autobrecciated carapace indicates that this was probably locally extrusive onto the sea floor. The overlying tuffs and their rhyolitic clasts, which mark the top of the LRTF, are geochemically distinct from the underlying tuff sequence but closely match the composition of the rhyolite. This supports their emplacement as a distinct magmatic event within the evolution of the Snowdon caldera.

Following resurgence and reworking of the main caldera-related tuffs, the Snowdon eruptive centre was dominated by basaltic volcanic activity. This activity is represented at Cwm Idwal by the Bedded Pyroclastic Formation which erupted into a shallow-marine environment, as is indicated by the presence of basaltic pillow lavas and reworking of the upper basic tuffs and lavas.

Conclusions

The Cwm Idwal GCR site provides magnificent exposures of volcanic products from both the 1st and 2nd eruptive cycles which occurred in Snowdonia in Caradoc times. The heterogeneous sequence at Cwm Idwal records the dramatic influence of major caldera-related explosive volcanism on the sedimentation patterns in a marine environment. Combined volcanological and sedimentological studies reveal significant changes in sediment supply and modification of the tectonic environment in response to volcanic activity. In addition the Cwm Idwal site provides excellent examples of emplacement features of welded ash-flow tuffs in a shallow-marine setting and emphasizes the importance of reworking and the widespread redistribution of volcanic material in such an environment.

CURIG HILL
(SH 722 580–736 593)

M. Smith

Introduction

Following the cessation of volcanic activity related to the 1st Eruptive Cycle, northern and central Snowdonia underwent large-scale extension in later Caradoc times to form an elongate NW-trending trough or graben structure. The sediments and lesser volumes of acidic and basic tuffs that infilled this trough provide valuable information on the environment of deposition and volcanic activity prior to the initiation of the 2nd Eruptive Cycle of acid ash-flow tuff volcanism in North Wales.

The GCR site in the vicinity of Curig Hill records an impressive heterogeneous sequence of marine sedimentary rocks, interlayered with tuffaceous sedimentary rocks and distal acid and basic tuffs. This distinctive association, which occurs immediately above the Capel Curig Volcanic Formation, passes conformably up into a distal outflow tuff derived from the Snowdon Centre (the Lower Rhyolitic Tuff Formation or LRTF). This tuff is in turn overlain by tuffs and tuffaceous sedimentary rocks derived from the Crafnant Centre (the Lower Crafnant Volcanic Formation or LCVF). The site is also important in being one of the few examples of a well-exposed section through a basic vent or tuff cone.

Originally, the area was mapped in 1848, with the first geological maps and sections published between 1851 and 1854, and described by Ramsay (1881). It was mapped at the 1:10 560 scale by Williams (1922) and later by the Geological Survey in 1968–70. Detailed descriptions are presented in Howells *et al.* (1978) with later revisions and re-interpretation of the stratigraphy in Howells *et al.* (1991). Published geochemical data for the LCVF indicate predominantly rhyolitic compositions with individual tuffs distinguished by their Zr/TiO_2 ratios (Howells *et al.*, 1991).

Description

The Curig Hill GCR site lies on the limb of a paired fold structure with moderate dips predominantly to the NE (Figure 6.52). The western (and lowest) parts of the succession are exposed immediately north of the A5 at Plas Curig and comprise greyish-green, well-bedded sandstones of the Cwm Eigiau Formation, interbedded with acid tuff, tuffaceous sedimentary rocks and basic tuffs. Pre-tectonic deformation of the strata is common and prominent in a 2 m-thick sandstone containing slump structures including overfolds and oversteepened foresets. Fossiliferous beds dominated by shelly faunas occurring in layers up to 10 cm thick are also common locally; north of Curig Hill these include the brachiopod *Plaesiomys multifida*, indicating a Soudleyan age. Elsewhere along strike, the presence of Longvillian faunas near the top of the section suggests that the Soudleyan–Longvillian boundary probably lies within the sequence.

The interbedded acid tuffs are fine grained and composed of devitrified, recrystallized fragmentary shards and dust, with or without a mudstone matrix. Locally, with increasing additions of sedimentary debris, the tuffs grade into tuffaceous sandstones and siltstones. Generally up to 5 m thick, the tuffaceous sandstones commonly show cross-bedding in the tops of units and washouts. In places they are disturbed by soft-sediment deformation, for example north of the Capel Curig Youth Hostel (at 7258 5811).

Basic tuffs form two distinctive layers within the succession. The lower crops out as a wedge-shaped intrusive mass some 200 m wide, forming the mass of Curig Hill immediately to the north of the Bryn Tyrch Hotel (724 581) (Figures 6.52 and 6.53). Well-cleaved, poorly sorted and rarely graded, the basic tuffs contain abundant volcanic blocks and lapilli, with bedding defined by grain-size variations and clast or block concentrations. Petrographically the tuffs are composed mainly of aggregates of chlorite, carbonate and iron oxide.

Grain-size analysis reveals that towards the summit of the hill there is a gradual increase in the size of the blocks and lapilli. This is associated with the development of a slumped, agglomerate zone characterized by blocks of tuff up to 1 m in diameter and penecontemporaneous minor faults. In addition, there are important variations in the dip patterns around the hill. In the west concentric inward (centroclinal) dips decrease from 80° to 50° towards the zone of slumped agglomerate. In contrast, above a planar discordance, which trends N–S within the eastern part of the tuff pile, dips of between 25° to 30° are more constantly to the east. The tuffs here are finer grained and include lenses of fine-grained reworked tuff and tuffaceous sedimentary rocks and pass conformably up into younger sandstones typical of the Cwm Eigiau Formation.

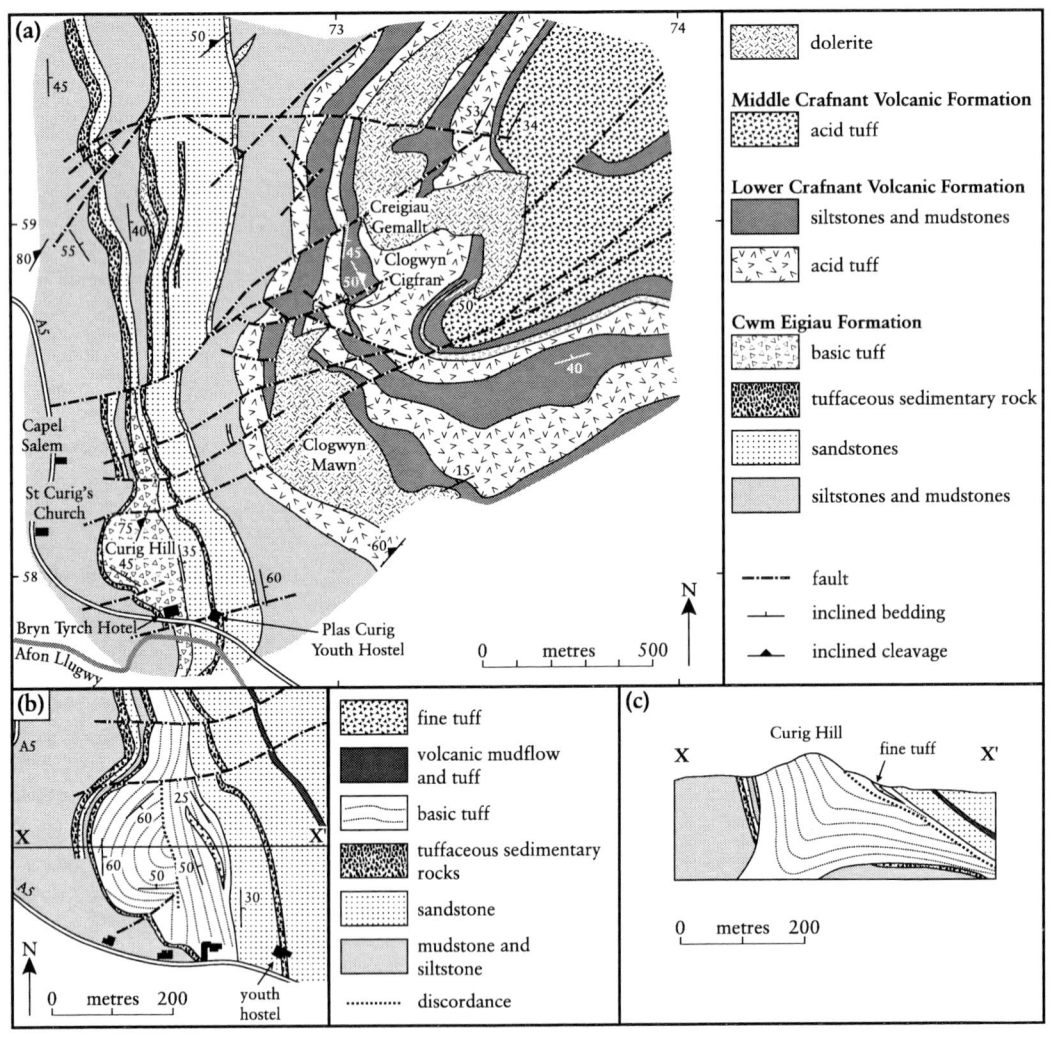

Figure 6.52 (a) Map of the Capel Curig area (after BGS 1:10 000 Sheet SH75NE). Insets (b), (c) show sketch map and section of the basaltic vent at Curig Hill (after Howells *et al.*, 1991).

Figure 6.53 Base of bedded basalt agglomerate, Curig Hill 'vent' with acid tuff forming the lower feature near the wall. (Photo: BGS no. L1868.)

The upper basic tuff layer lies near the top of the sandstone succession at 7275 5790. It is composed of two distinct horizons: a lower laharic mudflow, less than 2 m thick, includes angular clasts of sedimentary rock and acid tuff showing a crude alignment parallel to the regional bedding; and an upper basic tuff, 2–3 m thick, which marks the contact between sandstones and overlying mudstones and siltstones. Well-exposed around 7275 5793, the tuff contains clasts of chloritized and altered basalt and basic pumice or scoria in a fine-grained matrix.

The higher parts of the GCR site, from Clogwyn Mawr across to Creigiau Gemallt, are dominated by a succession of NE-dipping acidic ash-flow tuffs separated by siltstones and mudstones and intruded by a dolerite sill.

The overlying volcanic rocks are poorly cleaved, often flinty, vitric, non-welded acid tuffs with variable proportions of crystals and lithic clasts. The lower tuff (No. 1 of the LCVF of Howells *et al.* (1973, 1978)) is equivalent to part of the most easterly outflow tuff of the LRTF (Howells *et al.*, 1991). Up to 56 m thick at 7270 5760, the tuff displays a distinctive upwards-fining sequence from a basal zone rich in crystals and lithic clasts, through a sparsely porphyritic middle part with small pumice clasts, to a fine-grained crystal-depleted top. The basal zone includes clasts of siltstone, brachiopod and trilobite fragments, and rare ooliths. The middle zone, between 7 and 21 m above the base, is regularly bedded with thin (up to 3 cm) well-cleaved silty layers and passes up into massive columnar-jointed tuff with clasts of pumice and rhyolite. Above are interbedded siltstones and mudstones.

The middle tuff (No. 2 of the LCVF) is well exposed on Clogwyn Cigfran at 7295 5873, where it is underlain by a rusty brown-weathering feldspar-phyric dolerite sill. The tuff is uniform and massive with visible pumice clasts, feldspar crystals and rare siliceous nodules. The upper tuff (No. 3 of the LCVF), up to 40 m thick, is a more heterogeneous unit, distinguished by the absence of xenocrysts and a wide range in shard sizes. Forming the eastern slopes of Clogwyn Cigfran (e.g. at 7321 5879), it includes clast-rich vitric tuff interlayered with tuffaceous siltstones. Lithic clasts include andesite, hyaloclastite, basic and acid tuff, pumice and siltstone. Bedding is demonstrated at 21 m above the base by a thin agglomeratic bed with rounded clasts.

Interpretation

The strata in the Curig Hill GCR site area lie within the middle part of the interval between the two major Caradoc eruptive cycles in Snowdonia and have been interpreted as marking a marine transgression, with the progressive development of deeper water environments from shallow marine to offshore down a south-facing palaeoslope. Sedimentological studies in adjacent areas have interpreted the sandstones in the lower part of the succession as having formed within fluctuating inner and outer shelf regimes subject to periodic storm events (Orton, 1988). Within this environment the acid tuffs and tuffaceous sedimentary rocks represent distal ash fall-out, subsequently reworked in the marine environment and disrupted by soft-sediment deformation. The coarser-grade tuffs probably represent secondary emplacement by transport as high-density debris flows and slurries of pyroclastic debris.

The centroclinal dips, lateral wedging and grain-size variation in the lower basic tuffs of Curig Hill were interpreted by Howells *et al.* (1978, 1991) to represent the upper levels of an intrusive funnel-shaped volcanic vent which probably fed a tuff cone on the ancient surface. The cone superstructure was reworked and the sediments were redeposited as the fine-grained bedded volcaniclastic sediments at the top of the section. The limited contamination of the adjacent sediment with basaltic debris suggests that such eruptions were minor and that debris dispersal was limited.

The lowest tuff of the overlying Lower Crafnant Volcanic Formation is the sole distal representative of a rhyolitic ash-flow tuff (Lower Rhyolitic Tuff Formation) derived from the Snowdon Centre. The fine-grained top of vitric dust probably represents the elutriation of devitrified and recrystallized volcanic dust and ash material from the head of the pyroclastic flow. The overlying tuffs and tuffaceous sedimentary rocks are the products of rhyolitic ash-flow eruptions with limited ingestion of substrate sediment and were emplaced in a deep-water marine environment. Later remobilization of these tuffs was periodic and localized.

Conclusions

The Curig Hill GCR site preserves an important section recording the reworking of volcanic deposits and sedimentation between the two major eruptive cycles in Snowdonia. A complex heterogeneous sequence of marine sedimentary and volcaniclastic deposits indicates the progressive subsidence and development of a major trough. The products of initial basaltic eruptions, in shallow shelf settings, were progressively reworked and buried by finer-grained marine sediments from a more distant source as the trough deepened. Distant minor volcanic activity released small volume ash-fall deposits. The start of the 2nd Eruptive Cycle is heralded by the Lower Crafnant Volcanic Formation and its emplacement into quiescent deep marine conditions.

SARNAU
(SH 779 587–776 600)

M. Smith

Introduction

The Sarnau GCR site provides key exposures of the eruptive products from the most north-easterly and youngest of the three major eruptive centres that were active during the 2nd Eruptive Cycle in Caradoc times in Snowdonia. In contrast to the two other centres at Snowdon and Llwyd Mawr, the Crafnant Centre is completely buried by younger strata and its eruptive history and palaeoenvironment can only be inferred from its outflow products. The exposed strata at the Sarnau site include the Middle and Upper Crafnant volcanic formations which represent acid ash-flow tuffs derived from this centre, and their emplacement into dark marine mudstones and siltstones suggests a deep water submarine eruption.

Originally mapped by the Geological Survey in 1848, the area was included in a regional study by Davies (1936) but was not studied in detail until the resurvey on the 1:10 560 scale by the Geological Survey in 1968–70. The site area is covered by the geological sheets SH75 (Capel Curig and Betws-y-Coed), 1:25 000 scale (1976), and 109 (Bangor), 1:50 000 scale (1985). The strata, of Longvillian age, are described in Howells *et al.* (1978, 1991) and comprise ash-flow tuffs and reworked tuffs of rhyolitic to rhyodacitic composition. They are probably contemporaneous with the Bedded Pyroclastic and Upper Rhyolitic Tuff formations in central northern Snowdonia and form part of the Snowdon Volcanic Group.

Description

The Sarnau site (Figure 6.54) lies in the wooded uplands NNW of Betws-y-Coed, between Llyn Sarnau and Mynydd Bwlch-yr-haiarn. Moderately dipping to the north and east, the strata lie on the SE limb of an anticline, one of a series of NE-trending open fold structures in the area. A complex array of steeply dipping brittle faults form prominent topographical features in the area and are the focus of lead and zinc mineralization (Howells *et al.*, 1978).

The Middle Crafnant Volcanic Formation, which elsewhere reaches a total thickness of 90 m, forms the lowest stratigraphical unit in the area. At Sarnau the formation is characterized by an ordered sequence of thin, primary acid ash-flow tuffs, interlayered with flaggy, evenly bedded, remobilized tuffs, tuffaceous to volcaniclastic sedimentary rocks, and black pyritic mudstones and siltstones. These form a series of well-developed scarp and dip features (e.g. at 7730 5880). The interlayered sedimentary rocks comprise mainly dark-grey structureless mudstones and siltstones with infrequent ribs of turbiditic sandstone. The mudstones and siltstones contain varying proportions of iron oxide, sericite and chlorite, with scattered fragments of feldspar crystals, locally up to 0.5 mm in length. The thin sandstones are normally graded with scoured and loaded bases often marked by concentrations of pyrite. Flame structures, indicating subsequent deformation of the sequence while semi-lithified, include lobes of sandstone completely surrounded by flames of the underlying mudstone.

The various units range from primary tuffs through tuffaceous to volcaniclastic sedimentary rocks and display a wide range of grain size from ash to breccia. The thin flaggy turbiditic tuffs with recrystallized cuspate shards set in a fine-grained vitroclastic and micaceous matrix show fine cross-lamination and grading and may contain fragmentary fossiliferous debris including crinoid columns and graptolites (Howells *et al.*, 1978). The coarser tuff-breccias are well exposed in the forest track (at 7705 5910) where they form massive beds up to 1.3 m thick with blocks of angular to subrounded, indurated siltstone, mudstone (Figure 6.55) and tuffaceous and volcaniclastic sedimentary rocks, and are invariably associated with thin fine-grained siliceous air-fall tuffs. Some of the blocks show a faint internal planar lamination that parallels the irregular periphery of the block indicating that the blocks were probably unlithified at the

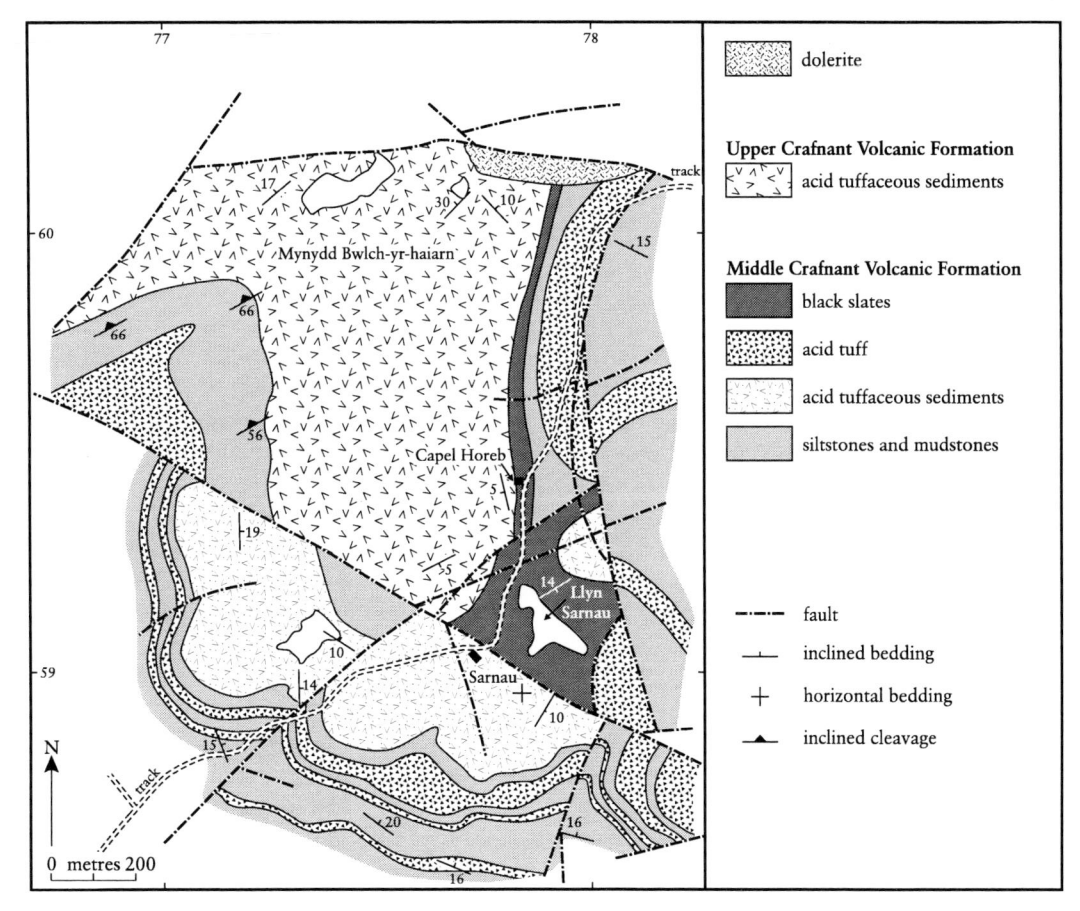

Figure 6.54 Map of the Sarnau GCR site area, after BGS 1:10 560 sheets SH75SE (1977) and SH76NE (1979).

Figure 6.55 Acid ash-flow tuff with blocks and clasts of mudstone, Middle Crafnant Volcanic Formation, Sarnau (SH 7721 5891). (Photo: BGS no. L2905.)

time they were incorporated into the breccia.

The reworked tuffs, comprising varying admixtures of pyroclastic and sedimentary components, form distinctive striped parallel-bedded sequences of fine-grained blue-grey mudstones and paler siltstones. Gradations from fine-grained tuffaceous mudstone to mudstone and siltstone are common and locally the basal contacts may be deformed by loading. Other sedimentary features include convolute lamination, penecontemporaneous microfaulting, and flame structures (e.g. 7745 5884). Within the striped sequences coarser beds, with recognizable crystals and glass shards in thin section, form units up to 1.5 m thick and grade in places into tuffs. They consist of small rounded albite crystals, up to 1.2 mm, in a matrix of quartz, feldspar, chlorite and fragments of carbonaceous material.

The Upper Crafnant Volcanic Formation comprises a massive heterogeneous unsorted tuffaceous sedimentary unit up to 70 m thick. It is separated from the underlying formation in the Sarnau area by blue-black cleaved mudstones. The formation is well exposed in the northern part of the site in the region of Mynydd Bwlch-yr-haiarn but is poorly bedded and only forms subdued scarp features. The tuffaceous rocks are generally structureless, strongly cleaved due to the high proportion of mudstone, and consist of cuspate shards with fragmentary feldspar crystals and pumice admixed with fine aggregates of chlorite, sericite and carbonaceous material. The proportions of constituents are highly variable. A typical exposure is in a small quarry adjacent to the forestry track at 7729 6019.

Interpretation

By its nature, a submarine eruptive centre will be more difficult to identify in the rock record due to subsequent burial and obscuring of contemporaneous tectonic features. Nevertheless, using regional variations in the lithological characteristics and the degree of internal disruption, Howells *et al.* (1978, 1991) interpreted the Middle and Upper Crafnant volcanic formations as representing outflow tuffs derived from a major caldera centre tentatively located to the north and east (the Crafnant Centre).

The tuffs exposed at the Sarnau GCR site include primary ash-flows and block-and-ash deposits disrupted during transport to form slurries of unstable pyroclastic debris. The thin siliceous beds associated with the massive coarse tuffs and breccias are considered to represent the settling out of the fine ash elutriated into the water column during ash-flow transport. These were also subject to reworking and were resedimented as tuff-turbidites represented by the striped beds. The frequent restriction of convolute lamination to these thin beds may be the result of emplacement of tuff on partly lithified water-saturated tuffaceous sediment which deformed thixotropically, possibly in response to seismic shocks (Howells *et al.*, 1978).

Within the Upper Crafnant volcaniclastic deposits, the general lack of sorting and internal bedding suggests that they were deposited from high-density turbid flows of pyroclastic debris that incorporated sedimentary material from a semi-lithified substrate during transport. As Howells *et al.* (1991) noted, however, such a process might not be expected to produce such a well-mixed sequence, and an alternative model of an underwater explosive eruption through unlithified mud could also be considered.

Conclusions

The GCR site at Sarnau represents a typical series of primary and reworked acid ash-flow tuffs emplaced within a deep water marine environment. It provides a well-exposed section through tuffs erupted from the Crafnant Centre, the most north-easterly of the three main Ordovician caldera centres which developed during the 2nd Eruptive Cycle in Snowdonia in Caradoc times.

FFESTINIOG GRANITE QUARRY (SH 696 453)

M. Smith

Introduction

Within the Ordovician (Caradoc) cycle of volcanism in Snowdonia there are only a small number of subvolcanic intrusions that are not spatially related to the main eruptive centres. Among the largest are the granite plutons of Tan y Grisiau and Mynydd Mawr, which lie several kilometres outside the main zone of Caradoc caldera structures, intrusions and related extrusions as defined by Howells *et al.* (1991). They were emplaced mainly into Lower Ordovician (Tremadoc) or Cambrian (Merioneth) sandstones and siltstones. Sections in the Ffestiniog Granite Quarry provide a rare example of the contact relationships and autometasomatic effects that occurred in the granite roof during volatile streaming and crystallization. The area was originally described by Jennings and Williams (1891) and later revised by Bromley (1963), whose maps and descriptions are incorporated into the recent resurvey (Howells and Smith, 1997).

The Tan y Grisiau Granite has an outcrop area of *c.* 4 km² and is intruded into sandstones and siltstones of the Dol-cyn-afon Formation (Figure 6.56). The outcrop pattern is that of a truncated ellipsoid but the extent of the hornfels and aureole and associated geophysical anomalies show a much larger body at depth (Howells and Smith, 1997). The gravity and magnetic data have been interpreted to reflect a steep-sided, sub-vertical body, elongated some 10 km to the NE and 5 km to the SW of the main exposures with a NNW-dipping roof (Cornwell *et al.*, 1980; Campbell *et al.*, 1985). The roof zone of the granite is best exposed in a small quarry, now largely infilled, situated along the north-eastern margin of the granite outcrop.

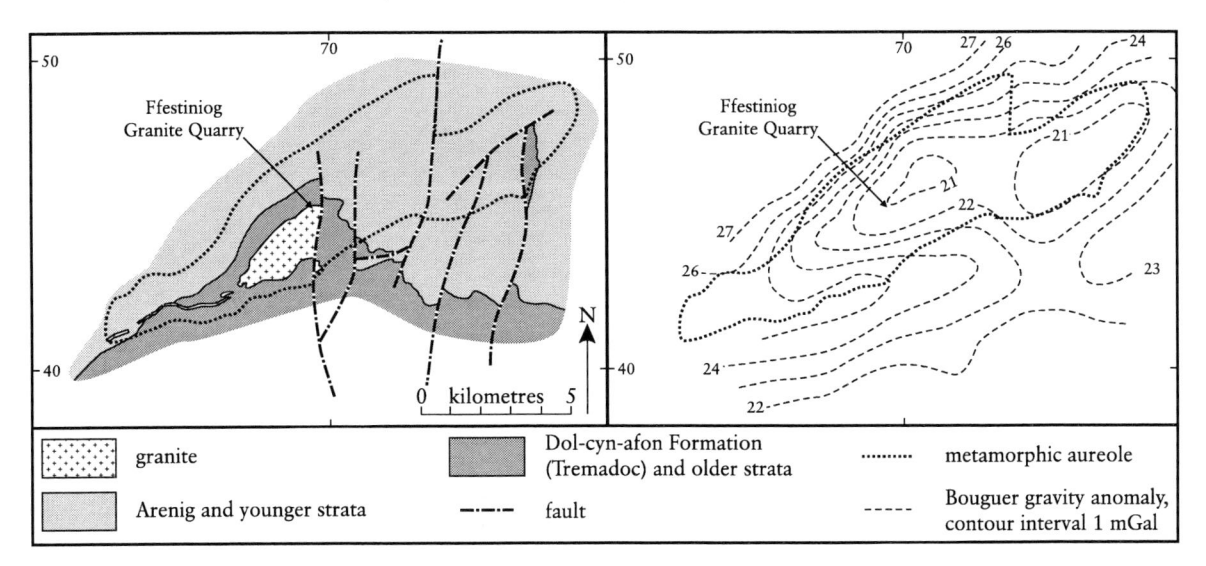

Figure 6.56 Map of the Ffestiniog area showing the surface outcrop and limit of metamorphic aureole of the Tan y Grisiau Granite and the associated Bouguer gravity anomaly.

Figure 6.57 The roof zone of the Tan y Grisiau granite intrusion, Ffestiniog Granite Quarry. (Photo: reproduced from Roberts, 1979.)

Description

The Ffestiniog Granite Quarry GCR site exposes the upper contact of the Tan y Grisiau Granite, here dipping 40° to the NW. The contact is clearly discordant (Figure 6.57) and separates Tremadoc siltstones and mudstones in the upper part of the quarry face from a distinctive roof facies of the granite.

In the lower quarry face (now largely obscured) the granite is a grey-green, homogeneous, fine-grained equigranular mosaic of plagioclase (albite–oligoclase), perthite and quartz with dark clots (0.5–1.0 cm in diameter) of chlorite after biotite. Common accessories include magnetite, zircon and allanite with traces of titanite, monazite, fluorite and epidote. In thin section, granophyric intergrowths of quartz and alkali feldspar are common and feldspars are altered to sericite. Bromley (1963) also recorded the presence of the blue-green amphibole ferrohastingsite. Towards the contact the granite becomes finer grained and vesicular, perthite is altered to muscovite, and plagioclase to aggregates of albite, quartz and calcite. This zone, heavily veined with graphic pegmatites and granite apophyses, contains rounded metasomatized xenoliths with whitish reaction rims (Bromley, 1964). Mineral assemblages in the marginal zone include biotite, almandine garnet and cordierite. Cavities, vugs and irregular thin pipes within this zone, and well exposed in a small quarry to the west of, and below, the main quarry (Roberts, 1979), contain allanite, pyrophyllite, quartz and traces of molybdenite.

The overlying country rocks belong to the Upper Sandstone Member of the Dol-cyn-afon Formation and comprise well-bedded medium- to fine-grained sandstones and siltstones. Immediately adjacent to the granite these strata are a very fine-grained, flinty, yellow-grey, unfoliated hornfels with irregular planar light and dark banding. In thin section a granoblastic texture of quartz with decussate muscovite is dotted with irregular dark blebs of chlorite and magnetite. Rare euhedral laths of sericite and chlorite may represent pseudomorphs after andalusite. Thin veins of albitic feldspar, often following early joints, and associated with enhanced levels of sericitic alteration in the adjacent wall rocks, are interpreted as former pathways for escaping volatiles.

At 10–15 m from the contact, the above rocks pass gradationally up into coarser grained, foliated, spotted hornfelses comprising the assemblage albite-epidote-chlorite-sericite-quartz. The spots, between 1 to 10 mm in diameter, consist of radial or concentric aggregates of sericite, quartz and penninitic chlorite with inclusions of fine-grained magnetite. A thin rim of leucoxene and microcrystalline chlorite often mantles the spots. Where the spots are weakly altered or deformed they are identifiable as pseudomorphs after porphyroblastic andalusite and less commonly cordierite. The presence of relict andalusite, cordierite and minor amounts of biotite and hornblende indicate original hornblende hornfels facies rocks subsequently retrogressed to the albite-epidote hornfels facies.

Interpretation

Compositionally, the Tan y Grisiau Granite has a rhyolitic to rhyodacitic trace element signature and Howells *et al.* (1991) noted a close comparison with the main phases of rhyolite dome emplacement in the Lower Rhyolitic Tuff Formation caldera. The granite appears to have been emplaced by stoping and contains a prominent marginal zone choked in places with partially digested xenoliths associated with pegmatite and tourmalinized breccias, and cut by veins of micrographic aplite and granophyre. Deformation within much of the granite is weak to non-existent and the lack of a penetrative foliation or mineral lineation supports a passive emplacement model. Locally, however, discrete narrow (1–2 m wide) zones of intense strain have been recorded (Smith, 1988). These zones, often associated with sheared dolerite dykes, are characterized by an intense foliation, a marked reduction in grain size and mylonitic fabrics; they indicate that the granite was deformed during the Caledonian Orogeny. Deformation is more clearly expressed in the overlying aureole rocks which are characterized by the development of thermal spots in sedimentary rocks up to 1 km stratigraphically above the granite roof. The spots, composed mainly of radial aggregates of sericite, quartz and chlorite, make excellent strain markers and have been studied by a number of workers (Bromley, 1963; Coward and Siddans, 1979; Smith, 1988; see review by Scott, 1992).

The extensive hornfels aureole and associated geophysical anomalies clearly show that the outcrop in the Tan y Grisiau area represents only a small part (less than 10%) of a large granite body (Figure 6.56). Subsurface, the granite has a sheet-like form, elongated NE–SW, with a NW-dipping roof. During emplacement and crystallization, streaming of silica-rich volatiles altered the original magmatic assemblage, producing a distinctive vesicular mineralized facies within the roof zone and retrogression of the surrounding hornfels. This autometasomatism is expressed elsewhere along the north-western margin of the granite with enhanced radiogenic values in areas of allanite mineralization and cavity development associated with tourmalinized breccias (Bromley, 1969).

The timing of emplacement of the Tan y Grisiau Granite has long been the subject of debate. On the basis of petrography, mineralization and deformation within the aureole, it has been assigned either to the Caradoc or to a late stage in the Caledonian Orogeny (see Bromley, 1969 for review). To the SW of the main outcrop, granophyric apophyses and sheets intrude Arenig strata, and transgress the mid-Caradoc unconformity (Smith *et al.*, 1995) and the disrupted strata within the Rhyd mélange (Bromley, 1969; Smith, 1988; Howells and Smith, 1997). The extent of the hornfels (thermal spotting) aureole (Bromley, 1963, 1969) further indicates that rocks up to Costonian–Harnagian in age (the Moelwyn Volcanic Formation) are affected whereas younger rocks are not. Thus, the granite is stratigraphically constrained to the Caradoc. This, together with its geochemical correlations with the Lower Rhyolitic Tuff Formation, suggests that it is a subvolcanic intrusion associated with the 2nd Eruptive Cycle of caldera volcanism in North Wales (see the Snowdon Massif GCR site report).

Magnetic studies by Piper *et al.* (1995) indicate that the tilt-adjusted remnance directions are pre-Silurian and that magnetization occurred after deformation in late Ordovician times. K-Ar isotopic determinations by Thomas *et al.* (1966) suggest an age of *c.* 408 Ma, and recent Rb-Sr determinations by Evans (1990) provide an age of 384 ± 10 Ma. In common with many other younger Caledonian granites in North Wales these are considered to be reset ages affected by low-grade metamorphism and deuteric alteration during the Acadian Event (Evans, 1991). The emplacement age is probably concordant with the Mynydd Mawr Granite (438 ± 4 Ma) and the Bwlch y Cwyion hornfels (454 ± 20 Ma) which are the only two North Wales granites currently known to have escaped isotopic resetting (Evans, 1991).

The granite is broadly parallel to the main Caradoc volcanic rift structure in northern Snowdonia (Howells *et al.*, 1991), and is on the northern margin of the Harlech Dome, so it may well have been focused along a pre-existing basement fracture (Smith, 1988).

Conclusions

The Tan y Grisiau Granite represents a large, elongate, subvolcanic intrusion within the Ordovician (Caradoc) marginal basin of Wales. The Ffestiniog Granite Quarry GCR site provides excellent exposures of the granite as well as preserving an important section through its heavily veined and mineralized roof zone and into the overlying hornfelsed sedimentary rocks.

PANDY
(SJ 195 362, 199 360 AND 197 356)

P.J. Brenchley

Introduction

The Pandy GCR site consists of three outcrop areas in the Ceiriog Valley near Pandy, Denbighshire, where three tuffs, belonging to the Cwm Clwyd, Swch Gorge and Pandy tuff formations, are well exposed. Two of these tuffs are products of explosive silicic volcanism, and comprise pumiceous ash-fall, ash-flow, and pyroclastic surge deposits, while the third is mainly composed of volcaniclastic sandstones, deposited at the coastal fringe of subaerial ash-flow tuffs. The tuffs clearly show a range of depositional processes. The three tuffs relate to activity at volcanic centres widely separated across North Wales, contemporaneous with the major caldera volcanism of Snowdonia (Howells *et al.*, 1991; Bevins *et al.*, 1992) (see, for example, the Snowdon Massif GCR site). The centres were apparently short-lived and produced discrete subaerial pyroclastic deposits within a predominantly marine succession, implying contemporaneous emergence in some way linked to the volcanism.

Research on the volcanic rocks in the Berwyn has occurred in two main phases. The first involved the mapping and preliminary description of the volcanic rocks. Ramsay (1866), in the Geological Survey Memoir, reported the continuity of 'ash bands' along the northern flank of the Berwyn, while subsequently the outcrops and petrography of the igneous rocks were described in several papers in the early part of the 20th century, most notably by Cope and Lomas (1904) and Cope (1910, 1915). Cope (1910) included the first attempt to reconstruct volcanic events from the sequence of deposits. Groom and Lake (1908) presented a detailed account of the Glyn Ceiriog area that included perceptive descriptions of the pyroclastic deposits, recognizing that the flinty felsitic parts of the Pandy Tuff had a vitroclastic texture and were not intrusive as others had thought previously. They also noted the 'bogen'-like texture of what is now recognized as a welded tuff. Resurvey by the Geological Survey of part of the eastern Berwyn area (Wedd *et al.*, 1927, 1929) showed errors in the previous maps of the Swch Gorge and Pandy tuffs and established their true

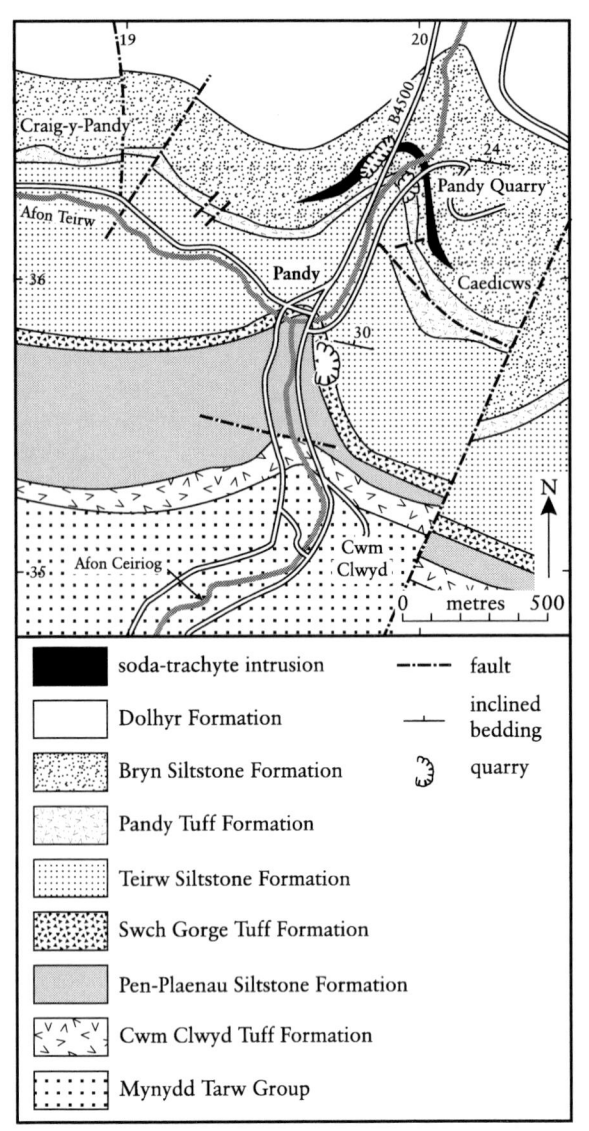

Figure 6.58 Map of the Pandy area.

continuity along the northern flanks of the Berwyn. Subsequently, in the second phase of investigation, detailed studies of the pyroclastic beds focused on the depositional processes and environments (Brenchley, 1964, 1969, 1972).

The Pandy GCR site is critical to understanding the Caradoc volcanic history of the Berwyn area of North Wales. It is particularly important because the pyroclastic deposits exhibit magnificently many of the characteristic features of the explosive silicic volcanism that was widespread across North Wales. The significant localities are within a small area, the tuffs are well exposed and are relatively unaffected by deformation or metamorphism, so that they retain remarkably clear details of primary vitroclastic textures. Additionally the contacts between the Pandy Tuff and the marine sedimentary rocks above and below, and the internal relationships between the constituent welded and unwelded ash-flow tuffs, contribute to an understanding of the nature and origin of transiently emergent volcanic islands. The site therefore contributes to a broader view of silicic volcanism in the context of the Ordovician marginal basin of Wales (Kokelaar *et al.*, 1984b).

Description

Strata of Caradoc age, including the three main tuff formations, crop out along the north flank of the Berwyn and are intersected by the Ceiriog Valley near Pandy, where the succession is well exposed on the valley sides (Figure 6.58). The Caradoc sedimentary strata that separate the tuff formations are predominantly silty mudstones with interbedded sandstones, interpreted as having formed in shallow subtidal environments in water depths of less than 25 m (Brenchley and Pickerill, 1980). The tuffs are probably all of Soudleyan age, though the Pandy Tuff could be Lower Longvillian (Brenchley, 1978).

The Cwm Clwyd Tuff is a sequence of bedded tuffs that are well exposed in crags on the eastern slopes of the Ceiriog Valley (Figure 6.58). It comprises alternations of thick-bedded lithic-crystal-pumice and thin-bedded pumiceous tuffs with partings typically 1–2 cm apart. The thick-bedded tuffs are tabular and may be massive, normally graded or, more rarely, inversely graded. Normal grading is most common in the coarser lithic lapilli-tuffs which show a reduction in grain size and increase in the amount of pumice towards their tops. The juvenile component of these tuffs is mainly pumice shreds and up to 12% quartz and 40% albite crystals, together with a variable content of accidental lithic clasts of pink rhyolite up to lapilli grade. The thin-bedded vitric tuffs are tabular, parallel-laminated or cross-laminated with dips of up to 10°. The vitric fraction in the tuffs is generally fine-grained tubular pumice, but bicuspate or tricuspate glass shards predominate in some beds. Accretionary lapilli occur in beds a few centimetres thick, interbedded within the sequence at several levels. The lapilli, which are flattened in the plane of the bedding, typically have a rim of fine vitric ash enclosing a pumice core. At the top of the sequence there is a massive unit of vitric-crystal tuff, 4 m thick, with irregular-shaped pumice lapilli weathered to form cavities. One laterally impersistent breccia containing blocks of tuff similar to the enclosing sequence is interpreted as a lahar deposit infilling a small channel and this, together with a few small channels containing a concentration of lithic clasts, is the only record of erosion and reworking of the tuffs.

The Swch Gorge Tuff, well exposed in quarries and crags on the eastern slopes of the Ceiriog Valley (Figure 6.58), comprises 40 m of volcaniclastic sandstones. The succession consists of thick-bedded, generally tabular sandstones up to 1 m thick that commonly appear massive although some show large-scale cross-stratification. There are thin shale partings between some beds, while large flakes of mudstone, some showing imbrication, are present in some of the sandstones. Desiccation cracks are known from a loose block. The sandstones are medium grained and are composed predominantly of feldspar and grains of volcanic rock with laths of feldspar, suggesting that they are from a lava source and not derived from the ash-flow tuffs that form the thicker Swch Gorge Tuff sequence farther west (Brenchley, 1969). Rare graptolites have been recorded from within the Swch Gorge Tuff at Pandy (Groom and Lake, 1908) and a varied trilobite–brachiopod fauna occurs in the thinner-bedded sandstones at the top of the sequence (Brenchley, 1978).

The Pandy Tuff is exposed in crags on both the eastern and western slopes of the Ceiriog Valley and in quarries at Caedicws, Craig-y-Pandy and near Pandy (Figure 6.58), where a particularly flinty facies of the tuff was exploited for the production of china. The sequence through the Pandy Tuff is best exposed in the Pandy Quarry where it consists of 6 m of massive lithic-crystal-vitric tuff overlain by 21 m of massive vitric-pumice lapilli-tuffs that have unflattened pumice in the basal metre, show flattened pumice fiamme and a eutaxitic texture through the succeeding 8 m, and pass transitionally into a further 12 m of unwelded tuffs (Brenchley, 1964). In the top few metres of the tuff there is a marked increase in the abundance of uncompacted pumice clasts. The overlying sedimentary rocks lie in erosional hollows in the tuffs and contain a shallow-marine brachiopod–bryozoan–bivalve fauna (Harper and Brenchley, 1993). The quarries at Caedicws and Craig-y-Pandy show a similar sequence, but detail is partly obscured by large silicic nodules. Columnar-jointing of the welded ash-flow tuff is

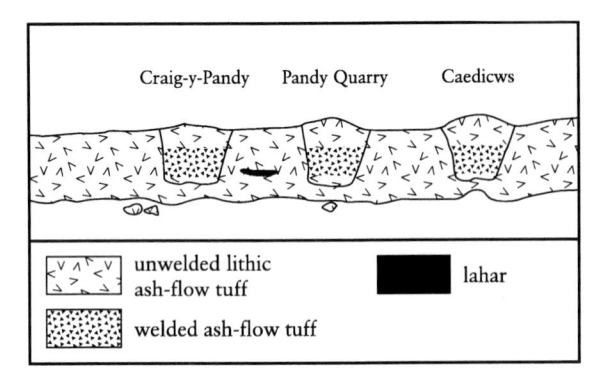

Figure 6.59 Relationship between welded and unwelded ash-flow tuffs in the Pandy area.

particularly well preserved at Caedicws.

In the crags laterally adjacent to the quarries and throughout most of the outcrop of the Pandy Tuff, welded tuffs are absent and the sequence consists of massive lithic-crystal-vitric tuffs, characteristically with coarse accidental lithic clasts at the base (coarse-tail grading) that range up to a few centimetres in diameter and protrude from the rock face. No internal divisions have been recognized within this development of the tuff, except in crags below the eastern end of Craig-y-Pandy where there is one horizon 6 m above the base where the clasts are concentrated to form a bed that thins laterally from 3 m to 4 cm over a distance of 12 m. The base of the Pandy Tuff is sharp but commonly irregular where the tuff has loaded into the underlying marine sediments, which locally contain detached blocks of the tuff several metres long. The marine mudstones extend upwards as tongues into the tuff, and sandstone beds are deformed into ball-and-pillow structures.

Interpretation

The site exhibits diverse pyroclastic deposits, formed by ash-fall and ash-flow processes representing accumulations related to separate volcanic centres in a generally marine, marginal basin setting. The Cwm Clwyd Tuff, with its predominance of fine pumice and beds of accretionary lapilli, records plinian or phreatoplinian eruptions that deposited more than 40 m of ash subaerially to a distance of at least 7 km from the volcanic centre. Beds have low-angle cross-stratification, in some instances occurring in stacked sets, implying deposition from currents. The low dip in most of the laminae and the variable direction of foreset dips suggest that the associated bedforms were low-amplitude, long-wavelength, sinuous dunes constructed by pyroclastic surges. The massive, thick-bedded lithic-crystal-pumice tuffs and the graded lithic-crystal-pumice lapilli-tuffs were interpreted as ash-fall tuffs by Brenchley (1972) because of their tabular nature and the vertical separation of the lithic clasts from the pumice in the graded beds. However, neither of these criteria is diagnostic and many of the beds could equally well have been deposited from pyroclastic surges. The range of bed types, including massive, inversely graded, graded and parallel-laminated or cross-laminated tuffs, could all be associated with downcurrent changes in the flow characteristics and depositional mechanisms of pyroclastic currents (Chough and Sohn, 1990). There is, however, a close relationship between ash-fall deposits and pyroclastic density currents, because ash columns may collapse (Cas and Wright, 1987) and evolve into different or varying concentrations and it is likely that the Cwm Clwyd Tuff records both ash-fall and pyroclastic surge deposits.

In contrast to the Cwm Clwyd Tuff, the Swch Gorge Tuff, in the Ceiriog Valley, is mainly reworked rather than primary. A high proportion of lithic grains of lava appear to have been derived from a source outside the immediate area, or from a local source that was entirely destroyed by erosion. Additionally there are a few thin interbedded marine mudstones. Laterally, towards the east, the volcaniclastic sandstones are interbedded with ash-flow tuffs and within 15 km the succession thickens and is wholly composed of ash-flow tuffs. The lateral association with welded ash-flow tuffs and the shallow subtidal sediments above and below (Brenchley and Pickerill, 1980) place the volcaniclastic sediments in a shallow-marine context. This is supported by the brachiopod–trilobite fauna at the top of the tuff that belonged to a *Dinorthis* community living in estimated water depths of less than 10 m (Pickerill and Brenchley, 1979). The presence of flat mudstone clasts within the sandstones and the record of desiccation cracks suggest an intertidal setting. Some of the thick-bedded sandstones are cross-stratified, suggesting that dune bedforms were present, but other beds are either massive or planar laminated and some are separated by a mud parting. The welded ash-flow tuffs of the Swch Gorge Tuff probably formed islands of low relief (Brenchley, 1969) and the volcaniclastic sandstones in the Ceiriog Valley appear to represent an intertidal coastal fringe with dunes and sand flats intermittently covered by mud layers.

The Pandy Tuff represents a third type of pyroclastic accumulation, which is almost wholly composed of ash-flow tuffs. Throughout most of the length of its outcrop the tuff is a massive unwelded ash-flow tuff with a distinctive content of accidental lithic lapilli. The presence locally of a lenticular laharic deposit within the tuff suggests that it was formed from more than one flow. The deformation of the underlying substrate shows that the tuff was deposited on unconsolidated marine sediments. The absence of any signs of marine reworking suggests that if the pyroclastic flow deposited its contents in a marine environment, accretion above sea level was very rapid. Alternatively the sea floor may have been uplifted tectonically.

The welded ash-flow tuffs, confined to three separate short lengths of outcrop, form a single cooling unit and have a concentration of pumice at their top reflecting segregation during transport and eventual deposition from the waning current. Their lateral margins are not exposed, but it appears that the welded ash-flow tuffs abut against the unwelded ash-flow tuffs (Figure 6.59). They are interpreted as ash-flow deposits that infilled canyons incised into the non-welded ash.

Conclusions

The Pandy GCR site incorporates three contrasting manifestations of silicic volcanism in a shallow-marine setting towards the edge of the Ordovician marginal basin of Wales. Each volcanic formation is both underlain and overlain by shallow-marine sedimentary rocks, but two of the three tuffs were deposited subaerially, the third on the coastal fringe of a volcanic island. The Cwm Clwyd Tuff was formed from the products of explosive steam- and gas-charged (plinian/phreatoplinian) eruptions. Deposition was probably from pyroclastic flows and by ash-fall, but the relative frequency of the two depositional processes is uncertain. The Swch Gorge Tuff is formed of volcaniclastic sandstones deposited in an intertidal environment with shallow-marine dunes and sand flats, on the coastal fringe of a volcanic island. The Pandy Tuff is composed of ash-flow tuffs that appear to have accumulated fast enough in a marine environment to form a volcanic island or were deposited on a tectonically uplifted surface. The initial sheets of unwelded tuff were dissected by canyons that became the conduits for subsequent ash-flows which formed welded tuffs.

The importance of the site is that it exhibits particularly well some important aspects of Ordovician (Caradoc) silicic volcanism in a small, accessible and well-exposed area. The site is particularly relevant to an understanding of the genesis of transiently emergent volcanic accumulations in general, and contributes more particularly to an overall understanding of silicic volcanic processes in the Ordovician marginal basin of Wales.

TRWYN-Y-GORLECH TO YR EIFL QUARRIES (SH 348 455–363 461)

T.P. Young and W. Gibbons

Introduction

The Trwyn-y-Gorlech to Yr Eifl quarries GCR site (Figure 6.60) encompasses a superbly exposed transect across the Garnfor Multiple Intrusion, one of the most interesting of a number of major Ordovician intermediate and acidic intrusions that crop out along the north limb of the Llŷn Syncline. An outer microgranodiorite is exposed on the coast at Trwyn-y-Gorlech in the west and in the Yr Eifl quarries to the east. The Yr Eifl quarries also expose the 'Blue Rock', a dark porphyritic hypersthene-bearing microgranodiorite. The centre of the GCR site, the mountainside of Garnfor, exposes an inner, microtonalitic intrusion that is also exposed in the smaller Garnfor Quarry.

Tremlett (1962) interpreted the field relationships of the Garnfor Intrusion as indicative of a 'Caledonian' age (end Silurian to Early Devonian). In a subsequent article (Tremlett, 1972), he described the geochemistry of these rocks, but his interpretation relied heavily on the distribution of elements now generally believed to be mobile during alteration and low-grade metamorphism (see Merriman *et al.*, 1986). Croudace (1982) revised the geochemical interpretation of these intrusions using 'immobile' elements and mineral chemistry, and demonstrated a genetic link between the granitoid intrusions and the Moelypenmaen lavas (see the Moelypenmaen GCR site report), thus establish-

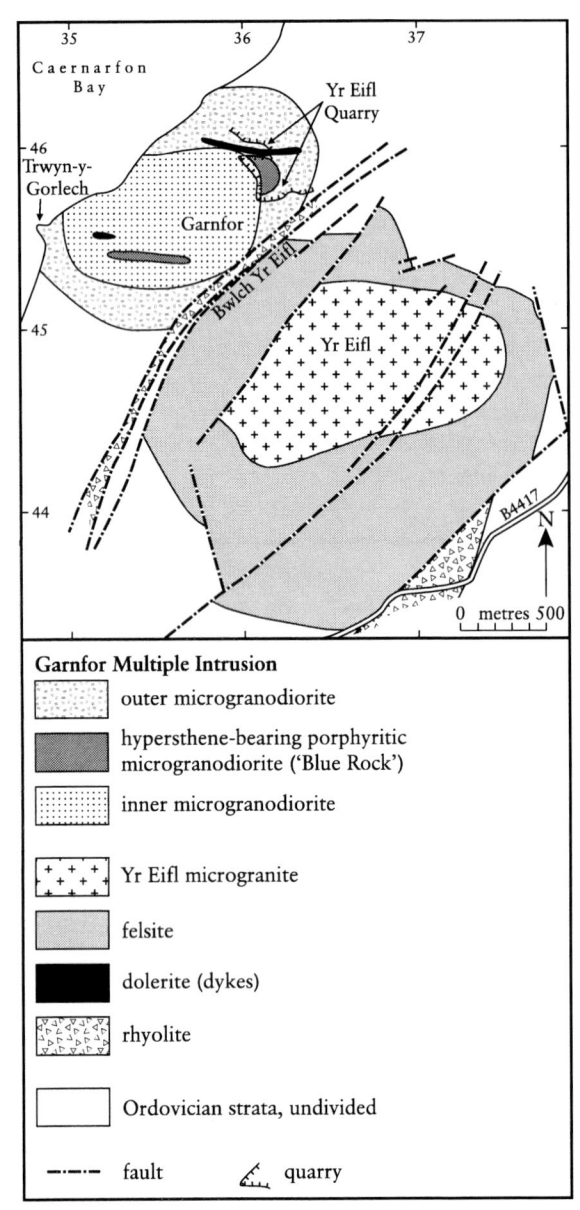

Figure 6.60 Map of the Garnfor Multiple Intrusion, north Llŷn (adapted from Tremlett, 1962).

ing the Ordovician age of the intrusions. Leat and Thorpe (1986) reworked the data of Croudace (1982) and argued that the intrusions form part of a peralkaline evolutionary series, together with the Moelypenmaen lavas.

Description

The Garnfor rocks comprise an outer intrusion of white or pink microgranodiorite and an inner intrusion of microtonalite of grey vitreous appearance, with a finer groundmass than the outer intrusion (both intrusions were described as 'granodiorite porphyry' by Tremlett, 1962). The inner intrusion becomes more mafic, with more enclaves towards its centre and shows a chilled margin at its contact with the outer intrusion. Both inner and outer intrusions are cut by smaller bodies of 'blue rock', a microgranodiorite similar to the outer intrusion but with a bluish-grey groundmass. These field relationships established the relative succession of the intrusions, which were subsequently cut by several dolerite dykes orientated NW–SE.

In petrographical detail, the outer intrusion is porphyritic with cumulophyric clots and individual phenocrysts of plagioclase showing chemical zonation from An_{58} to An_{28}, and commonly rimmed by orthoclase. The groundmass includes plagioclase with interstitial quartz, hornblende, biotite and magnetite. The inner intrusion shows similar core-to-rim plagioclase zonation (An_{48-20}), with a mafic mineral assemblage comprising magnetite and ilmenite, rare biotite, and scattered pyroxene crystals which become more abundant inwards.

Interpretation

Croudace (1982) and Leat and Thorpe (1986) referred the intrusions and volcanic rocks of Llŷn to the 'Lleyn volcanic complex'. Young *et al.* (in press) have suggested the presence of two distinct volcanic centres on Llŷn, one around Llanbedrog and the other farther north in the Nefyn district. The Llanbedrog centre, which produced the Llanbedrog Volcanic Group (Woolstonian), is largely restricted to the area south of the Efailnewydd Fault, whereas the products of the more northerly centre, which is slightly older (?Soudleyan–Longvillian), crop out along the northern limb of the Llŷn Syncline. Both volcanic sequences show an evolution from early trachybasaltic volcanism, through trachydacites, culminating in the eruption of rhyolitic tuffs. It would appear likely therefore that the evolution of the igneous rocks reported by Croudace (1982) and by Leat and Thorpe (1986) did not occur as a single event, but happened on two occasions. The distribution of the large granitoid intrusions, such as the Garnfor Multiple Intrusion, along the northern coast of Llŷn suggests a closer spatial relationship with the more northerly centre than with the younger Llanbedrog eruptive rocks.

Investigation of the geochemistry of the Garnfor intrusions by Croudace (1982) demonstrated that the rare-earth elements show a negative correlation with SiO_2, and it was argued that this was likely to have been produced through the fractionation of apatite. The analyses of the Garnfor intrusions plot in the field of trachydacite on the TAS (total alkalis vs silica) diagram and in the field of trachyandesite on the Zr/TiO_2 vs Nb/Y diagram (Winchester and Floyd, 1977). Tremlett (1962) argued that the range of compositions in the intrusions suggest that magma mixing or contamination was involved, with more basic material progressively contaminating the acid magma. Croudace (1982) demonstrated that the range of compositions was more likely to have been generated through fractional crystallization. Hence, the sequence of lithologies described by Tremlett (1962) derives from the progressive tapping of a magma chamber and the 'xenoliths' are in fact co-magmatic enclaves.

Conclusions

The well-exposed Garnfor Multiple Intrusion provides evidence for processes operating in an alkaline magma chamber. The geographical location of the intrusion suggests that it is likely to have been associated with a major centre of alkaline volcanism in northern Llŷn that is of probable Soudleyan–Longvillian age. As such it represents one of the best exposures in North Wales of an arc-related subvolcanic intrusion complex involving multiple intrusive events. It is especially interesting in revealing an acid-to-basic sequence of intrusive events, suggesting the progressive evacuation of a fractionating and possibly layered magma chamber at depth.

PENRHYN BODEILAS (SH 318 422)

T.P. Young and W. Gibbons

Introduction

The Penrhyn Bodeilas GCR site (Figure 6.61) preserves one of the best-exposed and most accessible of several intermediate to acid Ordovician intrusions that crop out east of Nefyn in northern Llŷn. The intrusion is particularly interesting in being rich in co-magmatic enclaves, and is considered to represent a subvolcanic intrusion.

Description

The Penrhyn Bodeilas Granodiorite Intrusion is exceptionally well exposed in coastal outcrops around the headland of the same name. The intrusion is a coarse-grained, greyish coloured granodiorite. It contains crystals of plagioclase 3–4 mm in length and of intermediate composition (An_{32}). Similarly sized clots of mafic minerals (hornblende, some clinopyroxene, chlorite and magnetite) occur together within a fine-grained (1 mm) groundmass, mostly of plagioclase together with quartz-feldspar intergrowths. Enclaves include examples of both basic and intermediate composition ('dolerite' and 'andesite' of Tremlett, 1962) (Figure 6.62). A late-stage, more evolved magmatic component is represented by thin aplitic veins, most of which are steep and strike NNE–SSW. Some of these late-stage aplites show chilled margins against the main body of the intrusion (Tremlett, 1962).

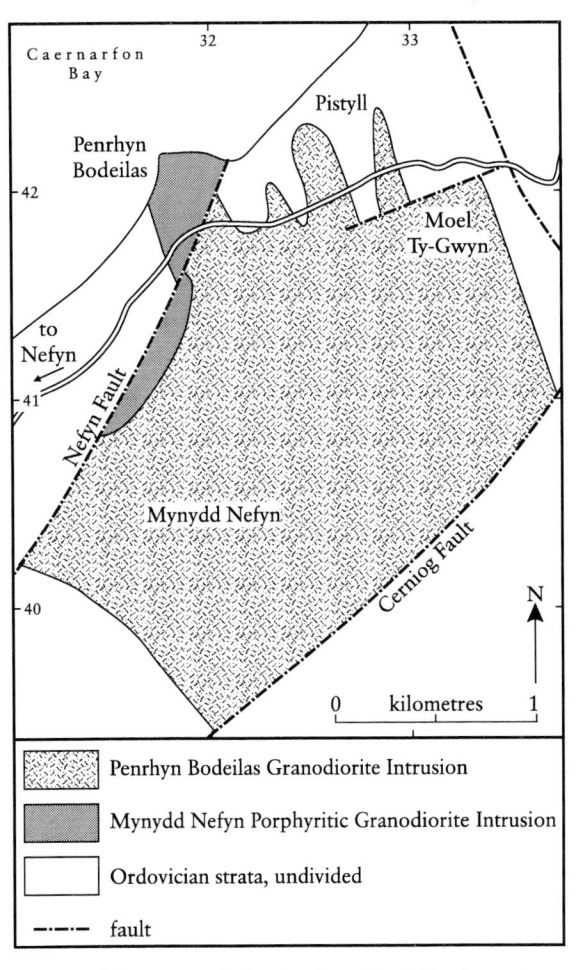

Figure 6.61 Map of the Penrhyn Bodeilas Intrusion, north Llŷn (adapted from Tremlett, 1962).

Figure 6.62 Co-magmatic mafic enclaves in the Penrhyn Bodeilas Intrusion, Penrhyn Bodeilas. (Photo: R.E. Bevins.)

Interpretation

The Penrhyn Bodeilas Granodiorite Intrusion is interpreted as one of a suite of Caradoc age subvolcanic intrusions. It remains unclear whether this intrusion was directly related to the magmatism associated with either of the two Caradoc age magmatic centres in Llŷn (Young *et al.*, in press). The geographical position of the intrusion is marginal to the area of distribution of the Llanbedrog Volcanic Group (Woolstonian), but it lies closer to the more northerly centre. Volcanic rocks from this northern centre include the Upper Lodge and Allt Fawr Rhyolitic Tuff formations both of which are interpreted as ?Soudleyan–Longvillian in age (Young *et al.*, in press). Analyses presented by Croudace (1982) plot in the trachyandesite field on the Zr/TiO_2 vs Nb/Y diagram (Winchester and Floyd, 1977).

Conclusions

The Penrhyn Bodeilas Granodiorite Intrusion is a well-exposed example of a high-level subvolcanic slightly alkaline intrusion belonging to one of the magmatic centres that developed on Llŷn in Caradoc times. It is of particular interest in containing a suite of abundant and compositionally variable co-magmatic enclaves, and a well-developed late aplitic facies preserved as a swarm of steeply inclined dykes.

MOELYPENMAEN (SH 338 386)

T.P. Young and W. Gibbons

Introduction

The excellent exposures of trachyandesitic rocks at Moelypenmaen (Figure 6.63) represent one of the best-preserved examples of intermediate igneous rocks produced during the widespread Caradoc magmatism in North Wales. Moelypenmaen is a low rocky hill on the northern limb of the Llŷn Syncline NW of Pwllheli. The majority of the rocks exposed on the hill are basaltic trachyandesites referred by Young *et al.* (in press) to the Penmaen Formation of the Llanbedrog Volcanic Group (not named after this locality, but after the hill of Penmaen on the western outskirts of Pwllheli). In addition, the southern side of the hill has small exposures of both the Foel Ddu Rhyodacite Formation and the base of the Carneddol Rhyolitic Tuff Formation, which are younger, more evolved components of the Llanbedrog Volcanic Group.

Matley and Heard (1930) described the succession at Moelypenmaen as 'oligoclase-keratophyre' and 'pyroxene-soda-trachyte' (corresponding respectively to the Penmaen and Foel Ddu formations of the current nomenclature). The rocks dip steeply to the north, and it was not until areas farther south in the Llŷn Syncline were mapped (Matley, 1938) that it was realized that the beds are overturned and actually young to the south. Tremlett (1962, 1969, 1972) described the dominant rock type at Moelypenmaen as andesite (part of his 'Main Andesitic Series').

Moelypenmaen is of greatest significance because of the geochemical studies undertaken by Tremlett (1969), Croudace (1982) and Leat and Thorpe (1986), which revealed the intermediate chemical character of these rocks. Intermediate rocks are comparatively rare among the products of the Caradoc igneous centres of central Snowdonia (see the Snowdon Massif and Cwm Idwal GCR site reports), but are volumetrically much more important in volcanic centres preserved in central and north-western Llŷn.

Description

The bulk of the succession exposed at Moelypenmaen comprises a massive unit of basaltic trachyandesites at least 200 m thick, the uppermost part of which shows amygdales orientated at a high angle to bedding. No evidence to suggest individual flow units within the body of these lavas has been recorded. The rocks have andesine and less abundant augite phenocrysts within a dark-green chloritic groundmass with flow-aligned plagioclase microlites; amygdales reach up to 1.3 mm and are mostly infilled with chlorite.

The southern part of the site includes a second basaltic trachyandesite flow, 30 m thick and separated from the main body of trachyandesites by approximately 50 m of sandstones, which are locally conglomeratic and fossiliferous. These are in turn overlain by the Foel Ddu Rhyodacite Formation (represented by small exposures of a thin, red-coloured flow-foliated lava) and the base of the crystal-rich tuffs of the Nant y Gledrydd Member of the Carneddol Rhyolitic Tuff Formation.

Mapping by Young *et al.* (in press and unpublished data) suggests that the Moelypenmaen locality is very close to the eastern limit of the distribution of the Llanbedrog Volcanic Group on the northern limb of the Llŷn Syncline. The Foel Ddu Rhyodacite Formation is reduced to only a few metres at Moelypenmaen, and the Penmaen Formation, despite being at least 200 m thick at Moelypenmaen, is absent at Pont Penprys, only 1200 m farther east. Tuffs of the Carneddol Rhyolitic Tuff Formation do continue to the east, although they are much reduced in thickness.

Interpretation

Tremlett (1962, 1969, 1972) proposed that most of the major granitoid intrusions of northern Llŷn are 'Caledonian' in age (end-Silurian to Early Devonian), largely on the basis of structural arguments. Croudace (1982) considered that the Moelypenmaen andesite lavas were generated by approximately 70% fractional crystallization of a primitive tholeiitic magma, and that the peralkaline microgranites and granophyres probably represent residual melts (less than 10%) of the same, or similar magmas. The implication of this interpretation is that the Llŷn granitoids must also be Ordovician in age. Leat and Thorpe (1986) refined the model further and argued a direct link between the Moelypenmaen rocks ('trachybasalts and probable mugearites' in their terminology) and the evolution of the peralkaline granitoids.

The bulk of the Moelypenmaen exposures are

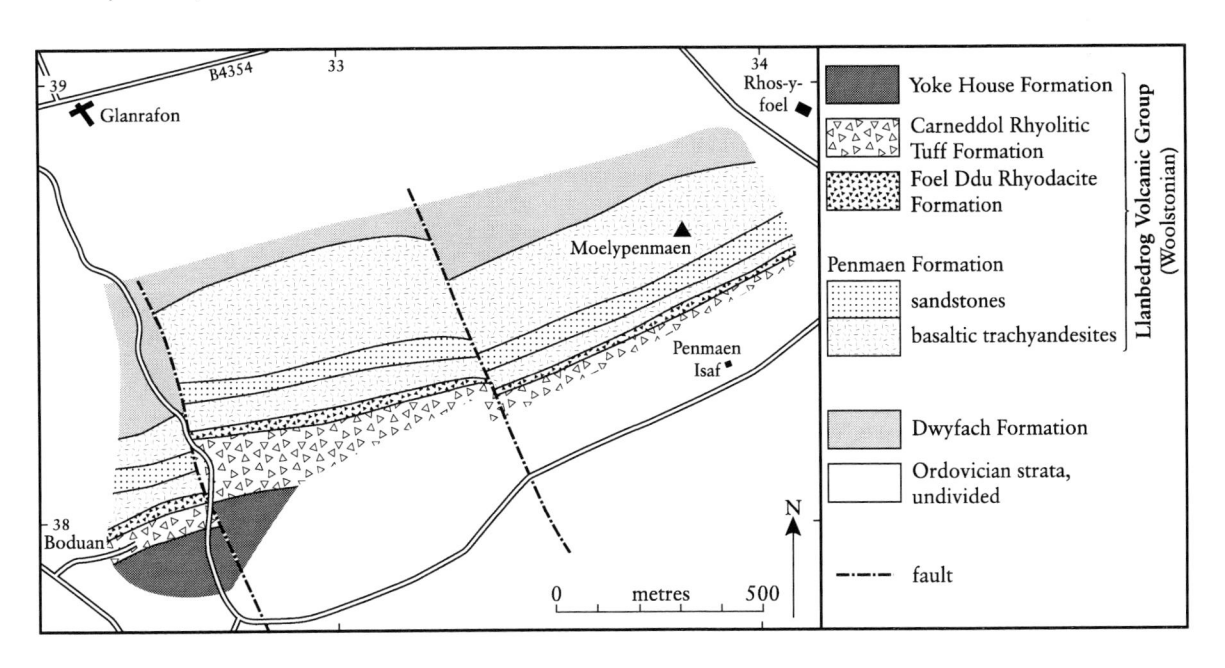

Figure 6.63 Map of the Moelypenmaen area, north Llŷn.

now recognized to represent the earlier, less evolved products of magma emanating from a major volcanic centre near Llanbedrog (Young *et al.*, in press). However this site is of additional interest in preserving more evolved representatives derived from the same centre, namely the Foel Ddu Rhyodacitic Formation and the Carneddol Rhyolitic Tuff Formation.

Conclusions

The well-exposed basaltic trachyandesites of Moelypenmaen are excellent representatives of intermediate rocks in the transitionally alkaline volcanic centres of Llŷn. Such rocks contrast with the subalkaline centres of central Snowdonia where intermediate rocks are comparatively rare. The existence of a range of overlying volcanic lithologies from andesite to rhyolite makes this site especially representative of the magmatic centre that produced the Llanbedrog Volcanic Group and associated intrusions.

LLANBEDROG (SH 337 307)

T.P. Young and W. Gibbons

Introduction

Llanbedrog Head preserves one of the best-exposed and topographically most prominent of a series of late Ordovician (Woolstonian) subvolcanic intrusions that crop out on Llŷn. The GCR site (Figure 6.64) occupies much of the area of Mynydd Tir-y-cwmwd (Llanbedrog Head), on the southern coast of Llŷn, to the west of Pwllheli. The microgranite has been quarried in the past from a series of quarries along the southern margin of the headland.

The Mynydd Tir-y-cwmwd Porphyritic Granophyric Microgranite (Young *et al.*, in press) is important for its close geochemical affinity with the Carneddol Rhyolitic Tuff Formation of the Llanbedrog Volcanic Group of central Llŷn. It is the only large subvolcanic intrusion on the southern limb of the Llŷn Syncline. For a long time, this intrusion has been believed to be associated with the local extrusive rocks; Matley (1938) described the Mynydd Tir-y-cwmwd intrusion as a subvolcanic plug of 'granite porphyry' and Fitch (1967) identified an unconformity within the local volcanic succession, which he attributed to the effects of the intrusion of the Mynydd Tir-y-cwmwd intrusion (his 'Llanbedrog Granophyre').

The intrusion is one of the few on Llŷn for which an Ordovician age has been accepted by almost all previous authors. Tremlett (1969) interpreted the Mynydd Tir-y-cwmwd Porphyritic Granophyric Microgranite as contemporary with the Caradoc volcanic succession, despite interpreting many other intrusions, such as the Garnfor Multiple Intrusion (see the Trwyn-y-Gorlech to Yr Eifl quarries GCR site report) as having a younger, end-Silurian to Early Devonian age. Croudace (1982) re-investigated the granitoid intrusions of Llŷn and interpreted them all as being of Ordovician age.

Description

The Mynydd Tir-y-cwmwd Porphyritic Granophyric Microgranite intrusion forms Llanbedrog Head, and is elliptical in plan (1500 m E–W and 900 m N–S). The intrusion has major radial and concentric joint sets which are well exposed in the quarries around the southern margin. The joint pattern suggests that the present level of erosion is close to the exhumed top of the intrusion. At one point in the SW of the intrusion a small fault downthrows part of the roof, so that baked mudrocks are exposed in contact with the microgranite. The country rocks (the 'Tal-y-fan argillites' of Fitch, 1967) are undivided mudstones of the Nant Ffrancon Subgroup (Young *et al.*, in press), which are probably of early Llanvirn age. Abundant, pendent didymograptid graptolites can be found in these cleaved mudstones close to the intrusion.

The microgranite is dominated by perthite phenocrysts (up to 4 mm) with lesser plagioclase (0.7 mm) and minor green biotite. Quartz is abundant in the equigranular, unfoliated groundmass, and abundant granophyric intergrowths commonly nucleate around feldspar phenocrysts.

Interpretation

Fitch (1967) elaborated on the earlier interpretation of the Llanbedrog area by Matley (1938), identifying an unconformity between his 'Mynytho Volcanic Group' and 'Llanbedrog Ignimbrite Group', which he attributed to the topographical effects of the intrusion of the

Llanbedrog

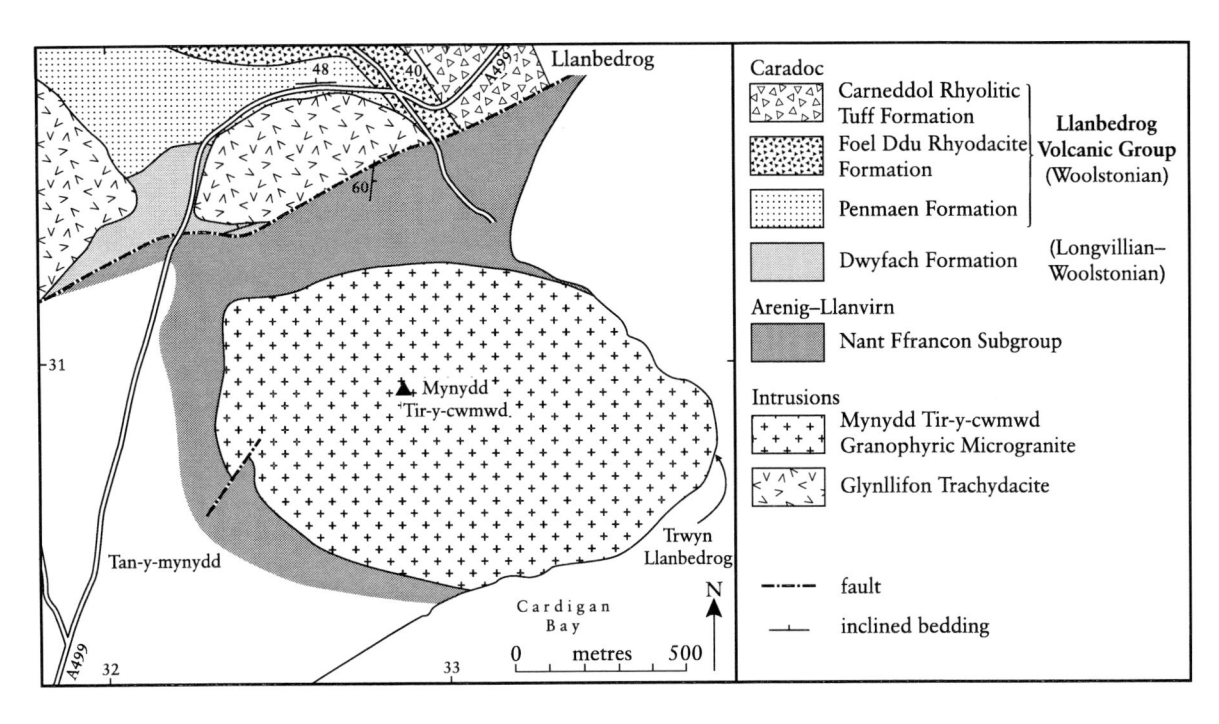

Figure 6.64 Map of the Llanbedrog area, south Llŷn.

Mynydd Tir-y-cwmwd intrusion (his 'Llanbedrog Granophyre'). The stratigraphical position of Fitch's unconformity is equivalent to the base of the Carneddol Rhyolitic Tuff Formation in the stratigraphical scheme of Young *et al.* (in press), although the remapping of the area did not prove the existence of an unconformity but instead emphasized the importance of folding and faulting to explain the present-day outcrops.

In addition to the geographical proximity there is a strong petrographical and geochemical case for linking the intrusion with the Carneddol Rhyolitic Tuff Formation. The perthite phenocrysts and granophyric intergrowths of the microgranite can be matched with similar components in the Carneddol Rhyolitic Tuff Formation, particularly in its lower part (the Nant y Gledrydd Member). Both the intrusion and the rhyolitic tuff are Zr-poor and show a closely similar trace element signature that is especially evident on plots of Y vs Zr, and Nb vs Th.

The Mynydd Tir-y-cwmwd intrusion lies close to the southern corner of a triangular region identified by Young *et al.* (in press) as the site of a late Ordovician 'Llanbedrog' volcanic centre. This centre is characterized by dramatic thickness variations in the Carneddol Rhyolitic Tuff Formation, the youngest component of the Llanbedrog Volcanic Group and the one which is genetically associated with the Mynydd Tir-y-cwmwd intrusion.

Conclusions

The Llanbedrog GCR site provides excellent exposures of the Mynydd Tir-y-cwmwd Porphyritic Granophyric Microgranite, interpreted as a subvolcanic intrusion associated with the volcanic centre from which the Carneddol Rhyolitic Tuff Formation of the Llanbedrog Volcanic Group was erupted in Caradoc (Woolstonian) time. Both petrographical and geochemical evidence strongly support a common source of magma for the intrusion and for the major ash-flow tuff eruptions from the Llanbedrog volcanic centre. Extensive exposures of the intrusion, particularly around the southern side of the headland, show good evidence for the chilling of the magma against the country rocks and for the joints produced by contraction of the solidifying magma. The intrusion presents excellent opportunities for the study of the emplacement of a high-level granitic intrusion paired with an adjacent, overlying, silicic ash-flow tuff.

FOEL GRON
(SH 301 309)

T.P. Young and W. Gibbons

Introduction

The Foel Gron Granophyric Microgranite is the most southern of the peralkaline intrusions of the Nanhoron Suite (Young *et al.*, in press) which crop out in central Llŷn, and is closely related to the Nanhoron Microgranite (see the Nanhoron GCR site report). The peralkaline nature of the Nanhoron Suite is extremely unusual among the Ordovician igneous rocks of North Wales and the Foel Gron intrusion is the most evolved component of that suite. It is interpreted as being associated with the Llanbedrog Volcanic Group and represents one of the most evolved lithologies associated with that alkaline igneous centre. The intrusion is elliptical in plan and is intruded into mudstones of the Nant Ffrancon Subgroup (Llanvirn).

The first field description of the Foel Gron intrusion was by Matley (1938), with subsequent descriptions and interpretations of its geochemistry by Tremlett (1972), Croudace (1982) and Young *et al.* (in press).

Description

The Nanhoron Suite comprises three strongly peralkaline intrusions, the Nanhoron Granophyric Microgranite, the Mynytho Common Riebeckite Microgranite and the Foel Gron Granophyric Microgranite (Figure 6.65). The fine-grained, pale microgranite of the Foel Gron intrusion is very slightly elliptical in plan (330 m N–S, 280 m E–W). It was initially referred to as an aplite (Matley, 1938), until Tremlett (1972) suggested that the term microgranite is more appropriate. It comprises an equigranular groundmass of anhedral quartz (0.4 mm), subhedral oligoclase (0.4 mm), subhedral alkali feldspar laths (up to 1.4 mm), altered biotite, and 1 mm clusters of quartz and alkali feldspar in granophyric intergrowths. No contacts with the country rocks are exposed, although cleaved mudstones of the upper part of the Nant Ffrancon Subgroup are exposed in close proximity to the microgranite at various points, but particularly on the NE side of the intrusion.

Interpretation

The elliptical outcrop of the Foel Gron Granophyric Microgranite led Matley (1938) to interpret it as a subvolcanic plug. The intrusion is one of several steeply inclined bodies distributed along a N–S line that has been interpreted by Young *et al.* (in press) as defining one margin of the Llanbedrog centre, which was active during Caradoc (Woolstonian) time.

Regional geochemical variations show increasing fractionation from north to south in the Nanhoron Suite. Rocks of the Foel Gron intrusion are geochemically the most evolved members of the group, with samples plotting in the comendite/pantellerite field of the Nb/Y vs Zr/TiO_2 diagram (Winchester and Floyd, 1977). They have very high concentrations of the incompatible elements Y, Zr, Th and Nb and the rare-earth elements (REE), with very low P_2O_5 contents. Chondrite-normalized REE data show steep profiles of light REE enrichment for the Nanhoron Suite, with the Foel Gron Granophyric Microgranite showing the most dramatic values. The steepening of the profiles may be due to zircon removal, which would preferentially deplete the heavy REE. The REE profiles show marked negative Eu anomalies, suggestive of extensive plagioclase fractionation, whereas the negative Ce anomalies are extreme and may be due to fractionation of monazite.

The relationship of the Foel Gron Granophyric Microgranite (and the Nanhoron Suite as a whole) to the other acidic components of the Llanbedrog volcanic centre is uncertain but interesting. The rocks associated with the centre show an evolutionary series from trachybasalts and trachyandesites through to trachydacites and rhyodacites, all showing progressive enrichment in incompatible elements. However, the more rhyolitic compositions show a marked division into Zr-depleted rocks (the Carneddol Rhyolitic Tuff Formation and the Mynydd Tir-y-cwmwd intrusion) and Zr-enriched rocks (the Nanhoron Suite). Whatever the reason for this striking geochemical subdivision, these Zr values emphasize the geochemically extremely fractionated nature of the Foel Gron Granophyric Microgranite.

Conclusions

The Foel Gron Granophyric Microgranite is the most evolved component of the Llanbedrog volcanic centre and is the most evolved Ordovician intrusion in North Wales. This intrusion is part of a suite, interpreted as having been emplaced along a major volcanotectonic structure close to, or defining, the western boundary of the Llanbedrog caldera margin. The existence of such a peralkaline rock in the area is an important demonstration of the alkaline nature of the Llanbedrog Volcanic Group, which is in strong contrast with the mostly subalkaline character of Ordovician igneous activity elsewhere in North Wales, especially in the eruptive centres of Snowdonia.

NANHORON QUARRY
(SH 287 329)

T.P. Young and W. Gibbons

Introduction

The Nanhoron Quarry GCR site preserves a rare exposure of the contact between the Nanhoron Granophyric Microgranite and its envelope of lower Ordovician sedimentary rocks. The intrusion is part of the Nanhoron Suite, a group of late Ordovician (Woolstonian) peralkaline intrusions aligned N–S along the western margin of the Llanbedrog volcanic centre. Nanhoron Quarry is a small working quarry, situated on the NW side of Nanhoron and just to the north of a major NE–SW fault (Figure 6.65). Although long-established, the quarry has expanded in recent years to satisfy an increased local demand for its product, providing new evidence concerning the contacts of the microgranite.

Description

The Nanhoron Granophyric Microgranite is exposed in Nanhoron Quarry (see Figure 6.65) and in a small quarry to the north, near Penbodlas. The host rocks are mudstones of the Nant Ffrancon Subgroup (Llanvirn). The microgranite is non-porphyritic and comprises anhedral quartz (up to 6 mm) and alkali feldspar (up to 1.0 mm), with less common subhedral oligoclase (up to 0.2 mm). A granophyric texture is variably overprinted by secondary alteration. The margin of the intrusion is fine grained, pervasively devitrified, weakly porphyritic (rare quartz, oligoclase and alkali

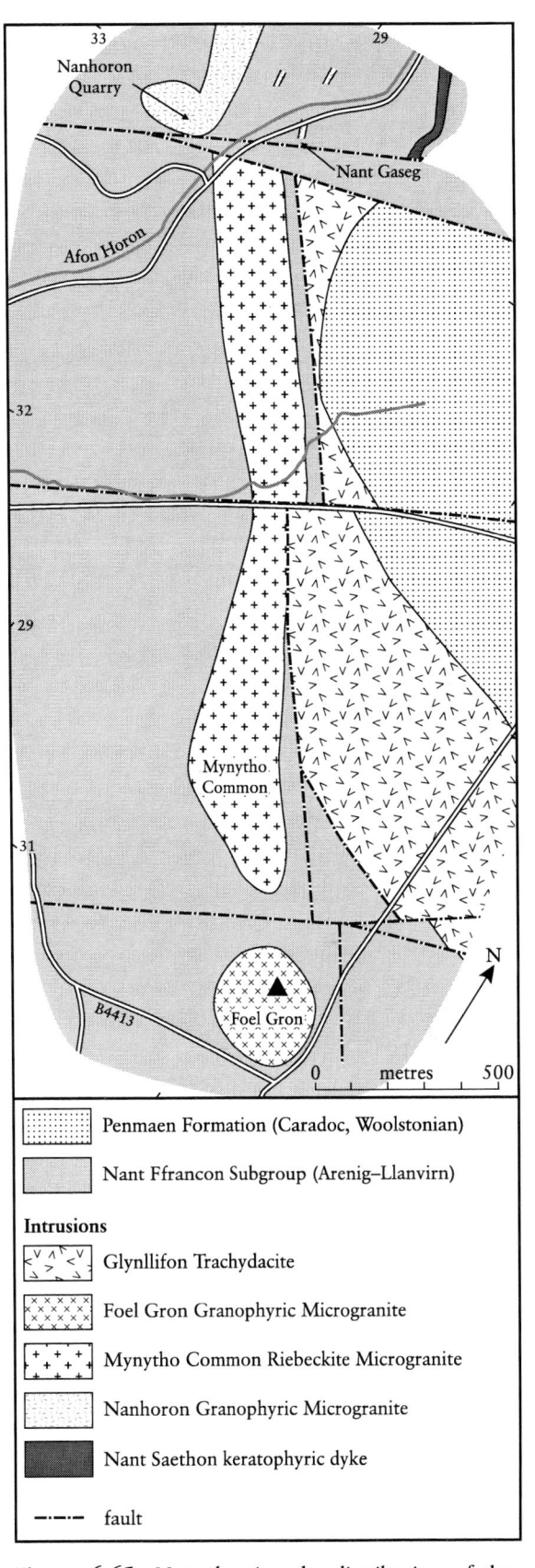

Figure 6.65 Map showing the distribution of the Nanhoron Suite of intrusions, south Llŷn.

feldspar phenocrysts up to 0.3 mm), and crowded with spherulites.

The main quarry provides excellent exposures of the core of the intrusion. In addition, a chilled margin can be traced running NW–SE close to the entrance to the main quarry and passing behind the south-west face. This margin can be followed into the lower, abandoned, section of the quarry, where it swings towards the north-east before apparently passing just to the east of the eastern face (see Figure 6.65). In the northern corner of the main quarry, a faulted contact with mudstones lies close to the centre of the arc defined by the chilled contact, suggesting that this faulted contact may constitute the core of a fold. Unfortunately the margins of the intrusion are not traceable outside the quarry, therefore the overall form of the intrusion is not known. Further exposures in the small quarry near Penbodlas suggest that the intrusion extends northwards, probably in continuity with the eastern limb of the folded structure in Nanhoron Quarry.

Interpretation

The geochemical features of the Nanhoron Suite are described in the account of the Foel Gron GCR site (p. 332). The Nanhoron Granophyric Microgranite is the least evolved component of the suite and is the most northern of a north–south line of cogenetic intrusions interpreted as defining the western margin of the Llanbedrog caldera.

Conclusions

Nanhoron Quarry provides a large fresh exposure of the Nanhoron Granophyric Microgranite, the least evolved member of the Nanhoron Suite of intrusions, exposed along the presumed western margin of the Llanbedrog volcanic centre on Llŷn. Old workings provide evidence for an original curving contact around the east of the main quarry. The quarry also shows faulted contacts of the intrusion against mudstones in its northern corner. These features present new data on structure in an otherwise very poorly exposed tract of ground and provide important constraints on the evolution of a major phase of alkaline igneous activity of Caradoc (Woolstonian) age.

MYNYDD PENARFYNYDD (SH 214 259–236 273)

T.P. Young and W. Gibbons

Introduction

The Mynydd Penarfynydd GCR site encompasses one of the best examples of a layered basic intrusion in the southern part of the British Caledonides. The site area centres on the prominent headland of Mynydd Penarfynydd, on the southern coast of Llŷn, which provides excellent coastal exposures through the intrusion (Figures 6.66 and 6.67). Much of the outcrop of the intrusion occurs within the GCR site boundary, but it continues inland to the north-east along the eastern side of Mynydd Rhiw for several kilometres, where it is only poorly exposed. The intrusion has great significance for the role it has played in influencing ideas on magma evolution during Ordovician igneous activity on Llŷn.

The Mynydd Penarfynydd Layered Intrusion was emplaced into Ordovician (Llanvirn) sedimentary rocks, close to the north-western margin of a sedimentary basin aligned along the Menai Straits Fault-zone. Most accounts have suggested that the intrusion is of Llanvirn age, pre-dating the major late Ordovician phase of igneous activity on Llŷn. Young *et al.* (in press), however, have proposed that the intrusion is associated with younger (Caradoc) Ordovician activity, being contemporary with the extrusive basaltic volcanism seen in the Nod Glas Formation.

Originally the Mynydd Penarfynydd intrusion was described as a greenstone (Sharpe, 1846; Ramsay, 1866; Tawney, 1880; Bonney, 1881; Teall, 1888). The earliest detailed description of the intrusion was provided by Harker (1888, 1889) and the area was subsequently mapped by Matley (1932). The intrusion was the subject of two major studies in the 1960s and 1970s giving rise to several publications on the geology and mineralogy by Hawkins (1965, 1970), and on the geochemistry by Cattermole (1976). More recent re-examination of the intrusion, including new geochemical data, has been undertaken during mapping of both the Aberdaron and Pwllheli 1:50 000 Geological Survey sheets (Gibbons and McCarroll, 1993; Young *et al.*, in press).

Mynydd Penarfynydd

Description

The Mynydd Penarfynydd Layered Intrusion is an easterly-dipping transgressive picritic and gabbroic sill within Llanvirn sedimentary rocks (Hawkins, 1970) of the Nant Ffrancon Sub-group. At Mynydd Penarfynydd, it is intruded into the undifferentiated mudstones of the sub-group, at least 200 m above the top of the Trygarn Formation, but it lies entirely within the Trygarn Formation to the north. On Mynydd y Craig the sill is over 150 m thick, and the intrusion thins gradually over a distance of 3 km to the north, before terminating abruptly NW of Tyddyn Corn.

The intrusion best demonstrates its layered nature on Mynydd Penarfynydd, within the GCR site area. The exposed section has been fully described by Hawkins (1970) and by Gibbons and McCarroll (1993). The lower 100 m of the sill is dominated by picrite (Figure 6.67). The base of the picrites shows a complex chilled zone about 10 m thick comprising a basaltic margin up to 2 m thick (Zone A of Hawkins, 1970) and about 8 m of fine-grained hornblende gabbros (Zone B of Hawkins). The thick picrites (Zone C of Hawkins) show a variable degree of layering. The picrites are overlain by banded leucogabbros, 9 m thick (Zone D of Hawkins). Hornblende olivine-gabbros, 13 m thick, (Zone E of Hawkins) appear to lie erosively on the leucogabbros, and are rich in augitic clinopyroxene and magnetite. Above this lies a unit of banded melagabbros (Figure 6.68). Higher in the intrusion, the rocks become progressively more differentiated, through a zone of secondarily-altered feldspathic ilmenitic gabbroic and dioritic lithologies, to the granophyric rocks which form the highest part of the exposed intrusion (Zones F, G, H and I of Hawkins). One small outcrop of the roof of the intrusion occurs in a small un-named cove (at 2226 2620).

On Mynydd y Graig the gabbro typically contains brown intercumulus hornblende, commonly enclosing augitic clinopyroxene, altered plagioclase and up to 10% opaque minerals (magnetite, ilmenite, pyrite and pyrrhotite). The texture and mineralogy are similar to the hornblende-cumulate gabbros (Zone F of Hawkins) exposed on Mynydd Penarfynydd, although layering is much less well developed. On Mynydd Rhiw (outside the GCR site) the intrusion is less well exposed, but shows a similar range of lithologies, although the gabbro is commonly

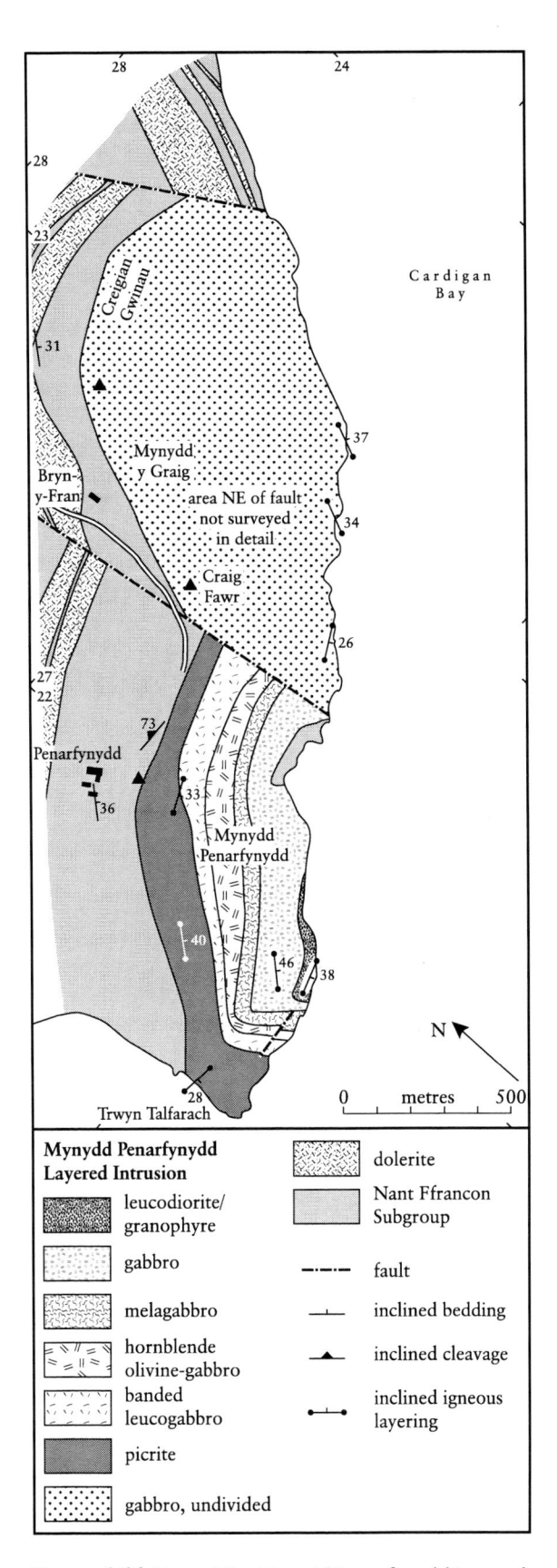

Figure 6.66 Map of the Mynydd Penarfynydd Layered Intrusion, south Llŷn.

Figure 6.67 Picrite within the Mynydd Penarfynydd Layered Intrusion, north of Trwyn Talarfach (SH 2152 2580) Weathering of the cumulate texture has produced the distinctive honeycomb pattern. (Photo: W. Gibbons.)

more pegmatitic. In the most northerly outcrops of the intrusion, Zone F probably constitutes the floor of the intrusion, and Hawkins (1970) suggested that the upper zones (G, H and I) are probably absent or not well developed.

Interpretation

Hawkins (1970) investigated the geochemistry of the intrusion and argued for an alkaline, rather than tholeiitic, character on the basis of the succession of rock types and the suite of minerals present. He did accept, however, that the absence of analcime and feldspathoids and the great range of rock types are features more associated with tholeiitic layered intrusions. Cattermole (1976) suggested that the intrusion was derived through fractional crystallization of a hydrated alkali olivine basaltic magma, with the marked geochemical variations produced by the intrusion of separate influxes of magma. Young *et al.* (in press) presented new geochemical data suggesting that the Mynydd Penarfynydd Layered Intrusion is subalkaline and tholeiitic in character, rather than alkaline.

Young *et al.* (in press) described the Mynydd Penarfynydd gabbros as characterized generally by very low concentrations of incompatible elements (Zr, Nb, Y) and high total Fe, MgO, Ni, Cr and V, features that are especially marked in the picritic cumulates, reflecting olivine and/or pyroxene accumulation. Two samples of melagabbro have relatively low Ni and Cr and markedly higher concentrations of TiO_2 and V, indicative of high modal proportions of an Fe-Ti oxide. On the Zr/TiO_2 vs Nb/Y diagram Mynydd Penarfynydd samples plot in the subalkaline basalt and andesite/basalt fields. Accordingly, Young *et al.* (in press) interpreted the intrusion as a suite of tholeiitic picrites and gabbros. The geochemical characteristics of the intrusion are in marked contrast to the transitionally alkaline Upper Lodge and Llanbedrog volcanic groups (see, for example, the Foel Gron and Trwyn-y-Gorlech to Yr Eifl quarries GCR site reports).

Mynydd Penarfynydd

Figure 6.68 Banded melagabbros from the Mynydd Penarfynydd Layered Intrusion at Trwyn Talfarach (SH 2173 2580). (Photo: W. Gibbons.)

The Mynydd Penarfynydd intrusion has commonly been interpreted as being of Llanvirn age. However, the style of intrusion contrasts markedly with the nearby dolerite and basalt sills. Although the latter are locally transgressive, they only attain a highest stratigraphical level of intrusion close to the top of the Trygarn Formation (Lower Llanvirn) and commonly exhibit features (fine-grained apophyses, peperitic textures and pillows) suggestive of high-level intrusion into wet, unconsolidated, sediments. It is intruded, at the south-western limit of exposure, at a level 200 m above the top of the Trygarn Formation, but shows no indication of a shallow level of intrusion. The Mynydd Penarfynydd intrusion has therefore been reinterpreted as being of significantly younger, Caradoc age (Gibbons and McCarroll, 1993; Young *et al.*, in press).

The subalkaline geochemical character distinguishes the Mynydd Penarfynydd intrusion from the major mid-Caradoc volcanic centres in western Llŷn, but it is similar to the final phase of Caradoc igneous activity in western Llŷn, which is seen to be associated with the deposition of the Nod Glas Formation, of *Dicranograptus clingani* Biozone age. The only other major basic intrusion in the area is the Carreg yr Imbill Intrusion at Pwllheli, and the general geochemical characters of the two intrusions are very similar. The position of the Carreg yr Imbill Intrusion on the line of the Efailnewydd Fault and the existence, at only slightly higher stratigraphical levels, of strongly transgressive tholeiitic minor intrusions associated with the Nod Glas Formation extrusive basaltic volcanism, argues in favour of the association of the Carreg yr Imbill Intrusion (and by extension the Mynydd Penarfynydd intrusion) with these late Caradoc basalts. However, the Nod Glas Formation is not preserved close to the Penarfynydd area, where the highest stratigraphical levels preserved are Llanvirn.

Conclusions

The Mynydd Penarfynydd Layered Intrusion is a classic example of a layered basic sill, one of only two Ordovician layered intrusions in Wales, with relatively easy access to the base of the sill exposed in sea cliffs. The superbly defined layering, and the range of rock types present, from picrites through gabbros to more differentiated, intermediate lithologies, makes this site one of the best exposures of intrusive igneous rock in the southern Caledonides of the British Isles. The age of the intrusion is not known with certainty, but it is now considered to be associated with the late Caradoc basaltic volcanism seen in the Nod Glas Formation on Llŷn. This period is of great significance for it marks a major change in the style and chemistry of volcanism after cessation of activity at the major volcanic centres at Llanbedrog and Snowdon.

SKOMER ISLAND
(SM 722 088–731 091, 727 086, 728 084, 738 092, AND 747 091)

R.E. Bevins

Introduction

Skomer Island, off the coast of Pembrokeshire, is composed almost entirely of volcanic and related rocks of Llandovery age (Figure 6.69). The volcanic rocks are chiefly basic in composition, with rarer intermediate and silicic rocks. They are mainly of extrusive origin, or relate to the intrusion of magma at a high level; a small proportion are pyroclastic. Rarer sedimentary rocks within the sequence reflect periods of volcanic quiescence. These sedimentary beds provide important palaeogeographical evidence for the setting in which the volcanism took place. The Skomer Island GCR site is of national importance in that it provides excellent exposures of the youngest major volcanic episode in the southern part of the British Caledonides.

The volcanic rocks are also exposed farther west on Grassholm, on the shoals known as the Hats and Barrels, and on the Smalls (Figure 6.69). To the east they are exposed on Middleholm (Midland Island) (Figure 6.70), on the Deer Park Peninsula, where crucial evidence for the age of the Skomer Volcanic Group is seen at Renney Slip (see the Deer Park GCR site report), and extending as far as St. Ishmael's. In all there is an E–W extent of some 43 km of the Skomer volcanic rocks. Clearly, this sequence represents the remnants of a major volcanic field which developed in the southern part of the Welsh Basin during early Silurian times.

The earliest, brief, accounts of the volcanic and related rocks of Skomer Island were presented by Howard and Small (1896a, 1896b, 1897), while the first major report was that by Thomas (1911). Further reference to the sequence appeared in the Geological Survey memoir to the district around Milford Haven (Cantrill *et al.*, 1916). In all of these accounts, the age of the Skomer volcanic rocks was thought to be Ordovician (Arenig). It was the detailed stratigraphical investigation of Ziegler *et al.* (1969) that established the true age of the sequence as Silurian (Llandovery); they re-examined the Skomer volcanic rocks and their associated sedimentary rocks, concluding that in fact they grade laterally and pass vertically into rocks of early Silurian age. Poorly preserved ostracodes from Middleholm (Midland Island) also argue against an Arenig age for the Skomer volcanic rocks, as does a consideration of structural relationships at Musselwick, on the Marloes Peninsula, where the Skomer volcanic rocks are apparently downthrown against sedimentary rocks of Llandeilo age (Ziegler *et al.*, 1969).

Thomas (1911) provided preliminary geochemical data for the Skomer volcanic rocks; these data were elaborated on by Ziegler *et al.* (1969) who provided, in particular, analyses of a unit termed the 'Skomer Ignimbrite'. Hughes (1977) and Fitton *et al.* (1982) presented additional geochemical data, while the most recent geochemical investigations are those of Thorpe *et al.* (1989). These last authors presented new major and trace element analyses for basic and acidic volcanic rocks from the area, concluding that in the main they are related by crystal fractionation. In addition, they reported that the silicic rocks could be divided into two unrelated groups, a high-Zr peralkaline group and a low-Zr group.

Description

Ziegler *et al.* (1969) adopted the term Skomer Volcanic Group for the thick volcanic succession exposed on Skomer Island and in the adjacent region, although no formal stratigraphical subdivision of the Skomer volcanic rocks has been established. In the absence of such, the sequence is described on the basis of field exposures, from the exposed base to the exposed top. Descriptions are predominantly of the

Skomer Island

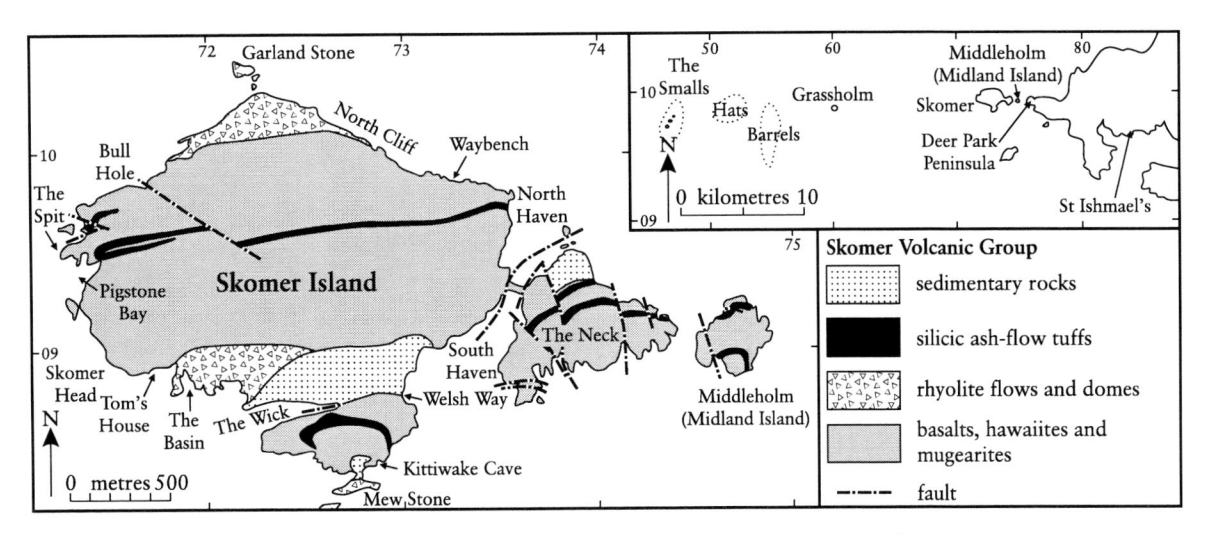

Figure 6.69 Map of Skomer Island (after Ziegler *et al.*, 1969).

coastal outcrops, in particular those along the eastern and southern cliffs, exposure inland being relatively scant and of inferior quality.

On Skomer Island, the effusive and tuffaceous rocks form up to 760 m of the exposed section, with interbedded sedimentary rocks representing an additional 140 m. The various units dip consistently to the SSE at around 20° to 30°. The oldest rocks exposed, therefore, are the silicic ash-flow tuffs of the Garland Stone and adjacent areas on the north coast, while the youngest are the rhyolites of the Mew Stone in the south.

The Garland Stone rocks were called 'soda-rhyolites' and 'soda-trachytes' by Thomas (1911). There is probably more than one unit in this section, with a composite thickness of around 130 m. In the field they are typically dark-grey in colour, locally show banding and are spherulitic in parts. They contain plagioclase feldspar (albite) crystals, along with minor Fe-Ti oxides, set in a fine-grained quartzo-feldspathic matrix. In places, shards, and pumiceous and

Figure 6.70 Oblique aerial view of Skomer Island from the NW, with Middleholm (Midland Island) and the Deer Park Penisula behind. Both islands are made up chiefly of basalts, hawaiites and mugearites of the Skomer Volcanic Group. (Photo: S. Howells.)

Figure 6.71 Spherulites (up to 10 cm across) in The Basin Rhyolite, Skomer Volcanic Group, The Basin, Skomer Island. (Photo: R.E. Bevins.)

lithic lapilli are present, set in a matrix which shows a well-developed eutaxitic texture.

To the south, as far as Tom's House on the west coast and South Haven on the east coast, a thick sequence (up to 385 m in total) of grey to greenish-grey hawaiite to mugearite lava flows is intruded by thin doleritic sheets. Both flows and sheets are particularly well exposed along the west coast of the island, for example in the vicinity of Skomer Head. The lavas form thin units, typically 5 m in thickness, separated by thin red scoriaceous layers. Very rarely, the flows show pillowed forms, for example on the north coast to the SE of the Garland Stone. In thin section, the lavas show variable degrees of alteration, some showing only very minor development of sericite in plagioclase coupled with groundmass recrystallization. They are sparsely porphyritic, with plagioclase, olivine, clinopyroxene and Fe-Ti microphenocrysts, set in a fine-grained groundmass chiefly composed of plagioclase and commonly showing a flow texture. Some flows show glomeroporphyritic clusters of olivine and plagioclase.

Within the above section thin layers of silicic rock occur, for example at Bull Hole, to the north of The Spit, and at Pigstone Bay. The layer at Pigstone Bay is up to 7.5 m thick at its maximum development; the rocks are fragmental, and contain randomly orientated pumice lapilli, whole and broken crystals and pseudomorphs after glass shards. Minor amounts of accessory minerals are present, including monazite and zircon. Fragmental, silicic rocks are also exposed on the eastern side of the island, in the vicinity of Waybench, which contain coarse lithic fragments, some of which are vesicular, associated with broken feldspar crystals.

To the east of Tom's House, extending as far as The Wick, a spectacular silicic sequence known as The Basin Rhyolite is excellently exposed. These rhyolitic rocks, which reach a maximum thickness of 77 m but thin dramatically over a distance of only 400 m, show magnificent flow-banding and flow-folding, but are perhaps best renowned for the extreme development of nodules, described originally by Rutley (1885b). The nodules, which show spherulitic forms, are best developed in the vicinity of The Basin, where they reach up to 25 cm in diameter and commonly occur in layers (Figure 6.71). The rhyolites show alternating light and dark bands on the millimetre scale, with small spherulites developed preferentially in the darker bands. In addition to the predominant spherulitic texture, the rhyolites sporadically show a snowflake texture. The rocks are almost entirely aphyric, with only very rare lithic lapilli and even rarer plagioclase microphenocrysts. Monazite and xenotime are present in very minor amounts, typically occurring as crystals 2–30 microns across.

To the east of The Basin, occupying the low-lying ground of The Wick eastwards to Welsh Way, is an intraformational sedimentary sequence described by Bridges (1976). To the south, and overlying these sedimentary rocks, are a further 74 m of hawaiite and mugearite lavas; a spectacular section through these lavas is seen to the south of The Wick. Within this succession of lavas is a distinctive red silicic tuff, up to 6 m thick, known as the Skomer Ignimbrite. This unit can be traced across the peninsula, cropping out on the east coast at Kittiwake Cove. It can then be traced across the Neck, onto Middleholm and as far as the Deer Park Peninsula, providing an important stratigraphical marker (Ziegler *et al.*, 1969). Microscopically, it shows elongate pumice fragments, lithic clasts and fiamme (up to 5 cm in length) in a recrystallized matrix with a well-developed eutaxitic texture.

The top of the section on Skomer Island is represented by the flow-banded silicic rocks of the Mew Stone. These rhyolites are up to 55 m thick, display excellent flow-folds on the northern face and show a crude columnar-jointing in the upper part of the exposed section. They are sparsely porphyritic and show well-developed perlitic textures, particularly at the base.

Interpretation

The Skomer volcanic sequence is dominated by relatively thin, uniform lava flows showing considerable lateral extent and reddened tops and bases. Ziegler *et al.* (1969) considered these to be subaerial flows, which were non-explosive in character, and did not result in the generation of any significant topography. Locally, however, lavas were either erupted under or emplaced into water, as evidenced by the rare occurrences of pillowed flows. Such flows are found near the top and at the base of the sequence and so there is no simple story of the gradual emergence or submergence of a volcanic island.

The silicic rocks show a variety of forms related to contrasting modes of eruption and emplacement. The rhyolitic lava exposed around The Basin thins dramatically over a short distance, and is thought to represent the remnants of a steep-sided extrusive flow-banded and flow-folded obsidian dome. The rhyolitic rocks exposed in the south, at the Mew Stone, are also thought to represent a thick, extrusive flow, although whether it represents a single flow or a compound set of flows is not certain. In contrast, the silicic tuffs exposed, for example, at Pigstone Bay and on the headland to the south of The Wick are relatively thin (up to 7.5 m maximum thickness) but are laterally extensive. These tuffs are recognized as being ash-flow tuffs, and show both welded and non-welded varieties. The silicic rocks exposed in the north of the island, in the vicinity of the Garland Stone, appear composite and possibly comprise both rhyolitic lava flows and ash-flow tuffs. All of the silicic rocks were apparently emplaced in a subaerial environment, although Ziegler *et al.* (1969) considered that The Basin Rhyolite formed a volcanic island which was subject to erosion in a coastal environment.

Thorpe *et al.* (1989) presented the most recent geochemical data for basic, intermediate and silicic volcanic rocks from the Skomer Volcanic Group. They concluded that the basic to intermediate lavas are hawaiites and mugearites belonging to an alkaline series. Two groups of rhyolites were discriminated, however, a low-Zr group and a high-Zr (peralkaline) group. Silicic rocks from Pigstone Bay and The Basin Rhyolite belong to the low-Zr group, while silicic rocks from the Garland Stone and the Skomer Ignimbrite belong to the high-Zr group. Thorpe *et al.* (1989) considered that rocks of the high-Zr group were derived from the hawaiites and mugearites as a result of low-pressure fractional crystallization, and relate to a basalt–hawaiite–mugearite–comendite series such as is seen in within-plate oceanic and continental settings. The low-Zr group, however, they considered to be unrelated to the other volcanic rocks exposed in the area, although recent unpublished work contradicts that view, linking their generation to the high-Zr group by crystal fractionation involving minor mineral phases, in particular monazite and xenotime (R.E. Bevins and G.J. Lees, unpublished data). The consensus is that the parental magmas were derived from a within-plate ocean-island mantle source which had been modified by earlier subduction-related events.

Conclusions

The Skomer Island site is of national importance in that it provides excellent exposures of the youngest major volcanic episode in the southern part of the British Caledonides. The volcanic rocks show geochemical features which suggest that their source rocks were influenced by earlier Caledonian subduction events.

A range of volcanic rocks, ranging from basic through intermediate to acidic compositions, are excellently exposed on Skomer Island, especially in the rugged cliffed coastline. The basic to intermediate rocks originated chiefly as subaerial flows, although rare pillowed flows show the local occurrence of subaqueous flows. The silicic rocks appear to be entirely of subaerial origin, and were generated in part as extrusive flows resulting in steep-sided domes, and in part from explosive eruptions, leading to the generation of ash-flow tuffs. The age of these various volcanic rocks is provided by their relationship to fossil-bearing sedimentary rocks on the mainland, which indicates a Llandovery age.

Geochemically, two distinct groups of silicic rocks have been determined, a low-Zr group and a high-Zr group. The latter are thought to be related to the basic to intermediate lavas through crystal fractionation, while the low-Zr group rocks have been considered to be unrelated to any of the other volcanic rocks exposed, although recent work suggests a link to the high-Zr group through crystal fractionation. The volcanic rocks are alkaline in character, and were apparently derived from a within-plate mantle source.

DEER PARK
(SM 756 091–760 088)

R.E. Bevins

Introduction

Lavas, pyroclastic rocks and high-level intrusive rocks of the Skomer Volcanic Group, of Llandovery age, crop out intermittently over 40 km of a near-strike section in south Pembrokeshire. Offshore, the volcanic rocks form The Smalls in the west, the reef known as the Hats and Barrels, and Grassholm, as well as much of Skomer Island, and Middleholm (Midland Island); onshore, they crop out across the Marloes Peninsula as far east as St. Ishmael's (Figure 6.69), although exposures are scattered and poor. The thickest and best exposed development of the volcanic rocks is seen on Skomer Island (see the Skomer Island GCR site report), while exposures in the rugged coastline of the Deer Park Peninsula, at the Deer Park GCR site, provide critical evidence for the age of the group, which represents the most important episode of Silurian volcanism in the southern Caledonides of the British Isles. Broad details of the Skomer Volcanic Group are described in the Skomer Island GCR site description.

Volcanic and related rocks are well exposed all around the coast of the headland known as Deer Park, from Martin's Haven to Renney Slip (Figure 6.72). Access, however, is for the most part difficult; the easiest access is in the vicinity of Wooltack Point and at Renney Slip.

Description

Basic to intermediate lavas are the most common volcanic rocks at Deer Park, as indeed they are on Skomer Island. Some 45 m of thin (c. 5 m-thick), generally massive flows crop out on the headland, in places showing reddened tops and bases. These flows have been correlated with those exposed on Skomer Island in the vicinity of North Haven and South Haven (Ziegler *et al.*, 1969), and are chiefly hawaiites and mugearites (Thorpe *et al.*, 1989). Locally, as at Jeffry's Haven, the lavas are pillowed. Petrographically, they are identical to the lavas exposed on Skomer Island, being fine grained, greenish-grey in outcrop and commonly vesicular. Under the microscope they are seen to be sparsely porphyritic, containing plagioclase, clinopyroxene, Fe-Ti oxide and rare olivine microphenocrysts in a fine-grained matrix dominated by plagioclase microlites, some showing flow alignment.

Silicic rocks are rare in the Deer Park section. However, a 5.5 m-thick ash-flow tuff exposed at Jeffry's Haven, has been correlated with the 'Skomer Ignimbrite' which crops out at the southern end of the Skomer Island section (see the Skomer Island GCR site report) and on Middleholm (Midland Island). This ash-flow tuff forms an important stratigraphical marker within the Skomer Volcanic Group. Ziegler *et al.* (1969) presented major element chemical analyses of the ash-flow tuff from this locality, but no trace element analyses are available; hence it is not possible to ascribe this unit to either of the

Deer Park

two silicic groups identified by Thorpe *et al.* (1989).

The key section for establishing the age of the Skomer Volcanic Group occurs to the south of Anvil Bay, although much of the sequence in this area is made up of sedimentary rocks. Here a 7 m-thick basalt flow, which is vesicular both at its base and top, is exposed at Limpet Rocks and extends eastwards across the promontory dividing Anvil Bay from Renney Slip. The sequence is much faulted around the bay; sedimentary beds immediately overlying the basalt are best exposed in the steep dip faces forming the north side of Renney Slip. Faunas contained within this sedimentary sequence are indicative of an early Upper Llandovery (late Aeronian, C_{1-2}) age (Walmsley and Bassett, 1976; Bassett, 1982). The unconformity at the top of the Skomer Volcanic Group is exposed on the SE side of Renney Slip.

Interpretation

The basic to intermediate lavas exposed across the site area are lateral equivalents of the lavas exposed on Skomer Island, and form a part of the Skomer Volcanic Group. Predominantly they were erupted in a subaerial environment, although locally they were either erupted or emplaced subaqueously. Silicic rocks are rare across the site area, restricted to the single occurrence of an ash-flow tuff to the south of Jeffry's Haven. This unit has been correlated with similar rocks exposed towards the top of the sequence exposed on Skomer Island; however recent work has identified a number of ash-flow tuffs on Skomer Island, at various stratigraphical levels and with contrasting chemistries (R.E. Bevins and G.J. Lees, unpublished data), and hence the correlation is in need of re-appraisal.

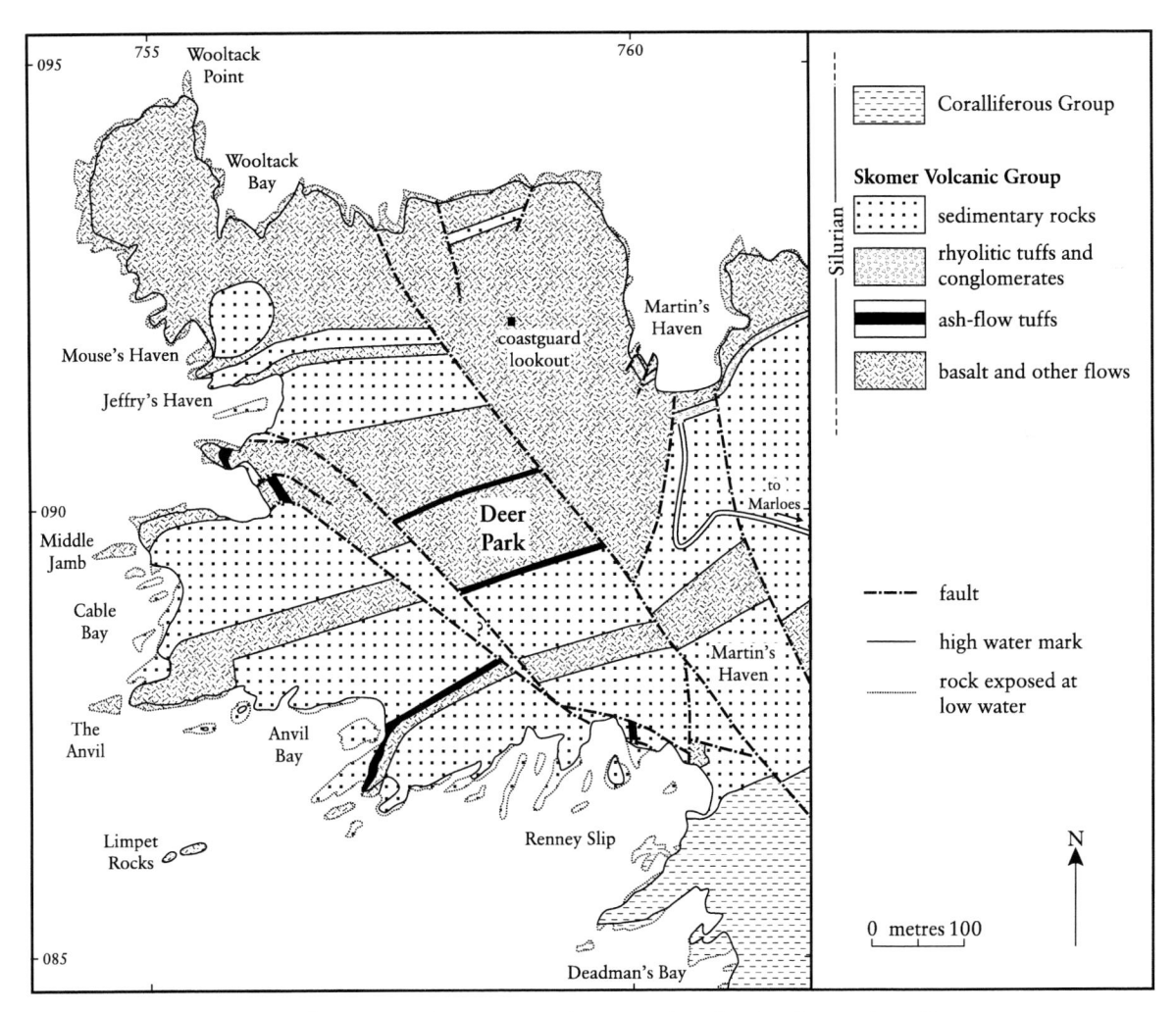

Figure 6.72 Map of the Deer Park Peninsula (after Ziegler *et al.*, 1969).

The volcanic rocks of the Skomer Volcanic Group are part of a basalt–hawaiite–mugearite–comendite series. Thorpe *et al.* (1989) presented the most recent interpretation of the geochemistry of the volcanic rocks of the group (see the Skomer Island GCR site report).

Conclusions

The Deer Park site exposes lavas and pyroclastic rocks of the Skomer Volcanic Group, along with associated sedimentary rocks. This sequence represents the most important episode of Silurian volcanism in the southern part of the British Caledonides. The Skomer Volcanic Group is best exposed on Skomer Island itself (see the Skomer Island GCR site report); however, coastal outcrops in the Deer Park area provide critical evidence for the true age of activity, based on fossils indicative of the Upper Llandovery. This volcanic sequence therefore represents the youngest volcanic episode of importance in the southern Caledonides of the British Isles.

Chapter 7

Late Ordovician to mid-Silurian alkaline intrusions of the North-west Highlands of Scotland

Introduction

INTRODUCTION

I. Parsons

A group of small alkaline igneous plutonic complexes and a suite of associated dykes and sills of late Caledonian age occur in a belt roughly parallel to the Moine Thrust in the NW Highlands of Scotland (Figure 7.1). This alkaline magmatism represents the NW edge of the otherwise overwhelmingly calc-alkaline Caledonian Igneous Province north of the Highland Boundary Fault. Radiometric ages show that the alkaline activity was protracted, spanning the period 456–426 Ma (Table 7.1). The rocks were thus emplaced concurrently with the earlier members of the 'newer granite' suites, although most of these intrusions are a little younger, with a peak of activity around 410–400 Ma (Brown, 1991). The earliest alkaline pluton, the Glen Dessarry syenite (Richardson, 1968), was penetratively deformed by late orogenic events, while the younger intrusions are deformed only locally. The Loch Loyal intrusions were emplaced after the regional metamorphism of the enclosing Moine and interleaved Lewisian rocks. The intrusions in Assynt occur in all the thrust sheets, including a minor development in the Moine, and also in the unmoved Foreland, the only incursion of Caledonian magmatism into that structural unit. The reader must bear in mind that what is now a relatively compact area of alkaline rocks in Assynt includes rocks which must have been emplaced in a belt extending perhaps of the order of 50 km towards what is now the ESE, that has been shortened by movements on the lower thrusts.

Alkaline magmas were available during several other phases of igneous activity in the British Isles (for example in the Carboniferous magmatism in the Midland Valley, and in the British Tertiary Igneous Province) but no highly evolved plutons occur. Many of the rock types in the NW Highlands are therefore unique in a British context, and all are rare in a world context. The ultrapotassic rocks of the Loch Borralan intrusion (Figure 7.1), in particular, have extreme compositions matched at only a handful of localities worldwide. Like most alkaline suites, that of the Scottish Caledonian Province has its own distinctive character. Reviews of the alkaline suite have been provided by Sutherland (1982) and Brown (1991).

Because of the early recognition of their

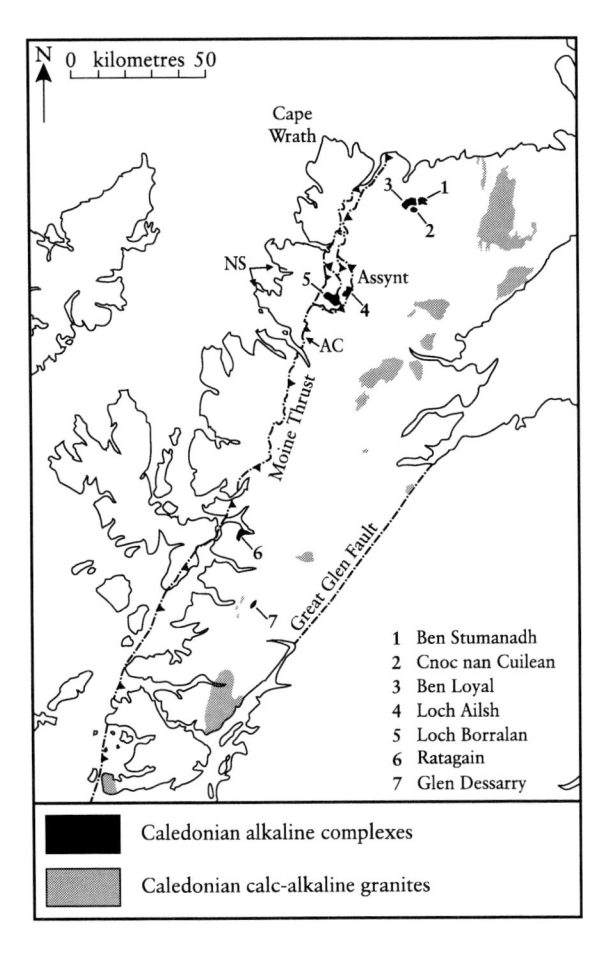

Figure 7.1 Map of NW Scotland showing localities of alkaline intrusions, aligned roughly parallel to the Moine Thrust. Many alkaline dykes and sills occur in the Assynt district and also near Ullapool in the Achall Culmination (AC). GCR sites exemplifying nepheline-syenite dykes in the Foreland are indicated by NS. Caledonian calc-alkaline granites NW of the Great Glen are also shown. The Ratagain intrusion is largely calc-alkaline in character but has minor syenitic members (after Halliday *et al.*, 1987, fig. 1).

unusual mineralogy, the Caledonian alkaline rocks were prominent in the evolution of ideas in igneous petrology in the early part of the 20th century. The province provided type localities of many rock types named by these early workers (Table 7.2), some of which have remained in widespread use. Adoption of a modern terminology for the suite makes much of this classic early work very hard to follow and in the text the old names have been used, but with their newer equivalents, and rather clumsy mineral-based names, added where appropriate. Credit for the first chemical recognition of the alkaline character of the rocks appears to be due to Heddle

Table 7.1 Inter-relationship of alkaline igneous activity and major tectonic events in the Moine thrust zone (after Halliday et al., 1987).

FORELAND	SOLE THRUST SHEET	BEN MORE NAPPE	MOINE NAPPE	AGE (Ma)
Peak of illite metamorphism in Foreland sediments				c. 408[1]
			Ross of Mull Granite cuts Moine thrust plane	414±4[2]
Nepheline-syenite dykes	*Late undeformed pegmatites in Loch Borralan*		*Cnoc-nan-Cùilean intrusion*	426±9[5]
	Penetrative deformation of pseudoleucite rocks at Loch Borralan. Crush Zones in quartz-syenites		*Final movements on the MTP*	
		Late crushing in Loch Ailsh		
		'Nordmarkite' sills near the MTP		
		Loch Borralan intrusion		430±4[3]
Canisp Porphyry	Main movements on the STP, folding BMTP?			
		Main movements on the BMTP	Moine mylonites and 'D1' Main movements on MTP	
		'Grorudite' dykes		
		Mylonites and greenschist-facies metamorphism in Loch Ailsh		
		Sgonnan Mór folds and fabric		
		Loch Ailsh intrusion		439±4[5]
	'Hornblende-porphyrite' and vogesite sills and dykes			
			'D3' of Glen Dessarry Moine. Deformation of syenite	
			Glen Dessarry intrusion	456±5[4]

Events in italic were essentially synchronous. MTP: Moine thrust plane. BMTP: Ben More thrust plane. STP: Sole thrust plane. The radiometric ages are from the following sources: 1. Johnson et al., 1985. 2. Halliday et al., 1979a. 3. Van Breemen et al., 1979a. 4. Van Breeman et al., 1979b. 5. Halliday et al., 1987.

(1883a; see Teall, 1900, p. 26) who analysed an albitite from Assynt. Murchison and Cunningham made earlier references to 'syenite' and described Ben Loyal (Ben Laoghal) in some detail (see Heddle, 1883b). These early workers regarded syenite as differing from granite only by 'a mineralogical accident'. The first descriptions of the alkaline rocks with a modern ring are those of Horne and Teall (1892) and Teall (1900), who introduced the name 'borolanite' for the exotic melanite-garnet nepheline-pseudoleucite-syenites for which the Loch Borralan intrusion is most famous.

For more than thirty years, Shand (1906–1939) maintained the province at the forefront of the developing science of igneous petrology by his introduction of the important concept of 'silica saturation' and his assertion that the silica-undersaturated character of some alkaline rocks (in his view a good example being the Loch Borralan intrusion) was a result of the extraction of silica from granitic magmas by reactions with limestones, precipitating calcium silicate minerals and releasing carbon dioxide. The idea was taken up by Daly (1914) and became known as the 'Daly–Shand hypothesis'. The hypothesis has fallen out of favour, not least because it is now recognized that the carbonate rocks often associated with nepheline-syenites are themselves intrusive igneous carbonatites. The discovery, as recently as 1988 (Young et al., 1994), of a carbonatite body slightly outside the

Table 7.2 Glossary of uncommon or varietal rock names employed for members of the alkaline suite in the NW Highlands.

Rock name	First use in NW Highland's literature	Modern equivalent(s)	Petrography and mineralogy	Comments
Assyntite	LB. Shand (1910) NW of Cnoc na Sroine	Sodalite nepheline-syenite	Trachytic texture; alkali feldspar, interstitial nepheline, both enclosing sodalite, with biotite, magnetite and titanite	Obsolete name. An exotic rock but poorly exposed
Borolanite	LB. Horne and Teall (1892) from SE end of intrusion	Melanite-biotite (pseudoleucite-) nepheline-syenite	Alkali feldspar-nepheline intergrowths (both in pseudoleucite and matrix), well-formed melanite and biotite. Pseudoleucite not always present	The original name is still occasionally used informally
'Canisp Porphyry'	MI. Adopted by Sabine (1953) from early usage	Porphyritic quartz-microsyenite	Alkali and plagioclase feldspar phenocrysts in a groundmass of turbid feldspar and quartz	Forms major sill complex
Cromaltite	LB. Shand (1910) from Bad na h-Achlaise. After Cromalt Hills	Melanite-biotite pyroxenite	Diopsidic pyroxene and ilmenomagnetite enclosed by biotite and replacive melanite	Obsolete name. Similar pyroxenites without melanite at LA
Grorudite	MI. Sabine (1953)	Peralkaline rhyolite Comendite	Alkali feldspar and aegirine phenocrysts in fine quartz-feldspar matrix full of aegirine needles	Dykes. Equivalents are strictly volcanic
Hornblende porphyrite	MI. Sabine (1953) following Bonney (1883)	Hornblende microdiorite Spessartite	Phenocrysts of hornblende and plagioclase, sometimes biotite, in fine feldspathic groundmass	Many sills. Calc-alkaline
Ledmorite	LB. Shand (1910), from Ledmore River	Melanite-augite nepheline-syenite Melanocratic nepheline-syenite	Equigranular, medium grained with closely intergrown melanite, diopsidic augite, biotite. Alkali feldspar intergrowths with nepheline	Name occasionally used informally
Nordmarkite	LA. Phemister (1926), after Nordmarken, Norway	Quartz-syenite	Leucocratic syenites made of alkali feldspar and interstitial quartz with variable aegirine-augite and/or alkali amphibole	Main rock of BL. Also occurs as deformed sills
Perthosite	LA. Phemister (1926), main syenite unit	Alkali feldspar-syenite	Nearly monomineralic alkali feldspar rock. Name refers to microperthitic texture	Name still widely used
Pulaskite	LA. Phemister (1926) after Pulaski Co, Arkansas	Pyroxene syenite Melasyenite	Similar to 'nordmarkites' and 'perthosites' but with more aegirine-augite. Some variants have melanite at LA, with minor nepheline and melanite at LB	Type example is nepheline-bearing so use at LA is incorrect
Shonkinite	LA. Phemister (1926) after Shonkin Sag, Montana	Pyroxene (nepheline-) melasyenite	At LA diopside and biotite, sometimes hornblende occur in glomeroporphyritic clusters set in alkali feldspar. Nepheline-bearing at LB	Nepheline usual but not essential. Associated with ledmorites at LB
Sövite	LB. Young et al. (1994)	Calcite carbonatite	Porphyritic sövite has large calcite rhombs set in finer calcite matrix. Phlogopite sövite has small phlogopite crystals together with apatite set in calcite matrix	Small body with xenoliths from LB outside southern contact
Vogesite	MI. Sabine (1953) after Vosges mountains	Vogesite Hornblende-rich lamprophyre	Hornblende phenocrysts set in fine-grained matrix of euhedral plagioclase, alkali feldspar, hornblende and minor quartz. Diopside occurs as glomeroporphyritic clots and rare phenocrysts	Many sills. Calc-alkaline
Vullinite	LB. Shand (1910), from Allt a'Mhuillin	None	Fine-grained, sometimes schistose rock, with altered plagioclase, alkali feldspar, plagioclase, diopside, hornblende and biotite	Obsolete name. Shand considered it probably metamorphic

LB: Loch Borralan intrusion; LA: Loch Ailsh intrusion; BL: Ben Loyal intrusion; MI: Minor Intrusion.
Rock names in **bold** were named from type examples in Assynt. Historical details are from Holmes (1920) and Brögger (1921). Note that many of the old varietal rock names are used in the text, between quotation marks, for clarity when referring to earlier publications.

Loch Borralan intrusion, rounds-off a long diversion in petrological thinking, and shows the potential for continuing field research in the province.

The ultimate source of the alkaline magmas, and the processes that have affected them on their rise through the crust and during their final crystallization, remain topics of intense research. The modern view of alkaline magmatism is that it is initiated by small degrees of partial melting in the Earth's mantle, which has sometimes been subject, before it melts, to a metasomatic process that enriches it in alkalis and certain other elements characteristic of alkaline magmas. These elements, particularly potassium, titanium, phosphorus, barium, strontium, uranium, thorium and the rare-earth elements, are normally present in very low concentrations in mantle rocks but reach high concentrations in alkaline magmas. The carrier that introduces these elements may be melts related to the carbonatites that are commonly associated with alkaline silicate magmas. Whatever the ultimate sources, basic alkaline parental magmas fractionate strongly as they ascend to give rise to a vast range of alkaline igneous rocks. The relative importance of variation arising during mantle metasomatism, partial melting of the mantle, crystal–liquid fractionation during uprise through mantle and crust, and reactions with wall-rocks, remain contentious, and no doubt vary from one instance of alkaline magmatism to another, accounting for the extraordinary diversity in the final consolidated products. The field relationships described in this chapter provide evidence of differentiation prior to emplacement, fractionation during final solidification, reactions with country rocks, and subsequent metasomatic reactions during cooling. It is necessary to take account of all these processes when attempting to deduce the ultimate sources of the magmas using sophisticated geochemical and isotopic techniques.

The structural setting of the Scottish alkaline suite is somewhat unusual in that its emplacement overlaps, both in time and space, a period of intense crustal shortening. Worldwide, the greatest upwellings of alkaline magma are in environments of major crustal extension, often preceded by large-scale doming, such as in the present-day East African rift system. Much early discussion on the rocks of the NW Highlands centred on this established correlation. For example, van Breemen *et al.* (1979a) suggested that the Scottish alkaline magmatism was related to arching on the scale of the entire NW Highlands Moine outcrop. They noted that the alkaline rocks formed a zone at the edge of the Caledonian mobile belt (Figure 7.1), close to or on, the rigid older crust of the stable Foreland, which could withstand differential stresses provided either by orogenic compression or in compensation for isostatic sag produced by the weight of the thrust sheets.

More recently it has become accepted that alkaline magmatism can be associated with small degrees of melting in the deeper parts of subduction zones, or with regions of the mantle that contain relics of earlier, now inactive subduction zones. This type of igneous association, known as shoshonitic magmatism, includes members with calc-alkaline affinities (like the granites that dominate Caledonian igneous activity) but includes some members with a strongly potassic, silica-undersaturated character, like the pseudoleucite-syenites in the Loch Borralan intrusion. Other members may be oversaturated and strongly sodic, like the late quartz-syenites at Loch Borralan or the 'grorudite' (comendite) suite of dykes in the Ben More thrust sheet. Shoshonites themselves are basaltic rocks unusually rich in potassium so that sanidine occurs as rims on plagioclase phenocrysts and in the groundmass.

Recent contributions dealing with the ultimate origins of the NW Highlands alkaline magmas are based largely on the trace element and isotopic chemistry of the rocks and can be touched on only briefly here. Thompson and Fowler (1986) were the first to apply the association between rocks of shoshonitic affinities and subduction to the Caledonian alkaline suite. Thirlwall (1981a) had postulated the existence of a NW-dipping subduction zone beneath the Scottish Caledonides from the chemistry of Old Red Sandstone lavas. Basing their work primarily on the trace element chemistry of the leucocratic syenite members of the major intrusions, Thompson and Fowler suggested that the parental magmas of the Caledonian alkaline rocks were ultrabasic shoshonitic magmas developed by deep melting of the asthenosphere with included slabs of crustal rocks, perhaps carried down as far as the seismic discontinuity at 670 km by this subduction zone. Fowler (1988b) later showed that basic members of the Glen Dessarry pluton had been contaminated by reactions with the Moine envelope rocks and

Introduction

later (1992) used isotopic data to support the shoshonite hypothesis for Glen Dessarry, invoking a two-stage fractionation model and a multicomponent mantle source.

North-west-dipping subduction was also invoked by Halliday *et al.* (1987) but they pointed out that the alkaline magmatism stayed in a single narrow zone, albeit made even narrower by thrusting, over a period of 30 Ma while the region was a convergent plate margin. They also pointed out that there is a progressive increase in the alkaline characteristics of even the Caledonian granitoids towards the NW (Halliday *et al.*, 1985). They considered that these factors rule out a deep, well-mixed, asthenospheric source and pointed to a source in the lithospheric mantle, which had been subject to metasomatic enrichment in the elements characteristic of alkaline magmatism. They considered that the thermal state of the lithosphere, on the edge of the orogen, exercised the main control on the magmatism, with small-degree partial fusion of ancient, cold and dry lithosphere underlying the Lewisian gneisses of the Foreland to produce the alkaline melts. For the most western, most potassic and chemically by far the most extreme complex, Loch Borralan, they invoked special, potassium-rich subcontinental mantle, with subduction seen as the trigger for melting. Thirlwall and Burnard (1990) carried out a chemical and isotopic study of this intrusion, and concluded that all its rock types were primarily generated by strong fractional crystallization of mantle-derived, subduction-related shoshonitic magmas closely similar to those that produced the late Silurian Lorn lavas to the south of the Great Glen Fault. The magmas producing the oversaturated syenites were modified, prior to emplacement, by reactions with Lewisian crust. On geochemical grounds Thirlwall and Burnard ruled out the derivation from old, stable lithosphere favoured by Halliday *et al.* (1987), but they were not able to reach a conclusion as to whether the source was in the deep lithospheric mantle or the asthenosphere. In conclusion, it is fair to say that there is much to be done to explain the origins of Britain's most exotic suite of rocks, and in the writer's view solutions will only come when field relationships, petrography, mineralogy and the modern geochemical approach are made to work more closely together.

The alkaline rocks of the NW Highlands also have great importance because of their structural and geochronological implications. They provide evidence of the order and scale of movements in the Moine thrust zone, and provide the only exact time markers for events in this internationally famous major structure. A number of workers have discussed the possibility that the igneous rocks in Assynt were responsible for the embayment in the Moine Thrust known as the Assynt Culmination. This interweaving of igneous activity and structural events is an unusual and outstanding characteristic of the province and a number of the sites described in this chapter have been chosen to illustrate not only petrographical types but also critical structural relationships.

Although the early map-makers (Peach *et al.*, 1907) thought that all the igneous activity in Assynt occurred prior to the movements on the great thrust planes, it was Bailey and McCallien (1934) who first pointed out evidence that thrust movements actually overlapped the emplacement of the igneous rocks. They noted undeformed pegmatites cutting pseudoleucite-syenites in the Loch Borralan complex in which the normally rounded pseudoleucites had been flattened. They suggested that the flattening occurred as a result of thrust movements. The origin of this flattening is still controversial, but contemporaneity of igneous activity and thrusting was firmly established in an important paper by Sabine (1953), in which he recognized that particular types of minor alkaline intrusion occurring as dykes and sills were restricted to particular structural units in the thrust region.

Figure 7.2 is a simplified map of the structure of Assynt, showing the main thrust planes. The development of nomenclature of the thrusts is reviewed in Johnson and Parsons (1979). There are many minor planes of movement, in particular the arrays of high-angle reverse faults joining low-angle thrusts originally known as imbricate structure but called duplex structure by modern workers. In places the sills of Assynt are repeated by such structures. For modern treatments of the structural setting of the igneous rocks in Assynt the reader is recommended to read Johnson and Parsons (1979), Elliott and Johnson (1980) and Coward (1985). The uppermost thrust is the Moine Thrust itself, bringing the metamorphic Moine Supergroup of the Moine Nappe, mylonitized at the base, over the Lewisian, Torridonian and Cambro-Ordovician rocks of the Ben More Nappe, itself carried westward on the Ben More Thrust. North of Inchnadamph the rocks below the Ben More

Thrust were moved on a lower thrust, the Glencoul Thrust, carrying the Glencoul thrust sheet, but west of Inchnadamph the Ben More and Glencoul thrusts join. Sabine (1953) noted that a type of peralkaline felsite dyke, for which he used the name 'grorudite', occurred only in the Glencoul and Ben More thrust sheets and suggested that this meant that both sheets had behaved as a single tectonic unit, for which he proposed the name Assynt Nappe. He suggested that, south of Inchnadamph, the combined Ben More and Glen Coul thrusts be called the Assynt thrust plane. However Sabine's terminology has not been adopted by recent workers and the early terminology of Peach and Horne (in Peach *et al.*, 1907) is retained in this chapter. Sabine located a solitary 'grorudite' in rocks of the Cam Loch Klippe (Figure 7.2, locality 2), confirming the classic interpretation of Peach and Horne that these rocks were outliers of the Ben More thrust sheet. The lowest thrust sheet, the Sole Nappe, brings thrust rocks over the undisturbed Foreland region.

Sabine noted many other striking restrictions on the distribution of the alkaline dykes and sills in Assynt. These can be used to make important deductions concerning the order of events in the thrust belt, and together with radiometric ages obtained on the plutons, to provide brackets for the ages of the main thrust movements. Table 7.1 is the most recent interpretation. Although some workers (e.g. Macgregor and Phemister, 1937) regarded the thrust sequence as propagating upwards, lower thrusts being truncated by higher ones, the modern view is that thrusts normally evolve downwards, with older thrusts riding piggy-back on younger lower ones. This picture is compatible with the distribution of igneous rocks in Assynt, although all workers agree that late movements on the Moine thrust plane must have occurred.

A final important point is the possible role of the alkaline rocks in Assynt in giving rise to the Assynt Culmination. This embayment in the Moine Thrust (Figure 7.2) is actually a broad upwarp or bulge in the Moine thrust plane, and between it and the almost perfectly planar Sole Thrust is a thick lenticular complex of thrust rocks. Several early workers (Phemister, 1926; Bailey, 1935; Sabine, 1953) suggested that there was a causal connection between this thickening and the presence of the minor and major bodies of igneous rocks, and their view was supported by Elliott and Johnson (1980). The connection

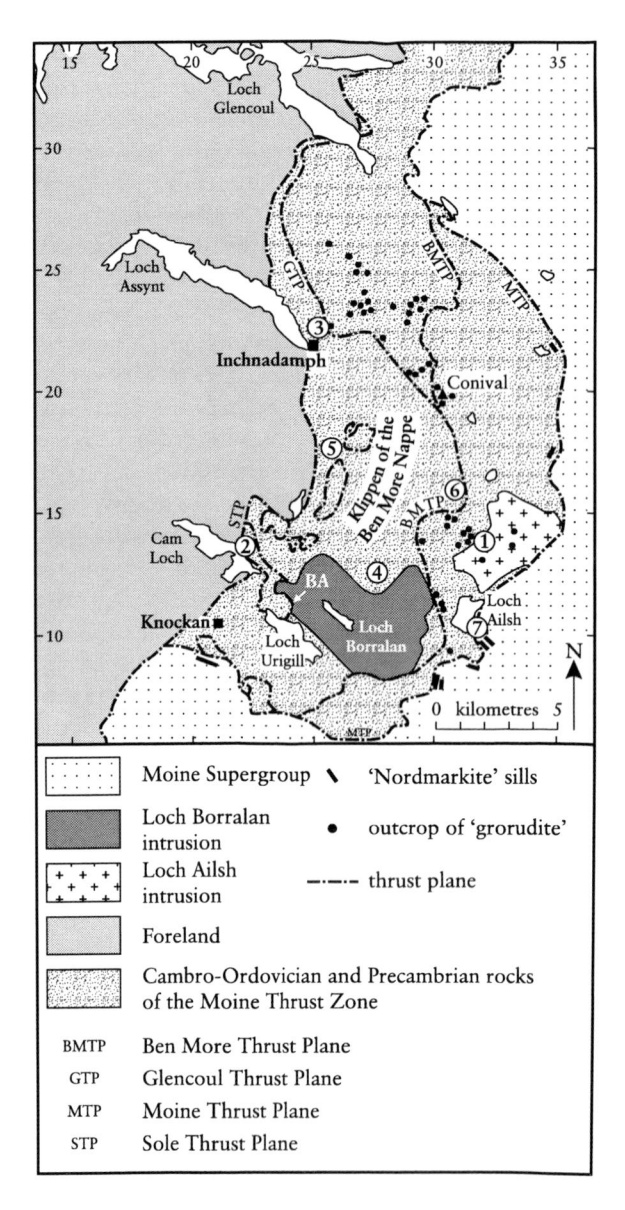

Figure 7.2 Map of the Assynt district showing the major thrusts, the two major alkaline intrusions, and the distribution of two of the six types of minor intrusive rocks. BA is the critical locality, at Bad na h-Achlaise, where nepheline-syenites and pyroxenites of the Loch Borralan intrusion are intruded into one of the klippen (the Cam Loch Klippe) of the Ben More Nappe. GCR sites in the thrust zone related to minor intrusive rocks are shown by circled numbers. 'Grorudite': 1, Glen Oykel South; 2, Creag na h-Innse Ruaidhe. 'Hornblende porphyrite': 3, Cnoc an Droighinn; 4, Luban Croma. 'Vogesite': 5, Allt nan Uamh; 6, Glen Oykel North (diatreme). 'Nordmarkite': 7, Allt na Cailliche. (After Sabine, 1953 and Johnson and Parsons, 1979, fig. 3.)

would not be a direct one, because the thickness and volume of the igneous rocks is far less than the thickness and volume of the culmination. But it is plausible that the igneous rocks and their thermal metamorphic aureoles increased the resistance to slip of the Cambro-Ordovician horizons on which most of the thrusts moved, roughening the slip surfaces, and causing duplication and doming.

Three of the sites selected for this Geological Conservation Review are relatively large and complex, and an extended description of each is given. These are the major intrusions at *Loch Borralan* and *Loch Ailsh* in Assynt, and the group of intrusions around *Loch Loyal*, 40 km to the NE (Figure 7.1). Localities have been chosen within these sites that illustrate the great range of unusual rock types, internal and external contact relationships, and structural implications.

The remaining 12 sites are much smaller and of a different character. Examples have been selected of each rock type represented in the extensive suite of dykes and sills that occur in Assynt, on the grounds of typical character, accessibility, and where appropriate, their relationship with the major structures in the thrust belt. The exposures are individually important because of the relatively rare rock types represented, and also important as a suite, because of their petrogenetic, structural and chronological implications. A common introduction to each of the rock types is provided, setting out their petrography and structural implications. Peralkaline rhyolites ('comendites', the 'grorudites' of Sabine, 1953) are described cutting the Loch Ailsh syenite in *Glen Oykel, south*, illustrating an important relative intrusive age relationship, and from the Cam Loch Klippe at *Craig na h-Innse Ruaidhe*, east of the Cam Loch, exemplifying the restriction of this rock type to the Ben More thrust sheet. Porphyritic quartz-microsyenite (the Canisp Porphyry), is described from a large and physiographically important site on *Beinn Garbh*, from the *Laird's Pool* GCR site near Lochinver (which demonstrates its extension into the Foreland), and from a structurally important site west of Loch Awe (*Cnoc an Leathaid Bhuidhe*). Calc-alkaline hornblende microdiorite sills (so-called 'hornblende porphyrites') are extremely common in Assynt and examples of these are described from intense swarms on *Cnoc an Droighinn*, east of Inchnadamph and *Luban Croma*, north of the Loch Borralan intrusion. More mafic but otherwise similar hornblende lamprophyres (vogesites) are also common and are described from *Allt nan Uamh* and from *Glen Oykel, north* where the vogesite is associated with a remarkable diatreme with a carbonate matrix. A set of sills of quartz-microsyenite ('nordmarkite') occurs near the Moine thrust plane (usually just above) and an example from *Allt na Cailliche*, SE of Loch Ailsh, is described. Finally, two examples of melanite nepheline-microsyenite ('ledmorite') cutting Torridonian and Lewisian rocks on the west coast, at *Camas Eilean Ghlais* and *An Fharaid Mhór* show that the source of the most strongly alkaline magmatism was beneath the Lewisian Foreland, and also have important implications for late movements on the Sole Thrust.

ALKALINE PLUTONIC COMPLEXES

LOCH BORRALAN INTRUSION (NC 235 110–277 081–297 085–306 107–298 140–260 150–235 150)
I. Parsons

Introduction

The Loch Borralan intrusion, in the SW corner of the Assynt region (Figure 7.2) is of international importance for petrological reasons, and of great regional significance for structural reasons. It is the only plutonic complex composed largely of silica-undersaturated (i.e. feldspathoid-bearing) igneous rocks in the British Isles, and many of its members are exceptionally alkaline. It provides Britain's only example of truly ultrapotassic magmatism. The most potassic members, with as much as 15 wt% K_2O, are among the most K_2O-rich rocks encountered on Earth. The site includes the only British example of carbonatite.

The unusual character of the rocks was recognized in the 19th century and the intrusion has held an important place in the international development of igneous petrology. The first specific account of the intrusion was by Horne and Teall (1892) who described the pyroxene-melanite nepheline-pseudoleucite-syenite, which they called 'borolanite'. (There have been several spellings; the current spelling for the Loch is 'Borralan' but 'Borolan' has precedence for the rock name). Additional rocks were described by Teall (1900). The intrusion was mapped, and an account given, by Peach *et al.* (1907), and S. J. Shand carried out pioneering detailed petrographical and analytical work in the following years (1906, 1909, 1910 and 1939). Rock names introduced by these early workers, 'borolanite', 'ledmorite', 'cromaltite' and 'assyntite' (Table 7.2) found some worldwide application as similar rock types were found at other localities, but are now used only very occasionally.

Shand suggested that the Borralan rocks owed their silica-deficient character to disilication reactions between a magma of broadly granitic composition and the Cambro-Ordovician dolomitic limestones that form much of the envelope (Figure 7.4), providing support for the 'Daly–Shand hypothesis', which was developed because of a common association observed between nepheline-bearing rocks and 'limestone'. It is, however, now known that the carbonate rocks that often occur in association with nepheline-syenites are carbonatites, and this association has recently been confirmed just outside the Loch Borralan intrusion in excavations made by Scottish Natural Heritage, near Loch Urigill (Young *et al.*, 1994). Modern petrologists see the source of both magmas as lying in the Earth's upper mantle; modern thinking on the ultimate origins of the Caledonian alkaline rocks is given in the chapter introduction.

Shand (1910) also postulated that the intrusion has the form of a gradationally stratified laccolith, a model accepted by Bowen (1928) who reproduced Shand's section through the supposed laccolith, and argued that the pseudoleucite-bearing rocks at the base of the laccolith formed as a result of crystal settling under the influence of gravity leaving a silica-oversaturated liquid which crystallized to produce quartz-syenites. He justified this interpretation on the basis of new insights gained from his experimental petrological studies. Tilley (1957) rejected Bowen's hypothesis on the grounds that leucite would not crystallize at likely high water-vapour pressures. He provided some new analyses of 'ledmorites' from Loch Borralan, and compared them with dykes in the Foreland, from Camas Eilean Ghlais, Coigach and from Achmelvich, far from dolomitic limestones of the Durness Group, complementing analyses provided by Sabine (1952, 1953). This connection, between distinctive alkaline dyke rocks in the Foreland, and the Loch Borralan pluton in the thrust zone, is of considerable regional structural and temporal significance.

Subsequent interpretations of the internal structure of the intrusion have questioned the gradational character of the boundaries. The complex has a broadly concentric form (Figure 7.4) but exposure is extremely poor in many critical areas. The diversity of rock types (Table 7.2), many of them very rare on a world scale, is exceptional, but the boundaries of most units are not exposed. The most recent detailed treatment of the internal relationships is that of Woolley (1970, 1973) and his terminology is used here (Table 7.2). The map (Figure 7.4) is based on that of Woolley, modified slightly by Johnson and Parsons (1979) and by later work (Parsons and McKirdy, 1983; Notholt *et al.*, 1985; Young *et al.*, 1994). Woolley divided the complex into an earlier suite of silica-undersaturated (nepheline- and/or pseudoleucite-bearing) rocks, most members of which are relatively mafic or ultramafic, and a relatively leucocratic later suite of silica-saturated or oversaturated rocks. Woolley accepted the views of Harker (in Tilley, 1957) and Macgregor and Phemister (1937) that the boundary between the two suites is an intrusive one. He considered that the early suite has a laccolithic form, but that the later suite has the form of a thick plug-like body, punching through the earlier units. Since Woolley's work, some re-interpretation has become necessary because of an extensive drilling programme in the vicinity of the SW margin (Matthews and Woolley, 1977; Notholt *et al.*, 1985; Shaw *et al.*, 1992). An excavation made by the Nature Conservancy Council (Parsons and McKirdy, 1983) in Bad na h-Achlaise demonstrated the intrusive character of the ultramafic members of the complex, ruling out an in-situ skarn origin postulated by Johnson and Parsons (1979). A new marble quarry, near Ledbeg, provides exposures of 'borolanite' cutting both highly altered Durness Group dolomitic limestone, with spectacular contact metamorphic and metasomatic effects, and quartzite. This still-to-be-described new locality offers outstanding opportunities for research in metamorphic reactions, and provides superb teaching opportunities.

The structural relationships of the Loch Borralan intrusion are of considerable geochronological importance (Halliday *et al.*, 1987), as summarized in Table 7.1. Van

Breemen *et al.* (1979a) obtained U-Pb ages from zircons separated from four units of the intrusion and concluded that it was emplaced over a relatively short period of time at 430 ± 4 Ma, making it a little younger than the 439 ± 4 Ma obtained by Halliday *et al.* (1987) for the nearby Loch Ailsh intrusion, which is petrographically similar to the late syenite suite at Loch Borralan. This rules out a suggestion made by Bailey (1935) that the Loch Ailsh intrusion is an easterly extension of the Loch Borralan intrusion viewed through a 'window' in the Ben More thrust sheet.

There is considerable discussion in the literature concerning the structural relationships between the Loch Borralan intrusion and the Ben More Thrust (sometimes called the Assynt Thrust in this part of Assynt, see Johnson and Parsons, 1979) and the Sole Thrust which crops out slightly to the west of the intrusion and must dip beneath it. Several structurally critical areas of the intrusion are described below in a separate section. The relationship of the igneous rocks to the thrust movements is an important issue, because several workers (Bailey and McCallien, 1934 and Woolley, 1970) have suggested that early members of the intrusive complex were emplaced before or during movements on the Ben More Thrust, and hence the complex can provide a very exact date for this episode of movement in the Moine thrust zone. However, more recent work (Parsons and McKirdy, 1983) has shown that the ultramafic and nepheline-syenite members were emplaced after the movements on the Ben More thrust plane, and new exposures in the marble quarry at Ledbeg show 'borolanite' cutting quartzites which have moved on the Ben More thrust plane. The evidence now seems to suggest that emplacement of all the units of the Loch Borralan complex occurred after the movements on the Ben More Thrust had ceased. The Loch Ailsh mass was undoubtedly emplaced in the rocks of the Ben More Nappe before the movements on the thrust plane, so that the 439 ± 4 and 430 ± 4 Ma ages on the two intrusions provide an important bracket on the time of the main movements on the Ben More Thrust. A K-Ar age of 394 ± 8 Ma obtained on a mica from Loch Borralan by Brown *et al.* (1968; recalculated with more recent decay constants by van Breemen *et al.*, 1979a) has been interpreted to mean that some 30 Ma elapsed before the temperature in the pile of nappes fell below an Ar blocking temperature of *c.* 300°C (van Breemen *et al.*, 1979a).

The relationship of the Loch Borralan complex to the Sole Thrust cannot be established directly as the two are not seen in contact. The intrusion does not become more deformed as it approaches the Sole Thrust, which crops out about 1 km to the west, in contrast with its behaviour as the Moine Thrust is approached. The alignment of nepheline-syenite ('ledmorite') dykes in the Foreland (Sabine, 1952, 1953) with the Loch Borralan nepheline-syenites suggests that little or no horizontal displacement of the mass has occurred since emplacement. Halliday *et al.* (1987) accepted the implication that the Loch Borralan mass was emplaced after the main movements on the Sole Thrust. However, from structural mapping, Coward (1985) concluded that the Loch Borralan intrusion has been moved at least 30 km on the Sole Thrust since emplacement. These apparent contradictions between structural interpretations in Assynt and the chronology of igneous events remain unresolved.

The Borralan site also provides examples of contact metamorphism and metasomatism of which the most spectacular examples are those seen in the marble quarry NE of Ledbeg (248 136). Other examples of marble occur in the Ledbeg river around Ledbeg. Examples of alkali metasomatism (fenitization) of quartzite have been described from various localities by Woolley *et al.* (1972), Rock (1977), Martin *et al.* (1978) and Parsons and McKirdy (1983).

Description

The intrusion covers an area of around 26 km^2, and comprises several low hills culminating in Cnoc-na-Sroine at 398 m, surrounded by an area of low, largely peat-covered ground in which most of the more unusual rock types crop out (Figure 7.3). The higher ground is mainly of leucocratic feldspar-rich syenites, quartz-bearing at the top, while the lower ground comprises undersaturated, usually rather mafic syenites (Figure 7.4). At the southern margin pyroxenites occur in isolated exposures; geophysical work (Parsons, 1965a) and drilling (Matthews and Woolley, 1977; Notholt *et al.*, 1985; Shaw *et al.*, 1992) have shown that these are part of an extensive sub-vertical sheet between syenite and altered dolomitic limestones. The igneous rocks

Figure 7.3 'Ledmorite' (melanite-augite nepheline-syenite) exposures at Ledmore in the Loch Borralan intrusion, looking west, with Cùl Mòr (849 m) behind. Cùl Mòr is in the Foreland and is composed of Torridonian and Cambrian sandstones. The Sole Thrust runs beyond the low hills in the middle distance. (Photo: I. Parsons.)

are in contact along three sides with Cambro-Ordovician Durness Group limestones, while the northern margin is against Cambrian quartzite.

The description below follows Woolley's (1970) division of the complex into two suites, with an intrusive junction between them. The early suite comprises pyroxenites, nepheline-syenites and pseudoleucite-syenites, while the later suite is feldspathic syenites ('perthosites') and quartz-syenites. The two suites are not now believed to be related by in-situ fractionation processes. The early suite appears to have a sheet-like (laccolithic) form, while the later suite appears to have the form of a plug punching through the earlier rocks. The variety of rocks in the early suite is extremely large, exposure is very poor, and few exposures reveal contact relationships. The most useful descriptions of field relationships are those of Woolley (1970, 1973), extended by the drilling work mentioned above and by the excavations reported by Parsons and McKirdy (1983) and Young et al. (1994).

Early suite

Ultramafic rocks

Biotite-magnetite pyroxenites, with and without melanite (called 'cromalite' by Shand, 1910), and hornblendites crop out only in the low ground in the SW of the intrusion. Despite their extent demonstrated by geophysical means there are only poor exposures. A map of the western corner of the Loch Borralan complex, showing all the exposures and the extent of the pyroxenites deduced from magnetic and gravity anomalies and proved by drilling (Matthews and Woolley, 1977; Notholt et al., 1985; Shaw et al., 1992), shown in Figure 7.5. The only natural exposures of pyroxenite are in the Ledmore River 200 m downstream from the footbridge (246 119) where coarse-grained biotite pyroxenite, with a few outstanding ribs of syenite, can be seen, and in Bad na h-Achlaise (245 115). The main evidence for the 'stratified laccolith' structure for the intrusion favoured by Shand (1910, 1939) was the relatively low topographical position of these rocks and the higher position of the leucocratic syenites. However, the considerable magnetic anomaly associated with the hidden ultramafic rocks (Parsons, 1965a; Matthews and Woolley, 1977) clearly indicates that the pyroxenites form a sub-vertical screen between syenite and Durness Group dolomitic limestone. The main pyroxenite body forms a dyke-like mass dipping at approximately 70° to the NE, and plunging beneath less basic rocks towards the SE. The exposures at Bad na h-Achlaise have little magnetic expression and are of limited vertical extent and floored by skarn rocks. Drilling in the main pyroxenite body showed that the pyroxenites are interleaved with screens of heterogeneous melanite-rich syenite, pyroxene syenite, and more leucocratic nepheline-syenite. Only three small natural exposures of these types occur, to the NW of Bad na h-Achlaise, but recently excavations nearby have provided more substantial exposures showing this inter-relationship (Parsons and McKirdy, 1983).

The origin of the pyroxenites is controversial. Matthews and Woolley (1977) favoured the Bowen (1928) hypothesis that the pyroxenites are cumulate rocks from the base of the sheet forming the 'early suite', and postulated that they have been brought to their present attitude by faulting or by squeezing of a partly consolidated layered sequence, but the writer (in Johnson and Parsons, 1979) suggested that they are a metasomatic assemblage at the junction between the syenites and dolomitic limestones. The incorrectness of the latter view was demonstrated by the Nature Conservancy Council excavations near Bad na h-Achlaise reported by Parsons and McKirdy (1983) which clearly show the intrusive character of the pyroxenites into quartzites of the Cam Loch Klippe carried over the Durness Group rocks by the Ben More Thrust (Figure 7.5). These excavations are important, complementing the very limited but historically important exposures in Bad na h-Achlaise itself, and providing the only accessible evidence of the character of the rock types found by the very extensive drilling programme (Figure 7.5). The exposures in and near this small hollow (Bad na h-Achlaise means 'place of the armpit') were recorded in detail by Shand (1910). Melanite pyroxenite ('cromaltite') occurs in the stream where it is cut by a 'two foot dyke' of 'aegirite pegmatoid' (aegirine-nepheline-alkali feldspar pegmatite). A 'dyke' of pyroxene-melanite syenite with 'flesh coloured feldspar' occurs below locality 2 on Figure 7.5, and a body of carbonate-bearing pyroxenite forms a knob on the east side of Bad na h-Achlaise. Shand considered this rock to be 'half-fused sediment which has absorbed a certain proportion of silicates from the intruded 'cromaltite', and later (1930) considered the exposure to provide 'clear evidence ... supporting the Daly–Shand hypothesis'. However, Phemister (1931) studied this exposure and concluded that it was carbonated melanite pyroxenite, and that the progressive replacement of the silicate minerals by carbonate could be demonstrated.

The excavations (1–3 on Figure 7.5) resolve several features of the geology around Bad na h-Achlaise. Locality 1 provides a small exposure of deeply weathered, coarse-grained, melanite-biotite pyroxenite ('cromaltite') with a few thin feldspathic veinlets overlain by a thicker southward-inclined sheet of syenite. Locality 2 was originally a small exposure of leucocratic syenite pegmatite, but the excavation provides a 19 m section through heterogeneous, variably feldspathic, pyroxene-hornblende-melanite syenites, showing a faint, nearly vertical layering, surrounded by pyroxenite. The syenites are cut by a zoned syenite pegmatite with striking 25 cm euhedral, black feldspars. Excavation 3 shows important relationships and is an enlargement of a small exposure of Cambrian quartzite cut by a red syenite pegmatite. Woolley (1970) correlated this syenite with similar sheets cutting the pyroxenites in Bad na h-Achlaise. He suggested that the quartzite (which is part of the Cam Loch Klippe) was brought into place on the Ben More Thrust after emplacement of the pyroxenites and the main body of undersaturated syenite, but before injection of the pegmatite, and that the igneous activity therefore bracketed the movements on the Ben More Thrust. However exposure 3 (a field sketch is given in Parsons and McKirdy, 1983) shows no evidence of a thrust relationship between igneous rocks and quartzite, and at the western end pyroxenite is clearly intrusive into the quartzite, which is fenitized, with rosettes and veins of asbestiform, pale-blue amphibole. The movements on the Ben More Thrust were therefore complete before all the igneous rocks of this part of the complex were emplaced, and the hypothesis of Parsons (in Johnson and Parsons, 1979) that the pyroxenites are an in-situ skarn is disproved.

The commercial interest in the pyroxenites was initially because of their high magnetite content but the more recent drilling work was to evaluate the phosphate potential of the apatite-bearing pyroxenite. While not currently economic, the body constitutes the most significant phosphate resource yet found in the United Kingdom (Notholt *et al.*, 1985).

Nepheline-syenites

In addition to the exposures of syenite at Bad na h-Achlaise (previous section) members of the less mafic part of the early suite crop out extensively in the Ledmore River at Ledmore (Figure 7.3, 247 121) and on the A837 (244 132). These are Shand's (1910) 'ledmorites', mesocratic melanite-pyroxene nepheline-syenites. There are few other exposures (Figure 7.5) and little hint of the petrologically very exotic rocks which extend over at least 3 km from near Ledmore to the SE (Figure 7.4).

In their drillcore material, Notholt and Highley (1981) recognized two generations of syenites intrusive into the pyroxenites: (1) leucocratic, pink syenite veins usually a few centimetres in thickness; (2) two types of more mafic syenite: (a) melanite garnet-bearing, sometimes with as much as 50% garnet; (b) pyroxene syenites, sometimes garnetiferous, showing both intrusive and gradational relationships to the pyroxenites. All of these types can be seen in the very poor exposures near Bad na h-Achlaise although their spatial relationships cannot be established. A pile of large boulders extracted during the building of a forestry road can be inspected in a shallow quarry at the track side just to the north of Bad na h-Achlaise. These include several varieties of nepheline-syenite showing cross-cutting relationships, with some intimate vein networks. There are some pyroxenite xenoliths in syenite, a relationship found in boreholes by Notholt and Highley (1981). They provide good evidence of the fractionation of several magma-types before emplacement. These exposures (and excavations) of plutonic nepheline-syenites are unique in the British Isles.

Pseudoleucite-syenite and associated rocks
This suite, which includes the intrusion's best known rock type 'borolanite', (Horne and Teall, 1892), mainly crops out in the eastern part of the complex, and is particularly well exposed around the Allt a' Mhuillin (Figure 7.6). The most important exposures are in the quarry east of the Allt a' Mhuillin (287 097), and in the Allt a' Mhuillin gorge. These are the best exposed localities for 'borolanite', which is a pyroxene-melanite nepheline-syenite with conspicuous white spots, which are generally believed to be nepheline-alkali feldspar pseudomorphs after leucite. These 'pseudoleucites' have varying degrees of ellipticity within the quarry, and in the lower part of the Allt a' Mhuillin gorge are flattened into white streaks giving the rock a schistose appearance. This flattening has controversial implications concerning the timing of the thrust movements, noted in the introduction to the Loch Borralan complex and in a later section. The 'borolanites' are cut by a set of undeformed pegmatite veins containing an assemblage unique in Britain: feldspar, nepheline, biotite, melanite, magnetite, titanite, allanite, zeolites and a blue, sulphatic cancrinite (vishnevite) described by Stewart (1941). 'Borolanites' with white spots more convincingly of the icositetrahedral pseudoleucite shape are best found on the 358 m hill east of Loch a' Mheallain (291 108).

Other silica-undersaturated rocks in the eastern part of the Borralan complex form an extremely diverse suite. They are exposed sporadically in the ground east of Allt a' Mhuillin, for which Woolley (1973) gives an accurate map defining three main types arranged in eastward-dipping sheets (Figure 7.6). At the top, beneath a roof of Durness Group dolomitic limestone forming the eastern margin is a strongly potassic 'muscovite group'. Some of these rocks reach 15 wt% K_2O (Woolley, 1973) and are among the most extreme potassic igneous rocks known on Earth. Below this group are a suite of biotite–magnetite rocks, and then the 'borolanites'. Nepheline and K-feldspar are the felsic minerals in each case. Thinner layers of pyroxene- and hornblende-rich rocks (including 'shonkinites', Table 7.2) were encountered below the Allt a' Mhuillin quarry, in boreholes that went to nearly 50 m. Woolley called these unexposed rocks the 'lower suite'. There are numerous xenoliths of a more mafic melanite-pyroxene-biotite syenite in the 'borolanites', well seen on the north wall of the quarry, and ascribed to an earlier, disrupted phase of the intrusion by Macgregor and Phemister (1937). The rocks at the bottom of the deeper boreholes are reddened, and red syenite veins appear, suggesting proximity to the later syenite intrusion.

The lowest exposures in the Allt a' Mhuillin gorge are of highly deformed 'borolanite', but upstream there are layers of pyroxene-rich 'shonkinite', chemically similar to 'ledmorite' (Woolley, 1973). A fine-grained alkali feldspar-biotite-albite rock called 'vullinite' by Shand (1910) occurs above the gorge. Shand considered it to be a metamorphosed sediment but Macgregor and Phemister (1937) thought it was a metamorphosed earlier igneous rock. Woolley (1973) considered that the lower and pseudoleucite suites had a generally sheet-like form, and that the pseudoleucite-bearing rocks were emplaced after the lower suite. Within the 'borolanites' a roughly contact-parallel boundary (Figure 7.6) separates lower melanite-bearing and upper melanite-free zones, which Woolley suggested could result from in-situ settling of melanite (and also pyroxene). He also suggested that the extraordinary upward increase in potassium to sodium ratio in the pseudoleucite suite could be explained by the settling of sodium-bearing leucite. The origin of the variants found in the eastern part of the Borralan complex is a far from resolved problem and their unique chemistry makes the conservation of this part of the complex a matter of considerable importance.

Rocks of 'borolanite' type occur in the thrust-defined body known as the Loyne mass, at the NW extremity of the intrusion (Figure 7.4), where they have a roof and floor in dolomitic limestone. The Loyne mass is a lenticular 'horse', defined above by the Ben More Thrust and below by the Ledbeg Thrust (Johnson and Parsons, 1979; Elliott and Johnson, 1980). Spectacular examples of 'borolanite' sheets cutting marbles can be seen in the recently opened quarry at Ledbeg. A diversity of igneous rocks are visible in this quarry, some of which are extremely rich in melanite garnet. Melanite syenites with white spots resembling pseudoleucite were recorded by Notholt and Highley (1981) in a borehole near dolomitic limestone exposures (at 256 098), but these, and the phlogopite- and serpentine-carbonate rocks that they cut, are never exposed.

Carbonatite

The igneous carbonate rock, carbonatite, was discovered as blocks of orange-brown carbonate rock on the beach at Loch Urigill (at 247 105) (Figure 7.5), where they are still visible (Young *et al.*, 1994). This is the only British example of this important rock type except possibly some thin carbonate dykes found in association with albitites cutting Moine and Dalradian rocks in the Great Glen near Inverness (Garson *et al.*, 1984). The carbonate-rock blocks contain xenoliths of syenite and biotite pyroxenite with pronounced reaction rims. Subsequently a white sövite (coarse-grained calcite-carbonatite) was found cutting the Durness Group dolomitic

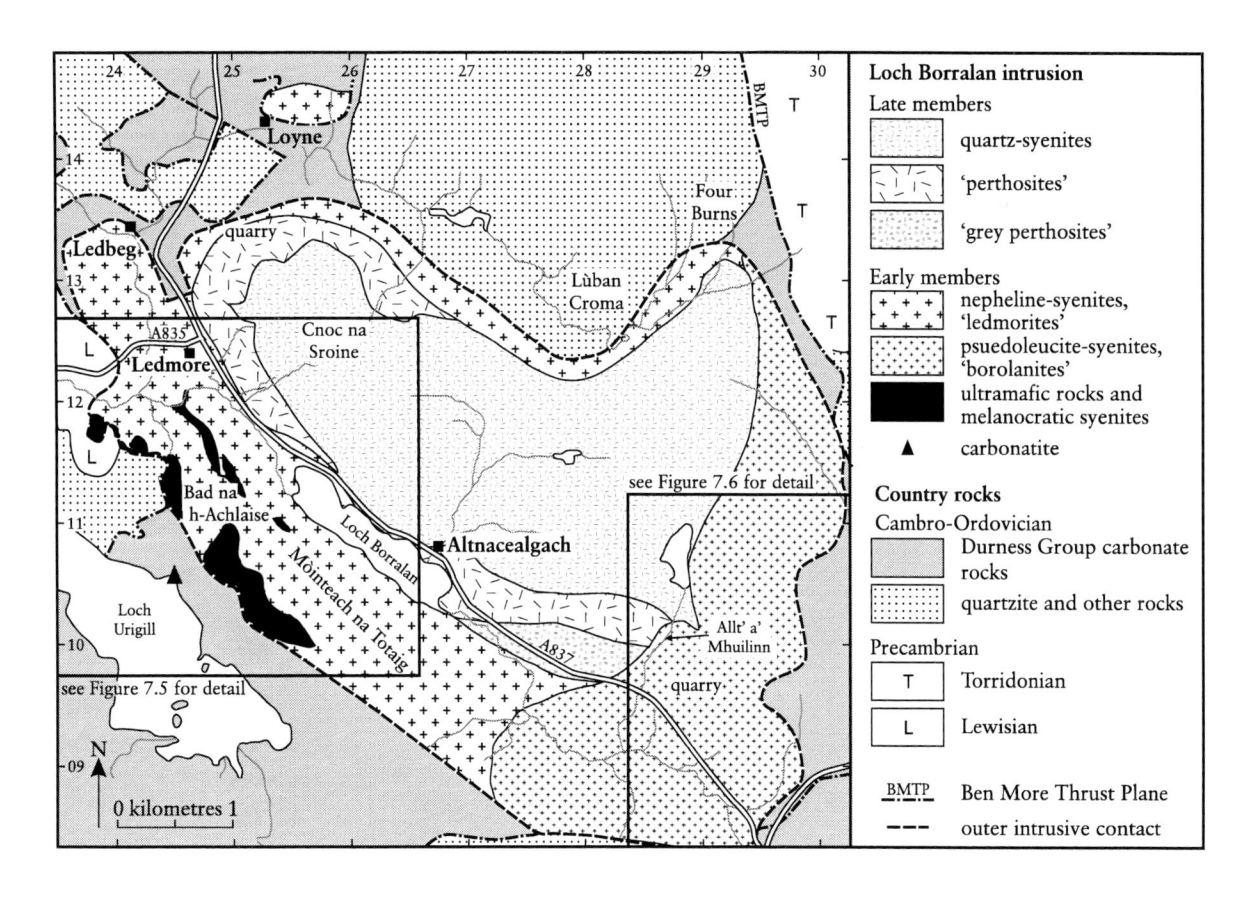

Figure 7.4 Map of the Loch Borralan intrusion and its envelope rocks (modified after Johnson and Parsons, 1979).

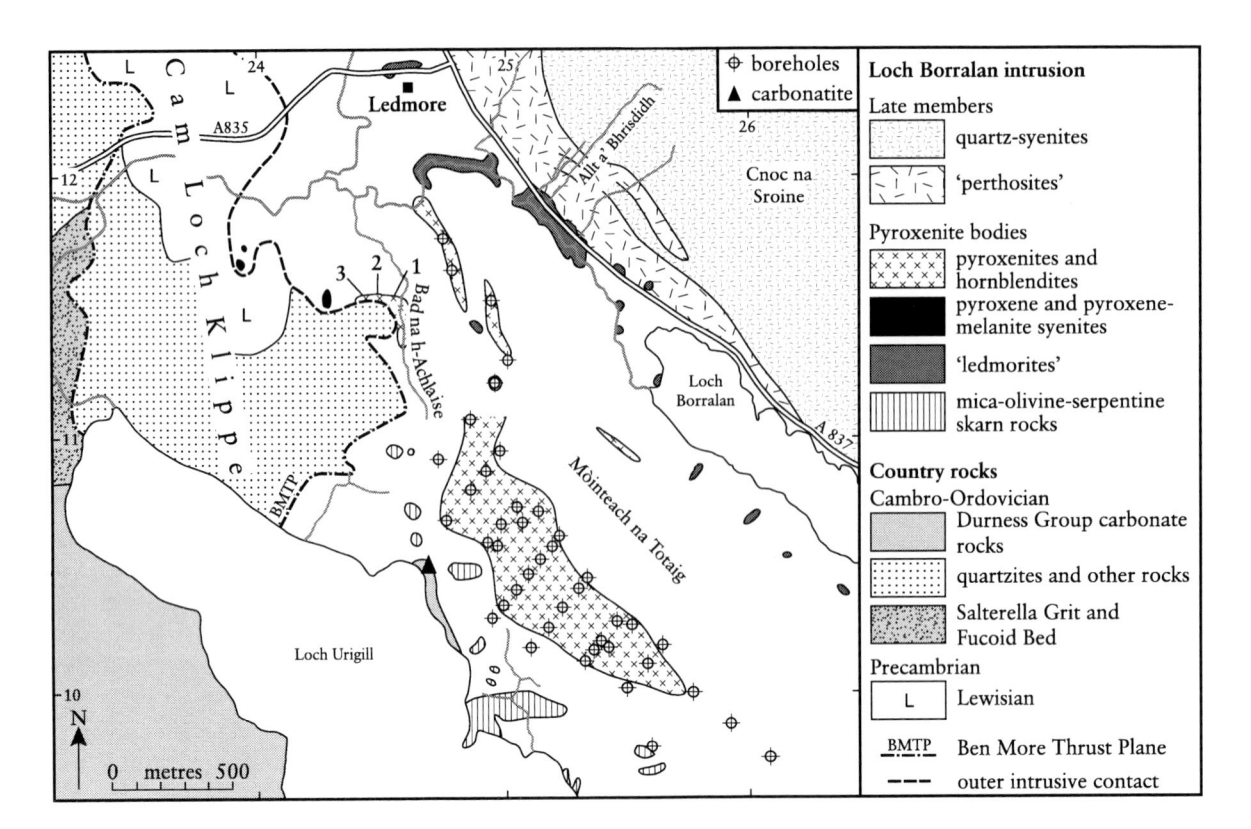

Figure 7.5 Map of the western part of the Loch Borralan intrusion. Units within the Cam Loch Klippe and on the western side of Loch Urigill are interpolated from exposures, as are the alkali feldspar-syenites of Cnoc na Sroine. The central part of the map shows actual exposures, boreholes and the extent of the pyroxenite bodies interpolated from them and from magnetic anomalies. The unornamented area in the central part of the map is a complex, largely unexposed assemblage of leucocratic nepheline-syenites, ledmorites and pyroxenites. Localities 1 to 3 are discussed in the text. (Compiled from Parsons and McKirdy, 1983, fig. 1; the Geological Survey special sheet for Assynt, 1923; Woolley, 1970, fig. 1; Notholt et al., 1985, fig. 3; Young et al., 1994, fig. 1.)

limestone a few metres to the north. This exposure was subsequently enlarged by Scottish Natural Heritage using an excavator (Threadgould et al., 1994); a drawing of the new exposure is given in Young et al. (1994). The carbonatite contains numerous xenoliths of both nepheline-syenites and pyroxenites which can be matched in the intrusion and xenoliths of Durness Group dolomitic limestone that have been rotated during the emplacement of the carbonatite magma. The carbonatite body is 400 m outside the contacts of the Loch Borralan intrusion but it is very likely to be part of the magmatism that gave rise to the Loch Borralan mass. The association of carbonatite with nepheline-syenites and diopsidic pyroxenites is recognized worldwide. Chemical evidence that the rocks are carbonatites of deep origin, rather than locally mobilized Durness Group carbonates, comes from their distinctive trace element, carbon and oxygen isotope signatures (Young et al., 1994). The overall extent of the carbonatite in this area cannot be established from the topography owing to the poor exposure.

Four varieties of carbonatite have been found, three *in situ*. These are porphyritic sövite, phlogopite sövite, and sövite breccia. The fourth variety, a foliated silicocarbonatite, has been found only as a 30 cm block in the drift. Considerable internal heterogeneity is a common feature of carbonatites, which often involve several generations of brecciation and incorporation into later phases of injection. The most striking rock is the phlogopite sövite, which owes its orange colour to myriads of small phlogopite plates included in a matrix of large calcite crystals. The rock also contains rosettes of apatite. The porphyritic sövite is white in colour, and is made of coarse calcite crystals. On one face of the exposure it is layered, with 2 cm-thick bands of the relatively rare mineral chondrodite (a hydrated magnesium silicate) separated by 25 cm-thick layers of normal sövite. The sövite breccia is a matrix-supported breccia of brown carbonatite fragments in a coarsely crystalline, brown sövite matrix resembling the phlogopite sövite.

Late suite

The silica-saturated and oversaturated alkali feldspar-syenites of the late suite are relatively well exposed on the southern slopes of Cnoc-na-Sroine. The rocks are less exotic and controversial than those of the earlier suite, and are similar to the 'perthosites' and melanite syenites in the Loch Ailsh intrusion. The top of Cnoc-na-Sroine is formed of quartz-syenites ('nordmarkites') which, with around 12 vol.% quartz, are more quartzose than the quartz-syenites at Loch Ailsh and a little richer in potassium relative to sodium (Parsons, 1972). Shand (1910) considered that the quartz-syenites grade downwards continuously into quartz-free syenites, with or without melanite, and eventually into the melanite syenites ('ledmorites') in the Ledmore River.

Woolley (1970) divided the syenites into a downward succession of quartz-syenites, 'perthosites' (with or without melanite) and 'grey perthosites'. The former two variants form the bulk of Cnoc-na-Sroine, and their relationships are best seen in the Allt a' Bhrisdidh (Figure 7.5, 252 119) where the quartz-syenites can be shown to be interleaved with the 'perthosites', with intrusive junctions (Woolley, 1970). The 'grey perthosite' variant is seen only in the low ground west of the Allt a' Mhuillin gorge (Figure 7.4, around 283 099).

Junctions between early and late suites

Exposures illustrating the relationships of early and late suites are of importance because of the bearing they have on the genesis of the rocks of the complex as a whole, and because of the possibility that the two suites were emplaced respectively before and after the main movements on the Ben More thrust plane (as suggested by Woolley, 1970, but disputed by Elliott and Johnson, 1980).

Shand (1910) believed that all boundaries in

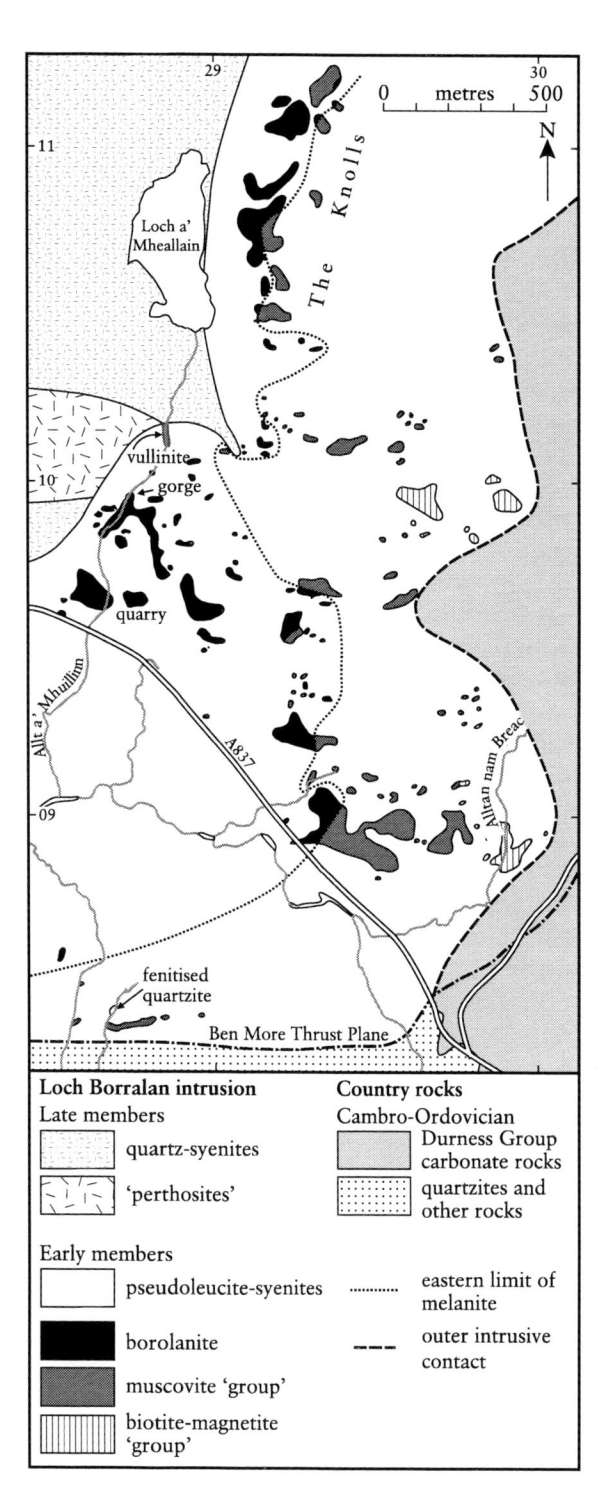

Figure 7.6 Exposure map of the geology of the pseudoleucite-bearing 'borolanites' and associated rocks of the SE part of the Loch Borralan intrusion. (After Woolley, 1973, fig. 2.)

the complex are gradational, but Woolley (1970) agreed with Macgregor and Phemister (1937) that the later suite is intrusive into the earlier one, with sharp boundaries. Critical relationships are seen only at two localities. The most important junction (Woolley, 1970) is in the lower part of the Allt a' Bhrisdidh (253 119; see Figure 7.5). At its confluence with the Ledmore River this stream flows in medium-grained, brownish 'ledmorite', but upstream this rock becomes darker in colour, and veined and speckled by pink feldspar. About 50 m north of the A837 a fairly sharp but irregular contact can be seen between a pink leucocratic syenite and somewhat foliated mesocratic rock. This contact is overall nearly vertical, and only pink syenite occurs above. Cross-cutting relationships between pink syenite veins and 'ledmorite' can be seen in streams nearer to Ledmore, and in road cuttings SE of Ledbeg, but the correlation of these syenite veins with the main mass of Cnoc-na-Sroine is not certain.

A second critical junction (Woolley, 1970) is in poorly exposed ground about 0.5 km NE of the deep section of the Allt a' Mhuilinn gorge (290 130; see Figure 7.6). Here a tongue of quartz-syenite can be mapped extending into the 'borolanites'. Sharp intrusive contacts can be demonstrated and the later syenite becomes finer grained towards the contact. If penetrative deformation exhibited by the 'borolanites' in the nearby Allt a' Mhuillin quarry is tectonic, this junction is strong evidence that the emplacement of the complex overlapped the thrust movements.

Localities important for structural reasons

Three localities have had particular importance for structural and geochronological reasons. They provide the best evidence for the temporal relationship between the igneous activity and the thrusting, and the measured ages of rocks in the Loch Borralan and Loch Ailsh intrusions provide the best estimate of the timing of movements in the Moine thrust belt in general (Table 7.1). Evidence for minor sub-horizontal movements is provided by the presence of locally developed cleavage in almost all units of the Loch Borralan intrusion, but this is not thought to be associated with large-scale movements on the major thrusts. Overlap of igneous and tectonic activity was postulated first by Bailey and McCallien (1934) on the basis of relationships seen in the quarry in 'borolanite' east of the Allt a' Mhuillin (287 097) where undeformed pegmatites cut 'borolanite', which appears to be deformed. Here, and in the Allt a' Mhuillin itself, the normally equidimensional pseudoleucites are flattened, first into ellipses and ultimately into white streaks (best seen in the Allt a' Mhuillin gorge, 286 098). Woolley (1973) has suggested that the flattening is due to penetrative deformation associated with the Ben More thrust plane, which presumably lay not far above the present exposures, while Elliott and Johnson (1980) have argued that the flattening is due to 'igneous' displacements during emplacement. Whatever the character of the deformation in the 'borolanites', however, it is certain that the later igneous rocks at this locality escaped deformation, because clearly undeformed pegmatites cut rocks with flattened pseudoleucites. The pegmatites form a network on the wall of the 'borolanite' quarry, and a large zircon from this pegmatite formed part of the dating study of van Breemen *et al.* (1979a).

In Bad na h-Achlaise (Figure 7.5) and in the series of exposures extending 200 m to the west, there was for a time thought to be further evidence that the early igneous activity pre-dated the main movements on the Ben More Thrust. This series of exposures was originally interpreted as showing quartzite overthrust on to syenites and pyroxenites, with residual late pegmatites cutting into the quartzite as well as passing through the pyroxenites (Woolley, 1970; Johnson and Parsons, 1979). However, the excavations reported by Parsons and McKirdy (1983) clearly show that this quartzite mass is actually intruded by massive syenite, syenite pegmatite and pyroxenite. If the interpretation of this quartzite as part of the Cam Loch Klippe (as shown on the original Geological Survey maps of 1892) is correct then movements on the Ben More Thrust *pre-date* the emplacement of the entire suite of nepheline-syenites and pyroxenites at Borralan. A less attractive hypothesis is that the quartzite is a xenolith that has been carried up from a normal stratigraphical position beneath the Durness limestone, which crops out at Loch Urigill to the south, and which has come to rest, fortuitously, adjacent to the thrust rocks of the Cam Loch Klippe. Because the timing of emplacement of the pseudoleucite suite relative to the nepheline-syenites and pyroxenites is at present unknown, it is still possible that the emplacement of the Loch Borralan mass overlapped the movements on the Ben More thrust plane, but this requires that the emplacement of the pseudoleucite suite pre-dates all other units of the complex, a view accepted by Woolley (1970). However, in the marble quarry at Ledbeg, 'borolanites' cut quartzites in the Cam Loch klippen, so here both pseudoleucite-bearing rocks and silica-oversaturated members of the intrusion were emplaced after the thrust movements.

Large-scale evidence that the intrusion punches through the Ben More thrust plane can be obtained by consideration of the relationships between geology and topography at the west end of Cnoc-na-Sroine. To the east of Ledbeg (Figure 7.4, around 252 140), in the low ground between the steep slopes formed of late-suite leucosyenites and the A837, are a number of exposures that show Cambrian quartzites carried by the Ben More Thrust on to Durness Group dolomitic limestones, and forming one of the Cam Loch klippen (Peach et al., 1907). Peach and his co-workers considered that the Ben More Thrust passes above the summit of Cnoc-na-Sroine and then dives steeply down to the west. Woolley (1970) rejected this interpretation and instead postulated that the late syenite suite punches through the thrust, and this interpretation is consistent with the relationships at Bad na h-Achlaise.

External contacts

Contacts of the intrusion against country rocks are very badly exposed. On the A837 near Ledbeg (244 133), red early-suite melanite nepheline-syenites are seen enclosing xenoliths of recrystallized dolomitic limestone, with pyroxene selvages, the syenite being strongly deformed at the margins of the xenoliths. Farther north similar limestone–syenite relationships are seen, and the syenites are cut by a leucocratic melanite syenite dyke which Woolley (1970) equates with the later suite of syenites.

An important group of exposures occur to the north of Loyne (around 253 145). The Loyne mass is a separate thrust wedge (or 'horse' in the terminology of Elliott and Johnson, 1980) beneath the Ben More Thrust. The mass shows a complete section through igneous rocks with both roof and floor exposed. Dolomitic limestones on top of the igneous rocks are tens of metres thick, but only small patches of limestone are exposed beneath. To the south, exceptionally well-exposed, but very complex, contacts are seen in the new marble quarry near Ledbeg. The marbles are cut by massive sheets of 'borolanite', with the production of beautiful serpentine marbles, with strikingly banded reaction zones, and clear evidence for the contemporaneous presence of both 'borolanite' and mobilized carbonate liquids. Igneous sheets cutting limestones are also exposed in the 'Four Burns' area in the NE of the complex, (around 293 132; see also Figure 7.16). The main body of the intrusion is probably at a shallow depth below this locality, since Woolley (1970) believed that the upper contact of the intrusion dips east beneath the limestones at about 5°.

Contacts between igneous rocks and the envelope are nowhere exposed along the SW and E edges of the intrusion. Only the drilling work (Matthews and Woolley, 1977; Notholt et al., 1985; Shaw et al., 1992) has revealed the extensive zone of metamorphic calc-silicate rocks that forms the contact of the igneous pyroxenite bodies beneath the peat on Mòinteach na Totaig. At the NW end of the mixed pyroxenite–nepheline-syenite zone there are a few contacts of syenite with quartzite, of which the excavated example from Bad na h-Achlaise is the most instructive. Fenitization of quartzite exposures from near here has been described by Woolley et al. (1972), Rock (1977) who described fenitization of a block in drift, and Martin et al. (1978).

Interpretation

The poor exposure of the Loch Borralan intrusion, its exceptional petrological diversity, and its complex tectonic setting, make interpretation of field and petrogenetic relationships extremely difficult. The excavations and drilling that have taken place since the exposure mapping of Shand (1910) and Woolley (1970) have invariably led to major re-assessments, and the recent report of carbonatite (Young et al., 1994) shows that even exposural evidence has not yet been fully exploited. The reader should have an open mind when assessing the following brief interpretation.

The original interpretation of the whole intrusion as a continuously stratified laccolith, with the various rock types related by crystal settling, has not stood the test of time. The leucocratic, silica-saturated and oversaturated members have an intrusive relationship to the earlier, generally more mafic, undersaturated suite, which may be demonstrated in Allt a' Bhrisdidh and near Allt a' Mhuillin. Both suites show clear internal evidence of the emplacement of pulses of magma of different composition, presumably fractionated before emplacement. Nepheline-syenites, melanite syenites and pyroxene syenites around the critical exposures at Bad na h-Achlaise were certainly emplaced in several phases. The mafic melanite syenites and 'ledmorites' are part of this suite, and all types show complex cross-cutting relationships. The more leucocratic syenites cut the pyroxene syenites and both types cut the pyroxenites and the skarn rocks. The western edge of the intrusion appears to be a complex interleaving of all these rock types but even the considerable drilling programme does not reveal the overall structure.

Woolley's (1973) mapping of the main area of pseudoleucite-bearing rocks in the SE part of the intrusion suggests that the rocks there are stratified and possibly fractionated *in situ*. The structural relationship between the main pseudoleucite suite and the nepheline-syenites that are now known (through drilling only) to occupy a large part of the SW margin of the intrusion is not clear, although there are several localities in the western part of the intrusion (Loyne, Ledbeg quarry and the hidden contacts on Mòinteach na Totaig) where 'borolanites' occur, interestingly always in close association with limestones. Evidence that the early pseudoleucite-bearing suite around Allt a' Mhuillin was emplaced early, before the 'ledmorites' and nepheline-syenites, as suggested by Woolley (1970), hinges on the interpretation placed on the flattening of the pseudoleucites. As the pyroxenite and nepheline-syenite members of the early suite are clearly intrusive into rocks of the Ben More Nappe at Bad na h-Achlaise, but are undeformed (Parsons and McKirdy, 1983), the pseudoleucite-bearing assemblage must be earlier. But if the fabric in the pseudoleucite-bearing rocks at Aultivullin is related to their mode of emplacement, their relative emplacement age is equivocal. There is an urgent need to investigate the structural relationship between these exotic rocks and the remainder of the intrusion, particularly the newly exposed 'borolanites' at Ledbeg.

The large, steeply dipping mass of biotite-magnetite pyroxenite under the peat of Mòinteach na Totaig is earlier than at least three generations of the nepheline-syenites, which occur, largely unseen, to the NE. The pyroxenites are known by drilling to be interleaved with skarn rocks, but were undoubtedly magmatic as is demonstrated by their intrusive relationships at Bad na h-Achlaise, where the pyroxenites are intrusive into quartzites that were moved previously into position on the Ben More Thrust. The high temperatures implied by their bulk mineralogy remain problematical. The cumulate origin favoured by Matthews and Woolley (1977) requires faulting or emplacement as a crystal mush to explain both their near-vertical form and structural level. Their evidence for a cumulate origin is the presence, in borehole material, of alternating, sharply defined pyroxene- and hornblende-rich layers, a few centimetres thick, in which the hornblende shows a preferred orientation. The 'cumulate' textures are not unequivocally due to crystal settling, although it is a possible interpretation, and it is not clear how the layering would survive the proposed squeezing of a crystal mush. Furthermore, the rocks are similar (apart from the presence of garnet) to the pyroxenites in the nearby Loch Ailsh complex which have a dyke-like form between syenite and dolomitic limestone. Although the pyroxenites are intrusive rocks, their intimate association with a major calc-silicate and magnesium-silicate skarn body at the margin of the silicate rocks is at least suggestive of an origin involving reactions between silicate magma and the dolomitic limestones. Young *et al.* (1994) provided Rare Earth Element (REE) plots of a range of rocks from Loch Borralan. The Bad na h-Achlaise pyroxenites and a diopside-rich skarn rock from a borehole nearby have very similar patterns, as do 'borolanites' and 'ledmorites'. In contrast, the leucocratic nepheline-syenites and the carbonatite show much greater enrichment in the light REE. Although the alkaline magmas undoubtedly originated in the Earth's mantle, the visitor to the Loch Borralan and Loch Ailsh intrusions should be open-minded about the origin of the pyroxenites. Perhaps, to this extent, Shand's ideas live on.

The affiliation of the recently discovered carbonatite to the Loch Borralan intrusion is demonstrated by the syenite and pyroxenite xenoliths it contains, and its true character as a carbonatite by its trace and rare-earth element contents and patterns, and its carbon and oxygen isotopes (Young *et al.*, 1994). The mineralogy and internal heterogeneity are characteristic of carbonatites. While it is perhaps surprising that these rocks went unnoticed until 1988, the association of carbonatite with nepheline-syenite magmatism is seen worldwide and the occurrence itself is unsurprising. Diopsidic pyroxenites are also commonly associated with carbonatites. The shape of the Loch Urigill carbonatite is unknown but it cannot be more than approximately 100 m in diameter.

The late suite of melanite alkali feldspar-syenites and quartz-syenites is internally much simpler than the early suite. Cnoc-na-Sroine shows an upward progression from 'perthosites', often with melanite, to quartz-syenites at the top, complicated only by the presence of some quartz-syenite sheets at lower levels. The contact relationships with the early suite give little information on the overall shape but it is perhaps a stock-like body at least 275 m thick (Woolley, 1970). These rocks are generally similar to the alkali feldspar-syenites at Loch Ailsh, but are chemically subtly different (Parsons, 1972) and were emplaced significantly later, after the movements on the Ben More Thrust (Halliday *et al.*, 1987), so that their current proximity may hide an initial separation of perhaps several tens of km. The overall shape of the early suite seems to be sheet-like (see Woolley, 1980) as is its internal structure in the eastern part of the complex. Exposed contacts with country rocks are extremely rare, but the quartzite–pyroxenite contact at Bad na h-Achlaise and the dolomitic limestone–'borolanite' contacts in the Ledmore marble quarry are particularly important. Huge volumes of skarn rocks occur beneath Mòinteach na Totaig but are unexposed.

Conclusions

The Loch Borralan intrusion is the only plutonic igneous complex composed of silica-undersaturated rocks in the British Isles, and it contains several rock types that are extremely rare on a worldwide scale. Some of its members are among the most potassium-rich rocks on Earth. Very recently, a small body of igneous carbonate rock (carbonatite), with syenite xenoliths, has been discovered just outside the main intrusion. This too is a unique occurrence of this rock type in the British Isles. Exposure around Loch Borralan is notoriously bad but the intrusion has an important historical position in the development of igneous petrology through its contribution to the concept of silica saturation. It also has historical prominence because of the idea that reactions between limestone and silicate magma (desilication) were essential to the formation of feldspathoid-bearing rocks (the 'Daly–Shand hypothesis'), and the suggestion that the intrusion was a single, internally stratified, gravitationally differentiated laccolith. Neither hypothesis has stood the test of time; the modern view of such magmatism is that it has its origins in the Earth's mantle, but the detailed geotectonic setting of the Borralan mass and its associated rocks is still a matter of debate.

The current structural view is that the intrusion was emplaced in two major episodes. An early suite, consisting of ultramafic and feldspathoid-bearing rocks involving several (at least five, and perhaps several more) pulses of already differentiated magmas, is extremely complex. It includes the celebrated pseudo-leucite-bearing 'borolanites', has a sheet-like, laccolithic form and may have partly differentiated *in situ*. The ultramafic rocks (biotite-magnetite pyroxenites) contain Britain's largest reserves of phosphate (as apatite) and form a steep-sided, extended lenticular dyke-like body, cut by several generations of feldspathoidal syenite, and interleaved with diopside-, phlogopite- and forsterite-rich rocks produced by reactions with Durness Group dolomitic limestones. A new quarry, at Ledmore, provides outstanding exposures illustrating reactions between 'borolanites' and carbonate rocks.

The later suite is composed of alkali feldspar-syenites ('perthosites') and quartz-syenites. It appears to punch through the early suite and may have a stock-like form. It becomes more quartz-rich upwards but sheets of quartz-syenite cut 'perthosite' lower in the mass. These rocks are mineralogically quite similar to, although chemically distinct from, the syenites in the neighbouring Loch Ailsh intrusion.

The intrusion provides an important time-marker for movements in the Moine thrust zone. Most (and perhaps all) of the early suite were emplaced after the main movements on the Ben More thrust plane. Thus its U-Pb age of 430 ± 4 Ma provides a minimum age for these movements, while the just-significantly-different age of 439 ± 4 Ma for the neighbouring Loch Ailsh complex provides a maximum. It is possible that a flattening fabric affecting the 'borolanites' was produced during movements on the Ben More Thrust. This interpretation is controversial, but if correct it implies that the 'borolanites' result from the earliest phase of emplacement, and that the movements on the Ben More Thrust were very close to 430 Ma. Evidence that the Loch Borralan complex post-dates large-scale movements on the Sole Thrust comes from its alignment with nepheline-syenite dykes in the Foreland, which also place the source of this extreme magmatism firmly in the mantle underlying the Lewisian gneisses.

LOCH AILSH INTRUSION
(NC 330 115–360 150–
330 160–310 140–310 125)

I. Parsons

Introduction

The Loch Ailsh intrusion (Figures 7.7, 7.8), lies at the eastern margin of the Assynt culmination, immediately below the Moine Thrust, which brings metasedimentary rocks of the Moine Supergroup over its eastern edge (Figure 7.2). It is largely composed of syenite, an unusual rock type, both in a worldwide and a British context, and the Loch Ailsh syenites have an unusually high ratio of sodium to potassium. Its main rock types are similar to the late syenite suite in the nearby Loch Borralan intrusion but it does not include the very strongly alkaline silica-undersaturated rocks of the latter intrusion, is mineralogically much less diverse and has figured less in the geological literature. Although it is rather better exposed than the Loch Borralan complex, many critical relationships are nevertheless obscured by peat. The Loch Ailsh intrusion is the world type-locality for the nearly mono-mineralic alkali feldspar rock 'perthosite', and provides evidence for the fractionation, prior to emplacement, of a series of syenitic magmas that become more leucocratic and more peralkaline with time. It is also particularly interesting because of the direct evidence it affords concerning the incorporation of material from Cambro-Ordovician dolomitic limestones, and it provides an instructive range of contact metamorphic rocks from various sedimentary lithologies.

The intrusion was first described by Peach *et al.* (1907), who considered that it rests on a thrust (which they called the Sgonnan Beag Thrust). The petrology and internal structure were first described, in considerable detail, by Phemister (1926). Following a fashion of the time, he considered the intrusion to be a stratified laccolith, with a floor of dense pyroxenites. This interpretation did not survive later geophysical work (Parsons, 1965a), but the overall shape of the intrusion is still enigmatic. The age relationships of the Loch Ailsh and Loch Borralan intrusions are critical in understanding the relative and absolute timing of thrust movements in Assynt.

The igneous rocks are mainly sodium-rich alkali feldspar-syenites, for the most leucocratic of which Phemister (1926) coined the name 'perthosite'. Phemister subdivided the syenites into numerous named varieties (modern equivalents are given in parentheses): 'perthosite' (leucocratic alkali feldspar-syenite); aegirine-melanite syenite; 'nordmarkite' (quartz-syenite); 'pulaskite' (pyroxene syenite); riebeckite syenite; 'shonkinite' (pyroxene-rich syenite). Phemister's use of 'shonkinite' is not followed by more recent terminology, which requires the presence of nepheline. In contrast with Loch Borralan, feldspathoids are not found in the Loch Ailsh pluton. In addition, he recognized ultramafic biotite pyroxenites and hornblendites, similar (apart from the absence of garnet from the Loch Ailsh examples) to the 'cromaltites' described by Shand (1910) in the Loch Borralan complex, and drew attention to the similarity between them and the rock type 'jacupirangite' discovered in other alkaline complexes.

The pyroxene syenites occur chiefly as xenoliths enclosed in the leucosyenites, and form a discontinuous roof to the earlier syenite units, while the pyroxenites and hornblendites form a substantial, although poorly exposed vertical marginal body. These rocks occur between the more felsic intrusive rocks and Durness Group dolomitic limestones along the eastern margin, in a similar structural setting to the equivalent 'cromaltites' at Loch Borralan.

Halliday *et al.* (1987) obtained a U-Pb age of 439 ± 4 Ma on zircons from two samples of Loch Ailsh syenite, a little older than the age of 430 ± 4 Ma obtained for the Loch Borralan intrusion using the same method. Although the western contact of the Loch Ailsh mass is now only about 1 km east of the contact of the Loch Borralan intrusion (Figure 7.2), at the time of its emplacement the Loch Ailsh intrusion may have been several tens of kilometres farther to what is now the SE, because the Ben More Thrust lies in the ground between. There is good evidence

Loch Ailsh intrusion

Figure 7.7 Loch Ailsh and the upper valley of the River Oykel from the south. The snow-covered ridge is Ben More Assynt (998 m), with Conival (987 m) at the extreme left. The Loch Ailsh intrusion extends from just north of the loch to the base of the eastern end of this ridge. The dark, rocky hill in the left middle distance is Black Rock, formed of syenite S3 ('perthosite'). The rough ground immediately behind the cottage is Durness Group carbonate rocks, while the low cliff in the foreground is an exposure of Moine metasedimentary rocks. (Photo: I. Parsons.)

that most units of the Loch Borralan intrusion were emplaced after the main movements on the Ben More thrust plane, because alkaline dykes of the 'grorudite' suite cut the Loch Ailsh intrusion and rocks above the Ben More Thrust only.

There is some disagreement over the age relationships between the Loch Ailsh intrusion and structures within the enclosing Ben More Nappe. Milne (1978) suggested on the basis of careful mapping that the intrusion was emplaced later than the earliest phase of deformation in Assynt, the Sgonnan Mór folding. However, Halliday *et al.* (1987) suggested that greenschist facies recrystallization in some xenolithic pyroxene syenites in the Loch Ailsh intrusion could be correlated with the Sgonnan Mór phase of folding.

The contact relationships of the Loch Ailsh intrusion, and its three-dimensional shape, are not easily defined. On Sgonnan Beag mylonitized syenite is seen against Cambrian quartzite (the Sgonnan Beag Thrust of Peach *et al.*, 1907) but the plane of movement at the exposure dips steeply in a southerly direction rather than NE, beneath the intrusion. In the ground north of Loch Sail on Ruathair, in the unnamed stream that Phemister (1926) called the 'Metamorphic Burn' (333 153), the intrusion appears to finger into Cambrian sedimentary rocks dipping to the SE, showing that it is emplaced in the rocks of the Ben More Nappe. The range of contact metamorphosed Cambrian sedimentary lithologies exposed in this stream are an instructive and valuable feature of the intrusion. In the SE of the intrusion, in the Allt Cathair Bhàn (324 122), an interpretation of the very large magnetic anomalies caused by the high magnetite content of the pyroxenites (Parsons, 1965a), showed that the contact of these rocks against Durness Group dolomitic limestones is steep.

In places, rocks of the Loch Ailsh intrusion are considerably deformed by late movements in the Moine thrust zone. Zones of mylonite occur at several localities and are well seen in the River Oykel (325 127) and at the SW corner of Black Rock (318 135). Coward (1985) considered that there was no evidence for late movements on the Moine Thrust itself in eastern Assynt. The geophysical work (Parsons, 1965a) showed that the eastern edge of the Loch Ailsh complex very probably passes under the Moine Thrust, although there are no exposures to confirm this, and it is not possible to say whether the thrust truncated the intrusion or merely acted as a roof. Feldspar in the Loch Ailsh syenites is often very turbid and coarsely exsolved, perhaps because of deuteric alteration beneath an advancing, warm Moine Nappe, and the absence of alkaline hypabyssal rocks from the Moine also suggests late movements on the Moine Thrust itself.

Description

The intrusion is about 10 km² in area (Figure 7.8), extending from within 750 m of the northern shore of Loch Ailsh and on either side of the Oykel valley for some 3 km. It forms several low hills (Figure 7.7), Black Rock on the eastern shoulder of Sgonnan Mór and the ridge of Sail an Ruathair, and extends up to roughly the 350 m contour on the lower slopes of Meall an Aonaich. Most of the intrusion is composed of syenites, very rich in feldspar and poor in ferromagnesian minerals (leucosyenites). Three intrusive phases can be recognized, called S1–S3 by Parsons (1965b). The syenites are all silica-saturated or oversaturated; there are minor amounts of quartz in some rocks, but nepheline has not been found. The earlier syenite units, S1 and S2, contain appreciable amounts (up to 20%) of pyroxene or alkali amphibole, but the last phase, S3, is largely the near-monomineralic alkali feldspar-syenite 'perthosite'. Contact relationships show that S1 was emplaced first, and chemically it is less evolved than S2 and S3, a feature best shown by the progressive increase in the aegirine content of the pyroxenes (Parsons, 1979). The age relationships between S2 and S3 can be seen in the centre of the complex, where the junction includes an extensive zone of mixing. The central part of S3, forming the 408 m summit of the Sail an Ruathair ridge (333 143), which Phemister (1926) suggested perhaps represented a feeder for the intrusion, is slightly richer in mafic minerals than the bulk of S3 and contains melanite garnet, titanite and thoroughly sodic clinopyroxenes (aegirine-hedenbergite), showing it to be the most evolved part of the complex (Parsons, 1979).

Pyroxene syenites ('shonkinites' of Phemister, 1926) occur at localities in the River Oykel (326 127), and in the Black Rock Burn (318 133). Phemister believed that these are entirely igneous, forming part of a lower, more mafic zone in a stratified laccolithic body, but Parsons (1968) provided chemical and textural evidence that they are partly formed of mafic material contributed by contact metamorphosed Cambro-Ordovician dolomitic limestone xenoliths. The pyroxene syenites appear to map out on the upper surface of the earlier S2 phase of syenite intrusion, suggesting that they represent a fragmented roof to this earlier phase of injection.

Ultramafic biotite-magnetite pyroxenites and hornblendites, crop out in isolated localities along the eastern margin of the complex, along the Allt Cathair Bhàn (324 122). These rocks were taken by Phemister (1926) to represent the base of a laccolith, much as Shand (1909) had suggested for the 'cromaltites' at Loch Borralan. However, the profile of the extremely large magnetic anomalies associated with the pyroxenites (Parsons, 1965a) can be explained only by a set of sub-vertical screens of ultramafic rock, also, like Loch Borralan, interposed between syenite and Durness Group limestone. The similar structural setting of the pyroxenites in these two otherwise distinctly different intrusions, and the mineralogy of the pyroxenites, led Parsons (1979) to suggest that the pyroxenite bodies are large metasomatic skarns, formed by reactions between syenite and dolomitic limestone, but at Loch Borralan later excavations clearly showed that pyroxenites were intrusive into quartzites and by analogy with Loch Borralan, the igneous origin of the very similar Loch Ailsh pyroxenites is no longer in doubt. Nevertheless the emplacement of such a rock mass, very rich in the Ca-Mg pyroxene, diopside, as a magma, is an unresolved petrogenetic problem and the reader is referred to the discussion of this in the description of the Loch Borralan GCR site.

The isolated hillock known as Sròn Sgaile in the NE corner of the complex (348 148) is composed of ultramafic hornblendic rocks at the base passing up into a more leucocratic feldspar-hornblende rock at the top. Phemister (1926) thought that this represents a section through the lower zone of his postulated stratified laccolith. The rocks have distinctive textures, particularly the presence of large plates of a green mica with a 'sieve' texture, enclosing feldspar and hornblende, and are cut by a striking network of syenitic veins with conspicuous ferromagnesian minerals. Although these rocks are undoubtedly part of the Loch Ailsh intrusion, exposure around them is so poor that their

Loch Ailsh intrusion

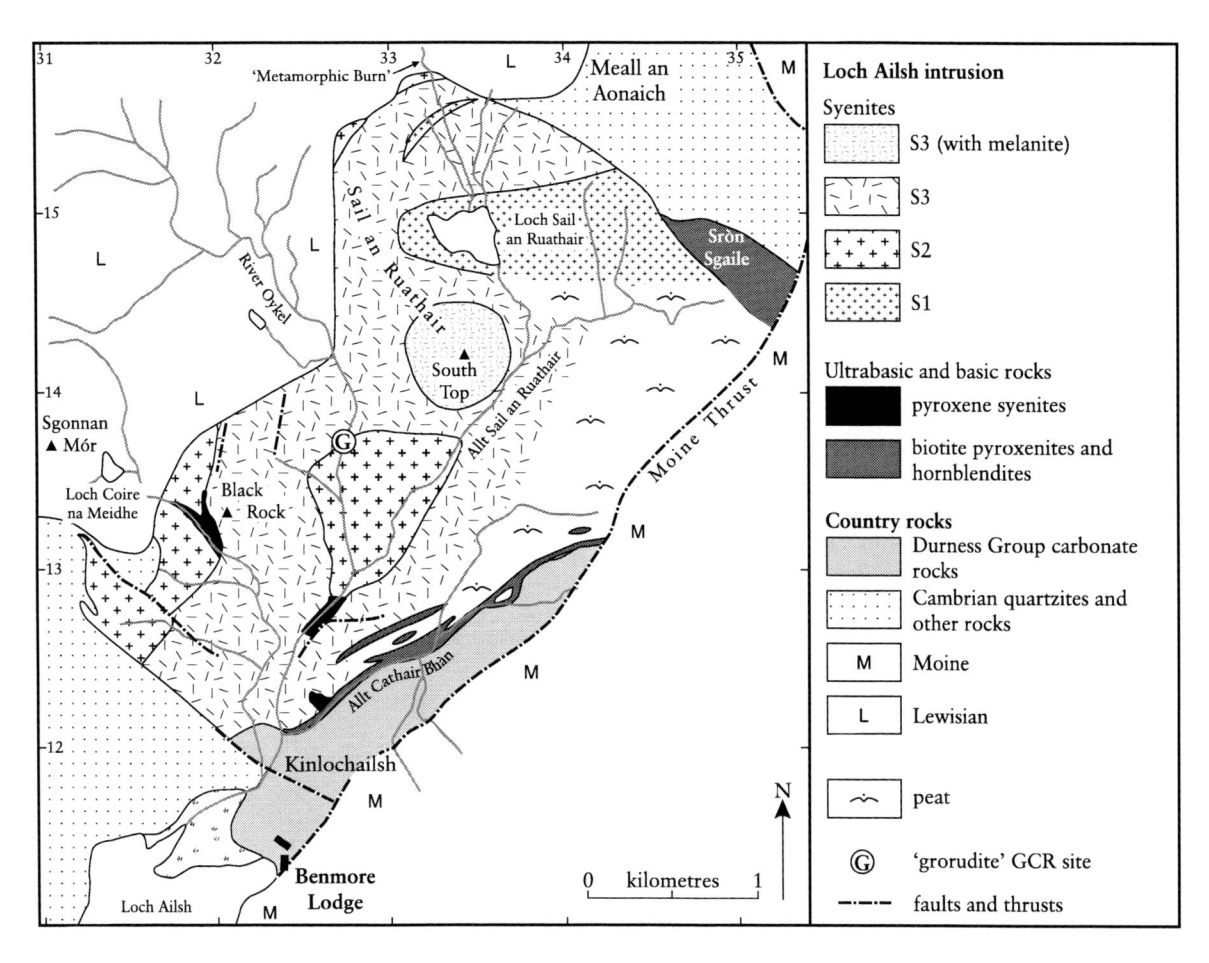

Figure 7.8 Map of the Loch Ailsh intrusion. The extent of the pyroxenites in the Allt Cathair Bhan is based largely on magnetic anomalies. (After Johnson and Parsons, 1979, fig. 15.)

structural relationships remain enigmatic. A magnetometer survey did not reveal a connection with the pyroxenites of Cathair Bhàn (Parsons, 1965a).

Syenites

The leucocratic syenites were described in detail by Parsons (1965b). Units S2 and S3 are well exposed throughout the southern part of the complex. S1 is less well exposed and crops out in Coire Sail an Ruathair. The most informative areas are those that show the inter-relationships between the units, and three critical exposures are of particular interest. At the base of the cliffs beneath the northern summit of the Sail an Ruathair ridge the junction between S1 and S3 can be examined (331 152) (Figure 7.9). S1 is coarse grained, red in colour, and has a distinct igneous lamination, shown by alignment of the slightly flattened feldspars. The S1 unit forms a dome-shaped body that does not penetrate S3 on the west side of the Sail an Ruathair ridge. The later S3 veins the earlier syenite, and encloses it as xenoliths. It is possible to demonstrate from rotation of the igneous lamination that the xenoliths have been rotated by the forceful injection of S3. This feature also shows that the lamination is igneous, and not tectonic, in origin. A screen of red syenite resembling S1 is enclosed in S3 in the central portion of the exposed section of the 'Metamorphic Burn' (see below), an interpretation confirmed by detailed work on the feldspars (Parsons, 1965b). This mass has sharp contacts, but the S3 in this section of the stream contains disseminated red feldspars which are no doubt xenocrysts from S1.

S2–S3 relationships can be seen in the central part of the intrusion around the confluence of the Allt Sail an Ruathair and the River Oykel (327

Figure 7.9 Loch Sail an Ruathair and the ridge of Sail an Ruathair in the northern part of the Loch Ailsh intrusion, from the east. The sketch shows the position of the upper contact of a dome of the early syenite, S1, overlain by the perthosite member, S3. (Photo: I. Parsons.)

130). Here there is an extensive zone of mixing between the two units, and a zone of pink xenocrysts can be mapped in the brown or grey S3 around an inner dome of S2 (Parsons, 1965b). The southern edge of this mixed zone, around the large waterfall in the River Oykel, (326 127) includes pyroxene syenite xenoliths (see next section). Rather similar relationships can be well seen in the upper section of the Black Rock Burn (318 133). Here, red xenoliths of S2 can be seen in S3, again in a zone including pyroxene syenite xenoliths. At the base of the SW cliffs on Black Rock itself (318 134) the suite is involved in a minor thrust plane, and streaking-out of xenoliths can be observed. As elsewhere in the intrusion, S2 underlies S3, but here the upper surface of S2 dips to the SE, and S2 extends on to the flanks of Sgonnan Mór.

Mineralogically, S1, S2 and S3 form an evolutionary series. The mafic mineral in most rocks is a pyroxene, those in S1 being diopsidic (calcium- and magnesium-rich), while those in S3 can have nearly 50 molecular % of the sodium-iron pyroxene component, aegirine; S2 is intermediate (Parsons, 1979). There are slight parallel changes in alkali feldspar composition, those in S1 being exceptionally rich in the albite molecule (*c.* 75 molecular %), those in S3 richer in orthoclase (*c.* 65 molecular % albite). Some facies, particularly of S2, contain a strongly pleochroic riebeckitic alkali amphibole instead of, or in addition to, pyroxene. Melanite garnet, often zoned and intergrown with titanite, appears only in the part of S3 that forms the southern summit on the Sail an Ruathair ridge (Figure 7.8). The boundaries of this variety appear to be gradational, but it contains the most evolved pyroxenes in the complex, suggesting that it was the final part of the intrusion to solidify. This part of the intrusion sometimes contains very small amounts of quartz and also muscovite. The presence of melanite together with quartz is unusual: melanite is usually present together with feldspathoids.

Pyroxene syenites – 'shonkinites'

The most important exposure of these rocks is in the River Oykel (326 127). Here, the pyroxene syenites are in the form of blocks, characteristically less than 1 m across, enclosed in leucosyenite, and the whole xenolith complex is cut by a network of leucosyenite veins. These veins are of two types, grey and red, which presumably correlate with the S2 and S3 syenite generations. The pyroxene syenite xenolith zone is enclosed by the gradational contact zone between S2 and S3. The xenoliths are therefore enclosed by mixed syenite forming the upper surface of a dome of S2 (Figure 7.8). The blocks are extremely heterogeneous (Figure 7.10, see Parsons, 1968); there are 'ultramafic clots' on the microscopic scale and up to a few centimetres across, composed almost entirely of pyroxene and/or biotite and amphibole, with either sharp or diffuse margins, enclosed in the pyroxene syenites, which are mostly diopside-biotite-alkali feldspar rocks. The 'ultramafic clots' are texturally very different, particularly in their fine grain-size, to the pyroxenites of Allt Cathair Bhàn and they do not provide direct evidence for the disrupted lower ultramafic zone to the Loch Ailsh pluton postulated by Phemister (1926). On the other hand they have great textural and mineralogical similarities to rocks occurring at syenite–limestone contacts elsewhere in the intrusion. Parsons (1968) illustrated the similarities and proposed that the pyroxene syenites represent the remains of a metasedimentary roof to S2.

Pyroxene syenites also occur as xenoliths in syenite in the Black Rock Burn (319 132 to 316 134). The S2 unit forms an inclined surface extending on to the flank of Sgonnan Mór and the pyroxene syenites are enclosed by both S2 and S3 near their interface. Xenoliths also occur under the conspicuous overhang at the SW corner of Black Rock (318 134) where they and their vein-networks are stretched and in places mylonitized in a minor thrust plane. At three isolated localities in Black Rock Burn (Figure 7.10) altered limestone xenoliths occur among the pyroxene syenites. Parsons (1968) considered this to be strong support for his hypothesis that the 'shonkinites' are not wholly igneous rocks.

A screen, about 10 m thick, of laminated pyroxene syenite, similar to those seen in the southern part of the intrusion, occurs about halfway up the exposed section of the 'Metamorphic Burn'. As at the other localities, the rock occurs on the upper surface of a body of earlier syenite, in this case S1. There are numerous altered limestone xenoliths both above and below this locality. A 5 m-wide screen of dark-green pyroxenite, and some metre-scale smaller xenoliths, occur 100 m higher up the stream. These resemble the ultramafic clots found elsewhere in the pyroxene syenites. The lowermost screen (stratigraphically highest) is attached directly to a mass of metasedimentary rock representing the Fucoid Beds, whereas the uppermost xenolith is resting against a large mass of quartzite.

Ultramafic rocks

A suite of unusual diopside pyroxenites and hornblendites is exposed at isolated localities along the eastern edge of the complex in Allt Cathair Bhàn (Figure 7.8). Although Phemister (1926) suggested that the exposures of ultramafic rocks are the sole representatives of the lowest zone of a stratified laccolith, large magnetic anomalies (Parsons, 1965a) show that under the peat the pyroxenites form a continuous, dyke-shaped body running from Kinlochailsh, at least to a point 2 km to the NE, where the magnetic anomaly dies out, probably because the rocks of

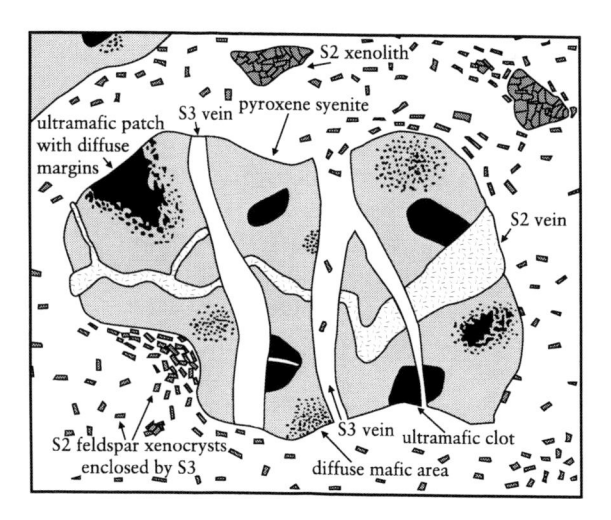

Figure 7.10 Sketch illustrating the relationships between a pyroxene syenite xenolith and feldspathic syenites in the Loch Ailsh intrusion, as seen in the River Oykel and Black Rock Burn areas. A typical xenolith would be about 1 m in length. (After Parsons, 1968, fig. 2.)

the Loch Ailsh intrusion pass under the Moine Thrust. Smaller, discontinuous lenses of ultramafic rocks are implied by the magnetic anomalies to occur in the leucosyenites to the west, and this has been confirmed by excavation.

The ultramafic rocks are deep blue-green biotite pyroxenites. They are often somewhat sheared, when the pyroxene is converted to a green-brown hornblende. The pyroxenes are close to pure diopside in composition (Parsons, 1979). Like the equivalent 'cromaltites' in the Loch Borralan intrusion, apatite is abundant, as is ilmenomagnetite, and the latter mineral leads to the very large local magnetic anomalies found over the pyroxenite members of both intrusions. The only worthwhile exposures are at 326 123, where pyroxene syenites occur to the west, and 328 123 and 334 127, where veins and screens of syenite can be seen cutting the ultramafic rocks. Parsons (1965a) presented a computer model of the profile of the magnetic anomalies found at the latter locality in which a vertical body of magnetic ultramafic rocks some 75 m thick is divided by screens of non-magnetic syenite; the depth to the base of the body must be at least 90 m. The nearest adjacent rocks to the west are mostly feldspathic syenites except for some pyroxene syenites at the south end of Cathair Bhàn. To the east are exposures of Durness Group dolomitic limestones, in the screes beneath which diopside marbles can be found. Just east of Kinlochailsh (323 120) beautiful green serpentine marbles can be found. It seems that the pyroxenites form a narrow, continuous screen against these dolomitic limestones and it was this relationship, together with the very Ca- and Mg-rich diopsidic pyroxenes, which led Parsons (1968, 1979) to initially favour the hypothesis that the pyroxenite were formed by in-situ syenite–dolomitic limestone reactions. Whether or not the pyroxenites were emplaced as a mobile magma (as was the case in the Loch Borralan complex) cannot be established from the limited exposure available in the Loch Ailsh complex. The cross-cutting veins of syenite at 334 127 suggest that emplacement of the pyroxenites pre-dated at least the S3 phase of syenite emplacement, although the possibility of rheomorphic back-veining must be considered.

Metamorphic xenolithic rocks

An excellent suite of metamorphic xenolithic rocks occur in an unnamed stream (called the 'Metamorphic Burn' by Phemister, 1926) which flows into Loch Sail an Ruathair (335 151–333 158). The exposures are important because of the evidence they afford of reactions between syenite magma and sedimentary rocks, which bears upon the origin of pyroxene syenites elsewhere in the complex, and they provide a useful instructional suite of contact metamorphic and alkaline metasomatic rocks. The igneous (and igneous-looking) rocks exposed in the stream are mostly brown 'perthosite' (S3) enclosing screens of earlier syenites, which may be correlated with S1 and S2 on the basis of colour and texture, and inclusions of pyroxene syenite and pyroxenite similar to those seen in the southern part of the intrusion. Interspersed with these are screens and xenoliths of the Cambro-Ordovician succession in correct stratigraphical order, but with Durness Group dolomitic limestones at the base of the slope, and quartzites at the top, implying a set of tongues of metasedimentary rocks more-or-less in place but dipping SE.

A log of the stream bed, starting at the lowest exposures, which occur about 400 m above Loch Sail an Ruathair, was given by Parsons (1968). At the base there are many originally dolomitic xenoliths, now diopsidic and phlogopitic calc-silicate rocks. Mafic patches can be seen in the enclosing syenites, and at certain localities individual pyroxenes can be observed apparently in the process of incorporation into the syenite. At 170 m (measured along the ground from the lowest exposures) an unusual white, melanite garnet-bearing syenite seems to be related to a large limestone xenolith, and just above here is a thick screen of red syenite (S1). At 270 m a 3 m body of fine, flinty dark-green rock with conspicuous pink feldspars and dark minerals, in contact with white or dark-grey quartzite, represents the Salterella Grit of the Cambro-Ordovician succession. Slightly above, 20 m of baked grey shale, with black streaks, represents the Fucoid Beds. Immediately in contact is a 2 m mass of dark-green pyroxenite, with micaceous patches (a 'shonkinite' of Phemister, 1926). There are also two small xenoliths of altered dolomitic limestones; which must have been moved out of their stratigraphical position by the magma. From the 308 m point upwards the syenite (S3) contains massive quartzite xenoliths that include developments of alkali amphibole, an example of the metasomatic process of fenitization. Conspicuous red feldspar xenocrysts sometimes have their long axes aligned, dipping downstream, and there are xenoliths of red syenite (probably S2). The last exposures seen before the stream flows through drift are of a pink, riebeckite-bearing syenite, probably S2.

Interpretation

The overall form of the Loch Ailsh intrusion is difficult to establish because of poor exposure in the vicinity of the contacts. Evidence that it formed a thrust sheet in its own right, as proposed by the early Survey workers (Peach *et al.*, 1907; Phemister, 1926), is not strong and it seems likely that the intrusion was emplaced in the Ben More Nappe either prior to the first folding phase in Assynt (as postulated by Halliday *et al.*, 1987) or shortly after (as proposed by Milne, 1978). This relatively early emplacement has been confirmed by the radiometric age of 439 ± 4 Ma obtained by Halliday *et al.* (1987). The ages provided by the alkaline rocks in Assynt are crucial for dating movements in the Moine thrust belt. The Loch Ailsh rocks in places are deformed (mylonitized) by late movements on the Moine Thrust, and there is geophysical evidence that the eastern contact passes under the Moine. The interfingering of syenite with a largely undisturbed sequence of altered Cambro-Ordovician sedimentary rocks, in the 'Metamorphic Burn' is consistent with a relatively gentle style of emplacement into the rocks of the Ben More Nappe. There is no convincing evidence that the intrusion is a stratified laccolith, as proposed by Peach *et al.* (1907) and Phemister (1926), although the late syenite unit S3 appears to overlie one or other of the earlier S1 and S2 units over much of the intrusion suggesting that S3 has a sheet-like form.

The eastern contact, along which a screen of pyroxenites is interposed between syenite and Durness Group dolomitic limestones, is certainly sub-vertical but the magnetic anomalies that lead to this conclusion give little information on the vertical extent of the intrusion. Like the similar pyroxenites along the southern margin of the Loch Borralan intrusion the origin of these rocks is enigmatic. The pyroxenites in the latter intrusion are definitely intrusive into quartzites and therefore certainly existed as a magma, but there are no exposures at Loch Ailsh that demonstrate that the pyroxenites have an intrusive character. It is curious that in both intrusions the pyroxenites occur only where silicate rocks and dolomitic limestone are in contact. There is no easy explanation for the extended sinuous form of the pyroxenite body if it is entirely intrusive. If it is earlier than the syenites as the cross-cutting veins of syenite superficially suggest, then it is possible that it is an incomplete section of an earlier, arcuate intrusion; alternatively if the syenite veins result from rheomorphism then it is a partial ring dyke. Whatever its structural relationships, the high temperature mineralogy requires that the pyroxenites were emplaced as a crystal mush; there is no direct evidence that they are mobilized cumulate rocks formed *in situ* from the syenite magma, as suggested by Matthews and Woolley (1977) for Loch Borralan, although this is a possible mode of origin.

The syenitic rocks were emplaced in three pulses. They were fractionated before arrival in their final resting place, and become chemically more evolved and peralkaline with time. Both the earlier members (S1 and S2) form domeshaped bodies overlain by a final unit mostly composed of very leucocratic alkali feldspar-syenite ('perthosite', S3). The slightly more mafic, aegirine- and melanite-bearing variant forming the South Top of Sail an Ruathair is the most highly evolved member. More melanocratic, pyroxene syenites occur as xenolithic blocks at various localities; these seem to appear discontinuously on the upper surface of the earlier syenite units and perhaps represent a disrupted roof, a view supported by the sporadic appearance of metasedimentary xenoliths at the same level. The pyroxene syenites have textural and chemical similarities to syenites demonstrably (in the 'Metamorphic Burn') modified by partial assimilation of altered dolomitic limestone. While on the one hand some of the pyroxene syenites have thoroughly igneous textures, there is also strong textural evidence of assimilation of material of metasedimentary origin. Perhaps the pyroxene syenites represent an early phase of igneous activity itself modified by reactions with the sedimentary envelope. Major element chemistry, mineral chemistry and even trace-element chemistry are equivocal on this subject (Parsons, 1979; Young *et al.*, 1994).

Conclusions

The Loch Ailsh intrusion includes a suite of sodic syenites unique in the British Isles, which provide evidence for igneous fractionation processes before emplacement, and include the world type locality for 'perthosite'. The pyroxenites of the eastern margins are extremely enigmatic rocks whose extent and subsurface shape have been elucidated by geophysical means. A suite of intermediate, pyroxene syenites include rocks that have a thoroughly igneous appearance as well as types that have certainly formed by reactions between the syenite magma and dolomitic sedimentary rocks from the envelope. There are very instructive exposures illustrating contact metamorphism, alkali metasomatism (fenitization) and assimilation of a large range of Cambro-Ordovician sedimentary rocks. The structural relationships of the Loch Ailsh body, and its known age of 439 Ma, provide an important age-marker for movements in the Moine thrust zone.

LOCH LOYAL SYENITE COMPLEX (NC 610 440–670 500–670 520–560 510–560 470–590 440)

I. Parsons

Introduction

The scenically magnificent Loch Loyal intrusions (Figure 7.11) form the largest area of alkaline rocks in Britain, and contain the only extensive body of the quartz-syenite type, 'nordmarkite'. There are three centres, emplaced in metamorphic Moine and Lewisian country rocks, but unaffected by Caledonian deformation (Table 7.1). The largest intrusion, Ben Loyal itself, is now thought to be separated from two smaller satellites, Ben Stumanadh and Cnoc nan Cùilean, by a major NE–SW dextral oblique fault (Holdsworth and Strachan, in press), called by these authors the Loch Loyal Fault. This may mean that the Ben Loyal body represents a deeper level of erosion through a single intrusion of which the Ben Stumanadh and Cnoc nan Cùilean bodies are upward apophyses.

The Ben Loyal intrusion is the only leucosyenite in the NW Highlands to be truly peralkaline, showing consistent normative *acmite* (Robertson and Parsons, 1974). It has an interesting internal structural subdivision into a two-feldspar (subsolvus) outer syenite and a chemically identical, one-feldspar (hypersolvus) core syenite (Robertson and Parsons, 1974). The Cnoc nan Cùilean intrusion has a distinctly different chemical character, in particular higher K_2O (Robertson and Parsons, 1974) and a high radiometric anomaly (Gallagher *et al.*, 1971). The Ben Stumanadh intrusion is chemically and petrographically similar to the Ben Loyal intrusion. All three intrusions contain numerous xenoliths of Moine and Lewisian country rocks.

The Loch Loyal intrusions are emplaced in Moine psammites or in Lewisian gneisses that are interleaved with them and which were reworked during the metamorphism of the Moine. A suggestion by Robertson and Parsons (1974) that the country rock structures were re-orientated on a large scale by the intrusions has not been supported by recent structural work (Holdsworth and Strachan, in press), and the current view is that a localized change in strike of the country rock is a result of SE-plunging folds pre-dating the emplacement of the intrusions. All the Loch Loyal intrusions were emplaced after the metamorphism of the enclosing Moine. Unlike the Assynt complexes they are not penetratively deformed or mylonitized. A U-Pb age of 426 ± 9 Ma obtained by Halliday *et al.* (1987) on zircon from Cnoc nan Cùilean is, within errors, the same as the age (430 ± 4 Ma) of the Loch Borralan complex in Assynt (Table 7.1).

Heddle (1883b) records descriptions of the Ben Loyal syenite by Murchison and Cunningham and provides entertaining detailed descriptions of the field relationships and mineralogy of the Ben Loyal mass. Read (1931) noted the similarity of the Ben Loyal syenites to the quartz-syenites in the Assynt area, and therefore suggested that they were 'comagmatic'. Since the Loch Loyal syenites are entirely non-metamorphic, while the Assynt rocks are involved in the Moine thrust belt, he came to the important conclusion that metamorphism of the Moine pre-dated the post-Cambrian movements. This was an important deduction in the days before widespread use of radiometric dating.

Description

Ben Loyal intrusion

The outcrop of the intrusion has the form of a half-circle, in area *c.* 16 km², with a circular boundary in the NW and straight boundary in the SE (Figure 7.12) which Holdsworth and Strachan (in press) interpret to be the major Loch Loyal Fault. The intrusions form high

Loch Loyal syenite complex

Figure 7.11 Ben Loyal (764 m) from the north. The quartz-syenite peaks rise dramatically out of the surrounding moorland underlain by Moine metasedimentary rocks. (Photo: I. Parsons.)

ground above the surrounding Moine, and the NW flank of Ben Loyal provides some of the most striking mountain scenery in Scotland, with excellent exposure of syenites on its imposing summits (Figure 7.11). King (1942) thought that both the Ben Loyal and Cnoc nan Cùilean intrusions have the form of irregular cones, with apices pointing downwards, but Phemister (1948) thought of the Ben Loyal intrusion as a sheet or laccolith dipping towards the SE. Robertson and Parsons (1974) considered that overall the intrusion has steep outward dips in the NW and W, citing exposures in Allt a' Chalbach Coire (568 495) and in the gorge of Allt Fhionnaich (564 476). At both these localities the relationships are complicated by syenite sheets, which may be concordant with the Moine or steeply dipping and which contain lenses of Moine rocks in places. Moine rocks can be seen overlying syenite, and there is a gradual decrease downstream in the amount of syenite. The outward dip is steeper in the Allt a' Chalbach Coire than in the Allt Fhionnaich and on the west slope of Sgòr Fhionnaich the contact is nearly vertical. However, Robertson and Parsons' interpretation of the attitude of the contact is disputed by Holdsworth and Strachan (in press) who consider that the contact dips east and SE beneath the pluton, parallel to the compositional layering of the Moine. In the SE, between Ben Loyal and Cnoc nan Cùilean, the syenite contact probably dips gently SE beneath a thickening Moine cover. Inclusions of Moine occur at various points on the southern slopes and the largest, at Bealach Clais nan Ceap (590 490), is exposed over some 800 × 300 m. Other large inclusions crop out on the northern and eastern slopes of Ben Hiel (595 502, 599 503, 604 497). Smaller lensoid Moine fragments characteristically between 5 and 15 cm long are common in many of the marginal areas and usually lie in the plane of the lamination in the syenites. Some of these inclusions are sharply defined, with the same mineralogy as the regional Moine rocks, but others show extensive feldspathization and are represented by diffuse 'ghosts'.

Robertson and Parsons (1974) considered that the intrusion had produced widespread deformation effects. The dip of the regional foliation and the pronounced lineation and quartz-rodding of the Moine rocks change markedly in the vicinity of the intrusion. The regional strike of the Moine is generally NE–SW, with a dip of 20–30° to the SE, but within 2 km of the syenite

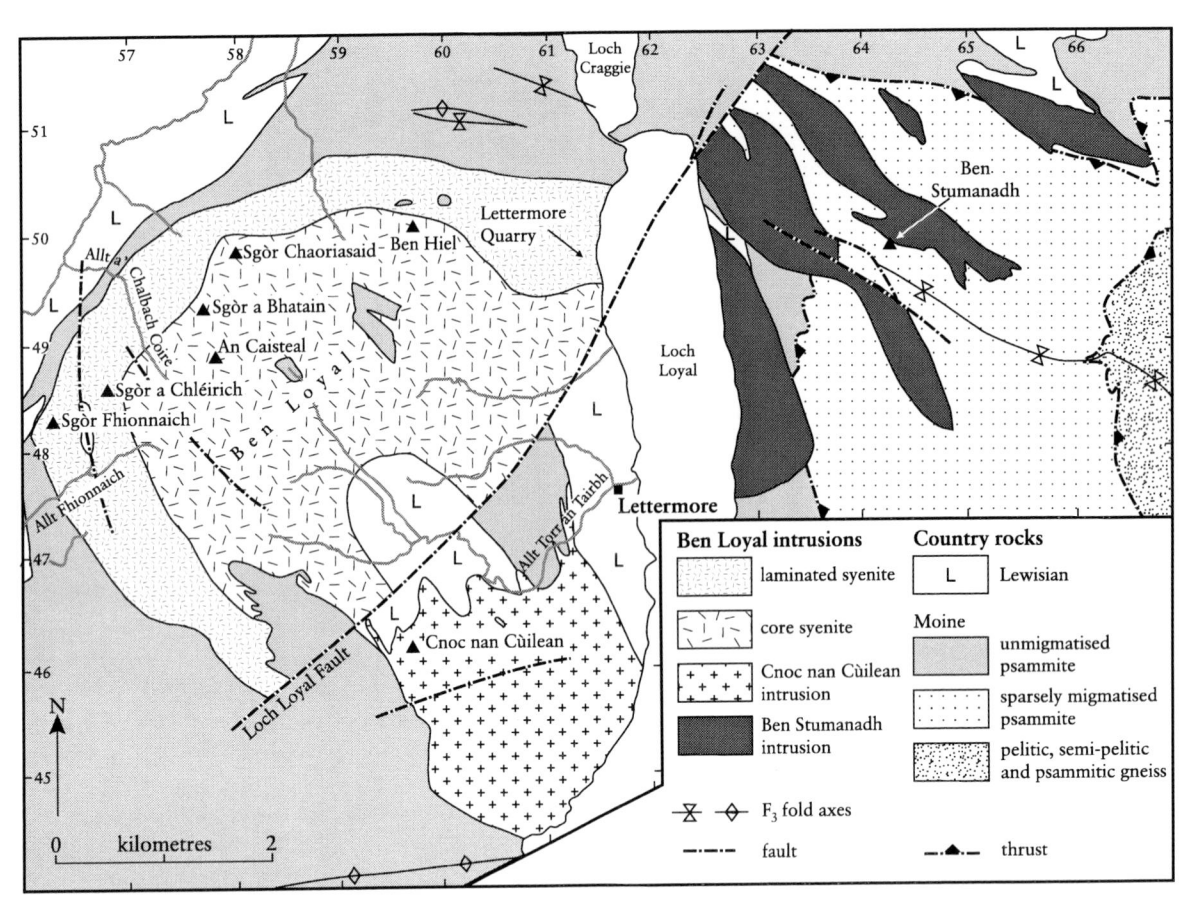

Figure 7.12 Map of the Loch Loyal syenite intrusions and their envelope rocks (compiled from Holdsworth and Strachan, in press; and Robertson and Parsons, 1974, fig. 1).

it begins to swing into parallelism with the margin of the intrusion. Dips are always towards the intrusion and increase towards the contact to 40–60°. Very close to the contact dips decrease again locally and the schists become crumpled. The most recent view of the change of strike in the envelope rocks (Holdsworth and Strachan, in press) is that the intrusion has been emplaced in a zone of large-scale SE-plunging folds of local F3 age, attributable to differential displacements on underlying ductile thrusts before the emplacement of the syenite.

In contrast with the chemically similar pink and grey syenites of Assynt, the Ben Loyal syenites are usually white or cream in colour, although late faulting may lead to development of pink variants. Two distinct variants can be distinguished: an outer, laminated syenite, and a relatively structureless core syenite (Gallagher *et al.*, 1971; Robertson and Parsons, 1974). The boundary between the two variants is gradational over several hundred metres, and faint laminations are sometimes seen even in the core syenites. Chemically, the two variants are indistinguishable (Robertson and Parsons, 1974) and this shows that, despite the quite frequently encountered Moine xenoliths in the laminated syenites, chemical effects of assimilation are unimportant. The lamination dips inward usually at 20–40° and is brought out by parallel prismatic amphibole and pyroxene crystals, and by tabular feldspars. The minerals show minor 'swirl' effects and this, together with the lack of evidence of assimilation, led Robertson and Parsons (1974) to reject the idea that the lamination was a 'ghost' Moine stratigraphy. Instead they suggested that it was caused by movements in a crystal mush during the last stages of consolidation. At the same time as the lamination developed the feldspar assemblage in the syenite changed by strain-facilitated exsolution and marginal recrystallization from a one-feldspar, hypersolvus syenite into a two-feldspar, subsolvus assemblage. The core of the intrusion escaped this flow deformation and preserves the original hypersolvus assemblage. This is an interesting and important textural change in felsic rocks, first suggested by Tuttle and Bowen (1958), in their classic memoir on the origin of granite, and it is not seen in the other Scottish syenites. Indeed two-feldspar syenites seem to be rather uncommon rocks, on a worldwide basis.

The Ben Loyal quartz-syenite has a thoroughly peralkaline character and consistently shows normative *acmite* (the pyroxene component $NaFeSi_2O_6$) in analyses (Robertson and Parsons, 1974). This is not true of the majority of the syenites in Assynt. Both core and marginal variants are extremely consistent in quartz content, with 5–10% normative *quartz*. The most abundant coloured mineral is a bright-green aegirine-augite, but in the laminated marginal syenites a green amphibole of the eckermannite–arfvedsonite series may be present and is dominant in parts. Both syenite units, but particularly the marginal variant, contain vugs lined by a yellow, powdery mineral identified by von Knorring and Dearnley (1959) as a rare-earth-bearing monazite group mineral. These cavities also contain montmorillonite, harmotome and stilbite. Boulders of pegmatite with green amazonite (a variety of K-feldspar), thorite, galena, titanite and topaz were reported by Heddle (1883b, 1901) from the boulder-scree slopes in the NW of the Ben Loyal mass.

A quarry at Lettermore (612 498) provides ready access to good exposures of fresh, very slightly laminated marginal syenite, with rare small schist xenoliths. Druses with the yellow rare-earth-rich mineral coating are common. A much more remote area, around Allt Fhionnaich (564 475), Sgòr a' Chleirich (568 485) and Allt a' Chalbhach Coire (572 488) can be used to demonstrate all the main features of the complex and its contact relationships. It is an area of quite exceptional scenic grandeur. On the peaks of Sgòr Fhionnaich and Sgòr a' Chleirich the inward-dipping lamination of the marginal variants can be readily mapped (Robertson and Parsons, 1974). The unlaminated core variant appears on the NE slopes of Sgòr a' Chleirich and in the upper Allt a' Chalbach Coire (573 483).

Cnoc nan Cùilean intrusion

This pluton has an oval exposure over an area of about 3 km² to the south of the main Ben Loyal mass and forms an imposing conical hill rising above the Moine rocks. On the early maps of the Geological Survey it is shown connected to the Ben Loyal intrusion in the poorly exposed ground at the head of Allt Torr an Tairbh, but mapping and geophysical work reported by Robertson and Parsons (1974) suggested that the two intrusions are separated by an area of Moine and Lewisian rocks. This separation has been confirmed by the recent mapping of Holdsworth and Strachan (in press) who see the Cnoc nan Cùilean body as separated from the main Ben Loyal mass by the major Loch Loyal Fault (Figure 7.12). That the two intrusions are separate is further supported by their chemical differences. Normal Cnoc nan Cùilean syenites, which are usually pink in colour, have little or no normative *quartz*, have more normative *orthoclase* than the Ben Loyal syenites, and no normative *acmite*. They are also richer in mafic minerals and have a larger radiometric anomaly because of high concentrations of thorite (Gallagher *et al.*, 1971). Mapping by McErlean (1993) reported by Holdsworth and Strachan (in press) has revealed that the Cnoc nan Cùilean intrusion was emplaced as a series of NW-trending sheets, and that the syenites have internal foliations similar to those in the other Loch Loyal intrusions. They therefore suggest that the 426 ± 9 Ma U-Pb age obtained for the Cnoc nan Cùilean syenite can be used reliably to date the emplacement of all the Loch Loyal syenites as a whole.

King (1942) presented a map of the Cnoc nan Cùilean body that showed a pronounced concentration of basic xenolithic inclusions around the margins. Such inclusions are not uncommon in the interior of the mass as well. King noted that there is a considerable gradation in appearance of these basic inclusions, from those that are obviously schistose and were clearly originally Moine rocks, to much more highly recrystallized, structureless inclusions. He presented a detailed account of the metasomatism of these inclusions, which was a highly topical field at what was the time of the 'granite controversy'. Although the scale of metasomatism envisaged was probably greater than would now be accepted, the study nonetheless is a very valuable account of the progressive metasomatism of these Moine inclusions, and the textural variety can readily be appraised in the field. Excellent examples of the xenolithic rocks can be obtained in the stream section of Allt Tòrr an Tairbh (612 473–609 469) where the sheet-like form of the syenite can also be seen. Less clear exposures can also be seen in the cliffs above Loch Loyal Lodge (around 615 465), and on top of the ridge, areas of 'normal' relatively xenolith-free syenite can be seen.

Ben Stumanadh intrusion

This set of intrusions is of less general petrological interest than the other Loch Loyal intrusions but is structurally important. Robertson and Parsons (1974) showed that the rocks are chemically similar to those of Ben Loyal and suggested that they might be easterly protrusions. More recent mapping (Holdsworth and Strachan, in press) has led to the suggestion that the Ben Stumanadh syenite is separated from that of Ben Loyal by the Loch Loyal Fault. The new mapping shows a set of at least five, major, steeply dipping, NW-trending sheets of syenite (Figure 7.12) which appear to be coalescing to the NW. There are also many, thinner, steeply dipping sheets that occur on all scales from centimetres to several hundreds of metres thick. The sheets are emplaced sub-parallel to the strike of the foliation of the country rocks. Contacts are usually sharp although in places feldspathization of the Moine can be demonstrated, and the junctions can be gradational. Hornfelsing of country rocks is visible for tens of metres from contacts, and the development of fibrolitic sillimanite in the Moine semipelites around the summit and to the NE of Ben Stumanadh (e.g. 647 508) may be due to a thermal metamorphic overprint. There are deformational structures in the Moine which seem to be associated with the emplacement of the sheets. Folds in the country rocks and offsets of bedding in host psammites at the margins of foliated members of the Ben Stumanadh sheets (e.g. on Ben Stumanadh itself, 649 502) indicate a steep NW-trending dextral sense of shear parallel to sheet walls during emplacement.

Despite their chemical similarity to the Ben Loyal syenites, the Ben Stumanadh rocks are dark-brown to pink in colour, the colour variation perhaps depending on late faulting (Holdsworth and Strachan, in press). They are usually two-feldspar syenites (like the outer unit in the Ben Loyal mass) with strongly aligned aegirine-augite and arfvedsonitic amphibole. Up to 16% quartz may be present, locally occurring in graphic intergrowth with feldspar. Miarolitic cavities similar to those in the Ben Loyal syenites also occur. Large parts of the Ben Stumanadh intrusions display cataclastic textures associated with low temperature brittle faulting. Exposures showing both the overall structure, petrography and contact relationships are conveniently found in the wooded slopes of Sròn Ruadh (627 507) east of the northern end of Loch Loyal.

Interpretation

The Loch Loyal syenites are petrographically relatively simple compared with the Assynt plutons, only the Cnoc nan Cùilean intrusion showing obvious petrographical variety in the form of mafic inclusions. These are in all probability variably metasomatized xenoliths of Moine and Lewisian envelope rocks. Chemically, the Ben Loyal and Ben Stumanadh syenites are similar, although there are textural and colour differences that perhaps reflect more intensive, late brittle deformation, and associated alteration, in the Ben Stumanadh rocks. The Cnoc nan Cùilean intrusion is chemically distinctive, which perhaps suggests multiple emplacement of two magmas, fractionated at depth.

Internally, the Ben Loyal syenite can be roughly sub-divided into chemically indistinguishable outer laminated and inner structureless syenites. There is general agreement that the lamination is non-metamorphic and formed during the late stages of consolidation, very probably during a 'ballooning' diapiric form of emplacement in which the core syenites were emplaced last. There is a degree of controversy about the attitude of the contacts on the north and west sides of the intrusion, but the body appears to dip beneath country rocks to the SE, and to be truncated by a large fault. The Cnoc nan Cùilean and Ben Stumanadh syenites occur to the SE of this fault. A good case can be made that the Ben Stumanadh syenites, which have the form of a series of steeply dipping sheets, represent a high-level section, brought down by the Loch Loyal Fault, equivalent to the eroded upper part of the Ben Loyal intrusion. The Cnoc nan Cùilean body also has a sheeted internal structure, but the distinctive mineralogy (little or no quartz, and significant enrichment in normative *orthoclase* relative to the other feldspar components when compared with the Ben Loyal syenites) and abundant metasomatized schist and gneiss xenoliths, set it apart from Ben Loyal and suggest that it represents a separate phase of intrusion. The chemistry of the syenites at Cnoc nan Cùilean suggests that they formed from a slightly less evolved magma than those of Ben Loyal, but care must be taken in this interpretation because of the clear evidence for assimilation of country rocks.

The Ben Loyal syenites are consistently peralkaline and quartz bearing. The marginal, laminated syenites are unusual in being two-feldspar rocks, in contrast with the one-feldspar, hypersolvus core syenites. The development of a two-feldspar assemblage probably occurred by a continuous process of exsolution and recrystallization during the mild deformation accompanying the ballooning period of emplacement. McErlean (1993) has suggested that serrated grain boundaries in syenites from Ben Loyal may be due to a solid-state overprint representing a late deformational phase. However, such textures are seen in undeformed syenites elsewhere and in the writer's view textures like those illustrated by Robertson and Parsons (1974, plate 2) do not necessarily imply externally imposed deformation. The yellow rare-earth monazite-group mineral, often present in miarolitic cavities, is a thoroughly alkaline characteristic, and was no doubt deposited from a late-stage aqueous fluid phase. The Cnoc nan Cùilean and Ben Stumanadh intrusions lack normative *acmite* in bulk analyses, although at Ben Stumanadh the presence of alkali pyroxenes shows the similarity to the Ben Loyal syenites. A U-Pb age for the Cnoc nan Cùilean intrusion of 426 ± 9 Ma is probably representative of the Loch Loyal intrusions as a whole. Within errors it is the same as the age of the Loch Borralan complex in Assynt. The Loch Loyal intrusions were certainly emplaced after the regional metamorphism of the Moine in Sutherland and Caithness, which also pre-dated the emplacement of the perhaps slightly older (439 ± 4 Ma) Loch Ailsh intrusion, which is cut by the Moine Thrust in Assynt.

Conclusions

The Ben Loyal intrusion is the grandest expression of alkaline magmatism in the British Isles. It is composed of a peralkaline quartz-syenite ('nordmarkite') and its alkaline character is underlined by the common presence of an unusual rare-earth mineral. The laminated, outer unit of the intrusion is a two-feldspar syenite (the only British example) which shows important microtextural changes in its transitional relationship to the chemically identical, unlaminated, one-feldspar core syenite. The fabric of the outer unit was produced during emplacement as a 'ballooning' diapir and is the only proven example of such a style of intrusion in the NW Highlands alkaline suite.

The two satellite intrusions, of Ben Stumanadh and Cnoc nan Cùilean, have different characters. They are separated from the Ben Loyal intrusion by the major Loch Loyal Fault. The Ben Stumanadh syenites are similar to the outer unit of Ben Loyal and were emplaced as a set of steeply dipping, NW-trending sheets in Moine psammites. A good case can be made that they represent a downfaulted portion of an upper section of the Ben Loyal mass, and the two intrusions thus provide important insights into emplacement mechanisms of plutons. The Cnoc nan Cùilean satellite has an internal sheeted structure but is chemically less evolved than the other two intrusions, suggesting that the Loch Loyal magmatism proceeded in at least two pulses, the magmas fractionating at depth prior to rising to their present level. The presence of numerous metasomatized Moine and Lewisian xenoliths in the Cnoc nan Cùilean intrusion, many of which show signs of assimilation, also sets this satellite apart from the Ben Loyal mass, in which xenoliths are relatively rare. Early descriptions of metasomatic reactions seen in the Cnoc nan Cùilean xenoliths have a historical place in discussions on the origin of granite (the 'granite controversy'). The Loch Loyal intrusions were emplaced after the metamorphism of the Moine envelope rocks, the only thoroughly alkaline rocks in this tectonic setting. A U-Pb age of 426 ± 9 Ma for the Cnoc nan Cùilean intrusion probably applies to the entire group of intrusions, providing an important regional time-marker. It is, within errors, the same as the age of the Loch Borralan intrusion in Assynt. In Assynt there are late displacements post-dating the Loch Borralan and Loch Ailsh intrusions, and it seems probable that the Loch Loyal intrusions will have been displaced towards the west by these late movements on the Moine Thrust.

ALKALINE MINOR INTRUSIVE ROCKS

I. Parsons

INTRODUCTION

The extensive suite of dykes and sills of alkaline

and related rocks in the Assynt district are petrologically unusual in a British context, and are important representatives of the Caledonian alkaline magmatism. They extend over a considerable distance, from north of the Assynt district at Loch More (NC 330 350), to south of the Assynt Culmination in the structure known as the Achall Culmination, near Ullapool (NH 144 953, Figure 7.1). Although they are less well-known than the alkaline plutons, a glance at the Geological Survey special sheet of the Assynt district shows at once that they constitute a major part of the magmatism in Assynt. Although many of the rock types are not strictly alkaline their association in time and space with the alkaline magmatism is clear. They have very important structural and geochronological implications for understanding the evolution of the Moine thrust zone, into which many members were emplaced. The value of the minor intrusive rocks considered as a suite is greater than that of individual sites in isolation, and their relationship to the major alkaline plutons and to the individual thrust sheets, or nappes, is of critical importance. The profound implications of the alkaline rocks for the magmatic and tectonic evolution of the NW Highlands are discussed in the introduction to this chapter, and the age relationships are summarized in Table 7.1. The distribution of some of the minor intrusive rock types is shown on Figure 7.2. The reader is reminded that we are dealing with a region of very considerable crustal shortening, perhaps in excess of 100 km normal to the thrust belt if the Moine Thrust itself is included (Elliott and Johnson, 1980), and that the region involved in the alkaline magmatism must have extended from the unmoved Foreland (where nepheline-syenite dykes are found on the Atlantic coast) to a point many tens of kilometres to what is now the ESE.

The petrography, chemistry and distribution of the dykes and sills was described by Sabine (1953) who also recognized the importance of the suite as structural markers. He recognized six main petrographical types of minor intrusive rock, together with some localized varieties. The sites are grouped below according to the six rock types using the varietal and local rock names adopted by Sabine (1953), with the more usual modern equivalents given where appropriate. The different types are as follows.

1. 'grorudite' (peralkaline rhyolite, comendite)
2. The Canisp Porphyry (porphyritic quartz-microsyenite)
3. 'hornblende porphyrite' (hornblende microdiorite, spessartite)
4. vogesite (hornblende-rich lamprophyre)
5. 'nordmarkite' (quartz-microsyenite)
6. 'ledmorite' (melanite nepheline-microsyenite)

A common introduction is provided here for each of the rock types, rather than for each individual site. The sites described are widely distributed (Figures 7.2, 7.13) and were chosen to provide examples with particularly significant structural relationships or which are relatively accessible examples of the different rock types making up the minor intrusive suite. There is one example of a probable diatreme (called a 'volcanic vent' by the early Geological Survey mappers) in Assynt. This is associated with a vogesite sill and is included in that section.

1. 'GRORUDITE' (PERALKALINE RHYOLITE, COMENDITE)

Introduction

The 'grorudites' are fine-grained acid (SiO_2 = 72–76%) peralkaline rocks for which aegirine rhyolite or comendite would be appropriate modern names. They are compositionally the equivalent of peralkaline granites and have no counterpart among the Caledonian plutonic rocks. They are therefore important in that they extend the compositional range of the Scottish Caledonian magmatism to include both strongly silica under- and oversaturated alkaline rocks, a dichotomy found in many much larger alkaline provinces. Thompson and Fowler (1986) pointed out the chemical similarity between an Assynt 'grorudite' and a comendite from the shoshonitic volcano of Lipari in the Aeolian Arc. The freshest examples may be pale-green in colour but in general they are pink or reddish-brown. The rocks are usually very fine grained and the matrix is aphanitic, although small pink alkali feldspar phenocrysts are sometimes visible. In section the rocks are characterized by phenocrysts of alkali feldspar and the sodium-iron pyroxene, aegirine, set in a quartz–feldspar matrix crowded with tiny aegirine needles. Various sub-varieties were recognized by Phemister (1926) and Sabine (1953). The reader is referred to these works for petrographical detail. 'Grorudites' are common in Assynt, particularly to the east and north of Inchnadamph (where they may be seen on the Cnoc an Droighinn GCR site (263 226), and cutting and close to the Loch Ailsh syenite intrusion (Figure 7.2). They usually occur as thin dykes (up to 1 m) but also, although less commonly, as sills, often describing a rather sinuous path through the country rocks. The examples are chosen to demonstrate the structural implications of the 'grorudite' suite.

GLEN OYKEL SOUTH
(NC 327 136)

Description

The site, in slabs in the bed of the River Oykel *c.* 750 m north of its confluence with the Allt Sail an Ruathair (Figure 7.8), shows an unusually thick (*c.* 4 m) dyke of reddish-brown 'grorudite', trending E–W and cutting the S2 syenite member of the Loch Ailsh intrusion. The dyke very probably corresponds with a similar large dyke in the Allt Sail an Ruathair. This dyke is a member of a swarm that cuts both the Loch Ailsh intrusion and its envelope on Sgonnan Mór, where the dykes run undeviated across the Sgonnan Mór Syncline (Milne, 1978) and end abruptly at the Ben More thrust plane. These age relationships are discussed by Elliott and Johnson (1980).

Interpretation

The 'grorudites' are restricted to the Ben More Nappe (Figure 7.2). As this example cuts the Loch Ailsh pluton the syenites must pre-date the 'grorudites'. The pluton must therefore have moved from what is now the east on the Ben More thrust plane. As the Loch Ailsh syenite was emplaced at 439 ± 4 Ma, and the Loch Borralan syenite was emplaced after the movements on the Ben More Thrust at 430 ± 4 Ma these dates bracket both the age of movements on the Ben More Thrust and the time of emplacement of the 'grorudites'. Conclusions concerning the relationship of the Loch Borralan complex to the Ben More Thrust were reached from observed cross-cutting relationships (see the Loch Borralan GCR site report, Interpretation), but the interpretation is supported by the absence of 'grorudites' from the Loch Borralan mass. The presence of a swarm of 'grorudite' dykes cutting both the Loch Ailsh mass, and nearby Lewisian gneiss on Sgonnan Mór, was considered by Bailey (1935) to throw doubt on the existence of the Sgonnan Mór Thrust postulated by Peach *et al.* (1907).

Conclusions

The Glen Oykel South GCR site provides a thick example of a 'grorudite' dyke, an unusual rock type that is unique in the British Caledonides. At this locality it cuts the Loch Ailsh syenite, demonstrating that the Loch Ailsh intrusion was emplaced in the Ben More Nappe prior to movements on the Ben More thrust plane. The cross-cutting relationships are extremely important for the understanding of the timing, absolute and relative, of events in the Moine thrust zone.

CREAG NA H-INNSE RUAIDHE
(NC 224 140)

Description

A fine-grained, red 'grorudite' dyke, about 1 m thick, striking at outcrop NNE–SSW, cuts the Cambrian 'false-bedded quartzite' in a GCR site that is also notified for structural reasons, the Cam Loch Klippe (Figure 7.2, locality 2; and see the GCR volume *Lewisian, Torridonian and Moine Rocks of Scotland*). The Lewisian–Cambrian unconformity occurs about 200 m to the SE, and is in part inverted in the lower limb of a fold that is truncated below the 'grorudite' exposure by the Ben More thrust plane.

Interpretation

The interpretation of Peach *et al.* (1907) of the structure of the Moine thrust zone in Assynt is a classic of British, and indeed world, geology. They interpreted an area of Lewisian rocks overlain unconformably by Cambrian quartzites to the east of the Cam Loch as forming an outlier, or klippe, of the Ben More thrust sheet (Figure 7.2). This interpretation has been supported by modern re-interpretations of the geology of the western part of the thrust belt in Assynt (Elliott and Johnson, 1980; Coward, 1985) although the latter author considers the eastern edge of the klippe to be a fault, not a thrust. As noted for the Glen Oykel South GCR site, 'grorudites' occur only in the rocks above the Ben More thrust plane, and the elegant and robust inter-

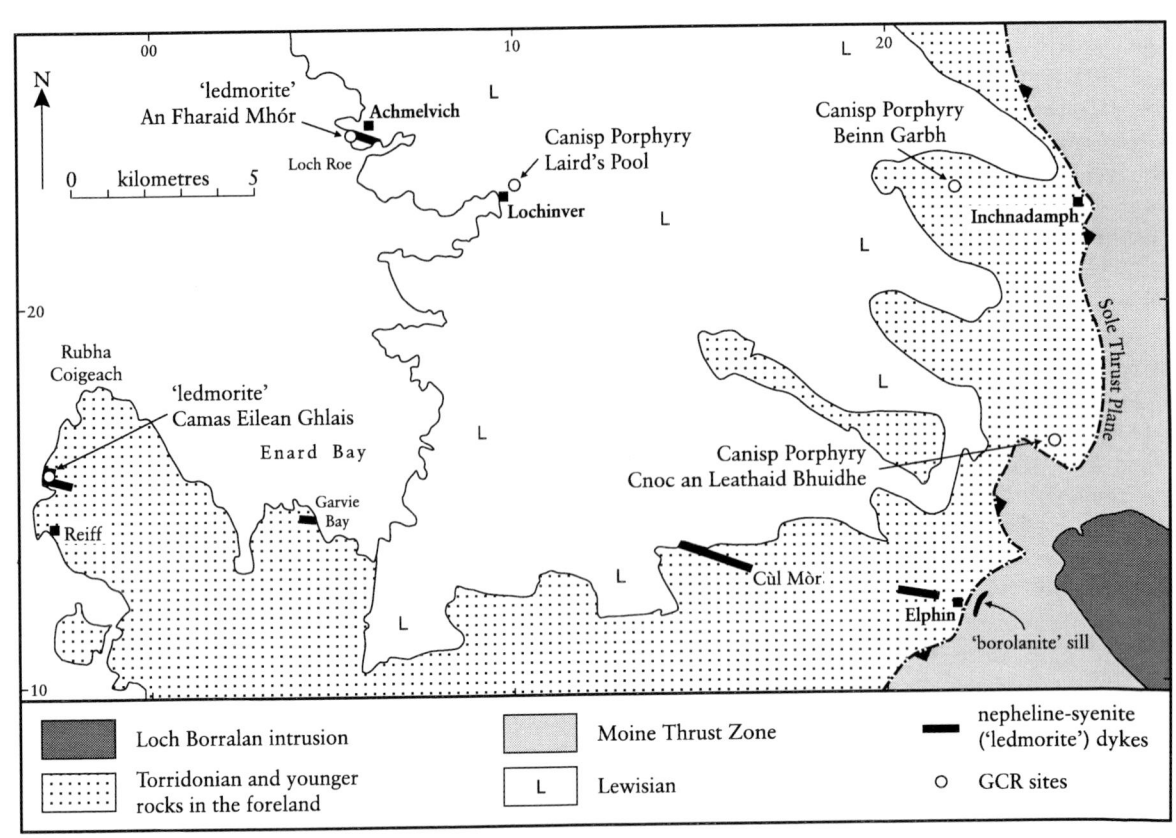

Figure 7.13 Map of western Assynt showing distribution of nepheline-syenite ('ledmorite') dykes in the Foreland and their relationship to the Loch Borralan nepheline-syenites in the Moine thrust zone. GCR sites exemplifying the 'ledmorite' dykes and the Canisp Porphyry are also shown. The full extent of the Canisp Porphyry around Beinn Garbh is shown on Figure 7.15.

pretation of Peach *et al.* (1907) is supported by the 'grorudite' at Creag na h-Innse Ruadh.

Conclusions

The Creag na h-Innse Ruaidhe GCR site demonstrates the presence of 'grorudite' in one of the outlying thrust slices (klippen) of the Ben More Nappe, providing support for the internationally well-known and historically important structural interpretation.

2. THE CANISP PORPHYRY (POR-PHYRITIC QUARTZ-MICROSYEN-ITE)

Introduction

The 'Canisp Porphyry' is one of the most striking igneous rocks in Assynt both because of its appearance in hand specimen and because of the way its sills dominate the skyline in the profiles of the well-known peaks of Canisp, Suilven and Beinn Garbh (Figure 7.14). The rock has the composition of a quartz-syenite although it is very sodic and the norm is unusually *albite*-rich. Its composition is similar to the upper quartz-syenites of Cnoc-na-Sroine in the Loch Borralan intrusion. In hand specimen the typical Canisp Porphyry is reddish-brown in colour with a fine-gained aphanitic groundmass containing well-shaped alkali feldspar phenocrysts up to 20 mm in length. These were analysed and figured by Heddle (1881) who described the rock as 'one of the most striking porphyrys of Scotland'. Paler, creamy phenocrysts are albitic plagioclase. The fine-grained groundmass consists of turbid K-feldspar, plagioclase and quartz.

The Canisp Porphyry crops out over an extensive area (Figure 7.15), one sill forming a plateau on the summit of Beinn Garbh (Figure 7.14) and then following the dip-slope of the Cambrian quartzites down its eastern flank towards Inchnadamph (Beinn Garbh GCR site). On Suilven and Canisp it forms sills, five in number on Canisp, in the Torridonian sandstones. Sabine (1953) provided a section suggesting correlations between the sills on the three mountains. It also occurs as dykes cutting Lewisian gneisses to the west, of which the most distant exposure is 12 km away from Beinn Garbh, near Lochinver, at the Laird's Pool GCR site (Figure 7.13). As Sabine (1953) noted, in view of its widespread distribution to the west of the Sole Thrust, the restriction of the Canisp Porphyry to the Foreland only is rather remarkable. In two places, on the lowest eastern slopes of Beinn Garbh, around the stream Cam Alltan (244 205) and near Loch Awe, on the Cnoc an Leathaid Bhuidhe GCR site (Figure 7.13), the Canisp Porphyry approaches close to the Sole Thrust, and although it is never seen to be truncated by the Sole it is certainly never seen in the rocks to the east.

The usual interpretation placed on the absence of Canisp Porphyry from the thrust zone (Parsons, 1979; Halliday et al., 1987) is that it represents the earliest phase of magmatism in Assynt, so that its emplacement was complete before the thrust-sheets arrived. It could also be roughly synchronous with the emplacement of the Loch Ailsh pluton and the 'grorudites', albeit they were emplaced many kilometres to the east.

Whatever the age relationships, there is no doubt that the source of the silica-oversaturated Canisp Porphyry magma was below the Foreland Lewisian, like that giving rise to the silica-undersaturated nepheline-syenite dykes that occur at the An Fharaid Mhór and Camas Eilean Ghlais GCR sites (Figure 7.13).

BEINN GARBH
(NC 227 222)

Description

The view of Beinn Garbh (540 m) from the north, across Loch Assynt, is one of the most celebrated in British geology (Figure 7.14), showing the striking contrast between unbedded Lewisian gneisses at the base, horizontal, well-bedded Torridonian sandstones resting on them, and both being overstepped by Cambrian quartzites dipping to the east at about 15°, the so-called 'double unconformity'. The implications of the different dips has intrigued generations of students. The dips are in fact brought out by two parallel major sills, 6–20 m in thickness, of Canisp Porphyry on Beinn Garbh (Figures 7.14, 7.15), of which the upper forms an extensive plateau on the flat-lying

Figure 7.14 Beinn Garbh (540 m, left) and Canisp (846 m, right) from Loch Assynt. The plateau of Beinn Garbh and the steps in the skyline of Canisp are formed of sills of Canisp Porphyry. (Photo: I. Parsons.)

Torridonian sandstones and an extensive easterly-dipping exposure on the dip-slope of the Cambrian quartzites to the east, extending almost to the Sole Thrust south of Inchnadamph. There are no sills in the Lewisian.

Interpretation

The sills have clearly followed the bedding of the Torridonian and Cambrian rocks, and change dip as they cross the unconformity. The relatively slow-weathering Canisp Porphyry forms a conspicuous plateau on Beinn Garbh, and Sabine (1953, fig. 4) illustrates how the sills can be correlated with conspicuous topographical steps on Canisp.

Conclusions

Beinn Garbh is a visually outstanding GCR site providing the most extensive exposures of a unique and celebrated hypabyssal alkaline rock type, and excellent examples of the influence of rock type on topography and scenery. The outcrop of the Canisp Porphyry shows the structural control of the emplacement of the sills in the vicinity of the famous 'double unconformity' of Cambrian on both Torridonian and Lewisian rocks.

THE LAIRD'S POOL, LOCHINVER (NC 103 235)

Description

A dyke, about 4 m thick, of a red porphyritic rock with conspicuous pink K-feldspar phenocrysts up to 5 mm long, crosses the River Inver at the Laird's Pool (Figure 7.13), cutting Lewisian gneisses, and striking about 100°. The site can be reached easily along the maintained path on the south side of the river.

Interpretation

In view of the appearance and orientation of this dyke there seems every reason to correlate the rock with the Canisp Porphyry. It therefore represents the most westerly expression of Canisp Porphyry magmatism at the surface and, like the nepheline-syenite dykes farther west at Achmelvich (see the An Fharaid Mhór GCR site report, Figure 7.13) clearly shows that the alkaline magmatism was fundamentally a product of the Lewisian Foreland and the rocks beneath. The main focus of alkaline magmatism in Assynt is therefore incidentally related to the thrust belt rather than being in some way genetically connected to it. The Canisp Porphyry cuts Lewisian rocks at a number of localities in the Foreland (see Geological Survey special sheet) where it forms dykes; in the overlying sedimentary rocks it almost invariably forms sills.

Conclusions

The Laird's Pool GCR site is the most westerly example of Canisp Porphyry. Here it occurs as a dyke cutting Lewisian gneiss, which provides evidence for the widespread character of Canisp Porphyry magmatism and its relationship with the rocks of the Foreland.

CNOC AN LEATHAID BHUIDHE (NC 235 154)

Description

The exposure at this locality is poor and the main interest is in the structural implications. The Geological Survey (Peach *et al.*, 1892) mapped a sill of Canisp Porphyry cutting the Pipe Rock on the slopes of Cnoc an Leathaid Bhuidhe to the west of Loch Awe (Figure 7.15). The outcrop extends towards the Sole Thrust but is obscured in poorly exposed ground to the west of Loch Awe. There is no sign of the sill to the east of Loch Awe, where Salterella Grit, Fucoid Beds and Durness Group dolomitic limestones both above and below the Sole Thrust are exposed. Unfortunately, in the critical area, the loch itself and surrounding bog intervene so that possible truncation of the dyke by the Sole Thrust cannot be proved.

Interpretation

The relationship of this Canisp Porphyry sill to the surrounding rocks suggests, but does not prove, that its emplacement preceded all movements on the Sole Thrust. However, the very widespread development of Canisp Porphyry in the Foreland, and its complete absence from the thrust sheets to the east, strongly suggests that this phase of magmatism occurred early in the history of the alkaline magmatism in Assynt.

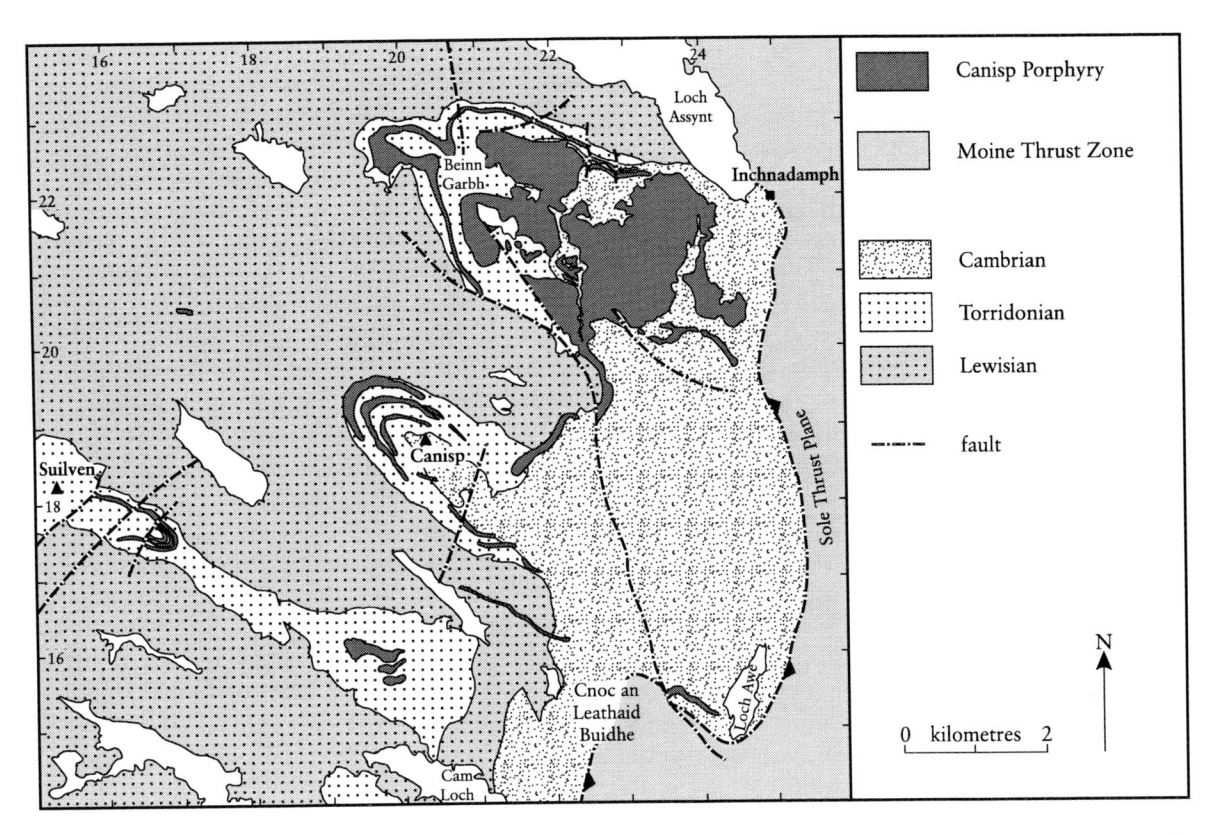

Figure 7.15 Distribution of sills and dykes of Canisp Porphyry in the Foreland. The dyke at the Laird's Pool, Lochinver, is farther to the west (see Figure 7.13). Only faults that affect Canisp Porphyry are shown. (After the Geological Survey special sheet for Assynt, 1923.)

Conclusions

The Cnoc an Leathaid Bhuidhe GCR site covers ground in which a sill of Canisp Porphyry approaches most closely the Sole Thrust from the west (in the Foreland), but is not exposed in the thrust zone to the east, providing evidence that Canisp Porphyry magmatism preceded movements on the Sole Thrust. Taken together with the absence of the porphyry from the thrust belt, this suggests that its emplacement pre-dated all thrust movements in Assynt.

3. 'HORNBLENDE PORPHYRITE' (MICRODIORITE, SPESSARTITE)

Introduction

Hornblende-bearing microdiorites are very abundant in Assynt and were first described by Bonney (1883). The majority form sills, although in places they cut across the bedding of the sedimentary rocks, and true dykes occur only in the Lewisian gneiss. The sills are usually of the order of 1 m thick, although one intrusion into Pipe Rock on the ridge of Breabag is over 30 m thick. They have a fine-grained feldspathic groundmass containing feldspar phenocrysts up to 6 mm in size and 3 mm hornblende phenocrysts. Phenocrysts of biotite are visible in places. In section the hornblende is commonly strongly colour-zoned. The feldspar phenocrysts are usually plagioclase, but alkali feldspar also occurs. The groundmass is usually made up of small prisms of euhedral or subhedral K-feldspar and sodic plagioclase enclosed in quartz. As in many lamprophyric rocks many of the constituent minerals are commonly very altered. While the Assynt microdiorites and lamprophyres are not strictly alkaline rocks, they undoubtedly overlapped the alkaline magmatism in both space and time and their role in the petrogenesis of the syenitic rocks must be considered.

Members of this suite are common in the Sole, Glencoul and Ben More thrust sheets but they do not occur in the Foreland or in the

Moine Nappe. This is yet another example of how the igneous rocks show that the thrust sheets have distinct characters. Sabine (1953, fig. 5) provided a map of the distribution of 'hornblende porphyrites' throughout Assynt. The relationships suggest that the 'hornblende porphyrites' were emplaced early in the igneous history of Assynt prior to all the thrust movements and perhaps contemporaneously with the Canisp Porphyry in the Foreland (Halliday et al., 1987, table 1). Of course, the upper-crustal shortening implied by the thrust movements means that the 'hornblende porphyrites', at the time of their emplacement were many kilometres to what is now the east of the Canisp Porphyry, and must have been emplaced over a wider area than now exposed because of later shortening on the Ben More thrust plane. The 'grorudites', discussed above, which are restricted to the Ben More Nappe, must have been emplaced prior to the main movements on the Ben More Thrust, in an area to the east which only overlapped partially that into which the 'hornblende porphyrites' were injected.

Because of their widespread distribution in Assynt, Read (1931) studied the cataclastic deformation of the 'hornblende porphyrites', following an early suggestion by Geikie (1888) that they show a progressive deformation eastward, mirroring the increase in metamorphism of the rocks generally in going from the Foreland to the Moine. Sabine (1953) revisited this problem and showed that cataclastic deformation of the 'hornblende porphyrites' is variable within the thrust belt and is most closely associated with proximity to one or other of the major thrusts.

There are many localities for viewing 'hornblende porphyrites' in the thrust belt. The GCR sites were chosen as representing intense swarms, one in an area of great structural complexity close to Inchnadamph, the other in an area with a considerable variety of hypabyssal rocks for which a map was provided by Sabine (1953).

CNOC AN DROIGHINN
(NC 263 226)

Description

This is a classic region of structural complexity in Assynt, overlooking Inchnadamph (Figure 7.2), in which the Glencoul Thrust brings Cambrian quartzites over thick duplexes (imbricated lenses) of Durness Group dolomitic limestones. The reader is referred to figure 15 of Elliott and Johnson (1980), and the accompanying text, for a modern structural interpretation. Numerous 'hornblende porphyrite' sills are structurally repeated in the duplexes beneath the Glencoul Thrust, but sills were also demonstrably emplaced repeatedly at several levels in the quartzites above the thrust. They may be seen in various stages of cataclastic deformation ranging from highly sheared to unsheared. Sabine's map (1953, figure 5) records the relative amount of deformation he observed. 'Grorudite' dykes and sills also occur in this site, but only above the Glencoul thrust plane.

Interpretation and conclusions

Cnoc an Droighinn is a structurally complex GCR site close to Inchnadamph, providing a major concentration of 'hornblende porphyrite' (and 'grorudite') sills cutting various Cambro-Ordovician rock types. Some sills are repeated structurally, others are emplaced at different levels in the succession. The 'hornblende porphyrites' are clearly pre-deformational and, despite structural repetition on the lower part of Cnoc an Droighinn, they were emplaced repeatedly at various levels in the quartzites above the Glencoul Thrust. The area is an object lesson in the difficulties of achieving correlations of igneous bodies in a region which, while largely non-metamorphic, is nonetheless structurally complex. The distribution of the 'hornblende porphyrites' provides a very useful marker of relative structural displacements in the thrust belt.

LUBAN CROMA
(NC 281 135)

Description

The site is a relatively accessible and representative part of a large area of barren, well-exposed ground to the north of the Loch Borralan intrusion (Figure 7.16). It is largely composed of Cambrian Pipe Rock, with some Fucoid Beds, cut by a large number of sills mainly of 'hornblende porphyrite'. Sabine (1953) provided a map of this part of Assynt which forms the basis of Figure 7.16. This shows the distribution of 'hornblende porphyrites', vogesites (see below), nepheline-syenite dykes representing outliers of the Loch Borralan mass, and a localized hypabyssal rock, which Sabine called the 'Breabag Porphyrite'. The latter occur as sills, 3–9 m thick, cropping out in a narrow belt about 5 km × 1.5 km between the Ledbeg River and the ridge of Breabag (Figure 7.16). All save one example (perhaps not *in situ*, according to Sabine) of this petrographical type occur in the Sole thrust sheet, i.e. below the Ben More Thrust. In hand specimen they have superficial similarities to fine-grained Canisp Porphyry but in section they prove to be microdiorites, with glomeroporphyritic aggregates of feldspar and phenocrysts of hornblende, set in a matrix of K-feldspar with a little quartz. Sabine's map gives a good impression of the variety of hypabyssal intrusive rocks found in this part of Assynt.

Interpretation and conclusions

The Luban Croma GCR site represents the wide variety of minor intrusions that can be demonstrated in this part of Assynt, including 'hornblende porphyrites' and a spatially restricted variety known as the 'Breabag Porphyrites'. It is possible that this variant provides a relative age-marker, because, like the 'grorudites', they must have been emplaced prior to movements on the Ben More thrust plane.

4. VOGESITE (HORNBLENDE-RICH LAMPROPHYRE)

Introduction

Rocks of the vogesite suite occur only in the Moine thrust belt, but are widespread in Assynt. They are found from Beinn Lice (NC 330 350) near Loch More in the north, to near Ullapool, 45 km to the SSW. A map showing the location of vogesites near Ullapool is provided by Sabine (1953, fig. 6); here they occur not far below the Moine thrust plane north and south of the Ullapool River at NH 147 958 and around NH 142 940, in an embayment structure in the Moine known as the Achall Culmination (Figure 7.1) (Elliott and Johnson, 1980), which is somewhat like a miniature version of Assynt. The structural implications of the vogesite suite in Assynt are essentially the same as those of the 'hornblende porphyrites' to which they have a similar distribution, although they tend to be most abundant in the Sole thrust sheet. Although vogesites are often altered, fresh exam-

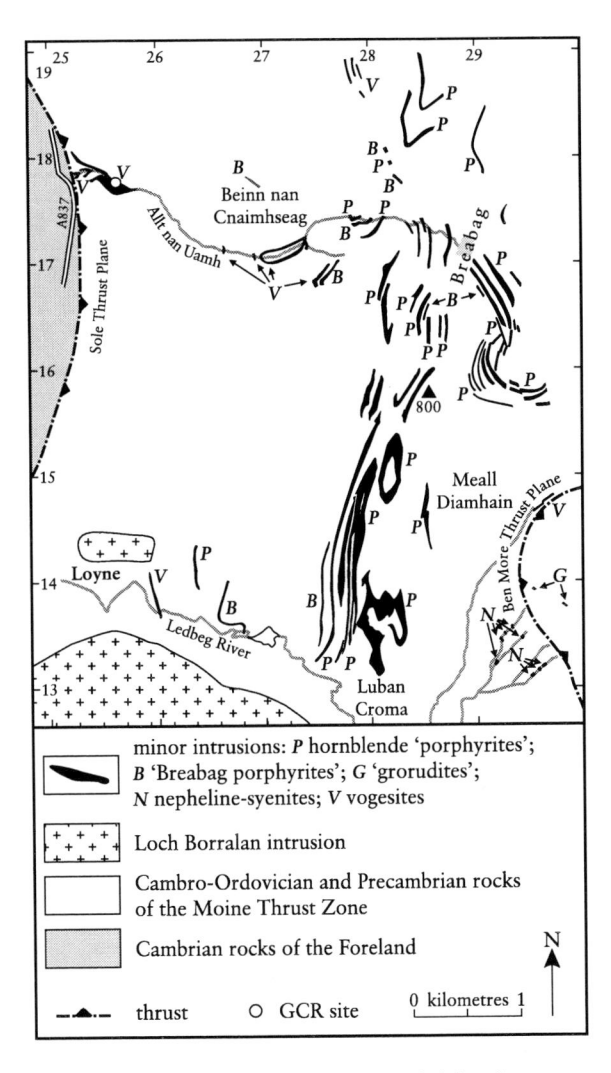

Figure 7.16 Distribution of sills and dykes between the Luban Croma and Allt nan Uamh sites, north of the Loch Borralan intrusion. (After Sabine, 1953, fig. 8.)

ples include some of the most attractive rocks among the minor intrusions. Petrographically they are relatives of the 'hornblende porphyrites' but with more hornblende and in some instances diopsidic pyroxene. The hornblende phenocrysts, which may be very abundant, are characteristically 3 mm in length and are set in a finer-grained matrix of euhedral plagioclase feldspar (albite-oligoclase), alkali feldspar, hornblende and a small amount of interstitial quartz. Some contain diopsidic pyroxene in glomeroporphyritic clots 3 mm across, or as euhedral phenocrysts.

The vogesites almost all occur in the form of sills, with a few dykes mainly in the Lewisian gneiss. They are almost all emplaced in Cambro-

Ordovician rocks, particularly in the Durness Group. The thickest are over 20 m but there are also thin sheets less than 1 m thick. Some sheets can be traced laterally for more than 3 km, for example in the limestone cliffs above Stronchrubie (NC 252 192), south of Inchnadamph. An historically interesting suggestion was made by Teall in 1886, pre-empting later arguments concerning the role of limestone in alkaline rock genesis in general, and the petrogenesis of pyroxene syenites in the Loch Ailsh intrusion. Teall noted that the pyroxene in the vogesites is nearly pure diopside (calcium-magnesium silicate) and that the pyroxene-bearing rocks usually intrude dolomitic limestones (dolomite is calcium-magnesium carbonate). He therefore suggested that the presence of the calcium-magnesium silicate mineral might be 'due to the absorption by the igneous magma of a certain amount of the dolomitic limestone into which the rock has been intruded'. Sabine (1953) revisited this problem and, from a study of 141 vogesite specimens, concluded that there is indeed a high correlation between the presence of diopside in a vogesite and injection into the Durness Group, although pyroxene does not occur exclusively in such rocks. He did not find field evidence for vogesite–dolomite reactions, however. Nonetheless, the chemical composition of the vogesites is similar to that of pyroxene syenites in the Loch Ailsh intrusion. There is some strong field and petrographical evidence for the importance of magma–dolomitic limestone reactions in the genesis of at least some of the pyroxene syenites (Parsons, 1968).

ALLT NAN UAMH (NC 256 179)

Description

A thick sill of a fresh vogesite with a high proportion of large prismatic hornblendes as much as 10 mm long, set in pink feldspar, occurs close to the A837 and can be reached by a maintained path in the beautiful valley of the Allt nan Uamh, best known for its bone-caves. The locality, slightly above the fish-farm, is shown on Figure 7.16. It has hardened the Salterella Grit which forms an attractive waterfall (Figure 7.17). The sill dips gently to the east and is about 20 m thick. According to Sabine (1953) this is one of the two thickest vogesite sills in Assynt. The other is on the A837 just north of Inchnadamph.

Figure 7.17 Waterfall in Salterella Grit hardened by vogesite sill below, Allt nan Uamh (see Figure 7.16). (Photo: I. Parsons.)

Glen Oykel north

Conclusions

The sill exposed in the Allt nan Uamh GCR site is a readily accessible, coarse-grained example of the most widespread hypabyssal intrusive rock type in Assynt, excellent for teaching purposes as an example of the lamprophyre family.

GLEN OYKEL NORTH (NC 312 161)

Description

This locality is in the floor of the Oykel valley about 1 km upstream of the western contact of the Loch Ailsh intrusion (Figure 7.2). Peach (in Peach *et al.*, 1907, p. 435) treated it in some detail and provided a charming 'ground plan' reproduced here as Figure 7.18. It is a roughly circular area about 22 m × 15 m, composed of variably sized jumbled blocks of hardened limestone, which Peach ascribed chiefly to the Eilean Dubh Formation of the Durness Group, in a carbonate matrix, surrounded by Pipe Rock and by a 3 to 5 m-thick vogesite sill, which extends southwards in the Pipe Rock. The breccia is arranged in layers of coarser- and finer-sized fragments, dipping steeply towards the edges of

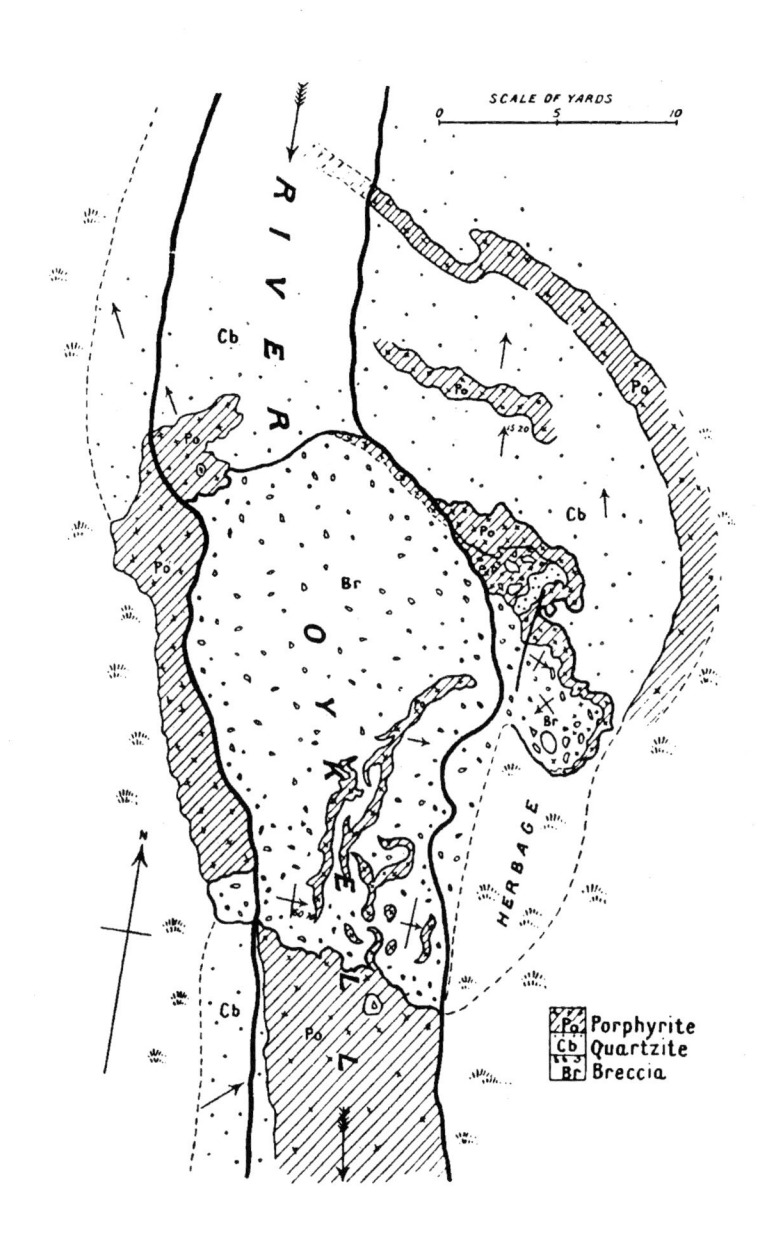

Figure 7.18 Facsimile of B.N. Peach's 'Ground Plan of possible Volcanic Vent, River Oykel, about three miles above Loch Ailsh' (from Peach *et al.*, 1907).

the depression in which the breccia lies. The breccia body has steep, clean-cut walls. The vogesite sill is emplaced in the Pipe Rock, which dips to the NE at 15–20°. According to Peach, the sill is later than the breccia, sends tongues into it, and has flowed between the limestone blocks, in some cases detaching them. A large mass of quartzite is also suspended in the vogesite. Peach noted that the vogestite is 'vesicular and slaggy' and Sabine (1953, p. 157) noted that it is very heavily pyritized where it intrudes the marble, more so than any other of the post-Cambrian sills in Assynt.

Interpretation

It is possible that Peach's interpretation of the general character of this locality is correct, although 'diatreme' would be a better modern description than 'volcanic vent'. If so it is the only example of transport by gas in the entire NW Highlands alkaline suite. The relationship of the vogesite to the diatreme is not clear. Sabine considered that the unusual pyritization lends weight to Peach's hypothesis that the breccia is volcanic in origin, although it is not clear how this equates with Peach's interpretation of the age relationships. Neither is it clear why the vogesite should have the vesicular character noted by Peach, if it is later than the breccia.

Peach discusses the age relationships of the pipe and the implications of its infill of limestone blocks. Because he was confident that the vogesite and 'hornblende porphyrite' sills in Assynt were emplaced prior to all the thrust movements, he concluded that the limestone blocks must have descended into the vent from above, estimating a minimum descent of about 70 m, the distance from the base of the Eilean Dubh Formation to the top of the Pipe Rock. On this basis the pipe formed prior to the thrusting in Assynt. On the other hand, if the vogesite is earlier than the breccia, and the shape of the body, the pyritization and the unusual vesicular character of the vogesite suggest that this may be the case, the breccia could have been emplaced later than the thrusting and could have incorporated limestone from hidden thrust sheets or from unmoved limestone beneath the Sole Thrust. The possibility, suggested by the presence of a carbonatite body near Loch Urigill (see the Loch Borralan GCR site report), that the matrix of the breccia might have been a hot carbonatite magma, can be ruled out because of recent measurements of carbon and oxygen isotopes. Both breccia fragments and matrix have identical isotopic compositions, characteristic of sedimentary carbonate rocks (K.M. Goodenough, pers. comm., 1997).

Conclusions

At the Glen Oykel North GCR site a pyritized vogesite sill is intimately associated with a jumbled mass of marble blocks, both being intruded into Pipe Rock. The structure was interpreted as a 'volcanic vent' by Peach et al. (1907). No other example of such a vent (or diatreme) is known in the NW Highlands alkaline suite. This is an intriguing locality of considerable historical and current importance. One hypothesis is best stated in Peach's own words:

'The appearances here observable afford plausible ground for believing that this orifice in the quartzite was a true volcanic vent, whence only gases may have escaped, and which was filled up by the descent of fragments from the walls of limestone above.'

On the other hand, if the diatreme is later than the vogesite, the limestone may have come from beneath the Sole Thrust. The carbonate matrix of the diatreme is not of igneous origin (i.e. it is not carbonatite).

5. 'NORDMARKITE' (QUARTZ-MICROSYENITE)

Introduction

This rock type occurs as sills in Assynt (Figure 7.2) and near Ullapool in the Achall structure (Figure 7.1). Their distribution is much more restricted than other types in the thrust belt because they are invariably found emplaced very close to the Moine Thrust itself, either in the Moine rocks just above the thrust plane, or in Cambro-Ordovician strata immediately below. They are thus unique in that their emplacement was apparently localized by the thrust structure itself, and in being emplaced in part in the Moine Nappe. The thickest sill, some 10 m thick, is emplaced just above the Moine Thrust at Druim Poll Eòghainn (NC 212 090) south of Knockan. Sabine (1953) reported that one of the least altered examples comes from a sill on Maol Calaisceig east of Ullapool (around

NH 144 944) and he provides a map (Sabine, 1953, fig. 6) and a detailed petrographical description. A detailed structural map of the Achall Culmination is in Elliott and Johnson (1980, fig. 22). They point out that the sill described by Sabine is slightly oblique to the foliation of the Moine rocks, so that its proximity to the Moine Thrust is to some extent fortuitous. Nevertheless the 'nordmarkites' do in general seem to be concentrated near the Moine thrust plane. The complete absence of any other members of the alkaline suite in the Moine Nappe is extremely striking and implies that the 'nordmarkites' were emplaced late in the tectonic history of Assynt (Halliday et al., 1987, table 1). As many examples are considerably deformed, late movements on the Moine Thrust must have occurred. This is an important conclusion for the tectonic reconstruction of the Moine thrust zone.

The 'nordmarkite' suite is varied petrographically and there is considerable range in quartz content and content of mafic minerals. Thus some are strictly syenites or pyroxene syenites. They are usually composed of variably fractured alkali feldspar crystals, easily visible in hand specimen, set in a fine-grained matrix of alkali feldspar, variable amounts of quartz, and chlorite. The rocks are strongly alkaline and similar in composition to the Canisp Porphyry in the Foreland, but the structural relationships of these two rock types suggests that the Canisp Porphyry was emplaced early and the 'nordmarkite' sills late. The quartz-syenites of Cnoc-na-Sroine in the Loch Borralan intrusion are also chemically similar and possibly almost contemporaneous (Halliday et al., 1987, table 1). Sabine discusses the possibility that the 'nordmarkites' were metamorphosed in the main regional metamorphism of the Moine, but favours the more modern view that the metamorphic changes occurred during later movements localized on the Moine Thrust.

ALLT NA CAILLICHE
(NC 320 102)

Description

The Allt na Cailliche (Figure 7.2) flows into the SE corner of Loch Ailsh and can be reached easily using forestry roads. The first exposures, just above the gravel fan, are of mylonitized, almost flinty Moine rocks. Upstream from this point pink 'nordmarkite' can be seen, forming a waterfall. In the gorge above, exposures in the stream bed are of foliated and sheared 'nordmarkite' with conspicuous pink feldspars. The rock is quite mafic. The Geological Survey mapped three sills near the Moine Thrust in the Allt na Cailliche, and also at several points near the thrust plane between there and the A837.

Interpretation and conclusions

This moderately deformed and unusual alkaline rock type was clearly emplaced as a sill just above the plane of the Moine Thrust, which must lie just below the lowest exposures in the Allt na Cailliche. This close proximity to the thrust characterizes all of the known exposures of deformed 'nordmarkites', and it seems highly unlikely that the association is fortuitous. One can conclude that the emplacement of the 'nordmarkites' was controlled by the thrust plane, occurred late in the evolution of the thrust zone, and that the moderate deformation and recrystallization of the 'nordmarkites' was caused by late movements on the Moine thrust plane.

6. 'LEDMORITE' (MELANITE NEPHELINE-MICROSYENITE)

Introduction

A dyke of silica-undersaturated syenite was recorded by Horne and Teall (1892) cutting Torridonian sandstones 28 km to the west of the Sole Thrust at Camas Eilean Ghlais (Figure 7.13) near Rubha Coigeach, NW of Achiltibuie. They pointed out its similarity to the rock type 'borolanite' that they had recently discovered and named in Assynt. A second example of a silica-undersaturated dyke, from near Achmelvich (Figure 7.13), was found in the Survey's collections by Sabine (1952). He suggested that the affinities of both rocks were with the 'ledmorites' because they lacked the pseudoleucite spots that characterize 'borolanite'. The two dykes are similar in mineralogy, although the example from Coigach is coarser grained and more altered. They are largely composed of an aggregate of orthoclase and nepheline (altered to natrolite) enclosing euhedral crystals of melanite garnet, prisms of aegirine and rare deep-brown biotite. Thomsonite occurs in interstitial clusters.

These dykes are extremely important, in both a historical and a modern context. Both extend well to the west of the thrust belt and, as Sabine (1953) pointed out, are far from any extensive exposures of limestone that might have been involved in the assimilation reactions central to the 'Daly–Shand' hypothesis. The nearest Durness Group carbonate rocks would have been well above the level of the dykes if carried to the west by the thrusts. Although one might argue that the dykes were emplaced laterally from the Loch Borralan magma chamber, Sabine's arguments contributed to the eventual abandonment of the limestone assimilation hypothesis. In a modern context, the dykes show that the source of this, the most alkaline magmatism in the British Isles, lies in the high-grade Lewisian gneisses of the Foreland or in the mantle beneath.

The dykes also have structural implications. Both strike towards the Loch Borralan intrusion in its present position in the thrust zone. Although horizontal movement of the Borralan mass of a few kilometres on the Sole Thrust cannot be ruled out, it certainly seems unlikely that a large displacement has occurred. The timing of major late movements on the Sole Thrust is a contentious issue, because Elliott and Johnson (1980) and Coward (1985), on the basis of detailed mapping, suggested that the Borralan intrusion has been considerably displaced by late movements on this thrust. Coward calculated a minimum displacement of 30 km. Halliday *et al.* (1987), however, accept the evidence provided by the 'ledmorite' dykes, and consider that the main movements on the Sole Thrust must have pre-dated the emplacement of the Loch Borralan complex.

CAMAS EILEAN GHLAIS
(NB 967 157)
POTENTIAL GCR SITE

Description

There are two sub-parallel dykes at Camas Eilean Ghlais (Figure 7.13), one presumably being a splay of its neighbour, striking approximately WNW. The larger can be found reasonably easily above the cliffs on the north side of this beautiful bay although, as a reddish-brown dyke cutting tilted Torridonian sandstones, it does not stand out very obviously. The presence of a high content of pink feldspars up to 2 mm in length is distinctive, however. Sabine (1953) provided a petrographical description and reported that the freshest specimens come from the more northern of the pair of dykes. An analysis was given by Horne and Teall (1892).

Sabine (1953) correlated the dykes at Camas Eilean Ghlais with a dyke at Garvie Bay (NC 039 139) and with a dyke that passes up through Lewisian into Torridonian on the NW flank of Cùl Mór (shown on the Geological Survey special sheet for Assynt at NC 140 129) (Figure 7.13). The dyke can be traced almost to Elphin (to NC 204 115) where it is less than 1 km west of a sheet of 'borolanite', an outlier of the Loch Borralan intrusion, in Durness Group carbonate rocks occurring above and to the east of the Sole Thrust. The dyke rock is very altered at this locality.

Interpretation and conclusions

The exposures in the Camas Eilean Ghlais site are at the western extremity of one of the two nepheline-syenite dykes that occur in the Foreland. These exposures of a unique dyke rock, here cutting Torridonian sandstone, demonstrate the spatial extent of the alkaline province, and tie the silica-undersaturated magmatism in Assynt firmly to the Foreland and the underlying mantle. The dyke can be traced sporadically to within 3 km of the main part of the Loch Borralan intrusion, in which it closely matches rocks of the 'ledmorite' type. Although continuity with the Loch Borralan intrusion cannot be proved directly, the alignment of this petrographically distinctive dyke with compositionally similar rocks east of the Sole Thrust, places important restrictions on the scale of displacements on this plane subsequent to the emplacement of the silica-undersaturated alkaline rocks.

AN FHARAID MHÓR
(NC 060 244)
POTENTIAL GCR SITE

Description

Sabine (1952) cut a section of a dyke-rock, from the Geological Survey's collections, that had been found cutting Lewisian gneisses near Achmelvich, WNW of Lochinver (Figure 7.13). Peach *et al.* (1907) thought it was equivalent to the Canisp Porphyry found at the Laird's Pool, Lochinver (Figure 7.13), but Sabine discovered that it is a nepheline-bearing rock similar to the dykes at Camas Eilean Ghlais. Sabine (1952) provides a sketch-map of the locality, and a detailed map of the Achmelvich peninsula, showing several slightly sinuous 'ledmorite' dykes striking approximately 110°, is given by Barber *et al.* (1978). The dykes occur on the An Fharaid Mhór-Clachtoll Lewisian GCR site and an example can be found on the shore of Loch Roe (060 244). A second dyke can be reached more easily by descending a gully in the cliffs on the western side of An Fharaid Mhór (053 244), where it is about 1–2 m thick. This dyke is a fine-grained, aphanitic, pale- chocolate-brown rock that is very distinct from the enclosing gneisses. It weathers in and cannot be traced in the higher ground on An Fharaid Mhór.

Interpretation and conclusions

The An Fharaid Mhór site provides an accessible second site for this important type of nepheline-syenite ('ledmorite') dyke in the Foreland, in this case cutting Lewisian gneiss. Although rocks that may be correlated with these dykes have not been discovered between this locality and the Loch Borralan intrusion, the measured strike would extrapolate only slightly to the north of the present position of the Loch Borralan 'ledmorites'.

Chapter 8

Late Silurian and Devonian granitic intrusions of Scotland

Introduction

INTRODUCTION

A. J. Highton

The late stages of the Caledonian Orogeny (*c.* 430–390 Ma) saw widespread voluminous granitic magmatism of essentially calc-alkaline characteristics in the Caledonian–Appalachian mountain belt. In Newfoundland, late Silurian pluton emplacement marks the final stages of collision between Laurentia and Avalonia (Bevier and Whalen, 1990; Whalen *et al.*, 1992). A similar event is manifest in the Greenland–Scottish sector. Within Scotland, the Caledonian Igneous Province encompasses both the orthotectonic zone from Shetland to the Highland Border and the paratectonic slate belt of the Midland Valley and Southern Uplands. The Scottish late Caledonian 'Newer Granites' (*sensu* Read, 1961), the principal subject of this chapter, are most abundant in the orthotectonic zone. With the possible exception of a few lamprophyre intrusions, e.g. on Iona (Rock and Hunter, 1987), and a few alkaline dykes described in Chapter 7, Caledonian magmatism extends little beyond the orogenic front into the Foreland (Figure 8.1). The late Silurian to Early Devonian magmatism post-dates the tectono-metamorphic event associated with the oblique convergence of Laurentia and Baltica and the closure of the northern Iapetus Ocean along the Iapetus Suture (Soper *et al.*, 1992). It spans the period of late orogenic uplift and extensional collapse (Watson, 1984), and coincides with the cessation of major sinistral strike-slip along the orogenic margin. There is some evidence in Scotland, however, for late Early Devonian ('Acadian') reactivation and contemporaneous magmatic activity (cf. Hutton and McErlean, 1991; Soper *et al.*, 1992), more commonly manifest in the slate belts of northern England (see Chapter 4).

Siluro-Devonian magmatism in Scotland, encompassing plutonic bodies, attendant dyke-swarms and volcanic rocks, is represented by predominantly high-K calc-alkaline rocks, and some have shoshonitic (high-K and high-Mg) affinities (Simpson *et al.*, 1979; Halliday, 1984; Stephens and Halliday, 1984; Plant, 1986). There is a significant background input of mantle-derived magmas, e.g. calc-alkaline lamprophyre and appinite suite intrusions. Although the plutons have historically been referred to as 'granites' they include a wide range of rock types from diorite through to monzogranite, with granodiorite predominant overall. These are derived mainly from a lower crustal source with some mantle component, and exhibit essentially 'I-type' characteristics (*sensu* Chappell and White, 1992), locally transitional to more alkaline 'A-type' in more evolved intrusions, e.g. Cairngorm. The chemistries of the Galloway Suite plutons within the Southern Uplands, e.g. Fleet, Criffel and Cheviot, bear some resemblance to the granites of the Lake District (see Chapter 4). Their weakly negative ϵNd, high δ^{18}O and ϵSr (Halliday, 1984) are comparable to patterns from the flysch sediments in the local Lower Palaeozoic sequences, and this argues in favour of an 'S-type' origin (*sensu* Chappell and White, 1992) through melting of this thickened young crust. (See the introduction to the Lotus quarries to Drungans Burn GCR site report for a full discussion of I-type and S-type characteristics.)

Stephens and Halliday (1984) divided the late Caledonian granites of the Grampian Highlands, Midland Valley and Southern Uplands terranes on geochemical and isotopic criteria into three suites: Argyll, Cairngorm and South of Scotland. This usefully demonstrates petrochemical provincialism within the orogen. The plutons of the Argyll Suite, and those within the Northern Highlands Terrane, map out an unusual granitic province with high Ba and high Sr characteristics and older crustal Nd signatures (Halliday, 1984; Stephens and Halliday, 1984; Thirlwall, 1989; Tarney and Jones, 1994). Intrusions of the South of Scotland and Cairngorm suites share relatively low Ba and low Sr characteristics (cf. Tarney and Jones, 1994), more typical of Palaeozoic magmatism worldwide. Tarney and Jones (1994) argue that these characteristics are derived from the subcontinental lithospheric mantle (SCLM) component rather than the continental crust. This suggests a significant change in the characteristics of the lithospheric mantle below the orthotectonic zone. Canning *et al.* (1996), from a study of geochemical differences within late Caledonian minette dykes, argue that the boundary between these mantle provinces coincides with the Great Glen Fault. However, the changes in chemistry and isotopic signatures of the plutonic rocks occur at the NE-trending 'mid-Grampian line' of Halliday (1984), within the Grampian Terrane and some 50 km to the SE of the Great Glen Fault. Our present understanding of the history of the Great Glen Fault would argue against any major significance of

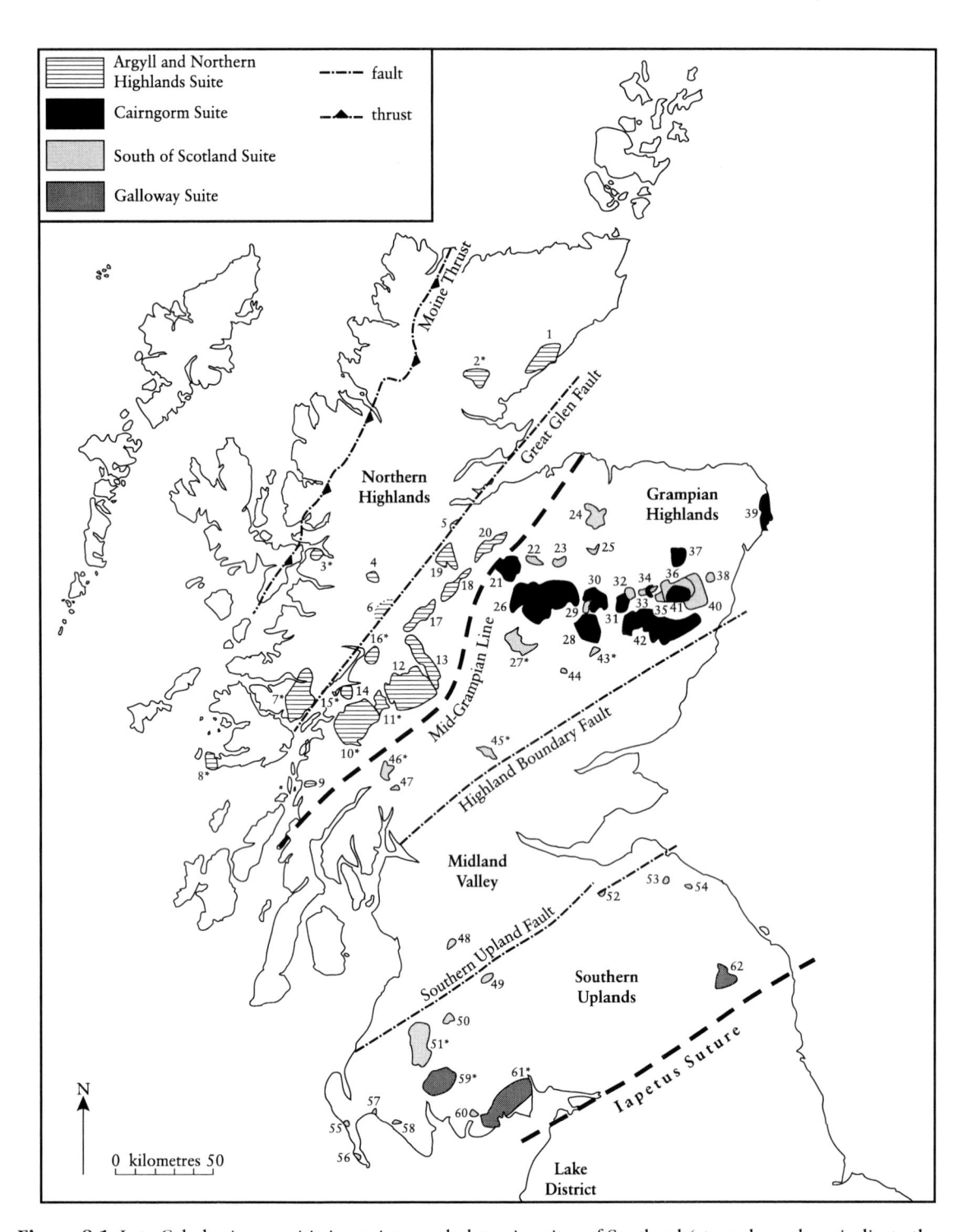

Figure 8.1 Late Caledonian granitic intrusions and plutonic suites of Scotland (starred numbers indicate those intrusions with GCR sites, named it italic type below):
1, Helmsdale; 2, Rogart (*Loch Airighe Bheg*); 3, Ratagain (*Glen More*); 4, Cluanie; 5, Abriachan; 6, Glen Garry; 7, Strontian (*Loch Sunart*); 8, Ross of Mull (*Cnoc Mor to Rubh' Ardalanish* and *Knockvologan to Eilean a'Chlamain*); 9, Kilmelford; 10, Etive (*Bonawe to Cadderlie Burn* and *Cruachan Reservoir*); 11, Glencoe fault intrusion (*Stob Mhic Mhartuin* and *Loch Achtriochtan*, Chapter 9); 12, Rannoch Moor; 13, Strath Ossian; 14, Ballachullish; 15, Duror of Appin (*Ardsheal Hill and Peninsula* and *Kentallen*); 16, Ben Nevis (*Ben Nevis and Allt a'Mhuilinn*, Chapter 9); 17, Corrieyairack; 18, Allt Crom; 19, Foyers; 20, Findhorn; 21, Monadhliath; 22, Boat of Garten; 23, Dorback; 24, Ben Rinnes; 25, Glen Livet; 26, Cairngorm; 27, Glen Tilt (*Forest Lodge*); 28, Lochnagar; 29, Craig Nardie; 30, Glen Gairn/Coilacreach; 31, Ballater; 32, Logie Coldstone; 33, Tomnaverie; 34, Cromar; 35, Torphins; 36, Balblair; 37, Bennachie; 38, Clinterty; 39, Peterhead; 40, Crathes/Gask; 41, Hill of Fare; 42, Mount Battock; 43, Glen Doll (*Red Craig*); 44, Glen Shee; 45, Comrie (*Funtullich* and *Craig More*); 46, Garabal Hill (*Garabal Hill to Lochan Strath Dubh-uisge*); 47, Arrochar; 48, Distinkhorn; 49, Spango; 50, Cairnsmore of Carsphairn; 51, Loch Doon (*Loch Dee*); 52, Broad Law; 53, Priestlaw; 54, Cockburns Law; 55, Cairngarroch Bay; 56, Portencorkrie; 57, Glenluce; 58, Mochrum Fell; 59, Fleet (*Clatteringshaws Dam Quarry* and *Lea Larks*); 60, Black Stockarton Moor; 61, Criffel (*Lotus Quarries to Drungans Burn* and *Millour and Airdrie Hill*); 62, Cheviot.

Introduction

this structure as a terrane boundary or re-activated older structure (Soper *et al.*, 1992; Stewart *et al.*, 1997). Similarly, there are no significant changes across any of the other major transcurrent shears, e.g. Highland Boundary and Southern Upland faults, suggesting that the tectonic blocks south of the mid-Grampian Line all have a similar SCLM.

Plutons within the orthotectonic Highland terranes present the widest range of ages (425–395 Ma) and emplacement levels seen in the orogen, from subvolcanic, e.g. Etive, Ben Nevis, through to mid-crustal *c.* 14 km, e.g. Strontian, Foyers, Findhorn. Some of the larger plutonic bodies, e.g. Etive and Strontian, consist of several intrusions with differing source components, emplaced at different crustal levels at resolvably different times. South of the Highland Boundary Fault, pluton emplacement is upper crustal, within the Palaeozoic sedimentary pile. It is clear that the major NE-trending fault shears throughout the orogen, but particularly within the orthotectonic zone, acted as magma conduits (Watson, 1984) and controlled pluton emplacement kinematics (Hutton, 1987, 1988a; Leake, 1990).

The period of magma generation is contemporaneous with rapid crustal uplift, erosion and development of flanking molasse basins, and this complex inter-relationship is well demonstrated in the Grampian Highlands. Here, mid-crustal *c.* 415 Ma plutons, e.g. Foyers and Findhorn, emplaced at depths of *c.* 12–14 km (Tyler and Ashworth, 1983), were unroofed by the time of deposition of the Middle Devonian piedmont sediments in the Great Glen area. In Glen Coe, clasts of granodiorite and microdiorite within Lower Devonian conglomerates, point to unroofing pre-dating volcanicity and the intrusion of the Etive pluton at *c.* 400–395 Ma (Bailey, 1960; Kynaston and Hill, 1908).

The suites identified by Stephens and Halliday (1984) provide a classification scheme independent of differences in granitic type or structural environment. Subsequently, data has become available for many more intrusions (such as in the NE Grampian Highlands) and the scheme is open to some modification on the basis of work by Tarney and Jones (1994). Hence, although the scheme is adopted here in broad terms, there are some groups of intrusions that are classified differently in Stephens and Halliday (1984), in Stephenson and Gould (1995) and in this volume. The Argyll Suite has been expanded to include the calk-alkaline plutons of the Northern Highlands. It is recognized, however, that some Northern Highland plutons are in part transitional into the slightly older alkaline suite described in Chapter 7 (cf. Fowler and Henney, 1996). The slightly younger plutons of the Southern Uplands, which have 'S-type' characteristics (Fleet, Criffel and Cheviot) were excluded from the scheme of Stephens and Halliday (1984) and are referred to here as the Galloway Suite.

Argyll and Northern Highlands Suite

This suite includes all late Caledonian plutons from the Grampian Highlands NW of the geochemically defined 'mid-Grampian Line' and from the Northern Highlands (Figure 8.1). With restoration of movement along the Great Glen Fault of *c.* 105 km, the suite defines a NE-trending belt some 60 km wide. The suite comprises predominantly hornblende–biotite granodiorite and biotite granodiorite plutons, with relatively minor diorite (some appinitic) and monzogranitic components. The plutons are strongly metaluminous (CaO + Na_2O + K_2O > Al_2O_3 > Na_2O + K_2O) with high Na, Sr and Ba and low Th, Nb and Rb. All have characteristically very low ϵNd values, -10 to +3, generally high ϵSr, -7 to +58 and $\delta^{18}O$ values in the range 7.2 to 10.7 ‰ (Halliday, 1984). Towards the western margin of the belt, the Ratagain intrusion has some elemental characteristics transitional between metaluminous calc-alkaline rocks and the alkaline rocks of the NW Scottish Highlands (see Chapter 7), e.g. very high Ba and Sr.

The Lochaber District, SE of the Great Glen Fault, is one of the few areas within the Caledonian Igneous Province, where extrusive products may be demonstrably linked to high-level plutons. Intrusive rocks at Ben Nevis and Glen Coe represent early phases of pluton emplacement into collapsing caldera systems. The Ben Nevis and Allt a' Mhuilluin GCR site (Chapter 9) provides evidence for high-level pluton emplacement and the derivation, in part, of the overlying volcanic pile from the underlying granitic magma body. While the availability of granitic magma during cauldron subsidence along a series of ring fractures beneath the Glen Coe centre is demonstrated within the Stob Mhic Mhartuin and Loch Achtriochtan GCR sites (Chapter 9), this magma is not related to that of the preceding volcanic rocks. The Etive pluton represents the final stages of significant magma intrusion in the Lochaber area at *c.* 400 Ma. Here, the Cruachan Reservoir GCR site demonstrates a ring-fracture system, and drop-down of a caldera fragment, subsequently exploited as a magma conduit by the Quarry Intrusion. The Bonawe to Cadderlie Burn GCR site is an instructive cross section through the multiple pulse main granodioritic-monzodioritic facies of the pluton. Assimilation of country rock enclaves and mingling of contemporaneous basic magma, are a feature of both sites. At the time of writing the GCR review did not include plutons from SE of the Great Glen Fault emplaced at mid-crustal levels between 10 and 14 km, for example Allt Crom (Key *et al.*, 1997), Ballachulish (Pattison and Harte, 1985; Weiss and Troll, 1989), Corrieyairack (Key *et al.*, 1997), Findhorn (Piasecki, 1975), Foyers (Marston, 1971), Rannoch Moor (Leighton, 1985) and Strath Ossian (Clayburn, 1981).

North of the Great Glen Fault, the GCR sites largely reveal aspects of deeper emplacement levels within the Proterozoic crust. Plutons such as Strontian, Ratagain and Rogart are composite intrusions, with basic, granodioritic and granitic magma components derived from differing source regions. Together with the Ross of Mull pluton, they demonstrate the availability of contemporaneous hydrated basic magma during emplacement. Within the Loch Sunart GCR site, shoshonitic meladiorites and mafic-rich enclaves are present in all facies of the Strontian pluton. The mafic enclaves here and in the Glen More and Loch Airighe Bheg GCR sites of the Ratagain and Rogart plutons probably represent disrupted synplutonic dykes. Within the central region of the Ross of Mull pluton a wide range of hybrid rocks are found in the Knockvologan to Eilean a' Chalmain GCR site. These are testament to mixing and mingling of the basic and granodioritic magmas, and give rise to an apparent reverse zonation within the pluton. The Cnoc Mor to Rubh Ardalanish GCR site contains examples of granitic hybrid rocks derived from the assimilation of country rock material. Isotopic studies of many of these intrusions consistently identify both crustal and mantle signatures in the main pluton facies (Halliday *et al.*, 1984). A subcontinental lithospheric mantle source is favoured for the basic enclaves (cf. Holden *et al.*, 1991) and the syenitic component of the Ratagain pluton (Thompson and Fowler, 1986).

High-temperature shear and magmatic state deformational fabrics are common features of the mid-crustal Argyll and Northern Highlands Suite intrusions from both sides of the Great Glen Fault. The Loch Sunart, Glen More and Loch Airighe Bheg GCR sites, within the Strontian, Ratagain and Rogart plutons respectively, provide fine examples of this emplacement deformation (Hutton, 1988b; Hutton *et al.*, 1993; Hutton and McErlean, 1991; Soper, 1963). In the high-level plutons SE of the fault, these fabrics are present but less distinctive, e.g. the Bonawe to Cadderlie Burn GCR site of the Etive pluton. These most likely derive from deformation within the linked NE–SW shear fracture systems during the waning stages of Caledonian deformation within the Northern Highlands and Grampian Highlands terranes.

Cairngorm Suite

At the time of writing, the GCR did not include examples from this important suite of intrusions, but for completeness a brief description is given here. The Cairngorm Suite, consists of late Caledonian (*c.* 400 Ma) voluminous granitic plutons, occurring within the northern Grampian Highlands (Figure 8.1). A large gravity low extends between the Monadhliath and Mount Battock masses. This suggests the presence of substantial volumes of low density rocks in the crust (Rollin, 1984), forming an easterly trending batholith at depth (the East Grampian Batholith of Plant *et al.*, 1990). The granite masses at the surface probably represent cupolas (Cornwell and McDonald, 1994). Aeromagnetic anomalies, commonly annular in form, e.g. at Cairngorm and Lochnagar, suggest the presence of magnetic lithologies. These may be either mafic-rich cumulates at depth (Brown and Locke, 1979) or magnetic rocks within the plutons not seen at the surface (Cornwell and McDonald, 1994). The plutons probably represent high emplacement levels, *c.* 5–8 km (Harrison and Hutchinson, 1987).

The intrusions comprise mainly biotite monzogranite, such as Monadhliath (Highton, 1998), Cairngorm (Harrison, 1986), Mount Battock (Webb and Brown, 1984b), Lochnagar (Oldershaw, 1974; Rennie, 1983), Ballater (Webb and Brown, 1984b), with minor granodiorite, but include a wide range of primary textural variants from microgranite to coarse-grained, K-feldspar megacrystic granite. Secondary magmatic tex-

Introduction

tures, such as xenocrystic aplitic microgranites, pegmatites and vuggy cavities (often mineral lined), are a consequence of either volatile fluxing, pressure quenching or fluidization in the plutons (Highton, 1999). High temperature late magmatic hydrothermal alteration is often extensive, e.g. Cairngorm and Mount Battock (Harrison, 1986, 1987a).

Intrusions within the suite are largely 'I-type' ($^{87}Sr/^{86}Sr \sim 0.706$), highly evolved, and quite distinct from the other suites, with low Ti, P, Ba, Sr and K/Rb, and high Rb, Nb, Th and U. The isotopic signatures bear some similarity to the Argyll Suite, having low ϵNd values, –8 to –1, generally high ϵSr, +24 to +33 and $\delta^{18}O$ values of c. 8.2 to 11.1 ‰ (Halliday, 1984; Stephens and Halliday, 1984). The origins of this Cairngorm Suite are problematical, with the large volume granite plutons that comprise the East Grampian Batholith having highly evolved characteristics more typical of Sn-U granites (Plant et al., 1990). Most are high heat-producing granites, e.g. Monadhliath and Cairngorm, and coincide with thermal anomalies in the crust (Webb and Brown, 1984b; Atherton and Plant, 1985). Some plutons show transitional 'A-type' characteristics, with moderate enrichments in B, Nb, F, Li, Sn and W, e.g. Cairngorm (O'Brien, 1985; Harrison, 1986, 1987a) and Monadhliath (Highton, 1999). This is more likely a reflection of the highly evolved nature of these granites, rather than having a genetic significance. With the exception of Glen Gairn/Coilacreich (Webb et al., 1992) most lack significant metalliferous mineralization.

South of Scotland Suite

This suite encompasses all remaining Siluro-Devonian plutons intruded into the Neoproterozoic crust SE of the 'mid-Grampian Line' and the Palaeozoic sequences of the Midland Valley and Southern Uplands. Most plutons of this suite within the Grampian Highlands terrane have been termed the 'South Grampians Suite' by Stephenson and Gould (1995). The suite may also include a group of diorite–granodiorite plutons of central Aberdeenshire that pre-date the Cairngorm Suite and have been assigned by Gould (1997) to a separate 'Crathes Suite', e.g. Crathes, Balblair and Torphins.

The intrusions in the Grampian Highlands Terrane, like those of the Argyll and Northern Highlands Suite, have close spatial associations with major NE-trending shear faults such as the Loch Tay and Glen Fyne faults, e.g. the Glen Tilt and Garabal Hill–Glen Fyne plutons. Most are composite multiple-pulse intrusions often dominated by granodiorite and moderately evolved monzogranites, but with significant gabbroic and pyroxene meladiorite (appinitic) facies rocks. Characteristically, they are strongly metaluminous ($CaO + Na_2O + K_2O > Al_2O_3 > Na_2O + K_2O$) with relatively higher K and Th, but lower Zr, La, Ce, Ba, Sr and Rb than the Argyll and Northern Highlands Suite intrusions. The transition element values in the basic to intermediate rocks, in particular V and Cr, are also commonly higher than in the Argyll and Northern Highlands Suite. All granodioritic components have low ϵNd values, -6 to -1, generally high ϵSr, +1 to +54 and $\delta^{18}O$ values of 7.5 to 10.5‰ (Halliday, 1984; Stephens and Halliday, 1984). All lack an inherited zircon component.

The depth and/or mode of emplacement of many intrusions is little documented. However, one of the finest examples of a high-grade contact metamorphic aureole within the Caledonian Igneous Province is found adjacent to the chilled margin of the Comrie pluton within the Craig More GCR site, indicating high T–low P emplacement conditions of c. 750°C and 2.5 kbar (Pattison and Tracy, 1991; Pattison and Harte, 1985). Comparable emplacement parameters have been obtained from the aureoles of quartz-diorite intrusions adjacent to, but pre-dating the Lochnagar pluton of the Cairngorm Suite (Goodman and Lappin, 1996).

Representative intrusions of this suite chosen for the GCR are the Glen Tilt, Garabal Hill–Glen Fyne, Glen Doll and Comrie plutons from the Grampian Highlands Terrane, together with Loch Doon from the Southern Uplands. Their designated GCR sites provide details of pluton construction rarely seen in other Caledonian intrusions, and the Forest Lodge GCR site in Glen Tilt is of international historical importance, as it was here in 1785 that James Hutton demonstrated the magmatic origin of granite. These are compositionally diverse (ultramafic to monzogranitic), zoned multiple pulse plutons, consisting of an early appinitic diorite and later granodiorite or monzogranite facies. The basic parts of these intrusions have historically attracted much attention and pioneering petrochemical research (Deer, 1938a, b, 1950, 1953; Nockolds, 1941; Nockolds and Mitchell, 1948).

Late Silurian and Devonian granitic intrusions of Scotland

The classical study by Nockolds (1941) of the Garabal Hill–Glen Fyne pluton (see below) brought to prominence the concept of zoned granitic bodies evolving through differentiation by fractional crystallization from a dioritic parental magma. Within the Garabal Hill to Lochan Dubh-Uisge GCR site, the granodioritic rocks were considered to be differentiates from a parental pyroxene-mica diorite magma, while the basic components represented crystal cumulates. Although now considered too simplistic, this model was to drive most early petrochemical studies of granite plutons worldwide. More recent interpretations point to different sources for the basic and granodioritic components. The predominantly dioritic Comrie pluton is compositionally zoned. Heterogeneities and merging contacts within the Funtullich GCR site, argue for multiple pulse injection and penecontemporaneous inward crystal accretion. The later microgranite facies, like many core facies, is highly evolved and unrelated to the earlier diorite–granodiorite facies. The Red Craig GCR site provides a fine example of wall-rock interaction and contamination at the SE margin of the predominantly dioritic Glen Doll intrusion.

The Loch Dee GCR site represents the Loch Doon pluton, which is the largest of the South of Scotland Suite plutons and was the first pluton to be defined by the Geological Survey. The pluton is the finest example within the Caledonian Igneous Province of a compositionally zoned intrusion, from a margin of pyroxene-mica diorite through granodiorite to central monzogranite. Some smaller intrusions within the Southern Uplands share these characteristics, for example Cairnsmore of Cairsphairn (Deer, 1935; Tindle *et al.*, 1988) Portencorkrie (Stone, 1995), Cairngarroch (Allen *et al.*, 1981), Priestlaw (Shand, 1989) and Cockburn Law (Shand, 1989). Other intrusions, e.g. Broad Law, Spango, Glenluce, Bengairn and Kirkowan, are single-component intrusions or show only minor compositional variation. The diorite–granodiorite Distinkhorn pluton, dated at 412 ± 5 Ma (Thirlwall, 1988), and small satellite bodies at Hart Hill, Glen Garr and Tincorn Hill are the only significant intrusions found within the Midland Valley Terrane. These and minor diorite stocks at Fore Burn close to the Southern Upland Fault, at Lyne Water in the Pentland Hills, and in the western Ochil Hills, intrude and hornfels the Lower Old Red Sandstone sedimentary rocks and lavas.

Galloway Suite

This group of intrusions consists of the Fleet, Crifell and Cheviot plutons and smaller intrusions at Portencorkrie (Stone, 1995) and Kirkmabreck (Blyth, 1955). They are younger than those of the South of Scotland Suite (Thirlwall, 1988) and have a close affinity with intrusions of the Lake District and south-east Ireland. The north-western limit of these intrusions is marked by the NE-trending Moniaive Shear zone. This structure coincides with a major geophysical discontinuity in the subcontinental lithospheric mantle and lower crustal rocks beneath the Southern Uplands (Kimbell and Stone, 1995). A basement discontinuity is also suggested by a marked contrast in the isotopic characteristics of intrusions on either side of the shear zone (Shand, 1989; Thirlwall, 1989). The inception and propogation of the shear zone through the Palaeozoic cover, during early Wenlock times, is thought to reflect re-activation of the basement discontinuity (Stone *et al.*, 1997).

All of the plutons include members with S-type peraluminous characteristics (Al_2O_3 > $CaO + Na_2O + K_2O$) which are usually two-mica granites. $^{87}Sr/^{86}Sr$ ratios are typically 0.705 to 0.707, with $\delta^{18}O$ values in the range 8 to 12‰, while ϵNd values suggest derivation from Silurian turbiditic sedimentary sequences (Halliday *et al.*, 1980; Halliday, 1984). The generation of these magmas was probably a consequence of underthrusting of the Southern Uplands by the leading edge of the Eastern Avalonia continent, a view supported by Pb isotope data (Thirlwall, 1989).

The Fleet pluton is the youngest, at *c.* 390 Ma, and the most evolved of these Southern Upland intrusions. The less evolved outer pophyritic biotite granite facies of the Clatteringshaws Dam Quarry GCR site passes inwards into a weakly porphyritic biotite–muscovite granite. The garnet-bearing aphyric muscovite microgranite core facies of the Lea Larks GCR site is the most highly evolved of the Scottish late Caledonian granites. Variations in the orientation of a weak to moderately developed ductile fabric argues for intrusion during sinistral movement on the Moniaive Shear zone. The zoned Criffel pluton has played an important role in the modelling of pluton emplacement and magma chamber dynamics. Both diapirism and stoping have been proposed by different investigators and

Introduction

much of the crucial evidence is displayed in the country rocks and outer granodiorite facies of the Millour and Airdrie Hill GCR site. The pluton has an unusual compositional make-up, being concentrically zoned from a metaluminous 'I-type' outer facies to a core with peraluminous 'S-type' characteristics. This is a unique example, not only within the Caledonian Igneous Province, and must invoke open-system multiple injection from differing sources. This paradox is well illustrated in the traverse through the Lotus Hill to Drungans GCR site.

The monzogranites of the Cheviot pluton (Carruthers *et al.*, 1932; Jhingran, 1942) are of similar age to the Cheviot lavas at *c*. 396 Ma (Thirlwall, 1988) and are considered to be co-magmatic with them (Thirlwall, 1979).

Shetland Suite

The Shetland plutons, not yet represented in the GCR at the time of writing, are divisible into eastern and western 'suites' separated by the Walls Boundary Fault (Flinn, 1988 and references therein). The timing of their emplacement is generally poorly constrained, with only K/Ar mineral ages reported.

The apparently older (*c*. 400 Ma) eastern suite, including the Graven, Brae and Aith-Spiggie complexes and the Hildesay Granite, mostly comprise granodiorite-dominated intrusions with diorite, appinitic diorite and minor basic components. The epidote-bearing main granodiorite facies of the Aith-Spiggie complex (Flinn, 1988) is unusual, although this may merely reflect late-magmatic alteration. All members of the 'suite' have geochemical characteristics similar to plutons of the Argyll and Northern Highlands Suite in the orthotectonic zone of the mainland (M. P. Atherton and co-workers, pers. comm., 1997). The minor intrusions to the east of the Walls Boundary Fault (predominatly microdiorites and calc-alkaline lamprophyres, with some porphyritic microgranodiorites) fall into two 'suites', late-tectonothermal and post-tectonic. The former are consistently foliated, with metamorphic mineral assemblages, and pre-date the plutonic complexes. The latter are generally synplutonic and are rarely deformed (Mykura, 1976).

To the west of the Walls Boundary Fault the plutonic complexes appear to be younger, with reported K/Ar ages in the range 370–360 Ma. This western suite is dominated by the Northmaven complex (Miller and Flinn, 1966; Mykura and Phemister, 1976; Phemister, 1979) and the Sandsting complex (Mykura and Phemister, 1976). These complexes are a mixture of granophyre and granite plutons (e.g. Ronas Hill, Muckle Roe and Sandsting) with some hornblende-bearing plutons (for example Mangaster Voe). Hybridization through intermingling of basic and acid components is common. All are characteristically low Ba and Sr intrusions (M. P. Atherton and co-workers, pers. comm., 1997). The Sandsting complex clearly cuts and hornfelses Middle Devonian sedimentary rocks of the Walls Formation and is closely associated in space and time with a late phase of compressive deformation and metamorphism (see the Ness of Clousta to The Brigs GCR site report, Chapter 9). This Mid- to Late Devonian deformation, which post-dates the main extensional event(s) responsible for the development of the Orcadian Basin, was probably the last phase of Caledonian folding in Britain. Both complexes are cut by zones of shearing, fracturing and hydrothermal alteration (including scapolitization and zeolitization) associated with the Walls Boundary Fault.

Appinite Suite

Appinites and their Northern Highland equivalents, the Ach'uaine Hybrids, are probably the most enigmatic intrusions within the Caledonian Igneous Province (Read, 1961; Wright and Bowes, 1979; Hamidullah and Bowes, 1987). The Ardsheal Hill and Peninsula GCR site is the type locality for 'appinites', described by Bailey and Maufe (1916) as medium- to coarse-grained rocks with essential prismatic hornblende in a groundmass of sodic plagioclase, K-feldspar and quartz. These, Bailey and Maufe (1916) and Bailey (1960) regarded as the plutonic equivalents of the hornblende-bearing lamprophyres – spessartites and vogesites; a view confirmed by more recent geochemical studies of lamprophyres and subvolcanic vents in the the Southern Uplands (Rock *et al.*, 1986a; Henney, 1991). However, the term Appinite Suite has become a miscellaneous term for intrusions with a heterogeneous range of ultramafic (e.g. olivine-pyroxene hornblendite), melanocratic basic and intermediate (such as olivine monzonite), dioritic and granodioritic rocks. Most are shoshonitic in affinity and probably of mantle derivation (Fowler, 1988a; Henney, 1991;

Fowler and Henney, 1996), with elevated Sr, Ba, Ni, Cr and light rare earth elements (LREE). Appinitic intrusions have a wide distribution throughout the Scottish and Irish Caledonian Igneous Province, and typically have a close spatial relationship with the late Caledonian granitic plutons. More significant, however, is a clustering close to major faults or concentration along NW-trending lineaments (cf. Fowler, 1988b). Deeper-level intrusions often carry fabrics indicative of emplacement into a ductile shear-dominated environment (Phillips and May, 1996), although some deeper-level intrusions with breccias are known (Peacock *et al.*, 1992; May and Highton, 1997).

Invariably appinitic intrusions pre-date, or are in part contemporaneous with the late Caledonian granitic plutons, e.g. the Ballachulish, Garabal Hill and Arrochar plutons, and they have comparable emplacements ages of *c.* 430–425 Ma (Rogers and Dunning, 1991). However, they only rarely form a significant proportion of the xenolith population in such plutons, for example at Corrieyairack (Key *et al.*, 1997), Ratagain and Rogart. Appinitic intrusions were commonly preceded by breccia pipes, and acted as open-system feeders within the subvolcanic systems (Wright and Bowes, 1968; Rock *et al.*, 1986a; Platten, 1982; Henney, 1991). Numerous and varied individual intrusions occur within the Ardsheal Hill and Peninsula GCR site, where breccia pipes are intimately associated with and commonly infiltrated by both primitive and evolved magmas. A detailed example at the Kentallen GCR site illustrates the relationships between various late Caledonian events, involving an appinitic intrusion (the type 'kentallenite'), the Ballachulish pluton, several dyke sets, hydrothermal veins and faulting associated with the Great Glen Fault.

In the Northern Highlands Terrane, the Ach'uaine Hybrid intrusions of Sutherland (Read *et al.*, 1925) are a similar heterogeneous suite, predominantly of appinitic meladiorites, but ranging from ultrabasic and basic rocks through to syenite and granite. These occur as enclaves and intrusions within and peripheral to the Ratagain and Rogart plutons with good examples in the Glen More and Loch Airighe Bheg GCR sites, respectively. These appinitic rocks present evidence for the mingling of mantle-derived lamprophyric and contemporaneous syenitic magmas (Fowler and Henney, 1996).

Minor intrusions

Compositionally and temporally diverse suites of predominantly calc-alkaline, and shoshonitic minor intrusions are present in all terranes (Richey, 1938; Sabine, 1953; Smith, 1979; Cameron and Stephenson, 1985; Barnes *et al.*, 1986; Rock *et al.,* 1988; Henney, 1991; Swarbrick, 1992; Stephenson and Gould, 1995). Emplacement is clearly episodic, with several syn- to late-orogenic regional suites. Many are genetically related to individual plutons or acted as feeders for extrusive rocks. Others, particularly the calc-alkaline lamprophyres, simply reflect the regional background magmatism within the province (Rock *et al.*, 1988). Minor intrusions are a feature of many of the GCR sites described in this chapter and in Chapter 9, but no GCR sites have been designated specifically for their representation. Thus a brief review of the suites is presented below.

In general the background magmatism throughout the province is lamprophyric, with both hornblende- and mica-phyric types of calc-alkaline lamprophyre present (spessartite, vogesite, kersantite and minette). Overall, mica lamprophyres such as the minettes are more abundant in the Northern Highlands, while in the Southern Uplands spessartites are more common. The more primitive lamprophyric magmas generally provide a window into the nature and melt characteristics of the subcontinental lithospheric-mantle source area (Canning *et al.*, 1996; Fowler and Henney, 1996).

Deformed and/or metamorphosed intrusions are common to both the ortho- and paratectonic zones (Dearnley, 1967; Winchester, 1976; Smith, 1979; Barnes *et al.*, 1986; May and Highton, 1997). In the Highland terranes, these mainly sheet intrusions post-date the *c.* 470 Ma Caledonian tectonothermal peak and the *c.* 450 Ma late orogenic granite-pegmatite complexes, e.g. Glenmoriston, Arkaig, Kyllachy and Strathspey. Most pre-date emplacement of the *c.* 425–415 Ma, mid-crustal plutons of the Argyll and Northern Highlands Suite (cf. Smith, 1979). Later sheared intrusions, (syn- to post- 420 Ma plutons) are prevalent throughout the orthotectonic zone. Shear fabrics in some intrusions result from deformation penecontemporaneous with pluton emplacement (cf. Hutton and McErlean, 1991). This is illustrated in the Glen More River section of the Glen More GCR site, where shearing in some intrusions is a function of regional strike-slip movement. South-east of the Highland Boundary Fault, early lamprophyre dykes post-date the main deformation of the Palaeozoic succession but pre-date pluton emplacement. Folding and shearing of these intrusions occurred during compressional deformation towards the end of terrane accretion (Barnes *et al.*, 1986).

North-east- or ENE-trending dyke-swarms are particularly prominent within the upper crustal/subvolcanic environment. Some, for example the Ben Nevis, Etive, Distinkhorn and Doon swarms, are closely associated with plutons. The Cruachan Reservoir and Bonawe to Cadderlie Burn GCR sites, described in this chapter, and the Buchaile Etive Beag and Ben Nevis and Allt a Mhuillin GCR sites of Chapter 9, demonstrate the linkage between dyke-swarms and the multipulse Etive and Ben Nevis plutons, respectively. Swarms are less prominent adjacent to deeper-level plutons, but examples include the swarm of kersantites and spessartites that cut the Ross of Mull pluton locally, as seen in the Knockvologan to Eilean a' Chalmain and Cnoc Mor to Rubh' Ardalanish GCR sites (Rock and Hunter, 1987). A small composite swarm of spessartite to quartz-micromonzonite intrusions accompanies the Ratagain pluton (Glen More GCR site) (May *et al.*, 1993). Other small swarms associated with the Comrie pluton (Craig More and Funtullich GCR sites), Garabal Hill–Glen Fyne igneous complex (Garabal Hill to Lochan Dubh-Uisge GCR site) and Arrochar plutons, mostly comprise microdiorite and spessartite. To the south of the Highland Boundary Fault, widespread but not abundant minor intrusions within the Midland Valley are mostly microdiorites with some kersantites and minor acid variants (Swarbrick, 1992). Intrusions mostly take the form of dykes, although large sills and bosses occur in the Ochil and Sidlaw hills. Here, dyke-swarms are closely associated with the volcanic sequences, a feature reflected in their chemistries.

Lamprophyric intrusions are a significant component of the late Caledonian magmatism in the Southern Uplands (Rock *et al.*, 1986b; Henney, 1991; Shand *et al.*, 1994). Although a significant number of swarms are centred on the granitic plutons such as Loch Doon (Loch Dee GCR site), regional swarms are widespread. Most comprise hornblende-bearing lamprophyres and porphyritic microgranodiorites. A major ENE-trending swarm of largely mica-rich lamprophyres (mostly kersantites and minor minettes), with spessartites, microdiorites and a minor balsaltic component, extends from St Abb's through Dumfries-shire, the Mull of Galloway and into Northern Ireland (Read, 1926; Reynolds, 1931; Macdonald *et al.*, 1986; Rock *et al.*, 1986b). Near Kircudbright, the majority of the lamprophyre intrusions are contemporaneous with vents (see the Shoulder O' Craig GCR site report, Chapter 9) and the subvolcanic Black Stockarton Moor complex (Rock *et al.*, 1986a; Rock *et al.*, 1986b), and they pre-date the Criffel pluton.

LOCH AIRIGHE BHEG (NC 703 025)

N. J. Soper

Introduction

The Rogart igneous complex extends over about 115 km^2 in SE Sutherland (Figure 8.2). It has been described by Read *et al.* (1925, 1926) and Soper (1963), and consists of a zoned quartz-monzodiorite–granodiorite–granite pluton, together with a peripheral zone of migmatization. The complex was emplaced into metasedimentary rocks of the Altnaharra Formation in the Morar Group of the Moine Supergroup after the main, presumed Caledonian, deformation. It is overlain by Devonian strata of Old Red Sandstone facies. It therefore belongs to the late Caledonian 'Newer Granites' and is a component of the Argyll and Northern Highlands Suite.

The scale of the migmatitic envelope of the Rogart complex is unique in British Caledonian granites, and three GCR sites have been chosen to illustrate its features, Creag na Croiche, Aberscross Burn–Kinnauld and Brora Gorge. These are to be described in the *Lewisian, Torridonian and Moine Rocks of Scotland* GCR volume. The Loch Airighe Bheg GCR site described here displays features of the intrusive part of the complex, the pluton, together with hybridized appinitic rocks that occur as xenoliths.

The Appinite Suite of alkalic ultramafic, intermediate and felsic intrusive rocks is intimately associated with late Caledonian plutons throughout the Scottish Highlands. The type area of the suite is described in this GCR volume (see the Ardsheal Hill and Peninsula GCR site report). Appinitic intrusions are widely developed in the Moine rocks of the Northern Highlands where they form isolated bosses, often a hundred metres or so across. They were described by Read *et al.* (1925) as 'hybrids of Ach'uaine type', from a locality some 10 km SW of this GCR site (NH 624 952), where they range in composition from ultramafic olivine-pyroxene-amphibole-biotite rocks to granodiorite. Read interpreted the diversity of the suite as a result of hybridization of granitic magma with either ultrabasic magma or solid rock. More recent geochemical studies have invoked the contamination of mantle-derived, K-rich (shoshonitic) basic magmas (Fowler, 1988a; Fowler and Henney, 1996).

Description

The GCR site is located near the south-eastern margin of the Rogart pluton (Figure 8.2). This intrusion consists of hornblende-biotite granodiorite that grades into a marginal quartz-monzodiorite and is cut by a body of granite; the quartz-monzodiorite and granite are exposed at the site, but not the granodiorite.

The quartz-monzodiorite crops out immediately east of the loch and at Dalmore Quarry 600 m to the NE (709 029). At the latter locality it is composed of plagioclase An_{23} (50%), quartz (18%), K-feldspar (12%), biotite (10%) and amphibole (8%), together with minor titanite, allanite, zircon and apatite. It is cut by aplitic microgranite, granite pegmatite and rare microgranodiorite veins (Figure 8.3). The quartz-monzodiorite has a foliation defined by the preferred orientation of biotite, amphibole and to some extent plagioclase. This foliation is subvertical and here strikes NE, roughly parallel to the contact of the outer facies with its migmatitic envelope. Surfaces parallel to the foliation show a weak linear fabric, defined mainly by the alignment of amphibole, which plunges gently north-

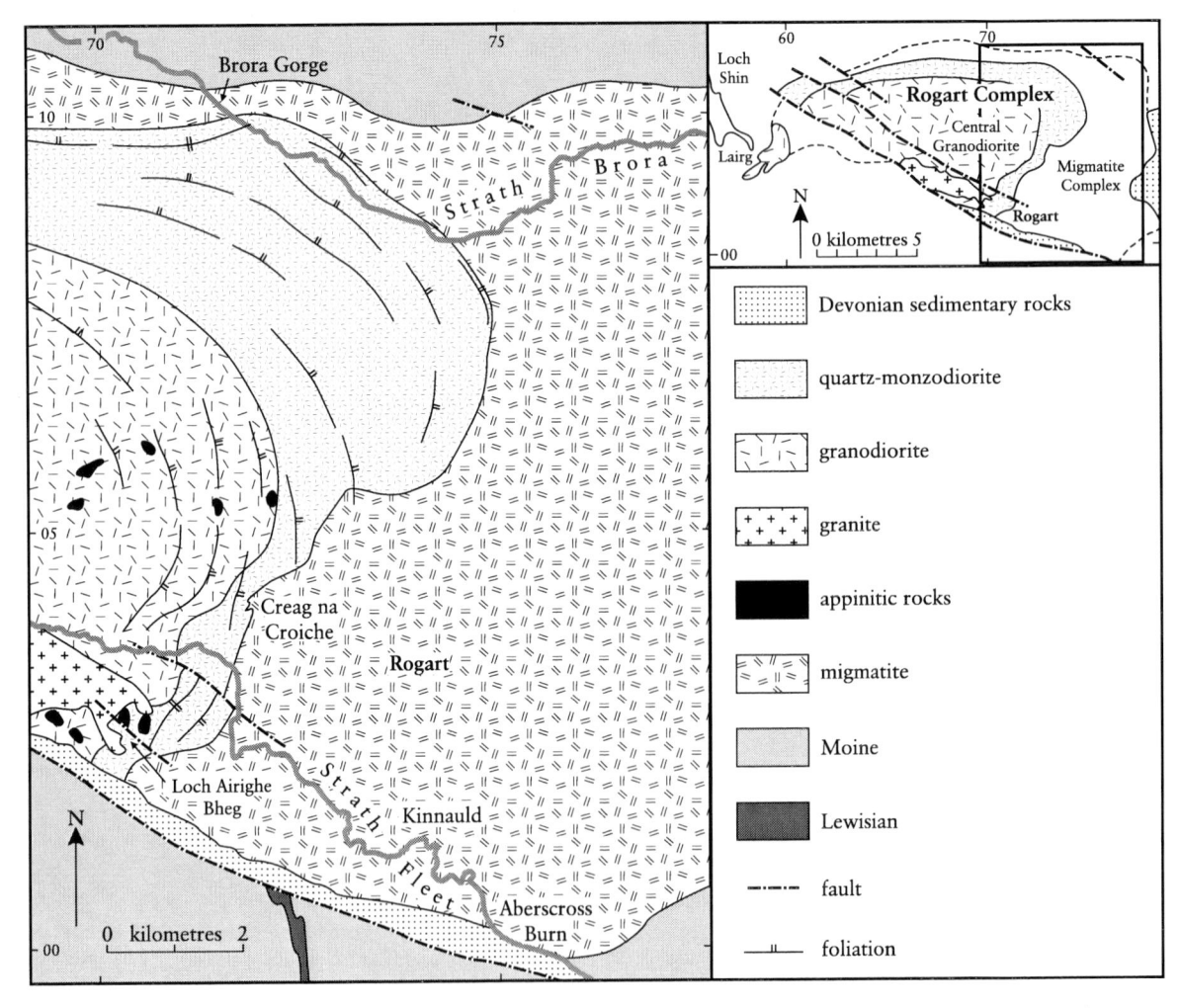

Figure 8.2 Map of the eastern part of the Rogart pluton, including the Loch Airighe Bheg GCR site. The inset shows the whole Rogart complex.

Loch Airighe Bheg

Figure 8.3 Poorly foliated outer quartz-monzodiorite of the Rogart pluton cut by veins of aplitic microgranite (NC 704 025). (Photo: Susan Hall.)

eastwards.

The later granite, exposed in the northern part of the GCR site, consists essentially of plagioclase An_{12} (50%), quartz (26%), K-feldspar (18%) and biotite (6%). It is unfoliated and although its contact with the quartz-monzodiorite is not exposed here, elsewhere there is a narrow gradation between the facies.

Crags immediately to the north and NW of the loch are composed of a wide variety of appinitic lithologies, ranging from ultramafic types composed of pyroxene, amphibole, biotite and minor feldspar to meladiorite and melanocratic syenogranite; the commonest type is made of amphibole and biotite with patches of pink K-feldspar. Angular to sub-rounded appinitic xenoliths are enclosed by the host quartz-monzodiorite, with biotite-rich selvages developed locally at the contact (Figure 8.4). Larger appinitic masses are veined by hornblende-rich quartz-monzodiorite and contain patches of hornblende-K-feldspar pegmatite.

Interpretation

In a structural study of the Rogart complex, Soper (1963) interpreted the foliation and sub-horizontal linear fabric in the outer quartz-monzodiorite as the result of ballooning during its final emplacement. Evidence that the pluton deformed and eventually punched through its own migmatite envelope is illustrated by the sites in the *Lewisian, Torridonian and Moine Rocks of Scotland* GCR volume.

Late Silurian and Devonian granitic intrusions of Scotland

Figure 8.4 Xenoliths of appinitic rock in quartz-monzodiorite of the Rogart pluton (NC 703 026). Dark biotite-rich selvages are visible at the margin of the xenolith above the compass. (Photo: Susan Hall.)

Read (1961) classified the 'Newer Granites' into 'forceful' and 'permitted'. The 'forceful' intrusions were emplaced by shouldering aside the country rocks and were thought to be deeper and older than the 'permitted' high-level, subvolcanic granites emplaced by brittle mechanisms: the Rogart pluton was cited as an example of the forceful type. While modern views on pluton emplacement emphasize tectonic controls on space creation (Hutton, 1988a), in the Highlands it is evident that concordant, foliated intrusions such as Rogart were emplaced in a more ductile environment than nearby cross-cutting plutons such as Helmsdale (Figure 8.1). The K-Ar age of the Rogart complex is about 420 Ma (Brown *et al.*, 1968), similar to K-Ar ages generally obtained from the Moine rocks of the Northern Highlands, suggesting that the complex was emplaced during the waning stages of regional metamorphism and cooled together with the country rocks during late orogenic uplift and erosion. The K-Ar age of the Helmsdale pluton is about 400 Ma (Brown *et al.*, 1968), and so this intrusion is likely to have been emplaced after substantial erosion had taken place. Thus Read's view that the 'forceful' Highland granites were deeper and older than the 'permitted' types is essentially true.

The 'Ach'uaine hybrids' are regarded as a component of the late Caledonian 'Newer Granite' magmatism (Read *et al.*, 1925). Fowler (1988a) interpreted them as differentiates of relatively primitive, K-rich mantle-derived magma that crystallized under hydrous conditions with some crustal contamination. Fowler and Henney (1996), however, invoked the mixing of shoshonitic magma with contemporaneous syenitic magma. The appinitic masses at Loch na Airighe Bheg are xenoliths within the quartz-monzodiorite, not intrusions, so their emplacement pre-dates that of the Rogart pluton at the level now exposed. No modern petrogenetic investigation of these rocks has been undertaken, so it is uncertain to what extent their compositional diversity is a result of hybridization with the Rogart magma.

Glen More

Conclusions

The Rogart complex is a member of the Argyll and Northern Highlands Suite of late Caledonian intrusions, emplaced into metasedimentary rocks of the Moine Supergroup. It consists of a quartz-monzodiorite–granodiorite–granite pluton flanked to the east and north by a concordant aureole and migmatite envelope. At the Loch Airighe Bheg GCR site several components of the pluton are exposed. The quartz-monzodiorite is foliated roughly parallel to its nearby contact with the migmatized envelope and also carries a weak sub-horizontal alignment of amphibole. This is thought to be due to ballooning of the quartz-monzodiorite–granodiorite magma during its emplacement. The later inner granite is unfoliated and apparently discordant.

The most interesting feature of the site is the presence of numerous appinitic xenoliths within the quartz-monzodiorite. These belong to a regional suite of mantle-derived, K-rich mafic intrusions known as the Ach'uaine hybrids. The xenoliths show a wide range of structural, textural and mineralogical relationships with the surrounding quartz-monzodiorite, which would repay further investigation to establish to what extent the compositional diversity of the appinitic rocks at this locality is a result of hybridization with the host quartz-monzodiorite.

GLEN MORE
(NG 861 201–917 188)

W. E. Stephens

Introduction

The Ratagain pluton is distinctive among the late Caledonian intrusions of Scotland in having compositional features transitional between the alkaline intrusions of Assynt and the NW Foreland (see Chapter 7) and the more common metaluminous calc-alkaline intrusions of the Argyll and Northern Highlands Suite. In this sense it forms a link between these very different but near-contemporaneous periods of magmatism. The pluton shows considerable petrological variety in a small area (c. 17 km²), and also has some notable compositional characteristics, having among the highest known Sr and Ba abundances for such rock types anywhere. The presence of mafic (meladiorite) bodies and a degree of mingling between mafic masses and felsic magmas are well displayed, in common with several other plutons lying between the Great Glen Fault and the Moine Thrust. It is one of the few plutons in Scotland that hosts gold-bearing veins.

The pluton was first described by the Geological Survey (Peach *et al.*, 1910), and a detailed petrological account was presented by Nicholls (1951a, b). An extension to the intrusion was recognized by Dhonau (1964), and the main pluton was characterized geochemically and isotopically by Halliday *et al.* (1984) and Hutton *et al.* (1993). A new map of the pluton, taking advantage of many new exposures associated with local forestry activities, has revised the various petrological facies and their distribution (Hutton *et al.*, 1993). Emplacement of the pluton in relation to movements on regional faults systems was the subject of a further study by Hutton and McErlean (1991).

The pluton was emplaced at 425 ± 3 Ma (U-Pb baddelyite age from the pyroxene-mica diorite facies; Rogers and Dunning, 1991). A Rb-Sr mineral-whole rock isochron age of 415 ± 5 Ma (Turnell, 1985) is now regarded as too young and is taken to reflect the fairly rapid cooling history of the complex. These data indicate that the pluton was emplaced more-or-less contemporaneously with late members of the Assynt alkaline suite such as the Ben Loyal syenite (van Breemen *et al.*, 1979a; Halliday *et al.*, 1987).

In recent years this pluton has contributed to the debate over the role of subduction in the origin of the Caledonian granites, the compositions of some components having been correlated with those of shoshonitic lavas, which tend to be associated with the deepest parts of subduction zones (Thompson and Fowler, 1986).

The Glen More GCR site contains all the important members of the pluton, as well as, at Braeside, the only outcrops of an olivine-gabbro component. The site includes the hillside of Moyle Wood in which various relationships between earlier and later members of the pluton are well displayed. Good examples of the small mafic bodies previously described as 'appinites' are well exposed in the Glen More river.

Description

The Ratagain pluton comprises principally diorites and quartz-monzonites (Figure 8.5). The

Late Silurian and Devonian granitic intrusions of Scotland

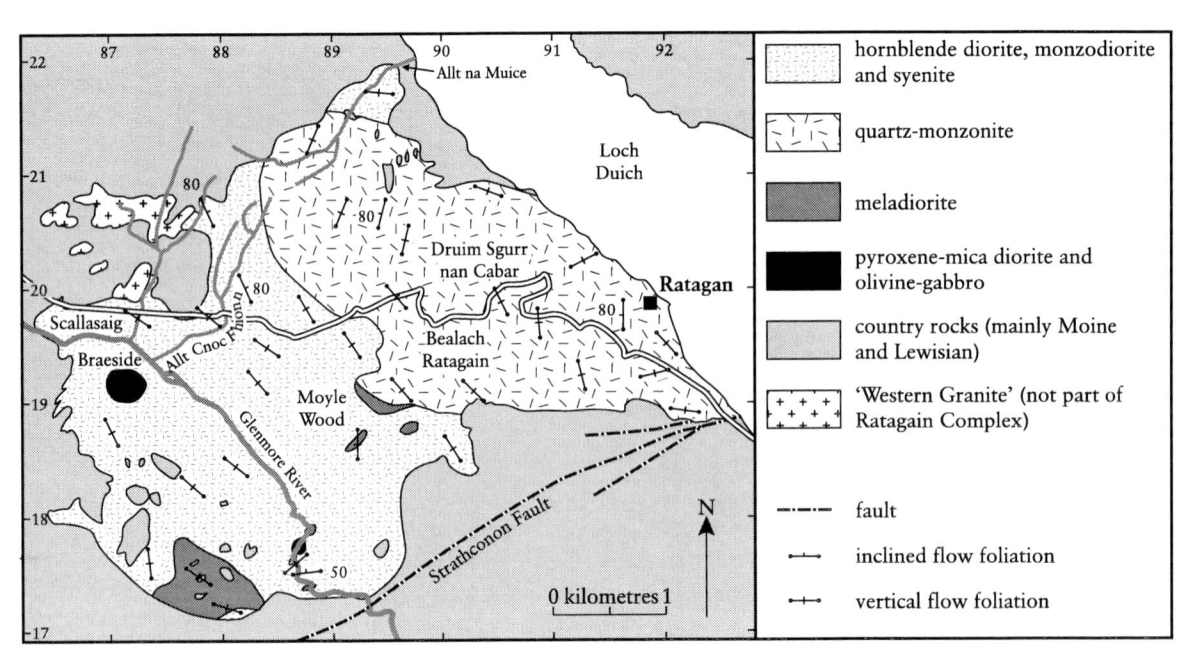

Figure 8.5 Map of the Ratagain pluton, adapted from Hutton *et al.* (1993).

diorites tend to occupy the low ground around Glen More and the valley sides of Moyle Wood. These are cut by later quartz-monzonites which form the hill of Druim Sgurr nan Cabar, above the Bealach Ratagain, and its easterly slopes down to Loch Duich. Overall, the pluton has shallow-dipping walls, where they can be mapped, but the main internal contact between the diorite and the quartz-monzonite is rather steep. In the Glen More region, the present topographical surface appears to be nearly parallel to the roof of the diorite.

The main diorite mass is highly xenolithic with abundant mafic-rich enclaves and some metasedimentary inclusions, all of which are well exposed in the Glen More river and in some tributaries such as the Allt Cnoc Fhionn. The diorites are usually medium grained with hornblende, biotite and plagioclase as the principal minerals, with accessory titanite and minor celestine. Compared with other diorites in the late Caledonian suites, these display considerable textural and mineralogical heterogeneity. One poor exposure of considerable petrogenetic importance in southern Glen More at Braeside is of pyroxene-mica diorite associated with olivine-gabbro (Figure 8.5).

Included within the diorite are large rafts of metasedimentary rock as well as mafic bodies originally described as 'appinites' (Figure 8.6). However these 'appinites' contrast strongly with the type appinites of Appin in Argyll in which amphibole is generally idiomorphic (see the Ardsheal Hill and Peninsula GCR site report). Amphibole is only rarely idiomorphic in the Ratagain rocks and biotite is more abundant than in the type appinites; hence the term meladiorite is preferred. These rocks are best displayed in forestry road cuttings in Glen More and in the woods to the NE. Some of the meladiorite masses may have been intruded in a solid state as pipe-like bodies facilitated by felsic magma. This is suggested by an exposure in a small quarry within Moyle Wood, where pillow-like meladiorite masses are vertically aligned within a monzonitic matrix. While most of these meladiorite bodies have igneous textures at least one example in the Glen More river is banded with layers containing abundant titanite poikilitically enclosed within large alkali feldspars. Petrographical and geochemical evidence indicates that some of these mafic masses may be the result of interaction between diorite and local calcareous metasedimentary rocks, as originally suggested by Nicholls (1951a).

The pink- or buff-coloured quartz-monzonites were intruded late (Figure 8.5), and it is difficult to find sharp contacts with the earlier diorites, most being transitional over about 200 m. An approximate boundary zone can be traced in the

Figure 8.6 Net-veined meladiorite ('appinite') from the Ratagain pluton. (Photo: W.E. Stephens.)

cuttings of the Glenelg–Ratagain road and in the forestry tracks in the northern parts of Moyle Wood. Away from the contact, veins of felsic monzonites intrude brittle fractures in the diorites. The quartz-monzonite adjacent to the contact is rich in mafic minerals, bearing conspicuous hornblende-rich aggregates, and containing large meladiorite inclusions (up to 0.5 m) in places. Farther into the quartz-monzonites, i.e. towards the interior of the pluton, the abundance of quartz appears to increase, although this has not been quantified.

As discussed later, much has been made of the syenitic rocks within the Ratagain pluton (e.g. Thompson and Fowler, 1986). The petrological map of Nicholls (1951a) shows extensive outcrops of syenite in the Glen More and Moyle Wood areas of this GCR site. Exposures of syenitic rocks can indeed be found, for instance at the junction of the Allt Cnoc Fhionn and the main road, but all are small (metres scale) and enclosed within the more evolved diorites. Thus the syenite is regarded as a local facies rather than as a major member of this pluton (Hutton *et al.*, 1993). Nicholls' petrological map also indicates outcrops of 'Western Granite' to the NE of the Glen More river at Scallasaig, which were interpreted as the earliest member of the pluton (Nicholls, 1951a). Mapping by Hutton *et al.* (1993) and associated geochemical studies have shown that this granite is not part of the main Ratagain pluton.

Interpretation

Strongly alkaline granites (*sensu lato*) are unusual and are normally associated with extensional tectonic environments, yet the contemporaneous alkaline syenitic rocks of Assynt (some 100 km to the north of Ratagain) are clearly associated with regional movements on the Moine Thrust, so precluding an extensional origin (see Chapter 7). Thompson and Fowler (1986) highlighted the similarity of the Ratagain syenites with those of Glen Dessarry, Loch Borralan and Loch Ailsh and argued that they were derived from deep asthenospheric mantle sources in a subduction-related setting. However, most recent studies have shown the syenites to be a relatively insignificant facies of the Ratagain pluton and hence they cannot be representative of the mass as a whole (Hutton *et al.*, 1993). An isotopic study of the whole pluton by Halliday *et al.* (1984) established that both mantle and crustal sources were involved in the genesis of the magmas, but detected no subduction-related characteristics.

Compositionally the whole pluton has some most unusual features for a late Caledonian granitic intrusion. As well as unusually high levels of alkalis, especially Na_2O, the trace elements Sr (1000–5000 ppm), Ba (1000–6000 ppm), and Ce (representing the light rare earth elements, 70–400 ppm) are extremely high. These enrichments are not closely correlated, with higher Sr tending to be found in the diorites and Ba being enriched in the monzonites. The origin of this extreme enrichment in incompatible trace elements is still not resolved (Hutton *et al.*, 1993). Such trace element characteristics in intermediate magmas are uncommon in typical subduction regimes, but are known from post-subduction and ridge-subduction systems (Saunders *et al.*, 1987).

The variety of igneous rocks in the pluton is greatest in the Glen More area, virtually spanning the whole range. Olivine-gabbros and pyroxene-mica diorites near the outer contact at Braeside have quenched magmatic textures and thus provide evidence of potential parental basic magmas to at least some facies within the pluton (Hutton *et al.*, 1993). The more basic rock types at Ratagain, including the pyroxene-mica diorite and meladiorite have unambiguous mantle isotopic signatures, which vary sufficiently to suggest that the mantle sources were heterogeneous. The more evolved rock types, including the quartz-monzonites and some diorites, have isotopic signatures that indicate interaction with crustal sources, though not the local metasedimentary rocks (Halliday *et al.*, 1984). The syenitic rocks, which Thompson and Fowler (1986) regarded as K-rich shoshonites with a mantle origin, are in fact more sodic than potassic and are probably local variants of the diorites.

Nicholls (1951a) explained the variety of rocks in the pluton in terms of the co-existence of a calc-alkaline magma of 'Newer Granite' affinity and an alkaline magma of Assynt and Ben Loyal affinity, which underwent extensive hybridization, both at depth and after emplacement. Field evidence in Moyle Wood and in the road cuttings suggests that some hybridization has occurred between the diorites and the quartz-monzonites, and it is likely that the quartz-monzonites have also undergone some fractional crystallization.

A further unusual feature of this pluton in the context of the late Caledonian granitic suites is the rather oxidized condition of the magmas, with the presence of sulphates (celestine and baryte) in the igneous rocks. This condition in monzonites is known to favour the occurrence of gold (Cameron and Hattori, 1987), and indeed small amounts of gold mineralization have been described from late veins in the pluton (Alderton, 1986, 1988).

Conclusions

The Glen More GCR site contains all the major facies of the Ratagain pluton, as well as providing constraints on some of the important field relationships between these members. This single, rather small intrusion is important for its unusual transitional alkaline composition and its implications for the plate tectonic environment of the NW Highlands during late Caledonian times. It has been constructed from magmas derived from a wide range of mantle and crustal sources and could be important in providing a better understanding of the relationships between thrust tectonics and the tapping of magmas from their source regions. The pluton is most unusual, not just in the Caledonian rocks of Britain, but worldwide, in having extreme enrichments of the trace elements Sr and Ba. There is no close analogue anywhere in the world and further studies will contribute to an understanding of this rare type of geochemical enrichment.

LOCH SUNART
(NM 776 607–872 593)

A. J. Highton

Introduction

The Strontian pluton (MacGregor and Kennedy, 1932; Sabine, 1963) occurs on the NW side of the Great Glen Fault and is assigned to the Argyll and Northern Highlands Suite of late

Caledonian granitic intrusions on the basis of its geochemical and isotopic characteristics (Halliday, 1984). The pluton falls into the category of 'forceful' intrusions, thought to have emplaced by diapirism (Read, 1961). As a means of pluton emplacement this mechanism is questionable, and alternative solutions have been presented (Hutton, 1988b). The Loch Sunart GCR site presents a cross section through the northern part of the Strontian pluton. Significant features include evidence for intrusion of basic magma contemporaneous with pluton emplacement, and fabrics resulting from syn-emplacement deformation.

The pluton extends over an area of some 200 km² in a N–S-trending outcrop from the NW shores of Loch Linnhe to the southern slopes of Meall a' Ghruith (822 653). It comprises:

1. an outer hornblende-biotite granodiorite facies, with porphyritic and non-porphyritic variants ('tonalite' and 'granodiorite' of early workers)
2. an inner biotite granodiorite ('biotite granite' or 'adamellite' of early workers) that extends eastwards as a vein complex cross-cutting the metasedimentary envelope (Figure 8.7).

These are referred to respectively as the Loch Sunart granodiorite and Glen Sanda granodiorite facies (Paterson *et al.*, 1992a, b). Mafic enclaves, including some large bodies of appinitic meladiorite, are common in both facies (Holden, 1987). Although previously dated at 435 ± 10 Ma (Pidgeon and Aftalion, 1978), recent zircon studies give an emplacement age of 425 ± 3 Ma for the hornblende-biotite granodiorite (Rogers and Dunning, 1991) and 418 ± 1 Ma for the biotite granodiorite facies (Paterson *et al.*, 1993). The latter facies, however, has a significant inherited zircon component (Paterson *et al.*, 1992a, b).

The envelope consists of middle to upper amphibolite facies metasedimentary rocks of the Glenfinnan and Loch Eil groups of the Moine Supergroup. A 3 km-wide high-grade, sillimanite-bearing thermal aureole encloses the pluton, from which Tyler and Ashworth (1982) derived a pressure estimate of 4 kbar. Along its northern contact, the pluton truncates the outcrop of the Precambrian West Highland Granite Gneiss (Barr *et al.*, 1985; Friend *et al.*, 1997), while the Great Glen Fault terminates the south-eastern boundary of the intrusion. Kennedy (1946) regarded

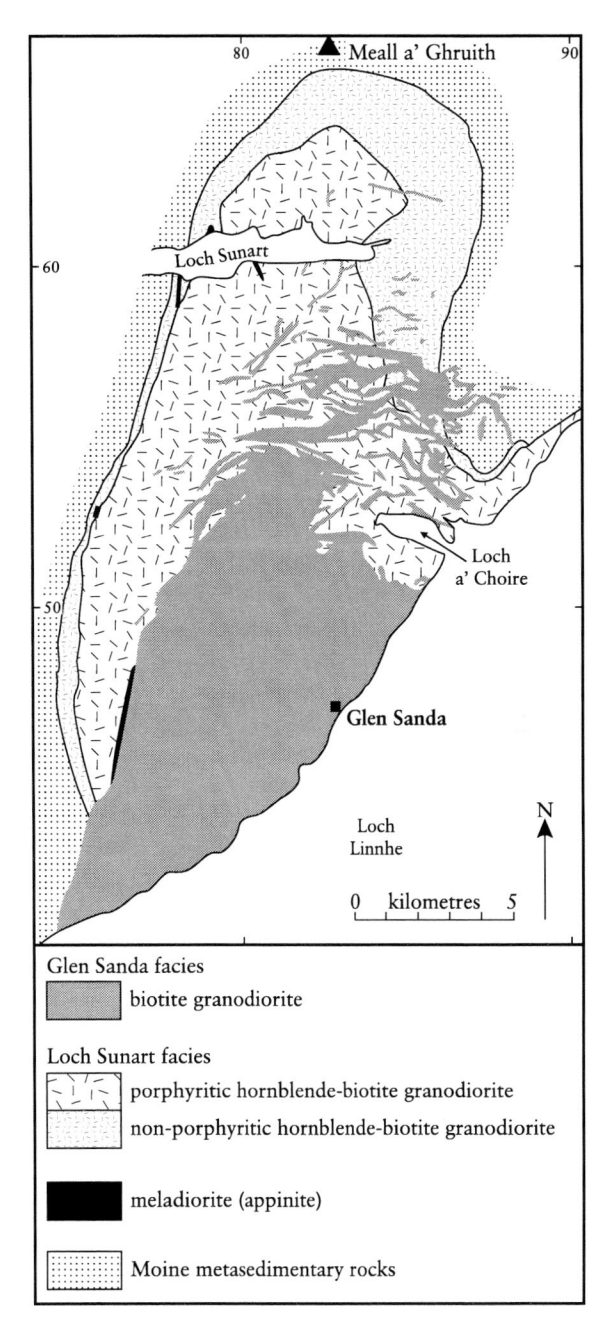

Figure 8.7 Map showing the distribution of facies within the Strontian pluton, adapted from Sabine (1963).

the Strontian outcrops and those at Foyers on the SE side of the fault as part of the same pluton, separated by a 105 km sinistral displacement. There are significant similarities in terms of lithologies, enclave populations, emplacement and synplutonic deformational histories, and both were intruded at corresponding crustal levels of *c.* 13 km (Tyler and Ashworth, 1983).

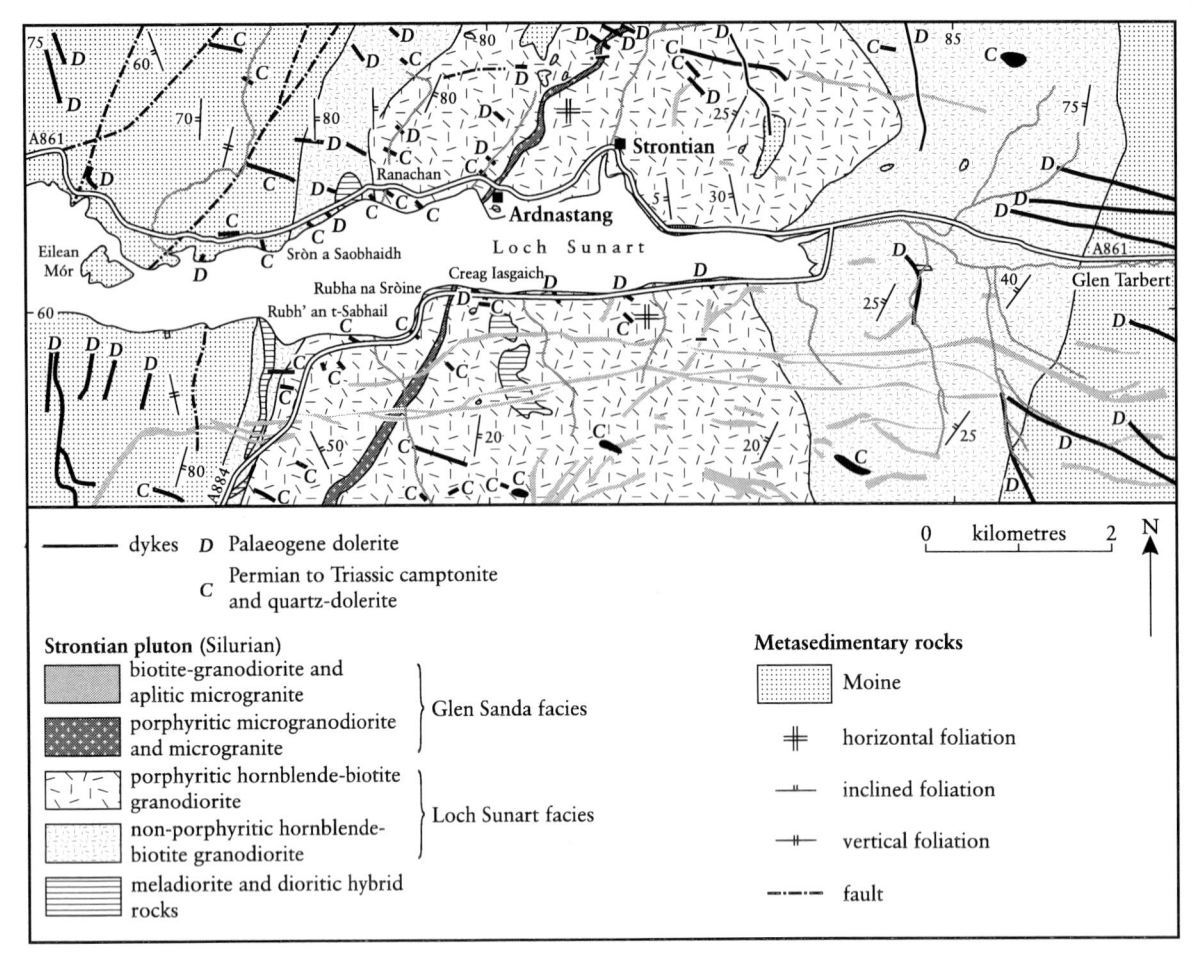

Figure 8.8 Map of the area around the Loch Sunart GCR site, Strontian pluton, adapted from BGS sheets 52E and 53.

However, geochemical and isotopic evidence (Marston, 1971; Pankhurst, 1979; Hamilton *et al.*, 1983; Halliday, 1984), geophysical constraints (Ahmad, 1967; Torsvik, 1984) and structural interpretations (Munro, 1973) have since demonstrated that this correlation is unlikely.

The pluton and country rocks are cross cut by Permian age ENE-trending camptonite dykes and quartz-dolerite plugs (Rock, 1983), and by later basic intrusions of Palaeogene age.

Description

The GCR site comprises a west to east traverse, including road cuttings, shoreline exposures along both sides of Loch Sunart, and the hills to the south of Glen Tarbert (Figure 8.8). It lies within the Loch Sunart facies of the Strontian pluton, which contains abundant mafic enclaves, and is cross-cut by veins of the Glen Sanda facies.

Margin and envelope

To the east of Sròn na Saobhaidh, the western contact of the pluton is steeply inclined against interlayered psammites and semipelites. Here, extensive recrystallization and a cordierite–K-feldspar assemblage represent the highest grade within the aureole. Sillimanite forms felts or less commonly coarse crystals, with cordierite, garnet and K-feldspar overgrowing the regional gneissose foliation. A migmatitic overprint, in the form of granitic segregation, disrupts the regional metamorphic fabrics within 500 m of the contact, but is absent from rocks west of Eilean Mór.

Loch Sunart facies

At the outer margin of the pluton, this hornblende-biotite granodiorite is a medium- to coarse-grained, strong to moderately foliated

Loch Sunart

Figure 8.9 Mafic microgranular enclaves (MME) in porphyritic biotite granodiorite of the Strontian pluton, Rubh' an Torr-mholaich, Loch Sunart (NM 8133 6015). (Photo: BGS no. C4000.)

non-porphyritic variant. Adjacent to the western contact aplitic microgranite veins and partially assimilated country rock xenoliths are numerous. The western contact between the outer non-porphyritic and inner porphyritic variant is transitional over a few metres, marked by the incoming of feldspar mesocrysts. On the south shore of Loch Sunart this boundary is partly obscured by a meladioritic intrusion, that is chilled against the non-porphyritic facies. The inner granodiorite is characterized by variably abundant pink microperthitic K-feldspar megacrysts (up to 2 cm long) and plagioclase phenocryts (up to 1 cm long). Both lithologies contain prominent phenocryst of pink-brown titanite up to 0.5 cm long.

A weak to strong foliation defined by the alignment of the ferromagnesian minerals and plagioclase, is present throughout most of the site. The foliation dips inwards, flattening towards the centre of the intrusion. It decreases in intensity away from the margins, and in the porphyritic facies, is often overgrown by the K-feldspar megacrysts. Abundant mafic-rich enclaves have regular ellipsoidal or lenticular shapes, flattened in the plane of the foliation. However, there is no evidence of any significant deformation or recrystallization of either the fabric-defining minerals or the interstitial quartz and K-feldspar in the granodiorites.

Meladioritic bodies and mafic microgranular enclaves

Of principal interest in the GCR site are numerous mafic-rich enclaves, prevalent within the Loch Sunart facies. These take the form of large meladiorite bodies, smaller microdioritic inclusions and amphibole-rich clots. Three large lenticular steep-sided bodies crop out at Ranachan, Rubh' an t-Sabhail and Rubha na Sròine (Figure 8.8). Their contacts with the host granodiorite have irregular lobate forms, with globular mafic detachments. Fine-grained, (?)chilled, margins are common, in which the ferromagnesian minerals have skeletal or crudely radial forms (7941 6016; 7852 6107). All show compositional zoning outwards from coarse-grained meladiorite to finer-grained heterogeneous, variably plagioclase-phyric, hybrid leucodiorites. The latter often enclose mafic-rich fragments. The hybrid marginal rocks are commonly either veined by the host granodiorite or net-veined by microgranitic pegmatite segregations. In hybrid rocks at the margin of the Rubh' an t-Sabhail body, the pervasive mineral fabric is refracted from the host granodiorite into the appinitic xenolith.

Mafic microgranular enclaves are abundant, forming trains of ellipsoidal fragments aligned parallel to the foliation fabric in the granodiorite host (Figure 8.9). The enclaves are predominantly hornblende-plagioclase quartz-diorites, and rarely porphyritic tonalites. The non-porphyritic granodiorite contains numerous amphibole-titanite-rich clots, up to 8 mm in diameter. Compositional zoning is common from marginal intergrowths of hornblende, biotite and plagioclase, enclosing largely monomineralic cores of actinolitic amphibole.

Minor intrusions of porphyritic microgranodiorite/granite

Rocks of the Loch Sunart facies and meladiorite bodies are cut by NE-trending sheets and/or dykes of compositionally heterogeneous porphyritic microgranodiorite and microgranite (Figure 8.8). A *c.* 20 m-wide intrusion crops out on both shores of Loch Sunart, to the west of Ardnastang and at Creag Iasgaich on the south shore. The intrusion contains both microdiorite and foliated mafic-rich microgranular enclaves, in varying stages of assimilation. Feldspar mesocrysts overgrow contacts between the hybrid rocks and enclaves.

Glen Sanda facies

This facies is not well represented within the GCR site (Figure 8.8), but sheets and dykes of a pink-grey, medium-grained biotite granodiorite extend northwards as far as Loch Sunart and Glen Tarbert. These sheets are mostly parallel sided, with sharp angular contacts; they contain xenoliths of hornblende-biotite granodiorite and lack the foliation that is ubiquitous in the earlier facies. They are variably feldspar-phyric, with pale-pink euhedral phenocrysts of plagioclase (up to 5 mm) enclosed by irregular white rims of albite and poikilitic mesocrysts of K-feldspar (up to 8 mm). Biotite is the predominant mafic mineral (with hornblende rare), but forms less than 10% of the rock.

Interpretation

The Strontian pluton contains two distinct facies, a hornblende-biotite granodiorite and a biotite granodiorite. A two-stage emplacement model has been suggested, with intrusion of the Loch Sunart followed by the Glen Sanda body (Munro, 1965, 1973). Sabine (1963) described the pluton as funnel shaped. Munro (1965) ascribed this form to initial intrusion of a stock-like mass centred in the southern part of the pluton. On reaching the level of emplacement, the magma body expanded laterally in a northerly direction, forcibly distending into the country rock envelope. The internal foliation in the Loch Sunart body was interpreted as a magmatic flow pattern, formed during forceful intrusion (MacGregor and Kennedy, 1932; Sabine, 1963; Munro, 1965). The Glen Sanda intrusion was seen as a later stock, with an apparently brittle mode of intrusion with stoping and sheeting, reflecting emplacement at a higher crustal level than the Loch Sunart intrusion (Munro, 1965). A considerable time gap was invoked to accommodate the apparent uplift, and current geochronology separates the intrusions by approximately 7 Ma. Given the estimated emplacement level of the Loch Sunart body at 14 km, an uplift rate in excess of 1 km per Ma would be necessary to accommodate the high level of intrusion implied for the Glen Sanda body.

More recent studies suggest that the internal foliation in the Loch Sunart facies is a pre-full crystallization fabric, rather than magmatic flow, with the highest strains occurring towards the margins of the pluton (Hutton, 1988a, b). Minerals both defining the fabric and filling the interstices show little evidence of recrystallization. Hence, the foliation in the northern part of the pluton is not a high temperature solid-state tectonic fabric, but reflects the impositon of strain upon an inward accreting crystal framework. Crystal plastic strain fabrics are present in the hornblende-biotite granodiorite elsewhere in the pluton, adjacent to the western boundary with the biotite granodiorite intrusion (Hutton, 1988b). This is an indication of continuing imposition of strain after local consolidation. The variation in orientation of the foliation within the northern part of the pluton (well illustrated within the GCR site), coupled with asymetric vein shear-sense indicators (seen elsewhere), was interpreted by Hutton (1988b) as consistent with a southerly directed listric extension at the time of emplacement, rather than with diapirism or ballooning. Space created during this deformation, at the extensional termination of a dextral shear splay of the Great Glen Fault, allowed the intrusion of the later biotite granodiorite (Hutton, 1988a, b).

Mafic-rich enclaves are common inclusions within 'I-type' granitic intrusions, and are abundant within the Strontian pluton. They have been interpreted as disrupted precursor 'appinite' and microdiorite intrusions (MacGregor and Kennedy, 1932; Sabine, 1963). However, recent studies suggest that the enclaves represent synplutonic basaltic intrusions and autoliths, which have mantle isotopic signatures (Holden, 1987; Holden *et al.*, 1987; Holden *et al.*, 1991; Stephens *et al.*, 1991). Interaction of these hydrous basic melts with the host granitic magma has given rise to the dioritic hybrids (Holden *et al.*, 1987). The lobate margins to most of the enclaves are indicative of liquid–liquid contacts, with the fine-grained edges representing quenching against a lower temperature host granitic magma. The liquid–chill contacts with differing granitic facies implies episodic intrusion of basic magma throughout pluton emplacement.

Castro and Stephens (1992) suggested that the amphibole-rich clots formed as reaction products between pyroxene and the host magma. The source of the pyroxene is equivocal. They may either represent phenocrysts from the basic magma dispersed during mingling with granodioritic magma, or they may have been derived from the source as restite. The granoblastic textures within the clots might favour a restite origin.

Conclusions

The Loch Sunart GCR site represents one of the finest examples worldwide of mid-crustal pluton emplacement. Fabrics and textures of the pluton clearly demonstrate the effect of active shearing during intrusion and crystallization of magma, which here was probably contemporaneous with movement within the Great Glen Fault system. The site is equally important in demonstrating succinctly the interaction of contemporaneous basic and granodioritic magmas during pluton emplacement, and also the incorporation of residual unmelted material from the source area (restite). The occurrence of basic rocks in all the granodioritic facies points to the continuous availibility of basic magma throughout the emplacement history of the pluton.

CNOC MOR TO RUBH' ARDALANISH (NM 367 186–360 160)

A. J. Highton

Introduction

The Ross of Mull pluton

The Ross of Mull pluton extends over an area of some 140 km², much of which lies offshore, forming skerries along the eastern coast of Iona

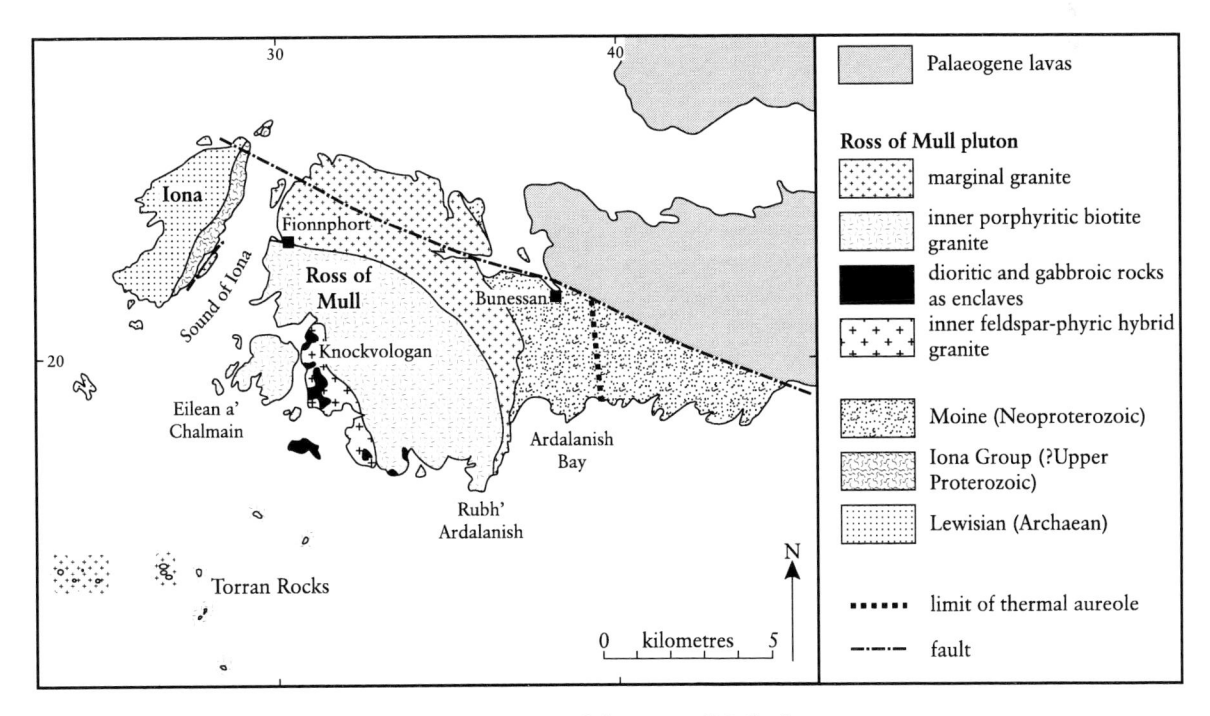

Figure 8.10 Map of the Ross of Mull pluton.

and the Torran Rocks (Cunningham-Craig *et al.*, 1911; Bailey and Anderson, 1925; Barber *et al.*, 1979). The pluton is mainly granitic, with a crudely concentric reverse zonation. Three discrete facies are identifiable within the land outcrop (on the basis of variations in biotite content), all with gradational internal contacts (Figure 8.10):

1. an outer non-porphyritic equigranular biotite granite, which is the most evolved part of the pluton; a xenolithic, two-mica contaminated variant forms small bodies along the pluton margin;
2. an inner pink K-feldspar-phyric biotite granite;
3. a heterogeneous hybridized variant of '2' that ranges from biotite granite to biotite granodiorite, and contains hybrid rocks and basic enclaves.

Little is known of the geology offshore, although the granites off eastern Iona and the Torran Rocks are similar to the outer facies.

The Ross of Mull granites have been quarried extensively in the past, particularly the outer non-porphyritic granite along the coast north of Fionnphort (Faithfull, 1995). The stone has been used for bridges, docks, lighthouses and other buildings throughout the world, as well as for ornamental stone. Noteable examples include Ardnamurchan Lighthouse, Westminster Bridge, New York and Liverpool docks, Glasgow University, Manchester Town Hall and the Albert Memorial.

The granitic components are of high-K calc-alkaline affinity, and are mostly metaluminous to weakly peraluminous, with compositions in the range 67–78% SiO_2. The rocks of the basic enclaves have SiO_2 contents of 45–55%, and have shoshonitic affinities. Halliday *et al.* (1979a) obtained a mineral/whole-rock Rb-Sr age of 414 ± 3 Ma from the outer biotite granite, that probably dates cooling rather than crystallization of the pluton.

The pluton is hosted mainly by metasedimentary rocks of the Moine Supergroup (Riley, 1966). These comprise the Assapol and Shiaba groups, thought to be lateral equivalents of the Glenfinnan and Morar groups of the mainland successions (Holdsworth *et al.*, 1987). Country rock xenoliths are present in all the facies and a relict stratigraphy is traceable through the pluton. To the west, the granite is in contact with low grade ?Upper Proterozoic metasedimentary rocks of the Iona Group. These overlie meta-igneous and metasedimentary rocks of the Archaean Lewisian Complex, with tectonically modified unconformity (Potts *et al.*, 1995).

The thermal overprint in the Moine rocks extends up to 3 km from the eastern margin of the pluton, but significant hornfelsing, marked by the incoming of andalusite, is found only within 500 m of the contact (Bailey and Anderson, 1925). Higher-grade assemblages containing fibrolite and sillimanite appear only at the pluton contact (Brearly, 1984). Regional metamorphic kyanite is metastable throughout much of the aureole (Bosworth, 1910; MacKenzie, 1949). Clough (in Bailey and Anderson, 1925) noted the occurrence of a small suite of hornfelsed microdioritic intrusions ('lamprophyres') close to the eastern margin, which pre-date granite emplacement. On Iona the aureole is generally less than 1 km wide.

Both the granite and the country rocks are cross-cut by a synplutonic suite of minor intrusions consisting of shoshonitic calc-alkaline lamprophyres (spessartite and kersantite) and porphyritic microgranodiorite. These intrusions are mostly sheets, with composite or multiple forms common. All are cut by, generally ESE-trending, camptonite and monchiquite dykes of Permian age (Beckinsale and Obradovich, 1973).

The Cnoc Mor to Rubh' Ardalanish GCR site demonstrates the form and nature of the eastern contact of the Ross of Mull pluton. The site provides a traverse through the inner aureole and marginal contact zone of the pluton. Units that comprise the central part of the pluton are represented by the Knockvologan to Eilean a' Chalmain GCR site.

Description

The Cnoc Mor to Rubh' Ardalanish GCR site encompasses the coastal outcrops along the Ardalanish peninsula on the southern coast of the Ross of Mull. Of special interest is the interaction of the granitic rocks with the Assapol Group country rocks during emplacement. Magmatic processing and assimilation of included metasedimentary material has resulted in a locally developed marginal hybrid granite. Here, enclaves and screens of Moine metasedimentary rocks are seen in various stages of incorporation into the host granite. Other notable features of the site are the aureole and

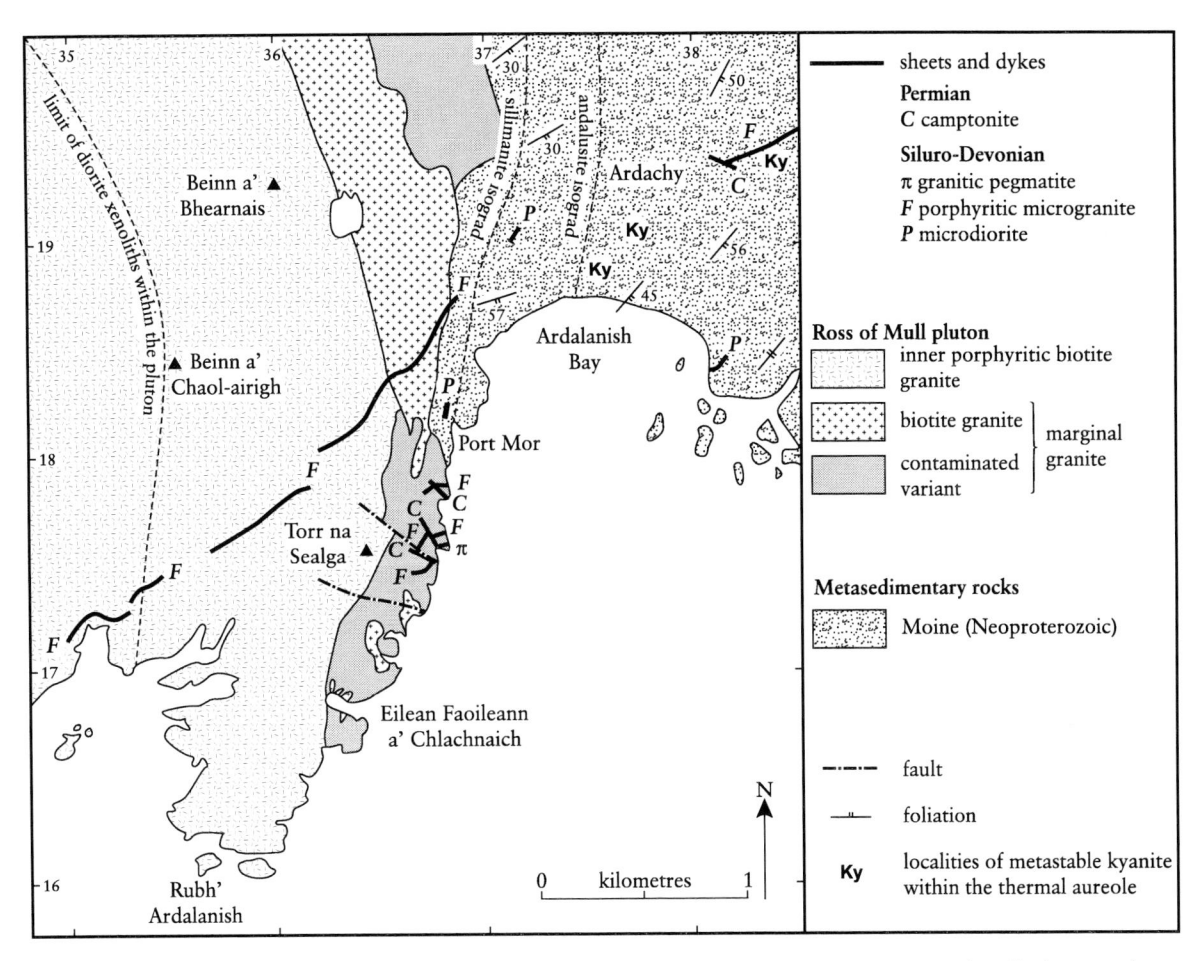

Figure 8.11 Map of the area around the Cnoc Mor to Rubh' Ardalanish GCR site, Ross of Mull pluton, adapted from BGS 1:50 000 Sheet 43 and unpublished work, University of Liverpool.

the syn-plutonic intrusions. Three principle units are recognized within the site (Figure 8.11):

1. country rocks
2. marginal granite, with locally developed contaminated variant
3. inner porphyritic biotite granite

Country rocks

To the east of Port Mor, the eastern margin of the granitic pluton is in contact with country rocks of the Assapol Striped and Banded Formation. These comprise variably interlayered beds of semipelite, psammite and minor pelite, with concordant ribs of calc-silicate rock. Both the metasedimentary rocks and the microdiorite intrusions within the envelope are hornfelsed. In the former, regional tectonic fabrics and mineral assemblages are barely recognizable. Knots of sillimanite (up to several centimetres long) and/or small porphyroblasts of andalusite are prominent in semipelitic and pelitic lithologies. Small dark crystals of cordierite are ubiquitous, and cordierite also replaces biotite in the more micaceous psammitic lithologies. Regional metamorphic muscovite is generally absent from the metasedimentary rocks of the inner aureole, although andalusite and/or fibrolite may be replaced by a late white mica. Within the microdiorite sheets recrystallization is extensive, with hornblende overgrown and replaced by a new red biotite.

Marginal granite

Of principal interest in this site is the occurrence of a zone of contamination, up to 200 m wide,

Figure 8.12 Irregular porphyritic microgranodiorite sheet cutting the marginal contaminated granite along the eastern margin of the Ross of Mull pluton. The slab-like xenoliths of the Moine country rock generally retain a pre-emplacement attitude in this marginal sheeted complex. Carraig Mhór (NM 3678 1762). (Photo: A.J. Highton.)

along the eastern margin of the pluton. This contaminated variant forms most of the outcrop of the marginal granite exposed along the eastern side of the peninsula between A' Bhualaidh and Eilean Faoileann a' Chlachanaich. The non-contaminated granite parent is seen only in exposures on Cnoc Mor, as a pale-pink, coarse-grained, equigranular biotite granite. The contaminated granite variant is marked by a decrease over several tens of metres in the pink colouration of the parent granite. Outcrops are commonly crowded with country rock xenoliths showing all stages of assimilation into the host. The contaminated granite is unusually quartz-rich and contains abundant mesocrysts of red-brown biotite, which contain inclusions of zircon and zoned dark-cored apatite. Muscovite is present but is not a primary mineral. Pods and small masses of granite pegmatite are locally common. A large sheet of coarsely crystalline pegmatite at 3677 1760, presents a fine example of macroscopic microcline perthite.

The contact with Moine rocks is essentially a sub-horizontal sheeted complex of interdigitating granite and country rock. The envelope is gradational over a few tens of metres, from country rocks containing granite sheets concordant with the prominent foliation, into granite with screens of metasedimentary rocks. The marginal granite carries abundant rafts, up to 250 m long (364 172) and slab-like xenolithic blocks (Figure 8.12). An outcrop (at 3673 1804) contains enclaves in varying states of assimilation from little altered angular slabs to rounded enclaves. The latter are enclosed by a reaction corona of leucocratic granodiorite. Locally the contaminated granite contains a nebulous mica fabric or schlieren of biotite-rich restite. These schlieren typically wrap around included blocks. K-feldspar megacrysts are common in many partially assimilated enclaves. These overgrow both the relict metamorphic fabrics in the enclaves and the contacts with the granite host.

Sheets of porphyritic microgranodiorite, generally dipping NW at 15–40°, cut all facies of the granite and country rocks. These irregular sheets, display necking and side-stepping typical of synplutonic minor intrusions (Figure 8.12).

Inner porphyritic biotite granite

Much of the site on the Ardalanish peninsula comprises a K-feldspar-phyric biotite granite facies. This component is generally a pink, coarse-grained biotite monzogranite, with abundant megacrysts, up to 5 cm long, of pink perthitic microcline. This is well exposed in low rounded outcrops of the Rubh' Ardalanish area. The boundary of the porphyritic biotite granite with the marginal granite is generally transitional over several tens of metres. Mica schlieren are, however, recognizable within the outer facies at some distance from the main outcrop of contaminated marginal variant (359 162). On A' Bhualaidh, the contact with the marginal granite is sharp.

Interpretation

This eastern margin of the Ross of Mull pluton shows features typical of the transition zone between roof and wall in mid to upper crustal granite intrusions (Fowler *et al.*, 1995), which helps to explain the relatively wide thermal aureole. In section, the contact is low to moderately inclined to the east, and is irregular with brittle-looking metre-scale angular steps. Pre-existing regional deformational structures in the envelope do not appear to control the overall form of the contact, and are truncated mostly at a high angle. However, low-angled sheets have exploited foliation planes along much of the margin. Here country rock blocks are typically angular slabs with minimal rotation from their original orientation in the envelope. Away from this sheeted margin, included blocks of country rock are commonly separated by granite with locally prominent biotite-rich schlieren and banding. A tectonic origin for this mica fabric is unlikely. Evidence of strain-induced recrystallization or ductile thinning of beds within included material is absent. Further, the synplutonic minor intrusions cut across internal structures in the pluton, but close to the margin take on a stepped form similar to that of the contact. Thus, the schlieren more likely derive from incomplete processing and incorporation of envelope material during intrusion, resulting in a biotite-rich restite fabric within a hybrid. Hence the areas of contaminated hybrid granite preserve clues to part of the pluton emplacement mechanism.

The emplacement process along this eastern margin was probably initiated by wall rock assimilation, but gave way to penetration and stoping of the envelope. Deflection of the schlieren fabric around included blocks points to foundering of blocks from the roof into a partially crystalline granite hybrid. The transitional boundary into the porphyritic biotite granite points to relative homogenization in the main part of the magma body. Preservation of the contaminated marginal rocks may well be fortuitous, reflecting local ponding within an irregularity in the roof or wall.

Conclusions

This GCR site is of national importance as a representative of the Ross of Mull pluton, which provides one of the most definitive examples of passive emplacement with assimilation of country rock in the Caledonian plutonic suites. Within the coastal outcrops on the western side of Ardalanish Bay, all stages are preserved, from the impregnation of the Moine country rocks by granite sheets, to the spalling off and sinking of slabs and blocks of country rock into the granite magma (stoping). Assimilation of metasedimentary material into the marginal granite resulted in the development of a contaminated hybrid granite, present at the pluton–country rock contact throughout most of the site. Localized ponding of fluids in the magma led to the crystallization of pegmatite. The complex sub-horizontal sheeted margin is unlikely to reflect the pluton form, but lies within the transition between the roof and wall of the intrusion.

KNOCKVOLOGAN TO EILEAN A'CHALMAIN (NM 309 175–309 204)

A. J. Highton

Introduction

The coastal exposures on the eastern side of Erraid sound, on Eilean Dubh and on Eilean a' Chalmain, which comprise this GCR site, encompass the central part of the Ross of Mull granitic

Late Silurian and Devonian granitic intrusions of Scotland

pluton (Figure 8.13). The general features of the pluton are described in the Cnoc Mor to Rubh' Ardalanish GCR site report. Here, heterogeneous hybrid granitic rocks contain chilled enclaves of gabbroic and dioritic rocks ('appinites'). These enclaves have high-K basalt to basaltic andesite compositions and themselves contain mafic microgranular enclaves and xenocrysts. The hybrid rocks grade over several hundreds of metres into the inner porphyritic biotite granite of the Ross of Mull pluton, which still contains mafic enclaves up to 3 km beyond the site. Of special interest are textures and lithologies resulting from the generation of these hybrid rocks through the incorporation of the basic to intermediate material into the granitic host. The site also preserves a 'ghost stratigraphy' of the Moine country rocks, which can be traced through the granitic rocks in the metasedimentary xenoliths.

Description

Although dioritic rocks form a small component of the Ross of Mull pluton, their abundance within this GCR site is considerable. The basic enclaves range in size from a few centimetres to several hundreds of metres and the largest discrete mass of basic rock comprises most of Eilean a' Chalmain. Rocks within this outcrop are heterogeneous, ranging from medium-grained diorite to coarse-grained appinitic monzodiorite characterized by poikilitic crystals of orthoclase and pale-coloured biotite. The outcrop is cut by numerous sheets and apophyses of biotite granite, but with little development of hybrid rocks.

Exposures on the small island to the north of Eilean Dubh (at 3064 1894) and on the west-facing shore of Erraid Sound (at 3077 2012) provide good examples of the mafic enclave-rich hybrid granitic rocks and of their heterogeneous textures. The hybrid host rock is predominantly a biotite-rich granite (± rare primary amphibole), with a significant accessory mineral component (titanite, allanite and magnetite). However, there is a range of compositions from mafic granodiorite to quartz-diorite. The enclaves range in size from a few centimetres to several metres across, and in composition from fine-grained hornblende gabbro or microgabbro to microdiorite. Their distribution is variable, but locally they comprise up to 10% of the granite. Larger enclaves are typically surrounded by

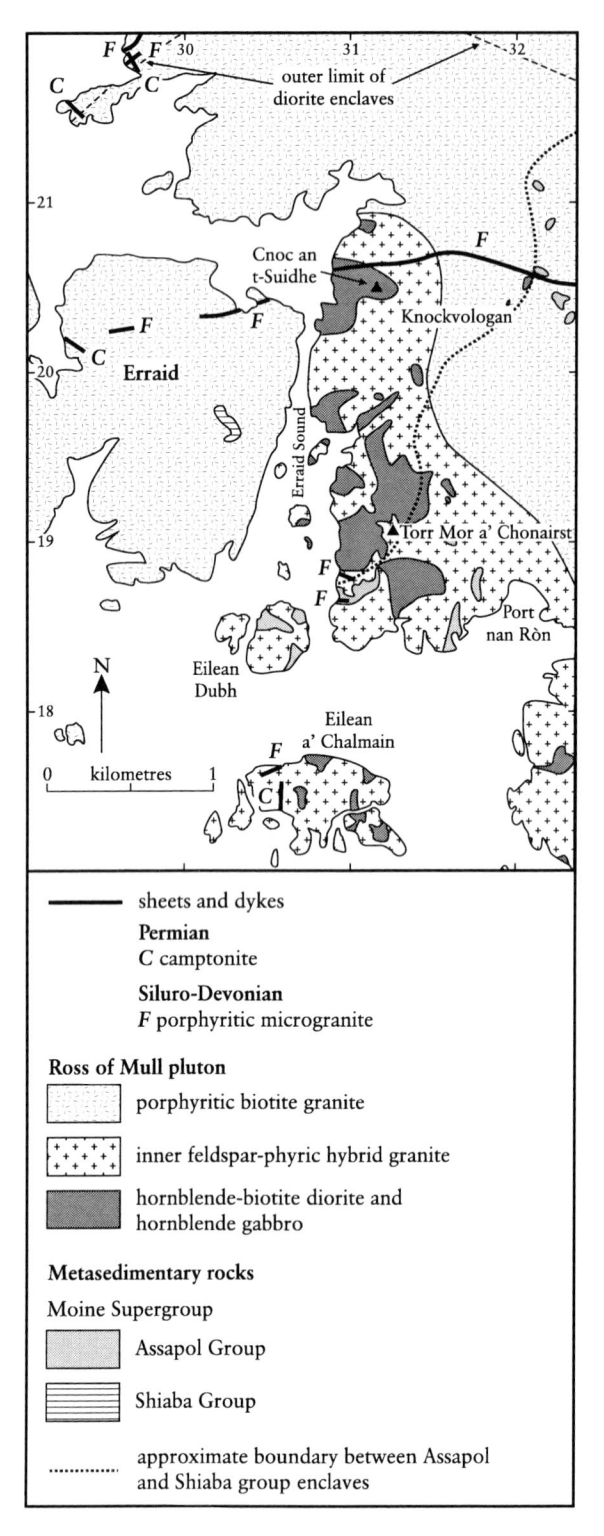

Figure 8.13 Map of the area around the Knockvologan to Eilean a' Chalmain GCR site, Ross of Mull pluton, adapted from BGS 1:50 000 Sheet 43 and unpublished work, University of Liverpool.

Figure 8.14 Small meladiorite ('appinite') enclaves enclosed by a hornblende-biotite granodiorite hybrid host, centre of the Ross of Mull pluton, west of Port nan Ròn (NM 3124 1858). (Photo: A.J. Highton.)

numerous smaller fragments (Figure 8.14). Most are rounded in form, and commonly show finely lobate contacts with their host. The contacts vary from sharp with fine-grained 'chilled' margins, to diffuse. Where contacts are diffuse, the enclaves are either enclosed by discrete coronas of hornblendic quartz-diorite or grade over several tens of centimetres into the host biotite-rich granite. Quartz-diorite with small hornblende-rich aggregates similar to the corona material also occurs as small rounded blobs within the hybrid granite. Veining or development of small areas of net-veined hybrid rocks are common. The veins, of fine-grained leuco-tonalite to microgranodiorite, cut both host and enclaves. Glomerocrysts of quartz and plagioclase are common in the hybrid rocks, while K-feldspar mesocrysts occur in all lithologies.

A weak to moderately developed mineral fabric is present in most of the igneous rocks of the site. This partially wraps the enclaves, which take on an ellipsoidal shape in areas where the fabric is strongly developed.

The metasedimentary inclusions in the granite define a 'ghost stratigraphy'. K-feldspar-rich arkosic psammites and finely laminated psammites of the Upper Shiaba Psammite crop out within the northern part of the GCR site, (e.g. 3095 1952), and on islands in the Erraid Sound, while striped psammites and semipelites of the Assapol Group occur elsewhere. The xenolith populations have a clearly definable distribution, with little intermixing. Structures and fabrics in the larger inclusions, for example at Cnoc an t-Suidhe, to the SW of Torr Mor a' Chonaist, at Port nan Ròn and on Eilean Dubh, are comparable in orientation to those in the country rocks. Hence, these are probably close to in-situ roof pendants. Examples of thermal overprinting, partial melting and assimilation into the granite are common. Within the inclusions of Assapol Group rocks to the west of Port nan Ròn, sillimanite forms knots that both overprint all fabrics, and mimetically replace the biotite that forms crenulation cleavages axial planar to minor folds in semipelitic lithologies.

Interpretation

The age of the mafic enclaves and meladiorite bodies within the Ross of Mull pluton has been interpreted in several ways. Cunningham-Craig *et al.* (1911), suggested that they represent a disrupted diorite complex intruded as a precursor to emplacement of the pluton. Recent studies (R. H. Hunter, University of Liverpool, pers. comm.) suggest that the lobate fine-grained 'chilled' margins to many enclaves result from quenching during the interaction of penecontemporaneously emplaced basic to intermediate and acid magmas. Physical mixing, with thermal equilibration between the magma types, resulted in the formation of the heterogeneous hybrid rocks. The presence of enclaves of similar composition to these hybrid rocks within the main porphyritic biotite granite indicates local dispersion and mingling of the hybrid liquids. The more basic magmas are probably contemporaneous with the suite of calc-alkaline lamprophyre and microdiorite dykes present within the envelope. However, the linear trains of enclaves recognized in other Caledonian plutons, e.g. Strontian (see the Loch Sunart GCR site report) have not been recorded.

The origin of the internal foliation is equivocal. The enclaves show little evidence of significant deformation indicative of high syn-emplacement strains. Hence the fabric may be magmatic in part.

The distribution pattern of the metasedimentary xenoliths follows predicted stratigraphical lines. The lack of fragmentary dispersal, with only local intermixing, and the lack of significant re-orientation suggests a passive emplacement mechanism into the envelope and close proximity to the roof of the pluton.

Conclusions

The Ross of Mull pluton is notable among the Caledonian plutons for the preservation of one of the finest examples of 'ghost' country rock stratigraphy within an intrusion. This demonstrates that the pluton was intruded through a process of passive emplacement with little disruption of the metasedimentary country rocks. The 'ghost stratigraphy' is best preserved within this GCR site, which imparts an international significance to the site and to the pluton as a whole. The hybrid rocks featured in this site provide an example of the co-existence of basic to intermediate and granitic magmas in the pluton. Features are typical of magma mixing and mingling, with dispersed rounded enclaves of both basic and hybrid material in the porphyritic biotite granite host. It is considered that the basic and granitic magmas were intruded at the same time and their interaction has given rise to zoning within the pluton, with more basic rocks passing outwards to more acid rocks (reverse zoning).

BONAWE TO CADDERLIE BURN (NN 008 336–038 385)

A. J. Highton

Introduction

The Etive pluton

Like many plutons of the late Caledonian Argyll and Northern Highlands Suite, the Etive pluton is composite, ranging from diorite through to monzogranite (Bailey and Maufe, 1916; Kynaston and Hill, 1908; Anderson, 1937; Batchelor, 1987). It was emplaced at *c.* 400 Ma (Pidgeon and Aftalion, 1978; Clayburn *et al.*, 1983) into metasedimentary and meta-igneous rocks of the Dalradian Supergroup. This large elliptical intrusion, covering an area of some 300 km², comprises four discrete intrusive phases. In order of emplacement these are the Quarry intrusion, the Cruachan facies, the Meall Odhar facies and the central, Starav facies (Figure 8.15).

An accompanying NE-trending swarm of syn-plutonic dykes, the Etive dyke-swarm, consists mainly of porphyritic microdioritic and microgranodioritic lithologies, and contains sub-suites that either cut or are truncated by the main granitic facies. The pluton is spatially associated with extrusive rocks of the Lorn plateau to the SW and the Glen Coe caldera volcano to the north (see Chapter 9), but it is unlikely to be the source of their magmas. Pressure estimates from the metamorphic aureole indicate a high crustal, subvolcanic, emplacement level at *c.* 3–6 km (Droop and Treloar, 1981).

The Bonawe to Cadderlie Burn GCR site

This site, along the western shore and succeeding hills of Loch Etive, includes the extensive quarries at Bonawe, which were worked historically for paving sets and latterly for hard rock aggregate. The site provides a broad traverse from the country rock envelope through most of the principle components of the Etive pluton

(Figure 8.16). This illustrates the range of lithologies and their sequence of emplacement, from the outer xenolith-rich monzodiorites of the Cruachan facies to the monzogranitic rocks of the inner, Starav facies (Anderson, 1937). The satellitic Quarry intrusion is not present here, but is represented by the Cruachan Reservoir GCR site. Synplutonic dykes are also present. The country rocks within the site are assigned to the Bonawe Succession, a possible correlative of the Easdale Slates within the lower Argyll Group of the Dalradian Supergroup (Litherland, 1980). These country rocks are the predominant enclaves in the outer part of the Cruachan facies, often occurring as large screens or roof pendants.

Description

Margin and envelope

On the south-western slopes of Beinn Duirinis above Bonawe, the irregular NW-trending contact of the pluton is traceable in almost continuous exposure, (e.g. 0065 3385). Along much of the western edge the pluton margin is a sheeted complex. Within the GCR site it is sharp and mainly sub-vertical but locally it dips inwards at a steep angle. To the SE, the country rocks of fine-grained semipelite and black slates of the Bonawe Succession form small outcrops on the shore at Bonawe (0065 3353). These rocks reached biotite grade during the Caledonian regional metamorphism. On emplacement of the pluton, there was extensive recrystallization of the country rocks, with macroscopic poikiloblasts of cordierite and andalusite overgrowing the tectonic fabrics. Close to the pluton contact (0068 3374) the host rocks became hornfelsed, with bedding and the regional tectonic fabrics largely obliterated.

Cruachan facies

Much of the GCR site between Bonawe and Cadderlie lies within the Cruachan facies (Figure 8.16). Of principal interest is a marginal, variably foliated, enclave-rich monzodioritic variant, that forms much of a 1 km-wide outcrop to the SE of Lag Choan (027 340). In the lower level of the current workings in Bonawe Quarry at 0215 3365, rafts and xenoliths of Bonawe Succession

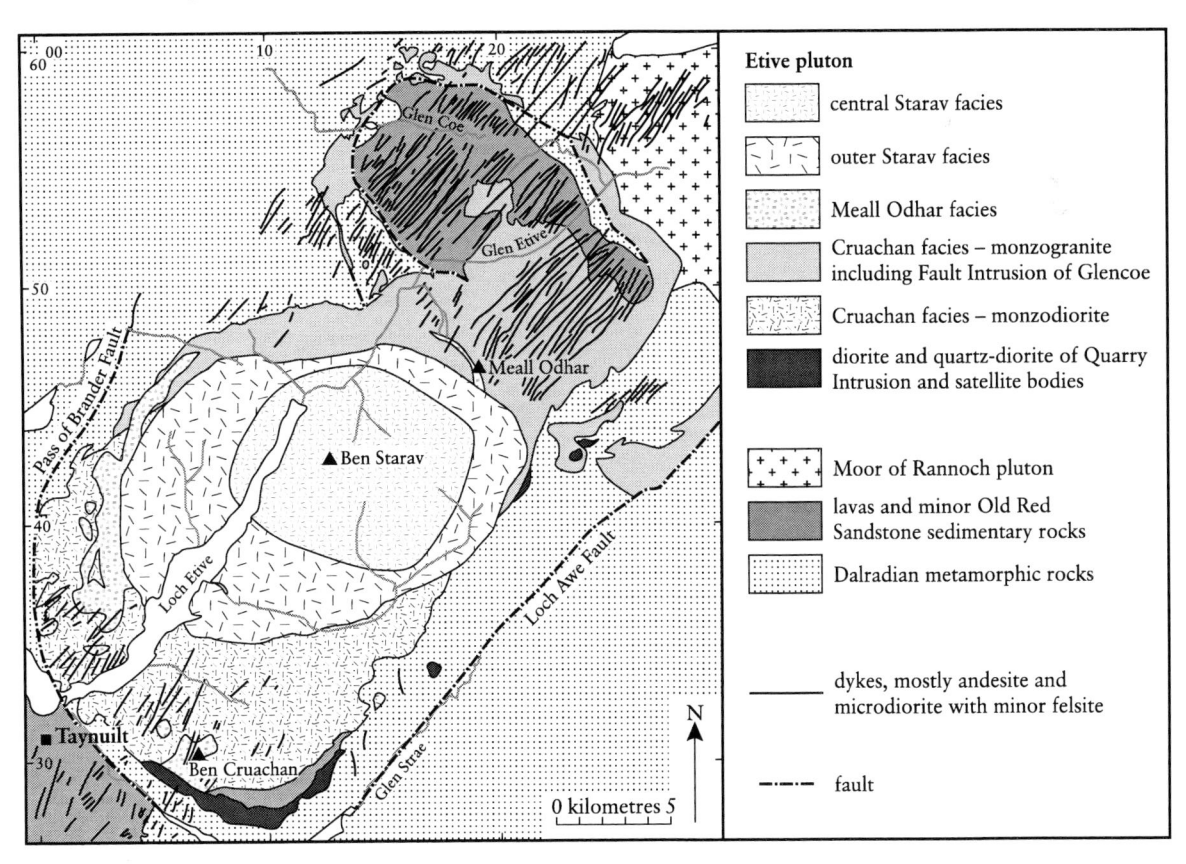

Figure 8.15 Map of the Etive and Glencoe complexes, after Anderson (1937) and Batchelor (1987).

Late Silurian and Devonian granitic intrusions of Scotland

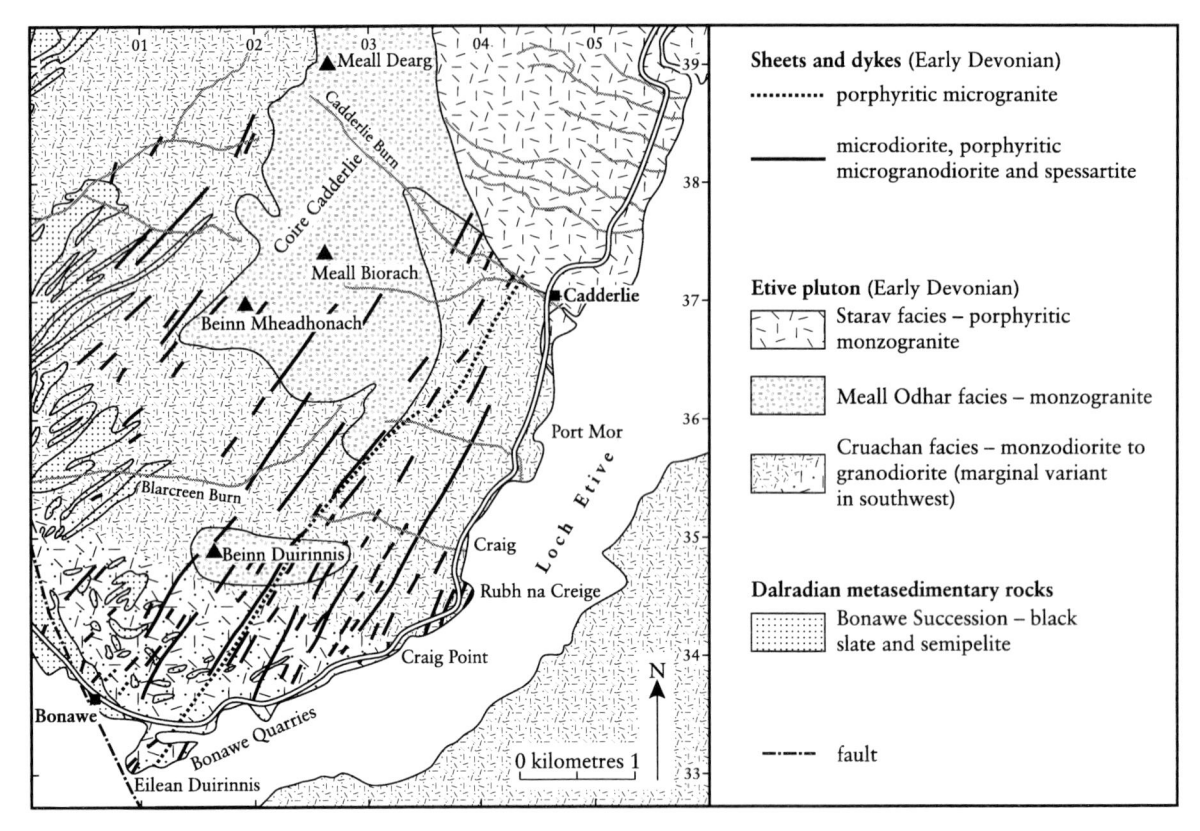

Figure 8.16 Map of the area around the Bonawe to Cadderlie Burn GCR site, Etive pluton, adapted from BGS 1:50 000 Sheet 45W.

rocks, weakly foliated porphyritic microdiorite, and mafic microgranular enclaves crowd the monzodiorite. The upper level workings (at 0214 3370) expose large spalled country rock blocks, several hundred metres long, which are probably roof pendants (Figure 8.17). Smaller inclusions range from angular to elliptical blocks, up to several tens of metres in diameter, with the latter flattened parallel to a foliation in the host rock. Evidence for assimilation is sparse, although thin selvages of leucogranite enclose some xenoliths. Mafic monzodioritic enclaves, with round to lobate fine-grained margins, form inclusion trains through this facies. Brittle deformation is ubiquitous close to the pluton margin, with zones of small-scale fracturing accompanied by cataclasite veining. Chlorite-carbonate slickensides are prominent on many joint surfaces.

In his summary of the 'Etive Complex', Anderson (1937) ascribed the distinctive palegrey, fine-grained, equigranular granitic rocks within the Bonawe Quarry (015 335) to a small marginal intrusion. This was purportedly separated from the main intrusion by a screen of metasedimentary rocks. This leucocratic biotite granodiorite is now regarded as a marginal phase of the Cruachan facies. Nebulous patches and net-veins of leucogranodiorite represent local heterogeneities.

The inner variant is a fine- to medium-grained hornblende-biotite monzodiorite. Although generally equigranular, textural and compositional heterogeneities include variations in grain size, mafic content and occurrence of feldspar megacrysts. On the shore of Loch Etive (0390 3460) and in the old quarry workings at Rubha na Creige (0390 3453), the rocks contain K-feldspar megacrysts (up to 1 cm long) and minor amphibole. Small mafic-rich enclaves are locally numerous.

A steep inwardly dipping, margin-parallel fabric, defined by aligned plagioclase and ferromagnesian minerals, occurs throughout the Cruachan facies. Within the marginal variant there is significant flattening of the enclaves (0215 0363). Although the foliation is generally less prominent inwards, rocks adjacent to the contact with the Starav facies in the Cadderlie Burn (0440 3709) contain a well-defined fabric.

Bonawe to Cadderlie Burn

Figure 8.17 Raft or roof pendant of hornfelsed Bonawe Succession (Dalradian) metasedimentary rocks (dark coloured) within the marginal variant of the Cruachan facies, Etive pluton. All are cross-cut by a c. 20 m-wide dyke of porphyritic microgranite (to the right of the photo). Quarry workings, Bonawe. (Photo: BGS no. MNS 4849.)

Meall Odhar facies

Within this GCR site, the Cruachan facies is cut by two large shallow-dipping intrusions comprising weak to moderately porphyritic, fine- to medium-grained pink granite, the Meall Odhar facies (Figure 8.16). A small body crops out on the summit of Beinn Duirinnis (021 347), while the slopes of Coire Cadderlie (035 385) and Meall Biorach (027 374) lie within a shallow, irregular, NE-dipping sheet. This intrusion is truncated by the Starav facies along a sharp sub-vertical contact (0383 3840). The monzodiorites of the Cruachan facies are also cut by irregular zones comprising anastomosing steeply inclined veins of a moderately porphyritic microgranite, as seen in the main face at Bonawe Quarry (015 335). At Craig Point (0307 3402) the veins coalesce into larger bodies. These are commonly separated by thin screens of monzodiorite, often only a few centimetres thick, (e.g. 0314 3398). At Craig Point, NE-dipping porphyritic microdiorite sheets, with chilled margins, cut both the Cruachan and Meall Odhar facies.

Starav facies

Only the outer porphyritic variant of the Starav facies lies within the GCR site (Figure 8.16). As seen on the NE flank of Meall Dearg (035 390), this variant is typically a coarse-grained pink-grey monzogranite, with conspicuous K-feldspar megacrysts up to 3 cm long. The rocks are hornblende- and biotite-bearing, with conspicuous phenocrysts of titanite, although the mafic content is generally less than 10%. Mafic-rich microgranular enclaves, up to 4 mm, comprising amphibole + biotite + opaque minerals ± pyroxene, are common. The K-feldspar megacrysts are numerous in the outer part of the intrusion, but decrease inwards in both size and abundance (Anderson, 1937). To the NW of Cadderlie (0414 3736), the marginal rocks are non-porphyritic up to 15 cm from the contact with the Cruachan facies. Elsewhere, the K-feldspar megacrysts overgrow a ubiquitous weak, steep inward-dipping margin-parallel foliation.

Minor Intrusions

The plutonic rocks of the GCR site are cut by numerous NE-trending dykes and some sheet-like intrusions (Figure 8.16). The swarm consists mainly of microdiorite and porphyritic microgranodiorite (formerly termed 'porphyrites'), but also includes meladiorite ('appinite'), spessartite, olivine kersantite and quartz-phyric microgranite (formerly termed 'quartz-porphyries'). Sub-suites include those that:

a. transgress all facies of the pluton;
b. cut all the pre-Starav facies;
c. are truncated by the Cruachan facies.

A pink quartz-phyric microgranite dyke cropping out in the Cadderlie Burn (0431 3716) provides the only example in the GCR site of an intrusion cutting the Starav facies, but does not extend much beyond the contact with the Cruachan facies. From here this dyke is traceable south-westwards to the main quarry at Bonawe, and on the southern flank of Beinn Duirinis (018 342) it cuts a microdiorite dyke (type b). Cross-cutting relationships within the minor intrusive suite are well seen in the new quarry workings (0214 3370). Here also, a c. 20 m-wide porphyritic microgranite intrusion (Figure 8.17) contains rounded microdiorite xenoliths. In most porphyritic intrusions the phenocrysts have a margin-parallel alignment. Examples of hornfelsed pre-Cruachan facies minor intrusions crop out on the south-facing slopes above Kenmore (0052 3393).

Interpretation

The Etive pluton is an excellent example of multiple pulse emplacement (Anderson, 1937; Frost and O'Nions, 1985; Batchelor, 1987). Overall, the pluton is one of the least evolved of the Argyll and Northern Highlands Suite intrusions, with initial $^{87}Sr/^{86}Sr$ values ranging from 0.7043 to 0.7068 (Clayburn et al., 1983; Frost and O'Nions, 1985). The isotopic data has been variously interpreted as providing: (a) evidence for either the recycling of lower continental crustal material during magmatic evolution, with significant assimilation of country rock (Frost and O'Nions, 1985); or (b) the incorporation of a substantial component of older continental crust (Hamilton et al., 1983) or a mantle component (Thirlwall, 1986). The Cruachan and Starav facies derive from different parental magmas; the latter having a 'juvenile' signature (Plant et al., 1985).

The Cruachan facies was historically thought to comprise two lobes, predominantly monzodioritic in the south and monzogranitic in the north (Anderson, 1937). These compositional differences have been attributed subsequently to tilting of the intrusion, exposing more evolved rocks to the north (Brown, 1975), although there may be two separate intrusions (Barritt, 1983). Compositional variations within the Cruachan facies are consistent with the intrusion of successive pulses from essentially the same magma source (Batchelor, 1987). This suggestion is also reinforced by compositional variations within those members of the Etive dyke-swarm, that are broadly contemporaneous with emplacement of the Cruachan facies. The presence of mafic-rich enclaves alludes to the availability of contemporaneous basaltic magma during emplacement of the Cruachan facies. Monzodioritic enclaves within the outer felsic variant, point to mingling of these basaltic magmas with their granitic host, leading to localized hybridization.

The irregular sheet-like intrusions of the Meall Odhar facies lie within the upper parts of the Cruachan facies while the vein complexes pervade differing levels. Batchelor (1987) suggested that the Meall Odhar facies intrusions are precursors to the emplacement of the Starav facies, although the conduit is no longer recognizable. However, rocks of the Meall Odhar facies are cut by minor intrusions consanguineous with the Cruachan facies magmas. Published geochemical analyses show that Meall Odhar facies rocks are consistently more evolved than the Starav facies (Rb/Sr 1.1–8.5 and 0.12–2.8, respectively; cf. Clayburn *et al.*, 1983; Batchelor, 1987), with significantly lower levels of Rb and Sr. This corroborates the suggestion of Clayburn *et al.* (1983), on isotopic evidence, that a new magma batch was introduced after the emplacement of the Meall Odhar facies. The occurrence of high-level residual fractionation melt bodies is a feature of other Caledonian plutons, e.g. Cairngorm and Monadhliath (Harrison, 1987a; Highton, 1999), and may provide a solution for the origin and distribution of the Meall Odhar facies. Hence it is unlikely that the Meall Odhar facies simply represents a tapping of contemporaneous Starav facies magmas.

Reverse compositional zonation in the pluton from outer felsic-rich to inner mafic-rich phases has been cited as evidence for cauldron subsidence (Batchelor, 1987; following Anderson, 1956). The abundance of xenoliths, rafts, screens or roof pendants is characteristic of carapace foundering into the magma body, with the form and size of xenolithic material demonstrating the varying stages of stoping. However, Jacques and Reavy (1994) interpret the margin-parallel foliation as a pre-full crystallization fabric. It is argued that this fabric is a consequence of high-level in-situ 'ballooning' contemporaneous with shearing on NE-trending faults. The flattening of enclaves and en echelon pull-aparts seen at Bonawe are consistent with shearing about a steeply inclined axis. Hence evidence of plastic strain during the later stages of crystallization would support the case for synmagmatic transpressional shear rather than simple block let down.

Conclusions

The Bonawe to Cadderlie Burn GCR site is of national and international importance for the cross section through the Etive pluton, which contains examples of differing mantle- or lower crustal-derived magmas. The site embraces some of the finest evidence of upper crustal, multiple pulse pluton emplacement and also illustrates important evidence for dykes that were intruded into the larger bodies of magma while they were still cooling. Fabric evidence at outcrop is consistent with intrusion via deep crustal fractures into an active shear environment. The magma conduit may well have developed at the confluence of long-lived basement fractures and late Caledonian shear zones. Space was created for pluton emplacement by means of fracturing of the metasedimentary envelope and foundering of large blocks into the magma (stoping).

CRUACHAN RESERVOIR (NN 077 285)

A. J. Highton

Introduction

To the north of the Cruachan hydroelectric power station, outcrops along the Pass of Brander and on the SE flanks of Ben Cruachan provide a traverse through the metasedimentary envelope and marginal facies of the Etive pluton. This large elliptical pluton comprises four discrete intrusive phases; in order of emplacement these are the Quarry intrusion, the Cruachan facies, the Meall Odhar facies and the central, Starav facies (Figure 8.15) (see the Bonawe to Cadderlie Burn GCR site report). The envelope and complex contact relationships along this southern margin of the pluton, first described by Kynaston and Hill (1908), are well displayed within this GCR site. The outer margin comprises rocks of the Cruachan and Meall Odhar facies, similar to those documented within the Bonawe to Cadderlie Burn site. The inner, Starav facies, which is well displayed at the Bonawe to Cadderlie Burn site, is not represented at the Cruachan Reservoir site. Here also, an apparently down-faulted block of andesitic lavas, the Beinn a' Bhuridh screen, separates an arcuate satellite body, the Quarry intrusion, from the main body of the pluton. Within the Quarry intrusion, assimilation of calcareous metasedimentary rock xenoliths has produced some unusual hybrid rocks (Nockolds, 1934).

The Cruachan Reservoir GCR site encompasses the Cruachan pump storage system, part of the Loch Awe hydroelectric scheme (Figures 8.18 and 8.19). Excavations for the dam construction and the modification of water courses reveal details of the contact relationships

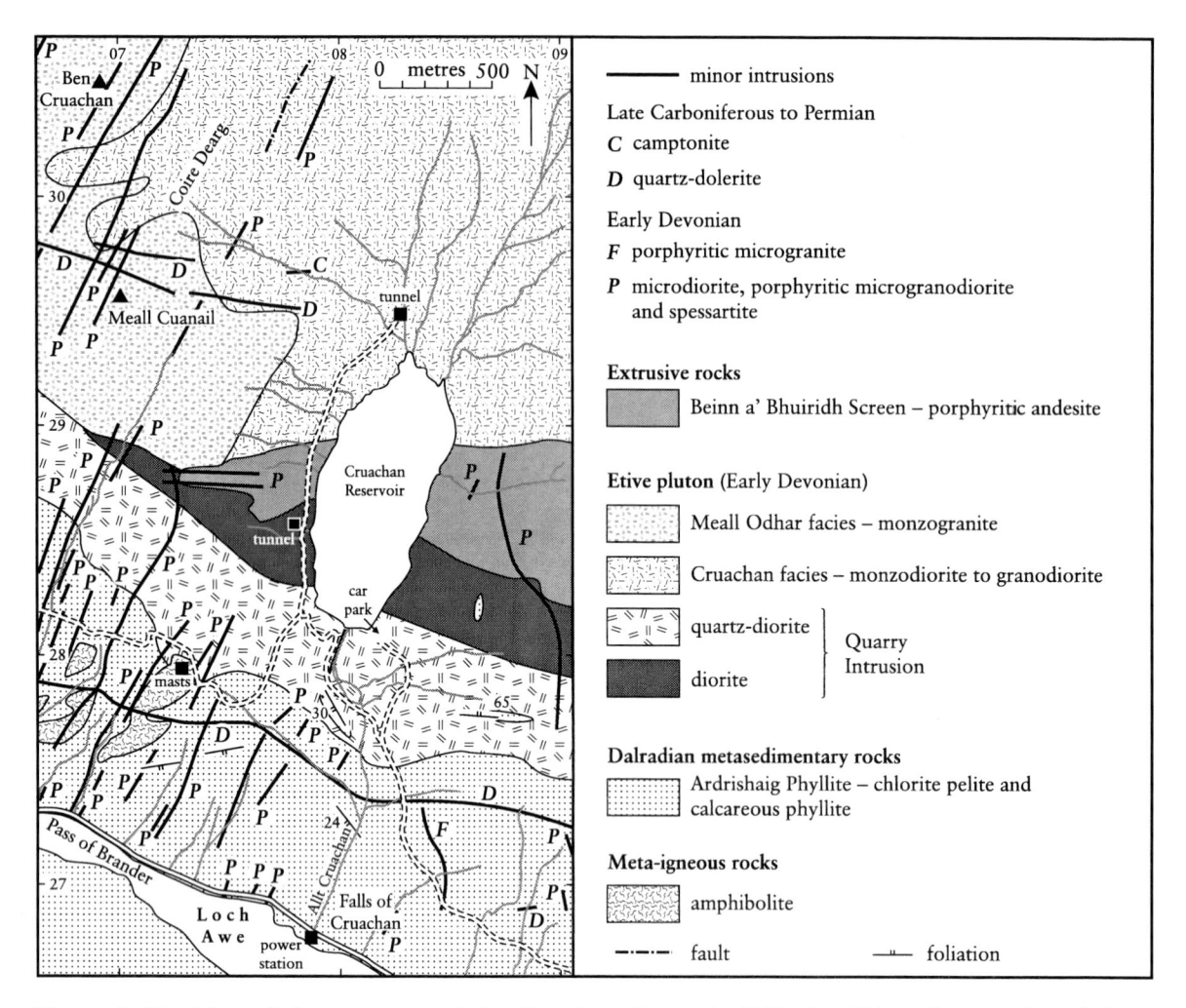

Figure 8.18 Map of the area around the Cruachan Reservoir GCR site, Etive pluton, adapted from BGS 1:50 000 Sheet 45E.

between the Etive pluton and its envelope. Detailed geological information was also obtained during the construction of underground tunnels, aqueducts and the power station (Knill, 1972).

Description

Margin and envelope

Much of the pluton envelope within the GCR site comprises metasedimentary rocks of the Ardrishaig Phyllite (Argyll Group, Dalradian Supergroup). These comprise finely interlayered black or purple chloritic phyllites and calcsilicate rocks, with thicker bands of impure metalimestone and calcareous quartzite. Close to the contact with the Quarry intrusion, the phyllites become hard splintery hornfelses, in which the regional deformational fabrics are poorly preserved. Biotite, as small porphyroblasts or aggregates, is the most common thermal aureole mineral, with rare cordierite spots. Within 200 m of the intrusion the calcareous rocks contain a contact metamorphic assemblage of palegreen diopsidic pyroxene, garnet, epidote and amphibole.

Quarry intrusion

This arcuate satellite intrusion consists of diorite and quartz-diorite variants. Both are represented within the GCR site (Figure 8.18). Only the quartz-diorite is in contact with the metasedimentary rocks of the envelope. The contact is discordant to the foliation in the host Ardrishaig Phyllite and has a steep outward dip, which is seen in the Allt Cruachan. Here, the quartz-diorite at the intrusion margin is fine grained but elsewhere, (for example at 0775 2813) and in the access road to the dam, it is commonly medium to coarse grained. It comprises phenocrysts of plagioclase, up to 1.5 cm long, and brown amphibole with interstitial quartz and K-feldspar. Country rock xenoliths in varying stages of assimilation are common, often enclosed by a heterogeneous quartz-rich hybrid granodiorite (0819 2804).

The diorite is finer grained than the quartz-diorite, with pyroxene phenocrysts, up to 8 mm long, as the predominant mafic mineral. A green amphibole occurs both as small phenocrysts and as aggregates, with biotite, after pyroxene. The contact between the two dioritic variants is transitional, for example along the shore of the reservoir (0785 2030). A margin-parallel pre-full crystallization fabric is present throughout the intrusion, but is most conspicuous in the coarser-grained rocks.

Beinn a' Bhuridh screen

The augite-hornblende andesites of the Beinn a' Bhuridh screen form the conspicuous crag (at 0775 2868), and crop out on the access track (at 0785 2870) and on the shore of the reservoir (Figure 8.18). A steeply dipping outer contact with the Quarry intrusion is exposed close to the tunnel entrance, west of the reservoir (0783 2865), and in the slopes above. The inner contact with marginal rocks of the main pluton (Cruachan facies) is not found within the site, and can only be inferred from elsewhere (cf. Anderson, 1937). However, the lavas are clearly hornfelsed and are commonly altered from dark-grey to a greenish colour. In outcrops to the west of the reservoir the lavas are commonly vesicular, the voids being infilled with quartz and/or chlorite. Here, fragments and small blocks of 'quartzite', up to 10 cm across, are also present.

Cruachan facies

Rocks of the Cruachan facies, which comprise the outer part of the main Etive pluton, form the outcrops close to the northern tail race entrance (0827 2946), on the NE-facing slope of Meall Cuanail (0775 2953), and in the burn emanating from Coire Dearg (0768 2972). Rocks of this facies cross-cut and vein rocks of the Quarry intrusion. The Cruachan facies comprises coarse-grained, pale pink-grey hornblende-biotite monzodioritic to granodioritic rocks, and is similar to the inner variant described from the Bonawe to Cadderlie Burn GCR site. Large crystals of biotite and small phenocrysts of green hornblende are the predominant mafic minerals in these rocks. The amphibole also occurs in mafic aggregates, up to 1 cm in diameter, with biotite, opaque minerals and rare pyroxene. Titanite is conspicuous as small deep-pink crystals. A pre-full crystallization fabric is ubiquitous, but is most intense at the pluton margin.

Within the tail race tunnel of the hydroelectric scheme, the monzodiorites of the Cruachan facies are cross-cut by veins of K-feldspar megacrystic hornblende-biotite granodiorite. This lithology may be a correlative of the marginal variant seen within the Bonawe to Cadderlie Burn GCR site. The mafic content of these rocks is variable, although mostly less than 15%. Biotite is predominant, while amphibole commonly occurs within small, 1–2 mm, mafic microgranular aggregates with titanite, magnetite and biotite.

Meall Odhar facies

This facies forms the twin summits of Ben Cruachan (Figure 8.19) and slab-like outcrops along the S-trending ridge from Meall Cuanail (072 288), where it imparts a characteristic pink colour to the ground. The facies comprises a fine- to medium-grained, equigranular, K-feldspar-phyric monzogranite. K-feldspar megacrysts, up to 1.5 cm long, vary from white to pale-pink where rocks are fresh, to deep-red in areas of alteration. Contacts with either the Cruachan facies or Quarry intrusion are not seen, although both are cut by thin sheets, or more commonly anastomosing veins, of this granite. In exposures on the flanks of Meall Cuanail, the granite contains large xenoliths of andesitic lava, similar to those of the Beinn a' Bhuridh screen, and of fine-grained acid (?rhyolitic) lava (Anderson, 1937).

Minor intrusions

The synplutonic intrusions of the Etive dyke-swarm consists mostly of steeply inclined NE-trending dykes, with a few shallow-dipping sheets (Figure 8.18). Branching and side-stepping along joint structures is a characteristic of many of these intrusions. The highest concen-

Late Silurian and Devonian granitic intrusions of Scotland

Figure 8.19 Aerial view of the Cruachan Reservoir GCR site, Etive pluton, looking WNW to Ben Cruachan. (Photo: BGS no. D 2571.)

tration of intrusions occurs in the country rocks and the swarm density decreases towards the centre of the pluton. Three distinct sub-suites are recognized as follows.

(a) Porphyritic andesites.
(b) Microdiorite, porphyritic micromonzodiorite/microgranodiorite (formerly 'porphyrites') and rare spessartite.
(c) Porphyritic microgranite (formerly 'quartz-porphyry') and aplitic microgranite (formerly 'felsite').

Members of sub-suite (a) occur only within the country rocks. These are dark grey-green, fine grained to aphanitic, and have been partially recrystallized by contact metamorphism. The more abundant microdiorite and microgranodioritic intrusions of sub-suite (b) may reach 15 and 35 m-thick respectively, but are mostly less than 6 m. The more basic intrusions are dark-green rocks, with phenocrysts of green or brown amphibole; the latter is commonly in association with augite, and plagioclase. The microgranodioritic rocks are a grey-buff colour, and are characterized by large oscillatory zoned crystals of plagioclase, 0.5–2 cm long. Plagioclase is the most abundant phenocryst in these rocks, with biotite predominant over amphibole. Within-dyke textural and compositional variations are common, often manifest as feldspathic net-veining.

On the basis of intrusive relationships the dyke swarm is further divisible into those that:

1. pre-date the Quarry intrusion;
2. are cut by the Cruachan facies or veins of monzodiorite/granodiorite;
3. cross-cut the Cruachan facies but are cut by the Meall Odhar facies or veins of similar age;
4. post-date all the main facies of the Etive pluton within the site.

Intrusions of sub-suite (c) consistently post-date all rocks of the pluton and the dykes of intermediate composition. A good example of this relationship is seen in the access road to the dam (0828 2735). All members of the Etive dyke-swarm are cut by ESE-trending quartz-dolerite and camptonite dykes of late Carboniferous to Permian age.

Interpretation

The origin of the Beinn a' Bhuridh screen is equivocal, although $^{87}Sr/^{86}Sr$ isotope values lie within the range of the Lorn Plateau lavas. However, contact metamorphic alteration is extensive, as seen by a $\delta^{18}O$ value of 6.3 (Frost and O'Nions, 1985), thus negating any meaningful comparison with other plutonic rocks or with Siluro-Devonian lavas. The arcuate form of the screen is compatible with Anderson's (1937) interpretation as a drop-down block within a ring fracture. There is no evidence for post-pluton emplacement ring fracturing within the GCR site, although elsewhere Anderson (1937) describes fault crush along the contact of the Quarry intrusion. Similarly, the shape of the satellitic Quarry intrusion suggests the exploitation of a pre-existing fracture system as a magma conduit.

Anderson (1937), following Nockolds (1934), attributed the turbid patches that disrupt oscillatory zoning in plagioclase crystals in the Quarry intrusion to contact metamorphism. However, this 'patchy zoning' is common in the Cruachan facies rocks, and is a feature of many plutons worldwide (cf. Vance, 1965). The origin of this texture is attributable to feldspar dissolution as a consequence of crystal-melt disequilibria during such processes as magma mixing (Wark and Watson, 1993). A pre-full crystallization texture is present in both the Quarry intrusion and the Cruachan facies. The absence of plastic strain fabrics and evidence of high temperature recrystallization, e.g. biotite overgrowth, argues against the dioritic Quarry intrusion being fully crystalline prior to pluton emplacement.

The multistage and multiple pulse emplacement history for the Etive pluton, described in the Bonawe to Cadderlie Burn GCR site report, is confirmed here. Both the Cruachan facies and the Quarry intrusion are cut by a large flat lying intrusion of the Meall Odhar facies at the highest preserved level within the pluton. Some members of the synplutonic dyke-swarm post-date all plutonic rocks within the GCR site. This points to the continued availability of Cruachan facies-type magmas after emplacement of Meall Odhar facies intrusions. This is demonstrated by the complex intrusive history of the Etive dyke-swarm, in which clearly defined sub-suites represent tapping of successive magma pulses during pluton emplacement. Cross-cutting relationships do not necessarily follow compositional maturity (i.e. acid cutting basic), but are time dependent, i.e. an acid intrusion of an earlier sub-suite will be cut by a basic component of a later sub-suite. The exception lies with the quartz-phyric and aplitic microgranites. These compositionally distinct intrusions post-date all basic to intermediate members of the swarm, and are likely to be penecontemporaneous with the Starav facies (not represented within this GCR site).

Conclusions

The Cruachan Reservoir GCR site is of national importance as a representative of the Etive pluton, in particular the satellitic Quarry intrusion, which is separated from the main pluton by a screen of andesitic lavas. The pluton is of major importance for its contribution to the understanding of multiple pulse emplacement mechanisms in the Caledonian plutonic suites. Of particular interest here are the numerous sub-suites of minor intrusions. Their emplacement relationships to facies within the Etive pluton, provide an unrivalled example of near-contemporaneous dyke intrusion into large bodies of magma that were still cooling. The high-level emplacement of the pluton is indicated by the presence of down-faulted volcanic rocks bound by a ring fracture. The arcuate form of the Quarry intrusion suggests exploitation of this fracture by the dioritic precursor magmas. Subsequent emplacement of the main body of the Etive pluton induced widespread baking and alteration of rocks within the envelope. Evidence for recrystallization of rocks within the satellite Quarry intrusion is, however, equivocal.

RED CRAIG
(NO 293 758)

S. Robertson

Introduction

Red Craig lies at the eastern margin of the Glen Doll pluton, a member of the South of Scotland Suite of late Caledonian intrusions. The pluton has long been recognized as providing well-exposed and easily accessible evidence of the interaction between component magmas of a basic to intermediate intrusion (Barrow and Cunningham-Craig, 1912). The Red Craig area exhibits transitions from quartz-diorite through quartz-monzodiorite to granite. The igneous rocks are predominantly xenolithic and provide excellent examples of the interaction between the intermediate part of the pluton and the host Dalradian metasedimentary rocks.

The Glen Doll pluton occupies approximately 12 km² astride the Glen Doll Fault in the upper part of Glen Clova. The pluton is dominated by intermediate rocks of dioritic to tonalitic composition, although with a significant component of gabbro (Jarvis, 1987; Mahmood, 1986). Local olivine-pyroxenite (Mahmood, 1986) was previously referred to as serpentinite or picrite (Barrow and Cunningham-Craig, 1912). Barrow and Cunningham-Craig originally described a 'narrow fringe of encircling granite'; Jarvis, however, only recognized a marginal facies of medium-grained xenolithic granite along much of the southern and eastern part of the pluton.

The pluton was emplaced into dominantly semipelitic metasedimentary rocks assigned to the Argyll and Southern Highland groups of the Dalradian Supergroup (Figure 8.20). Little has been reported on the contact metamorphic effects of the intrusion, although Barrow and Cunningham-Craig recognized a zone of alteration at least 70 m-wide at the eastern margin but failed to detect any mineralogical alteration of the regional sillimanite zone rocks under the microscope.

Description

The GCR site at Red Craig provides a section through the eastern part of the Glen Doll pluton. Medium- to coarse-grained diorite and quartz-diorite, some of which is xenolithic, is typical of a large part of the pluton and passes east into a heterogeneous marginal zone of xenolithic quartz-diorite with areas of quartz-monzodiorite, granodiorite and granite. Xenoliths are mostly of high-grade hornfelsed semipelite, some of which show evidence of partial melting and assimilation into the diorite. Appinitic meladiorites are developed locally as are sheets and dykes of fine-grained felsite and quartz-feldspar porphyry.

Diorites and quartz-diorites are well exposed in a small roadside quarry near Braedownie (2882 7572) and on hillslope exposures to the NE on Dùn Mòr (290 759). Farther north there are numerous exposures although access is more difficult because of forestry plantations. The diorite on the south slopes of Dùn Mòr is cut by both granite veins and felsite sheets. The granite veins are typically only 5 cm thick, sinuous and generally steeply inclined, whereas the felsites are vertical and up to 1 m thick. Both have sharp margins with the host diorite. East of Dùn Mòr, an appinitic meladiorite contains large (4 mm) euhedral hornblende megacrysts within a finer-grained quartz-diorite or quartz-monzodiorite groundmass. The contact relationships of this unit cannot be seen on the steep and loose slopes. Appinitic rocks also occur farther north towards the Cald Burn (295 770). Xenoliths of hornfelsed semipelite occur quite widely within the diorite although many are only a few centimetres long. However, a xenolith (at 294 766) is at least 100 m long.

The diorites are separated from the host Dalradian rocks to the east by a 500 m-wide heterogenous zone of intermediate to acid igneous

Red Craig

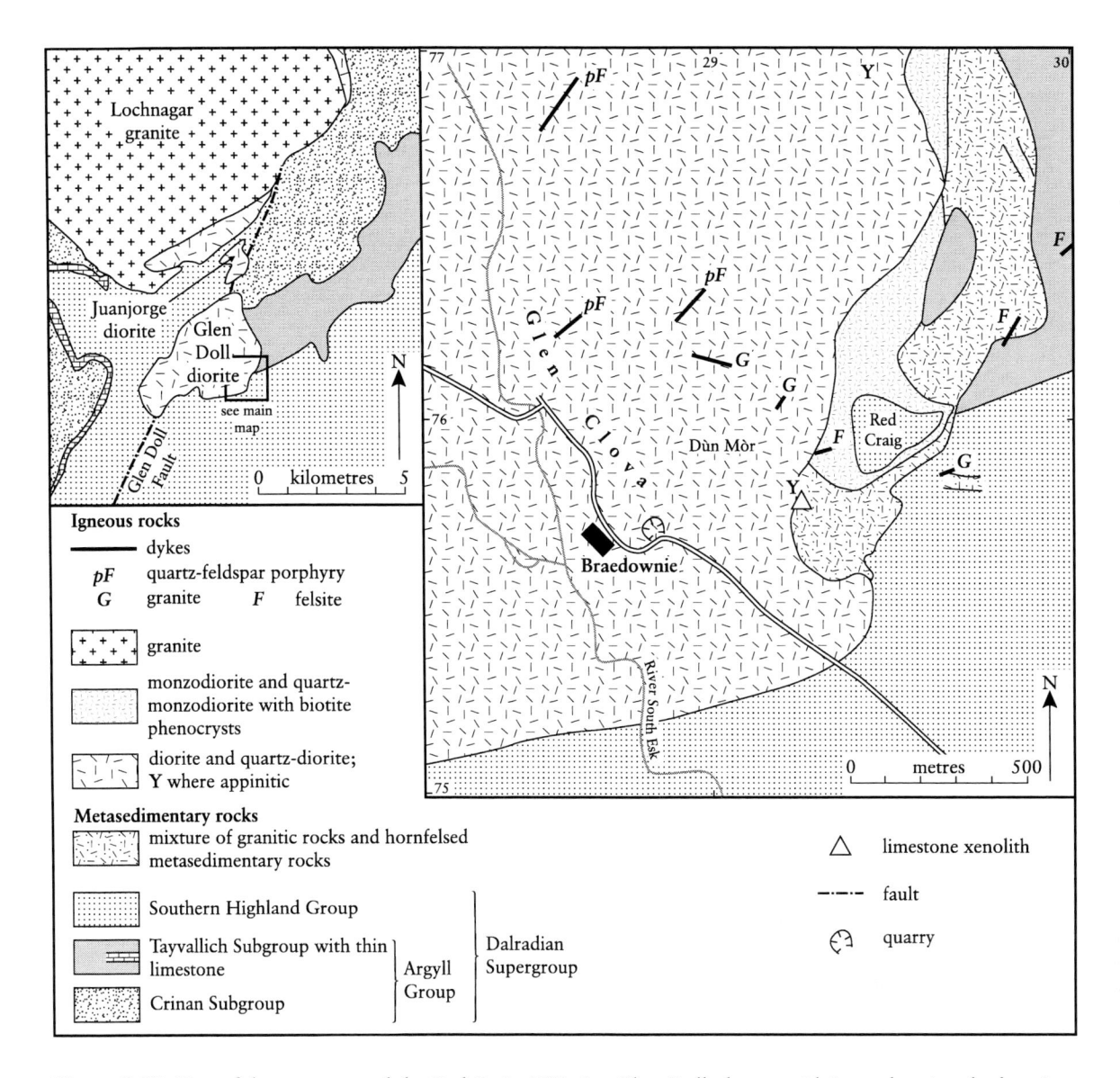

Figure 8.20 Map of the area around the Red Craig GCR site, Glen Doll pluton, with inset showing the location of the area with respect to the regional geology.

rocks and hornfelsed semipelite. Felsite and quartz-feldspar porphyry sheets and dykes ranging from a few centimetres to 15 m across and small bosses up to 50 m across occur widely. Within this zone, lithology, texture and grain size vary over short distances. Grey fine-grained diorite is net-veined by, and occurs as xenoliths within, medium-grained quartz-monzodiorite on the crags SW of Red Craig (2941 7558). Textural observations indicate that the fine-grained rocks were incorporated within and veined by the medium-grained rocks before crystallization was complete. Both are cut by veins and sheets of pink leucocratic granite up to 3 m thick. Craggy exposures to the SW of Red Craig reveal an eastward increase in interstitial to poikilitic perthitic orthoclase together with the appearance of biotite megacrysts, as quartz-diorite passes into quartz-monzodiorite. The quartz-monzodiorite forms a near continuous outcrop (Figure 8.20) and is typically grey and medium grained with biotite megacrysts up to 5 mm across comprising approximately 5% of the rock. In thin section, the quartz-monzodiorites show evidence of the addition of a potassic component to a parent diorite or quartz-diorite magma. The biotite megacrysts, together with perthitic orthoclase megacrysts that comprise up to 30% of the rock, poikilitically enclose areas with a dioritic texture that is typical of diorites to the west.

To the east, with increasing proportions of orthoclase and quartz, the quartz-monzodiorite grades into granodiorite and granite in which rafts and xenoliths of hornfelsed metasedimentary rocks are widespread (Figure 8.21). Many of the rocks in this zone are texturally heterogeneous; patches with dioritic texture are enclosed by areas with granitic features, including graphic intergrowths of quartz and alkali feldspar. The largest area of homogeneous granite is seen around 2967 7620, in a 150 m-long section beside a track, 200 m NE of Red Craig. The granite is pink to red, medium to coarse grained and rather weathered with some disseminated pyrite. The neighbouring metasedimentary rocks are cut by granite sheets, although more commonly granite contacts are gradational and marked by transition zones in which granite contains numerous xenoliths in various stages of assimilation.

Widespread xenoliths and larger rafts of hornfelsed metasedimentary rock form some of the most spectacular features of the GCR site. The largest crops out on, and immediately west of, Red Craig and covers an area of 250 m × 200 m. This is not a roof pendant since the inclination of the lithological layering is rotated with respect to the country rock envelope. Xenoliths and rafts are predominantly either hornfelsed semipelite or interlayered psammite and semipelite typical of the envelope rocks. The exception is a 1 m xenolith of metacarbonate rock at 2922 7584; the nearest metacarbonate rocks outside the pluton belong to the Loch Tay Limestone Formation, which is not recorded in the Glen Doll area. Clean exposures on cliffs south of Red Craig (294 757) show transitions over a few metres from hornfelsed semipelite, through a zone containing veins of granodiorite or monzodiorite, into a zone where the proportion of igneous material progressively increases and the semipelite becomes detached and re-orientated. This is followed by quartz-monzodiorite or granodiorite with heterogeneous grain size and texture, choked with randomly orientated xenoliths and schlieren in various stages of assimilation, ranging from a few millimetres to many centimetres long. Some xenoliths contain lenses or veins of pink granite, commonly less than 1 cm thick, which both permeate and cross-cut the lithological layering. They are mantled by biotite and encircled by leucocratic haloes with more orthoclase than the surrounding intrusive rocks (Figure 8.21). Locally, psammites are present as discrete xenoliths whereas the semipelitic interbeds have been largely assimilated and are preserved only as dismembered mafic-rich schlieren.

Semipelitic xenoliths, whether a few millimetres or tens of metres across, typically contain the assemblage: cordierite + perthitic orthoclase + plagioclase + sillimanite + biotite + quartz + minor spinel + minor corundum. Most are compact dark bluish-grey rocks, typified by rusty and in places gossanous weathering. Pyrite is abundant and occurs either in discrete lenses or with quartz segregations. Transitional contacts between the semipelites and igneous rocks show decreasing abundance of cordierite, sillimanite and spinel. Perthitic orthoclase occurs along with plagioclase, larger biotites and some quartz; the perthite poikilitically encloses biotite, aggregates of spinel and cordierite. Biotite is locally embayed by orthoclase and may include pinite after cordierite, suggesting the local breakdown of biotite.

Interpretation

Whole-rock geochemical and isotopic data have been interpreted as indicating heterogeneous crustal contamination of the parent magma to the Glen Doll diorite (Jarvis, 1987). Jarvis suggested that the partially assimilated rafted metasedimentary xenoliths 'provide an observable source of contamination'. The evidence from Red Craig indicates that most xenoliths are derived from the immediate envelope. However, some xenoliths are exotic, such as those of metacarbonate rock, which have probably been brought to the present level from deeper in the intrusion. None are roof pendants. Field evidence clearly demonstrates a link between the distribution of xenoliths and contamination of the quartz-diorite melt by granitic melt. The xenolith-rich zone also coincides with a heterogeneous, hybrid igneous assemblage, ranging from quartz-monzodiorite to granite and characterized by the presence of orthoclase and biotite megacrysts. On a small-scale there is a close association between the abundance of orthoclase within the igneous rocks and proximity to xenoliths. The occurrence of interfingering granitic segregations with graphic textures in the xenoliths indicates partial melting.

Red Craig

Figure 8.21 Disaggregated xenoliths showing evidence of partial melting and assimilation into the host quartz-monzodiorite of the Glen Doll pluton at Red Craig (NO 294 757). Leucocratic haloes are apparent around some xenoliths and schlieren. (Photo: BGS no. D 4550.)

Taken together, these features all suggest the contamination of magmas by partial melts derived from the xenoliths. The poikilitic or interstitial nature of the partial melt component, comprising orthoclase, biotite and quartz, indicates that it was introduced into the diorites after they had partly solidified. Minor granite veins may also be derived from melting of xenoliths although their sharper margins may indicate derivation from deeper within the pluton.

Conclusions

The assimilation of country rocks, resulting in contaminated xenolithic zones, is common in the Caledonian intrusive suites and is described from several GCR sites. High-grade hornfelses are observed adjacent to many dioritic intrusions and localized melting of country rocks to produce granitic magma has been inferred. The Red Craig GCR site provides an excellent illustration of all of these processes *in situ* and hence is a site of national and possibly international importance. Here a dioritic magma has been contaminated both with xenoliths and with a granite melt that was derived locally from within the xenoliths. The largely semipelitic xenoliths range from a few millimetres to more than 200 m across. They preserve high-grade contact metamorphic mineral assemblages and some have granitic segregations indicating partial melting. Many have transitional contacts with the host dioritic rocks where their marginal parts have been spalled off and assimilated by the magma.

FOREST LODGE
(NN 933 741)

D. Stephenson

Introduction

In the second half of the eighteenth century the controversy regarding the origin of rocks which we now regard as 'igneous' was at its height. On the one side were the *Neptunists*, inspired by the teaching of Abraham Werner in Saxony, who believed that rocks such as basalt and granite were 'sedimentary', having crystallized from a supposed primeval ocean, as rock-salt does from today's oceanic water. Granites were considered to have crystallized first, to form a thick layer around the Earth's 'nucleus'. The granites were overlain by other layers of crystalline rock, those that we now know as 'metamorphic', followed by layers of sedimentary rock formed as a result of erosion of the 'Primitive' crystalline rocks and subsequent deposition. Rocks resulting from observed volcanic eruptions were attributed to the local action of 'subterraneous fires' (generally thought to be due to the burning of coal seams), but these, it was believed, could only consolidate as a glass.

By the 1760s prehistoric volcanic features had been recognized, in particular in the Auvergne region of France, where flows of crystalline basalt had been traced back into undisputed volcanic craters. The *Vulcanists* had made their point and gradually, over the next fifty years, most Neptunists came to accept the magmatic origin of basalt. The origin of granite was more difficult to prove since, by its very nature according to any of the theories of the day, it could not be observed forming *in situ*. It was James Hutton (1726–1797), a prominent member of the Edinburgh scientific community, who was to lead the *Plutonists* in establishing that granite crystallized from a molten fluid and not from an aqueous solution.

In his 'Theory of the Earth', delivered in two lectures to the Royal Society of Edinburgh in the Spring of 1785 (Hutton, 1788), he argued that the relative insolubility in water of quartz and other component minerals of granite, precluded the Neptunist theory. He recognized the significance of the intergrowth texture between quartz and feldspar in a sample of coarse-grained graphic granite (Hutton, 1788, plate II) and concluded that granite might have 'risen in a fused condition from subterranean regions' and that the country rock should therefore be broken, distorted and veined.

In the late summer of 1785, Hutton embarked upon the first of several excursions to find field evidence in support of his philosophical theories; these must have been among the first ever geological field trips. From the examination of debris washed down by rivers in the Highlands, he had already concluded that a major junction existed somewhere in the region of Glen Tilt, between a largely granitic terrain to the north and *'alpine schistus'* to the south. He stayed at Forest Lodge, where he soon found the evidence that he was seeking, particularly in those exposures in the bed of the River Tilt, less than 1 km from the lodge, which now constitute the GCR site. In Hutton's words, 'the granite is here found breaking and displacing the strata in every conceivable manner, including the fragments of the broken strata, and interjected in every possible direction among the strata which appear'. He also inferred from the highly crystalline nature of the country rocks that they had been subjected to intense heat and that the granite was responsible; hence the granite must have been injected as a very hot liquid, i.e. a magma.

Over the next three years Hutton was to find and document further examples of granite veining, particularly in Galloway and on the Isle of Arran, and a brief account of the field evidence was included in a discussion of the origin of granite (Hutton, 1794). The Glen Tilt sites were revisited by several of Hutton's contemporaries (Playfair, 1802; Seymour, 1815; MacCulloch, 1816), but his own detailed descriptions were not published until 100 years after his death in volume 3 of 'Theory of the Earth', edited by Archibald Geikie (Hutton, 1899). By this time the magmatic origin of granite had, for the time being at least, become universally accepted and the Neptunist theory had lapsed into historical significance. In 1968 a collection of drawings came to light, many of which were clearly intended to accompany volume 3 of 'Theory of the Earth', but were 'lost' prior to its publication (Craig *et al.*, 1978). Several depict the Glen Tilt exposures and one is reproduced here (Figure 8.24).

The exposures described by Hutton occur on

Forest Lodge

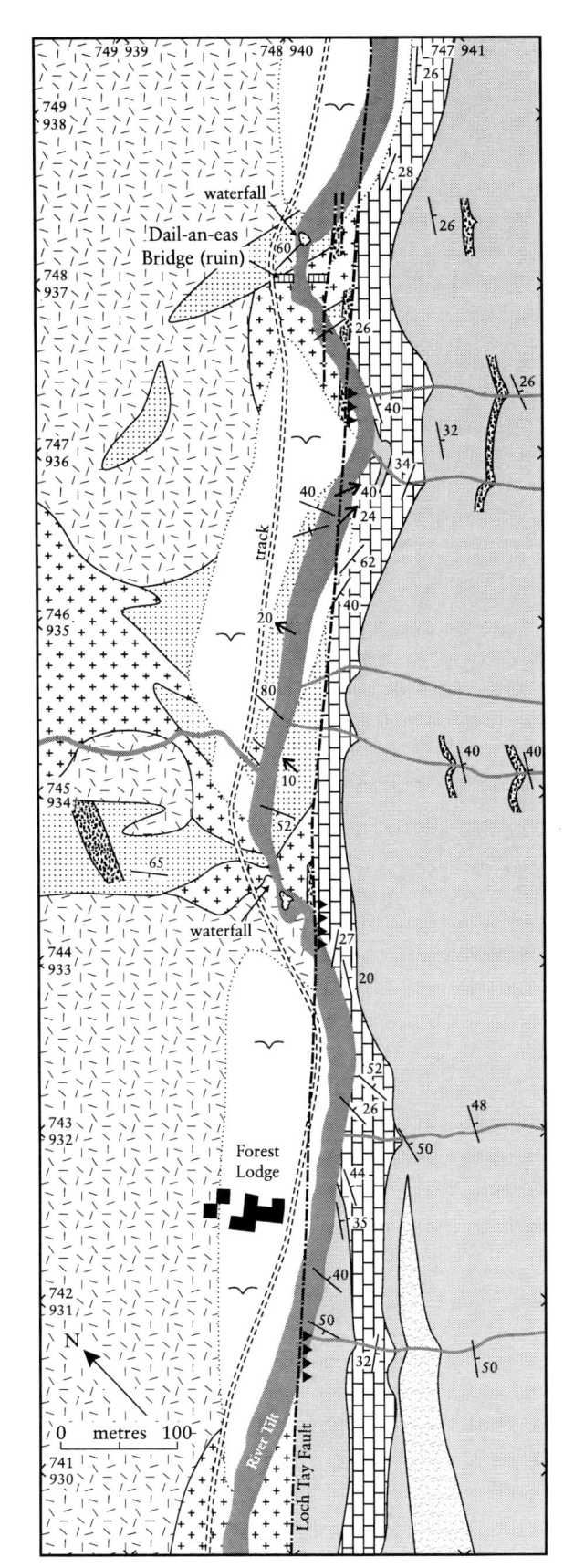

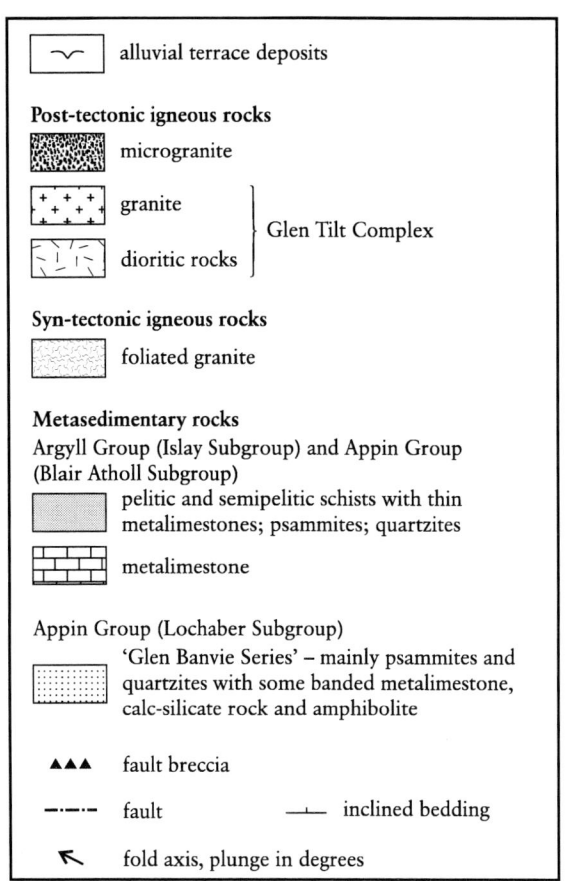

Figure 8.22 Map of the Forest Lodge GCR site, Glen Tilt igneous complex, based on unpublished British Geological Survey maps by D. Stephenson and B. Beddoe-Stephens. Note: Much of the area to the NW of the River Tilt is covered by head and scree deposits. Boundaries in this area are highly conjectural and the area shown as 'dioritic rocks' probably includes areas of granite and metasedimentary rocks.

the margins of the Glen Tilt igneous complex, which occupies most of the ground to the NW of Glen Tilt in the area around Forest Lodge. Here, complex and varied contact relationships between granite, diorite and the metasedimentary country rocks are well displayed. Since the visits of Hutton's followers in the early nineteenth century, the exposures have received surprisingly little attention. The area was mapped by the Geological Survey in the late nineteenth century, but the accompanying memoir makes little mention of the contacts (Barrow et al., 1913). Studies of the Glen Tilt complex by Deer (1938a, b, 1950, 1953) and Mahmood (1986) concentrate almost entirely upon the mineralogy and petrology of the granites, diorites and a wide range of contaminated and hybrid rocks, but provide few details of the field relationships. The site also lies along the line of the Loch Tay Fault, and the fault plane, with associated breccias and zones of silicification, is well exposed. The present account is based upon remapping by the British Geological Survey (Figure 8.22).

Description

The straight line of Glen Tilt at Forest Lodge is controlled by the Loch Tay Fault, one of the most important NE–SW-trending late Caledonian faults in the Grampian Highlands. Farther to the SW around Loch Tay, the fault has a net sinistral strike-slip movement of 7 km and a downthrow of at least 0.5 km to the SE (Treagus, 1991). In Glen Tilt the fault juxtaposes Grampian Group (and possibly lowest Appin Group) strata, cut by rocks of the Glen Tilt igneous complex, to the NW against upper Appin Group strata to the SE. There are no major intrusions and no widespread contact metamorphism to the SE of the fault. This would appear to be consistent with a significant downthrow to the SE, although there are no indications of the extent of strike-slip displacement in this area.

Where seen, the fault plane is near vertical and sharply defined, as below the waterfall at 935 743. It is commonly marked by zones of brecciation and silicification, such as below the waterfall at 938 746 and in the river due south of the lodge (932 740). The zone of brecciation is never very wide, considering the magnitude of the fault, and rarely extends for more than about 10 m on either side of the fault plane. Silicification can extend slightly farther and is particularly noticeable in limestone beds which take on a distinctive yellow-brown colour. In many places the fault plane is occupied by a dyke, no more than 1–2 m wide, of either microgranite or microdiorite, which in this area is invariably highly brecciated and silicified. Within 20 m of the fault plane minor buckle folds plunging generally between E and SE (e.g. around 937 745) are probably related to the strike-slip movements on the fault.

On the SE side of the fault, rocks assigned to the Blair Atholl Subgroup dip generally to the SE at 25–45°. At river level, most exposures are of metalimestones, which occur within a sequence of schistose to flaggy semipelites and pelites, with flaggy banded psammites and some massive quartzites (Smith, 1980). On the NW side of the fault in general, the metasedimentary rocks are psammites and quartzites of the Grampian Group, but in the area around the Glen Tilt igneous complex, lithologies include calcareous schists, para-amphibolites and impure metalimestones. This distinctive assemblage has been termed the Glen Banvie 'series' by Smith (1980); its stratigraphical affinities are uncertain, but it may be equivalent to part of the Lochaber Subgroup at the base of the Appin Group. In the Forest Lodge area the Glen Banvie 'series' rocks are commonly steeply inclined with a general NW–SE strike and tight upright minor folds that plunge north at 10–20°. It may be that they occupy the core of a large-scale late N–S fold, similar to those which control the outcrop pattern in the Schiehallion area and elsewhere in the Southern Highlands (Treagus, 1991). The Glen Tilt igneous complex occupies and largely replaces this proposed fold core.

The Glen Tilt igneous complex, a member of the South of Scotland Suite, comprises diorite, granodiorite, granite and a suite of microdioritic minor intrusions that together occupy an area of 12 × 6 km, extending north-westwards from the Loch Tay Fault in Glen Tilt. The north-western part is entirely granitic (the Beinn Dearg pluton), but the south-eastern part has a complex outcrop pattern of granite and diorite with large areas of metasedimentary rock. Exposures are discontinuous, especially on the lower valley sides around Forest Lodge, where scree and other superficial deposits cover much of the outcrop. Consequently, the form of the intrusions is difficult to determine; dioritic rocks form most of the area to the NE of the lodge, with many small outcrops of biotite granite around the lodge itself. Intrusive relationships, where seen, are equivocal. The granite is not chilled at any contacts, although locally it is slightly reduced in grain size against country rocks. The diorite generally has a finer-grained marginal facies. Most contacts between granite and diorite are sharp, but locally they are gradational with evidence of hybridization. The granite is cut by microdiorite dykes that seem to be related petrogenetically to the diorites (Beddoe-Stephens, 1999). However, marginal granite locally contains blocks of fine-grained basic rock in an intrusion breccia; marginal dioritic rocks are commonly veined by aplitic granite; and some marginal diorites develop small feldspar porphyroblasts. Critical evidence of the intrusive relationships between granitic and dioritic rocks and with the metasedimentary country rocks occurs in the River Tilt exposures, examined by Hutton and now part of the GCR site.

A sharp contact between diorite and granite is exposed at the top of a waterfall (935 743). The biotite granite, here mostly on the SE side of the river, is brick red and equigranular with conspicuous milky white quartz up to 5 mm in diameter that weathers proud to give a nodular appearance. Most of the outcrop close to the contact is an irregular mix of veins of granite, granodiorite, diorite and vein quartz with no consistent crosscutting relationships. The veining continues into psammites and quartzites above the waterfall where almost 50% of the exposure is vein material. The diorite on the NW side of the river is medium grained (2–3 mm), equigranular and is composed essentially of dark-green hornblende and dark-pink plagioclase. It is cut by thin, irregular quartzofeldspathic veins and contains small granitic pods. The diorite remains even textured right up to the contact, but the granite has a very irregular texture and is also very hard and splintery. Small inclusions of psammite, quartzite, calc-silicate rock and grey metalimestone occur in the granite close to the contact, but not in the diorite. An island within the waterfall is formed by a larger inclusion of banded quartzite or psammite cut by pods and thin veins of granite, and the NW edge of this inclusion is marked by an intrusion breccia of quartzite/psammite and some amphibolite in a dioritic matrix.

The waterfall beneath the ruined abutments of Dail-an-eas Bridge (939 747) (Figure 8.23) was the main focus of Hutton's observations and figures in several of his drawings (Figure 8.24). The abutments rest upon a coarse-grained brick-red biotite granite, crowded with xenoliths (up to 2 m), of psammite, quartzite, metalimestone and pale- to dark-green, banded para-amphibolite. The coarse-grained granite and the inclusions are cut by thin, irregular pink granite veins with sharp, angular contacts. Immediately upstream of the bridge, a sharp contact between granite and metasedimentary rocks is near vertical, planar, trends WNW–ESE and forms a gulley on the NW bank; it has possibly been modified by later faulting. Within 2–3 m of the contact the granite is slightly finer grained, with an irregular texture, and it is crowded with angular xenoliths. A few veins of red granite penetrate into the adjoining metasedimentary rocks, which dip steeply southwards and strike slightly oblique to the contact. The metasedimentary rocks consists of psammites, quartzites, metalimestone (one bed is 2.5 m thick) and banded, ribbed siliceous calc-silicate rocks with para-amphibolites. In thin section they show evidence of recrystallization due to hornfelsing and in several places skarns, with large (1 cm) pale-brown garnets, are developed at the junction between carbonate and calc-silicate rock. The veins that so impressed Hutton are mostly of irregular coarse-grained white quartzofeldspathic pegmatite, with some more regular veins and concordant sheets of white to grey microgranite or microgranodiorite. There are a few irregular veins of pink granite but, except for those at the contact (see above), these do not resemble the main granite body.

Interpretation

Although Hutton had correctly deduced that a junction between granite and 'schistus' would be found in Glen Tilt, he was unaware of the presence and significance of the Loch Tay Fault. So it was fortuitous that he found his evidence so easily; he may well have met with only a faulted contact with no veins, xenoliths or other evidence to indicate an intrusive relationship. The contacts of the intrusive complex with large masses of metasedimentary rock are well exposed only in the valley bottom, where the River Tilt departs from the line of the fault plane, albeit by less than 50 m, in places like Dail-an-eas. Fortunately, the zone of brecciation and alteration associated with the fault is too narrow to have affected these outcrops and obliterate the detail.

The presence of large outcrops of metasedi-

Figure 8.23 The waterfall at Dail-an-eas Bridge above Forest Lodge: view from the ruined bridge abutment. Siliceous metasedimentary rocks and a 2.5 m-thick bed of metalimestone (the latter surrounded by and tunnelled through by the waterfall) are cut by granitic veins of various types. The most prominent vein is also clearly visible in Hutton's 'lost drawing', reproduced in Figure 8.24. (Photo: D. Stephenson.)

mentary rock, within those of igneous rock, in the Forest Lodge area suggests that this is a marginal zone of the intrusive complex. Unfortunately the exposures are too discontinuous to identify this as either a roof zone with pendants, a lateral margin with screens, or an area of in-situ country rock heavily impregnated with veins and irregular apophyses from the main intrusions. The consistent NW–SE strike of the bedding and shallow northerly plunge of fold axes in most of the outcrops does however suggest that re-orientation as a result of the intrusions has been minimal. The form of the main intrusions is likewise impossible to determine. Many of the granite outcrops seem to be relatively small, isolated bodies, but they are all coarse grained and have petrographical similarities to a larger intrusion centred upon the ridge of Sron a' Chro, some 3 km to the west. The diorite is more widespread and continuous, dominating the higher, craggy slopes of Glen Tilt to the NE; it could be a highly irregular, steep-sided intrusion, but the outcrop pattern could be interpreted as a thick low-angled sheet with the Forest Lodge outcrops close to the lower margin.

Intrusive relationships between the granite and diorite are ambiguous. Earlier workers (Deer, 1938b; Mahmood, 1986) regarded the diorites as earlier. This is supported by the presence of granitic veins and feldspar porphyroblasts in the diorite. However, recent field and petrological studies (Beddoe-Stephens., 1999) suggest that the Sron a' Chro and related granite

Forest Lodge

Figure 8.24 Plan of exposures in the River Tilt on the upstream side of Dail-an-eas Bridge, drawn by John Clerk of Eldin during Hutton's visit in 1785 and published as one of Hutton's 'lost drawings' by Craig *et al.* (1978). The outline is inaccurate and the detail is somewhat exaggerrated and stylized, but it illustrates well the features that Hutton observed; in particular the 'granite', 'limestone', 'schistus' and 'mixed rocks' are clearly labelled on the original. Note that the drawing is reproduced upside down to facilitate comparison with the photograph of Figure 8.23; the bridge abutments are indicated by the blank rectangles (bottom right and bottom left). Reproduced by permission of Sir John Clerk.

bodies in the Forest Lodge area were the earliest phase of the complex and that, during subsequent intrusion of the diorite, melting, hybridization and remobilization occurred, with back-veining of aplitic and pegmatitic veins into the still hot diorite. During crystallization of the diorite, slightly fractionated melts were expelled to form the microdiorite dykes which cut the Sron a' Chro granites.

Within the GCR site, the overall impression is that the diorite cuts the granite. Whereas the granite is commonly crowded with metasedimentary xenoliths, none of diorite are recorded. The diorite has far fewer metasedimentary xenoliths, giving the impression of a later event and the rounded pods of granite that have been recorded could be partly digested xenoliths. The granite is slightly finer grained against the metasedimentary rocks, but not against the diorite. It develops a highly irregular texture against both, but is hard and splintery against the diorite suggesting secondary silicification during recrystallization. More importantly, whereas thin sections of the diorite show primary crystallization textures, those of the granite show evidence of strain-related recrystallization, implying thermal and deformational events associated with the diorite emplacement.

Relationships of the Loch Tay Fault to the intrusions are interesting. Treagus (1991) considers that the major NE–SW faults of the Grampian Highlands were initiated as dextral shears during a late phase of N–S-trending folding, which pre-dated the major intrusions. The Glen Tilt complex does seem to be located in the core of a major N–S late fold and an early shear or fracture could possibly have controlled the location of the SE margin or even have acted as a magma conduit. Clearly the major movements on the fault post-date the main intrusions of the complex, which are truncated and which do not appear on the downthrown SE side. Neither is there any major aureole development on the SE side. However, throughout most of its length in upper Glen Tilt, the fault plane contains minor intrusions of microgranitic and microdioritic rocks; most are highly brecciated with much secondary silicification, but others have suffered little or no brecciation. Clearly the fault was in existence soon after the emplacement of the Glen Tilt Complex, to control the site of minor intrusions. Some of these were then affected by subsequent brittle movement, to complete a long and complex cycle of inter-related faulting and intrusion in late Caledonian time.

Conclusions

This site is of international importance on historical grounds. It was here in 1785 that James Hutton first found and documented field evidence to support his theory that granite was intruded into country rocks in a hot, fluid state.

Although there are much clearer examples elsewhere of granite veins emanating from a pluton, including some that were visited subsequently by Hutton, the exposures around Forest Lodge were the first to be evaluated. They enabled Hutton to demonstrate that granite crystallized from 'matter made fluid by heat', i.e. magma; that this matter was able to flow within the Earth's crust, veining, disrupting and recrystallizing pre-existing rocks; and that consequently granite is not universally the earliest formed rock. This effectively ended the Neptunist arguments of Werner and his disciples and for well over a hundred years was accepted as a satisfactory explanation for the origin of all granites.

It was not until the 20th century that people began to address the problem of where and how granitic magma might be generated. Once again this led to a division into two camps, with the essentially Huttonian 'magmatists' opposed by 'transformationists' who believed that granite was generated *in situ* from pre-existing rocks, by fluid or gaseous 'fronts' which modified the composition, mineralogy and texture of the rock without generating a melt; the process of 'granitization'. This time, however, it eventually became accepted that both arguments have their merits and that granites have formed in many different ways; in the words of H. H. Read (1957), 'there are granites and granites'. Mechanisms such as broad-scale, in-situ, solid state diffusion as a product of ultrametamorphism and the precipitation of granitic pegmatites from silicic, hydrous fluids are now well documented, and fractional crystallization of a more basic magma can result in a granitic residual melt. But by far the greatest volumes of granite are now considered to have crystallized from magma that originated by partial or complete melting of crustal material; a true vindication of the Huttonian theory.

FUNTULLICH (NN 750 265)

W. E. Stephens

Introduction

The Comrie pluton (Figure 8.25) is a member of the South of Scotland Suite. It is somewhat unusual among the Scottish late Caledonian intrusions in its isolation from other plutons, and by the fact that the strike of its long axis runs counter to the main NE–SW Caledonian trend. The pluton is a fine example of a normally zoned pluton, with a dioritic outer member pierced by a granitic core. In most of the Scottish late Caledonian granite plutons diorite tends to be a minor component whereas more acid rocks are more abundant. The Comrie pluton is unusual in that diorite is the dominant lithology (see also the Craig More GCR site report). There is considerable heterogeneity among the diorites, and a sharp contact can be observed between the granitic rocks and the outer dioritic envelope. This is important, as in similar plutons elsewhere in Scotland the contacts are not usually exposed and gradational changes are often described. In this case the sharp contact and xenolithic inclusions clearly demonstrate the importance of multiple pulses in the construction of some of the late Caledonian plutons. An age of 408 ± 5 Ma has been obtained using Rb-Sr on whole rock-mineral pairs (Turnell, 1985).

Many of the late Caledonian 'granites' of Scotland are diorite–granodiorite–granite complexes in which the central and youngest magmatic member is located in the centre of the pluton. It has been long debated whether this disposition is the product of in-situ fractional crystallization, with early higher temperature minerals forming first near the cooler walls, or whether such plutons represent separate pulses of magma intruded either directly from the source of melting, or from an intermediate magma chamber. Many such plutons have been described from around the world and the form is known as normal zoning. Just how such plutons are constructed is the subject of considerable debate (e.g. Paterson *et al.*, 1996; Petford,

Funtullich

1996), and the question of whether plutons are constructed from single or multiple magmatic pulses is important in understanding the mechanisms involved. The Comrie pluton is one of the few good examples of multiple pulses in the British Caledonian igneous province.

The Funtullich GCR site displays many of the key features of the Comrie pluton in one small area. The main interest of this site is igneous, with all principal members of the pluton exposed and showing internal contact relationships. In addition, the contact effects of the pluton on the more psammitic lithologies of the Ben Ledi Grit may be examined and contrasted with the effects on the Aberfoyle Slate, seen in the Craig More GCR site. For reasons of outcrop quality and accessibility this site is widely used for instructional purposes, and the site is included in the excursion guide of MacGregor (1996).

Description

Diorites form much of the outer part of the pluton (see the Craig More GCR site report) but the central area, forming the high ground SW of Carn Chois, is formed of pink microgranite. An extension of this microgranite cuts through the diorites in the Funtullich area, and dioritic xenoliths are abundant close to this internal contact. A sheet of similar microgranite occurs in contact with the Ben Ledi Grit.

The diorites around Funtullich farm (Figure 8.26) are compositionally and texturally highly variable. A dark fine-grained variety seen near the road comprises hornblende and biotite with slightly altered plagioclase and minor quartz. Biotite generally occurs with opaques or in clusters with hornblende. Farther to the east of the road the rock becomes coarser grained with large crystals of hornblende (sometimes obviously after pyroxene) and plates of biotite within a matrix of altered plagioclase. Other varieties of diorite contain fresh pyroxenes, commonly both ortho- and clinopyroxene. These rocks range from 53–60% SiO_2, and geochemical characteristics suggest that some may be mafic-rich cumulates. Compositional heterogeneity in the diorites extends them into the granodiorite field, and some more evolved types reach about 65% SiO_2.

Intruding these various diorites is a mass of pink medium-grained granite (microgranite) which is best displayed in the nearest crags to the ENE of Funtullich. At the western end of

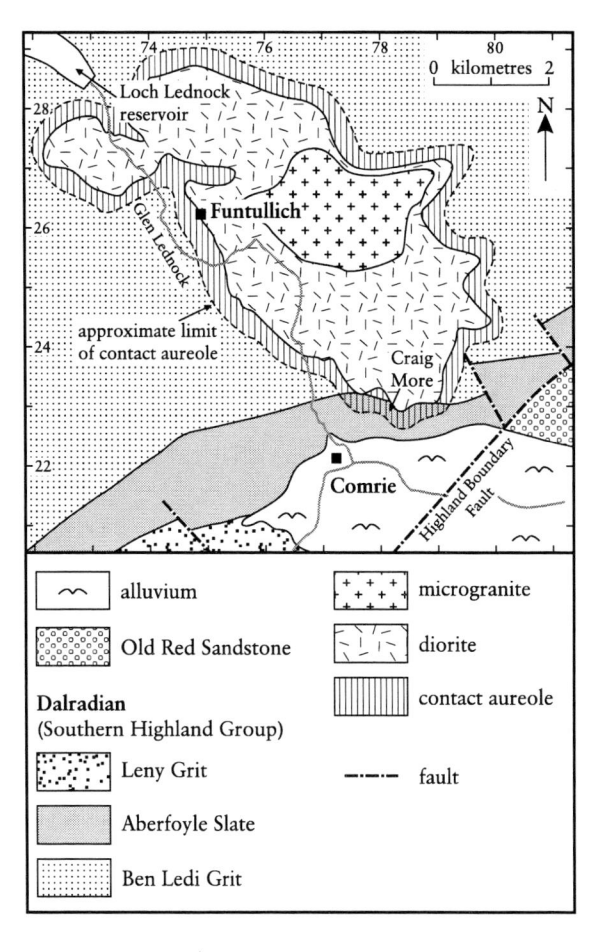

Figure 8.25 General location map of the Comrie pluton. Taken from a paper by Pattison and Tracy (1991), who drew it from data in Tilley (1924).

these crags is a coarse-grained diorite with pink aplitic veins, and a solitary tree marks the line of contact with the microgranite. Eastwards from this contact the microgranite contains abundant dark xenoliths, including some of dioritic composition. The mineralogy of the microgranite comprises principally plagioclase, orthoclase and quartz with biotite as the principal mafic mineral, together with aggregates of amphibole and biotite. This microgranite has an SiO_2 content of about 71% and it is depleted in most trace elements except Rb, relative to the outer diorites (Mahmood, 1986).

The outer contact of the pluton can be seen about 100 m NW of Funtullich, on the SW side of the road. Here a sheet of pink microgranite with prominent sub-horizontal joints is in contact with the Ben Ledi Grit. The contact can be located to within a metre or so, but here the psammitic rocks of the envelope show no obvious signs of contact metamorphism. The relationship of this sheet to the main granite mass is not known but it is of a similar facies and appears to be detached.

Interpretation

The Comrie pluton was constructed in two separate stages. An early diorite, intruded as a single (or possibly multiple) magma pulse(s) had crystallized before a pulse of more evolved granitic magma pierced the centre, picking up xenoliths of the diorite, as shown by the Funtullich outcrops.

The diorites began to crystallize at about 1100°C, but most crystallized at somewhat lower temperatures (see the Craig More GCR site report). Cumulates of mafic-rich minerals formed from these magmas, driving the magma(s) to more evolved dioritic and granodioritic compositions. Relationships between the different varieties of diorites (and granodiorites) are not known. Their heterogeneity appears to be due to gradual facies changes rather than to separate intrusive pulses, but critical field evidence is not available.

Using trace element evidence, Mahmood (1986) tested whether the microgranite could have been derived directly from the dioritic parental magma through fractional crystallization. The data show that this model is not tenable as no incompatible trace element (not even Rb) in the microgranite exceeds abundances found in the granodiorites. Hence, the central microgranite appears to have formed from the crystallization of a separate single pulse of magma.

Two samples from the diorites have yielded Sr isotope initial ratios of 0.7050 (Turnell, 1985). Such values can be interpreted as indicating a deep crust or upper mantle source. Whole rock-mineral separates on the same two diorite samples give concordant Rb-Sr ages of 408 Ma (Turnell, 1985), the probable minimum age of intrusion.

Conclusions

The Funtullich outcrops of the Comrie pluton afford an opportunity to view a wide range of rocks from diorite to granite in a small area, with an observable contact between a late granite and the host diorite, clearly establishing the age relationships. The outcrops display many of the essential field relationships that are required to understand the construction of zoned plutons, which commonly provide valuable information on the generation and evolution of deep crustal magmas.

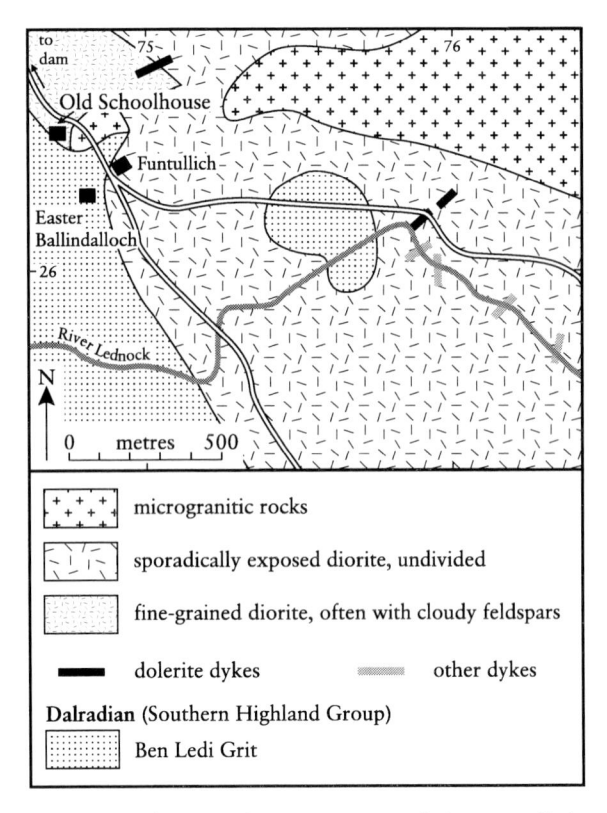

Figure 8.26 Map of the area around the Funtullich GCR site, Comrie pluton, adapted from MacGregor (1996).

CRAIG MORE (NN 787 227)

W. E. Stephens

Introduction

The Comrie pluton of Perthshire is best known for the classic study of a contact metamorphic aureole made by Tilley (1924), following Nichol (1863) who was the first to observe the hardening of the slates and attribute this to recrystallization close to the igneous contact. The intrusion of the pluton into Dalradian metasedimentary rocks with a variety of pelitic, semipelitic and psammitic lithologies and low regional metamorphic grade made it ideal for these pioneering studies. More recently the high-grade contact metamorphic rocks adjacent to the pluton have been studied in detail (Pattison and Harte, 1985) and a low-pressure environment of contact metamorphism has been established from the mineral assemblages. In a classification of contact metamorphic facies series (Pattison and Tracy, 1991), the aureole is cited as a type example of a particularly low-pressure facies that is notable for the prevalence of spinel and hypersthene and the absence of garnet in high-grade cordierite-bearing assemblages.

The Craig More GCR site has excellent exposures of pelitic rocks throughout the increasing contact metamorphic grade up to the contact with the diorite, and is included in the excursion guide by MacGregor (1996).

Description

The area between the Caravan Park at Old Lodge and the margin of the diorite just east of Craig More represents a complete section through the aureole (Figure 8.27). Here the country rocks are the Aberfoyle Slate of the Southern Highland Group of the Dalradian. The unaffected rocks near Old Lodge are greenish pelites said by Tilley (1924) to lie in the chlorite zone of regional metamorphism.

In a traverse from outside the aureole up to the igneous contact the following sequence of changes can be observed. About 400 m from the contact dark spots of phyllosilicates begin to appear. These spots are composed of aggregates of chlorite and muscovite which Pattison and Tracy (1991) suggest represent altered cordierite poikiloblasts. Within another 100 m extensive recrystallization is evident from the rapid loss of fissility along strike. In the final 120 m leading up to the contact on the side of Craig More, there is total loss of fissility and massive hornfelses occur containing assemblages with cordierite and K-feldspar. Muscovite has declined in abundance by this stage and the rock is very hard with a purplish colour on fresh surface, in marked contrast with the greenish-grey colour of the soft Aberfoyle Slate outside the aureole.

The contact of the hornfels with the intrusion is well displayed on the wooded slopes east of Craig More. Here a coarse-grained slightly pink granodioritic rock exposed on the lower ground is capped by the hornfels. A contact between chilled dark diorite and high-grade hornfels can be seen, although there is little visual distinction between these lithologies at the immediate contact. The contact appears to be at a shallow inclination, and it seems that the present exposure level is not far from the original pluton roof. White granitic veins cut the hornfelses and may be evidence of local anatexis.

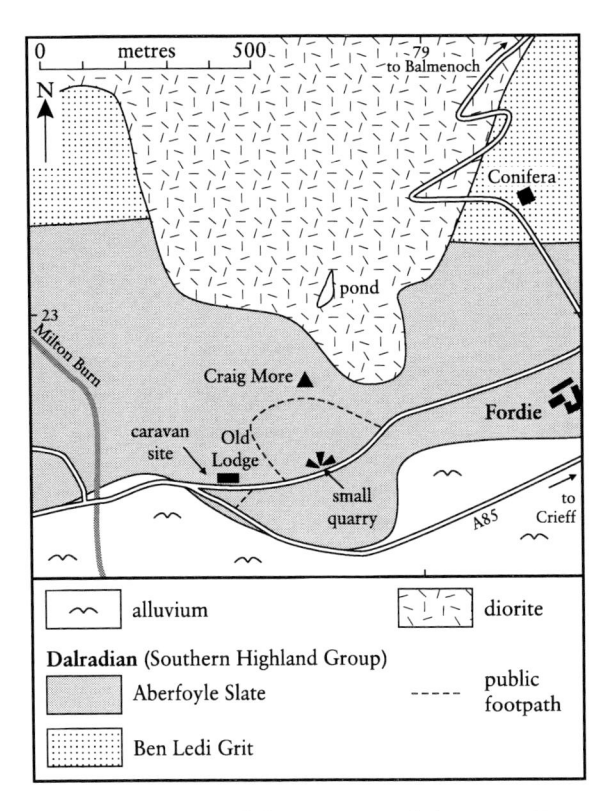

Figure 8.27 Map of the area around the Craig More GCR site, Comrie pluton, adapted from MacGregor (1996).

Microgranodioritic dykes (formerly termed 'porphyrites') radiate from the pluton and many can be seen cutting the aureole rocks. Sulphide is quite abundant locally in the major intrusions and this pluton was actively explored for mineralization in the 1980s.

Craig More is also a good vantage point to view the line of the Highland Boundary Fault which runs ENE–WSW about 1 km south of Comrie essentially along Glen Artney. Small movements on this fault are still felt and one of the first seismometers was built by villagers in Comrie.

Interpretation

The diorites of the Comrie pluton vary considerably in their mineralogy, but include two-pyroxene diorites. These are important because several geothermometric calibrations exist for coexisting ortho- and clinopyroxenes. Mahmood (1986) derived temperatures of 950–1100°C for two samples of diorite using various calibrations; it seems likely that these may be on the high side given more recent calibrations (as reviewed by Anderson, 1996). A large mass of dioritic magma, probably consisting largely of melt judging by the textures, was intruded probably in excess of 900°C. This generated the c. 400 m-wide metamorphic aureole around the pluton. Conditions inside the aureole probably reached melting temperatures locally, as the granophyric intergrowth texture of K-feldspar and quartz in some of the high-grade hornfelses suggests anatexis (Pattison and Tracy, 1991). Wilde (1995) has suggested that temperatures of about 750°C were reached in the immediate vicinity of the contact where local melts were injected as granitic veins. Analyses of these veins indicate that they are close to minimum melt compositions at 0.5 kbar. The detailed studies of the aureole mineral assemblages by Pattison and Harte (1985) did not place an estimate on the geobarometric conditions, except that the aureole was at lower pressures than the Ballachulish aureole, which they suggested tentatively was between 2.5 and 4 kbar.

The Craig More outcrops provide an outstanding example of a low-pressure aureole in pelitic rocks and, although there are still many uncertainties about the precise pressure and temperature conditions and other factors such as the role of water, progress in these areas is being facilitated by new technological and theoretical developments. The Comrie aureole will continue to be an important case study for metamorphic petrologists.

These outcrops also provide a good example of a Caledonian pluton exposed close to its original roof. The low apparent pressure of the aureole indicates emplacement at a shallow level in the crust, but the reason for the trend of the pluton across the regional strike is not known.

Conclusions

Country rocks close to the southern end of the Comrie pluton, in the vicinity of Craig More, exhibit one of the most spectacular contact metamorphic aureoles in Britain. The GCR site has historical importance for early studies of contact metamorphism and has been cited as a type example in a recent classification of contact aureoles. Pelites can be traced over a distance of about 400 m, from those showing no evidence of contact metamorphism, through rocks with a range of mineral assemblages, to rocks with evidence of melting. The aureole has several distinctive mineral assemblages, although the absence of garnet is notable. These assemblages have provided useful information on the conditions of metamorphism, up to contact melting at about 750°C adjacent to a body of dioritic magma, which was probably intruded at a temperature of at least 900°C.

Diorites and granodiorites are the main intrusive rocks at this site, and the pluton is exposed just beneath the level of the original roof.

GARABAL HILL TO LOCHAN STRATH DUBH-UISGE (NN 304 177–282 158)

W. E. Stephens

Introduction

The Garabal Hill–Glen Fyne igneous complex is well known worldwide from the pioneering study of Nockolds (1941). It is also celebrated for its wide and continuous range of rock types from peridotite to granite. The contribution of Nockolds, apart from publishing one of the first detailed geochemical studies of a plutonic complex, was to explain the whole sequence of rocks in terms of 'crystallization differentiation' of a pyroxene-mica diorite parental magma. This process generated basic and ultrabasic cumulate rocks and more acid differentiates along a liquid line of descent, essentially the same process as that now known as 'fractional crystallization'. This essentially closed system model for the petrogenesis of such complexes has been widely applied. Nockolds' paper influenced most studies of compositional diversity in granitic complexes worldwide for three subsequent decades and is still often cited as a classic study (e.g. Atherton, 1993).

The complex, which is a member of the South of Scotland Suite, was emplaced into Dalradian metasedimentary rocks of the Southern Highland Group at about 429 Ma. Compositional diversity is the most important feature of this complex, with four bodies of peridotite located within several varieties of diorite and gabbro. Peridotites are virtually absent from late

Garabal Hill to Lochan Strath Dubh-uisge

Caledonian granitic complexes elsewhere and their presence is important in understanding the petrogenesis of the more basic members of such complexes. The main mass of granodiorite is porphyritic, although megacrysts are lacking in the east. Appinitic rocks are closely associated with the granodiorites, and large numbers of other appinitic bodies are known from this area (Anderson, 1935a; Anderson and Tyrell, 1937).

The first account of the complex was published by Dakyns and Teall (1892) who were struck by the petrological variety, which they attributed to the differentiation at depth of a single parental magma. This was a rather advanced view for its day. This was followed by the Geological Survey memoirs (Gunn *et al.*, 1897; Hill, 1905), and a detailed study of the ultramafic and basic varieties of the complex (Wyllie and Scott, 1913). In the following eight decades or so, the only paper specifically to address the origin of this complex with new data was that of Nockolds (1941). Although Nockolds' model is often cited, it should be viewed with some caution in the light of modern studies. Most late Caledonian plutons that have been studied in any geochemical detail show evidence of significant open-system behaviour (e.g. Halliday *et al.*, 1980) and a compilation of the available but rather meagre Sr isotopic data for Garabal Hill (Summerhayes, 1966; Harmon and Halliday, 1980) suggests that the granodiorite probably has a different origin to the rest of the complex.

Description

Garabal Hill is situated to the north of Loch Lomond and gives its name to the more petrologically variable part of the complex; the larger and more homogeneous mass of granodiorite occupies the high ground to the SE of Glen Fyne (Figure 8.28). The more basic rocks of the complex are separated from the main plutonic mass by the Garabal Fault, which in turn is truncated by the Glen Fyne Fault. The exposed area of the entire complex is 32 km^2.

The wide range of petrology has been adequately described in the papers cited above, but a summary is necessary here to convey the key feature of extreme petrological variation. The descriptions are based on Nockolds (1941) supplemented with modal and electron microprobe data from Mahmood (1986).

Ultramafic rocks crop out as discrete bodies, 100 m to 1 km across (Figure 8.28), and include peridotites, pyroxenites and hornblendites. The fresh ultramafic rock is typically a black peridotite (wehrlite) with approximately equal amounts of olivine (about Fo$_{80}$) and clinopyroxene (augite–diopside), although the content of clinopyroxene can drop to about one-third. The peridotite is often at least partly serpentinized (Figure 8.29). Other types of peridotite, including dunite, lherzolite and plagioclase-bearing varieties are also present, usually showing internal contacts with the main peridotite. In the pyroxenites, olivine is typically absent and orthopyroxene becomes more important, although clinopyroxene is still the principal mineral. Clinopyroxene is commonly replaced by brown amphibole, and in extreme cases the rock has been described as a hornblendite by Nockolds.

Gabbroic rocks are especially important as they are the most basic component of the pluton with clear evidence of non-cumulate magmatic textures, and are thus potential candidates for a parental magma. The largest outcrop of gabbro stretches for about 1.4 km around Lochan Srath Dubh-uisge and, after being cut out for about a kilometre by granodiorite, also forms quite a large mass at the southern end of the complex (Figure 8.28). The typical gabbro has abundant clinopyroxene, partly or largely replaced by actinolitic amphibole or brown hornblende. Hypersthene-bearing varieties are known; biotite is present in the gabbros close to late veins of granodiorite.

Diorites were the favoured parental magma of Nockolds, specifically the pyroxene-mica diorite. This latter rock, found in the east of the complex (Figure 8.28), contains more clinopyroxene (about 17%) than orthopyroxene (7%), biotite (about 17%), and pseudomorphs after olivine. About half of the rock is plagioclase of about An$_{41}$ composition. As is typical in this complex, clinopyroxene is commonly replaced by amphibole. Hornblende-rich diorites are also present in the eastern part of the complex, and show marked variations in grain size which Nockolds described as coarse appinitic diorite (coarse-grained), to medium appinitic diorite (medium grained) to fine-grained quartz diorite. There is some debate as to whether the amphibole is primary or secondary in these rocks; although many amphibole crystals are idiomorphic in form and would appear to be primary, others clearly show evidence of replacement. Some diorites are also remarkably rich in xenoliths

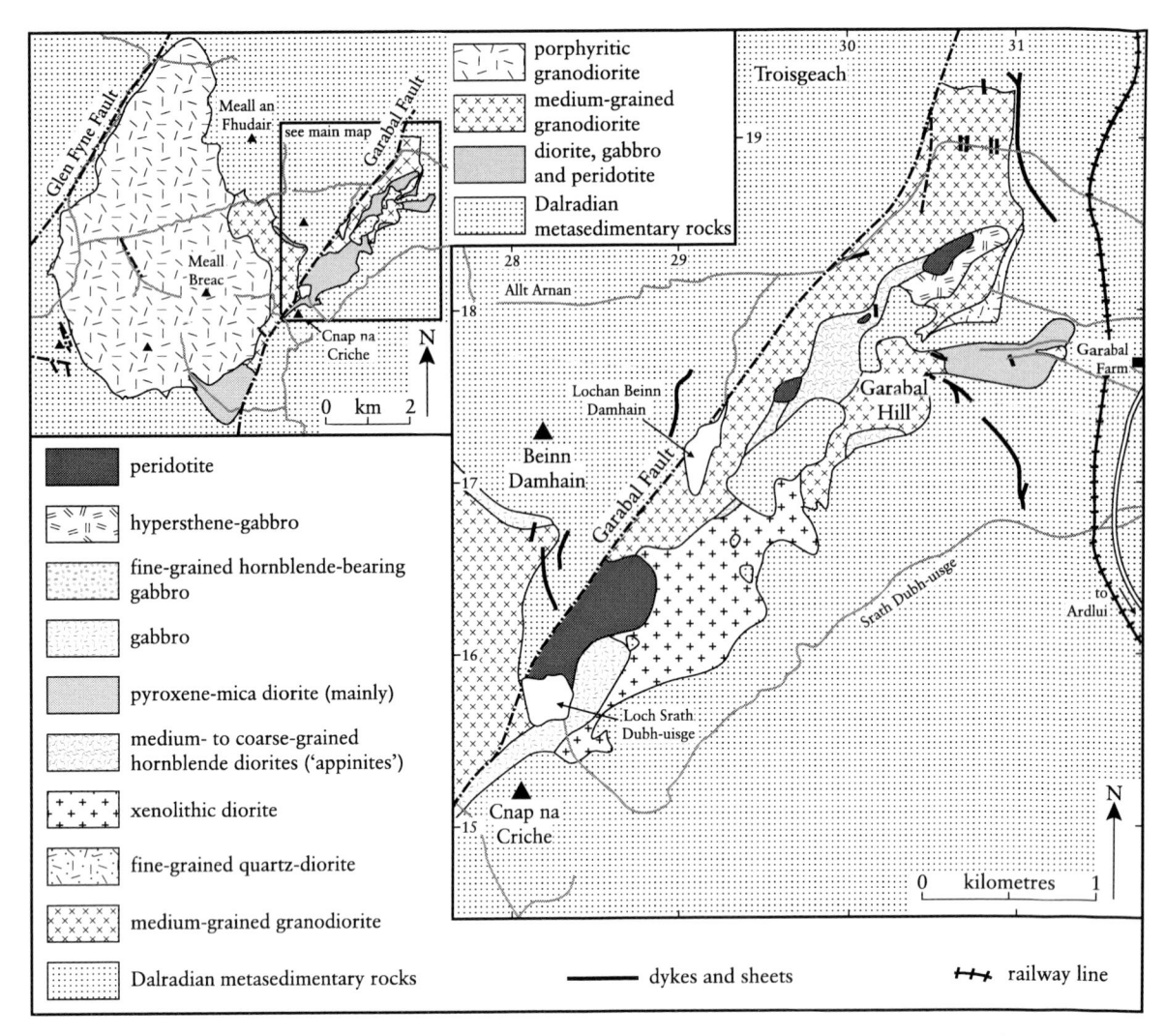

Figure 8.28 Map of the Garabal Hill–Glen Fyne igneous complex (after Nockolds, 1941).

(Nockolds' xenolithic diorite) which appear to be dominantly of basic igneous rocks.

The final intrusive phases consist of granodiorites. The medium-grained granodiorite is earlier than the porphyritic granodiorite, occurring both on the eastern margin of the main Glen Fyne intrusion and as a separate strip SE of the Garabal Fault. This facies contains hornblende and biotite with quartz, K-feldspar and plagioclase. A major feature is the abundance of xenocrysts of plagioclase, augite and amphibole, making it a very heterogeneous rock. Nockolds interpreted these xenocrysts as 'earlier igneous material' which he believed to have had a significantly modifying effect on the overall composition. The main porphyritic 'granodiorite' is a rather more homogeneous biotite granite, with a porphyritic character imparted by megacrysts of alkali feldspar.

Interpretation

Age relationships within the complex have been established by cross-cutting internal contacts, veins of late granodiorite, and xenolithic inclusions. It is clear on these criteria that the ultramafic rocks are early and are cut by the gabbros, which in turn provide xenoliths for the diorites, which are intruded by the granodiorites. Thus the age sequence appears closely to match the order of increasing acidity. The age of emplacement is also well constrained. Rogers and Dunning (1991) found the U-Pb age on an appinitic diorite to be 429 ± 2 Ma (zircon) and 422 ± 3 Ma (titanite). Combining these data with the Rb-Sr biotite-whole rock age of

Garabal Hill to Lochan Strath Dubh-uisge

406 ± 4 Ma (Summerhayes, 1966), it may be concluded that emplacement took place at about 429 Ma, followed by cooling through 600°C at 422 Ma, down to about 300°C at 406 Ma.

The geochemical study of the complex by Nockolds revealed a gradual change in composition from the peridotites (as low as 42% SiO_2) to the granites at 68% SiO_2, which led to his classic interpretation of closed system differentiation. Nockolds reasoned that rocks more basic than his proposed parental pyroxene-mica diorite magma must be the products of mineral accumulation, as evidenced by the abundant mafic minerals in the appinitic diorites, the gabbros and especially in the peridotites. The differentiated magmas went on to evolve into the granodiorites. Mahmood (1986) tested this hypothesis with numerical models of the compositions and found a gabbro parental magma to provide a better model for the observed compositions, although no simple closed-system fractionation model tested was capable of generating the granodiorite compositions. Mahmood also provided new trace element data that contrasted with the major oxides in not describing single smooth linear trends. The trace element data are therefore difficult to reconcile with a simple liquid line of descent, although they do provide evidence of cumulate processes (exceptional Ni- and Cr-enrichments in the peridotites). This evidence is not completely unambiguous, but it is likely that the granodiorites evolved quite separately from the main gabbro–diorite series. The data also suggest that the gabbros and diorites represent quite distinct magma series.

One of the very early applications of the Rb-Sr isotope technique to the study of granite plutons was performed on Garabal Hill (Summerhayes, 1966). By modern standards the precision of the age and initial ratio calculations is rather poor. Nevertheless the range of initial ratios

Figure 8.29 Serpentinized peridotites (wehrlite) at Lochan Strath Dubh-uisge, Garabal Hill–Glen Fyne igneous complex. (Photo: W.E. Stephens.)

measured is so large as to be significant in terms of the petrogenesis. Summerhayes found that the ($^{87}Sr/^{86}Sr$) initial values of all the uncontaminated components of the complex are the same within error, around 0.705. He noted, however, that several samples of the medium-grained granodiorite have higher values of this ratio (about 0.710) and concluded that these had the same parental magma as the main complex but were contaminated by local Dalradian metasedimentary rocks. This is consistent with the abundance of included xenoliths, as noted by Nockolds (1941). These are important petrogenetic conclusions but must be viewed in the context of modern Sr isotope studies in which variations of an order of magnitude smaller are taken to have petrogenetic significance. A higher precision isotopic study of a single sample from the medium-grained granodiorite, the area of contaminated granodiorites studied by Summerhayes, gave a ($^{87}Sr/^{86}Sr$)initial value of 0.7074 and $\delta^{18}O$ of 10.4‰ (Harmon and Halliday, 1980).

Isotope ratios are now used routinely to test whether complexes such as Garabal Hill evolved in a closed system (i.e. no addition of exotic material to the parental magma) or in an open system, in which case the contaminant is usually evident in changes to the isotope ratios. Whereas the old data of Summerhayes do not permit a rigorous test, the newer Sr and O isotope data are strong indicators of a significant involvement of evolved crustal material in the genesis of the granodiorite magmas, while the remainder of the complex appears to have the relatively primitive signature of young basic crust and/or mantle sources.

The trace element and limited isotopic data currently available do suggest therefore that Nockolds' model is inappropriate for this complex. However, there are other plutons (e.g. the Boggy Plain pluton of SE Australia, Wyborn *et al.*, 1987) where melts can be clearly shown to undergo crystal–liquid fractionation in the manner envisaged in 1941 for Garabal Hill–Glen Fyne, so that the wider conclusions of Nockolds' study remain valid to the present day.

In a further study, Garabal Hill was chosen as a good example in which to investigate the disappearance of Ca-poor pyroxenes from crystallizing calc-alkaline magmas (Cawthorn, 1976). Cawthorn concluded from petrographical evidence that the replacement of the assemblage calcic pyroxene–calcium-poor pyroxene–plagioclase by amphibole–plagioclase in the more evolved rocks is explicable in terms of increasing fugacity of water, during differentiation of a basic magma.

Conclusions

The Garabal Hill–Glen Fyne complex shows the greatest variety of 'granitic' rocks in any one complex in the Caledonian of the British Isles. An orderly sequence of intrusion from basic to acid magmas can be demonstrated and the complex is an important site for the testing and developing of modern ideas on the origins of granitic rocks.

A model in which a single dioritic parental magma underwent differentiation by the removal of crystals as they formed, causing the residual magma to change composition along a 'liquid line of descent' was proposed by Nockolds in 1941. This very detailed study was widely acclaimed and the ideas were applied to other plutons worldwide. The influence of the study has undoubtedly been seminal, but critical examination of new geochemical and isotopic data from the Garabal Hill–Glen Fyne complex suggests that the model is oversimplified and that here, more than one different magma and modification by crustal contamination may have been involved. However, the original concept, which subsequently became known as 'fractional crystallization', has been shown to be valid for other plutons elsewhere in the world.

LOCH DEE
(NX 458 761–466 847)

W. E. Stephens

Introduction

Loch Dee is located at the southern end of the Loch Doon pluton, which is one of the finest examples in Scotland of a zoned pluton showing inwards progression from a dioritic margin to an acid interior. The Loch Doon pluton stands out for the regularity of its zonation pattern over a very wide compositional range. This regularity has led to the suggestion that the zonation developed *in situ*. Another significant feature of the site is that the intermediate rocks include some of the best examples of hypersthene-mica diorite, the rock type considered to represent the parental magma for the Caledonian plutonic series in the influential model of Nockolds (1941) (see the Garabal Hill to Lochan Strath Dubh-uisge GCR site report).

The pluton was first recognized by the Geological Survey (Peach and Horne, 1899), with Teall (in the same volume) identifying the wide petrological range, which he attributed to differentiation in a deep-seated magma chamber prior to emplacement. The regularity and composite nature of the zonation within the pluton was demonstrated by Gardiner and Reynolds (1932) using a combination of mapping, petrography and density measurements, and the map of the complex has changed only in minor details with subsequent studies (although some of the rock nomenclature has been updated). Gardiner and Reynolds (1932) argued for separate intrusion of three successive magma pulses of increasing acidity.

In the 1950s the 'granitization' and 'basic front' transformationist theories of Doris Reynolds were applied to this pluton by McIntyre (1950), Rutledge (1952), and Higazy (1954) who argued on compositional grounds that the complex could be derived through solid-state metasomatic modification of the surrounding metasedimentary rocks. The magmatic versus replacement controversy over the origin of granites was largely settled in favour of the magmatists by the end of the 1950s, and a detailed study of the southern half of the complex (including the area around Loch Dee) by Ruddock (1969) returned to the magmatic view of petrogenesis for the pluton. He interpreted new whole-rock geochemical analyses as supporting the derivation of the complex by differentiation of a single parental magma of intermediate composition. Brown *et al.* (1979), also using a geochemical approach, concurred that a single parental magma was responsible for the complex. They identified the parental magma as monzodioritic, and postulated a two-stage fractional crystallization process. In contrast, Tindle and Pearce (1981), using an approach based largely on whole-rock trace element data, proposed that there had been two distinct parental magmas that evolved by fractional crystallization. These authors proposed that this fractional crystallization occurred *in situ* and the pluton has been widely cited subsequently as a leading example of in-situ fractional crystallization. Halliday *et al.* (1980) showed that the petrogenesis was even more complex, having identified distinctive isotopic signatures within and between the various plutonic members. Most marked is the variation in the oxygen isotope composition from the more primitive outer members to the more evolved interior which was attributed to increasing degrees of contamination of the magmas by metasedimentary rocks.

The pluton was emplaced into Ordovician metasedimentary rocks of low metamorphic grade at 408 ± 2 Ma, according to a mineral-whole rock Rb-Sr isochron (Halliday *et al.*, 1980) which is in agreement with U-Pb ages of 406 ± 2 and 410 ± 1 Ma obtained from single zircons (J. A. Evans in Floyd, 1997).

The Loch Doon pluton is important for understanding the origin of petrological zonation, there being differing views on the mechanism and the extent to which such processes can occur *in situ*. The Loch Dee site was selected for the GCR because it encloses all the important petrological variation in the southern part of the pluton. In this account the IUGS rock names of Le Maitre (1989) are applied using the modal analyses of Mahmood (1986), leading to significant differences with those of Gardiner and Reynolds (1932) and Brown *et al.* (1979).

Description

The hour-glass shaped Loch Doon pluton (Figure 8.30) is located between the Rhinns of Kells and the Merrick in Galloway. Much of the pluton forms low boggy ground between these hills, but the dioritic margins in the south and NW and the central ridge of granite form relatively positive features with moderately good exposure.

The outer contacts of the pluton are generally steep to vertical, but in the vicinity of the dioritic rocks of Loch Dee (Figure 8.30) contacts are much shallower and appear to represent a roof zone of the pluton (Stephens and Halliday, 1979). The evidence for this can be seen in small waterfalls that mark the contact zones between the diorite and country rocks in the White Laggan Burn and its tributaries. These all lie at similar topographical levels, suggesting a relatively flat, roof-like contact.

Petrological variation is concentric (Figure 8.30). There is an outer discontinuous zone of dark-coloured diorites, notably two-pyroxene-biotite diorite (fine- and medium-grained varieties with hypersthene, augite and minor olivine) and medium-grained hornblende-biotite diorites. Close to the contact south of Loch Dee these diorites are commonly quite xenolithic enclosing fragments of country rock. Good exposures of these xenolithic varieties, as well as the more normal fine-grained and medium-grained diorites, may be found in the road cuttings in the forestry plantations of this area.

The main mass of the pluton is granodiorite, with a tendency to form the low, poorly exposed ground. The rock is grey, medium- to coarse-grained, with abundant biotite and some hornblende as the principal mafic phases. The nature of the contact between the granodiorite and the dioritic facies is not clear; in places it appears to be quite sharp, but elsewhere a narrow transition zone is present. Passage to the central granite, which forms the ridge of high ground including Craiglee, Mulwharchar and Hoodens Hill, appears to be entirely gradational in the field, and no evidence of a sharp contact has been found. These white to buff-coloured granites have biotite as the only primary mafic phase and cordierite has been recorded in some microgranites at the very core of the pluton (Tindle and Pearce, 1981).

Interpretation

The main point of interest in this pluton is the exceptionally fine petrological variation, and an understanding of the origin of this variation requires combining field studies and geochemistry. The field evidence suggests a steep-sided plutonic body for the most part, except around the diorites which appear to form a roof zone. Three main facies make up the bulk of the pluton, namely diorite, granodiorite and granite. Each shows some internal variation, and the internal contact between the granodiorite and granite, in particular, is a rather broad transitional zone. The overall trend is one of increasing magmatic evolution towards the centre of the pluton.

In compositional terms, the pluton varies from the outer dioritic margin with about 57% SiO_2 (although rare samples of diorite in the NW have as little as 50% SiO_2) to the interior granite with about 72% SiO_2. The rocks are essentially metaluminous except for those with high SiO_2 levels, which are slightly peraluminous. Trace elements show no unusual enrichments or depletions relative to other South of Scotland Suite granitic rocks (Stephens and Halliday,

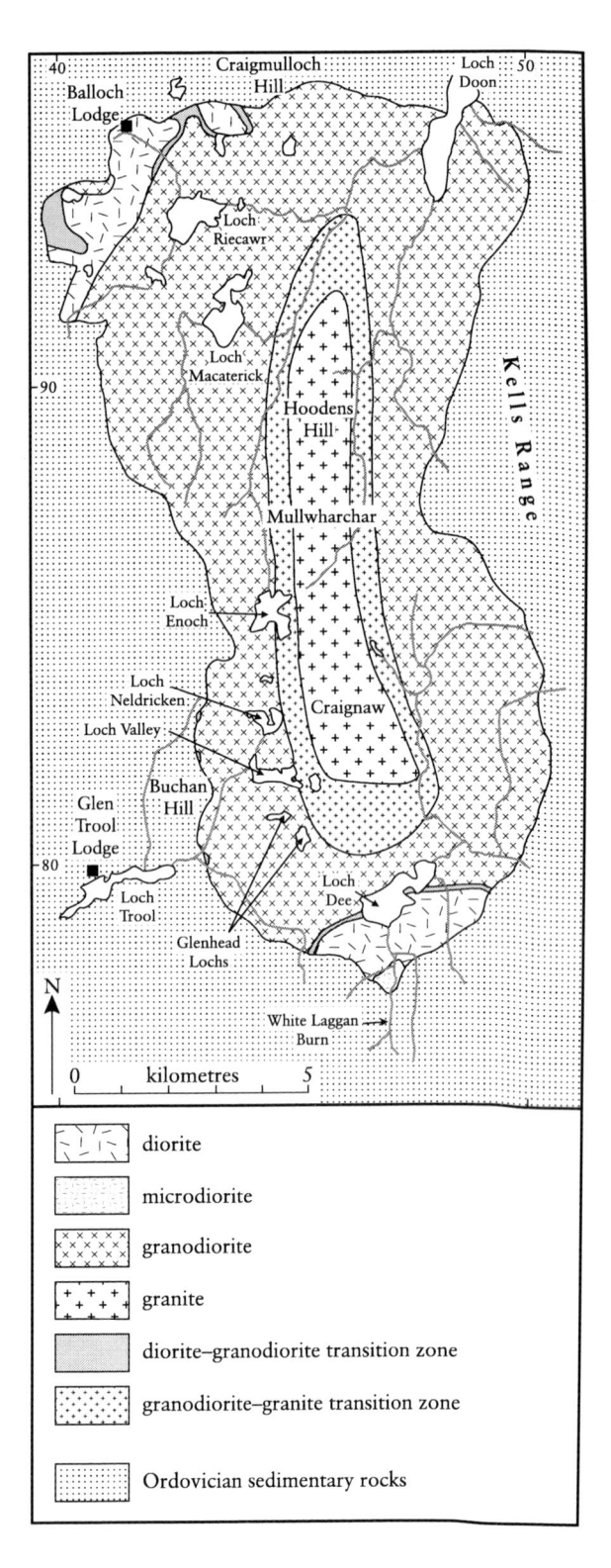

Figure 8.30 Map of the Loch Doon pluton, based largely on Gardiner and Reynolds (1932) with modifications from Ruddock (1969).

1984) and tend to vary smoothly through the sequence. Most trace elements decrease in abundance with increasing SiO_2 but Rb shows a continuous and marked increase as, to a lesser extent, does Th (Tindle and Pearce, 1981; Mahmood, 1986).

The concentric petrological variation is mirrored in the whole-rock isotopic composition. Initial $^{87}Sr/^{86}Sr$, for which most data are available, has the range of 0.7041–0.7051 in the diorites, 0.7052–0.7053 in the granodiorites, and increases from 0.7052 to 0.7059 in the granites (Halliday et al., 1980). Whilst these differences are small, they are significant and systematic. There is a similar systematic variation in the oxygen isotopes, although on a smaller dataset. $\delta^{18}O$ varies from 7.8 to 8.3‰ in the diorites, is 8.3‰ in a single analysed granodiorite sample, and jumps to 10.2–10.3‰ in the granites (Halliday et al., 1980). Nd isotopes may be more restricted with ϵNd values of –1 in the diorites and –1.4 in the granodiorites; no values have been determined for the granites (Halliday, 1984).

Most modern petrogenetic studies of this pluton have concluded that more than one magmatic pulse was involved in its construction, but the lack of clear internal contacts may indicate that at least some of the variation was formed by in-situ processes such as filter pressing and the accumulation of early formed crystals at the pluton walls (Tindle and Pearce, 1981). Such processes are isotopically closed systems and could not generate the observed isotopic variations described above; it would be necessary to involve some simultaneous assimilation of wall rocks in combination with fractional crystallization to account for such changes, as in the widely applied model of DePaolo (1981). It may be possible to apply this model satisfactorily to the geochemical variations in the Loch Doon pluton (there are no published attempts), but such a geochemical model would be inconsistent with the field evidence that the rocks requiring the greatest degree of assimilation are the innermost granites, those most isolated from the likely source of contamination. (It should be noted that the outcrops at the GCR site do not include the volumetrically less important 'Trend 1' and 'Trend 3' of Tindle and Pearce (1981) and thus have not been included in this discussion).

There is no model for the origin of the concentric zonation in the Loch Doon pluton that is consistent with all of the data. Isotopically, there is a requirement for two components in the magma generation, involving a primitive mantle-like component with another source which resembles the average of Southern Uplands Lower Palaeozoic crust (Halliday et al., 1980). Where and how these components interacted is not known, but there is also strong evidence for the operation of fractional crystallization as proposed by Stephens and Halliday (1979) and Tindle and Pearce (1981).

In regional terms the Loch Doon pluton is the most southerly large pluton of the South of Scotland Suite. The relatively unevolved I-type granites of this suite have little evidence of significant crustal contamination. Farther south, plutons of the Galloway Suite are more complex with evidence for greater incorporation of Lower Palaeozoic metasedimentary rocks during their genesis (Halliday et al., 1980) (see Clatteringshaws Dam Quarry, Lea Larks and Lotus quarries to Drungans Burn GCR site reports). One feature in common with many of the larger South of Scotland Suite granites is the presence of marginal pyroxene-mica diorites. Isotopic studies have shown that this facies is not the universal parental magma for the Caledonian granites as suggested by Nockolds (1941) and Nockolds and Mitchell (1948). Nevertheless it is important, as relatively anhydrous dioritic facies are not widely represented in granitic rocks of other orogens.

Conclusions

The Loch Dee GCR site was chosen as representing a key segment of the Loch Doon pluton, which has made important contributions to international studies of when and where compositional variation is developed in the history of a granitic pluton. Possible sites are in situ, by accumulation of crystals against the walls of the pluton and squeezing out of remaining magma towards the centre (filter pressing), or by fractionation in a feeder magma chamber beneath the pluton. Other possible factors include contamination during ascent, and heterogeneities in the original source. Presently no single model fully accounts for the variations within the Loch Doon pluton; it is expected that the Loch Dee GCR site will continue to provide new and valuable constraints in developing a fully consistent model, and that this will have implications for many plutons elsewhere in the world that show similar concentric petrological zonation.

CLATTERINGSHAWS DAM QUARRY (NX 548 754)

W. E. Stephens

Introduction

The Fleet pluton

The Cairnsmore of Fleet pluton (nowadays usually called the Fleet pluton) is one of three large and several small granitic bodies in the Southern Uplands of Scotland that comprise the Galloway Suite. The pluton lies within a broad zone of ductile shear, the Moniaive Shear zone, that affects Llandovery age sedimentary rocks of the Gala Group (Phillips *et al.*, 1995a). Weak to moderately developed pre-full crystallization fabrics, transitional into high-temperature fabrics within the pluton and shear-fabrics within the contact aurcole (Barnes *et al.*, 1995), argue the case for emplacement during the final stages of sinistral movement within the shear zone. The Orlock Bridge Fault, close to the NW margin of the pluton, is a late brittle structure that developed on the NW edge of the shear zone.

Most of the plutons of the Galloway Suite have similar mineralogical and compositional characteristics but the Fleet pluton is distinctly different. This uniqueness, coupled with its young age, has important implications for the petrogenesis of the relatively late Caledonian granites and the reconstruction of their plate tectonic environment. The pluton has two main granite facies, first established by Gardiner and Reynolds (1936), namely biotite granite and biotite-muscovite granite. The distribution of these facies was mapped by Parslow (1968) who further subdivided the biotite-muscovite granite facies into fine- and coarse-grained varieties, separated by a mappable sharp internal contact (Figure 8.31). Parslow (1971) also mapped the modal mineralogy and major oxide composition, and showed that some key parameters are concentrically zoned. The internal and external contacts were modelled from gravity anomalies, which suggest that the pluton continues to a depth of at least 11 km beneath the present surface (Parslow and Randall, 1973). In a regional study of the pluton and its environs Cook (1976) confirmed the general structure of the pluton and argued that extensive base metal mineralization in the south-western aureole is related to a shallow roof zone, in agreement with the suggestion from gravity anomalies of a western extension of the granite and a subsurface cupola (Parslow and Randall, 1973).

In a regional context the Fleet pluton is petrologically and geochemically unique (Stephens and Halliday, 1984). It has many of the characteristics of S-type granites (Chappell and White, 1974), indicating derivation of the magmas from a metasedimentary source, unlike the other late Caledonian granites in Scotland, which are all I-type and derived from meta-igneous sources (Stephens and Halliday, 1984). Isotopic compositions are generally similar to the Lower Palaeozoic host rocks suggesting that such a protolith may have provided a significant proportion of the granitic magmas from which the Fleet pluton crystallized (Halliday *et al.*, 1980).

The pluton has been dated by various methods. A U-Pb determination on zircon separates gave 390 ± 6 Ma (Pidgeon and Aftalion, 1978) within error of an Rb-Sr whole rock isochron age of 392 ± 2 Ma (Halliday *et al.*, 1980). A new unpublished zircon age is apparently similar (J. A. Evans, quoted by Barnes and Fettes, 1996). As the local metamorphic grade was low, these results establish the emplacement of this pluton at around 392 Ma, clearly Early Devonian, and the youngest reliably dated Caledonian pluton of mainland Scotland.

These data establish this pluton as somewhat anomalous in its context as a late Caledonian granite emplaced north of the Solway–Shannon line, which is usually taken as the line of the Iapetus Suture. Looking farther afield, the Fleet pluton is seen to have far stronger geochemical affinities with the plutons emplaced south of this line, that is those of the English Lake District and SE Ireland. It has been suggested that the source of the Fleet magmas was the same as those of the Lake District plutons (Stephens and Halliday, 1984) with the implication that magma genesis occurred after closure of the Iapetus Ocean and during underthrusting of the Southern Uplands by the leading edge of the southerly continent, and this is supported by more recent Pb isotope data (Thirlwall, 1989).

The importance of this pluton is its unique age and composition for a northerly late Caledonian granite, and its affinity with its southerly equivalents. The pluton is likely to provide important constraints on the timing and structure of the end-Iapetus closure and collision events. Two GCR sites have been selected, one to represent the more primitive outer por-

Clatteringshaws Dam Quarry

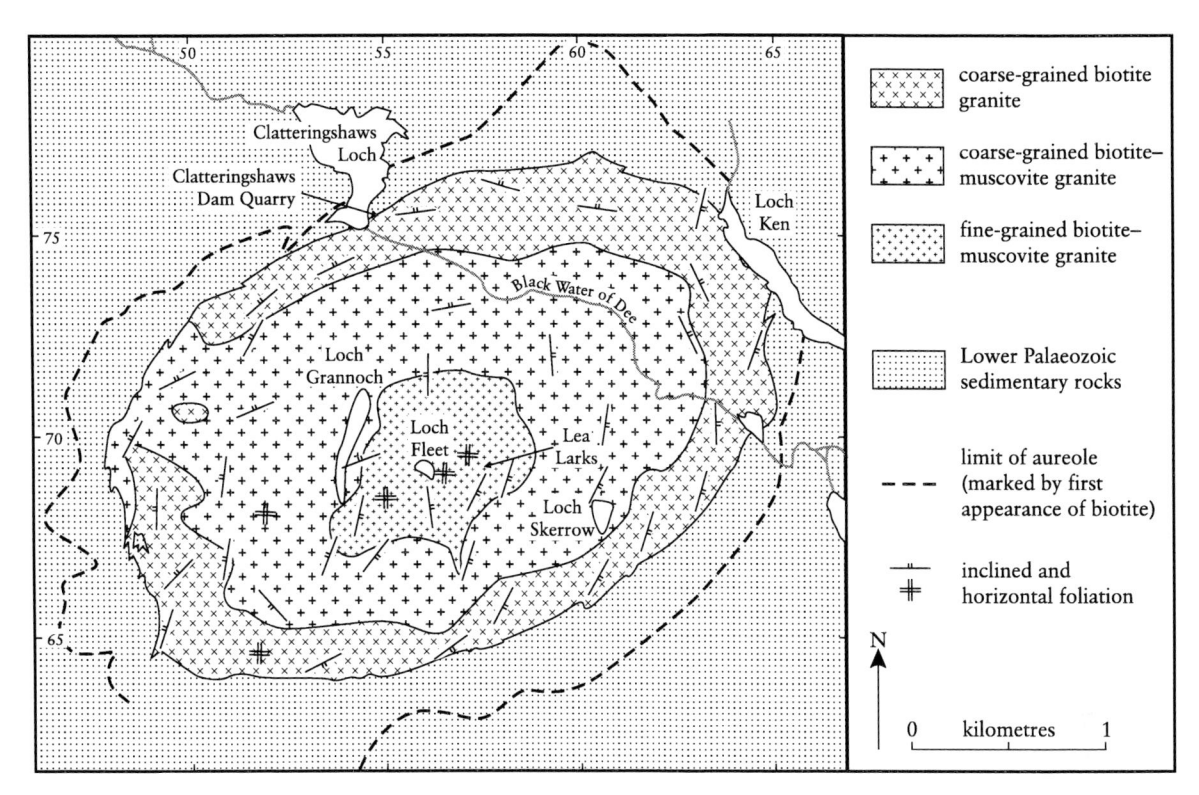

Figure 8.31 Map of the Fleet pluton, adapted from Parslow (1968), showing the locations of the Clatteringshaws Dam Quarry and Lea Larks GCR sites.

tions (Clatteringshaws Dam Quarry) and one to represent the more evolved central facies (Lea Larks).

Clatteringshaws Dam Quarry GCR site

The quarry and road cuttings at Clatteringshaws Dam (Barnes and Fettes, 1996) provide accessible and representative exposures of the marginal biotite granite facies of the pluton, which here is injected with aplite and pegmatite veins. The outer contact of the pluton with the metasedimentary rocks is also well exposed.

Description

The contact zone of the pluton is traversed by the road cuttings at Clatteringshaws Dam. Here the hornfelsed Silurian greywackes can be seen to dip northwards away from the pluton at about 40°, which is typical of that predicted by the gravity anomaly study (Parslow and Randall, 1973). The small quarry at Clatteringshaws Dam was cut in grey coarse-grained biotite granites, the outer facies of the zoned Fleet pluton, about 100 m from the outer contact (Figure 8.31).

This outer granite consists dominantly of alkali feldspar (microcline) and quartz. The alkali feldspars form quite large crystals (up to 5 cm) and are microperthitic; plagioclase is much less abundant and forms small crystals. Quartz forms large pools of strained polycrystalline aggregates. There is some biotite, usually chloritized, and some secondary muscovite replacing the alkali feldspars, but very few opaque minerals. The rock has a marked fabric with a strongly developed mortar texture in which strings of granulated quartz and mica develop around large 'islands' of alkali feldspar. This fabric gives the rock a foliation, obvious in the field, which can be traced throughout the pluton suggesting a relationship to its mode of emplacement. At this locality the foliation has a similar orientation to the foliation in the local hornfelsed country rocks.

Veins of aplite and pegmatite are common in both the quarry and the road cutting. The aplitic rocks are highly leucocratic bodies composed dominantly of alkali feldspar, plagioclase and quartz with minor chloritized biotite and accessory garnet. Analyses of these garnets show them to be unusually manganiferous (Macleod,

1992) and it is noteworthy that Tilley described similarly spessartine-rich garnets from garnet-rich lenticles within the Fleet aureole in his discussion of the paper by Gardiner and Reynolds (1936). The pegmatites are characterized by large alkali feldspars and quartz with some muscovite.

Interpretation

Studies of the composition of the pluton (Stephens and Halliday, 1979; Halliday et al., 1980; Stephens and Halliday, 1984) showed that it is the most evolved of all the Galloway Suite plutons in terms of major oxide composition and Sr isotopes. Despite being relatively evolved, the granites at Clatteringshaws Dam are among the least evolved of the pluton, with about 69% SiO_2 (in contrast to the much more evolved granite facies at the Lea Larks GCR site with 76% SiO_2). Like most of the rocks in the pluton the granites here are corundum normative, reflecting the peraluminous composition and the dominance of micas among the mafic minerals. However, at this locality the abundance of normative corundum is low, at less than 1%. This granite also has the least evolved isotopic compositions of this pluton, with a $^{87}Sr/^{86}Sr$ initial ratio of 0.7062 (Halliday et al., 1980) and ϵNd at this locality of –2.4 (Halliday, 1984). The oxygen isotope values for the whole pluton (including this site) are highly evolved, with $\delta^{18}O$ around 11‰, indicative of a source with a major sedimentary component (Halliday et al., 1980). A lead isotope study (Thirlwall, 1989) showed that the Fleet pluton is enriched in $^{207}Pb/^{204}Pb$, a signature more typical of Lake District granites such as Skiddaw than the Scottish late Caledonian granites.

In a regional survey of the Scottish late Caledonian granites, Stephens and Halliday (1984) found that the Fleet pluton, and to a lesser extent the Criffel pluton, do not fit neatly into their classification (see Chapter 8: Introduction) and they were omitted. Most of the late Caledonian plutons are I-type granites, whereas Fleet has many of the important characteristics of an S-type granite (Chappell and White, 1974) indicating that a major part of its protolith was metasedimentary. This conclusion is supported by the Sr and oxygen isotopic data, which are also consistent with derivation of the magmas in large part from the local Silurian metasedimentary rocks. These data establish the Fleet pluton as the only wholly S-type pluton among the late Caledonian granites of Scotland. Clearly the magmas were not derived *in situ* and the gravity data suggest that the pluton extends downwards at least 11 km (Parslow and Randall, 1973). A thickness of Lower Palaeozoic metasedimentary rocks rather greater than this depth has been modelled to underlie this area, and melting is likely to have occurred towards the bottom of this pile.

Given that the Fleet pluton is distinctive among more than 50 major granitic bodies in the Scottish Caledonian terranes (Brown, 1991), it is important to place the pluton in its regional context. Two-mica Caledonian granites with mild S-type characteristics and lacking zircon inheritance are known, but not from north of the Iapetus Suture. Plutons with many similar characteristics include the Skiddaw Granite in the Lake District, the buried Weardale and Wensleydale granites of the north of England, the Foxdale Granite on the Isle of Man, and the Leinster Granite of SE Ireland. This strong association with plutons on the other side of the Iapetus Suture led Stephens (1988) to suggest that they had a common, rather young, immature sedimentary source and that they shared an end-closure tectonic environment in which the same source rocks became available on either side of the suture. This implies underthrusting of the southerly continent beneath the leading edge of the northern continent, i.e. beneath the Fleet pluton, and this is consistent with the geophysical interpretations of Beamish and Smythe (1986).

Conclusions

The Fleet granite pluton is unique among all the late Caledonian granites of Scotland, as a consequence of the change in source of magmas near the leading edge of the northern (Laurentian) continental landmass. Geochemical evidence strongly suggests an origin of the magma in rocks similar to those that currently host the pluton, but the pluton's rather greater affinity with those in the English Lake District points to a more southerly source. These associations have important tectonic implications for the events that followed closure of the Iapetus Ocean and the collision of the continents either side of the Iapetus Suture. The site is thus of major national importance for the conclusions that have arisen from mineralogical and geochemical studies of the rocks present.

LEA LARKS
(NX 563 690)

W. E. Stephens

Introduction

This GCR site is located in the centre of the Fleet pluton; the rocks exposed are representative of the evolved central facies, which is very uncommon among the Scottish late Caledonian granites in being a peraluminous garnet-bearing two-mica granite. Such highly evolved garnet-bearing granites are rather unusual worldwide. The general significance of the Fleet pluton is discussed in the Clatteringshaws Dam GCR site. It represents the final event of Caledonian plutonism in mainland Scotland and is significant in providing important constraints on the end-stages of the Caledonian Orogeny and the closure of the Iapetus Ocean.

Zonation in the Fleet pluton takes the form of a small central facies of fine-grained granites within the main coarse-grained biotite granite and biotite-muscovite granite facies, as shown on Figure 8.31. There is a sharp contact between the inner and outer facies, and xenoliths of coarse-grained granite are found within the fine-grained, demonstrating that the central pulse is the last major intrusive event (Parslow, 1968). This central facies is the most evolved of the pluton and is particularly distinctive in compositional terms.

Small, apparently magmatic, garnets are present in the granites of this facies in the vicinity of this site. These are spessartines, similar to those described by Macleod (1992) in aplites from the marginal zones. Such garnets are usually associated with highly evolved granitic melts of appropriate bulk composition, probably as the result of extensive fractional crystallization (Speer and Becker, 1992).

The Lea Larks GCR site was selected as representative of the most evolved facies of the Fleet pluton. It is well exposed and accessible via forestry tracks. The rocks of this facies are rare, and have almost nothing in common petrogenetically with the other late Caledonian granites of Scotland.

Description

The white-weathering exposures at Lea Larks are of muscovite granite, an extreme variant of the 'fine-grained biotite-muscovite granite' facies of Parslow (1968), although technically the rock here is medium grained. It is a leucocratic granite, comprising quartz, alkali feldspar and plagioclase, with plates of muscovite up to 5 mm. Scarce garnet occurs as small, anhedral crystals (about 0.5 mm), which are not obvious in hand specimen. The rock is almost devoid of biotite save for some rare flakes that have been largely chloritized. There is little evidence of strain in these rocks and the textural features are essentially igneous apart from some late alteration of feldspars. The fabric is more isotropic than in the foliated granites of the main coarse-grained outer facies of the pluton.

Interpretation

Granites at Lea Larks are highly evolved, even in the broad context of the late Caledonian granites. They have about 76% SiO_2 and a normative composition of more than 95% quartz + orthoclase + albite, approaching the 2 kbar ternary minimum. About 2.5% normative corundum indicates a strongly peraluminous composition, which is reflected by the presence of muscovite. Most trace elements are strongly depleted relative to other late Caledonian granite plutons, and to the outer facies of this pluton. The sole exception is Rb, which reaches levels of over 500 ppm in some samples. Initial $^{87}Sr/^{86}Sr$ ratios of 0.7076–0.7109 are significantly higher than in the outer facies, but the oxygen isotopes are indistinguishable (Halliday *et al.*, 1980). Overall, the geochemical features suggest that the highly evolved composition was probably achieved by fractional crystallization, particularly through the removal of feldspars and accessory minerals.

The bulk composition of the Fleet pluton is peraluminous, and granites at the Lea Larks site are strongly so. Strongly peraluminous compositions suggest a pelitic source rock (Miller, 1985; White and Chappell, 1988) and the combined major oxide and isotopic composition of this facies are consistent with this interpretation (Halliday *et al.*,1980). However, such highly evolved compositions have also been attributed to fractional crystallization (Clarke, 1992); and Halliday *et al.* (1981), reviewing the origins of peraluminous compositions in granitic magmas, suggested that this may have involved subaluminous amphiboles. However, there is no evidence that amphibole ever crystallized from any of the magmas of the Fleet pluton and thus this mechanism can be discounted. It seems that the magma from which these muscovite granites formed was originally peraluminous, with S-type characteristics, and that it evolved through the fractional crystallization of feldspars.

Almandine garnets occur in earlier Caledonian granitic rocks of the Lake District (Firman, 1978b; see Chapter 4) and Connemara, Ireland (Bradshaw et al., 1969). However, spessartine occurs only in highly evolved granitic rocks; in the British Caledonides it has been recorded only from aplites and pegmatites, from the outer marginal facies of the Cairngorm pluton (Harrison, 1988) and from Lea Larks. The presence of garnet, albeit in rather small abundance, is important as an indication of extensive fractional crystallization of peraluminous granite magmas (Speer and Becker, 1992). Garnets in granites have also been interpreted as refractory relics from the melting of a pelitic protolith (Green, 1976). However, the composition of the spessartine garnets at Lea Larks and the evolved compositions of the whole rock are more consistent with a magmatic origin. In an analogous pluton, Speer and Becker (1992) show such garnets to be late magmatic, crystallizing at about 650°C.

The significance of the Fleet pluton in terms of its tectonic setting during end-Caledonian times has been discussed in the Clatteringshaws Dam GCR site. The Lea Larks site represents the most evolved facies of the zoned pluton.

Conclusions

The Lea Larks GCR site represents one of the most highly evolved forms of granite found in the late Caledonian granitic suites of Scotland. The granite at Lea Larks contains garnet and the chemical composition of the rock suggests a long history of fractional crystallization, leading to the generation of a highly evolved melt through the separation of crystals from the magma. These processes are important for understanding the mechanisms by which certain elements (including economically important metals) become enriched in evolved magmas, and ultimately in fluids derived from them.

The compositions of these rocks also reveal that the source of the magmas had a major component of crustal sedimentary rocks not unlike those which presently host the pluton but also with similarities to those of the Lake District, suggesting that these may be present at depth beneath the present Fleet pluton.

Further study of these outcrops will improve understanding of how highly aluminous magmas evolve in their late stages and may also provide tectonomagmatic constraints on models for the closure of the Iapetus Ocean and subsequent collisional events.

LOTUS QUARRIES TO DRUNGANS BURN
(NX 897 685–907 665)

W. E. Stephens

Introduction

The Criffel pluton

The Criffel pluton is an outstanding example of a granitic pluton that exhibits strong concentric zonation. This zonation has been well characterized in terms of petrology, geochemistry and structure. The origin of such zonation in granitic plutons has been the subject of controversy for many decades, and this pluton has contributed significantly to that debate. Criffel is also an important pluton because it lies close to the postulated suture line resulting from the closure of the Iapetus Ocean. It was emplaced soon after this closure, around 397 Ma ago (Halliday et al., 1980) into low-grade greywackes and pelites of Silurian (Llandovery to Wenlock) age.

The original survey of the pluton mapped the external boundaries of the pluton but did not distinguish different facies (Horne et al., 1896). Petrological zonation was first demonstrated by Phillips (1956) in developing the idea of Macgregor (1937) who showed that the western part of the pluton consists of multiple phases of granite. Phillips also produced the first detailed map of the whole pluton and its aureole, and this remains the most detailed geological study of the area. Stephens and Halliday (1980), Stephens et al. (1985) and Stephens (1992) focused on the petrological and petrogenetic aspects and refined the two-fold classification of Phillips. Five more-or-less concentric zones based on the dominant mafic mineralogy are now recognized (Figure 8.32); from the least to the most evolved these are clinopyroxene-biotite-hornblende (CBH) granodiorite, biotite-hornblende (BH) granodiorite, biotite (B) granite, muscovite-biotite (MB) granite, and biotite-muscovite (BM) granite. The pluton and host rocks are also notable for their mineralization (Gallagher *et al.*, 1971; Braithwaite and Knight, 1990), including uranium and other rare minerals.

Zonation of granitic plutons is a very common feature, but no consensus exists for its origin. Models to explain zonation include, among others, fractional crystallization, multi-pulse intrusions, magma mixing, variable degrees of restite separation, and contamination by assimilation. An isotopic study of zonation of the Criffel pluton demonstrated for the first time that closed-system process alone could not account for such zonation, at least in this pluton (Halliday *et al.*, 1980; Stephens and Halliday, 1980). Crystals and melts may separate in a closed system from a parental magma in various ways; by crystal settling or wall-rock accumulation, by the separation of entrained restite, or by filter-pressing processes. The system becomes open when the parental magma is contaminated in some way, either by assimilation of wall rock during ascent or emplacement, or by the mingling and ultimately mixing with a different magma. The only way of confidently distinguishing open and closed systems is by means of isotopic ratios; these parameters change little during closed-system processes but will disclose the open-system interaction of two magmas or magma + wall rock if these initially have significant isotopic differences. This methodology was first successfully applied to demonstrating the importance of open system processes at Criffel.

In terms of the widely used I- and S-type classification of granites (Chappell and White, 1974) Criffel is somewhat enigmatic. The early, outer hornblende-bearing facies (CBH and BH granodiorites) are undoubtedly I-types (derived from meta-igneous source rocks). The last, innermost member, the BM granite, has affinities with S-types (derived from metasedimentary source rocks), although it does not entirely fit in all of its features. The intermediate zones, the B and MB granites, are transitional in their characteristics. The pluton also provided mafic microgranular enclaves (igneous-textured xenolithic inclusions) for a study by Holden *et al.* (1987) which was the first to show that such enclaves can retain a Nd isotopic memory of a mantle or mantle-like source. The implications are either that mantle-derived material became involved in the crustal melting event (Holden *et al.*, 1987) or that primitive magmas became incorporated into the granitic magma around the time of emplacement, as has been demonstrated elsewhere (Castro *et al.*, 1990).

There are numerous analogous zoned plutons worldwide, especially in cordilleran batholiths such as the Sierra Nevada Batholith in California, and in almost every Phanerozoic orogenic belt. The importance of Criffel is not just that it was the first in which some of the important isotopic observations were made, but also because the nature of its zonation straddles the key I-type and S-type classification that has been widely applied since it was proposed in the mid-1970s (Chappell and White, 1974). Most plutons in Andean cordilleran settings belong to the I-type category and are believed to be the products of deep crustal melts (Chappell and Stephens, 1988) although some argue for a small or even a significant proportion of mantle material (DePaolo *et al.*, 1992). Either way, I-types represent the melting of an igneous protolith and are generally characterized by metaluminous bulk compositions and the presence of hornblende and titanite. In contrast, S-type granites are typical of continent–continent collision orogens and are common in the Himalayas and the Hercynian of Europe. These granites are the products of melting metasedimentary protoliths leading to very different compositional and mineralogical characteristics, including peraluminous bulk composition, reflected in the presence of peraluminous minerals such as cordierite and muscovite.

The studies of Phillips and co-workers on this pluton have been important in terms of emplacement mechanisms and magma chamber dynamics (Phillips, 1956; Phillips *et al.*, 1981, 1983; Holder, 1983). Several features such as enclave and mineral foliations, a steepening of the country rocks, and rotation of envelope fabrics into parallelism with the pluton margin, are suggestive of diapiric emplacement. However, as the amount of marginal deformation is very limited, Phillips *et al.* (1981) argued that emplacement was principally by stoping, and that convection in the magma was the cause of the primary mineral alignments and foliations. In discussion of their paper, Holder (1983) argued that the same features could be better

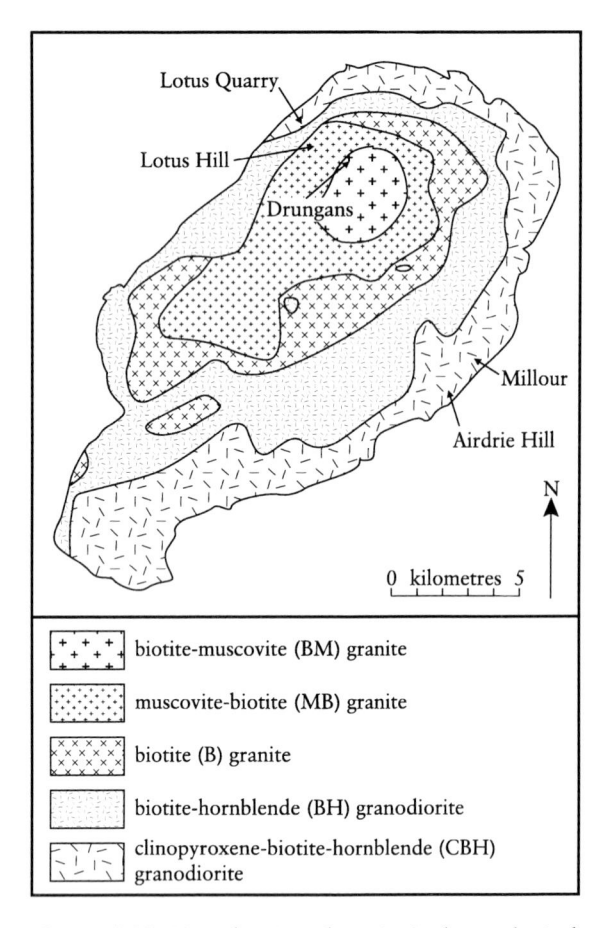

Figure 8.32 Map showing the principal petrological facies of the Criffel pluton.

Quarry is located within granodiorites close to the northern contact, and represents one end of a traverse in which there is a rapid petrological variation through the other facies to the two-mica granites to be found in the vicinity of Drungans, near the evolved centre of the pluton (Stephens, 1992) (Figure 8.32). This variation takes place over the whole pluton (Stephens and Halliday, 1980), but the gradient of change is steepest in this traverse; SiO_2 increases by about 10% over approximately 2 km, or roughly 1% SiO_2 every 200 m (Figure 8.33). This site has been selected as it offers the shortest traverse over all the major plutonic zones (i.e. steepest gradient of change) with reasonable exposure.

Description

The site is described from north to south, being the logical progression from least to most evolved members of the pluton (Figure 8.32). Exposure is quite good on the northern flanks of Lotus Hill to Lotus Quarry, but less good southwards towards Drungans Burn; the effects of forestry are to obscure some outcrops while creating others in road cuttings.

In and around Lotus Quarry the grey, foliated granodiorites of the CBH facies are exposed. As is the norm for this pluton the foliation dips outwards, in this case quite steeply at about 70° to the NW. The granodiorites are similar petrographically to those described more fully at the Millour and Airdrie Hill GCR site, and here also there are mafic enclaves lying in the foliation plane. The clinopyroxene is invariably included within amphibole, often in apparent reaction relationship. South of the quarry for about 300 m the granodiorite evolves to a clinopyroxene-free facies of biotite-hornblende granodiorites. The next zone inwards is normally the biotite granite facies but this is cut out in this area. Southwards, almost as far as Drungans, is the zone of muscovite-biotite granites. On Lotus Hill, a contact between this facies and the hornblende granodiorites can be located to within a few metres and the sharp junction is preserved in local boulders. This is important evidence that the pluton was emplaced as multiple pulses of magma.

Small knolls in the field west of Drungans cottage consist of the biotite-muscovite granite facies, but exposure is too poor to establish its relationship with the muscovite-biotite granite. The abundant large crystals of muscovite in the

explained by pluton ballooning. In an attempt to resolve the issue of the role of diapirism, Courrioux (1987) measured strain trajectories, strain gradients, and quartz fabrics over the whole pluton, which showed diapirism to have been the dominant emplacement mechanism for the second, inner magma pulse, and also possibly for the first.

The pluton is represented by two GCR sites: the Lotus Quarries to Drungans Burn site illustrates the main rock types and the concentric zoning; and the Millour to Airdrie Hill site exposes the outer contact and marginal zones, which contain abundant mafic-rich enclaves of structural and petrogenetic importance.

The Lotus Quarries to Drungans Burn GCR site

This site provides a section through the main components of the Criffel zoned pluton. Lotus

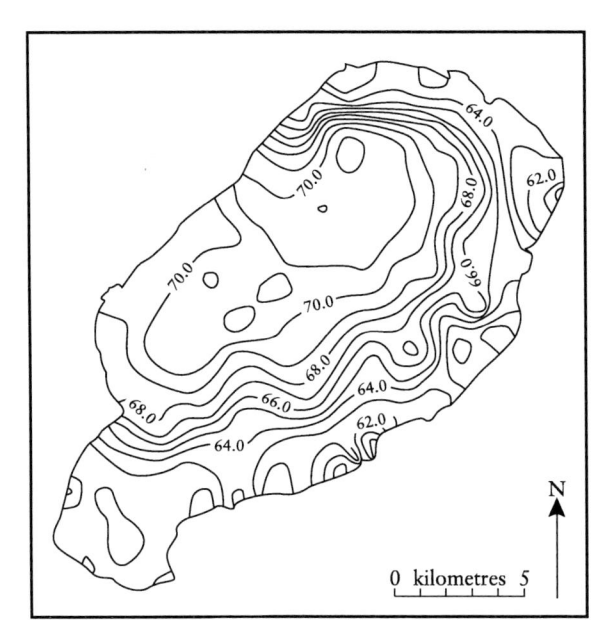

Figure 8.33 Contour map of SiO$_2$ variation in the Criffel pluton.

BM granite are magmatic on the basis of their igneous textures (Figure 8.35) and these are well seen in exposures at Drungans. About another 1.5 km SW of Drungans, on the northern end of Long Fell ridge, is the evolved core of the pluton in which muscovite is dominant with very little biotite present. In fact, biotite is normally extensively chloritized in this facies.

The contrast between the two ends of the traverse is illustrated in Figure 8.34 a, b, in which the grey, strongly foliated CBH granodiorite with mafic enclave is compared with the structurally more isotropic pink BM granite with abundant alkali feldspar and very few mafic minerals.

Interpretation

In order to understand the origins of zoning and the implications for the petrogenesis of such plutons it is essential to integrate the field data with the geochemistry. In this traverse the only evidence for an internal contact has been found between the MB granite and the BH granodiorite. Internal contacts between facies variations within the granodiorites (CBH to BH granodiorites) and within the granites (B to MB to BM granites) have never been located and it is presumed that these are gradational. However the sharp MB granite–BH granodiorite contact (which in places cuts out the B granite) shows that the granites were emplaced as a later magmatic pulse after the granodiorites had largely or entirely consolidated. This is consistent with the structural evidence for diapiric intrusion of the inner pulse discussed under the Millour and Airdrie Hill GCR site. The excision of the biotite granite leads to steepening of the compositional gradient in the region of the GCR site but elsewhere in the pluton even this boundary appears to be transitional, as described by Stephens and Halliday (1980).

Whole-rock geochemical variations along the traverse are best exemplified by SiO$_2$, which increases from about 62–72% (Figure 8.33), with very tightly bunched contours. At the same time, all other major oxides except K$_2$O decrease significantly, as do most trace elements except Rb which increases significantly. Most notable are the changes in isotopic ratios determined by Halliday *et al.* (1980) and Halliday (1984), with initial $^{87}Sr/^{86}Sr$ ranging from 0.7052 to 0.7073, ϵNd ranging from -0.6 to -3.1, and $\delta^{18}O$ ranging from 8.5 to 11.9‰ (all expressed as variations from outer to inner facies).

The varieties of outer granodiorite fit all the criteria for I-type granites (Chappell and White, 1974), including the presence of hornblende, metaluminous bulk compositions, and appropriate levels of the Sr, Nd and oxygen isotope ratios. The inner BM granites, however, depart from the I-type classification and have strong affinities with the S-types in most important respects, especially their peraluminous compositions and evolved isotopic compositions, including values of $\delta^{18}O$ well above 10‰. These granites are not unequivocally S-types, but nor are they I-types and are perhaps best described as transitional S-types.

The I-type to S-type zonation in a fairly regular concentric structure is unusual and its origin is not well understood. The outer I-type granodiorites are generally similar to I-types worldwide and have an origin dominantly in a relatively juvenile lower crust. The S-type characteristics of the most evolved inner granites can be correlated with the local Silurian greywackes, and it is likely that these have melted to provide the innermost magmas, although this must have happened at considerable depth given that there is no evidence for local melting. How these two melting events led to an organized zoned pluton is not clear; Stephens (1992) has suggested that the control was rheological, but it is possible that the inner pulse hybridized with the outer to

Figure 8.34 (a) Typical clinopyroxene-biotite-hornblende granodiorite, with enclave from the margin of the Criffel pluton. (b) Biotite-muscovite granite from the interior of the Criffel pluton. (Photos: W.E. Stephens.)

Figure 8.35 Photomicrograph of the biotite-muscovite granite from the Criffel pluton. (Photo: W.E. Stephens.)

produce the intermediate zones, although this would seem to be precluded by the fact that the outer granodiorites were largely crystallized when the granites were intruded (Courrioux, 1987).

Conclusions

The notion that originally homogeneous magmas differentiate to form diverse rock types is central to igneous petrology, and the products take many forms. In granites, such differentiation often takes the form of a zoned pluton, of which the Criffel pluton represents one unusual variety (I-type to S-type) which requires multiple pulses of magma derived dominantly from different sources (protoliths). The Lotus Hill to Drungans traverse is of international importance in this context. It provides a compact summary of this type of zonation and has contributed to detailed published studies of the observed variations as well as to some of the first discussions of the likely causes of such zonation. The contrasts in isotopic ratios between the zones make it especially likely that this pluton will be important in generating and testing new models for such zonation.

MILLOUR AND AIRDRIE HILL (NX 950 595)

W. E. Stephens

Introduction

The outcrops around Millour and Airdrie Hill (Figure 8.36), some 2–3 km south of the summit of Criffel, provide very good exposures of the outer contact and marginal zones of the Criffel pluton. A general introduction to the pluton and a discussion of the zoning is given in the Lotus Quarries to Drungans Burn GCR site description.

The marginal clinopyroxene-biotite-hornblende granodiorite facies contains abundant titanite and shows all the geochemical and isotopic characteristics of I-type granites (Stephens *et al.*, 1985). Features thought to be indicative of diapirism, including sharp contacts, steeply outward-dipping host rocks, mineral and enclave foliations with similar trends to the host rocks, and flattened enclaves, are well displayed. The evidence used in the stoping versus diapiric emplacement controversy (Phillips, 1956; Phillips *et al.*, 1981, 1983; Holder, 1983; Courrioux, 1987) can be examined here, as can many of the petrological features associated with the least evolved facies of the zoned pluton.

This site has been selected for the quality of the exposures and the range of features contained within a relatively small and accessible area. In particular, the site has among the best examples of discoidal enclaves in the UK, rivalling those of the classic Ardara pluton in Donegal (Pitcher and Berger, 1972; Holder, 1983; Pitcher, 1993).

Description

The outer contact of the pluton can be seen in the Kirkbean Burn in the SE corner of the site. The Silurian greywackes in contact with the granodiorites dip south-eastwards away from the contact at 50–60°. The contact may be traced along the break of slope in a south-westerly direction. The NE–SW strike of these contact rocks is more-or-less parallel to the regional strike but elsewhere, along the NE and SW contacts of the pluton, the regional strike is deflected into parallelism with the contact. For this reason, Phillips (1956) argued that the structures in the aureole are at least in part due to the emplacement of the pluton, and that the magma was emplaced forcefully (Phillips *et al.*, 1981). Veins of granitic material, rather more acidic than the local granodiorite, locally intrude along planar structures in the greywackes. These greywackes are recrystallized as a result of contact metamorphism, with significant growth of new biotite.

On the south and SE slopes of Millour and Airdrie Hill the clinopyroxene-hornblende-biotite granodiorite (Figure 8.34a) has a strong foliation, contains abundant mafic-rich enclaves (i.e. dark xenolithic inclusions) and is cut by microgranodiorite and microgranite dykes, and veins of aplite and pegmatite. The granodiorite consists of hornblende, commonly with cores of clinopyroxene, together with biotite and zoned plagioclase feldspar (andesine–oligoclase), some alkali feldspar and quartz. Titanite, zircon and apatite are the principal accessory minerals with some opaque minerals. The enclaves have the same mineralogy as the host granodiorites, but the mafic phases and plagioclase are more abundant. Country rock xenoliths can be seen on the southern slopes of Millour, close to the outer contact (i.e. near the break of slope), and these are aligned within the main foliation in the granodiorites. They decrease in number and size away from the contact and few are found beyond 100 m into the pluton.

Late stage veins of aplite are fairly common, usually just a few centimetres thick and whitish in colour, reflecting their dominant quartz and feldspar mineralogy. They cut across the enclaves and the main foliation in the granodiorites, but in places they show a deformation that is approximately parallel to the fabric in the host. Some pegmatites may also be found in these outcrops. Also present are dykes of 'porphyrite', a microgranodiorite with hornblende and/or plagioclase phenocrysts, and 'porphyry', a microgranite with quartz, plagioclase and/or biotite phenocrysts. These dykes are typically 20 cm to 1 m wide and trend approximately NW–SE towards the central granite.

The key feature of this site is the very strong foliation apparent in the granodiorite. This takes the form of a strong alignment of mafic enclaves combined with a parallel mineral alignment in the host. The enclaves appear as dark pod-shaped masses, typically 10 cm to 1 m in maximum length and disc-shaped in three dimensions. Measurements of the enclave dimensions by Courrioux (1987) show that the maximum:minimum length ratio varies from about 4 up to 20, indicating considerable deformation if the enclaves were initially equidimensional. The strong mineral foliation in the host rock is due to the alignment of feldspars, amphiboles and biotites. In thin section this is seen to be accompanied by strong deformation, with mortar texture of small quartz grains around large plagioclase crystals (protoclastic texture), bent plagioclase crystals and kinked biotites. These features indicate that the granodiorite suffered a high degree of deformation and, as all the quartz is seen to be deformed, the strain was taken up largely in the solid state. The foliation dips outwards at 45–65° to the SE or SSE, bracketing the dip of the country rocks.

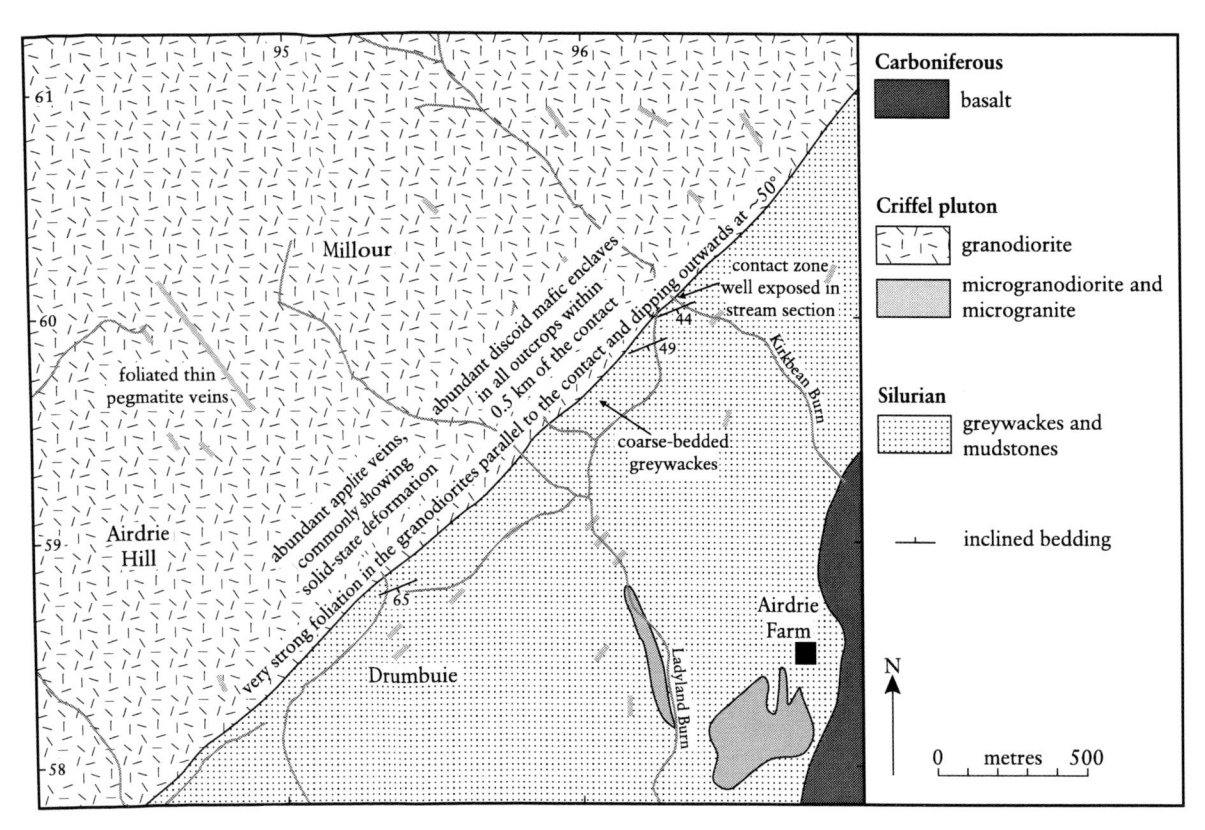

Figure 8.36 Map of the area around the Millour and Airdrie Hill GCR site, Criffel pluton, based on Phillips (1956), BGS 1:50 000 Sheet 5E (Dalbeattie) (1993) and observations by W.E. Stephens.

Interpretation

The field observations described above accord closely with the classic features of pluton emplacement by diapirism (i.e. magmas forcing their way into high levels of the crust by deforming their host rocks). The emplacement of granitic plutons is currently the subject of much debate, in particular concerning the role of diapirism at high crustal levels (Hutton, 1988a; Paterson et al., 1996; Petford, 1996). The present view is that diapirism needs heat to be effective; deformation of hot plastic rocks by magmas may be possible but it is mechanically impossible to deform cold brittle rocks in the same way. Thus, according to many authors, diapirism is confined to the hotter middle crust and is not appropriate to the emplacement of granites at a high crustal level. The evidence for the emplacement of at least part of the Criffel pluton by diapirism is compelling, yet several features place it in the mesozonal category of plutons emplaced within the upper crust. It was intruded at the end of Silurian time but was already unroofed and providing material to form local arkose deposits in the Late Devonian.

There is a partial reconciliation of this problem in the work of Courrioux (1987). Phillips (1956) and Phillips et al. (1981, 1983) argued that the alignment and increasingly discoidal form of the enclaves away from the pluton margin is the result of convective flow. However, Holder (1983) and Courrioux (1987) suggested that the flattening is a consequence of ballooning of the magma during emplacement. The enclaves therefore provide evidence of the distribution of strain over the magma body during emplacement. An increase in strain within the granodiorites towards the internal boundary with the granites was interpreted by Courrioux as the effect of the emplacement of the inner granite magma pulse on the outer, largely consolidated granodiorite. At this time the outer granodiorite envelope would certainly have been at a high temperature and thus potentially capable of being deformed by an intruding magma. It is less easy to reconcile the emplacement of the first granodiorite pulse by diaprism into cold Silurian host rocks, although the evidence of the steepening and rotation of these rocks indicates that they have been deformed by the intruding pluton. This pluton, and these outcrops in particular, provide an excellent location to study this lingering problem.

Samples from these outcrops have contributed to various geochemical and isotopic studies with the aim of understanding the origin of the magmas and their subsequent evolution into a zoned pluton, as described in the Lotus Quarries to Drungans Burn GCR site report. The granodiorites in these outcrops are among the most primitive of the whole pluton and may represent a magma that was parental to more evolved facies. The SiO_2 content locally is about 62%, whereas levels of 10% higher occur in the centre of the pluton.

In terms of understanding the origin of the magmas, the enclaves have made an important contribution. Several enclave–host sample pairs were collected for a geochemical and isotopic study of their chemical equilibration relationships. The results published by Holden *et al.* (1987), Holden *et al.* (1991) and Stephens *et al.* (1991) indicate that considerable disequilibrium is recorded in these enclaves. The enclaves are invariably more primitive and mantle-like in their Nd isotopic signatures than the host (ϵNd of *c.* –2 for the granodiorite, compared with ϵNd of *c.* 0 for the enclaves), indicating that their residence time in the main magma was limited to less than that at which full equilibrium would be achieved for all elements and isotopes. These authors provided the first firm indication that such enclaves may be the partially equilibrated remains of mantle-derived basic magmas that intruded the crust, bringing heat to assist the melting of deep crustal rocks to generate granodioritic magmas. This is not the only interpretation of the data, and the enclaves could represent exotic magmas mingled at the emplacement level, although there is no independent evidence for this at Criffel, or that they represent disequilibrium fragments of restite.

Conclusions

The Millour and Airdrie Hill GCR site exhibits, in a small area, all the main features of the early granodiorites of the Criffel pluton, which are of international importance. The structures preserved have allowed a partial reconstruction of the emplacement history of the pluton and this is contributing to the worldwide debate on how such large volumes of magma found space to occupy very near the Earth's surface. The compositions of the same rocks also provided the first evidence that the injection of mantle-derived magmas was probably the cause of deep crustal melting, which ultimately generated the magma for this pluton. This finding has since been applied by other researchers to plutons throughout the world.

ARDSHEAL HILL AND PENINSULA (NM 9611 5551–0000 5734)

I. M. Platten

Introduction

The small intrusions of the Appinite Suite occur typically in clusters. The Ardsheal GCR site occurs in the type area as part of the Duror of Appin cluster (Figure 8.37), which is 8 km in diameter but may extend farther beneath Loch Linnhe. Individual intrusions are numbered on Figure 8.37 to facilitate reference in the text. The Ardsheal Hill and Peninsula area was described initially by Bailey and Maufe (1916), Walker (1927) and Bailey (1960). The area was remapped by Bowes and Wright (1961, 1967) and McArthur (1971) who recorded 20 significant intrusions and breccia pipes. Petrological work has been presented by Bowes and McArthur (1976), Bowes *et al.* (1964), Hamidullah (1983), Hamidullah and Bowes (1987), Wright and Bowes (1979) and Platten (1991). Local copper mineralization was reported by Rice and Davies (1979) to be spatially related to the intrusions (intrusion 5). Parts of the cluster outside the site have also attracted attention. The well-known Kentallen intrusion (intrusion 1) is at the NE end of the cluster (see the Kentallen GCR site report) and aspects of the southern part of the cluster, particularly some breccia pipes, were documented by Platten (1982, 1984). The site is far more accessible than any of the other appinitic clusters in Scotland and is frequently visited by undergraduate and other field courses.

The country rocks are quartzites, metalimestones, phyllites and slates of the Dalradian, Appin Group that have been metamorphosed to greenschist facies before emplacement of the appinitic intrusions (Bowes and Wright, 1967;

Treagus and Treagus, 1971). The area is also cut by representatives of a number of dyke swarms (Bowes and Wright, 1967; see the Kentallen GCR site report). NE-trending lamprophyric microdiorites are inferred to be closely related to the appinitic rocks, whereas porphyritic microgranodiorites ('porphyrites') are considered to belong to the late Caledonian dyke swarms (Bailey, 1960). The north-eastern end of the site lies within the aureole of the late Caledonian Ballachulish pluton (Bowes and Wright, 1967; Pattison and Harte, 1985). The site is adjacent to the Great Glen Fault and is cut by associated faults and fracture systems.

Description

The intrusions and breccias are all small, ranging from 50 to 500 m across (Figure 8.37). Outlines vary from simple, near circular or oval shapes (intrusions 4, 5, 15 and 16), to irregular bodies with intricate outlines (intrusion 8). The most complex shapes result from the intersection of dioritic intrusions with earlier breccia pipes. Exposed contacts between intrusions and breccias and with the external country rocks are generally steep, hence the inferred pipe form. No evidence of upward closure has been found. Intrusions may penetrate locally between the clasts in the breccias, but generally the contacts are sharp.

The main igneous rock type is a pyroxene-bearing mesocratic diorite or monzodiorite with variable amounts of biotite (intrusions 2–8, 14 and 16) (Hamidullah and Bowes, 1987). The pyroxene is augite/salite, occurring either as large, 2–8 mm crystals, giving the rock a porphyritic aspect or as small grains of less than 1.0 mm. Orthopyroxene is absent or very rare; olivine is present as a minor component. Hornblende may occur as overgrowths on the pyroxenes or as co-existing elongate prisms. A porphyritic chilled facies of these diorites with pyroxene phenocrysts in a fine groundmass occurs in intrusions 8 and 14. However, chilled margins are commonly localized or completely absent, with coarse-grained rocks occurring at the contacts in spite of the small size of the intrusions.

Hornblende diorites with conspicuous euhedral, elongate or equant, prisms of pargasitic hornblende are a distinctive and diagnostic feature of the suite throughout the Scottish Caledonides. Small calcite-filled vugs are a common accessory element. These are the rocks that were originally called 'appinites' (Bailey and Maufe, 1916). They occur generally in four different settings.

1. Single, fairly uniform bodies of mesocratic or melanocratic appinitic diorite may form the bulk of a pipe-shaped intrusion, emplaced in breccia or in country rock (intrusions 11 and 15).
2. Small bodies of meladiorite and hornblendite occur at the margins of pyroxene- and hornblende-bearing diorite intrusions (intrusions 8 and 13). These include near-vertical layers of comb-textured hornblende that point to in-situ growth on the walls of the pipe. These bodies commonly show steep layering, marked by colour index or textural variation.
3. Small (< 20 m) bodies of very variable hornblende meladiorite and hornblendite with characteristically equant, euhedral to subhedral hornblendes developed by alteration of earlier dioritic rocks (intrusions 11 and 12).
4. Pegmatitic veins and patches of mesocratic or leucocratic hornblende diorite may be found in all of the three above situations.

Leucocratic hornblende and pyroxene quartz-diorites and monzonites, and biotite and hornblende granodiorites are present. Small druses and minor sulphides are not uncommon in these rocks. Some form discrete bodies with only minor amounts of dioritic or mafic material at their margins (intrusion 18). They also occur as pipes, dykes, patches and veins in the other igneous rocks (intrusions 5, 8 and 14). Some granodiorite dykes penetrate the country rocks for a short distance (intrusions 14 and 18).

Small bodies of melanocratic or ultramafic rock are a common regional feature of the Appinite Suite. At the Ardsheal GCR site, biotite pyroxenite and biotite peridotite are common as small bodies at the margins of pyroxene diorite intrusions (intrusion 14). Thin (1–20 mm) anorthosite layers occur at the inner margins of these ultramafic bodies (Figure 8.38a) and within the main pyroxene diorite intrusions (intrusion 14). Hornblendite and meladiorite occur as small bodies at the margins of both hornblende diorite and pyroxene diorite intrusions (intrusions 3 and 8).

The intrusions have narrow (40–100 m) contact metamorphic aureoles (Platten, 1982) showing fine-grained, flinty hornfels with cordierite

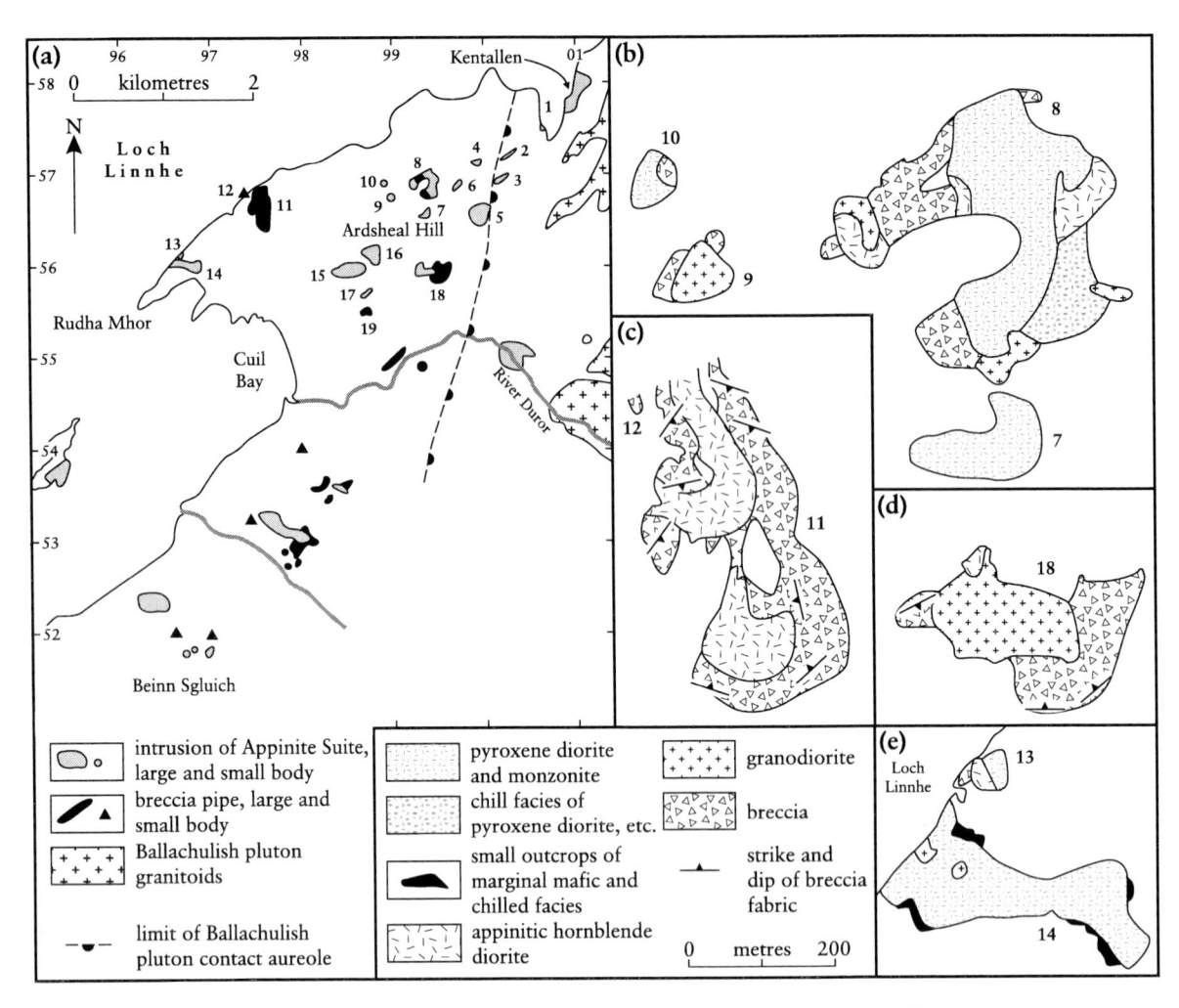

Figure 8.37 (a) The distribution of appinitic intrusions and breccia pipes in the Duror of Appin cluster. Intrusions within the Ardsheal Hill and Peninsula GCR site are numbered for reference in the text. (b), (c), (d) and (e) Examples of outcrop patterns within the Duror of Appin cluster. The intrusion numbers correspond to those given in Figure 8.37a. Figures b, d and e are redrawn with minor modification from Bowes and Wright (1967).

spots in pelites (intrusions 15 and 18). Localized partial melting and mobilization has been noted in pelites and feldspathic quartzites within a few metres of contacts with intrusions 14 and 18 (Platten, 1982). In the northern intrusions (1, 2, 3 and 5), later contact metamorphic effects of the Ballachulish pluton obscure the contact metamorphism due to the appinitic rocks (see the GCR site report).

Breccia pipes and rare dykes are associated with the appinitic intrusions (Bowes and Wright, 1961, 1967; Platten, 1982, 1984). These are characteristically composed of local country rock fragments that show evidence of rotation and local transport. Clasts are generally angular and most lie in the size range 0.05–1.0 m; sand-sized material is usually absent. Rounded clasts, mostly of quartzite, do occur as a minor component in some breccias (intrusion 11). Internal divisions may be mapped within some breccia pipes based on variations in clast size, rounding and lithology (intrusions 8 and 11). Breccias with abundant plate-shaped fragments show planar fabrics defined by parallel orientation of clasts (intrusions 11 and 18). In breccias dominated by phyllite clasts, most of the clasts are deformed and show face to face contacts leaving little pore space (intrusions 8 and 18). Breccias dominated by quartzite clasts show less deformation of clasts and some trace of pre-cement pore space may remain. This has a drusy quartz filling with minor pyrite and a carbonate mineral.

Ardsheal Hill and Peninsula

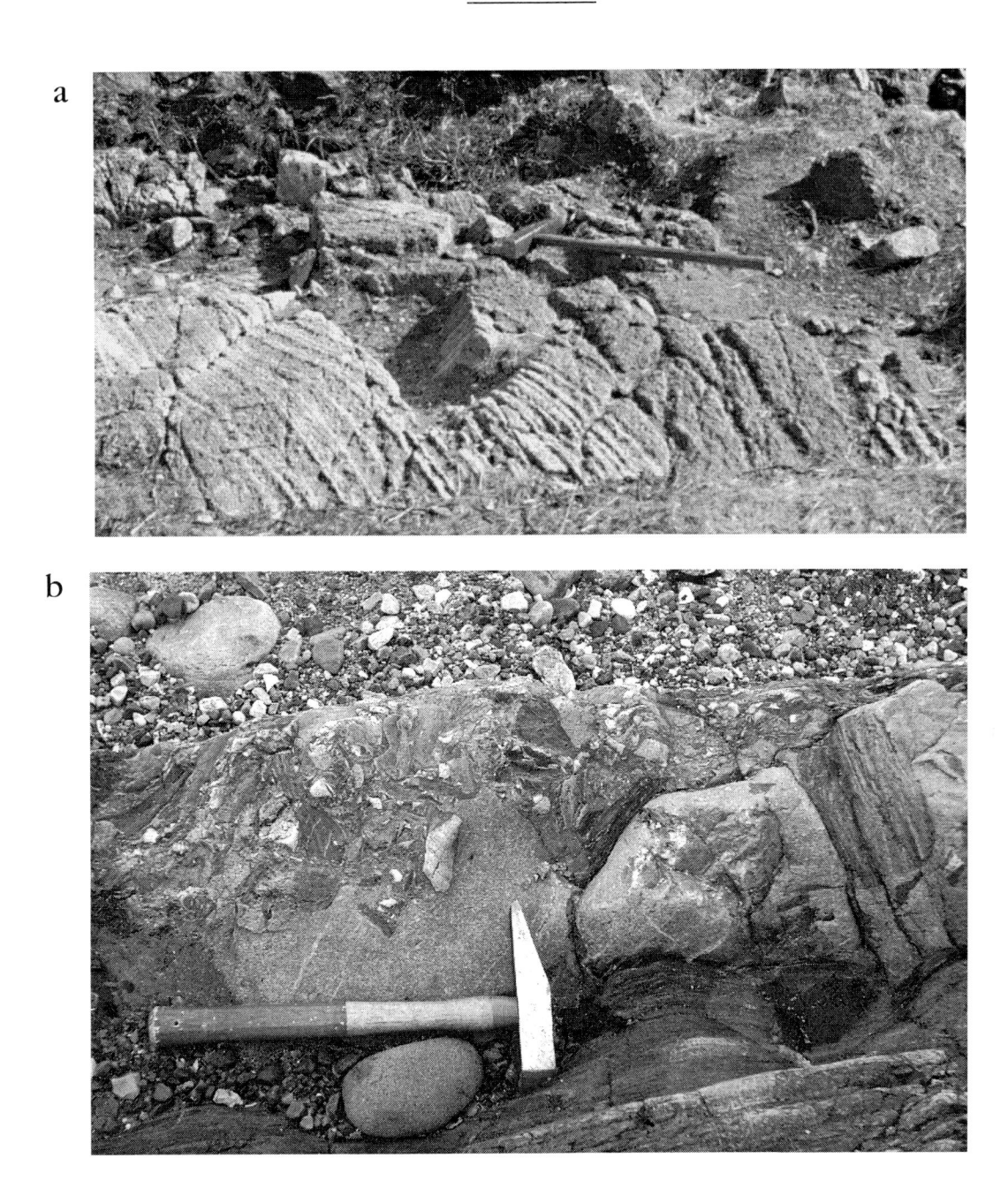

Figure 8.38 (a) Wall-parallel, steeply dipping layering of meladiorite and anorthosite at the SW margin of intrusion 14, in the Duror of Appin cluster (see Figure 8.37). (Photo: I.M. Platten.) (b) Composite dyke, Back Settlement, Ardsheal. The dyke, which cuts intrusion 12 to the NE, dips steeply towards the observer, with breccia in the footwall and lamprophyric microdiorite in the hanging wall. Magma has penetrated between breccia clasts at the irregular contact. (Photo: D. Stephenson.)

Interpretation

Bowes and Wright (1961, 1967) interpreted the breccias as forming as the result of explosions, largely based on the extensive fracturing of the Dalradian host rocks and the deformation of the breccia fragments. However the diorite intrusions, breccias, contact hornfels and later dykes of microdiorite and porphyritic microgranodiorite are also fractured and cut by drusy quartz veins. The timing of fracturing is discussed in the Kentallen GCR site description in which it is shown that the extensive fracturing is a very late event and thus unrelated to breccia emplacement. The breccia pipes at Ardsheal are now interpreted as having formed by the mechanisms proposed by Platten and Money (1987) for the Cruachan Cruinn breccias, near Crianlarich. Country rock collapse occurred above intrusions of volatile rich magma from which a vapour phase had separated. This produced narrow columns of fractured country rock which were fluidized and slightly transported as the vapour phase vented to the surface. Deformation of the clasts is interpreted as a post-breccia compaction driven by gravitational loading from overlying breccia and hydraulic loading by adjacent magma.

Emplacement of pipe-shaped intrusions resulted from the collapse of the country rock between the breccia pipes into an underlying, locally differentiated and de-gassed, magma chamber. Some very small intrusions (6, 7, 9 and 10) may have resulted from the removal, either upwards or downwards, of the clastic fill of breccia pipes. The relatively coarse-grained nature of the intrusions in even the smallest pipes and the evidence of local partial melting of wall rocks points to an active circulation of magma, probably with throughput to the surface. The cumulate marginal mafic and ultramafic rocks and the anorthositic layered rocks crystallized *in situ* on the walls from the convecting dioritic magmas. The more uniform cores, occupied by dioritic or evolved granodiorite, crystallized after convection ceased. Variation in the effectiveness of venting volatiles to the surface controlled the ferromagnesian mineralogy, pyroxene marking an open system and hornblende a closed system. Evolved leucocratic rocks intruded into the diorites may reflect the continued presence of magma bodies at depth below hot, but crystalline, diorite intrusions or simple filter-pressing of the lower parts of the diorite. A return to breccia formation is only known at one site (intrusion 12) where a breccia dyke cuts intrusive rocks (Figure 8.38b).

Conclusions

This is the type area of the Appinite Suite. It illustrates the extreme range of rock types (ultramafic, intermediate and acid) present in these highly differentiated intrusions, including the hornblende meladiorites with conspicuous euhedral hornblendes that characterize the suite. The site also provides excellent examples of the breccia pipes that are widely associated with the suite.

Igneous activity began with the formation of breccia pipes above underlying, volatile-saturated basic magma bodies. Primitive and more-evolved magmas were then emplaced, both into the breccia pipes and the country rock. Convection within the magma led to the accumulation of crystals against the walls of the pipes, which changed the composition of the remaining magma (crystal fractionation) to produce varied rock types. If the pipes remained sealed, volatiles were trapped and hydrous minerals crystallized (e.g. hornblende). These 'closed system' conditions alternated with 'open system' conditions in which the pipes functioned as feeders to surface volcanoes. Volatiles were released, possibly by explosive eruptions, and non-hydrous minerals crystallized in the remaining magma (e.g. pyroxene). Subsequently the cluster was cut by the late Caledonian Ballachulish pluton.

KENTALLEN
(NN 0091 5766–0135 5822)

I. M. Platten

Introduction

The Kentallen intrusion is a member of the Duror of Appin cluster of appinitic diorite intrusions (see the Ardsheal Hill and Peninsula GCR site report) and is the type locality of 'kentallenite', a melanocratic olivine monzonite with unusually high MgO (15%) and K_2O (2.5%). The rocks post-date deformation and metamorphism of the Dalradian country rocks but pre-date emplacement of the Ballachulish granite pluton (Bowes, 1962). The rock 'kentallenite' was described by Teall (1888, 1897) and its occurrence was described by Hill and Kynaston (1900), Bailey and Maufe (1916) and Bailey (1960). More recently the field relationships have been described by Bowes (1962), Bowes and Wright (1967) and Platten (1966) and the petrology by Westoll (1968), Wright and Bowes (1979) and Hamidullah and Bowes (1987). The area lies within the aureole of the Ballachulish pluton which has been described by Pattison and Harte (1985). The site was revisited for this review and new data were obtained about the Kentallen intrusion margin and the sequence of igneous, hydrothermal and structural events. The name 'kentallenite' has been formally replaced by olivine monzonite and should now only be used in the strictly local context.

Kentallen

Description

The Kentallen intrusion is about 0.6 × 0.3 km (Figure 8.39), typical of many of the Appinite Suite plugs. Short sections of near-vertical contact can be seen at two places and the general topographical relationships suggest that the entire northern contact must be steep. The northern and southern contacts are grossly discordant to the strike of the host rocks while the eastern margin is broadly concordant. The body is thus a steeply plunging pipe. The bulk of the observed intrusion is emplaced in semipelite and quartzite of the Appin Phyllite and Limestone Formation but the eastern margin is in contact with a dolomitic member. The intrusion exhibits a relatively uniform interior but varied marginal rocks and structures.

The interior of the intrusion is composed of a uniform melanocratic olivine monzonite ('kentallenite') (Figure 8.39) which is well exposed in the railway and road cuttings. The rock is composed of olivine, diopside, plagioclase, orthoclase, anorthoclase, phlogopite, magnetite and apatite. The olivine and diopside are coarse grained, most being 2–8 mm, but rare elongate olivine crystals may reach 20 mm. The crystals are euhedral to subhedral and are densely packed, approaching a grain-supported fabric. Plagioclase occurs as small interstitial tablets. Phlogopite crystals are large, 10 to 30 mm, and poikilitically enclose the plagioclase and mafic minerals. Alkali feldspar either overgrows plagioclase or forms poikilitic crystals. Rare examples of rootless veins and patches of coarse, leucocratic feldspar–biotite rock, superficially similar to the interstitial minerals of the olivine monzonite, are seen in displaced blocks on the coast. Xenoliths and enclaves are generally absent.

Changes occur in the main olivine monzonite towards the contacts. The phlogopite shows a marked decrease in grain size (from >10 mm to <1 mm) and a change from poikilitic to interstitial texture. The olivine is much less abundant and the large (>10 mm) olivine crystals are absent. Colour index is reduced and olivine and pyroxene crystals are well separated from each other. Some enclaves of a porphyritic facies with very fine-grained matrix are present locally, close to the outer margin. This finer-grained margin abuts directly against country rock in the east but in the north and south is separated from the country rocks by sheets of diverse, earlier contact facies rocks.

The early contact facies is best exposed on the coast just south of Sron Garbh (Figure 8.39). The earliest rock is a very fine-grained porphyritic 'kentallenite' with pyroxene and olivine phenocrysts in a very fine-grained matrix, the 'large lamprophyre body' of Bowes (1962). Traced south this shows a rapidly gradational, near-vertical contact with pyroxenite. The pyroxenite is a clinopyroxene-phlogopite-plagioclase rock that is coarser grained (5–10 mm pyroxene) than most other rocks in the intrusion. Olivine appears abruptly to the south and the rock becomes a phlogopite-clinopyroxene-bearing peridotite. The phlogopite crystals in these ultramafic rocks form a subpoikilitic framework and show evidence of cataclasis and kinking. This is succeeded southwards by pyroxene-olivine meladiorite and then the slightly chilled margin of the main 'kentallenite'. The meladiorite shows fine, millimetre-scale, vertical layering at and near the contact.

Two marginal intrusions are emplaced between the country rock and the early contact facies rocks in the Sron Garbh section. The first is a medium-grained pyroxene-hornblende-biotite diorite that carries conspicuous xenoliths of the Appin Quartzite. This diorite locally carries small numbers of pyroxene phenocrysts. The second is a pyroxene-biotite diorite with large (5 mm) pyroxenes that carries a rounded xenolith of chilled porphyritic 'kentallenite'.

The country rocks at the northern margin (Figures 8.39, 8.40 and 8.41) show a conspicuous antiform with an axial trace parallel to the contact. The fold style is concentric with multiple, angular hinges, local decollement planes, changes in profile along the fold axial surface and minor box folds. Domains of breccia consist of plate-like fragments of siliceous metasedimentary rock in a white microgranite matrix. Major and minor fold hinges show bedding-normal, wedge-shaped, gaping tensile fractures filled by microgranite. These veins are rootless. The southern limb of the antiform is the most disturbed with an extensive breccia of pelite and quartzite clasts. These breccias contain 4–8 mm-diameter altered pyroxenes, olivines and hornblendes similar to those found in the adjacent marginal intrusion.

Four dyke sets cut the Kentallen intrusion (Figure 8.39). Microdiorite dykes, Set Dk1, trending NNE and ENE and white feldsparphyric microgranodiorite dykes, Set Dk2, are cut by granite veins and show fine-grained biotite

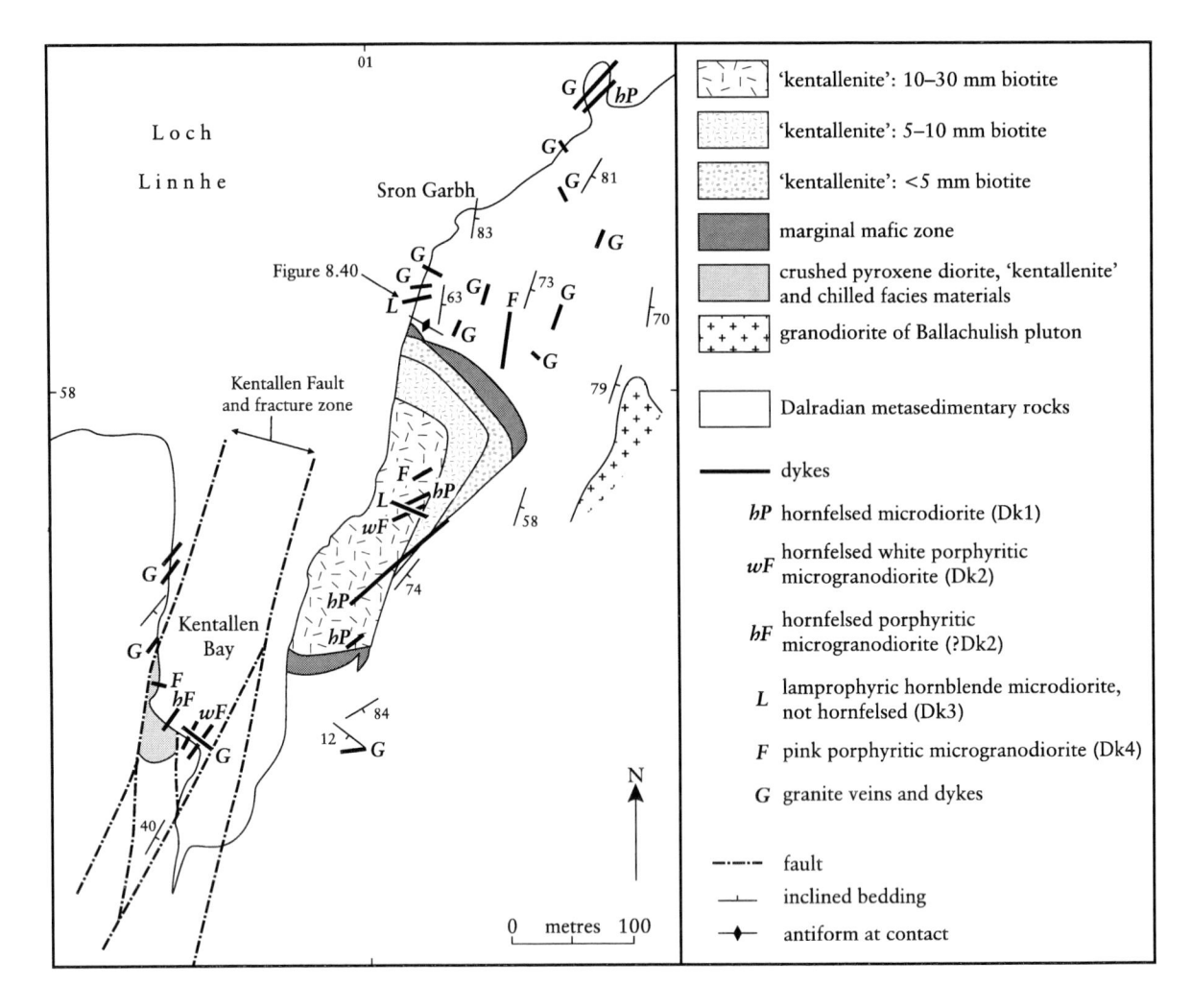

Figure 8.39 Map of the Kentallen appinitic intrusion.

and amphibole of hornfels origin in their matrix. An ENE-trending, hornblende microdiorite dyke, Set Dk3, with fresh, wholly igneous texture, cuts the hornfelsed microdiorites of Set Dk1 and one granite vein in the Sron Garbh section. A pink porphyritic microgranodiorite, Set Dk4, lacks secondary, hornfels-generated biotite and has a turbid altered appearance. It is cut by fracturing on the south side of Kentallen Bay that also affects the Kentallen intrusion. Relationships between Dk3 and Dk4 are not seen.

Four hydrothermal events have been recognized at Kentallen. (H1) Granular textured calcite veins occur as en echelon, concordant and discordant sheets in Dalradian rocks south of Sron Garbh. (H2) An early phase of hydrous alteration, with development of disseminated iron sulphides, is thought to have affected both the marginal rocks of the Kentallen intrusion and the early dykes of sets Dk1 and Dk2. These altered rocks have been extensively recrystallized by later contact metamorphism due to the Ballachulish pluton. (H3) The Kentallen intrusion contact in the Sron Garbh section is displaced by faults containing quartz veins. These lack an obvious drusy texture and contain <1% of pyrite and chalcopyrite. They are cut by granite veins and dykes of sets Dk3 and Dk4. (H4) The exposures on the south side of Kentallen Bay are cut by drusy-textured quartz-calcite-chalcopyrite veins associated with crushing and chlorite growth in the igneous rocks.

Two episodes of faulting are indicated at the site. NNE-trending faults lined with quartz (H3, above) cut the Kentallen intrusion and some Dk1 microdiorites in the Sron Garbh section. One is seen to be cut by a granite vein. These faults dip east at low to steep angles and have oblique-slip, normal and reversed movements. The second faulting event produced a sinistral offset of the margin of the Kentallen intrusion across Kentallen Bay (Figure 8.39) and some dextral and sinistral, oblique-slip, small faults dissecting the igneous rocks in the bay.

The Kentallen area lies within the inner aureole of the Ballachulish pluton (Pattison and Harte, 1985) and the northern part of the site shows good exposures of pelitic, semipelitic and calcareous hornfels. Calc-silicate rocks and granular calcite rocks are present, but poorly exposed, all along the eastern margin of the area. The pelitic rocks are coarse-grained cordierite hornfels with 2–10 mm poikiloblastic cordierite and local rootless veins of mobilized granitic material. Leucocratic biotite granodiorite dykes (trending NNE and ESE) with traces of yellow sulphides and lacking chilled margins cut the entire area. Contact metamorphic effects on fresh olivine monzonite are negligible, but hydrothermally altered igneous rocks develop contact metamorphic pyroxene, amphibole and biotite. The aureole obliterates most metamorphic effects associated with emplacement of the Kentallen intrusion.

Interpretation

The olivine monzonite ('kentallenite') is considered to be derived from magma represented by the porphyritic chill facies, which are similar to other chilled rocks in the Appinite Suite. Following the interpetation of similar rocks by Platten (1991), in-situ wall crystallization from a convecting magma formed the pyroxenite and peridotite cumulates of the marginal zones. Cataclasis of phlogopite in the pyroxenite and peridotite reflects the internal, gravitational deformation of these rocks while they formed weak layers with interstitial fluid that were only supported on one side. Temporary cessation of convection allowed crystallization of meladiorite at the walls. New, more mafic, magma was introduced into the pipe, displacing the existing magmas and triggering partial collapse and removal of the wall cumulates and initial chilled facies. This new magma quenched on the walls, trapping enclaves of penecontemporaneous fine-grained porphyritic rock and forming the relatively fine-grained outer facies of the main intrusion. Settlement and accumulation of olivine and pyroxene phenocrysts in the centre of the pipe led to the formation of the typical olivine monzonite ('kentallenite') with its densely packed olivine and pyroxene crystals. Compaction of this mass then expelled some interstitial residual liquid to form the leucocratic biotite monzonite segregations. The marginal intrusions at Sron Garbh may have been emplaced during collapse of the early contact facies rocks (Platten, 1983).

The time relationships of some events are re-interpreted here. The marginal fold at Sron Garbh is considered by Bowes and Wright (1967) to pre-date the Kentallen intrusion and to control its site of emplacement. The presence in the associated breccias of igneous material derived from the Kentallen intrusion and the rootless microgranite veins and patches in the fold hinges point to the fold forming in hot country rocks during emplacement of the Kentallen intrusion. The extensive fracturing south of Kentallen Bay was interpreted by Bowes and Wright (1967) as an important initial stage in the emplacement of the Kentallen intrusion. However, the recognition here that this fracturing affects the intrusion, and even later rocks, removes any link with emplacement. The fracturing is considered to be related to movements in the nearby Great Glen Fault-zone.

An extended sequence of late Caledonian magmatic, hydrothermal and structural events established at the site, can be summarized as follows.

1. Calcite veins (H1) are considered to pre-date the Kentallen intrusion; their texture results from contact metamorphism by the Ballachulish granite.
2. Kentallen intrusion emplaced in at least two stages and marginal fold formed.
3. Emplacement of Set Dk1 microdiorite and Set Dk2 leucocratic porphyritic microgranodiorite dykes.
4. Faulting and quartz vein emplacement (H3), post-dating microdiorite. Early Great Glen Fault movement.
5. Ballachulish granite emplacement and associated contact metamorphism of the products of events 1, 2 and 3 above.
6. Emplacement of Set Dk4 pink porphyritic microgranodiorite dykes.
7. Faulting, extensive shattering, chloritization and quartz-chalcopyrite-calcite vein emplacement (H4). The low temperature assemblages are considered to indicate that the frac-

Figure 8.40 Disharmonic minor folds at the margin of the Kentallen intrusion in the Sron Garbh section; view looking east. A 0.15 m-thick layer of quartzite on the left shows concentric folding (bottom) and fracture and pull-apart (top). A patch of breccia composed of 10 to 30 mm-long plates in a matrix of fine-grained microgranite occupies the space between differently folded adjacent layers (centre). A set Dk1 dyke cuts the structures at the top of the photograph.

tures post-date hornfelsing due to the Ballachulish granite. Late Great Glen Fault movement.

Certain events are only partly dated relative to this sequence. The extensive recrystallization of the Kentallen intrusion marginal zones and of the early Dk1 and Dk2 (event 3) dykes indicates that they had been reduced to a hydrous mineralogy that was unstable under prograde contact metamorphism during emplacement of the Ballachulish granite. This implied alteration (collectively termed H2) is most closely related in time to the emplacement of the Kentallen intrusion but is likely to post-date the main intrusive phases. It clearly pre-dates event 5 but time relationships with event 4 are unknown. Hornblende microdiorite dykes of Set Dk3 post-date event 5 but relationships with events 6 and 7 are unknown.

Conclusions

The Kentallen GCR site exposes a small intrusion of olivine monzonite, which is part of a large cluster of intrusions and breccia pipes in the type area of the Appinite Suite (see the Ardsheal Hill and Peninsula GCR site report). The olivine monzonite, which is unusually MgO and K_2O rich, is a distinctive lithology that was formerly termed 'kentallenite' after this, the type locality. The contact facies are particularly well exposed and the location, within the aureole of the Ballachulish pluton and close to the Great Glen Fault-zone, together with the wide range of igneous, hydrothermal and structural events that can be recognized, enable a complex sequence of time relationships to be determined.

The intrusion was emplaced in at least two stages, after the main deformation and metamorphism of the Dalradian country rocks,

Kentallen

Figure 8.41 Microdiorite dyke of set Dk1 cutting minor folds at the northern margin of the Kentallen intrusion; view looking east. The dyke shows axial concentration of mafic phenocrysts. The folded layer shows radial granitic veins forming a fan around the hinge (top). Traced downwards, this layer forms a box fold with few veins, but its core (centre) is mobilized and the bedding has been destroyed. The mobilized area forms a detachment surface isolating another fold closure (beside the scale), which shows an irregular mass of granite veins in the hinge and abrupt truncation of bedding on the upper limb. (Photos: I.M. Platten.)

but prior to two sets of dykes, hydrothermal alteration, early movement on the Great Glen Fault and subsequent emplacement of the Ballachulish pluton. This was followed by the intrusion of two further sets of dykes, extensive shattering, hydrothermal alteration and veining and later movements on the Great Glen Fault.

Chapter 9

Late Silurian and Devonian volcanic rocks of Scotland

Introduction

INTRODUCTION

D. Stephenson

The most voluminous phase of magmatism within the Caledonian Igneous Province of Scotland took place during the later stages of orogenesis, following the collision of Eastern Avalonia with Laurentia along the line of the Iapetus Suture. The 'Newer Granites' (described in Chapter 8), were emplaced during or following the ensuing late tectonic uplift, which ultimately gave rise to the Caledonian mountain chain. Overlapping in age with pluton emplacement was a widespread development of essentially calc-alkaline volcanic rocks and swarms of minor intrusions. On mainland Scotland the products of this volcanism are restricted to the upper Silurian and Lower Devonian, but in Orkney and Shetland they occur in the Middle Devonian, and show geochemical transitions away from calc-alkaline towards both tholeiitic (Shetland) and alkaline (Orkney) compositions. Both the plutonic and the volcanic rocks contributed greatly to the construction of the Caledonian mountain chain; many of the more dramatic mountains of the Scottish Highlands are carved out of the plutons, while the volcanic rocks also form many of the more prominent hill ranges of lowland Scotland.

The locations of all significant outcrops of late Caledonian volcanic rocks in Scotland are shown on Figure 9.1. The volcanic province extends south-westwards into Antrim, Tyrone and Roscommon in Ireland and similar volcanic rocks may have been exposed formerly to the east of the Grampian Highlands, contributing material to volcaniclastic sediments in the lower part of the Lower Old Red Sandstone succession of the northern Midland Valley. They are absent from the area NW of the Great Glen Fault, apart from in Orkney and Shetland, possibly because of significant lateral and/or vertical movements on this fault. All are associated with Old Red Sandstone sedimentary sequences of continental molasse facies. Most are now preserved in fault-bound basins and their original extent may have been far more extensive prior to erosion from the intervening uplifted areas. In some cases the volcanicity was intimately associated with high-level plutons. Volcanic rocks occur in small down-faulted blocks as a result of 'cauldron subsidence' at Glen Coe and within the Ben Nevis pluton. Elsewhere plutons have risen through volcanic sequences, as at Glen Etive and Cruachan where the Etive pluton cuts earlier lavas, and also far to the south in the Cheviot Hills, where an extensive lava field is intruded by a granite pluton.

The stratigraphical age of the volcanic rocks, and consequently of the continental Old Red Sandstone successions, was originally poorly defined, being based on a few fragmental fish, arthropod and plant remains (House *et al.*, 1977; Rolfe, 1980), although many more precise determinations are now available as a result of spore data (e.g. Richardson *et al.*, 1984; Marshall, 1991; Wellman, 1994). Published radiometric ages are of variable quality in these rocks, which are in general notorious for their secondary alteration, and many conflicting and confusing interpretations have been presented. All methods of radiometric age determination require critical interpretation of their analytical precision and petrological significance (i.e. as to which crystallization or alteration event is actually being dated). Even where radiometric dates appear to be confirmed by more than one independent method, good geological and palaeontological evidence should not be ignored, and in this volume this always takes precedence, where it is available (e.g. see Figures 1.2 and 9.2). However, radiometric age determinations by several methods have been reviewed objectively by Thirlwall (1988), who has shown that the radiometric ages of the volcanic rocks and closely related intrusions vary systematically across the strike of the Scottish Caledonides (Figure 9.2). In the Grampian Highlands the volcanicity ranges from 424 to 415 Ma (early to mid Ludlow), although related plutonism extends to a minimum of 408 Ma. In the Midland Valley the range of most volcanic rocks and intrusions is 415–410 Ma (late Ludlow to early Přídolí), although Thirlwall does not rule out the possibility of unexposed older igneous rocks in this area, coeval with those of the Highlands. In the north-western Southern Uplands, intrusions range from 413 to 407 Ma (late Ludlow to early Lochkovian), but intrusions and volcanic rocks in the south-eastern Southern Uplands and the Cheviot Hills, closest to the Iapetus Suture, are notably younger at 398–391 Ma (late Lochkovian to late Pragian), contemporaneous with the youngest granites in the Lake District.

These radiometric dates are very important in reconstructing the sequence of tectonic, magmatic and stratigraphical events in the late stages of the Caledonian Orogeny and, since the vol-

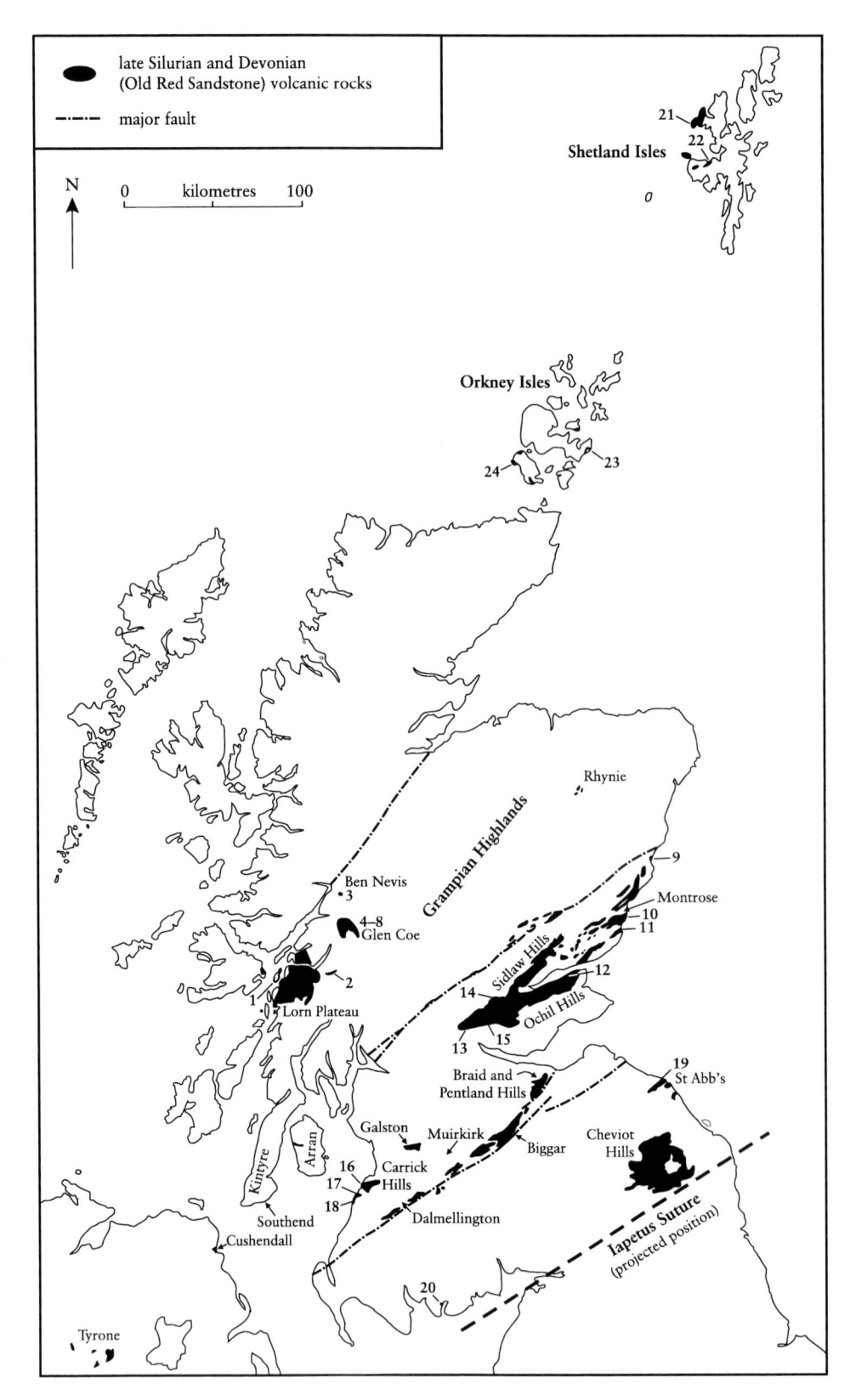

Figure 9.1 Location of late Silurian and Early- to Mid-Devonian age volcanic rocks of northern Britain. GCR sites: 1, South Kerrera; 2, Cruachan Reservoir (Chapter 8); 3, Ben Nevis and Allt a'Mhuilinn; 4, Bidean nam Bian; 5, Stob Dearg and Cam Ghleann; 6, Buachaille Etive Beag; 7, Stob Mhic Mhartuin; 8, Loch Achtriochtan; 9, Crawton Bay; 10, Scurdie Ness to Usan Harbour; 11, Black Rock to East Comb; 12, Balmerino to Wormit; 13, Sherriffmuir Road to Menstrie Burn; 14, Craig Rossie; 15, Tillicoultry; 16, Port Schuchan to Dunure Castle; 17, Culzean Harbour; 18, Turnberry Lighthouse to Port Murray; 19, Pettico Wick to St Abb's Harbour; 20, Shoulder O'Craig; 21, Eshaness Coast; 22, Ness of Clousta to The Brigs; 23, Point of Ayre; 24, Too of the Head.

Introduction

canic activity spans the Silurian–Devonian boundary, they have assumed great international significance in the construction of the Geological Time Scale. The most commonly accepted date for the base of the Devonian at the time of writing is around 409 Ma (e.g. Harland *et al.*, 1990, as used in this volume), which implies that most of the volcanicity, and consequently much of the Lower Old Red Sandstone sedimentary succession is of late Silurian rather than Early Devonian age. Thirlwall himself suggested 412 Ma on the basis of his dates on Scottish lavas, and a more recent rationalization of the British dates and spore evidence (in particular from Glen Coe and the northern Midland Valley) has resulted in a considerable revision of the estimated date of the Silurian–Devonian boundary to 417 Ma (Tucker and McKerrow, 1995). If this boundary date becomes accepted, as has been proposed by Gradstein and Ogg (1996), most of the volcanicity will be regarded as Early Devonian (Figure 9.2). The only exception will be that of the Grampian Highlands (Glencoe and Lorn), which has given slightly earlier dates of 424–421 Ma, in contradiction with the presence of Lower Devonian fish and arthropods, but within the range of recent spore estimates of latest Silurian to Early Devonian (Marshall, 1991; Wellman, 1994).

The late Silurian and Devonian age volcanic rocks of Scotland and the north of Ireland have been reviewed by Stillman and Francis (1979), Elliot (1982) and Fitton *et al.* (1982) and there have been several detailed studies of the petrology and geochemistry of lava sequences in individual areas (Taylor, 1972; Gandy, 1972, 1975; Groome and Hall, 1974; Slater, 1977; French *et al.*, 1979). However, it was the comparative study of lavas from across the whole province by Thirlwall (1979, 1981a, 1982) which led to a significant advance in the understanding of the origin and evolution of the magmas and a controversial unifying theory of their tectonic setting, related to the subduction of Iapetus oceanic lithosphere beneath the Laurentian continental margin.

The volcanic rocks comprise a high-K calc-alkaline suite, with particularly high K_2O (relative to silica) in such areas as the Sidlaw Hills, the western Ochil Hills and Lorn, suggesting affinities with the 'shoshonitic' suites of active continental margins. Compositions range from basalt to rhyolite, although in most of the more widespread sequences basaltic andesites and andesites predominate. The more acid magmas were erupted from local centres, commonly as pyroclastic material, and ignimbrites are common in several centres.

Most of the lavas are quartz-normative with SiO_2 ranging from 52 to 63% and alumina is generally high. High levels of Mg, Ni, Cr and V in the more basic rocks are taken by most authors to indicate that these are derived from primitive magmas that originated by partial melting of upper mantle material. However, the generally high levels of Ba, Sr, Rb, K, light rare earth elements (LREE) and other incompatible elements are difficult to derive by fractionation of mantle material, which generally has low levels of such elements. It is therefore necessary to invoke an additional source, such as partial melting of the lower continental crust (Groome and Hall, 1974) or subducted, hydrous, oceanic lithosphere (Taylor, 1972). Thirlwall (1982, 1983b, 1986) discussed the origin of the magmas in more detail, based upon a regional study of the trace element geochemistry, together with Sr, Nd and Pb isotopes. He invoked the mixing of melts from two primary sources: primitive mantle, giving the high Ni and Cr levels; and subducted Lower Palaeozoic greywackes, giving the high content of incompatible lithophile elements. The same study concluded that contamination by continental basement material is not likely and that most of the more evolved compositions were probably derived by closed-system fractional crystallization at high crustal levels. Most other authors have also invoked varying degrees of fractional crystallization to produce the more evolved compositions. For example high-pressure melting experiments on lavas from the Sidlaw Hills by Gandy (1975) suggested a multi-stage model involving fractionation during the ascent of a picritic magma to give a range of high-alumina basalts, followed by further low-pressure fractionation of olivine, plagioclase and ilmenite.

In his original study, Thirlwall (1979, 1981a) concentrated on the more primitive basalts and andesites with high Mg, Ni and Cr. He found that these rocks show pronounced spatial chemical variations, with Sr, Ba, K, P, LREE and the ratio La/Y in particular showing up to a six-fold increase north-westwards from the Southern Upland Fault across the Midland Valley and Grampian Highlands, perpendicular to the main Caledonian structural trend (Figure 9.3). The more evolved rocks show similar increases, but

Late Silurian and Devonian volcanic rocks of Scotland

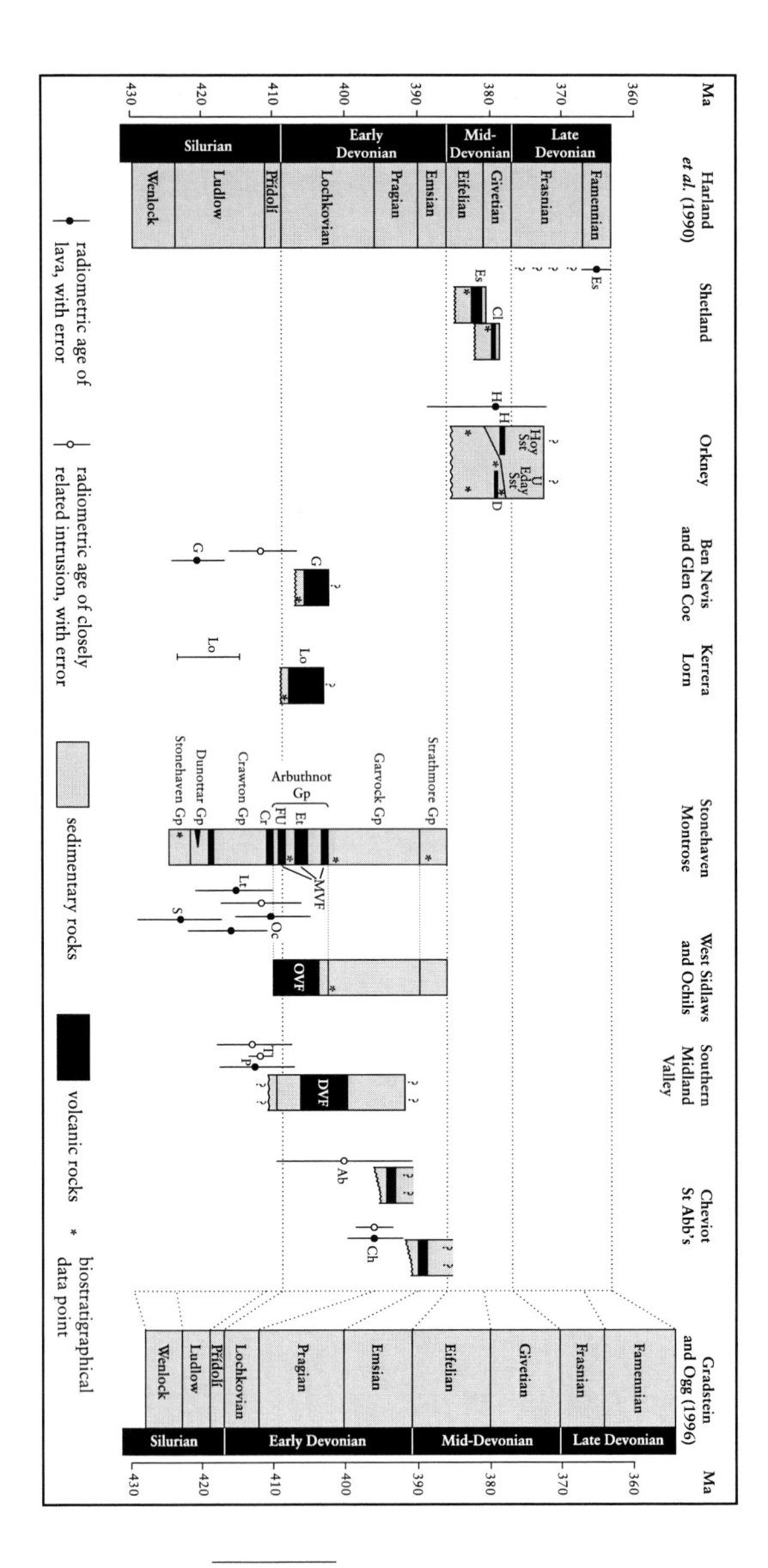

Figure 9.2 Stratigraphical relationships and ages of late-Silurian to Mid-Devonian age volcanic rocks of northern Britain. Biostratigraphical ages (where known) are given precedence and are plotted relative to the time-scale of Harland *et al.* (1990) (on the left). Note the consistent discrepancies between the biostratigraphical ages and the radiometric ages, which are not present if the time-scale of Gradstein and Ogg (1996) (on the right) is used. Where there is no biostratigraphical control (i.e. Southern Midland Valley, St Abb's and Cheviot), the volcanic sequences are projected from the radiometric dates. For example Cheviot at 396 Ma is early Emsian on the Gradstein and Ogg timescale, so it is plotted in the early Emsian position on the Harland *et al.* time scale.

Ab, St Abb's; Ch, Cheviot; Cl, Clousta; Cr, Crawton; D, Deerness; DVF, Duneaton Volcanic Formation; Es, Eshaness; Et, Ethie; FU, Ferryden and Usan; G, Glen Coe; H, Hoy; Lo, Lorn; Lt, Lintrathen; MVF, Montrose Volcanic Formation; Oc, Ochil Hills; OVF, Ochil Volcanic Formation; P, Pentland Hills; S, Sidlaw Hills; T, Tinto.

Introduction

their regional pattern is complicated by the effects of fractional crystallization. Broadly similar spatial chemical changes are shown by the late Caledonian plutonic suites (Stephens and Halliday, 1984; see Chapter 8). These spatial variations, coupled with the overall geochemical and petrographical features of the volcanic rocks, are typical of calc-alkaline suites associated with subduction at continental margins. Hence Thirlwall postulated the former presence of a WNW-dipping subduction zone beneath present-day Scotland that originated on the SE margin of the Laurentian continent during closure of the Iapetus Ocean. Compositional stratification in the mantle source, with depleted overlying enriched mantle, is considered to be the cause of the observed spatial variation, which reflects the depth of melting of the primitive source above the subduction zone (Thirlwall, 1982).

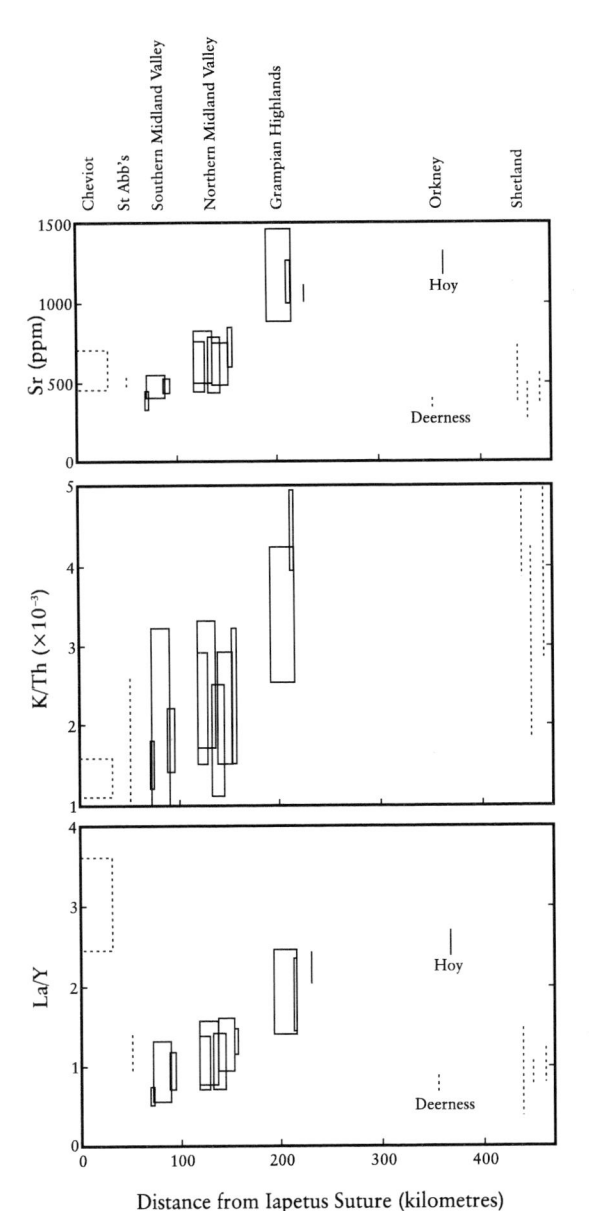

Although the subduction model is attractive in that it explains and unites many of the observed features of the late Caledonian igneous activity, it is beset by serious problems related to the overall timing of tectonic events and the distribution of some of the magmatism. For instance, it is now generally accepted from a wide range of evidence that the Iapetus Ocean had closed by the end of Wenlock time, and hence that active subduction had all but ceased, before the onset of volcanicity in the early Ludlow, and prior to the emplacement of most late Caledonian granitic plutons. A further problem is that the volcanic and plutonic rocks of the south-eastern Midland Valley and Southern Uplands are very close to the projected line of the Iapetus Suture relative to rocks of comparable composition in modern subduction settings.

The relationships of the later, Mid-Devonian volcanism in Orkney and Shetland to the main part of the volcanic province have been the subject of some debate. Thirlwall (1981a) and Fitton *et al.* (1982) argued that the trace elements and general transitional calc-alkaline to tholeiitic characteristics of the Shetland and some Orkney lavas suggest a relatively close proximity to the surface trace of the proposed subduction zone. Hence they proposed a curved suture, convex to the SE, which follows the regional Caledonian strike and passes close

Figure 9.3 Ranges of whole-rock Sr, K/Th and La/Y from late Silurian and Devonian volcanic rocks of northern Britain, plotted against projected distance from the Iapetus Suture of Phillips *et al.* (1976). Adapted from Fitton *et al.* (1982, fig. 9). The heights of rectangles and bars represent the range of values within geographically related outcrops; the width of the rectangles represents each outcrop width.
Southern Midland Valley: Pentland Hills and Lanarkshire, Ayrshire Coast, Straiton;
Northern Midland Valley: Ochil Hills, north Fife hills, Sidlaw Hills, Montrose, Highland Border;
Grampian Highlands: Ben Nevis, Glen Coe, Lorn Plateau;
Orkney: Deerness, Hoy;
Shetland: Eshaness, Papa Stour, Clousta;
Note the anomalous values (dotted lines) from the Cheviot and St Abbs outcrops, close to the proposed line of suture, and from the Deerness outcrops of Orkney and all of the Shetland outcrops (see text for discussion).

to the east of Shetland. However, Astin (1983) advocated caution in interpreting the Orkney and Shetland data in terms of subduction, and it has subsequently become accepted that the Middle Devonian sediments and volcanic rocks of this area developed in an extensional basin (McClay *et al.*, 1986; Enfield and Coward, 1987; Astin, 1990). The Shetland lavas and the Deerness lavas of Orkney do have geochemical characteristics that may be a relict of earlier subduction, but 'it cannot be stated with certainty that they are closely related to the rest of the province' (Thirlwall, 1981a); furthermore, the alkaline nature of the Hoy lavas could be seen as marking the start of the Carboniferous intraplate alkaline volcanism (Francis, 1988).

Since the subduction model depends upon the evidence of several suites of Caledonian igneous rocks, it has been discussed, with alternative interpretations, in Chapter 1 in the context of the tectonomagmatic setting of late Caledonian magmatism in general.

Grampian Highlands

The most extensive development of volcanic rocks in the Grampian Highlands is the Lorn Plateau sequence, which extends over some 300 km^2 and is up to 800 m thick, comprising mainly flows of basalt and andesite, with rare acid lavas and ignimbrites. The whole Lorn sequence is 'shoshonitic' (enriched in potassium and other incompatible elements) and the flows on the island of Kerrera are the most shoshonitic in the whole province. The South Kerrera GCR site illustrates a range of features in these lavas, which rest upon fluvial sedimentary rocks containing macro- and microfossils of latest Silurian to earliest Devonian age. Lavas of the Lorn sequence are also seen in the Cruachan Reservoir GCR site (Chapter 8), where they are preserved in a down-faulted screen within the margin of the Etive pluton.

Volcanic rocks are also preserved in cylindrical down-faulted blocks in the classic settings of Ben Nevis and Glen Coe, which illustrate the relationships between surface volcanicity, subvolcanic complexes and underlying granitic plutons. It was in these areas that the concept of 'cauldron subsidence' was first developed by E. B. Bailey and others of the Geological Survey almost a century ago. The model has subsequently been applied worldwide but these original sites, which provide spectacular three-dimensional exposures of the roots of central volcanos, have continued to stimulate teaching and research to the present day. The Ben Nevis and Allt a'Mhuilinn GCR site represents the Ben Nevis pluton (cf. Chapter 8), within which is a subsided block of sedimentary rocks and, largely volcaniclastic, andesitic volcanic rocks, some 650 m thick. The Glencoe volcanic sequence comprises over 1500 m aggregate thickness of fluvial, graben-controlled sedimentary rocks, andesitic lavas and high-level sills, rhyolitic lavas and ignimbrites, and a variety of other volcaniclastic rocks, all enclosed by a later ring fracture and granitic ring intrusion. The complex is represented by five GCR sites. The Bidean nam Bian site gives the most complete section through the volcanic sequence and exposes relationships with the underlying land surface of Dalradian metasedimentary rocks. At Stob Dearg and Cam Ghleann, sedimentary rocks below the volcanic sequence are particularly well developed and have yielded Lower Devonian plant remains and spores. Buachaille Etive Beag exhibits three thick pyroclastic flow units of ignimbrite, separated by erosional surfaces and fluvial sedimentary sequences. Two branches of the ring fault and the ring intrusion are exposed at Stob Mhic Mhartuin and contrasting expressions of these features, together with good exposures of the Dalradian rocks below the volcanic sequence are seen at Loch Achtriochtan.

Elsewhere in the Grampian Highlands there is only sparse evidence of volcanic activity. A 20 m-thick flow of andesite lava in the Lower Old Red Sandstone outlier at Rhynie is closely associated with siliceous sinters that include the internationally renowned plant-bearing 'Rhynie Chert' and significant gold mineralization (Rice and Trewin, 1988; Trewin and Rice, 1992). Single exposures of vesicular andesite in the Cabrach outlier and in the Gollachy Burn, near Buckie have been interpreted as lavas, but could equally be intrusive.

Highland Border

At several places along the Highland Border, Lower Old Red Sandstone rocks of the Strathmore sequence overlap the Highland Boundary Fault to give a series of small faulted outliers resting on Dalradian rocks (Armstrong and Paterson, 1970). Several of these outliers contain basaltic and andesitic lavas and a dis-

Introduction

tinctive dacitic ignimbrite, the 'Lintrathen Porphyry', forms a stratigraphical marker at the top of the Crawton Group on both sides of the fault (Paterson and Harris, 1969). This ignimbrite has been dated at around 416 Ma (Thirlwall, 1988) and is potentially a significant marker close to the Silurian–Devonian boundary. In Kintyre, Lower Old Red Sandstone conglomerates contain boulders of 'lava', possibly derived from the north, thin acid tuffs occur locally and three vents at Southend are very close to the projected position of the Highland Boundary Fault (Friend and Macdonald, 1968). An andesite lava, overlain by conglomerates containing lava pebbles, also occurs in the Lower Old Red Sandstone just south of the fault on Arran.

Northern Midland Valley

In the northern Midland Valley volcanic rocks are present throughout most of the lower part of the Lower Old Red Sandstone Strathmore succession (Figure 9.2), between the Highland Boundary and Ochil faults (Armstrong and Paterson, 1970). Volcaniclastic conglomerates first appear in the middle of the Stonehaven Group, possibly in the late Wenlock to early Ludlow (Marshall, 1991), above which they become a major part of the succession. The earliest lava occurs in the middle of the Dunottar Group and several flows of basalt comprise the Tremuda Bay Volcanic Formation at the top of this group. Isolated flows of andesite occur in the lower parts of the Crawton Group and the top of the group is marked by several flows of basalt and basaltic andesite that comprise the Crawton Volcanic Formation. The latter is well exposed in the Crawton Bay GCR site, where most flows are of the distinctive 'Crawton type' with large flow-orientated plagioclase phenocrysts.

The volcanic activity reached a peak during deposition of the Arbuthnott Group, which is regarded as early Lochkovian on palynological grounds and which has several radiometric age determinations around 412 Ma (Thirlwall, 1988; Tucker and McKerrow, 1995). The group contains two diachronous formations dominated by volcanic rocks. In the NE, the Montrose Volcanic Formation comprises several separate volcanic developments, all of which are thought to have emanated from a centre now covered by the North Sea. The Scurdie Ness to Usan Harbour GCR site represents the lower part of the formation, which comprises two sequences, informally termed the 'Ferryden lavas' (predominantly basaltic andesites) and the 'Usan lavas' (predominantly basalts). The Black Rock to East Comb GCR site exposes the basaltic andesites of the 'Ethie lavas' in the upper part of the formation. The Ochil Volcanic Formation forms the Ochil Hills (Francis *et al.*, 1970) and the Sidlaw Hills (Harry, 1956, 1958) and has a maximum thickness of over 2400 m. Olivine basalts and pyroxene andesites predominate throughout the formation, but minor trachyandesites, dacites and rhyodacites also occur. Pyroclastic rocks are thickest, coarsest and most common close to the Ochil Fault suggesting that the main centre of eruption lay to the south of the fault and is now concealed beneath younger strata. Two GCR sites provide sections through typical sequences of basaltic to andesitic lavas and intercalated volcaniclastic sedimentary rocks; Balmerino to Wormit in the eastern Ochils is a coastal section through a 350 m-thick sequence that has also provided radiometric and palaeontological age dating evidence, and Sheriffmuir Road to Menstrie Burn in the western Ochils exposes a 600 m-thick sequence in the dramatic Ochil Fault scarp. Rare acidic lavas are represented by the rhyodacite of the Craig Rossie GCR site and a suite of diorite stocks that cut the volcanic sequence within a wide thermal aureole show a variety of interesting contact relationships in the Tillicoultry GCR site.

The youngest volcanic activity in the northern Midland Valley is recorded by a few local flows of basalt and andesite in the upper part of the Garvock Group, south of Laurencekirk. The succeeding Emsian, Strathmore Group contains no primary volcanic rocks.

Southern Midland Valley

On the SE side of the Midland Valley volcanic rocks occur in a series of generally poorly exposed Lower Old Red Sandstone outcrops, separated by major NE–SW faults, in a 10 km-wide zone parallel to the Southern Upland Fault. At the NE end of this zone, the Braid Hills and Pentland Hills expose up to 1800 m of lavas and pyroclastic rocks with compositions ranging from olivine basalt to rhyolite (Mykura, 1960), that have been dated at *c.* 412 Ma by Thirlwall (1988). Farther to the SW major outcrops occur between West Linton and Douglas (the 'Biggar

Centre' of Geikie, 1897), SE of Muirkirk (the Duneaton Volcanic Formation; Phillips, 1994; Smith, 1995), and in the Straiton–Dalmellington area (Eyles *et al.*, 1949). Sequences in the SW are almost entirely of olivine basalt and basaltic andesite, although acid rocks are present in several large sills and laccoliths, such as that of Tinto hill (Read, 1927) which has yielded precise isotope dates of *c*. 412 Ma by both Sm-Nd and ^{39}Ar-^{40}Ar methods (Thirlwall, 1988). Intercalated volcaniclastic sedimentary rocks are common in the south-western outcrops but pyroclastic rocks and recognizable vents are rare.

Other outcrops of volcanic rocks in the Lower Old Red Sandstone occur to the NW of this zone near Galston and in the Carrick Hills. In the latter area a sequence of 300–450 m, consisting almost entirely of olivine basalts and andesites with intercalated fluvial and lacustrine sedimentary rocks, is particularly well exposed on the coast. Three GCR sites at Port Schuchan to Dunure Castle, Culzean Harbour and Turnberry Lighthouse to Port Murray exhibit a variety of sedimentary inclusions, infillings and intercalations within the igneous rocks, which suggest that the magmas were emplaced as shallow subvolcanic sills within unconsolidated wet sediment (Kokelaar, 1982). Such features have subsequently been recognized in many other parts of the Old Red Sandstone province, and also within Ordovician sequences (Chapters 4 and 6) where many parts of the lava sequences have now been re-interpreted as high-level sill complexes.

Southern Uplands

South-east of the Southern Upland Fault an extensive lava field in the Cheviot Hills is the most south-easterly expression of the Siluro-Devonian volcanicity (Carruthers *et al.*, 1932; Thirlwall, 1979, 1981a). A poorly exposed sequence of over 500 m, consisting almost entirely of acid andesites and dacites with only local pyroclastic interbeds, is located over the projected line of the Iapetus Suture. Some 25 km to the north, the somewhat smaller outcrop around St Abb's and Eyemouth is magnificently exposed on the coast at the Pettico Wick to St Abb's Harbour GCR site. Here, a 600 m-thick sequence of basalt and basaltic andesite lavas, with interbedded volcaniclastic sedimentary rocks, exhibits a variety of flow features, preserved by rapid burial in a high-energy volcano-sedimentary environment. Volcanic activity farther to the SW in the Southern Uplands is suggested by the presence of a number of subvolcanic vents in Kirkcudbrightshire that are represented by the Shoulder O'Craig GCR site. This coastal locality exposes an intrusion breccia, a basalt intrusion and a series of slightly younger lamprophyre dykes that are part of an important regional swarm (Rock *et al.*, 1986b; see Chapter 8, Introduction).

Shetland and Orkney

In Shetland and Orkney volcanic activity occurred during late Mid-Devonian time, significantly later than in mainland Scotland, and it is not certain whether it originated in the same tectonomagmatic setting. The most extensive outcrops are in west Shetland at Melby, Papa Stour and Eshaness, all of which are thought to be of similar late Eifelian age, as are thin tuffs in the Upper Stromness Flags of Hoy, Orkney. The Eshaness GCR site illustrates a sequence of lavas, shallow sills and pyroclastic rocks of olivine basalt to andesitic composition, and a rhyolitic ignimbrite, all exposed by spectacular coastal erosion. Volcanic rocks of Givetian age occur on the Walls Peninsula of Shetland (the Clousta volcanic rocks), and on Orkney at Hoy (the Hoy Volcanic Member) and at Deerness, Shapinsay and Copinsay (the Deerness Volcanic Member). The Clousta volcanic rocks are represented by the Ness of Clousta to the Brigs GCR site, which is notable for its well-bedded pyroclastic rocks, intercalated with alluvial deposits and well preserved beneath a penecontemporaneous basaltic shallow sill. The pyroclastic rocks have been interpreted as the products of phreatic and phreatomagmatic eruptions that emanated from maars and tuff-rings. The Hoy Volcanic Member comprises basaltic breccias, tuffs and lavas of nepheline-normative alkali olivine basalt and hawaiite; they are represented by the Too of the Head GCR site. Although some of the lavas of the Deerness Volcanic Member also show alkaline characteristics, this is regarded as a secondary effect. The member is represented by the Point of Ayre GCR site, where the margin of a basalt lava flow is exposed, together with an underlying air-fall tuff and lacustrine sediments that show synsedimentary deformation associated with the volcanicity.

SOUTH KERRERA
(NS 794 279–803 256)

G. Durant

Introduction

Volcanic and volcaniclastic rocks form part of a Lower Old Red Sandstone sequence that unconformably overlies Dalradian metasedimentary rocks on the island of Kerrera. The Kerrera volcanic rocks represent part of the extensive Lorn plateau lavas, which crop out over an area of 300 km² between Glencoe and the Oban district. The Lorn lava sequence is up to 800 m thick and comprises flows of basalt, pyroxene-, hornblende-, and biotite andesite with rare acid lavas and ignimbrites (Groome and Hall, 1974). Individual flows are 5 to 30 m thick but they cannot be traced for any distance inland because of poor exposure; only on the north side of Loch Etive can individual flows be traced for several kilometres. The Kerrera lavas are geochemically distinct in being the most shoshonitic (i.e. enriched in potassium and other incompatible elements), in the Old Red Sandstone volcanic suite (Groome and Hall, 1974; Thirlwall, 1979, 1981a). The volcanic rocks typically rest directly on Dalradian rocks although locally, as on Kerrera, there are sedimentary rocks at the base of the volcanic sequence.

The Lower Old Red Sandstone volcanic rocks crop out in two principal areas on Kerrera (Figure 9.4). The main outcrop is in the SW of the island, where it is well exposed around the coast immediately south of Port Phadruig (794 280) and just west of Port Dubh (793 263). Part of the sequence is also preserved as a narrow downfaulted outcrop which extends the length of the island, but which is particularly well exposed along the coast to the north of Rubha Seanach (8062 2620–8030 2563).

The sedimentary rocks underlying the lavas on Kerrera have yielded Early Devonian fish and arthropods (Kynaston and Hill, 1908; Lee and Bailey, 1925; Morton, 1979; Rolfe, 1980), but more precise spore data indicate a latest Silurian to earliest Devonian age (Marshall, 1991). Therefore the Lorn volcanic rocks have international significance as a time marker for the Silurian–Devonian boundary, with radiometric ages currently estimated in the range 424 to 415 Ma (Thirlwall, 1988).

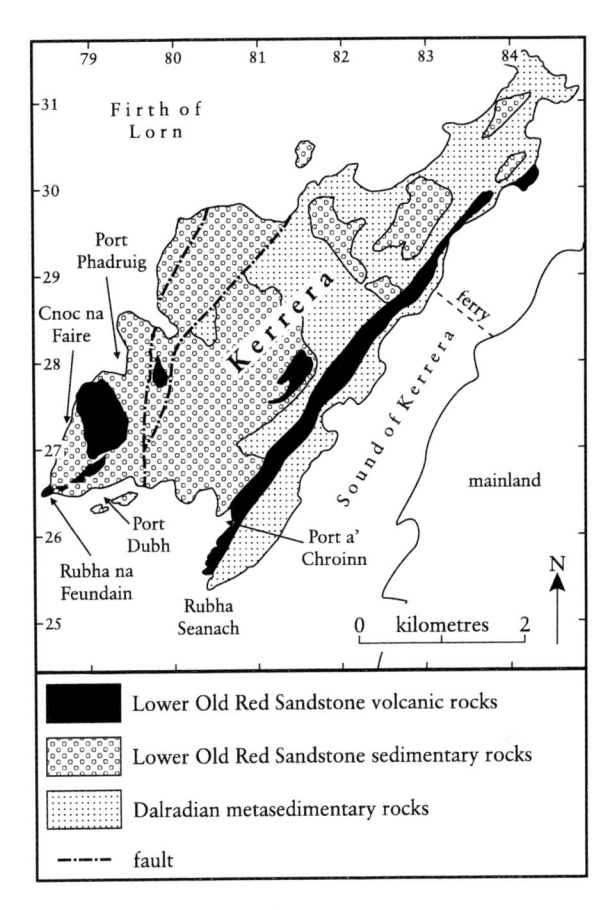

Figure 9.4 Map of Kerrera.

Description

The most complete section is that exposed along the western coast, where the volcanic rocks overlie a sequence of green conglomerates, green coarse-grained sandstones and well-laminated red fine-grained sandstones.

The base of the lava sequence is well exposed on the south side of Port Phadruig, where a 10 m-long lens of deep red jasper forms a conspicuous marker; large pillowed lobes of vesicular lava protrude into the jasper and selvages of jasper can be traced around one lobe. These relationships suggest that the jasper is silicified sediment, which was unconsolidated at the time of eruption of the basalt. A short way to the SW (7914 2788) the base is marked by a breccia characterized by blocks of vesicular basalt in a pale-pink matrix of fine-grained sandstone. The basalt fragments vary from being subrounded to angular and are up to 1 m across. Some show well-developed flow alignment of the vesicles which are now filled by calcite and quartz. The

Late Silurian and Devonian volcanic rocks of Scotland

Figure 9.5 Columnar jointing in andesite flow, south of Port Phadruig, Kerrera (NS 7888 2767). (Photo: BGS no. C 2647.)

blocks are monolithological, and some have been broken *in situ*, with little movement following disaggregation of the magma. Lobes of the succeeding lava protrude into this breccia.

The base of the lava sequence is also exposed in a raised sea stack beneath the main cliff NW of Cnoc na Faire (7884 2743). Here, the northerly dipping top surface of the underlying conglomerates and sandstones forms the wave-cut platform at the south end of the bay. In the sea stack, the lobate base of the lowest lava cuts down through a pale-green coarse-grained sandstone and a conglomerate into well-laminated red sandstone. Well-rounded pebbles from what was clearly an unconsolidated gravel have been caught up and are now included within the lava. The green sandstone includes fragments of vesicular basalt, the largest being 50 cm, and rip-up clasts of laminated red sandstone. The conglomerate contains well-rounded pebbles of volcanic rock and quartzite.

A breccia with angular to subrounded fragments of vesicular basalt in a limestone matrix forms a striking rock, particularly in the wave-polished boulders beneath the main cliff and on the beach. The basalt fragments, generally up to 10 cm but larger in some of the beach boulders, are supported within, and fairly evenly distributed through, the calcareous matrix. A second type of breccia has highly angular fragments of basalt in a white calcite matrix. The basalt fragments show little displacement during fragmentation and can easily be matched up to adjacent fragments. This type of breccia crops out on the wave-cut rock platform and is visible beneath and between the huge boulders. Basalt boulders on the beach contain amygdales and irregular cavities filled by quartz, calcite and agate.

In the main cliff itself (7892 2745), both columnar-jointed and more massive flows are present. Well-developed 20–30 cm-wide columns can be examined close to sea level in a down-faulted block of pale-grey compact basalt, which forms a marked 'whale-backed' outcrop, and also a short distance to the south (Figure 9.5). This flow is overlain by a massive extremely vesicular basalt, which is patchily autobrecciated and cut through by irregular curving joints, well seen close to a conspicuous gully formed by erosion of a Palaeogene dyke (7890 2772). Elsewhere, hexagonal columns up to 1 m across occur with salvages of sediment preserved locally within the joints.

In the extreme SW of the island the base of

the lava sequence is exposed in the low cliffs to the west of Port Dubh (7875 2668). Here the pillowed base of a vesicular basalt flow overlies sandstone and a coarse conglomerate. Enclaves of laminated sandstone occur between pillowed masses of basalt, which cut down through the bedding laminations of a pinkish-red sandstone to rest directly on top of the conglomerate. Individual pillowed lobes are flattened and are up to 2 m across. A short distance from Rubha na Feundain, just to the east of a conspicuous gully formed by the erosion of a Palaeogene dyke (7866 2671), the surface of a thin basalt flow shows distinctive pahoehoe texture, which indicates that this subaerial lava was locally flowing in what is now a south-westerly direction.

The hill forming Rubha na Feundain is made up of columnar-jointed orange-weathering biotite andesite. A steeply dipping contact can be traced around the base of the outcrop, defining an oval vent-like feature. The age of this plug is uncertain. It may be coeval with the Lower Old Red Sandstone lavas but it could be younger. A biotite-augite andesite cropping out on the nearby Bach Island (7780 2690) has been assigned to the Lower Old Red Sandstone volcanic sequence (Lee and Bailey, 1925).

In the down-faulted south-eastern inlier, the lavas vary from compact to rubbly and are much fractured due to the proximity of the faults. The tops and bases of individual flows are commonly brecciated, as is well seen in wave-polished exposures in the small bays at the NE and E sides of Port a'Chroinn.

The bay immediately north of Rubha Seanach has formed along the line of the NE–SW fault that separates Dalradian pelites and limestones from the down-faulted lavas. The lowest exposed part of the lava sequence, a breccia of volcanic rock fragments in a red sandstone matrix, is seen on the north side of the bay. This breccia is overlain by a sliver of red laminated sandstone and a lens of purplish coarse-grained sandstone which dip steeply to the west. The purplish sandstone is cut out by the lobate base of the overlying blocky lava which has a reddened top. The succeeding flow, a pale-grey weathering compact basalt, permeates the rubbly top of the underlying flow, forming irregular enclaves (8027 2566). A much reddened rubbly lava forms the small headland north of Rubha Seanach (8028 2577) and a short distance NE of this, a breccia composed of volcanic rock fragments in a green coarse-grained sandstone matrix crops out adjacent to a conspicuous Palaeogene dyke (8035 2589).

Interpretation

The basalt lavas of Kerrera are of a more uniform composition than those of similar age elsewhere in the Lorn Plateau where a greater thickness is preserved. The lavas commonly have slaggy tops and bases that are much brecciated due to eruption onto or into wet sediment. In considering the general inter-relationship of lava and sediment within the Lower Old Red Sandstone province, Geikie (1897) noted that 'a more striking proof of the subaqueous character of the eruptions could hardly be conceived'. Lee and Bailey (1925) noted that 'there is good reason to believe that the lavas from time to time, must have encountered expanses of water in their path' and that there must have been 'frequent recurrence of aqueous conditions during the accumulation of the volcanic rocks'. The re-interpretation of Lower Old Red Sandstone lavas as sills elsewhere in the province (Kokelaar, 1982; and see the Port Schuchan to Dunure, Culzean Harbour and Turnberry Lighthouse to Port Murray GCR site reports) has important bearing on the nature of the Kerrera lavas and some at least could be sills intruded in to wet sediment. However, the pahoehoe flow near Rubha na Feundain and the reddened tops of flows near Rubha Seanach argue for a subaerial origin for others. There can be no doubt that the lavas or sills encountered wet sediment in their path producing the breccias and that some of the sediment has been removed in the process. Rare veins of sandstone within columnar-jointed lavas could have originated in the manner suggested by Kokelaar for similar features in the Ayrshire coast Lower Old Red Sandstone sequence. The erosion of sediment at the base of lavas is demonstrated at a number of localities, particularly the raised sea stack below the cliffs NNW of Cnoc na Faire. However it is doubtful whether a flow could remove 20 m of sediment and hence the suggestion of Lee and Bailey (1925) that the lavas were erupted onto an eroded surface of sandstone and conglomerate with at least 20 m of relief requires further examination. It is clear that an adjacent volcanic area was being eroded prior to eruption of the main Kerrera lava sequence to provide the source of andesite, olivine basalt and highly vesicular basalt for the conglomerates that underly the main volcanic sequence on the island.

Conclusions

The Kerrera GCR site is of regional and national importance as a representative of the extensive Lorn Plateau volcanic sequence. These rocks are some of the most enriched in potassium and other 'incompatible' elements in the whole late Caledonian calc-alkaline volcanic suite and hence are believed to have originated in the deepest levels of a postulated subduction zone beneath the Laurentian continental margin (Thirlwall, 1981a).

Coarse conglomerates rich in volcanic debris are a significant component of the lower part of the sequence demonstrating that an adjacent volcanic landscape was being eroded before the eruptions of basalt that form the upper part of the sequence. Some of these lavas are characterized by columnar jointing and autobrecciation and some have pillowed bases demonstrating that they were erupted onto or into wet sediment. Other flows have reddened surfaces and one shows a pahoehoe-textured surface indicating subaerial eruption.

When taken in conjunction with the other geological features on Kerrera, particularly the spectacular unconformity at the base of the Lower Old Red Sandstone sequence, a visit to the south and west of Kerrera must rank as one of the finest excursions in Scottish geology.

BEN NEVIS AND ALLT A'MHUILINN (NN 140 757–167 713)

D. W. McGarvie

Introduction

The international importance of this GCR site is that it is a well-exposed Caledonian post-tectonic granitic intrusion, around 425 Ma old, within which is preserved a sequence of volcanic rocks. The presence of the volcanic rocks warrants inclusion of the site in this chapter, but the plutonic rocks are important representatives of the Argyll and Northern Highlands Suite described in Chapter 8.

Ben Nevis is Britain's highest mountain; its summit and spectacular north-facing cliffs (Figure 9.7) are part of a down-faulted block (*c.* 2.5 km in diameter) of volcanic and sedimen-

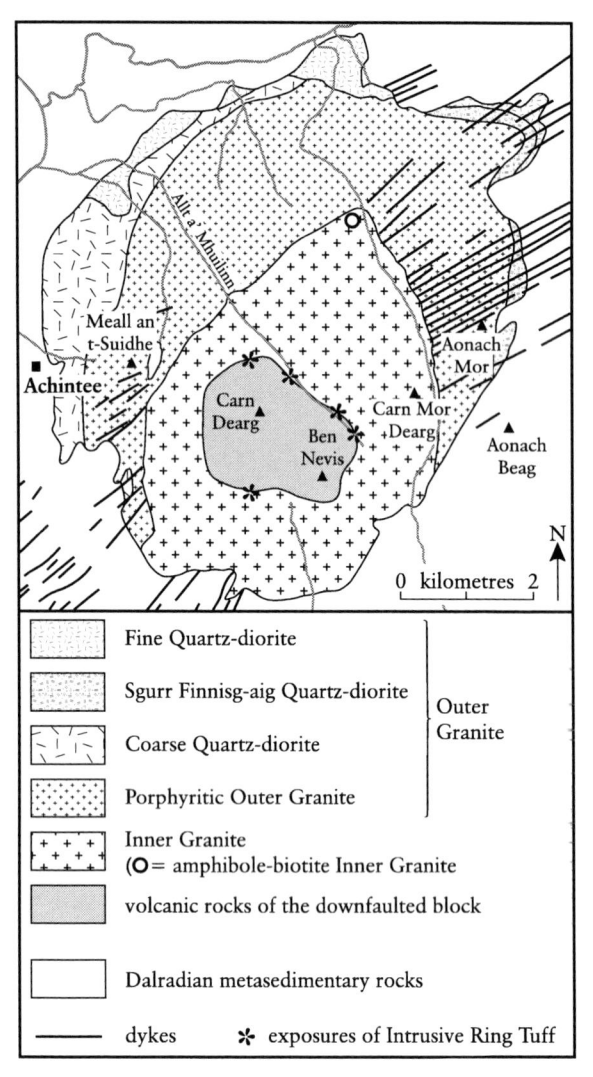

Figure 9.6 Map showing units comprising the Ben Nevis igneous complex. Note the Inner Granite (essentially a ring intrusion) completely surrounding the down-faulted block. The contact between the two is where the rhyolite and 'flinty crush-rock' (i.e. an ancient vent) is found. The Outer Granite consists of four distinct (and mappable) subunits, emplaced in a margin-to-core sequence. Note also that the dyke-swarm cuts through all subunits of the Outer Granite that lie in its path, but does not penetrate the Inner Granite or the downfaulted block. After Bailey (1960), Burt (1994) and Burt and Brown (1997).

tary rocks that subsided along an encircling ring fracture during a caldera-forming eruption. All traces of the surface caldera and the accompanying volcanic rocks have been removed by erosion, but the deeper structural feature (i.e. the cauldron subsidence) remains. The down-faulted block is completely surrounded by a granitic intrusion (the Inner Granite), which is in turn

Figure 9.7 View up the Allt a'Mhuilinn towards Coire Leis, showing the volcanic rocks of the downfaulted block (cliffs of Ben Nevis on the right), and the Inner Granite (low-lying ground on the left and the backwall of the coire in the distance. (Photomosaic: BGS nos. C 1794 and 1795.)

partly surrounded by an earlier granitic intrusion (the Outer Granite).

Geochemical relationships reveal a complex story of magma evolution involving interactions between different batches of magma and various petrogenetic processes (i.e. mantle melting, fractional crystallization, and crustal melting).

The GCR site includes terrain within 2 km of the summit of Ben Nevis, plus a narrow strip along the Allt a'Mhuilinn, which flows NW from the northern corrie, Coire Leis, giving a section across the plutonic units.

Description

The Ben Nevis igneous complex ($c.$ 42 km^2) is dominated by granitic rocks, with volcanic rocks restricted to an elliptical outcrop in the south ($c.$ 3.5 km^2). Early workers (Maufe, 1910; Bailey and Maufe, 1916) recognized three principal lithologies: (1) Outer Granite, (2) Inner Granite, and (3) a down-faulted block dominated by volcanic rocks. They noted that the Inner Granite is a homogeneous intrusion, whereas the Outer Granite consists of four distinct intrusive sub-units. Later workers (Anderson, 1935b; Haslam, 1965; Burt, 1994) remapped the contacts between the four Outer Granite sub-units. This description follows the work of Burt (1994), which is encapsulated in Figure 9.6.

The Ben Nevis granitic rocks are intruded into various garnet-grade Dalradian metasedimentary rocks, which generally strike NE–SW:

(youngest)	Ballachulish Slate (Ballachulish Subgroup)
	Ballachulish Limestone (Ballachulish Subgroup)
	Leven Schist (Lochaber Subgroup)
(oldest)	undifferentiated rocks of the Grampian Group

To the SE the granitic rocks cut the Fort William Slide which generally forms the junction between Grampian Group rocks and younger lithologies. This 'coincidental' association between a Siluro-Devonian volcano-plutonic complex and a major ductile slide is also evident at Glen Coe.

The Outer Granite

The Outer Granite has steep external contacts with the surrounding metasedimentary rocks (Maufe, 1910; Anderson, 1935b). The contact is usually sharp, but in the east there are zones showing considerable brecciation and veining. Veins from a few millimetres to several metres in width penetrate the metasedimentary rocks up to 100 m from the contact, while brecciation is confined to discrete zones within 20 m of the contact. The breccia matrix is granitic, and metasedimentary blocks within the breccia have moved only a few metres (Burt, 1994).

Burt (1994) noted that each of the four units of the Outer Granite (see Table 9.1) are themselves composed of multiple pulses of magma. This is more evident in the earlier units, where more rapid cooling has preserved boundaries between pulses. Later units that cooled more slowly (e.g. the Porphyritic Outer Granite) show considerable mingling between the pulses of magma.

All four units of the Outer Granite are cut by a prominent NE–SW dyke-swarm, which is particularly intense in the east. The average width of the dykes is 3 m (Anderson, 1935b), and Bailey (1960) reported that they vary in composition from $c.$ 58 to 75% SiO_2. In complete contrast, the dyke-swarm does not cut the Inner Granite, yet the Inner Granite can be observed cutting five dykes that intrude the adjacent Outer Granite (Maufe, 1910; Bailey, 1960).

The Inner Granite

The Inner Granite is a notably fine-grained granitic rock that is unusually rich in plagioclase, and this led Bailey (1960) to classify it as a 'trondhjemite' (= leucotonalite). The ring of Inner Granite has contacts with three different lithologies – outer contacts (steep and inclined outwards) with Dalradian rocks and the Outer Granite, and an inner contact (also steep and inclined outwards) encircling the down-faulted block (Maufe, 1910; Bailey, 1960; Burt, 1994).

In places, at the contact between the Inner Granite and the down-faulted block, there is an unusual fine-grained, pinkish to dark-grey rhyolite with contact-parallel flow-banding. Within this rhyolite is a 'flinty crush-rock' very similar to that at Glen Coe (see the Stob Mhic Mhartuin GCR site report). Burt and Brown (1997) described this rhyolite in detail, noting that the best exposure is along the Allt a' Mhuilinn $c.$ 250 m upstream of the climbing hut. The rhyolite contains abundant euhedral phenocrysts of plagioclase (up to 2 mm), plus some biotite, amphibole, and rare quartz. Rare xenoliths (rounded and less than 1 cm) include examples of Dalradian rocks, dacitic lava, and unmetamorphosed sedimentary rocks, all of which can be correlated with lithologies that occur within the down-faulted block. Rhyolite veins (from 1 to 40 cm in width) penetrate the down-faulted block for up to 500 m from the contact.

The down-faulted block

The down-faulted block is approximately 2.5 km in diameter and is surrounded on all sides by the Inner Granite. It has a basin-like structure, with the margins markedly buckled upwards (Maufe, 1910; Bailey, 1960). Burt (1994) has divided the down-faulted block into four distinct volcaniclastic formations (Table 9.2), which rest on Dalradian metasedimentary 'basement'. The exposed thickness is approximately 650 m.

Table 9.1 Nomenclature of the Outer and Inner granites of Ben Nevis by various workers. SiO_2 contents from Burt (1994).

Maufe (1910)	Anderson (1935)	Burt (1994)	SiO_2 (wt.%)
Outer Granite	Outer Quartz–diorite	Fine Quartz–diorite	58.0–62.2
Outer Granite		Sgurr Finnisg-aig Quartz-diorite	63.1
Outer Granite	Inner Quartz–diorite	Coarse Quartz–diorite	53.0–61.7
Outer Granite	Porphyritic Quartz–diorite	Porphyritic Outer Granite	63.7–70.9
Inner Granite	Inner Granite	Inner Granite	67.9–71.9

Three broad trends are evident: (1) the persistence of subaqueous conditions throughout the sequence; (2) the absence of volcaniclastic material in the early deposits, which contrasts with the dominance of such material in later deposits; (3) lateral thickness variations in all four formations.

Interpretation

Field relationships indicate an age sequence from Outer Granite to down-faulted block (i.e. volcanic activity) to Inner Granite. The prominent NE–SW dyke-swarm (which only cuts the Outer Granite) provides two important pieces of information: (1) that the regional stress field involved local dilation along a NW–SE axis; and (2) that the dykes were injected before intrusion of the Inner Granite. Although the dykes do not cut the down-faulted block, this may simply be a consequence of the down-faulted block descending from a level in the crust that did not favour dyke injection. However, it does indicate that dyke injection took place *before* subsidence of the down-faulted block, and it is conceivable that dyke injection was contemporaneous with the development of volcanism at Ben Nevis.

The variable nature of the Outer Granite has been commented upon by all workers (Maufe, 1910; Anderson, 1935b; Bailey, 1960; Haslam, 1968; Burt, 1994), who all recognized four mappable units within the discontinuous ring (although each drew slightly different boundaries between the units). Burt (1994) argued that the Outer Granite units were emplaced into the crust in a 'forceful ballooning style of intrusion' and that in places intrusion was accompanied by explosive release of volatiles, with the development of intrusion breccias. The preservation of internal contacts within each unit suggests that each consists of a number of separate pulses. These contacts are better preserved in the earlier units (e.g. the Fine Quartz-diorite) suggesting more rapid cooling in early units relative to later units. The margin-to-core sequence of intrusion partly explains this, with

Table 9.2 Succession in the down faulted block of Ben Nevis (after Burt, 1994).

Formation	Description	Interpretation
Summit Formation	Autobrecciated andesite-dominated; pervasive brecciation throughout andesite sheet; vesicle-poor; sills present; monolithological volcanic breccia beds are subordinate; lateral variations evident.	Proximal flows of largely degassed andesite lava, plus block-and-ash flows; probably erupted and deposited subaqueously (at least in part).
Ledge Route Formation	Moderately well-sorted volcanic (andesite-dominated) breccias; all strongly clast-supported; have deformed underlying fine-grained beds; lateral variations evident.	Proximal ash-fall deposit reworked by mass flow processes; fine-grained beds indicate quiescence and lacustrine conditions.
Coire na Ciste Formation	Massive unsorted volcanic breccias and block-and-ash flows; exotic clasts of welded ignimbrite and rhyolite lava; vesicle-poor andesite clasts; baked mudstone clasts; andesite lavas and sills; some quartzite-dominated breccias; lateral variations evident.	Volcaniclastic lahars and debris flows, andesite lavas and sills, and pyroclastic flow deposits; all deposited in subaqueous environment (i.e. lacustrine); fine-grained and laminated mudstones indicate periods of quiescence.
Allt a'Mhuilinn Formation	Unconformably overlies Dalradian lithologies; largely mudstone and siltstone (laminated, with rhythmic small-scale fining-up beds), with intercalations of non-volcaniclastic conglomerates (quartzite-dominated); lateral variations evident; no igneous materials present.	Freshwater lacustrine environment; mudstones and siltstones are low-volume fine-grained turbidites developed from bank collapse; conglomerates are subaqueous debris flows and lahars.
Dalradian rocks (Leven Schist?)	Pelites and semipelites; older ductile folding plus later brittle fracturing.	Part of the original land surface (bottom of lake bed).

early pulses encountering 'cold' Dalradian crust and later pulses encountering either pre-warmed crust or still hot earlier granitic magma. Anderson's (1935b) observations that the Outer Granite becomes more silicic with height (*c.* 1200 m of vertical exposure) is interesting, as this might be a relict of an original magma chamber stratification that existed prior to final cooling and solidification. This point has not been fully addressed by later workers and requires re-assessment.

The Inner Granite has steep, outward-dipping inner and outer contacts (Maufe, 1910; Burt, 1994), and appears to have been intruded in a passive (permitted) manner (Burt, 1994). It varies in composition from 67.9 to 71.9% SiO_2 which suggests that it might be the remnants of a zoned magma body.

The sedimentary rocks preserved in the down-faulted block appear to be similar to modern-day playa lake deposits and indicate that a freshwater lake existed (and persisted) at Ben Nevis. It is likely that a small sedimentary basin existed (or that the Ben Nevis basin was part of a larger structure), in which flash floods were a persistent feature in the warm and arid climate of the time (Burt, 1994). Early non-volcaniclastic sediments gradually became replaced by volcaniclastic sediments and by proximal volcanic rocks. The presence of Dalradian clasts in debris flows suggests that there was considerable relief in the area (Burt, 1994 estimates at least 300 m of relief), and there is good evidence of active erosion after emplacement of the various volcanic and sedimentary formations. It is notable that no volcanic rocks more evolved than andesites are found *in situ* – although 'exotic' clasts of dacite and rhyolite do occur, which may have come from neighbouring volcanic centres.

The down-faulted block indicates that substantial subsidence occurred (well over 650 m), and the upturned margins of the block suggest that there was frictional dragging as the block subsided (Maufe, 1910; Bailey, 1960). This conflicts with observations (and published cross sections) showing that the contact is outward-dipping, and further work is needed to resolve this paradox. (It is possible that inward-dipping ring faults are characteristic of near-surface environments, and that these become vertical and then outward-dipping at depth.) Subsidence was probably accompanied by venting of magma to the surface, and the fine-grained rhyolite at the junction of the down-faulted block and Inner Granite is interpreted by Burt and Brown (1997) as the remains of an ignimbrite conduit formed during caldera collapse. As such it is comparable to features developed in the Glencoe ring intrusion (see the Stob Mhic Mhartuin GCR site report) and it has been named the 'Ben Nevis Intrusive Ring Tuff' by Burt and Brown.

Large-scale subsidence along encircling ring fractures at evolved silicic centres is generally accompanied by pyroclastic eruptions at the surface (cf. Druitt and Sparks, 1984), and it appears that Ben Nevis supported an evolved magma system that vented to the surface during a cataclysmic eruption (Burt, 1994; Burt and Brown, 1997). It should be stressed that only the early products of the Ben Nevis volcano are preserved in the down-faulted block; later volcanic rocks (which may have included syncaldera dacites and rhyolites) have all been removed by erosion.

One of Burt's (1994) major contributions is a comprehensive geochemical survey of the igneous complex. The following points summarize his main findings.

1. Ben Nevis is a typical Argyll and Northern Highlands Suite calc-alkaline igneous complex.
2. High-K, calc-alkaline compositions of all igneous rocks (with some volcanic rocks being mildly alkaline).
3. Inner Granite and Outer Granite lie on separate geochemical trends, suggesting two distinct phases of intrusive activity; the Ben Nevis Intrusive Ring Tuff has geochemical similarities to a locally preserved amphibole-biotite granite marginal phase of the Inner Granite (Burt and Brown, 1997).
4. Volcanic rocks (andesite, dacite and rhyolite) spent little time in subsurface magma chamber(s) prior to their eruption.
5. Andesite compositions cluster between 63 and 67% SiO_2.
6. Early andesites show the largest amount of contamination with crustal rocks; later andesites are relatively uncontaminated (inferred from Nd and Sr isotope systematics).
7. Plutonic rocks reached the upper crust (*c.* 1 kb depth).
8. Trace element geochemistry indicates a major role for amphibole fractionation at depth (for all magmas).
9. Plagioclase fractionation largely controlled the geochemical evolution of magmas once

they were emplaced in the upper crust.
10. Strong temporal trend of magma system, generating variable early compositions and more homogeneous later compositions.
11. Appinite intrusion may be intimately related to the development of the igneous complex (i.e. appinites pre-, syn- and post-date the Coarse Quartz-diorite).

Incorporating these findings with the field evidence, Burt (1994) provided a synthesis of the magmatic evolution of Ben Nevis as follows.

1. Initiation of melting in SW Highland light rare-earth element (LREE)-enriched mantle.
2. Early magmas contaminated by interaction with lower crust and influenced by processes of assimilation and fractional crystallization; periodic intrusion of magmas to mid-crustal magma chamber(s); some early andesites and granitic rocks probably reached upper crust direct from lower crust without residing in mid-crustal magma chamber.
3. Fractional crystallization in developing mid-crustal magma chamber(s), with assimilation of Dalradian metasedimentary rocks.
4. Establishment of upper crustal magma chamber(s) where fractional crystallization became dominant, accompanied by localized contamination (by upper crustal Dalradian metasedimentary rocks) of chamber magmas and the magmatic conduits leading to the surface; these upper crustal magmas later solidified to form the various granitic rocks (which are believed to be the source rocks for much of the indigenous volcanic rocks within the down-faulted block).

Conclusions

At Ben Nevis the highest ground is occupied by volcanic rocks that sunk hundreds of metres into an underlying body of still-molten magma. This subsidence was accompanied by the eruption of magma to the surface through an encircling ring fracture. Erosion has removed all trace of the volcanic depression (caldera) that would have been created during this major subsidence event (cauldron subsidence), and the deposits of the accompanying eruption (pyroclastic flows). The volcanic rocks are preserved only because of large-scale subsidence of a down-faulted block, and provide compelling evidence that some of the igneous complexes associated with the Caledonian Orogeny reached high levels in the crust and supported flourishing volcanic superstructures. The international scientific importance of Ben Nevis is that it provides access to levels within an ancient igneous complex where relationships between volcanic rocks (erupted on the land surface), subvolcanic rocks (emplaced just beneath the surface), and plutonic rocks (emplaced deeper beneath the surface), can be investigated and interpreted.

THE GLENCOE VOLCANO – AN INTRODUCTION TO THE GCR SITES

D. W. McGarvie

Introduction

Glen Coe is famous worldwide as a classic example of cauldron subsidence – where a roughly cylindrical block of volcanic and basement rocks subsided into an underlying magma body. As this down-faulted inner block subsided, magma rose up around the margins and intruded the rocks outside in the form of a ring intrusion. A full wine bottle is a useful analogy: push down the cork (the down-faulted inner block) into the wine (underlying magma body), and as the cork is pushed in, wine will rise up (the ring intrusion) around the edges of the cork. One fortuitous outcome of cauldron subsidence is that the down-faulted rocks are preserved long after the surrounding volcanic products and features have been eroded away. In this part of Scotland, two spectacular and important examples of cauldron subsidence exist, at Glen Coe (five GCR sites) and Ben Nevis (one GCR site). Of these, Glen Coe contains the greater diversity of rock types and the greater structural complexity; spectacular erosion has revealed the original land surface of Dalradian metasedimentary rocks upon which the volcanic rocks were deposited, plus the three-dimensional shapes of many volcanic and intrusive units. These attributes have challenged geologists for over a century, and will continue to challenge the geologists of the future.

Cauldron subsidences are the subvolcanic expressions of volcanic calderas, where deeper erosion has revealed the rock types and structural features that accompanied surface caldera collapse and volcanism. The Glencoe volcanic rocks currently occupy an elliptical area approx-

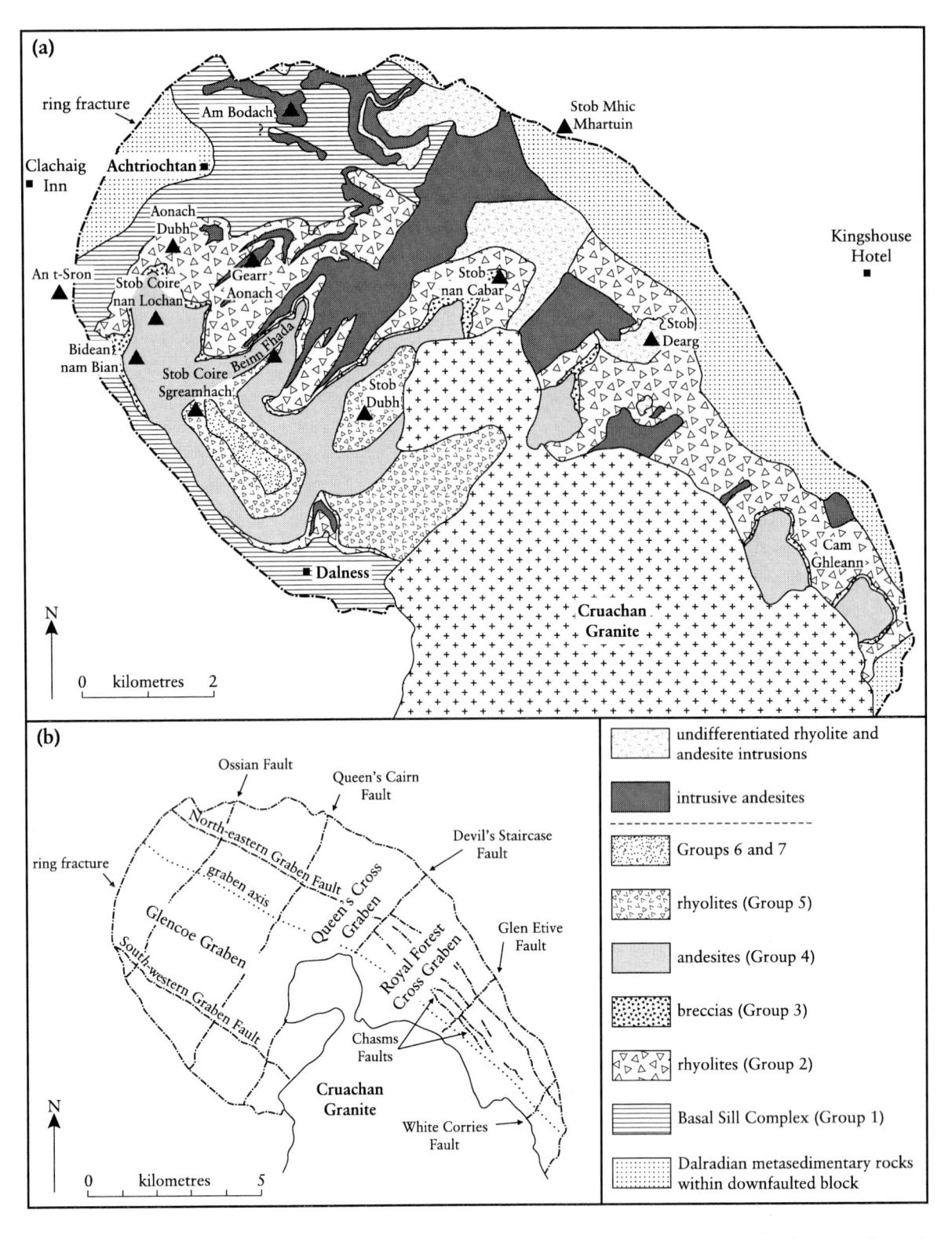

Figure 9.8 (a) Map of Glen Coe showing rocks enclosed by the ring fracture (i.e. within the down-faulted block); Dalradian metasedimentary 'basement'; groups 1 to 7 (with groups 6 and 7 shown together); and undifferentiated intrusive rocks (rhyolite and andesite). Group 3 rocks are sandwiched between groups 2 and 4 rocks throughout most of the area, and only substantial group 3 outcrops are shown. The Etive Dyke-Swarm, minor intrusions, and small outcrops are omitted. The ring intrusion is not shown (see the Stob Mhic Mhartuin GCR site report). Note the incursion of the younger Cruachan granite into the cauldron block from the south. Redrawn after Clough *et al.* (1909), Roberts (1966a), Roberts (1974), and Moore (1995).
(b) Map of the Glencoe cross-graben fault system preserved within the ring fracture (after Moore and Kokelaar, 1997).

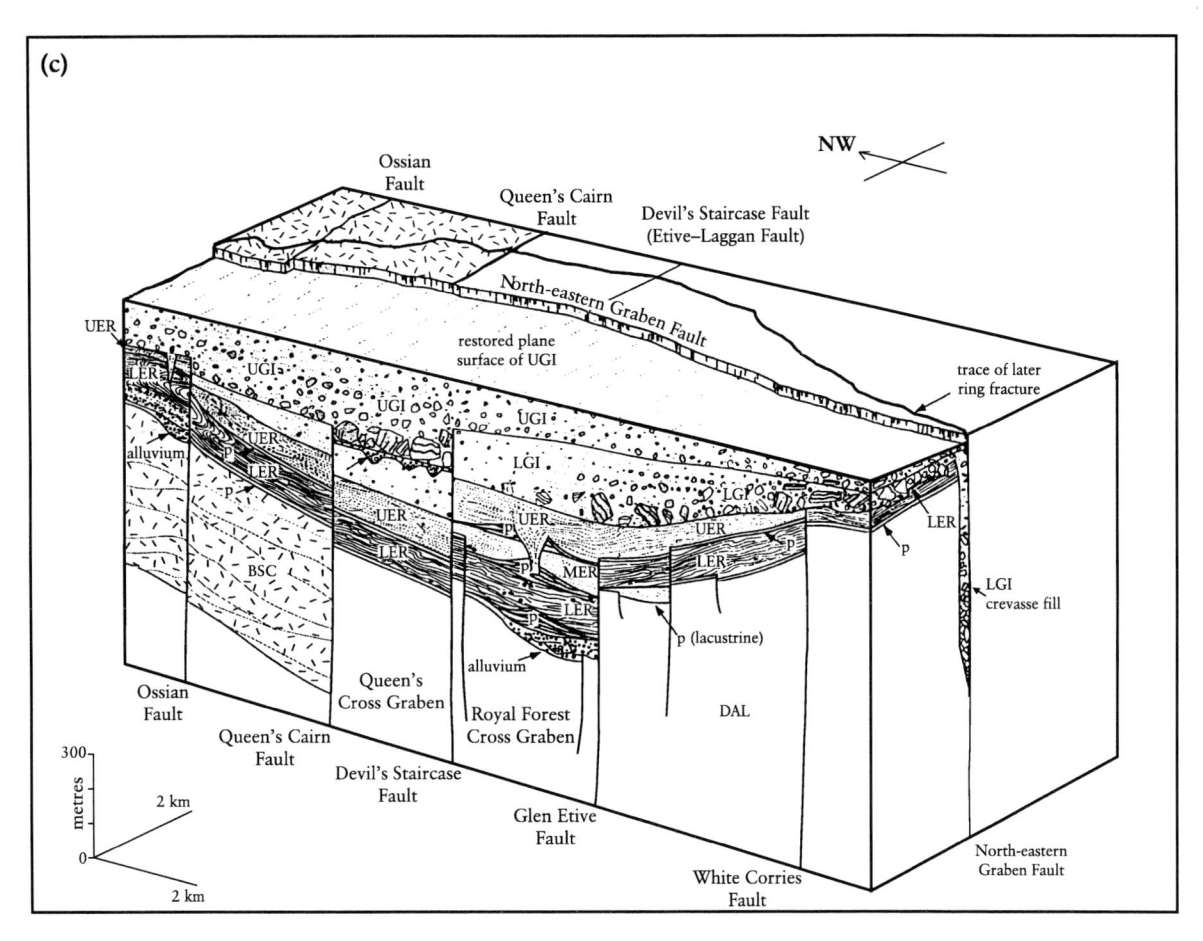

Figure 9.8 –*contd*. (c) Block diagram showing the 3D structure of the Glencoe caldera as interpreted by Moore and Kokelaar (1997). The sections have been restored to a horizontal plane surface presumed to have been formed by the eruption of the 'Upper Glencoe Ignimbrite' (top of Group 2). Thin deposits from this eruption extended north of the North-eastern Graben Fault, but are not shown. The long axis of the block diagram lies along the axis of the Glencoe Graben. DAL, Dalradian metasedimentary rocks; BSC, Basal Sill Complex; LER, MER and UER, Lower, Middle and Upper Etive rhyolites (lower Group 2); LGI and UGI, Lower and Upper Glencoe ignimbrites (upper Group 2); p, phreatomagmatic tuff.

imately 16 × 8 km, with the long axis orientated WNW–ESE (Figure 9.8). This is certainly not the original shape of the volcano; the injection of innumerable NE–SW dykes from the Etive Dyke-Swarm, well after cauldron subsidence, resulted in major dilation along the WNW–ESE axis. However, even with a generous allowance for this dilation, the original shape of the volcano was probably slightly elliptical.

The first major investigations at Glen Coe were undertaken by C. T. Clough and his co-workers when they mapped the area for the Geological Survey. Many subsequent investigators have commented on both the detail and the accuracy of their mapping, and on their perceptive and precise observations. The paper summarizing this work (Clough *et al*., 1909) and the minor modifications in later summaries (Bailey and Maufe, 1916; Bailey, 1960) are revered as classics. Here is a brief summary of their main findings.

1. All the volcanic and related rocks are contained within a ring fracture.
2. Marginal steepening of rocks within the inner block at the ring fracture.
3. Metasedimentary rocks within the ring fracture are of a lower grade than those outside.
4. Large-scale subsidence of a down-faulted inner block (cauldron subsidence).
5. Subdivision of volcanic (and related) rocks into seven groups.
6. General (regional) dip of rock units in a southerly direction (approximately 10–20°).

7. Well-preserved uneven palaeosurface of Dalradian metasedimentary rocks.
8. Early volcanic rocks are basalts to andesites; later rocks are rhyolites and andesites.
9. Lateral impersistence of all volcanic units.
10. Numerous sedimentary beds intercalated between volcanic units.
11. Presence of a ring intrusion outside the ring fracture.
12. Two generations of ring fractures and ring intrusions.
13. Presence of 'flinty-crush-rock' at contact of ring intrusion and ring fracture.
14. Igneous rocks show a calc-alkaline geochemical trend.
15. Recognition that the volcano had developed during the Early Devonian (fossil plant evidence – *Psilophyton* and *Pachytheca*).
16. Elliptical shape of volcano due largely to later dyke injection from adjacent igneous centre (Etive).
17. Later intrusion of Cruachan Granite into SE part of the volcano (after the cauldron subsidence).

Subsequently, various geologists looked at specific features of the volcano. The most pertinent publications are:

- Reynolds (1956): examined features of the ring fractures.
- Hardie (1963, 1968): investigated various breccia units.
- Roberts (1963, 1966a, 1966b, 1974): examined ring fracture and ring intrusion; recognized the presence of ignimbrites; recognized sub-groups within the major groups of Clough *et al.* (1909); developed a unifying model for the evolution of the volcano.
- Ferguson (1966): examined rhyolite flow structures.
- Taubeneck (1967): demonstrated overall inward-dip of ring fracture; recognized a major palaeoslope down to the west; and that abundant sedimentary rocks testified to incremental caldera collapse.
- Thirlwall (1979): analysed some Glen Coe volcanic rocks as part of a regional geochemical survey.
- Garnham (1988): research on ring fractures and associated intrusions.
- Moore (1995), Moore and Kokelaar (1997, 1998): re-mapping of the volcano; detailed investigation of early volcanism; developed radical new model for the early evolution of the volcano.

The two most comprehensive studies after Clough *et al.*, (1909) were by Roberts (1966a, 1966b, 1974) and Moore (1995). Roberts (1974) presented a unifying model for the evolution of the volcano, with two episodes of subsidence of a down-faulted inner block along inward-dipping ring fractures generating ignimbrite eruptions at the ring fractures (see also Reynolds, 1956). The non-erupted magma became the ring intrusion. While this unifying model was a useful attempt at placing Glen Coe into a world-wide framework of caldera formation accompanied by ring-fracture ignimbrite eruptions (established by Smith and Bailey, 1968), the model lacked geochemical confirmation and was based on the assumption that the ignimbrites of the caldera fill were genetically related to the ring intrusions. Moore (1995) has demonstrated that this model is incorrect.

Thirlwall (1979) classified the volcanic rocks of the SW Highlands (including Glen Coe rocks) as high-K calc-alkaline, and noted their geochemical similarity to Andean-type volcanic arc rocks. He concluded that geochemical parameters (variations in large ion lithophile (LIL) and high field strength (HFS) elements, and concentrations of Ni, Sr, and Cr) do not support fractional crystallization as a key process linking together the different magma types. Instead, production of magmas in the mantle (via partial melting, mixing, and contamination) was considered more likely. While Thirlwall's study enabled some generalizations to be made about magmatism in the SW Highlands and provided some analyses of Glen Coe volcanic rocks, the likelihood of magma mixing in many Glen Coe eruptions (discussed later) requires caution in interpreting the geochemical data. Analyses of a comprehensive and well-characterized suite of samples are not available at present, and consequently any comments on magma sources and evolution at Glen Coe would be highly speculative at best.

The recent work of Moore (1995) has brought a modern volcanological perspective to Glen Coe. This work focused on the early evolution of the volcano (groups 1 to 3 of Clough *et al.*, 1909). He further subdivided these three groups, and convincingly demonstrated that the early volcanic and structural evolution of the volcano was controlled by a rectilinear fault system; a series of major graben faults/hinges (WNW–ESE) and cross-graben faults/hinges (NNE–SSW). Moore's work has shown that the unifying model of Roberts (1974) is incorrect. Piecemeal subsidence along a rectilinear graben/cross-graben fault system accommodated all of the early caldera subsidence, without the involvement of the ring fracture (Moore and Kokelaar, 1997, 1998). The probable role of the encircling ring fracture and ring intrusion is discussed later.

Description

Clough *et al.* (1909) established seven volcanostratigraphical units in the Glencoe volcano (groups 1 to 7). Table 9.3 provides a comparison of the stratigraphy of Clough *et al.* (1909) and recent work by Moore (1995), plus approximate thicknesses.

The original land surface

The basement is composed of Dalradian metasedimentary rocks, generally phyllites in the west (Leven Schist Formation) and quartzites and semipelites in the east. The land surface developed on these rocks was heavily eroded and uneven. Clough *et al.* (1909) recorded lenticular masses of conglomerates that they considered to be deposits from flash floods, and also noted that the terrain in the east of the volcano was much steeper than that in the west. Taubeneck (1967) recorded an input of clastic sediment from the east, and noted a 'marked downward slope to the west' which he considered to be 'as much as 2000 feet'. Moore (1995) developed these findings further and described canyons and other fluvial channels etched into the basement.

Group 1: Basal Sill Complex

The lowest 'volcanic rocks' consist of approximately 17 separate sheets with an aggregate thickness of 450 m (Clough *et al.*, 1909; Bailey, 1960). Although Clough *et al.* (1909) described extensive brecciation of the sheet margins and the presence of red sandstones and shales in the matrix, it is only recently that these features have been recognized as peperites, which formed when magmas intruded wet sediments (Moore, 1995). Analyses in Bailey (1960) show that these sheets are basalts, basaltic andesites, and andesites. The rocks are black to dark-grey when fresh, with small black phenocrysts of augite and pseudomorphs after olivine (Bailey, 1960). Moore (1995) noted that the sill complex is deeply incised and eroded (with an unknown thickness of material removed), and that extensive alluvial deposits overly this angular unconformity.

Group 2: rhyolites

Clough *et al.* (1909) noted at least three separate rhyolite units in this group, and commented on their lack of lateral continuity; they also noted the presence of andesite sheets. After recognizing the presence of ignimbrites, Roberts (1966a, 1974) subdivided group 2 into the lower group 2 lavas and the upper group 2 ignimbrites, commenting on the striking variations in thickness of the rhyolite units.

Detailed mapping by Moore (1995) has enabled 11 sub-units to be identified (Table 9.3). These fall into two main units – the Etive Rhyolites and the Glencoe Ignimbrites – each of which represents a major eruptive cycle intimately related to graben-controlled caldera formation.

The first major eruptive cycle involved the eruption of three rhyolite lavas (producing the Lower, Middle, and Upper Etive rhyolites). Each began with a phreatomagmatic phase, which was followed by a flow-laminated rhyolite. A period of quiescence was marked by fluvial incision and sedimentation, and this sequence was repeated two further times. Each eruption was accompanied by subsidence (caldera collapse) of sufficient magnitude to enable fluvial systems to become re-established. Towards the end of this major eruptive cycle, andesite sheets (sills and possibly some lavas) were emplaced, which appear to be mixed magmas (with andesite dominant over rhyolite). Fluvial incision and sedimentation record further subsidence after andesite emplacement.

The second major eruptive cycle produced three ignimbrites (the Lower, Middle, and Upper Glencoe ignimbrites). Of these, the middle ignimbrite is the smallest and has a very limited outcrop, whereas the other two are larger in volume and are more widespread. The first two eruptions were accompanied by irregular

Table 9.3 Stratigraphy of the volcanic and associated sedimentary rocks preserved in the Glencoe cauldron subsidence.

	Group names of Clough *et al.* (1909)	Group names used in this account	Main units of Moore (1995)	Sub-units of Moore (1995)
Group 7 *c.*100 m thick	Andesites and rhyolites	Andesites and rhyolites	–	–
Group 6 *c.*20 m thick	Shales and sandstones	Shales and sandstones	–	–
Group 5 *c.*80 m thick	Rhyolites	Rhyolites	–	–
Group 4 *c.*280 m thick	Andesites	Andesites	–	–
Group 3 *c.*80 m thick	Agglomerates	Collapse breccias and alluvium	Collapse breccias and alluvial deposits	Glas Coire Alluvium Church Door Buttress Breccias Upper Queen's Cairn Breccias
Group 2 *c.*600 m thick	Rhyolites	Rhyolites	Glen Coe Ignimbrites	Upper Glen Coe Ignimbrite Lower Queen's Cairn Breccias Queen's Cairn Fan Middle Glen Coe Ignimbrite Lower Glen Coe Ignimbrite
			Etive Rhyolites	Upper Etive Rhyolite Crowberry Ridge Tuff Middle Etive Rhyolite Raven's Gully Tuff Lower Etive Rhyolite Kingshouse Tuff
Group 1 *c.*500 m thick	Augite andesites and basalts	Basal Sill Complex	**Pre-caldera Basal Andesite Sill Complex	

**Analyses in Bailey (1960) show that some of the sheets are in fact basalts and basaltic andesites.

caldera collapse. More substantial caldera collapse accompanied the eruption of the third ignimbrite, and two syncaldera breccia deposits were produced – the Lower Queen's Cairn Breccias and Upper Queen's Cairn Breccias. After this second major eruptive cycle a further series of andesite sheets was emplaced.

Group 3: collapse breccias and alluvium

Clough *et al.* (1909) noted that this group consists of a complex sequence of 'agglomerates', intercalated with various locally developed sedimentary units, all showing considerable thickness variations. They suggested that these 'agglomerates' might be detrital, and this view was shared by Roberts (1966a, 1974) and Taubeneck (1967), who considered them to be collapse breccias, fluvial deposits and lacustrine sediments. Moore (1995) concluded that there was widespread collapse of the caldera floor after eruption of the Upper Glencoe Ignimbrite, and that this subsidence is reflected in two contrasting types of deposit: fault-scarp collapse – the Church Door Buttress Breccias; and re-establishment of a fluvial system through the caldera, depositing alluvial and caldera lake sediments – the Glas Coire Alluvium.

Group 4: andesites

These are described by Clough *et al.* (1909) and Bailey (1960) as greenish-grey to black andesites typically containing small hornblende and feldspar microphenocrysts set in a flow-banded matrix. A number of sheets are present, attaining an aggregate thickness approaching 300 m; they crop out across much of the volcano. Thin rhyolite units, which may be ignimbrites, and sandstone and shale beds (up to 10 m thick) are intercalated within the andesites (Roberts, 1974).

Group 5: rhyolites

Clough *et al.* (1909) described this group as a black, vitreous, feldspar-phyric, flow-banded rhyolite up to 80 m thick, containing an abundance of lithic clasts (rhyolite, minor andesite, and infrequent quartzose schist) at its base. Both Roberts (1966a, 1974) and Taubeneck (1967) reinterpreted this sheet as an ignimbrite of rhyodacitic composition, and their descriptions hint at the presence of more than one eruptive unit. This rock may be a product of magma mixing, as Roberts (1966a) described flattened (i.e. once molten) inclusions (up to 25 cm) of a dark, porphyritic rock within the rhyolite.

Group 6: shales and sandstones

This group consists of a sequence of well-stratified greenish-grey shales and sandstones, of variable thickness up to a maximum of 20 m, which is exposed only around the southern shoulder of Beinn Fhada (Clough *et al.*, 1909). Taubeneck (1967) concluded that marked subsidence must have followed the eruption of the Group 5 ignimbrite(s) and attributed delicate lamination in the sediments to formation in a caldera lake.

Group 7: andesites and rhyolites

These are the youngest volcanic units preserved within the volcano. They consist of rhyolites and hornblende andesites (and possibly a basaltic andesite) that have accumulated in an irregular fashion, to a maximum thickness of 100 m (Clough *et al.*, 1909). They are exposed only in impersistent outcrops around the southern shoulder of Beinn Fhada. Taubeneck (1967) reported the presence of a thin dacitic to rhyodacitic ignimbrite, containing numerous fragments of volcanic rocks and quartzose basement.

Interpretation

The evolution of the Glencoe volcano can conveniently be divided into three phases: pre-caldera; graben-controlled caldera; and cauldron subsidence (ring-fracture-controlled caldera). Only the first two are represented in the preserved volcanic stratigraphy (Moore, 1995; Moore and Kokelaar, 1997, 1998).

Pre-caldera phase

1. Uneven palaeosurface developed on Dalradian metasedimentary rocks, sloping down to the west.
2. Input of fluvial material from the east, with fluvial system traversing the putative Glencoe caldera, exiting to the west.
3. Localized subsidence in the west, leading to development of a small sedimentary basin (containing red sandstones and shales).
4. Group 1. Sills of basalt to andesite composition injected into the wet sediments (the Basal Sill Complex).
5. Major erosion of the Basal Sill Complex, and deposition of alluvial deposits.

Graben-controlled caldera phase

6. Group 2. First cycle of rhyolitic volcanism (the Etive Rhyolites), with three subcycles of activity: each involving initial phreatomagmatic eruption, then rhyolite eruption, then caldera collapse, then re-established fluvial activity. First cycle culminated in eruption of andesite (mixed magma) sills and lavas.

7. Group 2. Second cycle of rhyolitic volcanism (the Glencoe Ignimbrites), similar in pattern to the first cycle, but explosive eruptions generated ignimbrites instead of lavas, and major collapse accompanied eruption of the last ignimbrite. Second cycle also culminated in eruption of andesite (mixed magma) sills and lavas.
8. Group 3. Major fault-scarp collapse generating breccias, alluvial deposits and lake sediments. Re-establishment of the fluvial system.
9. Group 4. Eruption of andesite sheets across the entire caldera. Presence of sedimentary beds and thin rhyolite units suggests periods of quiescence and collapse between eruptions. Thin rhyolite units may be ignimbrites.
10. Group 5. Ignimbrite eruption(s). Original lateral extent unknown. Caldera collapse. Erosion of upper parts of ignimbrite sheet.
11. Group 6. Establishment of caldera lake (and fluvial system?).
12. Group 7. Eruption of rhyolite and andesite, with at least one explosive rhyolite eruption producing an ignimbrite. Later eruptions of unknown volume and composition were subsequently removed by erosion (see 14 below).

Cauldron subsidence phase (ring-fracture-controlled caldera)

13. Cauldron subsidence (i.e. cataclysmic caldera collapse along ring fractures), probably accompanied by major ignimbrite eruption(s). Emplacement of the ring intrusion – which chilled against the ring fractures. Fine-grained facies (flinty crush-rock) found at the margins of the ring intrusion is possibly an intrusive tuff emplaced during surface venting that generated syncaldera ignimbrites.
14. Removal of an unknown thickness of volcanic and sedimentary rocks from above the preserved Group 7 rocks.

This late-stage, large-scale down-faulting of a cauldron block along encircling ring fractures (i.e. major caldera collapse) contrasts strongly with the piecemeal, incremental graben-controlled caldera subsidence described by Moore (1995) during Group 2 to Group 3 time. It follows that a substantial volume of magma was vented some time after Group 7 time, when the major (sub-circular) caldera was formed. The syncaldera volcanic rocks, which would have filled the large (*c.* 8 km diameter) caldera during this event, have been completely removed by erosion and consequently the volcanic rocks that remain (i.e. groups 1 to 7) represent only the early (non-cataclysmic) magmatic and structural evolution of the Glencoe volcano.

It is instructive to compare the Glencoe and the Ben Nevis igneous centres (see the Ben Nevis and Allt a'Mhuilinn GCR site report), as these are the only two centres in the region where caldera-forming eruptions are known to have taken place and where down-faulted blocks have preserved the products of central volcanoes. The following similarities are apparent.

- both volcanic centres developed on Dalradian 'basement',
- early pre-caldera volcanic sequences at both volcanic centres are dominated by basalts and andesites,
- there was substantial subsidence, deposition of fine-grained sediments, and lacustrine conditions at both centres,
- both centres display a sharp contact between the down-faulted blocks and the encircling ring intrusions,
- a fine-grained variant of the ring intrusion (rhyolite plus flinty crush-rock) is found at the contact with both down-faulted blocks,
- late-stage volcanic products (including possible syncaldera ignimbrites) have been removed by erosion.

However, the downfaulted block at Ben Nevis is only *c.* 2.5 km in diameter whereas at Glen Coe it is *c.* 8 km. The sizes of blocks that subside during caldera-forming ignimbrite eruptions are intimately related to magma chamber diameter (Smith and Bailey, 1968; Smith, 1979), and from this it can be concluded that the Glencoe upper crustal magma chamber was probably the larger of the two. There is thus a greater likelihood that larger volumes of more evolved (silicic) magma were produced at Glen Coe than at Ben Nevis. Furthermore, it is apparent that intrusive rocks are dominant at Ben Nevis (*c.* 90% of the complex) whereas at Glen Coe they are subordinate and are concentrated at the discontinuous ring intrusion. It is likely that Glen Coe and Ben

Nevis represent different erosion levels of broadly similar igneous centres, with Ben Nevis being more deeply eroded. The presence of major granite intrusions adjacent to Glen Coe (e.g. Starav, Etive, Cruachan, Moor of Rannoch) suggests that the Glencoe ring intrusion may be more extensive at depth, where it may look more like the Ben Nevis Inner Granite.

Two tectonic features may have an intimate association with the magmatism: (1) the major slides (Ballachulish and Sgurr a' Choise) that are cut by the volcanoes; and (2) the proximity of the Great Glen Fault. Jacques and Reavy (1994) commented on the possible influence of the Great Glen Fault on igneous centres that lie within 20 km of the fault. Moore (1995) recognized a system of main graben faults (and orthogonal cross-graben faults) at Glen Coe, with the cross-graben faults aligned parallel to the Great Glen Fault. Strike-slip movements on the Great Glen Fault could have developed localized transtension in the Glen Coe area (main graben faults), especially if a pre-existing crustal weakness or lineament was present. Evolved calc-alkaline magmatism in compressional regimes requires localized extension (see Pitcher, 1993), and localized transtension associated with movements of the Great Glen Fault is one possible mechanism for focusing and stimulating magmatism at Glen Coe.

It is apparent that, despite being a classic area of the geology of Great Britain, there are still numerous gaps in our understanding of the Glen Coe magmatism. At the time of writing no modern study has been made of groups 4 to 7, and it is not known how much material has been removed by erosion from above Group 7. Furthermore, it is possible that neighbouring igneous centres (e.g. Etive, Cruachan) supported flourishing volcanic systems (now removed by erosion), that could have contributed caldera fill material to Glen Coe, and that some of the rocks preserved could be from these sources. Other aspects needing further investigation include: how the volcano evolved geochemically; the role of magma mixing; eruption mechanisms of the rhyolites; why there is such extensive sill emplacement; relationships between local and regional magmatism; relationships between local and regional tectonics (especially the nearby Great Glen Fault system); the precise mechanism of cauldron subsidence; and emplacement of the ring intrusion.

BIDEAN NAM BIAN (NN 150 546)

D. W. McGarvie

Introduction

This GCR site is internationally important because it reveals a detailed section through the Glencoe volcano (see 'The Glencoe volcano – an introduction to the GCR sites', above). Spectacular erosion has fortuitously exposed the contact between the original land surface and the lower volcanic units, and has provided an easily accessible section through the volcanic pile preserved in the down-faulted inner block.

The site lies to the south of the River Coe and encloses the highest terrain in the area, including the mountains Bidean nam Bian (1150 m), Stob Coire Sgreamhach (1070 m) and Stob Coire nan Lochan (1115 m). It also includes parts of the 'three sisters' of Glen Coe – Aonach Dubh, Gearr Aonach, and Beinn Fhada.

Description

Figure 9.8a shows the surface outcrop of groups 1 to 7, within the whole of the Glen Coe down-faulted block. A general view of the Bidean nam Bian GCR site, looking south, is given in Figure 9.9, which shows the terrain from the River Coe up to Bidean nam Bian (part of the Loch Achtriochtan GCR site is seen to the right of the photograph). Figure 9.10 illustrates the outcrop of groups 1 to 4 on this photograph (after Bailey, 1960). The lower slopes consist of the Leven Schist Formation (Clough *et al.*, 1909) which is overlain by the Basal Sill Complex (Group 1). This in turn is overlain by the rhyolites of Group 2, which here consist of three thick units. These are the Lower Etive Rhyolite, the Upper Etive Rhyolite, and the Upper Glen Coe Ignimbrite (Moore, 1995). The Group 3 collapse breccias and alluvium deposits crop out on the shoulder of Aonach Dubh and beyond Stob Coire nam Beith, and then the Group 4 andesites crop out towards Bidean nam Bian. These are the youngest rocks seen on Figure 9.9 (Bailey, 1960). Outcrops of groups 1 and 2 on the west face of Aonach Dubh are shown in Figure 9.11.

Exposures revealing the unconformable contact between the Leven Schist Formation and the overlying Basal Sill Complex show an irregular

Figure 9.9 View up Coire nam Beitheach towards Bidean nam Bian, Glen Coe, showing the outcrop of groups 1 to 4, plus the ring fracture and ring intrusion. See Figure 9.10 and the text for details. (Photo: BGS no. B619.)

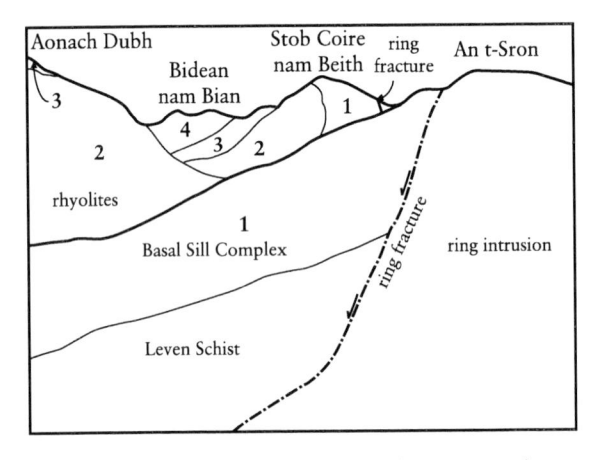

Figure 9.10 Interpretative sketch of Figure 9.9 showing the relationships between topography and lithologies at the Bidean nam Bian GCR site, after Clough *et al.* (1909). See text for details.

original land surface, with impersistent beds of conglomerate, sandstone, and shale above (Moore, 1995). Conglomerates are localized and appear to infill fluvial channels. In the stream draining Coire nam Beith, Clough *et al.*, (1909) recorded a *c*. 1 m-thick bed of purple sandy shale containing fragments of rock from the Leven Schist Formation.

Clough *et al.* (1909) and Bailey (1960) noted that the *c*. 17 sheets of igneous rock in Group 1 range from basalt through to andesite (the dominant lithology). The sheets exhibit flow-banding, and are largely non-vesicular. Brecciation of the upper and lower surfaces of sheets was noted by Clough *et al.* (1909), with sandy shale in the interstices (often retaining bedding). Recent investigations of these sheets (Moore, 1995) have shown that the entire sequence consists of sills with peperitic upper and lower surfaces (Figure 9.12); some sills pinch-out and bifurcate.

Clough *et al.*, (1909) commented on the uneven upper surface of the Group 1 rocks, and Moore (1995) has shown that this is a major erosional unconformity. Above this unconformity is a thin (less than 10 m thick) and irregular sequence of conglomerates and bedded sandstones infilling hollows and channels in the erosional surface.

Above the erosional unconformity are the

Bidean nam Bian

Figure 9.11 The west face of Aonach Dubh, Glen Coe. Scree-covered lower slopes are Leven Schist, which is overlain by intrusive sheets (up to 17) and sedimentary rocks of the Basal Sill Complex (Group 1). The summit region is capped by the thicker rhyolites of Group 2. (Photo: BGS no. B616.)

Group 2 rhyolites, which are described in detail by Moore (1995). A thin phreatomagmatic tuff layer up to 2 m thick and consisting of planar and cross-stratified tuffs with accretionary lapilli is overlain by the Lower Etive Rhyolite, a $c.$ 100 m-thick flow-laminated rhyolite with a low aspect ratio ($c.$ 1:140). A stratified tuff layer represents the first phase of the eruption, but the bulk of the flow consists of flow-laminated rhyolite, accompanied by a persistent upper autobreccia and a less well-developed lower autobreccia. Overlying this is the Upper Etive Rhyolite (20–30 m thick), a flow-laminated rhyolite with upper and lower autobreccias. This rhyolite also has a low aspect ratio (1:100). The third Group 2 rhyolite is the Upper Glen Coe Ignimbrite which is $c.$ 80 m thick here. Roberts (1966a, 1974) called this ignimbrite the upper Group 2 ignimbrite horizon.

Group 3 rocks (collapse breccias and alluvium) overlie the Group 2 rhyolites. Clough *et al.*, (1909) described these as 'agglomerates', but with a perceptive caveat, 'it is possible, indeed, that the deposit is mainly detrital in nature, and not, strictly speaking, a volcanic agglomerate at all'. Later workers (Roberts, 1966a; Taubeneck,

Figure 9.12 Peperite, clearly showing the brecciation of andesitic magma with fine-grained sandstones and shales forming the matrix between the andesite blocks. Beside the 'old road' near Achtriochtan farm, Glen Coe. This particular exposure is no longer visible; it was probably destroyed during construction of the new road, or has been covered by scree. (Photo: BGS no. C1154.)

1967) considered them to be breccias, a conclusion also reached by Moore (1995), who recorded various lithofacies distinguished by clast type; breccias dominated by either Group 1 or Group 2 clasts.

Group 4 rocks crop out around the summits of Stob Coire nan Lochan and Bidean nam Bian, and consist of a number of sheets (flows or sills?) of hornblende andesites approaching 280 m in total thickness (Clough et al., 1909; Bailey, 1960). These rocks contain small phenocrysts of plagioclase and hornblende set within a matrix that is frequently flow-banded. Outcrops of the next group (Group 5 rhyolites) occur far to the SE on the shoulder of Beinn Fhada. According to Roberts (1966a), this is a rhyodacite ignimbrite sheet c. 80 m thick, which is notable for its abundance of plagioclase phenocrysts (Clough et al., 1909). The sheet comprises a thin basal zone that contains abundant lithic clasts (rhyolite lava, hornblende andesite, quartzose schist – Clough et al., 1909) succeeded by a welded ignimbrite within which the proportion of flattened pumice fragments increases markedly towards the top of the sheet, which is brecciated (Roberts, 1966a).

The group 6 and 7 rocks are restricted to outcrops on the southern shoulder of Beinn Fhada (Clough et al., 1909). Surprisingly, no detailed account of these rocks has been published since this paper. The Group 6 rocks are well-bedded greenish-grey shales and sandstones that lie upon the eroded upper surface of the Group 5 rhyolites. Clough et al. (1909), and Bailey (1960) stated that there is much variation in thickness. The Group 7 rocks are andesites and rhyolites, and Clough et al. (1909) recorded rhyolites, hornblende andesites and one 'basic andesite' (basaltic andesite?), and made specific mention of their irregular accumulation. The oldest units crop out to the SE, while the youngest unit (a hornblende andesite containing large plagioclase phenocrysts) caps the southern summit of Beinn Fhada (Clough *et al.*, 1909). Taubeneck (1967) briefly mentioned the presence of an ignimbrite within Group 7, containing abundant 'inclusions of volcanic rocks and of quartzose basement rocks as much as an inch across'.

Interpretation

The palaeosurface developed on rocks of the Leven Schist Formation was heavily eroded prior to the inception of Lower Old Red Sandstone volcanism, and the evidence from sedimentary rocks lying within hollows and channels suggests that there was an active, E–W-running fluvial system. The presence of sandstones and shales suggests that a small sedimentary basin developed, which acted as a trap for finer-grained sediments. The localized development of this basin, and its later (but close) association with the Basal Sill Complex, possibly signifies localized subsidence, which was a precursor to magmatism in the area. (It is relevant to note that similar features are preserved in the neighbouring Ben Nevis cauldron subsidence – see the Ben Nevis and Allt a'Mhuilinn GCR site report.)

The emplacement of the Group 1 basaltic to andesitic sills into wet sediments within the basin resulted in explosive magma-water interactions at sill margins, with peperite development (Moore, 1995). If these sills were fed from a dyke-like body (cf. Francis, 1982), the magma source would have been beneath the area. In addition, most of the sheets are andesitic, and the higher viscosities of andesitic magmas (relative to basalt) would reinforce the argument for a local source. Accordingly, they are probably indigenous to Glen Coe, whereas Clough *et al.* (1909), Bailey (1960), and Roberts (1974) believed them to be outlying members of the Lorn lavas.

The erosional unconformity above the sill complex indicates that there was a substantial period of quiescence after the Group 1 magmatism, and Moore (1995) concluded that an unknown thickness of rocks had been removed. The Group 2 rhyolites mark the onset of the graben-controlled caldera phase of Glencoe volcanism, and they are therefore syncaldera eruptive rocks. Taubeneck (1967) speculated that incremental caldera collapse was intimately associated with Group 2 eruptions – a conclusion confirmed by the detailed mapping of Moore. The first Group 2 volcanic rock, a phreatomagmatic rhyolitic tuff, indicates that this volatile-rich early magma was fragmented even further by interactions with ground, surface or hydrothermal water at the vent. The two rhyolite lavas (Lower and Upper Etive rhyolites) are not typical of rhyolite lava flows. Moore commented on features that suggest surprisingly low viscosities for such high-silica magmas, concluding that (Hawaiian) fountaining at the vent produced a 'lava-like' deposit from a pyroclastic eruption column with a very high effusion rate (possibly with an initial high volatile content lost during degassing at the vent). Hausback (1987) reported similar features from a rhyodacite flow in Mexico, where he concluded that high volatile contents plus high eruption temperatures produced a lava flow varying in thickness from 120 m to 20 m, and of unusually low-viscosity $c.\ 10^5$ poise (note that 10^9 to 10^{11} poise is normal for rhyodacite magma). Clearly, further work is needed on the eruptive mechanisms of the Etive rhyolites. The pyroclastic origin of the third prominent rhyolite unit – the Upper Glen Coe Ignimbrite – was previously noted by Roberts (1966a, 1974) and Taubeneck (1967).

Graben-controlled caldera collapse was associated with all of the Group 2 rhyolite eruptions, and the presence of sedimentary rocks in eroded upper surfaces of each eruptive unit indicates two important features: that caldera collapse cancelled out any constructional features created by the Group 2 eruptions; and that periods of quiescence between eruptive episodes allowed re-establishment of fluvial systems (Taubeneck, 1967; Moore, 1995; Moore and Kokelaar, 1997, 1998).

The Group 3 clastic deposits record a major phase of fault-scarp collapse and re-establishment of fluvial/lacustrine conditions (Taubeneck, 1967; Roberts, 1974; Moore, 1995; Moore and Kokelaar, 1997, 1998). The origins of the Group 4 andesites overlying the Group 3 rocks are unknown, but their aggregate thickness (*c*. 280 m) indicates a major phase of andesitic magmatism after the two major cycles of syncaldera rhyolitic volcanism (Group 2). These andesites are probably indigenous to Glen Coe, and consequently they record a significant change in the underlying magma system, with less evolved compositions predominating and available for eruption; they could represent the 'dominant volume' magma of Smith (1979). These Group 4 andesites suggest that there was insufficient repose time for the magma system to produce new quantities of silicic magma. The Group 5 crystal-rich rhyodacite ignimbrite marks a return to the explosive eruption of evolved magma. The crystal-rich nature of the rhyodacite is unusual, as the Group 2 rhyolites are markedly crystal-poor. The presence of dark inclusions of more basic magma suggests a role for magma mixing, possibly as a trigger for the rhyodacite eruption. A thorough geochemical study of the rhyolites and andesites from groups 2 to 5 would clarify possible genetic relationships.

Little is known of the rocks of groups 6 and 7, although Taubeneck (1967) concluded that Group 6 rocks record a period of post-subsidence sediment accumulation in a caldera lake. The rhyolites and andesites of Group 7, including an ignimbrite, mark a phase of renewed volcanism in the caldera, with evolved compositions again being available for eruption. Post-Group 7 volcanic and sedimentary rocks, including any syncaldera ignimbrites erupted during the late-stage, cataclysmic caldera-forming (cauldron subsidence) event, have all been removed by erosion.

Conclusions

This extensive GCR site provides superb three-dimensional sections through the calc-alkaline Glencoe volcano, a well-preserved example of cauldron subsidence. Rocks preserved include those forming the ancient land surface (Dalradian metasedimentary rocks), and a graben-controlled caldera fill of volcanic and sedimentary rocks. An unknown thickness of material has been removed by erosion.

After an initial pre-caldera andesite-dominated phase (Group 1), there were two major cycles of syncaldera rhyolitic volcanism and caldera collapse (Group 2), which were followed by major fault-scarp collapse and a lengthy period of quiescence (Group 3). Renewed magmatism (Group 4) resulted in a substantial thickness of andesite sheets being emplaced, before a return to the explosive eruption of rhyodacite magma (Group 5). Sedimentary caldera-lake deposits (Group 6) indicate substantial graben-controlled caldera collapse after the volcanism of groups 4 and 5, while the rhyolites and andesites of Group 7 (the youngest rocks preserved) record further eruptions of evolved magma.

One final point is worthy of emphasis: the rocks preserved here record a consistent pattern of eruptive activity accompanied by incremental, graben-controlled caldera collapse. Subsidence kept pace with intracaldera volcanic and sedimentary fill during the early development of the Glencoe volcano. Late-stage, ring-fracture-controlled caldera collapse (cauldron subsidence) certainly took place, and this was directly responsible for preserving the rocks of groups 1 to 7, but unfortunately, the volcanic products of this late-stage, cataclysmic event have been removed by erosion.

STOB DEARG AND CAM GHLEANN (NN 224 547 AND 246 521)

D. W. McGarvie

Introduction

This site is the most easterly of the five GCR sites representing the Glencoe volcano. Relationships here between the volcanic rocks and the original land surface of Dalradian metasedimentary rocks contrast strongly with those in the west of the volcano, indicating important spatial differences in the early volcanic activity. Sedimentary rocks underlying the volcanic sequence at this site yielded the remains of Early Devonian plants during the initial survey, and spores obtained more recently have suggested a more precise biostratigraphical age.

The site comprises two areas: one NW of the River Etive consisting of the 1022 m-high peak of Stob Dearg (at the NE end of the Buachaille Etive Mor ridge) (Figure 9.13); and another SE of the River Etive, around Cam Ghleann.

Description

This GCR site is dominated by Group 2 rocks (see 'The Glencoe volcano – an introduction to the GCR sites', above), which lie directly on top of Dalradian metasedimentary rocks. The marked absence of Group 1 rocks at this site was commented on by Clough *et al.* (1909), Bailey (1960) and Moore (1995).

The Dalradian metasedimentary rocks are quartzites, quartzo-feldspathic psammites and semipelites, comprising the Eilde Flags and the Eilde Quartzite (Clough *et al.*, 1909). These are commonly brecciated and fragmented.

Stob Dearg and Cam Ghleann

Clough *et al.* (1909) and Bailey (1960) provided a detailed description of a complex sequence of psammite breccias, conglomerates and well-bedded quartzose sandstones, red laminated sandstones, and shales sandwiched between the metasedimentary rocks and the overlying Group 2 rhyolites near the foot of Stob Dearg (Table 9.4). From a dark shale bed beneath a well-known landmark (the Waterslide slab), remains of plants were collected in 1902 by Peach, Kynaston, and Tait of the Geological Survey, which were subsequently identified by Kidston and Lang (1924) as *Psilophyton* and *Pachytheca*. This provided an Early Devonian age for the sedimentary rocks beneath the volcanic succession. Spore assemblages collected more recently suggest a late early to early late Lochkovian age (Wellman, 1994). Although Hardie (1968) and Roberts (1974) believed that the outcrop from which these plant remains were taken is a detached block disturbed during explosive volcanic activity, they concluded that the block has moved only a short distance from its original location.

Group 2 is especially well developed at this site (Clough *et al.*, 1909; Roberts, 1974; Moore,

Figure 9.13 The NE face of Stob Dearg, Buachaille Etive Mor, Glen Coe. The lower, scree-covered slopes are of Dalradian metasedimentary 'basement'. The bulk of the mountain consists of rhyolite units (Group 2), with the summit area composed of a mass of intrusive rhyolite. The prominent slab in the lower left centre is the 'Waterslide slab' where fossil plant remains were collected. (Photo: D. Stephenson.)

1995). Using the more detailed subdivisions of Moore (1995), the following Group 2 units are recognized.

Stob Dearg area	Cam Ghleann area
Upper Glencoe Ignimbrite (top)	Upper Glencoe Ignimbrite (top)
Lower Glencoe Ignimbrite	Lower Glencoe Ignimbrite
Upper Etive Rhyolite	Upper Etive Rhyolite
Middle Etive Rhyolite	(missing)
Lower Etive Rhyolite	Lower Etive Rhyolite

In addition, Bailey (1960), Roberts (1974) and Moore (1995) report a mass of intrusive rhyolite which forms the summit region of Stob Dearg (Figures 9.13 and 9.14).

Other important lithologies present in the site are as follows (oldest first).

- Phreatomagmatic tuffs that constitute the first eruptive phase of each of the Etive rhyolite eruptions are particularly well developed at this site (note the absence of the Middle Etive Rhyolite and its underlying phreatomagmatic tuff at Cam Ghleann).
- Andesitic (mixed magma) flows and sills that intrude the Glencoe ignimbrites.
- A localized breccia wedge (psammite, quartzite, and semipelite clasts) at Cam Ghleann, which overlies the Lower Glencoe Ignimbrite.
- Group 3 sedimentary rocks (alluvium deposits only at this site).
- Group 4 andesites (Cam Ghleann only).

Interpretation

The quartzites, psammites, and semipelites exposed at this site are from a lower part of the Dalradian succession than the less resistant phyllitic lithologies (Leven Schist) dominant in the west of the volcano (Clough *et al.*, 1909). The overlying sedimentary rocks have been interpreted by Moore (1995) as a locally developed alluvial fan (the Kingshouse Fan), which crops out throughout this GCR site, but is best developed at Stob Dearg. Clasts have been derived from both talus and fluvial sources and the evidence of both clasts and sedimentary structures suggests an input into the area from the east (Taubeneck, 1967; Moore, 1995). Taubeneck also believed that the eastern parts of the pre-caldera land surface were elevated some 600 m higher than the western parts.

The absence of the Group 1 Basal Sill Complex suggests that the more-resistant rocks (plus more rugged topography) of the eastern parts of the putative volcano hindered formation of a sedimentary basin. It is possible that lavas similar to the basalt and andesite sheets of the sill complex were erupted in the east but, as Moore (1995) pointed out, the erosional unconformity seen in the west indicates removal of an unknown thickness of sheets, and the more rugged topography in the eastern part of the vol-

Table 9.4 Sequence of sedimentary rocks sandwiched between the Dalradian metasedimentary rocks and overlying volcanic rocks (from Bailey, 1960). Bed thicknesses are approximate.

Top of sequence

8. Bedded breccia often resembling conglomerate, with fragments of quartzite, micaceous schist, and some felsite – all in a matrix of gritty sandstone.
7. Red shales with cornstones. (3.5 m)
6. Purple shales. (1 m)
5. Greenish and black shales, showing alternations of coarser and more sandy layers with finer graded beds. (3 m)
4. Conglomerate, with angular and subangular boulders of quartzite (Eilde Quartzite?) and quartzose schists (Eilde Flags) in a green sandy matrix. (6 m)
3. Green shales, some red, and irregular beds of conglomerate. (5 m)
2. Fine greenish breccia containing quartzite fragments. (< 0.5 m)
1. Dalradian quartzite, much shattered at the surface.

Bottom of sequence

---

cano would have hastened their erosion. However, it is equally possible that no early basaltic to andesitic volcanism took place in the east of the volcano.

The dominance of Group 2 rocks, and their extreme development at this site (i.e. substantial thicknesses of individual units and well-developed tuff layers), strongly suggests that they are proximal deposits. Roberts (1974) noted the great thickness of Group 2 rhyolites here, and Moore (1995) suggested that many of the vents for the Group 2 eruptions were located in the central and eastern parts of the volcano, and that they were controlled by the rectilinear system of graben and cross-graben faults (Moore and Kokelaar, 1997, 1998). This was a significant departure from the model of Roberts (1974), who argued that the major rhyolite (ignimbrite) eruptions at Glen Coe were generated at ring fractures. Moore's evidence is convincing, and highlights the need for a re-evaluation of the role and importance of both the ring fracture and the ring intrusion.

Conclusions

The 'basement' of the down-faulted inner block in the east of the Glencoe volcano is a lower part of the Dalradian succession than in the west and comprises more resistant lithologies. Overlying this is an irregular succession of sedimentary rocks varying from shales to coarse conglomerates deposited in an alluvial fan. Fossil remains found in the shales indicate an Early Devonian (Lochkovian) age. A marked downslope to the west seems probable, as is indicated by sedimentary structures and by the provenance of boulders in the conglomerates. The absence of a sill complex within the basal sedimentary succession in the eastern part of the volcano reflects either complete removal by erosion, or a restriction of this early (Group 1) volcanism to the west of the volcano.

Rhyolites that represent the first major eruptive events of the Glencoe volcano (collectively termed Group 2) are extremely well developed at this site, and thickness variations indicate that feeder vents are nearby. Two eruptive cycles are recognized; all eruptive units comprising the largely effusive first cycle (the Etive Rhyolites) are exposed on Stob Dearg, while the two major units of the pyroclastic second cycle (the Glencoe Ignimbrites) crop out throughout the site. In addition, an intrusive mass of rhyolite occupies the upper third of Stob Dearg, while andesitic sills locally intrude the rhyolitic rocks. Overlying alluvial breccias (Group 3), indicate a re-establishment of a river system following the major eruptions, and at Cam Ghleann the succeeding andesites of Group 4 occur.

BUACHAILLE ETIVE BEAG (NN 201 554)

D. W. McGarvie

Introduction

Ignimbrites form distinctive deposits, and record episodes of explosive volcanism and pyroclastic flow generation. They are especially abundant at Glen Coe, and one of the best exposed and most accessible localities is this GCR site. The site consists of the north-eastern summit (Stob nan Cabar – c. 800 m) of the NE–SW mountain ridge of Buachaille Etive Beag (Figure 9.14). Here all three of the Glencoe ignimbrites (Group 2) occur together, and consequently the site preserves the complete sequence of events during this cycle of explosive volcanism.

Clough *et al.* (1909) and Bailey (1960) commented on the abundance of fragmental and welded rhyolite units at Stob nan Cabar, and classified them as the uppermost rhyolite members of Group 2. Roberts (1966a) re-examined the Group 2 rhyolites and identified ignimbrites at Stob nan Cabar with an aggregate thickness exceeding 300 m, the maximum thickness achieved by ignimbrites in Glen Coe. He called them the Upper Group 2 ignimbrite horizon, recognized various subdivisions and provided a description of the sequence (Roberts, 1966a, 1974). Following Moore (1995), the ignimbrites at Stob nan Cabar are now termed the Lower, Middle, and Upper Glencoe ignimbrites, and these are overlain by breccias of Group 3 (the Upper Queen's Cairn Breccias).

Description

The lower exposures at the base of Stob nan Cabar are undifferentiated rhyolite and andesite intrusive rocks that underlie the ignimbrites (Roberts, 1974). There are several NE- to NNE-trending dykes (3 cm to 2 m thick) in these lower exposures that show pyroclastic textures, pervasive lamination, and prominent flow structures. As some are very thin (*c.* 3 cm) and others pinch-out, they are probably indigenous to Glen Coe. Although Roberts (1966a) provided a description of the Stob nan Cabar ignimbrite sequence, Moore (1995) has added considerable detail that forms the basis for the following brief description.

1. Lower Glencoe Ignimbrite (*c.* 140 m thick). Oldest unit. Lower *c.* 40 m is a poorly welded, lithic coarse tuff that contains breccia lenses with clasts derived from the Etive rhyolites and quartzites (10–40% of the deposit). Upper *c.* 100 m is a poorly welded lithic coarse tuff grading upwards into a massive strongly welded tuff, with some inclusions of porphyritic andesite displaying ragged margins.
2. Middle Glencoe Ignimbrite (*c.* 20 m thick). Sharp contact with underlying strongly welded tuff. Bulk of deposit is poorly welded lithic coarse tuff containing abundant small fragments derived from the Etive rhyolites.
3. Queen's Cairn Fan (*c.* 10 m thick). Variable sequence of alluvial conglomerates, sandstones and tuffaceous siltstones that infill erosional features cut into the underlying ignimbrites.
4. Lower Queen's Cairn Breccias (*c.* 40 m thick). Angular clasts and blocks of Dalradian metasedimentary rocks (with subordinate andesite).
5. Upper Glencoe Ignimbrite (*c.* 180 m thick). The lower *c.* 80 m is a breccia deposit with a tuffaceous matrix, containing clasts derived mainly from Dalradian metasedimentary rocks. The next *c.* 30 m is a poorly welded lithic tuff, with clasts (Dalradian metasedimentary rocks plus andesite and rhyolite) up to 40 cm diameter, and pumice lapilli (around 10% of the deposit). The next *c.* 60 m is strongly welded lithic-rich tuff (10–20% lithic fragments). Pumice lapilli increase upwards to a maximum of 25%. The final *c.* 10 m comprises poorly welded stratified lithic tuff, with upper beds notably fine grained and well laminated.
6. Upper Queen's Cairn Breccias (*c.* 60 m thick). These are Group 3 rocks, and consist of a sequence of tuffaceous breccias, with clasts derived mainly from Dalradian metasedimentary lithologies (with minor quantities of volcanic clasts).

Interpretation

The *c.* 350 m thickness of ignimbrites at Stob nan Cabar represents the extreme development of the Group 2 Glencoe ignimbrites. Clough *et al.* (1909) mapped the Group 2 rhyolites, and their cross sections (op. cit., plate XXXIII) indicate that they were fully aware of their thickness variations across the volcano. Roberts (1966a, 1974) considered that differential subsidence of the caldera floor in this area had allowed the ignimbrites to pond. Moore (1995) developed this further and demonstrated that collapse had taken place along graben and cross-graben faults, and at the same time showed that caldera subsidence in this area accompanied the eruption of the Upper Glencoe Ignimbrite.

The Glencoe ignimbrites represent the second major cycle of syncaldera rhyolitic volcanism, and the widespread development of pyroclastic flows indicates that activity was generally more explosive than the first cycle (the Etive rhyolites). The following sequence of events was proposed by Moore (1995) and Moore and Kokelaar (1997, 1998). The Lower Glencoe Ignimbrite was produced by a single pyroclastic flow, which was followed by the much smaller Middle Glencoe Ignimbrite (also produced by a single pyroclastic flow). Post-eruption subsidence, a period of quiescence, alluvial fan development, and fluvial activity generated the sedimentary deposits of the Queen's Cairn Fan. The eruption of the Upper Glencoe Ignimbrite was preceded by large-scale foundering of the caldera floor and fault-scarp collapse, producing the Lower Queen's Cairn Breccias. The Upper Glencoe Ignimbrite was produced from a single pyroclastic flow, and grades upwards from a breccia-dominated lower zone through to a poorly welded lithic coarse tuff, which is strongly welded in its upper parts. The upper *c.* 10 m of fine-grained and laminated tuff may represent ash-fall tephra deposited after pyroclastic flow activity had ceased.

Based on relationships at this GCR site, Moore was able to discern a pattern in the lateral thickness variations of the three ignimbrites, and to demonstrate that topographical control on their spatial distribution was exerted by a long-standing, re-activated graben and cross-graben fault system. He also noted that the vents for these ignimbrites must have been nearby. Consequently, the unifying model of Roberts (1974), which invoked the Group 2 ignimbrites having

Stob Mhic Mhartuin

Figure 9.14 Buachaille Etive Beag, Glen Coe from the NE, looking towards Stob nan Cabar from Stob Mhic Mhartuin. (Photo: D.W. McGarvie.)

been vented at the ring fracture, must now be discarded.

Conclusions

At Stob nan Cabar there is an unusual thickness of rhyolitic ignimbrite, a rock type formed during the eruption of evolved magma when pyroclastic (hot, fragmental) flows are generated during explosive volcanic activity. The ignimbrites exposed here reach nearly 350 m in thickness, and within this total thickness three separate units are present, representing three separate phases of pyroclastic flow generation during one major eruptive cycle (the second cycle of rhyolitic eruptions responsible for Group 2 of the Glencoe volcanic succession). Ignimbrites infill depressions and hollows, and their extreme thickness in this area indicates substantial localized graben-controlled subsidence of the caldera floor accompanying the eruptions. The presence of sedimentary rocks intercalated between the second and third ignimbrites indicates a period of quiescence between eruptions.

STOB MHIC MHARTUIN (NN 208 576)

D. W. McGarvie

Introduction

The Glencoe volcano is of international importance because of the cauldron subsidence, which has preserved a sequence of Old Red Sandstone volcanic rocks within a downfaulted block, encircled by a ring fracture system and an irregular ring intrusion (see 'The Glencoe volcano – an introduction to the GCR sites', above). The ring fracture is the subject of this GCR site, which occupies the summit and surrounding area of Stob Mhic Mhartuin (706 m). Its importance lies in the good exposures of the ring fracture, where the relationships between the down-faulted inner block and the surrounding undisturbed metasedimentary rocks are especially clear. As with elsewhere at Glen Coe, there is an intimate association between the ring fracture and the ring intrusion. An unusual feature of this site is that there are two ring fractures, of different ages.

Description

Stob Mhic Mhartuin is regarded as the type locality for the ring fracture (Bailey, 1960), although it is in fact quite atypical in some aspects (Taubeneck, 1967). Clough *et al.* (1909), Bailey (1960), and Roberts (1966b) all provided detailed descriptions of the ring fracture and some of the key relationships are illustrated in Figure 9.16 (after Roberts, 1966b). Here it appears in two branches and contact metamorphic relationships led Clough *et al.* (1909) and Roberts (1966b) to conclude that the southern branch is the younger. This description concentrates on the southern (younger) ring fracture (Figure 9.15).

Immediately north of the ring fracture lies the ring intrusion, here a porphyritic microdiorite, which forms a near-continuous (but irregular) ring around the down-faulted inner block (Figure 9.17). The ring intrusion typically has a smooth and chilled inner contact against the ring fracture, which contrasts strongly with its irregular and variably chilled contact against the outlying metasedimentary rocks (Roberts, 1966b). Clough *et al.* (1909) placed great emphasis on a fine-grained lithology called 'flinty crush-rock' (described later), which lies at the smooth inner contact between the ring intrusion and the ring fracture (Figures 9.15 and 9.16). However, Roberts (1966b) reported that 'flinty crush-rock' also occurs at the outer, irregular contact, and argued that it is just a finer-grained facies of the ring intrusion.

The ring fracture strikes approximately NW–SE at this locality and dips outwards at 60° to the NE. The ring intrusion chills against the ring fracture, and there is a 2 cm-wide zone of dark, vitreous rock ('flinty crush-rock') between the chilled ring intrusion and the crushed metasedimentary rocks of the down-faulted inner block. The contact is surprisingly straight, sharp, and persistent. Within the 'flinty crush-rock' there are clear signs of flow, with lighter- and darker-coloured layers developed parallel to the contact (individual layers are traceable for a few metres), plus there are oval structures (long axes parallel to the contact) suggesting flow either down-dip or up-dip. There is a fairly sharp contact between the 'flinty crush-rock' and the adjacent lithology to the SW, which is a strongly flow-banded microbreccia with elongate and rounded fragments (smaller than 1 cm) set in a matrix of the 'flinty crush-rock'. At the contact with the 'flinty crush-rock' some whitish zones are present in the microbreccia, and in places the 'flinty crush-rock' appears to either cross-cut these zones or to have incorporated remnants of them within it. To the SW the flow-banded microbreccia grades (over a distance of 1 cm) into a zone of whitish rock that consists of crushed fragments of quartzite (Clough *et al.*, 1909). Within this crush zone there are two distinct lithologies: one adjacent to the flow-banded microbreccia, which is a mix of light and dark components, and one to the SW, which is characteristically white in colour. To the SW are relatively undisturbed quartzites of the down-faulted inner block with vertical bedding and NNW strike. However, within 40 m of the ring fracture there are crush zones (up to 10 cm wide) that have the same strike and dip as the ring fracture, and that are more abundant nearer to the ring fracture.

Interpretation

Roberts (1974) presented a unifying model of explosive eruptions at the ring fracture producing the ignimbrites that are now preserved within the down-faulted block. However, Moore (1995) has demonstrated that piecemeal subsidence associated with Group 2 ignimbrite eruptions took place within an early graben-controlled caldera, and not at the ring fracture, thus invalidating the model of Roberts. The cataclysmic event that led to late-stage cauldron subsidence (i.e. ring fracture-controlled caldera formation), contrasts with the earlier graben-controlled volcanism and caldera collapse. The ring intrusion probably represents a syncaldera, subvolcanic, non-erupted magma that was associated with cauldron subsidence on the encircling ring fracture(s).

Stob Mhic Mhartuin

Figure 9.15 The outward-dipping Glencoe ring fracture at Stob Mhic Mhartuin (down-faulted block on the right). The prominent dark band running up to the right is the 'flinty crush-rock', and above it (to its left) is the ring intrusion, here a porphyritic microdiorite, which is chilled at the contact. Between the 'flinty crush-rock' and the low ground on the right (Dalradian quartzites within the downfaulted block) are crushed and brecciated quartzites (see Figure 9.16). The outward-dipping orientation of the contact is probably due to rotation accompanying collapse of the downfaulted block. (Photo: BGS no. D1562.)

While there is good evidence for major subsidence of a down-faulted inner block along encircling ring fractures at Glen Coe (summarized by Clough *et al.*, 1909 and Bailey, 1960), there is uncertainty regarding the magnitude of the subsidence. Bailey (1960), Taubeneck (1967), and Roberts (1974) all placed considerable emphasis on the presence of sedimentary rocks intercalated throughout the volcanic pile, and concluded that episodic subsidence kept pace with the eruption of syncaldera volcanic rocks. Moore (1995) agreed with this conclusion, adding that the subsidence was graben-controlled. Furthermore, Moore has shown that the Group 2 ignimbrites were erupted from vents in the central area of the volcano, whilst Garnham (1988) found no geochemical correlation between the ring intrusion and volcanic rocks of the down-faulted inner block. Thus, any syncaldera volcanic rocks that accompanied *en bloc* subsidence of the down-faulted inner block have since been removed by erosion. If a major eruption accompanied cataclysmic caldera collapse (i.e. cauldron subsidence) – and this would be wholly consistent with ring-fracture-controlled caldera formation elsewhere (see Smith and Bailey, 1968; Smith, 1979) – the ring intrusion could be the remnants of a vent which gave rise to eruptions of ignimbrite.

At the ring fracture itself, subsidence was accompanied by comminution and crushing of the country rocks of the down-faulted inner block. There is a clear progression from crushed country rock quartzite in contact with undis-

Figure 9.16 Sketch (note variable scale) showing relationships in the Stob Mhic Mhartuin area between the downfaulted block, the ring intrusion, and the undisturbed metasedimentary rocks outside. Note the broad symmetry that is present, with 7 (ring intrusion) flanked by chilled margins (6 and 8), which are in turn flanked by 'flinty crush-rock' (5 and 9) – all of which probably constitute an ancient vent system. The 'flinty crush-rock' is flanked by a complex series of microbreccias to the west (2, 3 and 4), and by a much simpler microbreccia to the east (10), and these are themselves flanked by unbrecciated Dalradian metasedimentary rocks (1 – within the downfaulted block; and 11 – outside the downfaulted block). After Roberts (1966b), and Garnham (1988).
Key: 1, Bedded quartzites within downfaulted block; 2, Quartzites cut by veinlets of granulated quartzite; 3, Quartzite microbreccia; 4, Banded quartzite microbreccia with matrix of 'flinty crush-rock'; 5, 'Flinty crush-rock' (note sharp and straight contact with 6); 6, Chilled margin of ring intrusion; 7, Ring intrusion; 8, Chilled margin of ring intrusion; 9, 'Flinty crush-rock' (note sharp but irregular contact with 8); 10, Quartzite microbreccia; 11, Undisturbed Dalradian metasedimentary rocks outside the downfaulted block.

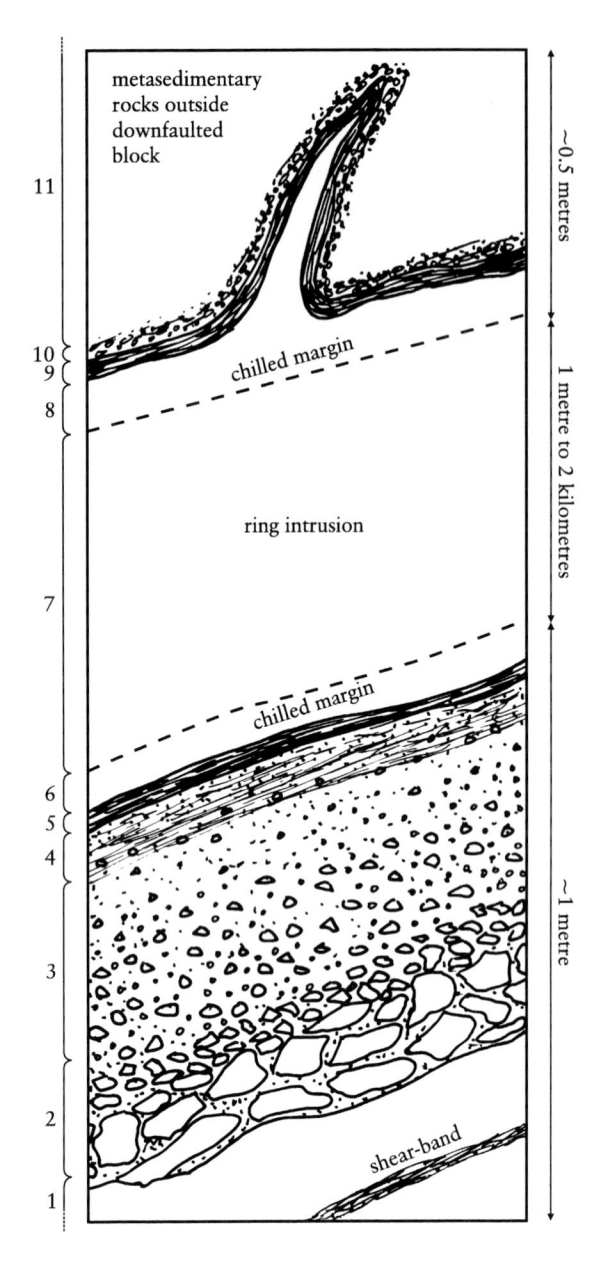

turbed vertically bedded quartzites of the downfaulted block, through to a bedded microbreccia incorporating progressively lesser amounts of crushed quartzite, through to the dark 'flinty crush-rock'. The incorporation of crushed quartzite into the 'flinty crush-rock' suggests that the 'flinty crush-rock' is younger. Roberts (1966b) presented convincing evidence that the 'flinty crush-rock' is simply a rapidly chilled variant of the ring intrusion, and this suggests that magmatic activity either accompanied or postdated down-faulting.

The original workers (Clough *et al.*, 1909; Bailey, 1960) regarded the 'flinty crush-rock' as a pseudotachylite (a melt produced by extreme friction). This interpretation was questioned by Reynolds (1956), Roberts (1966b), and Taubeneck (1967), who all proposed a magmatic origin. The veins and tongues of 'flinty crush-rock' that occur away from the ring fracture (Taubeneck, 1967) further support a magmatic origin. Clough *et al.* (1909) did comment on these offshoots of the 'flinty crush-rock', but attributed them to injection of the still-fluid pseudotachylite away from its source. It is significant that Roberts (1966b) also found 'flinty crush-rock' at the outer (irregular) contact, and concluded that the 'flinty crush-rock' is simply a marginal facies of the ring intrusion. Reynolds (1956), Roberts (1966b) and Taubeneck (1967) all regarded the 'flinty crush-rock' as an intrusive tuff. Recent work at Ben Nevis, where there is also 'flinty crush-rock' (Bailey and Maufe, 1916), has further strengthened the case for a magmatic origin. Burt and Brown (1997) concluded that the Ben Nevis 'flinty crush-rock' is an intrusive tuff, and that it represents the remnants of a vent which probably encircled the Ben Nevis volcano (see the Ben Nevis and Allt a'Mhuilinn GCR site report).

The presence of two separate ring fractures, of different ages, suggests two periods of subsidence, although Taubeneck (1967) has questioned this interpretation. The older ring fracture is not found encircling the entire down-faulted block, and its significance is therefore uncertain.

The ring fractures are outward-dipping at this locality, which is atypical. Both Taubeneck (1967) and Roberts (1974) demonstrated that the ring fractures are inward-dipping or vertical at the vast majority of localities, implying that the down-faulted inner block has the shape of an upward-opening cone. At calc-alkaline volcanic centres, major caldera collapse involving chamber roof collapse along concentric, inward-dipping ring fractures (cf. cauldron subsidence) invariably triggers the eruption of substantial volumes of ignimbrite (Druitt and Sparks, 1984). By analogy with caldera complexes elsewhere, it is likely that at least one such late-stage cataclysmic event took place at Glen Coe. Thus Roberts (1974) was partly correct in concluding that there had been a major ring fracture-controlled caldera event at Glen Coe. He did, however, misinterpret the role of the Group 2 ignimbrites, which were erupted during an earlier phase of small-scale, graben-controlled caldera formation, and not during late-stage cauldron subsidence (Moore, 1995).

Conclusions

Rocks exposed around the summit of Stob Mhic Mhartuin in Glen Coe preserve features developed during late-stage cauldron subsidence (cf. ring fracture-controlled caldera collapse), when a down-faulted inner block sank. Without this subsidence the volcanic rocks that form the rugged topography of Glen Coe would not have been preserved. The ring fracture, which separates the down-faulted inner block from the undisturbed rocks outside, is well exposed. By analogy with volcanic centres elsewhere, major subsidence of the down-faulted inner block along concentric, inward-dipping ring fractures, led to cataclysmic eruption of ignimbrites, much of which would have been deposited within the subsiding caldera. Subsequent erosion has removed these deposits, but the ring intrusion that partly encircles the ring fracture is probably the remnants of a concentric vent.

LOCH ACHTRIOCHTAN (NN 132 557 AND 140 575)

D. W. McGarvie

Introduction

This GCR site exhibits two contrasting topographic expressions of the ring fracture that encircles the down-faulted inner block of the Glencoe cauldron subsidence (Figure 9.17), together with good exposures of the metasedimentary rocks that underlie the volcanic rocks. Exposures within the site also reveal marginal tilting of rocks of the inner block immediately adjacent to the ring fracture, demonstrating that the subsidence was accompanied by drag.

The site comprises two areas: one to the north of the River Coe, extending from the banks of the river up to the western end of the Aonach Eagach ridge (Figure 9.18); and a second area to the south of the River Coe that includes the mountain spur of An t-Sron and the west side of Coire nam Beitheach (Figure 9.9).

Description

Rock types

This site is dominated by Group 1 rocks and the

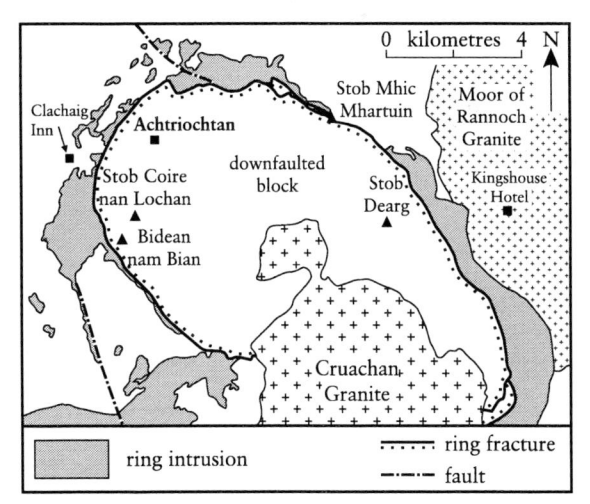

Figure 9.17 Sketch showing the ring fracture and ring intrusion system around Glen Coe. Relationships are less clear in the south where there are difficulties distinguishing the ring intrusion from neighbouring intrusions. Minor outcrops of the ring intrusion (which are numerous) have been omitted for clarity. Redrawn after Clough *et al.* (1909), Roberts (1974) and Garnham (1988).

underlying Dalradian metasedimentary rocks (Leven Schist). Outcrops of groups 2 and 3 occur high on the shoulders of Stob Coire nan Lochan and Stob Coire nam Beith.

The metasedimentary rocks that surround Glen Coe, and that form part of the down-faulted inner block, belong to the Grampian and Appin groups of the Dalradian Supergroup (Table 9.5). In Glen Coe the metasedimentary stratigraphy is complicated by the presence of two ductile slides or lags, the Ballachulish Slide and the Sgurr a'Choise Slide. This has created the structural sequence which is found in the River Coe just west of the Loch Achtriochtan GCR site and is as follows.

> Leven Schist
> Ballachulish Limestone
> *Sgurr a'Choise Slide*
> Ballachulish Slate
> *Ballachulish Slide*
> Leven Schist
> Glencoe Quartzite

At this site the Leven Schist (a 'phyllite') is predominant, and forms a roughly triangular outcrop within the ring fracture. Clough *et al.* (1909) and Bailey (1960) noted that the low metamorphic grade of the Leven Schist within the down-faulted block contrasts with the higher (i.e. garnet-bearing) metamorphic grade of the Leven Schist outside.

Between the Leven Schist and the Group 1 Basal Sill Complex is a variable and discontinuous sequence of sedimentary rocks varying from conglomerates (clasts up to 3 m) to finely bedded shales. Conglomerates are well exposed on the southern slopes of the Aonach Eagach in a prominent outcrop running along the base of the cliffs that trend SE from Coire Leith towards Achtriochtan farm. The bed (1–20 m thick) dips at *c.* 30° to the SE, and typically infills hollows in the irregular palaeosurface of Leven Schist. The clasts are set within a matrix consisting of either coarse sandstone or sandy shale; larger clasts are well-rounded, whereas smaller clasts tend to be more subangular (Clough *et al.*, 1909; Bailey 1960). The dominant clast type is quartzite, although minor schist, andesite, granite, quartz porphyry, and kentallenite are reported by Bailey (1960).

The Leven Schist outcrop forms fairly subdued topography, in contrast to the more-resistant rocks of the sill complex. Consequently, the rocks of the sill complex form prominent crags that enable its base to be traced with ease throughout the site. On the south side of the River Coe, sheets of the sill complex dip at approximately 15° to the SE.

The ring fracture

South of the River Coe is a striking (and rather atypical) topographical expression of the ring fracture. Much of the impressive and deeply incised gully known as The Chasm of An t-Sron follows the ring fracture, which passes just east of the summit of An t-Sron where it appears as a distinctive notch (Figure 9.9). On the southern slopes of An t-Sron the ring fracture forms a narrow (but pronounced) gully. The subdued topographical expression of the ring fracture just south of the River Coe, and its continuation north of the River Coe up to the summit of the Aonach Eagach ridge, is more typical (Figure 9.18). The line of the ring fracture over such terrain is not obvious from a distance, but it can be traced with ease when walking along the contact. Within this site the ring fracture changes direction abruptly: the strike is almost N–S from An t-Sron north to the River Coe, whereas it strikes NE to the north of the River Coe.

On the slopes of An t-Sron, south of the River Coe, the down-faulted rocks in contact with the ring fracture are Leven Schist up to the 350 m contour, with Group 1 rocks at higher elevations. Throughout the site, the ring fracture dips between 65° and 85° inwards (i.e. towards the centre of the down-faulted inner block), and flinty crush-rock (see the Stob Mhic Mhartuinn GCR site report) is commonly found at the contact between the ring intrusion and the ring fracture.

There is abundant field evidence of marginal steepening of rocks of the down-faulted inner block immediately adjacent to the ring fracture. For example, the primary foliation of the Leven Schist at An t-Sron increases from *c.* 35° some 20 m away from the ring fracture to *c.* 70° immediately adjacent to it. This marginal steepening is observed both in the metasedimentary rocks and in Group 1 rocks adjacent to the ring fracture.

The ring intrusion

At this site the ring intrusion is present outside

Loch Achtriochtan

Table 9.5 Metasedimentary rocks found in the Glen Coe area.

Group	Subgroup	Formation
Appin	Ballachulish	Appin Quartzite
		Ballachulish Slate
		Ballachulish Limestone
		Leven Schist
	Lochaber	Glencoe Quartzite
Grampian	–	Eilde Flags

the ring fracture, except where it is absent for short stretches north of the River Coe and at the summit of the Aonach Eagach ridge (Clough *et al.*, 1909). At An t-Sron, the ring intrusion has its greatest development anywhere in Glen Coe and forms a large intrusive mass that extends almost 2 km from the ring fracture (Clough *et al.*, 1909). Here, there is an abundance of accidental fragments, especially quartzite, in the ring intrusion. Most of the quartzite fragments are angular to subangular, and they vary from white to red in colour. The provenance of these fragments is unknown, but precise classification would clarify the emplacement mechanism of the ring intrusion.

Interpretation

The Basal Sill Complex (Group 1) is well developed at this site and at the adjacent Bidean nam Bian GCR site (where it is discussed in detail), in strong contrast with its absence from the eastern parts of Glen Coe (see the Stob Dearg and Cam Ghleann GCR site report). This suggests that the sedimentary basin intruded by the sills of the sill complex was restricted to the western part of Glen Coe and may have developed due to precursory subsidence related to Group 1 magmatism.

Clough *et al.* (1909) and Bailey (1960) interpreted the presence of garnet in the Leven Schist outside the down-faulted inner block as indicating a higher metamorphic grade than the non-garnet-bearing rocks of the inner block. Cauldron subsidence then brought rocks with similar lithologies yet contrasting metamorphic grade into juxtaposition. However, it is now known that subtle compositional differences can have a marked effect on index mineral development in metasedimentary rocks (e.g. Yardley, 1989). Consequently, it would be necessary to confirm that the Leven Schist lithologies in juxtaposition are compositionally identical before placing too much emphasis on this original interpretation.

The abrupt change in the strike of the ring fracture at this site was noted by Clough *et al.* (1909) and Bailey (1960). They mapped several persistent N–S-orientated shatter belts throughout Glen Coe (two are exposed on the south slopes of the Aonach Eagach ridge), and conjectured that the prominent N–S strike of the ring fracture at An t-Sron may coincide with one of these. Disruption of the ring intrusion at An t-Sron may also relate to this shattering. The shatter belts apparently pre-date cauldron subsidence and the intrusion of the Etive dyke swarm (Clough *et al.*, 1909), yet there is evidence that the shatter belts were also active after cauldron subsidence; Roberts (1974) noted that 'flinty crush-rock' at An t-Sron showed post-emplacement fracturing.

The marginal steepening of rocks adjacent to the ring fracture was interpreted by Clough *et al.* (1909) as resulting from frictional drag on the outer margins of the down-faulted inner block. Taubeneck (1967) agreed with this interpretation, and argued that it provides strong evidence for a predominant inward dip of the ring fracture and consequently an upward opening cone shape for the down-faulted inner block.

Conclusions

Outcrops at this site reveal spectacular features that accompanied late-stage cauldron subsidence within the Glencoe volcano. Metasedimentary rocks that formed the eroded landscape before volcanic activity began are preserved in the down-faulted inner block. The ring fracture that separates the down-faulted inner block from the undisturbed metamorphic

Figure 9.18 View across Loch Achtriochtan, Glen Coe towards the Aonach Eagach ridge on the skyline. The ring fracture enters the left edge of the photograph about halfway up and runs towards the low point of the ridge at its extreme left. This subdued expression of the ring fracture is typical, and contrasts strongly with that shown in Figure 9.9. The distinctive cliffs running down to the right are Group 1 rocks (Basal Sill Complex), which overlie Leven Schist (Dalradian basement). (Photo: BGS no. B624.)

rocks outside is also well exposed and, in the prominent gully of An t-Sron, it has a particularly impressive topographical expression. While the down-faulted inner block was subsiding, magma rose from depth and intruded the rocks outside the ring fracture. This magma is known as the ring intrusion, which at this site has its most extreme development in the mountain of An t-Sron (where it extends up to 2 km from the ring fracture).

CRAWTON BAY
(NO 880 797)

R. A. Smith

Introduction

These coastal exposures are the type locality for the Crawton Volcanic Formation, the youngest formation in the Crawton Group which comprises part of the Lower Old Red Sandstone succession in the Crawton Basin, a precursor to the Strathmore Basin (Figure 9.2). The formation consists of olivine-bearing basalts and basaltic andesites and interbedded conglomerates of late Silurian to Early Devonian age. The inter-relationships between the lava flows and intercalated sedimentary rocks of 'Highland origin' have been recorded from this vicinity since Geikie (1897). Subsequently Campbell (1913), Trewin (1987), Carroll (1994) and MacGregor (1996) have described the locality in detail. Of the four lava flows present in Crawton Bay (Figure 9.19), the lower three contain characteristic large, flow-orientated plagioclase phenocrysts (the 'Crawton type' of Campbell, 1913), whereas the uppermost one is aphyric. The distinctive 'Crawton type' lavas have been traced inland around the hinge of the Strathmore Syncline (Haughton, 1988). Some of the lavas have been analysed as part of a geochemical study of the British Lower Old Red Sandstone lavas (Thirlwall, 1979).

Crawton Bay

Description

Carroll (1994) estimated the Crawton Volcanic Formation to be 70 m thick. At Crawton Bay the rocks dip at about 13° to the WSW within the hinge zone of the Strathmore Syncline and are cut by minor normal faults that trend ENE. The parallel alignment of tabular feldspar phenocrysts up to 25 mm in length, gives the macroporphyritic basaltic andesites a platy structure. These flows generally have slaggy upper and lower surfaces. A columnar-jointed central portion commonly has potholes in the centres of the hexagonal columns as the rock close to the cooling joints is more resistant to wave erosion (Figure 9.20).

The base of the lowest flow, exposed near Trollochy (8805 7970), is more vesicular than the main part of the flow and contains disorientated feldspar phenocrysts. Lenses of laminated sandstone and mudstone underlying this flow have been disrupted in places and are baked by the lava (Trewin, 1987). The flow contains large amygdales filled with chalcedony, clear quartz, amethyst and calcite (MacGregor, 1996).

The second and third flows are separated by a few centimetres of sedimentary rock; the top of the second flow is marked by thin, impersistent red mudstones and blocks of altered lava, the reddening being due to a period of subaerial weathering (Trewin, 1987, fig. 1). Both the flows show a well-developed flow orientation of the feldspar laths. The top of the third flow is irregular, and in places the slaggy top was eroded prior to the deposition of the overlying conglomerate in the potholed surface.

The uppermost flow is a purplish massive basalt with scattered vesicles, which are locally over 10 cm in diameter. The vesicles are generally filled with calcite and quartz, but brick red stilbite is also recorded (Trewin, 1987).

Thick interbeds of clast-supported or matrix-supported conglomerate resting upon irregular eroded surfaces of lava are good evidence of penecontemporaneous erosion. These conglomerates consist of well-rounded pebbles of andesitic lava, psammite and quartzite with lesser amounts of metabasalt and greywacke within a matrix of poorly sorted volcaniclastic coarse-grained sandstone. The coarse fraction is of broadly 'Highland' provenance (i.e. Highland Border Complex and probably some Dalradian) although there is a component of locally derived lava.

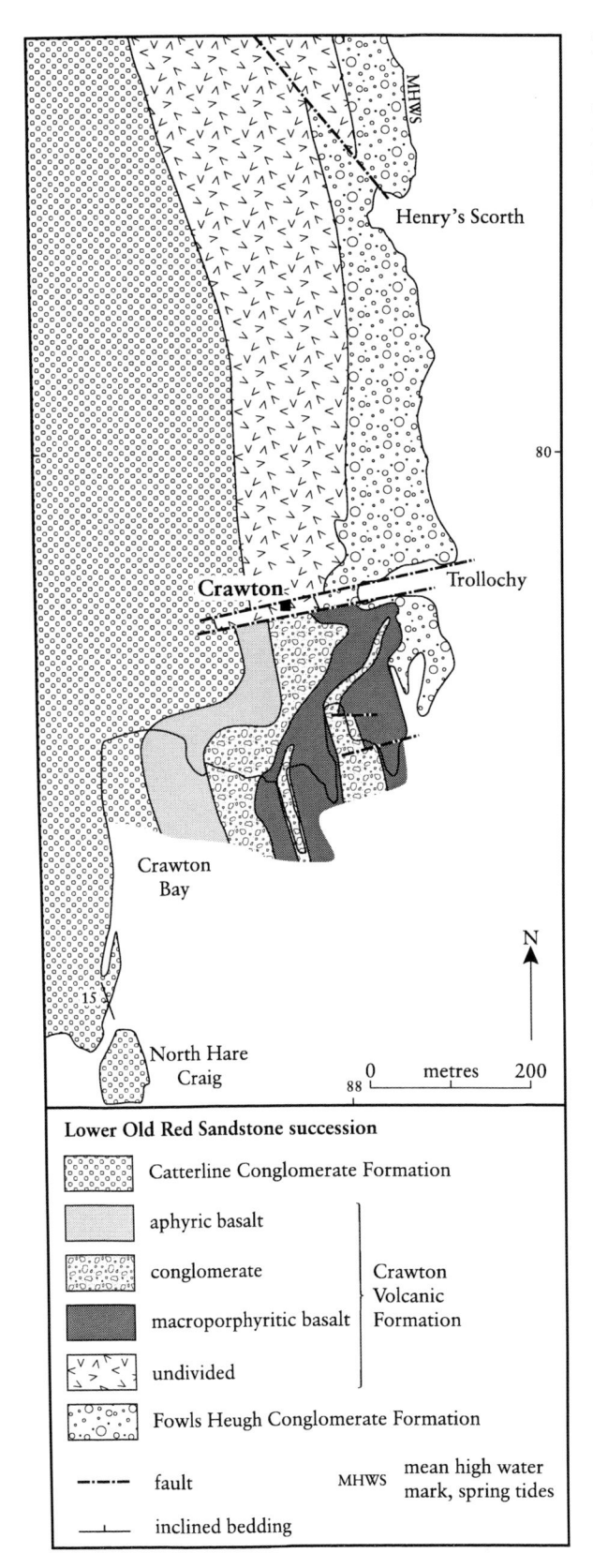

Figure 9.19 Map of the Crawton Bay Volcanic Formation at Crawton Bay.

Late Silurian and Devonian volcanic rocks of Scotland

Figure 9.20 Differential weathering in hexagonal columns of Crawton type basalt, Crawton Bay. (Photo: BGS no. D2454.)

Interpretation

Both types of lava present in the formation contain microphenocrysts of olivine and augite, with the 'Crawton type' also having megacrysts of plagioclase. The few available analyses have SiO_2 in the range 51–53% spanning the basalt/basaltic andesite division (Thirlwall, 1979), and they generally have high K_2O and related elements, making them shoshonitic *sensu* Le Maitre (1989). They are olivine-hypersthene normative, with moderately high alumina (16%), but the one analysed sample of 'Crawton type' shows substantial iron enrichment and hence has tholeiitic tendencies. The variation trends in trace elements may be accounted for by fractional crystallization of the phenocryst phases olivine, plagioclase and clinopyroxene, although some minor opaque oxide may have been involved.

The Crawton Volcanic Formation contains some of the oldest lava flows exposed within the Lower Old Red Sandstone in the northern Midland Valley, and Carroll (1994) considered the formation to be uppermost Silurian in age. This conclusion was reached by combining the evidence from the palynology of the overlying Arbuthnott Group (lowest Devonian according to Richardson *et al.*, 1984) and the age dating of Lower Old Red Sandstone lavas elsewhere in the northern Midland Valley (close to or before 410 Ma according to Thirlwall, 1988). The 'Lintrathen Porphyry', a dacitic ignimbrite in the Alyth area, north of the Highland Boundary Fault, dated at 415.5 ± 5.8 Ma (Thirlwall, 1988) has been correlated with the dacitic 'Glenbervie Porphyry' at the top of the Crawton Group, which gives a possible age for the Crawton Volcanic Formation. Because the Lintrathen Porphyry crops out to the north of the Highland Boundary Fault, Trench and Haughton (1990) considered that the scope for relative movement between the northern Midland Valley and Grampian terranes after Lower Old Red Sandstone deposition is only of the order of tens of kilometres.

Crawton Bay provides an excellent example of a palaeoenvironment in which subaerial/non-marine lavas poured out on land or into shallow water, cooled, cracked and became partially eroded, before being covered by clastic sediment mainly derived from an exotic source. In this case the Lower Old Red Sandstone lavas appear to have accumulated in a subsiding rift basin close to the Highland Boundary Fault, with the Grampian Terrane to the NW of the fault feeding in coarse clastic detritus. Further geochemical studies and age dating of both the lavas and the interbedded 'Highland' clasts could reveal details of the evolution and provenance of the succession.

Conclusions

The Crawton Bay GCR site is the best exposed and type section through the Crawton Volcanic Formation, which forms a significant marker at the top of the Crawton Group. These intercalated Lower Old Red Sandstone lavas and conglomerates have been studied since the end of 19th century because of the fine exposures showing the stratigraphical relationship between volcanic and sedimentary rocks in a marginal rifted basin. They provide important evidence of the environment of deposition of both the lavas and the sediments in the precursor to the Strathmore Basin, which developed within the northern Midland Valley; and also the volcanic setting that heralded the more widespread phase of Lower Old Red Sandstone lava eruption within both the Strathmore Basin and the adjacent Highlands.

SCURDIE NESS TO USAN HARBOUR (NO 734 567–726 545)

R. A. Smith

Introduction

The outcrops on the east coast of Angus, south of Montrose are the best exposures of the lower part of the Montrose Volcanic Formation, which comprises an older sequence of basaltic andesites (the 'Ferryden lavas') and younger basalts (the 'Usan lavas'). This formation is intercalated with clastic sedimentary rocks of the Arbuthnott Group, within the Lower Old Red Sandstone succession of the Strathmore Basin (Armstrong and Paterson, 1970). The thick sequence of lava flows is chiefly composed of olivine basalts and orthopyroxene-feldspar-phyric andesites with few intercalated conglomerates. Structures within the flows and their relationships to penecontemporaneous sedimentary rocks are well exhibited.

This excellent section was first described by Geikie (1897) and later in more detail by Jowett (1913) and Robson (1948). Pillows within the Usan lavas were described as the best examples Jowett had seen along this coast. Heddle (1901) recorded that calcite, chalcedony, agate and chloritic minerals occur within the amygdales. Some of the lavas within this section have been analysed as part of a petrochemical study of the British Lower Old Red Sandstone lavas (Thirlwall, 1979).

Description

The 'Ferryden lavas' are exposed at the north end of the section (733 567) around Scurdie Ness ('scurdie' is the local term for the lavas), and farther south along the coast (732 564) they are faulted against the younger 'Usan lavas' (Figure 9.21). The trace of a large open ENE-trending anticline, the Ochil-Sidlaw Anticline, transects Scurdie Ness, so that the section southwards along the coast passes from the oldest exposed lavas of the Montrose Volcanic Formation, up through the younger flows. The dip on the southern limb of the anticline is about 10° to the SSE. The Montrose Volcanic Formation in this area is estimated to be about 200 m thick with the lava pile thinning out to the NE and SW.

The Ferryden lavas around Scurdie Ness nearly all contain conspicuous feldspar phenocrysts. These lavas were termed enstatite-olivine basalts by Jowett (1913) and were classified as andesites by Robson (1948). Although the lavas are mainly dark-grey in colour, they locally weather greenish, purple or brownish-red. The flows rarely exceed 3 m in thickness and the base of each is irregular with a flinty, chilled zone. In places, the chilled lavas appear to have formed pillow shapes. The coarser-grained centre of the flow is capped by a purplish weathered slaggy, amygdaloidal top, which is commonly fissured. The amygdales and cavities at Scurdie Ness contain green chloritic infill and agates of various sizes. Veins of chalcedony and gypsum are also present. Lenses of hard, green or reddish-brown, fine-grained sandstone have infilled the fissures from an overlying bed. Subsequent lava flows have incorporated and disturbed the fine-grained, slightly calcareous sandstone not only baking it but causing it to buckle, break up and

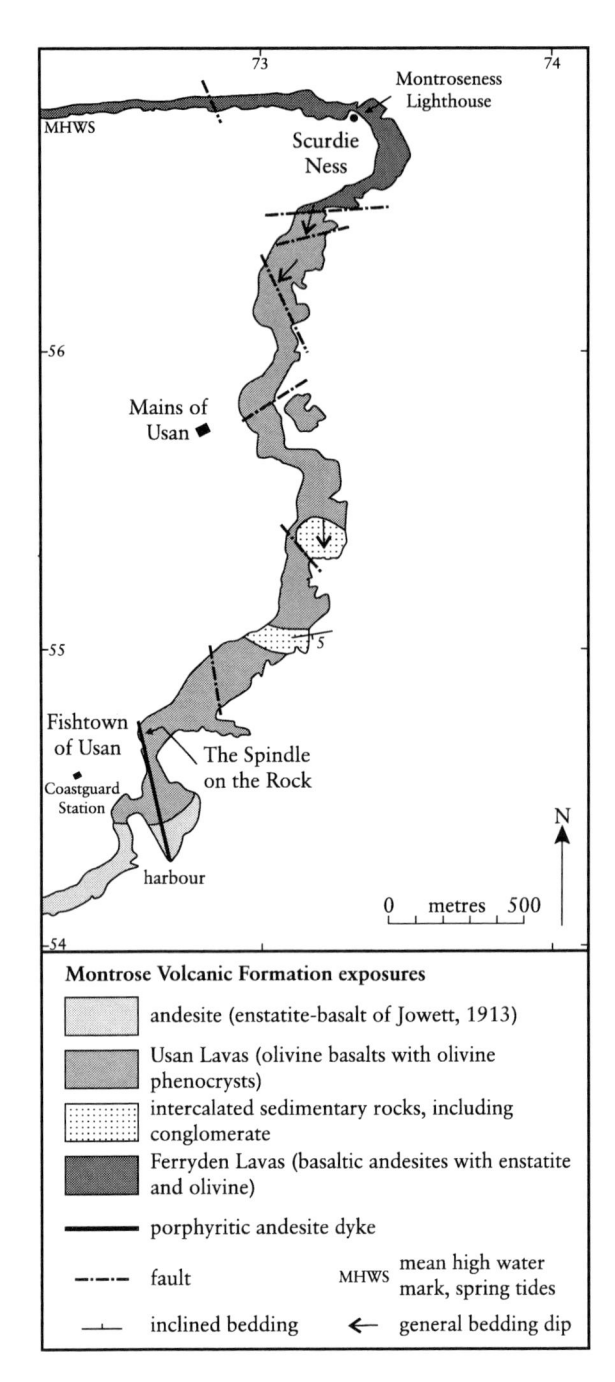

Late Silurian and Devonian volcanic rocks of Scotland

Figure 9.21 Map of coastal exposures of Montrose Volcanic Formation between Scurdie Ness and Usan Harbour, adapted from Jowett (1913).

Montrose Volcanic Formation exposures:
- andesite (enstatite-basalt of Jowett, 1913)
- Usan Lavas (olivine basalts with olivine phenocrysts)
- intercalated sedimentary rocks, including conglomerate
- Ferryden Lavas (basaltic andesites with enstatite and olivine)
- porphyritic andesite dyke
- fault
- inclined bedding
- MHWS mean high water mark, spring tides
- general bedding dip

extend from the fault southwards to Usan Harbour. They were described as olivine basalts by Jowett (1913) and as basalts by Robson (1948). These lavas are fine grained with reddish pseudomorphs after phenocrysts of olivine and they lack the conspicuous feldspar phenocrysts of the Ferryden lavas. Fresh 'Usan lava' is dark-grey but depending on the amount of weathering, ranges from shades of purple, brown and red into a lilac colour, particularly in the amygdaloidal tops. Fissures in the slaggy and brecciated tops are filled with sandstone, but locally the base of an overlying flow lies directly on sandstone-filled, fissured, compact lava, which suggests that any slaggy top or conglomerate originally intervening had been removed. Locally, there are intercalated lenses of reddish cross-bedded sandstone and conglomerate containing pebbles of locally derived lava (Figure 9.22).

Jowett (1913) mapped two thicker intercalations of conglomerate and sandstone within the Usan lava sequence. The northernmost, about 500 m SE of Mains of Usan (at 732 554), is a clast-supported, coarsening-upwards conglomerate (Figure 9.22) containing well-rounded boulders of volcanic rock, 0.6–1 m across, with a lens of red sandstone up to 0.45 m thick at its base. The boulders, some of which are feldspar-phyric lavas (i.e. not only the local Usan lavas), are set in a matrix of coarse-grained volcaniclastic sandstone. Spheroidal pillows with a concentric arrangement of amygdales occur in the lower part of the lava flow overlying this sedimentary intercalation. The other thick conglomeratic intercalation lies about 700 m SSE of Mains of Usan (at 730 550), and rests on an irregular channelled surface in the lava below. All the boulders in this conglomerate are volcanic rocks with feldspar phenocrysts (? Ferryden type lavas). On the coast, east of Fishtown of Usan, The Spindle on the Rock (725 547) is part of a NNW-trending, 2 m-wide dyke of porphyritic andesite (Jowett, 1913). On the rock-platform to the SE of this dyke, which constitutes the southernmost outcrops of the GCR site (727 544), Jowett mapped some of the lava flows as enstatite basalts; these are characterized by large plagioclase phenocrysts and a lack of olivine.

Interpretation

These lavas are inferred to be some of the earli-

even produce amygdales within the sediment itself (Jowett, 1913). Just north of the fault that separates the Ferryden from the Usan lavas, an intercalated lens of coarser sandstone is almost entirely composed of volcanic detritus.

The Usan lavas are well exposed on the wave-cut platform and in small sea stacks, which

Scurdie Ness to Usan Harbour

Figure 9.22 Bedded pebble and boulder conglomerate composed of lava clasts in coarsening up intercalation between lavas. Usan lavas, 500 m SE of Mains of Usan (at 732 554). (Photo: R.A. Smith.)

est of the Montrose Volcanic Centre, which lay to the NE, now under the North Sea (Geikie, 1897). The volcanic succession, exposed in the core of the Ochil–Sidlaw Anticline, must be at about its thickest in the Scurdie Ness to Usan Harbour area although its base is not seen. The lack of pyroclastic rocks within this sequence suggests that the eruptions from the Montrose Centre were essentially quiescent.

The Ferryden lavas are chiefly basaltic andesites, containing phenocrysts of labradorite and enstatite; olivine is restricted to the groundmass together with augite and andesine (Jowett, 1913; Robson, 1948). Some of the enstatite is intergrown with augite (?exsolution) but more commonly each phenocryst of enstatite has a border of granular augite. The Usan lavas were described by Jowett as olivine basalts with altered phenocrysts of olivine set in a fine-grained matrix of augite and feldspar. However, Robson (1948) recorded microphenocrysts of andesine and some groundmass feldspar of higher alkali content. A distinct flow-structure is indicated by alignment of feldspar laths and elongation of amygdales. The enstatite basalts without olivine (Jowett, 1913), in the upper part of the Usan lavas, were classed as andesites by Robson (1948) and attributed to fractionation of the more basic magmas.

Some lavas from this site have been analysed by Thirlwall (1979) as part of his 'Ferryden Member', which comprises all the lavas from Ferryden to Lunan Bay (i.e. both the Ferryden and Usan lavas). He divided the lavas into geochemical ranges using Al_2O_3 contents and Zr/Nb ratios. The alumina ranges from 15.8 to 16.1% for the Ferryden lavas as described above, and from 16.9 to 17.3% for the Usan lavas. He concluded from a study of the trace elements that the variation is unlikely to be the result of fractional crystallization.

The Scurdie Ness to Usan Harbour section shows in excellent three-dimensional detail the relationships between the lava flows of both the Ferryden and Usan lavas and the intercalated sedimentary rocks. The intermittent outpourings of fairly fluid gaseous lavas were punctuated by relatively short episodes of local erosion and deposition of fine sediment that accumulated in shallow waters. A slow but steady rate of subsidence is inferred within the Strathmore Basin, for although the lavas were hardly eroded before the deposition of the sediment, there was a considerable amount of penecontemporaneous oxidation (reddening) of the tops of lava flows. The fact that thicker, coarser conglomeratic beds are intercalated with the younger lavas might be due to increasing time intervals between flows as volcanic activity waned, or to changes in local topography as the lava pile built up. The local development of pillows at the bases of flows, as recorded by Jowett (1913), suggests that in places the lavas flowed into shallow water. On the contrary, Robson (1948) stated that pillow-structure is not known from these lavas, which he suggested precluded any suggestion that they formed in deep water. However, Robson agreed that the evidence from the fracturing and infilling of the lavas by sediment implied deposition in shallow water.

Interaction of the molten lavas with wet sediment can be studied in this section; as the wet sediment was ripped up and incorporated into the lava, it appears that the water was vaporized before the sediment was baked so that amygdales eventually developed in the sedimentary rock as well as in the lava. A study of the section could prove examples of fluidization by intrusive magmas as described from rocks of similar type and age from Ayrshire (Kokelaar, 1982)(see the Port Schuchan to Dunure Castle, Culzean Harbour and Turnberry Lighthouse to Port Murray GCR site reports).

Conclusions

The Scurdie Ness to Usan Harbour coastal section is of national importance, both as a representative of the lower part of the Lower Old Red Sandstone, Montrose Volcanic Formation and for its value in reconstructing the environment and evolution of the Montrose Volcanic Centre. The olivine basalt to basaltic andesite lava sequences are excellently exposed along this coast. The main distinction between the Ferryden and Usan lavas is that the former are enstatite-feldspar-phyric with only groundmass olivine, whereas the latter are mainly fine-grained olivine-phyric basalts. Detailed contact relationships with the sedimentary intercalations are of special interest in this section together with possible examples of lava–wet sediment interaction.

BLACK ROCK TO EAST COMB (NO 694 488–703 476)

R. A. Smith

Introduction

This section of the Angus coast exposes the 'Ethie lavas', the upper part of the Montrose Volcanic Formation in the Arbuthnott Group of the Old Red Sandstone succession in the Strathmore Basin (Armstrong and Paterson, 1970) (Figure 9.2). The three-dimensional exposure makes the section eminently suited to detailed study of sediment–lava contacts. These volcanic rocks and interbedded sandstones were first described by Fleming (1818). He used their complex relationships as an argument in favour of the Wernerian hypothesis, which considered that basalt was deposited from an aqueous fluid (see the Forest Lodge GCR site report). Subsequently the lavas were described by Geikie (1897) as sheets of andesite or 'porphyrite' erupted from the Montrose Volcanic Centre. Jowett (1913) produced a more detailed geological description and map (Figure 9.23). The lavas were placed at the base of the former 'Red Head Series' by Robson (1948) but he did not study them in detail. Although down-faulted Devono-Carboniferous rocks intervene between the 'Ethie lavas' and the lower part of the Montrose Volcanic Formation (see the Scurdie Ness to Usan Harbour GCR site report), the 'Ethie lavas' are clearly the younger and are succeeded by clastic red beds of the Red Head Formation of the Garvock Group. Some of the Ethie lavas have been analysed by Thirlwall (1979) as part of a regional geochemical study of Lower Old Red Sandstone lavas.

Description

The lavas are dull purplish-grey to green in colour and vary from compact to extremely amygdaloidal with large cavities. Some have roughly polygonal cooling cracks. Their most striking features are the detailed contacts with the intercalated sedimentary rocks. Fissures and large cavities in the lavas have been filled with pale-green and red sandstone and some of the slaggy lava tops, which can be 3–4 m thick, have interstices filled by horizontally stratified pale-green sandstone.

In his petrological study of the area, Jowett

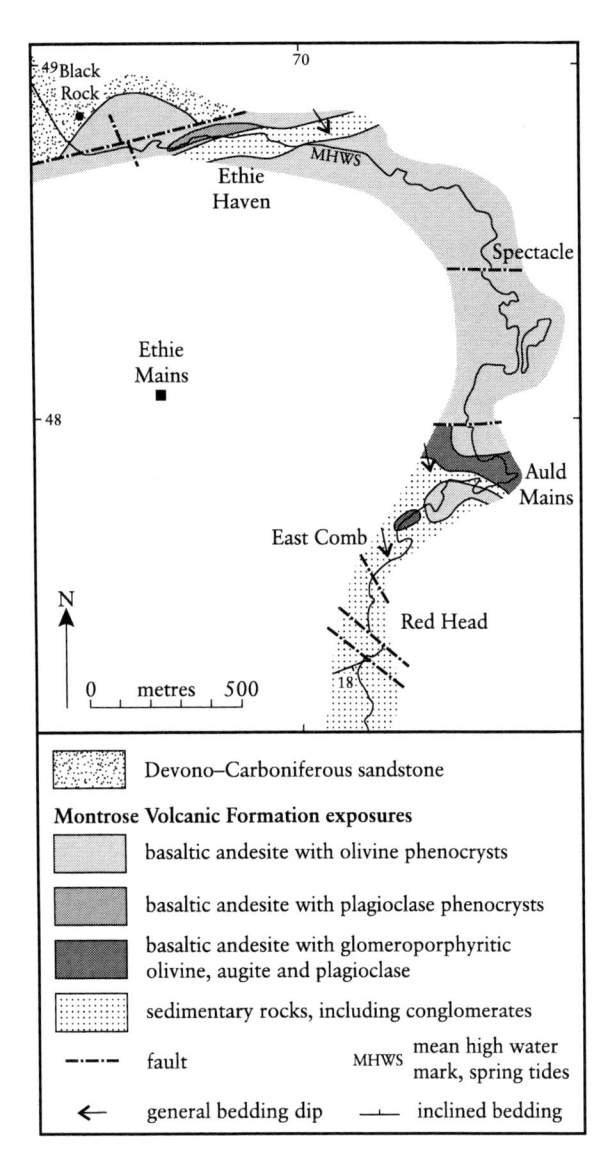

Figure 9.23 Map of coastal exposures of Montrose Volcanic Formation between Black Rock and East Comb, adapted from Jowett (1913).

(1913) described all the lavas as 'olivine basalts', most of which could not be distinguished in the field except for one feldspar-phyric type exposed in the lower part of the cliff at Ethie Haven (698 488) and which continues west for about 300 m. The typical 'Ethie lavas' are probably basaltic andesites; they contain phenocrysts of altered olivine and microphenocrysts of labradorite, augite and magnetite, set in a holocrystalline to partly glassy groundmass. All the olivine has been pseudomorphed by serpentine and haematite. Some of the youngest lavas in the succession, around Auld Mains (706 478) and in the bay to the SW (703 476), are less basic and contain glomeroporphyritic aggregates of olivine, augite and labradorite (Jowett, 1913). No orthopyroxene has been found in the Ethie lavas (cf. the Ferryden lavas of the Scurdie Ness to Usan Harbour GCR site). A flow structure is seen in some of the lavas due to the alignment of small feldspar laths.

The section is here described from north to south, which is generally up the sequence (Figure 9.23). Black Rock is a breccia within red sandstones of Devono-Carboniferous age that are presumed to be unconformable on the Lower Old Red Sandstone lavas. The latter are well exposed in the cliffs to the SE of a major fault trending ENE. Above the basaltic andesites and the feldspar-phyric basaltic andesite at Ethie Haven (699 487), there is a local unconformity overlain by a thick red conglomerate. Within the conglomerate are at least two thin flows of fine-grained basaltic andesite. The top surface of the conglomerate is irregular and it is overlain by fine-grained basaltic andesite.

Rugged coastal scenery has developed to the south, where the softer amygdaloidal zones of the basaltic andesites have been eroded away. The bases of some flows are pillowed and, locally, rounded blocks of highly amygdaloidal lava are set in scoriaceous lava or volcaniclastic breccio-conglomerate. The slaggy and brecciated parts of the flows are commonly paler coloured due to replacement by calcite. Thick beds of conglomerate, comprising lava fragments in a pale-green medium-grained sandstone matrix, are intercalated with the lavas near Spectacle (705 484). North of Kirk Loch (705 481), elongate, flow-orientated amygdales occur within the base of a lava flow overlying a volcaniclastic breccia (Figure 9.24).

South-west of Auld Mains (at 704 477), there are excellent exposures showing the relationship between a clast-supported conglomerate containing large rounded boulders of volcanic rock and an underlying, irregularly eroded lava surface. At Auld Mains (706 478), a resistant flow of lava infills and covers the uneven surface of the same conglomerate. The lower part of this lava is platy with partings parallel to the surface of the conglomerate. The compact part of the lava is at least 10 m thick and has rough columnar-jointing, but the slaggy top has been eroded away locally beneath the next conglomerate. The irregular slaggy top of the uppermost lava flow is overlain disconformably by volcaniclastic conglomerate and red sandstone of the

Figure 9.24 Base of lava with elongate amygdales above a lava breccia with a pale sandstone matrix. Ethie lavas, north of Kirk Loch (at 705 481). (Photo: R.A. Smith.)

Red Head Formation; this exposure has been figured by Geikie (1897).

Interpretation

The Black Rock to East Comb GCR site contains a well-exposed section through the lavas at the top of the Montrose Volcanic Formation which means that, except for minor outpourings in the overlying Garvock Group, these lavas are the youngest in the main Lower Old Red Sandstone lava sequences of the northern Midland Valley (Figure 9.2). Spores of Early Devonian age have been found in the intercalated sedimentary rocks of the Arbuthnott Group in Angus (Richardson *et al.*, 1984).

Thirlwall (1979) analysed phenocrysts from the Ethie lavas (his Ethie Haven Member); the olivines are Mg- and Ni-rich, the clinopyroxenes are low-Na diopsidic augites, and all but one of the analysed samples has bytownite–labradorite microphenocrysts. The analysed samples are quartz- or slightly olivine-normative, and the SiO_2 content ranges from 53.8–55% (Thirlwall, 1979). The lavas may therefore be classified as basaltic andesites. The 'Ethie lavas' have high MgO, Ni (*c.* 150 ppm) and Cr (350–540 ppm) contents (Thirlwall, 1979), suggesting that they are derived from primitive, mantle-derived magmas. The Ni and Cr contents are much higher than those for the Dunnichen and East Hills lavas in the Montrose Volcanic Formation to the west, and according to Thirlwall the two lava successions could be related by fractional crystallization with olivine and clinopyroxene as the major crystallizing phases.

The lavas are thought to have emanated from the Montrose Volcanic Centre, which lay to the NE, now under the North Sea, and to have accumulated in a slowly subsiding basin in which the interplay of successive lava flows with the intercalated sediments was complex. The fissured and scoriaceous zones permeated by fine-grained laminated siltstones could suggest that the sediment was washed in by shallow current action after the flow had cooled and cracked. However, the local development of pillow structures and the development of amygdales in the incorporated sediment suggests that the lavas flowed on to wet sediment. Hence, some of the lavas may have locally ploughed or intruded into wet sediment and caused fluidization of the sediment (cf. Kokelaar, 1982; and see the Port Schuchan to Dunure Castle, Culzean Harbour and Turnberry Lighthouse to Port Murray GCR site reports), which would account for the examples where the sediment has been homogenized and forced along a maze of narrow cracks. Because of the widespread reddening of the tops of the lavas and the local unconformities described at their tops, it is thought that most of the basaltic andesites are lava flows and not intrusive sheets into wet sediments, although this is an aspect of the geology that would require future study.

The intercalated conglomerates include boulders of volcanic rock and this indicates that during the longer periods between lava eruptions, strong currents deposited lenticular bodies in channels cut into the lava pile. The sandstones, siltstones and mudstones were probably deposited on the adjacent banks during periodic floods. South of Auld Mains, increasing amounts of mainly volcaniclastic sedimentary rocks occur within the succession, and above a local unconformity (at 702 477), the Red Head Formation begins. This clastic formation coarsens upwards towards the south, and the higher beds contain pebbles of quartzite (Jowett, 1913); it is therefore inferred that larger river systems became established, which eventually managed to bring in well-rounded quartzite pebbles from the Highlands to the north.

Conclusions

The Black Rock to East Comb coastal section is of national importance as a representative of the Ethie lavas, which are the youngest in the Lower Old Red Sandstone Montrose Volcanic Formation. The section provides an excellent opportunity to examine the relationships between the lavas and intercalated sedimentary rocks at the transition between the Arbuthnott and Garvock groups.

The lavas were probably erupted from a volcanic centre now under the North Sea. Most are probably basaltic andesites but the youngest ones may be less basic; as such they are typical of the calc-alkaline suite of Lower Old Red Sandstone lavas in the northern Midland Valley. They were erupted in a terrestrial shallow-water environment and have complex relationships with intercalated sedimentary rocks, partly due to the interaction of hot lava with wet sediment. The intercalated lenses of coarse conglomerate, composed mainly of volcanic detritus, indicate that initially the local drainage channels were sourced within the Strathmore Basin, but with the cessation of volcanic activity in this area, rivers draining the higher land to the north entered the basin.

BALMERINO TO WORMIT (NO 356 248)

M. A. E. Browne

Introduction

The coastal cliff sections and foreshore beside the Firth of Tay between Balmerino and Wormit provide the most complete section through the Lower Old Red Sandstone, Ochil Volcanic Formation in Fife (Figure 9.25). The GCR site illustrates the nature and environment of this volcanic activity and complements that of the Sheriffmuir Road to Menstrie Burn GCR site at the SW end of the Ochil Volcanic Formation outcrop. The eastern half of this section is described in a field guide by MacGregor (1996). A 350 m-thick sequence of basaltic to andesitic lavas and volcaniclastic sedimentary rocks is intercalated with sandstones and minor claystones of the Dundee Formation. Rocks from the section have provided radiometric ages and micro- and macro-palaeontological evidence, both indicating an Early Devonian (Lochkovian) age.

Description

The stratigraphy of the area has been described by Geikie (1902) and Armstrong *et al.* (1985). The lavas, usually less than 7–9 m thick, consist mostly of basalts and basaltic andesites, which are usually distinguishable only by geochemical analysis (Thirlwall, 1979). They vary from coarsely to finely feldspar-phyric and from fine- to medium-grained aphyric. The feldspar is labradorite–andesine and usually co-exists with small altered phenocrysts of one or more of forsteritic olivine, bronzitic orthopyroxene, titanomagnetite and diopsidic augite. Hornblende has not been reported and phenocryst biotite is restricted to the acid igneous rocks (Thirlwall, 1979). The lavas are commonly heavily altered and weather to a brown, greenish-grey (chloritic), or purplish and reddish-grey mottled appearance. When fresh they are grey

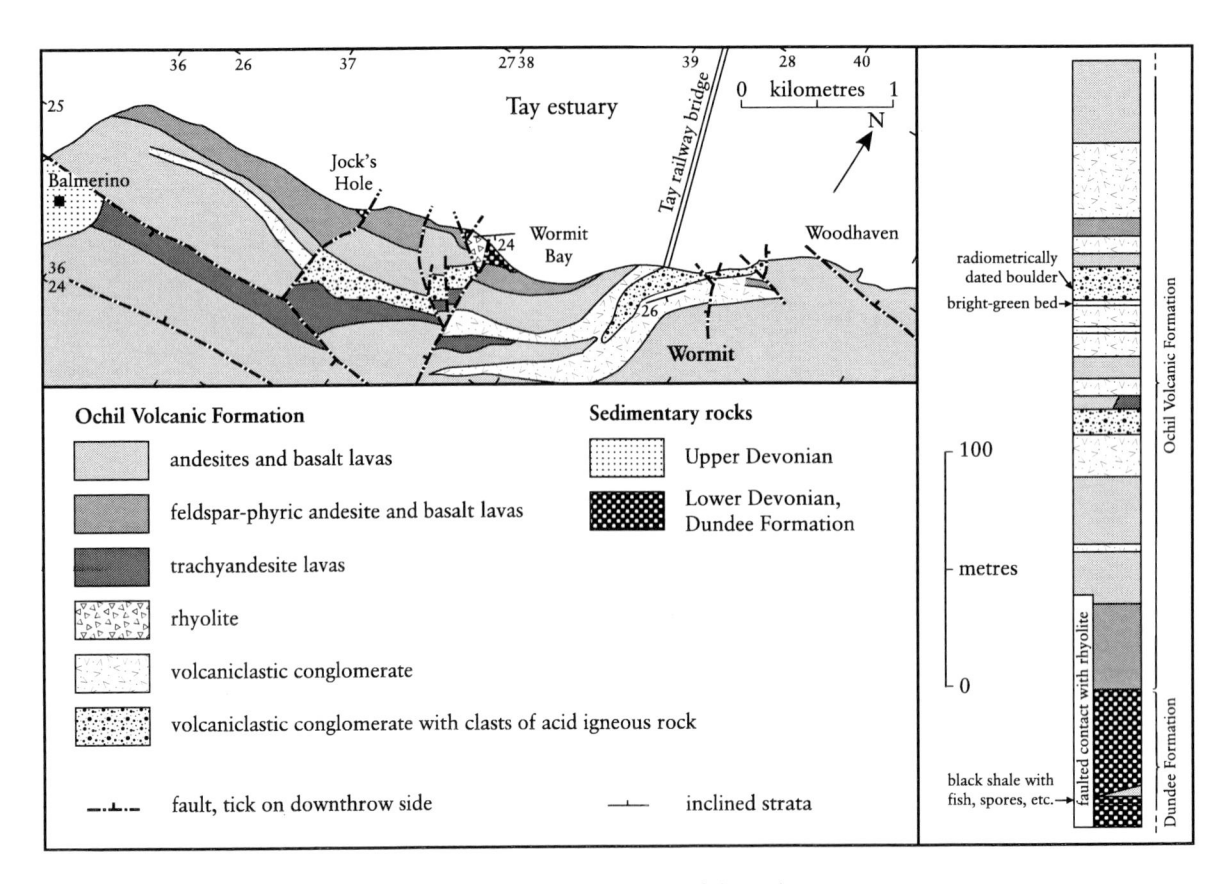

Figure 9.25 Map and generalized vertical section of the Balmerino to Wormit GCR site.

or less commonly brown. They show features such as autobrecciation, amygdales and infiltrated sediment in fissures and between blocks. Possible pillow lavas have been identified. Individual flow bases are not always easy to recognize in the absence of lateritic alteration. However, because the lava flows are interbedded with both volcaniclastic and non-volcaniclastic sedimentary rocks of the Dundee Formation (Arbuthnott Group), the section displays loadcast relationships between flow bases and underlying shales. Uneven, possibly eroded flow tops are infilled by sandstone in places, but elsewhere peperites formed by the interaction of hot magma and wet sediment are clearly seen (MacGregor, 1996) (Figure 9.26).

At Peasehill Point (3832 2580) highly altered and weathered andesite flows are faulted against a pale-grey, fawn and salmon-pink colour-banded rhyolite. On the western edge of the rhyolite outcrop, this thin flow-banding is near-parallel to the adjacent fault plane, but away from this contact it is highly irregular. The eastern contact of the rhyolite is with a rhyolitic breccia, which includes blocks of sandstone and shale. In places the rhyolite appears to intrude the breccia, which itself intrudes the sandstone country rock. Two possible origins have been suggested for the rhyolite. Harry (1956) suggested that it is a vent intrusion with the breccia pre-dating the rhyolite, whereas Geikie (1902) followed by Armstrong *et al.* (1985) suggested that it is a lava within a vent that has been faulted down into its present position.

Probably all the volcaniclastic rocks are sedimentary rather than pyroclastic. They consist predominantly of massive, thickly bedded, poorly sorted conglomerates and breccias in beds up to 9 m thick. The clasts in the conglomerates consist of basalt, basaltic andesite and rhyolitic lavas. The rhyolite clasts are probably not of local derivation and conglomerate units have been distinguished by their absence or presence and relative abundance. The clasts can be sub-angular to rounded and are usually less than 30 cm, but over 2 m across locally. The matrix is composed of grains of similar materials and quartz. Greenish-grey, cross-bedded, volcaniclastic sandstones are also present as impersistent lenses and beds.

Figure 9.26 Lens of well-laminated sandstone between igneous sheets on the foreshore west of the Tay Bridge at Wormit. The lower sheet has a sharp but irregular, apparently intrusive, contact with the sandstone; the upper sheet shows evidence of magma–wet sediment interaction, with fragments of igneous material separated by irregular veins of pale sandstone and peperitic texture well seen on the right. (Photo: M.A.E. Browne.)

The interbeddded Dundee Formation strata consist of cross-bedded, sometimes graded sandstones and finer-grained beds of siltstone and shale. The colours of these sedimentary rocks range from grey to green to red and also yellow. They are fluvial and lacustro-deltaic in origin and form part of the infill of the Strathmore Basin, which extends south-westward from Stonehaven across the whole of the Midland Valley of Scotland. Marshall *et al.* (1994) have described the vitrinite reflectivity of the carbonaceous shales at Wormit as a small part of a study of the Strathmore region. They concluded that maximum burial of 3–5 km, and therefore thermal maturity occurred during the late Carboniferous. This has major consequences for understanding the nature, depth and number of late Silurian to Early Devonian sedimentary basins in the eastern Midland Valley.

Interpretation

Thirlwall (1983b) analysed trace elements, including rare earths, and Sr, Nd and Pb isotopes of lavas within and around this site. He found all the samples from the site to be primitive basalts and andesites (high Mg, Ni and Cr) in contrast to samples farther south, which are more typical of modern calc-alkaline suites with less than 30 ppm Ni and Cr. He reported Rb-Sr age determinations on biotite and plagioclase separations from a rhyolite boulder in what he described as the lowest conglomerate in the local succession. The age was reported as 406.5 ± 5.6 Ma, later adjusted to 410.6 ± 5.6 Ma (Thirlwall, 1988).

The dark-grey, carbonaceous shales near Peasehill Point have yielded a fish fauna, including *Brachyacanthus*, *Ischnacanthus* and *Mesacanthus* (Westoll, 1951). The arthropods *Kampecaris* and *Pterygotus* and the plant *Parka decipiens* have also been found. Richardson *et al.* (1984) re-assessed the age of some Arbuthnott Group sedimentary rocks from the Strathmore region on the basis of a re-investigation of spore assemblages. At Wormit, samples associated with, but stratigraphically beneath the rocks dated by Thirlwall (1983b, 1988) as 410.6 Ma, yielded assemblages of Early (but not earliest) Devonian age belonging to the *micrornatus–newportensis* Zone (lower and middle subzones) of the Lochkovian Stage. The Wormit GCR site is possibly the only Lower Old Red Sandstone site where radiometric and biostratigraphical dating methods are so closely related. The radiometric dates are very close to the Silurian–Devonian boundary on most recently published time-scales and until recently may have been regarded as late Silurian. However, the palaeontological data indicate early Lochkovian (Gedinnian) and this is now accepted. Therefore this site not only provides a stratigraphical and radiometric date for the late Caledonian volcanicity in this part of the Midland Valley, but potentially it has even greater implications internationally for the dating of the Silurian–Devonian boundary.

Conclusions

The Balmerino to Wormit GCR site is a particularly important site both nationally and internationally. The former, because it represents a key 350 m-thick succession of the Ochil Volcanic Formation in the eastern Midland Valley, and the latter because of the intimate relationship it shows between the volcanic rocks and the interbedded fossiliferous claystones within the Dundee Formation (Arbuthnott Group). Studies of organic matter in the carbonaceous shales provide estimates of depth of burial and degree of maturity, which have contributed to the understanding of the development of the late Silurian to Early Devonian sedimentary basins in the eastern Midland Valley.

Geochemical studies have characterized the basalts and andesites as primitive calc-alkaline types related to subduction during the final closure of the Iapetus Ocean, and have contributed to more detailed suggestions regarding the magma source. Radiometric investigations on this site and elsewhere in the Midland Valley estimate the age of the late Caledonian magmatism as 415–410 Ma. This has been substantiated by spores which, together with the macrofauna, including fossil fish, show the succession to be early Lochkovian (Early Devonian). Hence there are possible international implications for the age of the Silurian–Devonian boundary.

SHERIFFMUIR ROAD TO MENSTRIE BURN
(NS 813 979–852 973)

M. A. E. Browne

Introduction

This well-exposed part of the Ochil Hills shows many of the characteristic features of the volcanic and volcaniclastic rocks of the Lower Old Red Sandstone, Ochil Volcanic Formation. The 600 m-thick sequence in the Ochil Fault scarp, from the summit of Dumyat down to the bottom of Menstrie Glen (Figure 9.28) consists of sub-aerial lava-flows interbedded with volcaniclastic rocks variously interpreted as being of pyroclastic and sedimentary origin. Detailed descriptions are given by Francis *et al.* (1970) and accounts of the petrology and geochemistry of lavas in the western Ochils in general are given by Taylor (1972) and Thirlwall (1979). A suite of dykes and sills adds to the interest of this site, which complements the Balmerino to Wormit GCR site at the NE end of the Ochil Volcanic Formation outcrop. The site also forms part of the classic Ochil Fault escarpment view, as seen from Stirling Castle and the Wallace Monument.

Description

A measured section of the volcanic succession, derived from this area, has been given by Francis *et al.* (1970, pp. 31–32) (Figure 9.27). It consists of at least 20 lavas, from 3 to 30 m thick, interbedded with volcaniclastic rocks in units up to 50 m thick. In the past the latter rocks have largely been described as of pyroclastic origin.

The lavas are mainly basalts, basaltic andesites and andesites with a few 'trachyandesites'. The basalts and andesites are normally only distinguished from each other by their geochemistry. They range from fine- to medium-grained and from micro- to macro-porphyritic in texture. Most are feldspar-phyric (labradorite–andesine) but clinopyroxene (diopsidic augite), orthopyroxene (bronzite) and olivine phenocrysts are also present in many rocks, and there are some hornblende-phyric andesites (Thirlwall, 1979). Thirlwall noted that the rocks described as trachyandesites by Francis *et al.* (1970) and classed as dacites by Taylor (1972) are not notably more alkaline than many other Ochil andesites. The term is, however, considered as a useful misnomer to describe these sparsely porphyritic and slightly more siliceous rocks. The lavas are usually grey when fresh but more commonly purplish where altered and weathered. They may be vesicular or amygdaloidal, and autobrecciation is common, but this is difficult to detect unless the voids between blocks or fissures contain infiltrated sediment. Flow texture and jointing are developed in some of the flows.

The volcaniclastic rocks, which form a significant part of this succession, consist of crudely and thickly bedded, massive, commonly matrix-

Sheriffmuir Road to Menstrie Burn

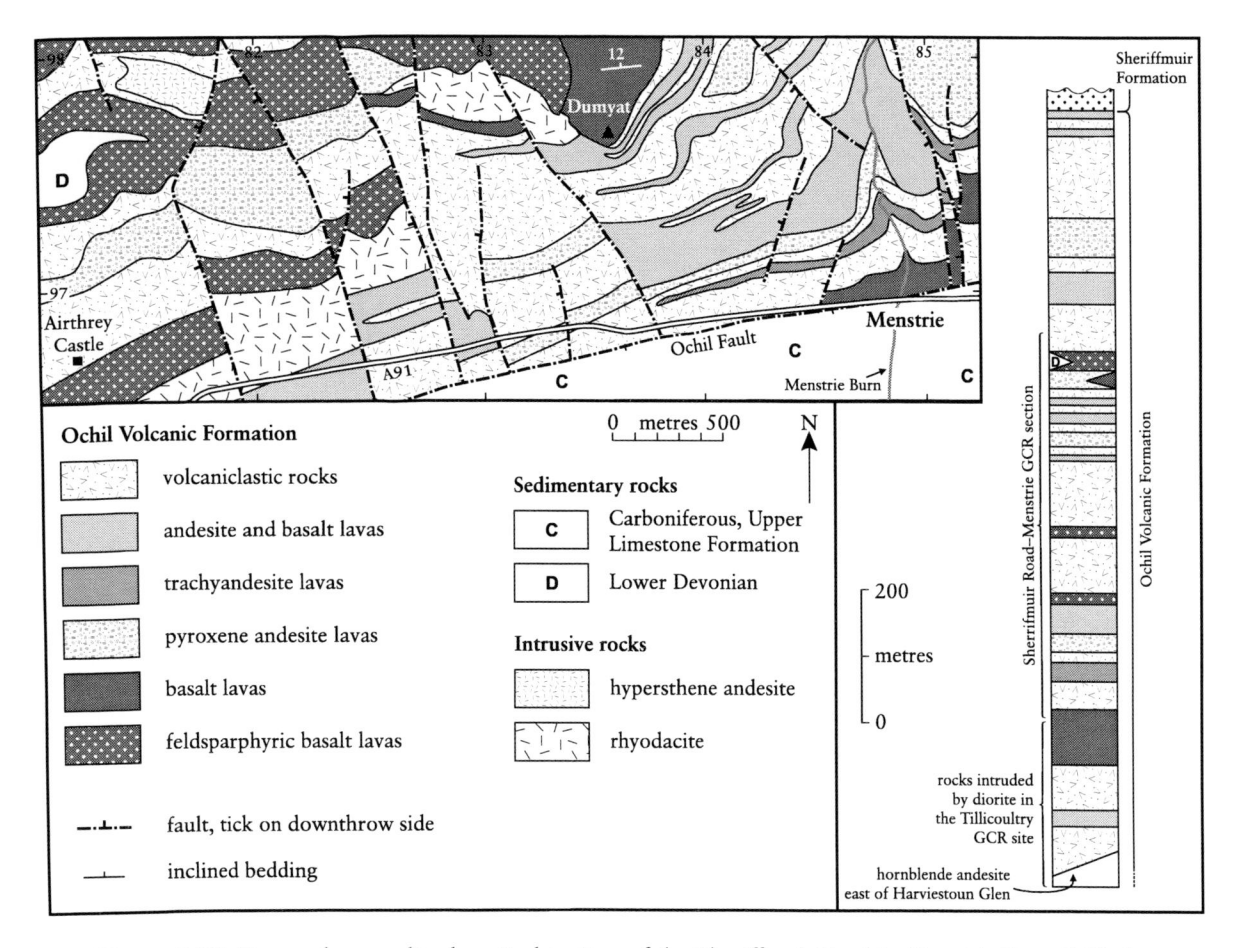

Figure 9.27 Map and generalized vertical section of the Sheriffmuir Road to Menstrie Burn GCR site.

supported breccias and conglomerates with only one example of volcaniclastic siltstone noted. The largest clasts are up to 1.2 m across, with an average size being 30 cm or less. They are extremely poorly sorted, with the matrix consisting mainly of volcanic detritus of sand to silt grade.

A sill of possible rhyodacite, over 100 m thick, forms pink-weathering crags and screes on the main fault scarp and in the foothills NW of Blairlogie. This intrusion forms an important marker as it is displaced upwards by faults to the higher slopes west of Dumyat. The rock in the centre of the sill contains scattered microphenocrysts of albitized plagioclase and altered amphibole and orthopyroxene in a cryptocrystalline matrix of quartz and alkali feldspar. The marginal aphyric rocks are pale and blotchy and show some resemblance to the autobrecciated lavas.

The succession is broken by a number of significant faults, mostly trending NNW. Locally these and related joints are mineralized (Dickie and Forster, 1976); the age of mineralization is possibly early Permian (Hall *et al.*, 1982).

Interpretation

According to Francis (1983), the more-proximal zones of central volcanoes of assumed heights of 1–3 km and diameter 40 km at the time of extinction, would not survive erosion to become buried. He suggested that volcanic outcrops of much greater linear extent are likely to represent overlapping or interdigitating products of more than one volcano. The extent of the linear outcrop forming the Ochil and Sidlaw hills far exceeds that of one edifice (Francis, 1983, fig. 22). The volcanic rocks preserved are suggestive of the lower proximal to distal zone of composite volcanoes (Williams and McBirney, 1979) with both moderately thick lavas and massive to thickly bedded volcaniclastic debris-flow (or

Late Silurian and Devonian volcanic rocks of Scotland

Figure 9.28 The Ochil Fault scarp above Menstrie. The scarp face exposes a 600 m-thick sequence through the Ochil Volcanic Formation, dipping gently to the north (right). Menstrie Glen is in the bottom right and the prominent summit is Dumyat. (Photo: BGS no. D2066.)

laharic) breccias and conglomerates predominant (rather than pyroclastic ash-fall or distal ash-flow tuffs and agglomerates). Only one possible vent structure has been identified in the Ochil Hills (now under the waters of the Upper Glendevon Reservoir) (Francis *et al.*, 1970), and it is uncertain whether the diorites of the Tillicoultry GCR site and elsewhere represent roots of volcanoes or late stage intrusions. In the above context, the Sheriffmuir Road to Menstrie Burn succession of basalts, andesites and volcaniclastic rocks represents at least one 600 m-thick fragment of a composite (or strato) volcano, the whole volcanic pile locally being over 2400 m thick in the Ochil Hills. There is little evidence for the alternative view that the lavas were erupted from fissures.

Thirlwall (1988) reported a clinopyroxene-whole rock Rb-Sr age of 416.1 ± 6.1 Ma from a lava a little to the north of this GCR site (NN 835 019).

Craig Rossie

Conclusions

The importance of this site lies in the thick sequence of the Ochil Volcanic Formation laid out largely on the scarp face of the Ochil Fault. The succession is of national importance in understanding the origin and architecture of a major volcanic pile and it complements that of the Balmerino to Wormit GCR site at the NE end of the Ochil Hills. On a regional scale, this calc-alkaline assemblage, with its associated geochemical and radiometric data, contributes to the understanding of late Caledonian subduction and closure of the Iapetus Ocean in northern Britain.

CRAIG ROSSIE (NN 980 125)

M. A. E. Browne

Introduction

The acid volcanic rocks around Craig Rossie in the Ochil Hills form only a small part of the outcrop of the Lower Old Red Sandstone, Ochil Volcanic Formation. In general this formation consists of a wide variety of basaltic and andesitic lavas interbedded with volcaniclastic rocks for which the GCR sites of Sheriffmuir Road to Menstrie Burn and Balmerino to Wormit provide adequate reference. In contrast, acidic flows are areally restricted and are confined to high stratigraphical levels within the volcanic pile. The Craig Rossie rhyodacite, on the NW limb of the Ochil Anticline, is well exposed in hillside exposures, in the cliffs of Craig Rossie itself and in a quarry at 9770 1277 (Figure 9.29). The flow is of local distribution and is believed to occupy a topographical low in the pre-existing volcanic landscape. It is a feldspar-biotite-quartz-phyric rock that shows well-developed flow-banding picked out by colour variation.

The hill of Craig Rossie is a good example of an escarpment with dip slope. The summit and the eastern escarpment (Figure 9.30) have been affected by major landslips, formed after the

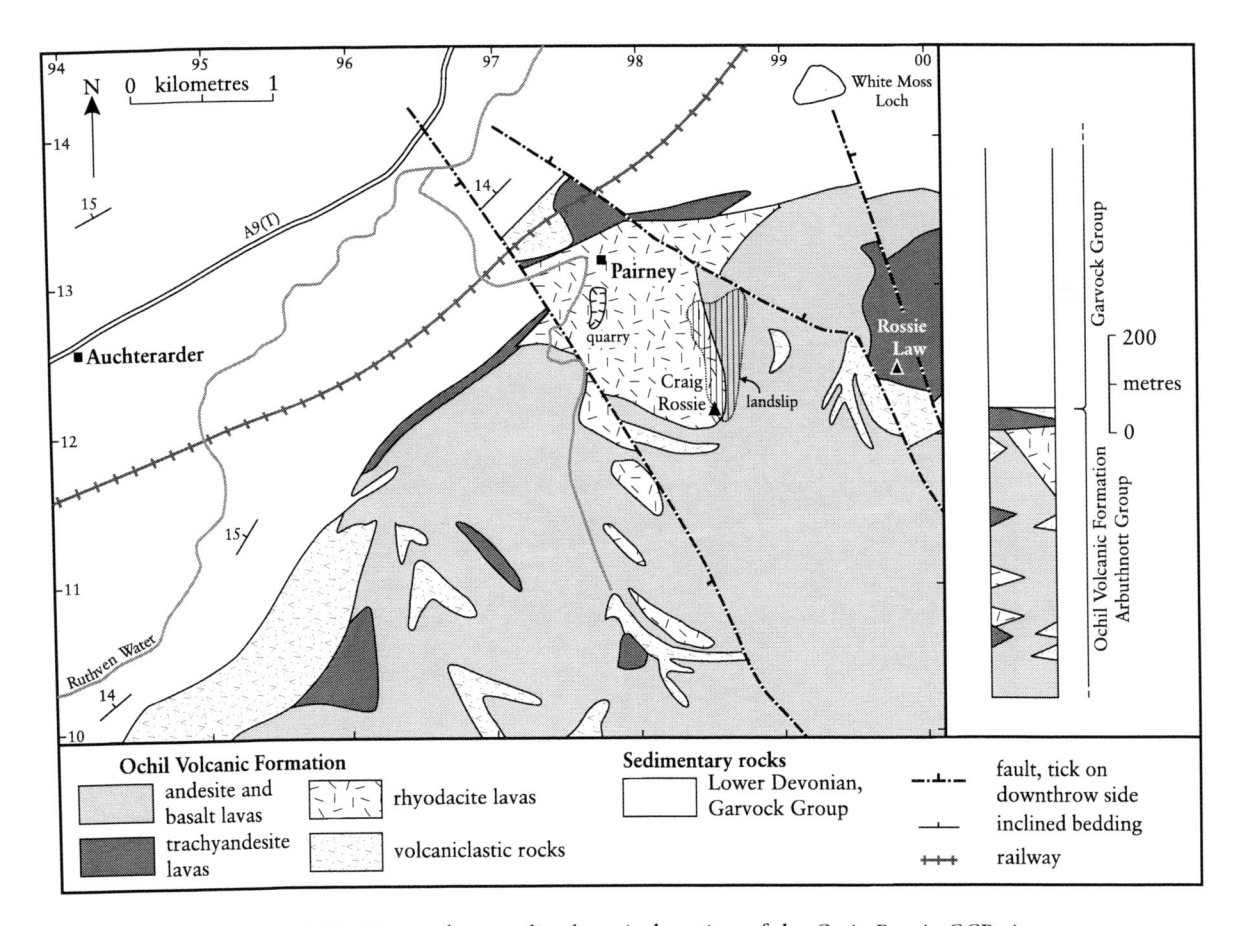

Figure 9.29 Map and generalized vertical section of the Craig Rossie GCR site.

retreat of the last ice sheet from the area less than 15 000 years ago.

Description

The Ochil Volcanic Formation in the Craig Rossie area generally consists of altered and weathered basalt and basaltic andesite lavas, which are commonly vesicular and autobrecciated (Geikie, 1900; Francis *et al.*, 1970; Taylor, 1972). These rocks are not conspicuously porphyritic but olivine-, feldspar-, clinopyroxene- and hypersthene-phyric types are present as well as aphyric forms. Interbedded in this succession are trachyandesites, rhyodacites and some volcaniclastic rocks. The Craig Rossie rhyodacite is immediately overlain by a trachyandesite which is the youngest flow in the Ochil Volcanic Formation of this area. A unit of volcaniclastic rocks of uncertain origin is present within the succession a little beneath the rhyodacite. It consists of gravel- to pebble-sized clasts of volcanic material, generally of poorly rounded to subrounded shape, in a massive, greenish-grey, poorly sorted sandy matrix. It is not known whether the matrix is entirely derived from volcanic rocks or whether it includes quartz grains derived from sedimentary rocks.

The Craig Rossie rhyodacite is brown, dull red or pink in colour. It contains phenocrysts, particularly of pale-coloured feldspar but also of quartz and biotite, in a fine-grained crystalline matrix. Its petrography and geochemistry are described by Francis *et al.* (1970) and by Taylor (1972). As far as is known, the rhyodacite is a single lava and no separate flow units have been recognized. Its most striking feature is the widespread occurrence of colour-contrasted flow-banding due to subtle mineralogical differences. This flow-banding developed as the cooling magma passed through a viscous phase, developing folds just prior to solidification. The folds

Figure 9.30 The east face of Craig Rossie; rhyodacite lava forms the summit and the lower crags are landslips. (Photo: BGS no. D2181.)

are usually open in form but locally become overfolded. Undeformed flow-banding is seldom parallel to the base of the flow. A feature of the rhyodacite in the quarry (9770 1277) is the presence of xenoliths, 5–10 cm across, of coarse-grained igneous material. These may be of gabbroic or dioritic rock by analogy with those found at the Rossie Braes Quarry (NO 249 120) near Auchtermuchty in Fife, where fragments of layered cumulate have also been found (Thirlwall, in Armstrong et al., 1985, p. 36). Other features noted include amygdales of calcite and red chalcedony near the supposed top of the rhyodacite and patches of fine-grained sedimentary rock, which have have been recorded as infilling fissures or spaces between autobrecciated blocks near the base.

The volcanic sequence at Craig Rossie dips to the NW at between 20 and 30°. It is cut by at least two significant NW-trending faults that throw down the base of the rhyodacite to the SW and NE respectively. Other smaller faults with similar trend are also present, together with related joints. Small crush zones associated with the faulting have voids infilled with chalcedony.

Interpretation

Armstrong (in Francis et al., 1970) concluded that the Craig Rossie rhyodacite occupies a depression in pre-existing lavas, is of lenticular form and thins out both eastwards and westwards. The evidence for the topography is to be found in the Pairney Burn and in an adjacent gully eroded along a NW–SE fault. Here the flow overlies an amygdaloidal basalt and the base can be traced. If the base is parallel to the flow-banding, the underlying basalt should occur along strike in a deeply cut glacial channel a short distance to the NE. Since the basalt does not occur here, the rhyodacite base must descend north-eastwards from the visible outcrops in a manner discordant to the flow-banding and thus presumably occupies a depression.

The rhyodacite forms a minor acidic part of the generally basalt and basaltic andesite assemblage of the Ochil Volcanic Formation. While noting the marked silica gap between the more basic rocks ($SiO_2 < 62\%$) and the rare 'rhyolites' ($SiO_2 > 74\%$), Thirlwall (1983b) believed that the acid rocks could be generated by fractional crystallization from some of the more mafic lavas, even though extensive liquid lines of descent are not present. In this context, it is not surprising that rhyodacitic flows are restricted to high stratigraphical positions in the volcanic pile.

Conclusions

The importance of this site lies in the excellent exposures and geomorphological expression of this rare, flow-banded, acidic lava within the primarily basaltic and andesitic Ochil Volcanic Formation. Craig Rossie thus provides a complement to the areally more extensive GCR sites between Balmerino and Wormit, and in the Sheriffmuir Road to Menstrie Burn area. The pre-existing topography below the flow base is an interesting feature of the site. The rhyodacite is important as a representative end member of fractional crystallization processes which may have operated in the late Caledonian magmas. Further study of the recently found coarse-grained xenoliths may provide information concerning possible parental material and this would add to the conservation value of this site.

TILLICOULTRY (NS 914 980, 918 982, 923 984 AND 931 987)

M. A. E. Browne

Introduction

The cliffs, ravines and hillside exposures associated with the Ochil Fault scarp around Tillicoultry display four diorite stocks that intrude related lavas and volcaniclastic rocks of the Lower Old Red Sandstone, Ochil Volcanic Formation within a 6 × 1 km thermal aureole (Figure 9.31). The stocks are cut by members of a contemporaneous radial dyke-swarm (Figure 9.32) and are truncated by the Ochil Fault. They are also cut by the late Carboniferous quartz-dolerite fault intrusion (Figure 9.33). The diorites are notable for their locally heterogeneous and hybrid character and show both sharp and diffuse contacts with the hornfelsed country rocks. The diffuse contacts, reflecting contamination of the magma by incorporation of xenolithic material, are well seen in the deeply incised Harviestoun Glen where a number of poorly defined enclaves of hornfelsed volcanic rock are apparently disposed in layers within the diorite.

Description

The diorite stocks are intruded into basalt and andesite lavas and volcaniclastic rocks of similar character to those in the nearby Sheriffmuir Road to Menstrie Burn GCR site. There are four masses of diorite; from west to east these are Castle Craig, Mill Glen, Wester Kirk Craig and Elistoun Hill. They are about 30–135 m apart and each is cut off to the south by the Ochil Fault. The Castle Craig and Mill Glen stocks are about 200 m in diameter, the latter with a 350 × 30 m westward-trending dyke-like extension, whereas those at Wester Kirk Craig and Elistoun Hill are 1050 m and 1250 m in diameter respectively. The diorites range in grain-size and colour from fine grained and grey to coarse grained and blue-grey or green with pink mottling. Transitions in grain-size occur on a metre scale and, although margins tend to be of fine-grained grey diorite, the latter's occurrence is not confined to contact zones. The heterogeneous nature of the diorites is further accentuated by the presence of numerous country-rock xenoliths, especially near to margins. As a consequence of the presence of xenoliths, there are only a few localities where the margins of the diorites are sharply defined. Where seen, the walls of the intrusions are vertical or highly inclined outwards. This suggests that the four masses may converge and unite at depth.

The Castle Craig diorite is fine- to medium-grained, with xenoliths of volcanic rock generally confined to the contact zone, some showing various stages of transition from hornfelsed lava to diorite. The Mill Glen rock is coarse grained and is mottled-pink and bluish-green. The dyke-like apophysis varies from fine- to coarse-grained, reflecting the presence of a chilled margin. The Wester Craig diorite is variable, being fine- to medium-grained in the north and west and coarser and of the mottled-pink form elsewhere. The Elistoun rock is also fine- to medium-grained in the west but is usually coarse grained elsewhere. It exhibits the best examples of rapid transition from hornfelsed lava into coarse-grained diorite.

The petrography of the diorites has been described by Francis *et al.* (1970). They are composed largely of plagioclase feldspar with varying proportions of ferromagnesian minerals (clinopyroxene, orthopyroxene, amphibole and biotite), and with sufficient quartz to classify them as quartz-diorites. The coarse-grained diorites may be subdivided into four types, three of which are present in the GCR site. A two-pyroxene diorite with little amphibole occurs within all of the Tillicoultry stocks and a variant with uralitic amphibole but little pyroxene occurs in several. A granodioritic variant, with

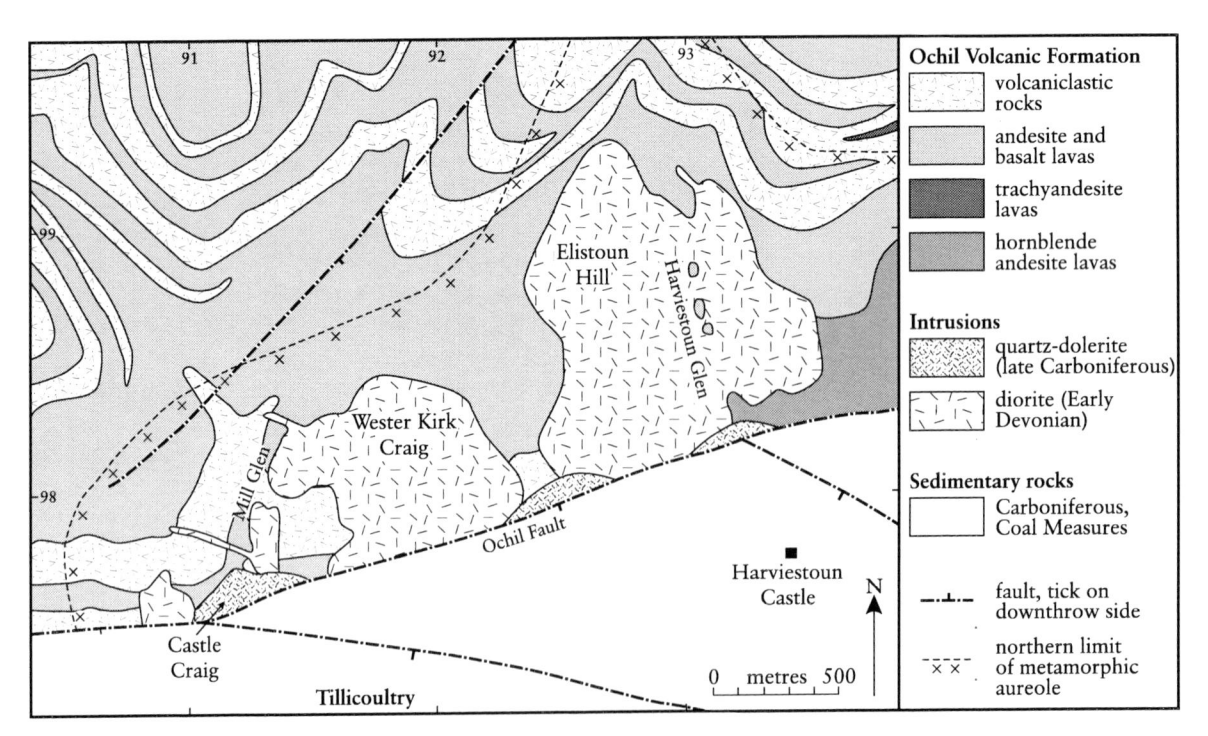

Figure 9.31 Map of the Tillicoultry GCR site. The general succession is similar to that shown on Figure 9.27.

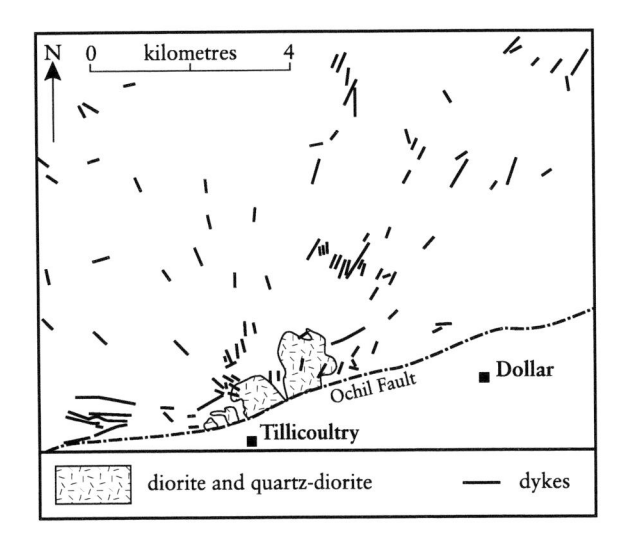

Figure 9.32 Sketch-map showing radial disposition of dykes relative to diorite stocks in the Ochil Hills, after Francis *et al.* (1970).

pink potash-feldspar and no pyroxene, is restricted to parts of the Mill Glen stock and its dyke-like apophysis. Francis described the fine-grained diorites as grey, saccharoidal rocks sharing many of the characteristics of both hornfelsed country rocks and coarse-grained diorites. In varieties most closely related to the hornfelsed lavas, plates of biotite are the only fresh ferromagnesian mineral. These fine-grained variants present some of the best evidence for the formation of some of the diorites by metamorphism of the country-rock lava. In general, the petrographical differences between the igneous and metamorphic/metasomatic varieties are that the former are relatively rich in quartz and potash-feldspar, whereas the latter tend to be richer in pyroxene.

Pink aplitic veins and segregation patches in the diorites and aureole are clearly a late-stage magmatic event. Diorites cut by the aplites show contact alteration but the veins themselves are not chilled and have gradational contacts with the host rock. Similar gradational contacts have been noted with some of the hornfelsed volcanic rocks.

Numerous small andesitic dykes ('porphyrites' and 'plagiophyres' on old maps) show a radial disposition relative to the diorite stocks (Francis *et al.*, 1970) (Figure 9.32). Many are similar in composition to the lavas, consisting of olivine basalts, pyroxene- and hornblende-phyric andesites, trachyandesites and albitized equivalents. Within the diorites, the dykes usually have irregular, gradational and unchilled contacts, indicating near-contemporaneity and their radial distribution probably reflects the stress pattern generated by emplacement of the stocks.

The diorite stocks all occur within a single extensive thermal metamorphic aureole that is always at least 300 m wide. The degree of alteration of the country rocks depends on their permeability and their distance from the diorite. The volcaniclastic rocks and autobrecciated lavas are generally more permeable and they alter to pale-pink, grey or even greenish-grey, partly amorphous rocks. The outlines of clasts become indistinct at an early stage of alteration and pink feldspar porphyroblasts develop. Amygdales in lavas recrystallize to ill-defined pink patches. Some lavas become hackly-looking and, when traced into the diorite, they pass through massive hornfels into metasomatically produced fine-grained diorite.

Interpretation

There seems little doubt that the diorite bodies are genetically linked to the extrusive rocks into which they are emplaced. They are probably of similar age, accepting the radiometric age of 411.4 ± 5.6 Ma for a related small olivine diorite body at Glenfarg, about 20 km to the ENE of Tillicoultry, and of 416.1 ± 6.1 Ma for an andesite lava flow at NN 835 019 just north of the Sheriffmuir Road to Menstrie Burn GCR site (Thirlwall, 1983b, 1988). These, together with other dates from the north Midland Valley, suggest that the late Caledonian volcanic activity in this area occurred during the limited interval of 415–410 Ma (see the Balmerino to Wormit GCR site report).

Large strips of vaguely defined hornfelsed country rock that crop out in the floor of Harviestoun Glen are possibly enclaves within the Elistoun diorite. These enclaves and their associated hybridization underlie non-hybridized, coarse-grained diorite, giving a sub-horizontal layered effect and Francis *et al.* (1970) recorded that the diorites here show differences in composition from one layer to another. Further evidence of hybridization (and metasomatism) cited by Francis *et al.* (1970) is the presence of 'ghost' scarp-featuring inherited from the subsumed lavas on Elistoun Hill. The 'ghosts' dip at about 10° to the NW in conformity

Figure 9.33 Castle Craigs Quarry, Tillicoultry. The quarry is along the line of the Ochil Fault plane; the back face exposes hornfelsed lavas and volcaniclastic conglomerates of the Ochil Volcanic Formation and the crags on the left are quartz-dolerite of the late-Carboniferous fault intrusion. (Photomosaic: M.A.E. Browne.)

with the regional dip of the rocks of the Ochil Volcanic Formation. Francis *et al.* also noted the way in which different topographical and weathering features may be used to map the generally narrow hybridized zones that delimit the outcrop of the diorites; the lava crags are more angular and the non-hybrid diorite crags are more rounded and massive.

Conclusions

The importance of this GCR site lies in the excellent exposures of the late Caledonian diorites that are intruded into the lavas and volcaniclastic sedimentary rocks of the Ochil Volcanic Formation. These include purely magmatic quartz-diorites and hybridized pyroxene-rich diorites. The changes caused by thermal metamorphism of the country rocks are also a feature. In particular the often diffuse contacts between the altered country rocks (hornfels) and the diorites, with ghost-like features within the hybridized diorites, contrasts with those where wholly magmatic diorite is in sharp contact with country rock. It may be concluded that the diorites were emplaced partly by simple intrusion and assimilation, with radial fracturing of the surrounding rocks, and partly by metasomatic replacement of country rocks.

PORT SCHUCHAN TO DUNURE CASTLE
(NS 247 152–252 159)

G. Durant

Introduction

Lower Old Red Sandstone volcanic rocks crop out along the Ayrshire coast at Dunure, Culzean and Turnberry where they occupy a total area of approximately 40 km². The Dunure coastal exposures are part of the Carrick Hills outcrop; those of Culzean and Turnberry are smaller down-faulted inliers (see the Culzean Harbour

Port Schuchan to Dunure Castle

and Turnberry Lighthouse to Port Murray GCR site reports) (Figure 9.34). The Ayrshire coast volcanic rocks conformably overlie the local Lower Old Red Sandstone succession of sandstones, conglomerates and cornstones and are in turn overlain unconformably by Upper Old Red Sandstone sedimentary rocks. The volcanic rocks are predominantly calc-alkaline andesites, suggested by Thirlwall (1981a) as being the product of subduction-related volcanism. The basal lavas are interbedded with the topmost beds of the sedimentary succession and sedimentary intercalations occur throughout the volcanic sequence suggesting that the volcanic rocks are the same general age as the sedimentary rocks.

Smith (1910) estimated that the volcanic rocks of the Dunure area are approximately 130 m thick in total, although a much higher figure of 300 m was given by Eyles *et al.* (1949). This section is important because of the complex relationship between volcanic rocks and intercalated sedimentary rocks, which has led to the suggestion that much of the sequence consists of subvolcanic sills rather than lava flows (Kokelaar, 1982). The volcanic units will therefore be referred to as 'sheets' in the following accounts. Petrographically the volcanic rocks are augite-, enstatite-, and pyroxene-olivine andesites with rare basalts near to the base of the sequence. The andesites are commonly markedly feldsparphyric and also highly vesicular. The vesicles are filled by a variety of secondary products, but it is the agates described by Smith (1910) that makes this section of the coast popular with collectors (see Heddle, 1901; Macpherson, 1989).

Description

A well-exposed sequence of andesite sheets with minor intercalations of sedimentary rock is exposed in the low cliffs, the sea stacks and on the wave-cut platform between Port Schuchan and Dunure. A porphyritic, vesicular andesite forms the wave-cut platform immediately west of Port Schuchan and two prominent sea stacks, the Two Sisters, are isolated outliers of the 7 m-thick overlying andesite sheet. At the base of the south-easterly of the Two Sisters (2468 1523), a highly distinctive pillow of vesicular andesite has cut into the finely laminated sandstone between the andesite sheets (Figure 9.35). The lamination within the sandstone remains close to that of the regional dip even when remnants are isolated by the pillowed andesite. Thin sandstone 'dykes' (up to 2 cm wide) cut through the massive central part of the andesite sheet which forms the stacks. The sandstone between the andesite sheets is intimately mixed with the volcanic rock in places; it can infill vesicles or occur within the andesite as irregular structureless patches, some of which are traceable back to larger more laminated patches. This is well displayed in outcrops near the top of the shore just to the east of the south-easterly of the Two Sisters stacks.

The 12 m-thick andesite sheet forming the distinctive cliff at Dunure Point (2473 1532) to the NE of the Two Sisters, is cut by numerous thin sandstone 'dykes', which follow broadly curving paths from the bottom to the top of the sheet.

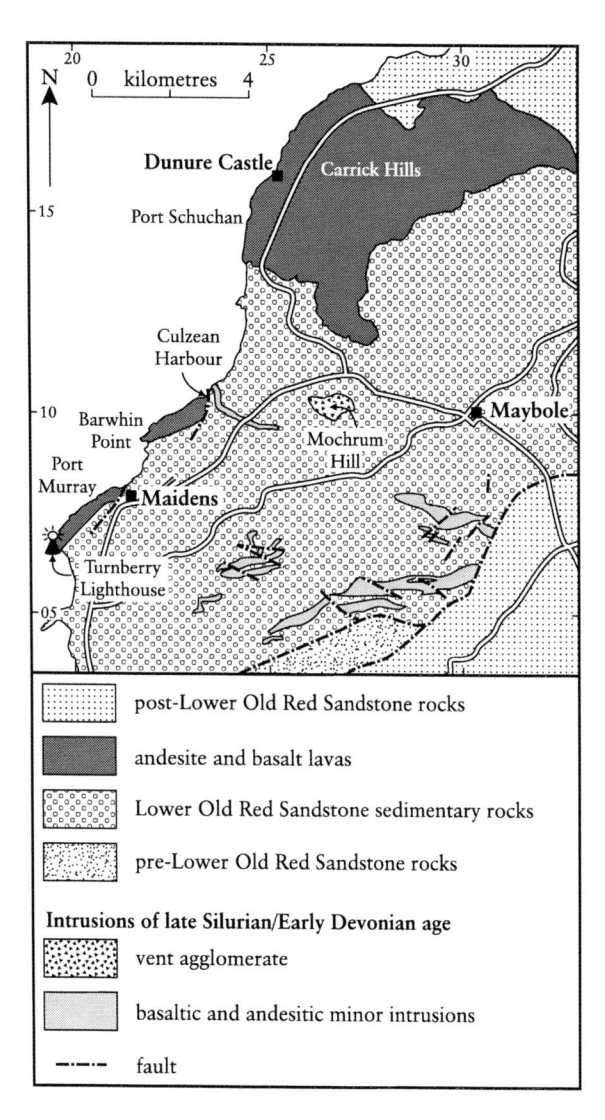

Figure 9.34 Map of the Ayrshire coast outcrops of Lower Old Red Sandstone volcanic rocks.

Where the andesite sheet is seen in horizontal section on the wave-cut platform, the sandstone 'dykes' can be seen to be part of a branching, pseudohexagonal network. The sandstone within the dykes has been hardened and often stands proud, being more resistant than the andesite. This is well displayed to the NE of Broad Crag (2500 1569) and on the shore just below Dunure Castle (2517 1582).

The bases and tops of individual sheets are often markedly amygdaloidal, the original vesicles having been infilled with quartz, agate and calcite. Minor amounts of galena, pyrite, manganese and baryte have been reported as associated minerals (Smith, 1910). Larger irregular masses of quartz and agate also occur within the andesites. A 75 × 50 cm oval mass of quartz occurs near the summit of Mackerel Rock (2471 1544), a low-lying stack north of the Two Sisters and a popular roost for sea-birds. Vertical veins of agate (up to 1 cm wide) also occur, a good example being seen in the stack below Dunure Castle.

Critical exposures for the re-interpretation of the andesite sheets as sill-like intrusions rather than lava flows occur in the low cliffs south of Dunure Castle. A good example of a typical contact between two andesite sheets occurs immediately to the east of Mackerel Rock (2482 1544). In this low cliff section the pillowed and highly vesicular upper contact of the lower andesite sheet encloses and is partially overlain by sandstone, which is laminated in places and structureless elsewhere. Pillows at the base of the upper andesite sheet cut down into the sandstone and sandstone 'dykes' traceable back to this sedimentary enclave run upwards through the upper sheet. This exposure has been figured by Kokelaar (1982, fig. 4B). A good example of the complex upper surface of one of the andesite sheets occurs in a low cliff at the top of the shore to the SE of Scart Rock (2494 1553). Protrusions of andesite from the main mass, penetrate upwards into laminated sandstone. The laminations within a diagonally orientated remnant wedge of sandstone are absent within 2 cm of the contact with andesite (see Kokelaar, 1982, fig. 4C). Such observations led Kokelaar (1982) to postulate a process of sediment removal by fluidization as the lobes of andesite burrowed into wet, unconsolidated sediment. The wet sediment into which the magma was

Figure 9.35 Pillowed lobe of vesicular andesite in laminated sandstone at the base of the south-easterly of the Two Sisters stacks, Port Schuchan (2468 1523). (Photo: G. Durant.)

intruded is now preserved in part as vesiculated sandstone. This feature can be seen on the shore just south of Dunure Castle (2518 1583).

Dunure Castle sits on top of a raised sea stack composed of enstatite andesite (Tyrrell, 1913) which shows a marked internal fabric parallel to the base of the flow. Loose wave-polished blocks beneath the castle show that this fabric is an internal feature caused by textural differences within the intrusion. Such internal textural differences and consequent differential erosion are also responsible for the pseudo-brecciated appearance of the andesite forming Scart Rock (2800 1560). At the base of Dunure Castle (2519 1586) numerous monolithological blocks of andesite occur within light-coloured calcareous sedimentary rock that form the enclaves between intrusive sheets. A fossil arthropod, *Kampercaris tuberulata*, was found by Smith (1909) in rocks close to Dunure Castle, and was subsequently described by Brade-Birks (1923). Rolfe (1980) places the discovery of these arthropod fossils and tracks in an overall evolutionary setting.

Interpretation

The relationship between the sedimentary rocks and the andesites has interested a number of authors over a considerable period since it was first noted in the Geological Survey Memoir for Sheet 14 (Geikie *et al.*, 1869). The survey authors noted vertical veins of sandstone traversing the 'lavas' and suggested 'that the veins were due to sand being washed into the irregular star-shaped cracks of cooled lava before the flow was covered by the next stream of molten matter'. Smith (1892) who discovered fossil arthropod tracks in the fine-grained sedimentary rocks was sceptical about this interpretation. The tracks figured by Smith (1909) are preserved in fine detail 'owing evidently to the fact that in the quiet recesses within the lavas, there would be no commotion to disturb the surfaces of the sedimentary laminae after the markings had been made on them'. A critical observation by Smith (1909) is that 'the lava has sometimes scoured away portions of the sedimentary beds. This is well demonstrated as sometimes a series of footprints will extend right up to the side of a lava ... where one row had been cut by it and the other left'. Heat blisters identified by Smith (1909) also demonstrate that the wet sediment was baked by the heat of the intruding magma.

Geikie (1897) viewed the lava–sediment relationships somewhat differently, believing that the sediments had entered the fissures from above but in a subaqueous environment. However, he also indicated thoughts of another possible explanation (op. cit., p. 283) 'the first and natural inference which a cursory examination suggests is that the molten rock has caught up and carried along pieces of already consolidated sandstone'. He countered this observation with another, 'that the lines of stratification in the sandstone, even in what appears to be detached fragments are marked by a general parallelism and lie in the same general plane with the surface of the bed of the lava in which the sandy material is enclosed'.

Tyrrell (1913) believed that the sediment infilled fissures in cooled lavas and hence that the lavas were subaerial. Eyles *et al.* (1949) also believed that the constancy of the alignment of the bedding of the sedimentary rock within the lavas is due to the sediment having been washed into position, 'the sediment infilling fissures was then greatly hardened possibly because the surrounding lava was still hot when the detritus was deposited in the cavities'.

Micaceous fine-grained sandstones occur as thin and impersistent intercalations between the igneous sheets and as a series of irregular vertical dykes and fissure fillings. They are commonly finely bedded and this bedding is mostly consistent with the regional dip. However, as was pointed out by Kokelaar (1982), the bedding is absent immediately adjacent to the volcanic rock and is now interpreted as being the original bedding of the sediment into which the magma burrowed, in so doing removing most of the unconsolidated wet sediment by fluidization. The bedded sedimentary rock that is observed in fissures within the andesite is therefore the last remaining vestige of a much greater mass of sediment that has been removed by the proposed fluidization process. This re-interpretation of the lava flows as high-level andesite sills emplaced into unconsolidated wet sediment (Kokelaar, 1982) means that the arthropod trackways found by Smith (1909) were probably formed in fine-grained sediment on the bottom of a shallow lake, prior to eruption, a suggestion which is in keeping with the findings of Pollard and Walker (1984), Pollard (1995) and E.F. Walker (1985) based on detailed examination of the fossil tracks.

The presence of multiple andesite sheets, requires a repetition of the conditions that resulted in the burrowing of magma into wet sediment rather than eruption at the surface. This argues for eruption into an actively subsiding sedimentary basin marked by a lake in which sediment accumulation took place in conditions quiet enough to preserve the arthropod trackways. The fine-grained nature of the sediment suggests that there was low relief around the margins of the lake, which was situated in a generally arid environment (Bluck, 1978b). In spite of the high degrees of vesiculation, the andesite magma seems to have been erupted relatively quietly, possibly from fissures. An earlier, possibly more explosive phase of volcanic activity is indicated by the presence of a breccio-conglomerate exposed at Barwhin Point at the SW end of the Culzean inlier (2185 0946), which contains volcanic rock fragments generally more siliceous than any found elsewhere within the overlying Ayrshire coast volcanic sequence (see the Culzean Harbour GCR site report).

Oxygen isotope studies of agates from both Dunure and Turnberry (Fallick *et al.*, 1985) support the idea of a low temperature (*c.* 50°C) origin for the agates from fluid having at least a component of meteoric water.

Conclusions

The Port Schuchan to Dunure coastal section has engaged the minds of geologists for more that 100 years. This section is of national and international importance because of the complex relationship between volcanic rocks and intercalated sediments. The volcanic rocks, predominantly pyroxene andesites, were originally interpreted as subaqueous flows (Geikie, 1897) or subaerial lavas (Tyrrell, 1913). A re-interpretation of the andesites as intrusions into wet, unconsolidated sediment (Kokelaar, 1982) relied on evidence from this section. Fossil arthropod trackways found in the laminated sedimentary rocks between the volcanic units furnish evidence of life in a Siluro-Devonian freshwater lake and Dunure agates, which form amygdales in the andesites, have been much prized by collectors since Victorian times. This is an enjoyable and instructive section, which has an important part to play in the understanding of Lower Old Red Sandstone times and volcanic processes.

CULZEAN HARBOUR (NS 231 102)

G. Durant

Introduction

A down-faulted inlier of Lower Old Red Sandstone andesites crops out along the coast by Culzean Castle (Figure 9.34). The andesites represent a southerly extension of the Carrick Hills volcanic sequence, which is also exposed at Dunure (Port Schuchan to Dunure GCR site) and Turnberry (Turnberry Lighthouse to Port Murray GCR site). A general description of the overall sequence is given in the Port Schuchan to Dunure site description. These coastal outcrops have been studied for over 100 years (e.g. Smith, 1892) and the spectacular setting of Culzean Castle on top of a 35 m-thick andesite intrusion has been a source of inspiration to artists and geologists alike (Figure 9.36). Evidence from this section has been important in the re-interpretation of some of the lavas as sill-like intrusions into wet sediment (Kokelaar, 1982).

Description

The base of the 35 m-thick andesite sheet which forms the main cliff beneath Culzean Castle is seen in the stack just to the north of the Culzean Harbour slipway (2311 1028). This enstatite andesite (Tyrrell, 1913) overlies a coarse debris flow deposit consisting of blocks of andesite and sandstone in a sandstone matrix (Figure 9.37, and see Kokelaar, 1982, fig. 4A). A highly vesicular basal pillowed zone cuts into the greenish grey laminated sandstone on top of the debris flow deposit. The laminations within the sandstone disappear close to the contact with the andesite and the sandstones are oxidized at the contact with andesite pillows (Kokelaar, 1982). The massive andesite is cut by numerous thin sandstone 'dykes' which can be traced back to a source in the sediment beneath the andesite. The pillowed top of the Culzean andesite intrusion contains pockets of structureless sandstone and is overlain by debris flow deposits similar to those beneath the sheet.

The strongly pillowed base of a vesicular andesite forms the low cliff to the south of the Culzean Harbour slipway (2304 1022) just to the north of Dolphin House which sits on the raised beach. Individual pillows are up to 2 m across.

Culzean Harbour

Pale greyish-green fine-grained laminated sandstone occupies the space between the pillows. The bedding laminations in the sandstone are constant throughout the outcrop and they are consistent with the regional dip. A similar relationship is well displayed in the southern wall of a raised sea-arch close to the southern end of the GCR site (2271 1000), where the pillowed lobes at the base of another andesite sheet protrude into sandstone.

Amygdales and veins of agate and quartz within feldspar-phyric andesite are well displayed on the wave-polished surface immediately beneath the small roundhouse close to the southern end of Dolphin House (2299 1018). Vesicular andesites from this section have been described by Geikie (1897) as being some of the most beautiful volcanic rocks in Scotland. Larger, irregular masses of agate and quartz are seen within andesites a short distance to the south where a 1 m-long mass of agate, quartz and calcite, with quartz crystals lining a drusy cavity is exposed in the low cliff. A coarse breccia containing vesicular andesite fragments represents the particularly fragmented part of one andesite intrusion. The space between the andesite blocks is here filled by coarsely crystalline pink calcite rather than by fine-grained sandstone as elsewhere along the section.

Some compositional variation of the volcanic rocks is present in the area around Culzean Castle. The cliffs immediately beneath the castle are formed of enstatite andesite and an intrusive sheet of augite andesite with olivine occurs at Port Carrick, 1 km along the shore SW of Culzean (Tyrrell, 1913). To the north of Culzean olivine basalt forms a broad dyke which runs inland in the direction of the Mochrum Hill vent (Eyles *et al.*, 1929). At Barwhin Point, at the SW end of the Culzean inlier, a breccio-conglomerate contains fragments of volcanic rock that are generally more siliceous than any others in the Ayrshire coast sequence.

Interpretation

A small number of andesite intrusions into fine-grained sediment are exposed in the cliffs along the coastal section around Culzean Castle. The many detailed features that result from such intrusions are described, interpreted and discussed in a historical context in the section on

Figure 9.36 Culzean Castle from an old engraving by W. Daniell (*c.* 1838). Despite some vertical exaggeration, the spectacular setting of the castle on a cliff formed from a 35 m-thick sill of andesite is well conveyed.

Figure 9.37 Pillowed base of an andesite sheet overlying volcaniclastic debris flow deposit at Culzean Harbour (NS 2311 1028). (Photo: G. Durant.)

the Port Schuchan to Dunure GCR site. In the Culzean section, the typical pillowed bases of andesite intrusive sheets are particularly well-displayed. The laminations in the sandstone between the andesite pillows appear to be generally undisturbed by the intrusion of the andesite magma and all follow the regional dip suggesting relatively passive intrusion of the magma. The debris flow deposits exposed north of the slipway (Figure 9.37) suggest that extrusive andesites existed close by and that these had been re-worked, or that shallow intrusive rocks were subject to penecontemporaneous erosion (Kokelaar, 1982). The exact site of eruption of the andesite magmas is not known but the volcanic vent at Mochrum Hill, 3.5 km east of Culzean, may be a source of some of the magma (Eyles *et al.*, 1949). An earlier, more siliceous and more explosive phase of activity may be represented by clasts in the breccio-conglomerate at Barwhin Point.

Conclusions

Like the other sites on the Ayrshire coast, the Culzean Harbour site is of national and international importance for the evidence critical to the re-interpretation of the Lower Old Red Sandstone lavas as intrusive sheets that have burrowed into wet, unconsolidated sediment. The highly vesicular lower contact of a thick andesite sheet below Culzean Castle shows characteristic well-developed pillow structure and the sheet incorporates sediment-filled veinlets and inclusions of fine sandstone. Laminated sandstone within pillowed andesite is well-exposed in the low cliff to the SW. The situation of this section in the grounds of Culzean Castle makes for a memorable visit.

TURNBERRY LIGHTHOUSE TO PORT MURRAY
(NS 196 072–207 081)

G. Durant

Introduction

A down-faulted inlier of Lower Old Red Sandstone andesites crops out along the shore between Turnberry Point and Port Murray just

Figure 9.38 Pillow structure in sparsely porphyritic pyroxene andesite south of John o'Groats Port (NS 2032 0780), Turnberry Lighthouse to Port Murray coast section. (Photo: G. Durant.)

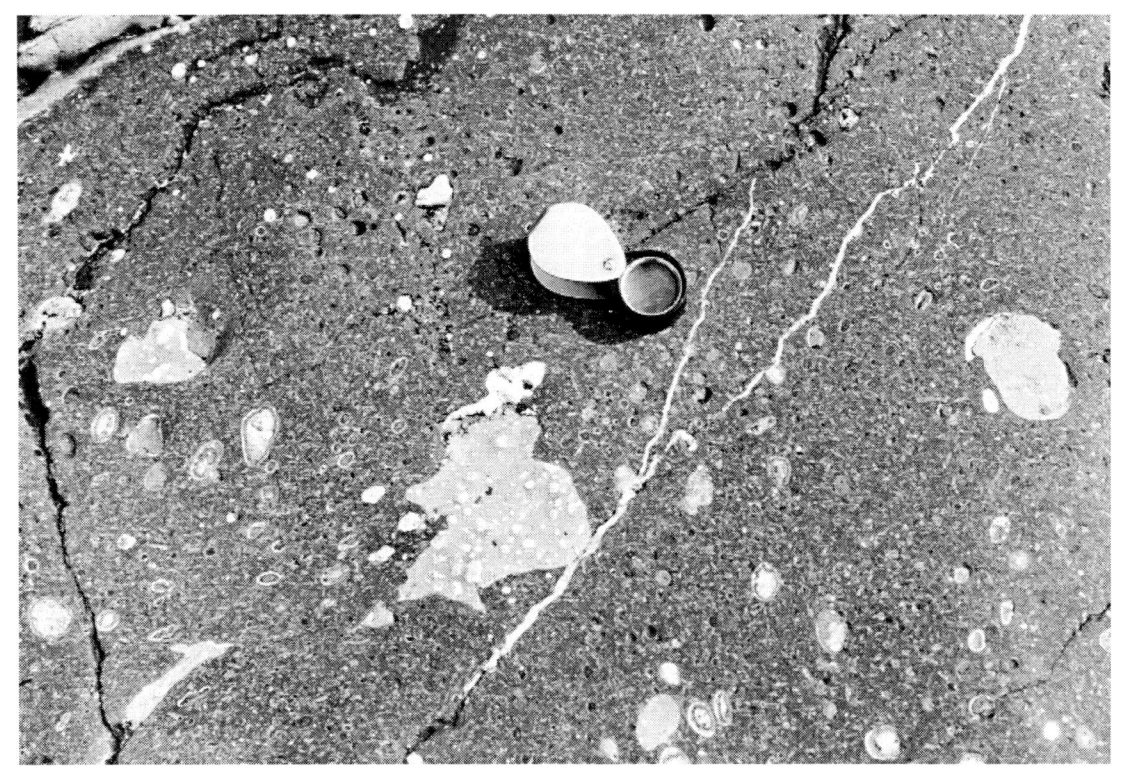

Figure 9.39 Inclusion of vesiculated sandstone in porphyritic andesite, Port Murray (NS 2058 0792). (Photo: G. Durant.)

south of Maidens Harbour (Figure 9.34). The lavas form the southernmost part of the Carrick Hills volcanic sequence and overlie a Lower Old Red Sandstone sedimentary sequence. A general description of the overall sequence is given in the Port Schuchan to Dunure GCR site description. Geikie (1897) identified 30 andesite flow-units within the Turnberry section that have been re-interpreted as sill-like intrusions into wet sediment (Kokelaar, 1982). The andesites crop out along the coast forming the wave-cut platform and low cliffs below the raised beach terrace. Inland exposures are masked by raised beach deposits.

Description

The lowest part of the section is exposed just south of Turnberry Lighthouse (1960 0718) where two andesite sheets occur, separated by a thin zone containing fine-grained sandstone. The pillowed base and brecciated top of one vesicular andesite intrusion is seen on the wave-cut rock platform in the bay immediately south of the lighthouse. The brecciated andesite fragments at the upper contact are locally dispersed within the sandstone, a structure which is known as peperite. The overlying andesite sheet forms the low cliff beneath the lighthouse. The base of this sheet is strongly pillowed and the andesite is partially autobrecciated within the intrusion.

The sequence of broadly similar andesites intruded into fine-grained sandstones continues north-eastwards along the coast. At the southern end of Broad Sands to the east of Castle Port (1978 0728), a strongly amygdaloidal andesite exposed towards the top of the beach, is particularly rich in agates. Sandstone dykes up to 20 cm thick conspicuously cut another andesite sheet exposed in the centre of Broad Sands Bay (1989 0738). The sandstone dykes are inter-branching and follow a meandering path through the andesite as seen in horizontal section.

Finely laminated red sandstone is exposed in the northern part of Broad Sands Bay and the northern edge of the bay is defined by a low cliff where an andesite sheet cuts into the sandstone (1993 0753). The andesite sheet has a partially pillowed base and there is local development of a breccia with 10 cm fragments of andesite now supported in a bleached pale-yellow sandstone. The lower 50 cm of the andesite has also been bleached to a pale-grey colour.

Good examples of sandstone-filled joints (sandstone 'dykes') and finely laminated sandstone inclusions within andesite are exposed on the wave-cut platform to the east of the low sea stack Yellow Craig (1997 0760). A rather decomposed enstatite andesite occurs at Cross Ports (Tyrrell, 1913). Just to the SW of John o'Groats Port the strongly vesiculated and pillowed base of a 7–8 m-thick andesite intrusion has been described and figured by Kokelaar (1982, fig. 2) as a typical example of the upper and lower contact features of a moderately thin andesite intrusion into wet unconsolidated sandstone. The base of the sheet, although broadly conformable with the underlying sandstones, is highly irregular with numerous lobate protrusions into sandstone and detached, rounded, more or less irregular pillows. Peperites occur in this basal zone, comprising dispersed andesite fragments within the structureless sandstone. The massive central part of the andesite intrusion is well jointed and some of the joints are filled with sandstone forming sandstone 'dykes'. The upper contact zone is marked by more in-situ hyaloclastite with numerous andesite fragments and detached pillows supported in a structureless sandstone matrix, above which laminated red sandstone is present.

At John o'Groats Port (2030 0777) andesite and sandstone are mixed in a complex way where an andesite sheet lenses out into sandstone (Geikie, 1897; Kokelaar 1982, figs 3A and 3C).

In the upper part of the Turnberry Lighthouse to Port Murray section, the andesite sheets commonly show pillowed zones. Well-developed pillow structure is seen in a sparsely porphyritic pyroxene andesite south of John o'Groats Port (Figure 9.38). The surface of the pillows exposed on the wave-cut platform has a metallic light-brown patina and the surface of individual pillows is crazed with a network of fine joints. A pyroxene andesite with olivine occurs at John o'Groats Port and two flows of augite andesite occur at Port Murray (Tyrrell, 1913). Just south of Port Murray (around 2058 0792) the lower part of an andesite sheet is highly vesicular and encloses a considerable amount of sandstone in irregular patches or as sandstone-filled amygdales. A 15 cm-wide circular inflation structure, conspicuous in the wave-polished rock surface at the top of the beach, is presumably the result of the expansion to steam of water enclosed within the sediment. Immediately adjacent to this an irregular fragment of sediment has been vesiculated following enclosure within the andesite (Figure 9.39, and see Kokelaar, 1982, fig. 3D). The vesicles in the pale-green sandstone fragment are now filled by calcite.

On the northern side of Port Murray the partially pillowed base of an andesite sheet is well exposed at low tide. Immediately to the east of the disused slipway near Maidens at the northern end of the GCR site (2087 0810) a granular-textured vesicular andesite intrusion shows well-developed hexagonal jointing. At the upper contact of this intrusion the joints have opened up to allow a significant amount of sediment to penetrate between joint blocks in the andesite (Figure 9.40). The andesite is also brecciated locally and a striking rock occurs where the angular blocks (up to 15 cm) of strongly vesiculated andesite are set in purple sandstone.

Interpretation

The detailed evidence for the re-interpretation of the andesites as intrusions into largely wet and unconsolidated sandstone has been presented by Kokelaar (1982) (see the Port Schuchan to Dunure GCR site report). Geikie (1897) records 30 separate intrusions along the Turnberry section. It is not clear whether these are the result of 30 separate intrusions or whether there have been a smaller number of multiple sill-like intrusions at different depths within a pile of unconsolidated sediment. It is difficult to correlate individual sheets and hence to quantify the precise number of intrusive events, due to the compositional similarity between intrusions, the variability of character of the andesite within a single intrusion, the rapid changes in thickness of sheets and the presence of minor faulting. However, Tyrrell (1913) reported sufficient petrological variation to suggest that at least some of the andesite sheets were emplaced as separate events. If there were indeed 30 separate intrusions, then the local Lower Old Red Sandstone basin of sedimentation, presumably marked by a lake at the surface, must have been continuously sinking, possibly due to rifting associated with magmatic activity, in order to accumulate sufficient unconsolidated sediment into which subsequent

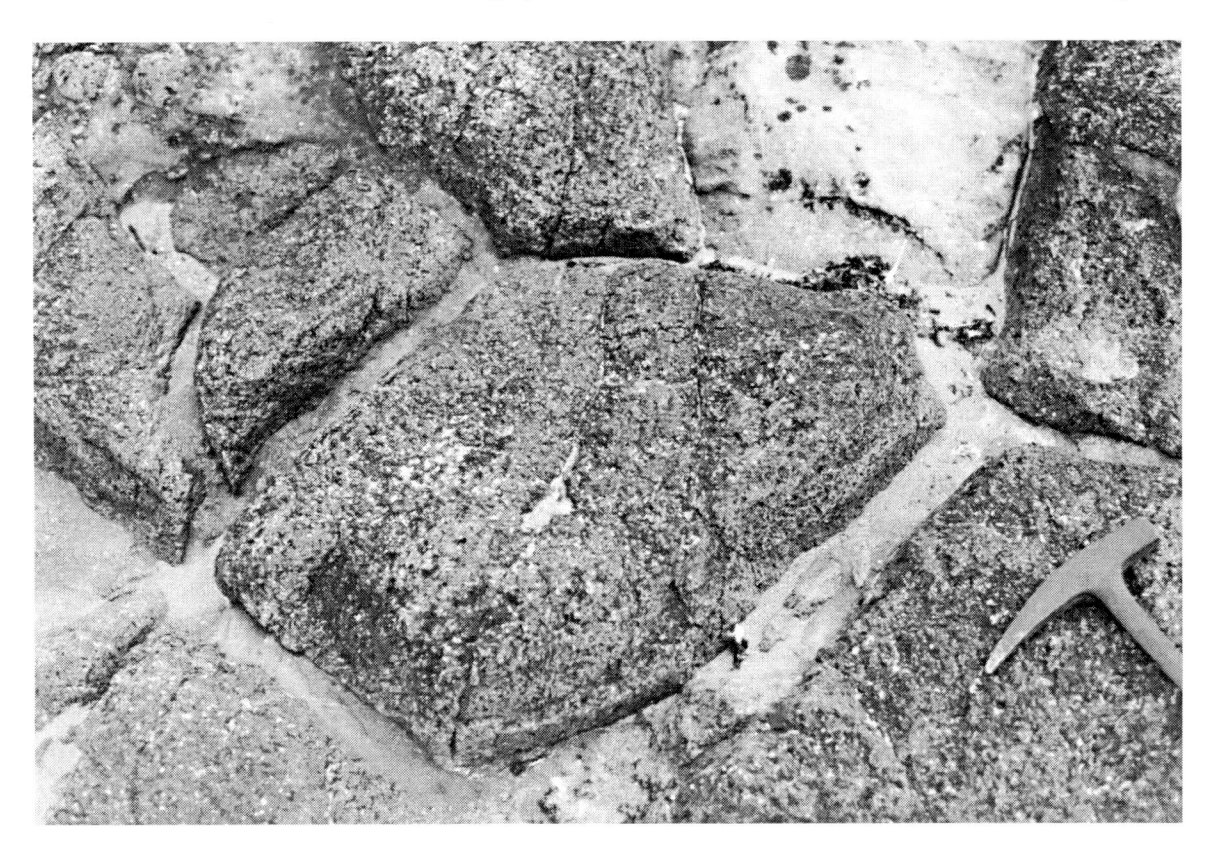

Figure 9.40 Sandstone infilling expanded cooling joints in andesite, Port Murray (NS 2087 0810). (Photo: G. Durant.)

sheets could be intruded.

As individual andesite sheets were intruded and cooled, a build-up of pressure beneath the intrusion would have caused the sediment to penetrate upwards into the cooling joints of the andesite as they opened. These joints are often curved and must have formed quite rapidly following intrusion. The upper parts of these sandstone 'dykes' are rarely seen and hence the relationship between the sandstone in the 'dykes' within an andesite sheet and that overlying the andesite intrusion cannot be fully evaluated. The fine-grained nature of the sediment and the local preservation of arthropod trackways (found by the author in Broad Sands Bay) argues for quiet depositional conditions.

Conclusions

Like the other GCR sites on the Ayrshire coast, this section is of national and international importance for the evidence demonstrating the high-level intrusion of magma into unlithified wet sediments. A sequence of 30 andesite 'lavas' in this historically significant section has been re-interpreted as a series of shallow intrusions. The basal and upper parts of the andesite sills are well-exposed along the section between Turnberry Lighthouse and Port Murray, and evidence critical to the re-interpretation is present.

PETTICO WICK TO ST ABB'S HARBOUR (NT 909 690–920 674)

D. Stephenson

Introduction

The volcanic rocks around St Abb's and Eyemouth are the best-exposed remnants of late Caledonian extrusive igneous rocks south of the Southern Upland Fault. The only other lava field, in the Cheviot some 25 km to the south, is more extensive but is less well exposed. A number of subvolcanic vents of comparable age and petrological affinity in Kirkcudbrightshire (see the Shoulder O'Craig GCR site report) provide evidence of volcanic activity farther to the SW. Although these volcanic rocks have overall calc-alkaline petrological characteristics, comparable with many modern orogenic suites, they do not conform to the systematic regional variations in geochemistry across the Midland Valley and Scottish Highlands, which have been attributed to the influence of a NW-dipping subduction zone (Thirlwall, 1981a, 1982; Fitton *et al.*, 1982).

Volcanic rocks occur within an outlier of Lower Old Red Sandstone rocks that extends inland from the coast between St Abb's and Eyemouth for some 10 km to the SW (Greig, 1988). The outlier rests unconformably upon an irregular topography of tightly folded greywackes of Llandovery age, and is overlain unconformably by the Upper Old Red Sandstone (Devono-Carboniferous in age). The Lower Old Red Sandstone affinity of the outlier is confirmed by a fragment of the arthropod, *Pterygotus* sp. (A. Geikie, 1863) and basal conglomerates and volcanic rocks are cut by a lamprophyre dyke dated at 400 Ma (Rock and Rundle, 1986). Other igneous intrusions in the immediate vicinity give similar Early Devonian radiometric ages. The sedimentary rocks of the outlier, which predominate to the south and south-west, almost all contain a volcaniclastic component, comprising fragments of fine-grained basaltic and andesitic rock. Reddish-brown sandstones predominate, but conglomerates, siltstones and mudstones occur locally and pedogenic cornstones have been recorded.

The volcanic rocks are concentrated in the coastal area, where two major and several minor vents have been recognized, and along the NW edge of the outlier. The GCR site comprises a 600 m-thick sequence of lavas and interbedded volcaniclastic sedimentary rocks that forms St Abb's Head, and a subvolcanic vent to the south, centred upon St Abb's village (Figure 9.41). They were first described by Archibald Geikie (1863) as part of the primary geological survey of Britain, and his account was subsequently expanded, together with a petrographical study by James Geikie (1887). A detailed account appears in the current memoir to Sheet 34 (Greig, 1988) and field guides based on this description have also been published by Greig (1975, 1992). The lavas are almost all autobrecciated and have all undergone extensive hydrothermal alteration, restricting the scope for geochemical investigation. Nevertheless, they were included in a geochemical study of Siluro-Devonian volcanic rocks in northern Britain by Thirlwall (1979) and this has been the basis for some speculation concerning magma genesis and tectonic implications.

Pettico Wick to St Abb's Harbour

Description

The volcanic succession of St Abb's Head is separated from Silurian sedimentary rocks to the SW by the NW-trending St Abb's Head Fault. This fault is marked by a low-lying valley, which at times of higher sea level would have cut off the headland from the mainland. The vertical fault plane, with a 2.5 m-wide breccia zone, is exposed at the cliff top at Hardencarrs Heugh (9176 6803). On the SW side of the fault, at Bell Hill (916 680), at least 120 m of conglomerates and sandstones rest unconformably on Silurian greywackes. The conglomerates contain pebbles of greywacke but, significantly, no volcanic rocks. These are the basal sediments of the local Lower Old Red Sandstone succession and immediately NE of the fault, at the base of the cliff of White Heugh (9185 6801), sandstones and siltstones are overlain by the lowest lava. A lamprophyre dyke (a minette with phenocrysts of biotite and pseudomorphs after olivine) which cuts sandstones on the east side of Bell Hill has been dated at 400 ± 9 Ma by K-Ar on separated biotite (Rock and Rundle, 1986), and a similar dyke cuts volcanic rocks 400 m to the NNE, at Horsecastle Bay.

The lavas at St Abb's Head are mostly fine grained and aphyric or olivine-phyric, with olivine content up to 12%; some are orthopyroxene-olivine-phyric. Phenocrysts are always pseudomorphed by secondary minerals. The groundmass generally consists of flow-aligned laths of plagioclase (labradorite to andesine when fresh), iron oxides and pseudomorphs after clinopyroxene. In some rocks biotite occurs as 'spongy' plates in the groundmass or associated with felsic segregations. Compositions are probably mostly in the range basalt to basaltic andesite, but in general the rocks are so altered hydrothermally that petrographical distinction is unreliable; rocks with over 5% olivine phenocrysts are generally considered to be basalts (Greig, 1988). Major element analyses are also unreliable, particularly because quartz is common in amygdales, as veins and as secondary silicification, exaggerating the silica content; all of the samples analysed by Thirlwall (1979) have silica in the range 54–57%, within the field of basaltic andesite. Feldspar-phyric lavas occur only on the east side of Kirk Hill, at the very top of the succession, but elsewhere in the outlier feldspar-phyric andesites are common and a single flow of dacite occurs near Eyemouth.

The lowest volcanic rocks, totalling at least 100 m in thickness, occur in a NW-younging sequence that extends from the top of the basal sedimentary rocks at White Heugh to Horsecastle Bay. Most of these lavas are red, aphyric, very finely vesicular and autobrecciated. The more massive parts of flows are up to 5 m thick and commonly exhibit good flow-jointing, mostly tabular and parallel to the flow margins but locally curved to bulbous suggestive of viscous flow. Most of the autobrecciation appears to have occurred *in situ* in blocky flows, but locally large tabular blocks of lava, tens of metres long, occur in poorly sorted, crudely bedded clastic units suggestive of down-slope mass movement accompanying the brecciation. This part of the succession is terminated by a NE-trending fault at Horsecastle Bay.

To the north of Horsecastle Bay, lavas and interflow sedimentary rocks dip generally to the SE at 30–40°. The sedimentary rocks, together with rubbly vesicular flow margins, are more easily weathered than the more massive flow centres and form hollows in a marked dip slope and scarp topography that is particularly well seen to the west and south of the lighthouse (Figure 9.42). Between the lowest exposed rocks at Pettico Wick and the highest at Kirk Hill, this part of the succession has an aggregate thickness of over 500 m; it has a different character to the basal flows south of Horsecastle Bay and is assumed to be entirely younger. Here, blocky lavas are up to 50 m thick. The upper and lower parts are vesicular or amygdaloidal, with much brecciation and hydrothermal alteration, and flow foliation is particularly well seen in the lower parts. The tops of flows are commonly marked by brick-red layers, 1–2 m thick. The more massive flow centres are mostly fine grained and purplish, with minute ferruginous pseudomorphs after original phenocrysts. Brecciation is commonly accompanied by ramifying veins and patches of bright-red homogeneous microbreccia with a finely crystalline matrix, which seems to have been auto-injected in a liquid state from the consolidating lava (e.g. in the road cutting around 9095 6905). The interflow volcaniclastic sedimentary rocks are commonly 1–2 m thick but units of up to 30 m occur locally. In places they overlie a planar eroded lava surface (e.g. by the Mire Loch dam (9145 6859)) but elsewhere they rest upon a perfectly preserved top surface of blocky lava and infill cavities to a depth of over 1 m (for

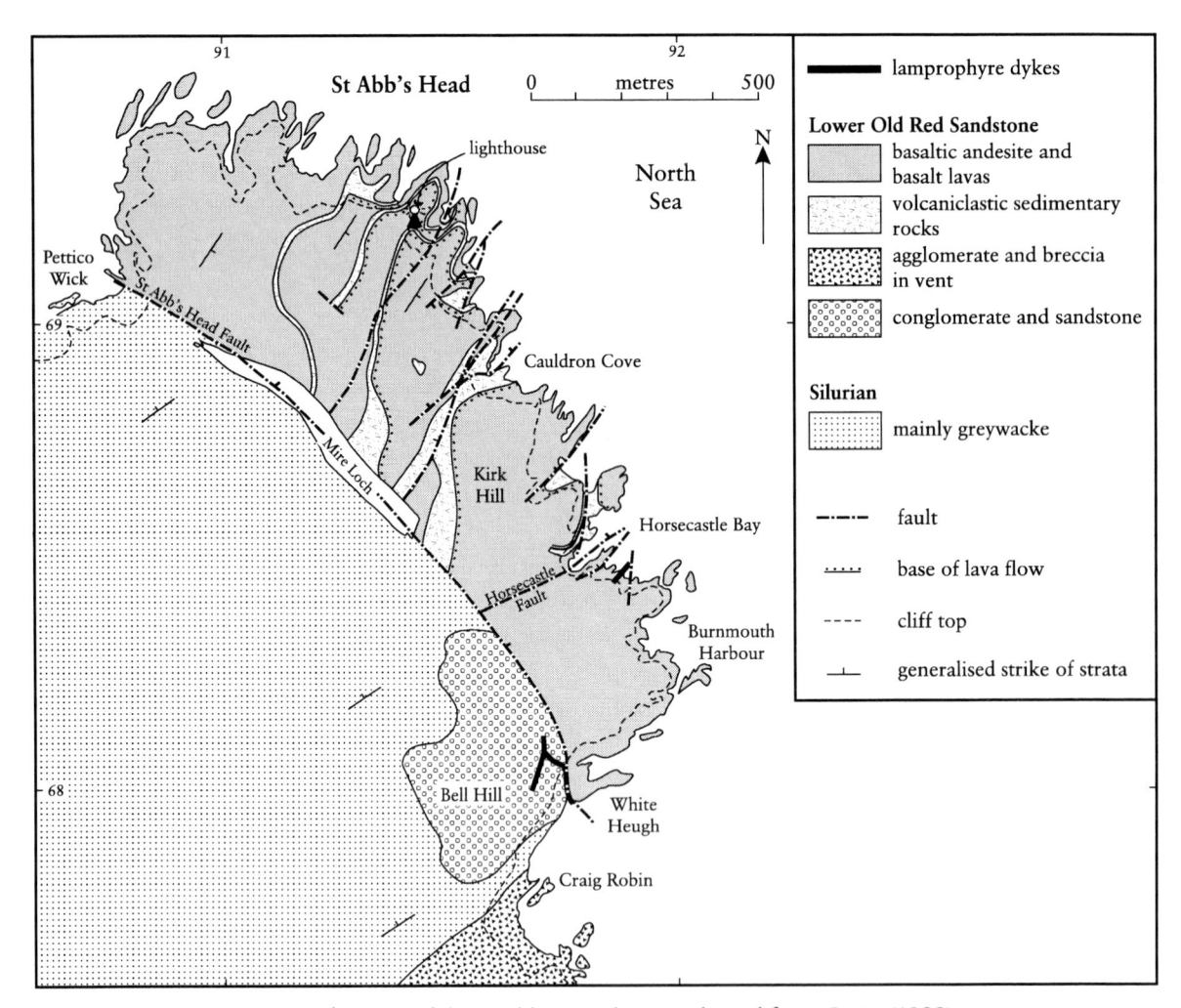

Figure 9.41 Map of the St Abb's Head area, adapted from Greig (1988).

example, below the lighthouse at 9137 6932 and at Cauldron Cove (9162 6889)). They are typically parallel bedded and poorly sorted, but with inverse grading in places. Clasts are angular and up to 20 mm in diameter; they consist entirely of fragments of reworked volcanic rock seemingly of local origin, although pale clasts may be more acid in composition and represent lavas that are no longer exposed. Although the volcaniclastic sedimentary rocks may contain some reworked pyroclasts and rare volcanic bombs, true bedded pyroclastic deposits are not represented in the St Abb's succession.

Shore and cliff exposures in the southern part of the GCR site are of breccia and agglomerate, consisting largely of unbedded, poorly sorted, very large blocks of igneous and volcaniclastic rock up to 2 m in diameter, many of which are moderately well rounded. Some large coherent masses of both lava and bedded volcaniclastic sedimentary rock occur, but their distribution and orientation is chaotic. In places a crude 'bedding' or size grading can be discerned, but this too has no obvious pattern. Many of the lava fragments are similar to those of the St Abb's Head succession, but there are also many that are feldspar-phyric. Some porphyritic andesites were considered by Greig (1988) to be intrusive into the fragmental rocks. At the NW margin of the agglomerates is a 100 m-wide zone of highly brecciated rocks with a marked, coarsely careous weathering that trends NE–SW through Craig Robin. To the north, this zone adjoins sheared Silurian greywackes and the marginal breccia contains blocks of both lava and sedimentary rock, including much yellowish weathering siltstone, in a finely comminuted matrix. Both the marginal breccia and the country rock are cut by

Figure 9.42 St Abb's Head from the north. The volcanic rocks of the headland are separated from Silurian greywackes, forming the smoother topography beyond, by the St Abb's Head Fault; the glacial channel which follows the fault contains the Mire Loch, which is just visible. Note the pronounced scarp and dip slope topography of the volcanic rocks, which dip to the left and away from the camera. (Photo: by kind permission of The National Trust for Scotland.)

abundant thin anastomosing veins of silica, carbonate and haematite. Although not exposed, the contact must be steep or vertical.

Interpretation

Outcrops of coarse breccia and agglomerate extend for almost 1 km on both sides of St Abb's village. Their unbedded, ill-sorted, chaotic nature, the rounding of the fragments and the steep, brecciated and veined margin are all indicative of a subvolcanic vent, as was first suggested by James Geikie (1887). A vent of similar size and nature occurs at Callercove Point between St Abb's and Eyemouth, and a cluster of small vents of a more basaltic nature around Hollow Craig, 5 km NW of St Abb's may also be related. These vents form a NW–SE trend, parallel to the coast, and it is tempting to suggest the presence of a volcanotectonic lineament which controlled the site of a group of small volcanoes and subsequently resulted in faulting on a similar trend. Detailed relationships between the vents and the volcanic sequences are not exposed. There is a general correspondence of rock types, but the St Abb's vent contains blocks of feldspar-phyric lava which are typical of the Eyemouth sequence rather than that now preserved at St Abb's Head.

The basal lavas, between White Heugh and Horsecastle Bay, are viscous autobrecciated flows that probably built up a moderate feature with sufficiently steep slopes to generate syn-eruptive mass-flow breccias on its flanks. Bedded volcaniclastic sediments were not deposited at this stage. Later eruptions, represented by the sequence between Pettico Wick and Kirk Hill, produced more massive, and possibly slightly more basic, blocky lavas up to 50 m thick. Between each eruption, weathering and oxidation of the flow tops occurred to a depth of

1–2 m. Planar-bedded, coarse-grained volcaniclastic sediments accumulated by reworking of debris as the volcanic rocks suffered rapid erosion. Reverse grading in many of these deposits suggests that they accumulated from high-energy flood deposits rather than by settling in areas of still water, and in several places blocky flow surfaces have been perfectly preserved by the rapid burial. If pyroclastic activity did accompany the eruptions of lava, the products must have been destroyed by erosion in a very short time, although obvious reworked pyroclasts, such as pumiceous or scoriaceous fragments are rare in the volcaniclastic sedimentary rocks now preserved.

The St Abb's and Eyemouth volcanoes probably had a limited lateral extent and the volcanic rocks give way to sequences entirely composed of volcaniclastic sedimentary rocks within a few kilometres to the south. Hence they were unlikely to have been connected directly with the Cheviot volcano and there is little compositional overlap between the two areas, although they were broadly contemporaneous.

The general high degree of alteration of the St Abb's and Eyemouth lavas and the consequent small number of samples analysed precludes any detailed geochemical discussion. Thirlwall (1979) describes them as calc-alkaline with relatively high Ni and Cr contents, implying a mantle component to the parental magma. However, they differ in detail from other Siluro-Devonian volcanic sequences and, although two groups of samples are identified, each with very distinct variation trends of many elements, neither of these can be attributed to simple fractional crystallization or progressive partial melting models. Like the Cheviot lavas, they do not fit into the spatial variation pattern of key elements such as Sr, Ba, P, K/Th and La/Y, which led Thirlwall (1981a, 1982) to suggest that the compositions of other Siluro-Devonian lavas relate to their position relative to a subducted slab of oceanic lithosphere. As with the major intrusions and dyke-swarms in the Southern Uplands Terrane (Rock *et al.*, 1986b), the chemistry of the lavas suggests magma generation at depths far greater than would have been possible, given their position relative to the proposed subduction zone. Both St Abb's and Cheviot are slightly later in age (400+ to 390 Ma) than the more northern lava sequences (424–410 Ma), the closure of the Iapetus Ocean, and the implied period of late-tectonic subduction (Thirlwall, 1988).

It is therefore possible that they relate to a deeper, slightly later, subduction zone; or in view of their very close proximity to the assumed final position of the Iapetus Suture, they may not be related to subduction at all (see Chapter 1).

Conclusions

This GCR site is of national importance as a representative of the most south-easterly Siluro-Devonian volcanic centres. The lavas were probably erupted some time after closure of the Iapetus Ocean and provide evidence of magma generation in the region of the suture.

The high sea cliffs of St Abb's Head, together with abundant craggy exposures and a marked dip slope and scarp topography inland, exhibit magnificent three-dimensional sections through a 600 m-thick lava sequence associated with a small calc-alkaline volcano. Block flows of basalt and basaltic andesite exhibit a variety of internal features including flow-jointing, vesiculation, auto-injection, autobrecciation and mass-flow brecciation. Volcaniclastic sediments, derived from the local lavas and deposited under high-energy flood conditions, have preserved the top surfaces of several flows. Very coarse agglomerates of a related subvolcanic vent are well exposed on the foreshore NW of St Abb's Harbour and the steep, brecciated and hydrothermally veined margin of the vent can be seen particularly well.

SHOULDER O'CRAIG (NX 663 491)

P. Stone

Introduction

The Shoulder O'Craig GCR site exposes an Early Devonian volcanic vent that cuts Silurian turbidite beds. An intrusion breccia represents the earliest intrusive phase and this is cut by a later intrusion of basalt. Both vent and country rocks are cut by a series of lamprophyre dykes thought to be only slightly younger than the vent itself. The Silurian strata are sandstone turbidites (greywackes) of the Carghidown Formation (Hawick Group) which, although not fossiliferous at Shoulder O'Craig, elsewhere in the region contain a graptolite fauna indicative of a late Llandovery age (White *et al.*, 1992). The tur-

Shoulder O'Craig

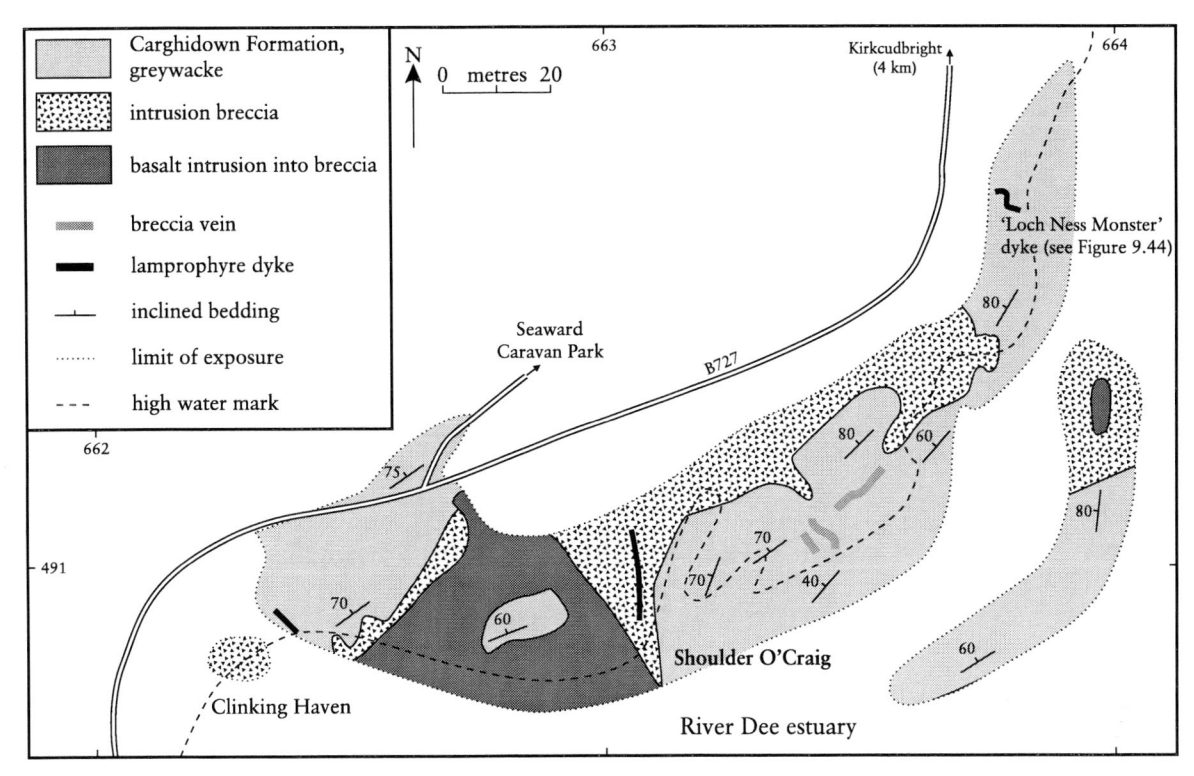

Figure 9.43 Map of the Shoulder O'Craig volcanic vent, after Rock *et al.* (1986a).

bidite beds were deformed and rotated to the vertical during the development of the Southern Uplands accretionary thrust belt (see Chapter 2, Introduction). The later stages of this deformation were accompanied by a range of minor intrusions and igneous activity continued after tectonism had ended, culminating in the emplacement of granitic plutons of the Galloway Suite at about 400 Ma (see Chapter 8). A number of late Silurian or Early Devonian volcanic vents are among the intrusive bodies seen. They appear to be entirely post-tectonic but for the most part are small and poorly exposed. The Shoulder O'Craig vent is one of the larger examples and its coastal locality provides excellent sea-cliff exposures illustrating the varied lithologies and textures within the vent itself, its relationship with the sedimentary country rock, and the morphology of slightly younger, but probably related, lamprophyre dykes. The site thus provides a rare opportunity to examine in detail the characteristics of a Caledonian volcano. A detailed description is provided by Rock *et al.* (1986a).

Description

The outline geology of the Shoulder O'Craig area is shown in Figure 9.43. The vent probably extends for a short distance inland beneath the caravan park. The country rock of the vent consists of beds grading upwards from sandstone to siltstone, each formed by deposition from a single turbidity current. In the immediate vicinity of the vent they strike approximately NE–SW and are vertical or dip steeply towards the NW, with sporadic zones of small-scale tight to isoclinal folding. A penetrative slaty cleavage is widely developed sub-parallel to bedding but terminates at the cross-cutting vent margin. This relationship establishes the vent as a post-tectonic intrusion.

The most striking aspect of the vent is the texture shown by the coarse breccia forming the earliest intrusive component. This is best examined on the wave-polished surfaces to the north of Clinking Haven where both matrix-rich and clast-rich varieties can be identified. The cliff sections provide more extensive exposure in three dimensions and confirm that the breccia consists principally of variably rounded to subangular clasts of sandstone, siltstone and rare basaltic or microdioritic lithologies set in a fine-grained matrix. The latter is pervasively altered to chlorite and carbonate but traces of a relict texture suggest an original igneous (basaltic?)

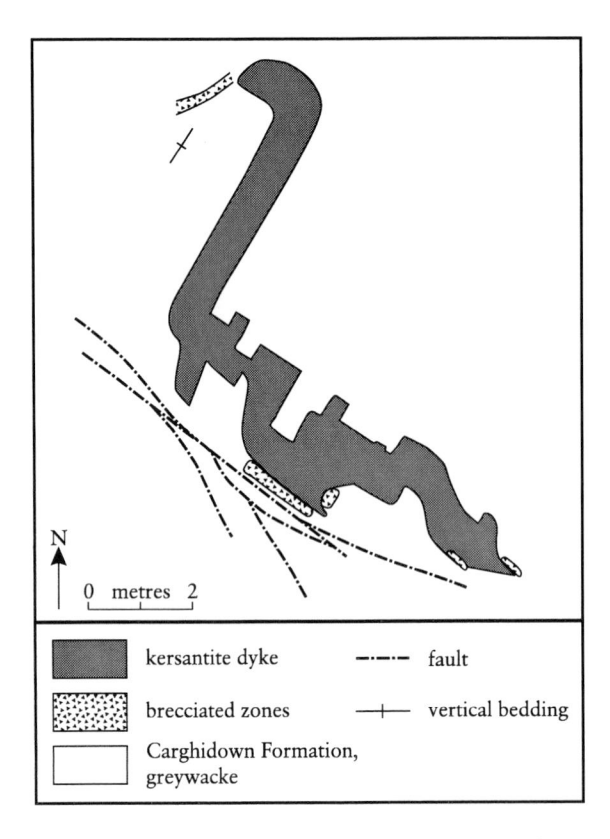

Figure 9.44 Sketch of the 'Loch Ness Monster' lamprophyre dyke, just north of the margin of the Shoulder O'Craig vent, after Rock *et al.* (1986a). (For location see Figure 9.43.)

composition. Most of the clasts seem likely to have been derived from the country rocks and in the cliff sections their size ranges up to rafts several metres in length. This part of the vent fill is an intrusion breccia (*sensu stricto*) although it has been generally referred to in the literature as a vent agglomerate (e.g. Rock *et al.*, 1986a). It is cut by at least two irregular and fractured masses of lamprophyric biotite-olivine basalt, the larger of which occupies much of the western end of the vent. The basalt is generally clast-free and its contacts with the adjacent intrusion breccia vary from sharp to diffuse and gradational. Oval, pillow-like textures and possible flow fractures may suggest intrusion in a semi-solid state (Rock *et al.*, 1986a). Breccia veins up to about a metre across cut the country rock in the vicinity of the vent and seem likely to be related to the intrusive episode.

Both vent and country rock are intruded by a suite of lamprophyre dykes, mostly kersantites in which large phenocrysts of biotite are contained in a melanocratic groundmass rich in plagioclase. The dykes range up to about 1.5 m wide but are commonly sinuous and highly irregular in shape. A particularly fine example occurs just beyond the NE extremity of the vent; its bizarre outcrop pattern has earned it the colloquial title 'the Loch Ness monster' (Figure 9.44). A fresh kersantite body cutting the intrusion breccia of the vent has given a K-Ar biotite age of 410 ± 10 Ma (Rock *et al.*, 1986b).

Interpretation

The Shoulder O'Craig vent contains two intrusive phases. The most abundant lithology is the vent-filling intrusion breccia, which consists largely of sandstone and siltstone country rock clasts carried into place within a fine-grained basaltic matrix. This has been intruded by a later basaltic mass containing very few xenoliths but enclosing one very large sandstone raft. Contacts between the basalt and the intrusion breccia are largely obscured but in places they appear to be gradational suggesting a continuum of intrusion rather than two separate episodes. Pillow-like textures and some possible flow-fractures could arise from emplacement of the basalt as a semi-solid mush (Rock *et al.*, 1986a). The basalt intrusion is altered, with chlorite generally replacing olivine, but some relict olivine remains together with a little biotite and augite. From this petrography and the abundances of trace elements such as Ti, Y and Zr, Rock *et al.* (1986a) classified this rock as a calc-alkaline basalt. The basaltic matrix of the earlier intrusion breccia is much more pervasively altered suggesting that this intrusive phase was more hydrous and volatile-rich. Breccia veins cutting the country rock close to the vent margin have been described as explosion breccias by Rock *et al.* (1986a) and may also relate to the earliest intrusive phase. However, they have an ambiguous relationship with the lamprophyre dykes which are demonstrably intrusive into the intrusion breccia. The dykes themselves are biotite-rich kersantites and most have highly irregular forms thought to reflect high volatile pressure during emplacement. They are an expression of deep-seated K-rich magmatism.

The calc-alkaline nature of the intrusions suggests subduction-related magmatism. However, that is difficult to reconcile with either of the proposed tectonic models for development of the Southern Uplands Terrane (see Chapter 1).

The problem has been discussed by Rock *et al.* (1986b) and two points are particularly pertinent:

1. On a regional scale, volcanic, subvolcanic and plutonic magmatism is juxtaposed in both space and time; volcanic vents, mantle-derived dykes and granite plutons were all intruded at about 410–400 Ma and are now seen at the same level of erosion.
2. The lamprophyres in particular are too K-rich and of too deep a mantle source for their close proximity to the putative trace of the Iapetus Suture, only some 30 km to the south beneath the Solway Firth.

In this context Shoulder O'Craig is the counterpart to the vent intrusion of similar age at the Pettico Wick to St Abb's Harbour GCR site in SE Scotland (see report); the similarities and contrasts between these two bodies are particularly instructive and have been discussed by Rock *et al.* (1986 a, b).

Conclusions

The Shoulder O'Craig locality provides the largest and best-exposed example in SW Scotland of a late Caledonian volcanic vent. The vent contains at least two components, an earlier intrusion breccia of country-rock sandstone clasts in a highly altered basaltic matrix, and a later basaltic plug-like intrusion. Possible explosion breccia forms veins cutting the country rock adjacent to the vent. Both vent and country rock are cut by lamprophyre dykes, which may assume highly irregular intrusive forms. A radiometric age of about 410 Ma from one lamprophyre dyke provides a minimum age for vent intrusion. The maximum age is constrained by the late Llandovery age (about 430 Ma) of the country rocks. These are turbidite sandstones and siltstones that were folded and cleaved prior to the emplacement of the intrusion breccia. Textures within the vent and its relationship with the country rock are exposed with unusual clarity.

The late Caledonian intrusive suite, of which the Shoulder O'Craig vent complex is a particularly fine example, is of regional tectonic significance in respect of subduction models for the closure of the Iapetus Ocean.

ESHANESS COAST (HU 217 807–211 775)

D. Stephenson

Introduction

The most extensive and thickest development of volcanic rocks in the Middle Old Red Sandstone of western Shetland is at Eshaness, where basaltic to andesitic lavas, andesitic pyroclastic rocks and a rhyolitic ignimbrite form a sequence some 500 m thick. The Eshaness Coast GCR site provides a section through most of this sequence, in spectacular sea cliffs that are renowned for their geomorphological features such as geos, blowholes, subterranean passages and cliff-top storm beaches (Figure 9.45). Volcanological highlights include well-preserved lava tops, textures due to contact with wet sediment, very coarse proximal pyroclastic breccias, a section through a welded ash-flow and a complex tuff with a variety of features suggestive of hydromagmatic eruption. The volcanic rocks were first described by Peach and Horne (1884) and subsequently in more detail by Finlay (1930). A succession modified from that established by Finlay forms the basis of the Geological Survey map by J. K. Allan (Wilson *et al.*, 1935) and of the summary by Mykura (1976). Some petrographical and geochemical details are given by Flinn *et al.* (1968) and Thirlwall (1979).

The Old Red Sandstone rocks of Shetland occur in three distinct structural blocks, differing in age, depositional and volcanological development, tectonic history and effects of igneous intrusion and low-grade metamorphism (Mykura, 1976). These blocks are separated by major N- to NNE-trending faults. The volcanic rocks of Eshaness, together with those of the island of Papa Stour and smaller outcrops at Melby on the western tip of the Walls Peninsula, all occur to the west of the Melby Fault (Figure 9.46) and hence are probably related temporally, if not magmatically (see Interpretation). The Eshaness outcrop, consisting almost entirely of volcanic rocks, is bound to the east by the probable northern extension of the Melby Fault, which juxtaposes the Northmaven plutonic complex. The rocks are folded into a shallow NNE-trending syncline, which plunges to the NNE in the northern part of the outcrop and to the SSW in the south. The GCR site (Figure 9.47) is entirely on the western limb of this syncline,

Figure 9.45 The cliffs of Eshaness, looking NE from the lighthouse. The nearest headland (the North Head of Caldersgeo) comprises andesitic pyroclastic breccias of unit 6 resting on andesites of unit 4. Most of the cliffs of the middle distance are andesites of unit 4; and the prominent dip-surface of the Grind of the Navir ignimbrite is just visible in the distance above the prominent stack (Moo Stack). (Photo: BGS no. D1660.)

which dips to the SE at 10–12°.

Description

The volcanic succession of Eshaness has been divided into nine units (Mykura, 1976), but the lowest two crop out only on the eastern limb of the syncline. Within the GCR site there is a continuous section from the ignimbrite of unit 3 in the north, to the pyroclastic rocks of unit 8 in the south (Figure 9.47).

1. and 2. The basal units, seen on the west side of Brae Wick (245 786) outside the GCR site, consist of reddish-purple micaceous sandstones and tuffaceous sandstones, overlain by olivine basalts and andesites with lenticular tuffaceous beds.
3. The well-jointed ignimbrite of unit 3 is responsible for the spectacular geomorphological feature of the Grind of the Navir (2127 8042). Here, large angular blocks have been, and still are being, excavated by the waves to form a natural passage and 'staircase' and then piled up on top of the cliff to form a high-level storm beach. Most of the outcrop consists of a relatively homogeneous pinkish-purple rhyolitic welded tuff with a well-developed eutaxitic fabric accentuated by flattening of the clasts. Broken and corroded crystals of pink alkali feldspar, commonly with a hollow core, are typically up to 10 mm long, but some are up to 150 mm; smaller fragments of collapsed pumice, shards of glass and rounded darker basic fragments are also abundant, with less common plagioclase, quartz and magnetite-rich aggregates that are presumably pseudomorphs after mafic minerals. All are etched out and well seen on weathered surfaces (Figure 9.48). The fine-grained grey matrix consists largely of devitrified glass with trails of opaque 'dust' and with elongate angular cavities. All parts of the rock are heavily altered, with the feldspars replaced by sericite and/or carbonate and pervasive secondary silicification throughout. The base of the ignimbrite is not exposed, but the top forms an extensive flat surface that dips inland behind the Grind of the Navir. A very sharp junction is well exposed between ignimbrite

with few clasts and a very fine-grained matrix and an overlying soft, yellow-brown-weathering tuff. The poorly sorted base of the tuff contains angular feldspar clasts up to 20 mm long similar to those of the ignimbrite, but it is generally fine grained above. Vesicles and cavities up to 20 mm long are flattened, but otherwise the tuff lacks the fabric of the underlying ignimbrite. Some of the cavities are filled by quartz, but others are hollow. The tuff is well bedded in parts and some beds have convolute flow structures with some brecciation. Coarser-grained beds up to 1 m thick contain large ragged fragments of very vesicular basic-looking rock.

4. Between Gruna Stack (213 802) and Drid Geo (209 790) are several sheets of aphyric andesite (Finlay recorded three), with distinctive vesicular, slaggy and autobrecciated tops. The vesicles are commonly elongated in the direction of flow. The central parts of the sheets have irregular to flaggy, flow-parallel jointing, commonly with a rather lenticular appearance. They are described as 'mugearites' on the Geological Survey map and by Mykura (1976), but the original designation as augite andesites is more appropriate for these essentially calc-alkaline rocks. Most are highly altered. At Brei Geo (2125 7975), the top surface of a sheet is spectacularly domed, the domes and hollows having an amplitude of 2–3 m. This surface can be examined in detail at the cliff top some 100 m to the north, where it is seen in contact with a remnant of bedded brown sandstone. Detached lobes and subangular patches of vesicular andesite, up to 20 cm across, cut across the bedding in the sandstone at a high angle and with a sharp contact (cf. peperite), suggesting intrusion of the magma into wet sediment. Undisturbed sandstone is not present between the sheets.

5. Andesites of unit 5 occur on the eastern limb of the syncline where they are highly silicified and oxidized. On the western limb they have been correlated on the map with a thin flow that rests upon green-, purple- and yellow-weathering clays developed on an amygdaloidal lava at the top of unit 4 (2083 7885). The thin flow, which is also yellow-brown weathering, is very flaggy at the base and has very strong flow-banding that is folded and convoluted in parts. Some layers are vesicular, with slightly elongate, lined vesicles 1–2 mm in diameter. Other layers contain angular pink fragments that could be silicified. The general impression is that this was a very viscous flow, more acid in composition than the underlying andesites.

6. Very coarse andesitic pyroclastic breccias, which are over 100 m thick in the GCR site, increase in thickness and overall clast size northwards (Finlay, 1930). Vertical sections are well seen around Calder's Geo (209 786) (Figure 9.45), but the breccias are best examined on the cliff top SW of the lighthouse (around 205 784), where all traces of soil and superficial deposits have been swept away by high-level wave action. The clasts are largely andesitic, but fragments of felsic rock, sandstone and metamorphic rocks are not uncommon. The larger blocks are up to 1 m in size and mostly angular. Some are slightly rounded, but there are no obvious bombs. The unit as a whole is very poorly sorted, but the relatively finer-grained beds are crudely bedded and crudely graded; more marked rounding of clasts suggests some reworking in parts.

7. The top surface of the pyroclastic breccias is remarkably planar and forms a prominent ledge around the north side of the headland 400 m SW of the lighthouse. It is overlain by a feldspar-phyric hypersthene andesite, which is homogeneous throughout most of its thickness with few vesicles and little development of rubble or visible hydrothermal alteration. For the most part it is massive and blocky, although the central part has spheroidal jointing, well seen at The Bruddans. At its base (2040 7825) it has a definite 5 cm chill. Its top surface forms the SE-dipping land surface on headlands to the north and south of The Cannon (a horizontal blowhole in the cliff). In contrast to the lower parts, the top of the flow has quite large elongate vesicles and some inclusions of sandstone, although it remains quite coarsely porphyritic, with no chill. There are abundant fissure fillings and some wider areas of poorly bedded yellowish-brown sandstone that seem to post-date the cooling of the lava. An overlying flow of pyroxene andesite, which crops out between the Bruddans and Stenness is less porphyritic and more scoriaceous with intense hydrothermal alteration in places. Finlay (1930) reported that this unit thickens towards the north where up to four flows occur.

8. A unit of very coarse andesitic pyroclastic breccias with subordinate interbedded sand-

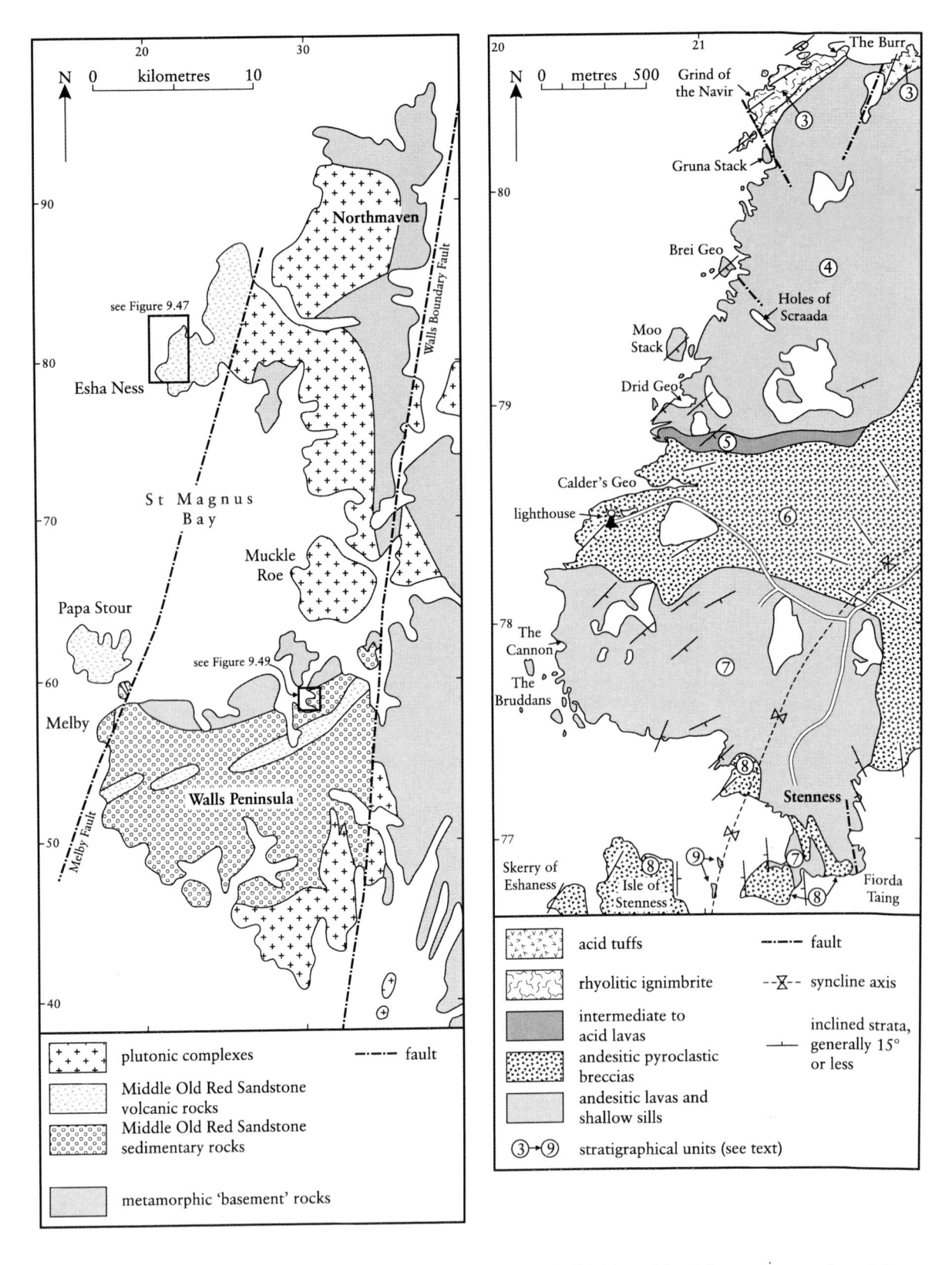

Figure 9.46 Location of Middle Old Red Sandstone volcanic rocks, major intrusions and major faults in western Shetland, after Mykura (1976).

Figure 9.47 Map of the Eshaness coast, adapted from Geological Survey 1:10 560 sheets Shetland 19 and Shetland 23 (1959).

stone and conglomerate crops out in the core of the syncline at Stenness and on islands just offshore. The junction with the underlying lava is irregular and fissures and hollows in the lava surface are filled by tuff and sandstone. Much of the unit is massive, with blocks up to 1 m in size, but bedding occurs in places.

9. The highest unit is a flow of vesicular fine-grained andesite, which forms two skerries, 500 m SW of Stenness, just outside the GCR site boundary.

Interpretation

Because of the lack of intercalated sedimentary rocks in the Eshaness succession, there is little direct evidence of the environment in which the volcanism took place. However, by analogy with successions west of the Melby Fault at Melby and on Foula (Mykura and Phemister, 1976; Mykura, 1976, 1991), it seems reasonable to assume an arid or semi-arid alluvial plain with temporary lakes. The sediment was derived from the W or WNW and the area was close to the NW margin of the main Orcadian Basin where alluvial fans may have been developed. According to Mykura, post Mid-Devonian dextral movement of 60–80 km on the Melby Fault has transposed these outcrops from much farther south than the other structural blocks of Shetland, and confident correlations have been made with successions on Orkney (see below).

The eruptions were almost entirely subaerial, and the lack of sedimentary intercalations at Eshaness could be interpreted as evidence that the volcanic rocks accumulated rapidly, with such sediment as did accumulate between eruptions being removed by subsequent flows. There is good evidence at Brei Geo for interaction of magma with wet sediment, suggesting that at least some magma was emplaced as high-level sills in thin unconsolidated sediments, possibly on a lake bed. However, it is difficult to imagine this as the dominant mechanism in view of the general lack of intercalated sediments. The lowest andesite of unit 7 has many sill-like features (planar base with preserved chill, homogeneity, lack of alteration and flow-brecciation, inclusions of sandstone near the top). If it was emplaced as a sill, its top surface must have been uncovered for the cooling cracks to be filled with sediment prior to the eruption of the next flow.

The very coarse pyroclastic breccias of units 6 and 8 are clearly the products of large-scale eruptions and are relatively proximal, although there is no indication of where the source may have been, apart from the observation that some units thicken and coarsen northwards (Finlay, 1930). In addition to juvenile material, the vents sampled both sandstones and metamorphic basement, which is consistent with a site close to the margin of the sedimentary basin. Some reworking is apparent within the pyroclastic units, but a lack of volcaniclastic sedimentary rocks within the sequence in general suggests that the volcanism did not result in a pronounced topography.

The rhyolitic rocks of the Grind of the Navir probably represent a number of distinct types of pyroclastic eruption. The ignimbrite exhibits classic features of a welded pyroclastic flow with broken crystals, collapsed pumice and a classic eutaxitic texture. Mixed lithofacies in the overlying, dominantly well-bedded tuffs suggest the involvment of several eruptive styles; the basal, very poorly sorted lapilli-tuff and the finer-grained vesicular tuffs are probably the result of hydromagmatic eruptions that may have included pyroclastic surges and ash-falls, whereas the beds with large, ragged vesicular clasts suggest a more dominantly magmatic, possibly strombolian type.

The complete succession at Eshaness includes a wide range of compositions and Thirlwall (1979) has identified basalts, andesites, dacites and rhyolites. Several geochemical and mineralogical features suggest that the rocks are best classified as transitional between calc-alkaline and tholeiitic, in marked contrast to the calc-alkaline suites that characterize the Old Red Sandstone volcanic province in general. Thirlwall (1979) also presented good evidence that the Eshaness sequence could have been derived by multistage low-pressure fractional crystallization from a parental magma close in composition to an olivine tholeiite and relatively low in incompatible elements, features which are also atypical of the province as a whole.

There is no direct evidence of the age of the Eshaness sequence, although Flinn *et al.* (1968) did obtain a Rb-Sr isochron age of 365 ± 2 Ma (recalculated from 373 Ma using new constants) from the Grind of the Navir ignimbrite; in view of the pervasive alteration, this age is probably a minimum (Thirlwall, 1983a). Several authors have proposed correlations on lithological

Figure 9.48 Ignimbrite of the Grind of the Navir, Eshaness coast. The larger clasts are of alkali feldspar, commonly with a hollow core; smaller clasts are mainly collapsed pumice, glass shards and basic fragments. (Photo: BGS no. D1662.)

grounds between the volcanic successions at Eshaness, Melby and Papa Stour (Finlay, 1930; Flinn *et al.*, 1968; Mykura, 1976), but Thirlwall (1979) identified geochemical differences. Although the Papa Stour rocks seem to be significantly distinct to have formed from a separate centre, he did conclude that the Eshaness and Melby sequences could be related. The volcanic rocks of Melby occur above the Melby fish beds, which have been reliably correlated with the middle Eifelian Sandwick Fish Bed of Orkney and palynological evidence has confirmed the Papa Stour and Melby volcanic rocks as late Eifelian (Marshall, 1988; Rogers *et al.*, 1989, fig. 2).

Despite their Mid-Devonian age, Thirlwall (1979, 1981a) attributed the Eshaness and other volcanic rocks of Shetland to the same late Caledonian, WNW-dipping subduction zone that was responsible for late Silurian and Early Devonian volcanic and plutonic activity in northern Britain. He pointed out that their geochemical characteristics are even closer to those of modern arcs than are those of the earlier volcanic rocks in the main part of the province, and attributed their transitional tholeiitic nature to a closer proximity to the surface trace of the subduction zone. Although the magmas do have features that could be related to a subducted slab of oceanic lithosphere, by Mid-Devonian time the tectonic environment was one of post-orogenic extensional basins. Indeed, most of the volcanic activity in Shetland and Orkney was coeval with, and hence was probably controlled by, extensional faulting in the Orcadian Basin (Astin, 1985, 1990; Enfield and Coward, 1987; McClay *et al.*, 1986; Rogers *et al.*, 1989).

Conclusions

The volcanic sequence at Eshaness is representative of several in the most westerly structural block of Old Red Sandstone outcrops in Shetland. Their late Eifelian age is significantly later than Old Red Sandstone volcanism elsewhere in northern Britain, but it is the earliest late Caledonian volcanism in Shetland and Orkney. Although the rocks have subduction-related characteristics, their eruption was probably related to a major phase of extensional faulting during the development of the Orcadian Basin.

The mainly andesitic and rhyolitic rocks of the GCR site have transitional calc-alkaline to tholeiitic petrological features and may be related by fractional crystallization. Proximal pyroclastic breccias are intercalated with subaerial lavas and some high-level sills intruded into wet sediment, although inter-volcanic sediments are rarely preserved in the sequence. The ignimbrite and overlying hydromagmatic tuffs at the Grind of the Navir constitute one of the best preserved records of continuous rhyolitic pyroclastic eruption in Britain, which would well merit further detailed study.

These and many other volcanological features, are well seen in magnificent sea cliffs that are also noted for their geomorphological structures.

NESS OF CLOUSTA TO THE BRIGS (HU 305 584)

D. Stephenson

Introduction

The Clousta volcanic rocks, which form an ENE-trending outcrop across the centre of the Walls Peninsula in western Shetland (Figure 9.46), comprise mainly basic and acid pyroclastic rocks, with some basaltic and andesitic lavas and shallow sills, rhyolitic lava domes and ignimbrites, and concordant intrusions of felsite. These are scattered as relatively thin and localized lenses within Middle Old Red Sandstone alluvial fan and lacustrine sequences. The Ness of Clousta to the Brigs GCR site (Figure 9.49) exhibits a variety of volcanic products but is particularly noteworthy for the evidence of interaction between magma and water-saturated, unconsolidated alluvial sediments, possibly giving rise to phraeatomagmatic explosions. The composition, internal structures and three-dimensional geometry of the pyroclastic accumulations in particular have been compared to those of maars and tuff-rings.

The volcanic rocks were first noted by Peach and Horne (1884) and were described briefly by Finlay (1930). The Walls Peninsula was mapped by the Geological Survey in the 1930s (Wilson *et al.*, 1935) and re-examined in detail in the 1960s, resulting in the current map and a detailed description of the volcanic rocks by Mykura (in Mykura and Phemister, 1976). A detailed, mainly sedimentological, study of the peninsula by Astin (1982) resulted in a radical re-appraisal of some of the volcanic rocks and their relationship to sedimentation. Some petrographical and geochemical details are given by Thirlwall (1979) and Astin (op. cit.).

The Old Red Sandstone rocks of Shetland occur in three distinct structural blocks separated by major N- to NNE-trending faults. The successions within each block differ in age, depositional and volcanological development, tectonic history and effects of igneous intrusion and low-grade metamorphism (Mykura, 1976). Most of the Old Red Sandstone of the Walls Peninsula occurs in the central block, bound to the west by the Melby Fault and to the east by the Walls Boundary Fault (Figure 9.46). Within this block, the Old Red Sandstone rocks rest unconformably on Precambrian metasedimentary rocks to the north and are intruded and hornfelsed in the south by the Sandsting plutonic complex (K-Ar mineral dates of 369 ± 10 and 360 ± 11 Ma by Snelling in Mykura and Phemister, 1976). They have been involved in two phases of intense folding with cleavage development, and have suffered low-grade regional metamorphism, locally up to low green-schist facies. Palynological data indicate a Givetian age (Rogers *et al.*, 1989, fig. 2), making the volcanic rocks younger than those of the western block (see the Eshaness GCR site report) and comparable in age to those of Orkney (see the Point of Ayre and Too of the Head GCR site reports).

In the Walls Peninsula, the Old Red Sandstone outcrop is divided by the ENE-trending Sulma Water Fault into areas of markedly different sedimentary facies that were assigned to two separate formations of different ages by Mykura and Phemister (1976). Astin (1982) recognized four diachronous sedimentary formations which, together with the Clousta volcanic rocks, comprise a single coherent sequence that can be correlated across the Sulma Water Fault. To the north of the fault, sedimentary rocks assigned to the Sandness Formation of Mykura and Phemister represent all four of Astin's forma-

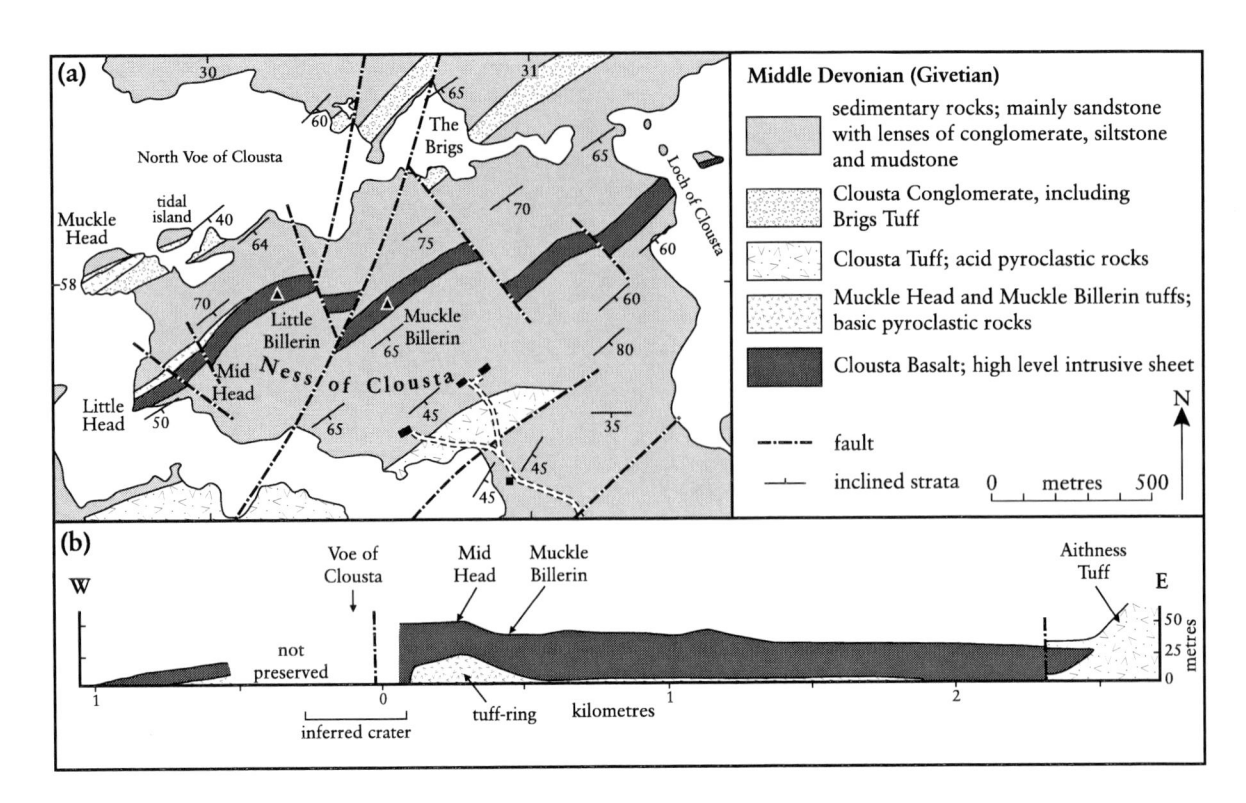

Figure 9.49 (a) Map of the Ness of Clousta to The Brigs GCR site, adapted from Geological Survey 1:10 560 sheet Shetland 42 (1967) and Astin (1982). (b) The Muckle Billerin Tuff and Clousta Basalt: reconstructed cross section based on measured sections, showing the different thickness either side of the Voe of Clousta. Vertical exaggeration ×2.5. (From Astin, 1982.)

tions. The strata dip generally to the SSE at moderate to high angles and the Clousta volcanic rocks are interbedded with the upper part of this sequence, adjacent to the fault.

Description

The GCR site (Figure 9.49) occurs towards the eastern end of the outcrop of Clousta volcanic rocks, which are here intercalated with medium- to coarse-grained sandstones locally with lenses of conglomerate and minor siltstones and mudstones, all assigned to the Vatslees Formation by Astin (1982). The Clousta Conglomerate, which is up to 70 m thick and can be traced for 5 km, forms a good topographical feature (Figure 9.50) and is a stratigraphical marker throughout the area. Within the GCR site there are two lenses of basic pyroclastic rocks, the Muckle Head Tuff and the Muckle Billerin Tuff; the latter is overlain directly by a sheet of basalt, termed the Clousta Basalt (the Muckle Billerin Basalt of Astin). Astin also identified a thin lens of acid volcaniclastic rocks within the Clousta Conglomerate, which he termed the Brigs Tuff.

These volcanic units are described in stratigraphical order, combining the observations of Mykura (in Mykura and Phemister, 1976) and Astin (1982) with those of the author.

The Muckle Head Tuff

Muckle Head is formed from a lens of poorly sorted basaltic lapilli-tuff up to 35 m thick, which thins markedly north-eastwards over a strike length of 350 m. Only 2 m are preserved west of the Voe of Clousta. The tuff rests upon conglomerate, which may be the basal part of the Clousta Conglomerate, capping a coarsening upward sequence of slumped alluvial sandstones. Large blocks of conglomerate and sandstone occur in the base of the tuff, but their size and abundance decrease upwards. Clasts of magmatic material, mostly of basic to intermediate composition form up to 70% of the tuff and become more dominant towards the top of the lens. Most are scoriaceous or vesicular, some are flattened and some are glassy and enclose small quartz grains. Clasts of fine-grained acid igneous material are rare. The matrix is com-

Ness of Clousta to the Brigs

Figure 9.50 View of the Ness of Clousta to The Brigs GCR site, looking east from Muckle Head. The Muckle Head Tuff forms the rocks of the foreground; the Brigs Tuff forms the extreme right of the tidal island beyond; the upper part of the Clousta Conglomerate, dipping to the right (SSE), forms the prominent feature crossing the tidal inlet in the middle distance; and the right skyline is the ridge of Muckle Billerin, formed by the Clousta Basalt overlying the Muckle Billerin Tuff. (Photo: D. Stephenson.)

posed mostly of quartz and feldspar derived from the underlying sediments, but garnet, epidote and titanite have also been recorded. The grains are commonly well rounded, but others are fractured and angular. The tuff is well bedded throughout its thickness and cross-bedding has been recorded. Overlying the tuff are well-bedded siltstones, which pass upwards into conglomerate. Astin has correlated these siltstones with a 2–3 m-thick lacustrine unit within the fluvial Clousta Conglomerate.

The Brigs Tuff

Immediately above the thin fine-grained lacustrine unit that divides the Clousta Conglomerate, is a sheet-like, coarse-grained volcaniclastic unit, up to 12 m thick, which thins only gradually to the ENE over a distance of 3 km. The unit consists largely of subrounded to angular lithic clasts of feldspar porphyry and less abundant flow-banded felsite in a finer-grained quartzofeldspathic matrix. Pumice, or other evidence of a magmatic eruption, is conspicuously absent. The lithic clasts are up to 40 cm across, there is little lateral variation in the size of clasts and, in general, the deposit is well sorted and largely clast supported. The unit is either massive or parallel-bedded and locally it has low-angle cross-bedding, with sets up to 20 cm thick that are well seen on the tidal island (at 2990 5811).

The Muckle Billerin Tuff

Poorly sorted basaltic pyroclastic breccia and lapilli-tuff are exposed on the northern side of Little Head (2976 5765) at the base of the Clousta Basalt and can be traced for about 2 km along the NW flank of the Muckle Billerin ridge. The lens is up to 24 m thick around Mid Head, but this proximal development only extends for about 450 m along strike. Farther to the NE there is only a thin distal fringe and to the SW, on the opposite side of the Voe of Clousta, there are 8 m (Figure 9.49b). The maximum clast size shows a systematic fining from blocks up to 30 cm around Mid Head to under 2 cm distally. The rock is composed of large amounts of quartzofeldspathic sand and basaltic to intermediate magmatic material, with less abundant blocks of acid igneous rock. The juvenile material varies in amount from 10–70%; it is commonly scoriaceous and vesicular and some clasts are flattened and welded, especially in proximal areas. The quartzofeldspathic and acid material is commonly fractured and angular and is more abundant in the proximal area. The whole deposit is very well bedded, dominantly parallel-bedded, but with some cross-bedding, low-angle discordances and shallow scour and fill structures. The set height of the cross-bedding varies systematically from up to 15 cm proximally to a few centimetres distally.

The Clousta Basalt

This sheet of basalt, which rests directly on the Muckle Billerin Tuff, is the most extensive of several in the eastern outcrops of the Clousta volcanic rocks. It forms a prominent, fault-stepped ridge extending for 2.5 km from Little Head to the shore of the Loch of Clousta and onwards, forming a string of small islands in the loch. The thickness varies from 25–40 m on the Ness of Clousta, but west of the Voe of Clousta, only 8–9 m are preserved. The basalt is aphyric and is pervasively altered, with small feldspar laths in a chloritic matrix that is replaced in parts by aggregates of green biotite, actinolite and epidote as a result of the regional metamorphism. It is vesicular throughout and has well-developed pipe amygdales at its base in places (for example on Muckle Billerin at 3064 5801). Partly remelted inclusions of tuff are also found in the base. The top surface of the sheet is well exposed on the east side of Little Head (2982 5763). Here, the contact is highly irregular, with bulbous protrusions and isolated globular to subangular pods of scoriaceous basalt in sharp contact with the overlying sediment (cf. peperite). Immediately overlying the basalt in places is a volcaniclastic coarse sandstone with quartz, feldspar and some dark igneous clasts that may be slightly flattened; it could therefore be a tuff. More generally, the contact is with purple siltstone and mudstone, the latter having large elongate vesicles. All the signs are that the basalt was intruded at a very shallow depth into the junction between tephra deposits and overlying unconsolidated wet sediments.

Interpretation

The volcanic rocks in the Clousta area were erupted on to the sands and gravels of braided river channels in an alluvial fan derived from metamorphic basement to the north (Astin, 1982). This fan bordered a shallow lake with beach ridges to the SW, which may have encroached north-eastwards at times, depositing finer-grained sediments such as those preserved in the middle of the Clousta Conglomerate. In this environment, volcanic activity is likely to have been phreatic or phreatomagmatic as a result of interaction of rising magma with groundwater or surface water, just below or at ground level. Astin has interpreted many of the pyroclastic deposits as the products of such eruptions.

The basic pyroclastic deposits of the Muckle Head and Muckle Billerin tuffs (and the Hollorin Tuff, 3 km to the WSW of the GCR site) have the composition, bedforms and geometry of phreatomagmatic deposits. The high content of detrital quartz and feldspar and the larger blocks of sandstone and conglomerate reflect the explosive excavation of a vent crater in the underlying alluvium. Indicators of lateral transport, such as cross-bedding, which characteristically decreases in set height away from the vent, coupled with finer-scale planar bedding are typical of pyroclastic surge deposits. But the high proportion of planar bedding suggests that much of the deposit resulted from ash-fall.

Astin reconstructed the original geometry of the basaltic tuffs from measured sections (Figure 9.49b). These formed very shallow cones, 700–1000 m in diameter, with approximate height to width ratios in the range 1:18 to 1:40. Allowing for possible incomplete preservation of the original height, these are comparable with those of modern tuff-rings (1:10 to 1:30; Heiken, 1971). The Muckle Head and Muckle Billerin tuffs each show their maximum thickness, maximum clast size and greatest proportion of sediment-derived clasts close to the Voe of Clousta. Although these features imply close proximity to the vents and possible craters, there is little direct evidence to indicate their sites. Astin did however point out that the lacustrine sediments that directly overlie the Muckle Head Tuff imply a horizontal surface. Hence the thickness variation of the tuff must have been accommodated in the substrata soon after eruption, possibly by slumping and subsidence on the site of the crater. He pointed to the slumped and chaotic sandstone beds below the tuff and steep normal faults restricted to the tuff and immediately underlying sediment, as further evidence for this mechanism. Only thin representatives of the tuffs, with limited lateral extent, are found on the SW side of the Voe of Clousta and the intervening sedimentary sequence is attenuated from about 200 m in the east, to less than 20 m in the west. Astin suggested that this is evidence for active syndepositional faulting on a N–S line along the Voe, which also acted as a magma conduit and controlled the positions of the vents and possible craters.

The Muckle Billerin Tuff is overlain directly by the Clousta Basalt, which has a similar lateral extent (Figure 9.49b). Astin (1982) interpreted this as a lava erupted immediately following the tephra, a transition that is commonly observed in modern phreatomagmatic eruptions as groundwater becomes excluded from the magma conduit. However, the peperitic features at Little Head and the vesiculation (= fluidization) of the overlying mudstones provide convincing evidence that the basalt was intruded into wet, unconsolidated, fine-grained sediments deposited on top of the tuff.

Acid pyroclastic rocks are a major feature of the Clousta volcanic rocks in general, forming large complex lenses such as the Clousta Tuff, west of the Voe of Clousta, and the Aithness Tuff to the east. These larger bodies are built almost entirely from ash-fall tuffs with a large magmatic component, much of it erupted in a plastic state and commonly welded. In the GCR site, only the Brigs Tuff is dominantly acid. This thin lens contains hardly any erupted magmatic material; angular clasts of feldspar porphyry and flow-banded felsite were interpreted by Astin as having originated from the break-up of small pre-existing lava domes or shallow intrusions, such as are found elsewhere among the Clousta volcanic rocks. Some clasts are quite rounded and may have come from the underlying alluvial gravels, along with the quartzofeldspathic sand that forms the matrix of the deposit. Clearly this was generated almost entirely by phreatic eruptions. The well-bedded and sorted nature suggests dominant ash-fall, but the cross-bedding indicates some pyroclastic surge. Measured sections suggest a height to width ratio of about 1:50, notably shallower than the basic tuff-rings and more comparable with modern day maars.

The compositions of the Clousta volcanic rocks are notably less varied than the volcanic sequences elsewhere in Shetland at Papa Stour, Melby and Eshaness (see the Eshaness Coast GCR site report), and Astin (1982) drew attention to the compositional gap between the basaltic and rare andesitic rocks and the more voluminous acid rocks. Thirlwall (1979) concentrated on analyses of the basic rocks, concluding that they have similar characteristics to those at Eshaness, transitional between calc-alkaline and tholeiitic, and were derived from similar parental magmas. Variation in these rocks was explained by low-pressure fractionation of olivine, clinopyroxene and plagioclase. Astin studied the acid rocks in more detail and concluded that, in view of the compositional gap and the presence of only K-feldspar, the acid rocks are more likely to have originated by partial melting of crustal rocks.

As with the other Mid-Devonian volcanic rocks of Shetland and Orkney, the Clousta volcanic rocks were erupted in an extensional basin setting, while retaining geochemical characteristics that are possibly attributable to earlier subduction (see the Eshaness Coast GCR site report). However, being located in the Walls structural block and slightly younger than the other sequences, they are more demonstrably related to the western Shetland plutonic complexes, both temporally and spatially (see Chapter 8: Introduction). These plutons are themselves closely related to the compressive deformation and metamorphism that affected the Old Red Sandstone rocks of the Walls block soon after deposition; Mykura and Phemister (1976) attributed the lack of deformation in rocks close to the Sandsting pluton to pre-deformation hornfelsing, but Astin (1982) implied that this hornfelsing resulted from early crystallization of the outer part of the pluton, which was followed by the main deformation and metamorphism as a result of continuing diapirism and isostatic rise of the plutons. So, it is possible that the volcanism, like the plutonism, may have been related to this Mid- to Late Devonian compressive event, the last phase of Caledonian folding in Britain, which post-dates the main extensional event(s) responsible for the development of the Orcadian Basin.

Conclusions

This GCR site represents the Clousta volcanic rocks of the central, Walls structural block of Old Red Sandstone outcrops in Shetland. Their Givetian age means that they, along with less extensive outcrops on Orkney, represent the youngest Caledonian volcanism in Britain. Although the rocks have subduction-related characteristics, they were erupted in an extensional basin setting, shortly before a reversion to compressive deformation and pluton emplacement.

The rocks were erupted onto an alluvial fan bordering on a lake margin, an environment that resulted in a preponderance of eruptions that involved the explosive gasification of ground and/or surface waters. Measured sections have enabled the three-dimensional form of the deposits to be determined which, together with the sedimentological and compositional features of the volcanic rocks, have suggested the presence of basic tuff-rings and an acid maar. Such features have not been described elsewhere in the Old Red Sandstone volcanic province of Britain and indeed are rarely well preserved in the geological record. An associated basaltic sheet has been intruded into wet unconsolidated sediments at a shallow depth and exhibits good textural features at its upper contact, comparable with those of many other sites in the province.

POINT OF AYRE (HY 590 038)

N. W. A. Odling

Introduction

On the Deerness Peninsula of eastern mainland Orkney, and on the neighbouring island of Shapinsay, volcanic rocks of the Deerness Volcanic Member (formerly known as the Eday volcanic rocks) occur within the Middle Devonian, Eday Flagstone Formation (Peach and Horne, 1880; Flett, 1898; Wilson *et al.*, 1935). Exposures of the volcanic rocks are generally very poor, but good exposures occur on the foreshore at Point of Ayre in the SE of Deerness. Here, the member consists of a vesicular basalt flow that was extruded on to wet lake sediments. The basalt has been substantially altered, which led to an original 'alkaline' classification (Kellock, 1969), but later geochemical studies have shown that it has similar calc-alkaline characteristics to other volcanic rocks of the Old Red Sandstone volcanic suite (Thirlwall, 1979; Fitton *et al.*, 1982). The lava has also provided significant palaeomagnetic information (Robinson, 1985).

Description

There are two outcrops of lava at Point of Ayre (Figure 9.51). The larger outcrop, in the north of the GCR Site, consists of the upper 7 m of an altered greenish-black basalt flow, which forms the foreshore and low cliffs above the Misker rocks. A second outcrop, on the upper shore 30 m to the SW, consists of the top 0.5 m of a vesicular lava. As the two exposures appear to be faulted against each other it is impossible to ascertain whether they represent different flows or different portions of the same flow. However, T. R. Astin (pers. comm., 1997) considers that the two exposures demonstrate a lateral reduction in thickness that occurred at the margin of a single flow.

The top 30 cm of the flow, exposed in low cliffs above high-water mark, contains numerous pipe amygdales up to 20 mm in diameter that are orientated perpendicular to the flow top. Although, most of the amygdales are filled with carbonate and zeolites, some now form hollow voids and so have either lost their filling or are simple vesicles. The pipes terminate at least 25 mm from the flow top and none connect with the upper surface. Below the amygdaloidal zone, the flow exhibits spectacular spheroidal weathering. Near the top of the flow the spheroidal masses are separated by a grid of sediment-filled fractures originating from the overlying Eday Flagstone Formation, indicating that the weathering occurred prior to burial. On the foreshore, cross-bedded tuffaceous sandstones fill hollows in the upper surface of the flow (Kellock, 1969, pl. 1B) and undisturbed laminations in the sediment can be traced to within 5 cm of the top of the igneous rock. The sediments filling the veins and immediately overlying the basalt show no evidence of thermal alteration.

Beneath the amygdaloidal zone the lava is composed of an intimate association of two types of basalt. The more abundant, massive component contains microphenocrysts of olivine and plagioclase set in a groundmass of lilac-coloured clinopyroxene, plagioclase, analcime and carbonate. The primary mineralogy has been subjected to extensive alteration, much of the olivine now being replaced by serpentine and bowlingite. The second component consists of pegmatitic veinlets and patches within the massive basalt. In the veins and patches, olivine up to 0.5 mm and feldspars up to 2 mm are ophitically enclosed by lilac-coloured clinopyroxenes up to 10 mm in diameter. The feldspars are mantled by clear rims of sanidine–anorthoclase and the olivines are substan-

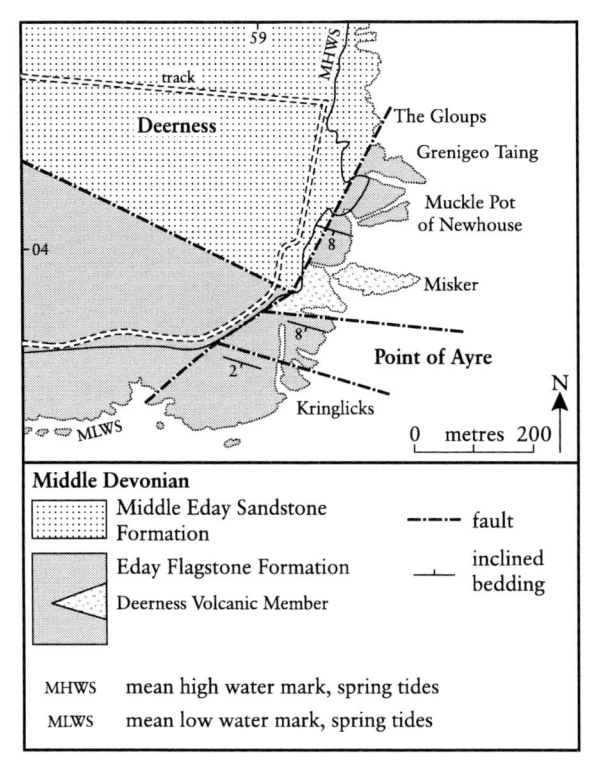

Figure 9.51 Map of the Point of Ayre GCR site, Orkney.

tially altered to serpentine. Analcime, carbonate and zeolite are present as interstitial patches.

Interpretation

The Point of Ayre basic rocks have been subject to a number of differing interpretations in the past (Peach and Horne, 1880; Flett, 1898; Wilson et al., 1935). However, more recently Kellock (1969) has suggested that the lack of thermal alteration and disruption of the overlying sediments, and the presence of sediment-filled veins are evidence that the flow became inundated by sediment only after it had cooled significantly. As the base of the flow(s) is not exposed it is impossible to tell what the nature of the underlying sediment was at the time of the basalt eruption. However, T. R. Astin (pers. comm., 1997) has reported synsedimentary, dish-shaped loading structures in the sediments below the flow and has suggested that these are the result of seismic tremors associated with the eruption of the lava. If so, the lava must have been erupted onto unconsolidated, possibly wet sediments.

The Deerness Volcanic Member occurs within the Eday Flagstone Formation, which is of Givetian age (Westoll, 1977; Marshall, 1996) and can be correlated throughout eastern Orkney (Astin, 1985). It has been suggested that the formation can be correlated on sedimentological grounds with the lower part of the Hoy Sandstone Formation of Hoy (D. Rogers, pers. comm. in Astin, 1990, p. 150; Marshall et al., 1996, p. 459) and hence it is possible that the Deerness Volcanic Member is near-coeval with the Hoy Volcanic Member. In a study of the palaeomagnetism of the 'Eday Group', Robinson (1985) found that most samples of sedimentary rock have been affected by a widespread remagnetization, possibly attributable to deep sub-unconformity weathering and oxidation in the Late Palaeozoic. The basalt of the Point of Ayre and sedimentary rocks in the contact zone, however, give a consistent remnant pole position of 8°N 167°E (present-day grid). Thus the Point of Ayre rocks are significant in preserving a late Mid-Devonian magnetic signature from the Orcadian Basin.

Kellock (1969) noted that the Deerness lava has many geochemical and petrographical features in common with the alkali basalts of the Carboniferous of the Midland Valley. However, Thirlwall (1979) (and in Fitton et al., 1982) noted that, although the major element bulk composition of these substantially altered rocks is that of an alkali basalt, their trace element signature is more allied to calc-alkaline rocks and is similar to the volcanic rocks of Shetland (see the Eshaness Coast GCR site report). It seems probable, therefore, that the apparent alkaline nature of these rocks is due to alteration, in particular the large amount of secondary analcime present in the groundmass. Thus, although the Deerness lava is possibly near-contemporaneous with the Hoy Volcanic Member (see the Too of the Head GCR site report), its inferred primary composition contrasts with the alkaline nature of the Hoy lava and shows that these two lavas cannot be related. The Deerness Volcanic Member comprises the youngest calc-alkaline rocks known in the Orcadian Basin, and hence marks the last possibly subduction-influenced magmatism in this area. The Hoy Volcanic Member shows geochemical features that are transitional between calc-alkaline and alkaline trends, which Francis (1988) considers to mark a change to an extensional tectonic regime. If so, these two GCR sites on Orkney mark an important time when magmatism ceased to be influenced by subducted Iapetus oceanic lithosphere, and became characteristic of the extensional tectonics that were to dominate Carboniferous times in Scotland.

Conclusions

The Point of Ayre GCR site is representative of the poorly exposed Eday volcanic rocks (the Deerness Volcanic Member). The basalt flow and surrounding sedimentary rocks are particularly interesting as they contain many features characteristic of lava extrusion in a subaerial environment. Although the alteration of the basalt has imparted an apparent alkaline character to the rock, the trace element geochemistry preserves evidence of an original calc-alkaline nature. These rocks therefore provide evidence of the last calc-alkaline volcanism associated with the closure of the Iapetus Ocean. Together with the Hoy Volcanic Member (see the Too of the Head GCR site report), the Eday volcanic rocks are significant in preserving a Mid-Devonian magnetic field, unmodified by the widespread late Palaeozoic remagnetization event which affected most of the associated sedimentary rocks.

TOO OF THE HEAD
(ND 184 992–196 990)

N. W. A. Odling

Introduction

The coastal exposures at Too of the Head, on the west side of Rackwick Bay, Isle of Hoy show the most extensive section through the Hoy Volcanic Member (Figures 9.52, 9.53). Here, the member comprises a lower volcaniclastic unit of ash-fall tuffs, agglomerates and tuffaceous sandstones and an upper basaltic lava. Elsewhere, one or other of the two units is commonly absent. The lower, volcaniclastic unit is only known from a number of outcrops north of the Bring Fault on Hoy and one small occurrence on the neighbouring coast of Mainland Orkney near Houton. The Hoy lava crops out most extensively in the north of Hoy, in particular at Too of the Head and at the base of the Old Man of Hoy sea stack. Only one limited outcrop occurs in the south of Hoy near the township of Melsetter. The Hoy Volcanic Member rests unconformably on an eroded surface of the previously folded and faulted Middle Old Red Sandstone, Lower Eday Sandstone Formation and is succeeded, apparently conformably, by the Lang Geo Sandstone Member of the Hoy Sandstone Formation.

The volcanic rocks of Hoy were described first by Peach and Horne (1880) and later by Flett (1898) and Wilson *et al.* (1935). More detailed descriptions, especially of the volcaniclastic rocks, are given by McAlpine (1979). The basalt from Too of the Head has been included in a geochronological study by Halliday *et al.* (1977, 1979b, 1982), and its geochemistry has been discussed by Thirlwall in relation to other Siluro-Devonian volcanic rocks of northern Britain (Thirlwall, 1979, and in Fitton *et al.*, 1982). The lava has also been the subject of a palaeomagnetic study (Storetvedt and Petersen, 1972; Storetvedt and Meland, 1985). The Hoy volcanic rocks are of particular interest because they are one of the youngest preserved representatives of the Old Red Sandstone volcanic suite and appear to represent a transitional phase between the dominantly calc-alkaline volcanism of Silurian and Devonian times and the alkaline

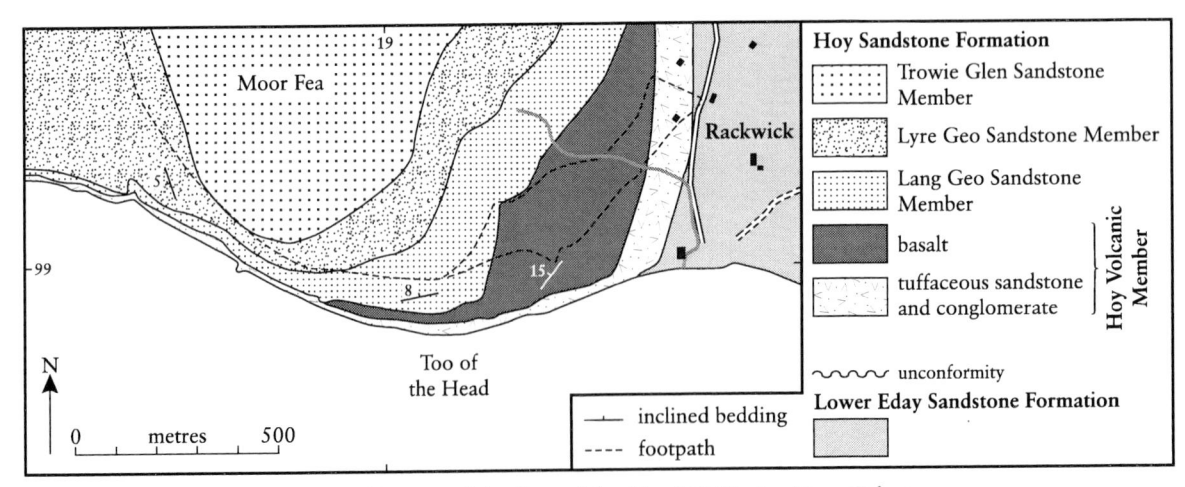

Figure 9.52 Map of the Too of the Head GCR site, Hoy, Orkney.

Too of the Head

Figure 9.53 The cliffs of Too of the Head on the west side of Rackwick Bay, Hoy. The pale rocks of the cliffs close to the buildings are composed of the Lower Eday Sandstone Formation. The lower and middle parts of the cliffs above and behind the buildings are composed of the basal volcaniclastic rocks and lava of the Hoy Volcanic Member. The paler rocks at the top of the cliff belong to the Lang Geo Sandstone Member of the Hoy Sandstone Formation. (Photo: BGS no. D1489.)

volcanicity characteristic of the Carboniferous in Scotland.

Description

The basal unit at Too of the Head consists of an ash-fall tuff that contains numerous angular blocks and lapilli of basalt and rounded volcanic bombs, and a brownish-red, locally cross-bedded tuffaceous sandstone. In the east of the GCR site, the unit is 20 m thick, but it thins to only a few metres and becomes finer grained in the west, below the headland of Moor Fea. The overlying columnar-jointed lava also thins markedly westwards from over 60 m at Rackwick, and it wedges out completely about one kilometre west of Rackwick Bay. The lava is a porphyritic basalt that contains phenocrysts, up to 4 mm, of euhedral to subhedral bytownite, euhedral or anhedral forsteritic olivine and anhedral sodic augite. All of the phenocryst phases are zoned and are variably resorbed. The groundmass consists of aligned laths of bytownite–labradorite with intergranular olivine, clinopyroxene, magnetite, K-feldspar and devitrified glass. The primary mineral assemblage is variably altered and analcime and calcite are significant secondary minerals.

Interpretation

As the outcrop of the Hoy Volcanic Member is discontinuous it is not known whether it is the result of a single eruption or is composed of several separate flows. The three-dimensional geometry of the volcaniclastic unit in the north of Hoy cannot be reconstructed, but the presence of large blocks and bombs at Too of the Head and the westward thinning implies that the eruption centre was located only a short distance away. A likely location for the centre was close to or along the WSW-trending Bring Fault, which cuts across the north of Hoy and was one of the major faults active during the formation of the Orcadian Basin.

As the unfolded Hoy Volcanic Member rests unconformably on an irregular surface of gently folded Middle Old Red Sandstone rocks, the volcanic rocks and the succeeding 'Hoy Sandstone' have formerly been regarded as Upper Old Red Sandstone (e.g. Mykura, 1976, 1991). However, it is now considered that the unconformity, although marked on Hoy, is of local extent only (Rogers *et al.*, 1989). It has further been suggested, on sedimentological grounds, that the 'Hoy Sandstone' is laterally equivalent to the 'Eday Group' of eastern Orkney (Rogers, pers. comm., in Astin, 1990, p. 150; Marshall *et al.*,

1996, p. 459). Although there is no palaeontological evidence from the Hoy Volcanic Member or the immediately overlying sandstones, the underlying strata on Hoy and the proposed laterally equivalent strata to the lower part of the 'Hoy Sandstone' in eastern Orkney, are both assigned to the Givetian on palynological evidence (Marshall, 1996). Hence it seems likely that the Hoy Volcanic Member is of Mid-Devonian age and possibly near-contemporaneous with the Eday volcanic rocks of eastern Orkney (see the Point of Ayre GCR site report). The geochronological study of the basalt of Too of the Head (Halliday *et al.*, 1977, 1979b, 1982) has yielded an Ar-Ar age of 379 ± 10 Ma, broadly consistent with this biostratigraphical age, although the uncertainty in the date and the altered state of the rocks does not allow precise correlation. However, it is clear that the Eday and Hoy volcanic rocks of Orkney are the youngest expressions of Old Red Sandstone volcanism in Britain.

The basalt at Too of the Head has been examined by Storetvedt and Petersen (1972) and Storetvedt and Meland (1985) as part of a palaeomagnetic study of the Devonian rocks of Hoy. Storetvedt and Petersen (1972) found that the lava contains a two-polarity magnetization structure consisting of a high-temperature remanence associated with spinel and a lower temperature remanence associated with haematite. They concluded that the spinel reflects the geomagnetic field at the time of eruption as it is a product of high-temperature alteration soon after the solidification of the lava. Analysis of the spinel remanence indicated a consistent remnant pole position of 23°N 146°E (present-day grid). This compares tolerably well with a pole position of 8°N 167°E obtained by Robinson (1985) for the near-contemporaneous Eday volcanic rocks of Mainland Orkney (see the Point of Ayre GCR site report). As the haematite was formed at a much lower temperature, it is likely that its remanence has recorded a significantly later geomagnetic field.

Thirlwall (1979) reported four analyses of the Hoy lavas, of alkali olivine basalt to hawaiite composition (48–52% SiO_2). Although there is variation in the compositions, he found no significant trace element correlations and concluded that the rocks cannot be related by simple fractional crystallization processes. The samples are unique within the Old Red Sandstone volcanic suite of northern Britain in having between 3 and 5% normative *nepheline*, which, because of the presence of fresh olivine in the rock, is believed to be a primary characteristic. Trace element concentrations and ratios are also typical of alkali basalts, in particular the high Nb, P and light rare earth elements. The clearly alkaline nature of the Hoy basalt sets it apart from the more calc-alkaline character of volcanic rocks from the rest of the province, although the relatively low TiO_2 is typical of arc-related, rather than continental alkali basalts elsewhere. Francis (1988) has suggested that this is the first evidence of a change from compressional, subduction-related tectonics to the extensional regime that was later to produce the voluminous alkaline volcanic rocks of Scotland during the Carboniferous.

Conclusions

The Too of the Head GCR site is of national importance as it contains the most complete section through the Hoy Volcanic Member. It is of international importance because the volcanic sequence provides a rare potential time-marker within the Devonian successions of Europe. A radiometric age of 379 ± 10 Ma from the lava is consistent with the Givetian age extrapolated from plant spores in the underlying strata and lateral correlation of the overlying strata on sedimentological grounds. Studies of the magnetic field preserved by the lava show that at this time the north magnetic pole was situated at 23°N 146°E (present-day grid). The markedly alkaline character of the Hoy lava contrasts with other volcanic rocks of the Middle Devonian of Orkney and Shetland and provides important evidence of the transition to the extensional tectonic regime that characterized Scotland during the Carboniferous.

References

In this reference list, the arrangement is alphabetical by author surname. Chronological order is used within each group of identical authors. Where there are references that include the first-named author with others, the sole-author works are listed chronologically first, followed by all the dual author references (alphabetically), followed by all the references with three or more authors are listed (alphabetically).

Ahmad, M. U. (1967) Some geophysical observations on the Great Glen Fault. *Nature,* **213**, 275–77.

Aitchison, J. C. (1998) A Lower Ordovician (Arenig) radiolarian fauna from the Ballantrae Complex, Scotland. *Scottish Journal of Geology,* **34**, 73–81.

Al Jawadi, A. F. (1987) Minor igneous intrusions of the Lake District; geochronology, geochemistry and petrology. Unpublished PhD thesis, University of Newcastle upon Tyne.

Alderton, D. H. M. (1986) Hessite and electrum from the Ratagain intrusion, northwest Scotland. *Mineralogical Magazine,* **50**, 179.

Alderton, D. H. M. (1988) Ag-Au-Te mineralisation in the Ratagain complex, northwest Scotland. *Transactions of the Institution of Mining and Metallurgy (Section B: Applied Earth Sciences),* **97**, 171–80.

Allan, W. C. (1970) The Morven–Cabrach basic intrusion. *Scottish Journal of Geology,* **6**, 53–72.

Allen, J. R. L. and Williams, B. P. J. (1981) Sedimentology and stratigraphy of the Townsend Tuff Bed (Lower Old Red Sandstone) in South Wales and the Welsh Borders. *Journal of the Geological Society of London,* **138**, 15–29.

Allen, P. M. (1987) The Solway line is not the Iapetus suture. *Geological Magazine,* **124**, 485–6.

Allen, P. M., Bide, P. J., Cooper, D. C., Parker, M. E. and Haslam, H. W. (1981) Copper-bearing rocks at Cairngarroch Bay, south-west Scotland. *Mineral Reconnaissance Programme Report, Institute of Geological Sciences,* No. **39**.

Allport, S. (1879) On the diorites of the Warwickshire coal-field. *Quarterly Journal of the Geological Society of London,* **35**, 637–42.

Allsop, J. M. (1987) Patterns of late Caledonian intrusive activity in eastern and northern England from geophysics, radiometric dating and basement geology. *Proceedings of the Yorkshire Geological Society,* **46**, 335–53.

Allsop, J. M. and Arthur, M. J. (1983) A possible extension of the South Leicestershire Diorite complex. *Report of the Institute of Geological Sciences,* No. **83/10**, 25–30.

Anderson, J. G. C. (1935a) The Arrochar intrusive complex. *Geological Magazine,* **72**, 263–83.

Anderson, J. G. C. (1935b) The marginal intrusions of Ben Nevis, the Coille Lianachain complex, and the Ben Nevis dyke swarm. *Transactions of the Geological Society of Glasgow,* **19**, 225–69.

Anderson, J. G. C. (1936) Age of the Girvan–Ballantrae serpentine. *Geological Magazine,* **73**, 535–45.

Anderson, J. G. C. (1937) The Etive granite complex. *Journal of the Geological Society of London,* **93**, 487–533.

References

Anderson, J. G. C. (1947) The geology of the Highland Border, Stonehaven to Arran. *Transactions of the Royal Society of Edinburgh*, **61**, 479–515.

Anderson, J. G. C. (1956) The Moinian and Dalradian rocks between Glen Roy and the Monadhliath mountains, Inverness-shire. *Transactions of the Royal Society of Edinburgh*, **63**, 15–36.

Anderson, J. G. C. and Pringle, J. (1944) The Arenig rocks of Arran, and their relationship to the Dalradian Series. *Geological Magazine*, **81**, 81–7.

Anderson, J. G. C. and Tyrell, G. W. (1937) Xenolithic minor intrusions in the Loch Lomond District. *Transactions of the Geological Society of Glasgow*, **19**, 373–84.

Anderson, J. L. (1996) Status of thermobarometry in granitic batholiths. *Transactions of the Royal Society of Edinburgh: Earth Sciences*, **87**, 125–38.

Anon. (1972) Ophiolites. *Geotimes*, **17**(12), 24–5.

Ansari, S. M. (1983) Petrology and petrochemistry of the Eskdale and adjacent intrusions (Cumbria) with special reference to mineralization. Unpublished PhD thesis, University of Nottingham.

Armstrong, H. A., Owen, A. W., Scrutton, C. T., Clarkson, E. N. K. and Taylor, C. M. (1996) Evolution of the Northern Belt, Southern Uplands: implications for the Southern Uplands controversy. *Journal of the Geological Society of London*, **153**, 197–205.

Armstrong, M. and Paterson, I. B. (1970) The Lower Old Red Sandstone of the Strathmore region. *Report of the Institute of Geological Sciences*, No. **70/12**.

Armstrong, M., Paterson, I. B. and Browne, M. A. E. (1985) Geology of the Perth and Dundee district. *Memoir of the British Geological Survey*, Sheets 48W, 48E and 49 (Scotland).

Ashcroft, W. A. and Munro, M. (1978) The structure of the eastern part of the Insch Mafic Intrusion, Aberdeenshire. *Scottish Journal of Geology*, **14**, 55–79.

Ashcroft, W. A., Kneller, B. C., Leslie, A. G. and Munro, M. (1984) Major shear zones and autochthonous Dalradian in the northeast Scottish Caledonides. *Nature*, **310**, 760–2.

Astin, T. R. (1982) The Devonian geology of the Walls Peninsula, Shetland. Unpublished PhD thesis, University of Cambridge.

Astin, T. R. (1983) Discussion on implications for Caledonian plate tectonic models of chemical data from volcanic rocks of the British Old Red Sandstone. *Journal of the Geological Society of London*, **140**, 315–18.

Astin, T. R. (1985) The palaeogeography of the Middle Devonian Lower Eday Sandstone, Orkney. *Scottish Journal of Geology*, **21**, 353–75.

Astin, T. R. (1990) The Devonian lacustrine sediments of Orkney, Scotland; implications for climatic cyclicity, basin structure and maturation history. *Journal of the Geological Society of London*, **147**, 141–51.

Atherton, M. P. (1993) Granite magmatism. *Journal of the Geological Society of London*, **150**, 1009–23.

Atherton, M. P. and Plant, J. A. (1985) High heat producing granites and the evolution of the Andean and Caledonian continental margins. In *High Heat Production (HHP) Granites, Hydrothermal Circulation and Ore Genesis*, Institution of Mining and Metallurgy, London, pp. 263–285.

Bailey, E. B. (1935) The Glencoul Nappe and the Assynt Culmination. *Geological Magazine*, **72**, 151–65.

Bailey, E. B. (1960) The geology of Ben Nevis and Glencoe and the surrounding country, 2nd edition. *Memoir of the Geological Survey of Great Britain*, Sheet 53 (Scotland).

Bailey, E. B. and Anderson, E. M. (1925) The Geology of Staffa, Iona and western Mull. *Memoir of the Geological Survey of Great Britain*, Sheet 43 (Scotland).

Bailey, E. B. and Maufe, H. B. (1916) The geology of Ben Nevis and Glen Coe and the surrounding country, 1st edn. *Memoir of the Geological Survey of Great Britain*, Sheet 53 (Scotland).

Bailey, E. B. and McCallien, W. J. (1934) Pre-Cambrian Association excursion to Scotland. *Geological Magazine*, **71**, 553–5.

Bailey, E. B. and McCallien, W. J. (1957) The Ballantrae Serpentine, Ayrshire. *Transactions of the Edinburgh Geological Society*, **17**, 33–53.

Baker, J. W. and Hughes, C. P. (1979) Summer (1973) field meeting in Central Wales, 31 August to 7 September 1973. Report by the organising directors. *Proceedings of the Geologists' Association*, **90**, 65–79.

Ball, T. K. and Merriman, R. J. (1989) The petrology and geochemistry of the Ordovician Llewelyn Volcanic Group, Snowdonia, North

References

Wales. *British Geological Survey Research Report*, No. **SG/89/1**.

Balsillie, D. (1932) The Ballantrae Igneous Complex, South Ayrshire. *Geological Magazine*, **69**, 107–31.

Balsillie, D. (1937) Further observations on the Ballantrae Igneous Complex, South Ayrshire. *Geological Magazine*, **74**, 20–33.

Barber, A. J., Beach, A., Park, R. G. Tarney, J., and Stewart A. D. (1978) The Lewisian and Torridonian rocks of North-West Scotland. *Geologists' Association Guide*, No. **21**.

Barber, P. L., Dobson, M. R. and Whittington, R. J. (1979) The geology of the Firth of Lorne, as determined by seismic and dive sampling methods. *Scottish Journal of Geology*, **15**, 217–30.

Barnes, R. P. and Fettes, D. J. (1996) Creetown and Cairnsmore of Fleet: igneous intrusion and tectonic deformation. In *Geology in South-west Scotland: an Excursion Guide* (ed. P. Stone), British Geological Survey, Keyworth, Nottingham, pp. 140–150.

Barnes, R. P., Lintern, B. C. and Stone, P. (1989) Timing and regional implications of deformation in the Southern Uplands of Scotland. *Journal of the Geological Society of London*, **146**, 905–8.

Barnes, R. P., Phillips, E. R. and Boland, M. P. (1995) The Orlock Bridge Fault in the Southern Uplands of SW Scotland: a terrane boundary? *Geological Magazine*, **132**, 523–9.

Barnes, R. P., Rock, N. M. S. and Gaskarth, J. W. (1986) The Caledonian dyke swarms in Southern Scotland: new field and petrological data for the Wigtown Peninsula, Galloway. *Geological Journal*, **21**, 101–25.

Barr, D., Roberts, A. M., Highton, A. J., Parson, L. M. and Harris, A. L. (1985) Structural setting and geochronological significance of the West Highland Granitic Gneiss, a deformed early granite within Proterozoic, Moine rocks of NW Scotland. *Journal of the Geological Society of London*, **142**, 663–76.

Barritt, S. D. (1983) The controls of radioelement distribution in the Etive and Cairngorm granites: implications for heat production. Unpublished PhD thesis, The Open University.

Barrow, G. and Cunningham Craig, E. H. (1912) The geology of the districts of Braemar, Ballater and Glen Clova. *Memoir of the Geological Survey of Great Britain*, Sheet 65 (Scotland).

Barrow, G., Hinxman, L. W. and Cunningham Craig, E. H. (1913) The geology of Upper Strathspey, Gaick and the Forest of Atholl. *Memoir of the Geological Survey of Great Britain*, Sheet 64 (Scotland).

Bartholomew, I. D. (1993) The interaction and geometries of diapiric uprise centres along mid-ocean ridges – evidence from mantle fabric studies of ophiolite complexes. In *Magmatic Processes and Plate Tectonics* (eds H. M. Prichard, T. Alabaster, N. B. W. Harris and C. R. Neary), *Geological Society Special Publication*, No. **76**, pp. 245–56.

Bassett, M. G. (1982) Silurian rocks of the Marloes and Pembroke peninsulas. In *Geological Excursions in Dyfed, South-West Wales* (ed. M. G. Bassett), National Museum of Wales, Cardiff, pp. 103–122.

Batchelor, R. A. (1987) Geochemical and petrological characteristics of the Etive granitoid complex, Argyll. *Scottish Journal of Geology*, **23**, 227–49.

Beamish, D. and Smythe, D. B. (1986) Geophysical images of the deep crust: the Iapetus suture. *Journal of the Geological Society of London*, **143**, 489–97.

Beavon, R. V. (1963) The succession and structure east of the Glaslyn river, North Wales. *Quarterly Journal of the Geological Society of London*, **119**, 479–512.

Beavon, R. V. (1980) A resurgent cauldron in the early Palaeozoic of Wales, U.K. *Journal of Volcanology and Geothermal Research*, **7**, 157–74.

Beckinsale, R. D. and Obradovich, J. D. (1973) Potassium-argon ages for minerals from the Ross of Mull, Argyllshire, Scotland. *Scottish Journal of Geology*, **9**, 147–56.

Beddoe-Stephens, B. (1990) Pressures and temperatures of Dalradian metamorphism and the andalusite-kyanite transformation in the northeast Grampians. *Scottish Journal of Geology*, **26**, 3–14.

Beddoe-Stephens, B. (1999) The Glen Tilt diorite: crystallization, petrogenesis and relation to granitic rocks. *Scottish Journal of Geology*, **35**, 157–78.

Bevier, M. L. and Whalen, J. B. (1990) Tectonic significance of Silurian magmatism in the Canadian Appalachians. *Geology*, **18**, 411–14.

Bevins, R. E. (1979) The geology of the Strumble Head–Fishguard region, Dyfed. Unpublished PhD thesis, University of Keele.

Bevins, R. E. (1982) Petrology and geochemistry

References

of the Fishguard Volcanic Complex, Wales. *Geological Journal*, **17**, 1–21.

Bevins, R. E. and Metcalfe, R. (1993) Ordovician igneous rocks of the Builth Inlier. In *Geological Excursions in Powys, Central Wales* (eds N. H. Woodcock and M. G. Bassett), National Museum of Wales, Cardiff, pp. 243–58.

Bevins, R. E. and Roach, R. A. (1979a) Early Ordovician volcanism in Dyfed, SW Wales. In *The Caledonides of the British Isles – Reviewed*, (eds A. L. Harris, C. H. Holland and B. E. Leake), Geological Society of London Special Publication, No. **8**, pp. 603–9.

Bevins, R. E. and Roach, R. A. (1979b) Pillow lava and isolated-pillow breccia of rhyodacitic composition from the Fishguard Volcanic Group, Lower Ordovician, SW Wales, U.K. *Journal of Geology*, **87**, 193–201.

Bevins, R. E. and Roach, R. A. (1982) Ordovician igneous activity in south-west Dyfed. In *Geological excursions in Dyfed, South-West Wales* (ed. M. G. Bassett), National Musuem of Wales, Cardiff, pp. 65–80.

Bevins, R. E., Bluck, B. J., Brenchley, P. J., Fortey, R. A., Hughes, C. P. *et al.* (1992) Ordovician. In *Atlas of Palaeogeography and Lithofacies* (eds J. C. W. Cope, J. K. Ingham and P. F. Rawson), The Geological Society of London, pp. 19–36.

Bevins, R. E., Kokelaar, B. P. and Dunkley, P. N. (1984) Petrology and geochemistry of lower to middle Ordovician igneous rocks in Wales: a volcanic arc to marginal basin transition. *Proceedings of the Geologists' Association*, **95**, 337–47.

Bevins, R. E., Lees, G. J. and Roach, R. A. (1991) Ordovician bimodal volcanism in SW Wales: geochemical evidence for petrogenesis of the silicic rocks. *Journal of the Geological Society of London*, **148**, 719–29.

Bevins, R. E., Lees, G. J. and Roach, R. A. (1992) Petrogenesis of Ordovician igneous rocks in the southern part of the Welsh Basin. *Geological Magazine*, **129**, 615–24.

Bevins, R. E., Lees, G. J., Roach, R. A., Rowbotham, G. and Floyd, P. A. (1994) Petrogenesis of the St David's Head Layered Intrusion, Wales: a complex history of multiple magma injection and *in situ* crystallization. *Transactions of the Royal Society of Edinburgh: Earth Sciences*, **85**, 91–121.

Bloxam, T. W. (1955) The origin of the Girvan–Ballantrae beerbachites. *Geological Magazine*, **92**, 329–37.

Bloxam, T. W. (1960) Pillow structure in spilitic lavas at Downan Point, Ballantrae. *Transactions of the Geological Society of Glasgow*, **24**, 19–26.

Bloxam, T. W. (1968) The petrology of Byne Hill, Ayrshire. *Transactions of the Royal Society of Edinburgh*, **68**, 105–22.

Bloxam, T. W. (1980) Amphibolite contact zones, amphibolite xenoliths, and blueschists associated with serpentinite in the Girvan–Ballantrae Complex, southwest Scotland. *Archive des Sciences*, **33**, 291–9.

Bloxam, T. W. (1982) Ordovician volcanism in Scotland. In *Igneous Rocks of the British Isles* (ed. D. S. Sutherland), Wiley, Chichester, pp. 51–63.

Bloxam, T. W. and Allen, J. B. (1960) Glaucophane-schist, eclogite and associated rocks from Knockormal in the Girvan–Ballantrae Complex, south Ayrshire. *Transactions of the Royal Society of Edinburgh*, **64**, 1–27.

Bluck, B. J. (1978a) Geology of a continental margin 1: the Ballantrae Complex. In *Crustal Evolution in Northwestern Britain and Adjacent Regions* (eds D. R. Bowes and B. E. Leake), Geological Journal Special Issue No. **10**, pp. 151–62.

Bluck, B. J. (1978b) Sedimentation in a late orogenic basin: the Old Red Sandstone of the Midland Valley of Scotland. In *Crustal Evolution in Northwestern Britain and Adjacent Regions* (eds D. R. Bowes and B. E. Leake), Geological Journal Special Issue No. **10**, pp. 249–78.

Bluck, B. J. (1982) Hyalotuff deltaic deposits in the Ballantrae ophiolite of SW Scotland: evidence for crustal position of the lava sequence. *Transactions of the Royal Society of Edinburgh: Earth Sciences*, **72**, 217–28.

Bluck, B. J. (1992) Balmaha, 110–29; Pinbain Block, 319–338; Bennane Head to Downan Point, 347–361. In *Geological Excursions around Glasgow and Girvan* (eds J. D. Lawson and D. S. Weedon), Geological Society of Glasgow.

Bluck, B. J. and Ingham, J. K. (1992) Dow Hill, Byne Hill and Ardmillan braes. In *Geological Excursions around Glasgow and Girvan* (eds J. D. Lawson and D. S. Weedon), Geological Society of Glasgow, pp. 366–77.

Bluck, B. J., Dempster, T. J. and Rogers, G.

References

(1997) Allochthonous metamorphic blocks on the Hebridean passive margin, Scotland. *Journal of the Geological Society of London*, **154**, 921–4.

Bluck, B. J., Gibbons, W. and Ingham, J. K. (1992) Terranes. In *Atlas of Palaeogeography and Lithofacies* (eds J. C. W. Cope, J. K. Ingham and P. F. Rawson), Geological Society of London Memoir No. 13, pp. 1–3.

Bluck, B. J., Halliday, A. N., Aftalion, M. and Macintyre, R. M. (1980) Age and origin of Ballantrae ophiolite and its significance to the Caledonian orogeny and Ordovician time scale. *Geology*, **8**, 492–5.

Bluck, B. J., Ingham, J. K., Curry, G. B. and Williams, A. (1984) The significance of a reliable age from some Highland Border rocks in Central Scotland. *Journal of the Geological Society of London*, **139**, 451–4.

Blyth, F. G. H. (1955) The Kirkmabreck Granodiorite, near Creetown, South Galloway. *Geological Magazine*, **92**, 321–8.

Blyth, F. G. H. (1969) Structures in the southern part of the Cabrach igneous area, Banffshire. *Proceedings of the Geologists' Association*, **80**, 63–79.

Bonney, T. G. (1878) On the serpentine and associated igneous rocks of the Ayrshire coast. *Quarterly Journal of the Geological Society of London*, **34**, 769–85.

Bonney, T. G. (1881) On a boulder of hornblende picrite near Pen-y-Carnisog, Anglesey. *Quarterly Journal of the Geological Society of London*, **38**, 137.

Bonney, T. G. (1883) Notes on a series of rocks from the North-West Highlands, collected by C. Callaway, Esq., D.Sc., F.G.S. *Quarterly Journal of the Geological Society of London*, **39**, 414–22.

Bosworth, T. O. (1910) Metamorphism around the Ross of Mull Granite. *Quarterly Journal of the Geological Society of London*, **66**, 376–401.

Bosworth, T. O. (1912) *The Keuper Marls Around Charnwood Forest*. Thornley and Son for Leicester Literary and Philosophical Society.

Bott, M. P. H. (1967) Geophysical investigations of the northern Pennine basement rocks. *Proceedings of the Yorkshire Geological Society*, **36**, 139–68.

Bott, M. H. P. (1974) The geological interpretation of a gravity survey of the English Lake District and the Vale of Eden. *Journal of the Geological Society of London*, **130**, 309–31.

Boulter, C. A. and Soper, N. J. (1973) Structural relationships of the Shap granite. *Proceedings of the Yorkshire Geological Society*, **39**, 365–9.

Bowen, N. L. (1928) *The Evolution of the Igneous Rocks*, Princeton University Press, (Reproduced by Dover Publications, 1956).

Bowes, D. R. (1962) Kentallenite–lamprophyre–granite age relations at Kentallen, Argyll. *Geological Magazine*, **99**, 119–22.

Bowes, D. R. and McArthur, A. C. (1976) Nature and genesis of the appinite suite. *Krystalinikum*, **12**, 31–46.

Bowes, D. R. and Wright, A. E. (1961) An explosion-breccia complex at Back Settlement, near Kentallen, Argyll. *Transactions of the Edinburgh Geological Society*, **18**, 293–313.

Bowes, D. R. and Wright, A. E. (1967) The explosion breccia pipes near Kentallen, Scotland and their geological setting. *Transactions of the Royal Society of Edinburgh*, **67**, 109–43.

Bowes, D. R., Kinloch, E. D. and Wright, A. E. (1964) Rhythmic amphibole overgrowths in appinites associated with explosion breccias in Argyll. *Mineralogical Magazine*, **33**, 963–73.

Boyd, R. and Munro, M. (1978) Deformation of the Belhelvie mass, Aberdeenshire. *Scottish Journal of Geology*, **14**, 29–44.

Brade-Birks, S. G. (1923) Notes on Myriapoda XXVIII. *Kampecaris tuberculata* n.sp. from the Old Red Sandstone of Ayrshire. *Proceedings of the Royal Physical Society, Edinburgh*, **20**, 277–80.

Bradshaw, R., Plant, A. G., Burke, K. C. and Leake, B. E. (1969) The Oughterard granite, Connemara, Co Galway. *Proceedings of the Royal Irish Academy*, **68**, 39–65.

Braithwaite, R. S. W. and Knight, J. R. (1990) Rare mineralisation near Dalbeattie, South Scotland. *Mineralogical Magazine*, **54**, 129–31.

Branney, M. J. (1988a) The subaerial setting of the Ordovician Borrowdale Volcanic Group, English Lake District. *Journal of the Geological Society of London*, **145**, 887–90.

Branney, M. J. (1988b) Subaerial explosive volcanism, intrusion, sedimentation, and collapse in the Borrowdale Volcanic Group, SW Langdale, English Lake District. Unpublished PhD thesis, University of Sheffield.

Branney, M. J. (1990a) Explosive volcanism and volcanotectonic subsidence at Crinkle Crags.

References

In *The Lake District* (ed. F. Moseley), The Geologists' Association, London, pp. 128–37.

Branney, M. J. (1990b) Subaerial pyroclastics of Side Pike. In *The Lake District* (ed. F. Moseley), The Geologists' Association, London, pp. 138–42.

Branney, M. J. (1991) Eruption and depositional facies of the Whorneyside Tuff Formation, English Lake District: An exceptionally large-magnitude phreatoplinian eruption. *Geological Society of America Bulletin*, **103**, 886–97.

Branney, M. J. (1995) Downsag and extension at calderas: new perspectives on collapse geometries from ice-melt, mining, and volcanoes. *Bulletin of Volcanology*, **57**, 304–18.

Branney, M. J. and Kokelaar, B. P. (1994a) Volcanotectonic faulting, soft-state deformation and rheomorphism of tuffs during development of a piecemeal caldera, English Lake District. *Geological Society of America Bulletin*, **106**, 507–30.

Branney, M. J. and Kokelaar, B. P. (1994b) Early caldera eruptions and piecemeal subsidence (Wrynose Pass to Crinkle Crags). In *Processes and controls of caldera collapse and related ignimbrite emplacements in Snowdonia (Wales), the Lake District (England) and Glencoe (Scotland), United Kingdom* (eds B. P. Kokelaar, M. J. Branney, I. Moore and M. F. Howells), Field Guide of the IAVCEI Commission on Explosive Volcanism Field Workshop, 18–29 May, 1994, pp. D1–7.

Branney, M. J. and Kokelaar, B. P. (1994c) Climactic eruptions and collapse of Scafell caldera (Crinkle Crags from Great Langdale). In *Processes and controls of caldera collapse and related ignimbrite emplacements in Snowdonia (Wales), the Lake District (England) and Glencoe (Scotland), United Kingdom* (eds B. P. Kokelaar, M. J. Branney, I. Moore, and M. F. Howells), Field Guide of the IAVCEI Commission on Explosive Volcanism Field Workshop, 18–29 May, 1994, pp. E1–4.

Branney, M. J. and Soper, N. J. (1988) Ordovician volcano-tectonics in the English Lake District. *Journal of the Geological Society of London*, **145**, 367–76.

Branney, M. J. and Sparks, R. S. J. (1990) Fiamme formed by diagenesis and burial-compaction in soils and subaqueous sediments. *Journal of the Geological Society of London*, **147**, 919–22.

Branney, M. J. and Suthren, R. J. (1988) High-level peperitic sills in the English Lake District: distinction from block lavas and implications for Borrowdale Volcanic Group stratigraphy. *Geological Journal*, **23**, 171–87.

Branney, M. J., Kokelaar, B. P. and McConnell, B. J. (1992) The Bad Step Tuff: a lava-like rheomorphic ignimbrite in a calc-alkaline piecemeal caldera, English Lake District. *Bulletin of Volcanology*, **54**, 187–99.

Branney, M. J., Davis, N. C., Kokelaar, B. P. and McConnell, B. J. (1993) The Airy's Bridge Formation in the Central Fells. *British Geological Survey Technical Report*, No. **WA/93/41**.

Branney, M. J., Kneller, B. C. and Kokelaar, B. P. (1990) Disordered turbidite facies (DTF): a product of continuous surging density flows (abstract). *13th International Sedimentological Congress, Nottingham, UK. No. 2*.

Brealy, A. J. (1984) A TEM study of the development of fibrolite sillimanite in a thermal aureole. *Journal of the Geological Society of London*, **141**, 190.

Brenchley, P. J. (1964) Ordovician ignimbrites in the Berwyn Hills, North Wales. *Geological Journal*, **4**, 43–54.

Brenchley, P. J. (1969) The relationship between Caradocian volcanicity and sedimentation in North Wales. In *The Pre-Cambrian and Lower Palaeozoic Rocks of Wales* (ed. A. Wood), University of Wales Press, Cardiff, pp. 181–202.

Brenchley, P. J. (1972) The Cwm Clwyd Tuff, North Wales: a palaeogeographic interpretation of some Ordovician ash-shower deposits. *Proceedings of the Yorkshire Geological Society*, **39**, 199–224.

Brenchley, P. J. (1978) The Caradocian rocks of north and west Berwyn Hills, North Wales. *Geological Journal*, **13**, 137–64.

Brenchley, P. J. and Pickerill, R. K. (1980) Shallow subtidal sediments of Soudleyan (Caradoc) age in the Berwyn Hills, North Wales, and their palaeogeographic context. *Proceedings of the Geologists' Association*, **91**, 177–94.

Briden, J. C. and Morris, W. A. (1973) Palaeomagnetic studies in the British Caledonides – III Igneous rocks of the Northern Lake District, England. *Journal of Geophysical Research*, **34**, 27–46.

Briden, J. C., Morris, W. A. and Piper, J. D. A. (1973) Palaeomagnetic studies in the British Caledonides – VI. Regional and global implications. *Geophysical Journal of the Royal*

References

Astronomical Society, **34**, 107–34.

Bridge, D. M., Carney, J. N., Lawley, R. S. and Rushton, A. W. A. (1988) The geology of the country around Coventry and Nuneaton. *Memoir of the British Geological Survey,* Sheet 169 (England and Wales).

Bridges, P. H. (1976) Late Silurian transgressive barrier sands, southwest Wales. *Sedimentology,* **23**, 347–62.

British Geological Survey (1996) Tectonic map of Britain, Ireland and adjacent areas, 1:500 000 (compilers: T. C. Pharoah, J. H. Morris, C. B. Long and P. D. Ryan). British Geological Survey, Keyworth, Nottingham.

Brögger, W. C. (1921) Die Eruptivgesteine des Kristianiagebietes. IV. Das Fengebiet in Telemark. *Norske Videnskapsselskjapets Skrifter. I. Math. Naturv. Klasse,* **9**, 1–408.

Bromley, A. V. (1963) The geology of the country around Blaenau Ffestiniog. Unpublished PhD thesis, University of Wales, Aberystwyth.

Bromley, A. V. (1964) Allanite in the Tan-y-Grisiau Microgranite, Merionethshire, North Wales. *American Mineralogist,* **49**, 1747–52.

Bromley, A. V. (1969) Acid plutonic igneous activity in the Ordovician of North Wales. In *The Pre-Cambrian and Lower Palaeozoic Rocks of Wales* (ed. A. Wood), University of Wales Press, Cardiff, pp. 387–408.

Brown, G. C. and Locke, C. A. (1979) Space-time variations in British Caledonian Granites: some geophysical correlations. *Earth and Planetary Science Letters,* **45**, 69–79.

Brown, G. C., Cassidy, J., Tindle, A. G. and Hughes, D. J. (1979) The Loch Doon granite: an example of granite petrogenesis in the British Caledonides. *Journal of the Geological Society of London,* **136**, 745–53.

Brown, G. C., Francis, E. H., Kennan, P. and Stillman, C. J. (1985) Caledonian igneous rocks of Britain and Ireland. In *The Nature and Timing of Orogenic Activity in the Caledonian Rocks of the British Isles* (ed. A. L. Harris), Geological Society of London Memoir No. 9, pp. 1–15.

Brown, J. F. (1975) Rb-Sr studies and related chemistry on the Caledonian calc-alkaline igneous rocks of NW Argyllshire. Unpublished PhD thesis, University of Oxford.

Brown, P. E. (1991) Caledonian and earlier magmatism. In *Geology of Scotland,* 3rd edn, (ed. G. Y. Craig), The Geological Society of London, pp. 229–95.

Brown, P. E., Miller, J. A. and Grasty, R. L. (1968) Isotopic ages of Late Caledonian granitic intrusions in the British Isles. *Proceedings of the Yorkshire Geological Society,* **36**, 51–276.

Brown, P. E., Miller, J. A. and Soper, N. J. (1964) Age of the principal intrusions of the Lake District. *Proceedings of the Yorkshire Geological Society,* **34**, 331–42.

Burt, R. M. (1994) The geology of Ben Nevis, Southwest Highlands, Scotland. Unpublished PhD thesis, University of St Andrews.

Burt, R. M. and Brown, P. E. (1997) The Ben Nevis Intrusive Ring Tuff, Scotland: re-interpretation of the 'flinty crush rock' as part of an ignimbrite conduit in the roots of an ancient caldera. *Scottish Journal of Geology,* **33**, 149–55.

Busby, J. P., Kimbell, G. S. and Pharoah, T. C. (1993) Integrated geophysical/geological modelling in southern Britain. *Geological Magazine,* **130**, 593–604.

Busrewil, M. T., Pankhurst, R. J. and Wadsworth, W. J. (1973) The igneous rocks of the Boganclogh area, NE Scotland. *Scottish Journal of Geology,* **9**, 165–76.

Busrewil, M. T., Pankhurst, R. J. and Wadsworth, W. J. (1975) The origin of the Kennethmont granite-diorite series. *Mineralogical Magazine,* **40**, 363–76.

Cameron, E. M. and Hattori, K. (1987) Archean gold mineralization and oxidized hydrothermal fluids. *Economic Geology,* **82**, 1177–91.

Cameron, I. B. and Stephenson, D. (1985) *British Regional Geology: The Midland Valley of Scotland,* (3rd edn), HMSO, London, for British Geological Survey.

Campbell, R. (1911) Preliminary note on the geology of south-eastern Kincardineshire. *Geological Magazine,* **8**, 63–9.

Campbell, R. (1913) The geology of south-eastern Kincardineshire. *Transactions of the Royal Society of Edinburgh,* **48**, 923–60.

Campbell, S. D. G. (1995) The Borrowdale Volcanic Group and related geology on 1:10 000 Sheets NY 32 SW and SE. *British Geological Survey Technical Report,* No. **WA/95/02**.

Campbell, S. D. G., Howells, M. F., Smith, M. and Reedman, A. J. (1988) A Caradoc failed-rift within the Ordovician Marginal Basin of Wales. *Geological Magazine,* **125**, 257–66.

Campbell, S. D. G., Reedman, A. J. and Howells, M. F. (1985) Regional variations in cleavage and fold development in North Wales. *Geological Journal,* **20**, 43–52.

References

Campbell, S. D. G., Reedman, A. J., Howells, M. F. and Mann, A. C. (1987) The emplacement of geochemically distinct groups of rhyolites during the evolution of the Lower Rhyolitic Tuff Formation caldera (Ordovician), N. Wales, U.K. *Geological Magazine,* **124,** 501–11.

Cannat, M. (1989) Late Caledonian northeastward ophiolite thrusting in the Shetland Islands, U.K. *Tectonophysics,* **169,** 257–70.

Canning, J. C., Henney, P. J., Morrison, M. A. and Gaskarth, J. W. (1996) Geochemistry of late Caledonian minettes from northern Britain: implications for the Caledonian sub-continental lithospheric mantle. *Mineralogical Magazine,* **60,** 221–36.

Cantrill, T. C., Dixon, E. E. L., Thomas, H. H. and Jones, O. T. (1916) The geology of the South Wales coalfield, Part 12: The country around Milford. *Memoir of the Geological Survey of Great Britain*, Sheet 227 (England and Wales).

Carney, J. N., Glover, B. W. and Pharaoh, T. C. (1992) Pre-conference field excursion: Precambrian and Lower Palaeozoic rocks of the English Midlands. *British Geological Survey Technical Report*, No. **WA/92/72**.

Carroll, S. (1994) Geology of the Inverbervie and Catterline District. *British Geological Survey Technical Report*, No. **WA/94/20**.

Carruthers, R. G., Burnett, G. A., Anderson, W. and Thomas, H. H. (1932) The geology of the Cheviot Hills. *Memoir of the Geological Survey of Great Britain*, Sheets 3 and 5 (England and Wales).

Cas, R. A. F. and Wright, J. V. (1987) *Volcanic Successions, Modern and Ancient: A Geological Approach to Processes, Products and Successions*, Allen and Unwin, London.

Cas, R. A. F. and Wright, J. V. (1991) Subaqueous pyroclastic flows and ignimbrites: an assessment. *Bulletin of Volcanology,* **53,** 357–80.

Castro, A., de la Rosa, J. D. and Stephens, W. E. (1990) Magma mixing in the subvolcanic environment: petrology of the Gerena interaction zone near Seville, Spain. *Contributions to Mineralogy and Petrology,* **106,** 9–26.

Castro, A. and Stephens, W. E. (1992) Amphibole polycrystalline clots in calc-alkaline granitic rocks and their enclaves. *Canadian Mineralogist,* **30,** 1093–112.

Cattermole, P. J. (1976) The crystallization and differentiation of a layered intrusion of hydrated alkali olivine-basalt parentage at Rhiw, North Wales. *Geological Journal,* **11,** 45–70.

Caunt, S. (1984) Geological aspects of the Threlkeld Microgranite, Cumbria. *Transactions of the Leeds Geological Association,* **10,** 89–100.

Cave, R. (1977) Geology of the Malmesbury district. *Memoir of the Geological Survey of Great Britain*, Sheet 251 (England and Wales).

Cawthorn, R. G. (1976) Calcium-poor pyroxene reaction relations in calc-alkaline magmas. *American Mineralogist,* **61,** 907–12.

Chappell, B. W. and Stephens, W. E. (1988) Origin of infracrustal (I-type) granite magmas. *Transactions of the Royal Society of Edinburgh: Earth Sciences,* **79,** 71–86.

Chappell, B. W. and White, A. J. R. (1974) Two contrasting granite types. *Pacific Geology,* **8,** 173–4.

Chappell, B. W. and White, A. J. R. (1992) I- and S-type granites in the Lachlan Fold Belt. *Transactions of the Royal Society of Edinburgh: Earth Sciences,* **83,** 1–26. (Also *Geological Society of America Special Paper*, **272**.)

Chough, S. K. and Sohn, Y. K. (1990) Depositional mechanics and sequences of base-surges, Songaksan tuff ring, Cheju Island, Korea. *Sedimentology,* **37,** 1115–35.

Church, W. R. and Gayer, R. A. (1973) The Ballantrae ophiolite. *Geological Magazine,* **110,** 497–510.

Clark, L. (1963) The geology and petrology of the Ennerdale Granophyre, its metamorphic aureole and associated mineralization. Unpublished PhD thesis, University of Leeds.

Clarke, D. B. (1992) *Granitoid Rocks*, Chapman and Hall, London.

Clarke, P. D. and Wadsworth, W. J. (1970) The Insch layered intrusion. *Scottish Journal of Geology,* **6,** 7–25.

Clayburn, J. A. P. (1981) Age and petrogenetic studies of some magmatic and metamorphic rocks in the Grampian Highlands. Unpublished PhD thesis, University of Oxford.

Clayburn, J. A. P., Harmon, R. S., Pankhurst, R. J. and Brown, J. F. (1983) Sr, O and Pb isotope evidence for the origin and evolution of the Etive Igneous Complex, Scotland. *Nature,* **303,** 492–7.

Clough, C. T., Maufe, H. B. and Bailey, E. B. (1909) The cauldron-subsidence of Glen Coe, and the associated igneous phenomena.

References

Quarterly Journal of the Geological Society of London, **65**, 611–78.

Cocks, L. R. M. and Fortey, R. A. (1982) Faunal evidence for oceanic separations in the Palaeozoic of Britain. *Journal of the Geological Society of London,* **139**, 467–80.

Coleman, R. G. (1977) *Ophiolites: Ancient Oceanic Lithosphere?* Springer-Verlag, New York.

Compston, W., McDougall, I. and Wyborn, D. (1982) Possible two-stage ^{87}Sr evolution in the Stockdale Rhyolite. *Earth and Planetary Science Letters,* **61**, 297–302.

Cook, D. R. (1976) The geology of the Cairnsmore of Fleet granite and its environs, southwest Scotland. Unpublished PhD thesis, University of St Andrews.

Cooper, A. H., Millward, D., Johnson, E. W. and Soper, N. J. (1993) The early Palaeozoic evolution of northwest England. *Geological Magazine,* **130**, 711–24.

Cooper A. H., Rushton, A. W. A., Molyneux, S. G., Hughes, R.A., Moore, R.M., *et al.* (1995) The stratigraphy, correlation, provenance and palaeogeography of the Skiddaw Group (Ordovician) in the English Lake District. *Geological Magazine,* **132**, 185–211.

Cooper, D. C., Lee, M. K., Fortey, N. J., Cooper, A. H., Rundle, C. C. *et al.* (1988) The Crummock Water aureole: a zone of metasomatism and source of ore metals in the English Lake District. *Journal of the Geological Society of London,* **145**, 523–40.

Cope, T. H. (1910) On the recognition of an agglomerate (Bala Volcanic Series). *Proceedings of the Liverpool Geological Society,* **11**, 37–46.

Cope, T. H. (1915) On the igneous and pyroclastic rocks of the Berwyn Hills (North Wales). *Proceedings of the Liverpool Geological Society* (Memorial Volume), 1–115.

Cope, T. H. and Lomas, J. (1904) On the igneous rocks of the Berwyns. *Report to the British Association (for 1903),* pp. 664–5.

Cornwell, J. D. and McDonald, A. J. W. (1994) Cairngorm and adjacent granites, Scotland. In *Granites, metallogeny, lineaments and rock-fluid interactions* (eds H. W. Haslam and J. A. Plant), *British Geological Survey Research Report* No. SP/94/1, pp. 31–54.

Cornwell, J. D., Patrick, D. J. and Tappin, R. J. (1980) Geophysical evidence for a concealed extension of the Tanygrisiau microgranite and its possible relation to mineralisation. *Mineral Reconnaissance Programme Report, Institute of Geological Sciences,* No. **38**.

Courrioux, G. (1987) Oblique diapirism: the Criffel granodiorite/granite zoned pluton (southwest Scotland). *Journal of Structural Geology,* **9**, 313–30.

Coward, M. P. (1985) The thrust structures of southern Assynt, Moine thrust zone. *Geological Magazine,* **122**, 596–607.

Coward, M. P. and Siddans, A. W. B. (1979) The tectonic evolution of the Welsh Caledonides. In *The Caledonides of the British Isles–Reviewed,* (eds A. L. Harris, C. H. Holland and B. E. Leake), Geological Society of London Special Publication No. 8, pp. 187–98.

Cox, A. H. (1915) The geology of the district between Abereiddy and Abercastle (Pembrokeshire). *Quarterly Journal of the Geological Society of London,* **71**, 273–342.

Cox, A. H. (1925) The geology of the Cadair Idris range (Merioneth). *Quarterly Journal of the Geological Society of London,* **81**, 539–94.

Cox, A. H. (1930) Preliminary note on the geological structure of Pen Caer and Strumble Head, Pembrokeshire. *Proceedings of the Geologists' Association,* **41**, 274–89.

Cox, A. H. and Wells, A. K. (1921) The Lower Palaeozoic rocks of the Arthog–Dolgelly district (Merionethshire). *Quarterly Journal of the Geological Society of London,* **76**, 254–324.

Cox, A. H. and Wells, A. K. (1927) The geology of the Dolgelly district, Merionethshire. *Proceedings of the Geologists' Association,* **38**, 265–318.

Cox, R. A., Dempster, T. J., Bell, B. R. and Rogers, G. (1996) Crystallisation of the Shap Granite: evidence from zoned K-feldspar megacrysts. *Journal of the Geological Society of London,* **153**, 625–35.

Craig, G. Y. (1991) *Geology of Scotland,* (3rd edn), The Geological Society, London.

Craig, G. Y., McIntyre, D. B. and Waterston, C. D. (1978) *James Hutton's Theory of the Earth: the Lost Drawings,* Scottish Academic Press, Edinburgh.

Croudace, I. W. (1982) The geochemistry and petrogenesis of the Lower Paleozoic granitoids of the Lleyn Peninsula, North Wales. *Geochimica et Cosmochimica Acta,* **46**, 609–22.

Cunningham-Craig E. H., Wright, W. B. and

References

Bailey, E. B. (1911) The geology of Colonsay and Oronsay with parts of Ross of Mull. *Memoir of the Geological Survey of Great Britain,* Sheet 35 with part of Sheet 27 (Scotland).

Curry, G. B., Bluck, B. J., Burton, C. J., Siveter, D. J. and Williams, A.. (1984) Age, evolution and tectonic history of the Highland Border Complex, Scotland. *Transactions of the Royal Society of Edinburgh: Earth Science*s, **75**, 113–33.

Curry, G. B., Ingham, B. K., Bluck, B. J. and Williams, A. (1982) The significance of a reliable Ordovician age for some Highland Border rocks in central Scotland. *Journal of the Geological Society of London,* **139**, 451–4.

Dakyns, I. and Teall, J. J. H. (1892) On plutonic rocks of Garabal Hill and Meall Breac. *Quarterly Journal of the Geological Society of London,* **48**, 104–21.

Dakyns, J. R. and Greenly, E. (1905) On the probable Peléan origin of the felsitic slates of Snowdon and their metamorphism. *Geological Magazine,* **42**, 541–9.

Dakyns, J. R., Tiddeman, R. H. and Goodchild, J. G. (1897) The geology of the country between Appleby, Ullswater and Haweswater. *Memoir of the Geological Survey of Great Britain,* Quarter Sheet 102SW (New Series Sheet 30, England and Wales).

Dalrymple, D. J. (1995) Contact anatexis of Dalradian metapelites from the Huntly-Knock area, Aberdeenshire, N.E. Scotland. Unpublished PhD thesis, University of Manchester.

Daly, R. A. (1914) *Igneous Rocks and Their Origin,* McGraw-Hill, New York.

Davies, D. A. B. (1936) Ordovician rocks of the Trefiw district (North Wales). *Quarterly Journal of the Geological Society of London,* **92**, 62–90.

Davies, J. R., Fletcher, C. J. N., Waters, R. A., Wilson, D., Woodhall, D. and Zalaziewicz, J. A. (1996) Geology of the country around Llanilar and Rhayader. *Memoir of the British Geological Survey,* Sheets 178 and 179 (England and Wales).

Davies, R. G. (1955) An investigation of the Cader Idris granophyre and its associated rocks. Unpublished PhD thesis, University of Wales, Aberystwyth.

Davies, R. G. (1956) The Pen-y-gader dolerite and its metasomatic effects on the Llyn-gader sediments. *Geological Magazine,* **93**, 153–72.

Davies, R. G. (1959) The Cadair Idris granophyre and its associated rocks. *Quarterly Journal of the Geological Society of London,* **115**, 189–216.

Davis, N. (1989) The relationship between ignimbrite eruption and caldera collapse in the Borrowdale Volcanic Group of the Central Fells, English Lake District. Unpublished PhD thesis, University of Sheffield.

Dearnley, R. (1967) Metamorphism of minor intrusions associated with the Newer Granites of the Western Highlands of Scotland. *Scottish Journal of Geology,* **3**, 449–57.

Deer, W. A. (1935) The Cairnsmore of Cairsphairn igneous complex. *Quarterly Journal of the Geological Society of London,* **91**, 47–76.

Deer, W. A. (1938a) The composition and paragenesis of the hornblendes of the Glen Tilt Complex, Perthshire. *Mineralogical Magazine,* **25**, 56–74.

Deer, W. A. (1938b) The diorites and associated rocks of the Glen Tilt Complex, Perthshire. I – The granitic and intermediate hybrid rocks. *Geological Magazine,* **75**, 174–84.

Deer, W. A. (1950) The diorites and associated rocks of the Glen Tilt Complex, Perthshire. II – Diorites and appinites. *Geological Magazine,* **87**, 181–95.

Deer, W. A. (1953) The diorites and associated rocks of the Glen Tilt Complex, Perthshire. III – Hornblende schist and hornblendite xenoliths in the granite and diorite. *Geological Magazine,* **90**, 27–35.

Dempster, T. J. and Bluck, B. J. (1991) The age and tectonic significance of the Bute amphibolite, Highland Border Complex, Scotland. *Geological Magazine,* **128**, 77–80.

DePaolo, D. J. (1981) Trace elements and isotopic effects of combined wallrock assimilation and fractional crystallisation. *Earth and Planetary Science Letters,* **53**, 189–202.

DePaolo, D. J., Peery, F. V. and Baldridge, W. S. (1992) Crustal versus mantle sources of granitic magmas: a two-parameter model based on Nd isotopic studies. *Transactions of the Royal Society of Edinburgh: Earth Sciences,* **83**, 439–46. (Also *Geological Society of America Special Paper,* **272**.)

Dewey, J. F. (1969) Evolution of the Appalchian/Caledonian orogen. *Nature,* **222**, 124–9.

References

Dewey, J. F. (1974) Continental margins and ophiolite obduction: Appalachian Caledonian System. In *The Geology of Continental Margins* (eds C. A. Burk and C. L. Drake), Springer-Verlag, New York, pp. 933–95.

Dewey, J. F. (1982) Plate tectonics and the evolution of the British Isles. *Journal of the Geological Society of London*, **139**, 371–414.

Dewey, J. F. and Shackleton, R. M. (1984) A model for the evolution of the Grampian tract in the early Caledonides and Appalachians. *Nature*, **312**, 115–21.

Dhonau, T. J. (1964) A possible extension of the Ratagain igneous complex, Wester Ross and Inverness. *Geological Magazine*, **101**, 37–9.

Dickie, D. M. and Forster, C. W. (1976) Mines and minerals of the Ochils. In *The Research Group of the Clackmannanshire Field Studies Society*, Clackmannanshire Field Studies Society.

Dodson, M. H., Miller, J. A. and York, D. (1961) Potassium-argon ages of the Dartmoor and Shap Granites using the total volume and isotopic dilution techniques of argon measurement. *Nature*, **190**, 800–82.

Downie, C. and Soper, N. J. (1972) Age of the Eycott Volcanic Group and its conformable relationship to the Skiddaw Slates in the English Lake District. *Geological Magazine*, **109**, 259–68.

Droop, G. T. R. and Charnley, N. (1985) Comparative geobarometry of pelitic hornfelses associated with the newer Gabbros: a preliminary study. *Journal of the Geological Society of London*, **142**, 53–62.

Droop, G. T. R. and Treloar, P. J. (1981) Pressures of metamorphism in the thermal aureole of the Etive Granite Complex. *Scottish Journal of Geology*, **17**, 85–102.

Druitt, T. H. and Sparks, R. S. J. (1984) On the formation of calderas during ignimbrite eruptions. *Nature*, **310**, 679–81.

Duff, P. McL. D. and Smith, A. J. (1992) *Geology of England and Wales*, The Geological Society, London.

Dunham, K. C. and Wilson, A. A. (1985) Geology of the northern Pennine orefield: Volume 2, Stainmore to Craven. *Economic Memoir of the British Geological Survey*, Sheet 40, 41, 50, and parts of 31, 32, 51, 60 and 61, (new series). British Geological Survey, Keyworth, Nottingham.

Dunham, K. C., Dunham, A. C., Hodge, B. L. and Johnson, G. A. L. (1965) Granite beneath Viséan sediments with mineralisation at Rookhope, north Pennines. *Quarterly Journal of the Geological Society of London*, **121**, 383–417.

Dwerryhouse, A. R. (1909) On some intrusive rocks in the neighbourhood of Eskdale (Cumberland). *Quarterly Journal of the Geological Society of London*, **65**, 55–80.

Eales, H. V. and Cawthorne, R. G. (1996) The Bushveld Complex. In *Layered Intrusions* (ed. R. G. Cawthorne), Elsevier, Amsterdam, pp. 181–229.

Eastwood T., Dixon E. E. L., Hollingworth, S. E., and Smith, B. (1931) The geology of the Whitehaven and Workington District. *Memoir of the Geological Survey of Great Britain*, Sheet 28 (England and Wales).

Eastwood, T., Gibson, W., Cantrill, T. C. and Whitehead, T. H. (1923) The geology of the country around Coventry. *Memoir of the Geological Survey of Great Britain*, Sheet 169 (England and Wales).

Eastwood, T., Hollingworth, S. E., Rose, W. C. C. and Trotter, F. M. (1968) Geology of the country around Cockermouth and Caldbeck. *Memoir of the Geological Survey of Great Britain*, Sheet 23 (England and Wales).

Elles, G. L. (1940) The stratigraphy and faunal succession in the Ordovician rocks of the Builth–Llandrindod inlier, Radnorshire. *Quarterly Journal of the Geological Society of London*, **95**, 383–445.

Elliot, R. B. (1982) The Old Red Sandstone continent: Devonian volcanism. In *Igneous Rocks of the British Isles* (ed. D. S. Sutherland), Wiley, Chichester, pp. 243–53.

Elliott, D. and Johnson, M. R. W. (1980) Structural evolution in the northern part of the Moine thrust belt, NW Scotland. *Transactions of the Royal Society of Edinburgh: Earth Sciences*, **71**, 69–96.

Ellis, N. V., Bowen, D. Q., Campbell, S., Knill J. L., McKirdy, A. P. et al. (1996) *An Introduction to the Geological Conservation Review*, GCR series, No. 1, JNCC, Peterborough, 131 pp.

Elsden, J. V. (1905) On the igneous rocks occurring between St David's Head and Strumble Head, Pembrokeshire. *Quarterly Journal of the Geological Society of London*, **61**, 579–607.

Elsden, J. V. (1908) The St David's Head 'Rock Series', Pembrokeshire. *Quarterly Journal of the Geological Society of London*, **64**,

References

273–96.

Enfield, M. A. and Coward, M. P. (1987) The structure of the West Orkney Basin, northern Scotland. *Journal of the Geological Society of London*, **144**, 871–84.

Evans, D. J., Rowley, W. J., Chadwick, R. A., Kimbell, G. S. and Millward, D. (1994) Seismic reflection data and the internal structure of the Lake District batholith, Cumbria, northern England. *Proceedings of the Yorkshire Geological Society*, **50**, 11–24.

Evans, D. J., Rowley, W. J., Chadwick, R. A. and Millward, D. (1993) Seismic reflections from within the Lake District batholith, Cumbria, northern England. *Journal of the Geological Society of London*, **150**, 1043–6.

Evans, J. A. (1990) Resetting of the Rb-Sr whole-rock system of an Ordovician microgranite during low-grade metamorphism. *Geological Magazine*, **126**, 675–9.

Evans, J. A. (1991) Resetting of Rb-Sr whole-rock ages during Acadian low-grade metamorphism in North Wales. *Journal of the Geological Society of London*, **148**, 703–10.

Ewart, A. (1962) Hydrothermal alteration in the Carrock Fell area, Cumberland, England. *Geological Magazine*, **99**, 1–8.

Eyles, V. A., Simpson, J. B. and MacGregor, A. G. (1929) The igneous geology of Central Ayrshire. *Transactions of the Geological Society of Glasgow*, **18**, 374–87.

Eyles, V. A., Simpson, J. B. and MacGregor, A. G. (1949) Geology of Central Ayrshire. *Memoir of the Geological Survey of Great Britain*, Sheet 14 (Scotland).

Faithfull, J. (1995) *The Ross of Mull Granite Quarries*, The New Iona Press, Iona.

Faller, A. M. and Briden, J. C. (1978) Palaeomagnetism of Lake District Rocks. In *The Geology of the Lake District* (ed. F. Moseley), The Yorkshire Geological Society, Leeds, pp. 17–24.

Fallick, A. E., Jocelyn, J., Donnelly, T., Guy, M. and Behan, C. (1985) Origin of agates in volcanic rocks from Scotland. *Nature*, **313**, 672–4.

Farrand, M. G. (1960) The distribution of some elements across four xenoliths. *Geological Magazine*, **97**, 488–93.

Ferguson, D. K. (1966) The structure of the Queen's Cairn rhyolite, Glen Coe, Argyllshire. *Scottish Journal of Geology*, **2**, 153–8.

Fettes, D. J. (1978) Caledonian and post-Caledonian metamorphism in the United Kingdom and Ireland. In *Metamorphic map of Europe (1:2 500 000): explanatory text* (eds H. J. Zwart, V. S. Sobolev and E. Niggli), UNESCO, pp. 75–83.

Fettes, D. J. and Munro, M. (1989) Age of the Blackwater mafic and ultramafic intrusion, Banffshire. *Scottish Journal of Geology*, **25**, 105–11.

Fettes, D. J., Leslie, A. G., Stephenson, D. and Kimbell, S. F. (1991) Disruption of Dalradian stratigraphy along the Portsoy Lineament from new geological and magnetic surveys. *Scottish Journal of Geology*, **27**, 57–73.

Finlay, T. M. (1930) The Old Red Sandstone of Shetland. Part II, North-western area. *Transactions of the Royal Society of Edinburgh*, **56**, 671–94.

Finlayson, A. M. (1910) The ore-bearing pegmatites of Carrock Fell, and the genetic significance of tungsten-ores. *Geological Magazine*, **7**, 19–28.

Firman, R. J. (1957) Fissure metasomatism in volcanic rocks adjacent to the Shap granite, Westmorland. *Quarterly Journal of the Geological Society of London*, **113**, 205–22.

Firman, R. J. (1978a) Epigenetic mineralisation. In *The Geology of the Lake District* (ed. F. Moseley), The Yorkshire Geological Society, Leeds, pp. 226–41.

Firman, R. J. (1978b) Intrusions. In *The Geology of the Lake District* (ed. F. Moseley), The Yorkshire Geological Society, Leeds, pp. 146–63.

Fitch, F. J. (1967) Ignimbrite volcanism in North Wales. *Bulletin Volcanologique*, **30**, 199–219.

Fitches, W. R. (1992) Introduction – a structural perspective. In *Caledonian Structures in Britain South of the Midland Valley*, (ed. J. E. Treagus), Geological Conservation Review Series, No. 3. Chapman and Hall, London, pp. 97–104.

Fitton, J. G. (1971) The petrogenesis of the calc-alkaline Borrowdale Volcanic Group, Northern England. Unpublished PhD thesis, University of Durham.

Fitton, J. G. (1972) The genetic significance of almandine-pyrope phenocrysts in the calc-alkaline Borrowdale Volcanic Group, Northern England. *Contributions to Mineralogy and Petrology*, **36**, 231–48.

Fitton, J. G. and Hughes, D. J. (1970) Volcanism and plate tectonics in the British Ordovician. *Earth and Planetary Science Letters*, **8**,

References

223–8.

Fitton, J. G., Thirlwall, M. F. and Hughes, D. J. (1982) Volcanism in the Caledonian orogenic belt of Britain. In *Andesites* (ed. R. S. Thorpe), Wiley, Chichester, pp. 611–36.

Fleming, J. (1818) On the mineralogy of the Redhead in Angusshire. *Memoirs of the Wernerian Natural History Society*, **2**, 339–69.

Fletcher, T. A. (1989) The geology, mineralisation (Ni, Cu, PGE) and precious-metal geochemistry of Caledonian mafic intrusions near Huntly, N.E. Scotland. Unpublished PhD thesis, University of Aberdeen.

Fletcher, T. A. and Rice, C. M. (1989) Geology, mineralization (Ni-Cu) and precious metal geochemistry of Caledonian mafic and ultramafic intrusions near Huntly, northeast Scotland. *Transactions of the Institution of Mining and Metallurgy (Section B: Applied Earth Science)*, **98**, B185–200.

Flett, J. S. (1898) The Old Red Sandstone of the Orkneys. *Transaction of the Royal Society of Edinburgh*, **39**, 383–424.

Flinn, D. (1958) On the nappe structure of north-east Shetland. *Quarterly Journal of the Geological Society of London*, **94**, 107–36.

Flinn, D. (1988) The Moine rocks of Shetland. In *Later Proterozoic Stratigraphy of the Northern Atlantic Regions* (ed. J. A. Winchester), Blackie, Glasgow, pp. 74–85.

Flinn, D. (1992) Late Caledonian northeastward ophiolite thrusting in the Shetland Islands, U.K. – Refutation. *Tectonophysics*, **216**, 387–9.

Flinn, D. (1993) New evidence that the high temperature hornblende-schists below the Shetland ophiolite include basic igneous rocks intruded during obduction of the cold ophiolite. *Scottish Journal of Geology*, **29**, 159–65.

Flinn, D. (1996) The Shetland Ophiolite Complex: field evidence for the intrusive emplacement of the 'cumulate' layers. *Scottish Journal of Geology*, **32**, 151–8.

Flinn, D., Miller, J. A., Evans, A. L. and Pringle, I. R. (1968) On the age of the sediments and contemporaneous volcanic rock of western Shetland. *Scottish Journal of Geology*, **4**, 10–19.

Flinn, D., Miller, J. A. and Roddam, D. (1991) The age of the Norwick hornblendic schists of Unst and Fetlar and the obduction of the Shetland ophiolite. *Scottish Journal of Geology*, **27**, 11–19.

Floyd, J. D. (1996) Lithostratigraphy of the Ordovician rocks in the Southern Uplands: Crawford Group, Moffat Shale Group, Leadhills Supergroup. *Transactions of the Royal Society of Edinburgh: Earth Sciences*, **86**, 153–65.

Floyd, J. D. (1997) Geology of the Carrick–Loch Doon district. *Memoir of the British Geological Survey*, Sheets 8W and 8E (Scotland).

Fortey, N. J., Merriman, R. J. and Huff, W. D. (1996) Silurian and late Ordovician K-bentonites as a record of late Caledonian volcanism in the British Isles. *Transactions of the Royal Society of Edinburgh: Earth Sciences*, **86**, 167–80.

Floyd, P. A., Exley, C. S. and Styles, M. T. (1993) *Igneous Rocks of South-West England*. Geological Conservation Review Series, No. 5, Chapman and Hall, London.

Fowler, M. B. (1988a) Ach'uaine hybrid appinite pipes: evidence for mantle-derived shoshonitic parent magmas in Caledonian granite gneiss. *Geology*, **16**, 1026–30.

Fowler, M. B. (1988b) Elemental evidence for crustal contamination of mantle-derived Caledonian syenite by metasediment anatexis and magma mixing. *Chemical Geology*, **69**, 1–16.

Fowler, M. B. (1992) Elemental and O-Sr-Nd isotope geochemistry of the Glen Dessarry syenite, NW Scotland. *Journal of the Geological Society of London*, **149**, 209–20.

Fowler, M. B. and Henney, P. J. (1996) Mixed Caledonian appinite magmas: implications for lamprophyre fractionation and high Ba-Sr granite genesis. *Contributions to Mineralogy and Petrology*, **126**, 199–215.

Fowler, T. K., Jr, Paterson, S. R., Crossland, A. and Yoshinobu, A. (1995) Pluton emplacement mechanisms: a view from the roof (abstract). In *The Origin of Granites and Related Rocks*, (eds M. Brown and P. M. Piccoli), *US Geological Survey Circular*, **1129**, 57.

Francis, E. H. (1982) Emplacement mechanism of late Carboniferous tholeiite sills in northern Britain. *Journal of the Geological Society of London*, **139**, 1–20.

Francis, E. H. (1983) Magma and sediment–II: Problems of interpreting palaeovolcanics buried in the stratigraphic column. *Journal of the Geological Society of London*, **140**, 165–84.

References

Francis, E. H. (1988) Mid-Devonian to early Permian volcanism: Old World. In *The Caledonian – Appalachian Orogen* (eds A. L. Harris and D. J. Fettes), *Geological Society Special Publication*, No. 38, pp. 573–84.

Francis, E. H. and Howells, M. F. (1973) Transgressive welded ash-flow tuffs among the Ordovician sediments of N.E. Snowdonia, N. Wales. *Journal of the Geological Society of London*, 129, 621–41.

Francis, E. H., Forsyth, I. H., Read, W. A. and Armstrong, M. (1970) The geology of the Stirling district. *Memoir of the Geological Survey of Great Britain*, Sheet 39 (Scotland).

Freeman, B., Klemperer, S. L. and Hobbs, R. W. (1988) The deep structure of Northern England and the Iapetus suture zone from BIRPS deep seismic reflection profiles. *Journal of the Geological Society of London*, 145, 727–40.

French, W. J., Hassan, M. D. and Westcott, J. (1979) The petrogenesis of Old Red Sandstone volcanic rocks of the western Ochils, Stirlingshire. In *The Caledonides of the British Isles – reviewed* (eds A. L. Harris, C. H. Holland and B. E. Leake). *Geological Society of London Special Publication*, No. 8.

Friend, C. R. L., Kinney, P. D., Rogers, G., Strachan, R. A. and Patterson, B. A. (1997) U-Pb zircon geochronological evidence for Neoproterozoic events in the Glenfinnan Group (Moine Supergroup): the formation of the Ardgour granite gneiss, north-west Scotland. *Contributions to Mineralogy and Petrology*, 128, 101–13.

Friend, P. F. and Macdonald, R. (1968) Volcanic sediments, stratigraphy and tectonic background of the Old Red Sandstone of Kintyre, W. Scotland. *Scottish Journal of Geology*, 4, 265–82.

Fritz, W. J., Howells, M. F., Reedman, A. J. and Campbell, S. D. G. (1990) Volcaniclastic sedimentation in an Ordovician subaqueous caldera, Lower Rhyolitic Tuff Formation, North Wales. *Bulletin of the Geological Society of America*, 96, 1246–56.

Frost, C. D. and O'Nions, R. K. (1985) Caledonian magma genesis and crustal recycling. *Journal of Petrology*, 26, 515–44.

Furnes, H. (1978) A comparative study of Caledonian volcanics in Wales and West Norway. Unpublished D.Phil. thesis, University of Oxford.

Gale, N. H., Beckinsale, R. D. and Wadge, A. J. (1979) A Rb-Sr whole-rock isochron for the Stockdale Rhyolite of the English Lake District and a revised Mid-Palaeozoic timescale. *Journal of the Geological Society of London*, 136, 235–42.

Gallagher, M. J., Michie, U. McL., Smith, R. T. and Haynes, L. (1971) New evidence of uranium and other mineralisation in Scotland. *Transactions of the Institution of Mining and Metallurgy (Section B: Applied Earth Sciences)*, 80, 150–73.

Gandy, M. K. (1972) The petrology and geochemistry of the Lower Old Red Sandstone lavas of the Sidlaw Hills, Perthshire. Unpublished PhD thesis, University of Edinburgh.

Gandy, M. K. (1975) The petrology of the Lower Old Red Sandstone lavas of the eastern Sidlaw Hills, Perthshire, Scotland. *Journal of Petrology*, 16, 189–211.

Gardiner, C. I. and Reynolds, S. H. (1932) The Loch Doon 'granite' area, Galloway. *Quarterly Journal of the Geological Society of London*, 88, 1–34.

Gardiner, C. I. and Reynolds, S. H. (1936) The Cairnsmore of Fleet granite and its metamorphic aureole. *Quarterly Journal of the Geological Society of London*, 92, 360–1.

Garnham, J. A. (1988) Ring-faulting and associated intrusions, Glencoe, Scotland. Unpublished PhD thesis, Imperial College, University of London.

Garson, M. S., Coates, J. S., Rock, N. M. S. and Deans, T. (1984) Fenites, breccia dykes, albitites and carbonatitic veins near the Great Glen Fault, Inverness, Scotland. *Journal of the Geological Society of London*, 141, 711–32.

Gass, I. G. (1980) The Troodos massif: its role in the unravelling of the ophiolite problem and its significance in the understanding of constructive plate margin processes. In *Ophiolites; Proceedings of the International Ophiolite Symposium, Cyprus, 1979* (ed. A. Panayiotou), Ministry of Agriculture and Natural Resources, Nicosia, pp. 23–35.

Gass, I. G., Neary, C. R., Prichard, H. M. and Bartholomew, I. D. (1982) The chromite of the Shetland ophiolite. A report for the Commission of the European Communities. Contract 043-79-1-MMP U.K.

Geikie, A. (1863) The geology of eastern Berwickshire. *Memoir of the Geological Survey of Great Britain*, Sheet 34 (Scotland).

References

Geikie, A. (1888) Report on the recent work of the Geological Survey in the North-West Highlands of Scotland, based on the field notes and maps of Messrs Peach, B. N., Horn, J., Gunn, W., Clough, C. T., Hinxman, L., and Cadell, H. M. *Quarterly Journal of the Geological Society of London*, **44**, 378–441.

Geikie, A. (1897) *The Ancient Volcanoes of Great Britain*, Macmillan, London.

Geikie, A. (1900) The geology of central and west Fife and Kinross. *Memoir of the Geological Survey of Great Britain*, Sheet 40 and parts of Sheets 32 and 48 (Scotland).

Geikie, A. (1902) The geology of east Fife. *Memoir of the Geological Survey of Great Britain*, Sheet 41 and parts of Sheets 40, 48 and 49 (Scotland).

Geikie, A., Geikie, J. and Peach, B. N. (1869) Ayrshire: southern district. *Memoir of the Geological Survey of Great Britain*, Sheet 14 (Scotland).

Geikie, J. (1887) Geology and petrology of St Abb's Head. *Proceedings of the Royal Society of Edinburgh*, **14**, 177–93.

Gibbons, W. and Gayer, R. A. (1985) British Caledonian terranes. In *The Tectonic Evolution of the Caledonian–Appalachian Orogen* (ed. R. A. Gayer), Vieweg, Wiesbaden, pp. 3–16.

Gibbons, W. and McCarroll, D. (1993) Geology of the country around Aberdaron, including Bardsey Island. *Memoir of the British Geological Survey*, Sheet 133 (England and Wales).

Gillen, C. (1987) Huntly, Elgin and Lossiemouth. In *Excursion Guide to the Geology of the Aberdeen Area* (eds N. H. Trewin, B. C. Kneller and C. Gillen), Scottish Academic Press, Edinburgh for the Geological Society of Aberdeen, pp. 149–59.

Gillen, C. and Trewin, N. H. (1987) Dunnottar to Stonehaven and the Highland Boundary Fault. In *Excursion Guide to the Geology of the Aberdeen Area* (eds N. H. Trewin, B. C. Kneller and C. Gillen), Scottish Academic Press, Edinburgh for the Geological Society of Aberdeen, pp. 265–273.

Gillespie, M. R. and Styles, M. T. (1999) BGS Rock Classification Scheme Volume 1: Classification of Igneous Rocks (2nd edn). *British Geological Survey Research Report*, No. **RR99-6**.

Goodman, S. and Lappin, M. A. (1996) The thermal aureole of the Lochnagar Complex: internal reactions and implications from thermal modelling. *Scottish Journal of Geology*, **32**, 159–72.

Gould, D. (1997) Geology of the country around Inverurie and Alford. *Memoir of the British Geological Survey*, Sheets 76W and 76E (Scotland).

Gradstein, F. M. and Ogg, J. G. (1996) A Phanerozoic time scale. *Episodes*, **19**, 3–5.

Grantham, D. R. (1928) The petrology of the Shap granite. *Proceedings of the Geologists' Association*, **39**, 299–331.

Green, J. F. N. (1913) *The Older Palaeozoic Succession of the Duddon Estuary*, Hayman, Christy and Lilly, London.

Green, J. F. N. (1915a) The garnets and streaky rocks of the English Lake District. *Mineralogical Magazine*, **17**, 207–17.

Green, J. F. N. (1915b) The structure of the eastern part of the Lake District. *Proceedings of the Geologists' Association*, **26**, 195–223.

Green, J. F. N. (1917) The age of the chief intrusions of the Lake District. *Proceedings of the Geologists' Association*, **28**, 1–30.

Green, J. F. N. (1920) The geological structure of the Lake District. *Proceedings of the Geologists' Association*, **31**, 109–26.

Green, T. H. (1976) Experimental generation of cordierite- or garnet-bearing liquids from a pelitic composition. *Geology*, **4**, 85–8.

Greig, D. C. (1975) St Abb's Head. In *The Geology of the Lothians and South-east Scotland: an Excursion Guide* (eds C. Y. Craig and P. M. D. Duff), Scottish Academic Press, Edinburgh, pp. 118–30.

Greig, D. C. (1988) Geology of the Eyemouth district. *Memoir of the British Geological Survey*, Sheet 34 (Scotland).

Greig, D. C. (1992) St Abb's Head. In *Scottish Borders Geology: an Excursion Guide* (eds A. D. McAdam, E. N. K. Clarkson and P. Stone), Scottish Academic Press, Edinburgh, pp. 31–40.

Gribble, C. D. (1968) The cordierite-bearing rocks of the Haddo House and Arnage districts, Aberdeen. *Contributions to Mineralogy and Petrology*, **17**, 315–30.

Groom, T. and Lake, P. (1908) The Bala and Llandovery rocks of Glyn Ceiriog (North Wales). *Quarterly Journal of the Geological Society of London*, **64**, 546–95.

Groome, D. R. and Hall, A. (1974) The geochemistry of the Devonian lavas of the northern Lorne Plateau, Scotland. *Mineralogical*

References

Magazine, **39**, 621–40.

Gunn, A. G. and Shaw, M. H. (1992) Platinum-group elements in the Huntly intrusion, Aberdeenshire, north-east Scotland. *British Geological Survey Technical Report,* No. **WF/92/4** (*BGS Mineral Reconnaissance Programme Report* No. **124**).

Gunn, A. G., Styles, M. T., Rollin, K. E. and Stephenson, D. (1996) The geology of the Succoth–Brown Hill mafic-ultramafic complex, near Huntly, Aberdeenshire. *Scottish Journal of Geology,* **32**, 33–49.

Gunn, W., Clough, C. T. and Hill, J. B. (1897) The geology of Cowal. *Memoir of the Geological Survey of Great Britain,* Sheet 29 (Scotland).

Gunn, W., Geikie, A. and Peach, B. N. (1903) The geology of north Arran, south Bute and the Cumbraes with parts of Ayrshire and Kintyre. *Memoir of the Geological Survey of Great Britain,* Sheet 21 (Scotland).

Hadfield, G. S. and Whiteside, H. C. (1936) The Borrowdale Series of High Rigg and the adjoining Low Rigg Microgranite. *Proceedings of the Geologists' Association,* **47**, 42–64.

Hall, I. H. S., Gallagher, M. J., Skilton, B. R. H. and Johnson, C. E. (1982) Investigation of polymetallic mineralisation in Lower Devonian volcanics near Alva, central Scotland. *Mineral Reconnaissance Programme Report, Institute of Geological Sciences,* No. **53**.

Halliday, A. N. (1984) Coupled Sm-Nd and U-Pb systematics in late Caldedonian granites and the basement under Northern Britain. *Nature,* **307**, 229–33.

Halliday, A. N., Aftalion, M., Parsons, I., Dickin, A. P. and Johnson, M. R. W. (1987) Syn-orogenic alkaline magmatism and its relationship to the Moine Thrust Zone and the thermal state of the lithosphere in NW Scotland. *Journal of the Geological Society of London,* **144**, 611–17.

Halliday, A. N., Aftalion, M., van Breemen, O. and Jocelyn, J. (1979a) Petrogenetic significance of Rb-Sr and U-Pb isotopic systems in the 400 Ma old British Isles granitoids and their hosts. In *The Caledonides of the British Isles – Reviewed* (eds A. L. Harris, C. H. Holland and B. E. Leake*), Geological Society of London Special Publication,* No. **8**, pp. 653–62.

Halliday, A. N., Dickin, A. P., Fallick, A. E., Stephens, W. E., Hutton, D. H. W. and Yardley, B. W. D. (1984) Open mantle and crust systems during ascent and emplacement of Late Caledonian alkali-rich magmas: a detailed multidisciplinary study of the Ratagain complex, NW Scotland. (Abstract) *Proceedings of I.S.E.M field conference on Open Magmatic Systems,* (August 1984, Taos, New Mexico), Institute for the Study of Earth and Man, Southern Methodist University, Dallas, Texas, pp. 175–6.

Halliday, A. N., McAlpine, A. and Mitchell, J. G. (1977) The age of the Hoy Lavas, Orkney. *Scottish Journal of Geology,* **13**, 43–52.

Halliday, A. N., McAlpine, A. and Mitchell, J. G. (1979b) The age of the Hoy Lavas, Orkney: Erratum. *Scottish Journal of Geology,* **15**, 79.

Halliday, A. N., McAlpine, A. and Mitchell, J. G. (1982) $^{40}Ar/^{39}Ar$ age of the Hoy lavas, Orkney. In *Numerical Dating in Stratigraphy* (ed. G. S. Odin), Wiley, Chichester, pp. 928–31.

Halliday, A. N., Stephens, W. E. and Harmon, R. S. (1980) Rb-Sr and O isotopic relationships in 3 zoned Caledonian granitic plutons, Southern Uplands, Scotland: evidence for varied sources and hybridisation of magmas. *Journal of the Geological Society of London,* **137**, 329–48.

Halliday, A. N., Stephens, W. E. and Harmon, R. S. (1981) Isotopic and chemical constraints on the development of peraluminous Caledonian and Acadian granites. *Canadian Mineralogist,* **19**, 205–16.

Halliday, A. N., Stephens, W. E., Hunter, R. H., Menzies, M. A., Dicken, A. P. and Hamilton, P. J. (1985) Isotopic and chemical constraints on the building of the deep Scottish lithosphere. *Scottish Journal of Geology,* **21**, 465–91.

Hallimond, A. F. (1930) On the magnetic disturbances in north Leicestershire. *Summary of Progress of the Geological Survey of Great Britain for 1929, Part II,* 1–23.

Hamidullah, S. (1983) Petrogenetic studies of the appinite suite, northwestern Scotland. Unpublished PhD thesis, University of Glasgow.

Hamidullah, S. and Bowes, D. R. (1987) Petrogenesis of the appinite suite, Appin District, Western Scotland. *Acta Universitas Carolinae – geologica,* **4**, 295–396.

Hamilton, P. J., Bluck, B. J. and Halliday, A. N. (1984) Sm-Nd ages from the Ballantrae complex, SW Scotland. *Transactions of the Royal Society of Edinburgh: Earth Sciences,* **75**,

183–7.

Hamilton, P. J., O'Nions, R. K. and Pankhurst, R. J. (1983) Isotopic evidence for the provenance of some Caledonian granites. *Nature*, **287**, 279–84.

Hancock, N. J. (1982) Stratigraphy, palaeogeography and structure of the East Mendips Silurian Inlier. *Proceedings of the Geologists' Association*, **93**, 247–61.

Hancox, E. G. (1934) The Haweswater dolerite, Westmorland. *Proceedings of the Liverpool Geological Society*, **16**, 173–97.

Hanson, R. E. and Schweickert, R. A. (1986) Stratigraphy of mid-Paleozoic island-arc rocks in part of the northern Sierra Nevada, Sierra and Nevada Counties, California. *Geological Society of America Bulletin*, **97**, 986–98.

Hardie, W. G. (1963) Explosion-breccias near Stob Mhic Mhartuin, Glen Coe, Argyll, and their bearing on the origin of the nearby flinty crush-rock. *Transactions of the Edinburgh Geological Society*, **19**, 426–38.

Hardie, W. G. (1968) Volcanic breccia and the Lower Old Red Sandstone unconformity, Glen Coe, Argyll. *Scottish Journal of Geology*, **4** (4), 291–9.

Harker, A. (1888) The eruptive rocks of Sarn. *Quarterly Journal of the Geological Society of London*, **44**, 442–61.

Harker, A. (1889) *The Bala Volcanic Series of Carnarvonshire and Associated Rocks*, Cambridge University Press, Cambridge.

Harker, A. (1894) Carrock Fell: a study in the variation of igneous rock-masses – Part I. The gabbro. *Quarterly Journal of the Geological Society of London*, **50**, 311–37.

Harker, A. (1895a) *Petrology for Students: an Introduction to the Study of Rocks under the Microscope*, Cambridge University Press, Cambridge.

Harker, A. (1895b) Carrock Fell: a study in the variation of igneous rock-masses – Part II. The Carrock Fell Granophyre. Part III. The Grainsgill Greisen. *Quarterly Journal of the Geological Society of London*, **51**, 125–48.

Harker, A. (1902) Notes on the igneous rocks of the English Lake District. *Proceedings of the Yorkshire Geological Society*, **14**, 487–93.

Harker, A. and Marr, J. E. (1891) The Shap Granite and associated rocks. *Quarterly Journal of the Geological Society of London*, **47**, 266–328.

Harker, A. and Marr, J. E. (1893) Supplementary notes on the metamorphic rocks around the Shap Granite. *Quarterly Journal of the Geological Society of London*, **49**, 359–71.

Harland, W. B., Cox, A. V., Craig, L. E., Smith, A. G. and Smith, D. G. (1990) *A Geologic Time Scale 1989*, Cambridge University Press, Cambridge.

Harmon, R. S. and Halliday, A. N. (1980) Oxygen and strontium isotope relationships in the British late Caledonian granites. *Nature*, **283**, 21–5.

Harmon, R. S., Halliday, A. N., Clayburn, J. A. P. and Stephens, W. E. (1984) Chemical and isotopic systematics of the Caledonian intrusions of Scotland and Northern England: a guide to magma source region and magma-crust interaction. *Philosophical Transactions of the Royal Society of London*, A **310**, 709–42.

Harper, D. A. T. and Brenchley, P. J. (1993) An endemic brachiopod fauna from the Middle Ordovician of North Wales. *Geological Journal*, **28**, 21–36.

Harris, P. and Dagger, G. W. (1987) The intrusion of the Carrock Fell Gabbro Series (Cumbria) as a sub-horizontal tabular body. *Proceedings of the Yorkshire Geological Society*, **46**, 371–80.

Harrison, T. N. (1986) The mode of emplacement of the Cairngorm granite. *Scottish Journal of Geology*, **22**, 303–14.

Harrison, T. N. (1987a) The evolution of the Eastern Grampians Granites. Unpublished PhD thesis, University of Aberdeen.

Harrison, T. N. (1987b) The granitoids of eastern Aberdeenshire. In *Excursion guide to the geology of the Aberdeen area* (eds N. H. Trewin, B. C. Kneller and C. Gillen), Scottish Academic Press, Edinburgh for The Geological Society of Aberdeen, pp. 243–50.

Harrison, T. N. (1988) Magmatic garnets in the Cairngorm granite, Scotland. *Mineralogical Magazine*, **52**, 659–67.

Harrison, T. N. and Hutchinson, J. (1987) The age and origin of the Eastern Grampians Newer Granites. *Scottish Journal of Geology*, **23**, 269–82.

Harry, W. T. (1956) The Old Red Sandstone Lavas of the western Sidlaw Hills, Perthshire. *Geological Magazine*, **93**, 43–56.

Harry, W. T. (1958) The Old Red Sandstone Lavas of the eastern Sidlaws. *Transactions of the Edinburgh Geological Society*, **17**, 105–12.

Harte, B., Booth, J. E., Dempster, T. J., Fettes, D. J., Mendum, J. R. and Watts, D. (1984)

References

Aspects of the post-depositional evolution of Dalradian and Highland Border Complex rocks in the Southern Highlands of Scotland. *Transactions of the Royal Society of Edinburgh: Earth Sciences,* **75**, 151–63.

Hartley, J. J. (1932) The volcanic and other igneous rocks of Great and Little Langdale, Westmorland; with notes on the tectonics of the district. *Proceedings of the Geologists' Association,* **43**, 32–69.

Haslam, H. W. (1968) The crystallisation of intermediate and acid magmas at Ben Nevis, Scotland. *Journal of Petrology,* **9**, 84–104.

Hatch, F. H. (1891) *An Introduction to the Study of Petrology: the Igneous Rocks,* Swan Sonnenschein, London.

Hatch, F. M., Wells, A. K. and Wells, M. K. (1971) *Petrology of the Igneous Rocks,* Murby, London.

Haughton, P. D. W. (1988) A cryptic Caledonian flysch terrane in Scotland. *Journal of the Geological Society of London,* **145**, 685–703.

Hausback, B. P. (1987) An extensive, hot, vapour-charged rhyodacite flow, Baja California, Mexico. In *The Emplacement of Silicic Domes and Lava Flows* (ed. J. H. Fink), *Geological Society of America Special Paper,* No. **212**, pp. 111–18.

Hawkins, T. R. W. (1965) A note on rhythmic layering in hornblende-picrites and dolerites at Rhiw, North Wales. *Geological Magazine,* **102**, 202.

Hawkins, T. R. W. (1970) Hornblende gabbros and picrites at Rhiw, Caernarvonshire. *Geological Journal,* **7**, 1–24.

Heddle, M. F. (1881) The geognosy and mineralogy of Scotland. Sutherland – Continued. *Mineralogical Magazine,* **4**, 197–254.

Heddle, M. F. (1883a) The geognosy and mineralogy of Scotland. Sutherland – Part IV. *Mineralogical Magazine,* **5**, 133–89.

Heddle, M. F. (1883b) The geognosy and mineralogy of Scotland. Sutherland – Part V. *Mineralogical Magazine,* **5**, 217–63.

Heddle, M. F. (1901) *The Mineralogy of Scotland.* D. Douglas, Edinburgh.

Heiken, G. H. (1971) Tuff rings: examples from Fort Rock–Christmas Lake Valley basin, south central Oregon. *Journal of Geophysical Research,* **76**, 5615–26.

Henderson, W. G. and Fortey, N. J. (1982) Highland Border rocks at Loch Lomond and Aberfoyle. *Scottish Journal of Geology,* **18**, 227–45.

Henderson, W. G. and Robertson, A. H. F. (1982) The Highland Border rocks and their relation to marginal basin development in the Scottish Caledonides. *Journal of the Geological Society of London,* **139**, 433–50.

Henney, P. J. (1991) The geochemistry and petrogenesis of the minor intrusive suite associated with the Late Caledonian Criffel–Dalbeattie pluton, SW Scotland. Unpublished PhD thesis, University of Aston, Birmingham.

Hepworth, B. C., Oliver, G. J. H. and McMurtry, M. J. (1982) Sedimentology, volcanism, structure and metamorphism of the northern margin of a Lower Palaeozoic accretionary complex; Bail Hill–Abington area of the Southern Uplands of Scotland. In *Trench-Forearc Geology: Sedimentation and Tectonics on Modern and Ancient Active Plate Margins* (ed. J. K. Leggett), *Geological Society Special Publication,* No. **10**, pp. 521–34.

Higazy, R. A. (1954) The trace elements of the plutonic complex of Loch Doon (southern Scotland) and their petrological significance. *Journal of Geology,* **62**, 172–81.

Highton, A. J. (1999) Geology of the Aviemore district. *Memoir of the British Geological Survey,* Sheet 74E (Scotland).

Hill, E. and Bonney, T. G. (1878) On the pre-Carboniferous rocks of the Charnwood Forest, Part II. *Quarterly Journal of the Geological Society of London,* **34**, 199–239.

Hill, J. B. (1905) The geology of Mid-Argyll. *Memoir of the Geological Survey of Great Britain,* Sheet 37 (Scotland).

Hill, J. B. and Kynaston, H. (1900) On Kentallenite and its relations to other igneous rocks in Argyllshire. *Quarterly Journal of the Geological Society of London,* **56**, 531–57.

Hinxman, L. W. and Wilson, J. S. G. (1890) Central Aberdeenshire. *Memoir of the Geological Survey of Great Britain,* Sheet 76 (Scotland).

Hitchen, C. S. (1934) The Skiddaw granite and its residual products. *Quarterly Journal of the Geological Society of London,* **90**, 158–200.

Holden, P. (1987) Source and equilibration studies of Scottish Caledonian xenolith suites. Unpublished PhD thesis, University of St Andrews.

Holden, P., Halliday, A. N. and Stephens, W. E. (1987) Neodymium and strontium isotope content of microdiorite enclaves point to mantle input to granitoid production. *Nature*, **330**, 53–6.

Holden, P., Halliday, A. N., Stephens, W. E. and Henney, P. J. (1991) Chemical and isotopic evidence for major mass transfer between mafic enclaves and felsic magma. *Chemical Geology*, **92**, 135–52.

Holder, M. T. (1983) Discussion on convection and crystallization in the Criffel Dalbeattie pluton. *Journal of the Geological Society of London*, **140**, 311–13.

Holdsworth, R. E. and Strachan, R. A. (in press) Geology of the Tongue district. *Memoir of the British Geological Survey*, Sheet 114E (Scotland).

Holdsworth, R. E., Harris, A. L. and Roberts, A. M. (1987) The stratigraphy, structure and regional significance of the Moine Rocks of Mull, Argyllshire, W. Scotland. *Geological Journal*, **22**, 83–107.

Holland, J. G. and Lambert, R. S. J. (1970) Weardale Granite. In *Geology of Durham County*. Transactions of the Natural History Society of Northumberland, Durham and Newcastle upon Tyne, **41**, 103–18.

Hollingworth, S. E. (1937) Carrock Fell and adjoining areas. Report of Field Meeting. *Proceedings of the Yorkshire Geological Society*, **23**, 208–18.

Holmes, A. (1920) *The nomenclature of petrology*. Murby, London.

Holmes, A. (1993) *Holmes' Principles of Physical Geology*, (ed. P. McL. D. Duff), (4th edn), Chapman and Hall, London.

Holub, F. V., Klápová, H., Bluck, B. J. and Bowes, D. R. (1984) Petrology and geochemistry of post-obduction dykes of the Ballantrae complex, SW Scotland. *Transactions of the Royal Society of Edinburgh: Earth Sciences*, **75**, 211–23.

Horne, J. and Teall, J. J. H. (1892) On borolanite – an igneous rock intrusive in the Cambrian limestone of Assynt, Sutherlandshire, and the Torridon sandstone of Ross-shire. *Transactions of the Royal Society of Edinburgh*, **37**, 163–78.

Horne, J., Peach, B. N. and Teall, J. J. H. (1896) Kirkcudbrightshire. *Memoir of the Geological Survey of Great Britain*, Sheet 5 (Scotland).

House, M. R., Richardson, J. B., Chaloner, W. G., Allen, J. R. L., Holland, C. H. and Westoll, T. S. (1977) A correlation of the Devonian rocks in the British Isles. *Special Report of the Geological Society of London*, **7**.

Howard, F. T. and Small, E. W. (1896a) Geological notes on Skomer Island. *Transactions of the Cardiff Naturalists' Society*, **28**, 55–60.

Howard, F. T. and Small, E. W. (1896b) The geology of Skomer Island. *Report of the British Association for the Advancement of Science*, 797–8.

Howard, F. T. and Small, E. W. (1897) Further notes on the geology of Skomer Island. *Transactions of the Cardiff Naturalists' Society*, **29**, 64–6.

Howells, M. F. and Leveridge, B. E. (1980) The Capel Curig Volcanic Formation. *Report of the Institute of Geological Sciences*, No. **80/6**.

Howells, M. F. and Smith, M. (1997) Geology of the country around Snowdon. *Memoir of the British Geological Survey*, Sheet 119 (England and Wales).

Howells, M. F., Campbell, S. D. G., Reedman, A. J. and Tunnicliff, S. P. (1987) An acidic fissure-controlled volcanic centre (Ordovician) at Yr Arddu, N. Wales. *Geological Journal*, **21**, 133–49.

Howells, M. F., Francis, E. H., Leveridge, B. E. and Evans, C. D. R. (1978) *Classical areas of British geology: Capel Curig and Betws y Coed: Description of 1:25 000 Sheet SH75*. HMSO, London, for Institute of Geological Sciences.

Howells, M. F., Leveridge, B. E., Addison, R., Evans, C. D. R. and Nutt, M. J. C. (1979) The Capel Curig Volcanic Formation, Snowdonia, North Wales; variations in ash-flow tuffs related to emplacement environment. In *The Caledonides of the British Isles – reviewed* (eds A. Harris, C. H. Holland and B. E. Leake). *Geological Society of London Special Publication*, No. **8**.

Howells, M. F., Leveridge, B. E., Addison, R. and Reedman, A. J. (1983) The lithostratigraphical subdivision of the Ordovician underlying the Snowdon and Crafnant volcanic groups, North Wales. *Report of the Institute of Geological Sciences*, No. **83/1**.

Howells, M. F., Leveridge, B. E. and Evans, C. D. R. (1973) Ordovician ash-flow tuffs in eastern Snowdonia. *Report of the Institute of Geological Sciences*, No. **73/3**.

Howells, M. F., Leveridge, B. E. and Reedman, A. J. (1981) *Snowdonia – Rocks and Fossils*, Unwin Paperbacks, London.

Howells, M. F., Reedman, A. J. and Campbell, S.

References

D. G. (1986) The submarine eruption and emplacement of the Lower Rhyolitic Tuff Formation (Ordovician), N. Wales. *Journal of the Geological Society of London*, **143**, 411–24.

Howells, M. F., Reedman, A. J. and Campbell, S. D. G. (1991) *Ordovician (Caradoc) Marginal Basin Volcanism in Snowdonia (Northwest Wales)*, HMSO, London, for the British Geological Survey.

Hudson, S. N. (1937) The volcanic rocks and minor intrusions of the Cross Fell Inlier, Cumberland and Westmorland. *Quarterly Journal of the Geological Society of London*, **93**, 368–405.

Hughes, C. P., Jenkins, C. J. and Rickards, R. B. (1982) Abereiddi Bay and the adjacent coast. In *Geological Excursions in Dyfed, South-West Wales* (ed. M. G. Bassett), National Museum of Wales, Cardiff, pp. 51–63.

Hughes, D. J. (1977) The petrochemistry of the Ordovician igneous rocks of the Welsh Basin. Unpublished PhD thesis, University of Manchester.

Hughes, R.A. and Fettes, D. J. (1994) Geology of the 1:10 000 Sheet NY 11 NW (Floutern Tarn). Part of the 1:50 000 Sheet 29 (Keswick). *British Geological Survey Technical Report*, No. **WA/94/69**.

Hughes, R. A. and Kokelaar, P. (1993) The timing of Ordovician magmatism in the English Lake District and Cross Fell inliers. *Geological Magazine*, **130**, 369–77.

Hughes, R. A., Evans, J. A., Noble, S. R. and Rundle, C. C. (1996) U-Pb geochronology of the Ennerdale and Eskdale intrusions supports sub-volcanic relationships with the Borrowdale Volcanic Group (Ordovician, English Lake District). *Journal of the Geological Society of London*, **153**, 33–8.

Hunter, R. H. (1980) The petrology and geochemistry of the Carrock Fell Gabbro-granophyre Complex, Cumbria. Unpublished PhD thesis, University of Durham.

Hunter, R. H. and Bowden, N. (1990) The Carrock Fell Igneous Complex. Itinerary 3. In *The Lake District* (ed. F. Moseley), The Geologists' Association, London, pp. 57–67.

Hutton, D. H. W. (1987) Strike slip terranes and a model for the evolution of the British and Irish Caledonides. *Geological Magazine*, **124**, 405–25.

Hutton, D. H. W. (1988a) Granite emplacement mechanisms and tectonic controls: inferences from deformation studies. *Transactions of the Royal Society of Edinburgh: Earth Sciences*, **79**, 245–55.

Hutton, D. H. W. (1988b) Igneous emplacement in a shear zone termination; the biotite granite at Strontian. *Geological Society of America Bulletin*, **100**, 1392–99.

Hutton, D. H. W. and McErlean, M. (1991) Silurian and early Devonian sinistral deformation of the Ratagain granite, Scotland: constraints on the age of Caledonian movements on the Great Glen fault system. *Journal of the Geological Society of London*, **148**, 1–4.

Hutton, D. H. W. and Reavy, R. J. (1992) Strike-slip tectonics and granite petrogenesis. *Tectonics*, **11**, 960–7.

Hutton, D. H. W., Stephens, W. E., Yardley, B., McErlean, M. and Halliday, A. N. (1993) Ratagain Plutonic complex. In *Geology of the Kintail district* (F. May, J. D. Peacock, D. I. Smith and A. J. Barber), *Memoir of the British Geological Survey*, Sheet 72W and part of 71E (Scotland), pp. 52–6.

Hutton, J. (1788) Theory of the Earth; or an Investigation of the Laws observable in the Composition, Dissolution, and Restoration of the Land upon the Globe. *Transactions of the Royal Society of Edinburgh*, **1**, 209–304.

Hutton, J. (1794) Observations on Granite. *Transactions of the Royal Society of Edinburgh*, **3**, 77–85.

Hutton, J. (1899) *Theory of the Earth with Proofs and Illustrations, III* (ed. A. Geikie), The Geological Society of London.

Ikin, N. P. (1983) Petrochemistry and tectonic significance of the Highland Border Suite mafic rocks. *Journal of the Geological Society of London*, **140**, 267–78.

Ikin, N. P. and Harmon, R. S. (1984) Tectonic history of the ophiolitic rocks of the Highland Border fracture zone, Scotland; stable isotope evidence from fluid rock interactions during obduction. *Tectonophysics*, **106**, 31–48.

Ingham, J. K., Curry, G. B. and Williams, A. (1985) Early Ordovician Dounans Limestone fauna, Highland Border Complex, Scotland. *Transactions of the Royal Society of Edinburgh: Earth Sciences*, **76**, 481–513.

Ingham, J. K., McNamara, K. J. and Rickards, R. B. (1978) The Upper Ordovician and Silurian rocks. In *The Geology of the Lake District* (ed. F. Moseley), The Yorkshire Geological Society, Leeds, pp. 121–45.

References

Institute of Geological Sciences (1963) *Summary of Progress for 1962*, HMSO, London, p. 57.

Jacques, J. M. and Reavy, R. J. (1994) Caledonian plutonism and major lineaments in the SW Scottish Highlands. *Journal of the Geological Society of London*, **151**, 955–69.

Jarvis, K. E. (1987) The petrogenesis and geochemistry of the diorite complexes south of Balmoral Forest, Angus. Unpublished PhD thesis, London Polytechnic University.

Jehu, T. J. and Campbell, R. (1917) The Highland Border rocks of the Aberfoyle district. *Transactions of the Royal Society of Edinburgh*, **52**, 175–212.

Jelínek, E., Souèek, J., Bluck, B. J., Bowes, D. R. and Treloar, P. J. (1980) Nature and significance of beerbachites in the Ballantrae ophiolite, SW Scotland. *Transactions of the Royal Society of Edinburgh: Earth Sciences*, **71**, 159–79.

Jennings, A. V. and Williams, G. J. (1891) Manod and the Moelwyns. *Quarterly Journal of the Geological Society of London*, **47**, 368–83.

Jhingran, A. G. (1942) The Cheviot granite. *Quarterly Journal of the Geological Society of London*, **78**, 241–54.

Johnson, E. W., Briggs, D. E. G., Suthren, R. J., Wright, J. L. and Tunnicliff, S. P. (1994) Non-marine arthropod traces from the subaerial Ordovician Borrowdale Volcanic Group, English Lake District. *Geological Magazine*, **131**, 395–406.

Johnson, E. W., Briggs, D. E. G. and Wright, J. L. (1996) Lake District pioneers – the earliest footprints on land. *Geology Today*, **12**, 147–51.

Johnson, M. R. W. and Harris, A. L. (1967) Dalradian–?Arenig relations in part of the Highland Border, Scotland and their significance in the chronology of the Caledonian Orogeny. *Scottish Journal of Geology*, **3**, 1–16.

Johnson, M. R. W. and Parsons, I. (1979) *Geological Excursion Guide to the Assynt District of Sutherland*, Edinburgh Geological Society, Edinburgh.

Johnson, M. R. W., Kelley, S. P., Oliver, G. J. H. and Winter, D. A. (1985) Thermal effects and timing of thrusting in the Moine Thrust zone. *Journal of the Geological Society of London*, **142**, 863–74.

Jones, F. (1927) A structural study of the Charnian rocks and of the igneous intrusions associated with them. *Transactions of the Leicester Literary and Philosophical Society*, **28**, 24–41.

Jones, O. T. and Pugh, W. J. (1941) The Ordovician rocks of the Builth District: a preliminary account. *Geological Magazine*, **78**, 185–91.

Jones, O. T. and Pugh, W. J. (1949) An early Ordovician shoreline in Radnorshire near Builth Wells. *Quarterly Journal of the Geological Society of London*, **105**, 65–99.

Jones, T. A. (1915) On the presence of tourmaline in Eskdale (Cumberland). *Proceedings of the Liverpool Geological Society*, **12**, 139.

Jowett, A. (1913) The volcanic rocks of the Forfarshire coast and the associated sediments. *Quarterly Journal of the Geological Society of London*, **69**, 459–83.

Kelemen, P. B., Shimizu, N. and Salters, V. J. M. (1995) Extraction of mid-ocean-ridge basalt from the upwelling mantle by focused flow of melt in dunite channels. *Nature*, **375**, 747–53.

Kellock, E. (1969) Alkaline basic igneous rocks in the Orkneys. *Scottish Journal of Geology*, **5**, 140–53.

Kemp, S. J. and Merriman, R. J. (1994) The petrology and geochemistry of Caradocian volcanics, Cadair Idris, Gwynedd. *British Geological Survey Technical Report*, No. **WG/93/6**.

Kennedy, W. Q. (1946) The Great Glen Fault. *Quarterly Journal of the Geological Society of London*, **102**, 41–76.

Key, R. M., May, F., Clark, G. C., Phillips, E. R. and Peacock, J. D. (1997) Geology of the Glen Roy district. *Memoir of the British Geological Survey*, Sheet 63W (Scotland).

Kidston, R. and Lang, W. H. (1924) Notes on fossil plants from the Lower Old Red Sandstone of Scotland. III. On two species of *Pachytheca* (*P. Media* and *P. fasciculata*) based on the character of algal filaments. *Transactions of the Royal Society of Edinburgh*, **53**, 604–14.

Kimbell, G. S. and Stone, P. (1995) Crustal magnetization variations across the Iapetus Suture. *Geological Magazine*, **132**, 599–609.

King, B. C. (1942) The Cnoc nan Cuilean area of the Ben Loyal igneous complex. *Quarterly Journal of the Geological Society of London*, **98**, 149–82.

King, R. J. (1959) The mineralization of the Mountsorrel Granodiorite. *Transactions of*

References

the Leicester Literary and Philosophical Society, **53**, 18–30.

King, R. J. (1968) Mineralization. In *The Geology of the East Midlands* (eds P. C. Sylvester-Bradley and T. D. Ford), Leicester University Press, Leicester, pp. 112–137.

Kneller, B. C. (1991) A foreland basin on the southern margin of Iapetus. *Journal of the Geological Society of London,* **148**, 207–10.

Kneller, B. C. and Aftalion, M. (1987) The isotopic and structural age of the Aberdeen Granite. *Journal of the Geological Society of London,* **144**, 717–21.

Kneller, B. C. and Leslie, A. G. (1984) Amphibolite facies metamorphism in shear zones in the Buchan area of NE Scotland. *Journal of Metamorphic Geology,* **2**, 83–94.

Kneller, B. C. and McConnell, B. J. (1993) The Seathwaite Fell Formation in the Central Fells. *British Geological Survey Technical Report,* No. **WA/93/43**.

Kneller, B. C., King, L. M. and Bell, A. M. (1993a) Foreland basin development and tectonics on the northwest margin of eastern Avalonia. *Geological Magazine,* **130**, 691–7.

Kneller, B. C., Kokelaar, B. P. and Davis, N. C. (1993b) The Lingmell Formation. *British Geological Survey Technical Report,* No. **WA/93/42**.

Kneller, B. C., Scott, R. W., Soper, N. J., Johnson, E. W. and Allen, P. M. (1994) Lithostratigraphy of the Windermere Supergroup, Northern England. *Geological Journal,* **29**, 219–40.

Knill, J. L. (1972) The engineering geology of the Cruachan underground power station. *Engineering Geology,* **6**, 289–312.

Kokelaar, B. P. (1977) The igneous history of the Rhobell Fawr area, Merionethshire, North Wales. Unpublished PhD thesis, University of Wales, Aberystwyth.

Kokelaar, B. P. (1979) Tremadoc to Llanvirn volcanism on the south-east side of the Harlech Dome (Rhobell Fawr), N. Wales. In *Caledonides of the British Isles – Reviewed,* (eds A. L. Harris, C. H. Holland and B. E. Leake), *Geological Society of London Special Publication,* No. **8**, pp. 591–6.

Kokelaar, B. P. (1982) Fluidization of wet sediments during the emplacement and cooling of various igneous bodies. *Journal of the Geological Society of London,* **139**, 21–34.

Kokelaar, B. P. (1986) Petrology and geochemistry of the Rhobell Volcanic Complex: amphibole-dominated fractionation at an early Ordovician arc volcano. *Journal of Petrology,* **27**, 887–914.

Kokelaar, B. P. (1988) Tectonic controls of Ordovician arc and marginal basin volcanism in Wales. *Journal of the Geological Society of London,* **145**, 759–75.

Kokelaar, B. P. (1992) Ordovician marine volcanic and sedimentary record of rifting and volcanotectonism: Snowdon, Wales, United Kingdom. *Geological Society of America Bulletin,* **104**, 1433–55.

Kokelaar, B. P. and Branney, M. J. (1994) Volcanotectonism and evolution of a shallow sedimentary basin during phreatomagmatic and magmatic eruptions: The Whorneyside and Airy's Bridge Formations at Sour Milk Gill. In *Processes and controls of caldera collapse and related ignimbrite emplacement in Snowdonia (Wales), the Lake District (England) and Glencoe (Scotland), United Kingdom* (eds B. P. Kokelaar, M. J. Branney, I. Moore and M. F. Howells) Field Guide of the IAVCEI Commission on Explosive Volcanism Field Workshop, 18–29 May, 1994, pp. H1–11.

Kokelaar, P. and Branney, M.J. (1999) *Inside silicic calderas (Snowdon, Scafell and Glencoe, UK): interactions of caldera development, tectonism and hydrovolcanism.* Field Guide for the IAVCEI Commission on Explosive Volcanism Field Workshop, 7–18 July, 1999.

Kokelaar, B. P., Bevins, R. E. and Roach, R. A. (1985) Submarine silicic volcanism and associated sedimentary and tectonic processes, Ramsey Island, S.W. Wales. *Journal of the Geological Society of London,* **142**, 591–613.

Kokelaar, B. P., Branney, M. J., McConnell, B. J., Kneller, B. C. and Smith, R. (1990) The English Lake District – an ancient continental arc. Volcanogenic sedimentation in ancient terrains. *International Association of Volcanology and Chemistry of the Earth's Interior /International Association of Sedimentologists Field Workshop Guide,* pp. 1–46.

Kokelaar B. P., Fitch, F. J. and Hooker, P. J. (1982) A new K-Ar age from uppermost Tremadoc rocks from North Wales. *Geological Magazine,* **119**, 207–11.

Kokelaar, B. P., Howells, M. F., Bevins, R. E. and Roach, R. A. (1984a) Volcanic and associated sedimentary and tectonic processes in the Ordovician marginal basin of Wales: a field guide. In *Volcanic and Associated sedimentary and Tectonic Processes in Modern and*

References

Ancient Marginal Basins (eds B. P. Kokelaar and M. F. Howells), *Geological Society Special Publication*, No. **16**, pp. 291–322.

Kokelaar, B. P., Howells, M. F., Bevins, R. E., Roach, R. A. and Dunkley, P. N. (1984b) The Ordovician marginal basin in Wales. In *Volcanic and Associated Sedimentary and Tectonic Processes in Modern and Ancient Marginal Basins* (eds B. P. Kokelaar and M. F. Howells), *Geological Society Special Publication*, No. **16**, pp. 245–69.

Kulp, J. L., Long, L. E., Griffin, C. E., Mills, A. A., Lambert, R. S. J. et al. (1960) Potassium-argon and rubidium-strontium ages of some granites from Britain and Eire. *Nature*, **185**, 495–7.

Kynaston, H. and Hill, J. B. (1908) The geology of the country near Oban and Dalmally. *Memoir of the Geological Survey of Great Britain*, Sheet 45 (Scotland).

Lake, P. and Reynolds, S. H. (1912) The geology of Mynydd y Gader, Dolgelly. *Quarterly Journal of the Geological Society of London*, **68**, 345–62.

Lambert, R. S. J. and Mills, A. A. (1961) Some critical points for the Palaeozoic time-scale from the British Isles. *Annals of the New York Academy of Sciences*, **91**, 378–89.

Lapworth, C. (1898) Sketch of the geology of the Birmingham District. *Proceedings of the Geologists' Association*, **15**, 313–89.

Lawrence, D. J. D., Webb, B. C., Young, B. and White, D. E. (1986) The geology of the late Ordovician and Silurian rocks (Windermere Group) in the area around Kentmere and Crook. *Report of the British Geological Survey*, Vol. **18**, No. 5.

Le Bas, M. J. (1968) Caledonian igneous rocks. In *The Geology of the East Midlands* (eds P. C. Sylvester-Bradley and T. D. Ford), Leicester University Press, Leicester, pp. 41–58.

Le Bas, M. J. (1972) Caledonian igneous rocks beneath Central and Eastern England. *Proceedings of the Yorkshire Geological Society*, **39**, 71–86.

Le Bas, M. J. (1982a) The Caledonian granites and diorites of England and Wales. In *Igneous Rocks of the British Isles* (ed. D. S. Sutherland), Wiley, Chichester, pp. 191–201.

Le Bas, M. J. (1982b) Geological evidence from Leicestershire on the crust of southern Britain. *Transactions of Leicester Literary and Philosophical Society*, **76**, 54–67.

Le Bas, M. J. (1993) The hidden mountains of Leicestershire. *Transactions of the Leicester Literary and Philosophical Society*, **87**, 33–5.

Le Maitre, R. W. (1989) *A Classification of Igneous Rocks and Glossary of Terms. Recommendations of the International Union of Geological Sciences Subcommission on the Systematics of Igneous Rocks*, Blackwell, Oxford.

Leake, B. E. (1990) Granite magmas: their sources, initiation and consequences of emplacement. *Journal of the Geological Society of London*, **147**, 579–89.

Leat, P. T. and Thorpe, R. S. (1986) Geochemistry of an Ordovician basalt-trachybasalt-subalkaline/peralkaline rhyolite association from the Lleyn Peninsula, North Wales, U.K. *Geological Journal*, **21**, 29–43.

Leat, P. T. and Thorpe, R. S. (1989) Snowdon Volcanic Group basalts date end of south-directed Caledonian subduction by Longvillian. *Journal of the Geological Society of London*, **146**, 965–70.

Leat, P. T., Jackson, S. E., Thorpe, R. S. and Stillman, C. J. (1986) Geochemistry of bimodal basalt-subalkaline-peralkaline rhyolite provinces associated with volcanogenic mineralization in the Southern British Caledonides. *Journal of the Geological Society of London*, **143**, 259–74.

Lee, G. W. and Bailey, E. B. (1925) The pre-Tertiary geology of Mull, Loch Aline and Oban. *Memoir of the Geological Survey of Great Britain*, Sheet 44 (Scotland).

Lee, M. K. (1986) A new gravity survey of the Lake District and three-dimensional model of the granite batholith. *Journal of the Geological Society of London*, **143**, 425–35.

Lee, M. K. (1989) Upper crustal structure of the Lake District from modelling and image processing of potential field data. *British Geological Survey Technical Report*, No. **WK/89/1**.

Lee, M. R. and Parsons, I. (1997) Compositional and microtextural zoning in alkali feldspars from the Shap granite and its geochemical implications. *Journal of the Geological Society of London*, **154**, 183–8.

Lee, M. R., Waldron, K. A. and Parsons, I. (1995) Exsolution and alteration microtextures in alkali feldspar phenocrysts from the Shap granite. *Mineralogical Magazine*, **59**, 63–78.

Leggett, J. K. (1987) The Southern Uplands as an accretionary prism: the importance of analogues in reconstructing palaeogeography.

References

Journal of the Geological Society of London, **144**, 737–52.

Leggett, J. K., McKerrow, W. S. and Eales, M. H. (1979) The Southern Uplands of Scotland: a Lower Palaeozoic accretionary prism. *Journal of the Geological Society of London,* **136**, 755–70.

Leighton, P. S. (1985) A petrological and geochemical reconnaissance of the Moor of Rannoch pluton, south-west Highlands, Scotland. Unpublished MSc thesis, University of St Andrews.

Leslie, A. G. (1987) Southern contact of the Insch mass, and the Bennachie Granite. In *Excursion Guide to the Geology of the Aberdeen Area* (eds N. H. Trewin, B. C. Kneller and C. Gillen), Scottish Academic Press, Edinburgh for the Geological Society of Aberdeen, pp. 179–84.

Lewis, A. D. and Bloxam, T. W. (1977) Petro-tectonic environments of the Girvan–Ballantrae lavas from rare-earth element distributions. *Scottish Journal of Geology,* **13**, 211–22.

Lipman, P. W. (1984) The roots of ash-flow calderas in western North America: windows into the tops of Granitic Batholiths. *Journal of Geophysical Research,* **89** B10, 8801–41.

Litherland, M. (1980) The stratigraphy of the Dalradian rocks around Loch Creran, Argyll. *Scottish Journal of Geology,* **16**, 105–23.

Locke, C. A. and Brown, G. C. (1978) Geophysical constraints on structure and emplacement of Shap granite. *Nature,* **272**, 526–8.

Lord, R. A., Prichard, H. M. and Neary, C. R. (1994) Magmatic platinum element concentrations and hydrothermal upgrading in the Shetland ophiolite complex. *Transactions of the Institution of Mining and Metallurgy (Section B: Applied Earth Sciences),* **103**, 87–106.

Lowe, E. E. (1926) *The Igneous Rocks of the Mountsorrel District,* Leicester Literary and Philosophical Society, Leicester.

McArthur, A. C. (1971) On the structural control of Caledonian igneous masses in the south-west Highlands of Scotland. Unpublished PhD thesis, University of Glasgow.

McAlpine, A. (1979) The Upper Old Red Sandstone of Orkney, Caithness and neighbouring areas. Unpublished PhD thesis, University of Newcastle upon Tyne.

McClay, K. R., Norton, M. G., Cony, P. and Davis, G. H. (1986) Collapse of the Caledonian Orogen and the Old Red Sandstone. *Nature,* **323**, 147–9.

MacCulloch, J. (1816) A geological description of Glen Tilt. *Transactions of the Geological Society of London,* **3**, 259–337.

Macdonald, R., Millward, D., Beddoe-Stephens, B. and Laybourn-Parry, J. (1988) The role of tholeiitic magmatism in the English Lake District: evidence from dykes in Eskdale. *Mineralogical Magazine,* **52**, 459–72.

Macdonald, R., Rock, N. M. S., Rundle, C. C. and Russell, O. J. (1986) Relationships between late Caledonian lamprophyric, syenitic, and granitic magmas in a differentiated dyke, southern Scotland. *Mineralogical Magazine,* **50**, 547–57.

Macdonald, R., Thorpe, R. S., Gasgarth, J. W. and Grindrod, A. R. (1985) Multi-component origin of Caledonian lamprophyres of northern England. *Mineralogical Magazine,* **49**, 485–94.

McErlean, M. A. (1993) Granitoid emplacement and deformation: a case study of the Thorr Pluton, Ireland, with contrasting examples from Scotland. Unpublished PhD thesis, University of Durham.

MacGregor, A. G. and Kennedy, W. Q. (1932) The Morvern–Strontian Granite. *Summer Programme of the Geological Survey for 1931,* Part II, pp. 105–19.

MacGregor, A. R. (1996) *Fife and Angus Geology – an Excursion Guide,* Pentland Press, Durham.

Macgregor, M. (1937) The western part of the Criffell–Dalbeattie igneous complex. *Quarterly Journal of the Geological Society of London,* **93**, 457–86.

Macgregor, M. and Phemister, J. (1937) *Geological Excursion Guide to the Assynt District of Sutherland,* Edinburgh Geological Society, Edinburgh.

McIntyre, D. B. (1950) The petrogenesis of the North West part of the Loch Doon plutonic complex, Galloway (abstract). *Abstracts of the Proceedings of the Geological Society of London,* 31–8.

McKerrow, W. S. and Soper, N. J. (1989) The Iapetus suture in the British Isles. *Geological Magazine,* **126**, 1–8.

McKerrow, W. S., Dewey, J. F. and Scotese, C. R. (1991) The Ordovician and Silurian development of the Iapetus Ocean. *Special Papers in Palaeontology,* **44**, 165–78.

MacKenzie, W. S. (1949) Kyanite-gneisses within

References

a thermal aureole. *Geological Magazine*, **86**, 251–5.

Macleod, G. (1992) Zoned manganiferous garnets of magmatic origin from the Southern Uplands of Scotland. *Mineralogical Magazine*, **56**, 115–16.

McLintock, W. F. and Phemister, J. (1931) A gravitational survey over a region of magnetic anomaly at Thrussington, Leicestershire, *Summary of Progress of the Geological Survey of Great Britain for 1930, Part II*, 74–90.

Macpherson, H. (1989) *Agates*, British Museum (Natural History) and the National Museums of Scotland.

McPhie, J., Doyle, M. and Allen, R. (1993) *Volcanic Textures: a Guide to the Interpretation of Textures in Volcanic Rocks*, Centre for Ore Deposit and Exploration Studies, University of Tasmania, Hobart.

Mahmood, L. A. (1986) Mineralogy, petrology and geochemistry of some zoned dioritic complexes in Scotland. Unpublished PhD thesis, University of St Andrews.

Manley, C. R. (1996) In situ formation of welded tuff-like textures in the carapace of a voluminous silicic lava flow, Owyhee County, S W Idaho. *Bulletin of Volcanology*, **57**, 672–86.

Marr, J. E. (1892) On the Wenlock and Ludlow Strata of the Lake District. *Geological Magazine*, **9**, 534–41.

Marr, J. E. (1900) Notes on the geology of the English Lake District. *Proceedings of the Geologists' Association*, **16**, 449–83.

Marr, J. E. (1916) *The Geology of the Lake District and the Scenery as influenced by Geological Structure*, Cambridge University Press, Cambridge.

Marshall, J. E. A. (1988) Devonian miospores from Papa Stour, Shetland. *Transactions of the Royal Society of Edinburgh: Earth Sciences*, **79**, 13–18.

Marshall, J. E. A. (1991) Palynology of the Stonehaven Group, Scotland: evidence for a mid Silurian age and its geological implications. *Geological Magazine*, **128**, 283–6.

Marshall, J. E. A. (1996) *Rhabdosporites langii, Gemnospora lemurata* and *Contagisporites optivus*: an origin for heterospory within the Progymnosperms. *Review of Palaeobotany and Palynology*, **93**, 159–89.

Marshall, J. E. A., Haughton, P. D. W. and Hillier, S. J. (1994) Vitrinite reflectivity and the structure and burial history of the Old Red Sandstone of the Midland Valley of Scotland. *Journal of the Geological Society of London*, **151**, 425–38.

Marshall, J. E. A., Rogers, D. A. and Whiteley, M. J. (1996) Devonian marine incursions into the Orcadian Basin. *Journal of the Geological Society of London*, **153**, 451–66.

Marston, R. J. (1971) The Foyers Granitic Complex, Inverness-shire, Scotland. *Quarterly Journal of the Geological Society of London*, **126**, 331–68.

Martin, R. F., Whitley, J. E. and Woolley, A. R. (1978) An investigation of rare-earth mobility: fenitized quartzites, Borralan Complex, N.W. Scotland. *Contributions to Mineralogy and Petrology*, **66**, 69–73.

Matley, C. A. (1932) The geology of the country around Mynydd Rhiw and Sarn, south-western Lleyn, Carnarvonshire. *Quarterly Journal of the Geological Society of London*, **88**, 238–73.

Matley, C. A. (1938) The geology of the country around Pwllheli, Llanbedrog and Madryn, southwest Carnarvonshire. *Quarterly Journal of the Geological Society of London*, **94**, 555–606.

Matley, C. A. and Heard, A. (1930) The geology of the country around Bodfean (South-western Carnarvonshire). *Quarterly Journal of the Geological Society of London*, **86**, 130–68.

Matthews, D. W. and Woolley, A. R. (1977) Layered ultramafic rocks within the Borralan complex, Scotland. *Scottish Journal of Geology*, **13**, 223–36.

Maufe, H. B. (1910) The geological structure of Ben Nevis. *Geological Survey of the United Kingdom, Summary of Progress for 1909*, pp. 80–9.

May, F. and Highton, A. J. (1997) Geology of the Invermoriston district. *Memoir of the British Geological Survey*, Sheet 73W (Scotland).

May, F., Peacock, J. D., Smith, D. I. and Barber, A. J. (1993) Geology of the Kintail district. *Memoir of the British Geological Survey*, Sheet 72W and part of 71E (Scotland).

Mellors, R. A. and Sparks, R. S. J. (1991) Spatter-rich pyroclastic flow deposits on Santorini, Greece. *Bulletin of Volcanology*, **53**, 327–42.

Meneisy, M. Y. and Miller, J. A. (1963) A geochronological study of the crystalline rocks of Charnwood Forest, England. *Geological Magazine*, **100**, 507–23.

Merriman, R. J. and Roberts, B. (1990)

References

Metabentonites in the Moffat Shale Group, Southern Uplands of Scotland: geochemical evidence of ensialic marginal basin volcanism. *Geological Magazine,* **127**, 259–71.

Merriman, R. J., Bevins, R. E. and Ball, T. K. (1986) Geochemical variations within the Tal y Fan intrusion: implications for element mobility during low-grade metamorphism. *Journal of Petrology,* **27**, 1409–36.

Merriman, R. J., Rex, D. C., Soper, N. J. and Peacor, D. R. (1995) The age of Acadian cleavage in northern England, UK: K-Ar and TEM analysis of a Silurian metabentonite. *Proceedings of the Yorkshire Geological Society,* **50**, 255–65.

Metcalfe, R. (1990) Fluid/rock interaction and metadomain formation during low grade metamorphism in the Welsh marginal basin. Unpublished PhD thesis, University of Bristol.

Miller, C. F. (1985) Are strongly peraluminous magmas derived from pelitic sedimentary sources? *Geology,* **93**, 673–89.

Miller, J. A. (1961) The potassium-argon ages of the Skiddaw and Eskdale Granites. *Geophysical Journal,* **6**, 391–3.

Miller, J. A. and Flinn, D. (1966) A survey of the age relations of Shetland rocks. *Geological Journal,* **5**, 95–116.

Millward, D. (1976) The Borrowdale Volcanics of the English Lake District: a study of the petrology, volcanicity and geochemistry of selected areas with particular reference to ignimbrites. Unpublished PhD thesis, University of Birmingham.

Millward, D. (1980) Three ignimbrites from the Borrowdale Volcanic Group. *Proceedings of the Yorkshire Geological Society,* **42**, 595–616.

Millward, D. and Lawrence, D. J. D. (1985) The Stockdale (Yarlside) Rhyolite – a rheomorphic ignimbrite? *Proceedings of the Yorkshire Geological Society,* **45**, 299–306.

Millward, D. and Molyneux, S. G. (1992) Field and biostratigraphic evidence for an unconformity at the base of the Eycott Volcanic Group in the English Lake District. *Geological Magazine,* **129**, 77–92.

Millward, D., Beddoe-Stephens, B., Williamson, I. T., Young, S. R. and Petterson, M. G. (1994) Lithostratigraphy of a concealed caldera-related ignimbrite sequence within the Borrowdale Volcanic Group of west Cumbria. *Proceedings of the Yorkshire Geological Society,* **50**, 25–36.

Millward, D., Johnson, E. W., Beddoe-Stephens, B., *et.al.* (in press) Geology of the Ambleside District. *Memoir of the British Geological Survey,* Sheet 38 (England and Wales).

Millward, D., Moseley, F. and Soper, N. J. (1978) The Eycott and Borrowdale volcanic rocks. In *The Geology of the Lake District* (ed. F. Moseley), The Yorkshire Geological Society, Leeds, pp. 99–120.

Milne, K. P. (1978) Folding and thrusting in the upper Glen Oykel area, Assynt. *Scottish Journal of Geology,* **14**, 141–6.

Mitchell, G. H. (1934) The Borrowdale Volcanic Series of the country between Longsleddale and Shap. *Quarterly Journal of the Geological Society of London,* **90**, 418–44.

Mitchell, G. H. (1940) The Borrowdale Volcanic Series of Coniston, Lancashire. *Quarterly Journal of the Geological Society of London,* **96**, 301–19.

Mitchell, G. H. (1956) The geological history of the Lake District. *Proceedings of the Yorkshire Geological Society,* **30**, 407–63.

Molyneux, S. G. (1988) Micropalaeontological evidence for the age of the Borrowdale Volcanic Group. *Geological Magazine,* **125**, 541–2.

Mongkoltip, P. and Ashworth, J. R. (1983) Quantitative estimation of an open-system symplectite-forming reaction: restricted diffusion of Al and Si in coronas around olivine. *Journal of Petrology,* **24**, 635–61.

Mongkoltip, P. and Ashworth, J. R. (1986) Amphibolitization of metagabbros in the Scottish Highlands. *Journal of Metamorphic Geology,* **4**, 261–83.

Moore, I. (1995) The early history of the Glencoe cauldron. Unpublished PhD thesis, University of Liverpool.

Moore, I. and Kokelaar, B. P. (1997) Tectonic influences in piecemeal caldera collapse at Glencoe volcano, Scotland. *Journal of the Geological Society of London,* **154**, 765–8.

Moore, I. and Kokelaar, B. P. (1998) Tectonically controlled piecemeal caldera collapse: A case study of Glencoe volcano, Scotland. *Geological Society of America Bulletin,* **110**, 1448–66

Morton, D. J. (1979) Palaeogeographical evolution of the Lower Old Red Sandstone basin in the western Midland Valley. *Scottish Journal of Geology,* **15**, 97–116.

Moseley, F. (1984) Lower Palaeozoic lithostrati-

References

graphical classification in the English Lake District. *Geological Journal*, **19**, 239–47.

Moseley, F. (1990) *Geology of the Lake District*, The Geologists' Association, London.

Moseley, F. and Millward, D. (1982) Ordovician volcanicity in the English Lake District. In *Igneous Rocks of the British Isles* (ed. D. S. Sutherland), Wiley, Chichester, pp. 93–111.

Munro, M. (1965) Some structural features of the Caledonian granitic complex at Strontian, Argyllshire. *Scottish Journal of Geology*, **1**, 152–75.

Munro, M. (1973) Structures in the south-eastern portion of the Strontian Granitic Complex, Argyllshire. *Scottish Journal of Geology*, **9**, 99–108.

Munro, M. (1984) Cumulate relations in the 'Younger Basic' masses of the Huntly–Portsoy area, Grampian Region. *Scottish Journal of Geology*, **20**, 343–59.

Munro, M. (1986) Geology of the country around Aberdeen. *Memoir of the British Geological Survey*, Sheet 77 (Scotland).

Munro, M. and Leslie, A. G. (1987) The 'Younger Basic' intrusions and their associated rocks. In *Excursion Guide to the Geology of the Aberdeen Area* (eds N. H. Trewin, B. C. Kneller and C. Gillen), Scottish Academic Press, Edinburgh for the Geological Society of Aberdeen, pp. 185–92.

Murchison, R. I. (1833) On the sedimentary deposits which occupy the western parts of Shropshire and Herefordshire, and are prolonged from NE to SW through Radnor, Brecknock and Carmarthen shires, with descriptions of the accompanying rocks of intrusive or igneous characters. *Proceedings of the Geological Society of London*, **1**, 470–7.

Murchison, R. I. (1839) *The Silurian System founded on geological researches in the counties of Salop, Hereford, Radnor, Montgomery, Carmarthen, Brecon, Pembroke, Monmouth, Gloucester, Worcester and Stafford with descriptions of the coal-fields and overlying formations* (in 2 vols), Murray, London.

Murchison, R. I. (1867) *Siluria*, 4th edn, Murray, London.

Murchison, R. I. (1872) *Siluria*, 5th edn, Murray, London.

Mykura, W. (1960) The Lower Old Red Sandstone igneous rocks of the Pentland Hills. *Bulletin of the Geological Survey of Great Britain*, **16**, 131–55.

Mykura, W. (1976) *British Regional Geology: Orkney and Shetland*, HMSO, Edinburgh, for Institute of Geological Sciences.

Mykura, W. (1991) Old Red Sandstone. In *Geology of Scotland*, 3rd edn (ed. G. Y. Craig), The Geological Society, London, pp. 297–344.

Mykura, W. and Phemister, J. (1976) The geology of western Shetland. *Memoir of the Geological Survey of Great Britain*, Sheet 127 and parts of 125, 126 and 128 (Scotland).

Nichol, J. (1863) On the geological structure of the Southern Grampians. *Quarterly Journal of the Geological Society of London*, **19**, 180–209.

Nicholls, G. D. (1951a) The Glenelg–Ratagain igneous complex. *Quarterly Journal of the Geological Society of London*, **106**, 302–21.

Nicholls, G. D. (1951b) An unusual pyroxene-rich xenolith in the diorite of the Glenelg–Ratagain igneous complex. *Geological Magazine*, **84**, 284–95.

Nicholls, G. D. (1958) Autometasomatism in the Lower Spilites of the Builth Volcanic Series. *Quarterly Journal of the Geological Society of London*, **114**, 137–62.

Noble, S. R., Tucker, R. D. and Pharaoh, T. C. (1993) Lower Palaeozoic and Precambrian igneous rocks from eastern England, and their bearing on late Ordovician closure of the Tornquist Sea: constraints from U-Pb and Nd isotopes. *Geological Magazine*, **130**, 835–46.

Nockolds, S. R. (1934) The contaminated tonalites of Loch Awe. *Quarterly Journal of the Geological Society of London*, **105**, 302–21.

Nockolds, S. R. (1941) The Garabal Hill–Glen Fyne Igneous Complex. *Quarterly Journal of the Geological Society of London*, **96**, 451–511.

Nockolds, S. R. and Mitchell, R. L. (1948) The geochemistry of some Caledonian plutonic rocks. A study in the relationship between the major and trace elements of igneous rocks and their minerals. *Transactions of the Royal Society of Edinburgh: Earth Sciences*, **61**, 533–75.

Notholt, A. K. and Highley, D. E. (1981) Investigation of phosphate potential of the Loch Borralan Igneous complex, Northwest Highlands, Scotland. *A report for the commission of the European communities,*

References

Contract No. 039-79-1. *Institute of Geological Sciences, open-file Report*, pp. 106.

Notholt, A. K., Highley, D. E. and Harding, R. R. (1985) Investigation of phosphate (apatite) potential of Loch Borralan igneous complex, northwest Highlands, Scotland. *Transactions of the Institution of Mining and Metallurgy (Section B: Applied Earth Sciences)*, **94**, 58–66.

Nutt, M. J. C. (1966) Field Meeting Report. *Proceedings of the Yorkshire Geological Society*, **35**, 429–33.

Nutt, M. J. C. (1970) The Borrowdale Volcanic Series and associated rocks around Haweswater, Westmorland. Unpublished PhD thesis, Queen Mary College, London.

Nutt, M. J. C. (1979) The Haweswater complex. In *The Caledonides of the British Isles – Reviewed* (ed. A. L. Harris, C. H. Holland and B. E. Leake), *Geological Society of London Special Publication*, No. 8, pp. 727–33.

O'Brien, C. (1985) The petrogenesis and geochemistry of the British Caledonian granites, with special reference to mineralized intrusions. Unpublished PhD thesis, University of Leicester.

O'Brien, C., Plant, J. A., Simpson, P. R. and Tarney, J. (1985) The geochemistry, metasomatism and petrogenesis of the granites of the English Lake District. *Journal of the Geological Society of London*, **142**, 1139–57.

O'Hanley, D. S. (1992) Solution to the volume problem in serpentinization. *Geology*, **20**, 705–8.

O'Hanley, D. S. (1996) *Serpentinization*, Oxford University Press, Oxford.

Old, R. A., Hamblin, R. J. O., Ambrose, K. and Warrington, G. (1991) Geology of country around Redditch. *Memoir of the British Geological Survey*, Sheet 183 (England and Wales).

Oldershaw, W. (1974) The Lochnagar granitic ring complex, Aberdeenshire. *Scottish Journal of Geology*, **10**, 297–309.

Oliver, R. L. (1954) Welded tuffs in the Borrowdale Volcanic Series, English Lake District, with a note on similar rocks in Wales. *Geological Magazine*, **91**, 473–83.

Oliver, R. L. (1956a) The origin of garnets in the Borrowdale Volcanic Series and associated rocks, English Lake District. *Geological Magazine*, **93**, 121–39.

Oliver, R. L. (1956b) The origin of garnets in the Borrowdale Volcanic Series. *Geological Magazine*, **93**, 516–17.

Oliver, R. L. (1961) The Borrowdale Volcanic and associated rocks of the Scafell area, English Lake District. *Quarterly Journal of the Geological Society of London*, **117**, 377–417.

O'Nions, R. K., Hamilton, P. J. and Hooker, P. J. (1983) A Nd isotope investigation of sediments related to crustal development in the British Isles. *Earth and Planetary Science Letters*, **63**, 229–40.

Orton, G. (1988) A spectrum of mid-Ordovician fan and braidplain deltaic sequences, North Wales: a consequence of varying fluvial input. In *Fan Deltas: Tectonic Setting, Sedimentology and Recognition* (eds W. Nemex and R. J. Steel), Blackie, Glasgow.

Pankhurst, R. J. (1969) Strontium isotope studies applied to petrogenesis in the basic igneous province of North-East Scotland. *Journal of Petrology*, **10**, 116–45.

Pankhurst, R. J. (1974) Rb-Sr whole-rock chronology of Caledonian events in north-east Scotland. *Bulletin of the Geological Society of America*, **85**, 345–50.

Pankhurst, R. J. (1979) Isotope and trace element evidence for the origin and evolution of Caledonian granites in the Scottish Highlands. In *Origin of Granitic Batholiths* (eds M. P. Atherton and J. Tarney), Shiva, Orpington, pp. 18–33.

Pankhurst, R. J. (1982) Geochronological tables for British igneous rocks. In *Igneous Rocks of the British Isles* (ed. D. S. Sutherland), Wiley, Chichester, pp. 575–81.

Pankhurst, R. J. and Sutherland, D. S. (1982) Caledonian granites and diorites of Scotland and Ireland. In *Igneous Rocks of the British Isles* (ed. D. S. Sutherland), Wiley, Chichester, pp. 149–90.

Parslow, G. R. (1968) The physical and structural features of the Cairnsmore of Fleet granite and its aureole. *Scottish Journal of Geology*, **4**, 91–108.

Parslow, G. R. (1971) Variations in mineralogy and major elements in the Cairnsmore of Fleet granite, S W Scotland. *Lithos*, **4**, 43–55.

Parslow, G. R. and Randall, B. A. O. (1973) A gravity survey of the Cairnsmore of Fleet granite and its environs. *Scottish Journal of Geology*, **9**, 219–31.

Parsons, I. (1965a) The sub-surface shape of the Loch Ailsh intrusion, Assynt, as deduced from magnetic anomalies across the contact, with a note on traverses across the Loch Borrolan

References

Complex. *Geological Magazine*, **102**, 46–58.

Parsons, I. (1965b) The feldspathic syenites of the Loch Ailsh intrusion, Assynt, Scotland. *Journal of Petrology*, **6**, 365–94.

Parsons, I. (1968) The origin of the basic and ultrabasic rocks of the Loch Ailsh alkaline intrusion, Assynt. *Scottish Journal of Geology*, **4**, 221–34.

Parsons, I. (1972) Comparative petrology of the leucocratic syenites of the Northwest Highlands of Scotland. *Geological Journal*, **8**, 71–82.

Parsons, I. (1979) The Assynt alkaline suite. In *The Caledonides of the British Isles – Reviewed* (eds A. L. Harris, C. H. Holland and B. E. Leake), *Geological Society of London Special Publication*, No. **8**, pp. 667–81.

Parsons, I. and McKirdy, A. P. (1983) The interrelationship of igneous activity and thrusting in Assynt: excavations at Loch Borralan. *Scottish Journal of Geology*, **19**, 59–67.

Paterson, B. A., Rogers, G. and Stephens, W. E. (1992a) Evidence for inherited Sm-Nd isotopes in granitoid zircons. *Contributions to Mineralogy and Petrology*, **111**, 378–90.

Paterson, B. A., Rogers, G., Stephens, W. E. and Hinton, R. W. (1993) The longevity of acid-basic magmatism associated with a major transcurrent fault (abstract). *Geological Society of America, Abstracts with programs*, **25**, (6), p. A42.

Paterson, B. A., Stephens, W. E., Rogers, G., Williams, I. S., Hinton, R. W. and Herd, D. A. (1992b) The nature of zircon inheritance in two granite plutons. *Transactions of the Royal Society of Edinburgh: Earth Sciences*, **83**, 459–71. (Also *Geological Society of America Special Paper*, **272**.)

Paterson, I. B. and Harris, A. L. (1969) Lower Old Red Sandstone ignimbrites from Dunkeld, Perthshire. *Report of the Institute of Geological Sciences*, No. **69/7**.

Paterson, S. R., Fowler, T. K. and Miller, R. B. (1996) Pluton emplacement in arcs: a crustal-scale exchange process. *Transactions of the Royal Society of Edinburgh: Earth Sciences*, **87**, 115–23.

Pattison, D. R. M. and Harte, B. (1985) A petrogenetic grid for pelites in the Ballachulish and other Scottish thermal aureoles. *Journal of the Geological Society of London*, **142**, 7–28.

Pattison, D. R. M. and Tracy, R. J. (1991) Phase equilibria and thermobarometry of metapelites. In *Contact Metamorphism* (ed. D. M. Kerrick). *Mineralogical Society of America, Reviews in Mineralogy*, **26**, 105–206.

Peach, B. N. and Horne, J. (1880) The Old Red Sandstone of Orkney. *Proceedings of the Royal Physical Society of Edinburgh*, **5**, 329–42.

Peach, B. N. and Horne, J. (1884) The old red volcanic rocks of Shetland. *Transactions of the Royal Society of Edinburgh*, **32**, 359–88.

Peach, B. N. and Horne, J. (1899) The Silurian rocks of Britain, 1: Scotland. *Memoir of the Geological Survey of the United Kingdom*.

Peach, B. N., Horne, J., Gunn, W., Clough, C. T., Hinxman, L. W. and Teall, J. J. H. (1907) The geological structure of the North-West Highlands of Scotland. *Memoir of the Geological Survey of the United Kingdom*.

Peach, B. N., Horne, J., Woodward, H. B. Clough, C. T., Harker, A. and Wedd, C. B. (1910) The geology of Glenelg, Lochalsh and South East part of Skye. *Memoir of the Geological Survey of Great Britain*, Sheet 71 (Scotland).

Peacock, J. D., Mendum, J. R. and Fettes, D. J. (1992) Geology of the Glen Affric district. *Memoir of the British Geological Survey*, Sheet 72E (Scotland).

Pearce, J. A. (1982) Trace element characteristics of lavas from destructive plate boundaries. In *Andesites* (ed. R. S. Thorpe), Wiley, Chichester, pp. 525–48.

Pearce, J. A. and Cann, J. R. (1973) Tectonic setting of basic volcanic rocks determined using trace element analyses. *Earth and Planetary Science Letters*, **19**, 290–300.

Pearce, J. A., Lippard, S. J. and Roberts, S. (1984) Characteristics and tectonic significance of supra-subduction zone ophiolites. In *Marginal Basin Geology: Volcanic and Associated Sedimentary and Tectonic Processes in Modern and Ancient Marginal Basins* (eds B. P. Kokelaar and M. F. Howells), *Geological Society of London Special Publication*, No. **16**, pp. 77–94.

Petford, N. (1996) Dykes or diapirs? *Transactions of the Royal Society of Edinburgh: Earth Sciences*, **87**, 105–14.

Petterson, M. G., Beddoe-Stephens, B., Millward, D. and Johnson, E. W. (1992) A pre-caldera plateau-andesite field in the Borrowdale Volcanic Group of the English Lake District. *Journal of the Geological Society of London*, **149**, 889–906.

References

Pharoah, T. C., Allsop, J. M., Holliday, D. W., Merriman, R. J., Kimbell, G. S. et al. (1997) The Moorby Microgranite: a deformed high level intrusion of Ordovician age in the concealed Caledonian basement of Lincolnshire. *Proceedings of the Yorkshire Geological Society*, **51**, 329–42.

Pharoah, T. C., Brewer, T. S., and Webb, P. C. (1993) Subduction related magmatism of late Ordovician age in eastern England. *Geological Magazine*, **130**, 647–56.

Pharoah, T. C., Merriman, R. J., Evans, J. A., Brewer, T. S., Webb, P. C. and Smith, N. J. P. (1991) Early Palaeozoic arc-related volcanism in the concealed Caledonides of southern Britain. *Annales de la Société Géologique de Belgique*, **114**, 63–91.

Pharoah, T. C., Merriman, R. J., Webb, P. C. and Beckinsale, R. D. (1987) The concealed Caledonides of eastern England: preliminary results of a multidisciplinary study. *Proceedings of the Yorkshire Geological Society*, **46**, 355–69.

Phemister, J. (1926) The alkaline igneous rocks of the Loch Ailsh district. In *The Geology of Strath Oykell and Lower Loch Shin* (eds H. H. Read, J. Phemister and G. Ross), *Memoir of the Geological Survey of Great Britain*, Sheet 102 (Scotland), pp. 22–111.

Phemister, J. (1931) On a carbonate-rock at Bad na h'Achlaise, Assynt, Sutherland. *Summary of Progress of the Geological Survey for 1930*, Part III, pp. 58–61.

Phemister, J. (1948) *British Regional Geology: The Northern Highlands*, HMSO, Edinburgh.

Phemister, J. (1964) Rodingitic assemblages in Fetlar, Shetland Islands, Scotland. In *Advancing Frontiers in Geology and Geophysics*, Osmania University Press, Hyderabad, pp. 279–95.

Phemister, J. (1979) The Old Red Sandstone intrusive complex of northern Northmaven, Shetland. *Report of the Institute of Geological Sciences*, No. **78/2**.

Phillips, E. R. (1994) Whole-rock geochemistry of the calc-alkaline Old Red Sandstone lavas, Sheet 15 (New Cumnock), Scotland. British Geological Survey Mineralogy and Petrology Brief Report, No. **WG/94/1**.

Phillips, E. R. and May, F. (1996) The Glas Bheinn Appinitic Complex, Glen Roy: a model for foliation development during emplacement. *Scottish Journal of Geology*, **32**, 9–21.

Phillips, E. R., Barnes, R. P., Boland, M. P., Fortey, N. J. and McMillan, A. A. (1995a) The Moniaive Shear Zone: a major zone of sinistral strike-slip deformation in the Southern Uplands of Scotland. *Scottish Journal of Geology*, **31**, 139–49.

Phillips, E. R., Barnes, R. P., Merriman, R. J. and Floyd, J. D. (1995b) The tectonic significance of Ordovician basic igneous rocks in the Southern Uplands, southwest Scotland. *Geological Magazine*, **132**, 549–56.

Phillips, E. R., Smith, R. A. and Floyd, J. D. (1999) The Bail Hill Volcanic Group: alkaline within-plate volcanism during Ordovician sedimentation in the Southern Uplands, Scotland. *Transactions of the Royal Society of Edinburgh: Earth Sciences*, **89**, 233–47.

Phillips, W. E. A., Stillman, C. J. and Murphy, T. (1976) A Caledonian plate tectonic model. *Journal of the Geological Society of London*, **132**, 579–609.

Phillips, W. J. (1956) The Criffel Dalbeattie granodiorite complex. *Journal of the Geological Society of London*, **112**, 221–40.

Phillips, W. J., Fuge, R. and Phillips, N. (1981) Convection and crystallisation in the Criffel Dalbeattie pluton. *Journal of the Geological Society of London*, **138**, 351–66.

Phillips, W. J., Fuge, R. and Phillips, N. (1983) Reply to: Discussion on convection and crystallization in the Criffell-Dalbeattie pluton. *Journal of the Geological Society of London*, **140**, 311–13.

Piasecki, M. A. J. (1975) Tectonic and metamorphic history of the Upper Findhorn, Inverness-shire, Scotland. *Scottish Journal of Geology*, **11**, 87–115.

Pickerill, R. K. and Brenchley, P. J. (1979) Caradoc marine communities of the south Berwyn Hills, North Wales. *Palaeontology*, **22**, 229–64.

Pickering, K. T. and Smith, A. G. (1995) Arcs and backarc basins in the Early Palaeozoic Iapetus Ocean. *The Island Arc*, **4**, 1–67.

Pickering, K. T., Bassett, M. G. and Siveter, D. J. (1988) Late Ordovician–early Silurian destruction of the Iapetus Ocean: Newfoundland, British Isles and Scandinavia: A discussion. *Transactions of the Royal Society of Edinburgh: Earth Sciences*, **79**, 361–82.

Pidgeon, R. T. and Aftalion, M. (1978) Cogenetic and inherited zircon U-Pb systems in granites: Palaeozoic granites of Scotland and England. In *Crustal Evolution in*

References

Northwestern Britain and Adjacent Areas (eds D. R. Bowes and B. E. Leake), *Geological Journal Special Issue*, No. **10**, pp. 183–220.

Piper, J. D. A. (1997) Palaeomagnetism of igneous rocks of the Lake District (Caledonian) terrane, northern England: Palaeozoic motions and deformation at a leading edge of Avalonia. *Geological Journal*, **32**, 211–46.

Piper, J. D. A., Nowell, D. A. G. and Crimes, T. P. (1995) Palaeomagnetism of the Tan y Grisiau granite, North Wales: evidence for a subvolcanic origin in Late Ordovician times. *Geological Journal*, **30**, 39–47.

Piper, J. D. A., Stephen, J. C. and Branney, M. J. (1997) Palaeomagnetism of the Borrowdale and Eycott volcanic groups, English Lake District: primary and secondary magnetization during a single late Ordovician polarity chron. *Geological Magazine*, **134**, 481–506.

Pitcher, W. S. (1993) *The Nature and Origin of Granite*, Chapman and Hall, London.

Pitcher, W. S. and Berger, A. R. (1972) *The Geology of Donegal: a Study of Granite Emplacement and Unroofing*, Wiley-Interscience, New York.

Plant, J. A. (1986) Models for granites and their mineralising systems in the British and Irish Caledonides. In *Geology and Genesis of Mineral Deposits in Ireland* (eds C. J. Andrew, R. W. A. Crowe, S. Finlay, W. M. Pennell and J. F. Payne), Irish Association for Economic Geology, Dublin, pp. 121–56.

Plant, J. A., Henney, P. J. and Simpson, P. R. (1990) The genesis of tin-uranium granites in the Scottish Caledonides: implications for metallogenesis. *Geological Journal*, **25**, 431–42.

Plant, J. A., O'Brien, C., Tarney, J. and Hurdley, J. (1985) Geochemical criteria for the recognition of High Heat Production Granites. In *High Heat Production (HHP) Granites, Hydrothermal Circulation and Ore Genesis*, Institution of Mining and Metallurgy, London, pp. 263–85.

Platten, I. M. (1966) The petrology of some Caledonian minor intrusions in Appin, Argyllshire. Unpublished PhD thesis, University of London.

Platten, I. M. (1982) Partial melting of feldspathic quartzite around late Caledonian minor intrusions in Appin, Scotland. *Geological Magazine*, **119**, 413–19.

Platten I. M. (1983) Partial melting of semipelite and the development of marginal breccias around a late Caledonian intrusion in the Grampian Highlands of Scotland. *Geological Magazine*, **120**, 37–49.

Platten, I. M. (1984) Fluidised mixtures of magma and rock in a late Caledonian breccia dyke and associated breccia pipes in Appin, Scotland. *Geological Journal*, **19**, 209–26.

Platten, I. M. (1991) Zoning and layering in diorites of the Scottish Caledonian Appinite suite. *Geological Journal*, **26**, 329–48.

Platten, I. M. and Money, M. S. (1987) Formation of late Caledonian subvolcanic breccia pipes at Cruachan Cruinn, Grampian Highlands, Scotland. *Transactions of the Royal Society of Edinburgh: Earth Sciences*, **78**, 85–103.

Playfair, J. (1802) *Illustrations of the Huttonian Theory of the Earth*. Caddell and Davies, London, and William Creech, Edinburgh.

Pollard, J. E. (1995) John Smith's discoveries of trace fossils from Old Red Sandstone and Carboniferous rocks of Southwest Scotland. In *John Smith of Dalry, Geologist, Antiquarian and Natural Historian*, Ayrshire Archaeological and Natural History Society, Darvel, pp. 30–39.

Pollard, J. E. and Walker, E. F. (1984) Reassessment of sediments and trace fossils from Old Red Sandstone (Lower Devonian) of Dunure, Scotland, described by John Smith (1909). *Geobios*, **17**, 567–76.

Ponnamperuma, C. and Pering, K. (1966) Possible abiogenic origin of some naturally occurring hydrocarbons. *Nature*, **209**, 982–4.

Potts, G. J., Hunter, R. H., Harris, A. L. and Fraser, F. M. (1995) Late-orogenic extensional tectonics at the NW margin of the Caledonides in Scotland. *Journal of the Geological Society of London*, **152**, 907–10.

Pratt, W. T., Woodhall, D. G., and Howells, M. F. (1995) Geology of the country around Cadair Idris. *Memoir of the British Geological Survey*, Sheet 149 (England and Wales).

Prichard, H. M. (1985) The Shetland ophiolite. In *The Caledonian Orogen – Scandinavia and Related Areas* (eds D. G. Gee and B. A. Sturt), Wiley, Chichester, pp. 1173–84.

Pringle, J. (1914) Geology of Ramsey Island, Pembrokeshire. *Reports of the British Association (for 1913)*, London, p. 151.

Pringle, J. (1915) Geology of Ramsey Island, Pembrokeshire. *Reports of the British Association (for 1914)*, London, p. 111.

Pringle, J. (1930) The geology of Ramsey Island

References

(Pembrokeshire). *Proceedings of the Geologists' Association*, **41**, 1–24.

Raine, P. (1998) Sedimentary processes and depositional environments in caldera lakes: Scafell (UK) and La Primavera (Mexico) calderas. Unpublished PhD thesis, University of Liverpool.

Ramsay, A. C. (1866) The geology of North Wales (1st edition). *Memoir of the Geological Survey of Great Britain*, Vol. **3**.

Ramsay, A. C. (1881) The geology of North Wales (2nd edition). *Memoir of the Geological Survey of Great Britain*, Vol. **3**.

Rast, N. (1969) The relationship between Ordovician structure and volcanicity in Wales. In *The Pre-Cambrian and Lower Palaeozoic Rocks of Wales* (ed. A. Wood), University of Wales Press, Cardiff, pp. 303–35.

Rast, N., Beavon, R. V. and Fitch, F. J. (1958) Sub-aerial volcanicity in Snowdonia. *Nature*, **181**, 508.

Rastall, R. H. (1906) The Buttermere and Ennerdale Granophyre. *Quarterly Journal of the Geological Society of London*, **62**, 253–74.

Rastall, R. H. (1940) Xenoliths at Threlkeld, Cumberland. *Proceedings of the Yorkshire Geological Society*, **24**, 223–32.

Rastall, R. H. and Wilcockson, W. H. (1915) Accessory minerals of the granitic rocks of the Lake District. *Quarterly Journal of the Geological Society of London*, **71**, 592–622.

Read, H. H. (1919) The two magmas of Strathbogie and Lower Banffshire. *Geological Magazine*, **56**, 364–371.

Read, H. H. (1923) The petrology of the Arnage district in Aberdeenshire: a study of assimilation. *Quarterly Journal of the Geological Society of London*, **79**, 446–84.

Read, H. H. (1926) Mica-lamprophyres of Wigtown. *Geological Magazine*, **63**, 422–9.

Read, H. H. (1927) The Tinto district. *Proceedings of the Geologists' Association*, **39**, 499–504.

Read, H. H. (1931) The geology of central Sutherland. *Memoir of the Geological Survey of Great Britain*, Sheets 108 and 109 (Scotland).

Read, H. H. (1934) The metamorphic geology of Unst in the Shetland Islands. *Quarterly Journal of the Geological Society of London*, **90**, 637–88.

Read, H. H. (1935) The gabbros and associated xenolithic complexes of the Haddo House district, Aberdeenshire. *Quarterly Journal of the Geological Society of London*, **91**, 591–635.

Read, H. H. (1956) The dislocated south-western margin of the Insch Igneous Mass, Aberdeenshire. *Proceedings of the Geologists' Association*, **67**, 73–86.

Read, H. H. (1957) *The Granite Controversy*, Murby, London.

Read, H. H. (1961) Aspects of the Caledonian magmatism in Britain. *Liverpool and Manchester Geological Journal*, **2**, 653–83.

Read, H. H. (1966) An orthonorite containing spinel xenoliths with late diaspore at Mill of Boddam, Insch, Aberdeenshire. *Proceedings of the Geologists' Association*, **77**, 65–77.

Read, H. H. and Farquhar, O. C. (1952) The geology of the Arnage district (Aberdeenshire); a re-interpretation. *Quarterly Journal of the Geological Society of London*, **107**, 423–40.

Read, H. H. and Haq, B. T. (1965) Notes, mainly geochemical, on the granite-diorite complex of the Insch igneous mass, with an addendum on the Aberdeenshire quartz-dolerites. *Proceedings of the Geologists' Association*, **76**, 13–19.

Read, H. H., Phemister, J. and Ross, G. (1926) The geology of Strath Oykell and Lower Loch Shin. *Memoir of the Geological Survey of Great Britain*, Sheet 102 (Scotland).

Read, H. H., Ross, G., Phemister, J. and Lee, G. W. (1925) The geology of the country around Golspie, Sutherlandshire. *Memoir of the Geological Survey of Great Britain*, Sheet 103 (Scotland).

Read, H. H., Sadashivaiah, M. S. and Haq, B. T. (1961) Differentiation in the olivine-gabbro of the Insch mass, Aberdeenshire. *Proceedings of the Geologists' Association*, **72**, 391–413.

Read, H. H., Sadashivaiah, M. S. and Haq, B. T. (1965) The hypersthene-gabbro of the Insch Complex, Aberdeenshire. *Proceedings of the Geologists' Association*, **76**, 1–11.

Reed, F. R. C. (1895) The geology of the country around Fishguard. *Quarterly Journal of the Geological Society of London*, **51**, 149–95.

Reedman, A. J., Colman, T. B., Campbell, S. D. G. and Howells, M. F. (1985) Volcanogenic mineralization related to the Snowdon Volcanic Group (Ordovician), North Wales. *Journal of the Geological Society of London*, **142**, 875–88.

Reedman, A. J., Howells, M. F., Orton, G. and

References

Campbell, S. D. G. (1987) The Pitts Head Tuff Formation: a subaerial to submarine welded ash-flow tuff of Ordovician age, North Wales. *Geological Magazine,* **124**, 427–39.

Rennie, F. W. (1983) The mineralogy and geochemistry of the Lochnagar complex. Unpublished PhD thesis, University of Aberdeen.

Reynolds, D. L. (1931) Dykes of the Ards Peninsula, County Down. *Geological Magazine,* **68**, 97–111.

Reynolds, D. L. (1956) Calderas and ring-complexes. *Nederlandsch Geologisch-Mijnbouwkundig Genootschap, Verhandelingen, Geologische Serie* **16**, 355–79.

Rice, C. M. and Davies, B. (1979) Copper mineralisation associated with an appinite pipe in Argyll, Scotland. *Transactions of the Institution of Mining and Metallurgy (Section B: Applied Earth Sciences),* **88**, 154–60.

Rice, C. M. and Trewin, N. H. (1988) A Lower Devonian gold bearing hot-spring system, Rhynie, Scotland. *Transactions of the Institution of Mining and Metallurgy (Section B: Applied Earth Sciences),* **97**, 141–4.

Richardson, J. B., Ford, J. H. and Parker, F. (1984) Miospores, correlation and age of some Scottish Lower Old Red Sandstone sediments from the Strathmore region (Fife and Angus). *Journal of Micropalaeontology,* **3**, 109–24.

Richardson, S. W. (1968) The petrology of the metamorphosed syenite in Glen Dessarry, Inverness-shire. *Quarterly Journal of the Geological Society of London,* **124**, 9–51.

Richey, J. E. (1938) The dykes of Scotland. *Transactions of the Royal Society of Edinburgh,* **13**, 393–435.

Riley, P. (1966) A re-investigation of the pre-Tertiary geology of the Ross of Mull, Argyll. Unpublished PhD thesis, University of Sheffield.

Roach, R. A. (1969) The composite nature of the St David's Head and Carn Llidi intrusions of North Pembrokeshire. In *The Pre-Cambrian and Lower Palaeozoic Rocks of Wales* (ed. A. Wood), University of Wales Press, Cardiff, pp. 409–33.

Roberts, B. (1969) The Llwyd Mawr Ignimbrite and its associated volcanic rocks. In *The Pre-Cambrian and Lower Palaeozoic Rocks of Wales* (ed. A. Wood), University of Wales Press, Cardiff, pp. 337–56.

Roberts, B. (1979) *The Geology of Snowdonia and Llŷn: an Outline and Field Guide,* Adam Hilger, Bristol.

Roberts, B. and Siddans, A. W. B. (1971) Fabric studies in the Llwyd Mawr Ignimbrite, Caernarvonshire, North Wales. *Tectonophysics,* **12**, 283–306.

Roberts, D. E. (1983) Metasomatism and the formation of greisen in Grainsgill, Cumbria, England. *Geological Journal,* **18**, 43–52.

Roberts, J. L. (1963) Source of the Glencoe ignimbrites. *Nature,* **199**, 201.

Roberts, J. L. (1966a) Ignimbrite eruptions in the volcanic history of the Glencoe cauldron subsidence. *Geological Journal,* **5**, 173–84.

Roberts, J. L. (1966b) The emplacement of the main Glencoe fault-intrusion at Stob Mhic Mhartuin. *Geological Magazine,* **103**, 299–316.

Roberts, J. L. (1974) The evolution of the Glencoe cauldron. *Scottish Journal of Geology,* **10**, 269–82.

Robertson, A. H. F. and Henderson, W. G. (1984) Geochemical evidence for the origin of igneous and sedimentary rocks of the Highland Border, Scotland. *Transactions of the Royal Society of Edinburgh: Earth Sciences,* **75**, 135–50.

Robertson, R. C. R. and Parsons, I. (1974) The Loch Loyal syenites. *Scottish Journal of Geology,* **10**, 129–46.

Robinson, M. A. (1985) Palaeomagnetism of volcanics and sediments of the Eday Group, Southern Orkney. *Scottish Journal of Geology,* **21**, 285–300.

Robson, D. A. (1948) The Old Red Sandstone Volcanic Suite of Eastern Forfarshire. *Transactions of the Edinburgh Geological Society,* **14**, 128–40.

Rock, N. M. S. (1977) A new occurrence of fenite in the Loch Borrolan alkaline complex, Assynt. *Mineralogical Magazine,* **41**, 529 and M7.

Rock, N. M. S. (1983) The Permo-Carboniferous camptonite-monchiquite dyke-suite of the Scottish Highlands and Islands. *Report of the Institute of Geological Sciences* No. 82/14.

Rock, N. M. S. (1987) The nature and origin of lamprophyres: an overview. In *Alkaline Igneous Rocks,* (eds J. G. Fitton and B. G. J. Upton), *Geological Society Special Publication,* No. 30, pp. 191–226.

Rock, N. M. S. and Hunter, R. H. (1987) Late

References

Caledonian dyke-swarms of northern Britain: spatial and temporal intimacy between lamprophyric and granitic magmatism around the Ross of Mull pluton, Inner Hebrides. *Geologische Rundschau*, **76**, 805–26.

Rock, N. M. S., Cooper, C. and Gaskarth, J. W. (1986a) Late Caledonian subvolcanic vents and associated dykes in the Kirkcudbright area, Galloway, south-west Scotland. *Proceedings of the Yorkshire Geological Society*, **46**, 29–37.

Rock, N. M. S., Gaskarth, J. W., Henney, P. and Shand, P. (1988) Late Caledonian dyke swarms of northern Britain: some preliminary petrogenetic and tectonic implications of their province-wide distribution and chemical variation. *Canadian Mineralogist*, **26**, 3–22.

Rock, N. M. S., Gaskarth, J. W. and Rundle, C. C. (1986b) Late Caledonian dyke-swarms in southern Scotland: a regional zone of primitive K-rich lamprophyres and associated vents. *Journal of Geology*, **94**, 505–21.

Rock, N. M. S. and Rundle, C. C. (1986) Lower Devonian age for the 'Great (basal) Conglomerate', Scottish Borders. *Scottish Journal of Geology*, **22**, 285–8.

Rogers, D. A., Marshall, J. E. A. and Astin, T. R. (1989) Devonian and later movements on the Great Glen fault system, Scotland. *Journal of the Geological Society of London*, **146**, 369–72.

Rogers, G. and Dunning, G. R. (1991) Geochronology of appinitic and related granitic magmatism in the W Highlands of Scotland: constraints on the timing of transcurrent fault movement. *Journal of the Geological Society of London*, **148**, 17–27.

Rogers, G., Paterson, B. A., Dempster, T. J. and Redwood, S. D. (1994) U-Pb geochronology of the 'Newer' gabbros, NE Grampians. In *Caledonian Terrane Relationships in Britain*, Symposium Abstracts, British Geological Survey, Keyworth, Nottingham.

Rolfe, W. D. I. (1980) Early invertebrate terrestrial faunas. In *Terrestrial Environment and the Origin of the Invertebrates* (ed. A. L. Panchen), Academic Press, London and New York, pp. 117–157.

Rollin, K. E. (1984) Gravity modelling of the Eastern Highlands granites in relation to heat flow studies. In *Investigation of the Geothermal Potential of the UK*, British Geological Survey, Keyworth, Nottingham.

Romano, M. and Spears, D. A. (1991) Bentonites from the Horton Formation (Upper Silurian) of Ribblesdale, Yorkshire. *Proceedings of the Yorkshire Geological Society*, **48**, 277–85.

Ruddock, I. (1969) The geochemistry of the southern half of the Loch Dee–Loch Doon pluton, South West Scotland. Unpublished PhD thesis, University of Newcastle upon Tyne.

Rundle, C. C. (1979) Ordovician intrusions in the English Lake District. *Journal of the Geological Society of London*, **136**, 29–38.

Rundle, C. C. (1981) The significance of isotopic dates from the English Lake District for the Ordovician–Silurian time-scale. *Journal of the Geological Society of London*, **138**, 569–72.

Rundle, C. C. (1982) The chronology of igneous intrusion in the English Lake District. Unpublished PhD thesis, University of London.

Rundle, C. C. (1992) Review and assessment of isotopic ages from the English Lake District. *British Geological Survey Technical Report*, No. **WA/92/38**.

Rushton, A. W. A., Stone, P. and Hughes, R. A. (1996) Biostratigraphical control of thrust models for the Southern Uplands of Scotland. *Transactions of the Royal Society of Edinburgh: Earth Sciences*, **86**, 137–52.

Rushton, A. W. A., Owen, A. W., Owens, R. M. and Prigmore, J. K. (1999) *British Cambrian to Ordovician Stratigraphy*. Geological Conservation Review Series, No. **18**, Joint Nature Conservation Committee, Peterborough.

Rushton, A. W. A., Stone, P., Smellie, J. L. and Tunnicliff, S. P. (1986) An early Arenig age for the Pinbain sequence, Ballantrae Complex. *Scottish Journal of Geology*, **22**, 41–54.

Rutledge, H. (1952) Contact phenomena of the southern part of the Loch Doon plutonic complex (abstract). *Abstracts of the Proceedings of the Geological Society of London*, **1484**, 60–6.

Rutley, F. (1885a) On strain in connexion with crystallisation and the development of perlitic structure. *Journal of the Geological Society of London*, **40**, 340–7.

Rutley, F. (1885b) Felsitic lavas of England and Wales. *Memoir of the Geological Survey of Great Britain*.

Ryan, P. D. and Dewey, J. F. (1991) A geological and tectonic cross-section of the Caledonides

References

of western Ireland. *Journal of the Geological Society of London,* **148**, 173–80.

Sabine, P. A. (1952) The ledmorite dyke of Achmelvich, near Lochinver, Sutherland. *Mineralogical Magazine,* **29**, 827–32.

Sabine, P. A. (1953) The petrography and geological significance of the post-Cambrian minor intrusions of Assynt and the adjoining districts of north-west Scotland. *Quarterly Journal of the Geological Society of London,* **109**, 137–71.

Sabine, P. A. (1963) The Strontian granite complex, Argyllshire. *Bulletin of the Geological Survey of Great Britain,* **20**, 6–42.

Sadashivaiah, M. S. (1954) The granite-diorite complex of the Insch Igneous Mass, Aberdeenshire. *Geological Magazine,* **91**, 286–92.

Saunders, A. D., Rogers, G., Marriner, G., Terrell, D. J. and Verma, S. P. (1987) Geochemistry of Cenozoic volcanic rocks, Baja California, Mexico: Implications for the petrogenesis of post-subduction magmas. *Journal of Volcanology and Geothermal Research,* **32**, 223–45.

Scarpati, C., Cole, P. and Perrotta, A. (1993) The Neapolitan Yellow Tuff – a large volume multiphase eruption from Campi Flegrei, southern Italy. *Bulletin of Volcanology,* **55**, 343–56.

Scott, R. (1992) Tan y Grisiau. In *Caledonian Structures in Britain South of the Midland Valley* (ed. J. E. Treagus). Geological Conservation Review Series, No. **3**, Chapman and Hall, London, pp. 118–20.

Seber, D., Barazangi, M., Ibenbrahim, A. and Demnati, A. (1996) Geophysical evidence for lithospheric delamination beneath the Alboran Sea and Rif-Betic mountains. *Nature,* **379**, 785–90.

Sedgwick, A. (1836) Introduction to the general structure of the Cumbrian Mountains, with a description of the great dislocations by which they have been separated from the neighbouring Carboniferous chains. *Transactions of the Geological Society of London,* **40**, 340–7.

Sedgwick, A. (1843) Outline of the geological structure of North Wales. *Proceedings of the Geological Society of London,* **4**, 212–24.

Seymour, J. W. (1815) An account of observations, made by Lord Webb Seymour and Professor Playfair, upon some geological appearances in Glen Tilt, and the adjacent country; drawn up by Lord Webb Seymour. *Transactions of the Royal Society of Edinburgh,* **7**, 303–75.

Shackleton, R. M. (1948) Overturned rhythmic banding in the Huntly gabbro of Aberdeenshire. *Geological Magazine,* **85**, 358–60.

Shackleton, R. M. (1954) The structural evolution of North Wales. *Liverpool and Manchester Geological Journal,* **2**, 216–52.

Shackleton, R. M. (1959) The stratigraphy of the Moel Hebog district between Snowdon and Tremadoc. *Liverpool and Manchester Geological Journal,* **2**, 216–52.

Shand, P. (1989) Late Caledonian magma genesis in Southern Scotland. Unpublished PhD thesis, University of Aston, Birmingham.

Shand, P., Gaskarth, J. W., Thirlwall, M. F. and Rock, N. M. S. (1994) Late Caledonian lamprophyre dyke swarms of south-eastern Scotland. *Mineralogy and Petrology,* **51**, 277–98.

Shand, S. J. (1906) Uber Borolanit und die gesteine des Cnoc-na-Sroine massivs im Nord Schottland. *Neues Jahrbuch fur Mineralogie,* **22**, 413–53.

Shand, S. J. (1909) On borolanite and its associates in Assynt (preliminary communication). *Transactions of the Edinburgh Geological Society,* **9**, 202–15.

Shand, S. J. (1910) On borolanite and its associates in Assynt (second communication). *Transactions of the Edinburgh Geological Society,* **9**, 376–416.

Shand, S. J. (1913) On saturated and unsaturated igneous rocks. *Geological Magazine,* **10**, 508–14.

Shand, S. J. (1930) Limestone and the origin of feldspathoidal rocks. *Geological Magazine,* **67**, 421–5.

Shand, S. J. (1939) The Loch Borolan Laccolith, North-West Scotland. *Journal of Geology,* **17**, 408–20.

Sharpe, D. (1846) Contributions to the geology of North Wales. *Quarterly Journal of the Geological Society of London,* **2**, 283–316.

Shaw, M. H., Gunn, A. G., Fletcher, T. A., Styles, M. T. and Perez, M. (1992) Data arising from drilling investigations in the Loch Borralan Intrusion, Sutherland, Scotland. *British Geological Survey, Open-File Report* No. **8**.

Shepherd, T. J. and Darbyshire, D. P. F. (1981) Fluid inclusion Rb-Sr isochrons for dating mineral deposits. *Nature,* **290**, 5807.

Shepherd, T. J., Beckinsale, R. D., Rundle, C. C. and Durham, J. (1976) Genesis of Carrock Fell tungsten deposits, Cumbria: fluid inclu-

References

sion and isotopic study. *Transactions of the Institution of Mining and Metallurgy,* **85**, B63–73.

Simpson, B. (1934) The petrology of the Eskdale (Cumberland) Granite. *Proceedings of the Geologists' Association,* **45**, 17–34.

Simpson, P. R., Brown, G. C., Plant, J. and Ostle, D. (1979) Uranium mineralisation and granite magmatism in the British Isles. *Philosophical Transactions of the Royal Society of London,* **291A**, 385–412.

Skillen, I. E. (1973) The igneous complex of Carrock Fell. *Proceedings of the Cumberland Geological Society,* **3**, 363–86.

Slater, G. B. (1977) A geochemical survey of the Lower Devonian volcanics of Scotland and northern England. Unpublished MSc thesis, University of Birmingham.

Smellie, J. L. (1984) Accretionary lapilli and highly vesiculated pumice in the Ballantrae ophiolite complex: ash-fall products of subaerial eruptions. *Report of the British Geological Survey,* Vol. **16**, No.1, 36–40.

Smellie, J. L. and Stone, P. (1984) 'Eclogite' in the Ballantrae Complex: a garnet-clinopyroxenite segregation in mantle harzburgite? *Scottish Journal of Geology,* **20**, 315–27.

Smellie, J. L. and Stone, P. (1992) Geochemical control on the evolutionary history of the Ballantrae Complex, SW Scotland, from comparisons with recent analogues. In *Ophiolites and their Modern Oceanic Analogues* (eds L. M. Parson, B. J. Murton and P. Browning), *Geological Society Special Publication,* No. 60, pp. 171–8.

Smellie, J. L., Stone, P. and Evans, J. A. (1995) Petrogenesis of boninites in the Ordovician Ballantrae Complex ophiolite, southwestern Scotland. *Journal of Volcanology and Geothermal Research,* **69**, 323–42.

Smith, D. I. (1979) Caledonian minor intrusions of the Northern Highlands of Scotland. In *The Caledonides of the British Isles – Reviewed* (eds A. L. Harris, C. Holland and B. E. Leake), *Geological Society of London Special Publication,* No. 8, pp. 683–97.

Smith, J. (1892) From the Doon to the Girvan Water, along the Carrick shore. *Transactions of the Geological Society of Glasgow,* **10**, 1–12.

Smith, J. (1909) *Upland Fauna of the Old Red Sandstone Formation of Carrick, Ayrshire.* Cross.

Smith, J. (1910) *Semi-Precious Stones of Carrick.* Cross.

Smith, M. (1988) The tectonic evolution of the Cambrian and Ordovician rocks in South Central Snowdonia. Unpublished PhD thesis, University of Wales, Aberystwyth.

Smith, M., Rushton, A. W. A. and Howells, M. F. (1995) New litho- and biostratigraphic evidence for a Mid-Ordovician hiatus in southern central Snowdonia, North Wales. *Geological Journal,* **30**, 145–56.

Smith, R. A. (1980) The geology of the Dalradian rocks around Blair Atholl, central Perthshire, Scotland. Unpublished PhD thesis, University of Liverpool.

Smith, R. A. (1995) The Siluro-Devonian evolution of the southern Midland Valley of Scotland. *Geological Magazine,* **132**, 503–13.

Smith, R. L. (1979) Ash-flow magmatism. In *Ash-Flow Tuffs* (eds C. E. Chapin and W. E. Elston), *Geological Society of America Special Paper,* No. **180**, pp. 5–27.

Smith, R. L. and Bailey, R. A. (1968) Resurgent cauldrons. In *Studies in Volcanology* (ed. R. R. Coates), *Geological Society of America Memoir,* No. **116**, pp. 613–62.

Smith, T. E. and Huang, C. H. (1995) Geochemistry and petrogenesis of the igneous rocks of the Builth–Llandrindod Ordovician Inlier, Wales. In *Magmatism in Relation to Diverse Tectonic Settings* (eds R. K. Srivastava and R. Chandra), A A Balkema, Rotterdam, pp. 261–81.

Soper, N. J. (1963) The structure of the Rogart igneous complex, Sutherland, Scotland. *Quaternary Journal of the Geological Society of London,* **11**, 445–78.

Soper, N. J. (1986) The Newer Granite problem: A geotectonic view. *Geological Magazine,* **123**, 227–36.

Soper, N. J. and Hutton, D. H. W. (1984) Late Caledonian sinistral displacements in Britain: Implications for a three-plate collision model. *Tectonics,* **3**, 781–94.

Soper, N. J. and Kneller, B. C. (1990) Cleaved microgranite dykes of the Shap swarm in the Silurian of NW England. *Geological Journal,* **25**, 161–70.

Soper, N. J. and Roberts, D. E. (1971) Age of cleavage in the Skiddaw Slates in relation to the Skiddaw aureole. *Geological Magazine,* **108**, 293–302.

Soper, N. J., Strachan, R. A., Holdsworth, R. E., Gayer, R. A. and O'Greiling, R. O. (1992) Sinistral transpression and the Silurian clo-

References

sure of Iapetus. *Journal of the Geological Society of London*, **149**, 871–80.

Soper, N. J., Webb, B. C. and Woodcock N. H. (1987) Late Caledonian (Acadian) transpression in north-west England: timing, geometry, and geotectonic significance. *Proceedings of the Yorkshire Geological Society*, **46**, 175–92.

Spears, D. A. (1961) The distribution of Alpha Radioactivity in a specimen of Shap Granite. *Geological Magazine*, **98**, 483–7.

Speer, J. A. and Becker, S. W. (1992) Evolution and magmatic subsolidus AFM mineral assemblages in granitoid rocks: biotite, muscovite, and garnet in the Cuffytown Creek pluton, South Carolina. *American Mineralogist*, **77**, 821–33.

Spray, J. G. (1988) Thrust related metamorphism beneath the Shetland Islands oceanic fragment, north-east Scotland. *Canadian Journal of Earth Sciences*, **25**, 1760–76.

Spray, J. G. and Dunning, G. R. (1991) A U/Pb age for the Shetland Islands oceanic fragment, Scottish Caledonides: evidence from anatectic plagiogranites in 'layer 3' shear zones. *Geological Magazine*, **128**, 667–71.

Spray, J. G. and Williams, G. D. (1980) The subophiolite metamorphic rocks of the Ballantrae Igneous Complex, SW Scotland. *Journal of the Geological Society of London*, **137**, 359–68.

Steinmann, G. (1927) Die ophiolithischen Zonen in dem mediterranen Kettengebirge. *14th International Geological Congress, Madrid*, **2**, 638–67.

Stephens, W. E. (1988) Granitoid plutonism in the Caledonian orogen of Europe. In *The Caledonian-Appalachian Orogen* (eds A. L. Harris and D. J. Fettes), *Geological Society Special Publication*, No. **38**, pp. 389–403.

Stephens, W. E. (1992) Spatial, compositional and rheological constraints on the origin of zoning in the Criffel pluton, Scotland. *Transactions of the Royal Society of Edinburgh: Earth Sciences*, **83**, 191–9.

Stephens, W. E. and Halliday, A. N. (1979) Compositional variation in the Galloway plutons. In *Origin of Granite Batholiths* (eds M. P. Atherton and J. Tarney), Shiva, Orpington, pp. 9–17.

Stephens, W. E. and Halliday, A. N. (1980) Discontinuities in the composition surface of a zoned pluton, Criffell, Scotland. *Geological Society of America Bulletin*, **91**, 165–70.

Stephens, W. E. and Halliday, A. N. (1984) Geochemical contrasts between late Caledonian granitoid plutons of northern, central and southern Scotland. *Transactions of the Royal Society of Edinburgh: Earth Sciences*, **75**, 259–73.

Stephens, W. E., Holden, P. and Henney, P. J. (1991) Microdioritic enclaves within Scottish Caledonian granitoids and their significance for crustal magmatism. In *Enclaves and Granite Petrology* (eds J. Didier and B. Barbarin), Developments in Petrology, No. **13**, Elsevier, Amsterdam, pp. 125–34.

Stephens, W. E., Whitley, J. E., Thirlwall, M. F. and Halliday, A. N. (1985) The Criffell zoned pluton: correlated behaviour of rare earth element abundances with isotopic systems. *Contributions to Mineralogy and Petrology*, **89**, 226–38.

Stephenson, D. and Gould, D. (1995) *British Regional Geology: the Grampian Highlands*, 4th edn, HMSO, London, for the British Geological Survey.

Stewart, F. H. (1941) On sulphatic cancrinite and analcime (eudnophite) from Loch Borralan, Assynt. *Mineralogical Magazine*, **26**, 1–8.

Stewart, F. H. (1946) The gabbroic complex of Belhelvie in Aberdeenshire. *Quarterly Journal of the Geological Society of London*, **102**, 465–98.

Stewart, F. H. and Johnson, M. R. W. (1960) The structural problem of the younger gabbros of north-east Scotland. *Transactions of the Edinburgh Geological Society*, **18**, 104–12.

Stewart, M., Strachan, R. A. and Holdsworth, R. E. (1997) Direct field evidence for sinistral displacement along the Great Glen Fault Zone: late Caledonian reactivation of a regional basement structure? *Journal of the Geological Society of London*, **154**, 135–9.

Stillman, C. J. and Francis, E. H. (1979) Caledonian volcanism in Britain and Ireland. In *The Caledonides of the British Isles – Reviewed* (eds A. L. Harris, C. H. Holland and B. E. Leake), *Geological Society of London Special Publication*, No. **8**.

Stone, P. (1995) Geology of the Rhins of Galloway. *Memoir of the British Geological Survey*, Sheets 1 and 3 (Scotland).

Stone, P. (1996) Girvan and Ballantrae: an obducted ophiolite. In *Geology in South-west Scotland: an Excursion Guide* (ed. P. Stone), British Geological Survey, Keyworth, Nottingham, pp. 69–79.

References

Stone, P. and Rushton, A. W. A. (1983) Graptolite faunas from the Ballantrae ophiolite complex and their structural implications. *Scottish Journal of Geology*, 19, 297–310.

Stone, P. and Smellie, J. L. (1988) *Classical areas of British Geology: the Ballantrae area: a description of the solid geology of parts of 1:25 000 Sheets NX 08, 18 and 19*. HMSO, London, for British Geological Survey.

Stone, P. and Smellie, J. L. (1990) The Ballantrae ophiolite, Scotland: an Ordovician island arc–marginal basin assemblage. In *Ophiolites: Oceanic Crustal Analogues* (eds J. Malpas, E. M. Moores, A. Panayiotou and C. Xenophontos), Geological Survey Department, Nicosia, Cyprus, pp. 535–546.

Stone, P., Floyd, J. D., Barnes, R. P. and Lintern, B. C. (1987) A sequential back-arc and foreland basin thrust duplex model for the Southern Uplands of Scotland. *Journal of the Geological Society of London*, 144, 753–64.

Stone, P., Green, P. M., Lintern, B. C., Simpson, P. R. and Plant, J. A. (1993) Regional geochemical variation across the Iapetus Suture zone: tectonic implications. *Scottish Journal of Geology*, 29, 113–21.

Stone, P., Kimbell, G. S. and Henney, P. J. (1997) Basement control on the location of strike-slip shear in the Southern Uplands of Scotland. *Journal of the Geological Society of London*, 154, 141–4.

Storetvedt, K. M. and Meland, A. H. (1985) Geological interpretation of palaeomagnetic results from Devonian rocks of Hoy, Orkney. *Scottish Journal of Geology*, 5, 337–52.

Storetvedt, K. M. and Petersen, N. (1972) Palaeomagnetic properties of the Middle-Upper Devonian volcanics of the Orkney Islands. *Earth and Planetary Science Letters*, 14, 269–78.

Styles, M. T. (1994) A petrological study of ultramafic rocks from the East Grampian region between Ballater and Huntly. *British Geological Survey Technical Report*, No. **WG/94/10/R**.

Summerhayes, C. P. (1966) A geochronological and strontium isotope study of the Garabal Hill–Glen Fyne igneous complex, Scotland. *Geological Magazine*, 103, 153–65.

Sutherland, D. S. (1982) Alkaline intrusions of north-western Scotland. In *Igneous Rocks of the British Isles* (ed. D. S. Sutherland), Wiley, Chichester, pp. 203–14.

Suthren, R. J. (1977) Volcanic and sedimentary facies of part of the Borrowdale Volcanic Group, Cumbria. Unpublished PhD thesis, University of Keele.

Suthren, R. J. and Davis, N. (1990) The Borrowdale Volcanic Group at Seathwaite, Borrowdale. In *Geologists' Association Guide No. 2: Geology of the Lake District* (compiled by F. Moseley, eds C. J. Lister and J. T. Greensmith) 2nd edn, The Geologists' Association, London, pp. 154–8.

Suthren, R. J. and Furnes, H. (1980) Origin of some bedded welded tuffs. *Bulletin Volcanologique*, 43, 61–71.

Swarbrick, A. (1992) Caledonian minor intrusions from the Midland Valley and Southern Uplands of Scotland: their geochemistry, isotopic characteristics and petrogenesis. Unpublished PhD thesis, University of Birmingham.

Sylvester-Bradley, P. C. and King, R. J. (1963) Evidence for abiogenic hydrocarbons. *Nature*, 198, 728–31.

Tanner, P. W. G. (1995) New evidence that the Lower Cambrian Leny Limestone at Callander, Perthshire, belongs to the Dalradian Supergroup, and a reassessment of the 'exotic' status of the Highland Border Complex. *Geological Magazine*, 132, 473–83.

Tarney, J. and Jones, C. E. (1994) Trace element geochemistry of orogenic igneous rocks and crustal growth models. *Journal of the Geological Society of London*, 151, 855–68.

Taubeneck, W. H. (1967) Notes on the Glen Coe cauldron subsidence, Argyllshire, Scotland. *Geological Society of America Bulletin*, 78, 1295–316.

Tawney, E. B. (1880) Woodwardian laboratory notes: North Wales rocks. *Geological Magazine*, 7, 207–15, 452–8.

Taylor, D. M. (1972) The geochemistry and petrology of the andesites and associated igneous rocks of an area of the Ochil Hills, near Dunning, Perthshire. Unpublished PhD thesis, University of Nottingham.

Taylor, J. H. (1934) The Mountsorrel granodiorite and associated igneous rocks. *Geological Magazine*, 71, 1–16.

Taylor, K. and Rushton, A. W. A. (1971) The pre-Westphalian geology of the Warwickshire Coalfield, with a description of three boreholes in the Merevale area. *Bulletin of the Geological Survey of Great Britain*, No. **35**.

Teale, C. T. and Spears, D. A. (1986) The miner-

References

alogy and origin of some Silurian bentonites, Welsh Borderland, U.K. *Sedimentology*, **33**, 757–65.

Teall, J. J. H. (1888) *British Petrography; with Special Reference to the Igneous Rocks*, Dulau, London.

Teall, J. J. H. (1897) Petrography of Kentallen District. *Annual report for 1896, Geological Survey of United Kingdom*, 21–3.

Teall, J. J. H. (1900) On nepheline-syenite and its associates in the North-west of Scotland. *Geological Magazine*, **7**, 385–92.

Thimmaiah, T. (1956) Mineralisation in the Caldbeck Fells area. Unpublished PhD thesis, University of London.

Thirlwall, M. F. (1979) The petrochemistry of the British Old Red Sandstone volcanic province. Unpublished PhD thesis, University of Edinburgh.

Thirlwall, M. F. (1981a) Implications for Caledonian plate tectonic models of chemical data from volcanic rocks of the British Old Red Sandstone. *Journal of the Geological Society of London*, **138**, 123–38.

Thirlwall, M. F. (1981b) Peralkaline rhyolites from the Ordovician Tweeddale Lavas, Peeblesshire, Scotland. *Geological Journal*, **16**, 41–4.

Thirlwall, M. F. (1982) Systematic variation in chemistry and Nd-Sr isotopes across a Caledonian calc-alkaline volcanic arc: implications for source materials. *Earth and Planetary Science Letters*, **58**, 27–50.

Thirlwall, M. F. (1983a) Reply to: Discussion on implications for Caledonian plate tectonic models of chemical data from volcanic rocks of the British Old Red Sandstone. *Journal of the Geological Society of London*, **140**, 315–18.

Thirlwall, M. F. (1983b) Isotope geochemistry and origin of calc-alkaline lavas from a Caledonian continental margin volcanic arc. *Journal of Volcanology and Geothermal Research*, **18**, 589–631.

Thirlwall, M. F. (1986) Lead isotope evidence for the nature of the mantle beneath Caledonian Scotland. *Earth and Planetary Science Letters*, **80**, 55–70.

Thirlwall, M. F. (1988) Geochronology of Late Caledonian magmatism in northern Britain. *Journal of the Geological Society of London*, **145**, 951–67.

Thirlwall, M. F. (1989) Movement on proposed terrane boundaries in northern Britain: constraints from Ordovician–Devonian igneous rocks. *Journal of the Geological Society of London*, **146**, 373–6.

Thirlwall, M. F. and Bluck, B. J. (1984) Sr-Nd isotope and geological evidence that the Ballantrae 'ophiolite', SW Scotland, is polygenetic. In *Ophiolites and Oceanic Lithosphere* (eds I. G. Gass, S. J. Lippard and A. W. Shelton), *Geological Society Special Publication*, No. **13**, pp. 215–30.

Thirlwall, M. F. and Burnard, P. (1990) Pb-Sr-Nd isotope and chemical study of the origin of undersaturated and oversaturated shoshonitic magmas from the Borralan pluton, Assynt, NW Scotland. *Journal of the Geological Society of London*, **147**, 259–69.

Thirlwall, M. F. and Fitton, J. G. (1983) Sm-Nd garnet age for the Ordovician Borrowdale Volcanic Group, English Lake District. *Journal of the Geological Society of London*, **140**, 511–18.

Thomas, G. E. and Thomas, T. M. (1956) The volcanic rocks of the area between Fishguard and Strumble Head, Pembrokeshire. *Quarterly Journal of the Geological Society of London*, **112**, 291–314.

Thomas, H. H. (1911) The Skomer volcanic Series (Pembrokeshire). *Quarterly Journal of the Geological Society of London*, **67**, 175–214.

Thomas, J. E., Dodson, M. H., Rex, D. C. and Ferrara, G. (1966) Caledonian magmatism in North Wales. *Nature*, **209**, 866–8.

Thompson, R. N. and Fowler, M. B. (1986) Subduction-related shoshonitic and ultrapotassic magmatism: a study of Siluro-Ordovician syenites from the Scottish Caledonides. *Contributions to Mineralogy and Petrology*, **94**, 507–22.

Thorpe, R. S., Gaskarth, J. W. and Henney, P. J. (1993a) Composite Ordovician lamprophyre (spessartite) intrusions around the Midlands Microcraton in central Britain. *Geological Magazine*, **130**, 657–63.

Thorpe, R. S., Leat, P. T., Bevins, R. E. and Hughes, D. J. (1989) Late-orogenic alkaline/subalkaline Silurian volcanism of the Skomer Volcanic Group in the Caledonides of south Wales. *Journal of the Geological Society of London*, **146**, 125–32.

Thorpe, R. S., Leat, P. T., Mann, A. C., Howells, M. F., Reedman, A. J. and Campbell, S. D. G. (1993b) Magmatic evolution of the Ordovician Snowdon Volcanic Centre, North

References

Wales (UK). *Journal of Petrology*, **34**, 711–41.

Threadgould, R., Parsons, I., and Young, B. N. (1994) The Assynt Carbonatite. *Earth Heritage*, **1**, 25–6.

Tilley, C. E. (1924) Contact-metamorphism in the Comrie area of the Perthshire Highlands. *Quarterly Journal of the Geological Society of London*, **80**, 22–71.

Tilley, C. E. (1957) Problems of alkali rock genesis. *Quarterly Journal of the Geological Society of London*, **113**, 323.

Tindle, A. G., McGarvie, D. W. and Webb, P. C. (1988) The role of hybridisation and crystal fractionation in the evolution of the Cairnsmore of Cairsphairn Intrusion, Southern Uplands of Scotland. *Journal of the Geological Society of London*, **145**, 11–21.

Tindle, A. G. and Pearce, J. A. (1981) Petrogenetic modelling of in situ fractional crystallisation in the zoned Loch Doon pluton, Scotland. *Contributions to Mineralogy and Petrology*, **78**, 196–207.

Torsvik, T. H. (1984) Palaeomagnetism of the Foyers and Strontian granites, Scotland. *Physics of the Earth and Planetary Interiors*, **36**, 163–77.

Torsvik, T. H. and Trench, A. (1991) The Ordovician history of the Iapetus Ocean in Britain: New palaeomagnetic constraints. *Journal of the Geological Society of London*, **148**, 423–5.

Torsvik, T. H., Smethurst, M. A., Meert, J. G., Van der Voo, R., McKerrow, W. S. *et al.* (1996) Continental break-up and collision in the Neoproterozoic and Palaeozoic – A tale of Baltica and Laurentia. *Earth-Science Reviews*, **40**, 229–58.

Torsvik, T. H., Smethurst, M. A., Van der Voo, R. Trench, A., Abrahamsen, N. and Halvorsen, E. (1992) Baltica: A synopsis of Vendian –Permian palaeomagnetic data and their palaeotectonic implications. *Earth-Science Reviews*, **33**, 133–52.

Treagus, J. E. (1991) Fault displacements in the Dalradian of the Central Highlands. *Scottish Journal of Geology*, **27**, 135–45.

Treagus, J. E. and Treagus, S. H. (1971) The structures of the Ardsheal peninsula, their age and regional significance. *Geological Journal*, **7**, 335–46.

Treloar, P. J., Bluck, B. J., Bowes, D. R. and Dudek, A. (1980) Hornblende-garnet metapyroxenite beneath serpentinite in the Ballantrae complex of SW Scotland and its bearing on the depth provenance of obducted oceanic lithosphere. *Transactions of the Royal Society of Edinburgh: Earth Sciences*, **71**, 201–12.

Tremlett, W. E. (1962) The geology of the Nefyn–Llanaelhaearn area of North Wales. *Liverpool and Manchester Geological Journal*, **3**, 157–76.

Tremlett, W. E. (1969) Caradocian volcanicity in the Lleyn Peninsula. In *The Pre-Cambrian and Lower Palaeozoic Rocks of Wales* (ed. A. Wood), University of Wales Press, Cardiff, pp. 357–385.

Tremlett, W. E. (1972) Some geochemical characteristics of Ordovician and Caledonian acid intrusions of Lleyn, North Wales. *Proceedings of the Yorkshire Geological Society*, **39**, 33–57.

Tremlett, W. E. (1973) Excursion 9: Balmaha. In *Excursion Guide to the Geology of the Glasgow District* (ed. B. J. Bluck), Geological Society of Glasgow, Glasgow, pp. 75–81.

Trench, A. and Haughton, P. D. W. (1990) Palaeomagnetic and geochemical evaluation of a terrane-linking ignimbrite: evidence for the relative position of the Grampian and Midland Valley terranes in late Silurian time. *Geological Magazine*, **127**, 241–57.

Trench, A. and Torsvik, T. H. (1992) The closure of the Iapetus Ocean and Tornquist Sea: new palaeomagnetic constraints. *Journal of the Geological Society of London*, **149**, 867–70.

Trewin, N. H. (1987) Crawton: Lavas and conglomerates of the Lower Old Red Sandstone. In *Excursion Guide to the Geology of the Aberdeen Area* (eds N. H. Trewin, B. C. Kneller and C. Gillen), Scottish Academic Press, Edinburgh for Geological Society of Aberdeen, pp. 259–64.

Trewin, N. H. and Rice, C. M. (1992) Stratigraphy and sedimentology of the Devonian Rhynie chert locality. *Scottish Journal of Geology*, **28**, 37–47.

Trotter, F. M., Hollingworth, S. E., Eastwood, T. and Rose, W. C. C. (1937) Gosforth District. *Memoir of the Geological Survey of Great Britain*, Sheet 37 (England and Wales).

Tucker, R. D. and McKerrow, W. S. (1995) Early Paleozoic chronology: a review in light of new U-Pb zircon ages from Newfoundland and Britain. *Canadian Journal of Earth Sciences*, **32**, 368–79.

Turnell, H. B. (1985) Palaeomagnetism and Rb-Sr ages of the Ratagain and Comrie intru-

References

sions. *Geophysical Journal of the Royal Astronomical Society*, **83**, 363–78.

Tuttle, O. F. and Bowen, N. L. (1958) Origin of granite in the light of experimental studies in the system $NaAlSi_3O_8$–$KAlSi_3O_8$–SiO_2–H_2O. *Memoir of the Geological Society of America*, **74**.

Tyler, I. M. and Ashworth, J. R. (1982) Sillimanite-potash feldspar assemblages in graphitic pelites, Strontian area, Scotland. *Contributions to Mineralogy and Petrology*, **81**, 18–29.

Tyler, I. M. and Ashworth, J. R. (1983) The metamorphic environment of the Foyers Granitic Complex. *Scottish Journal of Geology*, **19**, 271–85.

Tyrrell, G. W. (1913) A petrographical sketch of the Carrick Hills, Ayrshire. *Transactions of the Geological Society of Glasgow*, **15**, 64–83.

Tyrrell, G. W. (1928) The Geology of Arran. *Memoir of the Geological Survey, Scotland*.

Van Breemen, O., Aftalion, M. and Johnson, M. R. (1979a) Age of the Loch Borrolan complex, Assynt and late movements along the Moine Thrust Zone. *Journal of the Geological Society of London*, **16**, 489–95.

Van Breemen, O., Aftalion, M., Pankhurst, R. J. and Richardson, S. W. (1979b) Age of the Glen Dessarry syenite, Inverness-shire: diachronous Palaeozoic metamorphism across the Great Glen. *Scottish Journal of Geology*, **15**, 49–62.

Van de Kamp, P. C. (1969) The Silurian volcanic rocks of the Mendip Hills, Somerset; and the Tortworth area, Gloucestershire, England. *Geological Magazine*, **106**, 542–53.

Vance, J. A. (1965) Zoning in igneous plagioclase: patchy zoning. *Journal of Geology*, **73**, 636–51.

Vernon, R. H. (1986) K-feldspar megacrysts in Granites – phenocrysts not porphyroblasts. *Earth-Science Reviews*, **23**, 1–63.

Vistelius, A. B. (1969) O granitakh Shep (Westmorland, Angliya). (The Shap Granite, Westmorland). *Doklady Akademii Nauk SSSR*, **187**, 391–4, [In Russian–English translation].

Von Knorring, O. and Dearnley, R. (1959) A note on nordmarkite and an associated rare earth mineral from the Ben Loyal Syenite Complex, Sutherland. *Mineralogical Magazine*, **32**, 389–91.

Wadge, A. J. (1972) Sections through the Skiddaw–Borrowdale unconformity in eastern Lakeland. *Proceedings of the Yorkshire Geological Society*, **39**, 179–98.

Wadge, A. J. (1978) Classification and stratigraphical relationships of the Lower Ordovician Rocks. In *The Geology of the Lake District* (ed. F. Moseley), The Yorkshire Geological Society, Leeds, pp. 68–78.

Wadge, A. J., Gale, N. H., Beckinsale, R. D. and Rundle, C. C. (1978) A Rb-Sr isochron age for the Shap Granite. *Proceedings of the Yorkshire Geological Society*, **42**, 297–305.

Wadge, A. J., Harding, R. R. and Darbyshire, D. P. (1974) The rubidium–strontium age and field relationships of the Threlkeld Microgranite. *Proceedings of the Yorkshire Geological Society*, **40**, 211–22.

Wadsworth, W. J. (1970) The Aberdeenshire layered intrusion of north-east Scotland. *Geological Society of South Africa Special Publication*, **1**, 565–75.

Wadsworth, W. J. (1982) The basic plutons. In *Igneous Rocks of the British Isles* (ed. D. S. Sutherland), Wiley, Chichester, pp. 135–48.

Wadsworth, W. J. (1986) Silicate mineralogy of the later fractionation stages of the Insch intrusion, NE Scotland. *Mineralogical Magazine*, **50**, 583–95.

Wadsworth, W. J. (1988) Silicate mineralogy of the Middle Zone cumulates and associated gabbroic rocks from the Insch intrusion, NE Scotland. *Mineralogical Magazine*, **52**, 309–22.

Wadsworth, W. J. (1991) Silicate mineralogy of the Belhelvie cumulates, NE Scotland. *Mineralogical Magazine*, **55**, 113–19.

Walker, E. E. (1904) Notes on the garnet-bearing and associated rocks of the Borrowdale Volcanic Series. *Quarterly Journal of the Geological Society of London*, **60**, 70–104.

Walker, E. F. (1985) Arthropod ichnofauna of the Old Red Sandstone at Dunure and Montrose, Scotland. *Transactions of the Royal Society of Edinburgh: Earth Sciences*, **76**, 287–97.

Walker, F. (1927) The igneous geology of Ardsheal Hill, Argyllshire. *Transactions of the Royal Society of Edinburgh*, **55**, 147–57.

Walker, G. P. L. (1985) Origin of coarse lithic breccias near ignimbrite source vents. *Journal of Volcanology and Geothermal Research*, **25**, 157–71.

Walmsley, V. G. and Bassett, M. G. (1976) Biostratigraphy and correlation of the Coralliferous Group and Gray Sandstone Group (Silurian) of Pembrokeshire, Wales.

References

Proceedings of the Geologists' Association, **87**, 191–220.

Ward, J. C. (1875) Notes on the comparative microscopic rock structure of some ancient and modern volcanic rocks. *Quarterly Journal of the Geological Society of London*, **31**, 388–422.

Ward, J. C. (1876) The geology of the northern part of the English Lake District. *Memoir of the Geological Survey of Great Britain*, Quarter Sheet 101SE (New Series Sheet 29, England and Wales).

Ward, J. C. (1877) On the Lower Silurian Lavas of Eycott Hill, Cumberland. *Monthly Microscopical Journal (Transactions of the Royal Microscopical Society)*, **17**, 239–46.

Wark, D. A. and Watson, E. B. (1993) Plagioclase dissolution and origin of 'patchy' feldspars in igneous rocks. *Geological Society of America, Abstracts with Programs*, **25**, 259–60.

Watson, C. R. (1984) A volcanic vent at Colt Crag near Coniston, Cumbria. *Proceedings of the Cumberland Geological Society*, **4**, 237–44.

Watson, J. V. (1984) The ending of the Caledonian orogeny in Scotland. *Journal of the Geological Society of London*, **141**, 193–214.

Watts, W. W. (1947) *Geology of the ancient rocks of Charnwood Forest, Leicestershire*, Leicester Literary and Philosophical Society, Leicester.

Webb, P. C. and Brown, G. C. (1984a) The Lake District granites: heat production and related geochemistry. *Geothermal Resources Programme Report, British Geological Survey: Investigation of the geothermal potential of the UK*, No. **5**.

Webb, P. C. and Brown, G. C. (1984b) The Eastern Highlands granites: heat production and related geochemistry. *Geothermal Resources Programme Report, British Geological Survey: Investigations of the geothermal potential of the UK*, No. **6**.

Webb, P. C. and Brown, G. C. (1989) Geochemistry of pre-Mesozoic igneous rocks. In *Metallogenetic models and exploration criteria for buried carbonate-hosted ore deposits – a multidisciplinary study in eastern England* (eds J. A. Plant and D. G. Jones), The Institution of Mining and Metallurgy and the British Geological Survey, pp. 95–121.

Webb, P. C., Tindle, A. G. and Ixer, R. A. (1992) Wo-Sn-Mo-Bi-Ag mineralization associated with zinnwaldite granite from Glen Gairn, Scotland. *Transactions of the Institution of Mining and Metallurgy (Series B: Applied Earth Sciences)*, **101**, 59–72.

Wedd, C. B., King, W. B. R. and Wray, D. A. (1929) The geology of the country around Oswestry. *Memoir of the Geological Survey of Great Britain*, Sheet 137 (England and Wales).

Wedd, C. B., Smith, B. and Wills, L. J. (1927) The geology of the country around Wrexham, Part 1, Lower Palaeozoic and Lower Carboniferous rocks. *Memoir of the Geological Survey of Great Britain*, Sheet 121 (England and Wales).

Weedon, D. S. (1970) The ultrabasic/basic igneous rocks of the Huntly region. *Scottish Journal of Geology*, **6**, 26–40.

Weiss, S. and Troll, G. (1989) The Ballachulish igneous complex, Scotland: petrography, mineral chemistry and order of crystallization in the monzodiorite–quartz diorite suite and in the granite. *Journal of Petrology*, **30**, 1069–115.

Wellman, C. H. (1994) Palynology of the 'Lower Old Red Sandstone' at Glen Coe, Scotland. *Geological Magazine*, **131**, 563–6.

Wells, A. K. (1925) The geology of the Rhobell Fawr District (Merioneth). *Quarterly Journal of the Geological Society of London*, **81**, 463–538.

Westoll, N. D. S. (1968) The petrology of kentallenite. Unpublished PhD thesis, University of Edinburgh.

Westoll, T. S. (1951) The vertebrate-bearing strata of Scotland. *Report of the International Geological Congress, 18th session, Great Britian*, section **11**, 5–21.

Westoll, T. S. (1977) Northern Britain. In *A Correlation of Devonian Rocks of the British Isles* (M. R. House, J. B. Richardson, W. G. Chaloner, J. R. L. Allen, C. H. Holland and T. S. Westoll), *Special Report of the Geological Society of London*, No. **8**, pp. 66–93.

Whalen, J. B., Hegner, E., Jenner, G. A. and Longstaffe, F. J. (1992) A granite geochemical transect of the southern Canadian Appalchian Orogen (abstract). *Special Paper of the Geological Society of America*, **272**, 503.

Wheildon, J., King, G., Crook, C. N. and Thomas-Betts, A. (1984) The Lake District Granites: heat flow, heat production and model studies. *Geothermal Resources Programme Report, British Geological Survey: Inv-*

References

estigation of the geothermal potential of the UK, No.7.

White, A. J. R. and Chappell, B. W. (1988) Some supracrustal (S-type) granites of the Lachlan Fold Belt. *Transactions of the Royal Society of Edinburgh: Earth Sciences*, **79**, 169–81.

White, D. E., Barron, H. F., Barnes, R. P. and Lintern, B. C. (1992 [for 1991]) Biostratigraphy of late Llandovery (Telychian) and Wenlock turbidite sequences in the SW Southern Uplands, Scotland. *Transactions of the Royal Society of Edinburgh: Earth Sciences*, **82**, 297–322.

Whittle, G. (1936) The eastern end of the Insch igneous mass, Aberdeenshire. *Proceedings of the Liverpool Geological Society*, **17**, 64–95.

Wilde, S. A. (1995) Evidence for local anatexis at high-temperature diorite contacts of the Glen Lednock plutonic complex, Comrie, Scotland. *Petrologija*, **3**, 200–18.

Wilkinson, J. M. and Cann, J. R. (1974) Trace elements and tectonic relationships of basaltic rocks in the Ballantrae igneous complex, Ayrshire. *Geological Magazine*, **111**, 35–41.

Williams, D. (1930) The geology of the country between Nant Peris and Nant Ffrancon (Snowdonia). *Quarterly Journal of the Geological Society of London*, **96**, 191–233.

Williams H. (1922) The igneous rocks of the Capel Curig District (North Wales). *Proceedings of the Liverpool Geological Society*, **13**, 166–202.

Williams, H. (1927) The geology of Snowdon (North Wales). *Quarterly Journal of the Geological Society of London*, **87**, 346–431.

Williams, H. and McBirney, A. R. (1979) *Volcanology*, Freeman Cooper and Co., San Francisco

Williams, H. and Smyth, W. R. (1973) Metamorphic aureoles beneath ophiolite suites and Alpine peridotites: tectonic implications with west Newfoundland examples. *American Journal of Science*, **273**, 594–621.

Wilson, G. V., Edwards, W., Knox, J., Jones, J. C. B. and Stephens, J. V. (1935) The geology of the Orkneys. *Memoir of the Geological Survey, Scotland*.

Wilson, G. V., Robertson, T., Allan, J. K. and Buchan, S. (1935) Shetland. *Summary of Progress of the Geological Survey of Great Britain for 1934*, pp. 67–9.

Wilson, J. T. (1966) Did the Atlantic close and then re-open? *Nature*, **211**, 676–81.

Winchester, J. A. (1976) Different Moinian amphibolite suites in northern Ross-shire. *Scottish Journal of Geology*, **12**, 187–204.

Winchester, J. A. and Floyd, P. A. (1977) Geochemical discrimination of different magma series and their differentiation products using immobile elements. *Chemical Geology*, **20**, 325–43.

Woolley, A. R. (1970) The structural relationships of the Loch Borrolan Complex, Scotland. *Geological Journal*, **7**, 171–82.

Woolley, A. R. (1973) The pseudo-leucite Borolanites and associated rocks of the south-eastern tract of the Borralan complex, Scotland. *Bulletin of the British Museum (Natural History), Mineralogy*, **2**, 285–333.

Woolley, A. R., Symes, R. F. and Elliot, C. J. (1972) Metasomatized (fenitized) quartzites from the Borralan Complex, Scotland. *Mineralogical Magazine*, **38**, 819–36.

Wright, A. E. and Bowes, D. R. (1968) Formation of explosion breccias. *Bulletin Volcanologique*, **32**, 15–32.

Wright, A. E. and Bowes, D. R. (1979) Geochemistry of the Appinite Suite. In *The Caledonides of the British Isles – Reviewed* (eds A. L. Harris, C. H. Holland and B. E. Leake), *Geological Society of London Special Publication*, No. 8, pp. 599–704.

Wyborn, D., Owen, M., Compston, W. and McDougall, I. (1982) The Laidlaw Volcanics: a Late Silurian point on the geological timescale. *Earth and Planetary Science Letters*, **59**, 90–100.

Wyborn, D., Turner, B. S. and Chappell, B. W. (1987) The Boggy Plain Supersuite: a distinctive belt of I-type igneous rocks of potential economic significance in the Lachlan Fold Belt. *Australian Journal of Earth Sciences*, **34**, 21–43.

Wyllie, B. K. N. and Scott, A. (1913) The plutonic rocks of Garabal Hill. *Geological Magazine*, **50**, 499–508, 536–45.

Yardley, B. W. D. (1989) *An Introduction to Metamorphic Petrology*, Longman, Harlow.

Young, B. (1985) Greisens and related rocks associated with the Eskdale Granite, Cumbria. *British Geological Survey Report*, No. **PDA2/85/2**.

Young, B. (1987) *Glossary of the Minerals of the Lake District and Adjoining Area*, British Geological Survey, Newcastle upon Tyne.

Young, B., Ansari, S. M. and Firman, R. J. (1988) Field relationships, mineralogy and chemistry of the greisens and related rocks associ-

References

ated with the Eskdale Granite, Cumbria. *Proceedings of the Yorkshire Geological Society,* **47**, 109–23.

Young, B. N., Parsons, I. and Threadgould, R. (1994) Carbonatite near the Loch Borralan Intrusion, Assynt. *Journal of the Geological Society of London,* **151**, 945–54.

Young, T. P., Gibbons, W., and McCarroll, D. (in press) The geology of the country around Pwllheli. *Memoir of the British Geological Survey,* Sheet 134 (England and Wales).

Zhou, J. (1985) The timing of calc-alkaline magmatism in parts of the Alpine–Himalayan collision zone and its relevance to the interpretation of Caledonian magmatism. *Journal of the Geological Society of London,* **142**, 309–17.

Ziegler, A. M., McKerrow, W. S., Burne, R. V. and Baker, P. E. (1969) Correlation and environmental setting of the Skomer Volcanic Group, Pembrokeshire. *Proceedings of the Geologists' Association,* **80**.

Glossary

This glossary aims to provide simple explanations of all but the most elementary geological terms used in Chapter 1 and in the Introduction and Conclusions sections of site descriptions. It also includes many of the more important terms encountered in other sections of the volume. *The explanations are not intended to be comprehensive definitions, but concentrate instead on the way in which the terms are used in this volume.* Bold face indicates a further glossary entry.

Chronostratigraphical names are given in the correlation charts of Chapter 1 and the chapter introductions. For the names of minerals and non-igneous rock-types, the reader is referred to standard textbooks. The names of most common crystalline igneous rocks are better explained by means of classification diagrams (Figures G1 to G6, all simplified after Le Maitre, 1989 to include only rock names encountered in this volume). Names of igneous or igneous-related rocks that do not fit easily into these classification diagrams *are* included in the glossary, as are the names of most fragmental volcanic rocks, which require extended explanations commonly involving their mode of formation. Obsolete names and local names for distinctive rock-types are explained where they occur in the main text.

The classification and nomenclature of crystalline igneous rocks used in this volume follow the recommendations of the International Union of Geological Sciences (IUGS) Subcommission on the Systematics of Igneous Rocks (Le Maitre, 1989). Slight modifications follow the classification scheme of the British Geological Survey (BGS) (Gillespie and Styles, 1999), in which an attempt is made to distinguish 'root names' (i.e. largely those that figure on the main classification diagrams) from variants, mostly indicated by mineral qualifiers as prefixes to the root names. This is achieved through a strict use of hyphens:

- Compound root names, usually involving an *essential* mineral, are hyphenated (e.g. quartz-syenite, olivine-gabbro).
- Mineral qualifiers are hyphenated together (e.g. biotite-hypersthene andesite).
- Mineral qualifiers are *not* hyphenated to the root name, whether compound or not (e.g. biotite-hornblende trachyte, biotite quartz-trachyte, fayalite-augite nepheline-syenite).

Fragmental volcanic rocks are also classified and named according to the IUGS scheme, with minor modifications from the BGS scheme. Two points should be noted in particular:
The term 'volcaniclastic' is applied to all fragmental rocks that occur in a volcanic setting, including *both* rocks that have been fragmented by volcanic processes (i.e. pyroclastic rocks) *and* sedimentary rocks that comprise reworked fragments of volcanic rocks. The terms 'volcanogenic' and 'epiclastic', which are commonly used elsewhere in an inconsistent and confusing manner, are not used in the BGS scheme or in this volume.

The terms 'ignimbrite' and 'ash-flow tuff' are synonymous. In Britain their use tends to be regional, reflecting the preferences of recent workers. Both are used in this volume ('ash-flow tuff' in Wales and 'ignimbrite' elsewhere), so as to conform with previous literature in each area.

Glossary

A-type: refers to an igneous rock, usually a granite, with **alkaline** characteristics; an alkali granite.

Aa: **lava**, usually **basic**, typified by a spiny, clinkery scoriaceous surface.

Accretionary lapilli: concentrically layered, spherical, **lapilli**-sized volcanic **clasts** that form as moist aggregates of ash in eruption clouds.

Accretionary prism: a complex structural juxtaposition of inclined strata formed above an active **subduction** zone by the underthrusting of successively younger units of oceanic crustal rocks, which become attached to the leading edge of the overlying tectonic plate.

Acid: describes igneous rocks rich in silica (SiO_2 more than 63%).

Agglomerate: a **pyroclastic** rock with predominantly rounded **clasts** greater than 64 mm in diameter.

Albitization: replacement of a feldspar by the sodic plagioclase, albite.

Alkaline: describes igneous rocks that contain more sodium and/or potassium than is required to form feldspar and hence contain, or have the potential to contain (i.e. in the **norm**), other alkali-bearing minerals such as feldspathoids, alkali pyroxenes and alkali amphiboles.

Amygdale: a gas bubble cavity in an igneous rock that has been infilled later with minerals.

Aphyric: textural term, applied to igneous rocks that lack relatively large, conspicuous crystals (**phenocrysts**) compared with the grain size of the groundmass (or non-**porphyritic**).

Aplitic: describes relatively finer grained areas, or typically veins, within an igneous rock (contrast with **pegmatitic**).

Appinitic: describes a heterogeneous suite of coarse-grained **ultramafic**, **mafic** and **intermediate** igneous rocks, characterized by **shoshonitic** geochemical affinities and the presence of abundant hydrous minerals, particularly **euhedral** amphibole.

Ash-fall tuff: lithified **pyroclastic fall deposit** with grain size less than 2 mm in diameter.

Ash-flow tuff: equivalent to **ignimbrite**; term used typically in North America and, in this volume, in descriptions of volcanic rocks of Wales.

Assimilation: the addition of solid material such as country rock to a **magma**, changing its composition.

Asthenosphere: a weak layer within the Earth's **mantle** and immediately below the **lithosphere**.

Aureole: a zone around an igneous intrusion in which the texture, mineralogy and/or composition of the country rocks has been changed by heat and fluids from the intrusion.

Autobreccia: **breccia** caused by fragmentation of the chilled crust of **lava** or intrusion by continued flow of its fluid interior.

Basic: describes igneous rocks relatively rich in the 'bases' of early chemistry (MgO, FeO, CaO, Fe_2O_3); silica (SiO_2) is relatively low (nominally 45–52%).

Batholith: a very large discordant igneous intrusion or coalescing mass of related intrusions that extends to great depth in the Earth's crust.

Bentonite: a light coloured rock, mainly composed of clay minerals and colloidal silica, produced by devitrification and chemical alteration of glassy fine ash.

Block lava: lava, usually **intermediate** to **acid**, typified by a coarse, angular blocky surface.

Blueschist: a schistose rock containing blue sodic amphiboles, indicative of high pressure metamorphism.

Breccia: rock composed of angular broken fragments greater than 64 mm in diameter; can be **volcaniclastic**, sedimentary or fault related.

Caldera: a circular, basin-shaped depression, usually many times greater than the size of any individual volcanic vent, caused by collapse of the roof of an underlying **magma** chamber following an eruption; also refers to the underlying volcanic structure.

Calc-alkaline: describes a suite of **silica-oversaturated** igneous rocks, characterized chemically by the steady decrease in iron content relative to silica during evolution of the **magma**; typical of magmas generated during **orogenesis** at destructive plate margins.

Carbonatite: an igneous rock that contains more than 50% primary carbonate minerals.

Clast: a fragment in a rock.

Cleavage: plane of incipient parting in a rock, produced by the alignment of platy crystals such as mica in response to confining pressure during deformation.

Cognate xenolith: an inclusion in an igneous rock to which it is genetically related, for example as an earlier crystallized product of the same **magma**.

Glossary

Complex: used herein to refer to a large-scale spatially related assemblage of igneous rock units possibly, but not necessarily, with complicated igneous and/or tectonic relationships and of various ages and diverse origins.

Coulée: a thick viscous **lava** of limited length with blocky, very steep flow fronts; intermediate in shape between elongate **lava** flow and equidimensional **lava** dome.

Cumulate: an igneous rock formed by crystals that precipitated early from a **magma** and accumulated due to gravitational settling, current activity or other magmatic processes without modification by later crystallization.

Deuteric: describes reactions between primary minerals and the water-rich fluids that separate from the same body of **magma** at a late stage in its cooling history.

Diagenesis: the process of mineral growth and/or recrystallization leading to lithification of unconsolidated sediment to form rock.

Diapir: a dome-shaped body of **magma** or mobile rock that has risen through country rocks as a result of its lower density and/or greater plasticity.

Diatreme: a breccia-filled volcanic pipe formed by a gaseous explosion.

Distal: far from the source.

Dolerite: used herein as a synonym of microgabbro (see Figure G2).

Dyke: a tabular body of igneous rock, originally intruded as a vertical or steeply inclined sheet.

Dynamothermal: type of metamorphism involving directed pressure and shear stress as well as a wide range of confining pressures and temperatures.

Effusive: describes eruption as **lava** rather than as **pyroclasts**.

Enclave: an inclusion (**xenolith**) within an igneous rock, usually of some other igneous rock, which may or may not be related.

Euhedral: describes a mineral grain, such as a **phenocryst**, with well-formed crystal faces.

Eutaxitic: textural term describing elongate **fiamme** and glass shards, and produced through compaction and welding in an **ignimbrite/ash-flow tuff**; gradational to **parataxitic**.

Facies: the characteristic features of a rock unit, including rock type, mineralogy, texture and structure, which together reflect a particular sedimentary, igneous or metamorphic environment and/or process.

Felsic: describes light-coloured minerals (*fel*dspar/*fel*dspathoid and *si*lica) or an igneous rock containing abundant proportions of these minerals; the opposite of **mafic**.

Felsite: a field term for glassy and fine-grained **felsic** igneous rocks.

Fenitization: **metasomatism** by alkali-rich fluids.

Fiamme: dark, devitrified lenses in **welded tuff**, typically formed from the collapse of **pumice** during welding.

Flaser-banded: streaky layering with platy mineral aggregates surrounding lenticular bodies of granular material.

Fluidization: mobilization resulting from passage of a fluid (usually a gas) through a granular solid.

Foliation: the planar arrangement of components within a rock.

Foreland basin: a sedimentary basin developed by depression of a convergent continental margin due to the weight of sediment accumulating in front of the orogenic belt.

Fractional crystallization: process in which the early formed crystals in a **magma** are removed or otherwise prevented from equilibrating with the residual liquid, which consequently becomes progressively more evolved in composition (i.e. more fractionated).

Glomeroporphyritic: a **porphyritic** rock containing clusters of **phenocrysts**.

Graben: an elongate down-faulted crustal block, commonly with a marked topographical expression.

Granitization: the theory of the origin of granites by the chemical transformation of rock in its solid state by liquids and/or gases.

Granophyric texture: texture of an **acid** igneous rock in which quartz and alkali feldspar penetrate each other, having crystallized together.

Greenschist facies: the temperature and pressure conditions characteristic of hydrous low-grade regional metamorphism.

Greisen: a quartz-muscovite rock formed from the **hydrothermal** alteration of granite.

Hornfels: a well-baked, hard, splintery rock resulting from thermal (contact) metamorphism.

Hybridization: the intermixing of two or more **magmas**, which crystallize as a single rock, commonly having heterogeneous texture and complex mineralogy.

Glossary

Hydroclastic: describes fragmentation of **magma** or hot rock by its interaction with water; (see also **hydrovolcanic** and **phreatomagmatic**).

Hydromagmatic: processes driven by the interaction of **magma** with water.

Hydrothermal alteration: changes in mineralogy and chemistry in rocks resulting from the reaction of hot water with pre-existing minerals (cf. **metasomatism**).

Hydrovolcanic: volcanic processes driven by the interaction of **magma** with water.

Hypabyssal: describes an igneous intrusion, or its rock, emplaced at a depth intermediate between **plutonic** and volcanic.

Hypersolvus: describes granites and syenites in which a single type of alkali feldspar crystallized, rather than separate sodic and potassic feldspars.

I-type: refers to an igneous rock, usually a granite, that formed by the **partial melting** of some other igneous or meta-igneous rock, e.g. in the **mantle** or lower crust (contrast with **S-type**).

Incompatible elements: trace elements that are not readily accepted into the crystal structure of common rock-forming minerals during the crystallization of **magma** and hence are concentrated preferentially into the remaining liquid. They are also concentrated in the first liquids produced during **partial melting**.

Ignimbrite: the rock, typically silicic and pumiceous, formed by deposition from a **pyroclastic flow**; may partly or wholly comprise **welded tuff** (see also **ash-flow tuff**).

Intermediate: applied to an igneous rock that is transitional between **acid** and **basic** (i.e. SiO$_2$ between 52% and 63%).

Juvenile: applied to volcanic fragments that have been derived directly from **magma**.

Klippe: an isolated thrust-bound structural unit that is an erosional remnant of a large thrust sheet or **nappe**.

Laccolith: an igneous intrusion, roughly circular in plan and concordant with the structure of the country rock; generally has a flat floor, a shallow domed roof and a dyke-like feeder beneath its thickest point.

Lag breccia: coarse **breccia** of rock fragments, associated with **ignimbrite**; occurs typically near to the eruption site.

Lamprophyre: name used for a distinctive group of largely **hypabyssal** rocks characterized by abundant phenocrysts of **mafic** minerals, with **felsic** minerals confined to the groundmass. See Figure G5 for subdivisions.

Lapilli-tuff: pyroclastic rock predominantly comprising **clasts** with an average size of between 2 and 64 mm in diameter.

Lava: molten rock at the Earth's surface (contrast with **magma**).

Lava tube: a hollow space beneath the solidified surface of a **lava**, formed by the draining out of molten **lava** after the crust had formed.

Leucocratic: describes light-coloured igneous rocks containing few **mafic** minerals.

Lithosphere: the outer layer of the solid Earth, including the crust and upper part of the **mantle**, which forms tectonic plates above the **asthenosphere**.

Mafic: describes dark-coloured minerals, rich in *ma*gnesium and/or iron (*Fe*), **or** an igneous rock containing substantial proportions of these minerals, mainly amphibole, pyroxene or olivine; the opposite of **felsic**.

Magma: molten rock beneath the Earth's surface (contrast with **lava**).

Mantle: part of the interior of the Earth, beneath the crust and above the core.

Mass-flow: the transport, down slope under the force of gravity, of large, coherent masses of sediment, tephra or rock; commonly assisted by the incorporation of water, ice or air.

Megabreccia: a **breccia** of blocks so large that the brecciated nature of the rock may be obscured; commonly formed during collapse to form **calderas**.

Mélange: a chaotic rock unit, characterized by the lack of internal continuity of contacts between component blocks and including fragments of a wide range of composition and size.

Megacryst: any crystal (**phenocryst** or **xenocryst**) in a crystalline rock that is very much larger than the surrounding groundmass.

Melanocratic: describes dark coloured igneous rocks rich in **mafic** minerals.

Mesobreccia: breccia in which the **clasts** are visible within a single exposure; commonly used to describe tabular sheets in the upper and middle parts of **pyroclastic** deposits filling **calderas** (see also **megabreccia**).

Mesocratic: describes igneous rocks intermediate between **leucocratic** and **melanocratic** in colour.

Meta: prefix added to any rock name to indicate a metamorphosed variety e.g. metabasalt is a

Glossary

metamorphosed basalt.

Metaluminous: degree of alumina-saturation in igneous rocks in which the molecular proportion of Al_2O_3 is greater than that of Na_2O+K_2O, but less than that of Na_2O+K_2O+CaO.

Metasomatism: process involving fluids that introduce or remove chemical constituents from rock thus changing its chemical and mineralogical composition without melting.

Mid-ocean ridge: a continuous median mountain range within the oceans along which new oceanic crust is generated by volcanic activity.

Mid-ocean ridge basalt (MORB): type of **tholeiitic** basalt, generated at mid-ocean ridges. A world-wide, voluminous basalt type widely used as a fundamental standard for comparative geochemistry.

Migmatite: a partially melted rock generally consisting of light-coloured layers of igneous-looking **felsic** minerals and darker layers, richer in **mafic** minerals and having a metamorphic appearance.

Moho (=Mohorovicic Discontinuity): the boundary surface within the Earth below which there is an abrupt increase in seismic velocity; marks the base of the crust above the underlying **mantle**. Geophysical and petrological criteria define slightly different positions for the boundary.

Molasse basin: a sedimentary basin in an orogenic mountain belt within which thick sequences of coarse clastic sediments accumulate.

Nappe: a coherent body of rock, that has been moved a considerable distance away from its original location on a near-horizontal surface by thrusting or recumbent folding.

Norm: a recalculation of the chemical composition of an igneous rock to obtain a theoretical mineralogical ('normative') composition; useful for classification purposes and for comparison with experimental studies of **magma** crystallization.

Obduction: the over-riding/overthrusting of oceanic crust on to the leading edge of continental **lithosphere** during plate collision.

Olistostrome: a sedimentary deposit consisting of a chaotic mass of intimately mixed heterogeneous materials, commonly including very large blocks, and formed by submarine slumping of unconsolidated sediment.

Ophiolite: an ordered sequence of related **ultramafic** rocks, gabbros, **sheeted dykes** and basalt **lavas** that originated through the generation of oceanic crust.

Orogenesis: crustal thickening following the collision of tectonic plates and resulting from magmatism, folding, thrusting and accretion, leading to regional uplift and mountain building.

Outflow tuff: rock formed from **pyroclastic flows** that extend beyond the confines of a **caldera**.

Pahoehoe: basalt **lava** with a smooth, ropy surface.

Parataxitic: textural term, similar to **eutaxitic**, but where the **fiamme** and glass shards are extensively streaked out.

Partial melting: the incomplete melting of a rock to produce a **magma** that differs in composition from the parent rock.

Pegmatitic: textural description of an area within an igneous rock that is notably more coarsely crystalline and commonly forming veins and dykes (contrast with **aplitic**).

Peléan: a volcanic eruption characterized by gaseous ash clouds associated with the growth and collapse of volcanic domes.

Peperite: describes a **breccia** characterized by isolated blocks and lobes of igneous rock, commonly chilled and mixed with **fluidized** host sediment; typically present at the margins of high-level sills intruded into water-bearing sediment.

Peralkaline: degree of alumina-saturation in igneous rocks in which the molecular proportion of Al_2O_3 is less than that of Na_2O+K_2O.

Peraluminous: degree of alumina-saturation in igneous rocks in which the molecular proportion of Al_2O_3 is greater than that of Na_2O+K_2O.

Petrogenesis: the origin and evolution of rocks.

Petrography: the study of the mineralogy, texture and systematic classification of rocks, especially under the microscope.

Petrology: the study of the origin, occurrence, structure and history of rocks; includes **petrography** and **petrogenesis**.

Phenocryst: a crystal in an igneous rock that is larger than those of the groundmass, usually having crystallized at an earlier stage.

Phreatic: describes a volcanic eruption or explosion of steam, not involving **juvenile** material, that is caused by the expansion of ground water due to an underlying igneous

Glossary

heat source.

Phreatomagmatic: describes explosive volcanic activity caused by the contact of **magma** with large volumes of water, producing intensely fine ash and abundant steam.

Phreatoplinian: a rare type of explosive volcanic eruption and its deposits produced by **phreatomagmatic** processes (contrast with **plinian**).

-phyric: as in 'plagioclase-phyric', a **porphyritic** rock containing **phenocrysts** of plagioclase.

Pillow lava: subaqueously erupted **lava**, usually basaltic in composition, comprising an accumulation of smooth pillow shapes and **lava tubes** produced by rapid chilling.

Plinian: type of explosive volcanic eruption and its deposits; **magma** is fragmented through the release of magmatic gas and released at high velocity to form an eruption column that extends high into the Earth's atmosphere.

Pluton: an intrusion of igneous rock, emplaced at depth in the Earth's crust.

Plutonic: describes igneous rocks formed at depth in the Earth's crust.

Poikilitic: a texture of an igneous rock in which small crystals of one mineral are enclosed within a larger crystal of another mineral.

Porphyritic: textural term, for an igneous rock, in which larger crystals (**phenocrysts**) are set in a finer grained or glassy groundmass.

Porphyry: a field term for an igneous rock that contains **phenocrysts** within a fine-grained groundmass of indeterminate composition; usually preceded by a mineral qualifier indicating the type of **phenocryst** present; e.g. feldspar porphyry.

Protolith: the source rock from which an igneous rock was formed, most commonly by melting.

Proximal: near to the source.

Pumice: light-coloured **pyroclast** of generally **acid**, highly vesicular, glass foam.

Pyroclast: a fragment (**clast**) ejected from a volcano; ash, **lapilli** and block or bomb are pyroclasts that are respectively less than 2 mm, 2 to 64 mm and more than 64 mm in diameter.

Pyroclastic: describes unconsolidated deposits (**tephra**) and rocks that form directly by explosive ejection from a volcano.

Pyroclastic breccia: a rock comprising predominantly angular **pyroclasts** with an average size greater than 64 mm in diameter.

Pyroclastic fall deposit: **tephra** deposited by fall-out from a volcanic eruption cloud.

Pyroclastic flow: a volcanic avalanche; a hot density current comprising **pyroclasts** and gases, erupted as a consequence of the explosive disintegration of **magma** and/or hot rock; also describes the deposit from this eruption.

Pyroclastic surge: similar to a **pyroclastic flow** but turbulent and less dense.

Radiometric age: the age in years calculated from the decay of radioactive elements.

Restite: the material remaining after **partial melting**.

Rheomorphic: describes a very densely **welded tuff** that is characterized by folds and shears as evidence of the plastic deformation of the welding foliation by **mass flow**.

Rodingite: a rock that has suffered extensive calcium **metasomatism**; used here for veins rich in calcic pyroxene and garnet within serpentinite.

S-type: refers to an igneous rock, usually a granite, that formed by the **partial melting** of sedimentary or metasedimentary rocks (contrast with **I-type**).

Serpentinization: **hydrothermal alteration** of **ultramafic** rocks in which the **mafic** minerals are replaced by a range of hydrous secondary minerals, collectively known as serpentine.

Sheeted dykes: closely spaced **dykes** intruded parallel to each other; a major component of an **ophiolite**.

Shoshonitic: describes a suite of igneous rocks common to continental destructive plate margins with higher values of K_2O than **calc-alkaline** rocks.

Silica-saturation: a measure of the amount of silica available to form the major mineral components of an igneous rock, usually calculated from the **norm**. Silica-oversaturated rocks may contain free silica as quartz; silica-undersaturated rocks may contain feldspathoids in addition to feldspars.

Silicic: alternative term to **acid**.

Sill: a tabular body of igneous rock, originally intruded as a sub-horizontal sheet and generally concordant with the bedding or **foliation** in the country rocks.

Spherulite: spherical mass of acicular crystals, commonly feldspar, radiating from a central point; commonly found in glassy **silicic** volcanic rocks as a result of devitrification.

Glossary

Spilitization: the pervasive alteration of a basalt, commonly in a submarine environment; the dominant process is **albitization**, together with other hydrous mineralogical changes.

Stoping: the emplacement of **magma** by detaching pieces of country rock which either sink through or are **assimilated** by the magma.

Strombolian: type of volcanic eruption and its deposits characterized by continuous small explosive 'fountains' of fluid basaltic **lava** from a central crater.

Subduction: the process of one lithospheric plate descending beneath another during plate convergence.

Subsolvus: describes granites and syenites in which both sodic and potassic feldspars crystallized simultaneously.

Tephra: an unconsolidated accumulation of **pyroclasts**.

Terrane: a fault-bound body of oceanic or continental crust having a geological history that is distinct from that of contiguous bodies.

Tholeiitic: describes a suite of silica-oversaturated igneous rocks, characterized chemically by strong iron enrichment relative to magnesium during the early stages of evolution of the **magma**; formed in extensional within-plate settings, at constructive plate margins and in island arcs.

Transcurrent: a large-scale, steeply dipping fault or shear, along which the movement is predominantly horizontal.

Transpression: crustal shortening as a result of oblique compression across a **transcurrent** fault or shear zone.

Transtension: crustal extension as a result of oblique tension across a **transcurrent** fault or shear zone leading to localized rifts or basins.

Tuff: a rock comprising **pyroclasts** with average grain size less than 2 mm in diameter.

Tuff-breccia: a **pyroclastic** rock in which between 25 and 75% of the **pyroclasts** are greater than 64 mm in diameter.

Turbidite: a clastic rock formed through deposition from subaqueous sediment-laden density currents (turbidity currents) that move swiftly downslope under the influence of gravity.

Ultrabasic: describes an igneous rock with a silica content less than that of **basic** rocks (less than 45% SiO_2).

Ultramafic: describes an igneous rock in which dark-coloured minerals (amphibole, pyroxene, olivine) comprise more than 90% of the rock.

Ultrametamorphism: metamorphic processes at a temperature and pressure high enough to partially or completely fuse the affected rock and produce a rock with an igneous-looking texture.

Vesicle: a gas bubble cavity, usually in a **lava** or shallow intrusion.

Vitroclastic: describes a **pyroclastic** rock characterized by fragments of glass.

Volcaniclastic: generally applied to a clastic rock containing mainly material derived from volcanic activity, but without regard for its origin or environment of deposition (includes **pyroclastic** rocks and sedimentary rocks containing volcanic debris).

Volcanotectonic fault: fault along which the displacement occurred through subsurface movement of **magma** or during its eruption.

Welded tuff: a glass-rich **pyroclastic** rock in which the grains have been welded together because of heat and volatiles retained by the particles and the weight of the overlying material. (This is not synonymous with **ignimbrite** though many ancient ignimbrites are welded.)

Xenocryst: a crystal, like a **phenocryst**, but one that is foreign to the igneous rock in which it is found.

Xenolith: a rock fragment that is foreign to the igneous rock in which it is found.

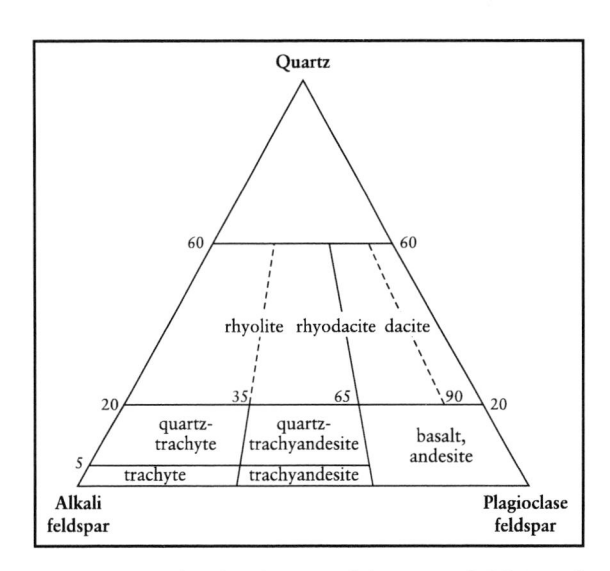

Figure G.1 The classification of fine-grained **felsic** and **mafic** crystalline igneous rocks. The distinction between basalt and andesite is based on the composition of the plagioclase feldspar present.

Glossary

Figure G.2 The classification of coarse-grained **felsic** and **mafic** crystalline igneous rocks. The distinction between gabbroic rocks and diorite is based upon the composition of the plagioclase feldspar present. Medium-grained rocks are named by attaching the prefix 'micro' e.g. microgranite.

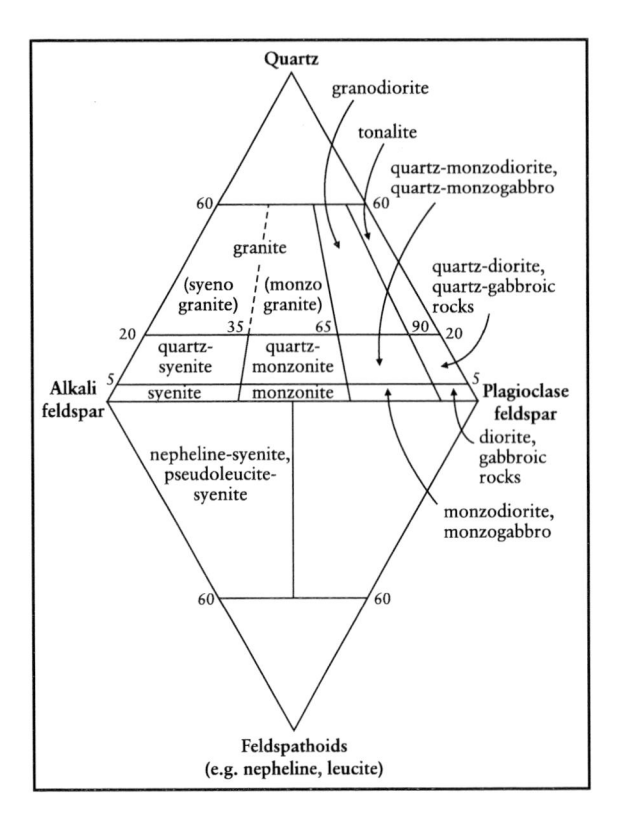

Figure G.3 The more-detailed classification of coarse-grained **mafic** crystalline igneous rocks, falling in the gabbroic rocks field of Figure G2. (a) Based upon the plagioclase, total pyroxene and olivine content, (b) based upon the plagioclase, orthopyroxene and clinopyroxene content, (c) figures (a) and (b) combine in three dimensions if necessary to form a tetrahedron.

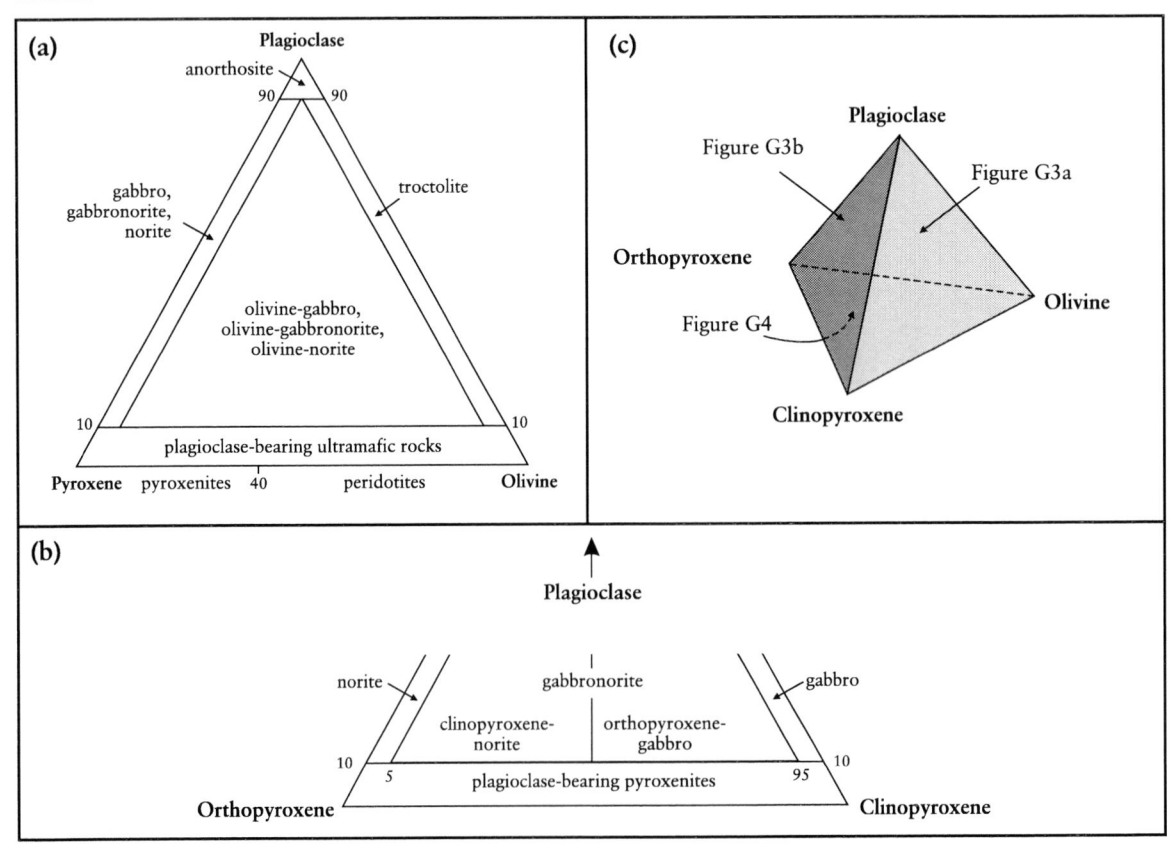

626

Glossary

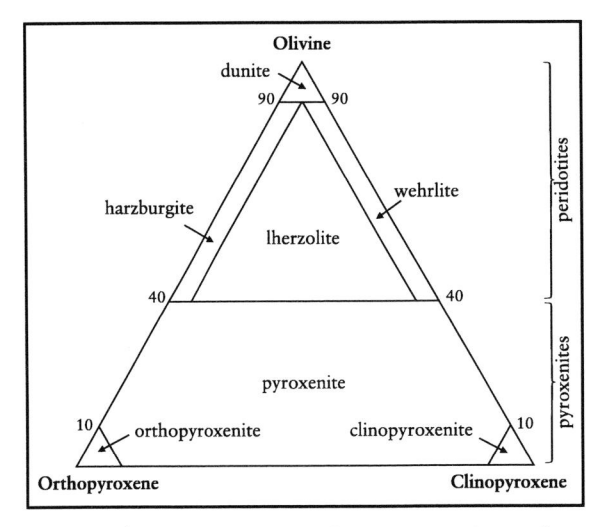

Figure G.4 The classification of coarse-grained crystalline ultramafic rocks. **Ultramafic** rocks also include 'hornblendite' for rocks with more than 90% hornblende.

Feldspar	Predominant mafic minerals	
	biotite, diopsidic augite, (±olivine)	hornblende, diopsidic augite, (±olivine)
more orthoclase than plagioclase	minette	vogesite
more plagioclase than orthoclase	kersantite	spessartite

Figure G.5 The classification of **lamprophyres** encountered in this volume.

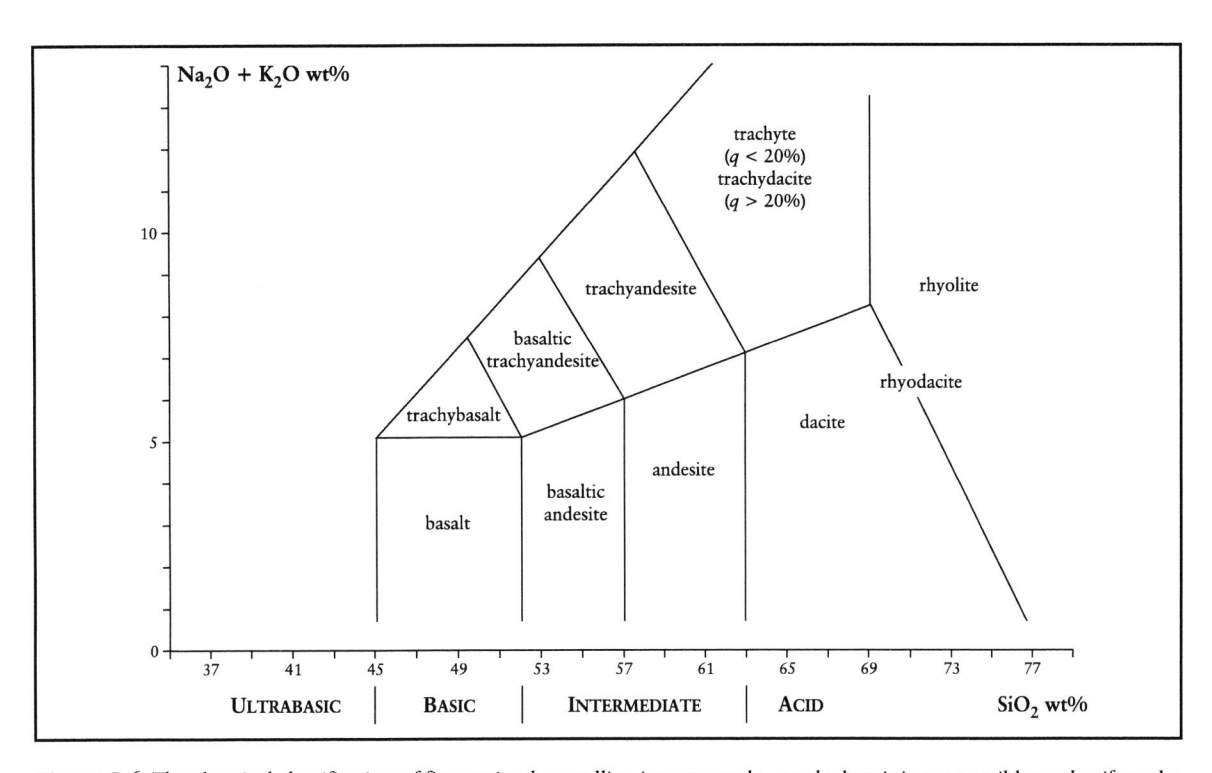

Figure G.6 The chemical classification of fine-grained crystalline igneous rocks, used when it is not possible to classify rocks according to their mineralogy due to very fine grain size. Heavily altered rocks such as are commonly encountered in the Caledonian Province can be difficult to classify chemically owing to the loss or addition of highly mobile elements such as sodium (Na) and potassium (K) and accompanying changes in silica (SiO_2).

Index

Note: Page numbers in **bold** and *italic* type refer to **tables** and *figures* respectively

Aber Mawr to Porth Lleuog GCR site **22**, *234*, 237, 252–6, *252*, *253*, *254*
Abereiddi Tuff Member, Llanrian Volcanic Formation 256–9, *258*
Aberfoyle 58, 61, 63, 64, 66
Aberfoyle Slate 445, 447
Ach'uaine Hybrids 403–9
Acmite 374, 377, 379
Airy's Bridge Formation 155, 160–1, *160*, *161*, 162, 163–5, 168, 179
Aithness Tuff 569
Alkaline intrusions (NW Highlands)
  deep subduction zone melting 350
  emplacement mechanisms 355, *356*, 361–5, 367, 373, 378–9
  fenitization (alkali metasomatism) 355, 357, 363, 372, 374
  glossary of rock names **349**
  influence on Assynt Culmination 352–3
  locality map *347*
  magma sources beneath Lewisian Foreland 350–1, 353, 366, 383, 384, 385, 392–3
Alkaline minor intrusions (NW Highlands) *352*, 353, 379–80, *387*
age relationships within suite **348**, 381
Breabag Porphyrites 387, *387*
Canisp Porphyry dykes and sills 353, 380, 382–5, *387*
grorudite dykes 350, 352–3, *352*, 367, 380–2, 386, *387*
hornblende porphyrites 385–6, 390
influence on Assynt Culmination 352–3
ledmorite dykes 354, *382*, 392, 393
locality map *347*
nordmarkite, relationship to Moine Thrust 390–1
pre-deformational 'hornblende porphyrite' sills 386
vogesite sills 387–8, *388*, 390
volcanic vent (diatreme) 389–90, *389*
Alkaline plutonic complexes *see* Loch Ailsh intrusion; Loch Borralan intrusion; Loch Loyal Syenite Complex
Allt Fawr Rhyolitic Tuff Formation 242, 328
Allt Felin Fawr Member, Carn Llundain Formation 237, 254–5, 256
Allt Lwyd Formation 270, 272–3
Allt na Cailliche GCR site **23**, *352*, 391
Allt nan Uamh GCR site **23**, *352*, *382*, 388–9, *388*
Alston Block *137*, 144
Altnaharra Formation 405
Amphibolites 78–9, 81
An Fharaid Mhor **24**, *382*, 392–3
Andesite 273, 275, 276, 432, 483, 486–7, 489, *490*, 496–7, 550
  augite-hornblende andesite 431
  basaltic andesite 483
  biotite andesite 491
  Glencoe 500, 501–2, 503–4, 506, 508, 509–10, 512, 513
    basaltic andesite 501
    hornblende andesite 503, 508
  Lake District 138–42, 145, 146–52, 155, 159, 161–4, 172, 174, 177, 179–83, *182*
    basaltic andesite 139, 146–9, 155, 176, 179–80
  Northern Midland Valley 487, 525–7, 534, 536–9, 541
    basaltic andesite 487, 522, 523–5, 527, 528, 529–31, 534
    pyroxene andesite 487
  Shetland 488, 559–60, *560*, 561–3, 563, 565

Index

Southern Midland Valley 488, 543–6, *544*, *548*, *549*, 550–2, *551*
   augite andesite 543, 547, 550
   enstatite andesite 543, 546, 547, 550
   pyroxene andesite *549*, 550
   pyroxene-olivine andesite 543
Southern Uplands 488
   basaltic andesite 553–4
Wales 275–6, 327, 329–30
Aplites 196–7, 206, 212, 225–7, *226*, 261, 263, 327, 457–8, 466, 541
Appin Group 440, 468
Appin Phyllite and Limestone Formation 473
Appinite Suite intrusions 397, 403–9, 468–77, *470*, 497
   Ballachulish pluton 469, 472
   breccia pipes 404, 469, 470–2, *470*, *471*
   Duror of Appin cluster 468, *470*, 472, *471*
   dyke swarms 469
   Garabal Hill–Glen Fyne complex 449, 451
Aran Volcanic Group 13, 237, 264, 269–73
Arbuthnott Group 487, 525, 528, 530, 531, 533–4
Ardsheal Hill and peninsula GCR site **25**, *398*, 403, 468–72, *470*, *471*
Armorica 11, 17
Arran 65–9
Ash-flow tuffs/ignimbrites 5
   Ben Nevis 486, 496
   breccias 172
   Glencoe Volcano 486, 500–1, 503–5, 507–8, 510, 512–15
   Highland Border 486–7
   Lake District 142, 153, 156, 159–63
      andesitic 159
      dacitic 148
      emplacement and environment 152–3, 159–60, 161, 162–3, 166, 168–71, 172–6, 179–80, 186–7
      rhyodacitic *175*
      silicic 138–9, 142, 155–64, *160*, 168–9, 172, 176, 178–80, 185–7
   rhyolitic 248, 316, 559, 560–1, 563, 565
   Shetland 488, 559–60, 563, 565
   subaqueous emplacement 259, 264, 269, 276–7, 283–4, 287
   Wales 5, 233, 236–41, 242–3
      Berwyn Hills 322–5, 323–5, *324*
      Builth Inlier 273, 274–9
      Cadair Idris 264–9, 270–2
      emplacement and environments 252, 255, 268, 272, 281, 283, *284*, 285, 286–7, *286*, 290, 312–13, 318
      Pembrokeshire 247–8, 249, 253–6, 259, 339–44
      Snowdonia 279–96, *279*, *295*, 297–312, 315, 316, 316–19
Askrigg Block *137*, 145
Assapol Group 418, 423
Aureoles
   Central England 219–21, 223, 225, 227–30
   dynamothermal 35, 57, 64, 78–81, *79*
   metamorphic 9
      Carrock Fell Complex 208
      Comrie pluton 401, 446–8
      Duror of Appin cluster 469–70
      Etive pluton 425, 430
      Kentallen 472, 475, 477
      Ochil Hills diorites 541–2
      Ross of Mull pluton 418–19, 421
      Shap granite 214
      Skiddaw pluton 144
      Strontian pluton 413–14
      Tan y Grisiau granite 319–20, *319*
Autobrecciated volcanic rocks
   felsite 178
   Glen Coe 507
   lavas 66, 151, 164, 166, 180, 181, 245, 312, 532, 541, 552–3
   pillow lavas 91, 93
   rhyolites 255, 256, 294, 312, 507
   silicic magmas 251
   Southern Uplands 552–3, 553, 555
   Wales 237, 245, 248, 251, 279

Bad Step Tuff 156, 163, 165, 166, 168, 170–1, 186
Bail Hill Complex 36
Balcreuchan Group 34–5, 73, 74–5, 79, 88, 91–6, 96–100
Balcreuchan Port to Port Vad GCR site **20**, 34, *34*, 35, 91–6, *92*, *93*, *94*, *95*
Ballachulish Slide 520
Ballantrae Ophiolite Complex 11, 13, 34–5, *34*, 69–100, *93*, *94*, *95*
   Byne Hill composite intrusion 35, 69–72, *70*
   chert–conglomerate–volcaniclastic rock assemblage 97, 98, *99*, 100
   locality map *34*
   Millenderdale gabbro dyke assemblage 35, 81–4, *83*
   see also Pinbain volcano-sedimentary block
Balmaha and Arrochymore Point GCR site **20**, 33, 34, 61–5, *62*, *63*
Balmedie Quarry GCR site **21**, *108*, 127–30, *128*, *129*
Balmerino to Wormit GCR site **25**, *482*, 487, 531–4, *532*, *533*
Baltica 9, 10–11, 15, 29, 107, 397
Barnt Green Volcanic Formation 219, 221
Basalt 96–7, *98*
   Glencoe 500–1, 503, 506, 512, 553
   Grampian Highlands 486, 487, 489–92
   Lake District 145, 157–9
   Midland Valley 483, 488, 531, 534, 536–9, 543
      olivine basalt 487, 522–6, 528, 547

Index

Shetland and Orkney 559–60, 565, 566, 568–9, 570–3
   olivine basalt 488
Southern Uplands 488, 553, 556
   biotite-olivine basalt 558
Wales 244–5, 249, 266–7, 273–9, 292–3, 296, 300–3, 306–7, 312, 343
Beckfoot Quarry GCR site **21**, *138*, 194–5, *195*
Bedded Pyroclastic Formation *240*, 241, 316
   Cwm Idwal 307–8, 310–12, 313
   Moel Hebog to Moel yr Ogof 291, 292–6, *295*
   Snowdon Massif 300–3, *304*, *305*, 306–7
Beinn Garbh GCR site **23**, *382*, 383–4, *383*
Belhelvie intrusion (NE Grampian highlands) *108*, 110
   compositional differences from Huntly–Knock intrusion 118–19
   cumulate zonation 111, *111*, 114
   shear zones 127–30, *128*
Ben Ledi Grit 445
Ben More Nappe 350, 351–2, *352*, 353, 355, 364, 373, 381–2
   relationship to grorudite dykes 350, 353, 367, 381–2
Ben More Thrust 351–2, 355, 357, 359, 362–6, 367, 381, 386, 387
Ben Nevis and Allt a'Mhuilinn GCR sites **24**, **25**, *398*, 399, 405, *482*, 486, 492–7, *492*, *493*, **494**
Ben Nevis granitic intrusion 399, 405, 492–7, *492*, *493*
   Dalradian 'basement' **494**, 496
   downfaulted volcanic rocks 493, *493*, 495, *495*, **495**, 496
   dyke-swarm *492*, 494
   'flinty crush-rock' *492*, 494

Inner and Outer granites 492–3, *492*, *493*, 494–6, **494**, 496
   Intrusive Ring Tuff 496
   magmatic evolution 497
   ring fracture 492, 496–7
Ben Nevis Intrusive Ring Tuff 496
Ben Stumanadh intrusion (Loch Loyal) *347*, 374, 378–9
Benan Conglomerate 70
Bennane Lea GCR site **20**, 34, *34*, 96–100, *98*, *99*
Bidean nam Bian GCR site **25**, *482*, 486, 505–10, *506*, *507*, *508*
Bin Quarry GCR site **20**, *108*, 115–19, *116*, *117*
Biostratigraphy
   Lake District 148, 186
   Scotland 481, *484*
      Ballantrae Ophiolite Complex 9, 35, 71, 73, 75, 77, 78, 91–3, 97, 100
      Grampian Highlands 489, 510–11, *511*, 513
      Highland Border Complex 58, 60, 61, 63, 65–6, 68–9
      Midland Valley 533–4, 545–6
      Orkney 574
      Shetland 564, 565
      Southern Uplands 100, 552, 556–7
   Wales 267, 270, 271–2
      Berwyn 323
      Builth Inlier 274
      Llŷn (Lleyn Peninsula) 288, 337
      Pembrokeshire 237, 256
      Skomer Volcanic Group 343
      Snowdonia 239, 280, 285, 297, 308, 313
Birker Fell Formation 148, 149, 152–3, 153, 155, 164
Black Rock to East Comb GCR site **25**, *482*, 487, 529–31, *529*, *530*
Blair Atholl Subgroup 440
Boganclogh intrusion *see* Insch–Boganclogh intrusion
Bonawe to Cadderlie Burn

GCR site **24**, *398*, 400, 405, 424–9, *425*, *426*, *427*, *430*, *432*
Bonawe Succesion (Dalradian) 425, *427*
Boninite lavas 35, 91, 94, 95
'Bonney's Dyke' (pegmatitic gabbro) 35, 72–4, *74*, 75–7, 78, 89
Borolanite (pyroxene-melanite nepheline-syenite) **349**, 354, 355, 358–9, *361*, 362–3, 364, 365
Borrowdale Volcanic Group 13, 137–8, 140–3, 145–6, *152*, 167, 174–6, 195
   basal breccia 149, 152, 153
   Birker Fell Formation 148, 149, 152–3, 153, 155, 164
   Haweswater mafic intrusions 205–7
   minor intrusions 145–6
   plateau-andesites 142, 149–51, 152–3, 159
   relations to Threlkeld microgranite 143, 187–90, *187*, *188*
   volcanotectonic structures 204
   xenoliths in Shap granite 212, 213
Bowness Knott GCR site **21**, *138*, 190–3, *191*, *192*
Braich tu du GCR site **22**, *234*, 239–40, *278*, 277–81, *279*
Braich tu du Volcanic Formation 239, 278–9, *279*, 280
Bramcrag Quarry GCR site **21**, *138*, 187–90, *187*, *188*
Breabag Porphyrites 387, *387*
Brigs Tuff 566, *567*, 569
Bryn Brith Member, Cregennen Formation 271
Buachaille Etive Beag GCR site **25**, 405, *482*, 486, 513–15, *515*
Buddon Hill GCR site **21**, 220, 224–7, *224*, *226*
Builth Inlier 13, 237, 273–9, *273*, *275*
   stratigraphy and lithology **274**
Builth Volcanic Group *275*

631

Index

Buttermere Formation 191
Byne Hill GCR site **20**, *34*, 35, 69–72, *70*, *73*
Byne Hill composite intrusion 35, 69–72, *70*
  origin of gabbro 70–2
  trondhjemite (leucotonalite) 69–72

Cadair Idris **22**, *234*, 237, 264–9, *265*, **266**, *266*
Cader Rhwydog Tuff 237, 253–4, *253*, *254*, 255
Cairnsmore of Fleet pluton *see* Fleet pluton
Calc-alkaline affinities
  extrusive rocks 140, 142, 207, 219, 236–7, 247, 485, 496, 531, 543, 552, 556, 570–2
  intrusions 219, 223, 224, 226–7, 397
  magmas 7, 11–12, 13, 15–16, 18, 19
  transitional to tholeiitic 15, 17, 236, 268, 481, 485, 563–5, 569
  see also Subduction
Caldera development 5, 142
  Ben Nevis 492–7
  collapse cycle 138, 153
  Crafnant Centre 241
  Craig y Garn 288, 290
  Falcon Crag area 149
  graben controlled subsidence
    Buachaille Etive Beag 514, 515
    Glencoe Volcano 498, 501–4, 505, 509–10, *510*, 513–17, 519, *519*
    Stob Mhic Mhartuin 515–17, 519, *519*
  Llwyd Mawr Centre 240
  Snowdon Centre 240–1
  Snowdonia 233, 239, 241, 290, 296, 300, 306
  unconformity on pre-caldera land surface (Glencoe) 497, 500, 501, 503, 509, 510–12, 520
  volcanotectonic faults 142, 154, 157, *157*, 159, 165, 166, 169–70, 171, *173*, 174
  see also Scafell Caldera
Caledonian Orogeny
  Acadian Event 11, 17, 140, 210, 211, 212, 214–15, 321, 397
  continental movements 3, 9–17, *14*, 19, 29–30, 35–6, 107, 137, 140, 219, 233, 397, 402, 456, 458, 483, 485
  igneous rocks, overview 3–19, *4*
  origin of Late Caledonian magmas 17–19, 71–2, 233, 237, 263, 438, 444, 452
  stratigraphical distribution of magmatic events *12*
  tectonic history 11–17
  tectonic settings and evolution 9–17
Caledonian–Appalachian Orogen 29, *29*
Cam Crags Member, Seathwaite Fell Formation 166
Cam Loch Klippe 352, *352*, 353, 357, *360*, 362–3
Cam Spout Tuff 155, 162, 163
Camas Eilean Ghlais **23**, *382*, 392
Cambrian Pipe Rock 386, 389–90
Canisp Porphyry (porphyritic quartz-microsyenite) 349, 353, 380, 382–5, *382*, *383*, *385*, 391
Capel Curig GCR site **22**, *234*, 240, 283–7, *286*, *287*
Capel Curig Anticline 285–7, *287*
Capel Curig Volcanic Formation 240, 280, 281–3, *282*, 285–7, *286*, *287*, 308, 313
Carbonatite 359–61, 364–5
Carghidown Formation 556–9, *560*
Carn Llundain Formation 237, 252–5
Carneddau and Llanelwedd GCR site **22**, *234*, 237, 273–7, *273*, **274**, *275*
Carneddol Rhyolitic Tuff Formation 242, 328, 329, 330
  links with Mynydd Tir-y-cwmwd Microgranite 330, 331, 332
Carreg yr Imbill Intrusion 337
Carrock Fell GCR site **21**, *138*, 198–204, *199*, *200*, *202*, *203*
Carrock Fell Complex 13–15, 144, 145, 198–204, *200*, *202*, 207, 208
  Carrock division 144, 198, 201–4, 208
  cogenetic relations with Eycott Volcanic Group 144, 198, 202–4
  emplacement orientations 203–4
  layered mafic rocks 144, 198, 199–201, 202, *203*, 204
  mineral lamination 199, 201
  Mosedale division 144, 198, 199–204, *203*, 208
Carrock Mine 207, 209
Castell Coch to Trwyncastell GCR site **22**, *234*, 237, *257*, 256–8, *258*, *259*
Cautley Volcanic Member 143
Cefn-hir Member, Cregennen Formation *270*, 271, 272
Chromite 30
  Highland Border Complex 63
  Shetland Ophiolite 37, 39–41, 48, 53
Clatteringshaws Dam Quarry GCR site **24**, *398*, 402, 456–9, *457*
Clinopyroxenite 85, 86
  garnet clinopyroxenite 35, 85–7
  olivine-clinopyroxenite 42
  wehrlite–clinopyroxenite 33, 41–2, 44–8, 50
Clousta Basalt 566–7, *566*, *567*, 568, 569, *569*
Clousta Conglomerate 566–7, *567*, 568
Clousta Tuff 569
Cnoc an Droighinn GCR site **23**, *352*, 386
Cnoc an Leathaid Bruidhe GCR site **23**, *382*, 384–5
Cnoc Mor to Rubh'Ardalanish

Index

GCR site **24**, *398*, 400, 405, 417–21, *417*, *419*, *420*
Cnoc nan Cuilean intrusion (Loch Loyal) *347*, 374, 377–8, 379
Columnar jointing
 ash-flow tuffs 162, 165, 176, 179, 180, 254, 279, *279*, 315, 323
 basalts and andesites 490–2, *490*, 523, *524*, 529
 composite dyke 159
 rhyolites 249–50
Comendite *see* Grorudite
Comrie pluton 9, 401–2, 405, 444–8, *445*, *446*, *447*
 geothermometric determinations 448
 metamorphic aureole 401, 446–8
 multiple intrusion pulses 444–5, 446
 normal zonation 444, 446
Coniston GCR site **21**, *138*, 176–80, *176*, *177*
Coniston Copper Mines 176
Conwy Rhyolite Formation 239
Cordierite norite 130–2
Cowie Formation (Stonehaven Group) 58
Craig Cau Formation 267, 268, 269
Craig Hall GCR site **21**, *108*, *109*, 132–3, *133*
Craig More GCR site **24**, *398*, 401, 446–8, *447*
Craig Rossie GCR site **25**, *482*, 487, 537–9, *537*, *538*
Craig y Garn GCR site **22**, *234*, 241, 287–90, *289*
Crawton Bay GCR site **25**, *482*, 487, 522–5, *523*, *524*
Crawton Volcanic Formation, Crawton Group 487, 522, 523–5, *524*
Creag na h-Innse Ruadhe GCR site **23**, *352*, 381–2
Cregennen Formation 265, 268, 271–2
Criffel pluton 402–3, 405, 460–8, *462*, *463*, *464*, *465*, *467*
 concentric zonation 460, 461, 465

deformation and foliation 466–8
Crinkle Tuffs 155–6, 162, 165, 166, 168–9
Croft Hill GCR site **21**, 221–3, *221*, *222*
Croft pluton (tonalite) 221–3, *221*, *222*, 224
 alteration and molybdenite mineralization 221, 223
 compositional zoning 221, 223
 magmatic development 107
 multiple intrusive phases 222–3
Cromaltite (melanite-biotite pyroxenite) 349, 356, 357, 364–6
Cruachan Reservoir GCR site **24**, *398*, 400, 405, 429–34, *430*, *432*, *482*
Crystal fractionation 6, 233, 242, 243, 483, 485
 Appinite Suite 472
 Ben Nevis 497
 Cadair Idris 268
 Capel Curig Volcanic Formation 285
 Carrock Fell Complex 144, 198, 203, 204
 Comrie pluton 444, 446
 Crawton Volcanic Formation 524
 Eshaness, Shetland 563, 565
 Eskdale pluton 195
 Fleet pluton 459–60
 Garabal Hill–Loch Fyne pluton 402, 448, 452
 Garnfor Multiple Intrusion 327
 Griff Hollow composite sill 230
 Insch intrusion 114, 121, 124
 Loch Borralan 358
 Loch Doon pluton 453, 455
 Moelypenmaen andesites 329
 Montrose Volcanic Formation 530
 Mynydd Penarfynydd Layered Intrusion 336
 Ochil Volcanic Formation 539

pre-emplacement 350, 351, 354, 356, 364, 366, 374, 379
 Ratagain pluton 412
 Rhobell Volcanic Complex 247
 St David's Head Intrusion 237, 263–4
 Skiddaw granite 210
 Skomer volcanic sequence 338, 341–3
 'Younger Basics' gabbro 110, 119, 121–4
Culzean Harbour GCR site **26**, *482*, 488, 546–8, *547*, *548*
Cumulates, layered 36–7, 144, 410, 472
 blocks in lavas 236, 243, 245, 246–7, *246*
 Grampian Highlands (Northeast) 110–11, 113–19, *117*, 121–2, 124, 127
 Lake District 199–204, 205–6, 208, 263
 Mynydd Penarfynydd intrusion 242, 335–8, *337*
 sedimentological features 115, *117*, 118–19, 121
 'stratigraphical' zoning 111–12, 118–19, 122–4, 127–8
 tectonic folding 114, 115, 119, 127
 textures 364
 ultramafic 110–12, 114–24, *117*, *120*, 128
Curig Hill GCR site **22**, *234*, 241, 242, 313–16, *314*, *315*
Cwm Clwyd Tuff Formation 322, 323, 324–5
Cwm Eigiau Formation
 Braich tu du 280
 Capel Curig 285
 Curig Hill 313–14
 Cwm Idwal 308–9, 312
 Llyn Dulyn 281, 283
 Moel Hebog to Moel yr Ogof 290–1
 Snowdon Massif 300–2, 306
 Yr Arddu 297
Cwm Idwal GCR site **22**, *234*, 241, 307–13, *307*, *308*, *309*, *311*

Index

Dacite 140, 142, 250–1, 273, 274, 275–7, 486–7, 488, 554
Dalradian 58–9, 61, 65–7, 68–9, 107, 112, 130–2, 424, 434, 447, 448
  Aberfoyle Slate 445, 447
  Ben Nevis 'basement' 493–4, 496
  Bonawe Succession 425, *427*
  Glencoe metasedimentary rocks 520, **521**
  Leven Schist 501, 505–6, *507*, 509, 520–1
  palaeo-landsurface 500, 501, 503, 505–6, 509–11, 569
  relations with Highland Border Complex 33–4
Daly–Shand hypothesis 7, 348, 354, 357, 365, 392
Deer Park GCR site **26**, *234*, 243, 342–4, *343*
Deerness Volcanic Member, Eday Flagstone Formation 488, 571–2, *572*
Dent Group 143, 178, 212
Diorite 144, 192, 193, 422
  Central England 219–21, 223, 225, 227–30
  ferrodiorite 144, 201, 203–4
  hornblende diorite 145, 220, 227–30
  microdiorite 206, 211, 213, 246
  Scotland 69, 72, 132, 401–2, 409–10, 419, 430, 434–6, 440–55, 469, 487, 539–42
    appinitic diorite 422, 449–50, 472
    appinitic meladiorite 413, 415, 428, 469
    ferrodiorite 125
    ferromonzodiorite 122–3
    hornblende diorite 353, 385–7, 469
    hornblende microdiorite 353, 385–7
    meladiorite 175, 410, *411*, 412, *423*, 434, 469, 473
    microdiorite 206, 405–6, 416, 422, 424, 426–8, 432–3, 469, 473–4, 516, *517*
    pyroxene meladiorite 401, 469
    pyroxene-biotite diorite 473
    pyroxene-mica diorite 412
    quartz-diorite 69, 416, 423, 430, 434–5, 469
Dol-cyn-afon Formation 319, 320
Dolerite 246, 272–3, 276, 293, 299, 326, 336, 340
  Cadair Idris 264, 267–8, 269, 272, 273
  Lake District 144, 145, 192, 205–6
  quartz-dolerite 261, 263, 539, *542*
  Snowdonia 280, 297, 299, 315, 321
Dounans Limestone, Aberfoyle 58, 61, 63
Downan Point Lava Formation 36, 100–3, *101*, *102*
Duddon Basin (Ulpha Syncline) 139, 142, 143, 204
Dundee Formation 531, 532–3, 534
Dungeon Ghyll Member, Seathwaite Fell Formation 166, 168, 170
Dunite/metadunite 33, 36–44, 46–8, 50–1, 53, 88, 90–1, *126*
Durness Group limestones 384, 386, 388
  Loch Ailsh 366, 367, *367*, 372
  Loch Borralan 354, 355, 356, 357, 358, 360, 362, 363, 365
Dyffryn Mymbyr Tuff 285, 286
Dykes 276, 293, 418, 466, 471, 473–5, 541, 543–6
  beerbachites 82, *83*, 84
  Ben Nevis Dyke Swarm *492*, 494–5
  Canisp Porphyry *382*, 384
  Carrock Fell Complex 144
  Comrie pluton 447
  Criffel pluton 466
  Crinkle Crags 157–9
  dyke swarms 16
    Appinite Suite 405–6, 468–9
    Ben Nevis *492*, 494–5
    Etive Dyke Swarm 405, 424–5, 428–9, 431–3, 499, 500, 521
    Ochil Hills diorites 539, 541, *541*
    Rhobell Fawr 246–7
    Shap granite 212, 214
  gabbro dykes (Millenderdale) 35, 81–4, *83*
  grorudite 350, 352–3, *352*, 367, 380–2, 386, *387*
  Kentallan intrusion 473–4
  Lake District 145–6, 157–9
  ledmorite 354, *382*, 392–3
  Millenderdale 81–4, *83*
  Mountsorrel complex 225, 227
  quasi-sheeted dykes 35, 37, 45–9, 50–1, 57, 81, 82
  spessartite 220
  see also Lamprophyre dykes; Sills
Dynamothermal metamorphism
  Ballantrae Ophiolite Complex 35, 78–81, *79*
  Shetland Ophiolite 54, 56, 57, 58, 64

Eastern Avalonia 10–11, 13, 15–16, 17, 18, 29, 36, 140, 219, 232, 397
Eclogite–blueschist assemblage (Knockormal area) 84–7
Eday Flagstone Formation 570–2
Eglwys Rhobell Formation 244, 245
Eilean Dubh Formation, Durness Group 389–90
Emplacement mechanisms
  Appinite Suite 471
  Bonney's Dyke 76
  continuing granite debate 6
  Criffel pluton 461–2, 466–8
  Etive pluton 426, 428, 429, 433, 434
  Glen Doll diorite 434, 437, *437*
  Highland Border Complex 68
  Kentallen intrusion 475–7
  Loch Ailsh intrusion 367, 373

Index

Loch Borralan intrusion 355, 356, 361–5
Loch Loyal Syenite 378–9
Ochil Hills diorites 540–2
Ross of Mull pluton 418, 420–1, 423–4
Strontian pluton 413, 416–17
Tan y Grisiau granite 319, 321
Enclaves *see* Xenoliths
Ennerdale intrusion *137*, 190–4, *191*, *192*
  associated basic rocks 192–3
  Crummock Water aureole 144
  minor intrusions 145, 192–3
  relations to Skiddaw Group 191–2, 193
Epidote schists 78–9, 81
Eshaness Coast GCR site **26**, *482*, 488, 559–65, *560*, *562*, *563*, *565*
Eskdale pluton 137, 143, 159, 190, 194–8, *195*, *196*, *197*
  xenoliths 196–7
Ethie lavas 487, 528, 529–31, *530*
Etive pluton 405, *425*, *426*, *427*, 429–34, *430*, *432*
  Beinn a'Bhuridh Screen 429, 431–2, *433*
  Cruachan facies 424, 425–9, *427*, 431, 433, *498*, 500
  dyke swarm 405, 424, 428–9, 431–3, 499, 521
  Lorn Plateau extrusive rocks 424, 433
  Meall Odhar facies 424, 427, 428–9, 431, 433
  Quarry Intrusion 429, 430–1, 433–4
  reverse zoning 429
  Starav facies 424, 426, 427, 428, 429
Etive Rhyolites 501, 503, 505, 507–8, 509, 512, 513–14, 519
Eycott Hill GCR site **21**, *138*, 145–9, *147*
Eycott Volcanic Group 13, 137–8, 139–40, 144, 146–9, *147*, 206–7
  'Eycott-type' basaltic andesite 146, *146*–9, *147*
  relations to Skiddaw Group 146, *146*, 148, 149, 207–8
  relationship with Carrock Fell Complex (Mosedale division) 198, 201–4

Falcon Crag GCR site **21**, *138*, 149–53, *150*, *151*, *152*
Faults
  Bennane Lea Fault 96, 99–100
  Bring Fault 572, 573
  Carrock End Fault 199
  Dove Cove Fault 100
  Drygill Fault 198
  Dungeon Ghyll Fault 170
  Efeilnewydd Fault 326, 337
  Garabal Fault 449
  Glen Doll Fault 434
  Glen Fyne Fault 401, 449
  Grave Gill Fault 170
  Great Glen Fault 397, 399, 413, 417, 475, 477, 505
  Gualann Fault 61
  Highland Boundary Fault 33, 58, 59–61, *59*, 61, 65, 405, 447, 487
  Isaac Gill Fault 157, *157*
  Langdale Fault 167, 168, 170–1
  Loch Loyal Fault 374, 377, 378, 379
  Loch Tay Fault 401, 440, 441, 443–4
  Melby Fault 559, 563, 565
  Menai Straits Fault-zone 334
  Moniaive shear-zone 16, 18, 402, 456
  Newmead Fault 276, 277
  Ochil Fault 487, 534, *536*, 537, 539, 540, *542*
  Orlock Bridge Fault 456
  Ramsay Fault 252
  Rhobell Fracture 244, 246, 247
  Roughton Gill Fault 144, 198
  St Abb's Head Fault 553, *556*
  Stinchar Valley Fault 100
  Sulma Water Fault 565
  volcanotectonic faults (Scafell Caldera) 142, 153, 157, *157*, 159, 165–6, 169–70, 171, *173*, 174
  Walls Boundary Fault 17, 403, 565
Felsites 178, 180, 183–6, *185*, 187, 193, 198
Ffestiniog Granite Quarry GCR site **22**, *234*, 241, 319–21, *319*, *320*
Fishguard Volcanic Group 236, 247–52, *249*, *250*
Fleet pluton 402–3, 456–60, *457*
  geophysics 456, 458
  Lake District affinities 456, 458, 460
Fluidization *see* Magma–wet sediment interactions
Foel Ddu Rhyodacite Formation 328, 329, 330
Foel Fras Volcanic Complex 239, 280
Foel Grach Basalt Formation 239, 279, 280
Foel Gron GCR site **23**, *234*, 242, 332–3
Foel Gron Granophyric Microgranite 332–3, *333*
Foliation
  Criffel pluton 466–8
  epidote schists and amphibolites 79
  Etive pluton 426
  Fleet pluton 457
  gabbro 35, 81–2, *83*, 84, 198
  Griff Hollow sill 228
  harzburgite 40
  Millenderdale 81–4, *83*
  Rogart pluton 406, 409
  Strontian pluton 415–16
Forest Lodge GCR site **24**, *398*, 401, 438–44, *439*, *442*, *443*
Fucoid Beds 372, 384, 386
Funtullich GCR site **24**, *398*, 402, 444–6, *445*, *446*

Gabbro 67, 70, 220, 335–6, *337*, 338, 449
  biotite quartz-gabbro 199, 201–2
  epidote-hornblende gabbro 45, 47, 48, 49
  ferrogabbro 110, 122, 201, 203, 208
  flasergabbro 48, 81, 128–9
  hornblende gabbro (xeno-

Index

liths) 70, 422
hypersthene gabbro 119,
  121–2, 124, 131, 133,
  199–202
microgabbro 144, 201–2,
  208
olivine-gabbro 70, 111, *117*,
  124, 261–2, 410, 412
pegmatitic gabbro 72–4, 76,
  76, 78, 118, 121
beerbachites 82, *83*, 84
foliated 35, 81–2, *83*, 84,
  129, 198
Grampian Highlands
  110–12, 115, *117*, 118–22,
  124, 128–31, 133
granular gabbro 112,
  119–22
syenogabbro 124
two-pyroxene gabbro
  111–12
Lake District 144, 205, 208
laminated 262–3
mylonite zones 127–8, 129,
  129, 130
olivine ferrogabbro *123*
Pembrokeshire 261–3, *261*,
  262
shear deformation 127–30
Gabbro–diorite complexes
  110–12, 114–19, 118–22,
  127–30
  Ballantrae Complex 35,
    69–72, 72–4, *74*, 76, 78,
    81–4
  Carrock Fell 198–204, *203*,
    208–9
  Carrock Fell *see* Carrock Fell
    Complex
  Millenderdale 35, 81–4, *83*
  Mynydd Panarfynydd
    Layered Intrusion 334–8,
    335, *336*, *337*
  St David's Head 259–64,
    261, *262*
  Shetland Ophiolite 33, 41,
    43–9, *43*, 46–54, *50*, *53*,
    69–72, 81–4
  see also Ballantrae Ophiolite
    Complex; Cumulates, lay-
    ered; Gabbro; Grampian
    Highlands
Gala Group 456
Games Loup GCR site **20**, 34,

34, 35, 87–91, *88*
Garabal Hill–Glen Fyne igneous complex 405, 448–52,
  450, *451*
  age sequence 450
  magma modified by contam-
    ination 452
  origin of granitic rocks 452
Garabal Hill to Lochan Strath
  Dubh uisge GCR site **24**, *398*,
  401–2, 405, 448–52, *450*, *451*
Garnet pyroxenite 78–81
Garnets, magmatic 113, 159
  Eskdale Pluton 196–8
  Fleet pluton 402, 457–9, *460*
  lava phenocrysts 140, 151
  microdiorite (Haweswater)
    206
  pyroclastic rocks 142, 156,
    160
  rodingite 52–3
  syenites 358, 370, 372
  Threlkeld microgranite 188,
    189
Garnfor Multiple Intrusion
  242, 325–7, *326*, 329–30
  comagmatic enclaves 326–7,
    327
  links with Moelypenmaen
    lavas 325
Garron Point to Slug Head
  GCR site **20**, 33, 58–61, *59*,
  60
Garth Tuff 281, 285–6
Garvock Group 487, 528, 530,
  531
GCR site locality maps *32*, *108*,
  109, *138*, *220*, *234*, *347*, *352*,
  382, *398*, *482*
GCR site selection process 7–9
  criteria 8, **19–26**
Glaramara Tuff 178, 181
Glaslyn Vent Complex 303, *304*
Glaucophane-crossite schist
  84–7
Glen Banvie 'series' 440
Glen Coe Ignimbrites 501,
  505, 507, 509, 512, 513–14
Glen Dessary syenite pluton
  13, 347, *347*, 351
  Moine envelope contamina-
    tion 350–1
Glen Doll diorite 434–8, *435*,
  437

appinitic minor intrusions
  434–5
magma contamination 431,
  436–8
xenolith melting and assimilation 434, 435, 436–8, *437*
Glen More GCR site **24**, *398*,
  400, 404, 409–12, *410*, *411*
Glen Oykel North GCR site **23**,
  352, 389–90
Glen Oykel South GCR site **23**,
  352, 381
Glen Tilt igneous complex
  438–44, *439*, *442*, *443*
  granitization 444
  intrusive contacts 440,
    441–2
  Neptunist and Vulcanist the-
    ories 438
  origins of granite 438, 444
Glencoe Volcano *425*, 486,
  497–522, *498*, *499*, **502**, *506*,
  508, *511*
  Basal Sill Complex 501, 503,
    505, *506*, *507*, 509, 512,
    520, 521, *522*
  cauldron subsidence 481,
    486, 497–501, *499*, 510,
    515–17, 519
  comparison with Ben Nevis
    504–5
  Etive Dyke Swarm 499, 521
  evolutionary phases 503–4
  'flinty crush-rock' 500, 504,
    516, 517–19, *517*, *518*,
    520, 521
  ring fracture and intrusion
    486, 497, 499–501, 504–5,
    506, 515–22, *517*, *518*,
    519, *522*
  volcanostratigraphy 501–3,
    502, 505–9, 510–13
Glencoul Thrust 351, *352*, 386
Glenfinnan Group 413
Gondwana 10, 11, 137, 233
Goodwick Volcanic Formation
  248, 249–50
Grainsgill GCR site **25**, *138*,
  207–10, *208*
Grampian Event 13, 107
Grampian Group 440
Grampian Highlands (North-
  east) 107–33, *108*
  Haddo House–Arnage intru-

Index

sion 130–2, *131*
Kennethmont intrusion *109*, 113, 132–3, *133*
locality map *108*
Maud intrusion 110, 112
mineralogical zoning 111–14, *111*, 118, 119, 122–4
Morven–Cabrach intrusion 110, 111, 112
'Newer Granites' and 'Older Granites' 112–13
'Older Basic' rocks 107–8, 112
'Younger Basic' intrusions 107–12
see also Belhelvie intrusion; Insch–Boganclogh intrusion
Grampian Highlands (Southwest) 486, 489–522
see also Ben Nevis; Glencoe Volcano; 'Younger Basic' intrusions
Granite 132–3, 319, 406–7, 492–7
ferromicrogranite 201, 204, 208
microgranite 201–4, 219, 225–6, 264, 268–9, 272–3, 330–4, 445
Lake District 137, 143–5, 159, 190–2, 194–8, 211–15, *213*
biotite granite 144, 207, 208–10
microgranite 143–4, 187–90, 192, 194–5
muscovite granite 143
Scotland 403–4, 436–46, 453–60
biotite granite 402, 418–20, 422–3, 450, 456–7, 459, 461–5
biotite-muscovite granite 402, 456, 459, 461–5, *465*
microgranite 416, 427, 428, 432–3, 445–6
muscovite granite 402, 459, *464*
muscovite-biotite granite 113, 461–5
Granitic intrusions 13, 16, 17,

18, *137*, *141*, 397–477, *398*
Ben Nevis 399, 405, 492–7, *492*, *493*
Cadair Idris Microgranite 264, 267–9
Comrie pluton 401–2, 405, 444–8
Cregennen Microgranite *270*, *271*, 272–3
Criffel pluton 402–3, 460–8, *462*, *463*, *464*, *465*, *467*
Croft pluton 221–3, *221*, *222*, *224*
emplacement mechanisms see Emplacement mechanisms
Ennerdale intrusion 144, 145, 190–4, *191*, *192*
Eskdale pluton 143, 159, 190, 194–8, *195*, *196*, *197*
Etive pluton 405, 424–34, *425*, *426*, *427*, *430*, *432*
Fleet pluton 402–3, 456–60, *457*
Foel Gron Granophyric Microgranite 242, 332–3, *333*
Garabal Hill–Glen Fyne complex 405, 448–52, *450*, *451*
Garnfor Multiple Intrusion 242, 325–7, *326*, 329–30
Glen Tilt complex (Forest Lodge) 438–44, *439*, *442*, *443*
Griff Hollow sill 227–30, *228*, *229*
I-types and S-types 107, 113, 192, 397, 401, 403
Argyll and Northern Highlands 399, 417
Southern Uplands 403, 455, 456, 458, 460, 463–5, 466
intrusive suites (Scotland)
Appinite Suite 397, 403–9, 468–77, *470*
Argyll and North Highlands Suite 397, 399–400, 405–34, 492
Cairngorm Suite 397, 400–1
Galloway Suite 397, 402–3, 456–68
Shetland Suite 403

South of Scotland Suite 397, 401–3, 434–455
Kennethmont complex *109*, 113, 132–3, *133*
Lake District batholith 137, 143–4, 190, 194–5
Loch Doon pluton 402, 452–5, *454*
Mountsorrel complex 219, 220, 224–7, *224*, *226*
Mynydd Mawr 241, 319, 321
Mynydd Tir-y-Cwmwd Microgranite 242, 330–1, *331*
Mynytho Common Riebeckite Microgranite 332
Nanhoron Granophyric Microgranite 242, 332–4
'Newer Granites' 397, 405–8, 412
orthotectonic and paratectonic zones 397–9, 404–5
Penrhyn Bodeilas Granodiorite 242, 327–8, *327*, *328*
petrological zoning 6
Comrie pluton 444
Criffel pluton 460–5
Croft pluton 221, 223
Etive pluton 429
Loch Doon pluton 453, 455, *454*
plutonic suites 397–406, 434–52, 452–77
Ratagain pluton 399, 400, 404, 404–5, 409–12, *410*, *411*
Rogart igneous complex 400, *401*, 404, 405–9, *407*, *408*
Ross of Mull pluton 400, 405, 417–424, *417*, *419*, *420*, *422*
Sandsting Complex 403, 565, 569
Shap granite 17, 140, 144, 211–15, *212*, *213*
Skiddaw granite 17, 140, 144, *199*, 204, 207–10, *208*
Strontian pluton 400, 412–17, *413*, *414*, *415*, *420*, 421, 424
Tan y Grisiau granite 241,

637

Index

319–21, *319*, *320*
tectonic settings 13, 16, 17, 18
Threlkeld microgranite 143, 187–90, *187*, *188*
Weardale granite 144–5
Wensleydale intrusion 145
zonation 452–6, 460–5, *462*
Granodiorite 143, 195, 196–8, 219, 220, 224–7, *226*, 327, 440
  biotite granodiorite 196, 197
  microgranodiorite 145, 196, 325–7, 420
  Scotland 401, 403, 449–52, 454–5, 462–3, 466
    biotite granodiorite 399, 413, *415*, 416, 418, 426, 461
    clinopyroxene-biotite-hornblende granodiorite 461–3, *464*, 465–8
    hornblende-biotite granodiorite 399, 406, 413–16, *423*, 431, 461, 469
    microgranodiorite 406, 416, 418, *420*, 424, 428, 432–3, 447, 469, 473–4
Griff Hollow GCR site **21**, 227–30, *228*, *229*
Griff Hollow composite diorite sill 227–30, *228*, *229*
  contact with country rock 227–8
  igneous layering 227, 230
  lamprophyre sheets and sills 227–8, 229
Grorudite (peralkaline rhyolite; comendite) **349**
  in Cam Loch Klippe 352, *352*
  cutting Loch Ailsh syenite 353, 367, 381
  dykes in Ben More thrust sheet 350, 380–2, *387*
  in Loch Borralan intrusive 381
  minor intrusives 380–2

Haddo House–Arnage intrusion 110, 112, 130–2, *131*
Ham Ness GCR site **19**, *32*, 33, 48–51, *49*

Hanging Stone Tuff 155, 162
Harestones Rhyolite 144, 186, 198, 204
Harrison Stickle Member, Seathwaite Fell Formation 168, 170–1
Harzburgite–metaharzburgite
  Ballantrae Complex 35, 69, 71, 72, 74, 89–91
  Shetland Ophiolite 31, 33, 36–41, *39*, 51, 53, *53*, 54
Hawaiite–mugearite 36, 233, 340–3, 488
Haweswater GCR site **21**, *138*, 205–7, *205*
Haweswater intrusions 205–7, *205*
  layered dolerite and gabbro 205–6
  relationship with Borrowdale Volcanic Group 207
Hawick Group 556
Highland Border Complex 11, 13, 33–4, 58–69, *59*, *60*, *62*, *66*, *67*
Highland Border Grits 64
Hill of Barra GCR site **20**, *108*, *109*, 113–15, *113*
Hill of Creagdearg GCR site **21**, *108*, *109*, 124–7, *125*
Hill of Johnston GCR site **20**, *108*, *109*, 122–4, *123*
Holehouse Gill Formation 140
Hollorin Tuff 568
Hornblende granofels 56–7
Hornblende porphyrite (hornblende microdiorite; spessartite) **349**, 353, 380, 385–7, *387*
Hornblendite 145, 366, 368, 371, 449, 469
Hoy Sandstone Formation 571, 572, *573*
Hoy Volcanic Member 488, 571–2, 573–4
Huntly–Knock intrusion 110, 112, 115–19, *116*, *117*, 130
Hutton, James 438, 441, 444
Hybridization 6, 71, 193, 403–10, 412, 415, 417–21, *420*, 422–4, 441, 539, 541–2
Hydromagmatic eruptions 563, 559, 565

Iapetus Ocean 7, 10–13, *14*, 17–18, 29–30, 35–6, 107, 137, 140, 252, 397, 458, 460, 559
  subduction of oceanic crust 107, 233, 236, 243, 247, 483, 485
Iapetus Suture 11, 15, 16–17, *17*, 19, 397, 456, 458, 460, 485, 488, 556
Idwal Syncline 277, 280, *307*, 308, *309*
Igneous geochemistry 233
  Ballantrae Ophiolite Complex 77, 90, 91, 92
  bimodal basic–silicic activity 233, 236–7, 242, 247, 259, 268
  discrimination of extrusive rocks 94, *95*
  granitic plutons (general) 397, 399, 401–3
  MORB characteristics 233, 268
  silica-undersaturated rocks 353–4, 356–65, 391–2
  silicic rocks, origins 233, 237, 263
  subduction-zone influence 230, 233, 237, 243, 247, 251, 342, 350, 483–5
  see also Calc-alkaline affinities; Granitic intrusions; individual igneous bodies; Subduction; Tholeiitic affinities
Ignimbrites see Ash-flow tuffs
Insch–Boganclogh intrusion *109*, 110–15, *110*, *111*, 118, 119–27, *120*, *125*, *126*, 130
  cumulate zonation 122, 124
  layered cumulates 114–15, 119–21, 122
  magnetic anomalies 114
  Red Rock Hills 122–4, *123*
  tectonically disturbed layering 114, 115, 119, 121
  ultramafic rocks 114–15, 124–5
Iona Group 418
Isotope signatures
  Comrie pluton 446
  Criffel pluton 461, 463, 465, 468

Index

Eskdale pluton 195
Etive pluton 428–9
Fleet pluton 456, 458
Garabal Hill–Glen Fyne igneous complex 449, 452
Haddo House–Arnage intrusion 132
Loch Doon pluton 453, 455
plutonic suites (Scotland) 399, 401, 402
Shap granite 214

Kennethmont granite–diorite complex *109*, 113, 132–3, *133*
Kentallen GCR site **25**, *398*, 404, 472–7, *471*, *474*, *476*, *477*
Kentallen intrusion 468, 472–7, *472*, *474*, *476*, *477*
  in aureole of Ballachulish pluton 472, 475, 477
  faulting episodes 474–5
  pipe-shaped intrusions 473
  sequence of time relationships 475–6, 477
Kentallenite (olivine monzonite) 404, 472–6
Kersantite 405, 418, 428, 558
Kingshouse Fan, Glencoe 512, **512**
Kirkley Bank Formation 184, 186
Knocklaugh GCR site **20**, *34*, 35, 78–81, *79*
Knockormal GCR site **20**, *34*, 35, 84–7, *85*
Knockvologan to Eilean a'Chalain GCR site **24**, *398*, 400, 405, 421–4, *422*, *423*

Lake District and northern England 137–215
  igneous stratigraphy *141*
  locality map *138*
  see also Carrock Fell Complex; Ennerdale intrusion; Eskdale pluton; Haweswater intrusions; Scafell Caldera; Shap granite; Side Pike complex; Skiddaw granite; Threlkeld microgranite

Lamprophyre 47, 48, 50, 220, 227–9, 403–7, 488, 556–9
  dykes, sheets and sills 353, 387–90, *387*, *389*, 403–4
  Griff Hollow composite sill 220, 227–8, 229
  Lake District 145
  Shetland Ophiolite 47, 48, 50
  Southern Uplands 488, 552–3, 556–9, *559*
  kersantite 405, 418, 428, 558
  minette 145, 404, 405, 553
  spessartite 145, 220, 227, 403–5, 404–5, 405, 418, 428, 432
  vogesite 353, 386, 387–90, *387*, *388*, *389*, 403–4
Lang Geo Sandstone Member 572, *573*
Langdale Pikes GCR site **21**, *138*, 167–71, *167*, *169*, *170*
Laurentia 9, 10–11, 13, 15–18, 29, 107, 137, 140, 233, 397
Lavas
  alkaline (Orkney) 573, 574
  andesitic see Andesite
  basaltic
    Ballantrae Ophiolite Complex 96–100
    Grampian Highlands 107
    Wales 233, 236, 237, 239–41, 243–5, 249, 264–8, 275–9, 293, 312
  boninite 35, 91, 92, 94, 96
  calc-alkaline
    Ben Nevis 496, 500, 505
    Lake District 140, 142, 153
    Wales 233, 236–7, 247, 274, 277
  cumulate blocks within 236, 243, 245, 246–7, *246*
  dacitic (Wales) 250–1, 274–6
  flow-breccias 101, 103, 151
  geochemical discrimination 94, *95*
  hawaiites and mugearites 36, 233, 340–3, 488
  Highland Border Complex 65–9
  hyaloclastites
    Ayrshire coast 75, 77, 550
    Wales 239, 240, 241, 247, 249, 267, 274–7, 293, 302–3
  Lorn Plateau within Etive pluton 424, 433
    South Kerrera 486, 489, 491–2
  MORB characteristics 40, 68, 233, 237, 251, 268
  oceanic island arc origin 13, 17–18, 35, 68, 72, 77–8, 88, 90–1, 94, 95
  pahoehoe and aa 146, 149, 155, 160, 277, 491, 492
  pargasite-bearing basalts 244–5, *245*, 247
  142, 149–51, 152–3, 159
  rhyodacite 250–1, 537–9, *539*
  rhyolitic see Rhyolite
  silicic (Wales) 242, 251, 252
  Southern Uplands (Ordovician) 13, 15–16, 17, 35–6, 100–3
  subaerial 77, 140, 149, 152, 246, 342, 491–2, 523–4, 528, 531, 534, 544, 553, 556
  submarine basalts 264
  submarine dacite lava (Wales) 250–1
  tholeiitic 13, 15, 481, 485, 524, 563–5, 569
    Ballantrae Ophiolite Complex 88–91, 92, 94, 95
    Grampian Highlands (North-east) 107
    Highland Border Complex 58–9, 67
    Lake District 140, 206–7
    Southern Uplands 101–2
    Wales 233, 236, 237, 268, 274, 277, 336
  within-plate oceanic island origin 16, 18
    Ballantrae Complex 35, 71, 77–8, 91–6, 99–100
    Highland Border Complex 67
    Southern Uplands 36, 103
Layered cumulates see Cumulates
Lea Larks GCR site **24**, *398*,

639

Index

402, 459–60, *457*
Ledmorite (melanite nepheline-microsyenite) **349**, 353, 354, 355, 356–8, *356*, 361–4, *382*, *387*, 391–3
Leucotonalite/trondhjemite 35, 69–72, 494
Leven Schist (Dalradian) 501, 505–6, *507*, 509, 520–1, *522*
Lewisian Complex 418
Lime Craig Quarry, Aberfoyle 63, 64
Lincomb Tarns Formation *176*, 177, 178–9
  columnar jointing 176, 179, 180
Lingcove Formation 155, 164
Lingmell Formation 163, 165, 168, 170, 179
Lingmoor Tuff 172, 174
Lintrathen Porphyry (dacitic ignimbrite) 486–7, 524
Llanbedrog GCR site **23**, *234*, 242, 330–1, *331*
Llanbedrog Volcanic Group 242, 326, 328–30, 330, 332–3, 336
Llanrian Volcanic Formation 256–9, *257*, *259*
Llanvirn Tarn Moor Formation 140
Llewelyn Volcanic Group 239–40, 278
Llyn Dulyn GCR site **22**, *234*, 240, 281–3, *282*, *284*
Llŷn (Lleyn Peninsula) 325–38
Llŷn Syncline 325, 328–9
Llyn y Gafr Volcanic Formation 265–7, 268, 272
Loch Achtriochtan GCR site **25**, *398*, 399, *482*, 486, 519–22, *522*
Loch Ailsh intrusion GCR site **23**, *352*, 353, 366–74, *367*, *369*, *370*, *371*
  age relationships to Ben More Nappe 366, 367
  assimilation of dolomitic limestones 366, 367
  contact metamorphism 366, 367, 372, 374
  deformation related to Moine Thrust 367–8, 373
  emplacement 367, 373
  external contacts 367
  'grorudite' dykes 381
  magnetic anomalies 367, 368, 371–2, 373
  multiple phases of syenite intrusion 368–72, 373
  syenite magma reactions with sediments 372–3
  ultramafic rocks 368, 371–2
  xenoliths 369–71, *371*, 372–3
    of altered limestone 371, 372
    of metamorphic rocks 372–3
    of pyroxene syenite 369–71, *371*
Loch Airighe Bheg GCR site **24**, *398*, 400, 404, 405–9, *406*, *407*, *408*
Loch Borralan intrusion GCR site **23**, *347*, *347*, 351, *352*, 353–66, *356*, *359*, *360*, *361*
  carbonatite 359–61, 364–5
  contact between early and late suites 354, 356, 361–2, 365
  cumulate textures 364
  Early Suite 354, 356–61, 364, 365
  external contacts 355, 363
  Late Suite 354, 356, 361, 363–4, 365, 366
  Ledbeg Thrust 359
  Loyne mass 359, 363
  pseudoleucite deformation 358, 362–3, 364, 365
  Rare Earth Element plots 364
  relationship to Ben More Thust 357, 362–3, 365–6, 367, 381
  skarn rocks 364, 365
  ultramafic rocks 355, 356–7, 365
  see also Alkaline intrusions; Loch Ailsh intrusion; Loch Loyal Syenite
Loch Dee GCR site **24**, *398*, 402, 452–5, *454*
Loch Doon pluton 402, 405, 452–5, *454*
Loch Eil Group 413
Loch Lomond Clastics (Highland Border Grits) 61, 64, 68
Loch Loyal Syenite Complex GCR site **23**, *347*, *347*, 353, 374–9, *375*, *376*
  ballooning diapiric emplacement 378–9
  Ben Loyal intrusion *347*, 374–7, *375*
  Ben Stumanadh intrusion *347*, 374, 378–9
  Cnoc nan Cuilean intrusion *347*, 374, 377–8, 379
  lamination 376, 377, 378–9
  Loch Loyal Fault 374, 377, 378, 379
  metasomatized Moine xenoliths 377–8, 379
  multiple phases of intrusion 379
  peralkaline character (Ben Loyal) 377, 379
  relationships to Moine envelope 374–6, 378–9
Loch Sunart GCR site **24**, *398*, 400, 404, 412–17, *413*, *414*, *415*
Lochaber Subgroup 440
Long Top Tuffs 155, 156, 164, 166, 168
Lorn Plateau Lavas 486, 489–92
Lotus Quarries to Drungans Burn GCR site **24**, *398*, 403, 460–5, *462*, *463*, *464*, *465*
Low Water Formation 177, 178, 179
Lower Crafnant Volcanic Formation 241, 313, 315–16
Lower Crystal Tuff Member, Llanrian Volcanic Formation 256, 258
Lower Eday Sandstone Formation 572, *573*
Lower Rhyolitic Tuff Formation 240, 241
  Cwm Idwal 307–8, 310–13, *311*
  geochemical correlation with Tan y Grisiau granite 321
  Moel Hebog to Moel yr Ogof 291, *293*, 296
  Snowdon Massif 300–5, *302*, 306
  Yr Arddu 297–300, *299*

640

Index

Luban Croma GCR site **23**, *352*, 386–7, *387*

Magma
  contamination 6, 130–1, 327, 350–1, 419–20, 431, 436–8, 452, 461, 496–7, 539
  mixing 6, 75, 132, 133, 213, 327, 422–4, 433, 461, 500, 510
  multiple phases of intrusion 401–2, 405
    Ben Nevis 494–6
    Criffel pluton 462, 465
    Croft pluton 223
    Etive pluton 424–9, 433
    Loch Ailsh intrusion 368–72, 379
    Loch Borralan 364–5
    Loch Doon pluton 453
    Loch Loyal intrusion 379
    St David's Head Intrusion 260, 263, 264
Magma sources 17–19, 350–1, 411–12
  sub-Lewisian Foreland 350–1, 353, 366, 383, 384, 385, 392–3
  within-plate oceanic island origin 16, 18, 35, 36, 67, 77–8, 82, 91–6, 99, 100, 103, 341
Magma–wet sediment interactions 5–6, 149
  Ayrshire coast 543–6, 548, 550–1, 550–2
  fluidization at contacts 5, 180–3, 183, 287, 528, 530–1, 544–5
  Glencoe Volcano 500–1, 503, *508*, 509
  Grampian Highlands 489, 491, 492
  Lake District 180–4, *183*
  Midland Valley 488
  minor intrusions 161, 180–2, *182*, *183*, 250, *251*, *254*
  Ochil Hills 532, *533*
  Shetland 559–60, 563, 565, 568, 569, 570
  Wales 248, 254, *254*, 255, *284*, 285, 287, 337

see also Peperites
Maud intrusion *108*, 110, 112
Mélange deposits 71, 72, 74, 77, 78, 86–7
Metagabbro *see* Gabbro
Metasomatism 71, 193, 207, 209–10, 213, 350–1, 355, 377–9, 541–2
Middle Crafnant Volcanic Formation 241, 316–19, *317*
Midlands Microcraton 11, 15, 16, 219–20
Millenderdale GCR site **20**, *34*, 35, 81–4, *83*
  gabbro dykes 35, 81–4, *83*
Millour and Airdrie Hill GCR site **25**, *398*, 403, 465–8, *467*
Mineralization 535
  Comrie pluton 447
  Croft pluton 221, 223
  Duror of Appin cluster 468
  Fleet pluton 456
  Huntly–Knock intrusion 118
  Kentallen 474
  Lake District 144, 176, 207, 209–10, 211, 212
  Mountsorrel complex 225, 226, 227
  Ratagain pluton 409
  Snowdon Massif 300, 317
  South Leicestershire diorites 221, 223
Minette 145, 404, 405, 553
Minor intrusions 236, 242, 246, 247, 249, 250
  basic 241, 246, 247–8, 293, 296
  calc-alkaline lamprophyres 397, 403, 404–5
  dacitic 274
  dolerites 264, 267–8, 272–3, 276, 280, 293, 297, 299, 315, 321, 326, 336
  Ennerdale 145, 192–3
  granite (Belhelvie) 129
  Lake District 145
  magma–wet sediment interactions 161, 180–2, *182*, 183, *183*, 250, *251*, *254*
  Midlands suite 220–1, 227–30
  rhyolitic 255, 256, 288–90, 291, 294, 296, 297–8, 300, 305, *305*, 306

silicic 240, 241
Strontian pluton 418, 420, 421, 424
see also Alkaline minor intrusions; Dykes; Sills
Moel Hebog to Moel yr Ogof GCR site **22**, *234*, 241, 290–6, *292*, *293*, *294*, *295*
Moelwyn Volcanic Formation 321
Moelypenmaen GCR site **23**, *234*, 242, 328–30, *328*, *329*
Moelypenmaen lavas 328–30, *329*
  links with Garnfor Multiple Intrusion 325–6, 329
  magmatic development 329–30
Moho
  geophysical 31, 41, 43–4, 45, 46, 47
  petrological 31, 33, 36, 37, 39, 40, 45, 47
Moine Supergroup 351, 366, 405, 409, 413, 418, *420*, 421
  relations to Ben Loyal intrusion 374–6
Moine Thrust 347, *347*, **348**, 351, *352*, 355
  Achall Culmination *347*, 380, 387, 390
  alkaline activity related to tectonic events **348**, 351
  Assynt Culmination 351, 352–3, *352*, 366, 380
  deformation of Loch Ailsh intrusion 367–8
  nordmarkite sills localized by 390–1
Montrose Volcanic Centre 525, 527, 528, 530
Montrose Volcanic Formation 487, 525–31, *526*, *527*, *529*, *530*
  Ferryden and Usan lavas 487, 525–7, *527*, 528
Monzodiorite 422, 425–7
  hornblende-biotite monzodiorite 426, 431
  quartz-monzodiorite 406–9, *407*, *408*, 434–6, *437*
Monzogranite 400, 401, 403, 421, 425, 428, 531
Monzonite 469

olivine monzonite (kentallenite) *123*, 472–7
quartz-monzonite 405, 409, 410–12
Morar Group 405
Morven–Cabrach intrusions 110, 111, 112
Mountsorrel complex (granodiorite) 219, 220, 224–7, *224*, *226*
   alteration and molybdenite mineralization 225, 226, 227
   calc-alkaline geochemistry 224, 226–7
   contact relations with host rock 224, 225–6, 227
   magmatic sequence 225–7
Mu Ness Klippe 48–51, *50*
Mu Ness peninsula 48–51, *49*
Muckle Billerin Tuff 566, *566*, 567–8, *567*, 569, *569*
Muckle Head Tuff 566–7, *567*, 568
Mugearite 36
Mynydd Penarfynydd GCR site **23**, *234*, 242–3, 334–8, *335*, *336*, *337*
Mynydd Penarfynydd Layered Intrusion 242–3, 334–8, *335*, *336*, *337*
   links with Nod Glas Formation 337–8
Mynydd y Gader dolerite 264–6, 267–8
Mynytho Common Riebeckite Microgranite 332

Nanhoron Quarry GCR site **23**, *234*, 242, 333–4, *333*
Nanhoron Granophyric Microgranite 242, 332–4
Nant Francon Subgroup 278, 283, 288, 330, 332, 333, 335
Nant y Gledryd Member, Carneddol Rhyolitic Tuff Formation 329, 331
Neolithic stone-axe factories 167, 168
Ness of Clousta to the Brigs GCR site **26**, *482*, 488, 565–71, *566*, *567*
Nod Glas Formation 334, 337–9

Nordmarkite (quartz-syenite) **349**, 350, 353, 356, 361–2, 365, 366, 374–9, 390–1
Norite *110*, 114, *120*, 128–30, 132
   cordierite norite 112, 130–2
   quartz-biotite norite 112, 125–7, 130–2
   quartz-norite 130–1
North Creake Borehole 219
North Glen Sannox GCR site **20**, 33, 34, 65–9, *66*, *67*
Norwick Hornblendic Schist 56–7

Obduction 3, 11–13, 29, 51, 54, 58, 64, 77, 80–1
   Ballantrae Complex 71, 80, 84
   ophiolites 35, 78, 80, 84
Ochil Volcanic Formation 487, 531–9, *535*, *537*, 539, *539*, 542, *542*
Ochil–Sidlaw Anticline 525, 527
Offrwm Volcanic Formation 265, 268, 270–2
Ogof Colomenod Conglomerate Member, Carn Llundain Formation 252–3, 255
Ogof Glyma Tuff 253, 256
Olistostromes
   Bennane Lea 96
   Cadair Idris 267
   Pinbain block 72, 74–5, *75*, 77, 78
Ophiolite complexes 7, 27–103
   amphibolite 78–9, 81
   chert 97, 98, *99*, 100–3
   eclogite 84–7
   geographical distribution 29, *29*
   glaucophane schist 84
   hornblendic rocks 54–6
   idealized ophiolite sequence 30–1, *30*
   jasper 63–4
   (meta)dunite 35, 36–7, 40–5, 47–8, 50, 91
   (meta)gabbro 33, 41, 43–8, *43*, 50–4, 57, 69–72, 81–4
   (meta)harzburgite 31, 33, 35, 37, 51, 53–4, 57, 64, 69–72, 74, 86
   Moho *30*, 31
   ultramafic nomenclature *31*
   see also Ballantrae Ophiolite Complex; Highland Border Complex; Moho; Shetland Ophiolite
Orcadian Basin 17, 403, 563, 564, 565, 571, 573
   extensional tectonic regime 485–6, 564–5, 569, 571–2, 574
Orthopyroxenite 118, 377, 529
Outwoods Shale Formation 227

Paddy End Member, Lickle Formation 178, 179
Palaeomagmatism 203, 321, 570–2, 572, 574
Pandy GCR site **22**, *234*, 242, 322–5, *322*, *324*
Pandy Tuff Formation 322–3, 323, 323–5
Pared y Cefn-hir GCR site **22**, *234*, 237, 269–73, *269*, *270*, *271*
Pavey Ark Member, Seathwaite Fell Formation 166, 168, 170–1, *170*
Pelagic sediments 60–1, 101, 103
Pen Caer GCR site **22**, *234*, 236–7, 247–52, *248*, *249*, *250*, *251*
Pen-y-gader dolerite 264, 267–8
Penmaen Formation 328, 329
Penrhyn Bodeilas GCR site **22**, *234*, 242, 327–8, *327*
Penrhyn Bodeilas Granodiorite Intrusion 242, 327–8, *327*, *328*
   comagmatic enclaves 327, 327–8
Penygadair Volcanic Formation *266*, 267, 268
Peperites
   Lake District 149, 151, 161, 162, 164, *173*, 174, 181, *182*, 183, *183*
   Scotland 501, 506, *508*, 509, 532, *533*, 550–1, 561, 568, 569

Index

Wales 255, 337
see also Magma–wet sediment interactions
Peridotite 110–11, 114, 115, 124–7, 448–9, 451, *451*, 469, 473, 475
  biotite peridotite 469
Perthosite (alkali feldspar-syenites) **349**, 355–6, 361, 365, 366, *367*, 368, *370*, 372, 374
Pets Quarry GCR site **21**, *138*, 180–3, *182*
Pettico Wick to St Abb's Harbour GCR site **26**, *482*, 488, 552–6, *554*, *556*
Phreatomagmatic/phreatic events and deposits
  Glen Coe volcano 501, 503, 504, 507, 509–10, 512
  Lake District 138, 153, 154–8, 159, 160–3, 164, 166, 170–2, 174
  Shetland 488, 565, 568, 569, 570
  Whorneyside Formation 155, 157, 159, 160–4, *160*, 166
Picrite 335–6, *336*, 338
Pillow lavas
  Scotland
    Ayrshire coast 544, *544*, 546–8, *548*, *549*, 550–2
    Ballantrae Ophiolite Complex 34, 72, 74, 77, 87–93, *93*, 95, 96–7
    Grampian Highlands 489, 491–2
    Highland Border Complex 33, 58–9, 61, 66
    Midland Valley 525–6, 528, 529–30, 532, 543–6, *544*
    Southern Uplands 100–3, *102*
  Wales 236, 237, 239, 243
    Builth Inlier 275
    Cadair Idris 266–7, *266*
    Pared y Cefn-hir 269, 271, 272
    Pembrokeshire 248–50, *249*
    Skomer Island 340, 341
    Snowdonia 290, 293, 296, 310, 312, 313
Pinbain volcanosedimentary block (Slockenray coast) 72–5, *73*, *76*, 77–8, 99
  genesis of volcanic rocks 77–8
  lava mixing 75, 77
  mélange deposits 71, 72, 74, 77, 78
  olistostrome 72, 74–5, *75*, 77, 78
Pitscurry and Legatesden quarries GCR site **20**, *108*, *109*, 119–22, *119*, *120*
Pitts Head Tuff Formation *240*, 241–2
  Braich tu du 280–1
  Craig y Garn 288, 290
  Cwm Idwal 308, 309, 312
  Moel Hebog to Moel y Ogof 291–2, *293*, 294, *294*, 296
  Snowdon Massif 302, 306
Point of Ayre GCR site **26**, *482*, 488, 570–2, *571*
Port Schuchan to Dunure Castle GCR site **26**, *482*, 488, 542–6, *543*, *544*
Porth Maen Melyn Volcanic Formation 248–9, *248*
Portsoy Lineament 107, 127
Pulaskite (pyroxene syenite) **349**, 357, 364, 366, 368, 370–1, *371*, 372, 373, 374
Punds to Wick of Hagdale GCR site **19**, 31, *32*, 33, 36–41, *38*
Pwll Bendro Member, Carn Llundain Formation 237, 253, 255
Pyroclastic rocks
  ash-fall tuffs 5, 138, 169, 172–3
    Lake District 142, 152, 153, 155, 161–3, 168, 170, 172–3
    Orkney 573
    Snowdonia 316, 317
    Wales 237, 242, 255, 322, 324–5
  breccias 30, 156, 165, 267, 291, 299, 310–12, 561–3, 563
    Lake District 156, 165, 168, 170–6
    Pembrokeshire 248
    Shetland 560–1, *562*, 563, 565
    Snowdonia 298–9, 303, 310, 312, 317
    Wales 267, 291, 299, 310–12
  co-ignimbrite ash-fall tuffs 169, 172–3
  co-ignimbrite lag breccias 156, 166, 167, *170*, 171, 312
  debris-flow deposits 266–7, 268, 271, 274–7, 292, 310, 316, *548*
  lapilli-tuffs
    Berwyn Hills 323
    Builth Inlier 274–6
    Lake District 140, 148, 152–3, 162, 165, 172, 178–80, 185
    Shetland 566–7
    Snowdonia 292, 297–8, 310
    Wales 253, 256–8, *258*, 292
  megabreccias 143, 171, 174, 176, 291, 296
  sediment-gravity flows 248, 249, 255
  siliceous nodules 156, 185, 340–1, *340*
    Lake District 156, 185
    Pembrokeshire 340–1, *340*
    Snowdonia 270, 280, 289, 291, *294*, 296, 299, *299*, 309
  surge deposits 142, 155, 160, 172, *175*, 242, 322, 324–5
  turbiditic tuffs 237, *254*, 255–6, 259, 266–7, 270–1, 274–6, 310
    Builth Inlier 274–6
    Lake District 140, 143, 153, 161, 164, 166, 170, 180, 181, 183
    Snowdonia 317–19
    Wales 237, *254*, 255–6, 259, 265–7, 270–1, 274–6, 310
  see also Ash-flow tuffs/ignimbrites; Bedded Pyroclastic Formation
Pyroxenite 35, 37, 40, 72, 74, 78, 449, 473, 475
  biotite pyroxenite 469

Index

garnet metapyroxenite 78–81
Loch Ailsh 366, 368, 373
  biotite pyroxenite 366, 372
  biotite-magnetite pyroxenite 368, 372
  diopside pyroxenite 371–2
Loch Borolan *352*, 355–7, 360, 362, 364
  biotite-magnetite pyroxenite (cromaltite) 356, 357, 364–6
  diopside pyroxenite 365

Quartz-albite (plagiogranite) veins 47, 48, 50
Queen's Cairn Breccias 513–14
Queen's Cairn Fan 514
Qui Ness to Pund Stacks GCR site **19**, *32*, 33, 45–8, *45*

Racks Tuff 281, 285–6
Radiometric ages 481–2, *484*
  Ballantrae Ophiolite Complex 34–5, 69–72, 77, 79, 80, 82, 87, 88, 90–2, 93
  Central England 219, 220, 221, 224, 227
  Grampian Highlands 107, 113, 129, 132, 133
  Granitic rocks of Scotland 403–4, 408, 409, 413, 418, 444, 446, 450–3, 456, 460
  Highland Border Complex 33, 487
  Lake District 140, *141*, 143, 144–5, 184, 186–7, 189–90, 193, 194–5, 198, 204, 209–11, 214
  Midland Valley 488, 489, 524, 533–4, 536, 541
  North-west Highlands 347, 354, 355, 365–6, 373, 374, 377, 379, 381
  Ochil Volcanic Formation 531, 534
  Orkney and Shetland 563–4, 574
  Shetland Ophiolite 31, 34, 48, 58
  Southern Uplands 100, 552, 553, 556, 558–9

Wales 244, 321
Ratagain pluton 400, 404, 405, 409–12, *410*, *411*
  'appinitic' inclusions 410, *411*
  hybridization 410, 412
  magma sources 411–12
  transitional calc-alkaline to tholeiitic aspects 399, 409, 411–12
Ray Crag and Crinkle Crags GCR site **21**, *138*, 142–3, 153–60, *154*, *156*, *157*, *158*
Red Craig GCR site **24**, *398*, 402, 434–8, *435*, *437*
Red Head Formation 528, 530–1
Red Rock Hills 122–4, *123*
Rest Gill Tuff 156–7, 162, 166, 168
Rheic Ocean 16, 19
Rhobell Fawr GCR site **22**, *234*, 236, 243–7, *244*, *245*, *246*
Rhobell Volcanic Complex 236, 243–7, *244*, *245*, *246*
Rhobell Volcanic Group 243–6, 272
Rhyodacite 248, 332, 487, 510, 535, 537–9
Rhyolite 204, 249, 251, 255–6, 329, 340–1, 494, 496, 532
  ash-flow tuffs 243, 248, 316, 559, 560–1, 563, 565
  autobrecciated lavas 255, 256, 294, 312, 507
  columnar jointing 249–50
  domes 244, 249, 251, 288–91, 294–6, 298–9, 305–6, 321
  Etive Rhyolites 501–3, 505, 507–8, 511–13, 519
  Glencoe 500, 501, 503–4, 505, 507–8, *507*, 509–10, *511*, 513
  Lake District 142, 145, 146, 159, 165–6, 178, 184–6
  lavas 233, 237, 239–41, 243, 248–9, 256, 278, 288–90, 303, 310, 312, 339–42
  minor intrusions 145, 157–9, 255–6, 288–91, 294, 296–8, 300, 305–6, *305*

Snowdonia 278, 288–91, 294, 296–9, 300, 303, 305–7, 310, 312, 321
  see also Grorudite
Rodingite (garnet-diopside-epidote-prehnite rock) 33, 51–4
Rogart igneous complex 400, *401*, 404, 405–9, *406*, *407*, *408*
  Ach'uaine Hybrids 403–9
  appinitic xenoliths 408, 409
  foliated quartz-monzodiorite pluton 406, 407–9, *408*
  migmatitic envelope 405, 407, 409
Rosneath Conglomerate 61
Ross of Mull pluton 400, 405, 417–24, *417*, *419*, *420*, *422*, *423*
  ghost stratigraphy 422, *423*, 424
  hybrid marginal granite, contaminated 418, 419–21, *420*, 422–3, 424
  magma mixing 422–4
  thermal aureole 418, 421
  xenoliths and enclaves 418, 420, 421, 422–3, *423*
Rosthwaite Fell GCR site **21**, *138*, 163–7, *164*, *165*
Rosthwaite Rhyolite 163, 165–6, 167

St David's Head GCR site **22**, *234*, 237, 259–64, *260*, *261*, *262*
St David's Head Intrusion 237, 259–4, *260*, *261*, *262*
  aplite veins 261, 263
  internal petrological contacts 263
  layering and lamination 261–3, *264*
  magmatic development 263–5
  xenoliths 261–2, 264
Salterella Grit 372, 384, 388, *388*
Sandness Formation 565–6
Sandstone dykes 543–5, 546, 550–2
Sarnau GCR site **22**, *234*, 241, 316–19, *317*, *318*

Index

Scafell Caldera 138, *139*, 142, 153–76, *156*
  co-ignimbrite lag breccias 156, 166, 167, *170*, 171
  lacustrine deposits (Langdale Pikes) 166–7, 168–9, 170–1
  piecemeal subsidence 138, 142, 153, *158*, 159, 163, 166–71, 176
  post caldera collapse magmatism 163, 165–6
  pre-caldera andesite sheets 155, 159, 164
  volcanotectonic faults 142, 154, *156*, 157, *157*, 159, 165–6, 169–70, 171, *173*, 174
  see also Caldera development; Whorneyside Formation
Scafell Syncline 139, *139*
Scurdie Ness to Usan Harbour GCR site **25**, *482*, 487, 525–8, *526*, *527*
Seathwaite Fell Formation 163, 166, 168, 170–1, 172, 174, 177, 178, 179–80, 181, *182*
Serpentinite
  Ballantrae Ophiolite Complex 69–74, 88–9, *89*, 90
    Northern Serpentinite Belt 34, 72, 78–9, 81, 84, 88, 90
    Southern Serpentinite Belt 34, 81, 82, 96, 97–8, 99
  Highland Border Complex 33, 59–60, 61, 62–5
  serpentinization volume problem 44
  Shetland Ophiolite 48, 51–4, 56
Sgavoch Rock GCR site **20**, *34*, 36, 100–3, *101*, *102*
Sgurr a'Choise Slide 520
Shap Fell Crags GCR site **25**, *138*, 211–15, *212*, *213*
Shap granite 17, *137*, 140, 144, 211–15, *212*
  megacrysts/phenocrysts 211–13, *213*
  mineral veins 211, 212

Shear deformation 59, 77, 107, 119, 121, 126, *128*, 403–5, 429
  Belhelvie intrusion 127–30
  Etive pluton 429
  Insch intrusion 119–22
  Moniaive shear-zone 16, 18, 402, 456
  mylonites 63, 127–30, *129*, 367
  Shetland Ophiolite 37, 46, 51–4
  Strontian pluton 413, 416–17
  Tan y Grisiau Granite 321
Sherriffmuir Road to Menstrie Burn GCR site **25**, *482*, 487, 534–7, *535*, *536*
Shetland Ophiolite 11, 30, 31–3, *32*, 36–58
  Lower Imbricate Zone 33
  Lower Nappe 31, 33, 36–51, *50*
  magnetic anomalies 43
  Middle Imbricate Zone 31, 33, 48, 51, 52, 54, 56–7
  rhythmic banding 37, *39*, 40, 43, 44, 45–6, 48
  rodingite 33, 51, 52–4
  'sheeted-dyke'-like intrusions 37, 45–9, *46*, 50–1, 57
  Upper Nappe 31–3, 48, 51–8
  see also Moho
Shiaba Group 418
Shineton Shales 220
Shonkinite (pyroxene-rich syenite) *349*, 358, 366, 368, 371, 372
Shoshonitic affinities 18, 397, 403–4, 483, 486, 489, 524
Shoulder O'Craig GCR site **26**, *482*, 488, 556–9, *558*, *559*
Side Pike GCR site **21**, *138*, 143, 171–6, *173*, *174*, *175*
Side Pike Complex 171–6, *173*
  Lingmoor Tuff 172, 174
  lithostratigraphy 172, *174*
  megabreccias 143, 171, 174, 176
  pyroclastic succession 172–5, *174*
  Side Pike ignimbrite 172–3, *175*

Sills
  andesite 177, 179, 180
  Canisp Porphyry 382–5
  dolerite 269, 270, 273, 315
  grorudite 380–2
  hornblende porphyrite 385–7, 389
  hornblende–microdiorite 353
  Lake District 139, 140, 142, 143, 149, 155, 161, 174, 176, 179, 180
  layered 334–8
  Midlands suite 220
  Mynydd Penarfynydd intrusion 335
  nordmarkite 390–1
  vogesite 388–9, *388*, 390
  Wales 255, 264, 265–70, 272–3, 276–7, 280, 288, 293, 294, 297, 300
  see also Dykes; Griff Hollow; Lamprophyre dykes; Minor intrusions
Site selection see GCR site selection
Skeo Taing to Clugan GCR site **19**, 31, *32*, 33, 41–4, *42*
Skiddaw granite 17, *137*, 140, 144, *199*, 204, 207–10, *208*
  mineralization and metasomatism 207, 208–10
  ore genesis 209–10
  relations to Carrock Fell Complex 207, 209, 210
Skiddaw Group 11, 137, 144, 145, 146, 187–9, 190–1, 199–201, 207–8, 214
Skomer Island GCR site **26**, *234*, 243, 338–42, *339*, *340*
Skomer Ignimbrite 341, 342
Skomer Volcanic Group 243, 338–44, *339*, *340*, *343*
Slockenray Coast GCR site **20**, 34, *34*, 35, 72–8, *73*, *76*
Snowdon Massif GCR site **22**, *234*, 241, 300–7, *301*, *302*, *304*, *305*
Snowdon Volcanic Group 239, 240–1, 287, 290, 300–8, 316
Snowdonia
  1st Eruptive Cycle 239, *240*, 277–8, 280–1, 283, 307, 313

645

Index

2nd Eruptive Cycle 239, *240*, 277–81, 287, 290–1, 296–7, 299–300, 307, 316, 321
  caldera development 233, 239, 241, 290, 296, 300, 306
  geological succession *240*
  volcanic vents 288, 291, 296, 303, 306, *315*, 316
Soft sediment deformation 97, 155, 159, 162, 166, 170–1, 178, 287, 303, 316
Sole Thrust 352, *352*, 353, 355, *356*, 366, 387, 390, 392
  relationship to Canisp Porphyry 383–5
Sour Milk Gill GCR site **21**, *138*, 143, 160–3, *160*, *161*
South Kerrera GCR site **25**, *482*, 486, 489–92, *489*, *490*
South Leicestershire diorites 219–20, 221, 223, 227, 230
Sövite (calcite carbonatite) **349**, 360–1
Spessartite 145, 220, 227, 403–5, *404–5*, *405*, 418, 428, 432
Spherulites 185, 193, 340–1, *340*
SSSI (Sites of Special Scientific Interest) 8
Stile End Formation 184, 186
Stockingford Shale Group 220, 227, 230
Stonesty Tuff 155
Stob Dearg and Cam Ghleann GCR site **25**, *482*, 486, 510–13, *511*
Stob Mhic Mhartuin GCR site **25**, *398*, 399, **482**, 486, 515–19, *517*, *518*
Stockdale Beck, Longsleddale GCR site **21**, *138*, 183–7, *184*, *185*
Strathmore Basin 522, 525, 528, 531
Strontian pluton 400, 412–17, *413*, *414*, *415*
  aureole 413–14
  foliation 415–16
  Glen Sanda Granodiorite facies 413, *413*, 416
  Loch Sunart Granodiorite facies 413, *413*, 414–15, 416
  mafic-rich enclaves 415–16, *415*, 417
Strumble Head Volcanic Formation 248–9
Subduction 3, 13, 35
  closure of Iapetus Suture 10–17, 107, 137
  influence on igneous geochemistry 230, 233, 237, 243, 247, 251, 342, 350, 483–5
  models 9–19, 485–6
  ophiolite generation 31
  shoshonitic magmatism 350–1, 409
Subduction related to calc-alkali magmas 13, 15–17, 18–19, 483–5
  Central England 219, 223, 224
  Grampian Highlands 492
  Shetland 564, 569
  Southern Uplands 543, 552, 556, 558–9
  Wales 247, 250
Submarine eruptions 247–52, 255–6, 259
Succoth–Brown Hill mafic complex 108, 127
Swch Gorge Tuff Formation 322, 323–5
Syenite 122–4, *123*, 355, 362, 368, 369–70, 372–9
  aegirine-melanite syenite 366
  feldspathic syenite (perthosite) 355–6, 361, 365, 366, *367*, 368, *370*, *371*, 372–4
  melanite nepheline-syenite (ledmorite) 353–5, 356–8, *356*, 361–4, *382*, *387*, 391–3
  melanite syenite 357, 361, 364, 372
  nepheline-pseudoleucite-syenite (borolanite) 353–5, 358–9, *361*, 362–5
  nepheline-syenite *352*, 355–7, 358, 360, 362, 364
  porphyritic quartz-microsyenite (Canisp Porphyry) 353, 380, 382–5, *382*
  pseudoleucite-syenite 350–1, 354, 356, 358–9, *361*, 362, 364–5
  pyroxene syenite (pulaskite; shonkinite) 357–8, 364, 366–8, 370–4, *371*, 388
  pyroxene-hornblende-melanite syenite 357–8
  quartz-syenite (nordmarkite) 110–11, 122, 124, 350, 353–4, 356, 361–2, 365–6, 374–9, *375*, 382, 390–1
  riebeckite syenite 366, 373
Syenodiorite 110–11

Tan y Grisiau granite 241, 319–21, *319*, *320*
  correlation with Lower Rhyolitic Tuff Formation 321
  geophysical anomalies 319, *319*, 321
  hornfels aureole 319–20
  stratigraphical age 321
The Laird's Pool, Lochinver GCR site **23**, *382*, 384
Tholeiitic affinities 3, 13, 15
  Ballantrae Complex 34, 91, 92
  basaltic rocks 36, 102, 198, 524
  Carrock Fell Complex 144
  extensional tectonic settings 107, 564–5
  ophiolite complexes 58–9, 68, 88–91, 92, 94, 96, 102, 107, 140
  pillow lavas 61, 91, 92, 101–2
  volcanic rocks
    Lake District 148
    Wales 233, 236, 237, 274, 277, 336
  Younger Basic rocks 110
Three Tarns Member, Seathwaite Fell Formation 166, 168
Threlkeld microgranite *137*, 143, 187–90, *187*, *188*
  gravity data 189
  relationship to Skiddaw and Borrowdale Volcanic

Index

groups 143, 187–9, *188*
Tillicoultry GCR site **25**, *482*, 487, 539–42, *540*, *541*, *542*
Tonalite 220, 222–3
    microtonalite 246, 326
Too of the Head GCR site **26**, *482*, 488, *572*, 572–4, *573*
Tornquist Sea 10, *14*, 15, 29, 219, 224, 233
Towie Wood **21**, *108*, 130–2, *131*
Townsend Tuff 243
Trachyandesite 332, 487, 534, 538
    basaltic trachyandesite 328, 329–30
Trachybasalt 332
Trachydacite 332
Treffgarne Volcanic Group 236
Tressa Ness to Colbinstoft GCR site **20**, *32*, 33, 51–4, *52*
Troctolite 111, 114, *114*–15, *115*, 118, *118*–19, *119*, 124
Trwyn-y-Gorlech to Yr Eifl quarries GCR site **22**, *234*, 242, 325–7, *326*
Trwyn yr Allt Tuff 253, 254, *254*, 256
Trygarn Formation 335, 337
Turnberry Lighthouse to Port Murray GCR site **26**, *482*, 488, 548–52, *549*, *550*, *551*
Tweeddale (Wrae) lavas 36
Ty'r Gawen Mudstone Formation 267, 268

Upper Crafnant Volcanic Formation 241, 316–19
Upper Lodge Formation 242, 328, 336
Upper Rhyolitic Tuff Formation 241, 300, 303–6, *305*, 306, 316
Upper Shiaba Psammite 423

Vaslees Formation 566
Virva GCR site **20**, 31, *32*, 54–8, *55*
Vogesite (hornblende lamprophyres) **349**, 353, 386, 387–90, *387*, *388*, *389*, 403–4
Volcanic centres
    Berwyn and Breidden Hills 237, 242, 322

Crafnant Centre 241, 313, 316–19
Dolgellau 237
Fishguard 236–7
Llanbedrog 326, 331–4
Llwyd Mawr Centre 241, 242, 280, *289*, 290, 291, 294, 296, 307, 309, 316
Llŷn (Lleyn Peninsula) 237, 242, 325–7, 330, 331
Montrose 525, 527, 528, 530
Ramsey Island 237, 252–8
St Abb's and Eyemouth 555–7
Snowdon Centre 237, 239–41, 286, 290–2, 296, 297, 299, 300–7, 313, 316
Volcanic rocks of Scotland (Late Silurian and Devonian) 481–575
    Grampian Highlands 399, 486, 489–522
    Highland Border 486–7
    locality map *482*
    Northern Midland Valley 487, 522–39
    palaeoenvironments 512–13, 524–5, 527–8, 531, 555–6, 563, 568–70
    radiometric dating 481–3
    Shetland and Orkney 485–6, 488, 559–75
    Southern Midland Valley 487–8, 542–53
    Southern Uplands 488, 552–9
    stratigraphical relations and ages *484*
    subduction model 485–6
Volcanic vents 166, *314*, *315*, 316, 403, 548, 552, 555, 556–9
    alkaline 389–90, *389*
    post-tectonic 557–60
    Snowdonia 288, 291, 296, 303, 306–7, *315*, 316
    Southern Uplands 552, 555–6, 557–9
    submarine silicic eruptions 252, 255–6, 259, 285, 287
    see also Lavas
Volcaniclastic deposits 245, 248–50, 322–3
    Ballantrae Ophiolite

Complex 74–5, 77, 90, *94*, 96–7, 98, 100
Berwyn Hills 322–4
Grampian Highlands 486, 495–6
Lake District 137–8, 161–4, 167, 171–2, 176–81, 186
    Borrowdale Volcanic Group 142, 151–3, 155
    Eycott Volcanic Group 140, 146–9
    turbiditic 180, 181, 183
    Whorneyside Formation 161–2
Midland Valley 487–8, 523, 526, 529, 531–2, 534–7, *542*
Orkney and Shetland 566–7, 572–3
reworked 138, 143, 149–51, 151, 186–7, 245, 250, 312–15, 317–19
Southern Uplands 488, 552, 553–7
Wales 245, 248–50, 267, 297, 302, 310–12, 317–19, 322, 323–5
Vord Hill Klippe 31, 33, 50, 51–5, 57
Vullinite (alkali feldspar-biotite-albite rock) **349**, 358

Waberthwaite Quarry GCR site **21**, *138*, 196–8, *196*, *197*
Wales 233–344
    geological successions *235*, *236*, *238*, *240*, *242*
    igneous rocks, locality map *234*
    volcanic episodes
        Arenig to Llandeilo 236–7
        Caradoc 237–43
        Llandovery and later Silurian 243
    in stratigraphical settings *235*, *236*, *238*, *240*, *242*
    Tremadoc 236
Wehrlite 35, 40, 41–2, 44, 45, 46–7, 48, 50, 86, 449, *451*
Wehrlite–clinopyroxenite 33, 41–2, 44–8
Welsh Basin 11, 13, 15, 16, 233, 247
Wensleydale intrusion 145

647

Index

Wet sediments *see* Magma–wet sediments
Whorneyside Formation 155, *157*, 159, 161–2, *161*, 163–4, 166
Windermere Supergroup 143, 144, *176*, 177, 204, 212
Wrae (Tweeddale) lavas 36
Wrengill Andesites 177, 179, 180

Xenoliths and enclaves
　assimilation of 379, 420–1, 426, 428–9, 431, 434, 435, 436–8, *437*, 455, 497
　Ben Nevis 494
　in carbonatite 359–60
　Carrock Fell gabbro 199–200
　Comrie pluton 444–5
　Craig Rossie 539
　Cregennen Microgranite *271*, 272
　Criffel pluton 466–8
　Croft pluton 222–3
　Eskdale Pluton 196–7
　Etive pluton 425–6, 428–9, 431
　Garnfor Multiple Intrusion 326–7
　Glen Doll diorites 434, 435, 436–8, *437*
　Haddo House–Arnage intrusion 130–1
　Kennethmont complex 132–3
　Loch Ailsh intrusion 369–73, *371*
　Loch Borralan intrusion 363
　Loch Loyal Syenite 374–5, 377–8, 379
　Mountsorrel complex 225–6
　Penrhyn Bodeilas Granodiorite 327–8, *328*
　Rogart pluton 404, 405, 408–9
　Ross of Mull pluton 418, 420–3, *423*
　St David's Head Intrusion 261–2, 264
　Shap granite 211–13, *213*
　Strontian pluton 415–16, *415*, 417
　Tan y Grisiau granite 321
　Threlkeld microgranite 189
　Tillicoultry 539, 540
　Towie Wood 130–2, *131*
　wehrlite–clinopyroxenite 40–2, 44–8
　Younger Basic intrusions 112, 130–2

Yarlside Volcanic Formation 143, 184–6, 187
'Younger Basic' intrusions (Grampian Highlands) 6, 107–13, *108*, *109*, 115–32
　associated granitic intrusions 112–13, 132–3
　cumulate layering 110–12, 113–19
　petrographical zoning 111–13, *111*, 114–15, 118–19, 122–4
　xenolithic complexes 130–2
Yr Arddu GCR site **22**, *234*, 241, 296–300, *298*, *299*
Yr Arddu Tuffs 241, 297–9, *298*